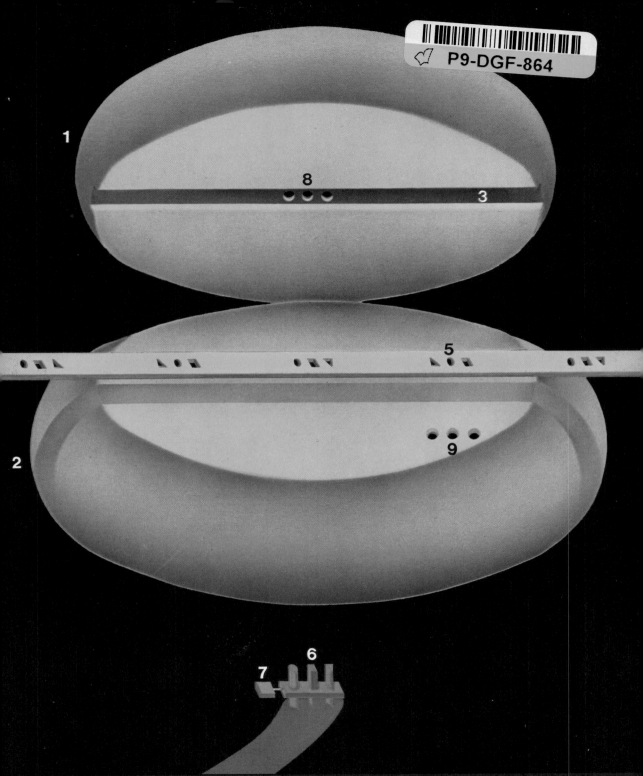

The ribosomes may be visualized as tiny computers which decipher genetic information and follow its directions to synthesize proteins. For simplicity's sake, the protein subunits are not shown in this end plate. The two ribosome subunits (1 and 2) are shown "open" according to the theory of the Russian biologist Alexander Spirin. Messenger RNA (4) flows along a hypothetical groove (3) and is depicted here in the form of a punched tape. Each group of three perforations (5) forms a codon, a term meaning a word in genetic language. Each codon represents a specific point at which carrier RNA (6) must attach a specific amino acid (7). There is a connection (8) in the upper half of the ribosome for the carrier, and another in the lower half for the protein chain in the process of formation (9).

(From Tibaldi, Ettore: The Ribosomes. Milan, Italy, RASSEGNA, Medical and Cultural Review (International Edition) XLVII, No. 1, 1970; illustrated by Silla Gardenghi.)

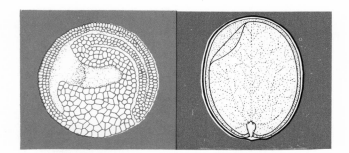

One of the major themes in current biological thinking
concerns the nature of the forces directing
development, represented here and on the cover by
an amphian gastrula (left) and a castorbean embryo (right).

W. B. SAUNDERS COMPANY · PHILADELPHIA · LONDON · TORONTO

CLAUDE A. VILLEE

Harvard University

VINCENT G. DETHIER

Princeton University

BIOLOGICAL
PRINCIPLES
and
PROCESSES

W. B. Saunders Company: West Washington Square
Philadelphia, Pa. 19105

12 Dyott Street
London, WC1A 1DB

833 Oxford Street
Toronto 18, Ontario

Biological Principles and Processes

ISBN 0-7216-9028-9

Print No.: 9 8 7 6 5 4 3

This book is dedicated, with affection and respect, to
Tyler Buchenau,
who has served with distinction as Biology Editor of
W. B. Saunders Company

PREFACE

Truly remarkable advances have been made during the past two decades in many aspects of biological science. This information explosion provides a formidable challenge to each instructor to present biological concepts in an interesting and understandable fashion to the beginning student. To be appreciated and to be fully understood these newer discoveries must be viewed against the background of the more classical aspects of biology. The aim of a course in general biology is to provide each student with an appreciation of the vast diversity of kinds of living things and their special adaptations, evolution and ecologic relations. It must, at the same time, emphasize the basic unity of life and the fundamental similarities in the problems faced and solved by all living things. The growing complexity of the biological sciences has made communication between specialists in the several areas of biology increasingly difficult; however, the advances in each area help the student to understand the problems in other areas.

The present text represents a somewhat different approach to the college level course in general biology, for it does not proceed with a phylum-by-phylum presentation of plants and animals, nor with a system-by-system description of man or the frog. It emphasizes biological principles and the functional aspects of biology. The term "functional" is used in its broadest sense to include actions and interactions at the molecular, cellular, organismal and population levels of organization in biology. In dealing with each principle or process we have chosen examples from bacterial, plant or animal sources, whichever seemed to provide the most appropriate basis for discussion. A host of different kinds of organisms have been utilized in the discussions of such general biological phenomena as the biosynthesis of proteins, the structural basis of membrane permeability, the transfer of energy and of information, the analysis of the biological basis of behavior, or the organization and interrelations of cells or communities.

Students occasionally complain that biology consists of an overwhelmingly large number of facts—the names of plants and animals, the names of macroscopic and microscopic structures, and the names of many kinds of biochemical compounds. By emphasizing principles and concepts, by decreasing the number of bits of information considered, and by relating each fact to the appropriate conceptual framework, we believe the student will be able to learn them more readily and retain them for a longer time. In developing the concepts we have chosen to examine a few examples in detail rather than attempt to mention all the possible examples of a given phenomenon. We have attempted to present those special areas of biology which are advancing most rapidly at present without ignoring those facts that were learned and theories that were formulated decades, even centuries, ago. This is not to say that the book is deliberately skimpy, for it is intended to provide enough knowledge of biology for the student who will continue with more advanced courses in the subject and to provide a general framework of biological information for those who will not continue in the subject.

There has never been general agreement among biologists as to the sequence in which the several topics in a general biology course should be taught. This is understandable, for reasonable arguments can be advanced for each of the many possible combinations and permutations. The several aspects of biology are so intimately interrelated that each can be grasped much more readily if all the other aspects have been learned previously. Since this cannot be done (except, perhaps, by a student repeating the course), each instructor must choose the sequence that seems optimal to him. It has been a tradition in many biology courses to consider genetics and evolution toward the end of the term, and any instructor who wants to can defer Part 2 until later, perhaps after Part 5 or Part 6. The various parts and chapters can be taken up in any of a variety of sequences.

After a brief historical perspective (Chap. 1) to set the scene, the major biological generalizations accepted today are introduced in Chapter 2. The section on Cellular and Molecular Biology presents the structural and ultrastructural features of cells and their chemical constituents, the structure of matter, and a brief review of some principles of chemistry. The concept of biologically useful energy is defined and elaborated on and the flow of energy through biological systems (bioenergetics) is traced from the sun through a chloroplast and a mitochondrion to its final utilization in an effector or in a biosynthetic reaction. The energy requirements and the pathways and control mechanisms for the biosynthesis of certain small and large molecules are examined as examples of that general phenomenon of life. The section ends with a discussion of information transfer, the gene concept and the synthesis of nucleic acids and proteins, affectionately called the "central dogma" of biology.

The second section, Genetics and Evolution, follows directly from the preceding discussion of the transfer of biological information from one generation of cells to the next. It opens with a presentation of classical transmission genetics, upon which is based the subsequent discussion of population genetics and the Hardy-Weinberg Principle. This, in turn, serves as the basis for the concept of differential reproduction and for a discussion of the principle of organic evolution.

The third section presents a brief picture of the major kinds of living things and their interrelations, which introduces the fourth section, concerned with the biology of organisms. Rather than the traditional catalogue of organ systems, this section is organized around the topics of biological membranes and osmotic regulation (Chap. 16), obtaining nutrients (Chap. 17), distribution systems in animals and plants (Chap. 18) and cellular homeostasis (Chap. 19).

Part 4 discusses the biological basis of behavior, beginning with the means by which cellular activities are coordinated and integrated by nervous and hormonal signals. The descriptions of receptors and effectors serve as an introduction to an extensive analysis of the behavior of organisms (Chap. 24). The section of reproductive biology includes the biology of development and the challenging problem of cellular differentiation, in addition to the process of reproduction.

The final section provides an introduction to the biology of populations and the problems of ecology and man's relation to the environment, which are so much in the forefront of discussion at present.

The appendix contains a taxonomic summary of the plant and animal kingdoms that includes the phyla, classes and some orders. It should serve as a convenient reference and should aid the student in recognizing the place in the taxonomic hierarchy of the organisms referred to in the text. Annotated suggestions for further reading are found at the end of each chapter.

We would like to express our deep appreciation to the many scientists, museums, biological supply houses and publishers who permitted us to use photographs or diagrams. Each is acknowledged in the figure's legend. We are especially pleased to acknowledge the careful and imaginative work of our illustrator, Frank Fancher, and his colleagues at New England Illustrators, Inc. We want to thank Hazel Cox, Sandra John, and Eleanor Cantrill, who typed the manuscript, and Janet Loring, who helped in preparing the index. Our special thanks are due to Tyler Buchenau, who as Biology Editor helped and encouraged us tremendously during the inception and early stages of the preparation of the manuscript, and to Carl May, his successor, who did yeoman service in assisting us during the final stages of the process. Finally we want to express our gratitude to our wives for their constant encouragement and patience, and to our families for their forbearance.

CLAUDE A. VILLEE

VINCENT G. DETHIER

CONTENTS

Part 1 CELL AND MOLECULAR BIOLOGY

Chapter 3

Chapter 4

Chapter 5

Part 3   THE KINDS OF ORGANISMS

Chapter 12

THE DIVERSITY OF LIFE: PROTISTA AND MONERA 319

Chapter 13

THE DIVERSITY OF LIFE: PLANTS 364

Chapter 14

Chapter 15

Part 4 THE BIOLOGY OF ORGANISMS

Chapter 16

Chapter 17

Chapter 18

Chapter 19

HOMEOSTASIS AND THE REMOVAL OF WASTE PRODUCTS 596

Part 5   BEHAVIOR AND ITS BIOLOGICAL BASIS

Chapter 20

CONTROL SYSTEMS: NEURAL INTEGRATION 617

Chapter 21

CONTROL SYSTEMS: HORMONAL INTEGRATION 651

Chapter 22

Chapter 23

Part 6   REPRODUCTIVE BIOLOGY

Part 7   POPULATION BIOLOGY: ECOLOGY

Chapter 27

ECOLOGY 893

INTRODUCTION: A HISTORICAL PERSPECTIVE

The first living things appeared on this planet between two and four billion years ago. The descendants of those early forms have diversified and ramified into the 1,000,000 kinds of animals and the 400,000 kinds of plants that are living today plus the very large number of kinds of plants and animals that appeared, flourished for a while and then became extinct. To deal with this vast array of living things, biologists have attempted to arrange them in natural groups of forms related by common ancestry and to understand the basic life processes that are common to them all.

In biology as in other sciences there has been a constant interplay between observations and the interpretation of those observations, and between facts and theories to explain the facts. This interplay is sometimes called "the scientific method." Biologists were chided at one time because it was held that, in contrast to physics and chemistry, biology was simply a descriptive science and had no core of theoretical concepts. In recent years the principles of biology have been emerging and our understanding of life processes has improved greatly. Indeed it is now possible to duplicate many facets of life in "cell free" systems, whose components are obtained by disrupting cells and separating the subcellular constituents, then reconstituting them in a test tube. Isolated chloroplasts can carry out the photosynthetic splitting of water to yield oxygen and reduced products which can react with carbon dioxide to form carbohydrates such as glucose. Isolated mitochondria and submitochondrial particles can carry out oxidative phosphorylation, the oxidation of citric or succinic acids with the formation of "energy-rich" phosphate bonds. Isolated actomyosin fibers will contract in a test tube if supplied with ATP, calcium and potassium. Isolated ribosomes will synthesize proteins from amino acids, and cell-free systems will synthesize specific kinds of DNA and RNA carrying specific kinds of biological information.

Today's complete biologist should be informed not only about cells, cristae and choanocytes, about plastids, pistils and phloem and about meiosis, motor endplates and müllerian ducts, but about concepts ranging from entropy, free energy and molecular conformation through the transfer of biological information, feedback controls and membrane transport to biomass, ecologic niche, evolution, biotic communities and imprinting.

Living things are islands of order in an environment of disorder. They must constantly expend energy to maintain and increase the ordered arrangement of their constituents, thus each is an energy transducer, converting one kind of energy, such as light, to another, such as the chemical energy of the bonds in the carbohydrates made by photosynthesis in green plants.

For decades biological facts and theories have had important applications in medicine, public health, agriculture and conservation. They are even more important now as man becomes concerned about the pollution of his environment and about what measures must be taken to ensure the future of this planet.

SECTION 1–1

LEVELS OF BIOLOGICAL ORGANIZATION

The science of biology is a vast collection of facts and theories about living things. To deal with this enormous array it has been customary to separate the study of plants (**botany**) from the study of animals (**zoology**) or to separate discussions of the structure of an organism (**morphology** or **anatomy**) from discussions of its function (**physiology**). Since plants and animals, despite their differences, have so many facets in common, and since it is difficult, if not impossible, to separate structure and function, to discuss the function of some organ system without describing its structure, it is perhaps more useful to divide biology into its several levels of organization.

The earliest biologists were concerned with studies of whole organisms, entire plants and animals, for to them the organism appeared to be the fundamental unit of life and biological activity. This aspect of biology, which we might term **organismic biology**, has certainly continued to be an important aspect of our current concern with living things and usually serves as an introduction to the other levels of biology.

The invention of the compound microscope and its application to the study of living things, beginning early in the seventeenth century, laid the groundwork for the enunciation in 1838 of the **cell theory** by Schleiden and Schwann. During the next century advances in the optics of the microscope and the development of improved methods for fixing, cutting and staining tissues made possible a rapid expansion of what has been called **cell biology.** The cell is the fundamental unit of both structure and function of living organisms, just as the atom is the fundamental unit of chemical structure and function.

The improvements in the electron microscope since its invention in 1938 and the development of methods for fixing tissues and cutting ultra-thin sections to be viewed in the electron microscope have revealed a whole new level of ultrastructural organization. Electron microscopy, together with x-ray diffraction and polarization optics, have advanced our understanding of the shape of the molecules that comprise living things and their orientation and aggregation into larger units such as membranes. The rapid development of chemical and physical methods for analyzing the sequence of amino acids in proteins, and the sequence of nucleotides in RNA and DNA has made possible an understanding of the genetic code and the synthesis of specific proteins. Our increasing knowledge of these aspects of living things, termed **molecular biology**, is elucidating the nature of the chemical and energy transformations that characterize the phenomena of life.

A fourth level of organization in biology is that of populations of living things and their relationship with their physical and biological environment. We are just beginning to appreciate the many ways in which populations of organisms interact with each other and with the physical environment. Particular kinds of plants and animals are not found at random over the earth but occur in interdependent communities of producer, consumer and decomposer organisms, together with certain nonliving components of the environment. Why communities are composed of certain organisms and not others, how they interact with each other, and how man can control them to his own advantage are major research problems in this field of *ecology.*

SECTION 1–2

THE BIOLOGICAL SCIENCES

The usual definition of biology as the "science of life" is meaningful only if we have some idea of what "life" and "science" mean. Life does not lend itself to a simple definition; the characteristics of living things— growth, movement, metabolism, reproduction and adaptation—will be discussed in Chapter 3. Biology is concerned with the myriad forms that living things may have, with their structure, function, evolution, development and relations to their environment. It has grown to be much too broad a science to be investigated by one man or to be treated thoroughly in a single textbook, and most biologists are specialists in some one of the biological sciences.

The *botanist* and *zoologist* study types of organisms and their relationships within the plant and animal kingdoms respectively. There are specialists who deal with one kind of living thing—ichthyologists, who study fish; mycologists, who study fungi; ornithologists, who study birds; and so on. The sciences of *anatomy, physiology* and *embryology* deal with the structure, function and development of an organism; these can be further subdivided according to the kind of organism investigated: e.g., animal physiology, mammalian physiology, human physiology. The *parasitologist* studies those forms of life that live in or on and at the expense of other forms, the *cytologist* investigates the structure, composition and function of cells, and the *histologist* inquires into the properties of tissues.

The science of *genetics* is concerned with the mode of transmission of the characteristics of one generation to another, and is closely related to the study of *evolution,* which attempts to discover how new species arise, as well as how the present forms evolved from previous ones. The study of the classification of plants and animals and their evolutionary relations is known as *taxonomy. Ecology* is the study of the relations of a group of organisms to its environment, including both the physical factors and other living organisms which provide food or shelter for it, or compete with or prey upon it.

SECTION 1–3

SOURCES OF SCIENTIFIC INFORMATION

Where, you may ask, do all the facts about biology described in this book come from? And how do we know they are true? The ultimate source of each fact, of course, is in some carefully controlled observation or experiment made by a biologist. In earlier times, some scientists kept their discoveries to themselves, but now there is a strong tradition that scientific discoveries are public property and should be freely published. It is not enough in a scientific publication for a man to say that he has discovered a certain fact; he must give all the relevant details by which the fact was discovered so that others can repeat the observation. It is this criterion of **repeatability** that makes us accept a certain observation or experiment as representing a true fact; observations that cannot be repeated by competent investigators are discarded.

When a biologist has made a discovery, he writes a report, called a "paper," in which he describes his methods in sufficient detail that another can repeat them, gives the results of his observations, discusses the conclusions to be drawn from them, perhaps formulates a theory to explain them, and indicates the place of these new facts in the present body of scientific knowledge. The knowledge that his discovery will be subjected to the keen scrutiny of his colleagues is a strong stimulus for carefully repeating the observations or experiments an adequate number of times before publishing them. He then submits his paper for publication in one of the professional journals in the particular field of his discovery (it is estimated that there are more than 7500 of them published over the world in the various fields of biology!) and it is read by one or more of the editors of the journal, all of whom are experts in the field. If it is approved, it is published and thus becomes part of "the literature" of the subject.

At one time, when there were fewer journals, it might have been possible for one man to read them each month as they appeared, but this is obviously impossible now. Journals such as *Biological Abstracts* assist the hard-pressed biologist by publishing, classified by fields, very short reports or **abstracts** of each paper published—giving the facts found, and a reference to the journal. A step beyond the publishing of abstracts is the journal *Current Contents*, which simply lists the titles and authors of the research reports appearing in each of several hundred journals, together with the name, volume and pages of the journal.

A considerable number of journals are devoted solely to reviewing the newer developments in particular fields; some of these are *Physiological Reviews*, *The Botanical Review*, *Quarterly Review of Biology*, *Annual Review of Physiology* and *Nutrition Reviews*. The new fact or theory thus becomes widely known through publication in a professional journal and by reference in abstract and review journals, and eventually may become a sentence or two in a textbook.

Other means for the dissemination of new knowledge are the annual meetings held by the professional societies of botanists, geneticists, physiologists and other specialists at which papers are read and discussed. Prominent among these are the meetings of the American Institute of Biological Sciences (AIBS) and the Federation of American Societies of Experimental Biology (FASEB). There are, from time to time, national and international gatherings, called symposia, of specialists in a given field to discuss the newer findings and the present status of the knowledge in that field. The discussions of these symposia are usually published as books.

SECTION 1–4

HISTORICAL PERSPECTIVES

Man's interest in plants and animals is probably somewhat older than the human race, for the ape-men and man apes that preceded us in evolution undoubtedly learned at an early time such practical things as which plants could be eaten, which animals were dangerous and which could be hunted for food and clothing, where these plants and animals are to be found, and so on. Some of prehistoric man's impressions of his contemporary animals have survived in the cave paintings found in France and Spain.

The civilizations of China, Mesopotamia and Egypt had a wealth of knowledge about plants and animals and had domesticated cattle, sheep, pigs, cats, geese and ducks. The Greek philosophers of the fifth and sixth centuries B.C. — Anaximander, Xenophanes, Empedocles and others — speculated on the origins of plants and animals. Aristotle (384–322 B.C.) was one of the greatest Greek philosophers and wrote on many topics, some of which dealt with biological problems. His *Historia animalium* contains a great deal of information about the animals of Greece and the nearby regions of Asia Minor. His descriptions of animals are quite good and are recognizable as those of specific animals living today. The breadth and depth of his biological interests are impressive — he made careful studies of the development of the chick and the breeding of sharks and bees. He elaborated the theory that plants and animals have gradually evolved up the "ladder of nature," stimulated by their drive to change from the simple and imperfect to the more complex and perfect. His contributions to logic, such as the development of the system of inductive reasoning from specific observations to a generalization that explains them all, have been of inestimable value to all branches of science.

The Greek physician, Galen (131–201 A.D.), was one of the first to do experiments and to dissect animals. The first experimental physiologist, he made some notable discoveries on the functions of the brain and nerves and demonstrated that arteries carry blood and not air. His descriptions of the human body were the unquestioned authority for some 1300 years, even though they contained some remarkable errors, since they were based on dissections of pigs and monkeys rather than of human bodies. Pliny (23–79 A.D.) made an encyclopedic compilation (37 volumes!) of the kinds of animals and where they lived; this was a remarkable mixture of fact and fiction.

The Renaissance in science began slowly with scholars such as Roger Bacon (1214–1294) and Albertus Magnus (1206–1280) who had interests in all branches of natural science and philosophy. The genius Leonardo da Vinci (1452–1519) was an anatomist and physiologist as well as a painter, engineer and inventor. Some of his many original observations in biology came to light only much later when his notebooks were deciphered.

By dissecting human bodies and making clear drawings of what he observed, Andreas Vesalius (1514–1564), a Belgian who was professor at the University of Padua in Italy, revealed many of the inaccuracies in Galen's descriptions of the human body. He published his observations and illustrations in *De Humani corporis fabrica* in 1543. His emphasis on the importance of relying on careful original observation rather than on Galenic authority laid the groundwork for modern anatomy but made him the object of much adverse criticism and he was finally forced to leave his post as professor.

A remarkable number of contributions to our understanding of circulatory physiology were made by the English physician, William Harvey (1578–1657), who received his medical training at the University of Padua, where Vesalius had taught. Harvey published in 1628 his *Exercitatio anatomica de motu cordis et sanguinis in animalibus*. Before then Galen's idea that blood is generated in the liver from food and passes just once to the organs, where it is used up, was accepted without question. The heart was believed to be nonmuscular and expanded passively by the blood flowing into it. Harvey described, from direct observations on animals, how first the atria and then the ventricles fill with blood and empty by muscular contraction. He showed experimentally that blood spurts from a severed artery in rhythm with the beating of the heart, and that when a vein is clamped it becomes full of blood on the side away from the heart and empty on the side toward the heart. He demonstrated the presence of valves in the veins that permit blood to flow toward the heart but not in the reverse direction. From these observations, he postulated that blood flows away from the heart in arteries and back to the heart in veins. By measuring the volume of blood delivered by each heart beat and the number of heart beats per minute, he could calculate the total volume of blood passing through the heart each minute. This proved to be so great that blood clearly could not be generated anew in the liver but must be recirculated, used over and over again. Harvey's was the first quantitative physiological argument. He inferred, though he could not see them, the presence of small vessels connecting arteries to veins by which the circular path of the blood is completed. Later in his life, he made and published a careful study of the development of the chick and postulated that mammals, like the chick, develop from an egg.

The development of the compound microscope by the Janssens in 1590 and by Galileo in 1610 provided the means for attacking many new types of biological problems. Among the first to use the microscope were Robert Hooke (1635–1703), Marcello Malpighi (1628–1694), Antonj van Leeuwenhoek (1632–1723) and Jan Swammerdam (1637–1680), who studied the structure of plant and animal tissues. With the aid of a microscope that magnified about 30 ×, Hooke reported his discovery of "cells" in a slice of cork. Leeuwenhoek, with lenses that magnified as much as 270 ×, described human sperm cells, bacteria, protozoa and the nuclei of blood cells. Malpighi was able to demonstrate the capillaries connecting arteries and veins. These observations of cells were made without any special theoretical insights; the latter appeared only early in the nineteenth century, with the formulation of the cell theory. Cell biology developed rapidly in the nineteenth century, aided by great improvements in the design of microscope lenses. The nucleus of the plant cell was described by Brown in 1833 and the nucleolus by Schleiden and Schwann in 1839. The cell theory was further advanced by Virchow's famous dictum (1855) that "all cells arise only by the division of previously existing cells." Chromosomes and the sequence of events in mitosis were described by Fleming (1880), and the more complex nuclear events of meiosis were clarified in the 1890's.

John Ray (1627–1705) and Karl Linnaeus (1707–1778) brought order into the classification of plants and animals and devised the binomial system (two names, genus and species) by which each kind of plant and animal is given a unique scientific name. This binomial system was first used consistently by Linnaeus in the tenth edition of his *Systema naturae* (1758).

Karl Ernst von Baer published in 1828 his book, *Developmental Studies of Animals*, which summarized the contributions to our understanding of development made by Fabricius, the professor of anatomy at Padua who taught William Harvey, by Harvey, Malpighi and Kaspar Wolff. In this book Baer made the generalization, known as **Baer's Law,** that the more general features, common to all the members of a group of animals, develop earlier in the embryo than the more special features, which distinguish the various members of the group. The features, for example, that characterize all vertebrates—brain, spinal cord, axial skeleton around a notochord, aortic arches and segmental muscles—develop earlier than the features that distinguish the several classes of vertebrates—limbs in tetrapods, feathers in birds and hair in mammals. The characteristics that distinguish families, genera and species appear last in development. The structure of the early embryo is very similar in all the members of a given large group (phylum) of the animal kingdom.

Baer's law was formulated long before Darwin's cogent arguments for the theory of evolution and was reinterpreted in the light of evolutionary theory by Müller and Haeckel. The latter named it the **biogenetic law**, briefly stated as "ontogeny recapitulates phylogeny." Organisms in the course of their development tend to pass through, to recapitulate, the same sequence of stages evident in the course of their evolution. The recapitulation is often greatly shortened and modified and many of the stages in the evolutionary sequence are omitted completely from the developmental (ontogenetic) sequence. The embryo recapitulates certain features of the *embryos* of his ancestors rather than their adult condition. This theory is useful in providing an explanation for such puzzling features of development as the appearance of pharyngeal pouches and mesonephric tubules in the higher vertebrates. Baer also established the theory of the germ layers and emphasized the need for comparative studies of development in different animals.

As soon as biologists had described the sequence of events in the development of any organism and had seen the remarkable regularity of the developmental processes, the question naturally arose, what forces guide the changes from a single cell, the fertilized egg, to the remarkably complex structures of the adult? During the seventeenth and eighteenth centuries, the theory of **preformation** was widely accepted as an explanation of development. This theory really explained development by denying that it occurred, for it supposed that the egg (or the sperm) contained all the essential features of the adult and that development was simply an unfolding of these preformed characteristics. If you open the pupa of a butterfly, all the parts of the adult can be seen, folded up and ready to unfold when the adult emerges from the pupal case. When the bud on a plant is dissected open, the leaves and the flowers and their parts may be exposed to view. By analogy it was argued that the egg contains all the parts of the future embryo but they are transparent, folded together, and very small, hence difficult to see. A lively argument of the time was between those who thought all the parts of the embryo were present in the egg, and the sperm were simply parasites present in the seminal fluid, and those who believed the parts of the embryo were present in the sperm, and the egg simply provided nutrition for the embryo developing from the sperm. The discovery by Bonnet in 1745 that certain insects such as aphids could develop parthenogenetically from unfertilized eggs argued strongly against the theory that the embryo is preformed in the sperm.

The opposite view, that there is a gradual differentiation of structure during development from a relatively structureless egg, termed **epigenesis**, was advanced by Kaspar Wolff in 1759. Wolff's careful studies of the development of the chick embryo could reveal no parts of the future embryo in the earliest stages of development. The egg did have some structure but it was quite different from that of the later embryo. Wolff concluded that the egg does not contain a preformed embryo but only the material from which the embryo would be formed. The argument between the preformation and epigenetic schools of embryologists has been revised and rephrased many times in the light of new discoveries. The advent of experimental embryology in the latter part of the nineteenth century helped resolve the question. If we separate the two cells that result from the first division of a fertilized egg and let each develop independently, the theory of preformation would predict that each would develop into an embryo lacking one half of its organs and parts. The theory of epigenesis would predict that each of the two cells would give rise to a complete embryo with a smaller than normal size. Wilhelm Roux attempted this experiment by destroying with a hot needle one of the two cells in the two-celled stage of the developing frog egg. The resulting embryo developed, at least initially, as though it were forming half of a complete embryo. The result was that predicted by the preformation theory, i.e., a defective embryo developed. However, it was found subsequently that the experimental method was at fault. In this experiment one of the two cells was killed but the dead material was not removed, and it was the presence of the damaged cell that caused the defective development of the undamaged cell. If the two cells are completely separated by tightening a loop of fine hair between them, then each gives rise to a normal whole embryo.

Following William Harvey, physiology was advanced by René Descartes (1596–1650), by Charles Bell (1774–1842) and by François Magendie (1783–1855), who made notable contributions to our understanding of the function of the brain and spinal nerves. Johannes Müller (1801–1858) studied the properties of nerves and capillaries, and through his textbook of physiology stimulated a great deal of interest and research in the field. Claude Bernard (1813–1878), one of the great advocates of experimental physiology, advanced our understanding of the functions of the liver, heart, brain and placenta. Henry Bowditch (1840–1911) discovered the "all-or-none" principle of the contraction of muscle and established the first laboratory for the teaching of physiology in the United States. Ernest Starling (1866–1927) made many contributions to the physiology of the circulatory system and the nature of lymph. Together with William Bayliss (1866–1924), he elucidated the hormonal control of the secretion of digestive enzymes by the pancreas.

The Scottish anatomist, John Hunter (1728–1793), and the French anatomist, Georges Cuvier (1769–1832), were among the pioneers in the study of similar structures in different animals, the field of **comparative anatomy**. Richard Owen (1804–1892) developed the concepts of homology and analogy. Cuvier was one of the first to study the structure of fossils as well as of living organisms and is credited with founding the science of **paleontology**. Cuvier, despite this, believed strongly in the unchanging nature of species and carried on bitter debates with Lamarck, who had proposed in 1809 a theory of evolution based on the concept of the inheritance of acquired characters.

Biology has expanded at a truly remarkable rate in the past century, and has witnessed the establishment of such subsciences as cytology, embryology, genetics, evolution, biochemistry, biophysics, endocrinology and ecology. The discoveries and advancing techniques of chemistry and physics have made possible new approaches to the problems of biology. A great many men have made contributions to the advancement of biology in the past century; only a few can be mentioned in each field: Mendel, deVries, Morgan, Bridges and Müller in genetics; Darwin, Dobzhansky, Wright and Goldschmidt in evolutionary theory; Harrison and Spemann in embryology; and Crick, Watson, Kornberg and Nirenberg in molecular biology.

SECTION 1–5

BIOLOGICAL NOMENCLATURE AND UNITS

The student of biology is confronted with an extensive list of terms for the kinds of plants and animals, their structures and functional mechanisms, and their interrelations. To be as precise as possible, and to have an internationally accepted system, it is customary to use Latin words where possible and to manufacture new words using Latin or Greek roots, casting the new words in Latin form, when describing a newly discovered structure or process. We have made an earnest effort to minimize the number of new terms introduced in this text, but many terms are, in fact, intrinsic parts of the concepts and principles under discussion and cannot be eliminated.

When dealing with cellular dimensions and the amounts of material present at the cellular level, units of an appropriately small size are necessary. Units of length include the micron (0.001 mm) and the Angström unit (1 micron = 10,000 Angström units). Weights are expressed in nanograms (10^{-9} gm) or picograms (10^{-12} gm) or in daltons. The **dalton** is the unit of molecular weight, the weight of a hydrogen atom. One molecule of water weighs 18 daltons, a molecule of hemoglobin, a middle-sized protein, weighs 64,500 daltons.

Some appreciation of the range of sizes of biological structures can be gained from Figure 1–1, in which the sizes of cells, viruses and molecules are arranged on a logarithmic scale.

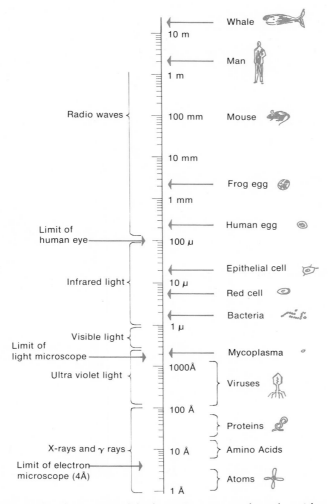

FIGURE 1–1   The dimensions of biology. To accommodate the wide range of sizes in the world of living things, the unit scale is drawn logarithmically.

SECTION 1–6

APPLICATIONS OF BIOLOGY

Some of the practical uses of a knowledge of biology will become apparent as the student reads on through this text—its applications in the fields of medicine and public health, in agriculture and conservation, its basic importance to the social studies, and its contributions to the formulation of a philosophy of life. There are esthetic values in a study of biology as well. A student cannot expect to learn all or even many of the names and characteristics of the vast variety of plants and animals, but a knowledge of the structure and functions of the major types will greatly increase the pleasure of a stroll in the woods or an excursion to the seashore. The average city-dweller gets only a small glimpse of the vast panorama of living things, for so many of them live in places where they are not easily seen—the sea, or parts of the earth that are not easily visited. Trips to botanical gardens, zoos, aquariums and museums will help give one an appreciation of the tremendous variety of living things.

It is impossible to describe the forms of life without reference to their habitats, the places in which they live. This brings us to one of the major unifying conceptual schemes of biology, that the living things of a given region are closely interrelated with each other and with the environment. The study of this is basic to sociology. The present forms of life are also related more or less closely by evolutionary descent. As we deal with each of the major life forms, the facts about them will be easier to understand and remember if we try to fit them into their place in the closely interwoven tapestry of life.

In our discussions of biological principles, we will focus our attention to a considerable part on man, to gain an appreciation of man's place in the biological world. It is only in man's somewhat biased opinion that he stands in the center of the universe, with other animals and plants existing only to serve him. In numbers, size, strength, endurance and adaptability, he is inferior to many animals, and in his adjustment to the environment— which, as we shall see, may be considered to be the most important biological attribute of any living organism—he often fails. However, in a survey study of general biology, both practical considerations and interest demand that our discussions focus on man.

SUGGESTIONS FOR FURTHER READING

Beveridge, W. I. B.: *The Art of Scientific Investigation.* New York, W. W. Norton Company, 1957.

Cohen, I. B.: *Science, Servant of Man.* Boston, Little Brown and Company, 1948.

Conant, J. B.: *Science and Common Sense.* New Haven, Yale University Press, 1951.
  Each of these three books discusses the scientific method and its application to the solution of research problems.

Conant, J. B.: *On Understanding Science.* New Haven, Yale University Press, 1947.
  A discussion of the role of science in society by one of the great figures in the American scientific community.

Feibelman, J. K.: *Testing Hypotheses by Experiment.* Perspect. Biol. Med. *4*:91, 1960.
  An excellent, brief discussion of scientific method and its application to specific research problems.

Gardner, Martin: *Fads and Fallacies in the Name of Science.* New York, Dover Publications, 1957.
  An interesting book which discusses many pseudosciences and, in revealing their shortcomings, helps one appreciate scientific evidence and standards.

Guthrie, Douglas: *A History of Medicine.* Philadelphia, J. B. Lippincott Company, 1946.
    This book describes the beginnings of anatomy, physiology and bacteriology.

Holton, Gerald: *The Making of Modern Science—Biographical Sketches.* Boston, Houghton Mifflin Company, 1971.
    A fascinating collection of biographical articles about scientists. The article on Francis Crick by Robert Olby is recommended reading in connection with *The Double Helix.*

Mazzeo, Joseph A.: *The Design of Life.* New York, Pantheon Books, Random House, 1967.
    A description and analysis of some of the major concepts in biology and their development from Aristotle to the present.

Nordenskiold, E.: *The History of Biology.* New York, Tudor Publishing Company, 1966.

Singer, Charles: *A History of Biology.* Rev. Ed. New York, Abelard-Schumann Limited, 1959.
    The histories of biology by Nordenskiold and Singer are well written and informative, providing historical accounts of the developments on which modern biology is based.

Sedgwick, W. T., Tyler, H. V. and Bigelow, R. P.: *A Short History of Science.* New York, The Macmillan Company, 1939.
    A well written account of the early development of the sciences in general.

Watson, J. D.: *The Double Helix.* New York, Atheneum Publishers, 1968.
    An interesting insight into the motivations and thought processes of one of the classic molecular biologists; "how to become a Nobel Laureate."

Wilson, E. Bright: *An Introduction to Scientific Research.* New York, McGraw-Hill Book Company, 1952.
    An excellent, nontechnical discussion of the methods of science and of some of the problems involved in scientific evidence and standards.

SOME MAJOR GENERALIZATIONS OF THE BIOLOGICAL SCIENCES

CHAPTER  2

PHYSICAL AND CHEMICAL PRINCIPLES GOVERN LIVING SYSTEMS

One of the basic tenets of modern biology is that all of the phenomena of life are governed by, and can be explained in terms of, chemical and physical principles. Until early in the present century most people, biologists and laymen alike, held that life processes differed in some fundamental way from those of nonliving systems. With the vast increase since then in our understanding of chemical and physical principles it has become clear that the myriad phenomena of life, although much more complex than nonliving systems, can be explained in chemical and physical terms without postulating some mysterious vital force. The properties of living cells and organisms that at one time seemed so mysterious, now appear to be quite straightforward. Many of the complex phenomena of living systems can be reproduced in the test tube under appropriate conditions. The corollary of this belief is, of course, that if we knew enough about the chemistry and physics of living systems we could re-create life in the test tube.

SECTION 2-2

THE CELL THEORY

Another fundamental biological generalization is the cell theory. The *cell theory*, as presently formulated, states that all living things, animals, plants and bacteria, are composed of cells and cell products, that new cells are formed by the division of preexisting cells, that there are fundamental similarities in the chemical constituents and metabolic activities of all cells, and that the activity of an organism as a whole is the sum of the activities and interactions of its independent cell units.

12

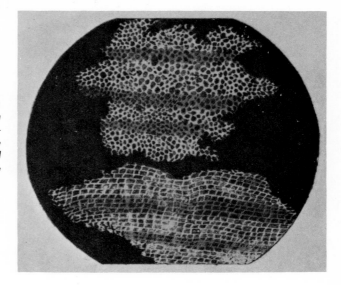

FIGURE 2-1 *Drawings by Robert Hooke of the microscopic structure of a thin slice of cork. From the book Micrographia, published 1665, in which Hooke described many of the objects he had viewed using the compound microscope he had constructed.*

Cells were first described by Robert Hooke, who looked at a piece of cork under one of the crude microscopes of the seventeenth century. What Hooke saw (Fig. 2-1) was in fact the cell walls of dead cells, and it was not until some two centuries later that it was realized that the important part of the cell was not its walls but the contents of the cell.

Like most broad and basic theories, the cell theory is not the product of any single person's research and thought. The German botanist. Matthias Schleiden, and the German zoologist, Theodor Schwann, are usually credited with this theory, for in 1838 they pointed out that animals and plants are aggregates of cells arranged according to definite laws. However, the French biologist, Dutrochet, had stated clearly in 1824 that "all organic tissues are actually globular cells of exceeding smallness, which appear to be united only by simple adhesive forces; thus all tissues, all animal organs are actually only a cellular tissue variously modified." Even before that, in 1809, Lamarck had stated that "no body can have life if its constituent parts are not cellular tissue or are not formed by cellular tissue." Dutrochet recognized that growth is the result of increases in the volumes of individual cells and of the addition of new little cells. The presence of the nucleus within the cell, now recognized as an almost universal feature of cells, was first described by Robert Brown in 1831. As in many other fields of science, Schleiden and Schwann, though not the first to enunciate a principle, stated it with such clarity and force that the idea gained general credence and was eventually accepted by the majority of the biologists at that time.

SECTION 2-3

BIOGENESIS

There appear to be no exceptions to the third generalization that all life comes only from living things. This in effect is an extension of the cell theory and was first stated clearly by Rudolf Virchow: "All living cells arise from preexisting living cells." In other words, there is no spontaneous creation of a cell from nonliving matter. The idea that even large organisms such as worms, frogs and rats could arise by **spontaneous generation** was widely held until the seventeenth century, until it was disproved by the experiments of Redi and Spallenzani (see p. 320).

The experiments of Pasteur, Tyndall and others just over a century ago finally provided convincing proof that microorganisms such as bacteria are also unable to originate from nonliving material by spontaneous generation. Previously the scientists who believed that organisms could not arise by spontaneous generation had shown that if nutrient medium is placed in a stoppered flask and the contents are boiled no organisms subsequently appear when the flask is cooled. The proponents of spontaneous generation replied that the boiling process had destroyed some essential life-forming nutrient and for this reason no spontaneous generation was observed. Pasteur's experiments involved putting nutrient broth into flasks with long **S**-shaped necks (Fig. 2–2). He boiled the broth in the flasks to kill any bacteria that might be present. Flasks with straight necks permitted dust particles and their adhering bacteria to settle into the broth; bacteria colonies appeared very quickly. However, although the **S**-shaped flasks were equally open to the air, bacteria could not penetrate to the nutrient broth, because they were trapped in the film of moisture in the **S**-shaped curve of the neck. This acted as a filter. The broths could be left in such flasks for weeks or months without any bacteria appearing in them. Pasteur showed further that if the **S**-shaped neck were broken off, the broth rapidly developed an extensive population of bacteria. By a

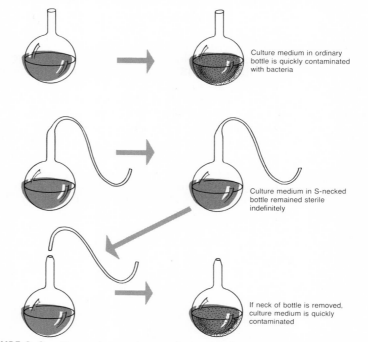

Culture medium in ordinary bottle is quickly contaminated with bacteria

Culture medium in S-necked bottle remained sterile indefinitely

If neck of bottle is removed, culture medium is quickly contaminated

FIGURE 2–2   *Pasteur's experiments disproving the spontaneous generation of microorganisms. Nutrient broth (sugar and yeast) was placed in flasks with long S-shaped necks and boiled to kill any bacteria present. Flasks with straight necks permitted bacteria to settle into the broth and it was quickly teeming with bacteria. The S-shaped neck did not permit bacteria to enter and, although the flask was open to the air, its contents did not become contaminated unless the neck was removed.*

series of such experiments, Pasteur showed that the bacteria that appeared in the nutrient broth did not develop by spontaneous generation. Rather, these bacteria were present in the air and were carried to the broth attached to dust particles.

In recent years it has been realized that although spontaneous generation does not occur at the present time it must have occurred some billions of years ago when life initially began on this planet. At that time, conditions in the atmosphere were quite different from those that exist at present. Although spontaneous generation is most unlikely to occur at the present time, it quite probably did occur under the markedly different environmental conditions of that primitive time.

SECTION 2–4

LIVING CELLS ARE ENERGY TRANSDUCERS

Living organisms and their constituent cells are not heat engines but are **transducers** which convert the chemical energy of foodstuffs—energy ultimately captured by green plants from sunlight—into electrical, mechanical, osmotic or other forms of useful work.

Each living cell is equipped with an efficient and complex series of devices for transforming energy. The radiant energy of sunlight is the major source of energy for all of the life forms on this planet. The first major transformation of energy on the planet earth is carried out by green plants. These transform the radiant energy of sunlight into chemical energy, which is stored as the energy of the bonds connecting the atoms of molecules such as glucose (Fig. 2–3). This first stage of energy transformation is photosynthesis. It is carried out by green plants using the pigment chlorophyll, which enables green plant cells to transform radiant energy into chemical energy. The chemical energy is used to synthesize carbohydrates and other molecules from carbon dioxide and water.

The second major stage in the flow of energy on this planet occurs in every cell, both plant and animal, in which respiration occurs. In this

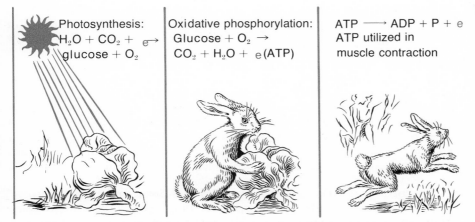

Photosynthesis:
$H_2O + CO_2 + e\rightarrow$
glucose + O_2

Oxidative phosphorylation:
Glucose + $O_2 \rightarrow$
$CO_2 + H_2O + e$(ATP)

$ATP \longrightarrow ADP + P + e$
ATP utilized in
muscle contraction

FIGURE 2–3   *The flow of energy from the sun through green plants to animals via photosynthesis, respiration and the utilization of ATP in the performance of work. The radiant energy of sunlight is converted into chemical energy, ATP, and finally is dissipated as heat.*

process, the chemical energy of carbohydrates and other molecules is transformed into a biologically useful kind of energy as these foodstuff molecules undergo oxidation. Cells metabolize foodstuffs such as glucose by a series of enzymic reactions, and the energy present in the chemical bonds of the foodstuffs is transformed and conserved as the energy of **adenosine triphosphate**, ATP.

In the third stage of energy transformation, the chemical energy recovered from foodstuffs in the form of ATP is used by cells to do a variety of kinds of work. The ATP is the source of energy used by cells to transmit nerve impulses, to cause muscle contraction, to carry out the synthesis of complex macromolecules from simpler smaller ones, and to carry on all of the myriad life functions. As these biological functions are performed, the energy eventually flows to the environment in the dissipated, useless form of heat. In none of these transformations of energy does the cell act as a heat engine. Neither glucose nor any other molecule is "burned" in the strict sense of the word.

The branch of physics that deals with energy and its transformations, **thermodynamics**, consists of a number of relatively simple basic principles which are universally applicable to chemical processes whether they occur in living or nonliving systems. It is well established that the laws of thermodynamics apply to the world of life as they do to the world of non-living systems.

Under experimentally controlled conditions, the amounts of energy entering and leaving any system may be measured and compared. It is always found that energy is neither created nor destroyed, but only transformed from one form to another. This is an expression of one of the fundamental laws of physics, the **law of the conservation of energy**. This law applies to living systems just as it does to nonliving ones.

THE THEORY OF ORGANIC EVOLUTION

The concept of organic evolution is one of the major generalizations of biology. That all of the many kinds of plants and animals now existing were not created *de novo* but have descended from previously existing, simpler organisms by gradual modifications which have accumulated in successive generations is one of the major concepts unifying all of biology (Fig. 2–4). Certain elements of this idea were implicit in the writings of Greek philosophers before the Christian era, from Thales to Aristotle. The theory of organic evolution was considered by a number of philosophers and naturalists from the fourteenth to the nineteenth century. However, not until Charles Darwin's publication in 1859 of *On The Origin of Species by Means of Natural Selection* was the theory brought to general attention. In his book Darwin presented a wealth of detailed evidence and cogent argument to show that organic evolution had occurred. He also presented a theory, termed **natural selection**, to explain how evolution may occur.

According to Darwin's **Theory of Natural Selection,** every group of animals or plants tends to undergo variation. More organisms of each kind are produced than can possibly obtain food and survive. There is a struggle for survival among the many individuals that are born, and those individuals that possess characters that give them some advantage in the struggle for existence will be more likely to survive than those without

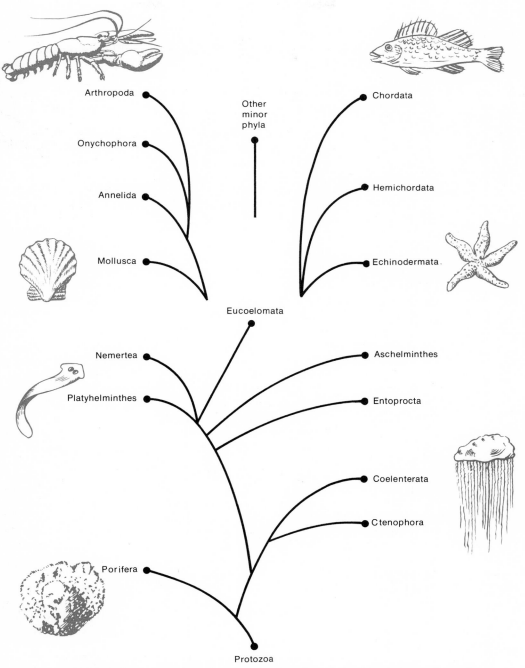

FIGURE 2–4   An evolutionary tree showing the relationships between the major groups of animals and plants.

them. The survivors will pass these advantageous characters on to their offspring, so that successful variations will be transmitted to future generations. The core of Darwin's theory was this concept of the struggle for existence, the survival of the fittest, and the inheritance of the advantageous characters by the offspring of the surviving individuals. This concept has had a central role in biological theory for the past century, and, with suitable amendments to bring it into line with subsequent discoveries in genetics and evolution, the theory is held by most present-day biologists.

A host of studies of the development of many kinds of animals and plants from fertilized egg to adult have led to the interesting generalization that organisms tend to repeat, in the course of their embryonic development, some of the corresponding stages of their evolutionary ancestors. This generalization, called the **Theory of Recapitulation**, was once interpreted as meaning that embryos have in succession the appearance of the *adult* forms of their ancestors. Most biologists would now prefer the statement that embryos recapitulate some of the *embryonic* forms of their ancestors. The human being, at successive stages in his development, resembles in certain respects first a fish embryo, then an amphibian embryo, then a reptilian embryo, and so on.

SECTION 2–6

THE GENE THEORY

It has been known for more than a century that each new organism originates from the union of an egg and a sperm (Fig. 2–5); however, the question of how the parental characters are transmitted to the offspring through these tenuous bits of living material continued to puzzle scientists for several more decades. One theory put forward by Charles Darwin was that each tissue or organ of the parent contributed some sort of model which he termed the "pangene". These pangenes were incorporated into the egg or sperm and thus were transmitted to the offspring. There they guided development so as to produce in the offspring a duplicate of the organ from which they came.

The theory of the "continuity of the germ plasm" was formulated by August Weismann in 1887. He stated his belief that the germ cells of an individual are derived from the parent germ cells and not from the body or somatic cells of the individual. He suggested that from the very first division of the fertilized egg one line of cells, the **germ plasm**, was distinct from the body cells or **somatoplasm**. The germ plasm was unaffected by the somatoplasm or by external influences. He realized, before chromosomes were discovered or before genes were known, that heredity involves the transfer of particular molecular constitutions from one generation to the next. A little reflection reveals an obvious corollary of Weismann's theory: acquired characteristics are not inherited. Only those changes which occur in the germ plasm and not in the somatoplasm can be transmitted to succeeding generations.

There are some invertebrate animals in which the continuity of the germ plasm from one generation to the next can be seen. Early in cleavage one cell can be distinguished as the precursor of the germ cells. Its descendants can be traced ultimately to their final location in the testis or ovary. In most animals, the distinction between germ plasm and somatoplasm is not clearly evident and germ cells appear to arise from unspe-

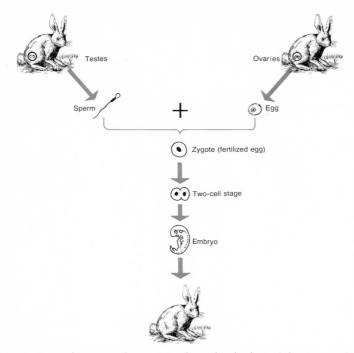

FIGURE 2–5   The union of an egg (with one haploid set of chromosomes) and sperm (with another haploid set of chromosomes) to form a zygote (with a diploid set of chromosomes) from which develops the new organism.

cialized somatic cells. As more has been learned about chromosomes and genes, it has become clear that genetic continuity from generation to generation resides primarily in the chromosomes, which are present in all cells, and not in some peculiar property of the germ line itself.

The generalizations about the mechanisms of inheritance are among the most exact and quantitative of biological theories. They permit us to make predictions regarding the probability that the offspring of two given parents will have a particular characteristic. These generalizations are called **Mendel's Laws**, for they were first enunciated by the Austrian abbot Gregor Mendel in 1866. He made these inferences from his careful breeding experiments with garden peas. The importance of Mendel's findings was not recognized until 1900, when the principles were independently rediscovered by three different investigators, Correns, de Vries and von Tschermak.

Mendel's First Law, called the **Law of Segregation**, states that genes, the units of heredity, exist in individuals as pairs. In the formation of gametes or germ cells, the two genes separate or segregate and pass to different gametes, so that each gamete has one and only one of each kind of gene. Mendel's Second Law, termed the **Law of Independent Assortment,** states that the segregation of each pair of genes in the process of gamete formation is independent of the segregation of the members of other pairs of genes. Thus the members of the pairs come to be assorted at random in the resulting gamete. Mendel's keen insight is remarkable, for he made these generalizations despite the fact that the details of chromosomes, meiosis and fertilization were not yet known. Later, when chromosomes were discovered and genetic and cytologic evidence was available, the modern concept was formulated that the units of inheritance are arranged in linear order in the chromosomes. This was stated by W. S. Sutton in 1902 and by T. H. Morgan in 1911.

GENETIC EQUILIBRIUM AND DIFFERENTIAL REPRODUCTION

The question that may puzzle a beginning biologist is why, if the genes for brown eyes are dominant to the genes for blue eyes, haven't all the blue-eyed genes and all the blue-eyed individuals disappeared? The answer lies partly in the fact that a recessive gene, such as the one for blue eyes, is not changed by having existed for a generation in the same cell with a brown-eyed gene. The remainder of the explanation lies in the fact that as long as there is no selection for either eye color, that is, as long as people with blue eyes are just as likely to marry and have as many children as people with brown eyes, successive generations will have the same proportions of blue- and brown-eyed people as the initial one. (Fig. 2–6).

The principle that a population of a given species of animals or plants in genetic equilibrium, i.e., in the absence of natural selection, tends to have the same proportion of organisms with a given characteristic in successive generations, was arrived at independently by the English mathematician, G. H. Hardy, and the German physician, G. Weinberg, in 1908. They pointed out that the frequency of the possible combinations of a pair of genes in a population may be calculated from the expansion of the binomial equation $(p\mathbf{A} + q\mathbf{a})^2$. When we consider all of the matings of all of the individuals in any given generation, a p number of $\mathbf{A}$-containing eggs and a q number of $\mathbf{a}$-containing eggs are fertilized by a p number of $\mathbf{A}$-containing sperm and a q number of $\mathbf{a}$-containing sperm: $(p\mathbf{A} + q\mathbf{a}) \times (p\mathbf{A} + q\mathbf{a})$. The proportion of offspring of all of these matings is described by the algebraic product: $p^2\mathbf{AA} + 2\,pq\,\mathbf{Aa} + q^2\mathbf{aa}$. Any population in which the distribution of a pair of alleles $\mathbf{A}$ and $\mathbf{a}$ conforms to the relation $p^2\mathbf{AA} + 2\,pq\,\mathbf{Aa} + q^2\mathbf{aa}$ is in *genetic equilibrium*. The proportions of these alleles in the members of successive generations will be the same, unless they are altered by selection, by mutation or by chance. This relationship, referred to as the ***Hardy-Weinberg Law***, has been of great importance in genetics, particularly in human genetics, for it is the basis of the statistical methods used in determining the mode of inheritance of a given trait in the absence of control and test matings.

This Hardy-Weinberg principle is also fundamental to the mathematical treatment of problems in evolution. Evolution by natural selection, stated in its simplest terms, means that individuals with certain genotypes, and therefore certain traits, have more surviving offspring in the next generation than do other individuals with contrasting genotypes. They contribute a proportionately greater percentage of genes to the gene pool of the next generation than do organisms with other genes. We now view the process of evolution as a gradual change in the gene frequencies of a population which occurs when the Hardy-Weinberg equilibrium is upset. This upset may result because mutations occur, because reproduction is nonrandom, i.e., because selection occurs, or because the population is

Generation 1

Generation 2

*FIGURE 2–6   The Hardy-Weinberg principle of genetic equilibrium. The gene frequency, i.e., the proportion of genes **A** and **a** in a population, tends to remain constant in successive generations unless altered by mutation, selection or chance.*

small and chance alone may determine the survival or the loss of a specific allele by a process termed **genetic drift**. This process of **differential reproduction** implies that the conditions of the Hardy-Weinberg equilibrium do not apply to that particular population. The individuals that produce more surviving offspring in the next generation are usually but not necessarily those that are best adapted to the given environment. Well-adapted individuals may be healthier, better able to obtain food and mates, and better able to care for their offspring, but the primary factor in evolution is how many of their offspring survive to be parents of the next generation.

SECTION 2–8

DNA IS THE MAJOR REPOSITORY OF GENETIC INFORMATION

By the early 1950's, Mirsky and Vendrely had shown that all the cells of the various tissues of a given organism have the same amount of DNA. The only exception to this are the gametes. The eggs and sperm have only one-half as much DNA per cell as do other cells of the same organism. The inference that DNA was the important part of the gene was obvious. Erwin Chargaff had carried out analyses of the relative amounts of purines and pyrimidines in DNAs from a variety of sources. His analyses showed that although DNAs from different sources may have quite different compositions, certain patterns are evident. The amount of adenine equals the amount of thymine and the amount of guanine equals the amount of cytosine. Studies by Maurice Wilkins in London had shown by x-ray crystallography that the DNA molecule was probably a helix, a giant coil. Using these facts, James Watson and Francis Crick in 1953 proposed a model structure for the DNA molecule (Fig. 2–7), which accounted for the known properties of the gene: its ability to replicate itself exactly, its ability to transmit information, and its ability to undergo mutation.

Watson and Crick suggested that the DNA molecule was a huge intertwined **double helix**. The Watson-Crick model of the DNA molecule pictures two polynucleotide chains wrapped helically around each other, with the sugar-phosphate residues forming a chain on the outside and the purines and pyrimidines on the inside of the helix. The two chains are held together by hydrogen bonds between specific pairs of purines and pyrimidines, e.g., between adenine and thymine as one pair and cytosine and guanine as another pair. Thus these two chains are complementary to each other; that is, the sequence of nucleotides in one chain dictates the sequence of nucleotides in the other. The two complementary strands have opposite polarity; that is, they run in opposite directions and have their terminal phosphate groups at opposite ends of the double helix. When Watson and Crick made an exactly scaled molecular model of this double helix, they found that the combinations adenine and thymine on the one hand and guanine and cytosine on the other hand would fit into the space available, whereas other combinations of purines and pyrimidines would not. The Watson-Crick model in addition provides an explanation to how DNA molecules may undergo replication. The two chains separate, each one brings about the formation of a new chain which is complementary to it, and thus two new double chains are established.

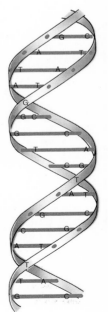

FIGURE 2–7   *The Watson-Crick model of the DNA molecule as a double helix, joined by hydrogen bonds between specific pairs of purines and pyrimidines.*

SECTION 2–9

CODING AND PROTEIN SYNTHESIS

Although the Watson-Crick model of the DNA molecule implied that genetic information is transmitted somehow by the sequence of its constituent nucleotides, the exact mechanism was unclear. Since there are only four types of nucleotides, A, T, C and G, in the DNA and twenty or more kinds of amino acids in the peptide chain, it was obvious that there could not be a one-to-one correlation between a nucleotide and an amino acid. Neither would two nucleotides to one amino acid provide a means of coding for 20 amino acids. The various combinations of four symbols taken two at a time would provide only 16 different combinations; however, a **triplet code** of three nucleotides for each amino acid would permit 64 different combinations of the four nucleotides taken three at a time. The mathematical and biological arguments for a triplet code were put forward by Francis Crick in 1961. An enormous amount of research since that time has substantiated his thesis that the genetic code is a triplet one with three adjacent nucleotide bases, termed a **codon**, specifying a particular amino acid (Fig. 2–8). Adjacent codons do not overlap. Each single base is part of a single codon.

A further generalization has emerged from these studies: the genetic code appears to be universal, that is, the codons in the DNA and RNA

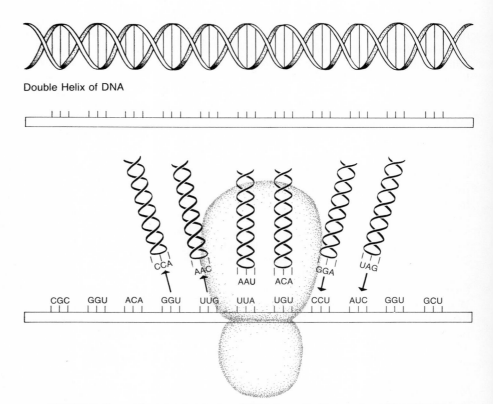

Double Helix of DNA

FIGURE 2–8   *The triplet code by which information is transferred from DNA in the nucleus via RNA to the ribosomes, where amino acids are assembled in an order dictated by the RNA codons, occurs, and then the synthesis of a specific peptide chain with a specific sequence of amino acids occurs.*

specify the same amino acid in all the organisms that have been studied from viruses to man. Experimental evidence to support Crick's hypothesis of a triplet code was quickly forthcoming from the experiments of Nirenberg and Matthei regarding the incorporation of specific labeled amino acids into proteins by purified enzyme systems under the direction of artificial polynucleotides of known composition. Thus the genetic code is composed of three-letter units or codons, the three nucleotides of which specify a given amino acid. The sequence of codons in a molecule of DNA in turn specifies the sequence of amino acids in the corresponding polypeptide chains.

In each cell generation the gene, the DNA chain, undergoes replication so that when the cell divides each of the two daughter cells receives an exact copy of the code. Also in each cell generation one or more transcriptions of the code may be made by which the genetic information is used to regulate the assembly of a specific enzyme or other protein. This is a two-step process. In the first step, the four-letter code of the nucleotides in the DNA of the genes is transcribed into a similar four-letter code composed of the linear sequence of four ribonucleotides, A, U, C and G. This RNA copy, called **messenger RNA**, is carried to the **ribosomes**, the submicroscopic structures in the cell on which amino acids are assembled to form enzymes and other proteins. In separate processes, the amino acids are activated and attached to another kind of RNA termed **transfer RNA.** This contains a sequence of three nucleotides, called an **anticodon** which binds the amino acid-transfer RNA combination to the appropriate codon on the messenger RNA.

The specificity of any protein, its physical and enzymatic properties, depends on the linear sequence of the amino acids that comprise it. There are some 20 different amino acids, and each protein molecule is composed of several hundred or more amino acids. All, or nearly all, of the various kinds of amino acids are present in each protein. At each step in the transfer of information from DNA to messenger RNA and in the alignment of messenger and transfer RNAs, the process depends upon the specific attraction between complementary pairs of purines and pyrimidines mediated by specific but rather weak **hydrogen bonds**. Thus the biosynthesis of any specific protein involves a specific template and the formation of specific hydrogen bonds between complementary purine and pyrimidine nucleotide pairs. We can summarize this as: the DNA gene, with a four-letter code, located in the chromosome in the nucleus of the cell → messenger RNA, with a four-letter code, made in the nucleus by transcribing (copying) the code of the genes → specific protein, an enzyme or other protein, with its specificity residing in the sequence of the amino acids that are present in its peptide chain. The amino acids comprise a 20-letter code and are assembled on the ribosomes from activated amino acids.

SECTION 2–10

METABOLIC PROCESSES ARE MEDIATED BY ENZYMES

One of the characteristics of all living things is their ability to metabolize, to carry on a great variety of chemical reactions. Our present-day generalizations about metabolism had their beginnings in 1780, when Lavoisier and LaPlace concluded that respiration is a form of combustion. They reached this conclusion from the results of simple experiments comparing the utilization of oxygen and the production of carbon dioxide by animals and by candles kept in bell jars (Fig. 2–9), even though their thinking was hindered by the then-current, erroneous "phlogiston" theory.

The concept that metabolism in all living organisms is mediated by specific organic catalysts, **enzymes**, synthesized by living cells, has gradually been crystallized since 1815, when Kirchhoff prepared an extract of wheat which would convert starch to sugar. A long argument between Liebig and Pasteur as to whether enzymes (or "ferments" as they were called then) themselves were living was resolved in Liebig's favor in 1897 when Eduard Büchner prepared a cell-free extract of yeast which would convert sugar to alcohol. Intensive research in enzymology has resulted in the isolation of many enzymes, the demonstration that they are all large protein molecules, and the generalization that each one controls a specific kind of chemical reaction because of the specific configuration of the enzyme molecule. The substance undergoing a chemical reaction (the **substrate**) unites with the enzyme to form a specific enzyme-substrate complex. In this way enzymes control the speed and specificity of essentially all the chemical reactions of living things.

The metabolic reactions of a wide variety of living things—animals, green plants, bacteria and molds—have been found to be remarkably similar in many respects. The continuation of life requires the expenditure of energy, and the ultimate source of the energy used by all living things is sunlight. This energy, captured by green plants in the process of photosynthesis, is made available to the plants in further reactions. Some of the energy may eventually be used by the animals that eat the plants, or by animals that eat the animals that ate the plants.

Metabolic processes are regulated so as to maintain the internal environment of the cell as constant as possible. This tendency toward constancy is called **homeostasis.** Changes in the external environment tend to produce comparable changes in the internal environment of the cell. Extreme changes in the internal environment lead to the death of the cell. Living organisms have a host of elaborate, complex devices that tend to resist the effects of these changes and thus keep the internal environment constant. Many of these devices involve the principle of **"feed-back" control** in which the accumulation of the product of a reaction leads to a decrease in its rate of production or a deficiency of the product leads to an increase in its rate of production. In the course of evolution, higher organisms have developed a greater degree of homeostatic control than the lower ones had.

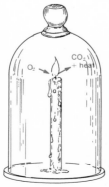

FIGURE 2–9   *A comparison of the utilization of oxygen and the production of carbon dioxide by a burning candle (combustion) and by an animal (respiration).*

SECTION 2–11

METABOLIC REACTIONS ARE UNDER GENIC CONTROL

One of the major biological generalizations is the "one gene—one enzyme—one reaction" hypothesis stated by George Beadle and Edward Tatum in 1941. According to this widely accepted theory, each biochemical reaction concerned with the development and maintenance of a particular organism is controlled by a particular enzyme, and the enzyme, in turn, is controlled by a single gene. A change (*mutation*) in the gene will result in an alteration or deficiency of the enzyme, a consequent alteration in the nature or rate of a particular metabolic step, and some particular change in the development of the organism. Thus this theory provides a basis for understanding the relationship of a given gene and its specific trait.

SECTION 2–12

CELLULAR DIFFERENTIATION RESULTS FROM THE DIFFERENTIAL ACTIVITY OF THE SAME SET OF GENES IN DIFFERENT CELLS

The mitotic process ensures the exact distribution of genes to each daughter cell in the organism; however, the various tissues of a multicellular organism do have quantitative and even qualitative differences in the pattern of enzymes and other proteins present. Thus the differences in the pattern of proteins found in different cells must arise by differences in the activity of the same set of genes in different cells (Fig. 2–10). The turning on or off of the synthesis of a specific protein could occur by some process regulating the transcription of DNA to form messenger RNA, by some process involving the combination of messenger RNA with the ribosome, or by some process in the transformation of the ultimate protein product. Each kind of messenger RNA has a half-life ranging from a few minutes in certain microorganisms to 12 or 16 hours in man and other mammals. Each molecule of RNA template probably can serve to direct the synthesis of many molecules of its protein, but eventually the template RNA is degraded and must be replaced. This provides a means by which a cell can alter the kind of protein that it synthesizes as new types of messenger RNA replace the previous ones. Thus the cell can respond to exogenous stimuli and produce new types of enzymes and other proteins. It has been suggested that the DNA of the genes that are not being transcribed at any given moment are "silent," bound to some kind of protein which makes the DNA unavailable for the transcription system. The process of cellular differentiation is not yet understood and is one of the major unsolved problems in present-day biology.

SECTION 2–13

VITAMINS ARE PRECURSORS OF COENZYMES

The discovery that substances other than salts, proteins, fats and carbohydrates are needed for adequate nutrition—substances called accessory food factors by F. G. Hopkins and *vitamins* by Casimir Funk in

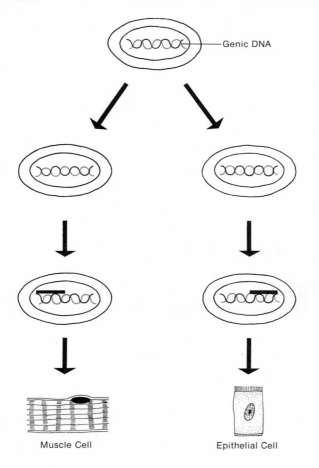

Genic DNA

Muscle Cell

Epithelial Cell

FIGURE 2–10   *The problem of cellular differentiation. The differences between the components of a muscle cell and of a nerve cell must arise in some way by differences in the activities of the same set of genes in the two kinds of cells. The diagram postulates that some genes are bound to protein and hence are "silent," whereas others are free and available for transcription to produce the specific protein characteristic of that cell type.*

1911—stimulated investigations into the role these substances play in metabolism and why they are needed in the diet of some organisms and not others. The generalization is now amply established that these substances are necessary for the metabolic processes of all organisms—bacteria, green plants and animals. Many organisms, however, are able to synthesize all of these substances that they require; the ones that cannot synthesize these materials must obtain them in their diet.

The specific roles in metabolism of many of these vitamins are now known. In each instance they are known to become a part of a larger molecule which functions as a **coenzyme**, a partner of the enzyme and substrate which is absolutely necessary for some particular reaction or reactions to occur. The deficiency diseases caused by the lack of the vitamins reflect the impaired metabolism caused by the deficient coenzyme.

SECTION 2–14

HORMONES REGULATE CELLULAR ACTIVITIES

The term **hormone** was originated by the British physiologist, E. H. Starling, in 1905 and was defined as "any substance normally produced in the cells of some part of the body and carried by the bloodstream to distant parts, which it affects for the good of the body as a whole." The science of endocrinology can be said to have begun in 1849, when, on the basis of experiments in which testes were transplanted from one bird to another, Berthold postulated that these male sex glands secrete some blood-borne substance essential for the differentiation of the male secondary sex characteristics. This substance, testosterone, was finally isolated and synthesized in 1935.

Our rapidly increasing knowledge of the many different hormones produced by vertebrate and invertebrate animals and by plants has led to the generalization that these special chemical substances, produced by some restricted region of an organism, diffuse or are transported to another region where they are effective in very low concentrations in regulating and coordinating the activities of the cells. Hormones thus provide for chemical control and coordination which complement and supplement the coordination controlled by the activities of the nervous system.

SECTION 2–15

INTERRELATIONS OF ORGANISMS AND ENVIRONMENT

The last major generalization we shall consider, and one of the major unifying concepts of biology today, comes from the field of ecology. From detailed studies of communities of plants and animals in a given area, the generalization has been made that all the living things in a given region are closely related with each other and with the environment. This generalization includes the idea that particular kinds of plants and animals are not found at random over the earth but occur in interdependent **communities** of producer, consumer and decomposer organisms together with certain nonliving components. These communities can be recognized and characterized by certain dominant members of the group, usually plants, which provide both food and shelter for many other forms. Why certain plants and animals comprise a given community, how they interact, and how man can control them to his own advantage are major research problems in ecology.

The list of biological principles given here is not intended to be exhaustive, but rather to emphasize the fundamental unity of biological science and the many ways in which living things are related and interdependent. These generalizations have been derived from careful observations and experiments made by many biologists over a long period of time. All of them have been tested repeatedly, and many have been revised as new information became available as the result of discoveries made with the aid of newer techniques such as the electron microscope, radioactive isotopes for tracer studies, and the many other physical and chemical methods which are being used in biological research. Future studies may result in further revisions of some of these principles.

SUGGESTIONS FOR FURTHER READING

Bonner, John T.: *The Ideas of Biology.* New York, Harper & Row, Publishers, 1962.
Several of the theories discussed in this chapter are presented more extensively in this book.

Gabriel, Mordecai L., and Fogel, Seymour: *Great Experiments in Biology.* Englewood Cliffs, New Jersey, Prentice-Hall, Inc., 1955.
> Presents some of the important ideas in biology by means of extensive quotations from original papers. The well chosen examples are accompanied by explanatory notes and by chronological tables of key developments in various areas of biology.

Schroedinger, E.: *What is Life?* New York, Cambridge University Press, 1944.
> A challenging look by a physicist at some of the basic concepts of biology.

Wightman, W. P. D.: *The Growth of Scientific Ideas.* New Haven, Connecticut, Yale University Press, 1951.
> Traces the development of ideas in science, with some especially interesting presentations of biological concepts.

CELL AND MOLECULAR BIOLOGY

CELLULAR STRUCTURE AND PROCESSES

Since biology is usually defined as the "science of living things" we must at the outset be able to differentiate "living" from "nonliving". We use the word **organism** to refer to any living thing, plant, animal or bacteria. It is relatively easy to see that a man, an oak tree, a rosebush, a lion and an earthworm are living, whereas rocks and stones are not. But whether such things as viruses are "alive" depends upon our precise definition of life.

The living material of essentially all organisms exists in discrete portions termed **cells**. These are the microscopic units of structure of the body, just as bricks may be the units of structure of a house. But cells are more than mere building blocks. Each one is an independent functional unit and the processes of the body are the sum of the coordinated functions of its cells. The cellular units may vary considerably in size, shape and function. Some of the smallest organisms have bodies made of a single cell. Others, such as a man or an oak tree, are composed of countless billions of cells fitted together.

In 1839, the Bohemian physiologist Purkinje introduced the term **protoplasm** for the living contents of the cell. In recent years, our increasing knowledge of cell structure and function has revealed that the living contents of the cell comprise an incredibly complex system of interdependent heterogeneous parts. The term "protoplasm" has no clear meaning in any chemical or physical sense, but it may still be used to refer to all of the organized constituents of a cell.

To get an idea of what protoplasm looks like, we may examine some simple organism, such as an amoeba or a slime mold, in which the living substance is naked and readily visible under the microscope. The living substance of such organisms is colorless, or perhaps faintly yellow, pink or green, and is translucent. It has a thick, viscid, syrupy consistency and would feel slimy to the touch. When seen in the light microscope it may appear to have granules or fibrils of denser material, droplets of fatty substances, or fluid-filled vacuoles, all suspended in the clear, continuous, semifluid "ground substance" (Fig. 3–1). However, when viewed in the electron microscope there is revealed a remarkable structural complexity in what had appeared in the light microscope to be a more or less homogeneous matrix (Fig. 3–1). The membranes of the cell and the various sub-

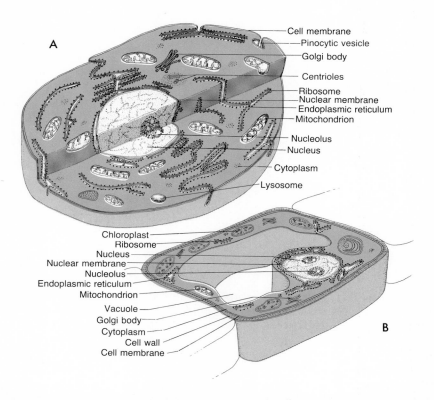

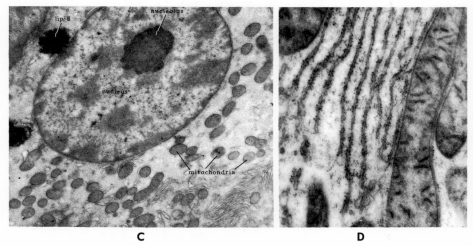

FIGURE 3–1   *The structure of cells. A, Diagram of a typical animal cell. B, Diagram of a typical plant cell. C, Electron micrograph of the nucleus and surrounding cytoplasm of a liver cell of a frog. Spaghetti-like strands of endoplasmic reticulum are visible in the lower right corner. Magnified, 16,500×. D, Electron micrograph of a rat liver cell shown at higher magnification. Granules of ribonucleoprotein are seen on the strands of endoplasmic reticulum and structures with double membranes are evident within the mitochondria in the upper left corner and on the right. Magnified, 65,000× (C and D, courtesy of Dr. Don W. Fawcett.)*

FIGURE 3–2　　　Codonella companella, a single-celled animal with a high degree of specialization of form and function within the single cell.

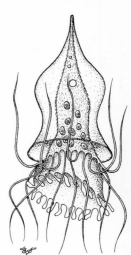

cellular constituents have been shown by x-ray diffraction analyses to have an even finer structure, one which can be related to the structure of the large molecules comprising them.

THE CHARACTERISTICS OF CELLS

Each cell contains a **nucleus** and is surrounded by a **plasma membrane.** Mammalian red blood cells and phloem sieve tube cells lose their nucleus as they mature, and skeletal muscles and many algae and fungi have several nuclei per cell, but these are rare exceptions to the general rule of one nucleus per cell. In the simplest plants and animals, all the living material is found within a single plasma membrane. Such organisms may be considered to be unicellular (i.e., single-celled) or acellular (i.e., with bodies not divided into cells). However, they may have a high degree of specialization of form and function within this single cell (Fig. 3–2). The cell may be quite large, larger than the whole body of other multicellular organisms. Thus it would be incorrect to infer that a single-celled organism is necessarily smaller or less complex than a many-celled one.

The cells of different plants and animals, and of different organs within a single plant or animal, present a bewildering variety of sizes, shapes, colors and internal structures, but all have certain features in common. Each cell is surrounded by a plasma membrane, and contains a nucleus, plus several kinds of **subcellular organelles**. These include mitochondria, granular endoplasmic reticulum, smooth endoplasmic reticulum, the Golgi complex, lysosomes and centrioles.

All organisms and their constituent cells have, to a greater or lesser extent, the properties of specific size and shape. They carry on metabolic reactions, they are able to move, they exhibit irritability, they grow, they reproduce, and they adapt to changes in the external environment. Although this list may seem specific and definite, the line between living and nonliving is rather tenuous. Viruses, for example, exhibit some but not all of the usual characteristics of living things. When we realize that we cannot really answer the question of whether they *are* living or nonliving but only the question of whether they should be *called* living or nonliving, the problem is put into proper perspective. Nonliving objects may show one or more of these properties but not all of them. Crystals in a saturated solution may "grow," a bit of metallic sodium will move rapidly over the surface of water, and a drop of oil floating in glycerol and alcohol may send out pseudopods and move like an amoeba.

Each kind of living organism is recognized by its characteristic shape and appearance; the adult of each kind of organism typically has a characteristic size. In contrast, nonliving things generally have much more variable shapes and sizes. Living things are not homogeneous but are made of different parts, each with special functions. Thus the bodies of living things are characterized by a specific complex organization. The structural and functional unit of all organisms is the cell. The cell in turn has a specific organization, and each type of cell has a characteristic size and shape by means of which it can be recognized.

The sum of all the chemical activities of the cell which provide for its growth, maintenance and repair is called **metabolism**. All cells are constantly changing by taking in new substances, altering them chemically in a variety of ways, building new cellular materials, and transforming the potential energy present in molecules of carbohydrates, fats and proteins into kinetic energy and heat as these substances are converted into simpler ones. This constant expenditure of energy is one of the unique and characteristic attributes of living things. Some types of cells, bacteria for example, have high metabolic rates. Other cells, such as seeds and spores, have rates of metabolism which are barely detectable even with the most sensitive instruments. Even within a particular species or person, metabolic rates may vary depending on factors such as age, sex, general health, the amount of endocrine secretion and pregnancy.

It is customary to divide metabolic processes into anabolic and catabolic ones. The term **anabolism** refers to those chemical processes in which simpler substances are combined to form more complex substances, resulting in the storage of energy, the production of new cellular materials, and growth. Catabolism refers to the breaking down of these complex substances resulting in the release of energy and the wearing out and using up of cellular materials. In fact, both types of processes occur continuously and are intimately interdependent and difficult to distinguish. Complex substances may be broken down and their parts recombined in new ways to form different complex substances. The interconversions of carbohydrates, proteins and fats that occur continuously in human cells are examples of a combination of catabolic and anabolic processes. Anabolic processes, in general, require energy, therefore some catabolic processes must occur to supply the energy to drive the reactions involved in building up the new molecules. Green plant cells have the ability to manufacture their own organic compounds out of inorganic materials in the soil and air. Animals must depend on plants ultimately for their food. Plant cells are simply better chemists than are animal cells.

Living things are characterized further by their ability to move. The movement of most animals is quite obvious. They wiggle, crawl, swim, run or fly. The movements of plants are much slower and less obvious, but they occur nonetheless. A few animals—sponges, corals, oysters and certain parasites—do not move from place to place, but most of these have cilia or flagella to move their surroundings past their bodies and thus bring food and other necessities of life to themselves. Movement may result from muscular contraction, from the beating of cilia or flagella, or from the slow oozing of a mass of cell substance termed **amoeboid motion** (Fig. 3–3). The streaming motion of the living material in the cells of the leaves of plants is termed **cyclosis**.

Living things respond to stimuli, to physical or chemical changes in their immediate surroundings. They are thus said to be irritable. Stimuli which are effective in evoking a response in most animals and plants are changes in the color, intensity or direction of light; changes in temperature, pressure or sound; and changes in the chemical composition of the earth, water or air surrounding the organism. Certain cells of the body may be highly specialized to respond to certain types of stimuli. The rods and cones in the retina of the eye respond to light, certain cells in the nose and in the taste buds of the tongue respond to chemical stimuli, and special cells in the skin respond to changes in the temperature or pressure. In lower animals, and in plants, no such specialized cells may be present, but the whole organism responds to stimuli (Fig. 3–4). Single-celled organisms will respond by moving toward or away from heat, cold, certain chemical substances, light or the touch of a microneedle.

The irritability of plant cells may not be as obvious as that of animal cells, but they are sensitive to changes in their environment. The streaming movement in plant cells may be speeded or stopped by changes in the

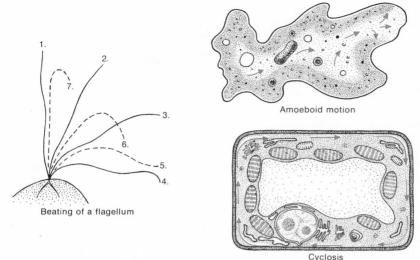

Amoeboid motion

Beating of a flagellum

Cyclosis

FIGURE 3-3   *Diagram illustrating several types of cellular movements. A, Ciliary motion. B, Cyclosis. C, Amoeboid motion.*

FIGURE 3-4   *A, The sensitive plant Mimosa pudica before being disturbed. B, The plant five seconds after being touched; note that the leaves have folded and drooped. (Courtesy of the General Biological Supply House, Chicago.)*

amount of light. A few plants, such as the Venus flytrap of the Carolina swamps have a remarkable sensitivity to touch and can catch insects (Fig. 3–5). Their leaves are hinged along the midrib, and the edges of the leaves are covered with hairs. The presence of an insect on the leaf stimulates it to fold, the edges come together, and the hairs interlock to prevent the escape of the prey. The leaf subsequently secretes a material which kills and digests the insect. The development of flytrapping is an adaptation which enables these plants to obtain part of the nitrogen they require for growth from the prey that they eat.

Cellular growth, defined as an increase in cellular mass, may result from an increase in the *size* of individual cells or from an increase in the *number* of cells, or both. An increase in cell size may occur simply by the uptake of water, but such swelling is generally not considered to be true growth. The term growth is restricted to those processes which increase the amount of living substance of the body measured by the amount of nitrogen or protein present. Why do you suppose nitrogen or protein is used as the yardstick rather than the amount of carbohydrate, fat, sulfur or sodium? Growth may be uniform in several parts of an organism, or it may be greater in some parts than in others so that the proportions of the body change as growth occurs. Many organisms will grow indefinitely. Other organisms have a definite growth period which terminates in an adult of a characteristic size. One of the remarkable aspects of the entire growth process is that each cell and each organ continues to function while undergoing growth.

If there is any one characteristic that can be said to be the *sine qua non* of life, it is the ability to reproduce. The simplest viruses do not metabolize, move or grow, yet because they can reproduce and undergo mutations, most biologists today would regard them as living. Since all life comes

FIGURE 3–5　　The Venus flytrap Dionaea. Top, a plant with leaves in various stages of closing in response to stimuli. Bottom, enlarged view of a single leaf showing central rib and the hairs along the edges of the leaf. (Courtesy of the Carolina Biological Supply Company.)

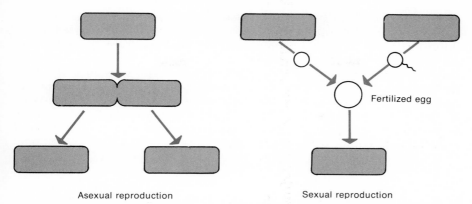

Asexual reproduction   Sexual reproduction

FIGURE 3–6   A comparison of asexual reproduction, in which one individual gives rise to two or more offspring, and sexual reproduction in which two parents each contribute a gamete which join to give rise to the offspring.

only from living things and not by spontaneous generation, this ability to reproduce is a most important characteristic of living things.

The process of reproduction may be as simple as the splitting of one individual into two (Fig. 3–6). In most organisms, however, it involves the production of specialized eggs and sperm which unite to form the fertilized egg or **zygote** from which the new organism develops. Reproduction in certain parasitic worms involves several quite different forms, each of which gives rise to the next in succession until the cycle is completed and the adult reappears.

The ability of a plant or animal to adapt to its environment is the characteristic which enables it to survive the exigencies of a changing world. Each species can become adapted by seeking out an environment to which it is suited or by undergoing modifications to make it better fitted to its present surroundings. Adaptation may involve immediate changes which depend upon the irritability of cells, or it may be the result of a long-term process of mutation and selection. It is clear that no single kind of plant or animal can possibly adapt to all the conceivable kinds of environment. Hence, there will be certain areas of the earth where it cannot survive. The list of factors that may limit the distribution of the species is almost endless: water, light, temperature, food, predators, competitors, parasites and so on.

SECTION 3–2

EXCHANGES OF MATERIALS BETWEEN CELL AND ENVIRONMENT

The outer surface of each cell is bounded by a delicate, elastic covering, an integral functional part of the cell termed the **plasma membrane**. This is of prime importance in regulating the contents of the cell, for all nutrients entering the cell and all waste products or secretions leaving it must pass through this membrane. It hinders the entrance of certain substances and facilitates the entrance of others. Cells are almost invariably surrounded by a watery medium. This might be the fresh or salt water in which a small organism lives, the tissue sap of a higher plant, or the plasma or extracellular fluid of the higher animals.

The plasma membrane behaves as though it has ultramicroscopic pores through which certain substances pass. The size of these pores determines the maximum size of the molecule that can pass through. Factors other than molecular size, such as the electric charge, if any, carried by the diffusing particles, the number of water molecules, if any, bound to the surface of the diffusing particle, and the solubility of the material in lipids, may also be important in determining whether the substance will pass through the membrane. The exact chemical and physical nature of the membrane is not yet completely clear, but it appears to be a three-layer sandwich about 120 Å units thick. The inner and outer layers, each some 30 Å units thick, are protein and surround a middle layer of phospholipid molecules 60 Å units thick (Fig. 3–7).

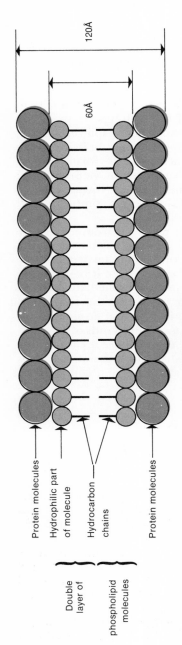

120Å

60Å

Protein molecules

Hydrophilic part
of molecule

Hydrocarbon
chains

Protein molecules

Double
layer of

phospholipid
molecules

FIGURE 3–7   Diagram representing one widely accepted hypothesis of the molecular architecture of biological membranes such as the plasma membrane surrounding the cell or the membrane around the mitochondria. An inner and outer layer of protein molecules some 30 Å thick surround a middle double layer of phospholipid molecules some 60 Å thick.

High resolution electron micrographs of the plasma membrane show a three-layered structure. It is of interest that the plasma membranes of animal, plant and bacterial cells and the membranes of a great many sub-cellular organelles all appear to have a similar three-layered structure. This pattern of protein-lipid-protein in membranes, termed the **unit membrane**, now appears to be a fundamental structural unit of widespread occurrence.

Nearly all plant cells (but not most animal cells) have a thick **cell wall** made of cellulose, which lies outside the plasma membrane. This cell wall is not living and is secreted by the cell substance. It is pierced in many places by tiny holes through which the contents of one cell may connect with those of the adjacent cells and through which materials may pass from one cell to the next. These tough, firm cell walls provide support for the plant body.

To understand the mechanisms by which materials are exchanged between cell and environment, we must appreciate first that all molecules in liquids and gases characteristically tend to move or diffuse in all directions until they are spread evenly throughout the available space. **Diffusion** may be defined as the movement of molecules from a region of high concentration to one of lower concentration, brought about by the kinetic energy of the molecules (Fig. 3–8). The rate of diffusion is a function of the size of the molecule and the temperature. The molecules that make up all kinds of substances, even solids, are constantly in motion. The chief difference between the three states of matter — solid, liquid and gas — is simply the freedom of movement of the molecules present. The molecules of a solid are relatively closely packed and the forces of attraction between molecules will allow them to vibrate but not to move around. In the liquid state the molecules are further apart, the intermolecular forces are weaker, and the molecules move about with considerable freedom. Finally, in the gaseous state, the molecules are so far apart that intermolecular forces are negligible and the movement of the molecules is restricted only by external barriers.

When a drop of water is examined under the microscope, the motion of the water molecules is not evident, but if a drop of India ink (which contains fine particles of carbon) is added, the carbon particles move about continually in aimless and zigzag paths. Each carbon particle is constantly being bumped by water molecules, and the recoil from these bumps gives the carbon particle its motion. This motion of small particles is termed **Brownian movement** after Robert Brown, an English botanist, who first observed it when he looked through the microscope at some tiny pollen grains in a drop of water.

In the process of diffusion, each individual molecule moves in a

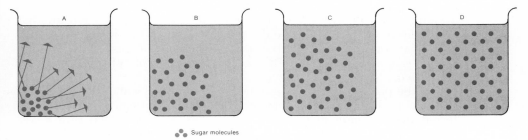

••• Sugar molecules

FIGURE 3–8   *The process of diffusion. When a small lump of sugar is dropped into a beaker of water its molecules dissolve (left) and begin to diffuse. The process of diffusion over a long period of time will result in an even distribution of sugar molecules throughout the water in the beaker (right).*

straight line until it bumps into something—another molecule or the side of the container—then it rebounds and moves in another direction. The molecules continue to move even when they have become uniformly distributed throughout a given space; however as fast as some molecules move, for example, from left to right, others move from right to left, so that an equilibrium is maintained. Any number of substances will diffuse independently of each other within the same solution. Individual molecules may move at a rate of several hundred meters per second, but each molecule can go only a fraction of a millimicron before bumping into another molecule and rebounding. Thus the progress of any given molecule in a straight line is quite slow. This can be demonstrated by placing a bit of dye at the bottom of a glass cylinder filled with water. As days and weeks go by, the colored substance will gradually move upwards, but it will take months before the dye is uniformly distributed throughout the cylinder. Thus, although diffusion occurs very rapidly over microdistances, it takes a long time for a molecule to travel distances measured in centimeters.

This fact has important biological implications, for it limits the number of molecules of oxygen and nutrients that will reach an organism by diffusion alone. Only a very small organism that needs relatively few molecules per second can survive if it sits in one place and lets molecules come to it by diffusion. A larger organism must either have some means of moving to a new region or some mechanism for stirring up its environment to bring molecules to it. As a third alternative, some organisms live where the environment is constantly moving past it—in a river or in the intertidal zone on the seashore. The larger land plants, shrubs and trees have solved the problem by developing a tremendously branched root system, thus obtaining their raw materials from a large area of the surrounding environment.

Whether or not a membrane will permit the molecules of a certain substance to pass through depends on its structure and on the size of the pores present. A membrane is said to be *permeable* if it will permit any substance to pass through, *impermeable* if it will permit no substance to pass, and *differentially permeable* if it will allow some but not all substances to diffuse through. Permeability, it must be emphasized, is a property of the membrane, not of the diffusing substance. All the membranes surrounding cells, nuclei, vacuoles and subcellular structures are differentially permeable.

The diffusion of a dissolved substance through a differentially permeable membrane is termed *dialysis*. To demonstrate the process of dialysis, one can make a pouch of collodion, cellophane or parchment and fill it with a sugar solution. The pouch is then placed in a beaker of water, and the sugar molecules will dialyze through the membrane, if the pores are large enough. After the passage of time, the concentration of sugar molecules in the water outside the pouch will equal that within the pouch. Molecules continue to diffuse but there is no net change in concentration, for the rate of diffusion into the pouch equals the rate of diffusion out of the pouch.

If the pores of the membrane of the pouch are somewhat smaller, so that it is permeable to water molecules but not to sugar molecules, a different phenomenon may be observed. The pouch is fitted with a cork stopper through which passes a glass tube. When this pouch, filled with sugar solution, is placed in a beaker of water, the sugar molecules will be unable to penetrate the membrane and thus remain inside the bag (Fig. 3–9). The water molecules, however, diffuse through the membrane into the pouch. The liquid inside the membrane is 5 per cent sugar, hence it

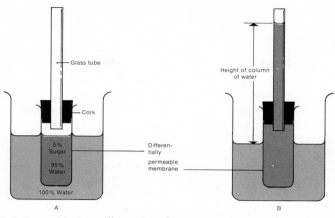

FIGURE 3–9   *Diagram illustrating the process of osmosis. A, When a 5-per cent sugar solution is placed in a sac made of a differentially permeable membrane such as cellophane and suspended in water, the water molecules diffuse into the sac causing the column of water in the glass tube to rise. The larger glucose molecules are unable to pass through the pores in the cellophane. B, When equilibrium is reached, the pressure of the column of water in the tube just equals, and is a measure of, the osmotic pressure of the sugar solution.*

is only 95 per cent water. The liquid outside the membrane is 100 per cent water; therefore, water molecules move from a region of higher concentration (100 per cent, outside the pouch) to a region of lower concentration (95 per cent, inside the pouch). The diffusion of water or solvent molecules through a membrane is termed **osmosis.**

As osmosis continues, water rises in the glass tube. If an amount of water equal to the amount originally inside the pouch were to pass through the membrane, the sugar solution would be diluted to be 2.5 per cent sugar and 97.5 per cent water. The concentration of water on the outside of the pouch would still be higher than that inside, and osmosis will continue. Eventually the water in the glass tube will rise to a height such that the weight of the water in the tube exerts a pressure just equal to that resulting from the tendency of the water molecules to enter the bag. Then there will be no further *net* change in the amount of water within the bag. Osmosis will continue to occur in both directions through the differentially permeable membrane with equal speed.

The pressure of the column of water is termed the **osmotic pressure** of the sugar solution. The osmotic pressure is brought about by the tendency of the water molecules to pass through the membrane and equalize the concentration of water molecules on the two sides. A more concentrated sugar solution would have a greater osmotic pressure, and thus would cause water to rise to a higher level in the tube. A 10 per cent sugar solution would cause water to rise approximately twice as high in the tube as a 5 per cent solution.

Dialysis and osmosis are simply two special forms of diffusion. Diffusion is the general term for the movement of molecules from a region of high concentration to one of lower concentration, resulting from their kinetic energy. Dialysis is the diffusion of dissolved molecules (solutes) through a differentially permeable membrane, and osmosis is the diffusion of solvent molecules through a differentially permeable membrane. In living systems, the solvent molecules are water.

The fluid compartment of every living cell contains dissolved salts, sugars and other substances that give that fluid a certain osmotic pressure. When this cell is placed in a fluid with the same osmotic pressure, there is no net movement of water molecules either into or out of the cell. There is no tendency for the cell either to swell or to shrink. This fluid is said to be **isotonic** or **isosmotic** to the fluid within the cell. Normally the blood plasma and all the body fluids are isotonic. They contain the same concentration of dissolved materials as the cells.

If the concentration of dissolved substances in the surrounding fluid is greater than the concentration within the cell, water tends to pass out of the cell and the cell shrinks. Such a fluid is termed **hypertonic** to the cell. If the surrounding fluid has less dissolved material than the cell, it is said to be **hypotonic**. Water will tend to enter the cell and cause it to swell. A solution of 0.9 per cent sodium chloride, sometimes termed "physiologic saline," is isotonic to the cells of human beings. Red blood cells placed in a solution of 0.6 per cent sodium chloride will swell and burst (Fig. 3–10), and those placed in a 1.3 per cent solution will shrink, but those placed in 0.9 per cent sodium chloride will neither swell nor shrink.

A cell placed in a solution that is not isotonic to it may adjust to the changed environment by undergoing a change in its water content (by swelling or shrinking), so that it finally reaches the same concentration of solutes as its surrounding fluid. Some cells have the ability to pump water or certain solutes in or out through the plasma membranes, and in this way can produce an osmotic pressure that differs from that of the surrounding medium. The protozoa living in pond water, which is very hypotonic, have evolved **contractile vacuoles**, which collect water from the interior of the cell and pump it to the outside. Plants living in fresh water also have the problem of dealing with the water that enters the cell by osmosis from the surrounding hypotonic environment. Plant cells have no contractile vacuole to pump out the water but have a firm cellulose wall that surrounds the cell and prevents undue swelling. As water enters, an internal pressure, called **turgor pressure**, is generated which counterbalances the osmotic pressure and prevents the entrance of additional water molecules. Turgor pressure is characteristic of plant cells in general and is responsible in part for the support of the plant body. A flower wilts when the turgor pressure in its cells has decreased owing to the lack of water.

Many organisms that live in the sea have phenomenal powers to accumulate selectively certain substances from the sea water. Seaweeds can accumulate iodine so that the concentration within the cell is 2,000,000 times that of the sea water. Tunicates, primitive chordates, can accumulate vanadium so that it is also some 2,000,000 times as concentrated within

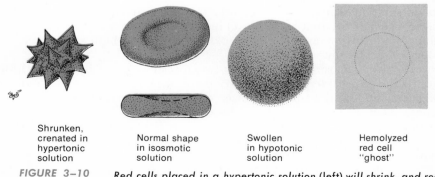

| Shrunken, crenated in hypertonic solution | Normal shape in isosmotic solution | Swollen in hypotonic solution | Hemolyzed red cell "ghost" |

FIGURE 3–10   *Red cells placed in a hypertonic solution (left) will shrink, and red cells placed in a hypotonic solution (right) will swell and burst. Red cells placed in an isosmotic solution (center) will neither swell nor shrink.*

the tunicate cells as it is in the sea water. The pumping of water or solutes in or out of the cell against the concentration gradient is physical work and requires the expenditure of energy. The cell is able to move molecules against the gradient only as long as it is alive and carrying on metabolic activities that yield energy. If the cell is treated with some metabolic poison, such as cyanide, it loses its ability to produce and maintain differences in concentration on the two sides of its plasma membrane.

SECTION 3–3

THE CELL NUCLEUS

Each cell contains a small, usually spherical or oval, organelle known as the **nucleus**. In some cells the nucleus has a relatively fixed position, usually somewhere near the center. In others it may move around freely and be found almost anywhere in the cell. The nucleus is an important center for the control of cellular processes. It contains the hereditary factors or genes responsible for controlling the traits of the cell and the organism, and it directly or indirectly controls many aspects of cellular activity. The nucleus is separated from the surrounding cytoplasm by a double layer of unit membranes, the **nuclear membrane**, which regulates the flow of materials into and out of the nucleus. The electron microscope reveals that the nuclear membrane is double-layered and that there are pores in the double membrane (Fig. 3–11) through which the nuclear contents are continuous with the cytoplasm and through which informational macromolecules may pass. The outer of the two layers of the nuclear membrane appears to be continuous with the membranes of the endoplasmic reticulum and the Golgi complex.

When a cell is killed by fixation in the proper chemicals and stained with appropriate dyes, several structures within the nucleus become visible. These are difficult to observe in the living cell with an ordinary light microscope, but are evident by the use of a phase microscope. Within the semifluid ground substance, termed the karyoplasm, are suspended a fixed number of extended, linear, thread-like bodies called **chromosomes**, composed of DNA and protein, and containing the hereditary units or genes. In a stained section of a nondividing cell (Fig. 3–12), the chromosomes usually appear as an irregular network of dark-staining strands and granules termed **chromatin**. Just prior to nuclear division these strands condense into compact, rod-shaped chromosomes which are subsequently distributed to the two daughter cells in exactly equal numbers. Each type of organism has a characteristic number of chromosomes present in each of its constituent cells. The fruit fly has eight chromosomes, sorghum has 10, the garden pea, 14, corn, 20, the toad, 22, the tomato, 24, the cherry, 32, the rat, 42, man, 46, the potato, 48, the goat, 60 and the duck, 80. The somatic cells of higher plants and animals each contain two of each kind of chromosome. A cell with two complete sets of chromosomes is said to be **diploid**. Sperm and egg cells, which have only one of each kind of chromosome, one full set of chromosomes, are said to be **haploid**. They have just half as many chromosomes as the somatic cells of that same species. When the egg is fertilized by the sperm the two haploid sets of chromosomes are joined and the diploid number is restored.

The **nucleolus**, a spherical body found within the nucleus, is extremely variable in most cells, appearing and disappearing, changing its form and structure. There may be more than one nucleolus in a nucleus, but the

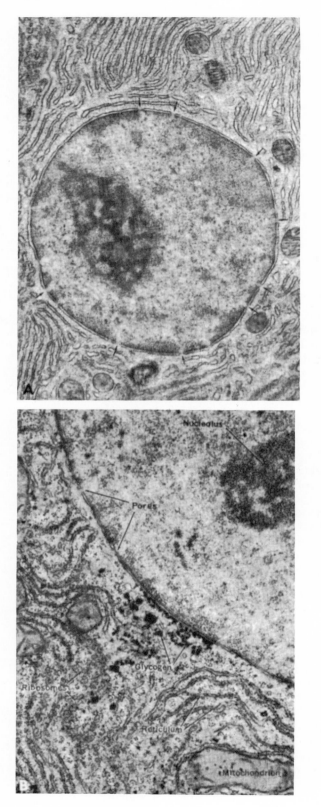

FIGURE 3–11 *Electron micrographs showing pores in the nuclear membrane. The endoplasmic reticulum is evident in both pictures; the lower one right shows ribosomes on the endoplasmic reticulum. A, magnified 20,000×; B, magnified 50,000×. (Courtesy of Drs. Don W. Fawcett and Keith R. Porter.)*

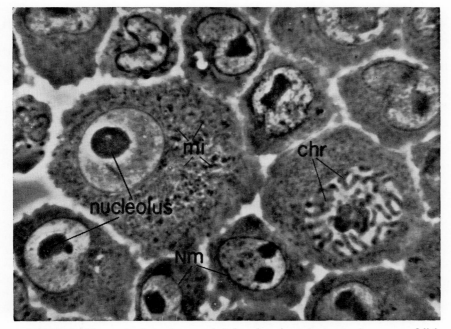

FIGURE 3–12   A photomicrograph taken by phase contrast microscopy of living cells from an ascites tumor. Chr, chromosomes; mi, mitochondria; Nm, nuclear membrane. (Courtesy of N. Takeda.)

cells of any given species of plant or of animal usually have a fixed number of nucleoli. The nucleoli disappear when a cell is about to divide, and reappear afterwards. They appear to play a role in the synthesis of the ribonucleic acid constituent of ribosomes. If the nucleolus is destroyed by carefully localized ultraviolet or x-irradiation cell division is inhibited. This does not occur in control experiments in which regions of the nucleus other than the nucleolus are irradiated.

But, you may object, what if the operation itself, not the loss of the nucleus, caused the ensuing death? We can decide this by a controlled experiment, in which we subject two groups of animals to the same operative trauma, but have them differ in the presence or absence of a nucleus. For example, we can stick a microneedle into some of the amoebas, perhaps push the needle around inside to simulate the operation of removing a nucleus, but then withdraw the needle, leaving the nucleus inside. An amoeba treated to such a "sham operation" will recover and will subsequently grow and divide, demonstrating that it was the removal of the nucleus, not the operation, that brought about the death of the first group of animals.

The role of the nucleus can be studied by removing it and observing the consequences. When we remove the nucleus of the single-celled amoeba with a microneedle, the cell will continue to live and move, but it cannot grow and will die after a few days. The nucleus, we conclude, is necessary for the metabolic processes, primarily the synthesis of nucleic acids and proteins that provide for growth and cell reproduction.

A classic series of experiments demonstrating the importance of the nucleus in controlling the growth of the cell was performed by Hämmerling, using the single-celled plant *Acetabularia mediterranea*. This marine alga, which may be as much as five centimeters long, superficially resembles a mushroom. It has roots and a stalk surmounted by a large, disc-shaped umbrella. The entire plant is a single cell and has but one nucleus located near the base of the stalk.

Hämmerling severed the stalk and found that the lower part could live and would regenerate an umbrella, recovering completely from the operation (Fig. 3–13). The upper umbrella portion, which lacked a nucleus, might live for a considerable time but eventually would die without being able to regenerate a lower part. In *Acetabularia*, as in the amoeba, the nucleus appears to be necessary for those metabolic processes underlying growth. Regeneration is, of course, a form of growth. In further experiments, Hämmerling severed the stalk above the nucleus and made a second cut just below the umbrella. The isolated section of stalk, when replaced in sea water, was able to regenerate a partial or complete umbrella. This might, at first, seem to show that a nucleus is not necessary for regeneration, however, when Hämmerling removed the second umbrella, the stalk was unable to form a third one. From such experiments, Hämmerling concluded that the nucleus produces some substance required for the formation of umbrellas which diffuses up the stalk and instigates the growth of an umbrella. In the experiments just described, enough of this umbrella material remained in the stalk after the initial cut to produce one new umbrella. After that had been exhausted in the formation of the first umbrella, no second regeneration was possible in the absence of a nucleus.

A second species of *Acetabularia* (*crenulata*) has a branched instead of a disc-shaped umbrella. Hämmerling grafted a piece of *crenulata* stalk that lacked a nucleus onto the base of a *mediterranea* plant, which contained a *mediterranea* nucleus. A new umbrella developed at the top of the stalk. However, the shape of the umbrella was dictated not by the species that supplied the stalk but by the species that supplied the base and the nucleus. The nucleus, through the activity of its genes, provided the specific information that controlled the type of umbrella that was regenerated. The nucleus can override the tendency of the stalk to form an umbrella characteristic of its own species. The nucleus controls the activities of the other portions of the cell because it has coded within its chromosomes the information needed for the synthesis of proteins and other materials on which cellular structure and function depend. Each time a cell divides, the entire set of instructions must be replicated, and a duplicate copy must be passed to each daughter cell.

SECTION 3–4

CENTRIOLES AND SPINDLES

The cells of animals and certain lower plants contain, adjacent to the nucleus, two small, dark-staining cylindrical bodies, the **centrioles**. These play a prominent role at the time of cell division, separating, migrating to opposite poles of the cell, and organizing the spindle between them. With the electron microscope, each centriole appears as a hollow cylinder with a wall in which are embedded nine parallel, longitudinally oriented groups of tubules, with three tubules in each group (Fig. 3–14). The cylinders of the two centrioles are typically oriented with their long axes perpendicular to each other.

When cell division begins, the centrioles move to opposite sides of

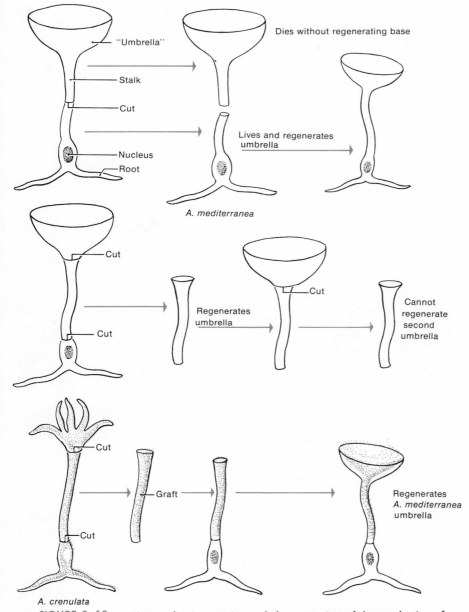

"Umbrella"

Stalk

Cut

Nucleus

Root

Dies without regenerating base

Lives and regenerates umbrella

A. mediterranea

Cut

Cut

Regenerates umbrella

Cut

Cannot regenerate second umbrella

Cut

Graft

Cut

Regenerates *A. mediterranea* umbrella

A. crenulata

FIGURE 3–13 *Hämmerling's experimental demonstration of the production of an umbrella-regenerating substance by the nucleus of the alga Acetabularia. Lower, when a stalk from* A. crenulata *is grafted on to the base (white) of an* A. mediterranea *plant, the stalk regenerates an umbrella with the shape characteristic of* A. mediterranea *plants.*

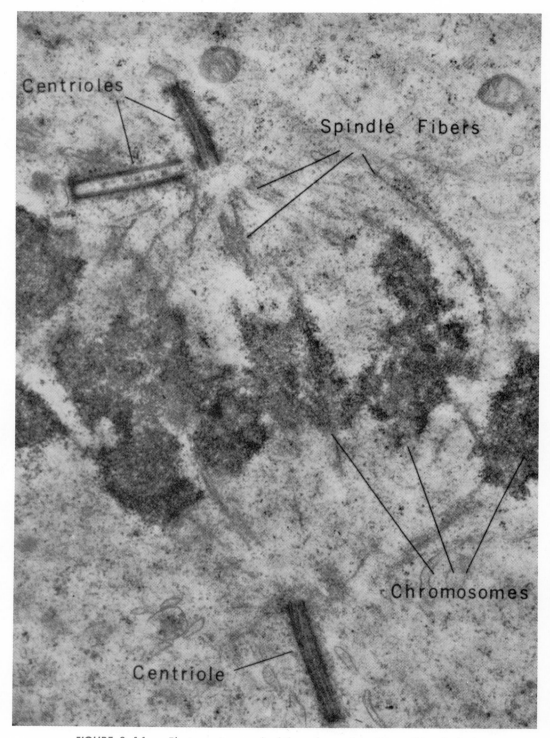

FIGURE 3-14   Electron micrograph of the polar region of the cell in mitosis showing one of the centrioles (Ce). Notice the tubular aspect of the spindle fibers which converge at the centriole. Two chromosomes and one centromere are evident at the bottom of
the figure. Magnification, 80,000×.

the cell. From each centriole there extends a cluster of ray-like filaments called an **aster**, and between the separating centrioles a **spindle** forms, composed of protein threads with properties similar to those of the contractile protein in muscle, **actomyosin**. The protein threads of this spindle are arranged like two cones placed together, base to base, narrow at the ends or poles near the centrioles and broad at the equator of the cell. The spindle fibers stretch from equator to pole and comprise a definite structure. One can introduce a fine needle into a cell and push the spindle around. Spindle fibers may be isolated by special techniques (Fig. 8–5) and have been found to contain protein, largely a single kind of protein, together with a small amount of RNA. Some of the spindle fibers are attached to the centromeres of the chromosomes and appear to push or pull the chromosomes to the poles during mitosis. When the spindle fibers are viewed under high magnification in the electron microscope they appear to be fine straight hollow tubules. During cell division they first lengthen and then shorten, but they do not appear to get thicker or thinner. This suggests that rather than stretching or contracting, new material is added to the fiber or removed from it as the spindle changes its size. When a moving spindle fiber was marked by burning a portion with ultraviolet light, the marked spot could be seen to move from a point near the equator to the pole and finally to disappear from the end of the fiber. This suggests that protein material is added at the equator, moves to the pole, and then is removed.

Those cells that bear cilia on their exposed surfaces have a structure at the base of each cilium, termed a **basal body**. The basal body strongly resembles the centriole in containing nine parallel tubules. Each cilium contains nine peripherally located, longitudinal filaments and, in addition, two centrally located ones. Like the centrioles, basal bodies can duplicate themselves.

SECTION 3–5

CYTOPLASMIC ORGANELLES: MITOCHONDRIA

The material within the plasma membrane but outside the nuclear membrane is termed **cytoplasm**. Under the light microscope this appears to be composed of a semifluid ground substance in which are suspended a variety of droplets, vacuoles, granules and rodlike or threadlike structures. The electron microscope reveals that the cytoplasm is an incredibly complex maze of membranes and spaces enclosed by membranes (Fig. 3–15). When a thin section of a cell is examined under the electron microscope, these membranes have the appearance of a profusion of spaghetti-like tubular strands termed **endoplasmic reticulum**. When viewed in three dimensions, these are sheetlike membranes filling most of the space in the cytoplasm. The remainder of the space is occupied by other specialized structures with specific functions—mitochondria, the Golgi apparatus, centrioles and plastids. All living cells, plant and animal alike, contain **mitochondria**, bodies ranging in size from 0.2 to 5 microns and in shape from spheres to rods and threads. The number of mitochondria per cell may range from just a few to more than a thousand. When living cells are examined, their mitochondria can be seen to move, to change size and shape, to fuse with other mitochondria to form bigger structures, or to cleave to form shorter ones. They are usually concentrated in the region of the cell with the highest rate of metabolism.

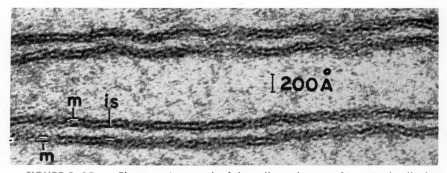

FIGURE 3–15   Electron micrograph of the cell membranes of intestinal cells showing the three-layered unit membrane. M, membrane; is, intercellular space. Magnification,
240,000×.

Some of the larger mitochondria are visible in the light microscope, but the details of their internal structure are revealed only in the electron microscope. Each mitochondrion is bounded by a double membrane, the outer layer of which forms a smooth outer boundary and the inner layer of which is folded repeatedly into parallel plates that extend into the center of the mitochondrial cavity (Fig. 3–16). These plates may meet and fuse with folds coming in from the opposite side. Each of the outer and inner membranes is a unit membrane consisting of a central double layer of phospholipid molecules and a layer of protein molecules on each side. The shelflike inner folds, called **cristae**, contain the enzymes of the electron transmitter system, which are of prime importance in converting the potential energy of foodstuffs into biologically useful energy for cellular activities. The semifluid material within the inner compartment, the matrix, contains certain enzymes of the Krebs citric acid cycle. The mitochondria, whose prime function is the release of biologically useful energy, have been aptly termed the "powerhouses" of the cell.

Biochemists are able to homogenize cells and separate mitochondria from other subcellular organelles by differential high speed centrifuga-

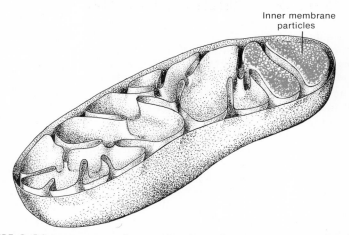

Inner membrane
particles

FIGURE 3–16   Diagram of a mitochondrion illustrating its internal structure and
the arrangement of its inner and outer membranes. The inner surface of the inner membrane
is covered with regularly spaced polygonal structures connected to the membrane by a
narrow stalk. These structures, termed "elementary particles" are believed to contain the
enzymes involved in oxidative phosphorylation. The elementary particles are indicated in
tint.

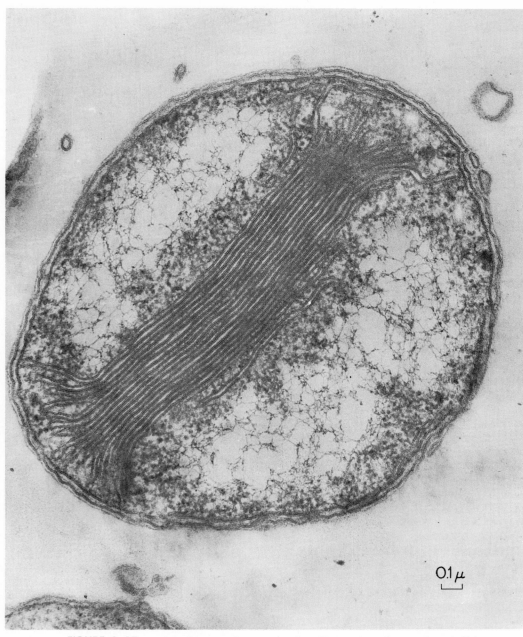

FIGURE 3–17   An electron micrograph of a thin section of a marine nitrifying bacterium, Nitrosocystis oceanus. Extending across the central portion of the cell are the parallel lamellae of the membranous organelle, which contain the enzymes of the electron transmitter system. On either side are light areas containing strands of DNA and near the periphery of the cell are ribosomes which appear as dark dots. Outside the plasma membrane is the cell wall composed of four dense layers. (Courtesy of Dr. S. W. Watson.)

tion. These purified mitochondria, when incubated *in vitro*, metabolize carbohydrates and fatty acids to carbon dioxide and water, utilizing oxygen and releasing energy rich phosphate compounds in the process. Mitochondria swell and contract as they carry out these metabolic functions.

Biologists have speculated about the evolutionary origin of mitochondria. The cells of bacteria do not contain mitochondria, but they do contain membranes in which are embedded the enzymes of the electron transmitter system. In some bacteria these membranes are just inside the plasma membrane. Other bacteria, such as certain marine forms, have a complex system of parallel membranous sheets that stretch across the central part of the cell (Fig. 3–17). The enzymes of the electron transmitter system are embedded in these membranes. One can speculate that as cells grew larger and more complex, these membranes underwent folding and finally pinched off to form discrete organelles, which were the precursors of the present-day mitochondria. Other investigators have speculated that entire bacterial cells, complete with their membranes of electron transmitter enzymes, invaded larger cells and took up a symbiotic existence as the mitochondria of the larger cells.

SECTION 3–6

CHLOROPLASTS

The cells of most plants contain **plastids**, small bodies involved in the synthesis or storage of foodstuffs. The most important plastids, **chloro-**

FIGURE 3–18   An electron micrograph of a chloroplast from a leaf of the tobacco plant Nicotiana rustica, showing the fine structure of the grana, magnified 30,000X. Note the alternate layers of protein and lipid in the grana within the chloroplast and the membranes separating the chloroplasts from the surrounding cytoplasm (Courtesy of Dr. E. T. Weier.)

plasts, contain the green pigment, chlorophyll, which imparts the green color to plants and is of paramount importance in photosynthesis in trapping the energy of sunlight. The chloroplasts of higher plants are typically disc-shaped structures, some 5 microns in diameter and 1 micron thick. Under the electron microscope, their internal structure is revealed as consisting of a lamellar arrangement of tightly stacked membranes (Fig. 3–18). Each cell has some 20 to 100 chloroplasts which can grow and divide to form daughter chloroplasts. Within each chloroplast are a large number of smaller bodies called **grana,** which contain the chlorophyll (Fig. 3–19).

   The chloroplast is not simply a bag of chlorophyll. Indeed, the capacity of chlorophyll to capture light energy depends upon its arrangement in the structure of the grana. A layer of chlorophyll molecules and a layer of phospholipid molecules are sandwiched between layers of proteins (Fig. 3–20). This arrangement spreads the chlorophyll molecules over a wide area, and the layered structure may be important in facilitating the transfer of energy from one molecule to an adjacent one during photosynthesis. The material surrounding each granum is termed the stroma. The several grana within a chloroplast are connected one with another by sheets of membranes which pass through the stroma.

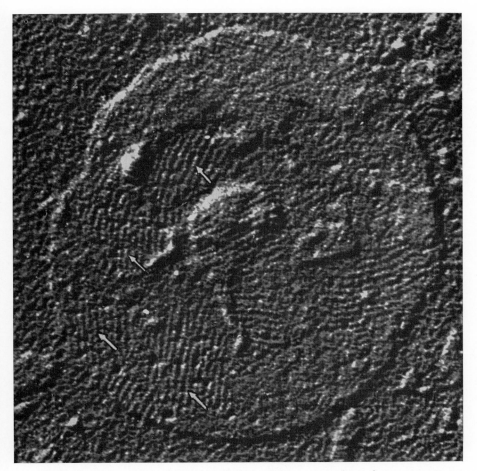

FIGURE 3–19   *Electron micrograph showing the inner surface of one compartment of grana. By shadow-casting, a uniform linear arrangement of spheroids 100 Å × 200 Å in diameter called quantosomes can be demonstrated (arrows). (Courtesy of M. Calvin.)*

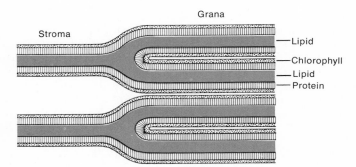

FIGURE 3–20   *Diagram of the lamellar structure of a granum. The protein layers are indicated in tint and chlorophyll is stippled. The lipid layers are indicated by vertical lines. (After Hodge, A. J., McLean, J. D., and Mercer, F. V.: Ultrastructure of the lamellae and grana in the chloroplasts of Zea mays L. J. Biophys. Biochem. 1:607, 1955.)*

Other colorless plastids, **leucoplasts**, serve as centers for the storage of starch and other materials. A third type of plastid, called **chromoplasts,** contain pigments and are responsible for the colors of flowers and fruits.

SECTION 3–7

RIBOSOMES: THE CELL'S PROTEIN-SYNTHESIZING MACHINERY

Cells, such as those of the pancreas, that are especially active in synthesizing proteins to be secreted, are crowded with the membranous labyrinth of the **endoplasmic reticulum** (Fig. 3–21). Other cells may have only a scanty supply of these membranes. Two types are found, granular or "rough" endoplasmic reticulum, to which are bound many ribosomes, and agranular or "smooth" endoplasmic reticulum, consisting of the membranes alone. Bound to the granular reticulum or free in the cell matrix are many **ribosomes,** small particles of ribonucleoprotein on which protein synthesis occurs. Both smooth and granular endoplasmic reticulum may be found in the same cell. The function of the agranular endoplasmic reticulum is not clear, but it may play some part in the process of cellular secretion. The tightly packed sheets of endoplasmic reticulum may form tubules some 50 to 100 millimicrons in diameter. In other portions of the cell, the cavities may be expanded, forming flattened sacs called **cisternae.** These membranes of the endoplasmic reticulum divide the cytoplasm into a multitude of compartments in which different groups of chemical reactions may occur. The endoplasmic reticulum serves a further function as a system for the transport of substrates and products through the cytoplasm to the outside of the cell and to the nucleus.

After mitochondria have been sedimented from homogenized cells by centrifugation, a heterogeneous group of smaller particles termed **microsomes** can be sedimented by centrifuging at about 100,000 times the force of gravity. Particles, termed ribosomes, can be separated from the rest of the microsomal fraction by treatment with appropriate detergents. Isolated ribosomes can then synthesize proteins *in vitro* if supplied with the appropriate instructions in the form of messenger RNA and with an assortment of amino acids, an energy source, and the other enzymes and transfer RNA required (cf. Chap. 7). Ribosomes are found in all kinds of cells from bacterial, plant and animal sources and thus are ubiquitous. Ribosomes

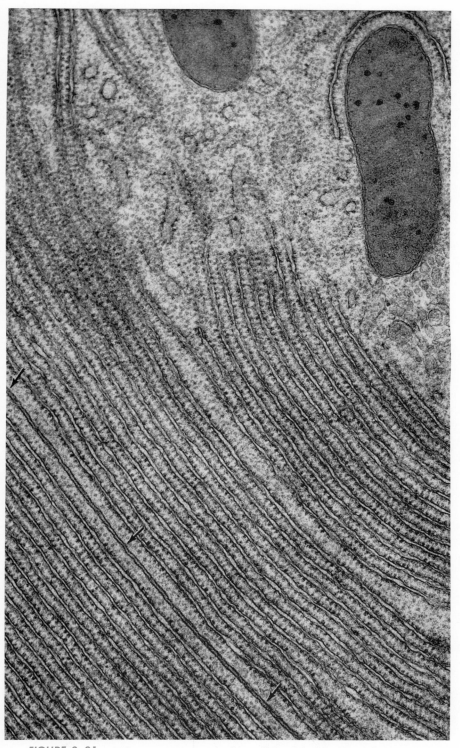

FIGURE 3–21   *Electron micrograph of a cell from the pancreas of a bat showing the tightly packed sheets of endoplasmic reticulum. The arrows indicate the plasma membranes of two adjacent cells. (From Fawcett, D. W.: The Cell. Philadelphia, W. B. Saunders Company, 1966.)*

contain RNA and protein and are composed of two nearly spherical sub-units, which are combined to form the active protein synthesizing unit (Fig. 3–22). Ribosomes are synthesized in the nucleus and pass to the cytoplasm where they are active in synthesizing proteins. Ribosomes may be bound to the membranes of the endoplasmic reticulum, or they may be free in the matrix of the cytoplasm. In many cells, clusters of five or six ribosomes, termed **polysomes** (Fig. 7–21), appear to be the functional unit that is effective in protein synthesis. It is estimated that a bacterial cell, such as *Escherichia coli*, contains some 6000 ribosomes and that the rabbit reticulocyte contains some 100,000 ribosomes. Ribosomes are remarkably uniform in their size, structure and composition, whether isolated from bacterial or mammalian sources. They contain protein and RNA in nearly equal proportions, with little or no lipid. The protein has a fairly high content of basic amino acids.

One of the subunits of the ribosomes has a molecular weight of about 1,300,000 and the other has a molecular weight of about 600,000. The protein components of ribosomes from different cells are remarkably similar in their amino acid composition; however, the nucleotide composition of the RNA of ribosomes from different species varies considerably.

The microsomal fraction contains, in addition to the ribosomes involved in the synthesis of polypeptide chains, several less well-characterized particles that contain enzymes involved in the metabolism of other types of compounds.

The **Golgi complex** is another cytoplasmic component present in almost all cells except mature sperm and red blood cells. It consists of an irregular network of canals lined with membranes and may have the appearance of small vesicles or vacuoles. The complex is usually located near the nucleus surrounding the centrioles. Its function in cellular activity is not known definitely, but it has long been believed to be important in the secretion of certain cell products. There is evidence to suggest that proteins made in the cisternae of the endoplasmic reticulum are sealed off in little packets of endoplasmic reticulum and travel to the Golgi complex. There they are repackaged in larger sacs made up of membranes from the Golgi com-

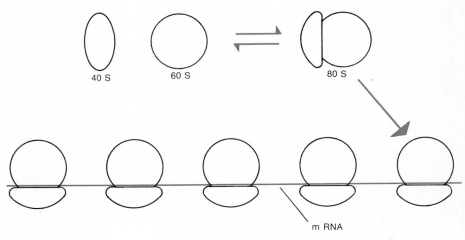

40 S   60 S   80 S

m RNA

Polysome

FIGURE 3–22 *Diagram illustrating the assembly of the two subunits 40 S and 60 S to form a ribosome, 80 S, the subcellular organelle on which proteins are synthesized. The 80 S ribosomes attach to a messenger RNA strand to form a polysome.*

plex. In these packets they travel to the plasma membrane, which fuses with the membrane of the vesicle, opening the vesicle and releasing its contents to the exterior of the cell. Under the electron microscope, the Golgi complex is seen to consist of parallel arrays of membranes without granules, which may be distended in certain regions to form small vesicles or vacuoles filled with cell products (Fig. 3–23). Some cytologists believe that the Golgi complex serves as a temporary storage place for substances produced in the endoplasmic reticulum. It is suggested that the canals of the Golgi complex are connected to the plasma membrane to provide for the secretion of these products.

The Golgi apparatus of plant cells may be involved in the secretion of the cellulose of the cell walls (Fig. 3–24). It usually appears as discrete bodies dispersed throughout the cell, each one of which consists of a stack of flattened vesicles that are slightly dilated at the edges.

Another group of intracellular organelles found in animal cells are **lysosomes**. About the size of mitochondria but somewhat less dense, lysosomes consist of a membrane-bounded structure containing a variety of enzymes that can hydrolyze the macromolecular constituents of the cell, proteins, polysaccharides and nucleic acids. These packets serve to segregate these enzymes in the intact cell and presumably prevent their digesting the contents of the cell. Rupture of the membrane of the lysosome releases these enzymes and accounts, at least in part, for the lysis of dead or dying cells and the resorption of cells such as those in the tail of a tadpole during metamorphosis. Since the lysosomes contain enzymes that can hydrolyze the major cellular constituents when the lysosomes rupture and release them, they have been termed "suicide bags" by the Belgium biochemist, Christian De Duve.

In addition to these living elements, the cytoplasm may contain **vacuoles**, bubble-like cavities filled with watery fluid and bordered by a vacuolar membrane similar in structure to the plasma membrane. Vacuoles are fairly common in the cells of plants and of lower animals but are rare in those of higher animals. Most protozoa have **food vacuoles**, containing food undergoing digestion, and **contractile vacuoles**, which serve to remove excess water from the cell. Cytoplasm, in addition, may contain granules of stored starch or protein, or droplets of oil.

Animal and plant cells differ in three major respects. Animal cells, but not the cells of higher plants, have a centriole; plant cells, but not animal cells, have plastids; and plant cells have a stiff cell wall of cellulose which prevents their changing position or shape, whereas animal cells usually have only the thin plasma membrane, and thus are able to alter their shape and move about. Most plant and animal cells are too small to be seen with the naked eye. Their diameters range from 1 to 100 microns and a speck 100 microns in diameter is near the lower limit of visibility. A few species of amoebas are cells as large as a millimeter or two in diameter. Some algae, such as *Acetabularia*, are single cells more than a centimeter long. Some of the largest single cells are the egg cells of fishes and birds. The egg cell of a large bird may be several centimeters in diameter. Only the yolk of the egg is the true egg cell; the white of the egg is noncellular material secreted by the hen's oviduct.

The ultimate size of the cell is limited by the physical fact that as a sphere increases in size, the volume increases as the cube of the radius whereas the surface increases only as the square of the radius. Cellular metabolism requires oxygen and nutrients that can enter only through the cell surface. Clearly there is a limit to the size of the cell above which the surface area is too small to supply its metabolic activities. The actual size of this limit in any given cell depends on the cell's shape and rate of

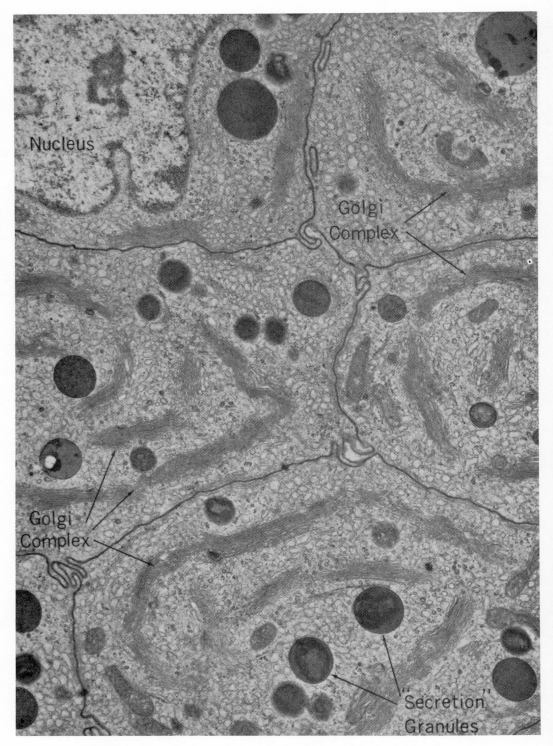

FIGURE 3–23   *Electron micrograph of rabbit epididymis showing the extensive Golgi complex, evident in the parallel arrays of the membranes. Magnification, 9500×. (From Fawcett, D. W.: The Cell. Philadelphia, W. B. Saunders Company, 1966.)*

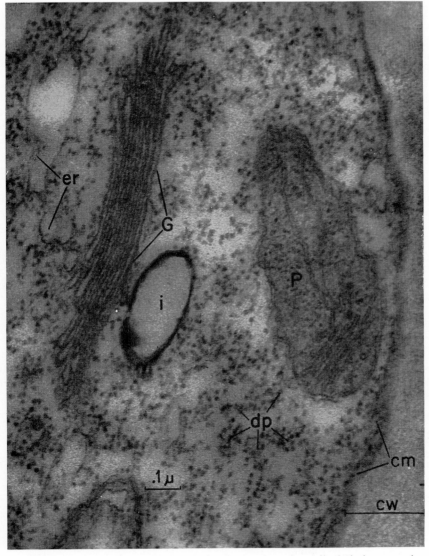

FIGURE 3–24   Electron micrograph of a portion of a cell of Elodea canadensis; er, endosplasmic reticulum, which is sparse and shows many ribosomal particles (some of which appear to be free in the cell and are not attached to the endoplasmic reticulum); G, Golgi body; P, chloroplast; cw, cell wall; cm, cell membrane. (Courtesy of P. Buvat.)

metabolism. When that limit is reached, the cell must either stop growing or divide.

TISSUES

One of the major trends in the evolution of both plants and animals has been that leading to specialization and division of labor of the constituent cells. The cells that make up the body of a tree or a man are not all alike; each is specialized to carry out certain functions. This specialization allows the cells to function more efficiently but also makes the parts of the body more interdependent: the injury or destruction of one part of the body may result in death of the whole. The advantages of specialization, however, outweigh the disadvantages.

A *tissue* may be defined as a group or layer of similarly specialized cells which together perform certain special functions. The study of the structure and arrangement of tissue is known as *histology*. Each kind of tissue is composed of cells which have a characteristic size, shape and arrangement. Tissues may consist of noncellular materials in addition to living cells. Blood and connective tissues, for example, contain some noncellular, nonliving material between the cells.

ANIMAL TISSUES

Biologists differ somewhat in their ideas of how the various types of tissue should be classified and, consequently, of how many types of tissue there are. We shall classify animal tissues in six groups: epithelial, connective, muscular, blood, nervous and reproductive.

Epithelial Tissues

Epithelial tissues are composed of cells which form a continuous layer or sheet covering the body surface or lining cavities within the body. They may have one or more of the following functions: protection, absorption, secretion and sensation. The epithelia of the body protect the underlying cells from mechanical injury, from harmful chemicals and bacteria, and from drying. The epithelia lining the digestive tract absorb nutrients and water into the body. Other epithelia secrete a wide variety of substances as waste products or for use elsewhere in the body. Finally, since the body is entirely covered by epithelium, it is obvious that all sensory stimuli must penetrate an epithelium to be received. Examples of epithelial tissues are the outer layer of the skin, the lining of the digestive tract, the lining of the windpipe and lungs, and the lining of the kidney tubules. Epithelial tissues are divided into six subclasses according to their shape and function.

Squamous epithelium is composed of flattened cells shaped like pancakes or flagstones (Fig. 3–25 *A*). Squamous epithelium is found on the surface of the skin and the lining of the mouth, esophagus and vagina. In man and the higher animals, there are usually several layers of these flat

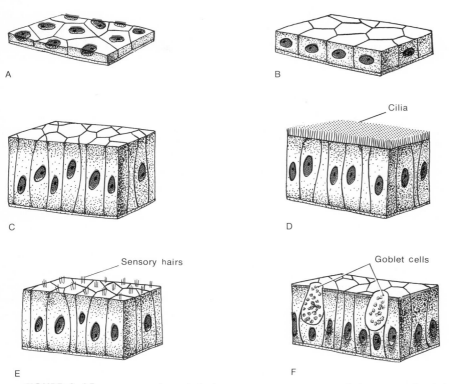

FIGURE 3–25   Types of epithelial tissue. A, Squamous epithelium; B, cuboidal epithelium; C, columnar epithelium; D, ciliated columnar epithelium; E, sensory epithelium (cells from the lining of the nose); F, glandular epithelium: two single-celled glands or "goblet cells" in the lining of the intestine are evident.

cells piled one on top of another, a condition called **stratified squamous epithelium.**

The tissue lining the kidney tubules, composed of cells that are cube-shaped, resembling dice, is known as **cuboidal epithelium** (Fig. 3–25 B).

The cells of **columnar epithelium** are elongated, resembling pillars or columns; the nucleus is usually located near the base of the cell (Fig. 3–25 C). The stomach and intestines are lined with columnar epithelium.

Columnar cells may have on their free surface small cytoplasmic projections called cilia (Fig. 3–25 D), which beat rhythmically and move materials in one direction. Most of the respiratory system is lined with **ciliated epithelium** whose cilia function to remove particles of dust and other foreign material.

Sensory epithelium is composed of cells specialized to receive stimuli (Fig. 3–25 E). The cells lining the nose—the olfactory epithelium, responsible for the sense of smell—are an example.

The cells of **glandular epithelium** (Fig. 3–25 F) are specialized to secrete substances such as hormones, digestive juices, milk, wax or perspiration. They are either columnar or cuboidal in shape.

Connective Tissue

Connective tissue, which includes bone, cartilage, tendons, ligaments and fibrous connective tissue, supports and holds together the other cells of the body. The cells of these tissues characteristically secrete a large amount of nonliving material, called *matrix*, and the nature and function of the particular connective tissue is determined largely by the nature of this intercellular matrix. The cells thus perform their functions indirectly by secreting a matrix which does the actual connecting and supporting.

In fibrous connective tissue the matrix is a thick, interlacing, matted network of microscopic fibers secreted by and surrounding the connective tissue cells (Fig. 3–26 *A*). Such tissue occurs throughout the body and holds skin to muscle, keeps glands in position, makes up the tough outer walls of the larger blood vessels, and forms a sheath around nerve fibers and muscle cells. Tendons and ligaments are specialized types of fibrous connective tissue. *Tendons* are not elastic but are flexible, cable-like cords that connect muscles to each other or to bone. *Ligaments* are somewhat elastic and connect one bone to another. There is an especially thick mat of connective tissue fibers just below the skin (Fig. 3–27). When this is treated chemically (tanned) it becomes leather.

Connective tissue fibers contain a protein called *collagen*. When the fibers are treated with hot water, collagen is converted into the soluble protein *gelatin*. Collagen and gelatin have nearly identical amino acid compositions. Because there is so much connective tissue present, about one third of all the protein in the human body is collagen. The collagen units comprising the fibers consist of a helix made of three peptide chains joined by hydrogen bonds.

The supporting skeleton of vertebrates is composed of cartilage or bone. *Cartilage* is the supporting skeleton in the embryonic stages of all

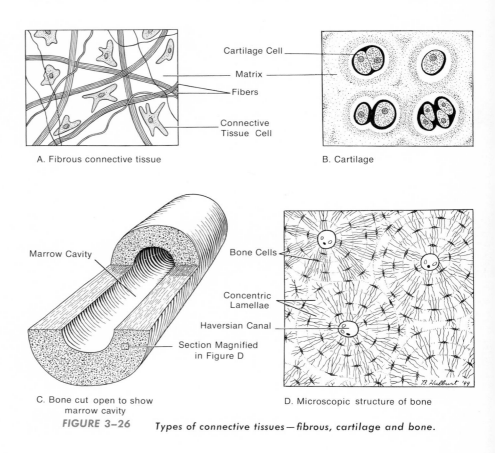

A. Fibrous connective tissue

Cartilage Cell

Matrix

Fibers

Connective Tissue Cell

B. Cartilage

Marrow Cavity

Bone Cells

Concentric Lamellae

Haversian Canal

Section Magnified in Figure D

C. Bone cut open to show marrow cavity

D. Microscopic structure of bone

FIGURE 3–26   *Types of connective tissues—fibrous, cartilage and bone.*

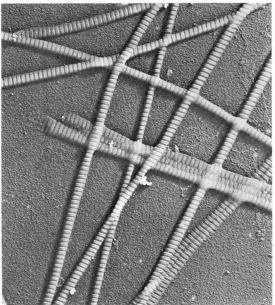

FIGURE 3–27   Electron micrograph of collagen fibrils teased from a preparation of calf skin. Notice the regular periodic striations of the fibrils which indicate its repeating structural unit. Magnification, 33,000×. (Courtesy of Dr. Jerome Gross.)

vertebrates, but is largely replaced in the adult by **bone** in all but the sharks and rays. In the human body, cartilage may be felt in the supporting structure of the ear flap (pinna) and in the tip of the nose. It is firm yet elastic. Cartilage cells secrete this hard, rubbery matrix around themselves and come to lie singly or in groups of two or four in small cavities in the homogeneous, continuous matrix (Fig. 3–26 B). The cartilage cells in the matrix remain alive; some of them secrete fibers which become embedded in the matrix and strengthen it.

Bone cells remain alive and secrete a bony matrix throughout a person's life. The matrix contains calcium salts (the mineral **hydroxyapatite**) and proteins, principally collagen. The calcium salts in the bony matrix make it very hard, and the collagen prevents the bone from being overly brittle. The dense, hard bony matrix enables bone to stiffen and support the body against the pull of gravity. Contrary to appearance, bone is not a solid structure. Most bones have a large marrow cavity in the center (Fig. 3–26 C), which may contain yellow marrow, mostly fat, or red marrow, the tissue in which red and certain white blood cells are produced.

Extending through the matrix of the bone are channels (**haversian canals**) in which lie blood vessels and nerves to supply and control the bone cells. The bone matrix is secreted in concentric rings (lamellae) around the canals, and the cells become embedded in cavities in these rings (Fig. 3–26 D). Bone cells are connected to each other and to the haversian canals by cellular extensions which lie in minute canals (canaliculi) in the matrix. The bone cells obtain oxygen and raw materials and eliminate wastes by way of these minute canals. Bones contain other cells which can dissolve and remove the bony substance; thus the shape of the bone can change in response to continued stresses and strains.

Muscular Tissue

The movements of most animals result from the contraction of elongate, cylindrical or spindle-shaped cells, each of which contains many

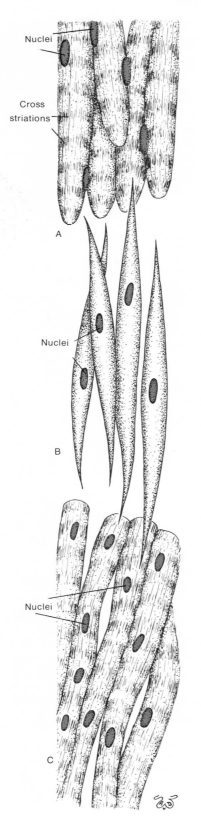

Nuclei

Cross
striations

A

Nuclei

B

Nuclei

C

FIGURE 3-28   Types of muscle tissues.
A, Skeletal muscle fibers, B, smooth muscle
fibers, C, cardiac muscle fibers.

TABLE 3–1   Comparison of the Types of Muscle Tissue

	Skeletal	Smooth	Cardiac
Location	Attached to skeleton	Walls of viscera, stomach, intestines, etc.	Wall of heart
Shape of fiber	Elongate, cylindrical, blunt ends	Elongate, spindle-shaped, pointed ends	Elongate, cylindrical, fibers branch and fuse
Number of nuclei per fiber	Many	One	Many
Position of nuclei	Peripheral	Central	Central
Cross striations	Present	Absent	Present
Speed of contraction	Most rapid	Slowest	Intermediate
Ability to remain contracted	Least	Greatest	Intermediate
Type of control	Voluntary	Involuntary	Involuntary

small, longitudinal, parallel, contractile fibers called **myofibrils**, composed of the proteins myosin and actin. Muscle cells can perform mechanical work only by contracting, by getting shorter and thicker; they cannot push. There are three distinct types of muscle in the human body: skeletal, smooth and cardiac (Fig. 3–28). Cardiac muscle is found in the walls of the heart and smooth muscle in the walls of the digestive tract and certain other internal organs. Skeletal muscle makes up the large muscle masses attached to the bones of the body. Skeletal and cardiac fibers are exceptions to the rule that cells have only one nucleus; each fiber has many nuclei. The nuclei of the skeletal muscle fibers are also unusual in their position: they lie peripherally, just under the cell membrane; presumably this is an adaptation to increase the efficiency of contraction. Skeletal muscle cells are extremely long—2 or 3 centimeters in length. Indeed, some investigators believe that muscle cells extend the entire length of the muscle.

Skeletal and cardiac fibers have alternate light and dark microscopic transverse stripes or striations. These stripes appear to be involved in contraction, for they change their relative sizes during contraction, the dark stripes remaining essentially constant and the light stripes decreasing in width. Striated muscles can contract very rapidly but cannot remain contracted; a striated muscle fiber must relax and rest momentarily before it can contract again. Skeletal muscle is sometimes called voluntary muscle, because it is under the control of the will. Cardiac and smooth muscles are called involuntary, because they cannot be regulated by the will. Table 3–1 summarizes the features that distinguish the three types of muscle tissue.

Blood

Blood includes the red and white blood cells and the liquid, noncellular part of the blood, the **plasma**. Many biologists classify blood with the connective tissues because they originate from similar cells.

The red cells (**erythrocytes**) of vertebrates contain the pigment, hemo-

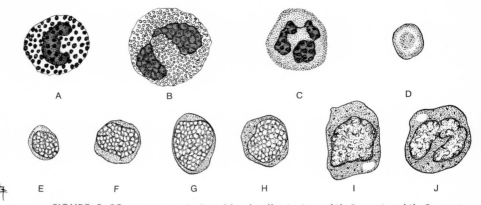

FIGURE 3–29   *Types of white blood cells. A, Basophil; B, eosinophil; C, neutro-phil; D, a red blood cell drawn to the same scale. E--H, a variety of lymphocytes; I and J, monocytes.*

globin, which has the property of combining easily and reversibly with oxygen. Oxygen combined as oxyhemoglobin is transported to the cells of the body in the red cells. Mammalian red cells are flattened biconcave discs without a nucleus; those of other vertebrates are more typical cells with an oval shape and a nucleus.

There are five different kinds of white blood cells—lymphocytes, monocytes, neutrophils, eosinophils and basophils (Fig. 3–29). White cells have no hemoglobin but can move around and engulf bacteria. They can slip through the walls of blood vessels and enter the tissues of the body to engulf bacteria there. The fluid part of the blood, plasma, transports many types of substances from one part of the body to the other. Some of the substances transported are in solution, others are bound to one or another of the plasma proteins. In certain invertebrate animals, the oxygen-carrying pigment is not localized in cells but is dissolved in the plasma and colors it red or blue. **Platelets** are small fragments broken off from large cells in the bone marrow. They play a role in the clotting of blood.

Nervous Tissue

Nervous tissue is made of cells, called **neurons**, specialized for conducting electrochemical nerve impulses. Each neuron has an enlarged **cell body**, which contains the nucleus and two or more thin, hairlike nerve fibers extending from the cell body (Fig. 3–30). The nerve fibers, made of cytoplasm and covered by a plasma membrane, vary in width from a few microns to 30 or 40 microns and in length from a millimeter or two to more than a meter. Those stretching from the spinal cord down the arm or leg in man may be a meter or more in length. The neurons are connected together in chains to pass impulses for long distances through the body.

Two types of nerve fibers, axons and dendrites, are differentiated on the basis of the direction in which they normally conduct a nerve impulse: **axons** conduct nerve impulses away from the cell body; **dendrites** conduct them toward the cell body. The junction between the axon of one neuron and the dendrite of the next is called a **synapse**. The axon and dendrite do not actually touch at the synapse; there is a small gap between the two. An impulse can travel across the synapse only from an axon to a dendrite;

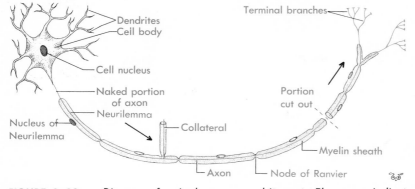

FIGURE 3-30   *Diagram of a single neuron and its parts. The arrows indicate the direction of the normal nerve impulse.*

the synapse serves as a valve to prevent the backflow of impulses. Neurons occur in many sizes and shapes but all have the same basic plan.

Reproductive Tissue

Reproductive tissue is composed of cells modified for the production of offspring—*egg* cells in females and *sperm* cells in males (Fig. 3-31). Egg cells are usually spherical or oval and are nonmotile. The eggs of most animals, but not of the higher mammals, contain a large amount of **yolk** which serves as food for the developing organism from the time of fer-

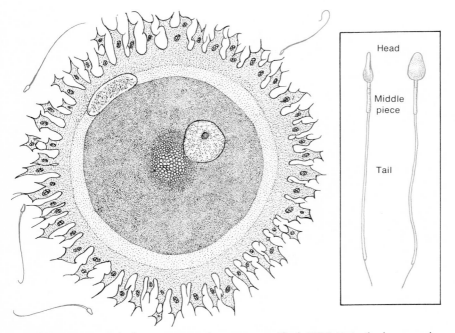

FIGURE 3-31   Left, human egg and sperm, magnified 400×. Note the large nucleus and nucleolus in the egg. The polar body is visible as a flattened oval body between the egg and the cells of the surrounding corona radiata. Right, side and top views of a sperm, magnified about 2000×.

tilization until it is able to obtain food in some other way. Sperm cells are much smaller than eggs; they lose most of their cytoplasm and develop a tail by which they propel themselves. A typical sperm consists of a **head,** which contains the nucleus, a **middle piece** and a **tail**. The shape of the sperm varies in different species of animals (Fig. 8–12). Because eggs and sperm develop from epithelial-like tissue in the ovaries and testes, some biologists classify them with those tissues.

SECTION 3–10

PLANT TISSUES

The cells of the higher plants are also organized and differentiated into tissues. Plant biologists recognize four main types of tissue: meristematic, protective, fundamental and conductive.

Meristematic Tissue

Meristematic tissues are composed of small, thin-walled cells with large nuclei and few or no vacuoles (Fig. 3–32). Their chief function is to grow, divide and differentiate into all the other types of tissue. An embryonic plant, as it begins development, is composed entirely of meristem; as it develops, most of the meristem becomes differentiated into other tissues. Even in an adult tree there are regions of meristem which provide for continued growth. Meristematic tissues are found in the rapidly growing parts of the plant—the tips of the roots and stems, and in the **cambium.** The meristem in the tips of roots and stems, called **apical meristem**, is responsible for the increase in length of roots and stems, and the meristem in the cambium, called **lateral meristem,** makes possible the increase in diameter of stems and roots.

Protective Tissue

Protective tissues consist of cells with thick walls that serve to protect the underlying thin-walled cells from drying out and from mechanical abrasions. The epidermis of leaves and the cork layers of stems and roots are examples of protective tissues. The epidermis of leaves secretes a waxy, waterproof material known as **cutin**, which decreases the loss of water from the leaf surface.

On the surface of leaves are specialized epidermal cells, termed **guard cells**, that occur in pairs around each tiny opening, termed a **stoma**, into the interior of the leaf. The turgor pressure in the guard cell regulates the size of the stoma and the rate at which oxygen, carbon dioxide and water vapor pass into or out of the leaf.

Some of the epidermal cells of the roots have outgrowths called **root hairs** that increase the absorptive surface for the intake of water and dissolved minerals from the soil. Stems and roots are covered by layers of cork cells, produced by a separate cork cambium, another lateral meristem. Cork cells are closely packed and their cell walls contain another waterproof material, **suberin**. Since the suberin prevents the entrance of water into the cork cells themselves, they are short-lived and all mature cork cells are dead.

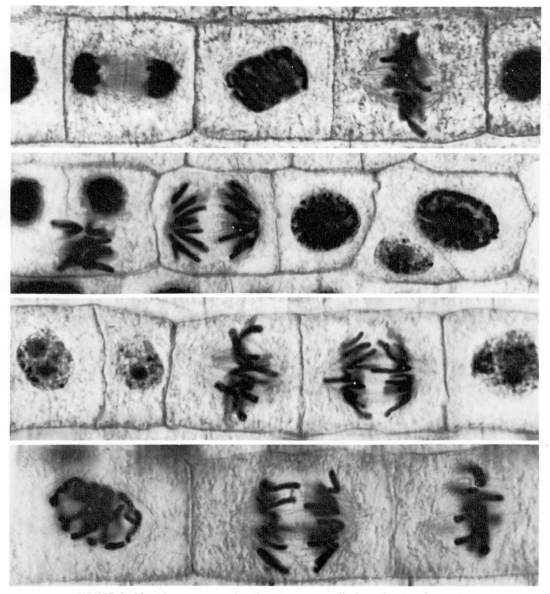

FIGURE 3–32   *Photomicrographs of meristematic cells from the tip of an onion root showing stages in mitotic division. Note the thin walls, large nuclei and absence of vacuoles characteristic of cells in the meristem. A, Prophase and anaphase; B, metaphase and anaphase; C, late telophase; D, telophase and anaphase. (Courtesy of the Carolina Biological Supply Company.)*

Fundamental Tissue

The fundamental tissues make up the great mass of the plant body, including the soft parts of the leaf, the pith and cortex of stems and roots, and the soft parts of flowers and fruits. Their chief functions are the production and storage of food. The simplest of the fundamental tissues, **parenchyma**, consists of cells with a thin wall and a thin layer of cytoplasm surrounding a central vacuole (Fig. 3–33). **Chlorenchyma** is a modified parenchyma containing chloroplasts in which photosynthesis occurs. The chlorenchyma cells are loosely packed and make up most of the interior of leaves and some stems. They are characterized by thin cell walls, large vacuoles, and the presence of chloroplasts.

In some fundamental tissue, the corners of the cells are thickened to provide the plant with support. Such tissue, called **collenchyma** (Fig. 3–33), occurs just beneath the epidermis of stems and leaf stalks. In still another type, known as **sclerenchyma** (Fig. 3–33), the entire cell wall becomes greatly thickened. These cells, which provide support and mechanical strength, are found in many stems and roots. They sometimes

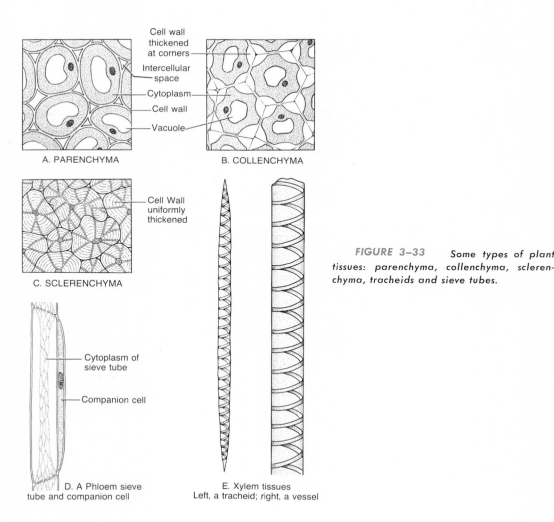

A. PARENCHYMA

B. COLLENCHYMA

C. SCLERENCHYMA

D. A Phloem sieve tube and companion cell

E. Xylem tissues
Left, a tracheid; right, a vessel

FIGURE 3–33   *Some types of plant tissues: parenchyma, collenchyma, sclerenchyma, tracheids and sieve tubes.*

take the form of long thin fibers. Spindle-shaped sclerenchyma cells called **bast fibers** are found in the phloem of many stems. Rounded sclerenchyma cells called **stone cells** are found in the hard shells of nuts.

Conductive Tissue

There are two types of conductive tissue in plants: **xylem,** which conducts water and dissolved salts, and **phloem,** which conducts dissolved organic materials such as glucose. In all higher plants, the first xylem cells to develop are long **tracheids**, with pointed ends and thickenings on the walls in a circular, spiral or pitted pattern (Fig. 3–33). Later, other cells join end-to-end to form **xylem vessels.** As the vessels develop, the end walls dissolve and the side walls thicken, leaving a long cellulose tube for the conduction of water. These vessels may be as much as 3 meters long. In both tracheids and vessels the cytoplasm eventually dies, leaving the tubes, which continue to function. The thickening, which involves the deposition of **lignin**, the substance responsible for the hard, woody nature of plant stems and roots, enables xylem to act as a supportive as well as a conductive tissue.

A similar end-to-end fusion of cells gives rise to the **sieve tubes** of the phloem (Fig. 3–33). The ends of the cells do not disappear, but remain as a perforated plate, the **sieve plate**. Unlike the tracheids and vessels of the xylem, mature sieve tubes remain alive and have an abundance of cytoplasm but lose their nucleus. Adjacent to the sieve tubes are nucleated "companion cells" (Fig. 3–33) which may serve to regulate the functions of the sieve tubes. The streaming movement or cyclosis of the cytoplasm of the sieve tubes is important in speeding up the transport of dissolved foods (translocation, cf. p. 555) by the sieve tubes. Sieve tubes are found in woody stems, in the soft bark just outside the cambium layer.

SECTION 3–11

THE FUNDAMENTAL PARTICLES OF MATTER

To get a more complete idea of the composition and properties of living matter, we must consider not only the readily visible macroscopic aspects of life, and those aspects visible under the microscope or electron microscope, but also those molecular patterns that lie far beyond the range of any microscope. This requires some understanding of certain basic principles of physics and chemistry.

Regardless of the form—gaseous, liquid or solid—that matter may assume, it is always composed of units called **atoms**. In nature there are 92 kinds of atoms, ranging from the smallest, hydrogen, to the largest, uranium. In addition to these, there are ten or more atoms larger than uranium that have been man made in a cyclotron or nuclear reactor. All atoms, natural and synthetic alike, are much smaller than the tiniest particle visible even in the electron microscope. In fact, no one has ever seen an atom; their structure and properties have been inferred from experiments made with many types of elaborate apparatus.

The atom is not the ultimate smallest unit of matter; atoms can be divided into even smaller particles organized around a central core, as our solar system of planets is organized around the sun. The exact number and kind of these particles and their arrangement in the atom are matters about which physicists are not yet in complete agreement. For our purposes we need consider only three simple types: *electrons*, which have a negative electric charge and an extremely small mass or weight; *protons*, which have a positive electric charge and are about 1800 times as heavy as electrons, and *neutrons*, which have no electrical charge but have essentially the same mass as protons.

The center of the atom, corresponding in position to the sun in our solar system, is the *nucleus* made up of protons and neutrons. It constitutes almost the entire mass of the atom. Just as the solar system is mostly empty space, so is the atom in which electrons move in circular or other paths in the empty space around the nucleus. Each type of atom has a characteristic number of electrons circling in orbits and characteristic numbers of protons and neutrons in its nucleus (Fig. 3–34). In all atoms, the number of protons in the nucleus equals the number of electrons circling around it, so that the atom as a whole is in a state of electrical neutrality. There are only a few kinds of particles other than protons, neutrons and electrons. The different kinds of matter are produced by differences in the number and arrangement of these basic particles. Living systems are composed of exactly the same kinds of atoms, with the same kind of atomic structure, as nonliving systems.

An *element* is a substance composed of atoms all of which have the same number of protons in the nucleus and therefore the same number of electrons circling in the orbits. Most elements have a tendency to unite with other elements to form compounds, but a few, such as gold, silver, iron and copper, occur in nature as such.

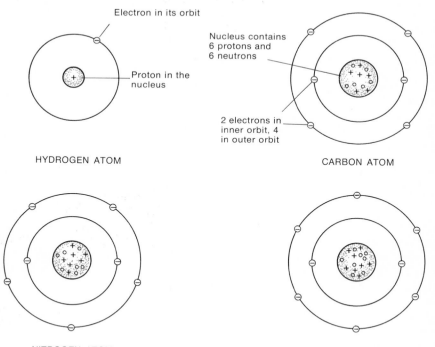

FIGURE 3–34   *Diagrams of the atomic structure of the four major elements found in living things. The nucleus of the nitrogen atom contains seven protons (+) and seven neutrons (○); its inner orbit has two electrons (−) and its outer orbit, five. The nucleus of the oxygen atom contains eight protons and eight neutrons, and there are two and six electrons, respectively, in its inner and outer orbits.*

The unique aliveness of living things does not reflect the presence of some rare or unique element. Four elements, carbon, oxygen, hydrogen and nitrogen, make up some 96 per cent of the material of the human body. Another four, calcium, phosphorus, potassium and sulfur, constitute another 3 per cent of the body weight. Minute amounts of iodine, iron, sodium, chlorine, magnesium, copper, manganese, cobalt, zinc and perhaps a few other elements complete the list. All of these elements, especially the first four, are abundant in the atmosphere, the earth's crust, and the sea. Life depends upon the complexity of the interrelations of these common abundant elements.

Elements are referred to by a symbol which is usually the initial letter of the element's name: O, oxygen; H, hydrogen; C, carbon; N, nitrogen. When several elements have the same initial letter, a second letter is added: C, carbon; Co, cobalt; Cl, chlorine; Cu, copper.

Most of the elements are composed of two or more kinds of atoms which differ in the number of neutrons in their nuclei. There are three kinds of hydrogen, five kinds of carbon, and 16 kinds of lead. The different types of atoms of an element are called *isotopes*, because they occupy the same place in the periodic table of the elements. All of the isotopes of any given element have the same number of electrons circling the nucleus. It is chiefly the number of electrons in the outermost shell which is responsible for the chemical properties of an element. Although the isotopes of a given element have the same chemical properties, they can be differentiated physically. Some are radioactive and can be detected and measured by an instrument such as a Geiger counter by the kind and amount of radiation they emit. Others can be differentiated by the slight difference in the mass of the atoms caused by an extra neutron in the nucleus. Substances containing ^{15}N, heavy nitrogen instead of ^{14}N, the usual isotope, or ^{2}H, heavy hydrogen (deuterium) instead of ^{1}H, will have a greater mass which can be detected with a *mass spectrometer.*

A tremendous insight has been gained into the details of the metabolic activities of cells by preparing substances such as sugar labeled with radioactive carbon, ^{11}C or ^{14}C, or heavy carbon, ^{13}C, in place of ordinary carbon, ^{12}C. The labeled substance is administered to a plant or animal, or cells are incubated in a solution containing it, and the labeled products resulting from the cell's or organism's normal metabolic processes are isolated and identified. By such experiments it has been possible to trace step by step the sequence of reactions undergone by a given compound and to determine the form in which the labeled atoms finally leave the cell or organism. The rate of formation of bone, for example, and the effect of vitamin D and parathyroid hormone on this process, can be studied with the aid of radioactive calcium, ^{45}Ca. Many biological problems that could not be attacked in any other way can be solved by this method.

The atoms of each element have a characteristic number of electrons. The distribution and behavior of these electrons, especially those in the outermost shell, determine the chemical properties of the atom. The simplest atom, hydrogen, has a single electron around the nucleus. In its most stable form (its ground state), the hydrogen atom has one electron moving around the nucleus in a spherical region; by suitable mathematical manipulations, one can calculate the most probable distance of the electron out from the nucleus. The electron does not circle the nucleus in a single orbit, as the earth circles the sun; instead it whirls around the nucleus, now close to it, then farther away.

Figure 3–35 depicts a hydrogen atom with a central nucleus (containing a proton) and a circular boundary. If the electronic charge were distributed over the entire region around the nucleus, 90 per cent would

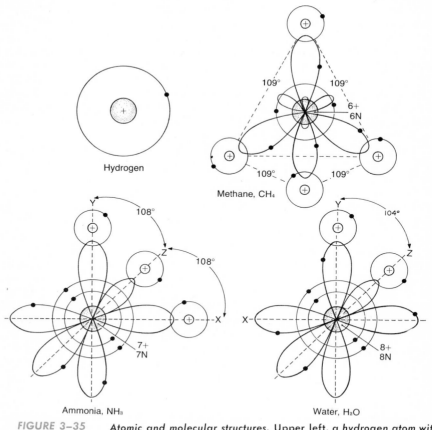

Hydrogen

Methane, CH₄

Ammonia, NH₃

Water, H₂O

FIGURE 3–35   *Atomic and molecular structures. Upper left, a hydrogen atom with a single proton in its nucleus and one electron distributed in a spherical atomic orbital. Upper right, a molecule of methane, composed of one carbon atom and four hydrogen atoms. The carbon atom has a nucleus containing six protons (+) and six neutrons (N) and six electrons surrounding the nucleus. Two of the electrons are paired and occupy an s orbital. The remaining four electrons occupy "hybrid" orbitals which overlap the s orbitals of the electrons of the hydrogen atom, thus forming covalent bonds. Lower left, a molecule of ammonia. The nitrogen atom has three p orbitals, each at right angles to the other two, which overlap the s orbitals of the three hydrogen atoms. Lower right, a molecule of water.*

be found within this circular boundary. This distribution of the electron about the nucleus is called the **atomic orbital**. All distributions of electrons which are spherical in shape are known as **s orbitals.** Other electrons may be distributed in a figure 8 pattern; such orbitals are called **p orbitals.** The electrons nearest the nucleus have a spherical distribution and those farther away have a figure 8 shape. Still other distributions are possible for electrons even more distant from the nucleus. The shape of the atomic orbital for an electron is determined by the energy of the electron.

The carbon atom, which has such a special role in the world of living things, has six electrons about its nucleus (Figs. 3–34 and 3–35). The two electrons nearest the nucleus are paired and occupy an s orbital. The four outer electrons have a distribution which is a mixture of s and p called a **hybrid orbital.** Any one of these four hybrid orbitals may overlap with the orbital of another atom, linking the two atoms together to form a molecule. This overlapping of two atomic orbitals is called a **covalent chemical bond.** An example of such a molecule is methane, CH₄, in which each hybrid orbital of the carbon atom overlaps with the s orbital of a hydrogen atom (Fig. 3–35). Carbon is said to have a valence of four, that is, it may form four chemical bonds. Hydrogen, with one unpaired electron, has a valence of one.

The nitrogen atom has seven electrons about its nucleus and seven protons plus seven neutrons in its nucleus. The first two electrons are paired and occupy an s orbital (called 1s) close to the nucleus. The next two electrons occupy a second s orbital (2s) farther out from the nucleus. Paired electrons in the 1s and 2s orbitals are very stable and do not contribute to the valence of an atom. Unpaired electrons in such orbitals, e.g., the electron in the 1s orbital of hydrogen, tend to overlap with unpaired electrons of other atoms and thus contribute to the valence. The three remaining electrons of the nitrogen atom are unpaired and occupy three different p orbitals farther out from the nucleus. These three unpaired electrons give nitrogen a valence of 3. If each unpaired electron overlapped a hydrogen 1s orbital, a molecule of ammonia, NH_3, would be formed (Fig. 3–35).

The oxygen atom has eight electrons around its nucleus. Two are paired and occupy the 1s orbital and another two occupy the 2s orbital. The spatial configuration of the p orbitals allows only three to exist, each at right angles to the other two and identified as p_x, p_y and p_z. Three of the remaining four electrons may occupy the p_x, p_y and p_z orbitals; the fourth electron must pair up with the electron in one of the p orbitals. Thus, there may be two paired electrons in the p_x orbital and one unpaired electron in the p_y and one unpaired electron in the p_z orbital. These two unpaired electrons give oxygen a valence of two. If each of the two unpaired electrons of the oxygen atom overlaps with a hydrogen 1s orbital, a molecule of water, H_2O, will be formed.

The number of electrons in the outermost (valence) shell varies from zero to eight in different atoms. If there are zero or eight electrons in the outer shell, the element is chemically inactive and will not readily combine with other elements. When there are fewer than eight electrons, the atom tends to lose or gain some in order to achieve an outer shell of eight. Since the number of protons in the atomic nucleus remains the same, the loss or gain of electrons results in an electrically charged atom called an *ion*.

Atoms with one, two or three electrons in the outer shell tend to lose them to other atoms and become positively charged because of the excess protons in the nucleus. Atoms with five, six or seven electrons in the outer shell tend to gain electrons from other atoms and thus become negatively charged because of the excess electrons. Atoms with four electrons in the outer shell (such as carbon) tend to share them with neighboring atoms. Both positively and negatively charged atoms are called ions. Because particles that have opposite electric charges attract each other, positive and negative ions tend to unite, forming an ionic or electrostatic bond.

SECTION 3–12

CHEMICAL COMPOUNDS

A chemical compound is a substance composed of two or more different kinds of atoms or ions joined together and thus can be decomposed into two or more simpler substances. The amounts of the elements in a given compound are always present in a definite proportion by weight. This reflects the fact that the atoms are attached to one another by chemical bonds in a precise way to form the compound. The assembly of atoms held together by chemical bonds is called a **molecule**. A molecule is the smallest particle of a compound that has the composition and properties of a larger part of it. A molecule is made of two or more atoms which may be the same as in a molecule of oxygen or nitrogen or they may be different elements. A molecule made of two or more different kinds of atoms is a chemical **compound**. The properties of a chemical compound are usually quite different from the properties of its component elements. Each water molecule contains two atoms of hydrogen and one atom of oxygen, and the chemical properties of water are quite different from those of either hydrogen or oxygen. Chemists state this fact by writing the chemical formula of water as H_2O. Thus the chemical formula gives, in the chemist's shorthand, the kinds of atoms that are present in the molecule and states their relative proportions.

A large portion of each cell is simply water. The percentage of water in different human tissues ranges from about 20 per cent in bone to 85 per cent in brain cells. About two thirds of our total body weight is water, and as much as 95 per cent of the jelly fish is water. Water serves several important functions in living systems. Most other chemicals present are dissolved in it. As we shall see, these substances must be dissolved in a watery medium in order to react one with another. Water dissolves the waste products of metabolism and assists in their removal from the cell and the organism. Water has a high heat capacity; that is, it has a great capacity for absorbing heat while undergoing minimal changes in its own temperature. This results from the fact that the neighboring water molecules in ice or in liquid water are held together by hydrogen bonds (Fig. 3–36) and some heat energy may be dissipated in breaking these hydrogen bonds. Water thus protects the cell against sudden thermal changes.

Water has the property of absorbing a great deal of heat as it changes from a liquid to a gas, thus enabling an organism to dissipate excess heat by evaporating water. For example, a football player weighing 100 kilograms might lose two kilograms of water from his body as perspiration in

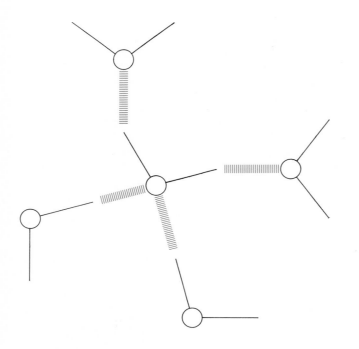

FIGURE 3–36   Hydrogen bonding of water molecules. Each water molecule tends to form hydrogen bonds with four neighboring water molecules. The hydrogen bonds are indicated by dashed lines.

the course of an hour's scrimmage. The heat of vaporization of water is 574 kcal per kg, hence $574 \times 2 = 1148$ kcal. If the water had not been vaporized, and if all the heat produced during the scrimmage had remained within his body, his body temperature would have risen 11.5°C or nearly 20°F. The characteristic high heat conductivity of water makes it possible for heat to be distributed evenly throughout the tissues of the body. Water serves a further indispensable function as a lubricant, and is present in body fluids wherever one organ rubs against another, and in the joints where one bone moves on another.

In contrast to a chemical compound such as water, a **mixture** is composed of two or more kinds of atoms or molecules that may be combined in varying proportions. Water and alcohol may be mixed in any ratio, and air is a mixture of varying amounts of oxygen and nitrogen plus small amounts of water vapor, carbon dioxide, argon and other gases. Although a pure compound will exhibit certain fixed and unchanging chemical and physical properties by means of which it can be identified, a mixture will have properties that vary with the relative abundance of its constituent parts.

The weight of any single atom or molecule is much too small to be expressed conveniently in terms of grams or micrograms. Instead these weights are expressed in terms of the **atomic weight** unit (dalton), approximately the weight of a proton or neutron. On this scale the weight of the lightest element, hydrogen, has an atomic weight of 1, carbon has an atomic weight of 12, and oxygen has an atomic weight of 16. The **molecular weight** is the sum of the atomic weights of the atoms in the molecule, thus the molecular weight of water is 18, or $(2 \times 1) + 16$. A molecule of the simple sugar glucose, composed of six carbon atoms, twelve hydrogen atoms, and six oxygen atoms, has the formula $C_6H_{12}O_6$ and a molecular weight of $(6 \times 12) + (12 \times 1) + (6 \times 16)$, or 180 daltons.

The constituent atoms of a molecule are joined together by forces called chemical bonds (Fig. 3–37). Ionic, covalent and hydrogen bonds are of importance in the molecules present in biological materials. Ionic and hydrogen bonds are relatively weak and easily broken. **Covalent bonds** are strong, and their formation is endergonic, that is, energy must be supplied to form them. Both the formation and cleavage of covalent bonds are carried out by enzymatic reactions within the cell.

Ionic bonds result from the attraction of particles with unlike charges, i.e., between an electropositive atom, such as a sodium ion, and an electronegative atom, such as a chloride ion. These unite to form crystalline sodium chloride or table salt.

Hydrogen bonds are very weak bonds formed when a hydrogen atom is shared between two atoms, one of which is usually oxygen. They tend to form between any hydrogen atom covalently bonded to oxygen or nitrogen and any strongly electronegative atom, usually oxygen or nitrogen in another molecule, or in another part of the same molecule. Hydrogen bonds have a specific length and a specific direction, which is of great importance in their role in determining the structure of macromolecules such as proteins and nucleic acids. Hydrogen bonds are geometrically quite precise. The water molecules in liquid water are held together in part by hydrogen bonds. This results from the fact that the oxygen atom and two hydrogen atoms in a water molecule form a triangle. The electrons belonging to the hydrogen atoms are more strongly attracted to the oxygen nucleus than to the hydrogen nuclei, and tend to be located nearer the oxygen atom. Because of this, the two hydrogen atoms have a small local positive charge, and the oxygen atom has a small local negative charge, although the water molecule as a whole is electrically neutral.

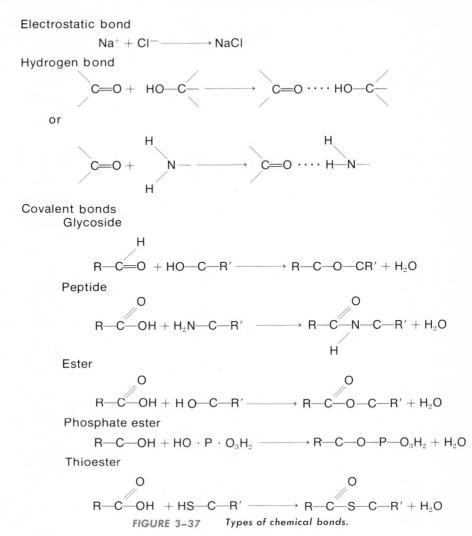

FIGURE 3-37   *Types of chemical bonds.*

Molecules which are positive at one end and negative at the other are said to be **polar**. Such molecules are usually soluble in water since the electrostatic attraction of the negative and positive charges tend to align the water molecules around them. When the positively charged hydrogen atom of one water molecule is next to an atom carrying an electronegative charge, such as the oxygen atom, in another water molecule, the attraction between them forms a hydrogen bond. The stronger covalent bonds are formed when two adjacent atoms share a pair of electrons. This is in contrast to the ionic bond in which one atom transfers an electron to another atom. Thus sodium transfers an electron to chlorine in the formation of a molecule of sodium chloride.

Each atom of hydrogen, as we have seen, has a nucleus and one electron in its orbit. When two atoms of hydrogen join to form a molecule of hydrogen, each atom shares its electron with the other atom so that, in a sense, each hydrogen has completed its first shell of two electrons. In this instance, the electrons are shared equally by the two atoms and there is no greater probability that the electrons will be nearer one nucleus than the other. Such a bond is said to be a **nonpolar covalent bond**. When two hydrogen atoms are covalently bonded to an oxygen atom to form a molecule of water, the oxygen atom shares electrons with two hydrogen atoms and completes its outer electron circle of eight. At the same time, each hydrogen atom completes its first shell of two electrons. This covalent bond is a little different from the one between two hydrogen atoms, because when a covalent bond is formed between two different elements, the shared electrons tend to be pulled more strongly to one element than to the other. Such a bond is called a **polar covalent bond.** As we have seen, in a water molecule the electrons tend to lie closer to the oxygen nucleus than to the hydrogen nucleus, giving the oxygen atom a partial negative charge and the hydrogen atoms, a partial positive charge. Covalent bonds may have all degrees of polarity, from ones in which the electrons are exactly shared as in the hydrogen molecule, to ones in which the electrons are much closer to one atom than to the other, and the bond is therefore quite polar. In a sense, an ionic bond is simply one extreme of this, in which the electrons are pulled completely from one atom to the other.

Since electrons do not remain in any one position but are constantly moving, the bond may exhibit resonance. It may be essentially covalent one instant and ionic another. For this reason a compound that is primarily covalent may be slightly ionized. To understand the properties of various molecules in living systems we must have a good grasp of this phenomenon of the polarity of bonds. Two other types of weak bonds, van der Waals bonds and hydrophobic bonds, are of particular importance in the structure of protein molecules, and will be considered in Chapter 7.

The covalent bonds of great importance in joining molecules together are **anhydro bonds**, formed by the removal of water between two molecules, i.e., by removing an OH group from one molecule and an H group from the other. In biosynthetic reactions, the bond is usually formed not by actually removing a molecule of water, but by substituting a phosphate group or some other group for an OH group on one molecule and then removing the phosphate and an H from the second molecule to liberate inorganic phosphate and form the anhydro bond.

The anhydro bonds of carbohydrates are **glycosidic bonds**, formed by removing an H from an alcohol group of one sugar and an OH from an aldehyde group of the other (Fig. 3–37). The anhydro bonds of proteins are **peptide bonds**, formed by removing an OH from the carboxyl group of one amino acid and an H from the amino group of another. The anhydro bonds of fats are **ester bonds**, formed by removing an OH from the carboxyl group of a fatty acid and an H from an alcohol group of glycerol. Other ester bonds of great biological importance are phosphate esters, formed by removing an H from phosphoric acid and an OH from a sugar, and thioesters, which involve the removal of an OH of the carboxyl group of an acid and an H from an SH group rather than an OH group. Nucleotides have a glycosidic bond between the sugar and the purine or pyrimidine and a phosphate ester bond joining the phosphate to the sugar. Coenzyme A, for example, forms thioesters (sulfur esters) with a variety of substances. Acetyl coenzyme A is the thioester of acetic acid and coenzyme A.

There is a chemical tradition to refer to molecules other than carbonates, containing the element carbon, as "organic" compounds, and to refer to all other ones as inorganic compounds. Inorganic compounds do play important roles in the physiology of living organisms.

The outer orbit of the carbon atom contains four electrons which can be shared in a number of different ways with adjacent atoms. Because of this, carbon can form a wider variety of compounds than any other element. It was believed at one time that organic compounds were in some way uniquely different from others and could be produced only by living matter. This hypothesis was disproved in 1828 when the German chemist Wöhler succeeded in synthesizing **urea**, an organic product found in the urine of many animals, from the inorganic compounds **ammonium sulfate** and **potassium cyanate**. Since that time many thousands of organic compounds have been prepared by chemical synthesis. Some of these compounds are quite complex molecules of great biological importance, such as vitamins, hormones, antibiotics and other drugs.

Among the inorganic compounds present in living systems are water, carbon dioxide, acids, bases and salts. An **acid** is a compound which releases hydrogen ions, H^+, when dissolved in water. Acids turn blue litmus paper to red, and have a sour taste. Hydrochloric (HCl) and sulfuric (H_2SO_4) are inorganic acids; lactic (from sour milk) and acetic (from vinegar) are two common organic acids. A **base** is a compound which releases hydroxyl ions (OH^-) when dissolved in water. Bases turn red litmus paper blue. Sodium hydroxide (NaOH) and ammonium hydroxide (NH_4OH) are common inorganic bases.

For convenience, the degree of acidity or alkalinity of a fluid, its hydrogen ion concentration, may be expressed in terms of pH, the negative logarithm of the hydrogen ion concentration. Most animal and plant cells are neither strongly acid nor alkaline, but contain an essentially neutral mixture of acidic and basic substances. The hydrogen ion concentration of such a solution (or of pure water) is about 10^{-7} molar, and thus its pH is 7.0. At pH 7.0 the concentrations of free H^+ ions and of free OH^- ions are exactly equal. Any considerable change in the pH of a cell is inconsistent with life. Since the scale is a logarithmic one, a solution with a pH of 6 has a hydrogen ion concentration 10 times greater than a solution with a pH of 7, and is much more acidic.

When an acid and a base are mixed, the hydrogen ion of the acid unites with the hydroxyl ion of the base to form a molecule of water (H_2O). The remainder of the acid (anion) combines with the rest of the base (cation) to form a **salt**. Hydrochloric acid, for example, reacts with sodium hydroxide to form water and sodium chloride, common table salt:

$$H^+Cl^- + Na^+OH^- \rightarrow H_2O + Na^+Cl^-$$

A salt may be defined as a compound in which the hydrogen atom of an acid is replaced by some metal.

When a salt, an acid or a base is dissolved in water it dissociates into its constituent ions. These charged particles can conduct an electric current, hence these substances are known as **electrolytes**. Sugars, alcohols, and the many other substances which do not separate into charged particles when dissolved, and therefore do not conduct an electric current, are called **nonelectrolytes.**

Cells and extracellular fluids contain a variety of mineral salts, of which sodium, potassium, calcium and magnesium are the chief cations (positively charged ions), and chloride, bicarbonate, phosphate and sulfate are the important anions (negatively charged ions). Although the body fluids of terrestrial animals differ considerably from sea water in their total salt content, they resemble it in general in the kinds of salts present and in their relative concentrations (Table 3–2). The total concentration of salt in the body fluids of most marine animals is equal to that in sea water, about 3.4 per cent. Vertebrates, whether terrestrial, fresh water or marine, have less than 1 per cent of salts in their body fluids. The body fluids of fresh water and terrestrial invertebrates contain 0.3 to 0.7 per

TABLE 3–2   The Ionic Composition of Sea Water and the Body Fluids of Various Animals, with Concentrations Expressed Relative to Sodium as 100

	Na	K	Ca	Mg	Cl	SO$_4$
Sea Water (Woods Hole)	100	2.74	2.79	13.94	136.8	7.10
Aurelia, coelenterate	100	2.90	2.15	10.18	113.1	5.15
Strongylocentrotus, echinoderm	100	2.30	2.28	11.21	116.1	5.71
Phascolosoma, sipunculid	100	10.07	2.78	—	114.1	—
Venus, mollusk	100	1.66	2.17	5.70	117.3	5.84
Carcinus, crustacean	100	2.32	2.51	3.70	105.2	3.90
Hydrophilus, insect	100	11.1	0.92	16.8	33.6	0.12
Lophius, fish	100	2.85	1.01	1.61	71.9	—
Frog, amphibian	100	2.40	1.92	1.15	71.4	—
Man, mammal	100	3.99	1.78	0.66	84.0	1.73

cent salts. Life processes require the presence of certain salts in relative concentrations that lie within certain limits.

The blood of man and other terrestrial vertebrates is not simply a dilute sea water, but differs in having relatively more potassium and less magnesium and chloride than sea water. Life probably originated in the sea and the cells of those early organisms were adapted to the relative concentration of the salts present (though that early sea probably had a lower total concentration of salts than the present sea). In the course of evolution, animals have evolved with body fluids having generally similar relative concentrations of salts, for any marked difference in the kinds of salts present would inhibit certain enzymes in the cells and place that kind of animal at a marked disadvantage in the competition for survival. Some animals have evolved kidneys and other excretory organs that selectively retain or secrete certain ions, thus leading to body fluids with somewhat different relative concentrations of salts. The concentration of each type of ion is determined by the relative rates of its uptake and excretion by the organism.

The concentration of the various salts is kept extremely constant under normal conditions, and any great deviation from the normal values causes marked effects on cell function, even death. A decreased concentration of calcium ions in the blood of mammals results in convulsions and death. Heart muscle can contract normally only in the presence of the proper balance of sodium, potassium and calcium ions. If a frog heart is removed from the body and placed in a pure sodium chloride solution, it soon stops beating in the relaxed condition. If placed in a solution of potassium chloride or a mixture of sodium and calcium chloride, it will stop in the contracted state. It will continue to beat, however, if placed in a solution containing the proper balance of these three salts. This frog heart method is so sensitive that it can be used to measure the concentration of calcium ions in solutions.

In addition to these specific effects of salts on certain cell functions, mineral salts are important in maintaining the osmotic relationships between the cell and its environment.

SECTION 3–13

BIOLOGICAL MOLECULES

The major types of organic compounds that cells synthesize and use include carbohydrates, proteins, lipids, nucleic acids and steroids. Some of these are required for cell structure, others serve to supply energy for its functioning, and still others are of prime importance in regulating chemical reactions within the cell. Carbohydrates and lipids are important sources of chemical energy for almost every form of life; proteins are structural elements but are of even greater importance as catalysts and regulators of cell processes. Nucleic acids are of prime importance in the storage and transfer of information used in the synthesis of specific proteins and other molecules. The types of substances and even their relative proportions are remarkably similar in cells from the various parts of the body and in cells from different organisms. A bit of human liver and the cell of an amoeba both contain about 80 per cent water, 12 per cent protein, 2 per cent nucleic acid, 5 per cent lipid, 1 per cent carbohydrate and a fraction of 1 per cent of sterols and other substances. Certain specialized cells, of course, have unique patterns of chemical constituents. The mammalian brain, for instance, is rich in certain kinds of lipids.

Carbohydrates

Carbohydrates are compounds containing carbon, hydrogen and oxygen atoms in the ratio of 1 C : 2 H : 1 O. Sugars, starches and celluloses are examples of carbohydrates. The simplest carbohydrates of biological importance are the single sugars with the formula $C_6H_{12}O_6$. **Glucose,** also called dextrose, and **fructose** are single sugars that differ slightly in the arrangement of the constituent atoms in the molecule. These different arrangements give the substances slightly different chemical properties. Compounds with identical molecular formulas but different arrangements of the atoms are termed **isomers.** These internal structures are represented by structural formulas in which the atoms are represented by their symbols — C, H, O, and so forth — and the chemical bonds, the forces that hold the atoms together, are indicated by connecting lines (Fig. 3–38). Hydrogen, as you can see, has one bond to connect to other atoms; oxygen has two, and carbon, four.

Carbon atoms can unite with each other as well as with other kinds of atoms and form an infinite variety of compounds. Carbon atoms linked together may form a long chain, as in a fatty acid, and a branched chain, as in certain amino acids; they may form rings, as in purines and pyrimidines, and complex rings (two-, three- and four-membered rings), as

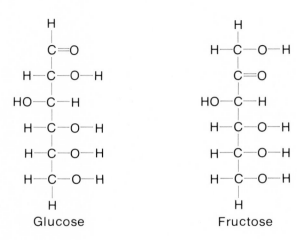

FIGURE 3–38   *Structural formulas of two simple sugars, glucose and fructose.*

Glucose

Fructose

in sterols and steroids. Molecules are in fact three-dimensional structures and not simple two-dimensional ones, as the formulas on a printed page would suggest. There are more complex ways of representing the third dimension of the molecule. Since the properties of the compound depend in part on its **conformation**, its three-dimensional structure, such conformational, three-dimensional formulas are helpful in understanding the intimate relations between molecular structure and biological function. The molecules of glucose and other single sugars in solution are not extended straight chains as shown in Figure 3–38, but are present as flat rings formed by the union of the carbon with one of the oxygens. Glucose for example has five carbons and one oxygen in its ring (Fig. 3–39).

Glucose is the only single sugar found in any quantity in the body. The other carbohydrates we eat are converted by the liver into glucose. Glucose is an indispensable component of blood. Normally it is present in the blood and tissues of mammals in a concentration of about 0.1 per cent by weight. No particular harm results from a simple increase in the amount of glucose in the body, but a reduced concentration leads to an increased irritability of certain brain cells, so that they respond to very slight stimuli. As a result of impulses from these cells to the muscles, twitches, convulsions, unconsciousness and death may ensue. Brain cells require glucose for their metabolism and a certain minimal concentration of glucose in the blood is necessary to supply this. An extremely complex mechanism, involving the nervous system, liver, pancreas, pituitary and adrenal glands, maintains the proper concentration of glucose in the blood.

Double sugars, all of which have the formula $C_{12}H_{22}O_{11}$, are made of two single sugars joined together by the removal of a molecule of water. Both cane and beet table sugars are **sucrose**, a combination of one molecule of glucose with one of fructose. Several other double sugars are found in biological systems; all have the formula $C_{12}H_{22}O_{11}$, but differ in the arrangement of their constituent atoms and hence in some of their chemical and physical properties. **Maltose**, or malt sugar, is composed of two molecules of glucose; **lactose**, or milk sugar, found in the milk of all mammals, is made of one molecule of glucose and one of galactose, a third kind of single sugar. Sugars differ markedly in their sweetness. Fructose is the sweetest of the common sugars; sucrose is intermediate, and lactose, the least sweet, is less than one tenth as sweet as fructose (Table 3–3). **Saccharin**, a synthetic sweetening agent, is much sweeter than any sugar and is used by dieters and diabetics who want to sweeten food without using sugar.

The largest carbohydrate molecules are starches and celluloses, composed of a large number of single sugars joined together either in a single long chain (**amylose**) or in a branched chain (**amylopectin**). Since the exact number of sugar molecules joined to make a starch molecule is unknown, and indeed may vary from one molecule to the next, the formula for starch may be written $(C_6H_{10}O_5)_x$, where x stands for the unknown, large number

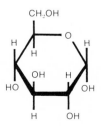

FIGURE 3–39   Structural formula of glucose depicted as a six-membered ring containing five carbon atoms and one oxygen atom.

TABLE 3–3　Relative Sweetness of Some of the Common Sugars

Sugar	Relative Sweetness (Sucrose 100)
Lactose	16.0
Galactose	32.1
Maltose	32.5
Glucose	74.3
Sucrose	100.0
Fructose	173.3
Saccharin	55,000

of single sugars which comprise the starch molecule. Starches and other polysaccharides are hydrolyzed by enzymes called **amylases** to shorter chains of single sugar units, and ultimately to free single sugars. The amylases catalyze reactions in which water molecules enter at the anhydro bond between adjacent units and result in the formation of free single sugars.

Starches vary in the number and kind of sugar molecules present and are common constituents of both plant and animal cells. Animal starch, called **glycogen**, differs from plant starch in being quite highly branched (Fig. 3–40) and more soluble in water. Carbohydrates are stored in plants as starches and in animals as glycogen; glucose could not be stored as such, for its molecules are small and would leak out of the cells. The larger, less soluble starch and glycogen molecules will not pass through the plasma membrane. In man and other higher animals, glycogen is stored especially in the liver and muscles. Liver glycogen is readily converted into glucose by four enzymes working sequentially and the glucose is then carried in the blood to other parts of the body.

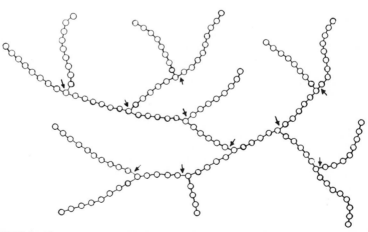

FIGURE 3–40　*A diagrammatic representation of a portion of a glycogen molecule showing its highly branched structure. Each circle represents a glucose molecule bound by a glycoside bond from its carbon 1 to carbon 4 of the adjacent glucose in the straight chain portion of the molecule, or from its carbon 1 to carbon 6 of the adjacent glucose at the branch points indicated by arrows.*

Most plants have a strong supporting outer wall of **cellulose**, an insoluble compound sugar (polysaccharide), which, like starch, is made of many glucose molecules. However, the chemical bonds between the successive glucose molecules of cellulose are β-glycosidic bonds, and are different from the ones in starch or glycogen. They are not split by the enzymes that digest starches. Paper, explosives, rayon, celluloid, certain plastics, photographic film and varnishes are some of the commercially important forms of cellulose.

Carbohydrates serve primarily as a readily available fuel to supply energy for metabolic processes in the cell. Glucose is ultimately metabolized to carbon dioxide and water and energy is released:

$$C_6H_{12}O_6 + 6O_2 \rightarrow 6H_2O + 6CO_2 + \text{energy}$$

Some carbohydrates are combined with proteins or lipids and serve as structural components of cells and cell walls. Ribose and deoxyribose are five-carbon sugars of great importance biologically as components of ribonucleic acid (RNA) and deoxyribonucleic acid (DNA).

Lipids (Fats)

True fats are also composed of carbon, hydrogen and oxygen, but have much less oxygen in proportion to the carbon and hydrogen than carbohydrates have. Fats have a greasy or oily consistency; some, such as beef tallow or bacon fat, are solid at ordinary temperatures, others, such as olive oil or cod liver oil, are liquid. Each molecule of fat is composed of one molecule of glycerol and three molecules of fatty acid; all such neutral fats, called **triglycerides**, contain glycerol but may differ in the kinds of fatty acids present. Fatty acids are long chains of carbon atoms with a carboxyl group at one end. All the fatty acids in nature have an even number of carbon atoms—palmitic has 16, and stearic, 18. Fatty acids with one or more double bonds are called "unsaturated." Oleic acid has 18 carbons and one double bond (and hence has two less hydrogen atoms than stearic). A fat common in beef tallow, tristearin, $C_{57}H_{110}O_6$, has three molecules of stearic acid and one of glycerol (Fig. 3–41). Fats composed of unsaturated fatty acids are usually liquids at room temperature, whereas saturated fats such as tristearin are solids.

Fats are important as fuels and as structural components of cells, especially cell membranes. Glycogen or starch is readily converted to glucose and metabolized to release energy quickly; the carbohydrates serve as short-term sources of energy. Fats yield more than twice as much energy per gram as do carbohydrates and thus are a more economical

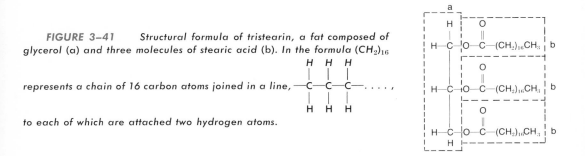

FIGURE 3–41   *Structural formula of tristearin, a fat composed of glycerol (a) and three molecules of stearic acid (b). In the formula* $(CH_2)_{16}$ *represents a chain of 16 carbon atoms joined in a line,* —C—C—C—. . . . , *to each of which are attached two hydrogen atoms.*

form for the storage of food reserves. Carbohydrates can be transformed by the body into fats and stored in this form—a restatement of the generally known fact that starches and sugars are "fattening." The reverse may also occur to some slight extent: Parts of the fat molecule may be converted into glucose and other carbohydrates. This has been shown by preparing fatty acids or glycerol labeled with radioactive or heavy carbon, feeding or injecting these into a rat or dog, and then isolating glucose from the blood or glycogen from the liver and demonstrating that these molecules now contain some of the labeled carbon atoms.

Fats are important structural elements of the body. The plasma membrane around each cell and the nuclear membrane contain fatty substances as important constituents, and the myelin sheath around the nerve fibers (p. 67) has a high lipid content. Animals store fat as globules within the cells of adipose tissue. The layer of adipose tissue just under the skin serves as an insulator against the loss of body heat. Women tend to have a thicker layer of adipose tissue than men do and should be more tolerant to cold. Whales, which live in cold water and have no insulating hair, have an especially thick layer of fat (blubber) just under the skin for this purpose. The subcutaneous fat in man keeps the skin firm in addition to restricting heat loss. The fat deposits are not long-term reserves of stored food used only in starvation, but are constantly being used up and re-formed. Studies with labeled fatty acids showed that mice replace half of their stored fats each week.

Besides the true fats, composed of glycerol and fatty acids, lipids include several related substances that contain components such as phosphorus, choline and sugars, in addition to fatty acids. The **phospholipids** are important constituents of the membranes of plant and animal cells in general and of nerve cells in particular. The fatty acid portion of the phospholipid molecule is hydrophobic, not soluble in water. The other portion, composed of glycerol, phosphate and a nitrogenous base such as choline, is ionized and readily water-soluble. For this reason, phospholipid molecules in a film tend to be oriented with the polar, water-soluble portion pointing one way and the nonpolar, fatty acid portion pointing the other. This configuration appears to underlie the three-layered unit membrane structure of proteins:phospholipid:proteins.

Plants contain certain red and yellow pigments called **carotenoids,** which are included with the lipids because they are insoluble in water and have an oily consistency. Carotenoids are found in all plant cells from the lowest to the highest. They play some role in phototropism, the orientation of plants toward light. One of the common carotenoids, **carotene,** is a molecule with a six-carbon ring at each end of a long chain of carbon atoms, linked together with alternating single and double bonds. When the carotene molecule is cut in half it yields a molecule of vitamin A. The chemical present in the cells of the retina of the eye that is sensitive to light is termed **retinene.** Retinene is a derivative of vitamin A and undergoes a chemical reaction in the presence of light, and thus is involved in the actual reception of light stimuli. It is remarkable to realize that photoreceptors or eyes have evolved independently in three different lines of animals—the mollusks, the vertebrates and the insects. These organisms have no common evolutionary ancestor, i.e., the three types of eyes have no common evolutionary source, yet each one of these eyes contains the same chemical compound, retinene, involved in the photoreception process. In contrast to a number of other instances, the fact that retinene is present in each of the eyes is not the result of a common evolutionary ancestry, but perhaps because of some unique fitness of this kind of molecule for the process of light reception.

Steroids

Steroids are complex molecules containing carbon atoms arranged in four interlocking rings, three of which contain six-carbon atoms each and the fourth of which contains five (Fig. 3–42). Some steroids of biological importance are the male and female sex hormones, the adrenal cortical hormones, bile salts, cholesterol and vitamin D. Cholesterol is an important structural component of nervous tissue and other tissues, and is the source of the steroid hormones. The steroid hormones are of prime importance in regulating certain phases of metabolism.

Proteins

Proteins, containing carbon, hydrogen, oxygen, nitrogen and usually sulfur and phosphorus constitute all enzymes, certain hormones, and many of the important structural components of the cell. Protein molecules are very large, containing thousands of atoms, and their structure may be extremely complex. **Hemoglobin,** a typical protein, is the red pigment responsible for the color of blood. Some idea of the complexity of the hemoglobin molecule can be gained from its formula: $C_{3032}H_{4816}O_{872}N_{780}S_8Fe_4$. (Fe is the symbol for iron.) Large as this molecule is, it is only a small-to-medium sized protein. A large fraction of the proteins present within cells are **enzymes**, substances which control the rates at which the many processes of the cell occur.

Protein molecules are made of simpler components, known as **amino acids**. Some 35 different amino acids have been found as the result of the chemical breakdown of proteins; the existence of about 25 of these has been confirmed by further investigations.

The 20-odd amino acids commonly found in proteins all contain an amino group ($-NH_2$) and a carboxyl group ($-COOH$) but differ in their side chains. Glycine, the simplest, has an H for its side chain, and alanine has a $-CH_3$ group (Fig. 3–43). The amino group enables the amino acid to act as a base and combine with acids; the acid group enables it to combine with bases. Amino acids and proteins in solution serve as **buffers**; i.e., they resist changes in acidity or alkalinity. Amino acids are linked together to form proteins by **peptide bonds** between the amino group of one and the

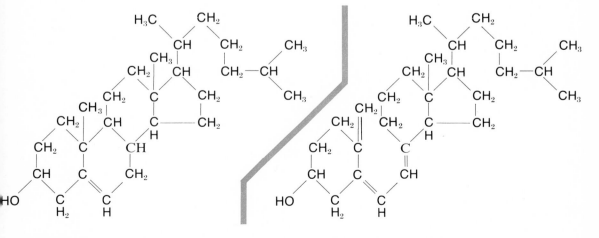

FIGURE 3–42   Structural formulas of cholesterol (left) and vitamin D_3 (right).

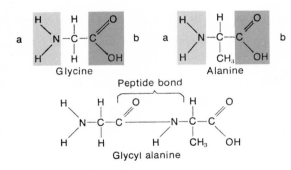

Glycine

Alanine

Peptide bond

Glycyl alanine

FIGURE 3–43   *The structural formulas of the amino acids glycine and alanine, showing (A) the amino group and (B) the acid or carboxyl group. These may be joined by a peptide bond to form glycylalanine by removing one molecule of water.*

carboxyl group of another (Fig. 3–43). Since each protein contains perhaps hundreds of amino acids combined in a certain proportion and in a particular sequence, an almost infinite variety of protein molecules is possible. Analytic methods have been developed for determining the exact sequence of amino acids in a protein molecule. *Insulin*, the hormone secreted by the pancreas and used in the treatment of diabetes, was the first protein whose structure was elucidated (Fig. 3–44). *Ribonuclease*, an enzyme secreted by the pancreas, was the first enzyme for which the exact order of its amino acids was known. Not all proteins contain all of the possible amino acids.

It is possible to distinguish several different levels of organization in the protein molecule. The first level is the so-called *primary structure* which depends upon the sequence of amino acids in the polypeptide chain. This sequence, as we shall see, is determined in turn by the sequence of

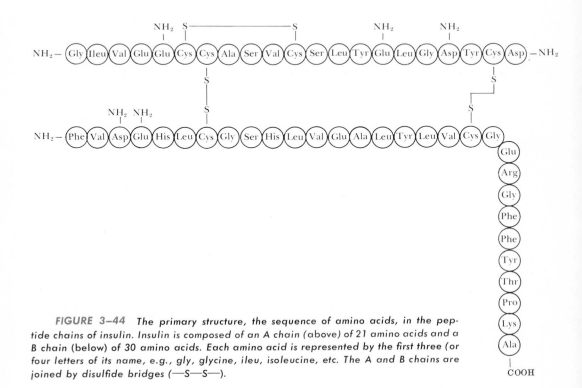

FIGURE 3–44   *The primary structure, the sequence of amino acids, in the peptide chains of insulin. Insulin is composed of an A chain (above) of 21 amino acids and a B chain (below) of 30 amino acids. Each amino acid is represented by the first three (or four letters of its name, e.g., gly, glycine, ileu, isoleucine, etc. The A and B chains are joined by disulfide bridges (—S—S—).*

nucleotides in the RNA and DNA of the nucleus of the cell. A second level of organization of protein molecules involves the coiling of the polypeptide chain into a helix or into some other regular configuration. The polypeptide chains ordinarily do not lie out flat in a protein molecule, but they undergo coiling to yield a three-dimensional structure. One of the common secondary structures in protein molecules is the α-*helix*, which involves a spiral formation of the basic polypeptide chain. The α-helix is a very uniform geometric structure with 3.6 amino acids occupying each turn of the helix. The helical structure is determined and maintained by the formation of hydrogen bonds between amino acid residues in successive turns of the spiral. A third level of structure of protein molecules involves the folding of the peptide chain upon itself to form globular proteins. Again, weak bonds such as hydrogen, ionic and hydrophobic bonds form between one part of the peptide chain and another part, so that the chain is folded in a specific fashion to give a specific overall structure of the protein molecule. Covalent bonds such as disulfide bonds ($-S-S-$) are important in the tertiary structure of many proteins. The biological activity of a protein depends in large part on the specific tertiary structure which is held together by these bonds. When a protein is heated or treated with any of a variety of chemicals, the tertiary structure is lost. The coiled peptide chains unfold to give a random configuration accompanied by a loss of the biological activity of the protein. This change is termed "denaturation."

Each cell contains hundreds of different proteins, and each kind of cell contains some proteins which are unique to it. There is evidence that every species of plant and animal has certain proteins which differ from those of every other species. The degree of difference in the proteins of two species depends upon the evolutionary relationship of the forms involved. Organisms less closely related by evolution have proteins which differ more markedly than those of closely related forms. Because of the interactions of unlike proteins, grafts of tissue taken from one species of animal usually will not grow when implanted into a host of different species, but degenerate and are sloughed off by the host. Indeed, even grafts made between members of the same species usually will not grow; but only grafts between genetically identical donors and hosts—identical twins or members of closely inbred strains.

Pure amino acids from proteins have a sweet taste. (Monosodium glutamate, or MSG, is the salt of glutamic acid, an amino acid of special importance in metabolism. It is widely used in cooking to add a "meaty" taste.) When proteins are eaten, they are hydrolyzed to amino acids before they are absorbed into the bloodstream. The amino acids are carried to all parts of the body to be made into new protein or to be metabolized for the release of energy. When a man eats "beef" proteins in a steak, they are digested and split to their constituent amino acids. The human tissues then rebuild the amino acids into "human" proteins.

Proteins play important roles as structural components of cells and as the functional constituents of enzymes and certain hormones; they may also serve as fuel for the liberation of energy. The amino acids first lose their amino group by an enzymatic reaction called **deamination**. The amino group reacts with other substances to form **urea**, and is excreted. The rest of the molecule may be changed, via a series of intermediate steps, into glucose and either used immediately as fuel or stored as glycogen or fat.

Information about the conversion of protein to carbohydrates and to fats has been derived from experiments with substances labeled with isotopes of carbon, hydrogen and nitrogen. In prolonged fasting, after the stores of glycogen and fats are exhausted, the cellular proteins may be used as fuel. The cells of the human body (and animal cells in general) can manufacture some, but not all, of the amino acids if the proper raw materials are present. Those which cannot be made by the animal body must be obtained directly or indirectly from plants as food or perhaps from the bacteria that live in our intestines.

Plants can synthesize all the amino acids from simpler substances. The ones which animals cannot synthesize, but must obtain in their food, are known as *essential amino acids*. It must be understood that these amino acids are no more "essential" as components of proteins than are any other ones; they are simply essential in the diet, since they cannot be synthesized.

Nucleic Acids

Nucleic acids are complex molecules, larger than most proteins, and contain carbon, oxygen, hydrogen, nitrogen and phosphorus. They were first isolated by Miescher in 1870 from the nuclei of pus cells and gained their name from the fact that they are acidic and were first identified in nuclei. At one time it was believed that there were but two kinds of nucleic acid — one containing ribose and called *ribose nucleic acid* or **RNA**, and one containing deoxyribose and called *deoxyribose nucleic acid* or **DNA**. It is now clear that there are many different kinds both of RNA and of DNA which differ in their structural details and in their metabolic functions. DNA occurs in the chromosomes in the nucleus of the cell and, in much smaller amounts, in mitochondria and chloroplasts. It is the primary repository for biological information. RNA is present in the nucleus, especially in the nucleolus, in the ribosomes, and in lesser amounts in other parts of the cell.

Nucleic acids are composed of units, called *nucleotides*, each of which contains a nitrogenous base, a five-carbon sugar and phosphoric acid. Two types of nitrogenous bases, purines and pyrimidines, are present in nucleic acids (Fig. 3–45). RNA contains the purines *adenine* and *guanine* and the pyrimidines *cytosine* and *uracil*, together with the pentose, ribose and phosphoric acid. DNA contains adenine and guanine, cytosine and the pyrimidine *thymine*, together with deoxyribose and phosphoric acid. The molecules of nucleic acids are made of linear chains of nucleotides, each of which is attached to the next by ester bonds between the sugar part of one and the phosphoric acid of the next (Fig. 3–46). The specificity of the nucleic acid resides in the specific sequence of the four kinds of nucleotides present in the chain; for example, CCGATTA might represent a segment of a DNA molecule, with C = cytosine, G = guanine, A = adenine and T = thymine.

An enormous mass of evidence now indicates that DNA is responsible for the specificity and chemical properties of the *genes,* the units of heredity. There are several kinds of RNA, each of which plays a specific role in the biosynthesis of specific proteins by the cell (cf. Chap. 7).

Nucleotides and Coenzymes

Related structurally to nucleic acids but with quite different roles in cellular function are several mono- and dinucleotides. Each is composed of phosphoric acid, ribose and a purine or pyrimidine base, like the units

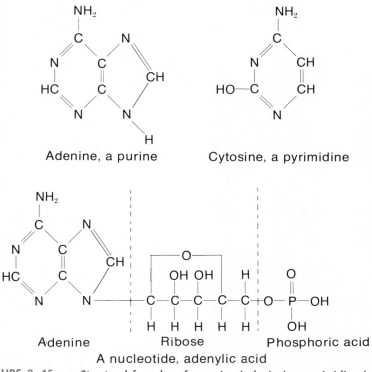

FIGURE 3–45   *Structural formulas of a purine (adenine), a pyrimidine (cytosine) and a nucleotide (adenylic acid).*

comprising the nucleic acids. Each of the bases may form a nucleoside triphosphate, with base, sugar and three phosphate groups linked in a row. **Adenosine triphosphate**, abbreviated **ATP**, composed of adenine, ribose and three phosphates, is of major importance as the "energy currency" of all cells. The two terminal phosphate groups are joined to the nucleotide by energy rich bonds, indicated by the $\sim$P symbol. The biologically useful energy of these bonds can be transferred to other molecules; most of the chemical energy of the cell is stored in these $\sim$P bonds of ATP, ready to be released when the phosphate group is transferred to another molecule. **Guanosine triphosphate, GTP**, is specifically required in certain steps in the synthesis of proteins; **uridine triphosphate**, UTP, is specifically required in certain steps in carbohydrate metabolism, e.g., glycogen synthesis; and **cytidine triphosphate**, CTP, is specifically required for the synthesis of fats and phospholipids. All four nucleotide triphosphates are necessary for the synthesis of RNA and all four deoxyribose nucleoside triphosphates are required for the synthesis of DNA.

Completing the list of nucleotides important in metabolic processes are the dinucleotides NAD, NADP and FAD. **Nicotinamide adenine dinucleotide**, abbreviated **NAD** (and also called diphosphopyridine nucleotide, DPN), consists of nicotinamide, ribose and phosphate attached to an adenine nucleotide, phosphate, ribose and adenine. NAD is of prime importance as a primary electron and hydrogen acceptor in cellular oxidation reactions. Enzymes called **dehydrogenases** remove electrons and hydrogen from molecules such as lactic acid and transfer them to NAD, which in turn passes them on to other electron acceptors. **Nicotinamide adenine dinucleotide phosphate**, abbreviated **NADP** (and also called triphos-

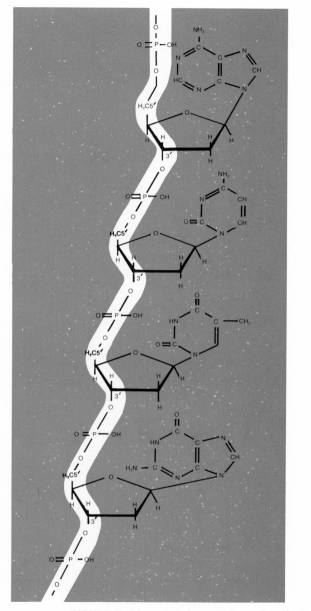

FIGURE 3–46   *A fragment of a polynucleotide chain showing how the phosphate groups form a bridge connecting the 3′ position of one deoxyribose to the 5′ position of the next deoxyribose to form a chain.*

phopyridine nucleotide, TPN), serves as electron and hydrogen acceptor for certain other enzymes. NADP is exactly like NAD except it has a third phosphate group attached to the ribose of the adenine nucleotide. *Flavin adenine dinucleotide, FAD*, consists of riboflavin-ribitol-phosphate-phosphate-ribose-adenine and serves as a hydrogen and electron acceptor for certain other dehydrogenases. Notice that these dinucleotides have vitamins—nicotinamide or riboflavin—as component parts. These molecules NAD, NADP and FAD are termed **coenzymes**; they are cofactors required for the functioning of certain enzyme systems but are only loosely bound to the enzyme molecule and are readily removed. When they have accepted electrons and hydrogens, they are changed from their oxidized form, e.g., NAD, to their reduced form, NADH. They are converted back to the oxidized form when they transfer their electrons to the next acceptor in the chain of respiratory enzymes.

SUGGESTIONS FOR FURTHER READING

Afzelius, Bjorn: *Anatomy of the Cell.* Chicago, University of Chicago Press, 1966.
  A brief discussion of the functional anatomy of the cell, emphasizing those areas of cell biology in which the functional organization is best known.

De Robertis, E. D. P., Nowinski, W. W., and Saez, F. A.: *Cell Biology.* 5th ed. Philadelphia, W. B. Saunders Company, 1970.
  An excellent text describing the cytological features of animal and plant cells.

Fawcett, Don: *The Cell.* Philadelphia, W. B. Saunders Company, 1966.
  A collection of superb electron micrographs of the several kinds of cells and cell organelles accompanied by a brief description of each.

Frey-Wyssling, A., and Mühlethaler, K.: *Ultrastructural Plant Cytology.* Amsterdam, Elsevier Publishing Company, 1965.
  A brief, well illustrated survey of the entire range of cell structures which attempts to integrate their associated biochemical features.

Hall, Thomas S.: *A Source Book in Animal Biology.* New York, McGraw-Hill Book Company, 1951.
  This book presents the development of the cell theory by means of long quotations from some of the original scientific papers.

Kennedy, Donald: *The Living Cell, Readings from the Scientific American.* San Francisco, W. H. Freeman and Company, 1965.
  A lively and readable collection of articles and photographs, which have appeared in the Scientific American, together with some new material.

Porter, K. R., and Bonneville, M. A.: *Fine Structure of Cells and Tissues.* 3d ed. Philadelphia, Lea & Febiger, 1968.
  A collection of superb electron micrographs of animal cells and tissues accompanied by text which is both concise and informative.

Swanson, C. P.: *The Cell.* 2nd ed. Englewood Cliffs, New Jersey, Prentice-Hall, Inc., 1964.
  A brief survey relating cell structure and function.

BIOENERGETICS

CHAPTER 4

The universe consists of matter and energy which are related by Einstein's famous equation $e = mc^2$, where e = energy, m = mass and c represents the velocity of light which is a constant. Einstein's equation provides the theoretical basis for the conversion of matter to energy that occurs in an atom bomb or nuclear reactor. In the more familiar every day world, matter and energy are separate and distinguishable. **Matter** occupies space, and has weight, and energy is the ability to produce a change or motion in matter—the ability to do work. **Energy**, defined as the capacity to do work, to produce a change in matter, may take the form of heat, light, electricity, motion or chemical energy. Physicists recognize **potential energy** as the capacity to do work owing to the position or state of a particle, and **kinetic energy** as the energy of motion. A boulder at the top of a hill has potential energy because of its position. As it rolls down the hill the potential energy is converted to kinetic energy.

ENERGY TRANSFORMATIONS

The never-ending flow of energy within a cell, from one cell to another, and from one organism to another organism is the essence of life itself. Living cells have complex and efficient systems for transforming one type of energy into another. Two important structures in energy transformations are the chloroplasts of green plants and the mitochondria present in both plant and animal cells. The study of energy transformations in living organisms is termed **bioenergetics.**

Three major types of energy transformation can be distinguished in the biological world (Fig. 4–1). In the first type, the radiant energy of sunlight is captured by the green pigment, chlorophyll, present in green plants and transformed by the process termed photosynthesis into chemical energy used to synthesize carbohydrates and other complex molecules from carbon dioxide and water. The radiant energy of sunlight, a form of kinetic energy, is transformed into a type of potential energy. The chemical

FIGURE 4–1   Energy trans-
formations in the biological world.
(1) The radiant energy of sunlight is
transformed in photosynthesis into
chemical energy in the bonds of
organic compounds. (2) The chemical
energy of organic compounds is
transformed during cellular respira-
tion into biologically useful energy,
the energy-rich phosphate bonds of
ATP and other compounds. (3) The
chemical energy of energy-rich phos-
phate bonds is utilized in cells to
do mechanical, electrical, osmotic
or chemical work. Finally energy flows
to the environment as heat in the
"entropy sink."

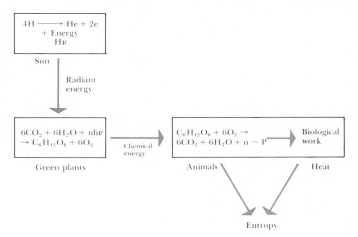

energy is stored in the molecules of carbohydrates and other foodstuffs
as the energy of the bonds connecting its constituent atoms. In a second
type of energy transformation, the chemical energy of carbohydrates and
other molecules is transformed by the process termed **cellular respiration**
into the biologically useful energy of energy-rich phosphate bonds. This
kind of energy transformation occurs in the mitochondrion. A third type
of energy transformation occurs when the chemical energy of these
energy-rich phosphate bonds is utilized by the cells to do work—the
mechanical work of muscular contraction, the electrical work of conducting
a nerve impulse, the osmotic work of moving molecules against a gradient,
or the chemical work of synthesizing molecules for growth. As these
transformations occur, energy finally flows to the environment in the dissipated, useless form of heat. Plants and animals have evolved some
remarkably effective **energy transducers** to carry out these processes,
together with efficient control systems to regulate them and enable the cell
to adjust to variations in environmental conditions.

   The branch of physics that deals with energy and its transformations,
thermodynamics, consists of certain relatively simple basic principles which
are universally applicable to chemical processes whether these are occurring
in living or nonliving systems. It is now well established that the laws of
thermodynamics apply to the world of life.

   Under experimentally controlled conditions, the amount of energy
entering and leaving any system may be measured and compared. It is
always found that energy is neither created nor destroyed but is only
transformed from one form into another. This is an expression of the first
law of thermodynamics, sometimes called the **law of the conservation of**
energy: the total energy of any system and its surroundings remains
constant. As any given system undergoes a change from its initial state to
its final state it may absorb energy from the surroundings or deliver energy
to the surroundings. The difference in the energy content of the system
in its initial and final state must be just equalled by a corresponding change
in the energy content of the surroundings. Heat is a convenient form in
which energy may be measured which is why the study of energy has been
called thermodynamics, i.e., heat dynamics. Nearly every physical or
chemical event is accompanied by the delivery of heat to the surroundings
or the absorption of heat from the surroundings. When a process occurs
with delivery of heat to the surroundings it is said to be **exothermic**. When

it occurs with the absorption of heat from the surroundings it is termed **endothermic**. Although heat is a simple and familiar means by which energy is transferred in many of the machines made by man it is not a useful way of transferring energy in biological systems, for the simple reason that living organisms are basically **isothermal** (equal temperature). There is no significant temperature difference between the different parts of the cell or between the different cells in a tissue. Stated in another way, cells do not act as heat engines. They have no means of allowing heat to flow from a warmer to a cooler body.

The **second law of thermodynamics** may be stated briefly as "the entropy of the universe increases." **Entropy** may be defined as a randomized state of energy that is unavailable to do work. The second law may be phrased that "physical and chemical processes proceed in such a way that the entropy of the system becomes maximal." Entropy then is a measure of randomness or disorder. In almost all energy transformations there is a loss of some heat to the surroundings, and since heat involves the random motion of molecules, such heat losses increase the entropy of the surroundings. Living organisms and their component cells are highly organized and thus have little entropy. They preserve this low entropy state by increasing the entropy of their surroundings. You increase the entropy of your surroundings when you eat a candy bar and convert its glucose to carbon dioxide and water and return them to the surroundings.

The force that drives all processes is the tendency to reach the position of maximum entropy. Heat is either given up or absorbed by the system from the surroundings to allow the system plus its surroundings to reach the state of maximum entropy. The changes of heat and entropy are related by a third dimension of energy termed **free energy**. Free energy may be visualized as that component of the total energy of a system which is available to do work under isothermal conditions. Entropy and free energy are related inversely; as entropy increases during an irreversible process, the amount of free energy decreases. All physical and chemical processes proceed with a decline in free energy until they reach an equilibrium in which the free energy of the system is at a minimum and the entropy is at a maximum. Free energy is useful energy, entropy is degraded useless energy.

CHEMICAL REACTIONS

A chemical reaction is a change involving the molecular structure of one or more substances. Matter is changed from one substance with its characteristic properties to another with new properties, and energy is released or absorbed. Hydrochloric acid reacts with the base sodium hydroxide to yield water and the salt, sodium chloride. In the process, energy is released as heat. The chemical properties of HCl and NaOH are quite different from those of H_2O and NaCl. In chemical shorthand, a plus sign connects the symbols of the reacting substances, HCl and NaOH, and the products of the reaction, H_2O and NaCl. An arrow indicates the direction of the reaction: $HCl + NaOH \rightarrow NaCl + H_2O +$ energy (heat). Note that there are on each side of the arrow the same number of each kind of atom. Atoms are neither destroyed nor created in a chemical reaction but simply change partners, which is an expression of the Law of Conservation of Matter.

Most chemical reactions are reversible; this is indicated by a double arrow, $\rightleftarrows$. The energy relations of the several chemicals involved, their relative concentrations, and their solubility are some of the factors that determine whether or not a reaction will occur and whether it will go from right to left or from left to right. For every reaction there is a constant, termed the thermodynamic equilibrium constant, K, which expresses the chemical equilibrium reached by the system. $K = (C) \times (D) \div (A) \times (B)$. This equation represents the equilibrium constant of the reaction $A + B \rightleftarrows C + D$. This equilibrium constant is unchanging and is determined by and is a function of the tendency of the reaction components to reach maximum entropy or minimum free energy for the system. Thus the equilibrium constant is related mathematically to the change in free energy of the components of the reaction: $\Delta G = -RT\ln K$. R is the gas constant, T the absolute temperature and lnK is the natural logarithm of the equilibrium constant. The symbol ΔG represents the standard free energy change; i.e., the gain or loss of free energy in calories as one mole* of reactant is converted to one mole of product. Examination of this equation reveals that when the equilibrium constant, K, is high, the standard free energy change, ΔG, is negative. Such a reaction will proceed with a decrease in free energy; however, when the equilibrium constant is very small the reaction does not go far in the direction of completion, and the free energy change is positive. It is necessary to put energy into the system to transform one mole of reactant into one mole of product. When the equilibrium constant is 1, then the change in free energy is zero and the reaction is freely reversible.

How does one go about measuring the equilibrium constant and free energy change of a biochemical reaction? Glucose-1-phosphate and glucose-6-phosphate are interconverted in the cell by a reaction catalyzed by the enzyme **phosphoglucomutase**. We set up a system with a carefully measured amount of glucose-1-phosphate, add an adequate amount of the enzyme, and measure the amounts of glucose-1- and glucose-6-phosphate present in the reaction mixture at succeeding times until no further change occurs, i.e., until equilibrium is reached. At equilibrium there will be 19 times as much glucose-6-phosphate as glucose-1-phosphate, and therefore the equilibrium constant, K, is 19. This number is substituted in the equation $\Delta G = -RT\ln K$, and we calculate that $\Delta G = -1745$ calories per mole. There is a decline in free energy of 1745 calories when one mole of glucose-1-phosphate is converted to one mole of glucose-6-phosphate at 25° C.

The rate at which a chemical reaction occurs is determined by a number of factors, one of which is temperature. Thus we must always state the temperature at which the reaction was carried out (25° C in this case). Each increase of 10° C approximately doubles the rate of most reactions. This is true of biological processes as well as reactions in a chemist's test tube, which indicates again that the chemical reactions of living things are fundamentally similar to those of nonliving ones. Whether or not an enzyme or any other catalyst is present in the system has no effect on either the equilibrium constant, K, or the free energy change, ΔG, undergone. Enzymes and other catalysts simply speed up the rate at which the system approaches equilibrium but do not change the equilibrium point itself. The unit of energy most widely used in biological systems is the **Calorie**, which is the amount of heat required to raise one kilogram of water one degree centigrade (strictly speaking, from 14.5° to 15.5° C).

---

*A mole or gram molecular weight of a compound is the amount of that substance equal to its molecular weight in grams. A mole of glucose is 180 grams of glucose.

Other forms of energy—radiant, chemical, electrical, the energy of motion or position—can be converted to heat and measured by their effect in raising the temperature of water. Reactions which have a high equilibrium constant, K, and a negative standard free energy change, ΔG, are said to be *exergonic*. A reaction with a very low equilibrium constant and therefore a positive standard free energy change is not spontaneous and will not go to completion under standard conditions, unless energy is supplied to it. Such processes are called *endergonic* processes. In biological systems, the endergonic processes must be coupled with some exergonic process so that the exergonic process delivers the required energy to the endergonic process. In such coupled systems, the endergonic process can occur only if the decline in free energy of the exergonic process to which it is coupled is larger than the gain in free energy of the endergonic process.

SECTION 4–3

CATALYSIS

Many of the substances that are rapidly metabolized by living cells are remarkably inert outside a cell. A glucose solution will keep indefinitely in a bottle if it is kept free of bacteria and molds. It must be subjected to high temperature, strong acids or bases before it will break down. Living cells unable to use such extreme conditions bring about chemical reactions by agents called enzymes, which belong to the class of substances known as catalysts.

A *catalyst* is a substance which regulates the speed at which a chemical reaction occurs without affecting the final equilibrium point attained and without being used up as a result of the reaction. Almost any substance may serve as a catalyst for some reaction. Water, for example, is an excellent catalyst to bring about the reaction of pure dry hydrogen gas and chlorine gas to form hydrogen chloride. Finely divided metals such as iron, nickel or platinum are used as catalysts in many industrial processes, such as the hydrogenation of cottonseed oil to make margarine. Since the catalyst is used over and over again, a very small amount of catalyst will speed up the reaction of large quantities of reactants.

There is an energy barrier to every chemical reaction, even an exergonic one with a strongly negative ΔG, which prevents the reaction from beginning. This energy barrier is termed the *activation energy*. In a population of molecules of any given kind, some have a relatively high energy content, others have a lower energy content, and the energy content of the molecules is normally distributed. Only those molecules with a relatively high energy content are likely to react to form the product. To make the reaction go faster we must raise the energy content of more of the population of molecules so that the activation energy barrier is overcome. This can be done by heating the mixture, for the heat absorbed by the molecules increases their internal energy and increases the likelihood that they will collide and react. Alternatively, the activation energy barrier can be overcome by adding a catalyst. This lowers the activation energy of the reaction and allows a larger fraction of the population of molecules to react at any one time (Fig. 4–2). The catalyst does this by forming an unstable intermediate complex with the substrate, which then decomposes to the product and frees the catalyst to react with a second molecule of reactant.

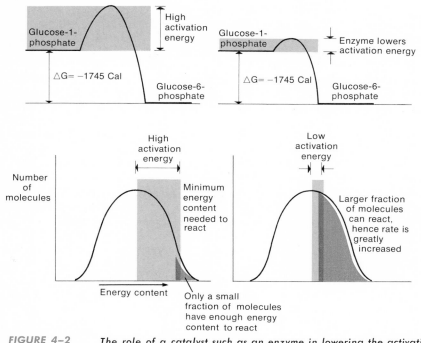

FIGURE 4–2   *The role of a catalyst such as an enzyme in lowering the activation energy of a reaction and in increasing the fraction of the population of molecules with an energy content great enough to react and form the product. The catalyst does this by forming an unstable intermediate complex with the substrate which then decomposes to the product and frees the catalyst to react with a second molecule of substrate.*

Enzymes are protein catalysts produced by living cells. They regulate the speed and specificity of the myriad chemical reactions occurring within the cells. Although enzymes are produced within a cell, they do not have to be inside a cell to be an effective catalyst. Many enzymes have been extracted from cells with their catalytic activity unimpaired. They can be purified and crystallized and their catalytic abilities can be studied. Reactions catalyzed by enzymes are basic to all of the phenomena of life: respiration, growth, muscle contraction, nerve conduction, photosynthesis, nitrogen fixation, deamination, digestion and so on. There is no need to postulate some mysterious vital force to account for these phenomena.

SECTION 4–4

THE PROPERTIES OF ENZYMES

All of the enzymes isolated and crystallized to date have been shown to be proteins. Enzymes are typically named by adding the suffix "ase" to the name of the substance acted upon, the **substrate**. For example, sucrose is split by the enzyme **sucrase** to give glucose and fructose. Enzymes are usually colorless, but they may be yellow, green, blue, brown or red. Most enzymes are soluble in water or in dilute salt solution, but some, the enzymes present in mitochondria for example, are bound together by lipoprotein, i.e., a phospholipid-protein complex, and are insoluble in water.

The catalytic ability of certain enzymes is truly remarkable. One molecule of the iron-containing enzyme *catalase*, for example, extracted from beef liver, will bring about the decomposition of 5,000,000 molecules of its substrate, *hydrogen peroxide*, per minute at 0° C. The number of molecules of substrate acted upon by a molecule of enzyme per minute is called the *turnover number* of the enzyme. The turnover number of catalase is thus 5,000,000. Most enzymes have high turnover numbers, and are very effective catalysts even though present in the cell in relatively minute amounts. Hydrogen peroxide is a poisonous substance produced as a by-product in a number of enzyme reactions. Catalase protects the cell by destroying the peroxide.

Hydrogen peroxide can be split by iron atoms alone, but only at a very slow rate. It would take 300 years for an iron atom to split the same number of molecules of H_2O_2 that a molecule of catalase (containing one iron atom) splits in one second. This is an example of the evolution of a catalyst and emphasizes one of the major characteristics of enzymes—they are very efficient catalysts.

Enzymes differ in their specificity, in the number of different substrates they will attack. A few enzymes are absolutely specific; *urease*, which decomposes urea to ammonia and carbon dioxide, will attack no other substance. Specific enzymes split each of the three common double sugars—sucrose, maltose and lactose. Other enzymes are relatively specific and will work upon only a few, closely related substances. *Peroxidase* will decompose several different peroxides, including hydrogen peroxide. Peroxidase, found in a wide variety of plant and animal tissues, can be demonstrated by mincing some raw potato and adding some hydrogen peroxide. A vigorous bubbling will ensue as peroxidase converts the peroxide to water and oxygen.

Finally, a few enzymes are specific only in requiring that the substrate have a certain kind of chemical bond. The *lipase* secreted by the pancreas will split the ester bonds connecting the glycerol and fatty acids of a wide variety of different fats.

Theoretically, enzyme-controlled reactions are reversible and the enzyme does not determine which way the reaction will go; it simply accelerates the rate at which the reaction reaches equilibrium. This *equilibrium point* depends upon complex thermodynamic principles, a discussion of which is beyond the scope of this book. Since reactions give off energy when going in one direction, it is obvious that the proper amount of energy in a usable form must be supplied to drive the reaction in the reverse direction.

To run an energy-requiring reaction, some energy-yielding reaction must occur at about the same time. In most biological systems, energy-yielding reactions result in the synthesis of "energy-rich" phosphate bonds, ~P, such as the terminal bonds of *adenosine triphosphate* (abbreviated ATP). The energy of these phosphate bonds can then be used by a cell to conduct a nerve impulse, contract muscle, synthesize proteins and so on. Biochemists use the term "*coupled reactions*" for two reactions which must take place together so that one can furnish energy, or one of the reactants, needed by the other.

Enzymes usually work in teams, with the product of one enzyme-controlled reaction serving as the substrate for the next. We can picture the inside of a cell as a factory with many different assembly lines (and disassembly lines) operating simultaneously. Each of the assembly lines is composed of a number of enzymes, each of which carries out one step such as changing molecule A into molecule B and then passes it along to the next enzyme, which converts molecule B into molecule C, and so on. From germinating barley seeds you can extract two enzymes that will convert starch to glucose. The first, **amylase**, hydrolyzes starch to maltose and the second, **maltase**, splits maltose to glucose. Eleven different enzymes, working consecutively, are required to convert glucose to lactic acid. The same series of 11 enzymes is found in human cells, in green leaves and in bacteria.

Some enzymes, such as **pepsin**, consist solely of protein. Others consist of two parts, one of which is protein (and called an **apoenzyme**) and the second (called a **coenzyme**) is a smaller organic molecule which typically contains phosphate. Coenzymes can be separated from their enzymes and, when analyzed, have proved to contain some vitamin as part of the molecule—thiamine, niacin, riboflavin, pyridoxine and so on. This finding has led to the generalization that all vitamins function as parts of coenzymes in body cells.

Neither the apoenzyme nor the coenzyme alone has any catalytic activity; only when the two are combined is activity present. Other enzymes may require for activity, in addition to a coenzyme, the presence of some ion. Several of the enzymes involved in the breakdown of glucose require **magnesium** (Mg^{++}). The amylase secreted by the salivary glands requires a **chloride** ion (Cl^-) for activity. Most, if not all, of the elements needed by plants and animals in very small amounts—the so-called **trace elements**, manganese, copper, cobalt, zinc, iron and so forth—function as such enzyme activators, usually as an integral part of the enzyme molecule.

In summary, enzymes, as catalysts, influence the rate but not the equilibrium of chemical reactions; they are very efficient catalysts; they have a high degree of specificity with respect to their substrates; they are subject to specific activators and inhibitors; and they direct the pathways of chemical reactions. Each enzyme is under the control of a specific gene.

SECTION 4–5

LOCATION OF ENZYMES IN THE CELL

Many of the enzymes present are simply dissolved in the cytoplasm of the cell. It is possible to make a water extract of ground liver that contains all the enzymes necessary to convert glucose to lactic acid. Other enzymes are tightly bound to certain cell bodies. The respiratory enzymes that catalyze the metabolism of lactic acid (and also substances derived from amino acids and fatty acids) to carbon dioxide and water are bound to the mitochondria; in fact, the mitochondrial membranes are in large part made of these enzymes. Other enzymes concerned especially with the synthesis of proteins are integral parts of smaller cytoplasmic particles, the ribosomes.

The location and functioning of enzymes within the cell can be studied histochemically. The tissue must be fixed and sliced by special methods which do not destroy enzyme activity. Then the proper chemical substrate for the enzyme is provided and, after a specified period of incubation, some substance is added which will form a colored compound with one of the products of the reaction mediated by the enzyme. The regions of the cell which have the greatest enzyme activity will have the largest amount of the colored substance (Fig. 4–3).

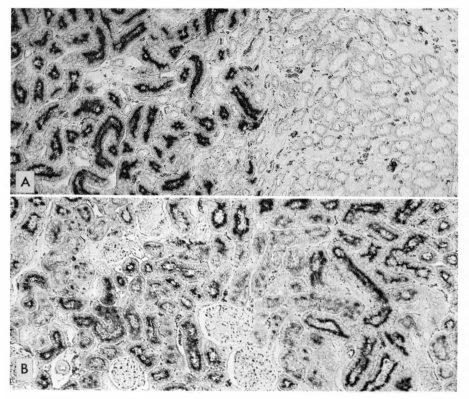

FIGURE 4–3   *Histochemical demonstration of the location of the enzyme alkaline phosphatase within the cells of the rat kidney. The tissue is carefully fixed and sectioned by methods which do not destroy the enzyme's activity. The tissue section is incubated at the proper pH with a naphthyl phosphate. Some hydrolysis of the naphthyl phosphate occurs wherever phosphatase is located. The naphthyl released by the action of the enzyme is coupled with a diazonium salt to form an intensely blue, insoluble azo dye which remains at the site of enzymatic activity. The photomicrograph reveals the sites at which the azo dye was deposited and thus indicates the location of phosphatase. The brush borders of the cells of the proximal convoluted tubules in the kidney cortex (A, left) shows thick deposits of azo dye, i.e., intense phosphatase activity. The tubules in the kidney medulla have no activity (A, right). (Courtesy of R. J. Barrnett.)*

SECTION 4–6

MODE OF ACTION OF ENZYMES

An enzyme can speed up only those reactions that would occur to some extent, however slight, in the absence of the enzyme. Many years ago the German chemist Emil Fischer suggested that the specific relationship of an enzyme to its substrate indicated that the two must fit together like a **lock and key.**

The idea that the enzyme combines with its substrate to form an intermediate **enzyme-substrate complex**, which subsequently decomposes to release the enzyme and the reaction products, was formulated mathematically by Leonor Michaelis nearly 50 years ago. By brilliant inductive reasoning, he assumed that such a complex does form, and then calculated how the speed of the reaction should be affected by varying the concentrations of enzyme and substrate. Exactly these relationships are observed experimentally, which is strong evidence that Michaelis' assump-

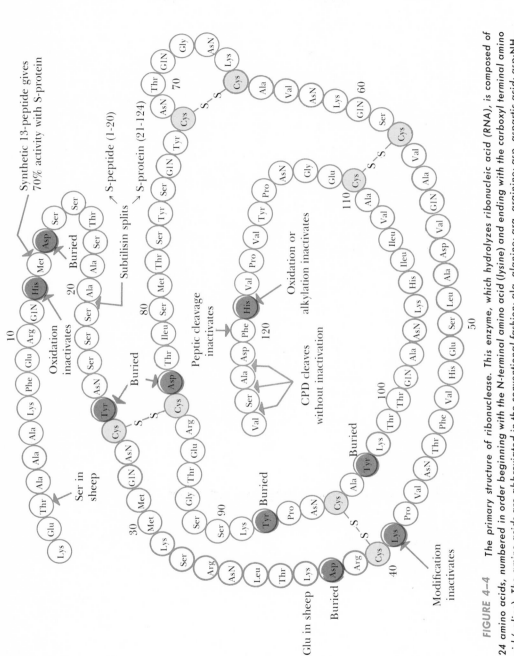

FIGURE 4–4   *The primary structure of ribonuclease. This enzyme, which hydrolyzes ribonucleic acid (RNA), is composed of 124 amino acids, numbered in order beginning with the N-terminal amino acid (lysine) and ending with the carboxyl terminal amino acid (valine). The amino acids are abbreviated in the conventional fashion: ala, alanine; arg, arginine; asp, aspartic acid; asp:NH₂, asparagine; cys, cysteine; glu, glutamic acid; glu:NH₂, glutamine; gly, glycine; his, histidine; ileu, isoleucine; leu, leucine; lys, lysine; met, methionine; phe, phenylalanine; pro, proline; ser, serine; thr, threonine; tyr, tyrosine; val, valine.*

tion, that an enzyme-substrate complex forms as an intermediate, is correct.

Direct evidence of the existence of an enzyme-substrate complex was obtained by David Keilin of Cambridge University and Britton Chance of the University of Pennsylvania. Chance isolated a brown-colored peroxidase from horseradishes. When he mixed this with hydrogen peroxide, a green-colored enzyme-substrate complex was formed. This was then changed to a second, pale red enzyme-substrate complex which finally split to give the original brown enzyme plus the breakdown products. By following the changes of color, Chance was able to calculate the rate of formation of the enzyme-substrate complex and its rate of breakdown.

Although it is clear that the substrate is much more reactive when part of an enzyme-substrate complex than when it is free, it is not clear why this should be true. One current theory postulates that the enzyme unites with the substrate at two or more points, and the substrate is held in a position which strains its molecular bonds and makes them more likely to break.

One approach to the study of enzyme action is to investigate the structure of the enzyme molecule itself. By certain analytic methods it is possible to determine the kinds of amino acids present in a given protein and their relative numbers. By more refined methods it is possible to determine the sequence of the amino acids in the peptide chain or chains comprising the protein. **Ribonuclease,** which consists of 124 amino acids in a single chain that is looped on itself like a pretzel, was the first enzyme for which it was possible to give the exact sequence of amino acids for the entire molecule (Fig. 4–4).

For most enzymes it is probable that only a relatively small part of the molecule combines with the substrate. This part is termed the **active site,** and another approach to the study of enzyme action is to determine the location of the active site, and the nature and sequence of the amino acids present in it. This information may throw light on the nature of the interaction between the active site and the substrate and provide an explanation of the mechanism by which the enzyme functions.

SECTION 4–7

FACTORS AFFECTING ENZYME ACTIVITY

Temperature

Enzymes are inactivated by heat—temperatures of 50 to 60° C rapidly inactivate most enzymes. The inactivation is irreversible, for activity is not regained upon cooling. This explains why most organisms are killed by a short exposure to high temperature: some of their enzymes are inactivated and they are unable to continue metabolism.

A few remarkable exceptions to this rule exist: there are some species of primitive plants, blue-green algae, that can survive in hot springs, such as the ones in Yellowstone National Park, where the temperature is almost 100° C. These algae are responsible for the brilliant colors in the terraces of the hot springs. Below the temperature at which enzymes are inactivated (about 40° C), the rates of most enzyme-controlled reactions, like other chemical reactions, are about doubled by each increase of 10° C.

Enzymes are not usually inactivated by freezing: their reactions go on very slowly or not at all at low temperatures, but their catalytic activity reappears when the temperature is raised to normal.

Acidity

Enzymes are sensitive to changes in pH, changes in the acidity and alkalinity of the reaction medium. **Pepsin**, the protein-digesting enzyme secreted by the stomach lining, is remarkable in that it will work only in a very acid medium, and works optimally at pH 2. **Trypsin,** a protein-splitting enzyme secreted by the pancreas, is an example of an enzyme that works optimally in an alkaline medium, at about pH 8.5. The majority of intracellular enzymes have pH optima near neutrality and will not work well in an acid or alkaline medium; stronger acids or bases will irreversibly inactivate them.

Concentration of Enzyme, Substrate and Cofactors

If the pH and temperature of an enzyme system are kept constant, and if an excess of substrate is present, the rate of the reaction is directly proportional to the amount of enzyme present (Fig. 4–5 B). This is used to measure the amount of some particular enzyme present in a tissue extract. If the pH, temperature and enzyme concentration of a system are kept constant, the initial rate of reaction is proportional to the amount of substrate present, up to a limiting value (Fig. 4–5 C). If the enzyme system requires a coenzyme or specific activator ion, the concentration of this substance may, under certain circumstances, determine the overall rate of the reaction.

Enzyme Poisons

Certain enzymes are particularly susceptible to poisons such as cyanide, iodoacetic acid, fluoride, lewisite and so forth—even a very low concentration of poison inactivates the enzymes. **Cytochrome oxidase**, one of the enzymes of the electron transmitter system (p. 141) is especially sensitive to cyanide, and a person dies of cyanide poisoning because his cytochrome enzymes have been inhibited. One of the enzymatic steps in the breakdown of glucose is inhibited by fluoride and another is inhibited by iodoacetic acid; biochemists have used these inhibitors as tools to investigate the properties and sequences of many different enzyme systems.

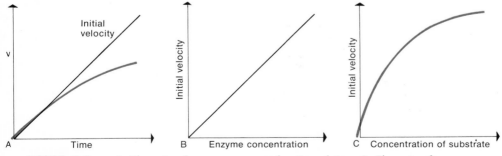

FIGURE 4–5   A, The rate of an enzyme as a function of time. B, The rate of an enzyme reaction as a function of the amount of enzyme added. Enough substrate and cofactors are added so that these do not limit the rate. C, The rate of an enzyme reaction as a function of the amount of substrate. Enough enzyme and cofactors are added so that these do not limit the rate of the reaction.

Enzymes themselves can act as poisons if they get into the wrong place. For example, as little as 1 milligram of crystalline trypsin will kill a rat if it is injected intravenously. Several types of snake, bee and scorpion venoms are harmful because they contain enzymes that destroy blood cells or other tissues.

SECTION 4–8

ENERGY FLOW IN LIVING SYSTEMS

Human beings, like other animals, derive their energy from the food-stuffs that they eat. Some of our food — peas, potatoes and passion fruit, for example — are derived directly from plants. Pork chops, roast beef, floun-der, steamed clams and lobster, are products of animals, but these animals in turn derived their food energy supply from plants. Ultimately all of the food and energy of the animal world comes from the plant world. Plants require for growth, water, carbon dioxide, nutrient salts and nitrogen, but, most important, they require an abundant supply of the radiant energy of sunlight. Sunlight is thus the ultimate source of all the biological energy on this planet. The radiant energy of sunlight arises from nuclear energy when, at the very high temperatures that occur in the interior of the sun, hydrogen atoms undergo transformation to helium atoms with the release of energy initially as gamma rays. The reaction is $4H \rightarrow He^4 + 2e + Hv$. H is Planck's constant and v is the wavelength of the gamma radiation. The gamma radiation reacts with electrons and the energy is ultimately emitted again as photons of light energy which pass out of the sun.

Only a small fraction of the light energy reaching the earth from the sun is trapped. Large areas of the earth have no plants, and plants can utilize in photosynthesis only some 3 per cent of the incident energy. The radiant energy is converted into the potential energy of the chemical bonds of the organic substances made by the plant. When an animal eats a plant, or when bacteria decompose it and these organic substances are oxidized, the energy liberated is just equal to the amount of energy used in syn-thesizing the substances (First Law of Thermodynamics), but some of the energy is converted to heat and not useful energy (Second Law of Thermo-dynamics). When the animal in turn is eaten by another animal, a further decrease in useful energy occurs as the second animal oxidizes the organic substances in the first to liberate energy to synthesize its own cellular constituents. Eventually all the radiant energy originally trapped by plants in photosynthesis is converted to heat and dissipated to outer space.

We can roughly estimate that the total amount of carbon that is fixed by all the plants on the earth and in the waters of the globe is some 200 billion tons each year. Land plants synthesize about one tenth of this total and the marine plants, mostly microscopic algae, synthesize the remainder. The formation of each mole of glucose — 180 grams — requires the input of 686,000 calories of radiant energy. If we make the simple assumption that the carbon is fixed in the form of glucose, then we can see that it would require biological energy amounting to 10^{19} calories per year. If we make appropriate corrections for friction losses, our estimate for the total biological energy flux rises another hundred fold, or perhaps more, to 10^{21} calories of solar energy captured per year. This, in turn, is only about one thousandth of the total solar energy, about 10^{24} calories per year, that is estimated to fall on the earth in the course of the year. It has been estimated that the activity of green plants leads to a renewal of all the carbon dioxide in the atmosphere and dissolved in the waters of the world every 300 years, and a renewal of all the oxygen in the atmosphere in about 2000 years.

LIGHT

To discuss photosynthesis (and other biological processes involving light—vision, phototropism and bioluminescence) we must first consider the properties of light itself. Light acts as though made of small packets of energy, called **photons** or light **quanta**, that have no electric charge and very little mass. Photons are a class of ultimate physical particles, like protons and electrons. Light is also characterized by its wave motion, and the different colors of light (different regions of the light spectrum) are identified by their wavelength or frequency.

Each photon has an energy content, E, equal to hc/λ, where c is the velocity of light (3×10^{10} cm per sec), λ is the wavelength of light and h is Planck's constant that relates frequency to energy. The energy content of a photon, the work it can do, is inversely proportional to wavelength. Light of shorter wavelength (violet) has a higher energy content than light of longer wavelength (red). The intensity of light equals the rate at which photons are being delivered. In photochemical reactions, each molecule, by absorbing one quantum of light, is excited to enter into a chemical reaction. One mole of the substance (6×10^{23} molecules) will be excited by 6×10^{23} quanta. The energy content of "one mole" of quanta, 6×10^{23} quanta, is defined as one **einstein.**

By multiplying the appropriate constants, we can find that the energy content of one einstein equals 2.854×10^7 gram calories divided by the wavelength in millimicrons. The energy content of one einstein of blue light (wavelength, 450 mμ) is 64,000 calories, that of red light (660 mμ) is 43,000 calories. The different colors of light represent quanta of different energy, those of blue light having greater energy than those of red light. All the radiations of the electromagnetic spectrum, from very short x-rays, through ultraviolet, visible and infrared light, to very long radio waves, represent a single phenomenon, differing only in wavelength and in the energy of their photons.

When light quanta strike a metal plate, electrons are ejected, the number of electrons ejected being proportional to the number of photons striking the plate (this is the principle of an exposure meter for photography). When a quantum strikes a molecule, an electron may be ejected or moved to an orbital farther from the nucleus (with a higher energy content). Energy must be put into the system to move an electron from an inner orbital to one farther out, because a negatively charged particle is being moved away from the positively charged nucleus. As we have learned (p. 75), there can be no more than two electrons in an s or a p orbital and the electrons must be paired—must spin in opposite directions. If one of these electrons absorbs a light quantum and moves to another orbital, the molecule is in an "excited" state (Fig. 4–6). To promote a reaction, light must be absorbed; there is a quantitative relationship between the number of quanta absorbed and the number of molecules activated. **Photochemical reactions** are characterized by the presence of intermediates, which are electronically excited atoms or molecules—ones in which an electron has been moved to an outer orbital of higher energy.

The transition of an electron to a new orbital leads to a redistribution

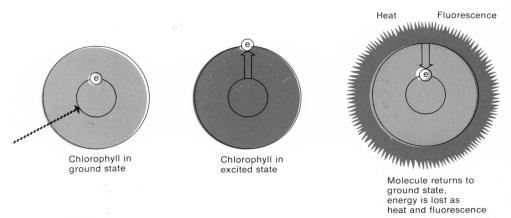

FIGURE 4–6   *Model of a photochemical reaction. A quantum of light energy strikes a molecule and its energy is used to move an electron to an orbital farther from the nucleus. When the electron returns to the inner orbital, the energy may be released as light energy of a different wavelength, a process termed fluorescence.*

of the electronic charge over the entire molecule. The chemical properties of a molecule depend on the energy and on charge density at different sites; thus the excitation of an electron results in the formation of a new molecule, one which is usually chemically more reactive than a molecule in the ground state.

The electrons of greatest significance in photochemical reactions are the **"lone-pair" electrons**, found in molecules with nitrogen and oxygen atoms, and not involved in binding. These two, occupying the same orbital, must be paired, i.e., have opposite spins. When one is activated by absorbing a photon and moves to an outer orbital of higher energy, the spins may be opposite (the singlet state) or alike (the triplet state).

The probability of a direct transition from the ground state to the triplet state is very small. Usually the **triplet state** is achieved by internal conversion from the singlet state of the same electronic configuration. The decay of an excited singlet state to the ground state is known as **fluorescence**; it is extremely rapid (10^{-8} seconds) and is independent of temperature. If the excited singlet state undergoes a transition to the related triplet state, and the latter then decays to the ground state, a different kind of radiation, **phosphorescence**, is emitted. This is much slower, lasting from 10^{-4} to 1 second.

The photochemical reactions most important biologically, however, do not involve the emission of light but are radiationless transitions from the triplet to the ground state in which the energy, instead of being emitted as light, is transferred to chemicals reacting in another system. The triplet state, with its relatively long half time, has a much greater probability of reacting chemically than the singlet state. Photosynthesis involves such a radiationless transition from the triplet to the ground state of the unique molecule **chlorophyll**.

The energy required to move an electron from one orbital to another depends on the difference in energy between the two. The energy in a photon must be used completely (one cannot use part of a quantum), hence only photons with the proper energy to move the electron to an allowed orbital will be absorbed. By shining lights of specific wavelengths through a solution of a substance, and measuring the fraction absorbed, one obtains the **absorption spectrum** of a given substance. If the absorption peak is at 660 mμ (in the red region), then about 43,000 calories per mole are required to move the electron from a filled inner orbital to an unfilled outer orbital. Each type of molecule has a characteristic absorption spectrum, and measuring the absorption spectrum can be useful in identifying some unknown substance isolated from a plant or animal cell.

SECTION 4–10

PHOTOSYNTHESIS

The essence of photosynthesis now appears to be the conversion of the radiant energy of sunlight into chemical energy in the form of ATP and reduced nicotinamide adenine dinucleotide phosphate (NADPH). Our theories of photosynthesis have undergone many changes as new evidence has come to light.

The first studies date back to 1630, when van Helmont, a Flemish botanist, showed that plants make their own organic materials and do not simply absorb them from the soil. He weighed a pot of soil and the willow tree planted in it and showed that the tree gained 164 pounds in five years but the soil weighed only two ounces less. Van Helmont concluded that the rest of the substance came from the water he had added; we now know that carbon dioxide removed from the air by the plant contributed largely to the plant material synthesized.

Joseph Priestley showed in 1772 that a sprig of mint would "restore" air that had been "injured" by the burning of a candle. Seven years later Jan Ingenhousz showed that vegetation could restore "bad air" only if the sun was shining, and that the ability of the plant to restore air was proportional to the clearness of the day and to the exposure of the plant to the sun. In the dark, plants gave off air "hurtful to animals."

The next major step in understanding photosynthesis came in 1804, when de Saussure weighed both air and plant before and after photosynthesis and showed that the increase in the dry weight of the plant was greater than the weight of carbon dioxide removed from the air. He concluded that the other substance contributing to the gain in weight was water. Thus, 160 years ago, the broad outlines of the photosynthetic process appeared to be: carbon dioxide plus water plus light energy yields oxygen and organic material.

Ingenhousz suggested that light functions in photosynthesis to split carbon dioxide to liberate oxygen and yield "carbon," which is used in forming plant substance. On this basis, living organisms were divided into green plants, which could use the radiant energy for the "assimilation" of carbon dioxide, and other organisms, without chlorophyll, which could not use radiant energy and could not assimilate CO_2.

This logical division of the living world was upset when Winogradsky discovered (1887) chemosynthetic bacteria, organisms without chlorophyll that could assimilate carbon dioxide—convert it to organic substances—in the dark. It was upset further by Engelmann's discovery (1883) of purple bacteria that carried out a kind of photosynthesis in which no oxygen was liberated. Although not fully appreciated at the time, the discovery of chemosynthetic bacteria that can carry on carbon dioxide assimilation in the dark revealed that this process is not peculiar to photosynthesis. Since 1940, by the use of labeled carbon as a tracer, experiments have shown that all cells, plant, bacterial and animal, can assimilate carbon dioxide—can incorporate it into organic molecules—but differ in the source of the energy that is required for this process.

Another major advance in our understanding of the photosynthetic process was made in 1905 when the British plant physiologist Blackman demonstrated that it includes two successive series of reactions, a rapid **light reaction** and a slower series of steps not affected by light, which he termed the **dark reaction**. Using light of high intensity, he found that photosynthesis proceeded as rapidly when the light was flashed alternately on and off for periods of a fraction of a second as when the light shone continuously, even though the photosynthetic system received less than half as much energy. Only when the length of the dark period was increased considerably was the rate of photosynthesis decreased. In further experiments, he showed that the rate of the dark reaction was markedly increased by increasing the temperature.

The next hypothesis regarding the central chemical process in photosynthesis was supplied by the experiments of C. B. van Niel, who showed in 1931 that bacterial photosynthesis could proceed anaerobically without the evolution of oxygen. He suggested that there is a basic similarity in bacterial and green plant photosynthesis; in the latter, light energy is used to carry out the photolysis of water, H_2O, to yield a reductant (H) which reacts in some way in the assimilation of carbon dioxide, and an oxidant (OH) which was postulated to be the precursor of molecular oxygen. In bacterial photosynthesis, the process is basically similar but a different hydrogen donor, H_2S or molecular hydrogen, is used and there is no evolution of oxygen.

It is now clear that all the reactions for the incorporation of CO_2 into organic materials can occur in the dark (the "dark reactions") and thus perhaps are not in the strict sense part of the process of photosynthesis. They fundamentally constitute a reversal of the reactions by which carbohydrates are broken down. The reactions dependent on light (the "light reactions") are those in which radiant energy is converted into chemical energy, and the first stable, chemically defined products of these reactions are ATP and NADPH.

Before proceeding with the discussion of the chemical reactions by which CO_2 assimilation occurs (which are now fairly well understood) and the reactions by which ATP and NADPH are formed (which are still not completely clear), let us consider the structure of the cell organelles and the properties of the pigments involved in this process.

Chloroplasts

When a bit of leaf is examined under the microscope it can be seen that the green pigment is not uniformly distributed in the cell but is confined to small bodies called **chloroplasts.** Each cell has some 20 to 100 chloroplasts, which can grow and divide to form daughter chloroplasts. The electron microscope reveals that the chloroplast, like the mitochondrion, has a double-layered outer membrane within which is a large number of smaller bodies, called **grana**, which contain the chlorophyll (Fig. 4–7).

Each granum is composed of layers of molecules arranged like a stack of coins. Layers of protein molecules alternate with layers containing chlorophyll, carotenes and other pigments, and special types of lipids (containing galactose or sulfur but only one fatty acid). These surface-active lipids are believed to be adsorbed between the layers and serve in stabilizing the lamellae composed of alternate layers of protein and pigments. As we shall see, this characteristic layered structure of the grana

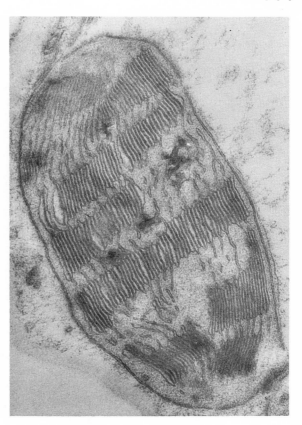

FIGURE 4-7   An electron micrograph of a chloroplast from a leaf of a tobacco plant, Nicotiana rustica, showing the fine structure of the grana 30,000×. (Courtesy of Dr. E. T. Weier.)

may be important in permitting the transfer of energy from one molecule to the adjacent one during the photosynthetic process.

Electron microscopy has revealed the presence in the lamellae of repeating unit structures made up of some 230 molecules of chlorophyll. These have been termed **quantosomes** and may prove to be the functional photosynthetic unit (Fig. 4–8). The material within the chloroplast and lying between the grana is called the **stroma**; it contains the enzymes which carry out the "dark" reactions. Isolated chloroplasts, when carefully prepared from leaves, are able to carry out *in vitro* the entire sequence of photosynthetic reactions.

Land plants absorb the water required for the photosynthetic process through their roots: aquatic plants receive it by diffusion from the surrounding medium. The carbon dioxide required diffuses into the plant by way of small holes, **stomata** (p. 121), in the surface of the leaves. Carbon dioxide is used up as a result of the photosynthetic process and its concentration in the cell is always slightly lower than that in the atmosphere. Oxygen is liberated in the process and diffuses out of the cell and finally out of the plant through the stomata. The sugars formed also tend to diffuse away from the site of formation to regions of lower concentration.

Plants need vast quantities of air to carry on photosynthesis, for air contains only 0.03 per cent carbon dioxide—10,000 cubic feet of air are needed to supply 3 cubic feet of carbon dioxide, enough to make about 110 grams (one quarter of a pound) of glucose. Plants generally grow better in air with a higher carbon dioxide content and some greenhouses are maintained with an atmosphere containing 1 to 5 per cent CO_2.

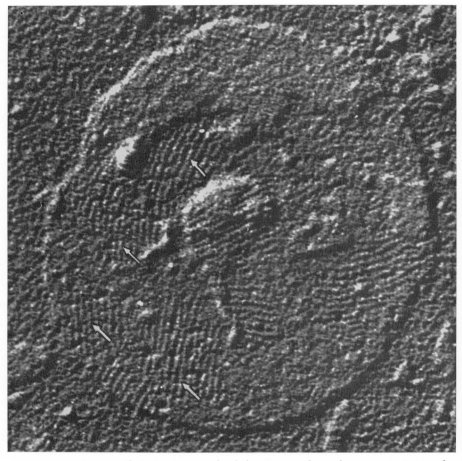

FIGURE 4–8   Electron micrograph of the inner surface of one compartment of a granum. Shadow-casting with a metal reveals a uniform linear arrangement of spheroidal particles 100 Å × 200 Å which are termed quantosomes (arrows). (Courtesy of M. Calvin.)

Chlorophyll and Other Photosensitive Pigments

The chlorophyll molecule, made of atoms of carbon and nitrogen joined in a complex ring (Fig. 4–9), is strikingly similar to the heme portion of the red pigment hemoglobin present in red blood cells, but contains an atom of magnesium instead of an atom of iron in the center of the ring, bound to two of the four nitrogen atoms. The chlorophyll molecule has a long tail composed of **phytol,** an alcohol containing a chain of 20 carbon atoms.

An examination of the chlorophyll molecule reveals that it is a **conjugated system,** with double bonds and single bonds alternating around the ring (indicated in light tint). Such a conjugated system provides for many rearrangements, many different patterns of single and double bonds in the ring structure. It is a **resonating system** which comprises many different ways of arranging the external electrons without moving any of the constituent atoms. The possibilities of resonance in the ring give the chlorophyll molecule considerable stability. Such systems of conjugated single and double bonds have mobile electrons, called **pi electrons,** which are associated not with a single atom or bond but with the conjugated system as a whole.

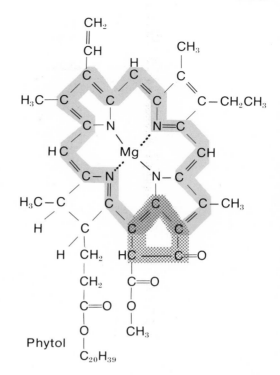

FIGURE 4-9 *Structural formula of chlorophyll. The light tone outlines the conjugated system of single and double bonds present in the molecule. The darker tone outlines a five-carbon ring which may function in transferring hydrogen atoms.*

Only a small amount of energy is needed to raise such pi electrons to an orbital of higher energy.

There are several types of chlorophyll, of which the two most important are called a and b. Chlorophyll b has one oxygen more and two hydrogens less than chlorophyll a. All green plants have chlorophyll a, but many algae and certain other plants lack chlorophyll b. Plants contain, in addition to chlorophyll, many pigments which give them their great variety of colors. Some of them play a role in absorbing light energy and transferring it to chlorophyll to be used in photosynthesis. Most plants have a deep orange pigment, **carotene,** which can be converted in the animal body to vitamin A, and a yellow pigment, **xanthophyll.** The red and blue-green algae contain other pigments, **phycocyanin** and **phycoerythrin,** which are more effective in photosynthesis in these red algae than are the chlorophylls present.

The sunlight reaching the surface of the earth has its maximal intensity in the blue-green and green portions of the spectrum, from 450 to 550 mμ, yet it is in this region that the chlorophyll molecule absorbs the smallest fraction of the incident light energy. Chlorophyll a and b both have absorption peaks in the violet region at about 440 mμ; chlorophyll b strongly absorbs light quanta in the red region at about 660 mμ and chlorophyll a at about 700 mμ in the far red region.

Despite its relative inability to absorb the wavelengths of sunlight with the greatest energy content, chlorophyll has a number of qualities that may account for its being selected in the course of evolution. In addition, it seems likely that other pigments in the grana such as the carotenoids, which absorb more strongly in the blue-green or green region, can transfer the energy absorbed to chlorophyll where it can be used in photosynthesis.

As we have seen, the absorption of a photon increases the energy of an electron, moving it from an inner orbital to one farther out from the nucleus. The amount of energy absorbed must just equal the difference in energy content of the old and new orbitals. An electron can occupy only a limited number of orbitals, and, unless the photon can raise the energy of the electron by just the right amount for a defined orbital level, the photon will not be absorbed. The unique capacity of chlorophyll to function in photosynthesis depends on its remarkable ability to absorb the energy of visible light with a high degree of efficiency and to transfer the energy to other molecules. Chlorophyll has a chemically reactive site (the five-carbon ring adjacent to the four rings made of a nitrogen and four carbon atoms, indicated in dark tint) which may function in transferring hydrogen atoms.

Carbon Dioxide Assimilation

The reactions by which carbon dioxide is incorporated into organic substances have now been worked out in detail by investigators who incubated suspensions of algae in the presence of carbon dioxide labeled with carbon-14. By incubating for varying lengths of time and isolating and identifying the products, they were able to show which substances became labeled first in these reactions. The "dark" reactions of carbohydrate synthesis occur in a cyclic sequence of (1) carboxylative, (2) reductive and (3) regenerative phases (Fig. 4–10).

The five-carbon sugar, ribulose-5-phosphate (with a phosphate group attached to carbon number 5), is phosphorylated by ATP to yield ribulose diphosphate. This is carboxylated by the addition of CO_2, presumably to an intermediate six-carbon substance, but this immediately splits by the addition of a molecule of water to give two molecules of phosphoglyceric acid. The phosphoglyceric acid is then reduced by an enzymic reaction which requires ATP and reduced nicotinamide adenine dinucleotide phosphate (NADPH) to give phosphoglyceraldehyde (a three-carbon sugar, triose). Two of these trioses can then condense to form one hexose molecule, which can be added onto a starch molecule and thus stored.

To achieve this point in the cycle, the photosynthetic process has taken up one molecule of CO_2 and has utilized three ATPs and four Hs (attached to two NADPH molecules). Ribulose phosphate can be regenerated from hexose phosphate by certain reactions of the pentose phosphate pathway (Fig. 4–11) and is then ready to pick up another molecule of carbon dioxide. None of these reactions, carboxylative, reductive or regenerative, is unique to photosynthetic cells. The only difference found so far is that the reductive reaction by which phosphoglyceric acid is converted to phosphoglyceraldehyde requires NADPH rather than the usual NADH.

The Light Reactions

The ATP and NADPH needed to drive these dark reactions is derived from the light reactions. It was thought at one time that the light reaction in the chloroplasts makes NADPH and that this might then be used by the mitochondria in the plant cell to synthesize ATP. However, the leaves that are most specialized for photosynthesis have cells containing very few mitochondria, and Arnon and his colleagues have demonstrated that isolated chloroplasts, free of mitochondria, could carry out photosynthesis and thus must be able to make ATP as well as NADPH.

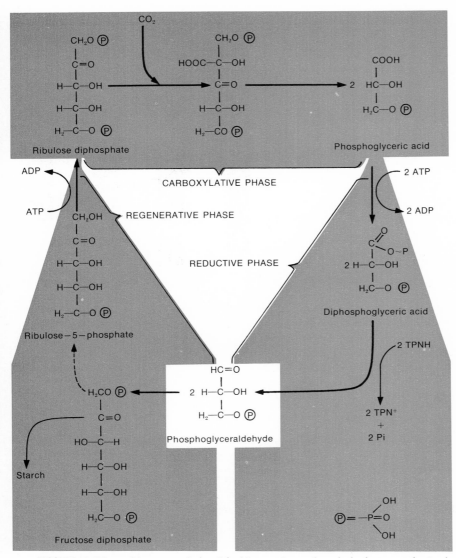

FIGURE 4–10 *Diagram of the "dark" reactions of carbohydrate synthesis, by which carbon dioxide is incorporated into sugars. The dotted line from fructose diphosphate to ribulose-5-phosphate indicates several reactions of pentose phosphate pathway. (Compare with Figure 4-11.)*

Working with isolated chloroplasts, Arnon demonstrated the complete separability of the light and dark reactions. He first illuminated chloroplasts in the absence of CO_2 but in the presence of ADP and NADP. Under these conditions the chloroplasts accumulated ATP and NADPH and evolved molecular oxygen. He then extracted the enzymes needed for the dark reactions of carbon dioxide assimilation, discarded the green part of the chloroplasts and, using the ATP and NADPH previously made in the light reaction, carried out CO_2 assimilation in the dark; i.e., the system made carbohydrates which could be isolated and identified. In other experiments these enzymes were supplied with NADPH and ATP made by animal cells and they were again able to carry out CO_2 assimilation in the dark.

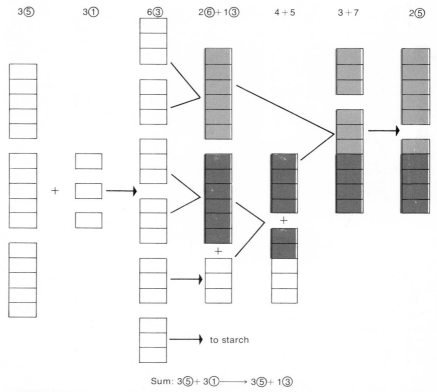

Sum: 3⑤ + 3① ⟶ 3⑤ + 1③

FIGURE 4-11 *Diagrammatic representation of the reactions by which three pentose molecules accept three molecules of carbon dioxide to yield six molecules of triose. One triose can be used in the synthesis of hexose and starch. The other five, by a further series of reactions in which the upper two or three carbons of the molecule are transferred as a unit, eventually regenerate three molecules of pentose so that the cycle can be repeated.*

Photosynthetic Phosphorylation

When Arnon illuminated isolated chloroplasts in the absence of CO_2 and pyridine nucleotide, but in the presence of large amounts of ADP and inorganic phosphate, the chloroplasts were able to utilize light energy to synthesize ATP from the ADP and inorganic phosphate. This process has been termed photosynthetic phosphorylation to distinguish it from the oxidative phosphorlyation that occurs in mitochondria (p. 141).

The chloroplast contains an electron transport chain that includes a flavoprotein, two or more cytochromes, an iron-containing protein called **ferredoxin**, which can undergo cyclic oxidation and reduction, and a substance related to ubiquinone called **plastoquinone.** The photosynthetic phosphorylation of ADP to produce ATP can proceed in the absence of oxygen. It has been demonstrated in chloroplasts from a variety of green plants and in the chlorophyll-containing chromatophores of several species of photosynthetic bacteria.

In oxidative phosphorylation, the energy to add the inorganic phosphate (P_i) to ADP is obtained when electrons are transferred from an electron donor at one energy level to an electron acceptor at another. This occurs in three steps, in each of which an energy-rich phosphate group ($\sim$P) is made and added to ADP to form ATP. The flow of electrons (in some of the steps the electrons are attached to protons to form hydrogen atoms) from one energy level to another releases energy. Each of these steps has been likened to a water wheel, turned by the "falling" electron, and driving the energy-requiring process of attaching an inorganic phosphate group to ADP. The process is fundamentally one of energy transfer—as the electrons move from one energy level to another, the energy is normally captured in the form of an energy-rich phosphate group. It is possible, however, to "uncouple" the phosphorylation process from the flow of electrons (p. 142). In oxidative phosphorylation the ultimate electron acceptor is oxygen and the electron donor is sugar or some other organic substance.

   Since photophosphorylation uses neither oxygen nor any substrate molecules, only light, the electron donors and acceptors must be within the chloroplast itself. Arnon proposed that the chlorophyll molecule can serve in both capacities. Light striking a chlorophyll molecule excites one of the electrons to an energy level high enough to eject it from the molecule (Fig. 4–12). The chlorophyll molecule, having lost an electron, is now ready to serve as an electron acceptor (**chlorophyll$^+$**). If it simply took the same electron back directly it would reemit the light energy as heat and fluorescence (pure chlorophyll does fluoresce in this way). However, if the electron ejected is taken up by ferredoxin, it can then return to the chlorophyll by a series of graded steps, somewhat like those in oxidative phosphorylation. The electrons pass from ferredoxin to the flavoprotein, to cytochrome and then to chlorophyll, yielding two ~P in the process as the energy of the electron is transferred via a phosphorylating enzyme system to ~P. The purpose of this flow of electrons out of the chlorophyll molecule through ferredoxin, flavoprotein and cytochrome back to chlorophyll is to conserve in a chemical form some of the energy taken in as radiant energy rather than have it lost as fluorescence and heat. Only the light-induced production of high energy electrons and the nature of the ultimate electron acceptor, chlorophyll, are peculiar to this photosynthetic process. This process of photosynthetic phosphorylation appears to be a primary reaction of photosynthesis, and it accounts for the produc-

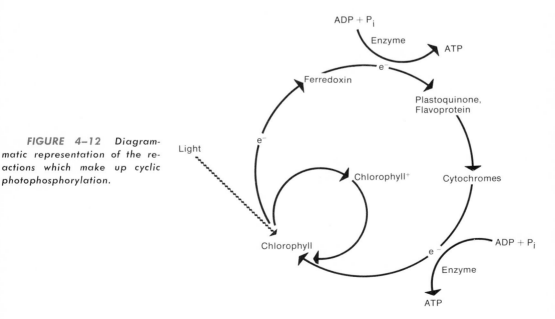

FIGURE 4–12 Diagrammatic representation of the reactions which make up cyclic photophosphorylation.

tion of some, at least, of the ATP needed in the dark reactions of CO_2 assimilation, but it does not account for the NADPH which is also needed nor for the splitting of water into hydrogen and oxygen.

Noncyclic Photophosphorylation

The production of NADPH occurs by a related system of electron transport, but one which is not cyclic and in which electrons from water are raised by absorbed light energy to a level at which reduced pyridine nucleotides can be formed. The excitation of chlorophyll a by light ejects high energy electrons which pass to ferredoxin and reduce it, and then pass to NADP, reducing it to NADPH. The electrons necessary to restore chlorophyll a$^+$ to the ground state are believed to come from the excitation of chlorophyll b. These high energy electrons pass to ferredoxin and then via flavoprotein and the cytochromes to chlorophyll a, in the course of which ADP is phosphorylated to ATP (Fig. 4–13). Both this **noncyclic photophosphorylation** and the cyclic process described previously are catalyzed by the same chain of electron carriers, but in cyclic photophosphorylation the electrons pass from chlorophyll a back to chlorophyll a whereas in noncyclic photophosphorylation the electrons pass from chlorophyll b to chlorophyll a.

The electrons required to restore chlorophyll b$^+$ to its ground state are believed to originate in OH$^-$ ions derived from the dissociation of water. Water molecules dissociate to a small extent into H$^+$ and OH$^-$ ions. The removal of electrons from OH$^-$ ions would yield (OH) radicals, which would subsequently combine to form water molecules and oxygen gas (Fig. 4–14).

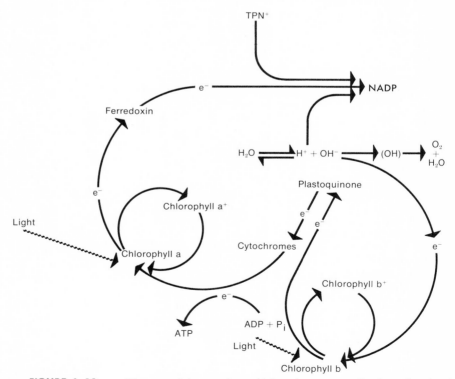

FIGURE 4–13   *Diagram of the reactions which make up noncyclic photophosphory-lation.*

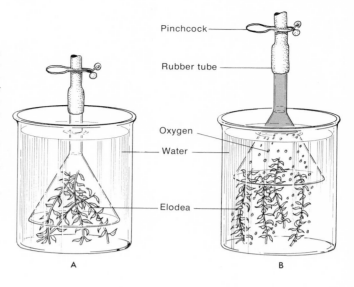

FIGURE 4–14   Experiment to show that oxygen is released during photosynthesis. Elodea or some other aquatic plant is placed in a beaker of water and covered by a glass funnel (A). The funnel is then raised (B) and the plant is exposed to light. The bubbles of gas given off collect in the stem of the funnel. To test the gas a glowing splinter of wood is placed over the stem of the funnel and the stopcock is opened. The glowing splinter will burst into flame, showing that the gas is oxygen.

This aspect of the theory fits with the evidence from experiments with water and CO_2 labeled with ^{18}O, that all of the oxygen gas evolved in photosynthesis comes from the O in H_2O and none comes from CO_2. Although the details of the reactions by which water is split are not yet known, it is clear (Fig. 4–13) that each complete sequence of reactions in noncyclic photophosphorylation, involving the excitation of one molecule of chlorophyll a and one molecule of chlorophyll b, would lead to the formation of one molecule of NADPH, two or more molecules of ATP from ADP and P_i, and the evolution of one atom of oxygen. This would require at least four quanta of light, two for each of the chlorophyll molecules.

By analogous reactions the chlorophyll of certain photosynthetic bacteria, such as the red sulfur bacterium, *Chromatium*, can be activated by light to raise electrons to a high energy level. These electrons can then either reduce pyridine nucleotides (as in green plants) or reduce nitrogen gas to ammonia or reduce protons to hydrogen gas.

It seems possible that cyclic photophosphorylation is the more primitive process and perhaps was the first to appear, when the earth's atmosphere contained little or no oxygen gas. It simply provided these primitive organisms with a way of synthesizing ATP in an anaerobic environment other than by the process of fermentation.

Perhaps the next evolutionary step was a noncyclic photophosphorylation, such as the one found in certain bacteria today, in which pyridine nucleotides are reduced (and thus made available for biosynthetic reactions) but the electron donors are molecules such as thiosulfate or succinate. The third step, achieved by the green plants, was the ability to use water molecules rather than thiosulfate or succinate as the electron donor in photophosphorylation. By evolving a system which could obtain electrons for reductive processes from water molecules, green plants were enabled to live nearly everywhere and were not restricted to places where special electron donors such as thiosulfate might be found.

Then, as plants spread and multiplied, they released into the atmosphere oxygen molecules that had previously been locked up in water; this made possible the further biochemical evolution of animals and other organisms that require molecular oxygen.

The overall efficiency of the photosynthetic process is quite high.

One can calculate that to carry out the chemical reduction of one mole of CO_2 to the level of carbohydrate would require about 120 kilocalories. The energy of the light effective in photosynthesis, red light with a wavelength of about 680 mμ, is about 41 or 42 kilocalories per mole. If the process were 100 per cent efficient, three quanta per molecule of CO_2 might suffice. Measurements made under a variety of experimental conditions give estimates ranging from four to 12 (most recent estimates average five or six) quanta per molecule of CO_2.

Thus the overall process of photosynthesis by which organic molecules are built up, represented by

$$6\ CO_2 + 6\ H_2O \rightarrow C_6H_{12}O_6 + 6\ O_2$$

is a very complex one, as complex as the reactions of cellular respiration by which they are subsequently broken down. In the light reactions, the remarkable properties of chlorophyll enable it to capture the radiant energy of light with a high degree of efficiency and, with the aid of ferredoxin, cytochromes and other compounds in the membranes of the grana, to produce NADPH, ATP and molecular oxygen.

In the subsequent dark reactions, enzymes in the stroma of the chloroplasts condense carbon dioxide with ribulose diphosphate to yield two molecules of phosphoglyceric acid, which are then converted to phosphoglyceraldehyde in a reaction driven by ATP and NADPH. The two phosphoglyceraldehydes condense to fructose diphosphate and a complex sequence of enzymatic transfers of two- and three-carbon units from one carbohydrate to another eventually regenerates ribulose phosphate and leads to the net synthesis of glucose.

SECTION 4–11

THE LEAF AND ITS FUNCTIONS

To carry on photosynthesis each green plant cell requires a continuing supply of water, carbon dioxide and radiant energy. Algae and other plants living in ponds or in the sea are abundantly supplied with all three. Land plants have evolved roots, stems, leaves, and special conducting tissues, xylem and phloem, to ensure the supply of raw materials and to make the photosynthetic process as efficient as possible. The roots anchor the plant and hold the stem upright, absorb water and minerals from the soil and conduct them to the stems and leaves.

Each leaf is a specialized nutritive organ whose function is to carry on photosynthesis. Leaves are generally broad and flat to present a maximum of surface to sunlight and to have a maximum surface area for the exchange of gases, oxygen, carbon dioxide and water vapor.

Leaves originate as a succession of lateral outgrowths, called *leaf primordia*, from the apical meristem at the tip of the stem. Each outgrowth undergoes cell division, growth and differentiation and finally a miniature, fully formed leaf is produced within the bud. In the spring, the leaves grow rapidly, forcing apart the bud scales and, largely by the absorption of water, unfold, enlarge and reach their full size. Many leaves have no meristematic tissue and thus do not live long—a few weeks in some desert plants, a few months for most trees, and up to three or four years for the needle-shaped evergreen leaves.

The leaf of a typical dicot consists of a stalk, the *petiole*, by which it is attached to the stem, and a broad *blade*, which may be one simple structure or a compound one, with two or more parts. The petiole may be short and in some species is completely lacking. Like a stem in cross section, it is composed of vascular bundles, attached at one end to those of the stem and at the other end to the midrib of the blade. Within the blade the bundles of conducting tissue, xylem and phloem fork repeatedly and form the **veins**.

A microscopic section through a leaf (Fig. 4–15) shows it to be composed of several types of cells. The outer cells, both top and bottom, make up a colorless, protective **epidermis** which secretes a waxy **cutin**. The epidermal cells—thin, tough, firm-walled and translucent—are well adapted to give protection to the underlying cells and decrease water loss yet admit light. Scattered over the epidermal surface are many small pores, called **stomata**, each surrounded by two **guard cells**. These cells, by changing their shape, can change the size of the aperture and so control the escape of water and the exchange of gases. In contrast to other epidermal cells, guard cells contain chloroplasts. There are 50 to 500 stomata per square millimeter of leaf, many more on the lower than on the upper surface in the leaves of most species.

The bean-shaped guard cells have thicker walls on the side toward the stoma than on the other sides. In general, the stomata open in the presence of light and close in the dark; the opening and closing are regulated by changes in the turgor pressure within the guard cells (Fig. 4–16). Increased turgor pressure causes their outer walls to bulge and the inner walls become curved so that they move apart, creating the stomatal opening between them. When the turgor pressure in the guard cells decreases, the elastic inner walls regain their original shape and the stoma is closed.

The mechanism that increases turgor pressure is complex, involving in part the production of glucose and other osmotically active substances by photosynthesis in the guard cells themselves. Light also initiates a sequence of enzymatic reactions that lead to the conversion of starch

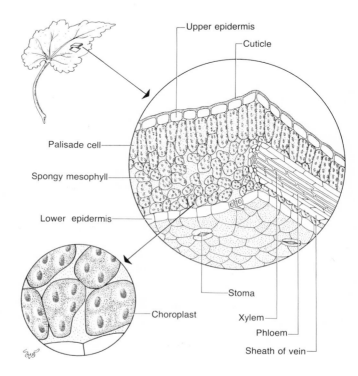

FIGURE   4–15   The micro-scopic structure of a leaf. A portion of a small vein is visible to the right. Inset: enlarged view of the meso-phyll cells containing chloroplasts.

Upper epidermis

Cuticle

Palisade cell

Spongy mesophyll

Lower epidermis

Choroplast

Stoma

Xylem

Phloem

Sheath of vein

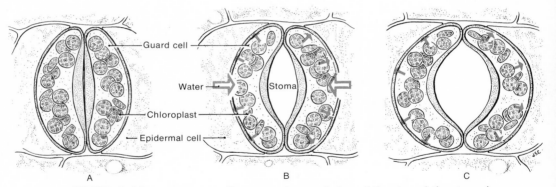

FIGURE 4-16   *Diagrams illustrating the regulation of the size of the stoma by the turgidity of the guard cells. A, Nearly closed condition. B, When osmotically active substances such as glucose are produced, water enters the guard cells, turgor pressure increases and the guard cells buckle, increasing the size of the stoma. C, Stoma open.*

stored in the guard cells (starch molecules are large, insoluble and osmotically inactive) to glucose, a small, soluble, osmotically active molecule. Experiments have shown that the turgor pressure in the guard cell can increase threefold or more in the light. Increasing the amount of light increases the rate of photosynthesis and the rate of utilization of carbon dioxide. The concentration of carbon dioxide in the leaf is decreased, increasing the pH of the system. The enzyme phosphorylase is stimulated by the increased pH to convert starch to glucose-1-phosphate. The glucose-1-phosphate is converted to glucose, increasing the concentration of glucose in the guard cell and increasing the osmosis of water into the cell, thereby raising the turgor pressure in the guard cells and opening the stoma! In the dark the process is reversed.

The open stomata permit the entrance of carbon dioxide and enable the leaf to carry on photosynthesis. In the absence of light, photosynthesis ceases in the guard cells as in all cells, turgor pressure decreases and the stomata close. If, on a hot, dry day, the amount of water supplied by the roots is too small, the guard cells will be unable to maintain the turgid state and hence will close, effectively conserving the decreased supply of water.

Most of the space between the upper and lower layers of the leaf epidermis is filled with thin-walled cells, called **mesophyll**, which are full of chloroplasts. The mesophyll layer near the upper epidermis is usually made of cylindrical cells, called **palisade cells**, closely packed together and so arranged that their long axes are perpendicular to the epidermal surface. The rest of the mesophyll cells are very loosely packed together, with large air spaces between them.

The veins of a leaf branch and rebranch repeatedly to form an extremely fine network, so that no mesophyll cell is far from a vein. Each vein contains both xylem and phloem tissues. The xylem is on the upper side of the vein, the phloem on the lower. In the smallest veins there are only a few xylem vessels and tracheids and a few phloem sieve tubes.

The leaves of a number of desert plants are thick and fleshy, and serve as storage places for water. The leaves of pond lilies and other aquatic plants have large air spaces to provide buoyancy. A few leaves, such as those of cabbages, store considerable amounts of food. The insect-trapping abilities of the leaves of plants such as the pitcher plant and the Venus flytrap have been discussed.

The fall of leaves in the autumn of the year is brought about by changes at the point where the petiole is attached to the stem. A special layer (called the **abscission layer**) of thin-walled cells, loosely joined together, extends across the base of the petiole, weakening the base of the leaf. The part next to the stem becomes corklike and forms a protective layer which will remain when the leaf falls off. When the abscission layer has formed, the petiole is held on only by the epidermis and the easily broken vascular bundles, so that a high wind will bring about the fall of the leaf. The change in color of the leaves is effected partly by the decomposition of the green chlorophyll, which exposes the yellow xanthophyll and orange carotene, previously hidden by the green pigment, and partly by the formation of red and purple pigments—**anthocyanins**—in the cell sap.

SUGGESTIONS FOR FURTHER READING

Baker, J. J. W., and Allen, G. E.: *Matter, Energy and Life*. Reading, Massachusetts, Addison–Wesley Publishing Company, Inc., 1965.
> A book for students who have had no previous chemistry or physics, which deals with such topics as the structure of matter, the formation of molecules, the course and mechanism of chemical reactions as well as with the chemistry of living systems.

Calvin, M. D., and Bassham, J. A.: *Photosynthesis of Carbon Compounds*. New York, W. A. Benjamin, Inc., 1962.
> An excellent source book for information about the chemical aspects of photosynthesis.

Lehninger, Albert L.: *Bioenergetics: The Molecular Basis of Biological Energy Transformations*, New York, W. A. Benjamin, Inc., 1965.
> A clear, well written discussion of the principles of thermodynamics and their application to living systems; an excellent account of the flow of energy in biological systems.

Rabinowitch, E. I., and Govindjee: *The Role of Chlorophyll in Photosynthesis*. Sci. Amer. *213*,74–83, July, 1965.
> A brief, informative presentation of the process by which the energy of sunlight is captured by chlorophyll.

Racker, E.: *Mechanisms in Bioenergetics*. New York, Academic Press, 1965.
> A detailed discussion of enzyme mechanisms and their role in energy transfer.

Rosenberg, Jerome L.: *Photosynthesis*. New York, Holt, Rinehart and Winston, 1965.
> An extensive and interesting treatment of the entire process of photosynthesis.

Steward, F. C.: *Plants at Work*. Reading, Massachusetts, Addison–Wesley Publishing Company, Inc., 1964.
> A fine discussion of photosynthesis, together with excellent presentations of many other aspects of plant physiology.

ENERGY TRANSFER: CELLULAR RESPIRATION

CHAPTER 5

BIOLOGICAL OXIDATION AND REDUCTION

The term *cellular respiration* refers to the enzymatic processes within each cell by which molecules of carbohydrates, fatty acids and amino acids are metabolized ultimately to carbon dioxide and water with the conservation of biologically useful energy. Many of the enzymes catalyzing these reactions are located in the walls and cristae of the mitochondria.

All of the phenomena of life — growth, movement, irritability, reproduction and others — require the expenditure of energy by the cell. Living cells are not heat engines and cannot use heat energy to drive these reactions. Instead they must use chemical energy, chiefly in the form of *energy-rich phosphate bonds*, abbreviated ~P. These bonds have a relatively high *free energy of hydrolysis,* ΔG; i.e., the difference in the energy content of the reactants and the products after the bond is split is relatively high. The free energy of hydrolysis is not localized in the covalent bond joining the phosphorus atom to the oxygen or nitrogen atom. Thus the term "energy-rich phosphate bond" is actually a misnomer, but it is so deeply ingrained by long usage that it is not likely to be changed.

All living cells obtain biologically useful energy by enzymic reactions in which electrons flow from one energy level to another. For most organisms, oxygen is the ultimate electron acceptor; oxygen reacts with the electrons and with hydrogen ions to form a molecule of water. Electrons are transferred to oxygen by a system of enzymes, localized within the mitochondria, called the *electron transmitter system.*

Electrons are removed from a molecule of some foodstuff and transferred (by the action of a specific enzyme) to some primary acceptor. Other enzymes transfer the electrons from the primary acceptor through the various components of the electron transmitter system and eventually combine them with oxygen (Fig. 5–1).

The chief source of energy-rich phosphate bonds, $\sim P$, in the cell is from the flow of electrons through the acceptors and the electron transmitter system. This flow of electrons has been termed the "electron cascade," and we might picture a series of waterfalls over which electrons flow, each fall driving a waterwheel, an enzymatic reaction by which the energy of the electron is captured in a biologically useful form — that of energy-rich compounds such as adenosine triphosphate, ATP.

ATP is the "energy currency" of the cell; all the energy-requiring reactions of cellular metabolism utilize ATP to drive the reaction. Energy-rich molecules do not pass freely from cell to cell, but are made at the site in which they are to be utilized. The energy-rich bonds of ATP that will drive the reactions of muscle contraction, for example, are produced right in the muscle cells.

Processes in which electrons (e^-) are removed from an atom or molecule are termed **oxidations**; the reverse process, the addition of electrons to an atom or molecule, is termed **reduction**. A simple example of oxidation and reduction is the reversible reaction

$$Fe^{++} \rightleftarrows Fe^{+++} + e^-$$

The reaction towards the right is an oxidation (the removal of an electron) and the reaction towards the left is a reduction (the addition of an elec-

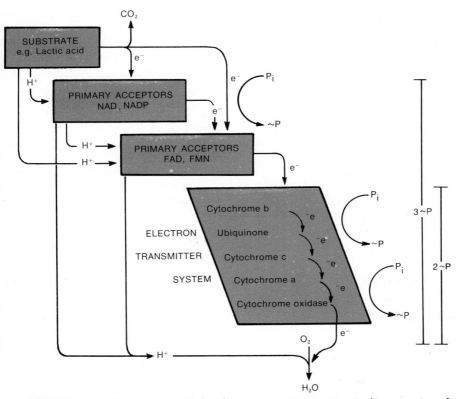

FIGURE 5–1   *The reactions of the electron transmitter system. In this succession of metabolic steps electrons are transferred from substrate to oxygen down an "electron cascade" and the energy is trapped in a biologically useful form as energy-rich phosphate bonds, $\sim P$.*

tron). Every oxidation reaction (in which an electron is given off) must be accompanied by a reduction, a reaction in which the electrons are accepted by another molecule; electrons do not exist in the free state.

The passage of electrons in the electron transmitter system is a series of oxidation and reduction reactions termed **biological oxidation**. When the energy of this flow of electrons is captured in the form of $\sim$P, the process is called **oxidative phosphorylation**. In most biological systems, two electrons and two protons (i.e., two hydrogen atoms) are removed together and the process is known as **dehydrogenation.**

The specific compounds of the electron transmitter system that are alternately oxidized and reduced are known as **cytochromes**. There are several of these and each one − a, b and c − is a protein molecule to which is bound a heme group similar to the one present in hemoglobin (p. 574). In the center of the heme group is an atom of iron, Fe, which is alternately oxidized and reduced − converted from Fe^{++} to Fe^{+++} and back − by giving off and taking up an electron:

$$Fe^{++} \rightleftharpoons Fe^{+++} + e^{-}$$

Another component of the electron transmitter system, called **ubiquinone** (it occurs everywhere!) or coenzyme Q, consists of a head, a six-membered carbon ring, which can take up and release electrons, and a very long tail. The tail is composed of ten repeating units, each of which consists of five carbon atoms. The repeating unit, called an **isoprenoid group**, is the basic unit of molecules of rubber, sterols and steroids.

All of the reactions of biological oxidation are mediated by enzymes and each enzyme is quite specific − it will catalyze the oxidation or reduction of certain compounds but not others. Only certain hydrogen atoms, ones having a certain spatial relationship to the rest of the molecule, can be removed (can undergo dehydrogenation) by a given enzyme.

SECTION 5–2

THE OXIDATION OF LACTIC ACID

As an example of biological oxidation, let us consider the oxidation of lactic acid (the acid of sour milk), an important intermediate in metabolism. The reaction catalyzed by the enzyme **lactate dehydrogenase** (reaction 1) is a dehydrogenation; the configuration

$$H-\overset{|}{\underset{|}{C}}-OH$$

is one from which two hydrogens can be removed enzymatically. In this

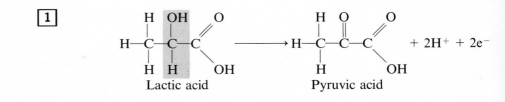

1   Lactic acid   Pyruvic acid

reaction, and in all dehydrogenations, the electrons given off are transferred to a **primary acceptor**. The primary electron acceptor in this reaction is the one called NAD or DPN (p. 91). The product of the dehydrogenation reaction, **pyruvic acid**, cannot itself undergo dehydrogenation directly. It does not have a structure suitable for attack by a dehydrogenase and must undergo further enzymatic action to attain a molecular configuration which can undergo dehydrogenation. The next reaction of pyruvic acid is a **decarboxylation**, the release of carbon dioxide (reaction 2).

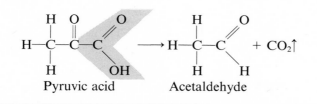

Pyruvic acid   Acetaldehyde

All the carbon dioxide in the air we breathe out is produced by similar decarboxylation reactions. Carbon dioxide is derived in biological systems only from carboxylic groups (—COOH) by the process of decarboxylation. The product of the decarboxylation reaction, **acetaldehyde**, has two carbons instead of three but is still not in a form suitable for dehydrogenation. It must undergo a **"make-ready" reaction,** a reaction that will result in an

$$H—\overset{|}{\underset{|}{C}}—OH$$

configuration suitable for dehydrogenation. The particular reaction used involves a large, complex organic molecule named **coenzyme A** and abbreviated CoA—SH. In this abbreviation, "—SH" represents the active end of the molecule, a **sulfhydryl group** composed of sulfur and hydrogen, and "CoA" represents all of the remainder of this complex molecule. The combination of coenzyme A with acetaldehyde results in a substance with an

$$H—\overset{|}{\underset{|}{C}}—OH$$

group that can be dehydrogenated to give acetyl coenzyme A (reaction 3).

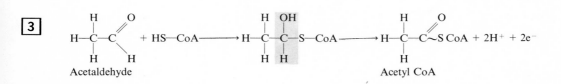

Acetaldehyde   Acetyl CoA

SECTION 5–3

THE OXIDATION OF SUCCINIC ACID

Another type of dehydrogenation, involving a different molecular configuration, is exemplified by the oxidation of succinic acid (reaction 4).

The dehydrogenation of molecules having a

group involves a **flavin** as hydrogen and electron acceptor. The product, **fumaric acid**, cannot be dehydrogenated directly, but undergoes a make-ready reaction in which a molecule of water is added (see reaction 5). The product of this reaction (**malic acid**) does have a configuration

$$ H-\overset{|}{\underset{|}{C}}-OH, $$

suitable for dehydrogenation with NAD.

Malic acid undergoes dehydrogenation by **malic dehydrogenase**, and the product of the reaction is **oxaloacetic acid** (see reaction 6). Oxaloacetic acid may undergo several different reactions, one of which is a decarboxylation to yield pyruvic acid. The further metabolism of pyruvic acid would proceed to acetyl coenzyme A by the reactions just discussed.

In the reactions by which carbohydrates, fats and proteins are oxidized, the cell requires only these three simple types of reactions—dehydrogenation, decarboxylation and make-ready. These may occur in dif-

4

$$ \begin{array}{c} \text{COOH} \\ | \\ \text{CH}_2 \\ | \\ \text{CH}_2 \\ | \\ \text{COOH} \end{array} \rightarrow \begin{array}{c} \text{COOH} \\ | \\ \text{CH} \\ \| \\ \text{CH} \\ | \\ \text{COOH} \end{array} + 2H + 2e^- $$

Succinic acid   Fumaric acid

5

$$ \begin{array}{c} \text{COOH} \\ | \\ \text{CH} \\ \| \\ \text{CH} \\ | \\ \text{COOH} \end{array} + H_2O \rightarrow \begin{array}{c} \text{COOH} \\ | \\ H-C-OH \\ | \\ \text{CH}_2 \\ | \\ \text{COOH} \end{array} $$

Malic acid

6

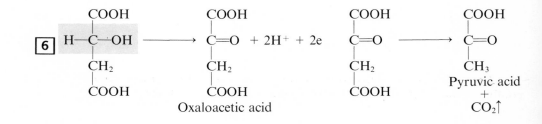

Oxaloacetic acid

ferent orders in different chains of reactions, as we have seen in the reactions of lactic acid and succinic acid. All of the dehydrogenation reactions are, by definition, oxidations, reactions in which electrons are removed from a molecule. The electrons cannot exist in the free state for any finite period of time but must be taken up immediately by other compounds, *electron acceptors.*

Two of the primary electron acceptors of the cell are pyridine nucleotides, *nicotinamide adenine dinucleotide* (abbreviated NAD) and *nicotinamide adenine dinucleotide phosphate* (abbreviated NADP). The functional end of both pyridine nucleotides is the vitamin, *nicotinamide.*

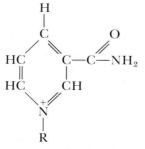

The nicotinamide ring accepts one hydrogen ion and two electrons from a molecule undergoing dehydrogenation (e.g., lactic acid) and becomes reduced nicotinamide adenine dinucleotide, NADH, releasing one proton. (See reaction 7).

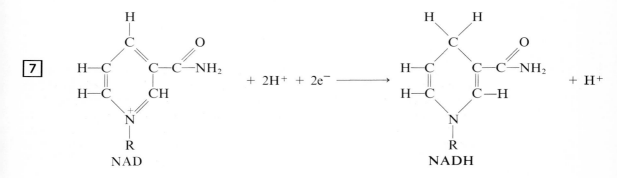

NAD and NADP serve as primary electron and hydrogen acceptors in dehydrogenation reactions involving substrates with the

configuration, such as the dehydrogenation of lactic or malic acid. The two pyridine nucleotides differ from one another in that NAD has two phosphate groups and NADP has three phosphate groups in the tail attached to the niacin ring ("R"). NAD may also be termed DPN, diphosphopyridine nucleotide, and NADP may also be termed TPN, triphosphopyridine nucleotide. Most dehydrogenases specifically require either NAD or NADP as the hydrogen acceptor and will not work with the other; some enzymes are less specific and will work with either one, though they usually work more rapidly with one than the other.

Another primary hydrogen acceptor, *flavin adenine dinucleotide* (abbreviated **FAD**), serves in reactions involving the

configuration, as in the dehydrogenation of succinic acid. In some reactions **flavin mononucleotide, FMN**, which consists of part of the FAD molecule, may serve in its place. The FAD of succinic dehydrogenase is bound very tightly to the protein part of the enzyme and cannot be removed easily. Such tightly bound cofactors are termed **prosthetic groups** of the enzyme. The pyridine nucleotide effective in the lactic dehydrogenase system, in contrast, is very loosely bound and is readily removed. Such loosely bound cofactors are termed **coenzymes.**

The reduced pyridine nucleotides, NADH or NADPH, cannot react with oxygen; their electrons must be passed through the intermediate acceptors of the electron transmitter system (the cytochromes) before they can react with oxygen. The flavin primary acceptors usually pass their electrons to the electron transmitter system, but some flavoproteins can react directly with oxygen. When this occurs, hydrogen peroxide, H_2O_2, is produced and no ~P is formed. An enzyme that can mediate the transfer of electrons *directly* to oxygen is termed an **oxidase**; one that mediates the removal of electrons from a substrate to a primary or intermediate acceptor is termed a **dehydrogenase** — e.g., lactic dehydrogenase, malic dehydrogenase or succinic dehydrogenase.

SECTION 5–4

THE CITRIC ACID CYCLE

Let us consider some further reactions of acetyl coenzyme A (which cannot undergo dehydrogenation directly), keeping in mind these three types of reactions: dehydrogenation, decarboxylation and "make-ready." Acetyl coenzyme A undergoes a make-ready reaction by combining with oxaloacetic acid (which contains four carbons) to yield citric acid (six carbons) plus free coenzyme A (reaction 8). Citric acid has neither an

$$H—\overset{\displaystyle |}{\underset{\displaystyle |}{C}}—OH \quad\quad nor\ a \quad\quad —\overset{\text{H H}}{\underset{\text{H H}}{C—C}}—$$

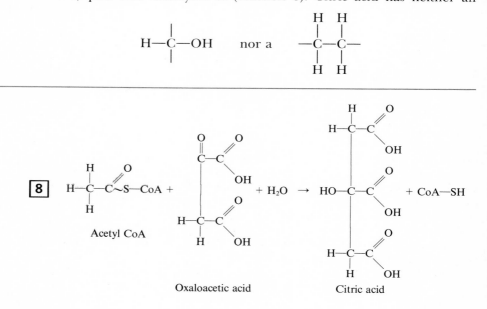

8   Acetyl CoA + Oxaloacetic acid + H_2O → Citric acid + CoA—SH

group and cannot undergo dehydrogenation. Two more make-ready reactions, involving the removal and addition of a molecule of water, yield **isocitric acid,** which can undergo dehydrogenation at its

$$H—\overset{|}{\underset{|}{C}}—OH$$

group (reaction 9). The hydrogen acceptor is a pyridine nucleotide, usually NADP, and the product is **oxalosuccinic acid.** This undergoes decarboxylation to yield **α-ketoglutaric acid** (reaction 10).

Inspection of the formula of α-ketoglutaric acid reveals that the lower part of the molecule is just like pyruvic acid, and α-ketoglutaric acid is metabolized by a set of reactions similar to those of pyruvic acid. A dehydrogenation, a decarboxylation and a make-ready reaction using coenzyme A combine to yield **succinyl coenzyme A** (reaction 11).

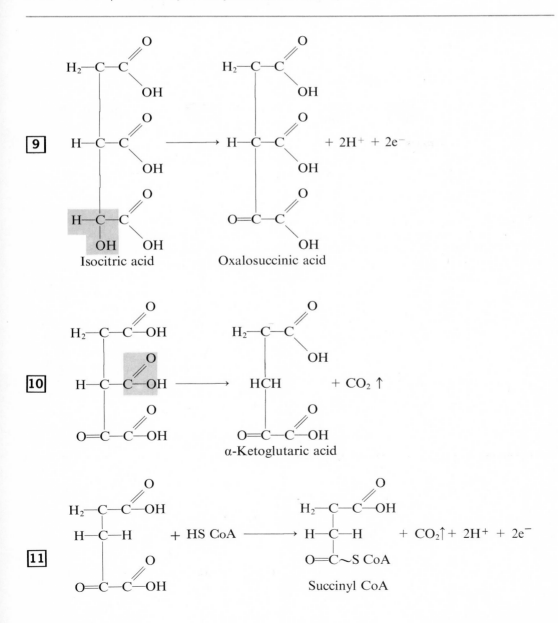

Isocitric acid

Oxalosuccinic acid

α-Ketoglutaric acid

Succinyl CoA

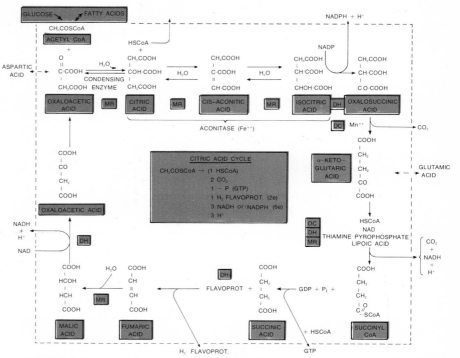

FIGURE 5–2   *The cyclic series of reactions, termed the Krebs citric acid cycle, by which the carbon chains of sugars, fatty acids and amino acids are metabolized to yield carbon dioxide. The reactions are of three types, designated DH, dehydrogenations; DC, decarboxylations; and MR, make-ready reactions. The overall reaction effected by one "turn" of the cycle is summarized in the center.*

Just as pyruvic acid is converted to the coenzyme A derivative of an acid with one less carbon atom, α-ketoglutaric acid (five carbons) is converted to succinyl coenzyme A (four carbons). In animal cells the metabolism of pyruvic acid to acetyl coenzyme A and the metabolism of α-ketoglutaric acid to succinyl coenzyme A are complex reactions involving as coenzymes, in addition to NAD and coenzyme A, *thiamine pyrophosphate* (a coenzyme containing another vitamin, *thiamine*) and *lipoic acid.*

The bond joining coenzyme A to succinic acid is an energy rich one, ~S, like the bond joining coenzyme A to acetic acid in acetyl coenzyme A. The energy of the bond of acetyl coenzyme A was utilized in bringing about the addition of the acetyl group to the oxaloacetic acid. The energy in the ~S bond of succinyl coenzyme A can be converted to an energy-rich phosphate bond, ~P, in ATP. The reaction of succinyl coenzyme A with inorganic phosphate yields succinyl phosphate and free coenzyme A (reaction 12).

The phosphate group is then transferred to ADP (via guanosine di-

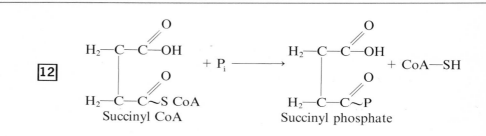

phosphate as an intermediate) to form ATP and free *succinic acid*. This is an example of an energy-rich bond synthesized at the substrate level—by reactions not involving the electron transmitter system (reaction 13).

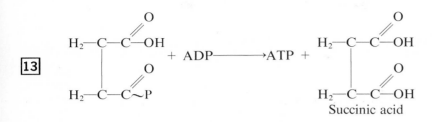

Only a small fraction of the energy-rich bonds are synthesized by reactions such as this. Normal cells metabolizing in an atmosphere containing oxygen synthesize most of their ATP by oxidative phosphorylation in the electron transmitter system.

We have discussed the reactions by which succinic acid is metabolized to fumaric, malic and oxaloacetic acids to complete a cycle of reactions (Fig. 5–2). This cycle of reactions was first described by the English biochemist, Sir Hans Krebs, and is usually called the **Krebs citric acid cycle.** In this cycle, a two-carbon unit, acetyl coenzyme A, combines with a four-carbon unit, oxaloacetic acid, to yield a six-carbon product, citric acid. By a series of dehydrogenations, decarboxylations and make-ready reactions, citric acid is metabolized back to oxaloacetic acid, which is then ready to combine with another molecule of acetyl coenzyme A. In the course of this cycle, two molecules of CO_2 are released, eight hydrogen atoms are removed, and one molecule of ~P is synthesized at the substrate level (reaction 14).

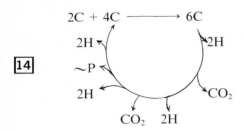

The Krebs citric acid cycle is the final common pathway by which the carbon chains of carbohydrates, fatty acids and amino acids are metabolized. All of these substances are fed into the cycle at one point or another. For example, pyruvic acid undergoes reactions which convert it to acetyl coenzyme A. Acetyl coenzyme A can also be produced from fatty acids by a series of reactions which cleave the long carbon chain into two-carbon units.

SECTION 5–5

FATTY ACID OXIDATION

Before it can be metabolized, a fatty acid must be "activated" by reacting with ATP and then with coenzyme A to yield the fatty acyl coenzyme A. **Palmitic acid** (16 carbons), for example, reacts to form palmityl coenzyme A (see reaction 15). Palmityl coenzyme A undergoes a dehydrogenation between the second and third carbons of the chain. The configuration undergoing dehydrogenation is

and, as you might now expect, the hydrogen acceptor is a **flavin** (see reaction 16).

The product of that reaction undergoes a make-ready reaction, the addition of a molecule of water (see reaction 17). This molecule has an

$$H—\overset{|}{\underset{|}{C}}—OH$$

configuration and can undergo dehydrogenation (reaction 18) with NAD as the hydrogen acceptor to yield

$$CH_3(CH_2)_{12}\overset{O}{\overset{\|}{C}}—CH_2\overset{O}{\overset{\|}{C}}{\sim}S\ CoA.$$

This in turn can undergo a make-ready reaction with coenzyme A to cleave off a two-carbon unit as **acetyl coenzyme A** and leave a carbon chain two carbons shorter (reaction 19). This is already activated; it contains

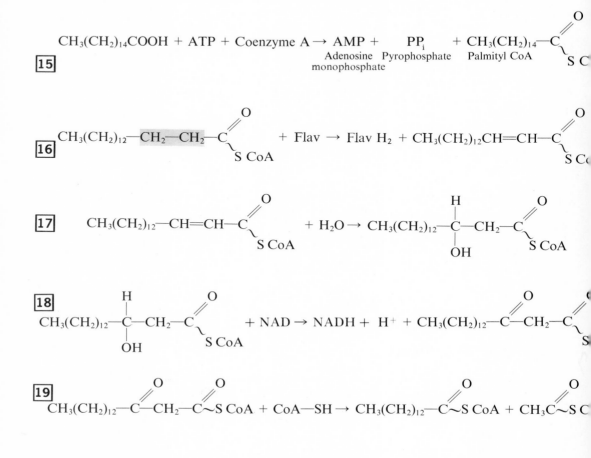

1. The activation reaction: $R \cdot CH_2 \cdot CH_2 \cdot COOH + ATP \longrightarrow R \cdot CH_2 \cdot CH_2 \cdot CO \cdot AMP + PyroPO_4$

$R \cdot CH_2 \cdot CH_2 \cdot CO \cdot AMP + HSCoA \longrightarrow R \cdot CH_2 \cdot CH_2 \cdot COSCoA + AMP$

---

2. Dehydrogenation: $CH_3(CH_2)_{10}CH_2 \cdot CH_2 \cdot CH_2 \cdot CH_2 \cdot COSCoA + Flavoprot. \rightleftharpoons CH_3(CH_2)_{10}CH_2 \cdot CH_2 \cdot CH = CHCOSCoA + H_2 Flavoprot.$

3. Make-ready: $CH_3(CH_2)_{10}CH_2 \cdot CH_2 \cdot CH = CHCOSCoA + H_2O \rightleftharpoons CH_3(CH_2)_{10}CH_2 \cdot CH_2 \cdot CHOH \cdot CH_2 \cdot COSCoA$

4. Dehydrogenation: $CH_3(CH_2)_{10}CH_2 \cdot CH_2 \cdot CHOH \cdot CH_2 \cdot COSCoA + NAD \rightleftharpoons CH_3(CH_2)_{10}CH_2 \cdot CH_2 \cdot CO \cdot CH_2 \cdot COSCoA + NADH + H^+$

5. Make-ready: $CH_3(CH_2)_{10}CH_2 \cdot CH_2 \cdot COCH_2 \cdot COSCoA + HSCoA \rightleftharpoons CH_3(CH_2)_{10}CH_2 \cdot CH_2 \cdot COSCoA + CH_3COSCoA$

    Acetyl   CoA

    To Reaction 2 and Repeat Sequence

Sum: $C_{16}H_{32}O_2 \longrightarrow 8 CH_2 COSCoA + 7H$ Flavoprot. $+ 7 NADH + 7H^+$

28 e

FIGURE 5–3   *The series of reactions by which a fatty acid is oxidized, two carbons at a time, to yield acetyl coenzyme A, reduced pyridine nucleotides and reduced flavoproteins.*

a coenzyme A on its carboxyl group, and thus is ready to be dehydrogenated by an enzyme which uses a flavin as hydrogen acceptor. This repeating series of reactions (which includes dehydrogenations and make-ready reactions, but not decarboxylations) cleaves a fatty acid chain two carbons at a time (Fig. 5–3). Seven such series of reactions split palmitic acid to eight molecules of acetyl coenzyme A (reaction 20).

SECTION 5–6

GLYCOLYSIS

The metabolic reactions of glucose, like those of other carbohydrates, convert the carbon chain to acetyl coenzyme A (Fig. 5–4). The series of glycolytic reactions begins (as with fatty acids) with one in which glucose is "activated." The reaction of glucose with ATP to yield glucose-6-phosphate and ADP is catalyzed by the enzyme **hexokinase** (see reaction 21). Only the terminal phosphate of ATP is transferred and adenosine diphosphate, ADP, remains. After this first "make-ready" reaction, additional make-ready reactions finally establish a configuration which can undergo dehydrogenation (Fig. 5–4). There is first a rearrangement to yield **fructose-6-phosphate** and the transfer of another phosphate from ATP to form **fructose-1,6-diphosphate** (fructose with phosphate groups at both carbon 1 and carbon 6) and ADP (reactions 22 and 23). Fructose-1,6-diphos-

---

[20] $CH_3(CH_2)_{14}-\overset{O}{\overset{\|}{C}}-OH \rightarrow 8CH_3\overset{O}{\overset{\|}{C}}\sim S\,CoA + 7Flav\,H_2 + 7\,NADH + 7H^+$

[21]   Glucose $+ ATP \xrightarrow[\text{hexokinase}]{}$ Glucose-6-phosphate $+ ADP$

[22]   Glucose-6-phosphate $\rightleftharpoons$ Fructose-6-phosphate

[23]   Fructose-6-phosphate $+ ATP \rightarrow$ Fructose-1,6-diphosphate $+ ADP$

phate is split by the enzyme **aldolase** into two three-carbon sugars, **gly-ceraldehyde-3-phosphate** and **dihydroxyacetone phosphate**. These two are interconverted by the enzyme triose phosphate isomerase (reaction 24).

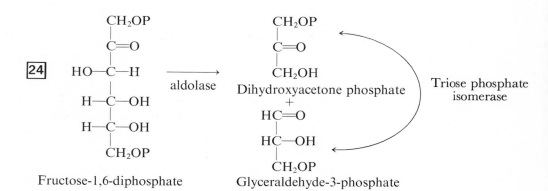

Fructose-1,6-diphosphate   Glyceraldehyde-3-phosphate

Glyceraldehyde-3-phosphate reacts with a compound containing an —SH group—not acetyl coenzyme A, as in certain reactions discussed previously, but an —SH group in an amino acid that is part of the enzyme molecule, part of the glyceraldehyde-3-phosphate dehydrogenase. The combination of the glyceraldehyde-3-phosphate with the —SH group yields an

$$H—\overset{|}{\underset{|}{C}}—OH$$

configuration that can undergo dehydrogenation with NAD as the hydrogen acceptor (reaction 25). The product, **phosphoglyceric acid** bound to the —SH group of the enzyme, then reacts with inorganic phosphate to yield, **1,3-diphosphoglyceric acid** and free enzyme —SH. The phosphate at carbon 1 is an energy-rich group, which can react with ADP to form ATP.

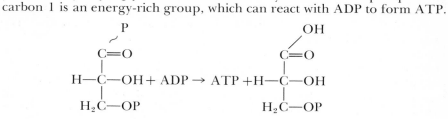

3-Phosphoglyceric acid

This, like the energy-rich phosphate of succinyl phosphate, is one made at the *substrate* level. The resulting 3-phosphoglyceric acid undergoes rearrangement to 2-phosphoglyceric acid. Next, in an unusual reaction, an energy-rich phosphate is generated by the removal of water, by a **dehydration**, rather than by the removal of two hydrogens, a **dehydrogenation** (reaction 26).

The product, phosphopyruvic acid, can transfer its phosphate group to ADP to yield ATP and free pyruvic acid. This is the second energy-rich phosphate bond generated at the substrate level in the metabolism of glucose to pyruvic acid. Each glucose molecule yields two molecules of glyceraldehyde-3-phosphate (the second by the conversion of dihydroxy-

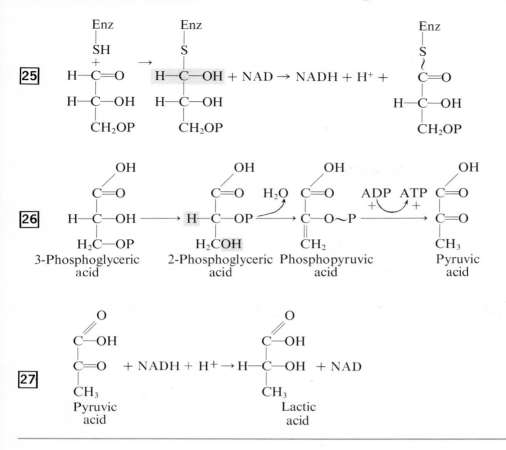

acetone phosphate) and hence a total of four energy-rich bonds are produced as glucose is metabolized to pyruvic acid. However, two energy-rich
phosphate bonds are utilized in the process—one to convert glucose to
glucose-6-phosphate, and the second to convert fructose-6-phosphate to
fructose-1,6-diphosphate. The net yield in the process is two $\sim$P (four
$\sim$P produced minus two $\sim$P used up in the reactions). Pyruvic acid is then
metabolized to acetyl coenzyme A by the reactions described previously.

SECTION 5–7

ANAEROBIC GLYCOLYSIS

Under anaerobic conditions, without oxygen to serve as the ultimate
electron acceptor, the reactions of the electron transmitter system cease
when all of the intermediate acceptors have been converted to the reduced
condition, when they have taken up all the electrons possible. Then the
metabolism of glucose will lead to the accumulation of pyruvic acid (since
it cannot be metabolized to acetyl coenzyme A) and pyruvic acid accepts
hydrogens from reduced pyridine nucleotides to yield lactic acid and
oxidized pyridine nucleotide, NAD (reaction 27). This is catalyzed by lactic
dehydrogenase operating in the opposite direction. The oxidized pyridine
nucleotide can then again accept hydrogens from glyceraldehyde-3-phos-
phate and be reduced to NADH and H$^+$. By the cyclic utilization of NAD
in these two reactions, glucose is utilized under anaerobic conditions and
lactic acid accumulates (reaction 28).

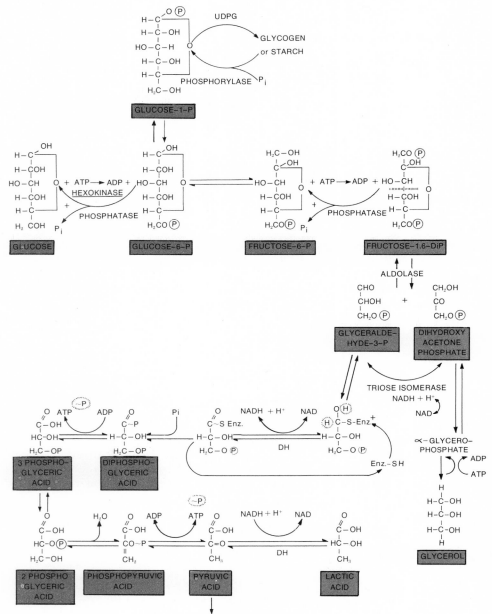

FIGURE 5–4   *Glycolysis, the series of reactions by which glucose and other sugars are metabolized to pyruvic acid. The reversible steps are indicated by double arrows. Other steps are essentially irreversible; two different enzymes catalyze the reaction in the two different directions. For example, glucose-6-phosphate is converted to glucose by glu-cose-6-phosphatase but hexokinase catalyzes the conversion of glucose to glucose-6-phosphate.*

The conversion of glucose to lactic acid will yield a net production of two ~P, as we have seen, and cells can obtain in this way a small amount of energy in the absence of oxygen. When glucose is simply split into two moles of lactic acid, the free energy change, $\triangle G$, is −52,000 calories per mole. However, when glucose is enzymatically converted to lactate by anaerobic glycolysis, in which there is a net phosphorylation of two moles of ADP to yield two moles of ATP, the decrease in free energy, $\triangle G$, is only −38,000 calories per mole. The other 14,000 calories are conserved in the two moles of ATP formed, some 7000 calories per mole.

The reactions by which glucose is metabolized in the absence of oxygen are identical to those by which it is metabolized in the presence of oxygen except for this very last step, at pyruvic acid. In the absence of oxygen, pyruvic acid is converted to lactic acid, which accumulates. In the presence of oxygen, pyruvic acid is metabolized to acetyl coenzyme A and, via the Krebs citric acid cycle, to carbon dioxide and water.

In yeast cells pyruvic acid is converted to acetaldehyde and this can accept hydrogens from reduced NAD to yield oxidized NAD and ethyl alcohol (reaction 29). If the cells of our bodies had this same enzyme we could, by exercising violently, so as to metabolize glucose anaerobically to pyruvate and acetaldehyde, produce ethyl alcohol within our cells and, perhaps, become intoxicated! Mammalian cells, however, metabolize pyruvic acid not to free acetaldehyde but, via a series of intermediates, to acetyl coenzyme A.

In anaerobic glycolysis the ultimate hydrogen acceptor is some substance other than oxygen—lactic acid in animal cells, ethanol in yeasts, and glycerol or butanol in certain bacteria.

SECTION 5–8

THE PENTOSE PHOSPHATE PATHWAY

Glucose-6-phosphate has an

$$H\text{—}\overset{\displaystyle |}{\underset{\displaystyle |}{C}}\text{—}OH$$

configuration at carbon 1 which can undergo dehydrogenation. The enzyme catalyzing this reaction, **glucose-6-phosphate dehydrogenase**, specifically requires NADP as its hydrogen acceptor. A make-ready reaction, the addition of water, converts the product to **6-phosphogluconic acid.** This can undergo dehydrogenation (it has an

$$H\text{—}\overset{\displaystyle |}{\underset{\displaystyle |}{C}}\text{—}OH$$

configuration) by a second enzyme which also specifically requires NADP as hydrogen acceptor. The product of this reaction undergoes decar-

---

28 Glyceraldehyde-3-phosphate + NAD → NADH + H$^+$ + 3-phosphoglyceric acid

lactic acid + NAD ← NADH + H$^+$ + pyruvic acid

29
$$H\text{—}C\text{=}O \quad \overset{\displaystyle |}{\underset{\displaystyle CH_3}{}} \;+\; NADH + H^+ \;\rightarrow\; H_2C\text{—}OH \quad \overset{\displaystyle |}{\underset{\displaystyle CH_3}{}} \;+\; NAD$$
Acetaldehyde   Ethyl alcohol

boxlation to yield a five-carbon sugar, ***ribulose-5-phosphate***, which is of major importance in reactions associated with photosynthesis (p. 114). It is also the source of the five-carbon sugars which are constituents of nucleotides and nucleic acids (reaction 30).

In further reactions, rearrangements occur in which either the upper two or the upper three carbons of the molecule are transferred as a unit to another sugar. By such reactions, sugars with three, four, five, six, or even seven carbon atoms are synthesized (Fig. 4–11). These reactions are termed the "pentose phosphate" pathway and provide an alternative pathway for the metabolism of glucose.

SECTION 5–9

AMINO ACID OXIDATION

Amino acids are oxidized by reactions in which the amino group is first removed, a process called ***deamination***, then the carbon chain is metabolized and eventually enters the Krebs citric acid cycle. The amino acid ***alanine***, for example, yields pyruvic acid when deaminated, ***glutamic acid*** yields α-ketoglutaric acid, and ***aspartic acid*** yields oxaloacetic acid (reaction 31).

These three amino acids enter the Krebs citric acid cycle directly. Other amino acids may require several reactions in addition to deamination to yield a substance which is a member of the Krebs cycle, but ultimately the carbon chains of all the amino acids are metabolized in this way.

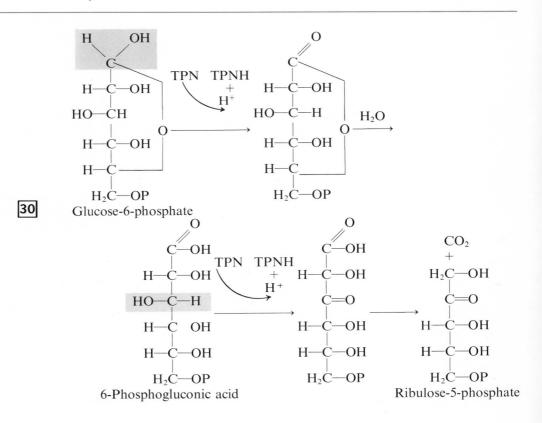

30

Glucose-6-phosphate

6-Phosphogluconic acid

Ribulose-5-phosphate

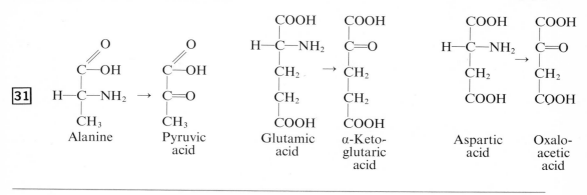

SECTION 5–10

THE ELECTRON TRANSMITTER SYSTEM

Thus far we have considered only the reactions by which electrons are removed from substrate molecules and transferred to the primary acceptors, either a pyridine nucleotide or a flavin. The major reactions by which biologically useful energy is conserved, however, occur when the electrons pass down the "electron cascade" in the electron transmitter system and provide the energy to drive the reactions of oxidative phosphorylation. The electrons entering the electron transmitter system from NADH have a relatively high energy content. As they flow along the chain of enzymes they lose much of their energy, some of which is conserved in the form of ATP.

The enzymes of the electron transmitter system are located within the substance of the mitochondrion in the mitochondrial membranes. They are immediately adjacent to one another and it seems likely that the electrons in fact flow through a solid phase rather than between enzymes that are in solution.

The enzyme **succinic dehydrogenase** is also located in the mitochondrial membranes but this is the only one of the enzymes of the Krebs cycle which is located there. The enzymes that convert pyruvic acid to acetyl coenzyme A and α-ketoglutaric acid to succinyl coenzyme A appear to be located in particles within the mitochondria which are evident in the electron microscope.

The components of the electron transmitter system are given in the order of their oxidation reduction potentials, which range from −0.32 volt for the pyridine nucleotides to +0.81 volt for oxygen (Fig. 5–5). However, it is not known whether a particular electron, in passing from pyridine nucleotide to oxygen, must go through each and every one of the intermediates or whether it might skip some of the steps. It probably has to pass through at least three different steps to account for the three ~P which are made for each pair of electrons that pass from pyridine nucleotide to oxygen.

It is not known whether the enzymes of the electron transmitter system are arranged in the mitochondrial membrane in some fixed mosaic pattern, which corresponds to the order in which electrons pass from one to the next in order of their oxidation-reduction potentials, or whether the enzymes simply lie close enough together (but not in a fixed pattern) so that electrons can pass from one to another with the proper oxidation-reduction potential.

Oxidative phosphorylation is measured by the rate at which inorganic

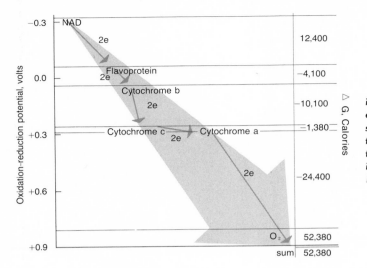

FIGURE 5–5   *The changes in potential and the changes in free energy that occur in the successive steps of the electron transmitter system. The energy released as electrons pass from one carrier to the next is trapped as energy-rich phosphate bonds, ~P.*

phosphate, P_i, is converted to ATP as NADH or some other substance undergoes oxidation. It is possible to extract mitochondria by homogenizing the cells and then separating the subcellular particles by centrifugation. Such mitochondria, when removed carefully from the cell, will still carry out oxidative phosphorylation. Indeed, it is possible to disrupt mitochondria by ultrasonic vibrations and obtain submitochondrial particles which will carry out oxidative phosphorylation.

In these purified systems NADH will be oxidized and oxygen will be utilized only if ADP is present to accept the energy-rich phosphates produced by the flow of electrons. The flow of electrons is *tightly coupled* to the phosphorylation process and will not occur unless phosphorylation can occur, too. This, in a sense, prevents waste, for electrons will not flow unless ~P can be formed.

It is known that oxidation is similarly coupled to phosphorylation when the reactions occur within the intact cell. Certain substances, one of which is the hormone **thyroxin**, can "uncouple" phosphorylation from oxidation so that the flow of electrons occurs without their energy being trapped as ~P; the energy is released as heat, instead. The details of the mechanism by which these reactions convert inorganic phosphate to energy-rich phosphate in the electron transmitter system are under intensive investigation at present, but they are not completely clear.

The flow of electrons all the way from pyridine nucleotide to oxygen, a total drop of 1.13 volts (from −0.32 to +0.81 volt), would yield 52,000 calories per pair of electrons if the process were 100 per cent efficient. This may be calculated from the formula $\triangle G = -nF \triangle E$, where $\triangle G$ is the change in free energy, n is the number of electrons (2), F is the Faraday (23,040 calories), and $\triangle E$ is the difference in the oxidation-reduction potentials of the reactants (1.13 volts). Under experimental conditions, most cells will produce at most three ~P per pair of electrons as the electrons pass from pyridine nucleotide to oxygen (Fig. 5–5). Each ~P is equivalent to about 7000 calories. The efficiency of the electron transmitter system may thus be calculated at 21,000/52,000 or about 40 per cent.

The amount of ATP present in any cell is usually rather small. In muscle cells, in which large amounts of energy may be expended in a short time during the contraction process, an additional substance, **creatine phosphate,** serves as a reservoir of ~P. The terminal phosphate of ATP is transferred by an enzyme, **creatine kinase,** to creatine to yield creatine phosphate and ADP (reaction 32). The phosphate bond of creatine phosphate is also an energy-rich bond. The $\sim$P of creatine phosphate must be transferred back to ADP, converting it to ATP, to be utilized in some reaction requiring energy, such as muscle contraction.

The fact that phosphorylation is tightly coupled to oxidation (electron flow) in the electron transmitter system provides the basis for a system of control which can regulate the rate of energy production and adjust it to the rate of energy utilization. In a resting muscle cell, oxidative phosphorylation will occur until all of the ADP has been converted to ATP. Then, since there are no more acceptors of $\sim$P, phosphorylation stops. Since oxidation is tightly coupled to phosphorylation, oxidation (i.e., the flow of electrons and utilization of oxygen) will also cease.

When the muscle contracts, the energy required is obtained by the splitting of the energy-rich, terminal phosphate of ATP:

$$ATP \rightarrow ADP + P_i + energy$$

The ADP formed can then serve as an acceptor of $\sim$P, phosphorylation begins, and the flow of electrons to oxygen occurs. Oxidative phosphorylation continues until all of the ADP has been converted to ATP. Electric generating systems have an analogous control device which adjusts the rate of production of electricity to the rate of utilization of electricity.

Some interesting calculations of the overall energy changes involved in metabolism in the human body have been made by E. G. Ball of Harvard University. The conversion of oxygen to water requires the participation of hydrogen ions and electrons; thus the total flow of electrons in the human body can be calculated and expressed in terms of amperes. From the oxygen consumption of an adult man at rest, 264 ml per minute, and the fact that each oxygen atom requires two hydrogen atoms and two electrons to form a molecule of water, Dr. Ball calculated that 2.86×10^{22} electrons are flowing from foodstuff via dehydrogenases and the cytochromes to oxygen each minute in all the body cells. Since an ampere equals 3.76×10^{20} electrons per minute, this current amounts to 76 amperes. This is quite a bit of current, for an ordinary 100 watt light bulb uses just a little less than 1 ampere.

The flow of electrons from substrate to oxygen involves a potential difference of 1.13 volts (from -0.32 v to $+0.81$ v). In electrical units, volts times amperes equals watts, and $1.13 \times 76 = 85.9$ watts.

The total expenditure of energy can also be calculated from the number of calories used per minute (about 1.27 at rest). Using the appropriate conversion factors this can be shown to be equivalent to about 88 watts, which agrees satisfactorily with the value calculated previously.

The body, then, utilizes energy at about the same rate as a 100 watt light bulb, but differs from it in having a much larger flow of electrons passing through a much smaller voltage change. How brightly are you glowing tonight?

The conversion of glucose to carbon dioxide and water in a calorimeter yields about 4 calories per gram. In describing the successive steps in the metabolism of glucose by cells, we noted that energy is released in the form of energy-rich phosphate bonds, a form in which it can be used

---

32   $$Creatine + A\overset{utilization}{\underset{synthesis}{\mathrm{T}}}P \rightleftharpoons Creatine \sim\!\!\underset{storage}{P} + ADP$$

TABLE 5–1

(1)	$C_6H_{12}O_6 + 2 \sim P \longrightarrow$	2 Pyruvate + 2 NADH + 4 $\sim$ P
(2)	2 Pyruvate $\longrightarrow$	2 CO_2 + 2 Acetyl CoA + 2 NADH
(3)	2 Acetyl CoA $\longrightarrow$	4 CO_2 + 6 NADH + 2 H_2FP + 2 $\sim$ P
(4) Sum	$C_6H_{12}O_6 \longrightarrow$	6 CO_2 + 10 NADH + 2 H_2FP + 4 $\sim$ P
(5)	$C_6H_{12}O_6 + 6 O_2 \longrightarrow$	6 CO_2 + 6 H_2O + 30 $\sim$ P + 4 $\sim$ P + 4 $\sim$ P

(From Villee, C. A.: *Biology.* 5th ed. Philadelphia, W. B. Saunders Company, 1967.)

to perform a variety of kinds of work. Let us now dissect the overall equation

$$C_6H_{12}O_6 + 6O_2 \rightarrow 6CO_2 + 6H_2O + \text{energy}$$

and review where useful energy is released (Table 5–1).

In glycolysis (reaction 1 of Table 5–1), glucose is activated by the addition of 2 $\sim$P and converted to 2 pyruvate + 2 NADH + 4 $\sim$P. Then the two pyruvates are metabolized (reaction 2) to 2 CO_2 + 2 acetyl CoA + 2 NADH. Finally, in the citric acid cycle (reaction 3), the 2 acetyl CoA are metabolized to 4 CO_2 + 6 NADH + 2 H_2FP + 2 $\sim$P.

These reactions can be added together (reaction 4) by eliminating items that are present on both sides of the arrows. Then, since the oxidation of NADH in the electron transmitter system yields 3 $\sim$P per mole, the 10 NADH = 30 $\sim$P. The oxidation of H_2FP yields 2 $\sim$P per mole and the 2 H_2FP = 4 $\sim$P. Summing these we see that the complete aerobic metabolism of one mole (180 gm) of glucose yields 38 $\sim$P (reaction 5). Each $\sim$P is equivalent to about 7000 calories and the 38 $\sim$P = 266,000 calories.

When a mole of glucose is burned in a calorimeter some 686,000 calories are released as heat. The metabolism of glucose in cells conserves 266,000/686,000 or about 42 per cent of the total energy as biologically useful energy, $\sim$P. The remainder of the energy is dissipated as heat.

SECTION 5–11

THE MOLECULAR ORGANIZATION OF MITOCHONDRIA

Isolated mitochondria can carry out the entire sequence of reactions in the Krebs citric acid cycle, the electron transmitter system and the oxidative phosphorylation of ADP to ATP. Mitochondrial shapes range from nearly spherical to elongated, sausage-like structures, but an average mitochondrion is an ellipsoid about 3μ long and a little less than 1μ in diameter. They can be seen with a high power microscope, especially in a stained section of a tissue, but the details of their internal structure are revealed only by the higher magnification possible in the electron microscope. Some very generously endowed large cells, such as the giant amoeba, *Chaos chaos*, have several hundred thousand and an average mammalian liver cell has about 1000 mitochondria. The mitochondrial protein accounts for about 20 per cent of the total proteins of the liver cell.

Each mitochondrion has two membranes, an outer smooth one, and an inner one that is folded repeatedly to form the cristae, or shelflike projections within the cavity of the mitochondrion (Fig. 5–6). In some mitochondria the cristae extend completely across the central cavity, from wall to wall. This inner membrane has a much greater area than the outer one.

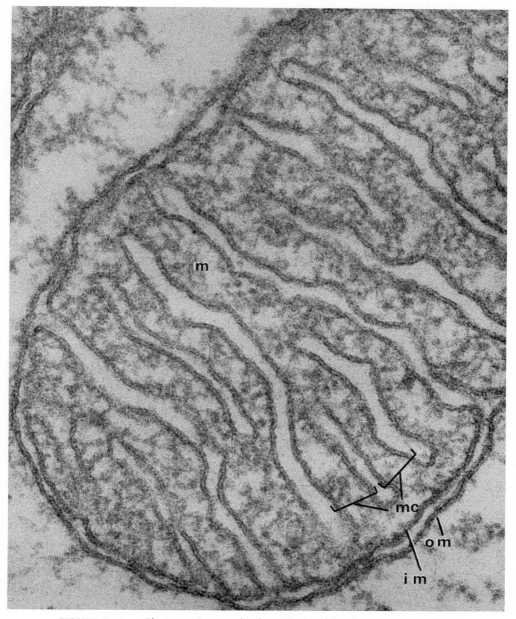

FIGURE 5–6   Electron micrograph of a mitochondrion from a pancreatic acinar cell. The double-layered unit membrane is evident in the smooth outer membrane (om) and in the inner membrane (im) which folds to form the cristae (mc) Magnified 207,000✕. (Courtesy of G. E. Palade; from De Robertis, E. D., Nowinski, W. W., and Saez, F. A.: Cell Biology. Philadelphia, W. B. Saunders Company, 1970.)

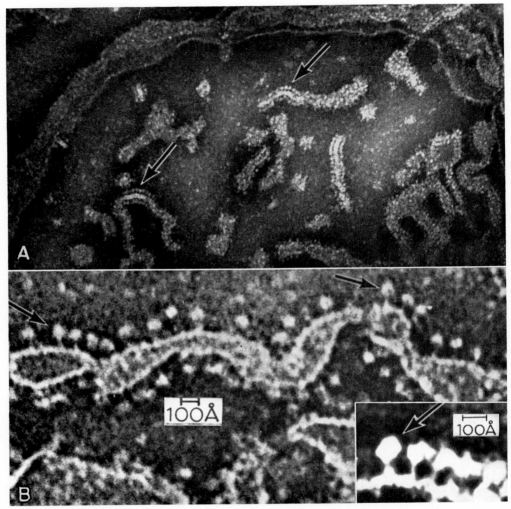

FIGURE 5–7   *Electron micrograph of a mitochondrion swollen in a hypotonic solution and negatively stained with phosphotungstate. A, Isolated cristae 85,000×. B, At higher magnification, 500,000×, the elementary particles attached by a stalk to the surface of the cristae are evident. Insert: At 650,000× the polygonal shape of the elementary particle and its slender stalk are clearly visible. (Courtesy of H. Fernandez-Moran; from De Robertis, E. D., Nowinski, W. W., and Saez, F. A.: Cell Biology. Philadelphia, W. B. Saunders Company, 1970.)*

Both are "unit membranes" composed of a core of a double layer of lipid molecules coated on both sides with a layer of protein molecules. The fluid material within the central cavity, within the inner membrane, is called the matrix.

The enzymes of the Krebs cycle have been found in the soluble matrix within the mitochondrion. The enzymes of the electron transmitter system are tightly bound to the inner membrane. There is some evidence to suggest that successive enzymes in the chain are actually located adjacent to each other in the membrane. Each group of these electron transmitter enzymes, termed a **respiratory assembly**, is one of the fundamental units of cellular activity. It has been estimated that the mitochondrion of the liver cell contains some 15,000 respiratory assemblies and that they make up about one quarter of the mass of the mitochondrial membranes. Thus the mitochondrial membrane is not just a protective skin but an important functional part of the mitochondrion.

High resolution electron micrographs of mitochondria, made by Humberto Fernandez-Moran at the University of Chicago, have demonstrated the presence of small particles on the outer surface of the outer membrane and the inner surface of the inner membrane (Fig. 5–7). The particles on the inner membrane typically have a spherical knoblike head, a cylindrical stalk and a base plate. These particles may be sites of the enzyme reactions carried out by mitochondria. The particles of the inner membrane contain the respiratory assemblies, the electron transmitter system and the enzymes of oxidative phosphorylation.

It seems clear that much of the cell's biologically useful energy, ATP, is generated by enzyme systems located in the inner membrane of the mitochondrion, yet most of the energy utilized by the cell is required for processes that take place outside of the mitochondrion. ATP is used in the synthesis of proteins, fats, carbohydrates, nucleic acids and other complex molecules, in the transport of substances across the plasma membrane, in the conduction of nerve impulses, and in the contraction of muscle fibers, all of which are reactions occurring largely or completely outside of the mitochondrion in other parts of the cell. We do not yet know how the ~P generated within the mitochondrion becomes available outside of the mitochondrion, for membranes such as those comprising the walls of the mitochondria are largely impermeable to large, charged molecules such as adenosine triphosphate.

Another problem of great concern to biochemists at the present time is that of the regulation of the rate of metabolism of the mitochondria. A single cell may have a thousand or more mitochondria and the function of each one must be controlled appropriately to generate the amount of energy required by the cell at any given moment. The rate at which ~P is utilized by a cell may vary over a remarkably wide range as the cell becomes active or quiescent. This has been measured in the muscle cells of a frog. One hundred grams of frog muscle utilizes 1.6 μmoles ~P per minute when quiescent and 3300 μmoles ~P per minute in a state of tetanus, that of continuous contraction. The rate of production of ATP in the cell is controlled in large part by the rate of utilization of ATP by the cell. The flow of electrons in the electron transmitter system is tightly coupled to phosphorylation, and oxidative phosphorylation can occur only when there is ADP to be converted to ATP. These facts provide the basis for the system by which a cell's utilization of ATP, which produces ADP, is used to regulate the rate at which ATP is produced. In addition, the structure and biological activity of some of the enzymes involved in glucose oxidation are affected by the concentration of ADP present. In this way an increased concentration of ADP can lead to increased activity of these enzymes and an increased production of ATP. The problem of the nature and interrelations of these various biological control systems is very much in the forefront in biology today.

SECTION 5–12

METABOLIC RATE AND TEMPERATURE

Only some 50 per cent of the energy of foodstuff molecules is conserved in the form of ATP by the metabolic reactions of the cell. Some of this energy is in the form of entropy and cannot be used, and the remainder is lost to the surroundings in the form of heat. All plants and most animals lose most of this heat energy to the surrounding environment. In other words, nearly half of the energy from their food molecules is lost to such organisms without providing any special benefit to them. This, as we shall see, is important in the concept of **food chains**. The body temperature of such animals will fluctuate with the environment and they are termed poikilothermic.

The rate of chemical reactions in living cells, as well as in the chemist's test tube, is a function of the temperature. Many if not most chemical reactions double with each increase of 10°C. Thus the metabolic rate of poikilothermic animals will increase as the temperature rises and decrease as the temperature falls in a regular fashion. Such animals can move, feed and grow in warm weather, but become inactive in cold weather.

A few animals, birds and mammals, have evolved physiological mechanisms which enable them to maintain their body temperature constant, despite wide fluctuations in the environmental temperature. Animals that have such built-in thermostats are termed **homoiothermic**. Mammals have body temperatures usually ranging between 36 and 39°C, and birds have temperatures slightly higher, ranging between 40 and 43°C. Homoiothermic animals, able to maintain a relatively constant body temperature, can maintain high rates of metabolism and can be active even when the temperature falls. By evolving insulation such as fat, fur or feathers, which retard the loss of heat to the environment, they use the heat that is conserved to maintain their body temperature and a high rate of metabolism.

A little thought should make clear why it is easier for an animal to maintain a temperature above that of his environment than to maintain a constant temperature below that of the environment. An animal with effective insulation can conserve heat by minimizing the flow of blood to the surface of its body, and it can increase the rate of production of heat by the muscular contraction of shivering. However, homoiothermic animals cannot withstand environmental temperatures more than a few degrees above their body temperature for more than a short time. The mechanisms for cooling an animal are not very effective. They can increase the flow of blood to the surface and thereby increase heat loss; they can cool the surface by evaporation, using perspiration in man, or panting in dogs, but these have very limited effectiveness in keeping an animal cool when the temperature of the environment exceeds the temperature of the animal's body.

Measurements of the metabolic rates of a great many homoiothermic animals have revealed the interesting fact that the basal metabolic rate of an organism is inversely proportional to its body size. In other words, the smaller the organism, the higher is the metabolic rate, expressed as cubic millimeters of oxygen used per gram of body weight per hour. The smaller the animal, the greater is the ratio of its surface to its volume, therefore it will have a greater relative heat loss to the environment per unit mass. This would necessitate a higher metabolic rate. This fact actually sets a lower limit to the size of an animal that can be a homoiotherm, because a very small size would require a greater food intake and respiration per minute than is physically possible. The smallest living mammal is a shrew that weighs only about 4 grams, and this poor thing must eat nearly its own body weight of food every day, or will starve to death in a few hours!

This inverse relationship between body size and metabolic rate raises serious problems for small animals during the cold season of the year. During the winter months, the rate of loss of heat to the environment increases by a considerable extent at the same time that the supply of food for the animal is usually much smaller than in the warmer months. Certain mammals, rodents, insectivores and bats, evade this problem to some extent by **hibernation**, i.e., by permitting their body temperature to fall below the normal level to just a degree or two above the temperature of the environment. The rates of all their metabolic reactions and their physiological functions are greatly decreased, and the animal passes the winter in a dormant state. Birds, with very few exceptions, do not hibernate during the winter, but either migrate to a warmer territory to avoid the cold, or they spend an increased fraction of their time looking for food. Bats can maintain their body temperature at a fairly high and constant level at night, but it drops nearly to that of their surroundings during the day when they are roosting. In this way, they can conserve the energy of the food taken in during the night.

Bears and other large mammals do not hibernate during the winter. They do become inactive, and spend much of their time sleeping and use up the extensive fat reserves they have laid down in the summer and the fall. Their body temperature, however, decreases by only a few degrees, and they are not dormant as are the small mammals.

SUGGESTIONS FOR FURTHER READING

Baldwin, Ernest: *Dynamic Aspects of Biochemistry.* 4th ed. New York, Cambridge University Press, 1964.
   A technical but extremely interesting discussion of the details of cellular metabolism by the late great British biochemist.

Dawkins, M. J. R., and Hull, D.: *The Production of Heat by Fat.* Sci. Amer. August (Offprint 1018) 1965.

McGilvery, R. W.: *Biochemistry.* Philadelphia, W. B. Saunders Company, 1970.
   A well written, detailed exposition of the molecular mechanisms underlying many biochemical processes.

Dixon, M., and Webb, E. C.: *Enzymes.* 2nd ed. London, Longmans, Green and Co., Ltd., 1964.
   An advanced discussion of enzymes and their properties, with a catalogue of the well known ones.

Mahler, H. R., and Cordes, E. H.: *Biological Chemistry.* New York, Harper & Row, Publishers, 1966.
   One of the standard texts of biochemistry; an excellent discussion of cellular respiration.

Weyer, E. M. (ed.): *Multiple Molecular Forms of Enzymes.* Ann. N. Y. Acad. Sci. *151:*1–747, 1968.
   Papers presented at a conference held in 1967 regarding the many facets of the subject of isozymes.

ENERGY TRANSFER: BIOSYNTHETIC PROCESSES

CHAPTER 6

The previous chapter dealt with those enzymatic processes in which molecules of foodstuffs are broken down and their energy is conserved in the biologically useful form of ~P, ATP. Such processes may be termed "catabolic" to distinguish them from anabolic processes in which larger molecules are built up from smaller ones. Both animal and plant cells have a wide range of anabolic reactions. Their enzymes catalyze a remarkable array of biosynthetic processes utilizing the energy of ATP, and using as raw materials some of the 5-, 4-, 3-, 2- and 1-carbon compounds that are intermediates in the metabolism of glucose, fatty acids and other compounds. The anabolic, building up, processes and the catabolic, breaking down, processes are interrelated and interdependent in a variety of ways, and it may be difficult to say whether some particular reaction is anabolic or catabolic. Indeed, a given reaction may be anabolic in one sequence and catabolic in another.

SECTION 6–1

BIOSYNTHETIC PRINCIPLES

This whole subject of intermediary metabolism, the sequences of enzymic reactions, by which an enormous variety of compounds is synthesized and broken down, is much too complex to be discussed in detail here. The enzyme that controls each reaction is a protein molecule whose structure is determined genetically by processes to be considered in the next chapter. The overall sequence of enzymic reactions by which any compound is synthesized usually includes one or more self-adjusting control mechanisms to regulate and integrate the rates of the several reactions in the sequence. Several basic principles of cellular biosynthesis may be distinguished:

1. Each cell, by and large, synthesizes its own proteins, nucleic acids, lipids, polysaccharides and other complex molecules. It does not receive them preformed from other cells. The molecules of glycogen present in

muscle, for example, are synthesized within the muscle cell and are not liver glycogen that has been transported there in the blood.

2. Each step in a biosynthetic process is catalyzed by a separate enzyme.

3. Although certain steps in a biosynthetic sequence will proceed without an external supply of energy, i.e., without using energy-rich phosphate, the overall synthesis of these complex molecules does require the input of chemical energy at various points along the way. Why should this be true? Can you relate this to your concept of entropy?

4. The synthetic processes utilize as raw materials relatively few substances, among which are acetyl coenzyme A, glycine, succinyl coenzyme A, ribose, pyruvate and glycerol.

5. Synthetic processes in general are not simply the reverse of the processes by which the molecule is degraded. The synthetic process includes one or more separate steps that differ from any step in the degradative process. Since these steps are catalyzed by different enzymes, this permits the synthesis and the degradation of the complex molecule to be governed by separate control mechanisms.

6. The process of biosynthesis includes not only the formation of the macromolecules from simple precursors, but the assembly of these macromolecules into the several kinds of membranes that comprise the outer boundary of the cell and the intracellular organelles. A cell that is growing rapidly must allocate a large fraction of its total energy output to biosynthetic processes, especially the biosynthesis of proteins. A rapidly growing bacterial cell may use as much as 90 per cent of its total biosynthetic energy for the synthesis of proteins. Even a cell that is not growing, not increasing in mass, uses a considerable portion of its total energy for the chemical work of biosynthesis. Each cell's constituent molecules are constantly being broken down and rebuilt. They are said to be in a dynamic state.

SECTION 6–2

THE DYNAMIC STATE OF CELLULAR CONSTITUENTS

Your body, or the body of a plant or animal, usually appears to be unchanging as days and weeks go by. It had been thought that the cells of the body and the component molecules of the cells are equally unchanging. Until some 30 years ago, it was believed that the molecules of nutrients that were not used to increase the total cellular mass were used within a short time to provide energy. It followed from this concept that you could distinguish two kinds of molecules: relatively unchanging, static ones, which made up the cellular "machinery," and other molecules which were rapidly metabolized and corresponded to cellular "fuel."

However, since that time investigators have fed rats or other experimental animals with amino acids, fats, carbohydrates, water or other molecules, each suitably labeled with some radioactive or heavy isotope. Such experiments have shown that the cellular constituents are in a constant state of flux. Labeled amino acids are rapidly incorporated into body proteins, and labeled fatty acids are rapidly incorporated into fat deposits, even though there is no increase in the total amount of protein or fat present in the body. The carbon atoms of sugar molecules may become part of an amino acid or fatty acid molecule.

The proteins and fats of the body, even the substance of the bones, are constantly and rapidly being synthesized and broken down. In the adult the rates of synthesis and degradation are essentially equal so that there is no change in the total mass of the body. The distinction between "machinery" molecules and "fuel" molecules becomes much less sharp, for some of the "machinery" molecules are constantly being broken down and used as fuel and some "fuel" molecules are used in the synthesis of new machinery.

One exception to this general rule of molecular flux is provided by the DNA molecules that make up the units of heredity, the **genes** in the nucleus of each cell. Experiments with labeled atoms have shown that the DNA molecules are remarkably stable, they are broken down and resynthesized only very slowly, if at all. The amount of DNA in the nucleus of each cell of an organism is constant. New DNA molecules must, of course, be synthesized each time a cell divides. The stability of the DNA molecules may be of importance in ensuring that genetic information is transmitted to succeeding generations with the greatest fidelity and the least possible errors. In marked contrast, the molecules of RNA are constantly undergoing degradation and synthesis, and the amount of RNA in each cell may vary within wide limits.

From the rate at which labeled atoms are incorporated into macromolecules, you can calculate that half of all the tissue proteins of man are broken down and rebuilt every 80 days. This is an average figure, for some proteins are replaced much more rapidly than others. The proteins of the liver and of blood serum are replaced very rapidly, half of them every 10 days. Some of the specific enzymes in the liver have a half time as short as two to four hours. Muscle proteins are replaced more slowly than the liver proteins, with a half time of about 180 days. In a very strict sense, then, you are not the same person, chemically speaking, that you were yesterday.

Some aspects of our current ideas of where specific enzyme systems are located within the cell are shown in Figure 6–1. The synthesis of DNA and of RNA occurs within the nucleus. The process of electron transport and oxidative phosphorylation occurs in the mitochondria. The synthesis of proteins occurs on the ribosomes located either free or on the endoplasmic reticulum. The activation of amino acids for protein synthesis, the process of glycolysis and many other reactions occur in the soluble cell sap. These latter enzymes apparently are not attached to any of the subcellular particles.

Many of the steps in biosynthetic processes involve the formation of an **anhydro bond,** e.g., the peptide bonds of proteins, the glycosidic bonds of polysaccharides, and the ester bonds of lipids and of nucleic acids; however, these bonds are not formed by reactions in which a molecule of water is removed. The biosynthesis of sucrose in the cane sugar plant, for example, does not proceed via

$$\text{Glucose} + \text{Fructose} \rightleftarrows \text{Sucrose} + H_2O$$

This reaction would require energy, some 5500 calories per mole, to go to the right, if all reactants were present in a concentration of 1 mole per liter. However, the concentration of glucose and fructose in the plant cell is probably less than 0.01 mole per liter, whereas the concentration of water is extremely high, some 55 moles per liter. Thus the equilibrium point of the reaction under these conditions would be very far to the left.

Instead, one or more of the reactants, either glucose or fructose, is activated by a reaction with ATP in which the terminal phosphate is enzymatically transferred to glucose with the conservation of some of the energy of the terminal phosphate bond. The glucose phosphate, with a higher energy content than free glucose, can react with fructose via another enzyme-catalyzed reaction to yield sucrose and inorganic phosphate.

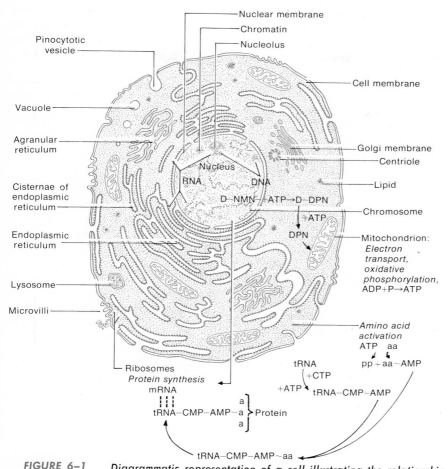

FIGURE 6-1   *Diagrammatic representation of a cell illustrating the relationship of certain enzymatic functions to subcellular structures.*

$$ATP + Glucose \rightarrow ADP + Glucose\text{-}1\text{-}phosphate$$
$$Glucose\text{-}1\text{-}phosphate + Fructose \rightarrow Sucrose + phosphate$$
$$Sum: ATP + Glucose + Fructose \rightarrow Sucrose + ADP + P_i$$

This reaction proceeds to the right because there is a net decrease in free energy. The 7000 calories of the $\sim P$ bond are used to supply the 5500 calories required to assemble the glucose and fructose into sucrose. The overall decrease in free energy is 1500 calories per mole. This is enough to drive the reaction to the right and bring about the synthesis of sucrose. Since water is not a product of this reaction, the high concentration of water in the cell does not inhibit it.

ATP has two energy-rich bonds and it has a high potential for group transfers. It can react in several different ways to transfer energy to another molecule and provide that molecule with the energy needed to carry out a further reaction. Depending on which of the energy-rich bonds of ATP is split, the result can be (a) the transfer of the terminal phosphate group and the release of ADP; (b) the transfer of the final two phosphate groups (called a pyrophosphate group) and the release of AMP, adenosine monophosphate; (c) the transfer of the adenosine monophosphate group with the release of pyrophosphate; or (d) the transfer of the entire adenosine group and the release of both a pyrophosphate from the terminal two phosphate groups, and an inorganic phosphate from the third phosphate group of ATP. These four types of reactions are shown in Figure 6–2.

The most common reaction is type (a). If the terminal phosphate group is transferred to water, the reaction is a hydrolysis of this terminal phosphate group, and the free energy change of this reaction is negative with a $\triangle F$ of 7000 calories per mole. The fact that the transfer of phosphate onto water is exergonic, it releases energy, explains the transfer potential of the terminal phosphate group of ATP. This terminal phosphate

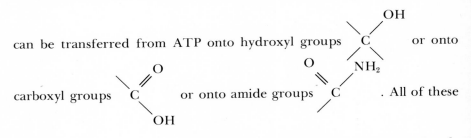

can be transferred from ATP onto hydroxyl groups, or onto carboxyl groups, or onto amide groups. All of these transfer reactions are catalyzed by enzymes called **kinases.** You are already familiar with hexokinase, which transfers a phosphate from ATP onto the

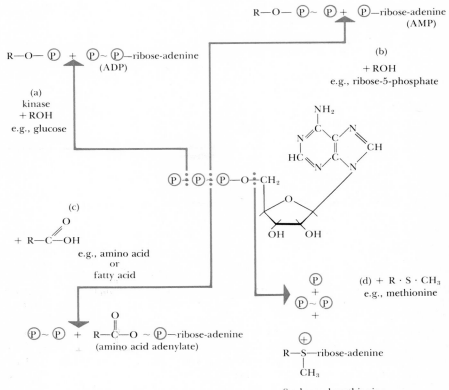

FIGURE 6–2   *Four major types of reactions of adenosine triphosphate: the transfer of phosphate* (A), *the transfer of pyrophosphate* (B), *the transfer of adenylic acid* (C) *and the transfer of an adenosyl group* (D).

hydroxyl group (OH) at carbon 6 of glucose. The phosphate donor for most kinases is ATP; however, in a few reactions, uridine triphosphate (UTP), cytidine triphosphate (CTP) or guanosine triphosphate (GTP) may act in the same way.

The second type of reaction, in which a **pyrophosphate** group is transferred, is much rarer than the first type. The reaction of ribose-5-phosphate with ATP provides an example of such a transfer. The two terminal phosphates of ATP are transferred to the ribose-5-phosphate to yield 5-phosphoribosyl-1-pyrophosphate (called PRPP) and free AMP. The PRPP molecule is an important reactant in the synthesis of both purine and pyrimidine nucleotides.

The third type of reaction, the transfer of AMP to another molecule with the release of pyrophosphate, is fairly common. The result is a compound (R~AMP) with potential for group transfer, an "activated" compound. This type of reaction occurs in the activation of amino acids to prepare them for protein synthesis, and in the activation of fatty acids to

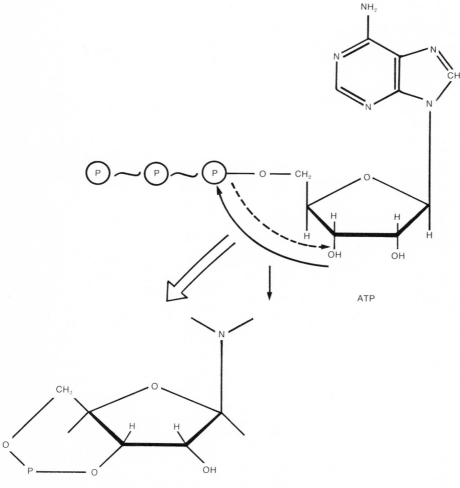

FIGURE 6–3   *The formation of 3′,5′-cyclic adenylic acid from adenosine triphosphate in the reaction catalyzed by adenyl cyclase.*

prepare them for metabolism. In the latter, the resulting fatty acid $\sim$AMP compound undergoes a second reaction with coenzyme A to yield free AMP and fatty acid $\sim$coenzyme A.

A curious variation of this type of transfer is seen in the synthesis of **cyclic adenylic acid**, adenosine 3,′5′-monophosphate, Figure 6–3. Cyclic adenylic acid is in the center of interest at the present time because of its role in the mechanism of action of many types of hormones. In its synthesis, ATP carries out an intramolecular transfer of AMP onto the 3′-hydroxyl group of the ribose with the elimination of pyrophosphate.

The fourth type of reaction, the transfer of an adenosyl group, with the release of both orthophosphate (P_i) and pyrophosphate (PP_i), occurs in the reaction of ATP with the amino acid **methionine.** The products are inorganic phosphate, pyrophosphate and an adenosine group attached to the sulfur of methionine, a compound called **S adenosyl methionine** (SAM). This reaction results in an activation of the methyl group of methionine so that it can be transferred from S adenosyl methionine to certain acceptor compounds.

SECTION 6–3

THE SYNTHESIS AND DEGRADATION OF GLYCOGEN

In general, the reactions by which a larger molecule is synthesized from smaller molecules by a cell differ from the reactions by which the larger molecule is broken down to smaller products. The synthetic process usually requires at one or more steps the utilization of energy-rich phosphate groups from ATP to drive the overall sequence of reactions. When the large molecule is broken down, the energy is generally not recovered as ATP, but is wasted as heat. The separation of the two pathways is advantageous to the cell because the synthetic pathway and the degradative pathway can be regulated independently, either by hormones or by other regulatory mechanisms.

Glycogen provides a simple example of this separation of synthetic and degradative pathways. In animal cells, glycogen is synthesized from glucose by steps in which the glucose is first converted to glucose-6-phosphate by the enzyme **hexokinase,** using ATP to supply the energy and the terminal phosphate. In a second step the glucose-6-phosphate is converted to glucose-1-phosphate by the enzyme **phosphoglucomutase** (Fig. 6–4). Then glucose-1-phosphate reacts with uridine triphosphate (UTP) to yield **uridine diphosphoglucose.** The uridine diphosphoglucose has a great enough transfer potential so that the glucose portion can be added onto a chain of glucose molecules to yield a chain that is one glucose unit longer. The enzyme that carries out this transfer is called **glycogen synthetase,** and as it repeats the reaction many times, the glycogen chain is lengthened.

Glycogen is broken down by a different process, termed **phosphorolysis,** in which the glycosidic bonds are cleaved by the addition of inorganic phosphoric acid (Fig. 6–4). The term is analogous to hydrolysis in which a bond is split by the addition of water. The process involves a transfer of the terminal glucose unit onto phosphoric acid, which serves as the acceptor molecule to form **glucose-1-phosphate.** The enzyme that catalyzes this reaction is called **phosphorylase.** The product of the phosphorolytic reaction is a sugar molecule that is phosphorylated and ready to enter either the glycolytic pathway or the pentose phosphate pathway. The next step, common to glycogen breakdown and glycogen synthesis, involves the con-

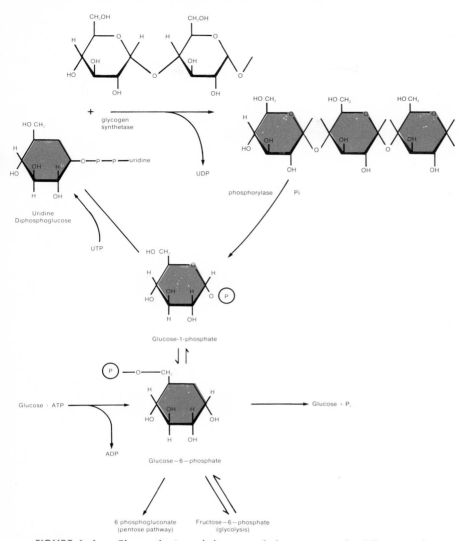

FIGURE 6-4   *The synthesis and cleavage of glycogen occur by different pathways. Glycogen is synthesized by reactions involving uridine triphosphate (UTP) and the formation of uridine diphosphoglucose (UDPG). Glycogen is broken down by the action of phosphorylase, which strips off individual glucose units as glucose-1-phosphate.*

version of glucose-1-phosphate to glucose-6-phosphate, catalyzed by phosphoglucomutase.

The enzyme phosphorylase is the site at which the breakdown of glycogen is controlled. Phosphorylase exists in two forms, an inactive phosphorylase b, and a very active phosphorylase a. The inactive phosphorylase b is transformed into the active phosphorylase a by a reaction in which a phosphate group is transferred from ATP to the OH on a serine, one of the amino acids in the protein chain of the enzyme. This reaction requires another enzyme, a specific kinase, for the transfer, and a special cofactor, **cyclic adenylic acid.** The initial step in the activation of phosphorylase requires the formation of cyclic adenylic acid from ATP by the enzyme **adenyl cyclase.** The adenyl cyclase is bound to membranes in the cell and is stimulated by either the hormone **epinephrine** or the hormone **glucagon.** It is through this rather complex mechanism that these two hormones bring about an increase in the level of blood sugar in the mammal when they are released.

Because of the fact that the synthesis and degradation of glycogen are catalyzed by different enzymes, it is possible for one of the enzymes to be absent while the other one is present. In one form of the rare hereditary disease in man called **glycogen storage disease,** the glycogen forming system is intact, but the phosphorylase for the breakdown of glycogen is absent. Because of this, the liver becomes charged with a vast excess of glycogen.

SECTION 6-4

THE STEPWISE ASSEMBLY OF A MOLECULE

The large molecules of nucleic acids and proteins are put together from their subunits by a continuous process in which the subunits are assembled on a **template.** The assembly of subunits is catalyzed by a single enzyme system. The details of this assembly of macromolecules on a template will be discussed in the next chapter.

A great many other molecules are put together laboriously, step by step, by a series of enzyme reactions working in sequence. As an example of this stepwise assembly of a molecule, let us consider the synthesis of **purines,** a process which has been studied by John Buchanan and by Robert Greenberg (Fig. 6-5). The purine ring consists of five carbon and four nitrogen atoms assembled into a double-ring system. The purines adenine and guanine are important constituents of nucleic acids and of a number of nucleotide coenzymes. It had been known that in man and in birds the purine ring is excreted in the form of **uric acid.** To study this process of purine synthesis, Buchanan fed a number of different substances labeled with ^{13}C to pigeons. He isolated the uric acid from their urine and took apart the uric acid molecule so that he could determine which of the carbons contained ^{13}C. From these studies he could infer that carbon 2 and carbon 8 were derived from formic acid, that carbon 6 came from carbon dioxide, and that carbons 4 and 5, together with nitrogen 7, came from the amino acid glycine. In subsequent experiments using purified enzyme systems, the specific reactions involved in each step were clarified.

The synthesis of the purine ring begins with the reaction of phosphoribosyl-pyrophosphate (p. 155) and glutamine to form 5-phosphoribosylamine. Glutamate and inorganic pyrophosphate are the other reaction products. The nitrogen that had been the amide group of glutamine is the first portion of the purine ring to be present. It eventually becomes nitrogen 9 of the ring. In the next step, glycine is added onto this framework. The reaction requires ATP, and the product is called glycine-amide-ribotide. In the next reaction, catalyzed by another enzyme, this glycine amide ribotide reacts with an active form of formic acid (formic acid bound to **tetrahydrofolic acid**) to add onto the framework the carbon which eventually becomes carbon 8 of the purine ring. Another reaction with glutamine and ATP transfers another nitrogen from glutamine onto this compound to yield 5-amino-imidazole ribotide. In the next reaction the ring is closed by an enzyme that requires ATP. The product is a five-membered ring containing three carbons and two nitrogens attached to ribose phosphate. Carbon dioxide is added next, and the carbon that is added becomes carbon 6 of the finished purine ring. In the next reaction, a nitrogen is transferred from the amino group of aspartic acid. This reaction requires ATP to drive it and the nitrogen added is the one that eventually becomes nitrogen 1 of the ring. Another transfer of an active formate group from

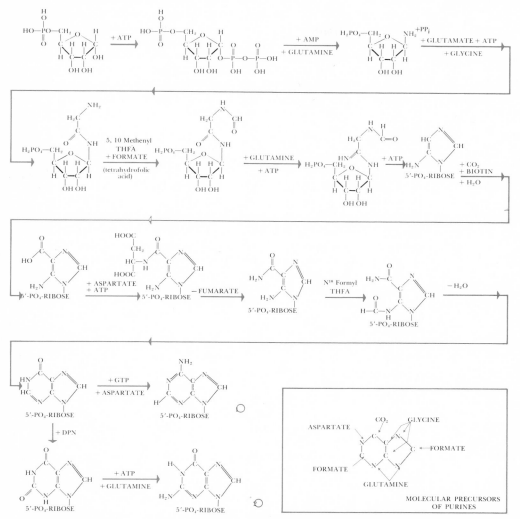

FIGURE 6–5   *A diagram of the enzymatic reactions by which the purine ring is synthesized. Lower right, the purine ring and the precursors of each of the atoms in the purine ring. The two major purine products are adenylic acid (1) and guanylic acid (2).*

formyl tetrahydrofolic acid adds the carbon which will become carbon 2. The last reaction involves the removal of a molecule of water to close the six-membered ring. The product contains a nine-membered purine ring (composed of a six-membered ring and a five-membered ring) and is called *inosinic acid.* This was first isolated from skeletal muscle. Inosinic acid is transformed by further reactions into adenylic acid or guanylic acid. Thus the synthesis of adenine or guanine requires a dozen or more different enzymatic steps, and some 5 moles of ATP per mole of purine to supply the energy needed to drive the reactions.

The synthesis of cholesterol is an even longer process, requiring perhaps two dozen enzymes to assemble this molecule composed of 27 carbon atoms arranged in four rings and a tail, from its subunits, which are two-carbon acetic acid molecules. These examples are cited to give you some appreciation of the complexity of the processes involved in synthesizing biological molecules, the number of enzymes required for each, and the expenditure of chemical energy in the form of ATP which must go on constantly to provide for a continued supply of these molecules. A man who rests in bed for 24 hours and eats nothing must still expend some 1600 calories to stay alive. Some of this is required to provide energy for such basic physiological functions as the beating of the heart and breathing, but another sizable portion of this is required for those biosynthetic processes which must continue day and night.

OTHER BIOSYNTHETIC PROCESSES

Bacteria, animals and plants each synthesize certain characteristic products. Green plants, by the process of photosynthesis, put together the elements of carbon dioxide and water and form many kinds of carbohydrates. Certain plants produce in addition a wide variety of pigments, perfumes and certain types of drugs. Bacteria and molds, among the most versatile chemists of all living things, can make everything from antibiotics to deadly poisons. A few fish, such as the electric eel, can produce shocking amounts of electricity, and certain bacteria, molds and animals can produce light enzymatically.

Although the firefly and the glowworm are the most conspicuous light-emitting organisms, a number of other animals and some bacteria and fungi have this same ability. Luminescent animals are found among the protozoa, sponges, coelenterates, ctenophores, nemerteans, annelids, crustaceans, centipedes, millipedes, beetles, echinoderms, mollusks, hemichordates, tunicates and fishes (Fig. 6–6). The ability to emit light has appeared independently a number of times in the course of evolution. Thus there appears to be no single evolutionary line of bioluminescent forms. It sometimes is difficult to establish that an organism is itself luminescent. In a number of instances the light has been found to be emitted not by the organism, but by bacteria living on the organism. Several unusual East Indian fish have light organs under their eyes in which live luminous bacteria (Fig. 6–7). The light organ contains special long, cylindrical cells well equipped with blood vessels to supply the bacteria with adequate amounts of oxygen. The bacteria emit light continu-

FIGURE 6–6   A school of luminescent squid, Watasenia scintillans. Drawn by Miss E. Grace White. (From Dahlgren, U.: Phosphorescent Animals and Plants. Nat. Hist. N.Y. 22:23, 1922.)

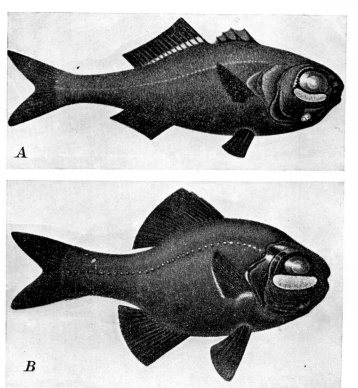

FIGURE 6–7   Two species of luminescent fish from the waters of the Malay Archipelago. A, Anomalops katoptron. B, Photoblepharon palpebratus. The half-moon-shaped luminescent organs just below the eyes are equipped with reflectors. (From Steche, O.: Ueber die Leuchtorgane von Anomalops katoptron und Photoblepharon palpebratus, zwei Oberflächenfischen aus dem malayischen Archipel. Ein Beitrag zur Morph. u. Physiol. der Leuchtorgane der Fische, Z. Wiss. Zool. 93:349, 1909.)

ously and the fish have a black membrane like an eyelid that can be drawn up over the light organ to turn off the light. An unsolved mystery is how the bacteria come to collect in the fish's light organ, as they must in each newly hatched fish.

Some species of shrimp have accessory lenses, reflectors and color filters that form a unit with the light-emitting organ. Thus the whole assembly resembles a lantern.

The amount of light produced by some luminescent animals is remarkable. Fireflies may produce as much light, expressed in terms of lumens per square centimeter, as do fluorescent lamps. Animals may emit lights of different colors—red, green, yellow or blue. Among the luminescent animals, the **railroad worm** of Uruguay, which is the larva of a beetle, is remarkable in that it can produce two different colors. It has a row of green lights along each side of the body, and a pair of red lights on its head. The light produced by luminescent organisms is entirely in the visible spectrum. No ultraviolet or infrared light is produced. Bioluminescence has been termed "cold light" because very little heat is given off along with the light.

The production of light is a reaction controlled by enzymes the details of which differ in different species. Bacteria and fungi will produce light continuously if oxygen is available. Most luminescent animals give off flashes of light only when their luminescent organs are stimulated. The names **luciferin** (the substrate) and **luciferase** (the enzyme) have been given to the two major components of the light-emitting system. The luciferin and luciferase from one species of animal may be quite different from those in another. The luciferins from the crustacean *Cypridina* and from the firefly *Photinus* have been isolated and crystallized and have been found to be chemically quite different.

The luciferin-luciferase reaction is a form of biological oxidation and can occur only in the presence of oxygen as an electron acceptor. When luciferin and luciferase are extracted with appropriate means from a firefly and mixed in a test tube with added magnesium and adenosine triphosphate, light is emitted. The production of light is an energy-requiring process, and the biologically useful energy to drive the reaction is supplied by ATP. Under certain conditions the amount of light emitted is proportional to the amount of ATP present and the system can be used to measure the amount of ATP in a tissue extract.

The bioluminescent reactions in the firefly begin with the reaction of luciferase (E) with the reduced form of luciferin (LH_2) and ATP to yield an intermediate complex of enzyme-luciferyl-adenosine monophosphate with the release of inorganic pyrophosphate (PP_i).

$$E + LH_2 + ATP \rightleftharpoons E\text{-}LH_2\text{-}AMP + PP_i$$

The adenosine monophosphate is attached to a carboxyl group of the luciferin. Then, in the presence of oxygen, light is emitted when $E\text{-}LH_2\text{-}AMP$ is oxidized to E-L-AMP, the combination of enzyme with oxyluciferin (L) and adenosine monophosphate (it seems most appropriate that L-AMP should be the abbreviation for a light-emitting substance!). Finally the E-L-AMP dissociates to yield free luciferase, luciferin and adenosine monophosphate. In this sequence the chemical energy of ATP is converted to light energy.

Two varieties of the fungus *Panus stipticus* are known—an American one which is luminescent, and a European one which is not. Crossing the two varieties reveals that the ability to luminesce is inherited by a single dominant gene.

What advantage the ability to emit light may confer on an organism is not clear. For deep sea animals living in perpetual darkness, light organs would conceivably be useful to enable members of a species to recognize each other, or to serve as a lure for prey or a warning to would-be predators. It is known that the light emitted by fireflies serves as a signal to bring the sexes together for mating. The light emitted by bacteria and fungi probably serves no useful purpose to the organism. Instead it is simply a by-product of oxidative metabolism, just as heat is a by-product of metabolism in other organisms.

SUGGESTIONS FOR FURTHER READING

Cohen, G. N.: *Biosynthesis of Small Molecules.* New York, Harper & Row, Publishers, 1967.
    An interesting discussion of the enzymatic reactions by which molecules are assembled from their constituent parts.

Greenberg, D. M. (ed.): *Metabolic Pathways.* New York, Academic Press, Vols. 1–3, 1967–1968.
    A reference book for the details of the reactions of intermediary metabolism.

Harvey, E. Newton: *Bioluminescence.* New York, Academic Press, 1952.
    A classic presentation of the field by one of the men who made many important discoveries in it.

Henderson, L. T.: *Fitness of the Environment.* New York, The Macmillan Company, 1913.
    A fine book to read to sample the flavor of its wit and the author's thesis that the environment had to have certain chemical and physical characteristics for life to develop.

Karlson, P.: *Introduction to Modern Biochemistry.* 2nd ed. New York, Academic Press, 1968.
    A concise exposition of the basic principles of biochemistry with some very helpful diagrams of metabolic pathways.

McElroy, W. D., and Seliger, H. H.: *Biological Luminescence.* Sci. Amer. *207:*76–89, 1962.
    A brief, up-to-date discussion of the process by which light is emitted by living organisms.

INFORMATION TRANSFER: THE GENE CONCEPT, THE CODE AND PROTEIN SYNTHESIS

Our modern concepts of genetics originated with the rediscovery of Mendel's Laws in 1900. Since that time, geneticists have been attempting to determine the physical structure and chemical composition of the hereditary units—the genes—and to discover the mechanisms by which they transfer biological information from one cell to another and control the development and maintenance of the organism. During the first half of this century, much was learned about the complexity of the molecular structure of proteins. As a result, nearly all biochemists assumed that any complex biological unit with such marked specificity as the gene must also be a protein. There was great difficulty, however, in explaining how if genes were protein molecules they could be duplicated precisely, as genes must be with each cell division.

SECTION 7–1

THE ORIGIN OF THE "CENTRAL DOGMA"

More than a century ago (1869), Friedrich Miescher isolated from the nuclei of pus cells a new class of chemicals which he called "nuclein." These substances, later called **nucleic acids**, were acidic in nature, uniquely rich in phosphorus, and contained carbon, oxygen, hydrogen and nitrogen. Subsequent experiments revealed that there are two types of nucleic acids—**deoxyribonucleic acid**, or **DNA**, present in the nucleus, and **ribonucleic acid**, or **RNA**, present in nucleus and cytoplasm. DNA was shown by P. A. Levene to be composed of four nitrogenous bases—two purines (**adenine** and **guanine**) and two pyrimidines (**cytosine** and **thymine**)—a five-carbon sugar, **deoxyribose**, and phosphate groups. Levene showed that the purine or pyrimidine base is attached to the sugar in a glycosidic linkage and the sugar is attached to the phosphate by an ester bond. The

163

combination of base-sugar-phosphate comprises the basic unit, termed a **nucleotide**, of nucleic acid. There are four kinds of nucleotides, one with each of the four kinds of nitrogenous bases—adenine, guanine, cytosine and thymine. Levene incorrectly concluded that all DNAs, of whatever source, are composed of *equal* amounts of these four nucleotides. Such simple molecules could not provide the basis for the biological specificity of the gene.

Our present belief that the nucleic acids, DNA and RNA, are the primary agents for the transfer of biological information arose gradually, culminating in 1953 in the proposal by James Watson and Francis Crick of a model of the DNA molecule that explained how it could transfer information and undergo replication. This proposal stimulated an enormous flood of research, and has led to the present **"central dogma"** of biology (Fig. 7–1): Genes are composed of DNA, and are located within the chromosomes. Each gene contains information coded in the form of a specific sequence of purine and pyrimidine nucleotides within its DNA molecule. The unit of genetic information, called a **codon**, is a group of three adjacent nucleotides that specify a single amino acid in a polypeptide chain. Thus the genetic code is a **"triplet"** code. The DNA molecule consists of two complementary chains of polynucleotides twisted about each other in a regular helix and joined by hydrogen bonds between specific pairs of purine and pyrimidine bases. The DNA molecule is replicated when the two strands of the helix separate, and each acts as a template for the formation of a new complementary strand. Each pair of

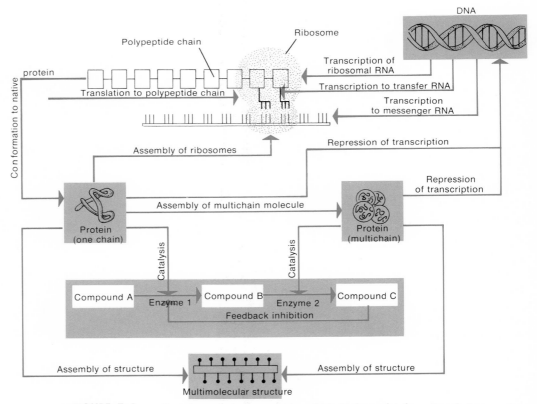

FIGURE 7–1   Overview of the process by which biological information is transferred from DNA via RNA to specific polypeptides. The peptide subunits are then assembled into multichain proteins.

strands, one old and one new, then twist together to form two daughter helices.

GENETIC INFORMATION IS TRANSMITTED BY DNA

The DNA of each gene has a sequence of nucleotide triplets which differs from that of every other gene. This information is transcribed from the DNA of the gene to a kind of RNA termed *"messenger" RNA,* a nucleotide sequence complementary to the genic DNA. Messenger RNA is synthesized in the nucleus and passes to the ribosomes in the endoplasmic reticulum. It combines with five to 10 ribosomes to form **polyribosomes** and serves as a template for the synthesis of an enzyme or some other specific protein.

For protein synthesis to occur, amino acids must be "activated" with ATP and then joined to specific adaptor molecules termed *"transfer" RNA.* Each kind of transfer RNA has a triplet code **(anticodon)** at some specific part of the molecule. The amino acid-transfer RNA complexes are arranged on the messenger RNA in an order dictated by the complementary nature of the nucleotide triplets in the messenger RNA codons and the transfer RNA anticodons. The information initially coded as a specific sequence of nucleotides in the DNA is **transcribed** as a specific sequence of nucleotides in messenger RNA, and is eventually **translated** into the specific order of the amino acids in the protein molecule.

Now let us trace some of the major steps in the origin of this concept. The first direct evidence that DNA can transmit genetic information came from the experiments of Avery and his coworkers with the *"transforming agent"* isolated from pneumococci and certain other bacteria. These experiments in turn were based upon those of Fred Griffiths, an English bacteriologist. In 1928 Griffiths was studying two different strains of pneumococci, a virulent "smooth" one with a polysaccharide capsule, and a nonvirulent "rough" one without the capsule. When he injected live "rough" bacteria into mice, the mice would survive. If he injected mice with live "smooth" bacteria, ones with a capsule, the mice died (Fig. 7–2). However, mice would survive if injected with heat-killed smooth bacteria. In a crucial experiment Griffiths injected mice with a mixture of live rough bacteria and heat-killed smooth bacteria. Although neither of these alone was harmful, the mixture of the two caused the death of the mice. He could recover live smooth bacteria from the dead mice. From these and other experiments, Griffiths concluded that the live rough bacteria had been transformed into live smooth bacteria by some material from the dead smooth cells. These bacteria, when grown in culture, reproduced smooth bacteria. Thus it appeared that some sort of material from the dead bacteria had entered the live rough bacteria and changed them into smooth ones. There are several types of pneumococci, and in further experiments Griffiths found that injecting live Type II rough pneumococci into mice, together with heat-killed smooth Type III pneumococci, killed the mice. He could isolate live, virulent, "smooth" or encapsulated pneumococci from the dead mice, but these were of Type III and not of Type II! Again the conclusion was clear that some sort of genetic material had passed from the dead Type III cells to the living Type II cells, and that this material changed them into Type III cells.

Methods were developed by which this bacterial transformation could

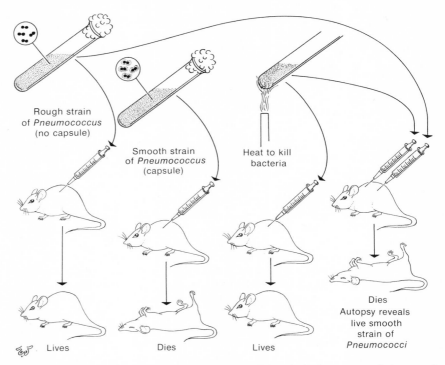

FIGURE 7-2   The experiments of Fred Griffiths which demonstrated the transfer of genetic information from dead, heat-killed bacteria to living bacteria of a different strain. Although neither the rough strain of Pneumococcus nor heat-killed smooth strain pneumococci would kill a mouse, a combination of the two did. Autopsy of the dead mouse showed the presence of living, smooth strain pneumococci.

be carried out in culture media rather than inside the bodies of mice. For example, live rough cells in a test tube could be transformed into smooth or encapsulated virulent bacteria by the fluid in which dead virulent cells had been dissolved. Finally in 1944 Avery and his colleagues at the Rockefeller Institute showed unequivocally that the transforming principle is DNA. The transforming agent lost activity when treated with deoxyribonuclease, but not when treated with a proteolytic enzyme such as trypsin or chymotrypsin. Avery isolated and purified enough of the material to show that it was DNA with a high molecular weight.

During the 1940's, A. E. Mirsky and Hans Ris, working at the Rockefeller Institute, and André Boivin and Roger Vendrely, working at the University of Strasbourg, independently showed that the amount of DNA per nucleus is constant in all of the body cells of a given organism. By making cell counts and chemical analyses, Mirsky and Vendrely showed that there is some 6×10^{-9} milligrams of DNA per nucleus in somatic cells, but only 3×10^{-9} milligrams of DNA per nucleus in egg cells or sperm cells (Table 7-1).

In tissues known to be **polyploid**, ones having more than two sets of chromosomes per nucleus, the amount of DNA per nucleus was found to be a corresponding multiple of the usual amount. For example, cells with four sets of chromosomes, termed **tetraploids**, were found to have 12×10^{-9} milligrams of DNA per nucleus. From the amount of DNA per cell, one can estimate the number of nucleotide pairs per cell, and thus the amount of genetic information present in each kind of cell (Table 7-2).

Only the amount of DNA and the amount of certain basic, positively

TABLE 7–1   Amount of Deoxyribonucleic Acid (DNA) per Nucleus in Animal Tissues, Expressed as mg. $\times$ 10^{-9}

Species	Sperm	Red Cell	Liver	Heart	Kidney	Pancreas	Spleen
Shad	0.91	1.97	2.01				
Carp	1.64	3.49	3.33				
Brown trout	2.67	5.79					
Toad	3.70	7.33					
Frog		15.0	15.7				
Chicken	1.26	2.49	2.66	2.45	2.20	2.61	2.55
Dog			5.5		5.3		
Rat			9.47	6.50	6.74	7.33	6.55
Ox	3.42		7.05		6.63	7.15	7.26
Man	3.25	7.30	10.36		8.6		

(From Villee, C. A.: *Biology.* 5th ed. Philadelphia, W. B. Saunders Company, 1967.)

charged proteins, termed **histones**, are relatively constant from one cell to the next. The amounts of other kinds of protein and of RNA vary considerably from cell to cell. Thus the fact that the amount of DNA, like the number of genes, is constant in all the cells of the body, and the fact that the amount of DNA in germ cells is only half the amount in somatic cells, is strong evidence that DNA is an essential part of the gene. However, because of the strong belief that only protein molecules have enough complexity to account for genetic specificity, this evidence was not accepted by most biochemists. Biologists assumed that the genetic material in the chromosomes must be protein, even though it was shown that the amount of structural protein in the chromosomes is not constant, but varies with the activity of the cell.

TABLE 7–2   The Amount of DNA per Cell in Animal and Plant Cells and in Virus Particles

	DNA mg. $\times$ 10^{-9} per cell	Nucleotide Pairs per cell
Mammals	6	5.5×10^9
Birds	2	2×10^9
Reptiles	5	4.5×10^9
Amphibia	7	6.5×10^9
Fish	2	2×10^9
Insects	0.17 — 12	0.16×10^9
Crustacea	3	2.8×10^9
Mullusks	1.2	1.1×10^9
Echinoderms	1.8	1.7×10^9
Sponges	0.1	0.1×10^9
Higher plants	2.5 — 40	2.3×10^9
Fungi	0.02 — 0.17	0.02×10^9
Algae	3	2.8×10^9
Bacteria	0.002 — 0.06	2×10^6
T_2 bacteriophage	0.00024	2.2×10^5
λ bacteriophage	0.00008	7×10^4
Papilloma virus	—	6×10^3

(From Villee, C. A.: *Biology.* 5th ed. Philadelphia, W. B. Saunders Company, 1967.)

Further evidence that DNA is the carrier of genetic information came from studies with bacteria, and with the viruses that infect bacteria, called **bacteriophages**. Bacteriophages are very small organisms visible only by electron microscopy. Each phage particle is shaped like a Ping-Pong paddle, with a head made of a DNA core surrounded by a thin protein membrane, and tail composed of membrane. The replication of bacterial viruses was studied by Delbruck and Luria, while working at California Institute of Technology. When bacterial cells are infected with phage particles and then are broken open and examined with the electron microscope after the infection has occurred, there is no trace at all of the bacteriophage during the first half of the period of infection; there is no trace of any of the Ping-Pong paddle-like particles. Then complete phage particles begin to appear and, in addition, bits of incomplete phage particles are found mixed in with them (Fig. 7–3). During the period of infection, the number of phage particles increases in a regular fashion, but the rate is linear rather than geometrical. Bacteria, for example, increase geometrically, i.e., 1, 2, 4, 8, 16, 32, 64 and so on. In contrast, the viruses appeared in a linear progression: 1, 2, 3, 4, 5, 6, 7 and so on, as though they were being assembled in a factory on an assembly line.

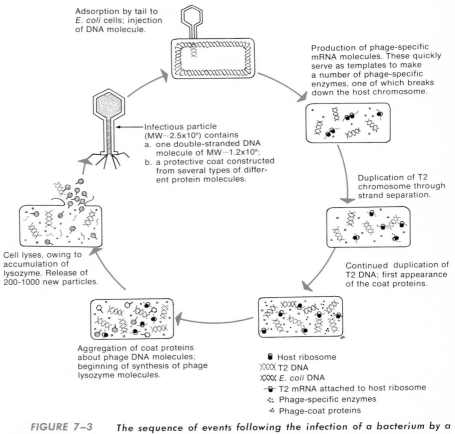

Adsorption by tail to *E. coli* cells; injection of DNA molecule.

Production of phage-specific mRNA molecules. These quickly serve as templates to make a number of phage-specific enzymes, one of which breaks down the host chromosome.

Infectious particle (MW~2.5x10⁸) contains
a. one double-stranded DNA molecule of MW~1.2x10⁸;
b. a protective coat constructed from several types of different protein molecules.

Duplication of T2 chromosome through strand separation.

Cell lyses, owing to accumulation of lysozyme. Release of 200-1000 new particles.

Continued duplication of T2 DNA; first appearance of the coat proteins.

Aggregation of coat proteins about phage DNA molecules; beginning of synthesis of phage lysozyme molecules.

Host ribosome
T2 DNA
E. coli DNA
T2 mRNA attached to host ribosome
Phage-specific enzymes
Phage-coat proteins

FIGURE 7–3   *The sequence of events following the infection of a bacterium by a T₂ bacteriophage particle. Only the DNA from the bacteriophage enters the bacterial cell, yet this provides all the information needed for the synthesis both of DNA and of new viral proteins.*

Studies by A. D. Hershey and Martha Chase at Cold Spring Harbor, New York, showed that only the DNA and not the protein of the virus enters the bacterial cell. Hershey and Chase carried out an experiment to determine whether the phage, as it infects the bacterium, injects DNA, proteins, or both into the bacterial cell (Fig. 7–4). They took advantage of the fact that DNA contains phosphorus, whereas protein does not. Protein contains some sulfur atoms, whereas DNA does not. By culturing bacteriophage on bacteria grown in a medium containing ^{32}P and ^{35}S, they were able to grow phage that contained ^{32}P in its DNA, and ^{35}S in its protein. The radioactive phage particles were recovered and purified, and nonradioactive bacteria were infected with them. After the infection had begun, the bacterial cells were agitated in a Waring blendor to remove the extra virus, then broken apart and analyzed. The remains of the virus contained ^{35}S, but the bacterial cells contained radioactive phosphorus and very little, if any, radioactive sulfur. This is evidence that the DNA of the phage entered the cell, whereas the protein coat of the phage remained outside, attached to the surface. This was evidence that only DNA had been injected into the phage, and that this DNA transmitted to the bacteria the genetic information that caused the bacteria to produce new phage. This viral DNA, injected into the bacterial cell by the phage, in some way commandeers the machinery of the bacterial cell that ordinarily makes new bacteria and programs it to make new bacteriophage material instead. If the bacteriophage was allowed to multiply within the bacteria and then escape, the new generation of bacteriophage contained ^{32}P but no ^{35}S.

Yet another type of evidence that DNA is the genetic material came from experiments with different strains of bacteria begun by Lederberg

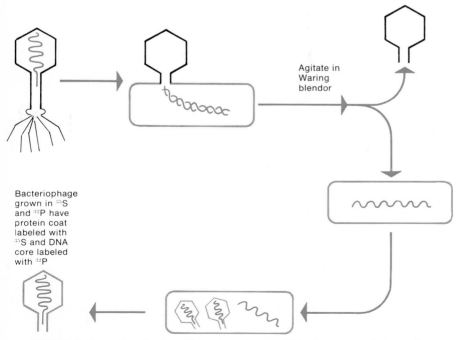

Bacteriophage grown in ^{35}S and ^{32}P have protein coat labeled with ^{35}S and DNA core labeled with ^{32}P

Agitate in Waring blendor

FIGURE 7–4   The Hershey-Chase experiment which demonstrated that only DNA from bacteriophage is injected into bacteria while the protein coat of the bacteriophage remains outside. All the genetic information needed for the synthesis of both new protein coat and new viral DNA is provided by the viral DNA.

and Tatum in 1946 (Fig. 7–5). One strain of bacteria contained a number of mutants which resulted in the loss of certain enzymes required for the synthesis of specific materials. This strain could survive if it were supplied with the products that it could not synthesize itself. Another strain of bacteria, with different mutants, required different nutritional materials to survive. In their key experiment, Lederberg and Tatum mixed these two strains together and grew them on a medium containing all of the nutrients required by both strains. After they had grown and reproduced, samples of the progeny were transferred to a simple culture medium containing no special nutrients. Most of the resulting bacteria could not survive, and died, but a few colonies did form. Subsequent analysis showed that these bacteria, like the so-called "wild type" bacteria, could grow and reproduce indefinitely without requiring special nutrients. Lederberg and Tatum suggested that genetic material, DNA, had been transferred from one strain of bacteria to the other so that the genetic material for the enzymes missing in one strain was transferred from the other. The process was probably analogous to that by which transforming material is passed from one kind of pneumococcus to another. This was a novel concept at the time, because it was believed that bacteria could not undergo genetic recombination; however, subsequently Lederberg and Tatum's hypothesis has been found to be correct. When the process was studied by electron microscopy, it was possible to show a bridge of cytoplasm forming between two bacterial cells (Fig. 7–6). Appropriate experiments demonstrated that the amount of DNA transferred across the cytoplasmic bridge is proportional to the amount of genetic information, measured on a genetic map, that is transferred from the donor to the recipient. Usually only a portion of the donor's DNA enters the recipient cell before the two mating cells separate. Samples of bacteria are placed in a Waring blendor at various times after the cytoplasmic bridge has formed; varying amounts of DNA and of genetic material are transferred across the bridge. The proportionality between the amount of DNA and the lengths of the genetic map transferred indicates that genetic information is contained in DNA.

Bacterial genes may be transferred passively from one bacterium to

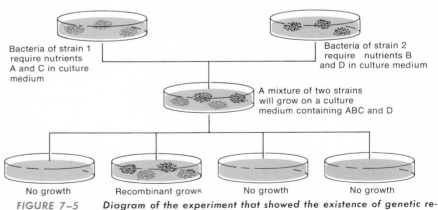

FIGURE 7–5   Diagram of the experiment that showed the existence of genetic recombination in bacteria. Bacteria of strain 1 required nutrients A and C, and bacteria of strain 2 required nutrients B and D in the culture medium. A mixture of the two strains will grow on a culture medium containing all four nutrients. Some of their offspring will grow on minimal culture media. These are bacteria in which genetic recombination has joined the wild type alleles of the mutants in one strain with the wild type alleles of the mutants in the other. The recombinant, like the original wild type, can grow without any added nutrients.

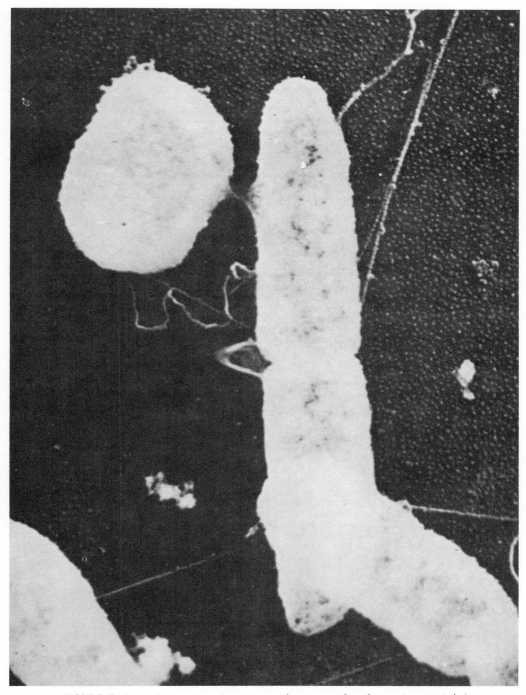

FIGURE 7–6   *Conjugating Bacteria conduct a transfer of genetic material. Long cell (right) is an Hfr "supermale" colon bacillus, which is attached by a short temporary bridge to a female colon bacillus (Electron micrograph, shown at a magnification of 100,000 diameters, by Thomas F. Anderson of the Institute for Cancer Research, Philadelphia.)*

another by a bacteriophage particle, a process that has been termed **transduction**. As a virus particle forms within the host cell, it may enclose and come to contain a small segment of the bacterial genetic material. When the phage is subsequently released, it becomes attached to a new bacterium and injects the segment of bacterial chromosome from the previous host into the new host, along with its own DNA. The segment of DNA may undergo "crossing over" with the new host's chromosome and thus incorporate genes from the previous host strain. Since only DNA is transferred in this way, this is further evidence confirming the hypothesis that genes are DNA.

It had been known for a long time that nucleic acids absorb ultraviolet light very strongly, with the maximum absorption at a wavelength of 260 millimicrons. It was also known that mutations could be produced by irradiating organisms with ultraviolet light. The wavelength of ultraviolet light most effective in producing mutations is also at 260 millimicrons. When you compare the number of mutations produced per unit of energy delivered and the wavelengths at which that energy is delivered, you obtain an **action spectrum** for the production of mutations. The very close correlation between this action spectrum for the production of mutations and the absorption spectrum for nucleic acids suggests that genes are composed of nucleic acids. Mutations are produced when the nucleic acids absorb energy, and the absorbed energy in some way changes the nucleic acid molecule to produce a new mutant gene.

Other evidence for the genetic role of nucleic acid came from the studies of plant viruses by Stanley and Fraenkel-Conrat at the University of California. They extracted from one plant virus by very gentle procedures both its nucleic acid and its proteins, which retained their biological properties; i.e., when protein and nucleic acid were mixed together they recombined and viral activity reappeared. When the reconstituted virus was applied to the leaf of the plant, the specific plant disease caused by the virus appeared. These investigators next extracted nucleic acid from one strain of virus and mixed it with protein extracted from another strain to produce a hybrid virus. This had the serological properties of the viral strain from which the protein was derived, but had the viral activity of the strain from which the nucleic acid came. The reconstituted virus produced the specific disease characteristic of the strain which supplied the nucleic acid. In further experiments Fraenkel-Conrat found that isolated nucleic acid without any protein has some viral activity, although somewhat less than that of nucleic acid stabilized by the viral protein.

SECTION 7-3

THE CONSTITUENTS OF DNA

The analyses of P.A. Levene had suggested that DNA, from whatever source, was composed of four nucleotides in equivalent amounts. During the 1940's Erwin Chargaff and his colleagues at Columbia University analyzed purified DNA from a variety of sources and showed clearly that the different nitrogenous bases do not occur in equal proportions. Chargaff found that although the proportion of these bases is the same in the DNA from all the cells of a given species, the DNAs from different species may differ markedly in the ratios of the constituent nucleotides. This suggested that the variations in the ratios of nitrogenous bases might represent a language. Although the ratios of purine and pyrimidine bases differed considerably in different samples of DNA, a pattern became apparent when these analyses were compared (Table 7–3). In all the samples the total amount of purines equaled the total amount of pyrimidines ($A + G = T + C$); the amount of adenine equaled the amount of thymine ($A = T$); and the amount of guanine equaled the amount of cytosine ($G = C$). DNA isolated from mammalian cells in general was rich in adenine and thymine and relatively poor in guanine and cytosine, whereas DNA isolated from bacterial sources was generally rich in guanine and cytosine and relatively poor in adenine and thymine. These findings constituted one of the important experimental bases on which the Watson-Crick model of DNA was eventually erected.

A further important clue about DNA structure came from Linus Pauling's studies of protein structure. Pauling had shown that there are several possible ways in which amino acid chains of a protein may be held together. One of the favorite molecular structures, termed an *α-helix*, can be visualized as a peptide chain wound around a cylinder. This permits the formation of hydrogen bonds between the amino acids on successive turns of the screw. Pauling had described this α-helix form of protein molecules in 1950, and had suggested at that time that the structure of DNA might also prove to be some sort of helix held together by hydrogen bonds.

The primary clues about the structure of the DNA molecule came from studies using x-ray diffraction, carried out in the laboratory of M. H. F. Wilkins. When a pure crystal of DNA is bombarded with x-rays, the x-rays are diffracted or bent in specific directions as they pass through the substance. The amount and nature of the bending of the x-rays depends on the structure of the molecule itself. The pattern of x-ray diffraction (Fig. 7–7), although incomprehensible to the novice, provides to the experienced eye a number of clues about the structure of the molecule. From such x-ray diffraction pictures, Wilkins inferred that the nucleotide bases (which are flat molecules) are stacked one on top of the other like a group of saucers. Wilkins' x-ray diffraction patterns showed three major periodicities in crystalline DNA, one of 3.4 Å, one of 20 Å, and one of 34 Å.

TABLE 7–3   Relative Amounts of Purines and Pyrimidines in Samples of DNA

Source	Adenine	Guanine	Cytosine	Thymine
Beef thymus	29.0	21.2	21.2	28.5
Beef liver	28.8	21.0	21.1	29.0
Beef sperm	28.7	22.2	22.2	27.2
Human thymus	30.9	19.9	19.8	29.4
Human liver	30.3	19.5	19.9	30.3
Human sperm	30.9	19.1	18.4	31.6
Hen red cells	28.8	20.5	21.5	29.2
Herring sperm	27.8	22.2	22.6	27.5
Wheat germ	26.5	23.5	23.0	27.0
Yeast	31.7	18.3	17.4	32.6
Vaccinia virus	29.5	20.6	20.0	29.9
Bacteriophage T_2	32.5	18.2	16.7	32.6

(From Villee, C. A.: *Biology.* 5th ed. Philadelphia, W. B. Saunders Company, 1967.)

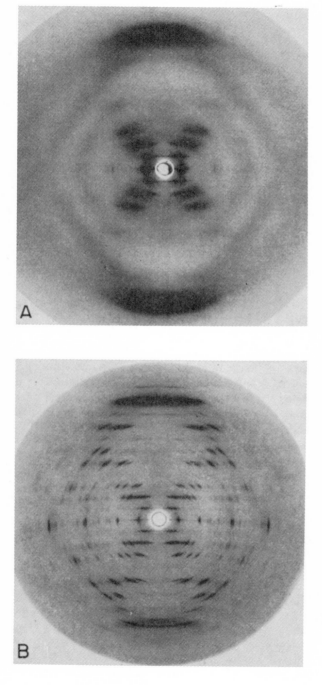

FIGURE 7–7   X-ray diffraction photographs of suitably hydrated fibers of DNA, showing the so-called B configuration. A, Pattern obtained using the sodium salt of DNA. B, Pattern obtained using the lithium salt of DNA. This pattern permits a most thorough analysis of DNA. The diagonal pattern of spots (reflections) stretching from 11 o'clock to 5 o'clock and from 1 o'clock to 7 o'clock provides evidence for the helical structure of DNA. The elongated horizontal reflections at the top and bottom of the photographs provide evidence that the purine and pyrimidine bases are stacked 3.4 Å apart and are perpendicular to the axis of the DNA molecule. (Courtesy of Biophysics Research Unit, Medical Research Council, King's College, London.)

SECTION 7-4

THE WATSON-CRICK MODEL OF DNA

On the basis of Chargaff's analytical results, and of Wilkins' x-ray diffraction patterns, Watson and Crick proposed in 1953 a model of the DNA molecule (Fig. 7–8), which has been very useful in providing a chemical explanation for many of its biological properties. The studies of Pauling and other chemists had provided a great deal of information about the exact distance between the atoms that are bonded together in a molecule, the angles between the bonds of a given atom, and the sizes of the atoms. Using this information, Watson and Crick began to build scale models of the component parts of DNA, and then fit them together to agree with the various experimental data. It had been known that the adjacent nucleotides in DNA are joined in a chain by phosphodiester bridges, which link the 5' carbon of the deoxyribose of one nucleotide with the 3' carbon of the deoxyribose of the next nucleotide. It seemed clear to

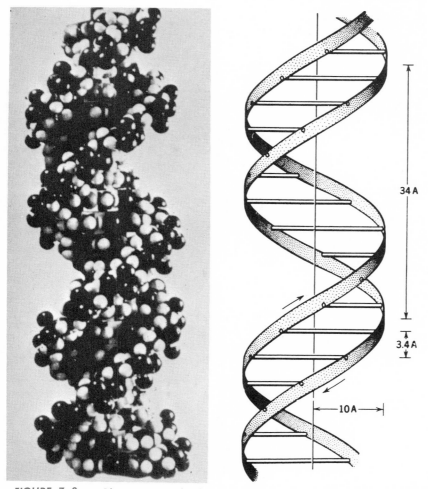

FIGURE 7–8   Photomicrograph of a molecular model of deoxyribonucleic acid. (Courtesy of M. H. F. Wilkins). Right, Schematic drawing of the two helices of this structure showing its major dimensions. The arrows indicate that the two strands of the double helix extend in opposite directions. (From Anfinsen, C. B.: Molecular Basis of Evolution, New York, John Wiley and Sons, 1968.)

Watson and Crick that the 3.4 Å periodicity found by Wilkins corresponded to the distance between successive nucleotides in the DNA chain. It further was a reasonable guess that the 20 Å periodicity corresponded to the width of the chain. To explain the 34 Å periodicity, they postulated, as Pauling had, that the chain was coiled in a helix. A *helix* is formed by winding a chain around a cylinder; in contrast, a *spiral* is formed by winding a chain around a cone. This 34 Å periodicity corresponds to the distance between successive turns of the helix. Clearly a chain can be wound around a cylinder either rather loosely or very tightly. This gives an indication of the steepness of the pitch of the screw in the helix. Since 34 is just 10 times the 3.4 Å distance between the successive nucleotides, it is clear that each full turn of the helix contained 10 nucleotides. From these data Watson and Crick could calculate the density of a chain of nucleotides coiled in a helix 20 Å wide, with turns that were 34 Å long. Such a chain would have a density only half as great as the known density of DNA. Clearly they had to postulate that there were two chains—a *double helix* of nucleotides—that made up a DNA molecule.

The next problem, of course, was to determine the spatial relationships between the two chains that made up the double helix. Having tried a number of arrangements with their scale model, they found that the best fit with all the data was given by one in which the two nucleotide helices were wound in opposite directions (Fig. 7–9), with the sugar phosphate chains on the outside of the coil and the purines and pyrimidines on the inside of the helix, held together by *hydrogen bonds* between bases on the opposite chains. It is these hydrogen bonds which hold the chains together and maintain the helix. A double helix can be visualized by imagining the form that would be obtained by taking a ladder and twisting it into a helical shape, keeping the rungs of the ladder perpendicular. The sugar and phosphate molecules of the nucleotide chains make up the railings of the ladder, and the rungs are formed by the nitrogenous bases held together by hydrogen bonds.

Further study of the possible models made it clear to Watson and Crick

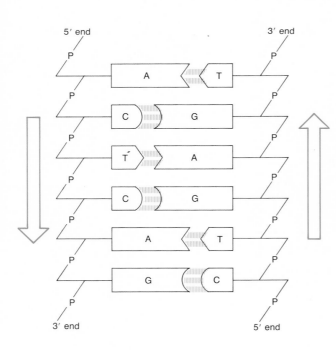

FIGURE 7–9   *Schematic diagram of a portion of a DNA molecule showing the two polynucleotide chains joined by hydrogen bonds. The chains are not flat as represented here, but are coiled around each other in helices (see Fig. 7–8). The two strands extend in opposite directions as indicated by the arrows.*

that each crossrung must contain one purine and pyrimidine. The space available with the 20 Å periodicity would accommodate one purine and one pyrimidine, but not two purines, which would be too large, and not two pyrimidines, which would not come close enough together to form proper hydrogen bonds. Further examination of the detailed model showed that although a combination of adenine and cytosine was the proper size to fit as a rung on the ladder, they could not be arranged in such a way that they would form proper hydrogen bonds. A similar consideration ruled out the pairing of guanine and thymine, however, adenine and thymine would form hydrogen bonds, and guanine and cytosine could form hydrogen bonds. The nature of the hydrogen bonds required that adenine pair with thymine, and that guanine pair with cytosine. This concept of specific base pairing provided a basis for Chargaff's rule that the amounts of adenine and thymine in any DNA molecule are always equal, and the amounts of guanine and cytosine are always equal. Two hydrogen bonds can form between adenine and thymine, and three hydrogen bonds between guanine and cytosine (Fig. 7–10). The specificity of the kind of hydrogen bond that can be formed assures that for every adenine in one chain, there will be a thymine in the other chain. Similarly for every guanine in the first chain there will be a cytosine in the second chain. Thus the two chains are *complementary* to each other; i.e., the sequence of nucleotides in one chain dictates the sequence of nucleotides in the other. The two strands run in opposite directions and have their terminal phosphate groups at opposite ends of the double helix. The most

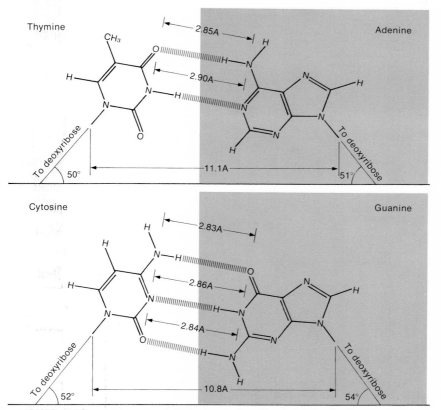

FIGURE 7–10   *Diagram of the hydrogen bonding between the base pairs adenine and thymine (above) and guanine and cytosine (below) in DNA. The A-T pair has two hydrogen bonds and the G-C pair has three.*

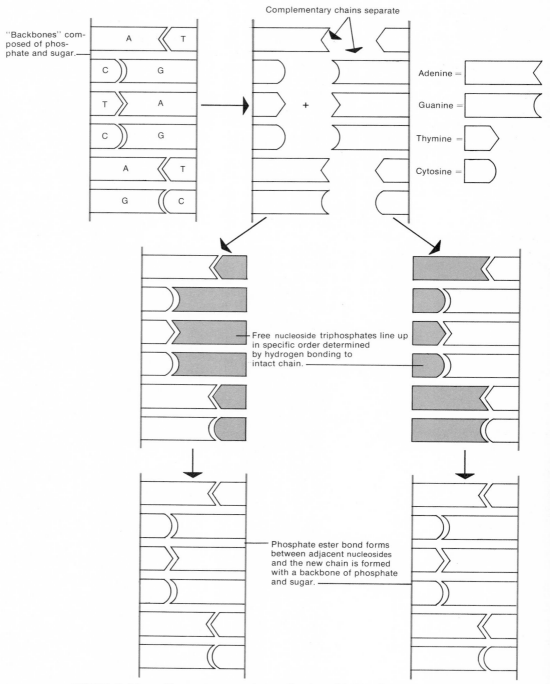

Complementary chains separate

"Backbones" composed of phosphate and sugar.

Adenine =

Guanine =

Thymine =

Cytosine =

Free nucleoside triphosphates line up in specific order determined by hydrogen bonding to intact chain.

Phosphate ester bond forms between adjacent nucleosides and the new chain is formed with a backbone of phosphate and sugar.

FIGURE 7–11   The semiconservative replication of DNA. The original base pairs separate and the two strands of the helix unwind. Each strand then serves as a template for the synthesis of a new complementary strand using nucleoside triphosphates as the building blocks.

distinctive properties of the genetic material are that it carries information and undergoes replication. The Watson-Crick model explains how DNA molecules may carry out these two functions. When a DNA molecule undergoes replication, the two chains separate and each one brings about the formation of a new chain, which is complementary to it. Thus two new chains are established (Fig. 7–11). The nucleotides in the new chain are assembled in a specific order, because each purine or pyrimidine in the original chain forms hydrogen bonds with the complementary pyrimidine or purine nucleotide triphosphate from the surrounding medium and lines them up in a complementary order. Phosphate ester bonds are formed by the reaction catalyzed by **DNA polymerase** to join the adjacent nucleotides in the chain, and a new polynucleotide chain results (Fig. 7–12). The new and the original chains then wind around each other, and two new DNA molecules are formed. Each chain, in other words, serves as a template or a mold against which a new partner chain is synthesized. The end result is two complete double chain molecules, each identical to the original double-chain molecule. A second prime function of DNA, in addition to its role in replication, is that the information contained in its specific sequence of nucleotides must be transcribed some time between cell division. The product of the transcription process, now termed messenger RNA, then combines with ribosomes to carry out the synthesis of enzymes and other specific proteins. Thus we can imagine how each gene can lead to the production of a specific enzyme. We shall return to this subject in another section.

SECTION 7–5

THE SYNTHESIS OF DNA: REPLICATION

Within a few years after its formulation, the Watson-Crick model received strong experimental support from several sources. Arthur Kornberg and his colleagues isolated (1957) the enzyme, DNA polymerase, from bacteria. This catalyzes the synthesis of DNA and requires as substrates the triphosphates of all four deoxyribonucleosides (abbreviated dATP, dGTP, dCTP and dTTP.) The reaction system further requires a magnesium ion (Mg^{++}) and a small amount of high molecular weight DNA polymer to serve as primer or "template" for the reaction. The product of the reaction is more DNA polymer and a molecule of pyrophosphate for each molecule of deoxyribonucleotide incorporated.

$$\left.\begin{array}{l} \text{dATP} \\ \text{dGTP} \\ \text{dCTP} \\ \text{dTTP} \end{array}\right\}_n \quad \xrightarrow[\text{DNA polymerase}]{\substack{\text{DNA} \\ Mg^{++}}} \quad \text{DNA} + n\text{PP}_i$$

The nucleoside triphosphate attacks the free 3'-hydroxyl of the last deoxyribose in the chain and forms an ester bond, freeing a molecule of pyrophosphate (Fig. 7–12). When the deoxyribonucleotide triphosphates were labeled with ^{14}C, the DNA polymer produced contained ^{14}C, permitting the inference that the labeled nucleotides had been incorporated into the DNA chain. By appropriate experiments with ^{14}C-labeled nucleotides, Kornberg could show that the ratios of A : T and of G : C in the DNA synthesized were the same as the corresponding ratios in the DNA used as primer. This suggested that the DNA produced is a copy of the primer DNA, as predicted by the Watson-Crick model.

The DNA polymerase from *Escherichia coli* will use template DNA

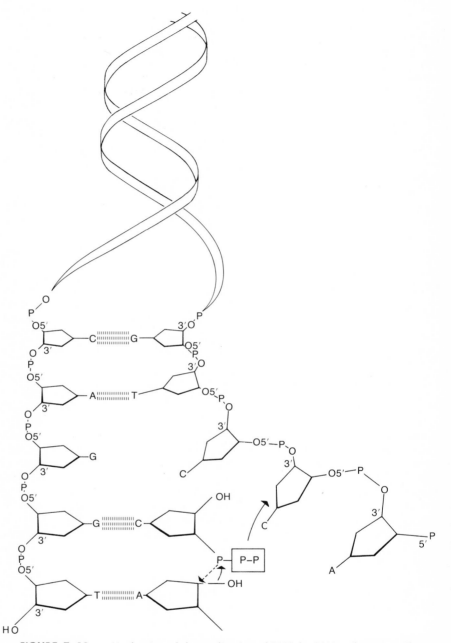

FIGURE 7-12   *Mechanism of the replication of DNA by DNA polymerase. The two strands of the DNA double helix are shown separating. The left one, which runs from the 5' phosphate to 3' OH, is being copied starting from the bottom. The newly synthesized chain begins with a 5' phosphate. In the new chain, as in the old, adenine forms base pairs with thymine and cytidine forms base pairs with guanine. The new phosphodiester bond is made between adjacent bases in the forming chain by an attack by the 3' OH group of the deoxyribose on the bond between the inner phosphate and the outer two phosphates of the adjacent trinucleotide. The two outer phosphates are split off as inorganic pyrophosphate.*

isolated from any of a wide variety of sources—bacteria, viruses, mammalian cells and plant cells—and will produce DNA with a nucleotide ratio comparable to that of the template used. Thus the sequence of nucleotides in the product is dictated by the sequence in the primer DNA, and not by the properties of the polymerase, nor by the ratio of the substrate molecules present in the reaction mixture. Using a more highly purified enzyme Kornberg was able in 1968 to synthesize biologically active viral DNA, using viral DNA as primer. The DNA produced would infect bacteria just like "live" viruses.

Khorana and his colleagues synthesized a number of deoxyribonucleotide polymers containing adenylic and cytidylic acids, and other polymers containing alternating thymidylic and guanylic acids. Neither of these polymers alone was able to serve as primer or template in the DNA polymerase system; however, a mixture of the two, which forms a synthetic double-stranded helix with conventional Watson-Crick pairing of the bases, can serve as a template.

The DNA template appears to have two functions in the DNA polymerase system. First, it provides 3'-OH groups which are free to serve as the growing end of the DNA polymer. Next, the DNA template provides coded information. A double-stranded molecule is required because each strand of the pair serves as a template for the extension of the complementary strand. It also serves as a primer for its own extension. The DNA-like polymer that is produced by the action of the DNA polymerase in the presence of a double-stranded template is also double-stranded. It has the same base composition as the template DNA. The ratios of the bases in the product are those predicted by the Watson-Crick model.

The synthesis of DNA occurs in the cells of higher organisms only during the interphase when chromosomes are in their extended form and are not readily visible. Thus if an enzyme system similar to the Kornberg DNA polymerase catalyzes the synthesis of DNA *in vivo*, there must be some sort of biological signal which will initiate DNA synthesis at this time and will turn it off at other times. It appears that both the enzyme, the DNA polymerase, and the substrates dATP, dGTP, dCTP and dTTP are present all of the time. The most likely explanation at present is that some sort of change in the DNA template initiates the synthesis of DNA at the appropriate time and then turns it off.

During the replication of DNA two new strands are formed, each of which is complementary to one of the existing DNA strands in the double-stranded helix. The double-stranded helix unwinds; one strand provides a template for one new strand, and the other original strand provides a template for the second new strand. This is called a "semiconservative" replication: the two original strands of DNA are retained or conserved in the product, one in each of the two daughter helices (Fig. 7–13).

The classic experiment of Meselson and Stahl used heavy nitrogen, ^{15}N, to distinguish "old" and "new" molecules of DNA and provided evidence that DNA replication is indeed carried out by a semiconservative process, at least in bacteria (Fig. 7–14). Bacteria grown for several generations in a medium containing heavy nitrogen had DNA (and RNA and protein) that was labeled with ^{15}N. When a sample of the DNA was isolated and centrifuged in a tube containing a cesium chloride density gradient, the DNA collected at a level in the tube which reflected the presence of the heavy nitrogen atoms in it.

The bacteria were then transferred from the ^{15}N medium to a medium containing ordinary nitrogen, ^{14}N, and were allowed to divide once in this medium. When some of the DNA from this generation of bacteria was

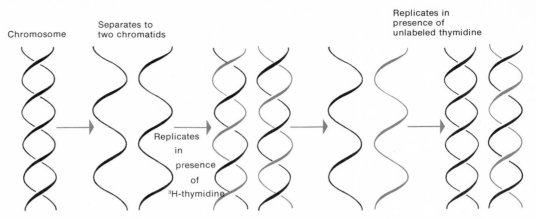

Chromosome | Separates to two chromatids | | Replicates in presence of unlabeled thymidine

Replicates
in
presence
of
³H-thymidine

FIGURE 7–13   When rapidly dividing cells were grown in the presence of ³H-thymidine and fixed for autoradiographic study, all the chromosomes were labeled (color) and radioactivity was equally distributed between the two chromatids of each chromosome (center). If these cells were not fixed, but allowed to divide once more in the presence of unlabeled thymidine and then fixed for study, only one of the two chromosomes was labeled (right). Thus each chromatid acts as a template for the formation of a new chromatid. The replication of chromosomes appears to be formally similar to the replication of the DNA double helix, but the relation between the cytologic and molecular events is not yet clear.

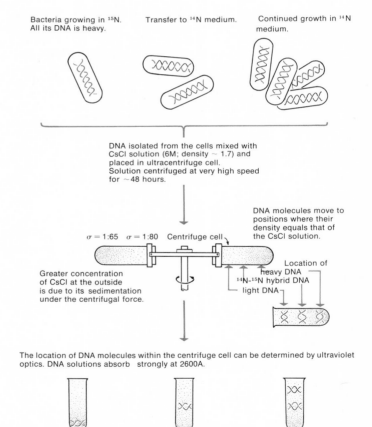

Bacteria growing in ¹⁵N. All its DNA is heavy.   Transfer to ¹⁴N medium.   Continued growth in ¹⁴N medium.

DNA isolated from the cells mixed with CsCl solution (6M; density ~ 1.7) and placed in ultracentrifuge cell. Solution centrifuged at very high speed for ~48 hours.

DNA molecules move to positions where their density equals that of the CsCl solution.

$\sigma = 1:65$ $\sigma = 1:80$ Centrifuge cell

Greater concentration of CsCl at the outside is due to its sedimentation under the centrifugal force.

Location of heavy DNA
¹⁴N-¹⁵N hybrid DNA
light DNA

The location of DNA molecules within the centrifuge cell can be determined by ultraviolet optics. DNA solutions absorb strongly at 2600A.

Before transfer to ¹⁴N.   One cell generation after transfer to ¹⁴N.   Two cell generations after transfer to ¹⁴N.

FIGURE 7–14   Diagram of the experiment of Meselson and Stahl which indicated that DNA is replicated by a semiconservative mechanism. The two original strands of DNA are retained in the product, one strand in each daughter helix.

182

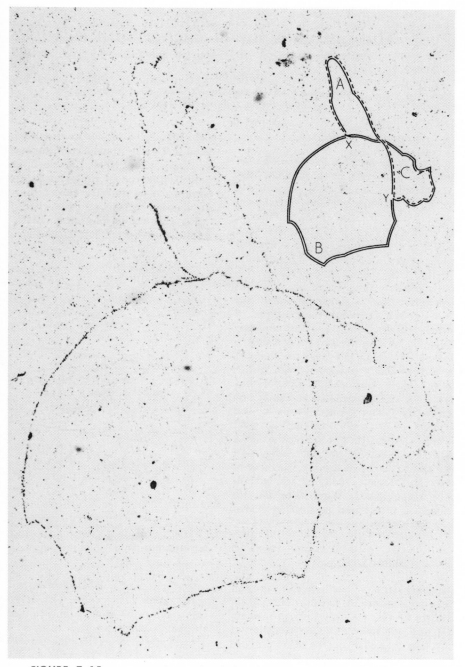

FIGURE 7–15   Autoradiograph of the chromosome of E. coli (strain K12 Hfr) labeled with tritiated thymidine for two generations and extracted by treatment of the cells with lysozyme. The inset shows the same chromosome in diagram. A predominantly half "hot" chromosome that has completed two-thirds of its second round of duplication is shown. Part of the still-unduplicated section is half marked with tracer (Y to C) and part is doubly marked with tracer (C to X). (From Cairns, J.: Cold Spring Harbor Symp. Quant. Biol. 28:44, 1963.)

isolated and centrifuged, all the DNA was lighter, with the density expected if it had just half as many ^{15}N atoms as the DNA of the parental generation. If the Watson-Crick theory is correct and replication is semiconservative, this result would be expected, because one strand of the double-stranded DNA in each organism would be labeled with ^{15}N and the other would contain only ^{14}N.

When these bacteria were allowed to divide a second time in the ^{14}N medium, each molecule of DNA in the progeny again received one parental strand and one new strand containing only ^{14}N. Some double-stranded DNA containing only ^{14}N was formed, and appeared as a light DNA on centrifugation. The parental strands containing ^{15}N made complementary strands containing ^{14}N, which were sedimented on centrifugation with a density characteristic of the half ^{15}N, half ^{14}N double-stranded state. Thus the original parental strands of DNA are not dispersed or split apart during the replication process but are conserved and passed to the next generation of cells. Each strand of the parental double helix is conserved in a different daughter cell, hence the process is termed "semiconservative."

If the replication of the strands begins as the strands begin to untwist, Y-shaped molecules of DNA should be evident during the replication process. Such Y-shaped regions have been found by autoradiography of chromosomes of *Escherichia coli* labeled with ^{3}H-thymidine (Fig. 7–15).

SECTION 7–6

THE GENETIC CODE

The Watson-Crick model of the DNA molecule implied that genetic information is transmitted by some specific sequence of its constituent nucleotides. Gamow appears to have been one of the first to suggest (1954) that the minimum coding relation between nucleotides and amino acids would be three nucleotides per amino acid. Four nucleotides taken two at a time provide for only 16 combinations ($4^2 = 16$), whereas four nucleotides taken three at a time provide for 64 combinations ($4^3 = 64$). At first glance, this would seem to provide for many more code symbols than are needed, since there are only 20 different amino acids. It was believed at one time that some of these 64 combinations were simply "nonsense" codes that did not specify any amino acid. However, there is now strong evidence that all but two of the 64 combinations do, in fact, code for one or another amino acid, and that as many as six different nucleotide triplets may specify the same amino acid.

The fundamental characteristics of the genetic code are now well established: it is a triplet code with three adjacent nucleotide bases, termed a codon, specifying each amino acid (Table 7-4). The Watson-Crick model of DNA assumed that information is coded in the sequence of the nucleotides in the helical strands, and subsequent experiments revealed that the code is triplet. For some years the question of whether the code was overlapping or not remained unsolved. For example, is the sequence CAG AUC GAC read only as CAG, AUC, GAC or can it also be read CAG, AGA, GAU, AUC, UCG, CGA, GAC? Is each nucleotide part of one codon or three? The amino acid sequences of each of the several mutant forms of the hemoglobin molecule have been analyzed. In each only a single amino acid in the peptide chain is substituted. In contrast, if the code were overlapping and a given nucleotide were part of three adjacent codons, we would expect three adjacent amino acids to be changed. An overlapping code would restrict the possible orders of amino acids in a

TABLE 7–4   The Genetic Code: The Sequence of Nucleotides in the Triplet Codons of Messenger RNA which Specify a Given Amino Acid

First position (5' end)	Second Position	Third Position (3' end)			
		U	C	A	G
U	U	Phe	Phe	Leu	Leu
	C	Ser	Ser	Ser	Ser
	A	Tyr	Tyr	Terminator	Terminator
	G	Cys	Cys	Cys?	Trp
C	U	Leu	Leu	Leu	Leu
	C	Pro	Pro	Pro	Pro
	A	His	His	Glu.NH$_2$	Glu.NH$_2$
	G	Arg	Arg	Arg	Arg
A	U	Ileu	Ileu	Ileu?	Met
	C	Thr	Thr	Thr	Thr
	A	Asp.NH$_2$	Asp.NH$_2$	Lys	Lys
	G	Ser	Ser	Arg	Arg
G	U	Val	Val	Val	Val
	C	Ala	Ala	Ala	Ala
	A	Asp	Asp	Glu	Glu
	G	Gly	Gly	Gly	Gly

(After Villee, C. A.: *Biology.* 5th ed. Philadelphia, W. B. Saunders Company, 1967.)

peptide. Thus the amino acid specified by CAG could be followed only by AG(X), by one of four amino acids. It is clear from analyses of amino acid sequences in peptides that there are no such restrictions on the sequences possible. Finally experiments with synthetic polynucleotides having known base sequences have shown conclusively that the code is not overlapping.

It is now generally believed that the code is commaless; it has no "punctuation." No punctuation is necessary since the code is read out beginning at a fixed point and the entire strand is read, three nucleotides at a time, until the read-out mechanism comes to a specific "termination" code, which signals the end of the message. The nature of the signal, "begin reading here," at least in bacteria, also appears to be specified by a specific sequence of bases.

There is now good reason to believe that the genetic code is universal, i.e., that a given codon specifies the same amino acid in all of the organisms that have been studied from viruses to man. Early in 1961, from a mathematical analysis of the coding problem, Crick concluded that three consecutive nucleotides in a strand of messenger RNA provide the code that determines the position of a single amino acid in a polypeptide chain. Experimental evidence to support this was quickly forthcoming from the laboratory of Nirenberg and Matthaei. Using purified enzyme systems they studied the incorporation of specific labeled amino acids into protein under the direction of artificial messenger RNAs of known composition. Nirenberg prepared a synthetic **polyuridylic acid** (UUUUU), using the enzyme polynucleotide phosphorylase. When this artificial messenger RNA was added to a system of purified enzymes for the synthesis of proteins, **phenylalanine,** and no other amino acid, was incorporated into protein; the polypeptide that resulted contained only phenylalanine. The inference that UUU is the code for phenylalanine was inescapable. Further similar experiments by Nirenberg and by Severo Ochoa showed that poly-adenylic acid provided the code for lysine and polycytidylic acid coded for proline. Making mixed nucleotide polymers (such as poly AC) and using them as artificial messengers made possible the assignment of many other nucleotide combinations to specific amino acids.

These experiments did not reveal the order of the nucleotides within the triplets, but this has been inferred from other kinds of experiments. Nirenberg and Leder discovered that even when no protein synthesis is occurring, specific amino acyl transfer RNA molecules will be attached to ribosomes when messenger RNA is present. Fortunately this effect does not require a long molecule of mRNA, which would be difficult to synthe-size. Indeed, synthetic messenger RNA molecules as short as trinucleo-tides will suffice to promote the binding of specific amino acyl transfer RNAs to the ribosomes. It is possible to synthesize trinucleotides of known sequence; using these the coding assignment of all 64 possible triplets has been determined. For example, GUU, but not UGU nor UUG, induces the binding of valine transfer RNA to ribosomes. UUG induces the bind-ing of leucine transfer RNA. Since GUU and UUG code for different amino acids, it follows that the reading of the code in the messenger RNA strand makes sense only in one direction.

Careful examination of the coding relationship shows that there is a pattern to the degeneracy in the code. Bernfield and Nirenberg found in 1965 that the binding of phenylalanine transfer RNA is induced by both UUU and UUC. The binding of serine transfer RNA is induced by both UCC and UCU, and proline transfer RNA is bound by CCC or CCU. In all of these the two alternative triplets are identical except for the sub-stitution of one pyrimidine for the other (C or U) at the 3' end. In other instances, adenine and guanine can be interchanged in the 3' position of the triplet. For example, the code for lysine may be either AAG or AAA. For a number of amino acids, the first two nucleotides of the codon are specific but any of the four nucleotides may be present in the 3' position. All of the four possible combinations will code for the same amino acid. In these, although the code may be read three nucleotides at a time, only the first two nucleotides appear to contain specific information. Only methionine and tryptophan have single triplet codes; all the other amino acids are specified by two to as many as six different nucleotide triplets.

That messenger RNA is read three nucleotides at a time was firmly established by experiments carried out by Khorana. He and his colleagues synthesized a poly UC messenger containing the regularly alternating base sequence UCUCUCUC. When this was used in a protein-synthesizing sys-tem, the resulting polypeptide contained a regular alternation of serine and leucine. Mathematical analysis of this result shows that the coding unit must contain an odd number of bases. Khorana then synthesized the nucleotide sequence AAGAAGAAGAAG. When this nucleotide polymer was used as template in a protein-synthesizing system, the result was either polylysine (AAG), polyglutamate (GAA) or polyarginine (AGA). Apparently the type of peptide that it synthesized depends on which nucleo-tide in the polynucleotide chain happens to be read first (Fig. 7–16). This effect, termed a *"frame shift" effect,* as in a movie camera, can be accounted for only if the chain is read in sequence, three nucleotides at a time, be-ginning from a fixed point. There are a few instances in which a specific nucleotide at the 5' end of the codon can be changed and still provide a code for the same amino acid. For example, both UUG and CUG provide a code for leucine. Arginine is coded by six nucleotide triplets, two of which differ in containing A instead of C at the 5' end. It appears that the middle nucleotide is the most informative one in the triplet, that the 5'

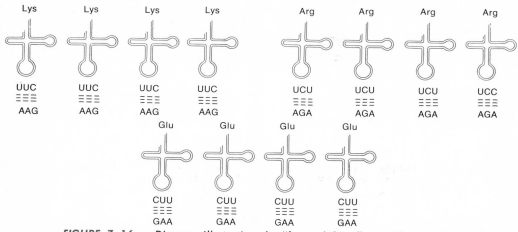

FIGURE 7–16   *Diagram illustrating the "frame-shift" effect in the translation of the synthetic polynucleotide AAGAAGAAGAAG. This led to the synthesis of polylysine, polyglutamate or polyarginine, depending on which nucleotide was read first. When read as AAG, polylysine was formed; when read as AGA polyarginine was formed; and when read as GAA polyglutamate was formed. The diagram emphasizes the three different codon-anticodon relationships possible.*

nucleotide is the next most informative, and the 3′ nucleotide is least specific.

The biological significance of the existence of a **degenerate code**, i.e., of alternate code words in protein synthesis, is not understood. We must keep in mind that what has been termed a code is, in fact, a complex physical chemical system that depends for its operation on the specificity of several reactions catalyzed by enzymes. The sequences of nucleotides in DNA will have no meaning except in relation to a specific set of amino acyl transfer RNAs, each of which is synthesized by a specific enzyme. The protein is then synthesized on the ribosome by yet another enzyme system operating on the enzyme-messenger RNA-ribosome-transfer RNA complex. The specificity of the system as a whole depends on the ability of the enzyme to recognize specific amino acyl transfer RNAs when these are present at the ribosomal binding site.

Although it is known that there may be more than one transfer RNA for a given amino acid, it is not yet clear whether the number of amino acyl transfer RNA molecules equals the number of codons for that specific amino acid. Do the six different codons for serine correspond to six different serine-transfer RNA molecules, or is there only a single serine tRNA that responds to any of the six mRNA codons? Crick's **"wobble" hypothesis** (Fig. 7–17) states that the first two bases of a transfer RNA anticodon undergo hydrogen bonding specifically with the first two bases of the messenger RNA codon in antiparallel. However, the third base can undergo unusual base pairing—i.e., it can "wobble." His theory predicts that the inosine residue of alanine transfer RNA may pair either with U, C or A in a messenger codon. This could account for three of the four codons for alanine, and would require only one additional transfer RNA to recognize the GCG code. The organisms that have been studied so far appear to have a set of at least 40 different transfer RNAs. The evidence available at present indicates that messenger RNA is read from the 5′ end toward the 3′ end, and that the 5′ end corresponds to the N-terminal amino acid of the polypeptide.

Two of the 64 codons, UAA and UAG, were termed **nonsense codons**, since they did not specify any amino acid. Brenner has suggested that these are "terminator" triplets, which signal the end of the polypeptide chain,

CGI   CGI   CGI
≡≡≡   . ≡≡≡   ≡≡≡
GCU   GCC   GCA

FIGURE 7–17   *Diagram illustrating Crick's "wobble" hypothesis. The first two bases of the transfer RNA anticodon (CG) form the expected specific hydrogen bonds with the first two bases of the messenger RNA codon in antiparallel; however, the third base of the anticodon (I) can base pair with U, C or A in a messenger codon. This theory accounts for three of the four known codons for alanine.*

and cause the protein chain to become detached from the ribosome; the codon UGA may also be a terminator triplet. Evidence regarding the universality of the code is based on analyses of relatively few types of organisms, but it does indicate that the code is universal. Synthetic poly U, for example, causes synthesis of polyphenylalanine in cell-free systems derived from bacteria or from rabbit reticulocytes. Perhaps even more convincing is the experimental finding that RNA isolated from tobacco mosaic virus and used as a template in the protein-synthesizing system derived from *E. coli* synthesizes proteins that are similar to native tobacco mosaic virus protein. More indirect but quite convincing evidence of universality can be obtained by looking carefully at the replacements of one amino acid for another found in the more than 60 mutant forms of human hemoglobin that have been analyzed. All of these replacements are ones that would be expected if there were a *single* base transformation between triplets assigned to amino acids on the basis of their behavior in the *E. coli* cell-free, protein-synthesizing system. The inference is that the code is the same in human beings and in bacteria.

One apparent exception to this generalization proved ultimately to be a striking confirmation of it. The amino acid sequence of a mutant form of hemoglobin, hemoglobin I, was initially reported as involving a substitution of aspartic acid for lysine. This would require a change in two of the three nucleotides in the codon, a "double mutation." This would be possible theoretically, but quite improbable statistically. A reinvestigation of the amino acid composition showed that the changed amino acid was glutamate, and a substitution of glutamate for lysine requires a change in only one of the three nucleotides in the codon.

Cytochrome c, a protein constituent of the electron transmitter system, has been isolated from several different organisms and the amino acid sequence of each peptide has been determined. The differences from one species to another are very small, and the amino acid substitutions are those expected if a single nucleotide were substituted for another in that codon. Similar analyses of a number of other proteins have given similar results, and this indirect evidence for the universality of the genetic code is very strong.

Since the DNA is a linear polynucleotide chain and a protein is a linear polypeptide, it was of interest to determine the relationship between these linear molecules. It is now clear that the sequence of amino acids in the peptide chain is dictated by the order of the corresponding nucleotide bases in codons in one of the two polynucleotide chains of the DNA molecule. Changing the sequence of nucleotides in the DNA molecule produces a corresponding change in the sequence of amino acids in peptides.

The DNA molecule and the resulting polypeptide chain are said to be *colinear*.

This concept of colinearity was implicit in the original Watson-Crick model of the DNA molecule. Direct evidence of colinearity came from analyses by Charles Yanofsky of the genetic control of the enzyme tryptophan synthetase in bacteria.

Tryptophan synthetase is composed of four subunits, two A chains and two B chains. The A chain, studied by Yanofsky, is a polypeptide containing 267 amino acids. Yanofsky carefully mapped the genetic locations of each of a large number of mutants with altered A chains. He then collected A-chain polypeptides from each of the mutant strains and analyzed them to determine which amino acid had been changed. His analyses showed that the relative position of the changed nucleotide within a gene, as determined by genetic analysis, corresponds to the relative position of the altered amino acid in the peptide chain of the enzyme molecule, as determined by direct chemical analysis of the peptide. Each amino acid substitution could be accounted for by a change in a single nucleotide in a codon. Glycine (GGA), for example, was replaced by glutamate (GAA) in the tryptophan synthetase of one mutant and by arginine (AGA) in another. Other examples of colinearity have been derived from analyses of the genetic control of hemoglobin synthesis in man.

SECTION 7–7

TYPES OF RNA: MESSENGER, RIBOSOMAL AND TRANSFER

RNA differs from DNA in containing ribose instead of deoxyribose, and uracil instead of thymine. In contrast to DNA, the purine and pyrimidine nucleotides are usually not present in RNA in complementary ratios. This fact indicates that RNA is not a double helix like DNA, but is single-stranded. The molecules of RNA are unbranched and contain the four kinds of ribonucleotides, A, G, C and U, linked by 3′, 5′ phosphodiester bonds.

The synthesis of proteins requires three kinds of RNA molecules: (1) *Messenger RNA* transmits genetic information from the DNA molecule in the nucleus to the cytoplasm; (2) *ribosomal RNA* makes up a large portion of the cytoplasmic particles called *ribosomes* on which protein synthesis occurs; and (3) *transfer RNA* acts as an *adaptor* to bring the proper amino acid into line in the growing polypeptide chain in the appropriate place. Messenger RNA is synthesized by a *DNA-dependent RNA polymerase* first found in the nuclei of rat liver, and identified subsequently as an important constituent of plant, bacterial and animal cells. The enzyme requires DNA as the template, and uses as substrate the triphosphates of the four ribonucleotides commonly found in RNA. The products are RNA and inorganic pyrophosphate. The reaction system can use single-stranded or native DNA or a synthetic deoxyribopolynucleotide as template.

Transfer RNA and ribosomal RNA are also produced by DNA-dependent RNA-synthesizing systems, and are transcribed from complementary deoxynucleotide sequences in DNA.

The RNA that is produced when carefully defined DNA templates are used is exactly that predicted by the kinds of base pairing permitted by the Watson-Crick model. The DNA template is double-stranded and contains two different but complementary template sequences which would have quite different genetic information. Only one DNA strand is apparently selected for transcription, and only one kind of messenger RNA is produced (Fig. 7–18). The molecular basis for the distinction between the two complementary DNA strands is unknown.

Electron microscopy has established that most cells contain an extensive system of tubules with thin membranes, termed **endoplasmic reticulum.** Associated with the endoplasmic reticulum, or floating freely in the cytoplasm, are small particles termed **ribosomes.** Ribosomes are about 50 per cent protein and 50 per cent RNA. When ribosomal RNA is extracted with phenol from the particles, two components can be identified with molecular weights of about 600,000 and 1,300,000, respectively.

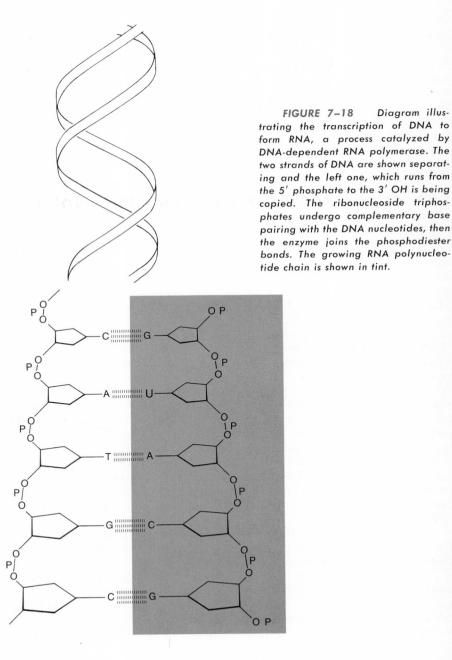

FIGURE 7–18 *Diagram illustrating the transcription of DNA to form RNA, a process catalyzed by DNA-dependent RNA polymerase. The two strands of DNA are shown separating and the left one, which runs from the 5' phosphate to the 3' OH is being copied. The ribonucleoside triphosphates undergo complementary base pairing with the DNA nucleotides, then the enzyme joins the phosphodiester bonds. The growing RNA polynucleotide chain is shown in tint.*

Molecules of transfer RNA are considerably smaller than the molecules of messenger or ribosomal RNA. Each functions as a specific adaptor in protein synthesis, binding to and identifying one specific amino acid; i.e., each of the 20 amino acids is attached to one (or more) specific kinds of transfer RNA. One portion of the nucleotide sequence in transfer RNA represents an **anticodon,** a nucleotide triplet complementary to the codon in messenger RNA that specifies that amino acid. Transfer RNA is unusual in containing not only the four usual ribonucleotides, adenylic, guanylic, cytidylic and uridylic, but also small amounts of unusual nucleotides such as **6-methylamino adenylic acid, dimethylguanylic acid** and **thymine ribotide.**

Transfer RNAs are polynucleotide chains of some 70 nucleotides. Each of the several kinds of transfer RNA has an identical sequence of nucleotides, CCA, at the 3' end to which the amino acid is attached. Each in addition has guanylic acid at the opposite, 5', end of the nucleotide chain. The chain is doubled back on itself, forming three or more loops of unpaired nucleotides; the folding is stabilized by hydrogen bonds between complementary bases in the intervening portions of the chain to form a double helix (Fig. 7–19). The loop nearest the amino acid acceptor (CCA) has seven nucleotides, with cytidine, pseudouridine and thymidine at positions 21, 22 and 23 from the CCA end. The triplet that is complementary to the codon is located in a seven-membered middle loop, and is preceded by uridine and followed by adenosine or a modified adenosine in the loop. Another unusual base, dimethylguanosine, is found eight positions before the anticodon, at the base of the larger (eight to twelve nucleotides) loop near the 5' end. The folding results in a constant distance between the anticodon and the amino acid in all of the tRNAs examined so far. The triplet anticodons found in these tRNAs agree very well (Table 7-5) with the complements predicted to the known mRNA codons for those amino acids.

The first transfer RNA to be completely analyzed was the transfer RNA for alanine derived from yeast cells. This dramatic achievement of Robert Holley and his colleagues was recognized with the awarding of a Nobel Prize to Holley in 1968. The alanine tRNA has 77 nucleotides arranged in a unique sequence. Nine of these are unusual bases, with one or more methyl groups which are added enzymatically after the nucleotides have been linked by phosphodiester bonds. Certain of these unusual bases cannot form conventional base pairs, and may serve to disrupt the

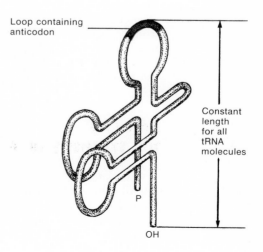

FIGURE 7–19   A diagram of the three dimensional clover leaf structure of transfer RNA. One loop contains the triplet anticodon which forms specific base pairs with the mRNA codon. The amino acid is attached to the terminal ribose at the 3' OH end, which has the sequence CCA of nucleotides. Each transfer RNA also has guanylic acid, G, at the 5' end (P). The pattern of folding permits a constant distance between anticodon and amino acid in all transfer RNAs examined.

Loop containing anticodon

Constant length for all tRNA molecules

P

OH

TABLE 7–5   Agreement Between the Anticodons Found in Isolated tRNAs and the Anticodons Predicted from the Genetic Code

Amino Acid	mRNA Codons $\longrightarrow$	Complements $\longleftarrow$	Observed Anticodons* $\longleftarrow$
Alanine	GpCpA	UpGpC	
	GpCpG	CpGpC	
	GpCpC	GpGpC	IpGpC
	GpCpU	ApGpC	
Phenylalanine	UpUpC	GpApA	OMGpApA
	UpUpU	ApApA	
Tyrosine	UpApU	ApUpA	
	UpApC	GpUpA	GpψpA
Serine	ApGpC	GpCpU	
	ApGpU	ApCpU	
	UpCpC	GpGpA	IpGpA
	UpCpU	ApGpA	
	UpCpA	UpGpA	
	UpCpG	CpGpA	
Valine	GpUpU	ApApC	
	GpUpC	GpApC	IpApC
	GpUpA	UpApC	
	GpUpG	CpApC	

*The anticodons contain some "unusual" bases, including I (inosine), OMG (2'-O-methylguanosine) and ψ (pseudouridine).

base pairing in other parts of the transfer RNA. This could expose specific chemical groups on the tRNA which form secondary bonds to messenger RNA or to the ribosome, or perhaps to the enzyme needed to attach the specific amino acid to its specific transfer RNA molecule. The exact sequence of nucleotides in several other transfer RNAs is now known. This implies that the sequence of the nucleotides in the genes that specify each of these particular kinds of transfer RNA is also known. The sequence in the genic DNA can be inferred from the Watson-Crick rules of specific base pairing. The gene specifying the transfer RNA for alanine would also be 77 nucleotides long. A complete, double-stranded DNA molecule with this sequence of 77 nucleotides was synthesized in 1970 by Khorana and was found to be transcribed to give alanine tRNA. This was the first gene to be synthesized.

SECTION 7–8

THE SYNTHESIS OF A SPECIFIC POLYPEPTIDE CHAIN

Before the subunits, the amino acids, can be assembled into a peptide chain, each amino acid must be activated by an enzyme-mediated reaction with ATP (Fig. 7–20). There is a separate specific activating enzyme for each amino acid. All of these enzymes are present in the cytoplasm and are not associated with any of the structural elements of the cell. The enzymes

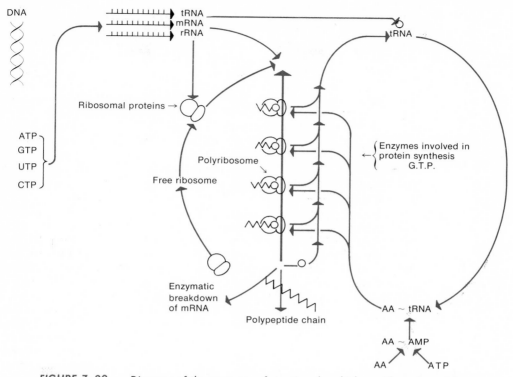

FIGURE 7–20   *Diagram of the sequence of reactions by which specific amino acids are activated and transferred to specific transfer RNAs (Aa~tRNA). On the ribosome the amino acid is transferred to a specific site in the growing polypeptide chain. (After Watson, J. D.: Molecular Biology of the Gene. New York, W. A. Benjamin, Inc., 1967.)*

catalyze the reaction of the amino acid with ATP to form the amino acid-adenylic acid compound (aa-AMP) and release inorganic pyrophosphate. The same enzyme next catalyzes the transfer of the amino acid from the adenylic acid (AMP) to the specific transfer RNA for that amino acid.

$$aa - AMP + tRNA \longrightarrow aatRNA + AMP$$

The amino acid is attached at the end of the transfer RNA, which contains the CCA nucleotides, to the ribose of the terminal adenylic acid. If these three nucleotides are removed the transfer RNA is unable to function. The classic experiments that clarified the nature of this two-step system were carried out by Paul Zamecnik and Mahlon Hoagland at Harvard Medical School.

Francis Crick had predicted on theoretical grounds that some sort of nucleic acid molecule must serve as an adaptor in the course of protein synthesis. He argued that since there is no simple steric correspondence between a polynucleotide chain and a polypeptide chain which would enable the nucleotide sequence to specify the amino acid sequence directly, the amino acids might be lined up in appropriate register by means of small RNA adaptor molecules. These adaptor molecules could assemble at specific places on the nucleic acid template by their complementary sequences of nucleotides. This hypothesis has now been validated by a wide variety of experiments. For example, Chapeville prepared a complex of cysteine with its specific transfer RNA (cysteinyl-tRNA^{cys}), and then converted the cysteine to alanine while it was still bound to the transfer RNA (alanyl-tRNA^{cys}). When this was added to a protein synthesizing system, a polypeptide was made which contained *alanine* at the sites in the peptide chain where cysteine should have been. Thus these experiments provided direct proof that the ordering of specific amino acids into their appropriate place is dictated by the specific transfer RNAs and not by the amino acid that is bound to it.

The amino acid bound to its specific transfer RNA is transferred to the ribosomes. The role of the ribosome is to provide the proper orientation of the amino acid-transfer RNA precursor, the messenger or template RNA and the growing polypeptide chain, so that the genetic code on the template or messenger RNA can be read accurately. There are some 15,000 ribosomes in a rapidly growing cell of *E. coli,* each with a molecular weight of nearly 3,000,000. These ribosomes account for nearly one third of the total mass of the cell. Only one polypeptide chain can be formed at a time on any given ribosome.

The template for the synthesis of a specific protein is supplied by the messenger RNA formed on one strand of the double helix of DNA. The messenger RNA leaves the nucleus and passes to the cytoplasm, where it becomes associated with the ribosome. Ribosomes from different kinds of cells may differ somewhat in their mass, in the composition of their RNA and in the ratio of RNA to protein, but there is a general similarity in their structures. Bacterial ribosomes have a molecular weight of about 2,600,000, and are composed of two portions—a larger one, molecular weight 1,800,000, and a smaller one, molecular weight 800,000. The ribosomes can be separated into these two subunits if they are placed in a solution with a low concentration of magnesium ion. When observed in the electron microscope, the subunit structure becomes apparent. The smaller subunit seems to sit on the flat surface of the larger subunit like a cap. The ribosomes of higher organisms tend to be somewhat larger than the bacterial ones, and are composed of two or four subunits. Each ribosome contains several kinds of proteins bound to the RNA. Ribosomes have GTPase activity, which appears to play an important but undetermined role in the transfer of amino acids from transfer RNA to the forming peptide chain. Protein synthesizing particles somewhat similar to ribosomes are also found in the nucleus, in the chloroplasts of plant cells and in mitochondria.

During protein synthesis the messenger RNA appears to move across the site on a ribosome at which protein synthesis occurs. In doing this it brings successive codons into a position on which the appropriate amino acid transfer RNA anticodons can be arranged. The transfer of the amino acid to the growing peptide chain of a ribosome requires guanosine triphosphate, one or more specific enzymes and glutathione.

Protein synthesis has been studied intensively in preparations of rabbit reticulocytes, which are engaged primarily in making just one protein—hemoglobin. Alexander Rich and his coworkers showed that the ribosomes most active in protein synthesis are those that interact in clusters of five or more (Fig. 7–21). These clusters, termed **polyribosomes,** are held together by the strand of messenger RNA. Peptide chains are formed by the sequential addition of amino acids beginning at the N-terminal end, the end having a free amino group. Studies of ribosomes in the electron microscope suggest that individual ribosomes become attached to one end of a polyribosome cluster, and gradually move along the messenger RNA strand as the polypeptide chain attached to it increases in length by the sequential addition of amino acids (Fig. 7–22). Thus each ribosome appears to ride along the extended messenger RNA molecule, "reading the message" as it goes. The ribosome is believed to play a part in bringing the transfer RNA molecule into line at the right position. After it com-

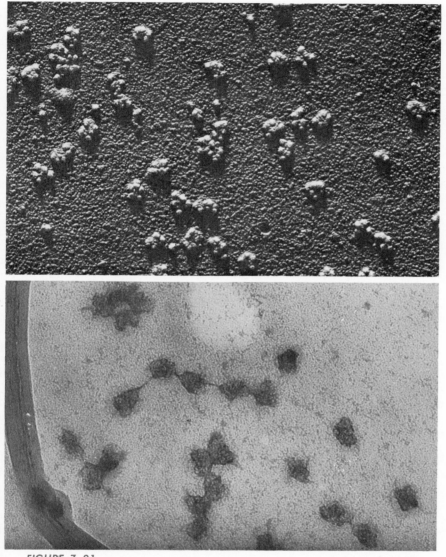

FIGURE 7–21   Electron micrographs of polyribosomes isolated from the reticulocytes of a rabbit. The upper preparation was shadowed with gold and the lower one was
stained with uranyl nitrate. The electron micrographs show that polyribosomes tend to
occur in clusters of four, five or six. The clusters are connected by a thin strand of mRNA.
(Courtesy of Dr. Alexander Rich.)

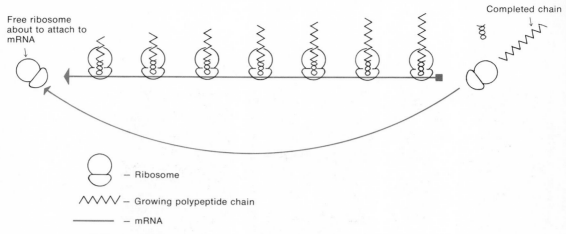

FIGURE 7-22   *Diagram of the mechanism by which a polypeptide chain may be synthesized on the ribosome. Each ribosome is believed to ride along the extended molecule of messenger RNA, reading and translating the genetic message. Amino acids are added in sequence at one end of the growing chain as dictated by the specific base pairing of the nucleotide triplets in the mRNA codon and the tRNA anticodon. (After Watson, J. D.: Molecular Biology of the Gene. New York, W. A. Benjamin, Inc., 1967.)*

pletes the reading of one molecule of messenger RNA and releases the polypeptide that it has synthesized, the ribosome appears to jump off the end of one messenger RNA chain, move to a new messenger RNA and begin reading it (Fig. 7–23). The growing peptide chain always remains attached to its original ribosome; there is no transfer of a peptide chain from one ribosome to another. Several ribosomes may be working simultaneously on a single strand of messenger RNA, each reading a different part of the message.

All the processes of gene replication, gene transcription and protein synthesis depend upon the formation of specific, though relatively weak, hydrogen bonds between specific pairs of purine and pyrimidine bases. The specificity of these bonds ensures the remarkable accuracy of the process; mistakes in base pairing occur less than one time in a thousand.

Experiments using cell-free protein-synthesizing systems from bacteria indicate that the messenger RNA becomes bound to the smaller of the two ribosomal subunits, whereas the forming peptide chain is attached to the larger of the two subunits. Each molecule of transfer RNA must first become attached to the smaller subunit so that it can be recognized and attached to the appropriate part of the messenger RNA, and then must be passed to the larger subunit for the formation of the peptide bond and the incorporation of the amino acid into the growing peptide chain.

From other experiments it has been possible to estimate the number of transfer RNA molecules bound to each ribosome during protein synthesis. Each ribosome has one transfer RNA linked to the growing polypeptide, one free transfer RNA which is not tightly bound and appears to be on its way out of the ribosome after contributing its amino acid to the polypeptide chain, and a third transfer RNA with its amino acid bound to it. The latter tRNA appears to be getting into position to contribute its amino acid to the growing chain. Since the messenger RNA may be a very long molecule, whereas the ribosome is no more than 240 Å in its longest dimension, it would follow that either the message must pass along the ribosome, or the ribosome must pass along the message. The mechanism for this is unknown.

Double helix of DNA

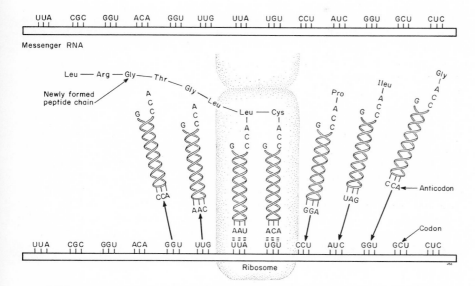

FIGURE 7-23   *Diagram of the mechanism by which protein synthesis is believed to occur on the ribosome. The diagram illustrates the relationship between the triplet code of the DNA helix, the complementary triplet code of messenger RNA and the complementary triplet code (anticodon) of transfer RNA. Molecules of tRNA charged with specific amino acids are depicted entering from the right, assuming their proper place on mRNA at the ribosome and then transferring the amino acid to the growing peptide chain. The transfer RNA molecules then (left) leave the ribosome to be recharged with amino acids for further reactions. The growing polypeptide chain remains attached to the original ribosome.*

Both the reading of the messenger RNA strand and the synthesis of the polypeptide chain are processes that proceed sequentially. Using cell-free systems prepared from rabbit reticulocytes it was shown that the chain is synthesized beginning at the amino-terminal end and proceeding sequentially to the carboxy-terminal end. To show this, tritium-labeled leucine was added to the hemoglobin-synthesizing system in pulses of different duration. With a very long pulse the labeled leucine was evenly distributed in all parts of the polypeptide; however, with short pulses, the labeled leucine was found only in those parts of the molecule synthesized last and these are at the C-terminal end. In the cell-free system the synthesis of the complete α chain of hemoglobin, which is 141 amino acids long, requires 1.5 minutes; i.e., about two amino acids are added each second.

It appears that two different enzymes are required to carry out the transfer of the amino acids from transfer RNA to the peptide linkage in the peptide chain. One enzyme binds the amino acid transfer RNA and the other is a synthetase that forms the peptide bond. The reaction which requires GTP is the binding of the tRNA to the ribosome rather than the synthesis of the peptide bond. GTP is used and the products are GDP and inorganic orthophosphate. In cell-free systems some 50 molecules of GTP are hydrolyzed for each amino acid transferred to a growing peptide. Presumably the protein-synthesizing system within the intact cell does not make such profligate use of its ~P!

The amino acid-transfer RNA molecule is first recognized by the complex of ribosome and messenger RNA by binding at a specific coding site, by the hydrogen bonding of its anticodon with a messenger RNA codon. It then displaces the previous C-terminal transfer RNA by forming a peptide bond between its amino group and the polypeptide terminal carboxyl group. The transfer RNA that is displaced, the one to which the peptide chain had been bound, is "spent" and then returns to the medium to be recharged for another cycle of peptide chain synthesis.

There is a special problem in incorporating the first amino acid of a new polypeptide chain, since there is no preexisting polypeptide transfer RNA complex to accept it at the combining site. It has been shown recently that the N-terminal amino acids of proteins are often substituted at their alpha amino end with an acetyl group or a formyl group. There is, in fact, a special enzyme which adds a formyl group to methionine after methionine has been linked to its transfer RNA. Such N-formyl methionine molecules are incorporated preferentially at the N-terminal end of peptide chains, whereas methionines without the formyl group are incorporated in intermediate positions in the chain. Two different types of methionine transfer RNA may be involved, only one of which can receive a formyl group.

Although we have learned a lot about the biosynthesis of proteins, there is much more yet to be learned. Even the best cell-free protein-synthesizing system operates at a rate no more than one one-hundredth of the rate within an intact, living cell.

SECTION 7–9

CHANGES IN GENES: MUTATIONS

Although genes are remarkably stable and are transmitted to succeeding generations with great fidelity, they do from time to time undergo changes called **mutations.** After a gene has mutated to a new form, this new form is stable and usually has no greater tendency than the original gene to mutate again. A mutation may be defined as any inherited change not due to segregation or to the normal recombination of unchanged genetic material. Mutations provide the diversity of genetic material which makes possible the study of the process of inheritance. Investigations of the mechanism of the mutation process have provided important clues as to the nature of the genetic material itself.

Some mutations, termed **chromosomal mutations,** are accompanied by a visible change in the structure of the chromosome. A small segment of the chromosome may be missing (a **deletion**), or may be represented twice in the chromosome (a **duplication**) (Fig. 7–24). A segment of one chromosome may be transferred to a new position on a new chromosome (a **translocation**) or a segment may be turned end for end and attached to its usual chromosome (an **inversion**). **Point mutations** or gene mutations involve small changes in molecular structure that are not evident under the micro-

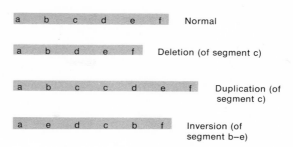

FIGURE 7–24   Diagram illustrating the types of mutations which involve changes in the structure of the chromosome.

scope. These gene mutations involve some change in the sequence of nucleotides within a particular section of the DNA molecule, usually the substitution of one nucleotide for another in a given codon.

From your knowledge of the DNA molecule, you might predict that replacing one of the purine or pyrimidine nucleotides by an analogue such as azaguanine or bromouracil would result in mutation. In several experiments in which such analogues were incorporated into bacteriophage, no mutations were evident. Because of the degeneracy of the genetic code (Table 7–4), a number of changes in base pairs could occur without changing the amino acid specified. In other experiments, the incorporation of bromouracil into DNA did lead to an increased rate of mutation. Other chemicals known to be mutagenic include nitrogen mustards, epoxides, nitrous acid and alkylating agents. These are all chemicals which can react with specific nucleotide bases in the DNA and change their nature. When an analogue is incorporated into DNA it may lead to mistakes in the pairing of nucleotides during subsequent replication processes. For example, whem bromouracil is incorporated into DNA in place of thymine it will pair with guanine rather than with adenine, the normal pairing partner of thymine. This would lead to the substitution of a GC pair of nucleotides at the point in the double helix previously occupied by an AT pair of nucleotides (Fig. 7–25).

The presence of mutagenic materials may increase the frequency of mistakes in nucleotide base pairing, which would result in the production of DNA molecules containing only natural bases, but with an altered base order. The change in normal base order is significant, because during subsequent replications, the altered base sequence will be reproduced by the normal process of DNA synthesis. The definition of mutation includes the restriction that the change that has been introduced into the DNA molecule must be propagated subsequently for an indefinite number of times.

Gene mutations generally result from errors in base pairing during the replication process; i.e., an AT base pair normally present may be replaced by GC, CG or TA pairs. The altered DNA will be transcribed to give an altered messenger RNA, and this will be translated into a peptide chain with one amino acid different from the normal kind of peptide. If the amino acid substitution occurs at or near the active site of the enzyme, the altered protein may have markedly decreased or altered enzymatic properties. However, if the amino acid substitution occurs elsewhere in the enzyme molecule it may have little or no effect on the properties of the enzyme, and may be undetected. The true number of gene mutations may indeed be much greater than the number observed.

If a single nucleotide pair were inserted into or deleted from the DNA molecule, it would shift the reading of the genetic message, alter all of the codons lying to the right of the substitution, and change completely the nature of the resulting peptide chain. Thus if the normal sequence is

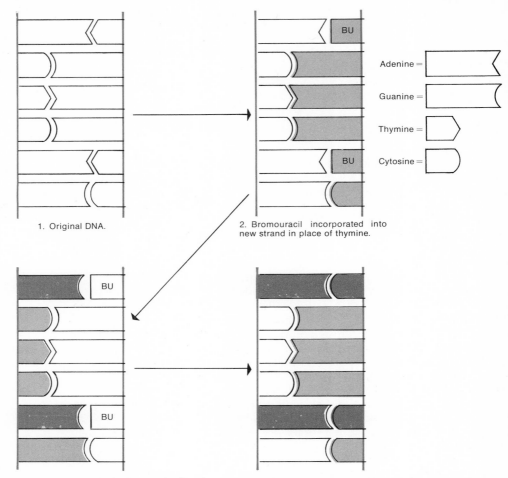

1. Original DNA.

2. Bromouracil incorporated into new strand in place of thymine.

Adenine =

Guanine =

Thymine =

Cytosine =

3. Strand with bromouracil leads to production of new strand with guanine paired to the bromouracil.

4. New, mutant DNA which contains no analogue bases, but has nucleotide sequence different from original, with GC pairs in place of AT.

FIGURE 7–25   *The sequence of events by which an analogue of a purine or pyrimidine might interfere with the replication process and cause a mutation, an altered sequence of nucleotides in the DNA. The nucleotides of the new chain at each replication are indicated in color. In this example two new GC pairs are indicated as replacing the original AT pairs. A single substitution of a GC pair for an AT pair would be sufficient to cause a mutation if it occurred in the triplet code at a point that changed the kind of amino acid specified, i.e., in the first or second base of the triplet or in the third member of certain triplets (see Table 7–4).*

CAGTTCATG (read CAG, TTC, ATG), the insertion of a G between the two Ts results in CAGTGTCATG, (read CAG, TGT, CAT, G . . .).

Gene mutations may be induced by x-rays, gamma rays, cosmic rays, ultraviolet rays and other types of radiation. How radiation may lead to changes in base pairs is not clear, but the radiant energy may react with water molecules to release short-lived, highly reactive free radicals that attack and react with specific bases. Mutations occur spontaneously at low but measurable rates which are characteristic of the species and of the gene. Some genes ("hot spots") are much more prone to undergo mutation than others. Spontaneous mutations may be caused by natural radiation such as cosmic rays or by errors in base pairing during replication. The rates of spontaneous mutations of different human genes range from 10^{-3} to

10^{-5} mutations per gene per generation. Since man has a total of some 25,000 genes, this means that the total mutation rate is on the order of one mutation per person per generation. Each of us, in other words, has some mutant gene that was not present in either of our parents.

SECTION 7–10

GENE-ENZYME RELATIONS

If each gene leads to the production of a specific enzyme by the method outlined in Section 7-8, we may next enquire how the presence or absence of a specific enzyme may affect the development of a specific trait. The expression of any structural or functional trait is the result of a number, perhaps a large number, of chemical reactions which occur in series, with the product of each serving as the substrate for the next: A→B→C→D. The dark color of most mammalian skin or hair is due to the pigment *melanin* (D), produced from dihydroxyphenylalanine (C), derived in turn from tyrosine (B) and phenylalanine (A). Each of these reactions is controlled by an enzyme. The conversion of dihydroxyphenylalanine to melanin is mediated by *tyrosinase. Albinism,* characterized by the absence of melanin, results from the absence of tyrosinase. The gene for albinism (*a*) does not produce the enzyme tyrosinase, but its normal allele (*A*) does.

The earliest attempts to connect the action of a specific gene with a specific enzymatic reaction were studies of the inheritance of flower colors in which the specific flower pigments were extracted and analyzed. Studies of the inheritance of coat color in mammals and of eye color in insects were also able to relate specific genes with specific enzymic reactions in the synthesis of these pigments. A major advance in this field was made in 1941 when George Beadle and Edward Tatum looked for mutations in the bread mold *Neurospora*—mutations which interfere with the reactions by which chemicals essential for its growth are produced. The wild type *Neurospora* requires as nutrients only sugar, salt, inorganic nitrogen and the vitamin biotin. A mixture of these makes up the so-called *minimal medium* for the growth of wild type *Neurospora.* Exposure of the conidia (haploid asexual spores) to x-rays or ultraviolet rays will produce mutations (Fig. 7–26). After irradiation the mold is transferred to a *complete medium,* an extract of yeast which contains all of the known amino acids, vitamins, purines, pyrimidines and so on. Any nutritional mutant produced by the irradiation is able to survive and reproduce when grown on this complete medium. It can subsequently be tested for its ability to grow on minimal medium. If the irradiated mold is unable to grow on minimal medium, the conclusion is that the mutant is unable to produce some compound essential for growth. By trial and error, by adding substances to the minimal medium in groups or singly, the required substance may be identified. Genetic tests show that the mutant strain produced by irradiation differs from the normal wild type by a single gene, and chemical tests show that the addition of a single chemical substance to the minimal medium will enable the mutant strain to grow normally.

Beadle and Tatum made the inference that each normal gene produces a single enzyme which regulates a single step in the biosynthesis of that particular chemical. The mutant gene does not produce the enzyme and therefore that strain of the organism must be supplied with the product of the reaction that is impaired. In certain instances it has been possible to extract the particular enzyme from the cells of normal *Neurospora,*

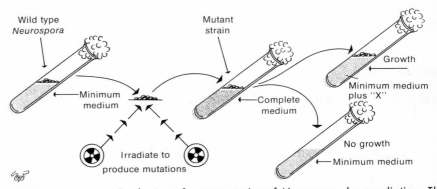

FIGURE 7-26   Production of mutant strains of Neurospora by x-radiation. The mutant strains produced can grow on complete medium but not on minimal medium; however, they can grow on minimal medium plus a single nutrient, X.

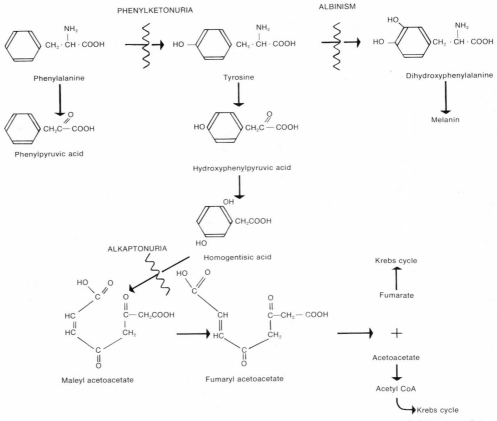

FIGURE 7-27   The sequence of enzymatic reactions by which phenylalanine and tyrosine are metabolized. A mutant which interferes with the production of one of the enzymes that catalyze one of these steps may result in an inborn error of metabolism such as phenylketonuria, albinism or alkaptonuria.

but not from cells of the mutant strain. The synthesis of each of these nutrients involves a number of different steps, each mediated in turn by a
separate gene-controlled enzyme. Indeed, biologists estimate the minimal
number of steps involved in the synthesis of a given substance from the
number of different mutants that will interfere with its production.

Similar one-to-one relationships of gene, enzyme and biochemical
reaction in man were described by the English physician, A. E. Garrod, in
1908. **Alkaptonuria** is an inherited condition in which a substance in the
patient's urine turns black when exposed to air. **Homogentisic acid,** a
normal intermediate in the metabolism of phenylalanine and tyrosine, is
excreted in the urine of alkaptonurics. The tissues of normal individuals
have an enzyme which oxidizes homogentisic acid so that it is ultimately
excreted as carbon dioxide and water (Fig. 7–27). Patients with alkaptonuria lack this enzyme because they lack the gene which controls its production. In these patients homogentisic acid accumulates in the tissues
and blood, and is excreted in the urine. Garrod coined the term *"inborn
errors of metabolism"* to describe alkaptonuria and comparable conditions such as **phenylketonuria** and albinism.

SECTION 7–11

THE OPERON CONCEPT: CONTROL OF PROTEIN SYNTHESIS

It appears from many kinds of experimental evidence that genes
normally do not operate to produce the maximum number of enzymes
all the time. Each gene appears to be "repressed" to a greater or lesser
extent under normal conditions. Then, in response to some sort of environmental demand for that particular enzyme, the gene becomes "derepressed" and leads to an increased production of the enzyme. When a
single gene is fully derepressed it can cause the synthesis of fantastically
large amounts of an enzyme. One enzyme may make up 5 to 8 or more per
cent of the total protein of the cell! If all enzymes were produced at a
similar very high rate, metabolic chaos would ensue. Thus the phenomena
of gene repression and derepression appear to be necessary to provide a
means of increasing or decreasing the rate of synthesis of some particular
enzyme in response to variations in the environmental requirement for
that enzyme.

In unicellular organisms and in the cells of multicellular organisms
there may be wide variations in the number of enzymes per cell and in the
amount of a given enzyme per cell. Thus there must be some mechanism
that controls how much of a particular enzyme is synthesized in a given cell
at a given moment. The rate of synthesis of a protein may be controlled
in part by the genetic apparatus and in part by factors from the external
environment. Most of the data relating to the control of protein synthesis
in cells has come from studies of microbial systems, especially in the bacterium *Escherichia coli.*

From the length of the chromosome of *E. coli* and the estimate that
the average gene contains about 1500 nucleotide pairs (and codes for a
polypeptide chain of 500 amino acids), it has been calculated that the
genes of *E. coli* code for 2000 to 4000 different polypeptides. Estimates
place the number of different enzymes required by an *E. coli* growing on
glucose at about 800. Some of these must be present in large amounts;
others are required in smaller quantities.

E. coli cells growing on glucose contain very little of the enzyme β-*galactosidase.* When grown on lactose as the sole carbon source, the cells require β-galactosidase to cleave lactose to glucose and galactose. Under these conditions β-galactosidase makes up some 3 per cent of the total protein of the cell; there are perhaps 3000 molecules of enzyme in each cell of *E. coli,* which is at least a thousandfold increase over the amount present in cells growing on glucose. Two other enzymes, a **galactoside permease** and a **galactoside transacetylase,** respond equally dramatically to the presence of lactose in the incubation medium. A substance such as lactose which elicits an increased amount of an enzyme is called an **inducer,** and enzymes that respond to inducers are termed **inducible enzymes.**

Cells of *E. coli* grown in a medium without any amino acids contain the whole spectrum of enzymes required to synthesize all of the 20 or so amino acids needed for the assembly of protein molecules. The introduction of one or more amino acids to the incubation medium greatly decreases the amount of the biosynthetic enzymes required for the production of that amino acid. Biosynthetic enzymes that are reduced in amount by the presence of the end product of a biosynthetic sequence (e.g., the amino acid) are called **repressible enzymes,** and the small molecule (the amino acid) that brings about the repression is called a **corepressor.**

Both induction and repression of enzymes are adaptive phenomena that appear to be of survival value to bacteria—it would be a waste of energy and material to synthesize a battery of enzymes not immediately required. The responses to inducers and repressors are not "all or none" like the responses of nerve conduction or muscle contraction, but allow for the formation of intermediate amounts of enzyme in response to intermediate conditions in the environment.

Differences in the amount of a specific protein in bacteria are generally due to variations in the rate of synthesis of that enzyme and are not due to altered stability or to a change in the rate of degradation. The rate of synthesis is controlled in turn by the amount of mRNA present per cell. Jacob and Monod have postulated that the amount of effective mRNA template for a given enzyme is controlled by a special kind of molecule, a protein called a **repressor,** which blocks the synthesis of messenger RNA. They have further postulated that these repressor proteins are coded for by special genes termed **regulator genes.** Gilbert and his colleagues at Harvard reported the isolation and characterization of the repressor for β-galactosidase from *E. coli.* This is a protein with a molecular weight of about 150,000. From their data they calculated that there are about 10 molecules of β-galactosidase repressor per cell.

Repressors have been postulated to block the synthesis of specific proteins by combining with a specific site on the DNA and blocking its transcription to form messenger RNA. Alternatively, the repressor might combine with the molecule of messenger RNA, prevent its being attached to the ribosome and thereby increase the probability of its enzymatic degradation.

If regulator genes made repressors all the time, the synthesis of messenger RNA would always be inhibited. Thus it is necessary to postulate further that repressors may exist in either active or inactive forms depending on whether the repressors are combined with specific small molecules, the inducers or corepressors. The attachment of an inducer inactivates the repressor. Thus, the combination of lactose with the repressor blocks the repressor and permits the synthesis of the enzyme.

The attachment of the corepressor, e.g., the amino acid, to the repressor increases the number of active repressor molecules. These bind to the DNA, decrease the transcription of the DNA and the number of messenger RNA molecules that are produced and code for the enzyme.

The repressor is thought to be bound to its specific inducer or corepressor by weak bonds such as hydrogen bonds (Fig. 7–28).

There is experimental evidence that two or more enzymes may vary in amount in a coordinate fashion, suggesting that they are under the control of the same repressor system. It has been inferred that a single repressor may control the formation of the mRNA for these "coordinately repressed" enzymes. Frequently, but not always, genetic mapping shows that the genes for these coordinately repressed enzymes are closely linked on the chromosome and that a single mRNA molecule may carry the message for the synthesis of two or more enzymes.

The genes whose codes are transcribed on a single mRNA molecule and which are under the control of a single repressor have been termed an **operon** by Jacob and Monod. Originally it was believed that all the genes controlled by the same repressor must lie closely adjacent in a chromosome, but more recently examples have been found of widely separated genes that appear to be coordinately repressed by the same repressor.

A further entity has been postulated to account for the control of the operon: the operator site. Operators are believed to lie adjacent to the genes in the operon and are thought to be the sites on the DNA to which active repressor molecules are bound, thereby inhibiting the synthe-

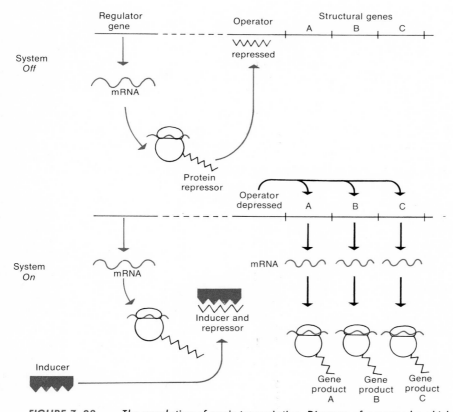

FIGURE 7–28   *The regulation of genic transcription. Diagram of a means by which a regulator gene may produce a messenger RNA which codes for a protein repressor. The repressor in turn inhibits an operator and thus prevents the transcription of the structural genes (above). Below, the messenger RNA produced by the regulator gene again produces a protein repressor but this is combined with an inducer. The operator gene is thereby derepressed, permitting the transcription of structural genes A, B and C and the formation of the protein gene products A, B and C.*

sis of mRNA by the genes in the adjacent operon. In the absence of active repressors, the genes in the operon are free to be transcribed, mRNA is formed and enzymes are produced on the ribosome. If the operator site is absent, the repressor cannot inhibit the synthesis of the specific mRNA and the corresponding enzyme is produced all the time. It is possible that the operator site in some way provides a code that directs the RNA polymerase to begin transcribing the DNA at that specific point at the end of the operon.

If there are such things as regulator genes, it has been pointed out that the transcription of the regulator gene must also be under regulation so that the synthesis of repressors is controlled. One cannot postulate an infinite sequence of repressors, each of which represses the synthesis of the next. Either a repressor can repress its own synthesis or repressors are synthesized without such control.

In summary, this overall control mechanism postulated by Jacob and Monod from their studies of β-galactosidase in *E. coli* states that in addition to the structural genes that provide the code for the synthesis of specific proteins there are "regulator genes" that code for the synthesis of "repressors"; that the repressors may be active or inactive depending on whether they are bound to small molecules, the inducers or corepressors; and that active repressors bind to operator sites in the DNA and turn off the transcription of the adjacent structural genes. The structural genes under the control of one repressor are termed an operon.

There is clear evidence that the synthesis of many proteins in *E. coli* and in other organisms is not influenced by substances such as corepressors and inducers in the external environment; either their operator sites are turned on all the time or their genes are not associated with an operator site.

INFORMATION TRANSFER OUTSIDE THE NUCLEUS

It has been known since early in this century that a few traits appear to be inherited not by the usual nuclear genes but by some mechanism restricted to the cytoplasm. Such traits are inherited exclusively from the maternal parent; the egg, but not the sperm, supplies cytoplasm to the zygote.

There is now clear evidence that not all biological information is transferred by the nuclear DNA; both mitochondria and plastids such as chloroplasts contain DNA (Fig. 7–29). About 2 per cent of the DNA of a liver cell is located in its mitochondria. Mitochondrial DNA may undergo mutations, leading to changes in the sequence of amino acids in the structural proteins of the mitochondria. Nass and Nass found fibrous structures in chick mitochondria that are removed by treatment with deoxyribonuclease. Mitochondria contain a DNA polymerase which carries out the replication of mitochondrial DNA; they also have a DNA- dependent RNA polymerase and can synthesize RNA.

The amount of DNA in the mitochondria from beef heart corresponds to a helix of molecular weight 3×10^7; however, the DNA in mitochondria exists as several pieces, each a closed circle of molecular weight 1×10^7. The replication of mitochondrial DNA is completely independent of the nuclear DNA; it is both synthesized and degraded more rapidly than

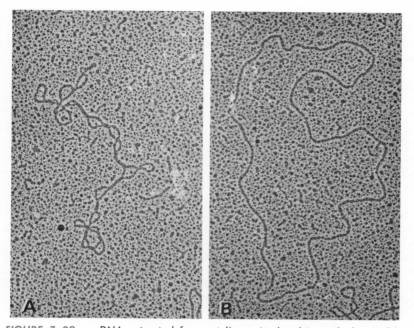

FIGURE 7–29   DNA extracted from rat liver mitochondria and observed by the
spreading technique. A, A DNA molecule in a twisted circle configuration. B, A DNA molecule in the open circle configuration. (From De Robertis, E. D., Nowinski, W. W., and Saez,
F. A.: Cell Biology, Philadelphia, W. B. Saunders Company, 1970; photograph courtesy
of B. Stephens.)

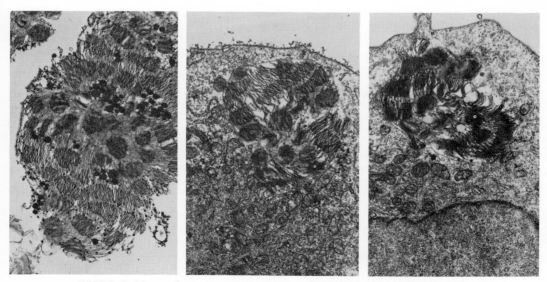

FIGURE 7–30   Electron micrographs of spinach chloroplasts that have undergone
phagocytosis by mouse cells grown in cell culture but have retained their structural integrity.
×12,000. (Courtesy of Dr. Margit Nass.)

nuclear DNA. The biological information in the mitochondrial DNA specifies some, but not all, of the mitochondrial proteins; the remainder are controlled by nuclear genes.

Plastids such as chloroplasts also contain DNA and RNA and, like the mitochondria, have the capacity for independent growth and division, and for the synthesis of specific proteins.

The single-celled alga, *Euglena,* which normally has chloroplasts, can survive without them if it is supplied with appropriate nutrients. *Euglenas* that are deprived of their chloroplasts never develop new ones; however, if they are again supplied with chloroplasts these bodies will undergo division and appear in all daughter cells. If the *Euglena* is supplied with structurally different chloroplasts from a mutant strain, these unusual chloroplasts will undergo division and appear in the daughter cells. This is cogent evidence that the control of chloroplast structure is at least in part under the control of genetic material (DNA) in the plastids themselves.

Mammalian cells grown in cell culture were shown by Margit Nass (1969) to be able to take up by phagocytosis chloroplasts prepared from spinach leaves. The chloroplasts survived for at least six cell generations (six days) and retained their structural integrity (Fig. 7–30). At the appropriate time they divided and were assorted at random to the daughter cells. The chloroplasts taken up could be reisolated from the mouse cells several days later and retained their ability to carry out at least some of the reactions of photosynthesis—the fixation of carbon dioxide and the photochemical reduction of dichlorophenolindophenol. The DNA from the reisolated chloroplasts was still in macromolecular form.

SUGGESTIONS FOR FURTHER READING

Beadle, George, and Beadle, Muriel: *The Language of Life.* Garden City, New York, Doubleday & Company, Inc., 1966.
    A popular, highly readable account of the chemistry of the gene.

Beermann, W., and Clever, U.: *Chromosome Puffs.* Sci. Amer. April (Offprint 180) 1964.
    A description and discussion of these interesting phenomena and their possible role in the biosynthesis of peptides.

Braun, W.: *Bacterial Genetics.* 2nd ed. Philadelphia, W. B. Saunders Company, 1965.
    A concise and well written survey of one of the most exciting frontiers of biology in the twentieth century.

Campbell, A.: *Episomes.* New York, Harper & Row, Publishers, 1969.
    A modern account of DNA molecules that occur free in the cytoplasm or are incorporated into chromosomes, their biological significance and their relationship to viruses.

Frisch, L. (ed.): *The Genetic Code.* Cold Spring Harbor, N. Y., Cold Spring Harbor Sympos. Quant. Biol. *31:* 1966.
    A rich source of facts and theories detailing the state of our knowledge of biochemical genetics as of June 1966.

Holley, R. W.: The Nucleotide Sequence of a Nucleic Acid. Sci. Amer. February 1966.
    An account of one of the major recent discoveries in genetics by the man who received a Nobel Prize for this work.

Ingram, V. M.: *The Biosynthesis of Macromolecules.* New York, W. A. Benjamin, Inc., 1966.
    A well written presentation of the molecular events in the synthesis of proteins and other large molecules.

Nirenberg, M. W.: *The Genetic Code:* III Sci. Amer. March (Offprint, 153) 1966.
    An updating of the triplet codes relating DNA, RNA and peptide chains.

Rich, A.: *Polyribosomes.* Sci. Amer. December (Offprint 171) 1963.
   A discussion of the relation of ribosomes, and strands of messenger RNA and their role in protein synthesis.

Watson, J. D.: *Molecular Biology of the Gene.* New York, W. A. Benjamin Inc., 1965.
   A detailed presentation of the genetics of *Escherichia coli* and the inferences that have been made from such studies.

PART
2

GENETICS AND EVOLUTION

TRANSMISSION GENETICS: MENDELISM

CHAPTER 8

GENETIC PRINCIPLES

Biological information is transferred from one generation to the next by the thousands of genes contained within the nucleus of each egg and sperm. Each contains the code for one particular kind of protein. The regularity of the process of inheritance and our ability to predict the nature of unborn offspring depend upon the fact that the hereditary units or genes are arranged in **chromosomes**. Each cell has two of each kind of chromosome and, therefore, two of each kind of gene. The great regularity of the mitotic process ensures that each daughter cell also will have two of each kind of chromosome and, therefore, two of each kind of gene. The exception to this rule occurs in the cell divisions in which mature eggs and sperm are formed. In these the members of the pairs of chromosomes separate and come to lie in different cells. As a result, mature eggs and sperm have only one of each kind of chromosome and one of each kind of gene. When the egg and sperm unite in fertilization, each contributes one of each kind of chromosome. This restores the pattern of paired chromosomes; hence, each cell again has two of each kind of gene. All of the phenomena of Mendelian genetics depend on these simple facts. The pairs of genes separate from each other during the formation of gametes. The set of genes supplied by the egg unites with the set of genes supplied by the sperm to give the pattern of genes present in the offspring and responsible for the development of its traits. Most organisms have many pairs of chromosomes (Man, for example, has 23 pairs), and the genes present in one pair of chromosomes separate from each other independently of the separation of genes in other pairs of chromosomes.

There may be a simple 1:1 relationship between a given gene and the trait which it controls, or one gene may participate in the control of several or many traits throughout the body, or many genes may cooperate to regulate the appearance of a single trait. As you learned in Chapter 7, each gene is a molecule of DNA in which biological information is stored as a triplet code in the sequence of nucleotides which comprise each strand

213

of the double helix of the DNA molecule. The information in each gene is "read out" and a specific protein is synthesized. The presence of the specific protein, an enzyme for example, provides the chemical basis for the trait.

SECTION 8–2

THE DEVELOPMENT OF GENETICS

Our basic understanding of the laws governing similarities and differences between individuals related by descent, the science of genetics, is a development of the 20th century. Yet, for many centuries before that men understood that "like begets like" and were able to breed sheep, cattle, horses, wheat, roses and other plants and animals with specific characteristics.

In the eighteenth and nineteenth centuries several attempts were made to discover how specific characteristics are transmitted from one generation to the next. An important discovery was made in 1760 by the German botanist Kölreuter when he crossed two species of tobacco by placing pollen from one species on the stigmas of the other. The plants, grown from the resulting seeds, had characteristics intermediate between those of the two parents. Kölreuter made the logical inference that parental characteristics are transmitted through both the pollen (sperm) and the ovule (egg). However, he and his contemporary plant and animal breeders were unable to discover the nature of the hereditary mechanism, in part because the cytologic basis was unknown, but primarily because they attempted to study the inheritance of *all* the characteristics of the plant or animal at one time.

Gregor Mendel, an Austrian abbot who bred pea plants in the garden of his monastery at Brunn, succeeded in discovering the basic laws of genetics where previous hybridizers had failed. He studied the inheritance of single contrasting characteristics; he counted and recorded the parents and offspring of each of his crosses. His knowledge of the principles of mathematics enabled him to interpret his data and led him to the hypothesis that each trait is determined by two genetic factors.

Mendel had several types of pea plants in his garden and kept records of the inheritance of seven clearly contrasting pairs of traits, such as yellow versus green seeds, round versus wrinkled seeds, green versus yellow pods, and so on. By crossbreeding and counting the types of offspring, Mendel was able to detect regularities in the pattern of inheritance that had escaped earlier breeders. Whenever he crossed plants with two different characteristics, such as yellow and green seeds, the plants in the next generation, the **F_1 generation,** were all like one of the two parents. The second or **F_2 generation** included individuals of both parental types. When he counted these, he found that the two types of individuals were present in the F_2 generation in a ratio of approximately 3:1 (Table 8–1). For example, (Fig. 8–1) when he crossed tall plants with short plants, all the members of the F_1 generation were tall. When two of these first generation tall plants were crossed, the F_2 generation included some tall and some short plants— 787 tall and 277 short. Clearly, in the first generation, the genetic factor (gene) for shortness was hidden or overcome by the gene for tallness. Mendal termed the gene for tallness "dominant" and the gene for shortness "recessive."

Having discovered that the crossing of two first generation plants led

TABLE 8–1.   An Abstract of the Data Obtained by Mendel from his Breeding Experiments with Garden Peas

Parental Characters	First Generation	Second Generation	Ratios
Yellow seeds × green seeds	all yellow	6022 yellow : 2001 green	3.01 : 1
Round seeds × wrinkled seeds	all round	5474 round : 1850 wrinkled	2.96 : 1
Green pods × yellow pods	all green	428 green : 152 yellow	2.82 : 1
Long stems × short stems	all long	787 long : 277 short	2.84 : 1
Axial flowers × terminal flowers	all axial	651 axial : 207 terminal	3.14 : 1
Inflated pods × constricted pods	all inflated	882 inflated : 299 constricted	2.95 : 1
Red flowers × white flowers	all red	705 red : 224 white	3.15 : 1

(From Villee, C. A.: *Biology.* 5th ed. Philadelphia, W. B. Saunders Company, 1967.)

to offspring in the second generation in a ratio of three with the dominant characteristic to one with the recessive characteristic, it occurred to Mendel that each plant must have two genetic factors whereas each egg and sperm has only one. The first generation tall plants also had two genetic factors—one for tallness and one for shortness—but the tall gene was "dominant" and these plants were tall. However, when these F_1 plants formed eggs or sperm, the gene for tallness separated from the gene for shortness so that half of the eggs and half of the sperm contained a "tall" gene and half a "short" gene. (The genes are not tall or short, but cause the plants to grow to different heights.) The random fertilization of eggs by sperm led to four possible combinations of genes—one with two talls, **TT;** one with two shorts, **tt;** and two with one tall and one short, **Tt** and **tT.** The tall gene (**T**) is dominant to short (**t**) and, therefore, three of the four kinds of offspring were tall plants, and only one was short. It is now the convention to use capital letters for dominant genes and lower case letters for recessive genes, e.g., **T** for the gene for tall plants, and **t** for the gene for short plants.

Mendel's mathematical abilities enabled him to recognize that a 3:1 ratio would be expected among the offspring if each plant had two factors for any given characteristic rather than a single one. This brilliant piece of reasoning was supported when chromosomes were seen and the details of mitosis, meiosis and fertilization became known.

Mendel reported his findings at a meeting of the Brunn Society for the Study of Natural Science and published his results in the transactions of that society. The importance of his findings was not appreciated by the other biologists of the time, and they were neglected for nearly 35 years.

In 1900, after the discovery of the details of mitosis, meiosis, and fertilization, three different investigators—Hugo DeVries in Holland, Karl Correns in Germany and Erich von Tschermak in Austria—conducted a series of experiments and independently rediscovered the laws of inheritance which had been described by Mendel. On finding Mendel's paper in which these laws had been clearly stated 35 years before, they gave him credit for his discoveries by naming two of the fundamental laws of inheritance after him.

In the first decade of the twentieth century experiments with a wide variety of plants and animals, together with observations of human inheritance, showed that these same basic principles govern inheritance in all of these different organisms. W. S. Sutton in the United States and Theodore Boveri in Germany showed that the genes which Mendel described are located in the chromosomes within the nucleus. Some investigators studied inheritance in mice, rabbits, cattle and chickens, but the

Tall plant   Short plant

FIGURE 8–1   A diagram illustrating one of the crosses carried out by Gregor Mendel. Crossing a tall pea plant with a short pea plant yielded offspring, all of which were tall. However, when these offspring were self-pollinated, the next generation included tall and short plants in a ratio of about 3:1.

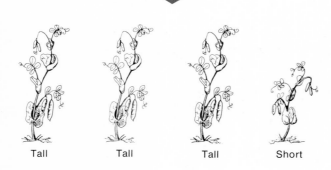

Tall   Tall   Tall   Short

favorite subject for genetic studies became the fruit fly, *Drosophila.* These small insects have a short life cycle of 10 to 14 days, are easily raised in the laboratory, and have only four pairs of chromosomes. In certain of their tissues the chromosomes become very large and the details of their structure can be studied with the microscope. Hundreds of inherited variations involving eye colors, wing shapes and bristle patterns were detected and studied. It was eventually possible to map the specific location of each gene on a specific chromosome. T. H. Morgan and his associates carried out extensive experiments which revealed the genetic basis for the determination of sex and provided the explanation for certain unusual patterns of inheritance in which a trait is linked with the sex of the individual, the so-called sex-linked traits.

A further great advance came in 1927 when H. J. Muller showed that genes could be changed, could undergo mutations, when fruit flies and other organisms were exposed to x-rays. This provided many new mutant genes with which to study heredity. The nature of the mutations gave clues as to the nature and the structure of the genes themselves. This era was followed in the 1940's by experiments to study the relationship of genes and enzymes. Investigators turned to the bread mold, *Neurospora*, in which a number of biochemical mutants, lacking some specific enzyme, could be produced artificially and studied. From the 1950's until the present, the most popular organisms for genetic studies have been the intestinal bacterium, *Escherichia coli*, and some of the bacterial viruses or bacteriophages which infect this bacterium. From 1900 to the present there has been a sustained interest in determining the inheritance of specific traits in man and in determining the inheritance of desirable and undesirable traits in domestic animals and plants. Armed with the growing knowledge of genetic principles, geneticists have been able to breed almost to order cattle that can survive in hot climates, cows that produce large amounts of milk with a high content of butterfat, chickens that lay large eggs with thin shells, corn and wheat plants that are highly resistant to specific diseases, and so on.

SECTION 8–3

CHROMOSOMES AND GENES

Each chromosome consists of a central thread, a **chromonema,** along which lie a series of beadlike structures, the **chromomeres.** Each chromosome has, at a fixed point along its length, a small, clear, circular zone called a **centromere** that controls the movement of the chromosome during cell division. As the chromosome becomes shorter and thicker just before cell division occurs, the centromere region becomes accentuated and appears as a constriction. Chromosomes are distinctly visible only at the time of cell division. At other times they appear as very long, thin, fine, dark-staining strands called **chromatin.** Chromosomes are present as highly extended but distinct physiologic and structural entities between successive cell divisions even though in most forms they are not visible.

When chromomeres were first described many biologists believed that they might be the **genes,** the hereditary factors which previous breeding experiments had shown to lie within the chromosome in a linear order. However, there is no one-to-one correspondence between chromomeres and genes; i.e., there is not a single gene for each chromomere. Some chromomeres contain several genes and some genes appear to be located between the chromomeres. The exact significance of these swellings along the chromonema is not clear.

Each cell of every organism of a given species contains a characteristic number of chromosomes. Each cell in the body of every human being has exactly 46 chromosomes (Fig. 8–2). Many other species of animals and plants also have 46. It is not the *number* of chromosomes that differentiates the various species of plants and animals but rather the nature of the hereditary factors in the chromosomes. A certain species of roundworm has only two chromosomes in each cell and some crabs have as many as 200. The highest chromosome number reported so far is about 1600 found in a radiolarian, a single-celled marine protozoan. Most species of animals and plants have chromosome numbers between 10 and 50. Numbers above and below this are comparatively rare.

Chromosomes always exist in pairs; there are invariably two of each kind in the somatic cells of higher plants and animals. Thus the 46 chromosomes of man consist of two of each of 23 different kinds. They differ in

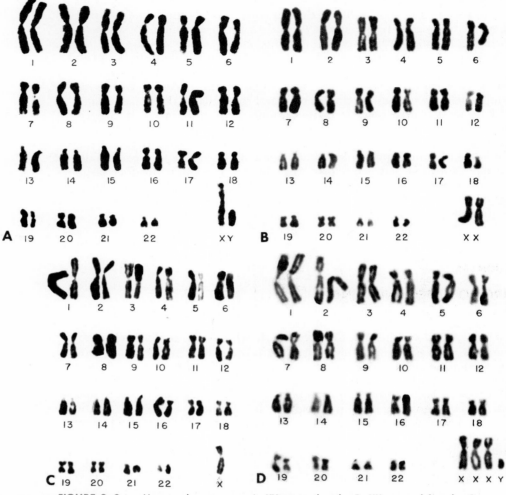

FIGURE 8–2   *Human chromosomes. A, XY, normal male; B, XX, normal female; C, XO condition—"gonadal dysgenesis" or Turner's syndrome; D, XXXY, an unusual example of Klinefelter's syndrome; the more usual individual with Klinefelter's syndrome has an XXY pattern of chromosomes. (Courtesy of Dr. Melvin Grumbach.)*

their length, shape, and the presence of knobs or constrictions along their length. In most species the chromosomes vary enough in these morphologic features so that cytologists can distinguish the different pairs.

SECTION 8–4

MITOSIS

The regularity of the process of cell division ensures that each daughter cell will receive exactly the same number and kind of chromosomes that the parent cell had. If a cell receives more or less than the proper number of chromosomes by some malfunctioning of the process of cell division, the resulting cell may show marked abnormalities and may be unable to survive. Although the chromosome appears to split longitudinally into two halves, the fact is that each original chromosome brings about the synthesis of an exact replica of itself immediately adjacent to it. The new chromosome is manufactured from raw materials present in the nucleus some time before the visible mitotic process begins. The old and new chromosomes are identical in structure and function and at first lie so close one to another that they appear to be one. As mitosis proceeds and the chromosomes contract, the line of cleavage between them becomes visible. In each human cell at the time of mitosis each of the 46 chromosomes has produced an exact replica of itself so that for a time there are 92 chromosomes in the cell. Then as cell division is completed, 46 go to one and 46 go to the other daughter cell. A rather complicated mechanism ensures an exactly equal division of the chromosomes between the two daughter cells.

The term mitosis in a strict sense refers to the division of the nucleus into two daughter nuclei and the term **cytokinesis** is applied to the division of the cytoplasm to form two daughter cells, each containing a daughter nucleus. Nuclear division and cytoplasmic division, although almost invariably well synchronized and coordinated, are separate and distinct processes. Each mitotic division is a continuous process, with each stage merging imperceptibly into the next one. For descriptive purposes mitosis may be divided into four stages: **prophase, metaphase, anaphase** and **telophase**, (Fig. 8–3). Between mitotic divisions a nucleus is said to be in the **interphase** or resting stage. The nucleus is "resting" only with respect to division, however, for during this time it may be very active metabolically. It is difficult to realize from a description, a diagram, or even from a prepared slide of cells undergoing mitosis (Fig. 8–4) just how active this process is. Time lapse movies of the process reveal graphically what an extremely active process mitosis is.

PROPHASE.   Prophase begins when the chromatin threads begin to condense and the chromosomes appear as a tangled mass of threads within the nucleus. Initially the chromosomes are stretched maximally and the individual chromomeres are clearly visible. In certain favorable instances the chromomeres differ enough in size and shape so that individual ones can be recognized. When the chromosomes subsequently contract, the chromomeres come to lie close together and individual ones can no longer be distinguished. Each chromosome had become doubled during the previous interphase, before mitosis began. Each part of the doubled

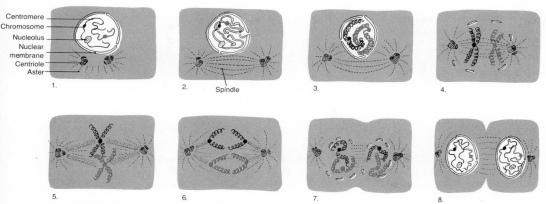

FIGURE 8–3   A diagram of the steps in mitotic division of a cell of an animal with a diploid number of four (haploid number = 2). 1, Resting stage; 2, early prophase: centriole divided; 3, and 4, later prophases; 5, metaphase; 6, and 7, early and late anaphases; 8, telophase: the nuclear membrane has reappeared and cytoplasmic division has begun.

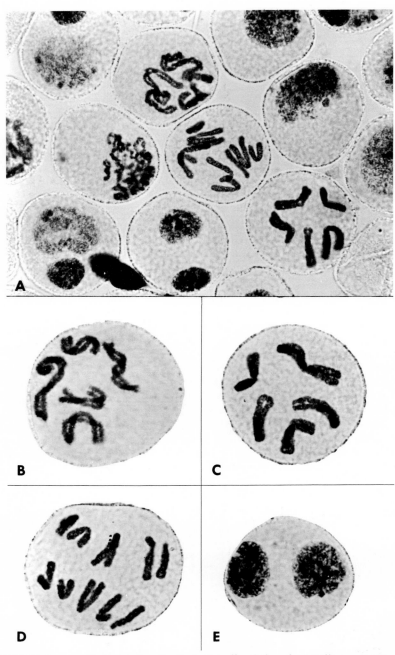

FIGURE 8–4   Photographs of mitosis in cells of the plant Trillium erectum. A, A field of normal microspore cells × 1400. B, Late prophase × 2300. C, Normal metaphase cell × 2300. D, Normal anaphase cell × 2300. E, Binucleate microspore (early) × 2000. (Courtesy of A. H. Sparrow and R. F. Smith.)

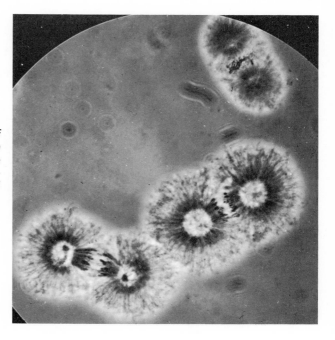

FIGURE 8–5   *Photomicrograph of isolated spindle fibers of dividing cells from a sea urchin embryo. A metaphase figure appears in the upper right and two anaphase figures below. (Courtesy of Dr. Daniel Mazia.)*

chromosome is called a **chromatid**; the two chromatids are held together at the centromere which remains single until the metaphase.

The cytoplasm of the cell contains a small granular structure called a **centriole**. At the beginning of prophase the centriole divides and the two daughter centrioles migrate to opposite sides of the cell. From each centriole there extends a cluster of raylike filaments called an **aster**. Between the separating centrioles a **spindle** forms, composed of protein with properties similar to those of the contractile proteins of muscle fibrils. The protein threads of the spindle are arranged like two cones placed together base to base, narrow at the ends or poles near the centrioles and broad at the center or equator of the cell. The spindle fibers stretch from equator to pole and constitute a definite structure; it is possible to introduce a fine needle into a cell and push the spindle around within the cell by means of a micromanipulator. Spindle fibers isolated from dividing cells (Fig. 8–5) contain protein, largely a single kind of protein, together with a small amount of RNA. While the centrioles have been separating and the spindle has been forming, the chromosomes in the nucleus have been contracting, getting shorter and thicker. Their double nature, which might not have been visible before, can now be clearly seen.

METAPHASE.   When the chromosomes have contracted fully and are short, dark-staining, rodlike bodies, the nuclear membrane disappears and the chromosomes line up across the equatorial plane of the spindle which has been forming around them (Fig. 8–3). Prophase is complete, and the short period during which the chromosomes are in the equatorial plane constitutes the metaphase. At this time the centromere divides and the two chromatids become completely separate **daughter chromosomes.** The division of the centromeres occurs simultaneously in all of the chromosomes, under the control of some as yet unknown mechanism. The daughter centromeres begin to move apart, marking the beginning of anaphase. In the division of human cells, prophase lasts from 30 to 60 minutes and metaphase from two to six minutes. The times vary considerably for different tissues and for different species.

ANAPHASE.   The chromosomes separate, one daughter chromosome going to each pole. The events from the time when the chromosomes first begin to move apart until they reach the poles constitute the anaphase, a period which lasts three to 15 minutes. The mechanism which moves the chromosomes to the poles is unknown. One theory suggests that the spindle fibers contract, in the presence of ATP, and pull the chromosomes to the poles. Spindles isolated from cells about to divide can be induced to contract when ATP is added. According to another theory, the spindle fibers act as guide rails along which the chromosomes glide in going to the poles. The motive power is believed to be supplied by substances which get between the daughter chromosomes, absorb water and swell, pushing the chromosomes apart. If it were not for the spindle fibers the chromosomes would be pushed at random in all directions, but with them all of one set of daughter chromosomes are gathered at one pole and all of the other set at the other pole. The chromosomes moving toward the poles usually assume a V shape with the centromere at the apex pointing toward the pole. It appears that whatever force moves the chromosome to the pole is applied at the centromere. Evidence from electron microscopy (Fig. 8–6) indicates that a spindle fiber attaches at the centromere. Chromosomes that lack a centromere, perhaps as a result of exposure to x-radiation, do not move at all in mitosis.

TELOPHASE.   When the chromosomes reach the poles the telophase begins. This period is roughly equal to the prophase and lasts 30 to 60 minutes. The chromosomes elongate and return to the resting condition in which only chromatin threads or granules are visible. A nuclear membrane forms around each daughter nucleus. This completes nuclear division **(karyokinesis),** and the division of the cell body **(cytokinesis)** follows. The division of animal cells is accomplished by a furrow which encircles the surface of the cell in the plane of the equator. The furrow gradually deepens and separates the cytoplasm into two daughter cells, each of which has a nucleus. In plants, division occurs by the formation of a **cell plate,** a partition which forms in the equatorial region of the spindle and grows laterally to the cell wall. The cell plate is secreted by the endoplasmic reticulum. Each daughter cell then forms a cell membrane on its side of the cell plate and the cellulose cell walls are finally formed on either side of the cell plate.

The frequency of mitosis varies widely in various tissues and in different species. In human red bone marrow, for example, where 10 million red blood cells are produced per second, 10 million mitoses must occur per second. In other tissues, such as those of the nervous system, mitoses occur very rarely. During the early development of an organism cell divisions take place extremely rapidly, every 30 minutes or so. The factors controlling the stage in development at which the mitotic rate slows and ceases are unknown. Cell divisions in the central nervous system largely cease in the first few months of life, whereas cell divisions in the red bone marrow, the lining of the digestive tract and the lining of the kidney tubules continue until the end of life.

CONTROL OF MITOSIS.   The factors that initiate mitosis are not known exactly, but the ratio between the volume of the nucleus and the volume of the cytoplasm appears to play a role. The increase in the size

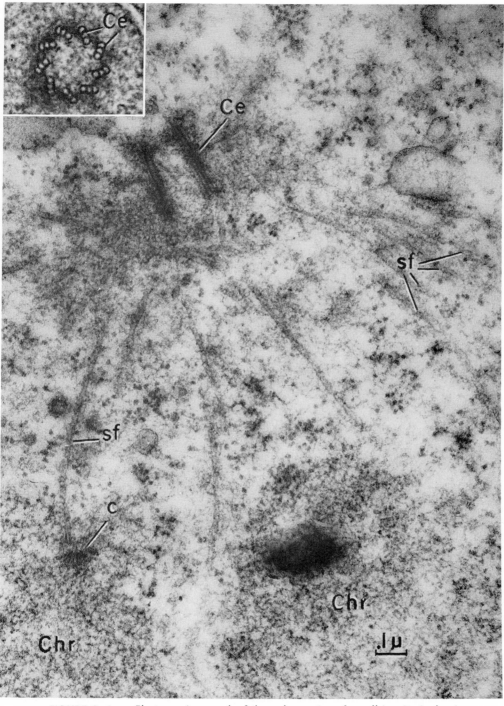

FIGURE 8–6   *Electron micrograph of the polar region of a cell in mitosis showing one of the centrioles (Ce). Notice the tubular aspect of the spindle fibers (sf) which converge upon the centriole. At the bottom of the figure are two chromosomes (Chr) and one centromere (c). × 80,000. Inset: a centriole cut transversely showing nine groups of three tubules each. (Courtesy of J. André.)*

of a cell involves the synthesis of proteins, nucleic acids, lipids and other cellular components. This requires the transport of substances back and forth through both nuclear and cell membranes. Although the volume of a sphere increases as the *cube* of its radius, the surface of a sphere increases only as the *square* of the radius. A cell is usually not a simple sphere, but comparable relationships hold and the volume of a cell increases more rapidly than the surface of the cell or nuclear membranes. At a critical point the surface of the nucleus becomes inadequate for the exchange of materials between nucleus and cytoplasm necessary to provide for further growth. The division of the cell greatly increases the surface of both nuclear and cell membranes without increasing the volume, and it is believed that this limiting factor in the nucleoplasmic ratio somehow initiates mitosis.

Some biologists have suggested that a hormone or hormone-like substance is involved in starting mitosis. The mitoses of various cells in a cleaving egg occur simultaneously for many cell divisions. Even in certain adult tissues mitosis may occur in waves, suggesting that some inductive substance is present. Haberlandt found evidence that dying cells may produce a substance that stimulates cell division. He cut a potato in half and examined the cut edges for mitoses. He found that if he cleaned off the cut edge few mitoses took place. If he did not clean off the cut edge, cell divisions were more frequent, and if he put a mash of cut-up potato cells on the cut edge many more occurred. From these experiments he concluded that cut-up and dying potato cells produce a "wound hormone" that stimulates cell division and the formation of scar tissue. Marshak and Walker prepared an extract of the nuclei of rat liver cells. This could be separated into two fractions, one of which increased and the other decreased the rate of division of liver cells.

BIOLOGICAL SIGNIFICANCE OF MITOSIS.   The process of mitosis ensures the precise and equal distribution of chromosomes to each of two daughter nuclei so that each cell in a multicellular organism has exactly the same number and kind of chromosomes as every other cell. The chromosomes contain genetic information coded in DNA and the regular and orderly mitotic process ensures that this genetic information is precisely distributed to each daughter nucleus; each cell has all the genetic information for every characteristic of the organism. Thus it is understandable why a single cell from a fully differentiated adult plant can, under suitable conditions in cell culture, develop into an entire plant. The mitotic process described above was for a diploid cell but the process is similar in haploid cells such as those of the gametophyte generation of plants.

SECTION 8–5

MEIOSIS

Constancy of the chromosome number in the cells of successive generations of organisms is ensured by the process of **meiosis**, which occurs during the formation of eggs or sperm, and in the formation of spores in plants. Meiosis is essentially a pair of cell divisions during which the chromosome number is reduced to one half, so that the gametes receive only half as many chromosomes as other cells in the body. When two gametes unite in fertilization, the fusion of their nuclei reconstitutes the diploid number of chromosomes. In meiosis the members of each pair of chromosomes separate and pass to different daughter cells. As a result of this, each gamete contains one and only one of each kind of chromosome; i.e., it contains one complete set of chromosomes. This is accomplished by the pairing or **synapsis** of the like chromosomes and a separation of the members of the pair, with one going to each pole. The like chromosomes which pair during meiosis are called **homologous chromosomes.** They are identical in size and shape, they have identical chromomeres along their lengths, and they contain similar hereditary factors or genes. A set of one of each kind of chromosome is called the **haploid** number; a set of two of each kind is called the **diploid** number. For man the haploid number is 23 and the diploid number is 46. Gametes—eggs and sperm—have the haploid number. Fertilized eggs (zygotes) and all the cells of the body developing from the zygote have the diploid number. A fertilized egg gets exactly half of its chromosomes and half of its genes from its mother and the other half from its father. Only the last two cell divisions, which result in mature, functional eggs or sperm, are meiotic; all other cell divisions are mitotic.

The process of meiosis consists of two cell divisions, the first and second meiotic divisions, which occur in succession (Figs. 8–7 and 8–8).

1a.

1b.

1c.

1d.

FIGURE 8–7 Diagrams illustrating the process of meiosis in an animal with a diploid number of four. 1a, Early prophase; 1b, later prophase, synapsis beginning; 1c, apparent doubling of the synapsed chromosomes to form tetrads; 1d, late prophase of first meiotic division. 2, Anaphase; and 3, telophase of first meiotic division. 4, Prophase of second meiotic division. 5, Metaphase; and 6, anaphase of second meiotic division. 7, Mature gametes, each of which contains the haploid number (two) of chromosomes, one of each kind of chromosome.

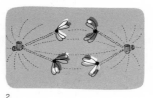

2.

3.

4.

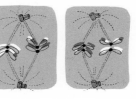

5.

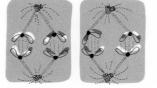

6.

7.

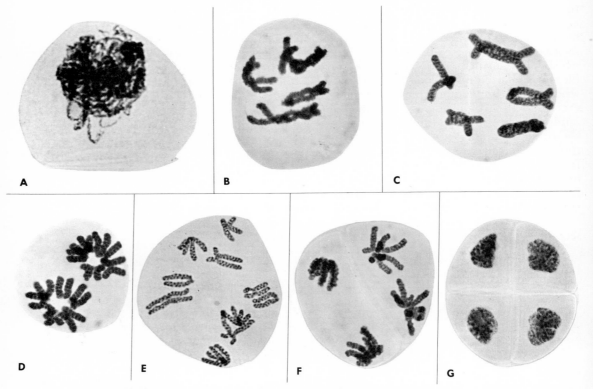

FIGURE 8–8 *Photographs of meiosis in the plant* Trillium erectum. *A, Early prophase of the first meiotic division. B, Later prophase of the first meiotic division. C, Metaphase. D, Anaphase of the first meiotic division. E, Metaphase of the second meiotic division. F, Anaphase of the second meiotic division. G, Four daughter cells. × 2000. (Courtesy of A. H. Sparrow and R. F. Smith.)*

Each of these includes prophase, metaphase, anaphase and telophase stages but there are important differences between mitosis and meiosis, especially in the prophase of the first meiotic division. The chromosomes first appear as long, thin threads and become shorter and thicker as in mitosis. The homologous chromosomes undergo synapsis while they are still elongate and thin. The homologous chromosomes pair longitudinally, coming to lie close together side by side along their entire length and twisted around each other. After synapsis or pairing has occurred, the chromosomes continue to shorten and thicken. Each one becomes visibly double, consisting of two chromatids as in mitosis. This doubling has occurred sometime before meiosis begins. At the end of the first meiotic prophase the chromosomes have doubled and undergone synapsis to yield a bundle of four homologous chromatids called a **tetrad.** Each pair of chromosomes gives rise to a bundle of four so the number of tetrads equals the haploid number of chromosomes. In human cells there are 23 tetrads (and a total of 92 chromatids) at this stage. The centromeres have not divided and there are only two centromeres for the four chromatids.

While these events are occurring, the centriole divides, the two centrioles go to opposite poles, a spindle forms between the centrioles, and the nuclear membrane dissolves. The tetrads line up around the equator of the spindle and the cell is said to be in metaphase. In the anaphase of the first meiotic division the daughter chromatids formed from each chromosome, still united by their centromere, separate and move toward opposite poles. Thus the homologous chromosomes of each pair, but not the daughter chromatids of each chromosome, are separated in anaphase I. This differs from mitotic anaphase in which the centromeres do divide and the daughter chromatids pass to opposite poles.

In the telophase of the first meiotic division in man there are 23 double chromosomes at each pole. Cytoplasmic division follows, but in most animals and plants there is no clear interphase between the two meiotic divisions. The chromosomes do not divide into daughter chromatids and there is no synthesis of DNA, as there is in the interphase between mitotic divisions. The chromosomes do not form chromatin threads; instead the centriole divides again, a new spindle forms in each cell (at right angles to the spindle of the first division), and the haploid number of double chromosomes lines up on the equator of the spindle. The telophase of the first meiotic division and the prophase of the second meiotic division are usually of rather short duration. The lining up of the double chromosomes on the equator of the spindle constitutes the metaphase of the second meiotic division. The metaphases of the first and second meiotic divisions can be distinguished, because in the first the chromosomes are arranged in bundles of four and in the second the chromosomes are arranged in bundles of two. No further splitting or doubling of chromosomes occurs in the second meiotic division; the centromeres divide and the daughter chromatids, now chromosomes, separate and move to opposite poles. Thus in the telophase of the second meiotic division in man, 23 chromosomes, one of each kind, arrive at each pole. The cytoplasm then divides, a nuclear membrane forms, and the chromosomes gradually elongate and become chromatin threads.

The two successive meiotic divisions yield four nuclei, each of which has one and only one of each kind of chromosome, a haploid set. The members of the homologous pairs of chromosomes are segregated into separate daughter cells. The four cells resulting from the two meiotic divisions are now mature gametes and do not undergo any further mitotic or meiotic divisions.

Essentially the same process occurs in the meiotic divisions in the testis which result in sperm and meiotic divisions in the ovary which result in eggs, but there are some differences in detail.

SECTION 8–6

SPERMATOGENESIS

The testis is made up of thousands of cylindrical sperm tubules, in each of which millions of sperm develop. The walls of these tubules are lined with primitive, unspecialized germ cells, **spermatogonia.** Throughout embryonic development and during childhood the spermatogonia divide mitotically, giving rise to additional spermatogonia to provide for the growth of the testis. After sexual maturity, some of the spermatogonia undergo **spermatogenesis,** the series of changes which results in mature sperm. Other spermatogonia continue to divide mitotically and produce more spermatogonia for later spermatogenesis. In most wild animals there is a definite breeding season, either in spring or fall, during which the testis increases in size and spermatogenesis occurs. Between breeding seasons the testis is small and contains only spermatogonia. In man and most domestic animals, spermatogenesis occurs throughout the year once sexual maturity is reached.

Spermatogenesis begins with the growth of the spermatogonia into

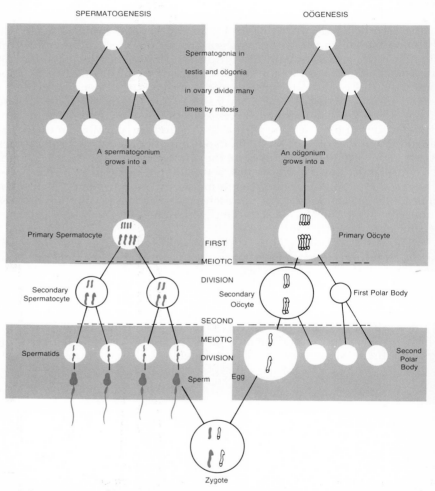

FIGURE 8-9   *Diagrammatic comparison of the formation of sperm and eggs in a hypothetical animal with a haploid number of two.*

larger cells known as **primary spermatocytes** (Fig. 8–9). These divide (first meiotic division) into two equal-sized cells, the **secondary spermatocytes**, which in turn undergo a second meiotic division to form four equal-sized **spermatids**. The spermatid, a spherical cell with a generous amount of cytoplasm, is a mature gamete with the haploid number of chromosomes. A complicated process of growth and change (though not cell division) converts the spermatid into a functional sperm. The nucleus shrinks in size and becomes the head of the sperm (Fig. 8–10), while the sperm sheds most of its cytoplasm. Some of the Golgi bodies congregate at the front end of the sperm and form a point (the **acrosome**), which may aid the sperm in puncturing the egg cell membrane.

The two centrioles of the spermatid move to a position just in back of the nucleus. A small depression appears on the surface of the nucleus and one of the centrioles, the proximal centriole, takes up a position in the depression at right angles to the axis of the sperm. The second or distal centriole, just behind the proximal centriole, gives rise to the **axial filament** of the sperm tail (Fig. 8–11). Like the axial filament of flagella, it consists of two longitudinal fibers in the middle and a ring of nine pairs or doublets of longitudinal fibers surrounding the two.

The mitochondria move to the point where head and tail meet and form a small middle piece which provides energy for the beating of the tail. Most of the cytoplasm of the spermatid is discarded; only a thin sheath remains, surrounding the mitochondria in the middle piece and the axial filament of the tail.

The spermatozoa of various animal species may be quite different. Nearly all sperm have a tail, but there are great variations in its size and shape, as well as in the characteristics of the head and middle piece (Fig. 8–12). The sperm of a few animals, e.g., the parasitic roundworm *Ascaris*,

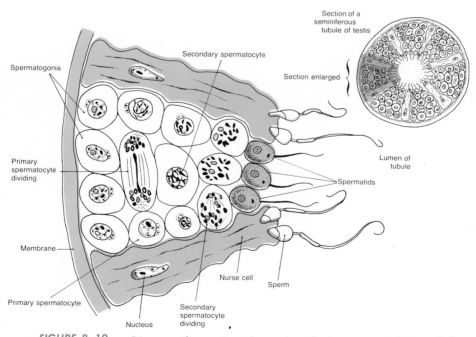

FIGURE 8–10   *Diagram of a portion of a section of a human seminiferous tubule to show the stages in spermatogenesis and in the transformation of a spermatid into a mature sperm.*

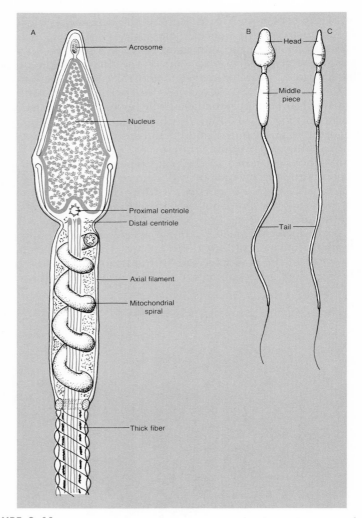

FIGURE 8-11   Diagram of the head and middle piece of a mammalian sperm greatly enlarged, as seen in the electron microscope. B, and C, Top and side views of a sperm seen by light microscopy.

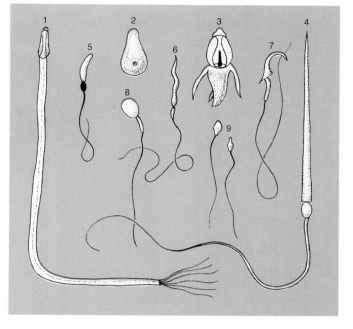

FIGURE 8–12 *Spermatozoa from different species of animals illustrating the wide differences in size and shape that may be observed. 1, Gastropod. 2, Ascaris. 3, Hermit crab. 4, Salamander. 5, Frog. 6, Chicken. 7, Rat. 8, Sheep. 9, Man.*

have no tail and move instead by ameboid motion. Crabs and lobsters have a curious tailless sperm with three pointed projections on the head, which stick to the surface of the egg, holding the sperm securely in place. The middle piece uncoils like a spring, and pushes the nucleus of the sperm into the egg cytoplasm, thus accomplishing fertilization.

SECTION 8–7

OÖGENESIS

The ova or eggs develop in the ovary from immature sex cells, oögonia. Early in development the oögonia undergo many successive mitotic divisions to form additional oögonia, all of which have the diploid number of chromosomes. In many animals, notably the vertebrates, the oögonia and oöcytes are surrounded by a layer of follicle cells derived from the germinal epithelium of the ovaries. In the human this occurs early in fetal development and by the third month the oögonia begin to develop into **primary oöcytes** (Fig. 8–13). When a human female is born, her two ovaries contain some 400,000 primary oöcytes, which have attained the prophase of the first meiotic division. These primary oöcytes remain in prophase for many years, until the woman reaches sexual maturity. Then as each follicle matures the first meiotic division resumes and is completed at about the time of ovulation (15 to 45 years after meiosis began!).

The events occurring in the nucleus—synapsis, the formation of tetrads and the separation of the homologous chromosomes—are the same as in spermatogenesis, but the division of the cytoplasm is unequal, resulting in one large cell, the **secondary oöcyte,** which contains the **yolk** and nearly all the cytoplasm, and one small cell, the first **polar body,** which consists of practically nothing but a nucleus (Fig. 8–9). It was named a polar

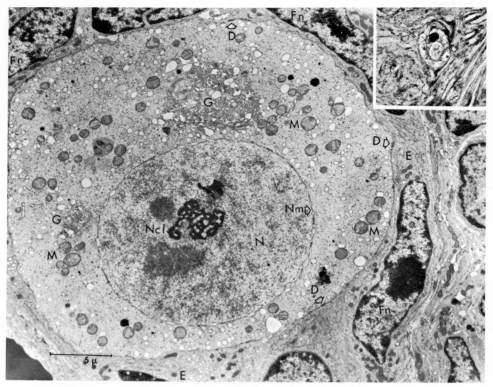

FIGURE 8–13   Electron micrograph of a young oöcyte of a guinea pig showing
Golgi material. Inset: Oöcyte in a similar stage, as seen by light microscopy. N, Nucleus;
Ncl, nucleolus; Nm, nuclear membrane; G, Golgi material; M, mitochondria; D, desmo-
somes, denser material connecting plasma membranes of oöcyte and follicle cells; Fn,
nuclei of follicle cells; E, endoplasmic reticulum in follicle cells. (Courtesy of Professor E.
Anderson. In Balinsky, B. I.: An Introduction to Embryology. Philadelphia, W. B. Saunders
Company, 1970.)

body before its significance was understood, because it appeared as a small
speck at the animal pole of the egg.

In the second meiotic division, which proceeds as the ovum enters the
fallopian tube, the secondary oöcyte divides again unequally into a large
oötid and a small second polar body, both of which have the haploid chromosome number. The first polar body may divide into two additional
second polar bodies. The oötid then becomes a mature ovum. The three
small polar bodies soon disintegrate, so that each primary oöcyte gives
rise to just one ovum, in contrast to the four sperm formed from each
primary spermatocyte. The unequal cytoplasmic division ensures that the
mature egg will have enough cytoplasm and stored yolk to survive, if fertilized. The primary oöcyte in a sense puts all its yolk in one ovum; the egg
has neatly solved the problem of reducing its chromosome number without
losing the cytoplasm and yolk needed for development after fertilization.

The union of one haploid set of chromosomes from the sperm with
another haploid set from the egg, which occurs in **fertilization,** reestab-
lishes the diploid chromosome number. Thus the fertilized egg or zygote,
and all the body cells which develop from it by mitosis, have the diploid
number of chromosomes. Each individual gets exactly half of his chromosomes and half of his genes from his mother and half from his father. Because of the nature of gene interaction, the offspring may resemble one
parent more than the other, but the two parents make equal contributions
to his inheritance.

SECTION 8-8

GENES AND ALLELES

The laws of heredity follow directly from the behavior of the chromosomes in mitosis, meiosis and fertilization. Within each chromosome are numerous hereditary factors or genes, each of which differs from the others and each of which controls the inheritance of one or more characteristics. Since the **genes** are located in the chromosomes and each cell has two of each kind of chromosome, it follows that each cell has two of each kind of gene. As the chromosomes separate during meiosis and recombine in fertilization, so, of course, must the paired genes separate and recombine. Each chromosome behaves genetically as though it were composed of a string of genes arranged in a linear order. The members of a homologous pair of chromosomes have genes arranged in similar order. The gene for each trait occurs at a particular point in the chromosome called a **locus.** When the chromosomes undergo synapsis during meiosis, the homologous chromosomes become attached point by point and presumably gene by gene.

The inheritance of any trait can be studied only when there are two contrasting conditions, such as the tall and short pea plants studied by Mendel, or brown and blue eye color in man, or brown versus black coat color in mice. Any individual may exhibit one or the other but not both of such contrasting conditions, which were originally termed *allelomorphic traits* or *alleles.* More recently the terms "gene" and "allele" have been used interchangeably. The gene **B** for brown eyes is said to be an allele of gene **b** for blue eyes. This usage of the term allele emphasizes that there are two (or more) alternative kinds of genes located at a specific point (locus) in the chromosome.

SECTION 8-9

MENDEL'S LAW OF SEGREGATION

Mendel began with one variety of tall pea plants and a second variety of short pea plants. He placed pollen from the male stamens of the tall variety on the female pistils of short plants. He also carried out the reciprocal cross, placing pollen from the stamens of short plants on the pistils of tall plants. In either cross each of the plants produced by these cross-fertilized flowers was tall. Mendel allowed the F_1 plants to undergo self-fertilization by letting pollen from one flower pass to the pistil of the same flower. When the resulting seeds were planted the offspring included 787 tall and 277 short plants.

When Mendel allowed the F_2 short plants to undergo self-fertilization, the resulting offspring were all short plants. When the tall plants were self-fertilized, roughly one third of them gave rise to seeds which produced only tall plants. The remaining two thirds of the tall plants when self-fertilized gave rise to seeds which produced tall and short plants in a ratio of three tall to one short. Thus one third of the tall plants in the F_2 generation were like the tall plants in the parental generation and two thirds were like the tall plants in the F_1 generation. Mendel showed that when tall plants are bred with short ones the gene for shortness is not expressed in the F_1 generation, but is expressed in a certain fraction of the F_2 generation of plants. To explain these observations Mendel suggested that: (1) Each pea plant contains a pair of genes which control the appearance of each characteristic, (2) Each plant receives one gene for each characteristic from its paternal parent and another from its maternal parent, and (3) Each plant transmits these genes as discrete, unchanging units to its progeny. Thus the short plants in the F_2 generation were just as short as the parental generation although in the intervening F_1 generation all the pea plants were tall. When eggs or sperm are formed, the pairs of genes separate during meiosis and are distributed, again as discrete units, to each gamete. This is usually referred to as Mendel's First Law, the *Law of Segregation.*

As another example of Mendel's First Law, let us consider the mating of a pure brown-colored male mouse (**bb**) and a pure black-colored female mouse (**BB**). During meiosis in the testes of the male, the two **bb** genes separate so that each sperm has only one **b** gene. In the formation of ova in the female the **BB** genes separate so that each ovum has only one **B** gene. The fertilization of the **B** egg by a **b** sperm produces offspring with the genetic formula **Bb**. These mice contain one gene for brown coat and one for black coat. What color would you expect them to be—dark brown, gray, or possibly brown with black spots? Like the gene for tallness in peas, the gene for black coat color is said to be *dominant* to the gene for brown coat. It will produce black body color even when only one dose of the black gene is present. The brown gene is said to be *recessive* to the black one and will produce brown coat color only when a double dose of the gene (**bb**) is present. Recessive genes may be defined as ones which will produce their effects only when an individual has two of them that are identical. Dominant genes, in contrast, are those which will produce their effects even when only one is present in an individual. The phenomenon of dominance supplies part of the explanation as to why an individual may resemble one of his two parents more than the other, despite the fact that both make equal contributions to his genetic constitution.

SECTION 8–10

HOMOZYGOUS AND HETEROZYGOUS ORGANISMS

An animal or plant with two genes exactly alike, two black (**BB**) or two brown (**bb**), is termed *homozygous* for the trait. An organism with one dominant and one recessive (**Bb**) is said to be *heterozygous* or hybrid. Using these terms we can now improve our definitions of dominant and recessive genes: A recessive gene is one which will produce its effect only when homozygous; a dominant gene is one which will produce its effect whether it is homozygous or heterozygous. When the F_1 black mice form gametes, the chromosome containing the **B** gene first synapses (pairs) with, and then separates from, the chromosome containing the **b** gene and the resulting sperm or egg has either a **B** gene or a **b** gene but never both. It follows that sperm and eggs containing **B** genes and ones with **b** genes are formed in equal numbers by heterozygous **Bb** mice. There are two types of eggs, two types of sperm and four possible combinations of egg and sperm in fertilization (Fig. 8–14). There is no special attraction nor repulsion between an egg and sperm that contain the same kind of gene, hence these four possible combinations are equally probable. The combinations of eggs and sperm can be determined by algebraic multiplication: ($\frac{1}{2}$ **B** + $\frac{1}{2}$**b**) eggs × ($\frac{1}{2}$**B** + $\frac{1}{2}$**b**) sperm.

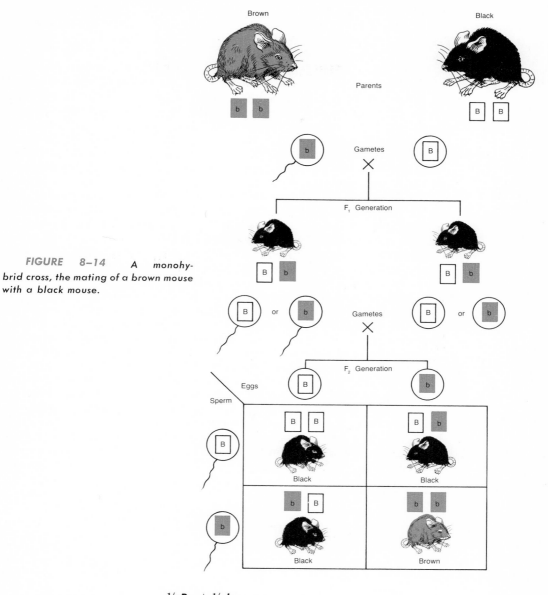

FIGURE   8-14   A monohybrid cross, the mating of a brown mouse with a black mouse.

$$\frac{1}{2}B \; + \frac{1}{2}b$$
$$\times \; \frac{1}{2}B \; + \frac{1}{2}b$$
$$\overline{\frac{1}{4}BB + \frac{1}{4}Bb}$$
$$\underline{\qquad\quad \frac{1}{4}Bb + \frac{1}{4}bb}$$
$$\frac{1}{4}BB + \frac{1}{2}Bb + \frac{1}{4}bb$$

You can also represent the possible combinations of eggs and sperm in a "checkerboard" or Punnett square, with the types of eggs represented along the top and the types of sperm indicated along the left side (Fig. 8–14). The squares are then filled in to show the combinations of genes in the offspring. Three fourths of all offspring will be **BB** or **Bb** and have black coat color, whereas one fourth will be **bb** and have brown coat color. This genetic mechanism originally postulated by Mendel to explain the results of his pea breeding experiments has been proved by many experiments to be generally true for breeding behavior in all kinds of animals and plants.

SECTION 8–11

PHENOTYPE AND GENOTYPE

The appearance of any individual with respect to a given inherited trait is known as its **phenotype.** The individual's genetic constitution, usually expressed in symbols, is called its **genotype.** In the cross of the mice the F_2 generation contains three black-coated mice to every one brown-coated mouse and, therefore, the phenotypic ratio is 3 black : 1 brown, whereas the genotypic ratio is 1 **BB** : 2 **Bb** : 1 **bb.** The phenotype may be some morphologic characteristic, such as color, size or shape, or a physiological characteristic, such as the presence or absence of a specific enzyme required to metabolize a specific substrate.

One third of the black mice in the F_2 generation of the mating of a black × brown are homozygous, **BB,** and the other two thirds are heterozygous, **Bb.** The situation is analogous to the tall pea plants in the F_2 generation, one third of which produced only tall plants when self-pollinated and two thirds of which produced a mixture of three tall to one short plants when self-fertilized. Since animals cannot, in general, be self-fertilized, how do you think the geneticist can distinguish the homozygous (**BB**) and heterozygous (**Bb**) black mice? He does this by a **test cross,** by mating each black mouse with homozygous brown (**bb**) mice. If all the resulting offspring are black, what inference would you make about the genotype of the black parent? If any of the offspring are brown, what conclusion would you draw regarding the genotype of the black parent? Can you be more certain about either of these two inferences?

Mendel did just such experiments and bred heterozygous tall pea plants with homozygous short ones. He predicted that the heterozygous parent would produce equal numbers of **T** and **t** gametes, whereas the homozygous short parent would produce only gametes containing **t**. This should lead to equal numbers of tall (**Tt**) and short (**tt**) individuals among the progeny. Thus, as a good hypothesis should, Mendel's hypothesis not only explained the known facts but enabled him to predict the results of other experiments.

SECTION 8–12

CALCULATING THE PROBABILITY OF GENETIC EVENTS

All genetic ratios are properly expressed in terms of probabilities. In the examples just discussed we stated that the offspring of the mating of two individuals heterozygous for the same gene pair would appear in the ratio of three with the dominant trait and one with the recessive trait. If the number of offspring is large enough this ratio will be very closely approximated, as Mendel's experiments demonstrated (Table 8–1). However, if the number of offspring is small, the ratio of the two types may be quite different from the expected 3 : 1. Why should this be? If there are only four offspring, any distribution from all four with the dominant trait to all four with the recessive trait might be found, although the latter would occur only very rarely. A better statement is that there are three chances in four ($3/4$) that any particular offspring of two heterozygous individuals will show the dominant trait and one chance in four ($1/4$) that it will show the recessive trait.

SECTION 8–13

PROBABILITIES: THE PRODUCT LAW

To understand this very important concept in genetics we must have a look at some of the principles of probability. One of the basic laws of probability, called the **Product Law,** states that the probability of two independent events occurring together is the product of the probabilities of each occurring separately. Using this principle we can calculate how frequently in a large series of matings, if each mating results in exactly four offspring, all four would show the recessive trait. How often in the mating of two heterozygous black mice, **Bb,** would one expect to get a litter of four brown mice? The probability that the first one will be brown is $1/4$. The probability that the second one will be brown is also $1/4$. The fertilization of each egg by a sperm is an independent event and we use the Product Law to calculate the combined probability of two (or more) events occurring together. Thus the probability that all four of the offspring of any given mating will be brown is $1/4 \times 1/4 \times 1/4 \times 1/4$ or $1/256$. In other words, there is one chance in 256 that all four mice will have brown coat color!

The probability that any given offspring will show the dominant trait, black coat color, is $3/4$. We can, in a similar fashion, calculate the probability that all four mice in the litter will be black. This is $3/4 \times 3/4 \times 3/4 \times 3/4$, or there are 81 chances in 256 that all four mice will have black coat color. You must not assume that if you actually mated 256 pairs of heterozygous mice and each pair had four offspring that you would be *guaranteed* to get one litter of four brown-coated offspring. A great many people have lost vast sums of money in gambling by making a similar mistake! If you have made 255 such matings without getting a set of four brown offspring, what is the probability of getting four brown mice on the 256th try? You might be misled into thinking it was bound to happen, but in fact there is still only one chance in 256 that it will occur, since each of these matings is an independent event.

How many different ways are there of getting three black mice and one brown mouse in a given mating? A look at Figure 8–15 shows that there are four ways: The first to be born could be brown and the next three would be black. Or, the second one born could be brown and the first, third and fourth black. The other two possibilities would be that the brown one would be the third or the fourth to be born. To calculate the probability that three of the four offspring will be black and one brown we must multiply the number of possible combinations by the probability of each type. There are $4 \times 3/4 \times 3/4 \times 3/4 \times 1/4$ or $108/256$ chances that there will be three black mice and one brown mouse in a litter of four. It may be surprising at first that the 3 : 1 ratio is actually obtained in less than half of the total number of litters of four. However, when we add up the total numbers of black and brown offspring from a large number of matings, the 3 : 1 ratio is more and more closely approximated.

How many different ways are there of getting two black and two brown mice in a litter of four? There are six ways, as shown in Figure 8–15. From this you can calculate the probability that two of the four offspring will be black and two brown is $6 \times 3/4 \times 3/4 \times 1/4 \times 1/4$, or $54/256$. You can see that there are four different combinations that give one black and three brown. The first born may be black and the rest brown, or the second will be black and all others brown, or the third or the fourth will be black. The probability of getting one black mouse and three brown mice is $4 \times 3/4 \times 1/4 \times 1/4 \times 1/4$ or $12/256$.

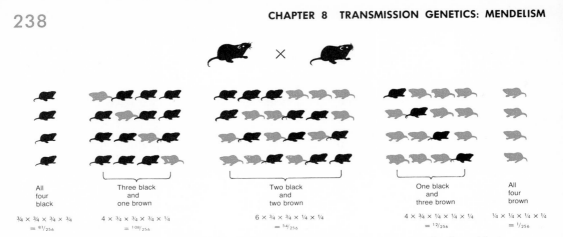

All four black	Three black and one brown	Two black and two brown	One black and three brown	All four brown
$\frac{3}{4} \times \frac{3}{4} \times \frac{3}{4} \times \frac{3}{4}$	$4 \times \frac{3}{4} \times \frac{3}{4} \times \frac{3}{4} \times \frac{1}{4}$	$6 \times \frac{3}{4} \times \frac{3}{4} \times \frac{1}{4} \times \frac{1}{4}$	$4 \times \frac{3}{4} \times \frac{1}{4} \times \frac{1}{4} \times \frac{1}{4}$	$\frac{1}{4} \times \frac{1}{4} \times \frac{1}{4} \times \frac{1}{4}$
$= \frac{81}{256}$	$= \frac{108}{256}$	$= \frac{54}{256}$	$= \frac{12}{256}$	$= \frac{1}{256}$

FIGURE 8–15   *The various ways in which one might obtain exactly three black mice and one brown mouse in a litter of four.*

All probabilities are expressed as fractions ranging from zero, expressing an impossibility, to one, expressing a certainty. Probabilities can be multiplied or added just like any other fraction. In this example there are no other possibilities; the four mice must be either four black, or three black and one brown, or two black and two brown, or one black and three brown, or four brown. The sum of the five probabilities will add up to one, $\frac{81}{256} + \frac{108}{256} + \frac{54}{256} + \frac{12}{256} + \frac{1}{256} = 1$.

The color of the iris of the human eye is inherited by several pairs of genes, but one pair is the primary factor differentiating brown eye color from blue. The gene for brown eye color, **B**, is dominant to the gene for blue, **b**. If two heterozygous brown-eyed people marry, what is the probability that they will have a blue-eyed child? Clearly there is one chance in four that any child of theirs will have blue eyes. Each mating is a separate, independent event and its result is not affected by the results of any previous matings. If these two brown-eyed parents have had three brown-eyed children and are expecting their fourth child, what is the probability that it will have blue eyes? Again, the unwary might guess that this one must have blue eyes, but in fact there is still only one chance in four of his having blue eyes and three chances in four that he will have brown eyes.

All genetic events are governed by the laws of probability. The prediction of any single event—e.g., predicting the characteristics of any single child—is highly uncertain; however, if the number of events is large enough, the laws of probability provide a reasonable prediction of the fraction of these events which will be of one type or the other.

The operation of the laws of probability may be demonstrated by tossing a coin or by rolling dice. The probability (usually expressed as **p**) of obtaining a head in any single toss of a coin is one chance out of two, expressed as $\frac{1}{2}$. The coin has two sides, each of which is equally likely to be up. Since dice have six sides, the probability that any one of these sides will come up in a given throw is one out of six or $\frac{1}{6}$. It is customary to express probabilities as a fraction obtained by dividing the number of "favorable events" by the number of total possible events. If you are engaged in some game of chance in which the favorable event involves turning up a three on a single die, the probability of this favorable event is one in six. What is the probability of drawing a heart from a full deck of cards? What is the probability of drawing an ace? What is the probability of drawing the ace of hearts? What is the probability of rolling a seven with a pair of dice?

The type of probability we have been discussing is called "***a priori***" *probability*, because the probability can be specified *in advance* from the nature of the event. You know the probability of getting heads or tails when flipping a coin simply from the nature of the act. These probabilities are independent of whether or not the coin is actually flipped.

The Product Rule of probability holds not only for two but for three or more independent events. We can calculate the probability of obtaining a head on the first toss, followed by a head on the second toss, followed by a head on the third toss, by multiplying $\frac{1}{2} \times \frac{1}{2} \times \frac{1}{2}$ to get $\frac{1}{8}$. In a similar way we can calculate the probability of choosing at random from a crowd an individual who is male, has blood group A and was born in September, if we know each of the individual probabilities. This would be approximately $0.5 \times 0.4 \times 0.084$, which is equal to 0.0168.

SECTION 8–14

PROBABILITIES: THE SUM RULE

A Second Law of Probability, the **Sum Rule**, states that the probability of one or another of two mutually exclusive events occurring is the sum of their separate probabilities. For example, in rolling dice the probability that the die will come up *either* two or five is $\frac{1}{6} + \frac{1}{6} = \frac{1}{3}$. This rule was used when we added the probabilities of getting all four black + three black and one brown + two black, two brown and so on and found that the sum of the probabilities of these five different, mutually exclusive, possibilities was one.

SECTION 8–15

THE LAW OF INDEPENDENT ASSORTMENT: DIHYBRID CROSSES

Exactly the same principles and procedures are involved in solving problems when one is concerned with the inheritance of two or more traits in the same group of individuals. When two pairs of genes are located in different, nonhomologous chromosomes, each pair is inherited independently of the other pair. In other words, the members of one pair separate during meiosis independently of the other pair. Mendel tested this by crossing pea plants differing in two characteristics. He cross-pollinated a pure breeding strain of pea plants bearing round seeds and yellow cotyledons with a pure breeding strain of pea plants with wrinkled seeds and green cotyledons.

All the offspring of this mating had the two dominant characteristics — round seeds and yellow cotyledons. When these yellow plants bearing round seeds were self-fertilized, they produced an F_2 generation in which round seeds and wrinkled seeds appeared in a ratio of $3 : 1$ and yellow cotyledons and green cotyledons appeared in a ratio of $3 : 1$. When we consider the inheritance of the two traits simultaneously (Fig. 8–16), we can see that nine of the 16 possible combinations in the F_2 generation are plants with round seeds and yellow cotyledons. Three of the 16 possible combinations are plants bearing round seeds and green cotyledons. Another three are plants with wrinkled seeds and yellow cotyledons. Finally there is one chance in 16 of getting a pea plant with wrinkled seeds and green cotyledons. Thus the segregation of the **Rr** genes for round and wrinkled seeds occurs independently of the segregation of the **Yy** genes for

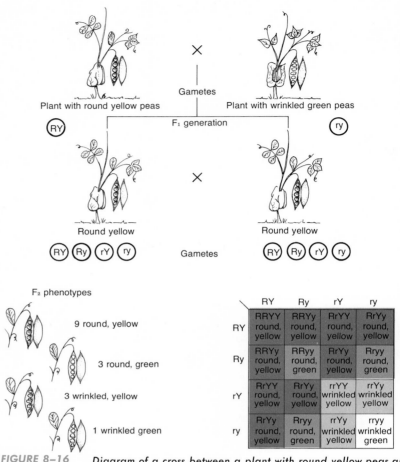

FIGURE 8–16   *Diagram of a cross between a plant with round yellow peas and a plant with wrinkled green peas, illustrating the principle of independent assortment of genes lying on different chromosomes.*

yellow and green cotyledons. Homologous pairs of chromosomes separate during meiosis, each pair independently of the other. In the F_1 generation, four kinds of gametes are produced in equal numbers: **RY**, **Ry**, **rY**, **ry**. The random mating of the four kinds of sperm with the four kinds of eggs produces the 16 possible combinations of offspring. This type of mating, involving individuals differing in two traits, is called a **dihybrid cross**.

Similarly, when one of a strain of mice with black short hair is mated with a mouse of another strain with brown long hair, the offspring all have black short hair (Fig. 8–17). Which is dominant, the gene for black or brown hair color? Is long hair dominant or recessive? The genotype of the F_1 individuals is **Bb Ss**; they are heterozygous for the two kinds of genes. These F_1 mice produce four types of gametes, **BS**, **Bs**, **bS**, **bs**. The random union of eggs and sperm in fertilization results in 16 possible combinations — nine mice with black short hair, three with black long hair, three with brown short hair, and one with brown long hair. This 9 : 3 : 3 : 1 ratio is characteristic of the second generation of a mating of two individuals that differ in two characters whose genes are located in nonhomologous chromosomes. The mating of such individuals illustrates Mendel's Second Law, the **Law of Independent Segregation**. This states that the members of

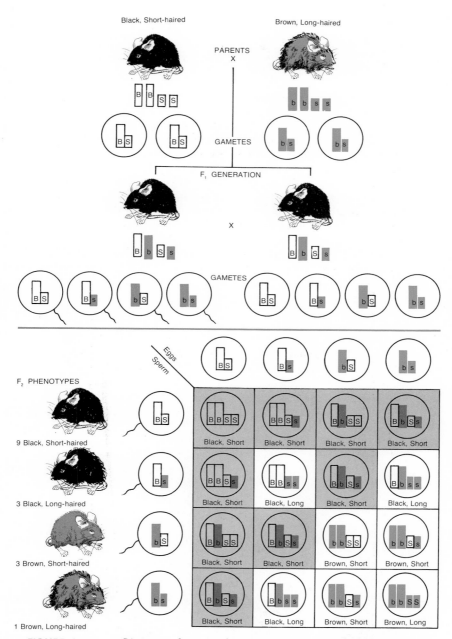

FIGURE 8–17   Diagram of a cross between a black, short-haired mouse and a brown, long-haired mouse, illustrating independent assortment in a dihybrid cross.

one pair of genes separate or segregate from each other in meiosis independently of the members of other pairs of genes and come to be assorted at random in the resulting gamete. This law does not apply if the two pairs of genes are located in the same pair of chromosomes.

SECTION 8–16

INCOMPLETE DOMINANCE

From studies of the inheritance of many traits in a wide variety of organisms it is clear that one member of a pair of genes may not be completely dominant to the other. Indeed it may be improper to use the words "dominant" and "recessive" in such instances. For example, red and white are common flower colors in Japanese Four O'Clocks. Each color breeds true when these plants are self-pollinated. What flower color might you expect in the offspring of a cross between a red flowering plant and one that bears white flowers? Without knowing which is dominant you might predict that all would have red flowers or all would have white flowers. This cross was first made by the German botanist, Karl Correns, who found that all the F_1 offspring have pink flowers! How can we explain that? Does this in any way prove that Mendel's assumptions about inheritance are wrong? Quite the contrary, for when two of these pink-flowered plants were

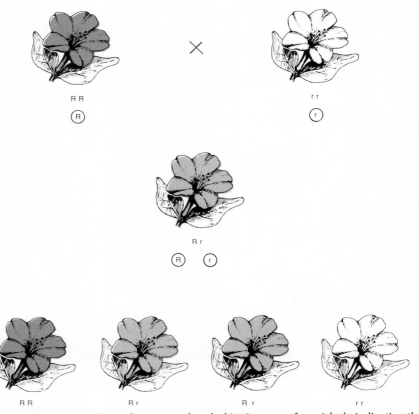

FIGURE 8–18   *A cross between red and white Japanese four o'clocks indicating the principle of incomplete dominance or codominance.*

crossed, offspring appeared in the ratio of one red-flowered to two pink-flowered to one white-flowered plant (Fig. 8–18). In this instance, as in other aspects of science, the finding of results that differ from those predicted simply prompts the scientist to reexamine and modify his assumptions to account for the new exceptional results. The pink-flowered plants are clearly the heterozygous individuals and neither the red gene nor the white gene is completely dominant. When the heterozygote has a phenotype which is intermediate between those of its two parents, the genes are said to show incomplete dominance or be **codominant**. In these crosses the genotypic and phenotypic ratios are identical.

Incomplete dominance is not unique to Japanese Four O'Clocks. Red- and white-flowered sweet pea plants also produce pink-flowered plants when crossed. In both cattle and horses reddish coat color is incompletely dominant to white coat color. The heterozygous individuals have roan-colored coats. If you saw a white mare nursing a roan-colored colt, what would you guess was the coat color of the colt's father? Is there more than one possible answer?

SECTION 8–17

GENETIC CARRIERS OF DISEASES

Careful investigation of the phenotypes of many gene pairs has revealed slight differences between the homozygous dominant and the heterozygous individual. In man many inherited diseases are transmitted by recessive genes, and it is important to be able to distinguish the homozygous normal individual from the heterozygous individual who may superficially appear to be normal but is a **carrier** for the trait. The mating of two carriers, two heterozygous individuals, would provide one chance in four for the appearance of a homozygous recessive individual showing the inherited disease. The human trait, **sickle cell anemia**, is inherited by a single pair of genes, but an individual must have two such genes to show the anemia. Thus the gene for sickle cell anemia may be said to be recessive to the normal gene. The red cells of a person with sickle cell anemia are shaped like a sickle or a half moon, whereas normal red cells are biconcave discs. The sickle cells contain hemoglobin molecules with a slightly different molecular structure from that found in normal red cells. The hemoglobin molecules in an individual with sickle cell anemia have the amino acid valine instead of glutamic acid at position 6, the sixth amino acid from the amino terminal end, in the β chain. The substitution of an amino acid with an uncharged side chain (valine) for one with a charged side chain (glutamate) makes the hemoglobin less soluble, and it tends to form crystals which break the red cell. The red cells of sickle cell anemics are more fragile than those of normal individuals. Individuals who are heterozygous for the sickling gene have a mixture of normal and abnormal hemoglobins in their red cells; about 45 per cent of the total hemoglobin has the abnormal chemical constitution. Their red cells do not usually undergo sickling but can be made to do so when the amount of oxygen in the blood is reduced. They are said to have sickle cell "trait." This provides a very simple test, using only a drop of blood, by which the heterozygote can be distinguished from the homozygous normal.

The human disease, **phenylketonuria**, is also inherited by a single pair of genes. The homozygous recessive individual lacks the enzyme needed to convert one amino acid, phenylalanine, to another, tyrosine. The phenyl-

alanine accumulates in his tissues and causes retarded mental development. The heterozygote appears perfectly normal, but when given a standard amount of phenylalanine in his diet, he may accumulate more in his
blood and excrete more in his urine. Unfortunately the results of such
tests are not always clear-cut in distinguishing between homozygous normal and heterozygous carrier individuals; the values for the two groups
overlap considerably. However, a very simple test, adding ferric chloride
to the urine in the diapers of the newborn, distinguishes the homozygous
phenylketonuric. He can then be kept on a diet containing a minimal
amount of phenylalanine. This is effective in minimizing the amount of
mental retardation as he develops.

SECTION 8–18

INTERACTIONS OF GENES

In the examples discussed so far, the relationship between a gene and
its phenotype has been direct, precise and exact. Each gene controls the
appearance of a single trait. However, some traits are controlled by the
interaction of several pairs of genes. One pair of genes may affect several
traits. One pair of genes may inhibit or reverse the effect of another pair of
genes. Finally a given gene may produce different effects when the environment is changed in some way. In all instances, the genes are inherited as
units, but as a result of their interaction the appearance of the trait may be
altered in a variety of different ways.

In poultry there are a number of genes that control the shape and size
of the comb on top of the head (Fig. 8–19). The gene for rose comb (**R**) is
dominant to that for single comb (**r**). Another pair of genes governs the inheritance of pea comb (**P**) and single comb (**p**). Thus, a chicken with a single
comb must have the genotype **pprr**. The genotype of a pea-combed fowl
is either **PPrr** or **Pprr**, and a rose-combed chicken is either **ppRR** or **ppRr**.
When a homozygous pea-combed fowl is mated with a homozygous rose-
combed fowl, the offspring have neither pea nor rose comb, but a completely different type called a walnut comb. The phenotype of walnut comb
is produced whenever a chicken has one or more **R** genes, plus one or
more **P** genes. What would you predict about the types of combs among the
offspring of the mating of two heterozygous walnut-combed fowls, **PpRr**?

SECTION 8–19

POLYGENIC INHERITANCE

Many of the characteristics of man, animals and plants cannot be
separated into distinct, contrasting traits and are not inherited by a single
pair of genes. Height, body form, intelligence and skin color in man, and
commercially important characteristics in animals and plants, such as the
production of milk or eggs, the size of fruits and so on, result from the
interaction of many different pairs of genes. The term **polygenic inheritance** is applied to the situation in which two or more independent pairs of
genes have similar and additive effects on the same characteristic. The inheritance of light and dark skin color in man was studied by Davenport in
Jamaica. He showed that two major pairs of genes are involved. He des-

Pea

Rose

Walnut

Single

FIGURE 8–19   *Diagram of some of the types of combs found in fowl.*

ignated these as **A-a** and **B-b**, with the capital letters representing genes producing dark skin. Thus, the genotype of the full Negro is **AABB** and that of a member of the white race is **aabb**. The offspring of a mating of **aabb** with **AABB** are all **AaBb** with a skin color—**mulatto**—intermediate between black and white. The mating of two such mulattos produces offspring with skin colors ranging from Negro to white.

Polygenic inheritance is characterized by an F_1 generation which is intermediate between the two parental types and shows little variation. The F_2 generation includes individuals which show a wide variation between the two parental types. The skin color of man is a rather simple example of polygenic inheritance, because only two major pairs of genes are involved. The inheritance of height in man is much more complex and involves 10 or more pairs of genes. In addition, the effects of the genes are modified by environmental factors such as the quantity and quality of the food. Surprising as it may seem, shortness is dominant to tallness. The more genes for shortness, i.e., the more "capital" letters in the genotype, the shorter an individual is. Because many pairs of genes are involved and because height is modified by a variety of environmental conditions, such as nutrition, the heights of adults range from about 135 cm. to 215 cm. If you measured the heights of 1000 adult American men taken at random, you would find that only a few are as tall as 215 cm or as short as 135 cm. The height of most would cluster around the mean, about 170 cm. When the number of individuals of each height is plotted against the height in centimeters and the points are connected (Fig. 8–20), the result is a bell-shaped curve which approximates a *curve of normal distribution*.

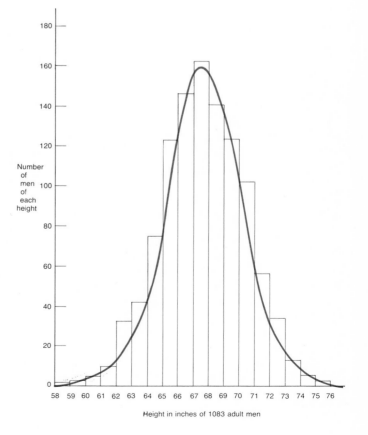

FIGURE 8–20   An example of a normal curve or curve of normal distribution, the heights of 1083 adult white males. The blocks indicate the actual number of men whose heights were within the unit range; for example, there were 163 men between 67 and 68 inches in height. The smooth curve is a normal curve of distribution based on the mean and standard deviation calculated from the raw data.

Height in inches of 1083 adult men

All living things show comparable variations in certain of their traits and these variations are usually distributed in a normal curve. If you measured the length of 1000 sea shells of a given species or counted the number of kernels per ear on 1000 ears of corn, or the number of pigs per litter in 1000 litters, or the weights of 1000 hens' eggs, you would find a normal curve of distribution in each instance. The variation resulting in this distribution may be caused by differences in hereditary factors, by differences in environmental conditions, or by some combination of the two. How do geneticists go about establishing a breed of cow that will give more milk or a strain of hens that will lay larger eggs or a strain of corn with more kernels per ear? By continued selection over many generations they gradually produce a true breeding strain with the commercially desirable trait. In other words, they continue their selective breeding until the strain becomes homozygous for all the polygenes for the desired trait. Obviously there is a limit to the effectiveness of breeding by selection, for when the strain becomes homozygous for all the genes involved, further selective breeding will not increase the desired quality.

SECTION 8-20

MULTIPLE ALLELES

In the examples so far we have dealt with situations in which at any given position, called a locus, in a chromosome there is one of only two alternative kinds of genes, the dominant and the recessive genes. However, at many, if not most, loci there may be additional possibilities, genes which produce phenotypes which differ from either the dominant or the recessive. The term **multiple alleles** is applied to three or more genes that can occupy a single locus, can fill the corresponding positions on a pair of homologous chromosomes. Each of the genes produces a distinctive phenotype. Any individual in a population may have any two of the possible genes but never more than two, and any gamete, of course, may have only one of them; however, in the population as a whole, three or more different alleles will occur.

The blood types of man, O, A, B and AB, are inherited by multiple alleles. Gene I^A provides the code for the synthesis of a specific protein, **agglutinogen A**, in the red cell. Gene I^B leads to the production of a different protein, agglutinogen B. Gene i produces no agglutinogen. Gene i is recessive to the other two, but neither gene I^A nor I^B is dominant to the other. The symbols I^A, I^B and i are used to emphasize that all three are alleles at the same locus. Individuals with genotypes $I^A I^A$ and $I^A i$ make up blood group A. Those with genotypes $I^B I^B$ or $I^B i$ comprise blood group B. Individuals with blood group O have the genotype ii. When both the I^A and I^B genes are present in the same individual, both agglutinogens A and B are produced in his red cells and the individual belongs to blood group AB (Table 8-2). These blood types are genetically determined and do not change during a person's lifetime. Determining the blood types of the individuals involved may be helpful in settling cases of disputed parentage. Such blood tests can never prove that a certain man *is* the father of a certain child, but only whether or not he *could be* its father. They may definitely prove that he could *not* be the father of a certain child. Could an AB man be the father of an O child? Could an O man be the father of an AB child? Could a type B child with a type A mother have a type A father? A type O father?

Nearly a dozen other sets of blood types, including the MN groups and

TABLE 8–2   The Human Blood Groups of the OAB Series

Blood Group	Agglutinogen in Red Cells	Agglutinin in Plasma	Can Give Blood to Groups	Can Receive Blood from Groups
O	None	a and b	O, A, B, AB	O
A	A	b	A, AB	O, A
B	B	a	B, AB	O, B
AB	A and B	None	AB	O, A, B, AB

(From Villee, C. A.: *Biology*. 5th ed. Philadelphia, W. B. Saunders Company, 1967.)

a series of Rh alleles, are inherited by other genes, independently of the ABO blood types. Determining all of these types in a given person may be useful in establishing relationships that could not be made certain by the ABO blood type alone.

SECTION 8–21

LINKAGE AND CROSSING OVER

Each species of animal or plant has many more pairs of genes than it has pairs of chromosomes. Obviously there must be many genes per chromosome. Man has 23 pairs of chromosomes, some large and some smaller (Fig. 8–2), but thousands of pairs of genes. The chromosomes are inherited as units—they pair and separate during meiosis as units; thus, all the genes in any given chromosome tend to be inherited together. If the chromosomal units never changed, the traits would always be inherited together and linkage would be absolute. However, during meiosis when the chromosomes are pairing and undergoing synapsis, homologous chromosomes may exchange entire segments of chromosomal material, a process called **crossing over** (Fig. 8–21). This exchanging of segments occurs at random along the length of the chromosome. Several exchanges may occur at different points along the same chromosome at a single meiotic division. It follows that the greater the distance is between any two genes in the chromosome, the greater will be the chance that an exchange of segments between them will occur.

In fruit flies the pair of genes **V** for normal wings and **v** for vestigial wings and the pair of genes **B** for gray body color and **b** for black body color are located in the same pair of chromosomes. They tend to be inherited together and are said to be linked. What would you predict the offspring would be like from a cross of a homozygous **VVBB** fly with a homozygous **vvbb** fly? They are all flies with gray bodies and normal wings and have the genotype **VvBb**. When one of these F₁ heterozygotes is crossed with a homozygous **vvbb** fly (Fig. 8–22), the offspring appear in a ratio which differs from that of the ordinary test cross for a dihybrid. If the two pairs of genes were not linked but were in different chromosomes, the offspring would appear in the ratio of ¼ gray-bodied normal-winged : ¼ black-bodied, normal-winged : ¼ gray-bodied, vestigial-winged : ¼ black-bodied vestigial-winged flies. If the genes were completely linked and no exchange of chromosomal segments occurred, then only the parental types—flies with gray bodies and normal wings and flies with black bodies and vestigial wings—would appear among the offspring, and these would be present in equal numbers (Fig. 8–22). However, there is an exchange of segments between the locus of gene **V** and the locus of gene **B**. Because of this crossing

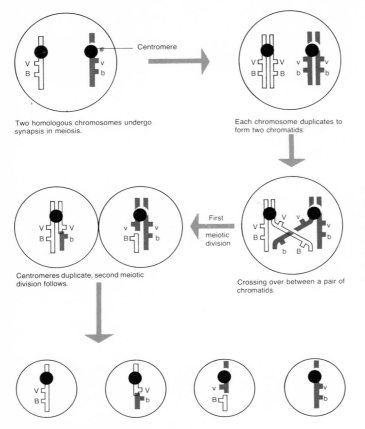

Two homologous chromosomes undergo synapsis in meiosis.

Each chromosome duplicates to form two chromatids.

Centromeres duplicate, second meiotic division follows.

First meiotic division

Crossing over between a pair of chromatids.

Four haploid gametes produced: here two crossover and two noncrossover gametes.

FIGURE 8–21 *Diagram illustrating crossing over, the exchange of segments between chromatids of homologous chromosomes. Crossing over permits recombination of genes (e.g., vB and Vb); the farther apart genes are located on a chromosome, the greater is the probability that crossing over between them will occur.*

over of part of the chromosomes some gray-bodied, vestigial-winged flies and some black-bodied, normal-winged flies (the crossover types) appear among the offspring (Fig. 8–22). Most of the offspring, of course, resemble the parents and are gray, normal or black, vestigial. In this particular instance, crossing over occurs between these two points in this chromosome in about one cell in every five or in 20 per cent of the total undergoing meiosis. In such crosses, about 40 per cent of the offspring are gray flies with normal wings. Another 40 per cent are black flies with vestigial wings. Ten percent are gray flies with vestigial wings, and 10 per cent are black flies with normal wings. The distance between two genes in a chromosome is measured in "cross-over units" which represent the percentage of crossing over that occurs between them. Thus, **V** and **B** are said to be 20 crossover units apart.

In a number of species the frequency of crossing over between specific genes has been measured. All of the experimental results are consistent with the hypothesis that genes are present in a linear order in the chromosomes. Thus, if the three genes A, B and C occur in a single chromosome, the amount of crossing over between A and C is either the sum of, or the difference between, the amounts of crossing over between A and B and B and C. For example, if the crossing over between A and B is five units and between B and C is three units, the crossing over between A and C will be found to be either eight units (if C lies to the right of B) or two units (if C lies between A and B) (Fig. 8–23). By putting together the results of a great

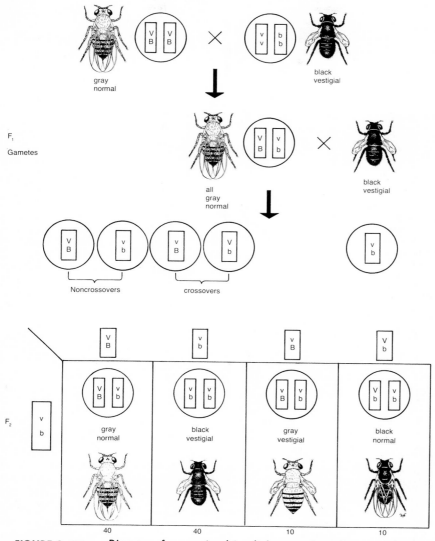

FIGURE 8-22   Diagram of a cross involving linkage and crossing over. The genes for vestigial versus normal wings and black versus grey body in fruit flies are linked; they are located in the same chromosome.

many such crosses, detailed maps of the location of specific genes on specific chromosomes have been made (Fig. 8–24).

Crossing over occurs at random and more than one crossover between two loci in a single chromosome may occur at a given time. You can observe among the offspring only the frequency of **recombinations** and not the frequency of crossovers. The frequency of crossing over will be slightly larger than the observed frequency of recombination, because the simultaneous occurrence of two crossovers between two particular genes will lead to the reconstitution of the original combination of genes in a particular chromosome.

FIGURE 8–23   Diagram illustrating the means of determining whether gene C lies between or to the right of genes A and B from the percentage of crossing over between each of the possible pairs.

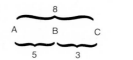
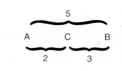

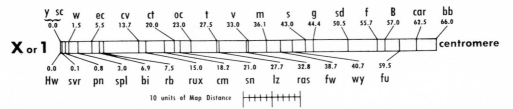

FIGURE 8–24   *Crossover map of the X chromosome of D. melanogaster. (From Herskowitz, I. H.: Genetics. 2nd ed. Boston, Little, Brown and Company, 1965.)*

KEY TO SYMBOLS

SYMBOL	NAME	SYMBOL	NAME
y	yellow body color	oc	ocelliless — ocelli absent; female sterile
Hw	Hairy-wing — extra bristles on wing veins, head, and thorax	t	tan body color
sc	scute — absence of certain bristles, especially scutellars	lz	lozenge — eyes narrow and glossy
svr	silver body color	ras	raspberry eye color
pn	prune eye color	v	vermilion eye color
w	white compound eyes and ocelli	m	miniature wings
spl	split bristles	fw	furrowed eyes
ec	echinus — large and rough textured eyes	wy	wavy wings
bi	bifid — proximal fusion of longitudinal wing veins	s	sable body color
		g	garnet eye color
rb	ruby eye color	sd	scalloped wing margins
cv	crossveinless — crossveins of wings absent	f	forked — bristles curled and twisted
rux	roughex — eyes small and rough	B	Bar — narrow eyes
cm	carmine eye color	fu	fused longitudinal wing veins; female sterile
ct	cut — scalloped wing edges	car	carnation eye color
sn	singed — bristles and hairs curled and twisted	bb	bobbed — short bristles

All the genes in a particular chromosome tend to be inherited together and comprise a **linkage group**. The number of linkage groups determined by genetic tests is always equal to the number of pairs of chromosomes. The most detailed chromosome maps are those for the bacterium, *Escherichia coli*, which has one circular chromosome (Fig. 8–25), and for fruit flies, which have four pairs of chromosomes. The chromosomes of corn, mice, *Neurospora* and certain other species of bacteria and viruses have been mapped in considerable detail.

Linkage provides an explanation for the common observation that certain traits in man and other organisms tend to be inherited together. Such traits are determined by genes which are located rather close together in a given chromosome. Crossing over provides another means by which genetic recombinations may arise. It plays a role in evolution by making possible new combinations of genetic units in the offspring.

SECTION 8–22

THE GENETIC DETERMINATION OF SEX

The **sex chromosomes** are an exception to the general rule that the members of a pair of chromosomes are identical in size and shape. The cells of the females of most species contain two identical sex chromosomes or **X** chromosomes. In males there is only one X chromosome and a smaller **Y** chromosome with which the X pairs during synapsis. Men have 22 pairs of

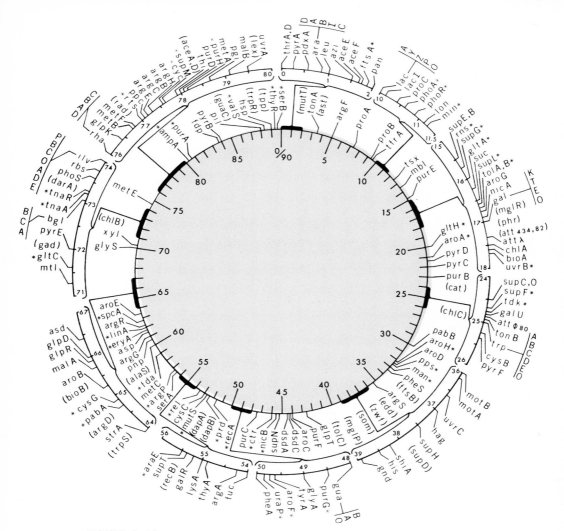

FIGURE 8–25   A map of the location of the genetic loci on the circular chromosome of Escherichia coli. (From Herskowitz, I. H.: Genetics. 2nd ed. Little, Brown and Company, 1965.)

ordinary chromosomes (autosomes) plus one X and one Y chromosome, and women have 22 pairs of autosomes plus two X chromosomes (Fig. 8–2). In some animals, butterflies and birds, males are XX and females are XY!

In man and other species where the male has one X and one Y chromosome, two kinds of sperm are produced. One half of the sperm contain an X chromosome and the other half contain a Y chromosome. Each egg, however, contains a single X chromosome. Fertilization of the X-bearing egg by an X-bearing sperm results in an XX or female offspring. The fertilization of an X-bearing egg by a Y-bearing sperm results in an XY or male offspring. Since there are approximately equal numbers of X- and Y-bearing sperm, the numbers of boys and girls born are nearly equal. In fact, some 106 boys are born for every 100 girls, and the ratio of male to female zygotes at conception is believed to be even higher. One possible explanation of this numerical difference is that the Y chromosome is somewhat smaller than the X chromosome. A sperm containing a Y chromosome might be slightly lighter and able to swim a little faster than an X-bearing sperm. Therefore, the sperm with the Y chromosome might be able to win the race to the egg slightly more than 50 per cent of the time.

Although there is a small excess of boys over girls at birth the number of boys dying in the first 10 years of life is somewhat greater than the number of girls dying. At about 10 years after birth, the number of males and females is equal. After that age the differential death rate continues and there are, at subsequent ages, fewer men than women.

SECTION 8–23

SEX-LINKED TRAITS

The human X chromosome contains many genes, whereas the Y chromosome contains very few, principally the genes for maleness. Any trait controlled by a gene located in the X chromosome is said to be sex-linked, because its inheritance is linked with the inheritance of sex. Each male offspring receives his single X chromosome from his mother. Therefore, he also gets all of his genes for sex-linked characters from his mother. Each female offspring receives one X chromosome from her mother and one from her father. Males have but a single X chromosome and but a single one of each kind of gene located in the X chromosome.

In fruit flies the genes for red versus white eye color are located in the X chromosome and, hence, are sex-linked. The Y chromosome has no gene for eye color. The male fruit fly, with only one gene for eye color, cannot be either homozygous or heterozygous but is termed **hemizygous** for the gene for eye color (and for any other gene in the X chromosome). The gene for red eye color, **R**, is dominant to the gene for white eye color, **r**. To avoid possible confusion, the genotype of the male is written with the Y present. The mating of a homozygous red-eyed female, **RR**, with a white-eyed male, **rY**, yields offspring, all of which have red eyes (Fig. 8–26). The female offspring are **Rr** and the male offspring are **RY**. However, mating a

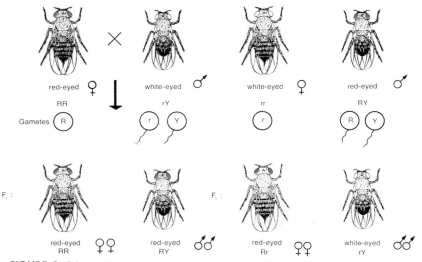

FIGURE 8–26   *The inheritance of eye color in the fruit fly, an example of a sex-linked trait. The gene for red versus white eye color is located in the X chromosome. Therefore the inheritance of this pair of eye colors is linked with the inheritance of sex.*

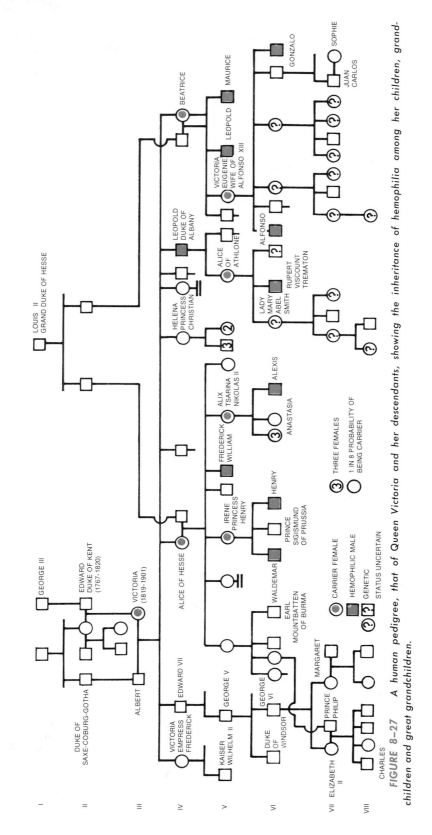

FIGURE 8-27 A human pedigree, that of Queen Victoria and her descendants, showing the inheritance of hemophilia among her children, grand-children and great grandchildren.

red-eyed male (**RY**) with a white-eyed female (**rr**) yields white-eyed male off-spring (**rY**) and red-eyed female offspring (**Rr**)!

In man the genes for **hemophilia** and **color blindness** are located in the X chromosome and the inheritance of these traits is sex-linked. Hemophilia is a disease in which blood does not clot properly—a hemophiliac will bleed profusely from even a small cut. One of the substances involved in the clotting mechanism, the so-called antihemophilic globulin, is lacking owing to the mutant gene of hemophilia. This gene is sex-linked, recessive and relatively rare, i.e., it is present in the population in a low frequency. The trait appears much more frequently in males than in females, because females require two of these genes for the trait to appear. Hemophilia is a very rare trait found only in a very few human males. It was unknown in human females until a woman with authentic hemophilia was described in 1951. Queen Victoria of England was heterozygous for the gene for hemophilia and passed this to several of her sons and grandsons (Fig. 8–27). What effects on the course of history can be traced to this fact?

One type of colorblindness, in which individuals are unable to distinguish between red and green, is another sex-linked recessive trait in man. About 4 per cent of all human males, but less than 1 per cent of human females, are affected with this type of colorblindness. As in the inheritance of hemophilia, females must have two of the genes to be colorblind whereas males need only one. A colorblind boy has a mother with normal vision and a colorblind father. From whom did he inherit his colorblindness? Could he have a colorblind sister?

SEX-LIMITED TRAITS

Certain other characteristics, termed **sex-limited**, are inherited by genes located not in the X chromosomes but in one of the pairs of autosomes. However, the appearance of the trait is altered or influenced by the sex of the animal, generally by the presence of male or female sex hormones. For example, in sheep the presence or absence of horns is determined by a single pair of genes. The gene **H** for the presence of horns is dominant in males but recessive in females, and its allele **h** for hornless is recessive in males but dominant in females. Thus, the genotype **HH** represents an animal with horns regardless of sex. The genotype **hh** represents an animal without horns regardless of sex, whereas the genotype **Hh** produces horns if the animal is a male and hornlessness if the animal is a female.

INBREEDING, OUTBREEDING AND HYBRID VIGOR

It it widely believed that **inbreeding**, the mating of two closely related individuals, such as brother and sister, is harmful and leads to the production of idiots and monsters. The process of inbreeding itself is not harmful. Inbreeding procedures are widely used by geneticists to improve strains of cattle, corn or cantaloupes. However, inbreeding greatly increases the chance for recessive genes to become homozygous and thus to express themselves phenotypically. All strains of organisms are heterozygous for many characteristics, and the recessive alleles of some of these genes produce undesirable traits when homozygous. If a stock is heterozygous for many recessive desirable traits, inbreeding can improve it; however, most people are heterozygous for recessive genes for undesirable traits. The inbreeding of humans leads to a high frequency of defects present at birth termed **congenital anomalies**.

The mating of totally unrelated individuals, termed **outbreeding**, frequently leads to offspring which are larger, stronger and longer lived than either parent was. This hybrid vigor is seen, for example, in the mule which results from the mating of a horse and a donkey. Almost all the corn grown commercially in the United States is a special hybrid strain developed by research geneticists of the Department of Agriculture from the mating of four different strains (Fig. 8–28). Hybrid vigor may be explained as follows: Each parental strain of corn is homozygous for certain recessive undesirable genes, but any two strains of corn are homozygous for different undesirable gene pairs. Each strain contains dominant genes which mask the undesirable recessive genes of the other strain. Thus, one strain might have the genotype **AAbbCCdd** and another strain **aaBBccDD**. The capital letters represent genes for dominant desirable traits and the lower case letters represent genes for the recessive undesirable traits. The hybrid offspring **AaBbCcDd** resulting from the mating of these two types would combine all of the desirable and none of the undesirable traits of the two pa-

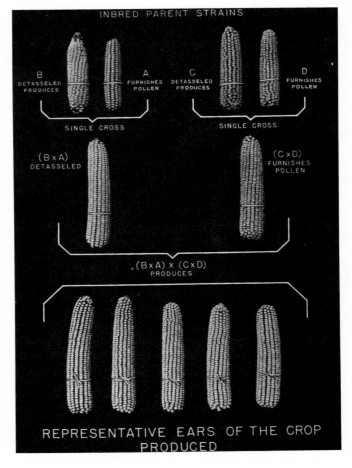

FIGURE 8–28   The mating of inbred strains of corn to produce the commercial variety with strong hybrid vigor. (Courtesy of the United States Department of Agriculture.)

rental strains. The actual genetic situation in hybrid corn is, of course, much more complicated than this.

SECTION 8–26

SOLVING GENETIC PROBLEMS

In genetics, as in mathematics, the student equipped with a clear understanding of a few basic principles can solve many kinds of problems. The basic genetic principles state that (1) inheritance is biparental; both parents contribute equally to the genetic constitution of the offspring, (2) genes are not altered when they exist together for a generation in a heterozygous individual, (3) each individual has two of each kind of gene, but each egg or sperm has only one of each kind of gene, (4) two pairs of genes located in different chromosomes are inherited independently, (5) two pairs of genes located in the same pair of chromosomes tend to be inherited together but will undergo an exchange when crossing over occurs, and (6) eggs and sperm unite at random; there is no attraction or repulsion between an egg and a sperm that contain identical genes.

In solving genetic problems it is helpful to: (1) write down the symbols used for each gene, (2) determine the genotypes of the parents, deducing them from the phenotypes of the parents and, if necessary, from the

Short-haired female A

L ?

Short-haired kitten

L ?

Short-haired male
Ll

Short-haired female B
Ll

Long-haired kitten
ll

FIGURE 8–29 *Deducing the parental genotypes from the phenotypes of the offspring: the inheritance of long hair versus short hair in cats. (See text for discussion.)*

Long-haired female C
ll

Long-haired kitten
ll

phenotypes of the offspring, (3) indicate all the possible types of eggs or sperm produced by each parent, (4) prepare the appropriate checkerboard and fill in the possible types of eggs on the top of the checkerboard and the types of sperm along the side, and (5) fill in the squares with the appropriate genotypes and read off the genotypic and the phenotypic ratios of the offspring.

This method of solving problems in genetics can be illustrated with the following example: The length of fur in cats is inherited; the gene for long hair (*l*), as in Persian cats, is recessive to the gene for the short hair (*L*) of the common tabby. Let us suppose that a short-haired male is bred to three different females, two of which (A and B) are short-haired and one of which (C) is long-haired (Fig. 8–29). Cat A gives birth to a short-haired kitten, but cats B and C each produce long-haired kittens. What offspring could be expected from further mating of this male with these three females?

Since the gene for long hair is recessive it is clear that all the long-haired cats must be homozygous *ll*. We can deduce then that cat C and the kittens produced by cats B and C have the genotype *ll*. All the short-haired cats have at least one *L* gene. The male has short hair but some of his offspring have long hair; hence, he must be heterozygous for this trait and have the genotype *Ll*. The fact that short-haired cat B gave birth to a long-haired kitten proves that she, too, is heterozygous and has the genotype *Ll*. From the data given, it is impossible to decide whether the short-haired cat A is homozygous *LL* or heterozygous *Ll*. How could a test cross with a long-haired male help in deciding this? Further mating of the short-haired male, *Ll*, with long-haired cat C, *ll*, would give half long-haired and half short-haired kittens. Further mating of the short-haired male with short-haired cat B would give three times as many short-haired kittens as long-haired ones.

The logical application of these simple principles enables each student to solve a wide variety of genetic problems and to predict the outcome of specific matings.

SUGGESTIONS FOR FURTHER READING

Hayes, W.: *The Genetics of Bacteria and Their Viruses.* 2nd ed. New York, John Wiley & Sons, Inc., 1968.
    One of the fundamental references in the field; giving a superb account of the genetics of these organisms.

Moore, K. L. (ed.): *The Sex Chromatin.* Philadelphia, W. B. Saunders Company, 1966.
    Our increasing understanding of the significance of the sex chromatin is summarized in this book.

Peters, James A. (ed.): *Classic Papers in Genetics.* Englewood Cliffs, New Jersey, Prentice-Hall, Inc., 1960.
    Papers by many of the scientists responsible for important developments in genetics — Mendel, Sutton, Morgan, Beadle and Tatum, Watson and Crick, Benzer and so on. Reading the original papers gives a feeling of immediacy that no other account can achieve.

Sturtevant, Alfred, H.: *A History of Genetics.* New York, Harper & Row, Publishers, 1965.
    A fine book for the informed layman. It gives an authentic account of the rise of classical genetics in which the author played a prominent role.

Whitehouse, H. L.: *Towards an Understanding of the Mechanism of Heredity.* New York, St. Martin's Press, 1965.
    An interesting book which discusses genetics in terms of its hypotheses and how they were tested.

THE GENETICS OF POPULATIONS

CHAPTER 9

The genetic problems considered in the previous chapter dealt generally with a single pair of parents of known genotypes and the types of offspring this pair may be expected to have. With this information you can predict the probability of getting each of the possible types of offspring. Other equally important problems in genetics concern the relative frequency of contrasting traits in an entire population of individuals. In such problems the geneticist attempts to measure the distribution of specific alleles in the whole population. Each pair of genes, such as **A** and **a**, is distributed in such a way that any member of the population may have the genotype **AA**, the genotype **Aa** or the genotype **aa**. If there is no selective advantage for either the **AA** genotype, the **aa** genotype or the **Aa** genotype, the frequency of these genes in successive generations of individuals will be exactly the same. As long as individuals with each of these genotypes are just as likely to mate and have offspring as individuals with the other genotypes, the three genotypes will be present in succeeding generations in the same proportion as in the initial generation.

SECTION 9–1

THE HARDY-WEINBERG PRINCIPLE

To see how this can be true, let us look at a specific example. Suppose we began with a generation in which the genotypes are present in the ratio of $\frac{1}{4}$ **AA** : $\frac{1}{2}$ **Aa** : $\frac{1}{4}$ **aa**. If each member of the population selects a mate at random, and if all the types of pairs produce on the average comparable numbers of offspring, what will be the ratio of the three genotypes in the succeeding generation? You can calculate this using the data given in Table 9–1, which presents the possible types of mating, the frequency of their random occurrence, and the kinds and types of offspring from each mating. The succeeding generation, it is clear, will consist of individuals with genotypes in the ratio of $\frac{1}{4}$ **AA** : $\frac{1}{2}$ **Aa** : $\frac{1}{4}$ **aa**.

Within a decade of the rediscovery of Mendel's laws, indeed a year

258

TABLE 9–1   The Offspring of the Random Mating of a Population Composed of $1/4$ AA, $1/2$ Aa and $1/4$ aa Individuals

Mating Male		Mating Female	Frequency	Offspring
AA	×	AA	$1/4 \times 1/4$	$1/16$ **AA**
AA	×	Aa	$1/4 \times 1/2$	$1/16$ **AA** $+ 1/16$ **Aa**
AA	×	aa	$1/4 \times 1/4$	$1/16$ **Aa**
Aa	×	AA	$1/2 \times 1/4$	$1/16$ **AA** $+ 1/16$ **Aa**
Aa	×	Aa	$1/2 \times 1/2$	$1/16$ **AA** $+ 1/8$ **Aa** $+ 1/16$ **aa**
Aa	×	aa	$1/2 \times 1/4$	$1/16$ **Aa** $+ 1/16$ **aa**
aa	×	AA	$1/4 \times 1/4$	$1/16$ **Aa**
aa	×	Aa	$1/4 \times 1/2$	$1/16$ **Aa** $+ 1/16$ **aa**
aa	×	aa	$1/4 \times 1/4$	$1/16$ **aa**
			Sum:	$4/16$ **AA** $+ 8/16$ **Aa** $+ 4/16$ **aa**

(From Villee, C. A.: *Biology.* 5th ed. Philadelphia, W. B. Saunders Company, 1967.)

before Johannson in 1909 proposed the word "gene" for Mendel's genetic factors, G. H. Hardy, an English mathematician, and G. Weinberg, a German physician, independently observed that the frequencies of the members of a pair of allelic genes in a population can be described by a mathematical equation. Multiplying the frequency of gene **A** plus the frequency of gene **a** (i.e., the relative numbers of sperm with gene **A** and with gene **a**) times the frequency of gene **A** plus the frequency of gene **a** (i.e., the relative numbers of eggs with gene **A** and with gene **a**) gives the product, the frequency of the several kinds of genotypes in the offspring.

This is exactly comparable to the operation used in the preceding chapter to predict the genotypes of the offspring of a specific mating, but now each gene is preceded by a fraction which represents its frequency in the population. Thus, $(1/2\mathbf{A} + 1/2\mathbf{a}) \times (1/2\mathbf{A} + 1/2\mathbf{a}) = 1/4\mathbf{AA} + 1/2\mathbf{Aa} + 1/4\mathbf{aa}$.

Since it is simpler to multiply decimals than fractions, it is common to express gene frequencies as decimals. Let us consider the pair of allelic genes **A** and **a,** present in a population in the ratio of 0.7**A** and 0.3**a**. In other words, 70 per cent of the eggs contain gene **A**, and 30 per cent of the eggs contain gene **a**. Similarly 70 per cent of the sperm contain gene **A** and 30 per cent of the sperm contain gene **a**. The random union of these eggs and sperm would result in offspring with genotypes in the ratio 0.49**AA** $+$ 0.42**Aa** $+$ 0.09**aa**, i.e., $(0.7\mathbf{A} + 0.3\mathbf{a}) \times (0.7\mathbf{A} + 0.3\mathbf{a})$.

SECTION 9–2

GENE POOLS AND GENOTYPES

The genetic constitution of a population of a given organism is termed the **gene pool**. Stated differently, all the genes of all the individuals in a population make up the gene pool. This may be contrasted to the **genotype**, which is the genetic constitution of a single *individual*. Any individual may have only two alleles of any given gene. In contrast, the gene pool of the population may contain any number of different alleles of a specific gene. The ABO blood groups (p. 246) are inherited by three alleles, I^A, I^B and i. In the population there are three alleles, but any given individual can have no more than two of the three.

The gene pools of different populations may differ in the ratios or proportions of the specific alleles. One population may have the alleles **A** and **a** in a ratio of 0.5 to 0.5. Another population of the same species may have the two alleles in the ratio 0.7**A** : 0.3**a**. If all the individuals in this second population have equal chances of surviving to adulthood and equal chances of producing gametes, then 70 per cent of the sperm produced by the entire population of males would have gene **A** and 30 per cent would have gene **a**. Similarly 70 per cent of the eggs produced by the entire population of females would contain gene **A** and 30 per cent would have gene **a**. The random union of the eggs and sperm would result in offspring in the ratio of 0.49**AA** + 0.42**Aa** + 0.09**aa**. Notice that the gene pool in the offspring is identical to the gene pool of the parents!

		sperm	
		.7A	.3a
	.7A	.49 AA	.21 Aa
eggs			
	.3a	.21 Aa	.09 aa

In a more general sense this can be represented as $(p\mathbf{A} + q\mathbf{a})^2$, where p represents the frequency of one allele (0.7 for gene **A** in this example) and q represents the frequency of the other allele (0.3 for gene **a** in this example). Multiplying $(p\mathbf{A} + q\mathbf{a}) \times (p\mathbf{A} + q\mathbf{a})$ gives $p^2\mathbf{AA} + 2\ pq\mathbf{Aa} + q^2\mathbf{aa}$. Since $p = 0.7$, $p^2 = 0.7 \times 0.7$ or 0.49, the frequency of **AA**. The value of $q = 0.3$ and $q^2 = 0.3 \times 0.3$ or 0.09, the frequency of **aa.** Finally, the value of $2\ pq = 2 \times 0.7 \times 0.3$ or 0.42, the frequency of **Aa.**

By similar calculations you can show that the next generation, and each succeeding generation, will contain an identical gene pool, 0.7**A** and 0.3**a**. The three kinds of genotypes will be present in the ratio 0.49**AA** : 0.42 **Aa** : 0.09 **aa** in succeeding generations provided that: (1) There are no mutations for **A** or **a**; (2) The three kinds of genotypic individuals have equal probabilities of surviving, mating and producing offspring, and there is no selection of mates according to these genotypes; and (3) The population of individuals is large enough so that chance cannot play a role in determining gene frequencies.

SECTION 9–3

THE GENETIC BASIS OF EVOLUTION

This principle that a population is genetically stable in succeeding generations is termed the **Hardy-Weinberg Principle**. Since the principle was established, J. B. S. Haldane, R. A. Fisher and Sewall Wright have developed mathematical methods for analyzing the inheritance of a given trait in a population. Subsequently it has become clear that the process of evolution, stated in the simplest terms, represents departure from the Hardy-Weinberg principle of genetic stability. Evolution involves changes in the gene pool of a population that result from mutations and selection. Thus an understanding of the Hardy-Weinberg Principle is of prime importance in understanding the mechanism of evolutionary change.

Hardy and Weinberg independently recognized that the frequencies of the members of a pair of genes can be described by the expansion of a binomial equation. If we let p represent the fraction of **A** genes and q represent the fraction of **a** genes in the entire population, then (since this allelic pair of genes must be either **A** or **a**) $p + q = 1$. If you know the value of either p or q, you can calculate the value of the other; e.g., $p = 1 - q$.

In a population of organisms mating at random, a p number of **A**-containing eggs and a q number of **a**-containing eggs will be fertilized by a p number of **A**-containing sperm and a q number of **a**-containing sperm: $(p\mathbf{A} + q\mathbf{a}) \times (p\mathbf{A} + q\mathbf{a})$. Completing this algebraic multiplication yields the product, p^2 **AA** + 2 pq **Aa** + q^2 **aa**. This describes the proportion of the types of offspring of these matings. If p, the frequency of gene **A**, is 0.5, then q, the frequency of gene **a**, is equal to $1 - p$ or $1 - 0.5$ or 0.5. From the formula you can calculate the frequency of genotype **AA**: $p^2 = 0.5^2$ or 0.25; the frequency of **Aa**: 2 $pq = 2 \times 0.5 \times 0.5$ or 0.5; and the frequency of **aa**: $q^2 = 0.5^2$ or 0.25.

Any population in which the distribution of alleles **A** and **a** conforms to the equation p^2 **AA** + 2 pq **Aa** + q^2 **aa** is in **genetic equilibrium**. This generalization implies that the proportion of these genes in successive generations of that population will be the same unless it is altered by mutation or by selection. This concept of genetic equilibrium in a population and of the changes in this equilibrium that may be produced by mutation and selection is the basis of our present concept of how the forces of natural selection operate in the process of evolution.

SECTION 9–4

ESTIMATING THE FREQUENCY OF GENETIC "CARRIERS"

Neither the value of p nor q, which are **gene frequencies**, can be measured directly. However, since the recessive phenotype can be distinguished you can determine q^2, the frequency of genotype **aa**. From this you can calculate the gene frequencies q (which is the square root of q^2) and p (which is $1 - q$). Finally you can calculate the frequencies of the other genotypes, p^2**AA** and 2 pq**Aa**. To calculate the number of individuals in a population that are genetic carriers for a given trait (i.e., are heterozygotes, **Aa**) you need to know only that it is inherited by a single pair of genes and the frequency with which the homozygous recessive individuals appear in the population.

SECTION 9–5

ALBINISM

Albinos are individuals with no pigment at all in their skin or hair. **Albinism** is an inherited trait in which the individual lacks a specific enzyme, **tyrosinase**. Tyrosinase catalyzes one of the reactions involved in the production of the dark pigment **melanin**. Albinism is inherited by a single pair of genes, and albinos, the homozygous recessive individuals, occur about once in 20,000 births. From this fact you can calculate that the frequency of **aa** individuals (q^2) is $\frac{1}{20,000}$. From the value of q^2 you can determine q by taking the square root of q^2. The square root of $\frac{1}{20,000}$ is about $\frac{1}{141}$. Since

$p = 1 - q$ or $1 - \dfrac{1}{141}$, $p = \dfrac{140}{141}$. You now have values for both p and q, and can calculate the value of $2\,pq$, which represents the frequency of the genetic "carrier" **Aa** individuals: $2 \times \dfrac{140}{141} \times \dfrac{1}{141} = \dfrac{1}{70}$. Surprising as it may seem, one person in 70 is a *carrier* of albinism, although only one person in 20,000 is homozygous and displays the trait. At first glance it may seem odd that there are so many carriers in a population that contains so few homozygous recessives, yet reflecting on the mathematical relations involved should lead you to realize that this must be true. When q is small (such as $\frac{1}{141}$), then q^2 will be very small, but $2\,pq$ will be much larger.

SECTION 9–6

THE TASTE TEST

As another example of the usefulness of the methods of **population genetics**, let us consider the inheritance of another human trait. Human beings differ in their ability to taste **phenylthiocarbamide** (PTC) and related compounds with the thiocarbamide group ($-$NCS), containing one atom each of nitrogen, carbon and sulfur. Some people find that PTC has a bitter taste, others report it to be completely tasteless. From the distribution of "tasters" and "nontasters" in specific families, L. H. Snyder assumed that the ability to taste PTC is inherited by a single pair of genes and that tasting (**T**) is dominant to nontasting (**t**). Snyder's genetic survey revealed that 70.2 per cent of the white population of the United States are tasters and 29.8 per cent are nontasters. From these values he could calculate (Table 9–2) that 12.4 per cent of the children from marriages of tasters with tasters will be nontasters. Further, he could calculate that in marriages of tasters with nontasters 35.4 per cent of the children should be nontasters. The percentages Snyder actually found were 12.3 per cent and 33.6 per cent respectively. Thus the observed percentages agreed closely with those predicted by the Hardy-Weinberg Principle. From this he concluded that his original assumption, that the tasting-nontasting trait is inherited by a single pair of genes, is correct. Nontasters, although comprising some 30 per cent of the white population, are very rare in populations of Negroes, Eskimos and American Indians. Thus the **T-t** gene pools are quite different in these different populations.

TABLE 9–2   The Inheritance of the Ability to Taste Phenylthiocarbamide

Marriage	Number of families	Offspring		Proportion nontasters	
		Tasters	Nontasters	Observed	Calculated*
taster × taster	425	929	130	0.123	0.124
taster × nontaster	289	483	278	0.336	0.354
nontaster × nontaster	86	5	218	0.979	1.0

*From the Hardy-Weinberg Law, it follows that among marriages of parents with unlike traits, such as taster× nontaster, the fraction of the offspring with the recessive trait is $q/1 + q$. Among marriages of parents with like traits, such as taster × taster, the fraction of offspring with the recessive trait, nontaster, is $(q/1 + q)^2$. Since 29.8 per cent of the population were nontasters, $q^2 = 0.298$, hence $q = 0.545$, $q/1 + q = 0.354$ and $(q/1 + q)^2 = 0.124$.

(From Villee, C. A.: Biology, 5th ed., Philadelphia, W. B. Saunders Company, 1967.)

SECTION 9–7

INHERITANCE OF M-N BLOOD GROUPS

As another example of population genetics, let us consider the blood groups, M, MN and N. These blood groups are independent of the ABO blood groups. Each of us is either A, B, O or AB, and also is either M, N or MN. The distribution of these blood groups in the white population of the United States has been found to be 29.16 per cent type M, 49.58 per cent type MN and 21.26 per cent type N. Let us test the theory that these traits are inherited by a single pair of genes with incomplete dominance, i.e., that MN represents the heterozygous individual. To do this we determine whether the distribution of phenotypes found would be expected on the basis of the Hardy-Weinberg Law of population genetics: p^2MM $+ 2\ pq$ MN $+ q^2$ NN.

We begin with the fraction of the population that is phenotypically N and assume that $q^2 = 0.2126$. From this we can calculate that $q = \sqrt{0.2126}$ or 0.46. The frequency, p, of the M gene is $1 - 0.46$ or 0.54. The square of this is $0.54^2 = 0.29$, which agrees well with the observed frequency of blood group M, 29.16 per cent. Furthermore, $2\ pq = 2 \times 0.54 \times 0.46 = 0.49$, which agrees with the 49.58 per cent observed to have blood type MN. The excellent agreement between the observed and predicted values supports the hypothesis that M and N blood types are inherited by a single pair of genes with neither one being dominant to the other. Such genes may be termed **codominants**. The heterozygous individual has both M and N antigens in his blood cells and is phenotypically MN. To test this hypothesis differently, repeat the calculations but begin with $p^2 = .2916$ and calculate p, then $q = 1 - p$, and so on.

Many of the diseases of mankind that are known to be inherited are determined by recessive genes. The undesirable trait—the disease condition—represents the homozygous recessive individual. When a man and a woman ask a geneticist whether they should have children, one of his primary concerns is whether both may be heterozygous for the same unwanted recessive trait. If they are, then there is one chance in four that any of their offspring will show this inherited disease.

SECTION 9–8

PHENYLKETONURIA (PKU)

Phenylketonuria is an inherited disease in which there is a deficiency of the enzyme which in normal individuals converts phenylalanine to tyrosine. Phenylalanine and phenylpyruvic acid accumulate in the blood stream and tissues and are excreted in the urine. The accumulation of phenylalanine in the nervous tissues interferes with their function; hence individuals with phenylketonuria have a mental deficiency associated with their disease. The gene frequency q for this trait in the United States population is about 0.005 and the gene frequency p of the normal allele is 0.995. The frequency of heterozygotes is about one in every 100 persons ($2\ pq = 2 \times 0.995 \times 0.005 = 0.01$). The heterozygous individual has a somewhat lower than normal ability to metabolize phenylalanine; however, the results of phenylalanine metabolism tests given to heterozygotes overlap those given to normal people so that it is not possible to distinguish between the two with confidence. If a combination of family history and the phenylalanine metabolism tests makes it likely that both husband and wife are heterozygous carriers for phenylketonuria (or for any other recessive trait) it becomes possible to predict the probability (one in four) that they will have a child showing the abnormality. It is also possible to predict the possibility that they will have one, two or three normal children. To do this one simply makes the appropriate expansion of the binomial and calculates the value of the appropriate term in the expansion.

To calculate, for example, the probability that a man and his wife, both heterozygous carriers for phenylketonuria, will have three normal children out of three you multiply $(p + q)^3$. In this, p represents the probability of a normal child, i.e., $\frac{3}{4}$, and q represents the probability of the phenylketonuric child, i.e., $\frac{1}{4}$, in a mating of two heterozygous individuals. Since we are calculating the probability of *three* normals in a total of three offspring we multiply $(p + q) \times (p + q) \times (p + q)$ or $(p + q)^3$. The algebraic product is $p^3 + 3\ p^2q + 3\ pq^2 + q^3$. The part of the product which represents the probability of the combination of three normal children is p^3. The next term, $3\ p^2q$, represents the probability of two normal children and one phenylketonuric child in a total of three children. To calculate the probability of three phenotypically normal children from such a couple we take $\left(\frac{3}{4}\right)^3$ $= \frac{27}{64}$. There are 27 chances out of 64 that a pair of heterozygous individuals could have three children, all of whom are normal. By similar reasoning and calculation you can see that q^3 is $\left(\frac{1}{4}\right)^3$ or $\frac{1}{64}$. In other words there is one chance in 64 that if the man and his wife have three children, all three would be phenylketonurics.

SECTION 9–9

GENETIC EQUILIBRIUM IN A POPULATION

The essence, then, of the Hardy-Weinberg Principle is that under specific conditions the frequencies of the several alleles of a given gene remain constant from one generation to the next. The first condition for genetic equilibrium is that either there must be no mutations or the rates of forward and reverse mutations must be in equilibrium. That is, either there is no change of gene **A** into gene **a**, or else the rate at which gene **A** mutates to gene **a** (forward mutation) is equal to the rate at which gene **a** mutates to gene **A** (reverse mutation). Genes undergo mutations continually; indeed, there is no way of preventing mutations from occurring. Further, the rates of forward and reverse mutations will rarely be equal. Thus there is usually a tendency, termed **mutation pressure**, for one of the alleles to increase in frequency and another allele to decrease in frequency. This mutation pressure may be countered by some other factor, such as selection. Even though mutations occur constantly, they occur at random. Mutations are seldom the major factor in producing changes in gene frequencies in a population. They increase genetic variability and ultimately provide the raw material of evolution, but mutations alone are unlikely to determine the nature or direction of evolutionary change.

SECTION 9–10

GENETIC DRIFT

To maintain its genetic equilibrium, a population must be large enough so that chance events are unlikely to change gene frequencies. The

Hardy-Weinberg Law is based on statistical concepts. Its operation requires that the sample size be large enough to minimize the possibility of chance deviations. In small (less than 100 members) isolated breeding populations of a given species, there is a relatively high probability that one allele or the other will be lost from the population by chance even though it may determine a trait that is of adaptive value. In such small populations there is a strong tendency for the population to become homozygous for one allele or the other. In contrast, large breeding populations have a tendency to remain more variable, and to include many heterozygous individuals. The production of random evolutionary changes by chance in small breeding populations is termed **genetic drift**. Genetic drift results in changes in the gene pool of a population and thus produces evolutionary change; however, such evolutionary changes are aimless, at random, and not adaptive. Genetic drift may explain the common observations that closely related species in different parts of the world frequently differ in curious, even bizarre, ways that have no adaptive value.

The maintenance of genetic equilibrium also requires that the population does not lose genes from its gene pool by the outward **migration** of certain members of the population. The population must also not receive new genes from immigrants from other populations. The populations of some species in nature are essentially isolated and do not undergo gene migration. Other populations do interbreed to some extent with neighboring populations and a considerable amount of gene migration occurs. New genes are introduced into the populations, genetic variability is increased, and this increased genetic variability may play a role in the evolution of that species or population.

SECTION 9–11

FACTORS CHANGING GENE FREQUENCIES: DIFFERENTIAL REPRODUCTION

To maintain genetic equilibrium, the members of the population must mate completely at random. The male must not select his mate because she has the same phenotype he has. Indeed, more than simply the selection of mates is involved, for **random reproduction** implies that all the many factors that contribute to success in reproduction, to success in producing viable offspring, are also operating at random and independently of the genotypes of the individuals involved. The selection of a mate, the fertility of the pair, the fraction of the resulting zygotes that complete development to birth, the survival of the young to reproductive age and their fertility are all factors that may influence the relative effectiveness of certain types or strains of organisms to perpetuate their kind. If individuals with certain genotypes are better able to raise large numbers of offspring to the age when they in turn reproduce, genetic equilibrium will not be maintained. Instead, the frequency of certain genes in the population will increase as a result of **differential reproduction.** This change in gene frequencies in the gene pool of a population caused by differential reproduction is the modern way of thinking about Darwin's concept of Natural Selection, the basis of evolution. If differential reproduction does not occur, the gene pool does not change and the Hardy-Weinberg concept of genetic equilibrium applies. If differential reproduction does occur (and it usually does) the gene pool changes. Evolutionary changes resulting from differential reproduction are characteristic of nearly all populations of organisms, including man.

SECTION 9–12

EVOLUTION: THE FAILURE TO MAINTAIN GENETIC EQUILIBRIUM

Evolution by natural selection, stated in its simplest terms, means that individuals with certain genotypes and certain traits have more surviving offspring in the next generation than other individuals do. They thus contribute a proportionately greater percentage of genes to the gene pool of the next generation than do organisms with other traits. The evolutionary changes resulting from differential reproduction tend to improve the average ability of members of the population to produce successive generations with genotypes like their parents. As we shall see, natural selection does not operate upon the phenotypes of single genes but upon the total phenotypic effect of the entire array of genes present. One group of organisms may survive despite some clearly disadvantageous character. Another group may be eliminated despite certain traits that are highly advantageous for survival. In evolution you don't necessarily get counted out because of a single bad trait, nor do you get a gold star for a single good adaptive trait. The organisms that ultimately survive and serve as parents of the next generation are ones whose total spectrum of qualities renders them a little better able to survive and reproduce their kind.

Thus, mutation is of importance in supplying the raw materials of evolution. Gene migration and genetic drift play generally minor roles in evolution; the major factor in evolution is differential reproduction. In the absence of mutation and selection, the gene frequencies in a population tend to remain constant in successive generations and the Hardy-Weinberg Principle of genetic equilibrium applies.

SUGGESTIONS FOR FURTHER READING

Carlson, E.: *The Gene: A Critical History.* Philadelphia, W. B. Saunders Company, 1966.
    This book traces the development of the concept of the gene.

Falconer, D. S.: *An Introduction to Quantitative Genetics.* Edinburgh, Oliver and Boyd, 1960.
    An advanced presentation for students who would like to pursue the subject of population genetics in greater detail.

Li, C. C.: *Population Genetics.* Chicago, Chicago University Press, 1955.
    A clear, readable presentation at a fairly advanced level of the quantitative aspects of the genetics of populations.

Mather, W. B.: *Principles of Quantitative Genetics.* Minneapolis, Minnesota, Burgess Publishing Company, 1964.
    A useful and up-to-date text of the genetics of populations.

Rasmuson, M.: *Genetics on the Population Level.* London, William Heinemann, Ltd., 1961.
    A recommended text for advanced students.

Spiess, E. B. (ed.): *Papers on Animal Population Genetics.* Boston, Little Brown and Company, 1962.
    A number of the classic papers dealing with the concepts of population genetics are reprinted in this volume.

Stern, C.: *Principles of Human Genetics.* 2nd ed. San Francisco, W. H. Freeman, Company, 1960.
    A standard text with an excellent treatment of the genetics of man and of populations.

Strickberger, M. W.: *Genetics.* New York, The Macmillan Company, 1968.
    One of the standard texts of genetics with a fine discussion at an elementary level of population genetics.

EVOLUTION

Chapter 10

The term evolution means a gradual, orderly change from one condition to another. Biological evolution refers to the tendency for populations of plants and animals to undergo changes in successive generations directed by natural selection. The various kinds of plants and animals now existing have descended from other, usually simpler, organisms by gradual modifications which have accumulated in successive generations.

Indeed no other theory of how the present array of organisms came into being is tenable. Our understanding of the concept of evolution can emerge logically and naturally from our understanding of genetics. This is not, however, how the concept of evolution arose historically. It came instead as a result of a large number of observations of similarities and differences in structures and functions of the various kinds of animals and plants in different parts of the world. It came from Charles Darwin's profound insight in putting together the various pieces of the puzzle of how these similarities and differences may have arisen. Darwin made his observations while serving as ship's naturalist on the *Beagle*, a small ship which had been sent on a five-year cruise around the world to gather data for oceanographic charts for the British Navy. It is a curious fact that Darwin's presence on the *Beagle* was very nearly accidental: he had begun a career in medicine and dropped this, and then turned to theology, but was not very interested in that subject either. He was fascinated by the natural sciences and had made a collection of beetles, mollusks and shells. While at Cambridge he spent some time studying both biology and geology, and became acquainted with Professor Henslow, the naturalist. Through his help, Darwin, shortly after leaving Cambridge and just 22 years old, was appointed to the position of naturalist on the *Beagle*. As the ship slowly went back and forth along the eastern and then the western coast of South America, Darwin frequently left the boat, and traveled to the interior of the country, where he fished, hunted, and collected fossils and specimens of living plants and animals. His careful observation led him to realize that the kinds of birds, animals and plants found, for example, on the east coast of South America, were quite different from those on the west coast. The *Beagle* made an extended visit to the Galápagos Islands off the western coast of South America, and Darwin was fascinated by the diversity of the giant tortoises and the finches that lived in each of the islands. Thinking about these observations eventually led him to reject the theory of special creation and to seek an alternative explanation for them.

The idea of natural selection occurred to Darwin shortly after his return to England in 1836, but he spent the next 20 years accumulating the vast body of facts which eventually became his book *On the Origin of Species by Means of Natural Selection*. In 1858 he received a manuscript from Alfred Russel Wallace, a young naturalist who was studying the distribution of plants and animals in the East Indies and the Malay Peninsula. In this paper Wallace set forth his concept of natural selection which was very similar to Darwin's and which he had reached independently. Wallace had been stimulated, as Darwin had been, by Malthus' book on population growth and pressure, and the struggle for existence. By mutual agreement, Darwin and Wallace presented a joint paper on their theory at the meeting of the Linnean Society in 1858, and Darwin's monumental book was published the next year.

SECTION 10–1

THE HISTORICAL DEVELOPMENT OF THE CONCEPT OF EVOLUTION

The idea that the present forms of life arose from earlier simpler ones was not new when Darwin published his *On the Origin of Species by Means of Natural Selection*. Elements of the theory of biological evolution can be found in the writings of early Greek philosophers such as Thales, Anaximander, Empedocles and Epicurus. Aristotle, a great biologist as well as a philosopher, developed an elaborate theory of gradually evolving life forms which he termed the "ladder of nature." He held the metaphysical belief that nature strives to change from the simple and imperfect to the more complex and perfect. The Roman poet, Lucretius, also gave an evolutionary explanation of the origin of plants and animals in his poem *De Rerum Natura*. With the Renaissance, interest in the natural sciences quickened, and from the fourteenth century on, more and more people found the concept of biological evolution reasonable. In the *On the Origin of Species by Means of Natural Selection* Darwin listed about 20 thinkers who had seriously considered the concept, among whom were his own grandfather, Erasmus Darwin, and the French scientist, Lamarck.

Even before the Renaissance, men had discovered odd fragments resembling bones, teeth and shells in the ground, and some of these were strangely unlike any living form. Indeed, some of the objects found in rocks high in the mountains resembled parts of marine animals! Leonardo da Vinci, a true universal genius, correctly interpreted these curious finds in the fifteenth century. Gradually others accepted his explanation that they were the remains of animals that had existed in previous ages but had become extinct. Such evidence of former life suggested the theory of **catastrophism** — the idea that fires and floods had periodically destroyed all living things and necessitated the repopulation of the world by successive acts of special creation.

Three Englishmen in the eighteenth and nineteenth centuries laid the foundations of modern geology. The first, James Hutton, developed (1785) the concept (**uniformitarianism**) that the geologic forces at work in the past were the same as those of the present. His study of the erosion of valleys by rivers and the formation of sedimentary deposits at the river mouths led him to the conclusion that the processes of erosion, sedimentation, disruption and uplift, acting over long periods of time, could account for the formation of the fossil-bearing rock strata. John Playfair published in 1802 another volume giving further explanation and examples of the idea of uniformitarianism in geologic processes. Sir Charles Lyell did much with his *Principles of Geology*, published in 1832, to establish this principle of uniformity. By demonstrating the validity of geologic evolution, he proved beyond a doubt that the earth is much older than a few thousand years—indeed old enough for the processes of biological evolution to have occurred. He was a close personal friend of Darwin's and had a great influence on his thinking.

Fifty years before *On the Origin of Species by Means of Natural Selection*, Jean Baptiste de Lamarck published his *Philosophie Zoologique*, which contained one of the earliest theories of biological evolution to be logically developed. Lamarck, like most biologists at that time, believed that organisms are guided by innate, mysterious forces which enable them to overcome handicaps in the environment. He believed that these adaptations, once made, are then transmitted from one generation to the next, i.e., that acquired characteristics can be inherited. In developing this notion, Lamarck went on to state that new organs arise in response to demands of the environment, and that their size is proportional to their "use or disuse." Again, such changes in size were believed to be inherited by succeeding generations. The Lamarckian concept of the inheritance of acquired characteristics is an attractive one and was used by Darwin to explain some aspects of evolutionary theory. It would explain the complete adaptation of many plants and animals to the environment; however, as our understanding of genetics has grown, Lamarck's theory has become unacceptable because it is clear that *acquired characteristics cannot be inherited*. Acquired characteristics are present in the *body* cells, whereas inherited traits are transmitted only by the *gametes*—the eggs and sperm.

SECTION 10–2

THE DARWIN-WALLACE THEORY OF NATURAL SELECTION

The explanation advanced by Darwin and Wallace as to the way evolution occurs may be summarized as follows:

(1) **Variation** is characteristic of every group of animals and plants. Darwin and Wallace assumed that variation was one of the innate properties of living things. Now we can distinguish inherited and noninherited variations. Only inherited variations, produced by mutations, are important in evolution. Only the natural selection of variations based on genetic differences has an effect on the character of successive generations of a population.

(2) More of each kind of organism begin to grow than can possibly obtain food, survive and reproduce. Yet since the number of members of each species in successive generations remains fairly constant under natural conditions, it must be assumed that most of the offspring in each generation perish. If all the offspring of any species remained alive and reproduced, they would soon crowd all other organisms from the earth.

(3) Since a larger number of individuals are born than can survive, there is a **competition** for food and space, a struggle for survival. This

might be an active kill-or-be-killed contest, or one less immediately apparent, but nonetheless real, such as the struggle of plants and animals to survive drought, cold, or other unfavorable environmental conditions.

(4) Those genetic variations that better equip an organism to survive in a given environment will favor their possessors over other organisms that are less well adapted. This idea of the *"survival of the fittest"* is the core of Darwin's and Wallace's theory of natural selection.

(5) The surviving individuals will give rise to the next generation and in this way the "successful" variations are transmitted to the next generation and to the succeeding ones.

This process would tend to provide successive generations of organisms with better adaptations to their environment. Indeed, as the environment changes, further adaptations would follow. The operation of natural selection over many years could lead ultimately to the development of descendants that are quite different from their ancestors — different enough to be recognized as a different kind of animal or plant. Certain members of the population with one group of variations might become adapted to environmental changes in one way, while other members with a different set of variations might become adapted in a different way. Thus two or more different kinds of organisms may arise from a single ancestral group. Darwin and Wallace recognized that animals and plants may exhibit variations which are neither a help nor a hindrance to them in their survival in a given environment. These will not be affected directly by natural selection, and the transmission of such neutral variations to succeeding generations will be governed by chance.

Darwin's theory of natural selection was so reasonable and so well supported by his arguments that most biologists soon accepted it. Some objected that the theory could not explain the presence of many apparently useless structures in an organism. Many of the differences between species may not be important for survival, but are simply incidental effects of genes that have invisible physiological effects of survival value. Other "useless" or nonadaptive differences may be controlled by genes that are closely linked in the chromosome to other genes controlling characteristics that are important for survival. Still other nonadaptive characteristics may become fixed in a population by chance, by the phenomenon termed "genetic drift."

Another early objection was that new characteristics would be "lost by dilution" as individuals possessing them bred with others that lacked them. We now know that although the phenotypic expression of a gene may be altered when it exists in combination with other genes, the gene itself is not altered or diluted. The genes emerge unchanged in successive generations.

The concepts of the "struggle for survival" and "survival of the fittest" were key points in the Darwin-Wallace theory of natural selection, but biologists have come to realize that actual physical struggle between animals for survival, or competition between plants for space, sun or water, is probably less important as an evolutionary force than Darwin had imagined.

SECTION 10–3

POPULATIONS AND GENE POOLS

The evolution of any kind of organism occurs over many generations during which individuals are born and die, but the population continues.

Thus the unit in evolution is not the individual, but rather a *population* of individuals. A population of similar individuals living within a defined area and interbreeding is termed a **deme** or a **genetic population.** The territorial limits of any given deme may be vague and difficult to define and the number of individuals in the deme may fluctuate widely from time to time. A deme commonly overlaps with one or more adjacent demes to some extent. The next larger unit of population in nature is the **species**, which is composed of a series of intergrading demes.

The relative frequencies of the genes in a population will remain constant from one generation to the next (1) if the population is large, (2) if there is no selection for or against any specific gene or allele, i.e., if mating occurs at random, (3) if no mutations occur, and (4) if there is no migration of individuals into or out of the population. The operation of the Hardy-Weinberg Principle will result in maintaining a given gene frequency in a population. The essential feature of the process of evolution is a gradual change in the gene frequencies of a population when this Hardy-Weinberg equilibrium is upset either because mutations occur, because reproduction is nonrandom, or because the population is small, so that the gene frequencies in successive generations will be determined by chance events.

The demes and species found in nature tend to remain unchanged for many generations. This fact implies that there has been no change in the genetic make-up of the demes, and no change in the environmental factors that determine survival. When the characteristics of the population do change, this reflects either changes in the genetic factors brought about by mutations or changes in the environmental factors which lead to the selective survival of one or another phenotype.

One of the basic concepts of population genetics, one of prime importance in understanding evolution, is that each population is characterized by a certain **gene pool.** Each individual in the population is genetically unique, and has a specific genotype. However, if we count all of the alleles of a given gene ($\mathbf{A_1A_2A_3}$ . . .), either in the entire population or in some statistically valid sample of the population, we can then calculate the fraction of the total pool represented by allele $\mathbf{A_1}$, by allele $\mathbf{A_2}$ and so on (Fig. 10–1). A population with a gene pool that is constant from one generation to the next is said to be in **genetic equilibrium.** When a population is in genetic equilibrium, the frequency of each allele in the population remains unchanged in successive generations.

In contrast, a population undergoing evolution is one in which the gene pool is changing from generation to generation. The gene pool of a population may be changed by mutations, by the introduction of genes from some outside population into this population, or by natural selection. Recombinations brought about by crossing over or by the assortment of chromosomes in meiosis may lead to new combinations of genes. The new phenotypes resulting from such recombinations may have some specific advantage or disadvantage for survival that would ultimately be reflected in a change in the genetic pool.

Evolution by natural selection implies that individuals with certain traits have more surviving offspring in the next generation. In this way they contribute a proportionately greater percentage of genes to the gene pool of the next generation than do organisms with other traits. New inherited variations arise primarily by mutation. If organisms with the new mutation survive and have on the average more offspring that survive than the organisms without that particular mutant allele, the gene pool of the population will gradually change. Thus the number of these mutant alleles in the population will increase in succeeding generations.

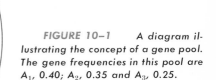

FIGURE 10–1   A diagram illustrating the concept of a gene pool. The gene frequencies in this pool are A_1, 0.40; A_2, 0.35 and A_3, 0.25.

This process, which is termed **differential reproduction** or nonrandom reproduction, implies that the conditions of the Hardy-Weinberg equilibrium do not apply in the population at that time. The individuals that produce more surviving offspring in the next generation, are usually, but not necessarily, those that are best adapted to survive in the given environment. Well adapted individuals may be healthier, better able to obtain food and mates, and better able to care for their offspring; however, of primary importance in evolution is how many of their offspring survive to be parents of the next succeeding generation.

The ultimate raw material of evolution is a **mutation** which establishes an alternative allele at a given locus and makes possible an alternative phenotype. Evolutionary changes are possible only when there are alternative phenotypes that may survive or perish. However, it is important to keep in mind that the process of selection does not operate gene by gene, but rather individual by individual, and on the basis of the effects of the individual's entire genetic system. The forces of natural selection operate on the entire individual and not on single traits.

When a mutation first appears, only one or a very few organisms in the population will bear the mutant gene, and these will breed with other members of the population from which the mutant arose. The change in the gene pool so that the mutant gene appears with greater and greater frequency in the population is a gradual process which may occur over many generations. In evolutionary changes in a large population, the success or lack of success of some new mutant gene will depend largely on its ability to confer on its possessors the capacity to leave a larger number of surviving individuals in the next generation.

Most, perhaps all, genes have many different effects on the phenotype (they are said to be **pleiotropic**). Some of the effects of a given gene may be advantageous for survival (termed **positive selection pressure**) and others may be disadvantageous (termed **negative selection pressure**). Whether the frequency of a given allele increases or decreases will depend on whether the sum of the positive selection pressures due to its advantageous effects is greater or less than the sum of the negative selection pressures due to its harmful effects.

In the reproduction of small populations chance alone may play a considerable role in determining the composition of the succeeding generation. The equilibrium of the gene pool of the population can be changed by chance processes rather than by natural selection. This role of chance in the evolution of small breeding populations has been described by Sewall Wright as *"genetic drift."* Within small interbreeding populations, heterozygous gene pairs tend to become homozygous for one allele or the other by chance rather than by selection. This may lead to the accumulation of certain disadvantageous characters and the subsequent elimination of the group possessing those characters.

The role that genetic drift actually plays in the evolution of organisms in nature has been the subject for debate among biologists but there seems little doubt that it does play at least a minor role. Certainly many animal and plant populations in nature are divided into subgroups small enough to be affected by the chance events underlying genetic drift. Genetic drift represents an exception to the Hardy-Weinberg Law underlying the tendency for a population to maintain its proportion of homozygous and heterozygous individuals. The Hardy-Weinberg Law is based on statistical events and, like all statistical laws, holds only when the number of individuals involved is large enough. Genetic drift may explain the common observation that closely related species in different parts of the world frequently differ in curious, even bizarre, ways which appear to have no particular adaptive value.

The role of chance in evolution is particularly evident when a species moves into a new area, for the number of individuals moving into that area is usually small. These first colonizers from which the entire new population develops rarely constitute a representative sample of the gene pool of the original population, but instead differ from the parent population in the frequencies of specific genes. These differences may be quite marked but the new colonizing generation tends to differ from the parent population in ways which are random rather than selective.

This effect is most apparent on islands and other areas of geographic isolation, and helps to account for the differences evident in island populations as compared to their mainland relatives. When a species is expanding continuously, the populations at the edge of the range, invading new areas, are likely to be small and differ genetically from the main body of the population. In all of these situations, when the breeding population is small, chance rather than selection may play a large role in determining the evolution of a particular group.

SECTION 10–4

DIFFERENTIAL REPRODUCTION

Populations become diverse by the action of the evolutionary forces of mutation, genetic drift, and the migration of genes from one population to another through migration and hybridization. Although these processes generally operate at random, a key feature of evolutionary change is the tendency of organisms to become adapted to survive and reproduce in a given environment. The evolutionary process itself is not random with respect to establishing adaptive features of the organism undergoing evolution.

The process of differential reproduction may not be random with respect to (1) union of male and female gametes, (2) the production of viable zygotes, or (3) the development and survival of the zygotes until they become adults and produce their own offspring. Nonrandom reproduction tends to produce nonrandom, directional changes in the genetic pool, which lead to nonrandom, directional evolution. In a strict sense, random mating implies that any male is as likely to mate with any one female as with any other, or more generally, that the gametes of any two individuals in a population will be equally likely to unite; however, in nature, mating is usually not completely at random.

There are well established behavior patterns of courtship and mating in many species which lead to the acceptance or refusal of one individual by another in mating. Such behavior patterns comprise one of the forces which may direct differential reproduction through **nonrandom mating**. For example, in many fishes or birds, some brightly colored part on the male (Fig. 10–2) serves as a stimulus to the female, which is necessary before copulation can be begun. Mutations which lead to the formation of bigger, brighter spots tend to make those males more attractive to females and may confer selective advantage on their possessors. Conversely, mutations which lead to the formation of smaller, duller spots would have a negative selection pressure. Darwin recognized this kind of evolutionary force, and termed it **sexual selection**, but it is simply one kind of natural selection, one factor that may result in differential reproduction.

Both the number of gametes produced by an individual and the proportion of the gametes that will unite with others to form zygotes, may be under genetic control. Both of these factors affect differential reproduction by what might be termed **"differential fecundity,"** differences in the number of viable zygotes that are produced in a given mating. Organisms in which the probability of survival of any given individual is low will usually have high fecundity to ensure survival of the species. This may be

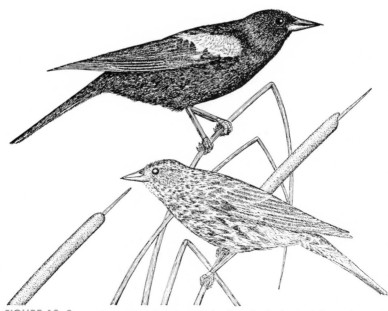

FIGURE 10–2   *The male (above) and the female (below) of the red-winged blackbird (Agelaius phoeniceus). Note the white spot on the male's wing. (From Orr, R. T.: Vertebrate Biology. 2nd ed. Philadelphia, W. B. Saunders Company, 1966.)*

an evolutionary advantage in one species but if the probability of survival of any given individual is high, an extremely high fecundity may actually reduce the chance of survival of the offspring by reducing the opportunity for parental care and feeding.

After a viable zygote has been formed by the union of two gametes, it must pass through a long period of development and growth before it becomes sexually mature and able to contribute offspring to the next generation. This phase of differential reproduction has had the greatest effect in determining the course of evolution. The differential success of organisms in a population in surviving to the reproductive age and contributing to the next generation's gene pool has been responsible for the more obvious features of organic evolution.

To survive to sexual maturity and to reproduce, each organism must be able to withstand a variety of physical factors in its environment, such as the amount of sunlight, moisture, temperature, gravity, light and darkness. In addition, each organism must live amidst other organisms, which leads to competition to eat and avoid being eaten. Plant species must compete for room in the soil and for sunlight, as well as for water and inorganic salts. Each plant is constantly threatened by animals that may eat it before it has had the opportunity to reach sexual maturity and release the spores or seeds that provide for the next generation. Animals are under similar pressure to avoid being eaten or killed and to find food for themselves. Any adaptation that improves an organism's ability to find food and avoid being eaten may play an important role in its differential reproduction. There are many ways by which the process of differential reproduction can be facilitated, and a correspondingly large number of ways in which natural selection may operate.

It is important to keep in mind that natural selection generally does not operate on the phenotypes of single genes, but rather on the phenotypic effect of the entire genetic system, or **genome.** One group of organisms may survive despite some clearly disadvantageous characteristic, while another group of organisms may be eliminated despite certain traits that appear to be highly advantageous for survival. The plants and animals that ultimately survive and are the parents of the next generation are those with qualities whose sum total renders them a little better able than their competitors to survive and reproduce their kind. Since the environment may change from time to time, the characteristics that are of adaptive value at one time may be useless or even deleterious at another.

SECTION 10–5

MUTATIONS—THE RAW MATERIALS OF EVOLUTION

The term "mutation" was coined by the Dutch botanist Hugo de Vries, one of the rediscoverers of Mendel's laws. De Vries carried out genetic experiments with the evening primrose and other plants which grew wild in Holland. When he transplanted these into his garden and crossed them, some of the resulting plants were unusual and differed markedly from the original wild plant. These unusual forms bred true in subsequent generations. For such sudden changes in the character of an organism, de Vries used the word "mutation." Darwin had described such sudden changes (called "sports" by earlier breeders) but believed that they occurred too rarely to be of importance in evolution. The vast number of genetic experiments with plants and animals carried on since 1900 have shown that mutations do occur constantly, and that the changes in the phenotype produced by such mutations may be of adaptive value and contribute to the survival of the organism. As the gene theory has developed, this word "mutation" has come to refer to a sudden, random, discontinuous change in the gene (or chromosome), although it is still used, to some extent, to refer to the new type of plant or animal. Hundreds of different mutations have been observed in the plants and animals most widely used in genetic experiments—corn and fruit flies. Among the mutations that have been observed in the fruit fly are changes in body color from yellow to brown, gray and black; eye colors ranging from red, brown and purple to white; wings that are curled, crumpled, shortened or completely absent; oddly shaped legs and bristles; and such remarkable changes as the development of a pair of legs on the forehead in place of the antennae (Fig. 10–3). The six-toed cats of Cape Cod and the short-

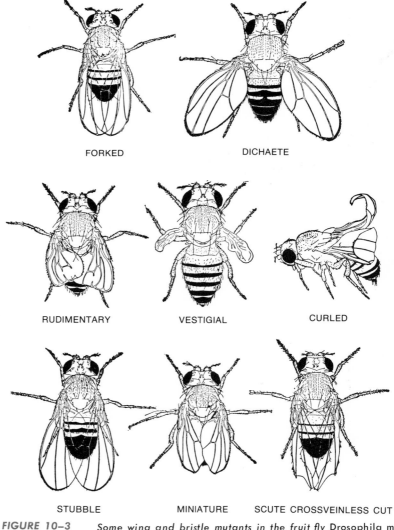

FORKED   DICHAETE

RUDIMENTARY   VESTIGIAL   CURLED

STUBBLE   MINIATURE   SCUTE CROSSVEINLESS CUT

FIGURE 10–3   *Some wing and bristle mutants in the fruit fly* Drosophila melanogaster. *(Drawn by E. M. Wallace; from Sturtevant, A. H., and Beadle, G. W.: An Introduction to Genetics. 2nd ed. Philadelphia, W. B. Saunders Company, 1939.)*

legged breed of Ancon sheep are examples of mutations among domestic animals.

At one time, there was heated discussion among biologists as to whether evolution results from natural selection or from mutations. It is now abundantly clear that evolution can take place only when mutations have occurred to provide alternative ways of overcoming the environment. The evolution of new kinds of plants and animals must involve both mutation and natural selection by differential reproduction.

Some mutations produce barely distinguishable changes in the structure or function of the organism in which they occur. Other mutations produce a major change early in development, and lead to multiple marked changes in the resulting body form or function. Usually when such a major change occurs in the control of some early stage of development, the result is a nonviable monster which dies almost immediately. A few such major changes may give rise to forms which are enabled by their mutation to occupy some new environment. These comprise what Richard Goldschmidt of the University of California called **"hopeful monsters."** He suggested, for example, that the ancestral type of bird, *Archaeopteryx*, evolved into the modern bird by a mutation which in a single step changed the shape of its tail. *Archaeopteryx* had a long, reptile-like tail that was covered with feathers. If a single mutation caused a shortening of the tail, then a "hopeful monster" with the fan-shaped arrangement of feathers characteristic of modern birds might have resulted (Fig. 10–4). This fan-shaped tail, better suited for flying than the former, long, reptile-like tail, would give its possessors a selective advantage during evolution. Such skeletal changes are known to occur as the result of a single mutation. The Manx cat, for example, owes its stubby tail to a mutation which causes the shortening and fusing of most of the tail vertebrae. A similar mutation in an ancestral *Archaeopteryx* might have led to the present fan-shaped arrangement of the tail in modern birds.

FIGURE 10–4   *A comparison of the structure of the tail of the primitive bird Archaeopteryx (right) and the tail of a modern bird (left).*

Types of Mutations

Some mutations, termed chromosomal mutations, are accompanied by a visible change in the structure of the chromosome, or by a change in the total number of chromosomes per cell. A single chromosome may be added to, or deleted from, the usual diploid set, or the entire set of chromosomes may be doubled or tripled, yielding organisms called **poly-ploids**. Polyploid plants and animals are usually larger and more robust than their diploid parents (Fig. 10–5). Changes in chromosome number are observed more frequently in plants than in animals, because the nature of the reproductive process in plants permits these altered chromosome numbers to be passed from one generation to the next. Some of the cultivated varieties of tomatoes, corn, wheat and other plants owe their vigor and the large size of their fruit to the fact that they are polyploids.

The structure of one or more chromosomes in the set may be altered by the removal of a small part of the chromosome (a **deletion**), by the doubling of a small part (**duplication**), or by the turning end for end of a small or large section of the chromosome (**inversion**). The interchange of segments between two nonhomologous chromosomes is called a **trans-location.** Each of these kinds of changes may produce phenotypic effects and may be inherited in a Mendelian fashion. Some of the traits that were thought at one time to be inherited by a true genic difference have been found to arise from chromosomal mutations. The mutations that are not accompanied by a visible change in chromosomal structure are termed "point" mutations or "gene" mutations. These involve some alteration in the nucleotide sequence within the DNA of the gene (p. 199).

Causes of Mutations

Both gene and chromosomal mutations can be produced artificially by α, β or gamma rays, x-rays, neutrons, heat, cold, and ultraviolet rays, or by chemicals such as nitrogen mustards, alkylating agents, and epoxides.

FIGURE 10–5   Photographs of diploid (left) and tetraploid (right) Easter lilies illustrating the differences in size typically seen in polyploid plants. (Courtesy of S. L. Emsweller, Bureau of Plant Industry, U. S. Department of Agriculture. Science in Farming, Yearbook of Agriculture, 1943–47.)

Cosmic and other natural rays bombarding the earth may cause some of the spontaneous mutations that are observed. Others may result simply from errors that occur during the replication of genes. Both chromosomal and point mutations occur in wild populations as well as in laboratory stocks, and they occur with a frequency great enough to account for the known rates of evolutionary change. Most of the mutations that occur both in laboratory stocks and in wild populations are detrimental, i.e., they cause phenotypic changes that decrease the probability of survival. The animals and plants that are living today have resulted from a long process of natural selection which preserved the most beneficial mutations. These present-day plants and animals are highly specialized and well adapted to their environment. Further mutations are much more likely to be disadvantageous than helpful.

SECTION 10–6

BALANCED POLYMORPHISM

Geneticists use the term **polymorphism** ("many shapes") to refer to two or more types of individuals that differ discontinuously — that is, without intergrading intermediate forms — in some genetically determined characteristic. The human blood groups O, A, B and AB (p. 246) provide a classic example of a polymorphism. Within a given population you might expect the selection process to yield a population completely homozygous for whichever member of a pair of alleles determines the trait with the greatest adaptive value. This does indeed happen in some evolutionary lines, but it is not the only possibility in differential reproduction. For example, the individuals heterozygous for the gene for sickle cell anemia (**Ss**) are somewhat more resistant to malaria than are homozygous normal individuals (**SS**). In Central Africa where malaria is endemic, there is strong positive selection pressure for the heterozygous individuals (**Ss**) because of their resistance to malaria, a strong negative selection pressure against the homozygous sickle cell anemics (**ss**) and a slight negative selection pressure against the homozygous normal (**SS**) individuals because of their lack of resistance to malaria. These separate and opposing forces act to maintain an equilibrium, a **"balanced polymorphism,"** of homozygotes and heterozygotes.

The existence of variation itself may be of adaptive value for a population, because a completely homozygous population would have no genetic substratum on which natural selection could act. A population which has a good prognosis for survival in the future is one that has maintained enough variation to permit further adaptive changes. Observations on wild populations of fruit flies and other organisms have shown that their genetic pools do change adaptively, even in response to such changes in the environment as the alternations of the seasons.

The heterozygous individual may have greater fitness for reproducing and surviving than either of the corresponding homozygous individuals. Obviously, however, the heterozygous state cannot be maintained in a population unless a certain number of the somewhat less fit homozygous individuals are also produced. The relative selective values of the heterozygous and homozygous states will determine the particular ratio of alleles in the gene pool that will result in the optimal proportion of heterozygotes and homozygotes.

SECTION 10–7

ADAPTIVE RADIATION

Because of the constant competition for food and living space, each group of organisms tends to spread out and occupy as many different habitats as possible. This process of evolution from a single ancestral species of a variety of forms which occupy somewhat different habitats, is termed "adaptive radiation" (Fig. 10–6). It is clearly advantageous in evolution in enabling the organisms to tap new sources of food or to escape from some of their enemies.

One of the classic examples of adaptive radiation is the evolution of placental mammals (Fig. 10–7). From a primitive, insect-eating, five-toed, short-legged creature that walked with the soles of its feet flat on the ground have evolved all the present-day types of placental mammals. These include dogs and deer adapted for terrestrial life in which running rapidly is important for survival; squirrels and primates adapted for life in the trees; bats equipped for flying; beavers and seals which maintain an amphibious existence; the completely aquatic whales, porpoises and sea cows; and the burrowing animals, moles, gophers and shrews. In each of these, the number and shape of the teeth, the length and number of leg bones, the number and attachment sites of muscles, the thickness and color of fur, the length and shape of the tail and so on have undergone changes that increase the adaptation of the animal to its particular environment.

A comparable major adaptive radiation of the reptiles occurred in an earlier geologic age. Adaptive radiation may take place on a very small scale, as represented by the variety of ground finches found today on the Galápagos Islands west of Ecuador. Some of these birds live on the ground and feed on seeds, others feed mainly on cactus, still others have taken to living in trees and eating insects. These have been accompanied by evolutionary changes in the size and structure of the beak. The essence of adaptive radiation, then, is the evolution from a single ancestral form of a variety of different forms, each of which is adapted and specialized in some unique way to survive in a particular habitat.

Adaptive radiation which gives rise to several different types of descendants, adapted in different ways to different environments, may be termed "divergent evolution." The opposite phenomenon, **convergent evolution**, also occurs fairly frequently; i.e., two or more quite unrelated groups may, in becoming adapted to a similar environment, develop characteristics that are more or less similar (Fig. 10–8). For example, wings have evolved not only in birds, but in mammals (bats), in reptiles (pterosaurs) and in insects. A very similar, streamlined shape, dorsal fins, tail fins and flipper-like fore and hind limbs have evolved in dolphins and

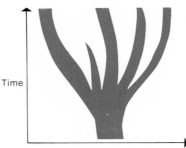

Time

Range of characteristic

FIGURE 10–6   *Diagram illustrating the increasing range of a characteristic as evolution occurs with time. Four different stocks are indicated as emerging by evolution from a single relatively homogenous stock.*

FIGURE 10–7   *Adaptive radiation. All the various mammals shown have evolved from the common ancestor depicted in the center. Each of the descendants has become adapted to a different type of habitat.*

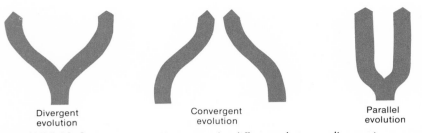

Divergent
evolution

Convergent
evolution

Parallel
evolution

FIGURE 10–8.   *Diagram illustrating the difference between divergent, convergent and parallel evolution. A single stock may branch to give two diverging stocks which become more and more different as evolution proceeds. In convergent evolution two stocks originally quite different come to resemble each other more and more as time passes, probably because they occupy a comparable habitat and become adapted to similar conditions. In parallel evolution a single stock branches into two which then evolve in parallel fashion for a long period of time as each stock independently responds to similar environmental influences.*

FIGURE 10–9   Convergent evolution. A, Shark, B, ichthyosaur (a fossil reptile) and C, dolphin (a mammal), all of which have a marked superficial similarity because of their adaptation to similar environments.

porpoises (which are mammals), the extinct ichthyosaurs (which were reptiles) and in both bony and cartilaginous fishes (Fig. 10–9). Moles and gophers have adapted to a burrowing life and have evolved similar fore and hind leg structures adapted for digging. The mole is an insectivore and the gopher is a rodent. The eye of the squid and the eye of a vertebrate such as the fish are also very similar in structure although very different in their embryonic origin.

SECTION 10–8

SPECIATION

The unit of classification for both plants and animals is the species. It is difficult to give a definition of this term which can be applied uniformly throughout the animal and plant kingdoms, but a **species** may be defined as a population of individuals with similar structural and functional characteristics, which have a common ancestry and in nature breed only with each other. A species is a collection of demes or populations within which interbreeding may occur—a group of populations with a common gene pool. It is implicit in this definition that there is no free flow of genes between two different species. One species is isolated from the next by reproductive barriers. These barriers are not absolute and the occasional hybrid that may occur between species does not alter their status as separate species.

Closely related species are grouped together in the next higher unit of classification, the **genus** (plural, genera). The scientific names of plants and animals consist of two words, the genus and the species, given in Latin. This system of naming organisms, called the **binomial system**, was first used consistently by the Swedish naturalist, Karl Linnaeus. Linnaeus, like many biologists before him, recognized that living things occur in discontinuous groups so that plants and animals can be assigned to separate, distinct kinds or species. Linnaeus catalogued and described plants in his *Species Plantarum* (1753) and animals in *Systema Naturae* (1758). This system of classification was based on structural similarities; but since these structural similarities are determined by the specific relationships of different gene pools, the modern classification of organisms based on evolutionary relationships is similar in many respects to the one set up by Linnaeus. Using the binomial system, the scientific name of the domestic cat, *Felis domestica*, applies to all the varieties of tame cats—Persian, Siamese, Manx, Abyssinian and so on. All of them belong to the same species, and they all are capable of interbreeding. Related species in the same genus are the lion, *Felis leo*, the tiger, *Felis tigris* and the leopard, *Felis pardus*. The dog, *Canis familiaris*, belongs to a different genus. The name of the genus is written first and is capitalized, and the name of the species is written second and is not capitalized.

Just as several species may be grouped to form a genus, a number of related genera may be grouped to constitute a **family**. Similarly, families may be grouped into **orders**, orders into **classes** and classes into **phyla** (singular, phylum). The phyla are the large major divisions of the plant and animal kingdoms just as the species are the basic small units of this evolutionary classification. The complete classification of man is phylum Chordata, subphylum Vertebrata, class Mammalia, subclass Eutheria, order Primates, family Hominidae, genus *Homo*, species *sapiens*.

Many plants and animals fall into easily recognizable, natural groups and their classification presents no difficulty. Others seem to lie on the borderline between two groups, with some characteristics in common with each, and are difficult to assign to one or the other. The number of the principal groups and the organisms included within each vary according to the basis of classification used and the judgment of the scientist making the classification (Fig. 10–10). Some taxonomists like to group things together in units that exist already. Others prefer to establish separate categories for forms that do not fall readily into one of the recognized classifications. For example, different taxonomists consider that there are from 10 to 36 animal phyla and from four to 12 plant phyla.

In providing an explanation for the origin of a new species, we must describe how the summation of unit evolutionary changes in a population may culminate eventually in the establishment of new species, genera, families and orders. This requires that *reproductive barriers* arise between the incipient species, as they are becoming established. When interbreeding between subgroups of a population becomes progressively less frequent and the resulting hybrids become progressively less fertile, the several groups eventually become different species. Any factor that decreases the amount of interbreeding between groups of organisms is termed an "*isolating mechanism*."

A very common type of isolation is *geographic*, the separation of groups of related organisms by some physical barrier, such as a river, desert, glacier, mountain or ocean. In a mountainous region the individual mountain ranges afford effective barriers between the valleys. Valleys only a short distance apart, but separated by ridges always covered with snow, typically have species of plants and animals that are peculiar to those valleys. Thus there are usually more different species in a given area in a mountainous region than on open plains. For example, in the mountains of western United States, there are 23 species and subspecies of rabbits, whereas in the much larger plains area of the Midwest and the East, there are only eight species of rabbits.

The isthmus of Panama provides another striking example of geographic isolation. On either side of the isthmus the phyla and classes of marine invertebrates are made up of different but closely related species. For some 16,000,000 years during the Tertiary period, there was no connection between North and South America, and marine animals could migrate freely between what is now the Gulf of Mexico and the Pacific Ocean. When the isthmus of Panama reemerged, the closely related groups of animals were isolated and the differences between the fauna in the two regions represent the subsequent accumulation of hereditary differences.

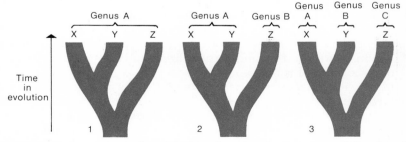

FIGURE 10–10   *A single original stock evolves in the course of time into three different groups of organisms—X, Y and Z. Different biologists looking at these three groups may either assign them all to the same genus (1) or to two genera (2) or to three different genera (3). Some biologists prefer taxonomic systems with a single large classification and many subdivisions. Others prefer separate subdivisions for each group that can be recognized. There is no single "correct" way of classifying organisms in any subdivision of the plant or animal kingdom.*

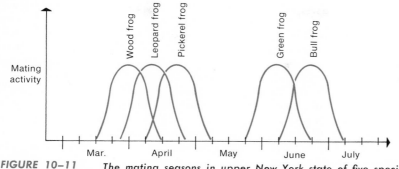

FIGURE 10–11   *The mating seasons in upper New York state of five species of* *frogs. Each species has a different period in which mating is most active. Where the mating* *seasons for several species overlap, the animals are usually separated because they occupy* *different breeding sites. (After Wallace, B., and Srb, A. M.: Adaptation. 2nd ed. Engle-* *wood Cliffs, New Jersey, Prentice-Hall, Inc. 1964.)*

Geographic isolation is usually not permanent, and the two previously isolated groups may come into contact again and resume interbreeding unless **genetic isolation** or interspecific sterility has arisen in the meantime. The various races of man have resulted from geographic isolation and the accumulation of chance mutations in different stocks. Since interracial sterility has not developed, the differences tend to disappear when geographic isolation is removed.

Genetic isolation results from one or more mutations that occur independently of mutations for structural or functional features. Genetic isolation may appear only after a long period of geographic isolation has produced striking differences between two groups of organisms, or it may originate within a single, otherwise homogeneous group of organisms. For example, in a species of fruit fly, *Drosophila pseudoobscura,* a mutation for genetic isolation has produced two groups of flies which, though externally indistinguishable, are completely sterile when cross-mated. These two groups are isolated genetically as effectively as if they lived on different continents. As generations pass and different mutations accumulate in each group by chance and by selection, the two groups will undoubtedly become visibly different. Two groups of organisms living in the same geographic area may be **ecologically isolated** if they occupy different habitats. Marine animals living in the intertidal zone are effectively isolated from other organisms living only a meter or so away below the low-tide mark. Ecologic isolation might result from the simple fact that two groups of organisms breed at somewhat different times of the year (Fig. 10–11).

SECTION 10–9

THE ORIGIN OF SPECIES BY HYBRIDIZATION

Although members of different species are usually not interfertile, occasionally members of two different but closely related species may interbreed to produce yet a third species by **hybridization.** Such phenomena make it difficult to establish a hard and fast definition of species. By hybridization, the best characters of each of the original species may be combined into a single descendant, thereby creating a new type better able to survive than either of its parents. If the new form combined the worst characters of both parents it would obviously be at a serious disadvantage and would be unlikely to survive.

When different species with different chromosome numbers are crossed, the offspring are usually sterile. The unlike chromosomes cannot pair in meiosis, and the resulting eggs and sperm do not receive the proper assortment of chromosomes, one of each kind. When, within such **interspecific hybrids**, the number of chromosomes is doubled, meiosis can take place in a normal fashion, and normal, fertile eggs and sperm will be produced. Thereafter the hybrid species will breed true, and indeed will not produce fertile offspring when bred with either of the parental species. Many related species of higher plants have chromosome numbers which are multiples of some basic number. The species of wheat include ones with 14, 28 and 42 chromosomes; there are species of roses with 14, 28, 42 and 56 chromosomes, and species of violets with every multiple of six from 12 to 54.

That such natural series arise by hybridization and doubling of the chromosomes is supported by laboratory experiments which yield similar series. One of the more famous of these experimental crosses was made by Karpechenko, who crossed the radish with a cabbage, hoping perhaps to get a plant with a cabbage top and a radish root. Radishes and cabbages belong to different genera, but both have 18 chromosomes. The resulting hybrid also had 18 chromosomes, nine from its radish parent and nine from its cabbage parent. Since the radish and cabbage chromosomes were unlike, they could not pair during meiosis, and the hybrid was almost completely sterile. By chance, however, a few of the eggs and pollen formed contained all 18 chromosomes, and a mating between two of these resulted in a plant with 36 chromosomes. This plant was fertile, for during meiosis the pairs of radish chromosomes underwent synapsis, and the pairs of cabbage chromosomes underwent synapsis. The hybrid exhibited some of the characteristics of each parent, and bred true for these characteristics. Unfortunately it had a radish-like top and a cabbage-like root. Since it could not be crossed readily with either of its parent species, it was, in effect, a new species produced by hybridization, followed by the doubling of the number of chromosomes.

A similar occurrence in nature has been documented in the marsh grasses. One of these, *Spartina townsendi*, first appeared more than 100 years ago in the harbor of Southampton, England, in company with two other species, *Spartina stricta* and *Spartina alterniflora*. The new species, *Spartina townsendi*, was much more vigorous than either of the parents and was soon widespread. It was especially valuable in collecting and holding soil and was transplanted to the Dutch dikes and to other parts of the world. Because it was intermediate in many characteristics between the two species with which it was first found, it was believed to have originated as a hybrid. When it was possible to examine the chromosome numbers, the hypothesis was confirmed, for *Spartina townsendi* was found to have 126 chromosomes, *Spartina stricta*, 56 and *Spartina alterniflora*, 70. Thus there is no doubt that the new species arose by hybridization and doubling of the chromosomes.

SECTION 10–10

PHYLOGENY

The evolutionary history of any group of organisms is termed its **phylogeny**. It is basic to many aspects of biological research to know which organisms are most closely related—i.e., which ones have common ancestors in the recent past, and which ones have common ancestors only in the more distant past. To establish the phylogenetic relationships of a group of organisms, each investigator must examine as many character-istics of each type as possible, looking for patterns of similarities and dis-similarities that may provide clues. Phylogeneticists originally were re-stricted largely to comparing morphologic characters — patterns of bones, muscles and nerves — but now a host of physiologic, biochemical, immuno-logic and cytologic characters can be examined and used to test the validity of the relationships inferred on the basis of morphologic characters. It is reassuring to find that the evolutionary relationships inferred on the basis of the newest, most sophisticated biochemical analyses of the types of pro-teins found in different species agree remarkably well with the evolu-tionary relationships established a century ago on the basis of gross morphologic similarities.

SECTION 10–11

THE EVIDENCE FOR EVOLUTION

The evidence that organic evolution has occurred is so overwhelming and comes from so many varied sources, that no one who is acquainted with it has any doubt that new species are derived from previous ones by descent with modification. For many animals and some plants there is an extensive **fossil record** that provides direct evidence of organic evolution, and gives the details of the phylogeny, the evolutionary relationship, of the several lines of descent. Most of the individual facts underlying each of the subdivisions of biological science acquire significance and make sense only when viewed against the background of evolution.

The term "fossil" (Latin *fossilium* — something dug up) refers not only to the bones, shells, teeth and other hard parts of an animal or plant body which have been preserved, but to any impression or trace left by some previous organism (Fig. 10–12). In view of the large number of fossils of animals and plants that have been found, it is sobering to realize that only a small fraction of all the organisms that ever lived have been preserved as fossils, and only a small fraction of those fossils have been dug up and studied to date. The science of **paleontology** deals with the finding, catalog-ing and interpreting of the abundant and diverse evidence of life in former times.

Footprints or trails made in soft mud which subsequently hardened are a common type of fossil. From such clues, one can make guesses about the structure and body proportions of the animals which made the foot-prints. Most of the vertebrate fossils are skeletal parts, from which it is possible to deduce the animal's posture and style of walking. From the bone scars which indicate the sites of muscle attachments, paleontologists can deduce the general position and size of the muscles, and from this, the contours of the body. From such studies, reconstructions have been made of how the animal may have looked in life. Such qualities as the texture and color of the fur or scales can only be guessed at.

One striking type of fossil is that in which the original hard parts or even the soft tissues of the organism have been replaced by minerals — a process called **petrifaction**. The minerals that replace the tissue may be iron pyrites, silica, calcium carbonate or some similar substance. The petri-fied muscles from a shark more than 300,000,000 years old were found. The petrifaction process had preserved them so well that not only could

FIGURE 10–12   One of the more famous examples of a fossil, the remains of Archaeopteryx, a tailed, toothed bird from the Jurassic Period. (Courtesy of the American Museum of Natural History.)

individual muscle fibers be detected, but their cross-striations could be observed in thin sections under the microscope.

Molds and casts are superficially similar to petrified fossils, but are produced in a different fashion. A **mold** is formed by the hardening of the material surrounding the buried organism, followed by the decay and removal of the organism by seepage of ground water. The molds may be filled subsequently with minerals, which in turn harden to form **casts**, replicas of the original structures. Small organisms such as insects and spiders may be preserved in amber, a fossil resin from pine trees. Originally the resin was the sap of the pine tree, soft enough to engulf the fragile insect and penetrate every part, then it gradually hardened and preserved the animal intact (Fig. 10–13). An unusual form of fossil is the preservation of an animal by freezing. Woolly mammoths more than 25,000 years old have been found in Siberia and Alaska, preserved in the frozen state. Their flesh was so well preserved that it was eaten by dogs.

The formation of a fossil requires that the organism be buried in some way. This may occur at the bottom of a pond or lake, or on land by the accumulation of wind-blown sand, soil or volcanic ash. The men and animals living in Pompei were preserved almost perfectly by the volcanic ash from the eruption of Vesuvius. Animals may be trapped and entombed in a bog or asphalt pit, such as the famous La Brea tar pits in Los Angeles, which have provided superb fossils of Pleistocene animals.

Even if the remarkably detailed fossil record did not exist, studies of the structure, function, development, genetics and chemistry of present-day plants and animals and their distribution over the earth's surface would provide overwhelming proof that biological evolution has occurred.

The classification of modern-day organisms into well defined groups is possible only because most of the intermediate forms have become extinct. If every animal and plant that ever lived were still living today, they would comprise a continuous series of forms intergrading from the very primi-

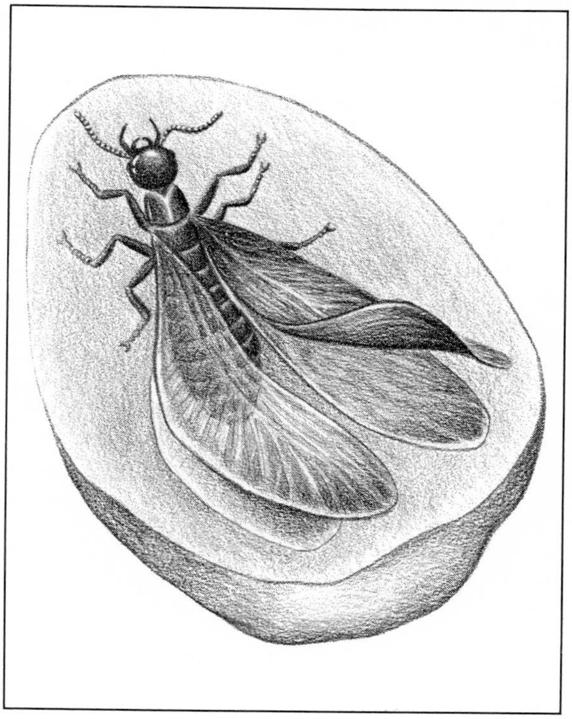

FIGURE 10–13   A termite embedded in amber. Such insects, dating from the middle Tertiary Period (about 38,000,000 years ago) have been preserved almost perfectly. (After a specimen, courtesy of A. E. Emerson.)

tive to the most complex, and it would be difficult to divide the living world into neat taxonomic categories. The species now living have been termed "islands in a sea of death" and have been compared to the terminal twigs of a tree, of which the trunk and main branches have disappeared. The problem of the taxonomist is to reconstruct the missing branches and put each twig on the proper branch.

The characteristics of living things are such that the organisms can be fitted into a hierarchical scheme of categories—species, genera, families, orders, classes and phyla. The most reasonable explanation for this is that this hierarchical scheme indicates evolutionary relationships. If the kinds of plants and animals were not related by evolutionary descent, their characteristics would be present in a confused, random pattern and no such hierarchy of forms could be established.

When one compares the structures of a given organ system among different animals or different plants, fundamentally similar patterns of structure become evident. Such comparisons are quite useful in helping to demonstrate evolutionary relationships; however, only structural patterns that are based on homologous organs are useful in inferring evolutionary relationships. *Homologous organs* are ones which are basically similar in structure, in their relationships to adjacent structures, in their embryonic development, and in their nerve and blood supply. For example, a seal's front flipper, a bat's wing, a cat's paw, a horse's front leg and the human hand and arm (Fig. 14–1), though superficially dissimilar and adapted for quite different functions, are homologous organs. Each consists of almost the same number of bones, muscles, nerves and blood vessels, arranged in the same pattern, and with very similar modes of development. The existence of such homologous organs, though each is adapted for quite different functions, is a strong argument for a common evolutionary origin.

Most plants and animals contain organs or parts of organs that are useless and degenerate, often undersized or lacking some essential part, compared to homologous structures in other related organisms. In the human body there are more than 100 such **vestigial organs**, including the appendix, the coccyx (the fused tail vertebrae), the wisdom teeth, the nictitating membrane of the eye, body hair and the muscles that move the ears (Fig. 10–14). Vestigial organs are the remnants of organs which were functional in some ancestral animal. Because of the change in the environment or mode of life of the species, the organ became unnecessary for survival and gradually lost its function. Mutations are constantly occurring which decrease the size and function of one organ or another. If such organs are necessary for survival, the organisms containing such mutations will, of course, tend to be eliminated. If the organs are not necessary for survival, they will become reduced in size, "vestigial," and eventually eliminated.

Evidence for evolutionary relationships can be inferred from functional and chemical comparisons as well as from morphological comparisons. Studies of the chemicals present in the blood have been particularly fruitful in yielding evidence of relationships. The degree of similarity between the plasma proteins of different animals may be tested by the **antigen-antibody technique**. An animal, such as a rabbit or sheep, is given repeated injections of the protein to be tested, such as human serum, which contains proteins foreign to the rabbit's blood. The plasma cells of the rabbit respond by producing antibodies specific for the antigenic proteins in the human blood. The antibodies are then obtained by withdrawing blood from the rabbit and allowing it to clot to obtain the serum. Even a dilute sample of the serum, when mixed with human blood, will result in a visible precipitation, caused by the combination of antigens and antibodies. The precipitation will be formed with nonhuman blood only if the serum is very much more concentrated, and even then the precipitation may be greatly delayed or nonexistent. By using a series of rabbits, each of which was injected with the blood of a different species, it is possible to ob-

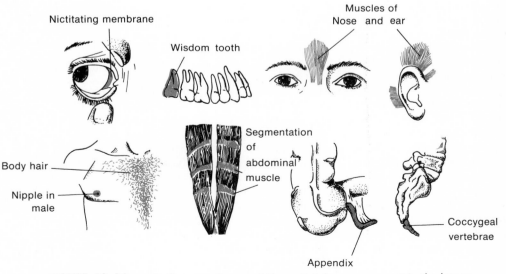

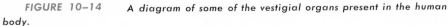

FIGURE 10–14   A diagram of some of the vestigial organs present in the human body.

tain a series of antibodies, each specific for the blood proteins of a particular species of animal.

Literally thousands of such tests involving different animals have revealed a basic similarity between the blood proteins of all the mammals, the degree of relationship being indicated by how much the antigen and antibody solutions can be diluted and still result in visible precipitation. Man's closest "blood relations," determined on the basis of these proteins, are the great apes, then, in order, the Old World monkeys, the New World prehensile-tailed monkeys and the tarsioids. Of all the types of primate blood, the lemur's results in the least precipitation when combined with antibodies specific for human serum. The biochemical relationships of a great number of different forms tested in this way correlate very closely with the relationships inferred from other types of evidence.

Investigations of the sequence of amino acids in the α and β chains of hemoglobin from different species have revealed great similarities, of course, and also specific differences, the pattern of which demonstrates the order in which the underlying mutations, the changes in the nucleotide base pairs, must have occurred in evolution. The evolutionary relationships inferred from these studies agree completely with those based on morphological studies. Other analyses of the sequence of amino acids in the protein portion of the cytochrome enzymes provide further concurring evidence of evolutionary relationships (Fig. 10–15). Thus evidence of evolutionary relations can be obtained by studying similarities and differences in molecular structure as well as by studies of gross structure.

Comparisons of the properties of specific enzymes from different organisms have revealed similarities and differences from which evolutionary relationships can be deduced. The rates of reaction of enzymes such as lactic dehydrogenases with the normal pyridine nucleotide coenzyme (NAD), relative to the rates with analogues of NAD, can be used to demonstrate evolutionary relationships. Enzymes from animals deemed to be closely related on the basis of anatomic or other evidence show very similar patterns of their rates of reaction, whereas enzymes from animals deemed to be only distantly related show very different patterns.

It might seem unlikely that an analysis of the urinary wastes of different species would provide evidence of evolutionary relationship, yet this is true. The kind of waste excreted depends upon the particular kinds of enzymes present, and the enzymes are determined by genes which have been selected in the course of evolution. The waste products of the metabolism of adenine and guanine are excreted by man and other primates as **uric acid**, by other mammals as **allantoin**, by amphibians and most fishes as **urea**, and by most invertebrates as **ammonia.** Vertebrate evolution has been marked by the successive loss of enzymes required for the stepwise breakdown of uric acid. Joseph Needham made the interesting observation that the chick embryo in the early stages of development excretes ammonia, then urea, and finally uric acid. The enzyme **uricase**, which catalyzes the first step in the degradation of uric acid, is present in the early chick embryo but disappears later. The adult frog excretes urea, but the tadpole excretes ammonia. These are biochemical examples of "recapitulation."

The importance of embryologic evidence for evolution had been stressed by Darwin and was reemphasized by Ernst Haeckel in 1866 when he developed his theory that embryos, in the course of development, repeat the evolutionary history of their ancestors in some abbreviated form. This idea, succinctly stated as "Ontogeny recapitulates phylogeny," stimulated research in embryology and focused attention on the general resemblance

	Human	Monkey	Pig, Bovine, sheep	Horse	Dog	Rabbit	Kangaroo	Chicken, turkey	Duck	Rattlesnake	Turtle	Tuna fish	Moth	Neurospora	Candida	Yeast
Human	0															
Monkey	1	0														
Pig, bovine, sheep	10	9	0													
Horse	12	11	3	0												
Dog	11	10	3	6	0											
Rabbit	9	8	4	6	5	0										
Kangaroo	10	11	6	7	7	6	0									
Chicken, turkey	13	12	9	11	10	8	12	0								
Duck	11	10	8	10	8	6	10	3	0							
Rattlesnake	14	15	20	22	21	18	21	19	17	0						
Turtle	15	14	9	11	9	9	11	8	7	22	0					
Tuna fish	21	21	17	19	18	17	18	17	17	26	18	0				
Moth	31	30	27	29	25	26	28	28	27	31	28	32	0			
Neurospora	48	47	46	46	46	46	49	47	46	47	49	48	47	0		
Candida	51	51	50	51	49	50	51	51	51	51	53	48	47	42	0	
Yeast	45	45	45	46	45	45	46	46	46	47	49	47	47	41	27	0

FIGURE 10–15   A diagram illustrating the differences in amino acid sequences in the cytochrome c's from a variety of different species of animals and plants. The numbers refer to the number of different amino acids in the cytochrome c's of the species compared. (From Dayoff, M. O., and Eck, R. V.: Atlas of Protein Sequence and Structure. Silver Springs, Maryland, National Biomedical Research Foundation, 1968.)

between embryonic development and the evolutionary process. It is now clear that the embryos of the higher animals resemble the *embryos* of lower forms, not the *adults*, as Haeckel had believed. The early stages of all vertebrate embryos are remarkably similar and it is not easy to differentiate a human embryo from the embryo of a pig, chick, frog or fish (Fig. 10–16).

In recapitulating its evolutionary history in a few days, weeks or months, the embryo eliminates some steps and alters and distorts others. In addition, some new characters have evolved which are adaptive and enable the embryo to survive. Early mammalian embryos have many characteristics in common with those of fish, amphibia and reptiles, but also have other structures which enable them to survive and develop within the mother's uterus rather than within an egg shell. These secondary traits may alter the original characters common to all vertebrates so that the basic resemblances are blurred. The concept of recapitulation must be used with due caution, but it can be helpful in understanding such curious and complex patterns of development as those of the vertebrate circulatory

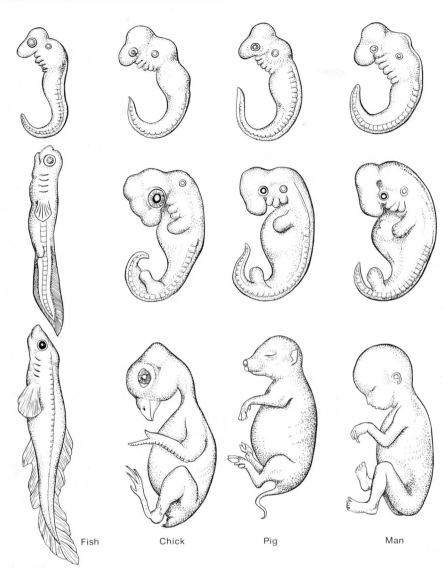

Fish   Chick   Pig   Man

FIGURE 10–16   *Successive stages in the embryonic development of the fish, chick, pig and man. Note that the earlier stages of development (top row) are remarkably similar and that differences become more marked as development proceeds.*

or excretory systems. It is also useful, when not taken too literally, in getting a broad picture of the whole of development. The fertilized egg may be compared to the single-celled flagellate ancestor of all animals, and the blastula to a colonial protozoan or some spherical multicellular form which may have been the ancestor of all the Metazoa. Haeckel believed that the ancestor of the coelenterates and all the higher animals was a gastrula-like animal, *Gastrea*, with two layers of cells and a central cavity connected by a blastopore to the outside.

After the gastrulation stage, development follows one of two main lines: in one (the echinoderms and chordates), the blastopore—the opening from the cavity of the gastrula—becomes the anus, or comes to lie near the anus; in the other (the annelids, mollusks, arthropods and others), the blastopore becomes the mouth, or comes to lie near the mouth.

In both lines, a third layer of cells—the mesoderm—develops between the ectoderm and endoderm. In the chordate-echinoderm line this de-velops, at least in part, as pouches or evaginations from the primitive diges-tive tract, while in the annelid line the mesoderm originates from special cells differentiated early in development.

Shortly after the appearance of the mesoderm, all chordate embryos develop a dorsal, hollow nerve cord as well as a notochord—the internal supporting rod for the body—and perforations in the pharynx (the gill slits). The early human embryo resembles a fish embryo, with gill slits, pairs of aortic arches, or blood vessels traversing the gill bars, a fishlike heart with a single atrium and ventricle, a primitive pronephros or fish kidney, and a tail, complete with muscles for wagging it. Later the human embryo resembles a reptilian embryo: Its gill slits close; the bones which make up each vertebra and which had been separate, as in fish embryos, fuse; a new kidney—the mesonephros—forms, and the pronephros dis-appears; and the atrium becomes divided into right and left chambers. Still later the human embryo develops a mammalian, four-chambered heart, and a third, completely new kidney (the metanephros), while the notochord regresses, and so on. During the seventh month of intrauterine development, the human embryo resembles—in being completely covered with hair and in the relative size of body and limbs—a baby ape more than it does an adult human.

The selection and breeding of domesticated animals and cultivated plants for the past several thousand years provide us with models of how some evolutionary forces may operate. All the varieties of present-day dogs are descended from one or a few related species of wild dog or wolf, and yet they vary tremendously in many characteristics. Compare, for example, the size of the chihuahua and the St. Bernard or Great Dane; the head shape of the bulldog and collie; the body proportions of the cocker, dachs-hund and Russian wolfhound. If these varieties were found in the wild, they would undoubtedly be assigned to different species, and perhaps even to different genera. But since all are known to come from common an-cestors and since all are interfertile, they are regarded as varieties or races of a single species.

The plant breeders who developed the present varieties of cultivated plants have similarly produced, by selection and interbreeding, a tremen-dous variety of plants from one or a few forms. The cliff cabbage, for ex-ample, which still grows wild in Europe, is the ancestor, not only of our cultivated cabbage, but of such dissimilar plants as cauliflower, kohlrabi, Brussels sprouts, broccoli and kale. The cultivated species of tobacco has been traced back to a cross between two species of wild tobacco, and corn has been traced to teosinte (a grasslike plant growing wild in the Andes and Mexico). Breeding experiments and observations indicate that species are not, as Linnaeus believed, unchangeable biological entities, each of which was created separately, but groups of organisms which have arisen from other species and which can give rise to still others.

The number and the detailed structure of the chromosomes of related species can be compared by cytologic methods. The patterns of inversions, deletions, duplications and translocations present in the chromosomes have provided useful evidence concerning the evolutionary history of fruit flies, jimson weeds, primroses and many other plants and animals.

Not all plants and animals are found in all parts of the world; they are not even found everywhere that they could survive, as one would expect if climate and topography were the only factors determining distribution. Central Africa, for example, has elephants, gorillas, chimpanzees, lions and antelopes, while Brazil, with a similar climate and other environmental con-ditions, has none of these, but does have prehensile-tailed monkeys, sloths and tapirs. The present distribution of organisms is understandable only on the basis of the evolutionary history of each species.

The *range* of a given species—i.e., the portion of the earth in which it is found—may be only a few square miles or, as with man, almost the entire world. In general, closely related species do not have identical ranges, nor are their ranges far apart; they are usually adjacent but separated by a barrier of some sort, such as a mountain or a desert. This generalization was formulated by David Starr Jordan and is known as *Jordan's Rule.* It follows from the role of isolation in the formation of species.

As one would expect, regions such as Australia and New Zealand, which have been separated from the rest of the world for a long time, have a flora and fauna which is peculiar to them. Australia has a population of monotremes and marsupials found nowhere else. During the Mesozoic, Australia was isolated from the rest of the world, so that its primitive mammals never had any competition from the better-adapted placental mammals, which eliminated the monotremes and most of the marsupials everywhere else. The primitive mammals gave rise to a variety of forms which were able to take advantage of the different habitats available.

The facts about the distribution of plants and animals constitute the science of *biogeography*, one of the basic tenets of which is that each species of animal and plant originated only once. The particular place where this occurred is known as its center of origin. The center of origin is not a single point, but the range of the population when the new species was formed. From its headquarters each species spreads out until halted by a barrier of some kind—physical, such as an ocean or mountain; environmental, such as an unfavorable climate; or biological, such as the absence of food or the presence of enemy organisms which prey upon it or compete with it for food or shelter.

SECTION 10–12

PRINCIPLES OF EVOLUTION

The experts in evolutionary theory may have differences of opinion regarding the precise molecular events that underlie mutations, the kinds of mutations involved in evolution, and the degree to which such factors as natural selection, isolation, genetic recombination, hybridization and the size of the breeding population affect the evolution of some particular organism. However, there are several fundamental facts about which they are agreed: Changes in the chromosomes and genes are the raw materials of evolution; some sort of isolation is necessary to establish a new species; and natural selection by differential reproduction is involved in the survival of some, but not all, of the mutations which occur. In addition, there are five principles of evolution to which nearly all biologists can subscribe: (1) Evolution occurs more rapidly at some times than at others. At the present time it is occurring rapidly, with many new forms appearing and many old ones becoming extinct. (2) Evolution does not proceed at the same rate among different types of organisms. At one extreme are the lampshells or Brachiopods, some species of which have remained unchanged for the last 500,000,000 years at least, for fossil shells found in rocks deposited at that time are identical with those of animals living today. In contrast, several species of man have appeared and become extinct in the past few hundred thousand years. In general, evolution occurs rapidly when a new species first appears, and then gradually slows down as the group becomes established. (3) New species do not evolve from the most advanced and specialized forms already living but from relatively simple, unspecialized forms.

For example, the mammals did not evolve from the large, specialized dinosaurs but from a group of rather small and unspecialized reptiles. (4) Evolution is not always from the simple to the complex. There are many examples of "regressive" evolution, in which a complex form has given rise to simpler ones. Most parasites have evolved from free-living ancestors which were more complex than the present forms; wingless birds, such as the cassowary, have descended from birds that could fly; many wingless insects have evolved from winged ones; the legless snakes came from reptiles with appendages; the whales with no hind legs evolved from mammals that had two pairs of legs. These are all reflections of the fact that gene mutations occur at random, and the resulting changes in phenotype do not necessarily increase the complexity or the "perfection" of the organism. If there is some advantage to a species in having a simpler structure, or in doing without some structure altogether, any mutations favoring the simpler condition will tend to accumulate by natural selection. (5) Evolution occurs by populations, not by individuals, by the processes of mutation, differential reproduction, natural selection and genetic drift.

SUGGESTIONS FOR FURTHER READING

Darwin, Charles: *On the Origin of Species by Means of Natural Selection*, or the *Preservation of Favored Races in the Struggle for Life*. Garden City, New York, Doubleday & Company, Inc., 1960.
> Every student of biology should at least browse through this book to catch its special flavor and to appreciate its extraordinary force.

Darwin, Charles: *The Voyage of the Beagle*. Garden City, New York, Natural History Library, Doubleday & Company, Inc., 1962.
> Darwin's chronicle of the expedition during which he made the observations that led him eventually to his theory of evolution. This is an excellent introduction to the man and his work.

Dobzhansky, Theodosius: *Evolution, Genetics and Man*. New York, John Wiley & Sons, Inc., 1955.
> This book discusses certain principles of evolution as they relate to human evolution.

Dobzhansky, Theodosius: *The Genetic Basis of Evolution*. Sci. Amer. 182:32, January 1950.
> Presents the Neo-Darwinian viewpoint of the importance of natural selection.

Ehrlich, P. R., R. W. Holm, and P. H. Raven (eds.): *Papers on Evolution*. Boston, Little, Brown and Company, 1969.
> A collection of research and review papers by many of the scientists presently making contributions to our understanding of evolutionary processes.

Ehrlich, P. R., and R. W. Holm.: *The Process of Evolution*. New York, McGraw-Hill, Inc., 1963.
> A clear, literate presentation of evolution, updating the theory of natural selection.

Grant, V.: *The Origin of Adaptations*. New York, Columbia University Press, 1963.
> An interesting presentation of this subject for the advanced student.

Goldschmidt, Richard B.: *The Material Basis of Evolution*. New Haven, Yale University Press, 1940.
> This text gives a detailed argument for the importance of large mutations in evolution.

Mayr, E.: *Animal Species and Evolution*. Cambridge, Massachusetts, Harvard University Press, 1963.
> One of the more detailed presentations of evolutionary theory, relating especially to the appearance of new species.

Stebbins, G. Ledyard: *Processes of Organic Evolution*. Englewood Cliffs, New Jersey, Prentice-Hall, Inc., 1966.
> A brief review of the entire field by one of the outstanding students of the evolution of plants.

Williams-Ellis, A.: *Darwin's Moon*. London and Edinburgh, Blackie and Sons, Ltd., 1966.
> Gives a vivid portrayal of Wallace's life and times and reemphasizes his solid contributions to advancing the theory of evolution.

THE KINDS OF ORGANISMS

BIOLOGICAL INTERRELATIONSHIPS

CHAPTER 11

As we turn from the more cellular and molecular aspects of biology to those concerning whole organisms, we will consider some of the enormous variety of plants and animals that exist and the various ways in which they are related. Each of these plants and animals at first appears to be quite different from the rest; each seems to be going its separate way at its own pace. However, all organisms, whether plant or animal, in fact have fundamentally similar requirements for survival. Each must obtain food for energy, find space in which to live, and survive to produce the next generation. In addition, each must interact with many other species in a variety of ways that enhance or detract from its chances to survive.

In the course of evolution, the various groups of plants and animals have developed somewhat different solutions to these problems. Many different forms have evolved, each of which is adapted to live in some particular kind of environment. Each has become adapted to a certain range of variations in the physical features of its environment—i.e., it has acquired a tolerance to certain variations in moisture, wind, sun, temperature, gravity and so on. Each has also become adapted to a certain spectrum in its biotic environment, to a certain constellation of plants and animals that live in the same general region. This study of the interrelations between living things and their environment—both their physical environment and their biological environment—is known as *ecology.* In recent years, the general public has begun to become aware of the awesome problems that have arisen from our past ignorance of, and disregard for, the principles of ecology. Our concern over the growing pollution of the air and water is encouraging, and suggests that eventually appropriate steps will be taken to alleviate these problems and prevent their recurrence; however, that is only a portion of our present ecologic predicament, which has a much broader basis. Man must eventually learn how to keep his own numbers in check and how to live on this planet without altering the physical and biotic environment in such a way as to endanger the continued existence of himself and the other 1,700,000 kinds of plants and animals.

Living organisms are related one to another in two main ways—by their evolutionary descent, which was discussed in the previous chapter,

299

and by their ecologic interrelationships. One organism may provide food or shelter for another, or produce some substance harmful to the second, or the two may compete for food or shelter. A detailed study of ecology requires a great breadth of knowledge of the structure and functions of a wide variety of plants and animals. In the latter part of this book, after we have considered some of the details of plant and animal physiology and the types of plants and animals, we will return to a discussion of ecology, which serves as one of the major unifying concepts in biology.

SECTION 11–1

DISTINCTIONS BETWEEN PLANTS, ANIMALS AND PROTISTS

The unit of classification for all types of organisms is the **species** (p. 283), which was defined as a population of similar individuals, alike in their structural and functional characteristics, which in nature breed only with each other, and which have a common ancestry. In the taxonomic hierarchy, species are grouped together into the next higher unit, the genus, genera are grouped into families, families into orders, orders into classes, and classes into phyla. The phyla are the major divisions of the biological kingdoms, and the members of each phylum differ in one or more fundamental and distinctive characteristics from the members of all other phyla.

Since the time of Aristotle, biologists have divided the world of living things into the two kingdoms, plants and animals. The word plant suggests trees, shrubs, flowers, grasses and vines, all of which are familiar objects in our everyday world. The word animal suggests cats, dogs, lions, tigers, birds, frogs and fish. Ferns, mosses, mushrooms and pond scums, although quite different, can be recognized as plants, and insects, lobsters, clams, worms and snails are definitely animals. A climb over a rocky coast at the seashore and an inspection of the organisms that cling to the rocks or live in the tide pools will reveal some organisms that might be difficult to identify as either animals or plants. In addition, many of the one-celled organisms visible under the microscope would not easily be assigned to the plant or animal kingdom.

Although there are many fundamental similarities between plants and animals, there are some obvious ways and some obscure ways in which they differ. The bodies of both plants and animals are made up of cells, which are the basic structural and functional units of the body. Both plant and animal cells have many metabolic processes, many enzymatic reactions, in common. The general structural plans of the cells of both plants and animals are similar; both contain a nucleus, in which lie the chromosomes, and cytoplasm that includes organelles such as mitochondria, ribosomes and Golgi bodies.

A first major difference between plant and animal cells lies in the fact that plant cells secrete a hard, outer **cell wall** of cellulose, which encloses the living cell and serves to support the plant body. Animal cells, in contrast, do not have an outer cell wall, and hence can change their shape. There are some plants which have no cellulose walls, and one group of animals, the primitive chordates (called tunicates or sea squirts), which do have cellulose walls surrounding each cell. Secondly, the growth of plants is generally indeterminant. Perennial plants keep on growing indefinitely because some of the cells remain in an actively growing state. Many tropical plants continue to grow throughout the year; those in the temperate regions grow primarily in the spring and summer. In contrast, the ultimate size of the body of most animals is established after a finite period of growth. Again there are exceptions to this rule, and animals such as alligators, turtles and lobsters continue to grow for a very long, if not indefinite, period.

Animals in general are able to move about, whereas plants in general remain fixed in one place. Their roots grow into the soil to obtain water and nutrients, and they obtain energy from the sun by exposing broad, flat surfaces—their leaves—to the sunlight. A little thought should bring to mind exceptions to both of these distinctions.

One of the more important differences between animals and plants is their means of obtaining nourishment. Animals, by and large, move about and obtain food by eating organisms in the environment, whereas plants are stationary and manufacture their own food. Plants characteristically have the green pigment, chlorophyll, which enables them to carry on photosynthesis. Light energy is absorbed and used to split water molecules, thereby providing the raw materials to reduce carbon dioxide to carbohydrates. There are exceptions to this rule, too. Bacteria and fungi are "plants" that lack chlorophyll. There are species or genetic strains of some of the higher plants that have lost the ability to synthesize chlorophyll and thus are white rather than green. Since they cannot synthesize their own food, they must obtain their nutrients from other organisms, as animals do.

Although the reproductive cycles of plants and animals are fundamentally quite different, there are enough exceptions to the generalizations that can be made about reproductive phenomena so that these criteria cannot be used to distinguish plants and animals. Indeed, there are no hard and fast rules for distinguishing plants and animals.

There is no difficulty in assigning wasps, worms and whales, mussels, mackerel and monkeys, or bees, boars and buffalo to the animal kingdom, or in assigning mosses, maples and mistletoe, and pines, palms and petunias to the plant kingdom. Sponges, sea anemones and corals may cause a little more difficulty, but can ultimately be assigned to the animal kingdom with certainty. Major difficulties arise with some of the smaller, primarily single-celled forms of life which resemble neither typical animals nor plants. Some of the single-celled forms, such as the **flagellates,** have an interesting mélange of animal and plant characteristics. They possess chlorophyll and can carry out photosynthesis, but they also have several animal-like characteristics, such as their long **flagella,** that enable them to swim actively. The bacteria and blue-green algae differ significantly from both plants and animals, and are especially difficult to classify.

About a century ago, the German biologist, Ernst Haeckel, suggested that a third kingdom, for which he proposed the name, the **Protista,** be established to include all the single-celled organisms that are intermediate in most respects between plants and animals. This suggestion, although appealing at first glance, has not won wide acceptance, because it complicates rather than simplifies classification. Some of the protists are very clearly plantlike and are closely related to other organisms that are definitely plants. Others are distinctly animal-like, and still others have intermediate characteristics. A few others have characteristics that are distinctly different from both plants and animals. Even the organisms included in the kingdom Protista by different biologists may differ. Some taxonomists include only unicellular forms in the Protista, whereas others include fungi and multicellular algae as well as bacteria and blue-green algae. Since the term Protista creates perhaps even more taxonomic problems than it solves, it has not been very widely accepted.

Other biologists have proposed a fourth kingdom, the **Monera,** composed of the bacteria and blue-green algae, which have in common many characteristics, such as the absence of a nuclear membrane. When we realize that the question is not what these organisms are but what we shall call them, the argument falls into appropriate perspective. It is not, in fact, particularly useful or meaningful to try to assign these single-celled organisms to either the plant or animal kingdom.

Our concepts of the evolutionary relationships between the major phyla of plants and animals are rather vague, because the evolutionary events occurred such a long time ago, and the fossil record of these early forms is very nearly blank. Thus the evolutionary relationships of viruses and bacteria to other organisms are unknown, there is little evidence regarding the relationships between the major kinds of algae and fungi, and the relationships of the several types of protozoa to multicellular animals is unclear.

SECTION 11–2

MODES OF NUTRITION

Organisms that synthesize their own food are termed **autotrophic** (self-nourishing). Autotrophs require only water, carbon dioxide, inorganic salts and an appropriate source of energy. Green plants and purple bacteria are **photosynthetic autotrophs** and derive the energy needed for the synthesis of organic molecules from sunlight. A few bacteria are **chemosynthetic autotrophs** and obtain energy by oxidizing inorganic substances such as ammonia or hydrogen sulfide. These bacteria have evolved special enzyme systems that catalyze the oxidation of these and couple the oxidations with the generation of energy-rich phosphates. For example, nitrite bacteria such as *Nitrosomonas* oxidize ammonia to nitrite. *Nitrobacter* and other nitrate bacteria oxidize nitrite to nitrate. Iron bacteria oxidize ferrous iron to ferric iron. Other species of chemosynthetic bacteria oxidize hydrogen sulfide to sulfate. The energy derived from these oxidations is transformed into biologically useful energy, ATP, and then can be used in the synthesis of all the organic materials needed for the maintenance of life and for growth. The nitrite and nitrate bacteria are present in enormous quantities and are important in the cyclic use of nitrogen, for together they convert ammonia into a form that can be readily used by green plants, nitrate.

Purple bacteria have specific pigments that can absorb the energy of sunlight and utilize it to fix carbon dioxide as carbohydrate. In these bacteria the photosynthetic process does not yield molecular oxygen. The organisms use hydrogen sulfide, molecular hydrogen or an organic compound such as succinate as the source of hydrogen for the reaction instead of the water molecules used by green plants in their process of photosynthesis.

In contrast to the autotrophs, heterotrophic organisms are unable to synthesize their own foodstuffs from inorganic materials. **Heterotrophs** must live at the expense of autotrophs or upon decaying organic matter. All animals, all fungi and most bacteria are heterotrophs.

Several types of heterotrophic nutrition can be distinguished. Food composed of entire organisms or bits of their bodies must be eaten, digested and ultimately absorbed by the process termed **holozoic nutrition.** Holozoic organisms must constantly find, catch and eat other organisms.

Frogs catch and eat flies; cats eat birds; hawks eat mice. Holozoic animals have evolved a variety of sensory, nervous and muscular structures to assist them in finding and catching food, and have evolved some sort of digestive system to convert the nutrients into molecules small enough to be absorbed. Insectivorous plants such as the Venus flytrap, sundew and pitcher plant supplement their photosynthetic capabilities by trapping and digesting insects and other small animals. From these the plants obtain amino acids and other nitrogenous compounds that increase their growth rate.

Herbivorous animals eat green plants and obtain their energy-rich compounds from the contents of the plant cells, compounds that had been made by the plant using energy derived from sunlight. Other holozoic animals eat other animals that ate plants and thus are called **carnivores** (meat-eaters). Many animals are omnivores and will eat either plant or animal material. All heterotrophic organisms get their energy-rich nutrients ultimately from autotrophic organisms such as green plants that used the radiant energy from the sun to make these compounds.

Yeasts, molds and most bacteria neither make their nutrients by autotrophic processes nor can they ingest solid food. They must absorb their required organic nutrients directly through the plasma membrane. This type of heterotrophic nutrition is called **saprophytic nutrition.** Saprophytes can grow only in places where the bodies of animals or plants are undergoing decomposition or where there are masses of plant and animal by-products.

Yeasts are excellent examples of saprophytic plants; they need only inorganic salts, oxygen and some form of sugar. From the last they can derive energy and using this energy they can make all the other substances they require—proteins, fats, nucleic acids, vitamins and so on. When there is an adequate supply of oxygen, the yeast obtains energy by oxidizing the glucose completely to carbon dioxide and water by way of the citric acid cycle. When the supply of oxygen is limited, they ferment glucose and form ethanol and carbon dioxide. The conversion of glucose to pyruvate by the glycolytic cycle and the further conversion of pyruvate to ethanol and carbon dioxide yields only about one twentieth as much energy as the complete oxidation of glucose, and therefore yeasts grow very slowly when oxygen is absent.

Yeasts are used in the manufacture of all alcoholic beverages. Indeed, the only practicable way of obtaining ethyl alcohol is by this action of yeast. Ethanol is used not only in alcoholic beverages, but in many industrial processes as a solvent or as a raw material for the production of plastics and synthetic rubbers. The yeast organisms have a remarkable resistance to the toxic effects of ethanol and continue to produce it until a concentration of 12 per cent ethanol is reached in the surrounding medium, at which point the yeast organisms are inhibited. To produce beverages such as brandy or whiskey with a higher alcoholic content, the wine or mash is distilled. When yeast is mixed with bread dough, it ferments some of the sugar present, converting it to ethanol and carbon dioxide. Most of the ethanol is dissipated during the baking process, but the bubbles of carbon dioxide trapped in the dough expand during the baking process and raise the dough, making the bread porous.

Other saprophytes may require a variety of organic compounds in addition to some kind of sugar as a source of energy. For example, the baker's molds, *Neurospora*, requires the vitamin biotin in addition to salts and sugar.

A third type of heterotrophic nutrition is termed **parasitism.** Parasites are plants or animals that live either in or on the living body of another plant or animal, termed the **host,** from which the parasite obtains its nourishment. Almost every living organism is the host for one or more parasites. A few plants, such as the mistletoe, are in part parasitic and in part autotrophic. The mistletoe has chlorophyll and makes some of its own food, but its roots grow into the stems of other plants and they absorb some of their nutrients from their hosts. Parasites may obtain nutrients either by ingesting and digesting solid particles or by absorbing organic molecules through their cell walls from the body fluids or tissues of the host. Some parasites cause little or no harm to the host, and the host may be totally unaware of their presence. Other parasites produce definite diseases, destroying host cells or producing toxic substances which interfere with the host's metabolic processes. The **pathogenic** (disease-producing) parasites of man and other animals include viruses, bacteria, fungi, protozoa and an assortment of worms. Most of the plant diseases are caused by parasitic fungi, but a few are due to viruses, worms or insects.

It is curious that parasites are usually restricted to one or a few species of host organisms. Most of the parasites that infect man will not infect other animals, or will infect only animals such as apes and monkeys that are closely related to man. Other human parasites have wider host ranges and will infect more distantly related mammals or birds. Saprophytes such as yeast or bread molds are readily grown in the laboratory for they require only inorganic salts, glucose and perhaps a vitamin or two. Furthermore, they can survive over a considerable range of temperatures. In contrast, parasitic bacteria usually require a temperature near that of their normal host and must be grown on a complex medium containing sugars, amino acids and vitamins. Indeed, some bacteria will grow only if provided with blood, liver or yeast extracts which contain one or more unknown growth factors. Finally, a few parasites, such as rickettsias and viruses, can be grown only in the presence of living cells. The poliomyelitis virus can be grown, for example, on tissue cultures of human cells or on the kidney of the rhesus monkey.

SECTION 11-3

ECOSYSTEMS

When any species of animal or plant is studied carefully in its natural habitat, it becomes clear that each species is not independent of its neighboring living things but is part of a system of interdependent and interacting parts which form a larger unit. Ecologists use the term **ecosystem** to indicate a natural unit of living and nonliving parts that interact to produce a stable system in which the exchange of materials between living and nonliving parts follows a circular path. An ecosystem may be as large as a lake or forest, or one of the cycles of the elements, or it may be as small as an aquarium jar containing tropical fish, green plants and snails. To qualify as an ecosystem, the unit must be a stable system in which the exchange of materials follows a circular path.

A classic example of an ecosystem compact enough to be investigated in detail is a small lake or pond (Fig. 11–1). The nonliving parts of the lake include the water, its dissolved gases such as oxygen and carbon dioxide, inorganic salts such as phosphates, nitrates and chlorides of sodium, potassium and calcium, and finally a variety of organic compounds. The living organisms may be subdivided into producers, consumers and

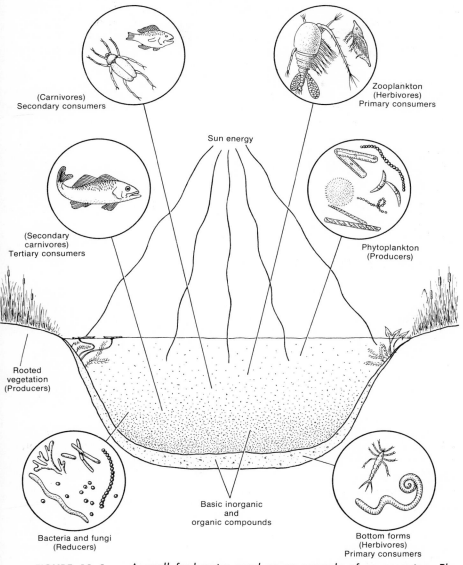

(Carnivores)
Secondary consumers

Zooplankton
(Herbivores)
Primary consumers

Sun energy

(Secondary
carnivores)
Tertiary consumers

Phytoplankton
(Producers)

Rooted
vegetation
(Producers)

Bacteria and fungi
(Reducers)

Basic inorganic
and
organic compounds

Bottom forms
(Herbivores)
Primary consumers

FIGURE 11-1   *A small fresh-water pond as an example of an ecosystem. The component parts of the system—the producer, consumer and decomposer or reducer organisms—are indicated, together with the nonliving parts of the ecosystem.*

decomposers, according to their specific role in keeping the ecosystem operating as a stable interacting unit. The **producer** organisms are the green plants that by photosynthesis manufacture organic compounds from simple inorganic substances. A typical small lake will have two kinds of producer organisms, larger plants that grow along the shore or float in shallow water, and microscopic floating plants, mostly algae, which are distributed throughout the water and are found as far down as light will penetrate. Such small plants are collectively called **phytoplankton**. When they are present in great abundance, they will give the water a greenish tinge, otherwise they may be invisible. The phytoplankton are usually more important food producers in the lake than are the larger plants.

The **consumer** organisms are heterotrophs such as insects, insect larvae, crustacea, fish and perhaps some freshwater clams. The animals that eat plants, the herbivores, are called primary consumers. The animals that eat the primary consumers are secondary consumers or carnivores. There may be a further group of tertiary consumers that eat the secondary consumers.

The ecosystem is completed by **decomposer** organisms—bacteria and fungi—which cleave the organic compounds of cells from dead producer and consumer organisms either into small organic molecules which they utilize themselves as saprophytes or into inorganic substances that can be used as raw materials by green plants. Even the largest and most complex ecosystems can be shown to consist of the same major types of organisms—producers, consumers and decomposers—plus a variety of nonliving components.

SECTION 11–4

HABITAT AND ECOLOGIC NICHE

In describing the ecologic relations of organisms it is important to distinguish between where an organism lives and what it does as part of its ecosystem. The terms habitat and ecologic niche refer to two concepts that are of prime importance in ecology. The **habitat** of an organism is the place where it lives. This is a physical area, some specific part of the earth's surface, air, soil or water. An organism's habitat may be as large as the entire ocean, or a vast expanse of prairie, or it may be as small and restricted as the underside of a rotten log or the intestine of a termite; however, a habitat is always a tangible, physically demarcated region. More than one animal or plant may live in a single habitat.

The **ecologic niche** is the status or role of an organism within the community or ecosystem. It depends upon the organism's structural adaptations, its physiological responses and its behavior. It may be helpful to think of the habitat as an organism's address, and the ecologic niche as its profession, biologically speaking. The ecologic niche is not a physically demarcated space, but an abstraction that includes all the physical, chemical, physiological and biotic factors that an organism requires to live. To describe an organism's ecologic niche, we must know what it eats, what eats it, its range of movement, and its effects on other organisms and on the nonliving parts of the surrounding ecosystem.

In the shallow waters at the edge of the lake or pond, you might find many different kinds of water bugs. They all live in the same place, and hence they all have the same habitat. Some of these water bugs, such as the backswimmer, *Notonecta*, are predators, catching and eating other animals of about the same size. Other water bugs, such as *Corixa*, feed on dead and decaying organisms. Each has a different role in the biological economy of the lake, and each occupies an entirely different ecologic niche.

A single species may occupy distinctly different niches in different regions, depending on the food supply that is available and on the number and kinds of its competitors. Some organisms, such as animals with distinctly different stages in their life history, occupy different ecologic niches in succession. The frog tadpole is a primary consumer and feeds on plants, but the adult frog is a secondary consumer, feeding on insects and other animals. In contrast, young river turtles are secondary consumers and eat snails, worms and insects, whereas the adult turtles are primary consumers and eat green plants such as tape grass.

THE CYCLIC USE OF MATTER

The total mass of all the organisms that have lived on the earth in the past billion and a half years is much greater than the mass of carbon and nitrogen atoms present on the planet. The Law of the Conservation of Matter assures us that matter has neither been created nor destroyed over this period. Obviously the carbon and nitrogen atoms must have been used over and over again in the formation of new generations of plants and animals. The earth neither receives any great amount of matter from other parts of the universe, nor does it lose significant amounts of matter to outer space. The atoms of each element — carbon, hydrogen, oxygen, nitrogen, phosphorus, sulfur and the rest — are taken from the environment, made a part of some cellular component, and finally, perhaps by a remarkably circuitous route involving their incorporation into a number of other organisms, are returned to the environment to be used again.

In the atmosphere over each acre of the earth's surface are about six tons of carbon as carbon dioxide. In a single year, an acre of luxuriant plant growth, such as sugar cane, can extract as much as 20 tons of carbon from the atmosphere and incorporate it into the plant body. If there were no way to renew the supply, the green plants would use up the entire supply of atmospheric carbon dioxide in a few centuries at most. The fixation of carbon dioxide by bacteria and animal cells is another, but quantitatively lesser, drain on the supply of carbon dioxide. Carbon dioxide is returned to the atmosphere by the decarboxylations that are part of cellular respiration (p. 147). Plant cells carry on respiration continuously. Green plant tissues are eaten by animals which, by cellular respiration, return more of the carbon atoms to the air as carbon dioxide. The process of respiration alone would not return enough carbon dioxide to the air to balance that withdrawn by photosynthesis. Carbon atoms would accumulate in the compounds making up the dead bodies of plants and animals. The carbon cycle is balanced by the decay bacteria and fungi which cleave the carbon compounds in the bodies of dead plants and animals and convert them to carbon dioxide (Fig. 11–2).

When the bodies of plants are compressed under water for long periods of time, they undergo chemical changes to form **peat,** then brown coal or **lignite,** and finally **coal.** Over a similar long time, the bodies of certain marine plants and animals may undergo changes to form **petroleum.** These processes remove some carbon from the carbon cycle temporarily but eventually either geologic changes or the mining and drilling activities of human beings bring the coal and oil to the surface to be burned to carbon dioxide and restored to the cycle. A large portion of the earth's carbon atoms are present in limestone and marble as carbonates. These rocks are gradually worn down, and the carbonates, in time, are added to the carbon cycle. However, other rocks are forming at the bottom of the sea from the sediments of dead animals and plants, and the amount of carbon in the carbon cycle at any given moment remains about the same.

The nitrogen for the synthesis of amino acids and proteins is taken up from the soil and water by plants as nitrates. The plant bodies may then be eaten by animals which use the amino acids from the plant proteins to synthesize their own amino acids, proteins, nucleic acids and other nitrogenous compounds. Decay bacteria will then convert these nitrogenous compounds into ammonia when the plants and animals die (Fig. 11–3). Animals also excrete several kinds of nitrogenous wastes — urea, uric acid, creatinine and ammonia — and decay bacteria convert these into am-

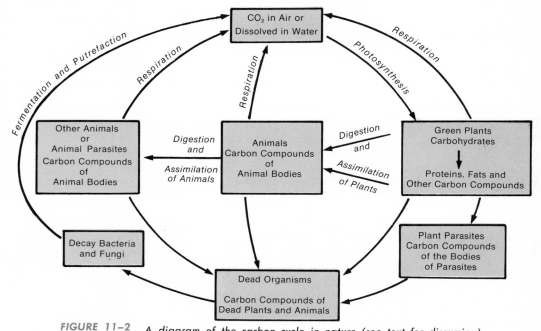

FIGURE 11-2   A diagram of the carbon cycle in nature (see text for discussion).

monia. The ammonia is converted by nitrite bacteria into nitrites, and these are metabolized into nitrates by nitrate bacteria. Some of the ammonia is converted to atmospheric nitrogen by denitrifying bacteria. Atmospheric nitrogen in turn can be converted to amino acids and other organic nitrogen compounds by certain algae and soil bacteria.

Other bacteria of the genus *Rhizobium*, although unable to fix atmospheric nitrogen alone, can do this when in combination with cells from the roots of peas, beans and other legumes. The bacteria invade the roots and stimulate the legume to form root nodules, which are a kind of benign tumor (Fig. 11–4). The combination of legume cell and bacterial cell is able to fix nitrogen, a process which neither can carry out alone. It is for this reason that legumes are often planted to restore the fertility of soil after other plants have been present for one or more growing seasons. These nodule bacteria may fix as much as 2.5 kilograms of nitrogen per acre per year, and soil bacteria as much as 3 kilograms per acre per year.

Atmospheric nitrogen can also be fixed by electrical energy supplied either by lightning or by man-made electricity. Although four fifths of the gas in the atmosphere is nitrogen, no animal and only a few plants can utilize molecular nitrogen. When the bodies of the nitrogen-fixing bacteria decay, the amino acids are metabolized to ammonia, and this is converted by nitrite and nitrate bacteria to nitrates.

In a similar fashion, the cyclic use of phosphorus and other elements can be traced. Both animals and plants take in inorganic phosphate and convert it to a variety of organic phosphates which are intermediates in the metabolism of carbohydrates, nucleic acids and fats. Phosphates in the dead bodies of plants and animals and in other organic wastes are being carried into the sediments at the bottom of the sea faster than they are being returned to land by the action of fish and marine birds. Sea birds play a role in returning phosphorus to the land cycle by depositing phosphate–rich **guano** on land. Man and other animals, by catching and eating fish, also

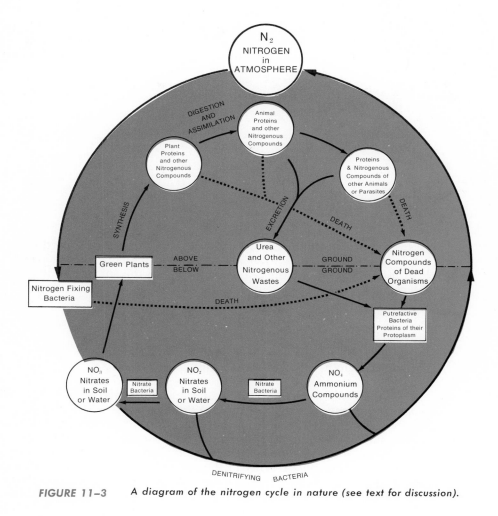

FIGURE 11-3   *A diagram of the nitrogen cycle in nature (see text for discussion).*

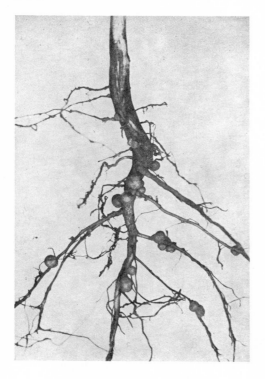

FIGURE 11–4   Roots of a soybean plant with root nodules formed by nitrogen-fixing bacteria. (From Weatherwax, P.: Botany. 3rd ed. Philadelphia, W. B. Saunders Company, 1956.

recover some phosphorus from the sea. Minerals are recovered from the sea bottom and become available for use once again when geologic up-heavals bring some of the sea bottom back to the surface and raise new mountains.

SECTION 11–6

INTERACTIONS BETWEEN SPECIES

The members of two different species of animals or plants may inter-act with each other in any one of several different ways. If each population is adversely affected by the presence of the other in its search for food, space, shelter or some other fundamental requirement for life, the inter-action is one of **competition.** If there is no interaction and neither popula-tion is affected by the presence of the other, the situation is termed **neutral-ism.** If each population is benefited by the presence of the other, but each can survive in the absence of the other, the relationship is termed **proto-cooperation.** When both species gain from an association and cannot sur-vive separately in nature, the association is termed **mutualism.** When one species is benefited and the second is not affected when the two exist to-gether, the relationship is termed **commensalism,** and **amensalism** refers to the relationship in which one species is inhibited by the second, but the second is unaffected by the first. One species may affect the second ad-versely, but cannot live without it. Such a relationship is termed **parasitism,** if one species lives in or on the body of the second. The relationship is termed **predation** if the first species catches, kills and feeds upon the second, the prey. The older term **symbiosis** (living together) is used by some biolo-gists as a synonym of mutualism, and by others in a wider sense to include mutualism, commensalism and even parasitism.

SECTION 11-7

COMPETITION

Two species may compete for the same space, food or light, or in escaping from predators or disease; this may be summarized as competition for the same ecologic niche. Competition may result in one species dying off or being forced to change its ecologic niche—to move to a different space or use a different food. Careful ecologic studies usually reveal that there is only one species in an ecologic niche (Gause's rule). One of the clearest examples of ecologic competition was provided by the classic experiments of Gause with populations of paramecia. When either of two closely related species, *Paramecium caudatum* or *Paramecium aurelia*, was cultured separately on a fixed amount of bacteria as food, it multiplied and finally reached a constant level (Fig. 11-5). But when both species were placed in the same culture vessel with a limited amount of food, only *Paramecium aurelia* was left at the end of 16 days. The *Paramecium aurelia* had not attacked the other species or secreted any harmful substance; it simply had been more successful in competing for the limited food supply. Studies in the field generally corroborate Gause's rule. Two fish-eating, cliff-nesting birds, the cormorant and the shag, which seemed at first glance to have survived despite occupying the same ecologic niche, were found upon analysis to have slightly different niches. The cormorant feeds on bottom-dwelling fish and shrimps, whereas the shag hunts fish and eels in the upper levels of the sea. Further study showed that these birds typically have slightly different nesting sites on the cliffs as well.

SECTION 11-8

BENEFICIAL ASSOCIATIONS

Commensalism, the living together of two species, one of which (the commensal) derives benefit from the association, whereas the other is

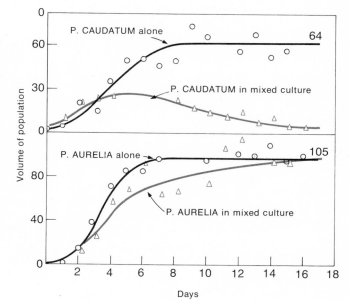

FIGURE 11-5   *An experiment to demonstrate the competition between two closely related species of paramecia which have identical ecologic niches. When grown separately in controlled cultures with a fixed supply of bacteria for food, both Paramecium caudatum and Paramecium aurelia show normal S-shaped growth curves (solid lines). When the two species are grown in the same culture, Paramecium caudatum is eliminated (colored line). (After Gause; from Allee, W. C. et al.: Principles of Animal Ecology. Philadelphia, W. B. Saunders Company, 1949.)*

unharmed by it, is especially common in the sea. Practically every worm burrow and shellfish contains some uninvited guests that take advantage of the shelter, and possibly of the abundant food, provided by the host organism but do it neither good nor harm. Certain flatworms live attached to the gills of the horseshoe crab and get their food from the scraps of the crab's meals. They obtain shelter and transportation from the host but apparently do it no harm. Many oysters and other bivalves have small crabs living in their mantle cavity, and there is a species of small fish that lives in the posterior end of the digestive tract of the sea cucumber!

If both species gain from an association but are able to survive without it, the association is termed **protocooperation.** Several kinds of crabs put coelenterates of one sort or another on top of their shells, presumably as camouflage. The coelenterates benefit from the association by getting bits of food when the crab captures and eats an animal. Neither crab nor coelenterate is absolutely dependent upon the other.

When both species gain from an association and are unable to survive separately, the association is termed **mutualism.** It is probable that associations begin as commensalism and then evolve through a stage of protocooperation to one of mutualism. A striking example of mutualism is provided by the relationship of termites and their intestinal flagellates. Termites have no enzymes to digest the cellulose of wood, yet that is their staple diet. Certain flagellate protozoa that live only in their intestines do have the enzymes (β-glucosidases) to digest cellulose to sugars. Although the flagellates require some of this sugar for their own metabolism, there is enough left over for the termite. Termites are unable to survive without their intestinal inhabitants; newly hatched termites instinctively lick the anus of another termite to get a supply of flagellates. Since a termite loses all of its flagellates along with most of its gut lining at each molt, termites must live in colonies so that a newly molted individual will be able to get flagellates from a neighbor. The flagellates are provided with plenty of food in a well-protected, relatively constant environment; they can, in fact, survive only in the intestines of termites.

SECTION 11-9

NEGATIVE INTERACTIONS

Commensalism, protocooperation and mutualism are types of positive interactions, ones in which one or both members of the associated pair derive benefit from the association, yet neither is harmed by it. Negative interactions between species—amensalism, parasitism and predation—are those in which one species is harmed by the association. If the second species is unaffected, the relationship between the two is termed **amensalism.** Organisms that produce antibiotics and the species inhibited by the antibiotic are examples of amensalism. The mold *Penicillium* produces the antibiotic penicillin, which inhibits the growth of a variety of bacteria, but the mold is unaffected by the bacteria. The clinical use of these bacteria-inhibiting agents has had the unexpected effect of increasing the incidence of fungus-induced diseases in man, which are normally kept in check by the presence of the bacteria. When the bacteria are killed off by the antibiotics, the pathogenic fungi seize the opportunity to multiply in the host.

It is incorrect to assume that host-parasite and predator-prey relationships are invariably harmful to the host or prey as a species. This is usually true when such relationships are first established, but the forces of natural selection tend, in time, to decrease the detrimental effects. If this did not occur, the parasite would eventually exterminate the host species and, unless it found a new species to parasitize, would die itself.

Studies of many examples of parasite-host and predator-prey associations show that, in general, when the associations have been established for a long time, evolutionarily speaking, the long-term effect on the host or prey species is not very detrimental. Conversely, newly acquired predators or parasites are usually quite damaging. The plant parasites and insect pests that are most troublesome to man and his crops are usually those which have recently been introduced into some new area and thus have a new group of organisms to attack.

The role of the predator-prey relationship in maintaining a balance between the number of predators and of prey is beautifully illustrated by the story of the Kaibab deer. The Kaibab plateau is located on the north side of the Grand Canyon of the Colorado River. In 1907 there were some 4000 deer living on the plateau, together with a considerable population of predators, mountain lions and wolves. When a concerted effort was made to "protect" the deer by killing off the predators, the deer population increased tremendously and by 1925 some 100,000 deer roamed the plateau, far more than the supply of vegetation could support. The deer ate everything in reach—grass, tree seedlings and shrubs—and there was marked damage to the vegetation. Over the next two winters, large numbers of the deer died of starvation, and the size of the herd fell to about 10,000. In the wild, the size of the predator population varies with the size of the population of the species which is preyed upon, with the swings in the size of the predator population lagging somewhat behind those of the prey.

SECTION 11–10

INTRASPECIFIC RELATIONS

In addition to the associations between the members of two different species just described, aggregations of animals or plants of a single species frequently occur. Some of these aggregations are temporary, for breeding; others are more permanent. Despite the fact that the crowding which accompanies dense aggregations of animals is ecologically undesirable and deleterious, both laboratory experiments and field observations show that such aggregations of individuals are able to survive when a single individual of the same species placed in the same environment would die. A herd of deer, with many noses and pairs of eyes, is less likely to be surprised by a predator than is a single deer. A pack of wolves hunting together is more likely to make a kill than is a lone wolf. The survival value of intraspecific aggregations is less obvious, but nonetheless real, in some of the lower animals. A group of insects is less likely to dry up and die in a dry environment than is a single insect, and a group of planaria is less likely to be killed by a given dose of ultraviolet light than is a single flatworm. When a dozen goldfish are placed in one bowl and a single one in a second bowl, and the same amount of a toxic agent such as colloidal silver is added to each bowl, the single fish will die, but the group will survive. The explanation for this has proved to be that the slime secreted by the group of fish is enough to precipitate much of the colloidal silver and render it nontoxic, whereas the amount secreted by a single fish is not.

Such animal aggregations do have survival value for the species. W. C. Allee has called this *"unconscious cooperation."* When genes governing a tendency toward aggregation arise in a species and prove to have survival value, natural selection will tend to preserve this inherited behavior pattern. The occurrence of many fish in schools, of birds in flocks and so on are examples of this "unconscious cooperation," which occurs very widely in the animal kingdom.

From such simple animal aggregations there may evolve complex animal societies, composed of specialized types of individuals, such as the colonies of bees, ants and termites. Man is another example of a social animal.

SECTION 11-11

FOOD CHAINS

Once a given species has reached and become established in a certain area, the number of organisms—or more precisely their total mass—is determined by the rate of flow of energy through the biological part of the ecosystem which includes them. Although carbon, nitrogen and the other elements are reused cyclically, energy is not and can be used only once by a given organism. It is ultimately degraded to heat and lost to the ecosystem.

The ultimate source of all the energy used by living things is sunlight, the energy of which is converted to a biologically useful form by the process of photosynthesis carried on by green plants. Only a small fraction, about 3 per cent, of the light energy striking the leaves of a green plant is transformed by photosynthesis into the potential energy of some food substance; the rest escapes as heat. This loss is not the result of inefficiency of the biochemical processes involved, but of the operation of the laws of thermodynamics. The Second Law of Thermodynamics may be stated as "Whenever energy is transformed from one form into another there is a decrease in the amount of useful energy; some energy is degraded into heat and dissipated." In other words, no transformation of energy can be 100 per cent efficient.

When an animal eats a plant, much of the energy is again dissipated as heat and only a fraction is used to synthesize the animal's tissues. When a second animal eats the first, there is a further loss of energy as heat, and so on. The transfer of energy from its ultimate source in plants through a series of organisms, each of which eats the preceding and is eaten by the following, is known as a **food chain.** The number of steps in a food chain is limited to perhaps four or five because of the great decrease in available energy at each step. The percentage of energy in the food consumed that is converted to new cellular materials, and thus is available as energy for the next organism in the food chain, is known as the percentage efficiency of energy transfer.

The flow of energy in ecosystems from sunlight through photosynthesis in autotrophic producers, through the tissues of herbivorous primary consumers to the tissues of carnivorous secondary consumers, determines the number and total weight, the **biomass,** of organisms at each level in the ecosystem. The flow of energy is greatly reduced at each successive level of nutrition because of the heat losses at each energy transformation, and this decreases the biomass in each level.

The first step in any food chain, the capture of light energy by photosynthesis and the production of carbohydrates and other energy-containing foods by plants, is relatively inefficient; only about 0.2 per cent of the light energy falling on a given area in the course of a year is stored as food. The efficiency of energy transfer when one animal eats a plant or another animal is higher, ranging from 5 to 20 per cent. Some animals eat but one kind of food and, therefore, are members of a single food chain. Other animals eat many different kinds of food and are not only members of different food chains but may occupy different positions in different food chains. An animal may be a primary consumer in one chain, eating green plants, but a secondary or tertiary consumer in other chains, eating herbivorous animals or other carnivores.

Man is the end of a number of food chains. For example, man eats a fish such as a black bass, which ate a smaller fish, which ate small crustacea which in turn ate algae. The ultimate size of the human population, or of the population of any animal, is limited by (1) the length of the food chain, (2) the percentage efficiency of energy transfer at each step in the chain, and (3) the amount of light energy falling on the earth. Since man can do nothing about increasing the amount of incident sunlight, and very little about the percentage efficiency of energy transfer, he can increase his supply of food energy only by shortening his food chain, i.e., by eating the primary producers, plants, rather than animals. In overcrowded countries such as India and China, men are largely vegetarians because this food chain is shortest, and a given area of land can in this way support the greatest number of people. Steak is a luxury in both ecologic and economic terms, but hamburger is just as much an ecologic luxury as steak is.

In addition to predator food chains, such as the man-black bass-minnow-crustacean one, there are parasite food chains. For example, mammals and birds are parasitized by fleas; in the fleas live protozoa which are, in turn, the hosts of bacteria. Since the bacteria might be parasitized by viruses, there could be a five-step parasite food chain.

A third type of food chain is one in which plant material is converted into dead organic matter, **detritus,** before being eaten by animals such as millipedes and earthworms on land, by marine worms and mollusks, or by bacteria and fungi. In a community of organisms in the shallow sea, about 30 per cent of the total energy flows via detritus chains, but in a forest community, with a large biomass of plants and a relatively small biomass of animals, as much as 90 per cent of energy flow may be via detritus pathways. In an intertidal salt marsh, where most of the animals — shellfish, snails and crabs — are detritus eaters, 90 per cent or more of the energy flow is via detritus chains.

Since, in any food chain, there is a loss of energy at each step, it follows that there is a smaller biomass in each successive step. H. T. Odum has calculated that 8100 kilograms of alfalfa plants are required to provide the food for 1000 kilograms of calves, which provide enough food to keep one 12-year-old, 48-kilogram boy alive for one year. Although boys eat many things other than veal, and calves other things besides alfalfa, these numbers illustrate the principle of a food chain. A food chain may be visualized as a **pyramid;** each step in the pyramid is much smaller than the one on which it feeds. Since the predators are usually larger than the ones on which they prey, the pyramid of numbers of individuals in each step of the chain is even more striking than the pyramid of the mass of individuals in successive steps: one boy requires 4.5 calves, which require 20,000,000 alfalfa plants.

SECTION 11–12

COMMUNITIES AND POPULATIONS

Each region of the earth—sea, lake, forest, prairie, tundra, desert—is inhabited by a characteristic assemblage of animals and plants which are interrelated in many and diverse ways as competitors, commensals, predators and so on. The members of each assemblage are not determined by chance but by the total effect of the many interacting physical and biotic factors of the environment. The ecologist refers to the organisms living in a given area as a **biotic community;** this is composed of populations, groups of individuals of any one kind of organism.

The intermeshings of the food chains in any biotic community are very complex and are sometimes called a food web, or "web of life." Some of the interrelated food chains of a deciduous forest in eastern North America are indicated in Figure 11–6. The basic principles of the ecologic relations of biotic communities have been elucidated by the study of somewhat simpler communities, such as the arctic tundra or desert. The producer organisms of the tundra are lichens, mosses and grasses. Reindeer and caribou feed on the lichens and are preyed upon by wolves and man. Grasses are eaten by the arctic hare and the lemming, which are eaten by the snowy owl and the arctic fox. The latter is preyed upon by man for its fur. During the brief arctic summer, the food web is enlarged by many insects and by migratory birds which feed upon them.

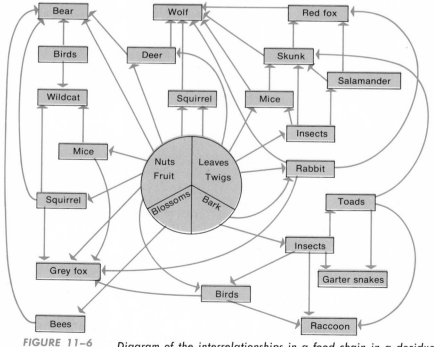

FIGURE 11–6   *Diagram of the interrelationships in a food chain in a deciduous forest in Illinois. (After Shelford, V. E.: Ecological Monographs 21:183–214, 1951.)*

THE DYNAMIC BALANCE OF NATURE

In discussing the concept of the dynamic state of body constituents (Chap. 5), we learned that the molecules of proteins, fats, carbohydrates and other constituents of animal and plant bodies are constantly being synthesized, broken down and resynthesized. A biotic community undergoes an analogous, constant reshuffling of its constituent parts, and the concept of the dynamic state of biotic communities is an important ecologic principle. Plant and animal populations are continuously subject to changes in their physical and biotic environments to which they must adapt or die. In addition, communities undergo a number of rhythmic changes — daily, tidal, lunar, seasonal and annual — in the activities or movements of the organisms which make them up and which result in periodic changes in the composition of the community as a whole. The size of a population may vary widely, but if it outruns its food supply, as the Kaibab deer did, or the lemmings in northern Europe do from time to time, equilibrium is restored by increasing the death rate. A population of organisms or a biotic community can be compared in many ways to a many–celled organism. The population can exhibit growth, specialization and interdependence of its parts, characteristic forms, and even a development from immaturity to maturity, old age and death.

It should be clear that the ecologic principles discussed so far apply to human populations as well as to populations of animals and plants. Human ecology deals with the dynamics of human populations and with the relationships of man to the physical and biotic factors which impinge upon him. If man can appreciate that human populations are part of larger units of biotic communities and ecosystems, he can deal much more intelligently with his own special problems. Man has wittingly and unwittingly exercised a great deal of control over his environment and has modified the communities and ecosystems of which he is a part. However, his control of the environment is far from complete, and man must, like other animals and plants, learn to adapt to those situations which he cannot change. By understanding and cooperating with the various cycles of Nature, man will have a better chance of surviving in the future than if he blindly attempts to change and control them.

The human population is clearly in danger of multiplying beyond the ability of the earth to support it. In the past several centuries, the population of the world has increased tremendously as new territories have been opened for exploitation and as methods of producing food have become more efficient. Most biologists and social scientists believe that the danger of human overpopulation is both great and imminent. It has been amply shown that the Malthusian Principle, that populations have an inherent ability to grow exponentially, is true for organisms generally, and the growth of the human population in the past 300 years follows an exponential curve. The productivity and carrying capacity of the earth for human beings can be maintained and perhaps increased somewhat, but eventually the human biomass must be brought into equilibrium with the space and food available. The human population is probably in greater danger of running out of drinkable water and breathable air than it is of exhausting its food supply! Some limitation of human reproduction is clearly inevitable. It remains to be seen whether man will do this voluntarily or involuntarily.

SUGGESTIONS FOR FURTHER READING

Gates, D. M.: *Energy Exchange in the Biosphere.* New York, Harper & Row Publishers, 1962.
The cycles of energy and matter in the biosphere.

Jackson, R. M., and Raw, F.: *Life in the Soil.* New York, St. Martin's Press, Inc., 1966.
Interesting discussions of the carbon, nitrogen and similar cycles.

Kormondy, E. J. (ed.): *Readings in Ecology.* Englewood Cliffs, New Jersey, Prentice-Hall, Inc., 1965.
The editor has selected with care a group of papers on ecologic subjects from the world literature and has added helpful annotations.

Odum, E. P.: *Fundamentals of Ecology.* 3rd ed. Philadelphia, W. B. Saunders Company, 1971.
The cycles of substances in nature and the various kinds of relations between plants, animals and their physical environment are discussed in detail in this standard text of ecology.

Smith, R. L.: *Ecology and Field Biology.* New York, Harper & Row, Publishers, 1965.
A useful reference work with an appendix describing some of the methods used in the study of plant and animal populations.

Stewart, W. D. P.: *Nitrogen Fixation in Plants.* London, The Athone Press, University of London, 1966.
In this concise volume, the author presents a balanced view of the expanding field of research in nitrogen fixation.

THE DIVERSITY OF LIFE: PROTISTA AND MONERA

CHAPTER 12

Single-celled plants and animals were probably first seen by Antonj van Leeuwenhoek (1632–1723), who was a draper in Delft, Holland. He was interested in the structure of the threads used in making linen and ground some crude microscope lenses to examine them. He discovered several ways to improve the grinding of lenses and made hand lenses which magnified about 150 diameters. Leeuwenhoek had broad interests and examined almost everything at hand—pond water, sea water, vinegar, saliva, semen and many other things—and described his findings in letters written to the Royal Society of London. The size, shape and characteristic motion of the organisms described in a letter written in 1683 leave no doubt that they were bacteria. Leeuwenhoek and other seventeenth and eighteenth century biologists called these creatures "animalicules" and supposed they were some kind of small worms or wormlike animals. It was not until 1845 that microscopists recognized that many of these small animals are unicellular in nature, and the phylum Protozoa was established to include the single-celled animals.

Recognition that bacteria may be important as agents of disease and decay came from the research of Louis Pasteur just a century ago. This in turn stimulated Robert Koch, Ferdinand Cohn, Joseph Lister and others, and the science of bacteriology blossomed in the latter part of the nineteenth century. Pasteur's studies of the diseases that brought about the souring of wine and beer revealed they were caused by microorganisms that entered the wine or beer from the air and brought about an undesirable type of fermentation yielding products other than ethanol. By gently heating the grape juice or beer mash, (a process now called **pasteurization**) he could kill the undesirable organisms and subsequently cool the juice and seed it with yeast to bring about the desired fermentation.

Another of Pasteur's contributions to biology was his unequivocal demonstration that bacteria cannot arise by spontaneous generation (p. 14). Lord Lister, an English surgeon, understood the significance of Pasteur's discoveries, and applied this germ theory to the procedures of surgical operations. He initiated antiseptic techniques in the operating room, dipping his operating instruments in carbolic acid and spraying the

319

scene of the operation with a germicide. In the past several decades, research with bacteria has been concerned mainly with their physiological, biochemical and genetic properties. Much of our present understanding of the molecular basis of life has been gained from studies of bacteria, especially of *Escherichia coli,* a harmless parasite of the intestines of man and other mammals.

The primitive plants which neither form embryos during development nor have vascular tissues are called **thallophytes.** More than 100,000 species of thallophytes are widely distributed in fresh and salt water, on land, and as parasites on other plants or animals. Although many of the members of this group are microscopic, single-celled plants, the thallophytes also include giant seaweeds (kelps) that may be more than 100 meters long. The body of such plants, called a **thallus,** may show some differentiation of parts, but it has no true roots, stem or leaves. The thallophytes include two general types of plants: the algae that have chlorophyll and are autotrophs, and the fungi that lack chlorophyll and are heterotrophs, living as saprophytes or parasites. In some instances, the separation of algae and fungi is artificial rather than natural, for it separates some organisms that are very much alike except in color. Indeed organisms like *Euglena* may lose chlorophyll and live as saprophytes if placed in the dark, but will regain the ability to synthesize chlorophyll if returned to the light. **Alga** is the latin word meaning seaweed, but although most seaweeds are algae, there are many algae that are not seaweeds, and some seaweeds that are not algae. Most algae live in water, either fresh or salt, but a few live on rock surfaces or on the bark of trees. The ones living in such relatively dry places usually remain dormant when water is absent. By virtue of their tremendous numbers, algae are important food producers. Almost all the photosynthesis in the oceans and most of that in fresh water is carried on by algae. Human beings do not commonly use algae as food, but a considerable fraction of human food is fish which eat either algae or other organisms that ate algae.

SECTION 12–1

THE ORIGIN OF LIFE

The current theories of mutation, natural selection and population dynamics provide a satisfactory explanation of how the animals and plants living today evolved from earlier forms by descent with modification. The question of the ultimate origin of living things on this planet has been given serious consideration by a number of biologists. Some have theorized that some kind of spore or germ may have been carried through space from some other planet to this one. This is unsatisfactory, not only because it begs the question of the ultimate source of the spores, but also because it seems most unlikely that any sort of living thing could survive the extreme cold and intense irradiation of interplanetary travel. Evidence for life in other parts of the cosmos came from the discovery in 1961 of what were identified as fossils of microscopic organisms, somewhat like algae, in meteorites, but this of course is no proof that *living* organisms could be transported through space.

The spontaneous origin of living things at the present time is believed to be extremely improbable. Francesco Redi's experiments showed, about 1680, that maggots do not arise *de novo* from decaying meat and laid to rest the old superstition that animals could appear by spontaneous generation.

Some 200 years later Louis Pasteur showed conclusively that microorganisms such as bacteria do not arise by spontaneous generation but come only from previously existing bacteria. Other investigators have shown since then that even the smallest organisms, the filtrable viruses, do not come from nonviral material by spontaneous generation. The multiplication of viruses requires the presence of previously existing viruses. Although the spontaneous generation of life at present is unlikely, it is most probable that billions of years ago, when chemical and physical conditions on the earth's surface were quite different from those at present, the first living things did arise from nonliving material.

This concept, that the first things did evolve from nonliving things, (and suggestions as to what the sequence of events may have been) have been put forward by Pflüger, J. B. S. Haldane, R. Beutner, and particularly by the Russian biochemist A. I. Oparin, in his book, *The Origin of Life* (1938). The earth originated some five billion years ago, either as a part broken off from the sun, or by the gradual condensation of interstellar dust. Most authorities now agree that the earth was very hot and molten when it was first formed and that conditions consistent with life appeared on the earth only perhaps three billion years ago. Twenty-two different amino acids were isolated recently from Precambrian rocks from South Africa that are at least 3.1 billion years old. At that time the earth's atmosphere contained essentially no free oxygen—all the oxygen atoms were combined as water, or as oxides. The primitive atmosphere was strongly reducing, composed of methane, ammonia and water originating by "outgassing" from the earth's interior.

A number of reactions are known by which organic substances can be synthesized from inorganic ones. Originally the carbon atoms in the earth's crust were present mainly as metallic carbides. These could react with water to form acetylene, which would subsequently polymerize to form compounds with long chains of carbon atoms. High-energy radiation such as cosmic rays can catalyze the synthesis of organic compounds. This was shown by Melvin Calvin's experiments in which solutions of carbon dioxide and water were irradiated in a cyclotron and formic, oxalic and succinic acids, which contain 1, 2 and 4 carbons respectively, were obtained. These, as you know, are intermediates in certain metabolic pathways of living organisms. Irradiating solutions of inorganic compounds with ultraviolet light, or passing electric charges through the solution to simulate lightning also produce organic compounds. Harold Urey and Stanley Miller in 1953 exposed a mixture of water vapor, methane, ammonia and hydrogen gases to electric discharges for a week and demonstrated the production of amino acids such as glycine and alanine, together with other complex organic compounds. The earth's atmosphere in prebiotic times probably contained water vapor, methane, ammonia and hydrogen gas from which irradiation could produce a tremendous variety of organic materials. Amino acids and other compounds could be produced in nature at the present time by lightning discharges or ultraviolet radiations; however, any organic compound produced in this way might undergo spontaneous oxidation or it would be taken up and degraded by molds, bacteria and other organisms.

The details of the chemical reactions that could give rise, without the intervention of living things, to carbohydrates, fats and amino acids have been worked out by Oparin and extended by Calvin and others. Most, if not all, of the reactions by which the more complex organic substances were formed probably occurred in the sea in which the inorganic precursors and the organic products of the reaction were dissolved and mixed. The sea became a sort of dilute broth in which these molecules collided, reacted and aggregated to form new molecules of increasing size and complexity (this might be called the "chicken soup" theory of evolution). As more has been learned of the role of specific hydrogen bonding and other weak intermolecular forces in the pairing of specific nucleotide bases and the effectiveness of these processes in the transfer of biological information, it has become clear that similar forces could have operated early in evolution before "living" organisms appeared.

The forces of intermolecular attraction and the tendency for certain molecules to form liquid crystals, provide us with means by which large complex specific molecules can be formed spontaneously. Oparin suggested that a kind of natural selection can operate in the evolution of these complex molecules before anything recognizable as life is present. As the molecules came together to form colloidal aggregates, these aggregates began to compete with one another for raw materials. Some of the aggregates which had some particularly favorable internal arrangement would acquire new molecules more rapidly than others, and would eventually become the dominant types.

Once some protein molecules had been formed and had achieved the ability to catalyze reactions, the rate of formation of additional molecules would be greatly speeded up. When combined with nucleic acids, these complex protein molecules should eventually acquire the ability to catalyze the synthesis of molecules like themselves. These hypothetical **autocatalytic** particles made of nucleic acids and proteins would have some of the properties of a virus or perhaps of a free gene. The next step in this theoretical development of a living organism might have been the addition of the ability of the autocatalytic particle to mutate, to undergo inherited changes. Then if a number of these free genes had joined to form a single, larger unit, the resulting organism would have been similar to certain present-day viruses. One major step in the evolution of these early prebiotic aggregates would have been the development of a protein-lipid **membrane** surrounding it which would permit the accumulation of some molecules and the exclusion of others. All the viruses known at present are parasites that can live only within the cells of higher animals and plants; however, if there were free living viruses—ones which do not produce a disease— they would be very difficult to detect.

The first living organisms, having arisen in a sea of organic molecules and in contact with an atmosphere lacking oxygen, presumably obtained energy by the fermentation of certain of these organic substances. The first organisms were almost certainly **heterotrophs,** and they could survive only as long as the supply of organic molecules that had been accumulated in the sea broth in the past lasted. Before the supply was exhausted, however, some of the heterotrophs evolved further and became autotrophs, able to make their own organic molecules by chemosynthesis or photosynthesis. One of the by-products of photosynthesis is gaseous oxygen; indeed, all the oxygen in the atmosphere has been produced and is still produced by photosynthesis. It can be estimated that all the oxygen in the earth's atmosphere is renewed by photosynthesis every 2000 years, and that all the carbon dioxide molecules pass through the photosynthetic process every 300 years. All the oxygen and carbon dioxide in the earth's atmosphere are the products of living organisms and have passed through living organisms over and over again during the course of evolution.

An explanation of how an autotroph may have evolved from one of these primitive, fermenting heterotrophs, was presented by N. H. Horowitz in 1945. Horowitz postulated that an organism would acquire, by successive gene mutations, the enzymes needed to synthesize complex substances from simple substances, but these enzymes would be acquired in the *reverse* order of the sequence in which they are ultimately used in normal metabolism. For example, let us suppose that our first primitive heterotroph required an organic compound, Z, for its growth. This substance, Z, and a vast variety of other organic compounds, Y, X, W, V, U and so forth, were present in the organic sea broth which was the environment of this heterotroph. They had been synthesized previously by the action of nonliving factors of the environment. The heterotroph would be able to survive as long as the supply of compound Z lasted. If a mutation occurred for a new enzyme enabling the heterotroph to synthesize Z from substance Y, the strain of heterotroph with this mutation would be able to survive when the supply of substance Z was exhausted. A second mutation which established an enzyme catalyzing a reaction by which substance Y could be made from substance X would again have survival value when the supply of Y was exhausted. Similar mutations, setting up enzymes enabling the organism to use successively simpler substances, W, V, U . . ., and eventually some inorganic substance, A, would result in an organism able to make substance Z, which it needs for growth, out of substance A. When, by other series of mutations, the organism could synthesize all its requirements from simple inorganic compounds, as the green plants can, it would have become an autotroph. Once the first simple autotrophs had evolved, the way was clear for the further evolution of the enormous variety of plants, bacteria, molds and animals that now inhabit the world.

From arguments such as these we are drawn to the conclusion that the origin of life as an orderly, natural process on this planet was not only possible, it was almost inevitable. Furthermore, with the vast number of planets in all the known galaxies of the universe, there must be many that have conditions which permit the origin of life. It is probable, then, that there are other planets—perhaps many other planets—on which life as we know it exists. Wherever the physical environment will support life, living things should, if given enough time, appear and ramify into a wide variety of types. Some of these may be quite unlike the ones on this planet, but others might be quite similar to those found here. Some might indeed be like ourselves. Living things on other planets might have a completely different kind of genetic code, or might be made up of elements other than carbon, hydrogen, oxygen and nitrogen.

It seems unlikely that we will ever know how life originated, whether it happened only once or many times, or whether it might happen again. The theory (1) that organic substances were formed from inorganic substances by the action of physical factors in the environment; (2) that they interacted to form more and more complex substances, finally enzymes, and then self-reproducing systems (free genes); (3) that these free genes diversified and united to form primitive, perhaps virus-like heterotrophs; (4) that lipid-protein membranes evolved to separate these prebiotic aggregates from the surrounding environment; and (5) that autotrophs then evolved from the primitive heterotrophs, has the virtue of being quite plausible. Many of the parts of this theory have been subjected to experimental verification.

SECTION 12–2

VIRUSES AND BACTERIOPHAGES

Viruses and bacteriophages are much smaller than bacteria, and indeed are scarcely larger than some very large single molecules of protein or nucleic acid. They are too small to be seen with the light microscope, and can be photographed only with an electron microscope. They resist classification as plants or animals, or even as protists or monerans. In one sense, viruses are not living organisms but large nucleoprotein particles which enter specific kinds of animal, plant or bacterial cells and multiply (or are multiplied) to form new virus particles. When viruses are outside the host cell they are metabolically inert; some viruses have been crystallized. Each is essentially a bit of genetic material, either DNA or RNA, enclosed within a protective coat of protein which permits it to pass from one cell to the next. The difficulty in deciding whether these forms should be considered living or nonliving simply reflects the difficulty in defining life itself. Viruses and bacteriophages exhibit some, but not all, of the usual characteristics of living things.

Filtrable viruses, ultramicroscopic particles small enough to pass through very fine-pored porcelain filters, were discovered by the Russian botanist, Iwanowski, in 1892. Iwanowski found that a disease of tobacco plants, called mosaic disease because the infected leaves had a spotted appearance, could be transmitted to healthy plants by daubing their leaves with the sap of diseased plants. The sap was effective even after it had been passed through filters fine enough to remove all bacteria.

Viruses are the infective agent in a wide variety of plant and animal diseases and in such human diseases as smallpox, rabies, poliomyelitis, measles, warts, fever blisters and the common cold. There is a strong possibility that viruses may be involved in the etiology of certain types of human cancer. One type of breast cancer in mice has been shown to be caused by a virus-like agent, but human cancers are not infectious, as one might expect a virus disease to be.

Viruses might be compared to the chromosomes of plants and animals, in that they can undergo duplication only within the complex environment of living cells. The many attempts to grow viruses in cell-free culture media, containing all of the known vitamins and amino acids, have been unsuccessful.

Viruses have very few, if any, of the metabolic properties of the cells of higher organisms. The entrance of a virus particle produces profound changes in the metabolic pattern of the host cell, which eventually lead to the production of new virus particles. Thus viruses do not really reproduce themselves, but are reproduced by the enzymatic machinery present in their host cells. Viruses are commonly cultured for experimental purposes by injecting them into fertilized hens' eggs (Fig. 12–1), or by adding them to cells growing in tissue culture. Iwanowski's tobacco mosaic virus was

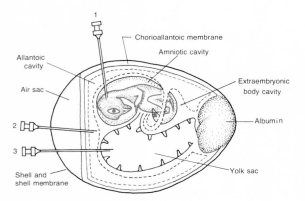

FIGURE 12–1 *Diagrammatic section through a developing chick embryo from 10 to 12 days old indicating how viruses can be inoculated into (1) the head of the embryo, (2) the allantoic cavity or (3) the yolk sac. (After Burrows, W. et al.: Textbook of Microbiology, 19th ed. Philadelphia, W. B. Saunders Co., 1968.)*

isolated and crystallized by W. M. Stanley in 1935. Since then many other viruses have been obtained as crystals. When these crystals are put back into the appropriate host, they multiply and produce the symptoms of the disease. The tobacco mosaic virus was separated into its component protein and nucleic acid parts by Stanley in 1956. He was then able to recombine these parts into an active virus. Fraenkel-Conrat subsequently showed that the nucleic acid portion of the virus is infectious when injected alone, although much less efficient as an infectious agent than the intact virus. Injecting the nucleic acid alone induced the tobacco plant to produce the specific protein of the viral coat, in addition to specific viral nucleic acid, so that the complete virus particle was reconstituted. Viruses generally consist of a core of nucleic acid surrounded by a coat of protein. The nucleic acid provides the specific genetic information of the virus; the protein coat serves to protect and stabilize the nucleic acid.

Viruses vary widely in size from the psittacosis virus, the cause of a disease transmitted by parrots and other birds (which is about 275 millimicrons in diameter), to the virus causing foot-and-mouth disease of cattle, which is only 10 millimicrons in diameter. Electron microscopy reveals that some of the viruses are spherical and others are rod-shaped (Fig. 12–2). Although individual virus particles cannot be seen in the light microscope, cells that are infected with viruses frequently contain **inclusion bodies** that are visible. These appear to be huge colonies of virus particles.

Each kind of virus usually attacks some specific part of the host's body. Apparently the virus particles can reproduce only in certain kinds of cells and not in all the cells of the body. The viruses of smallpox, measles and warts attack the skin. Those of poliomyelitis and rabies attack the brain and spinal cord, and those of yellow fever attack the liver. It is fortunate that many of the infections caused by viruses create a lasting immunity against reinfection. Thus inoculations for smallpox, rabies and yellow fever are highly successful.

The infection of one cell by one virus interferes with its subsequent infection by a second virus. The infected cell releases a substance which has been named **interferon** and identified as a protein about the size of hemoglobin. Interferon is released when the cell is infected with a live virus or with a heat-killed virus, and is produced in amounts which might indeed account for the subsequent resistance of the cell to viral infections. The synthesis of interferon may be stimulated by the presence within the cell of a foreign nucleic acid, either of viral or nonviral origin. Even synthetic polynucleotides such as poly UC are potent stimulators of the production of interferon. Mice infected with influenza virus had the greatest amount of virus in the lungs on the third day, after which the number of viruses decreased. The concentration of interferon in the lungs was greatest on days 3 to 5 and then slowly declined. Antibodies to the virus appeared only on day 7 and increased after that, which suggests that interferon may be more important than antibodies in accounting for the recovery of the animal from the acute infection.

Viruses that parasitize bacteria, called **bacteriophages,** were discovered in 1917 by the French scientist d'Herelle, who noticed that some invisible agent was destroying his cultures of dysentery bacilli. Bacteriophages are filtrable and will grow only within bacterial cells which they cause to swell and dissolve. Bacteriophages are found in nature wherever bacteria occur and they are especially abundant in the intestines of man and other animals. There are many varieties of bacteriophages, and usually one kind of phage will attack only one species or one strain of bacteria.

Electron micrographs show that phages are spherical or comma-

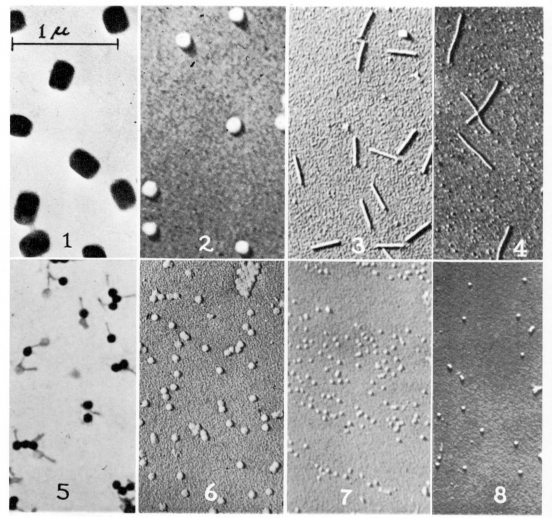

FIGURE 12–2   Electron micrograph of a variety of viruses. 1, Vaccinia virus (used in vaccinating for smallpox). 2, Influenza virus. 3, Tobacco mosaic virus. 4, Potato mosaic virus. 5, Bacteriophages. 6, Shope papilloma virus, 7, Southern bean mosaic virus. 8, Tomato bushy stunt virus. The viruses in plates 2, 3, 4, 6, 7 and 8 were shadowed with gold before being photographed in the electron microscope. (Courtesy of C. A. Knight.)

shaped or shaped like a Ping Pong paddle and are some 5 millimicrons in diameter (Fig. 12–3). The fact that bacteriophages destroy bacteria, of course, led biologists and medical scientists to attempt to use them to treat patients suffering from bacterial diseases such as dysentery and staphylococcus infections. None of these preparations of bacteriophage has had any significant effect and this, combined with experimental evidence that bacteriophages are ineffective in the presence of blood, pus or fecal material, have led to the abandonment of their use for therapeutic purposes.

To obtain bacteriophages, an emulsion of feces, soil or sewage is made and passed through a fine filter. If phages are present, a drop of this filtrate added to a turbid bacterial culture will cause the bacterial cells to swell and dissolve, and the culture becomes clear. When a drop of this clear culture is filtered and added in turn to a second bacterial culture, the

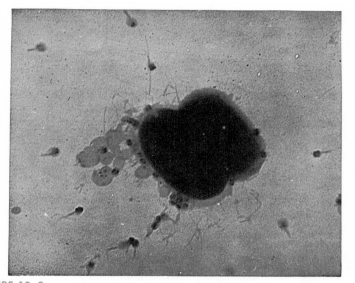

FIGURE 12–3   *An electron micrograph of an early stage in the destruction of two colon bacilli (the large dark bodies) by bacteriophages (the small particles shaped like Ping-Pong paddles). To the left of the bacilli are disc-shaped particles which are remnants of bacterial cells destroyed previously. Some of the bacteriophage particles have previously penetrated the outer layer of the bacterial cell. (Reduced in printing from an initial magnification of 30,000 ×; from Frobisher, M.: Fundamentals of Microbiology. 8th ed. Philadelphia, W. B. Saunders Company, 1968.)*

latter will also become clear. Serial transfers of the bacteriophage can be made in this way indefinitely.

The virus first attaches to the bacterial cell by means of its protein tail. The tail contains an enzyme which digests part of the bacterial cell wall and the DNA in the core of the virus is passed into the bacterial cell (Fig. 12–4). By labeling the protein of the virus with radioactive sulfur (^{35}S) and the nucleic acid with radioactive phosphorus (^{32}P), Hershey and Chase could show that only the DNA of the virus is injected into the bacterial cell. The protein coat of the head and tail remains outside. The inference that the DNA contains all the genetic information for the synthesis of the complete virus particle was established. The DNA directs the biosynthetic systems of the host cell to produce both viral DNA and viral protein. For 10 to 15 minutes after the infection of the bacterial cell, no virus particles can be detected within it. Over the next 10 to 15 minutes, increasing numbers of viral particles accumulate, and finally some 30 minutes after the initial infection, the bacterial cell bursts, and several hundred newly formed viral particles are released, ready to attack new bacterial cells.

When bacteria are treated with the enzyme **lysozyme** to remove most of the material in the cell wall, they can then be infected with purified DNA from a bacterial virus, as well as by intact viral particles. When these bacteria are subsequently lysed, *complete* viruses containing both DNA *and protein* are released. This is further evidence that all the information to make a complete virus is present in the DNA, and that the protein coat simply plays a role in penetrating the bacterial cell wall.

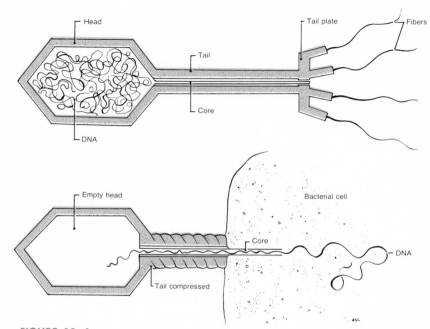

FIGURE 12–4   Upper, a diagram of the structure of T_2 bacteriophage revealed by electron microscopy. Lower, a diagram of the transfer of viral DNA into the bacterial host cell.

SECTION 12–3

RICKETTSIAS

The rickettsias are a group of organisms which are smaller than bacteria but larger than viruses. They are not filtrable, and are just barely visible under the light microscope. Some are spherical, others rod-shaped, and they vary in length from 300 to 2000 millimicrons. Their cellular structure is similar to that of bacteria. They were named after their discoverer, Howard Ricketts, who died in Mexico in 1910 of typhus fever while studying the organisms that cause it. Rickettsias, like viruses, will multiply only within living cells, and are obligate intracellular parasites. The single exception to this is a nonpathogenic parasite of the sheep tick.

Some 50 different kinds of rickettsias are harmless parasites in the intestinal tracts and salivary glands of insects such as lice, bed bugs and ticks. A few of these rickettsias, when transmitted to man by insect bites, will multiply inside human cells and produce the symptoms of disease. Only six kinds of rickettsias are known to produce human diseases; the principal rickettsial diseases of man are typhus fever and Rocky Mountain spotted fever. Rickettsias can be grown on chick embryos developing inside the egg shell to produce enough organisms to prepare a vaccine.

SECTION 12–4

BACTERIA

There are relatively few places in the world that are devoid of bacteria, for they can be found as much as 5 meters deep in the soil, in fresh and salt water, and even in the ice of glaciers. They are abundant in air, in liquids such as milk, and in and on the bodies of animals and plants, both living and dead. Bacterial cells range in size from less than 1 to 10 microns in length and from 0.2 to 1 micron in width. There are rod-like bacilli, spherical cocci and spiral forms of bacteria. Most bacterial species exist as single-celled forms, but some are found as filaments of loosely joined cells. Bacteria are classified largely by physiological and biochemical characteristics rather than by morphologic characters, since the various kinds of bacteria have generally similar shapes and cell structures.

The bacterial cell has a cell membrane and is covered by a strong, rigid cell wall which contains **diaminopimelic acid,** an amino acid found only in bacteria and in blue-green algae, and **muramic acid,** a derivative of glucose. With high resolution electron micrography, the cell walls of bacteria are revealed to be constructed of units 50 to 140 millimicrons in diameter, arranged in regular, hexagonal or rectangular patterns (Fig. 12–5). Most bacteria have a slimy **capsule** composed of polysaccharides which lies outside of the cell wall and serves as an additional protective layer. The cytoplasm within the bacterial cell is dense and contains granules of glycogen, protein and fat, but lacks mitochondria and an endoplasmic reticulum (Fig. 12–6). The ribosomes occur free in the cytoplasm and are not bound to an endoplasmic reticulum. The DNA is present in a distinct nuclear region, but no nuclear membrane separates it from the rest of the cytoplasm. Cocci usually have one nuclear region, but rod-like bacilli usually have two or more per cell. Many bacteria have whip-like cellular outgrowths called **flagella,** by means of which they can swim about. Bacterial flagella are curious in that they consist of a single fibril, whereas the flagella of higher organisms are composed of 11 fibrils arranged in a bundle, nine in the periphery and two in the center of the flagellum. Rod- and spiral-shaped bacteria tend to have flagella, whereas spherical ones do not. Some bacteria can travel as much as 2000 times their own length in an hour in this way.

Bacteria generally reproduce asexually by the division of the parental cell into two daughter cells. Cell division involves a mitotic process which is similar in many ways to that of higher forms. However, the duplication of the chromosome and the division of the nuclear region can get out of phase with the division of the rest of the cell, so that a given cell may have from one to four or even more nuclear regions. Bacterial cell division can occur with remarkable speed, and some species grown in an appropriately fortified and aerated culture medium can divide every 20 minutes. At this

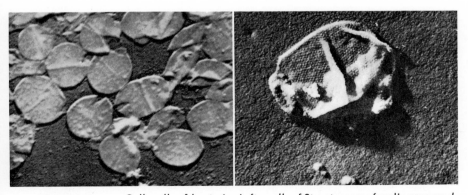

FIGURE 12–5   Cell walls of bacteria. Left, walls of Streptococcus fecalis prepared by grinding the cells; the splitting process permitted the cell contents to escape, magnified 12,000×. Right, a portion of the cell wall of Spirillum rubrum showing the regular pattern of the spherical bodies which comprise the cell wall, magnified 42,000×. (From Burrows, W.: Textbook of Microbiology. 19th ed. Philadelphia, W. B. Saunders Company, 1968.)

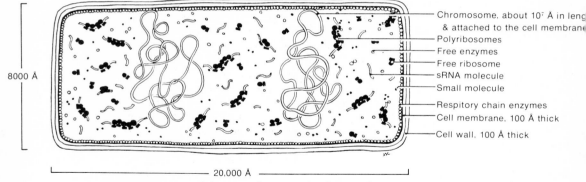

8000 Å

Chromosome, about 10⁷ Å in leng
& attached to the cell membrane
Polyribosomes
Free enzymes
Free ribosome
sRNA molecule
Small molecule

Respitory chain enzymes
Cell membrane, 100 Å thick
Cell wall, 100 Å thick

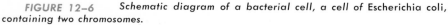

20,000 Å

FIGURE 12–6   Schematic diagram of a bacterial cell, a cell of Escherichia coli, containing two chromosomes.

rate, if nothing interfered, one bacterium would give rise to some 250,000 bacteria within six hours. This explains why the entrance of only a few pathogenic bacteria into a human being can result so quickly in the symptoms of disease. Fortunately bacteria cannot reproduce at this rate for a very long time, for they are soon checked by lack of food or by the accumulation of waste products.

Both cytologic and genetic evidence indicate that bacterial cells undergo something like sexual reproduction albeit rather rarely. Two cells fuse and hereditary material is transferred from one to the other. Genetic recombination occurs in those bacteria that have been carefully studied, and probably occurs in other species as well. One of the most intensively studied species, *Escherichia coli,* has been shown to have "sex." Some cells act as "males" and transfer genetic information by direct contact with "females." The ability to transfer genetic material is regulated by a fertility factor, F+, which can itself be transferred from a male to a female, thereby converting her into a male. The usual vegetative bacterial cell is haploid, and in sexual reproduction, part or all of the chromosome passes from the male cell to a female cell. The resulting cell is partly or completely diploid, and crossing over can then occur between the female chromosome and the male chromosome or fragment. This is followed by a process of segregation that yields haploid progeny cells.

When the environment of the bacterial cell becomes unfavorable, e.g., when it becomes very dry, many bacteria have the ability to become dormant. The cell loses water, shrinks a bit and remains quiescent until water is again available. Other species form **spores** to survive in extremely dry, hot or cold environments (Fig. 12–7). The formation of a spore is not a kind of reproduction, since only one spore is formed per cell. The total number of individuals does not increase as a result of the formation and subsequent hatching of the spore. During the formation of a spore the cell shrinks, rounds up within the former cell membrane, and secretes a new, thicker wall inside the old one. When the environmental conditions are again suitable for growth the spore can absorb water, break out of its inner cell, and become a typical bacterial cell. Spores of anthrax bacilli have been shown to be able to hatch out 30 years after they were formed.

Although a few bacteria are autotrophic and synthesize organic compounds by chemosynthetic or photosynthetic reactions, most bacteria are either saprophytes or parasites. The majority of bacteria, like animals and plants, are aerobic and utilize atmospheric oxygen in cellular respiration. A few bacteria can grow and multiply in the absence of gaseous oxygen. They obtain energy by the anaerobic metabolism of carbohydrates or

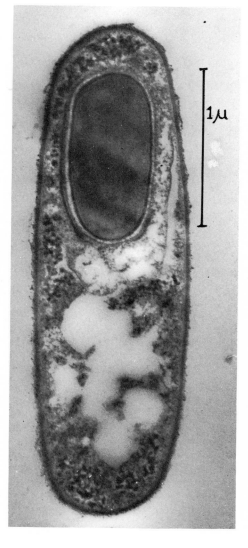

FIGURE 12–7   *Nearly mature spore within a cell of Bacillus cereus. Three spore coats can be distinguished. The large white bodies in the lower portion of the cell are vacuole-like inclusions. (Electron micrograph of ultrathin section by G. B. Chapman, J. Bact. 71:348-355, 1956.)*

amino acids, and, in the course of this, accumulate a variety of partially oxidized intermediates, such as ethanol, glycerol and lactic acid. A few bacteria, **obligate anaerobes,** can grow only in the absence of oxygen, and are killed in the presence of molecular oxygen.

The range of organic compounds that can be utilized by one or another kind of bacteria as a source of energy is impressive and includes sugars, amino acids, fats, urea, uric acid and other waste products. One strain of bacteria has become adapted to using penicillin, although many kinds are killed by this antibiotic. The anaerobic metabolism of carbohydrates is termed **fermentation,** and the anaerobic metabolism of proteins and amino acids is termed **putrefaction.** The foul smells associated with the decay of food and plant or animal bodies or wastes are due to nitrogen and sulfur-containing compounds formed in putrefaction. Bacteria play important roles in the carbon, nitrogen, and other cycles of nature, and, in fact, the substances produced by one kind of bacteria may be used as sources of energy by other kinds of bacteria. Without the bacteria and the fungi, all the available carbon and nitrogen atoms would eventually be tied up in the bodies of dead plants and animals, and life would cease for the lack of raw materials for the synthesis of new cellular components.

A **pure culture** of one species of bacteria can be set up by the method of **serial dilution.** A sample of soil, feces, blood or sputum is mixed with a large volume of nutrient medium and a small portion of this is removed and mixed with another large volume of medium. By repeating this process and making the serial dilutions rapidly enough so that the bacteria cannot reproduce between transfers, you eventually obtain a tube of nutrient medium with but a single bacterial cell, even though the initial sample contained many millions of bacterial cells. When the final tube is incubated for an appropriate period, and the single cell divides repeatedly, the resulting bacterial culture will have arisen by the reproduction of this single cell and will be a pure culture. Another method takes advantage of the fact that bacteria can move only a very short distance on a solid medium. A drop of feces or sputum is mixed with warm liquid **agar** (extracted from a seaweed) that is about to solidify. The mixture is poured in a thin layer on a flat glass dish with a cover. As the medium cools it solidifies and the bacteria are held in position. The bacteria are spread out on the agar by the mixing process, and when each subsequently multiplies it gives rise to a colony of daughter cells, all of which came from a single initial bacterium and thus is a pure culture. With a sterile wire loop some of the members of this colony are transferred to a fresh dish of agar, mixed and incubated again. In this way, any bacteria contaminating the original colony can be removed.

Bacteria are identified to some extent by their morphologic appearance and the color of the colonies, and by staining with the Gram stain devised by the Danish physician Christian Gram. Bacteria are also identified by their biochemical properties, by testing the kinds of substances they need to grow and the kinds of substances they produce. For example, the colon bacillus, *Escherichia coli,* is a normal inhabitant of the human colon, but typhoid bacilli and *Shigella,* a bacillus that produces food poisoning, are not normal inhabitants. The normal and the pathogenic bacilli cannot be distinguished morphologically, but the colon bacillus can ferment lactose, whereas the pathogenic forms cannot.

Man has used to good advantage the abilities of specific strains of bacteria to produce chemicals such as ethanol, acetic acid, butanol and acetone. Bacteria with specific enzymatic properties are widely used in the synthesis of drugs and other chemicals. Bacterial action is involved in the curing of tobacco, in preparing hides for tanning, and in treating the fibers from flax and hemp to make linen and rope. Bacteria also play a role in the production of butter, cheese, sauerkraut, rubber, cotton, silk, coffee and cocoa. Specific strains of decay bacteria play an important role in the functioning of sewage disposal plants. Sewage is allowed to pass slowly over beds of gravel and sand which contain bacteria. The action of the bacteria converts the raw sewage into solid material that can be dried and used as fertilizer. It removes and kills any disease-producing bacteria that may have been present, thus making the effluent safe as a source of drinking water.

Bacteria not only cause a variety of diseases in man and animals, but also are responsible for certain diseases of plants, such as the **fire blight** of apple and pear trees. The bacteria enter the tree through a wound or through a flower, multiply rapidly and kill the cells so quickly that the tree appears to be burned. The tree typically produces a sticky liquid around the infection which contains bacteria. Other trees become infected when drops of this liquid are transported by insects. Bacteria are responsible for the **soft rot** of a variety of vegetables, the **black rot** of cabbage and the growth of **crown galls** on other plants.

SECTION 12–5

THE THALLOPHYTES

Several different systems have been proposed for the classification of the thallophytes. One widely accepted classification, which we shall use, sets up seven phyla of **algae** and three phyla of **fungi.** An important criterion in classifying algae and fungi is their method of reproduction. With the possible exception of the blue-green algae, and perhaps certain of the bacteria, the members of all of these phyla reproduce sexually at some time in their life cycle. Sexual reproduction involves two fundamental processes: a special kind of nuclear division, **meiosis,** that reduces the chromosome number from diploid to haploid, and **fertilization,** the fusion of two haploid gametes to form a diploid zygote. Meiosis and the union of gametes are landmarks which divide the life cycle of a plant into two phases. The first, from gamete union to meiosis, is characterized by cells with a **diploid** (2n) number of chromosomes, and the second, from meiosis to the union of the gametes, is characterized by cells with a **haploid** (n) number of chromosomes (Fig. 12–8). In many algae and some fungi, the zygote nucleus divides by meiosis; thus only the zygote is a diploid cell and the haploid phase may be one-celled or many-celled. All the higher plants and some algae and fungi have a life cycle in which the zygote divides by mitosis to produce diploid daughter cells. The diploid phase in such plants includes the zygote and all the cells derived from it by mitosis. Eventually some of these cells undergo meiotic division to produce haploid spores, which in turn may divide mitotically to produce a many-celled haploid generation. The haploid generation completes the life cycle

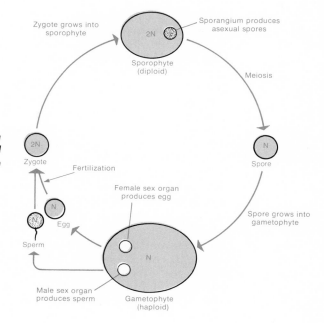

FIGURE 12–8   *Diagram illustrating the alternation of sporophyte (diploid) and gametophyte (haploid) generations in plants.*

Zygote grows into sporophyte

Sporangium produces asexual spores

2N

Sporophyte (diploid)

Meiosis

N

Spore

Spore grows into gametophyte

2N
Zygote

Fertilization

Female sex organ produces egg

N
Egg

Sperm

N

Gametophyte (haploid)

Male sex organ produces sperm

by producing gametes which unite to form zygotes. The diploid spore-producing generation is called the **sporophyte,** and the haploid gamete-producing generation is called the **gametophyte.** In most plants a haploid, gamete-producing generation alternates with a diploid, spore-producing generation. In the process of evolution from simple algae to a rose bush or apple tree, the basic pattern of the cycle has remained the same, but there have been tremendous changes in the relative size and nutritional relationships of the two generations and in the vegetative and reproductive organs.

SECTION 12–6

THE EVOLUTION OF SEX

Within a single living phylum, the green algae, are species that can be arranged in a series which illustrates how sexual reproduction might have evolved. The fact that these can be arranged in a series, however, proves neither that sex first evolved in the green algae nor that the evolution of sex took place in this series of steps.

The simplest green algae, such as *Protococcus,* reproduce only asexually, by simple cell division. In most other green algae, asexual reproduction occurs by the transformation of a vegetative cell into one or more specialized reproductive cells, called **zoospores,** each of which bears two or more flagella and is adapted for facilitating the dispersal of the species.

Chlamydomonas, found in pools, lakes and on damp soil, has a vegetative cell bearing flagella and protected by a heavy cellulose wall. Each cell has a single cup-shaped chloroplast containing a **pyrenoid** involved in the production of starch, an "eye" spot containing a red pigment and two contractile vacuoles near the base of the two flagella.

The cell may reproduce asexually by dividing to form two to eight zoospores within the cellulose wall. The rupturing of the parental cell wall sets these free to swim away, independent plants. In sexual reproduction the parent cell divides to form eight to 32 smaller gametes, which resemble the zoospores and adults but are smaller (Figure 12–9). Two of these gametes fuse, beginning at the end bearing flagella, to form a zygote which initially has four flagella, two contributed by each gamete. The zygote loses its flagella, rounds up, secretes a thick cell wall and is capable of surviving long periods in an unfavorable environment. When the environment is again favorable, the zygote undergoes meiosis to form four haploid cells, the cell wall cracks open and the liberated cells develop flagella and become independent plants.

Chlamydomonas illustrates a very primitive form of sexual reproduction, for the gametes are not specialized cells; they look exactly like the zoospores and adults. Most species of *Chlamydomonas* are **isogamous;** the two cells uniting are identical in size and structure. A few species are **heterogamous;** there are two kinds of gametes differing in size (both bear flagella), which unite to form a zygote.

The pond scum *Spirogyra* consists of long filaments of haploid cells arranged end to end. In the fall, when reproduction usually occurs, two filaments come to lie side by side and dome-shaped protuberances appear on the cells lying opposite (Fig. 12–10). These enlarge, fuse and form a conjugation tube connecting the two cells. One cell rounds up, oozes through the tube and joins the second. The nuclei of the two cells unite and fertilization is complete. The resulting zygote develops a thick cell

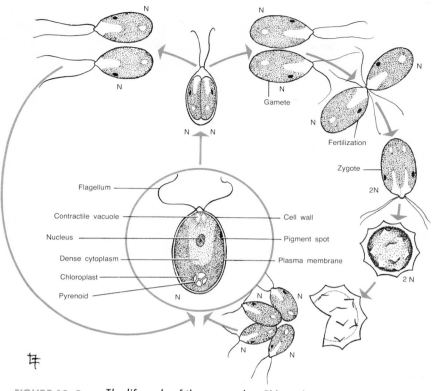

Flagellum
Contractile vacuole
Nucleus
Dense cytoplasm
Chloroplast
Pyrenoid

Cell wall
Pigment spot
Plasma membrane

Gamete
Fertilization
Zygote
2N
2N

FIGURE 12–9   *The life cycle of the green alga* Chlamydomonas, *asexual reproduction on the left, stages in sexual reproduction on the right. Inset: enlarged view of a single individual showing the body structures.*

wall and is able to survive during the winter. In the spring it divides meiotically to form four haploid nulei, three of which degenerate. The fourth remains and, after the thick wall breaks, divides mitotically to form a new haploid filament. Sexual reproduction in *Spirogyra* is primitive in that it involves unspecialized cells (any cell in the filament can fuse with one from a neighboring filament) and the two fusing cells are similar (isogamy).

Another filamentous fresh water green alga, *Ulothrix*, demonstrates what was possibly the next step in the evolution of sex. In this plant, each haploid vegetative cell in the filamentous chain contains a single collar-shaped chloroplast and several pyrenoids. One cell may divide to form four to eight zoospores, each bearing four flagella, which can be released and give rise to a new filament (Fig. 12–11). One of the cells of the filament may undergo several divisions to produce many small, similar gametes, resembling zoospores but with two instead of four flagella.

As in *Chlamydomonas*, two of these swimming forms fuse, forming a zygote which initially has four flagella. After swimming for a time, the zygote loses its flagella, secretes a thick cell wall and is capable of withstanding cold or drying. It later undergoes meiotic division and gives rise to four cells. These are finally liberated by the splitting of the old zygote wall and develop into new filaments. Thus, in *Ulothrix*, sexual reproduction is isogamous, involving fusion of two identical cells, but these cells are specialized, differing from the usual vegetative ones.

Another filamentous alga, *Oedogonium*, illustrates a third possible step in the evolution of sexual reproduction. The cells which fuse to form the zygote are unlike: One is a large, nonmotile, food-laden egg and the other

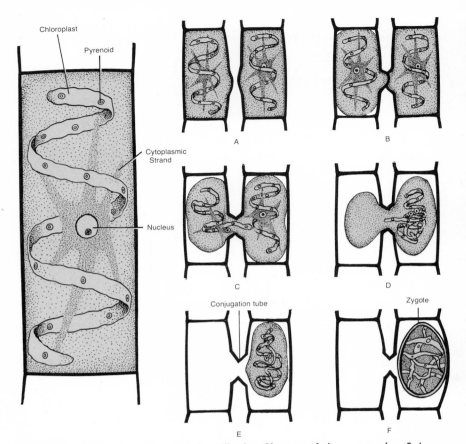

Chloroplast

Pyrenoid

Cytoplasmic
Strand

Nucleus

Conjugation tube

Zygote

A

B

C

D

E

F

FIGURE 12–10   Left, a single cell of a filament of the green alga Spirogyra, showing cell structures. Right A–F, stages in the sexual reproduction of Spirogyra (see text for discussion).

is a small, motile sperm (Fig. 12–12). Sexual reproduction by the fusion of unlike gametes, **heterogamy,** is characteristic of most higher plants. Any vegetative cell can differentiate into either an egg-forming cell **(oögonium)** or a sperm-forming cell **(antheridium).** The egg-forming cell is an enlarged, spherical cell which shrinks away from the hard cell wall to form a rounded, nonmotile, food-laden **egg.**

The sperm-forming cells are produced when a vegetative cell divides several times to produce a series of short, disc-shaped cells. Each of these antheridia divides to produce two small sperm, each of which has a circle of flagella at the anterior end (Fig. 12–12). The sperm swims to the egg, attracted by some chemical substance released by the egg. It enters the egg-containing cell through a crack and fuses with the egg. Both egg and sperm are haploid and their fusion results in a diploid zygote which secretes a thick cell wall and can survive extended periods of unfavorable environmental conditions. It eventually undergoes meiosis to form four haploid cells, each of which has a circle of flagella at the anterior end and resembles the asexual reproductive cells, the zoospores. The zoospores, whether formed sexually or asexually, can germinate and divide to form a new *Oedogonium* filament.

A final step in the evolution of sex is the production of specialized gametes only by special cells in the body—the **sex organs**—rather than by

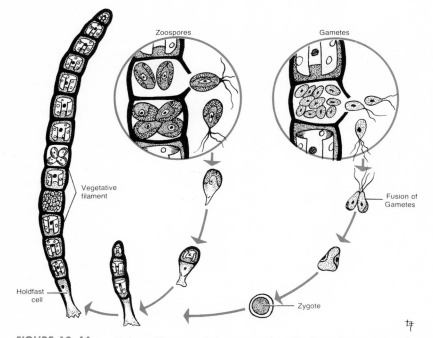

FIGURE 12–11   Left, a filament of the green alga Ulothrix. Center, an enlarged view of stages in asexual reproduction by zoospores. Right, sexual reproduction by the formation of gametes and the fusion of two gametes to form a zygote.

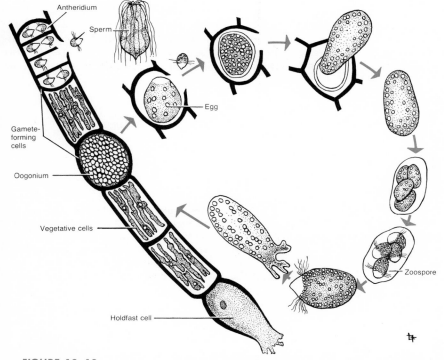

FIGURE 12–12   A filament of the green alga Oedogonium. Sexual reproduction in this alga involves the formation of differentiated eggs and sperm and their union in fertilization.

any vegetative cell in the organism as in *Ulothrix* and *Oedogonium*. *Volvox* is a colonial alga, a hollow ball of cells, each of which bears two flagella and is connected to its neighbors by fine cytoplasmic strands. Each colony may contain as many as 40,000 cells, most of which are alike and function vegetatively. Small motile sperm bearing two flagella are produced only in special sperm-producing organs or **antheridia** (this term is used generally for sperm-producing organs in higher plants as well). A single, large, nonmotile egg is produced within a special egg-producing organ or **oögonium** (Fig. 12–13).

The motile sperm are released and swim to the egg; their union results in a diploid zygote which forms a thick cell wall and can resist unfavorable conditions. During germination, meiosis occurs and haploid vegetative cells are formed. These, by many mitotic divisions, give rise to a new colony. In some species of *Volvox*, a single colony may have both antheridia and oögonia; in other species, a colony has one or the other but not both, and can be said to be "male" or "female." In these forms, sexual reproduction has evolved to a point where there is **sex differentiation.**

The series illustrates several trends in evolution, each of which is toward some sort of specialization. The trend from like gametes (isogamy) to unlike gametes (heterogamy) has obvious advantages for the survival of the species—the large number and motility of the sperm make them effective in seeking out the egg, and the large size and food stores of the egg provide nourishment for the zygote until it can become nutritionally independent. A second trend is toward the specialization of the cells of the colony or many-celled body so that some carry out only vegetative functions, others only reproductive ones. A third trend is toward differentiation of the sexes. In these primitive plants, asexual and sexual reproduction may occur in the same plant, depending on environmental conditions. A fourth trend is toward the retention of the fertilized nonmotile egg within the body of the parent plant. The higher algae and all the higher plants have a definite, regular alternation of a generation of plants reproducing sexually with a generation reproducing asexually by means of spores. The establishment of this pattern of **alternation of generations** is a fifth trend,

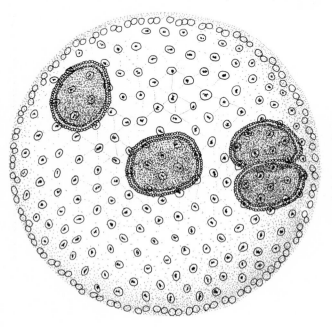

FIGURE 12–13 The colonial green alga Volvox. *Each colony may contain as many as 40,000 cells, most of which function only vegetatively. Small sperm cells are produced in antheridia and a single, large, nonmotile egg is produced within a special egg-producing oögonium.*

the beginnings of which occur in the green algae. The sea lettuce, *Ulva*, for example, consists of two kinds of plants which are identical in size and structure. One of these, however, is a diploid sporophyte which, by meiosis, produces haploid zoospores that develop into haploid gameto-phyte plants. This second type of plant produces gametes that fuse to form a diploid zygote which develops into the diploid sporophyte plant.

The **life cycle** of any species may be defined as the succession of devel-opmental stages and processes which occur between any given point in one organism's life span and that same point in the life span of its off-spring. For bacteria, blue-greens and *Protococcus*, which reproduce by splitting, the cycle is extremely simple. During most of its cycle a filamen-tous green alga such as *Ulothrix* consists of a colony of haploid cells which multiply asexually by mitosis. These divisions result in new vegetative cells in the filament, in haploid gametes or in asexual, haploid zoospores which divide to form new haploid colonies. The only diploid cells are the zygotes, for they divide meiotically to yield haploid vegetative cells.

In *Ulva* and the higher plants, the zygote divides mitotically to form a diploid sporophyte plant which in turn produces haploid spores and from them, haploid gametophytes. The resulting life cycle is more complex and involves a marked alternation of generations. Some parasitic plants such as the wheat rust and white pine blister rust have complex life cycles in-volving several kinds of spores and two host organisms.

SECTION 12–7

BLUE-GREEN ALGAE

The 2500 or so species of blue-green algae are probably the most primitive chlorophyll-containing plants that now exist. The oldest fossil plants found so far appear to have been blue-green algae. The chlorophyll of the blue-greens is not present in discrete chloroplasts but is scattered throughout the cytoplasm as small granules (Fig. 12–14). Blue-greens have a blue pigment, **phycocyanin**, in addition to chlorophyll, carotene and xanthophyll. The blue-greens store their carbohydrates as a unique kind of starch called **cyanophycean starch** and are covered by a sticky gelatinous outer sheath secreted by their cells.

Blue-green algae found in fresh water pools and ponds may occur in sufficient numbers to color the water and give it an unpleasant taste and smell. A few species live in hot springs and others live in the ocean. Mem-bers of the genus *Trichodesmium* contain a red pigment and occur in the Red Sea from time to time in such great numbers that they actually color the ocean. A few blue-greens live on the surface of soil and rocks in damp shady places. A few kinds of blue-green algae are unicellular, but most of them are found as long, many-celled filaments. None of these plants has flagella, but some of the filamentous ones are capable of a curious back and forth oscillatory movement.

SECTION 12–8

THE EUGLENOPHYTA

The unicellular alga *Euglena* and its relatives have a mixture of animal and plant characteristics and are difficult to classify. Botanists usually call them algae and assign them to a separate phylum, the Euglenophyta.

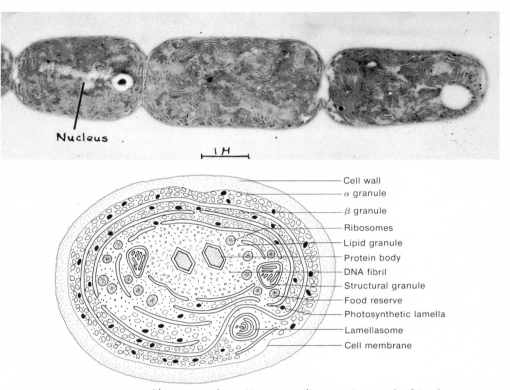

Nucleus

⊢ 1 μ ⊣

Cell wall
α granule
β granule
Ribosomes
Lipid granule
Protein body
DNA fibril
Structural granule
Food reserve
Photosynthetic lamella
Lamellasome
Cell membrane

FIGURE 12–14   Blue-green algae. Upper an electron micrograph of Anabaena. The low density material is nuclear substance, but no nuclear membrane is present. The membranous structures in the cytoplasm are the functional equivalents of chloroplasts. (Courtesy of G. B. Chapman; from Carpenter, P. L.: Microbiology. 2nd ed. Philadelphia, W. B. Saunders Company, 1967.) Lower, diagram of the structure of the cell of a blue-green alga. Compare with Figure 12–6; note the fundamental similarity of the structure of bacteria and blue-green algae.

Zoologists classify them as the order Euglenoidina, class Flagellata in the phylum Protozoa. Euglenoids are more advanced evolutionarily than the blue-green algae, for they have a definite, easily stained nucleus, and the chlorophyll is not scattered in granules but is localized in chloroplasts, as in the higher plants. All the euglenoids have one or two flagella by means of which they can swim actively (Fig. 12–15). They lack an outer cellulose cell wall but have a red-pigmented eyespot and a gullet near the base of the flagella. The pigment of the eyespot is **astaxanthin**, a substance found elsewhere only in crustaceans. It is the substance which gives a boiled lobster its red color. Animals usually store carbohydrates as glycogen, and plants store theirs as starch. The euglenoids store their carbohydrates as **paramylum**, chemically distinct from both starch and glycogen. The euglenoids with their curious mixture of plant and animal characters give us an idea of what early living things might have been like when the early autotrophs had evolved from heterotrophs but before plant and animals had evolved separately.

Euglenoid reproduction is usually asexual by simple cell division, but sexual reproduction has been observed in at least one genus. Although *Euglena* has plenty of chlorophyll, it apparently cannot survive solely by photosynthesis. It will not survive in a medium containing only inorganic salts but will flourish if small amounts of amino acids are added.

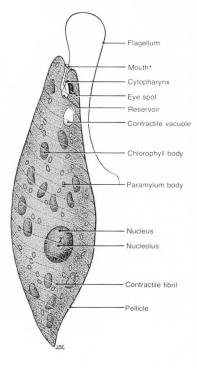

Flagellum

Mouth·

Cytopharynx

Eye spot

Reservoir

Contractile vacuole

Chlorophyll body

Paramylum body

Nucleus

Nucleolus

Contractile fibril

Pellicle

FIGURE 12–15   *Diagram of the structure of Euglena. These single-celled organisms are classified by botanists as members of the phylum Euglenophyta, whereas zoologists classify them as the order Euglenoidina within the class Flagellata.*

Some of the other species in the phylum are completely autotrophic. A few are completely saprophytic and some are holozoic, capturing and ingesting other organisms like typical animals.

SECTION 12–9

CHLOROPHYTA: THE GREEN ALGAE

Some 6000 species of green algae live in a wide variety of habitats ranging from salt to fresh water. Most botanists believe that higher plants evolved from some alga similar to the present-day green algae, for these algae have the greatest number of characters in common with the higher plants. The cell has a distinct nucleus with a nuclear membrane; its pigments, chlorophylls a and b, carotene and xanthophyll, are organized in chloroplasts; food is stored as starch; and a cellulose cell wall is present.

Some green algae are terrestrial and live on the moist, shady sides of trees, rocks and buildings. One species, adapted to living on snow and ice, has a red pigment in addition to chlorophyll and may grow in patches thick enough to give the snow a reddish tinge. The many-celled green algae living in fresh water include the pond scums which may grow very thickly in ponds and streams. Among the multicellular marine green algae living near the low-tide mark and in the upper 6 meters of water is the sea lettuce, with a body some 30 centimeters long, but only two cells thick. It resembles a crinkled sheet of green cellophane. Some of the tropical multicellular marine green algae have developed thickened plant bodies the size of a moss or small fern plant, with parts that superficially resemble roots, stems and leaves. Common nonmotile single-celled fresh water green algae, called **desmids**, have symmetrical, curved, spiny or lacy bodies with a constriction in the middle of the cell. When seen under the microscope they resemble snow flakes.

The **stoneworts,** found in fresh water ponds, are the most complicated of the green algae. These multicellular forms resemble miniature trees with structures that superficially look like, and serve the functions of, roots, stems, leaves and seeds. Despite this, they are not anatomically like their counterparts in the higher plants. Because they are more advanced than other green algae, some taxonomists place the 200 or so species of stoneworts in a separate phylum.

SECTION 12–10

CHRYSOPHYTA

The phylum Chrysophyta, or the golden brown algae, includes a variety of diverse types usually arranged in three classes. The **diatoms** are microscopic, usually single-celled forms, present in both fresh and salt water, and comprising an important food source for animals. Diatoms have cell walls containing silica, constructed in two overlapping halves which fit together like the two parts of a pillbox (Fig. 12–16). This siliceous wall is ornamented with extremely fine ridges, lines and pores which are characteristic for each species. The markings are either radially symmetrical or bilaterally symmetrical on either side of the long axis of the cell. Some are at the limit of resolution of the best light microscopes and are used as test objects to determine the quality of the lens. Diatoms are capable of a slow, gliding movement, apparently produced by the streaming of cytoplasm through the grooves on the surface of the cell wall. Diatoms have the brown pigment **fucoxanthin** and store their foods as the polysaccharide **leucosin,** and as **oil,** rather than as starch. It is generally held that petroleum is derived from the oil of diatoms that lived in previous geologic ages.

The remains of the silica-containing cell walls accumulate gradually as sediments in the bottom of the oceans. In later geologic times, uplifts may bring these sediments to the surface, and the diatomaceous earth is

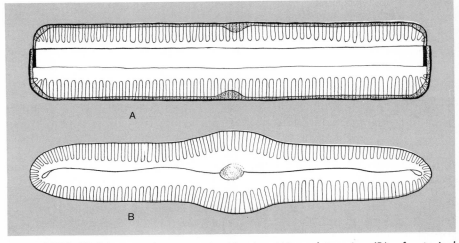

FIGURE 12–16   A diagrammatic side view (A), and top view (B), of a typical diatom, highly magnified. The shells have characteristic fine lines and the upper and lower shells fit together like the parts of a pillbox.

mined and used in making insulating bricks, as a filtering agent and as a fine abrasive. Several kinds of toothpaste contain diatomaceous earth. There are deposits of diatoms in California that are more than 300 meters thick.

It is estimated that some three quarters of all the organic material synthesized in the world is produced by diatoms and dinoflagellates. Diatoms can reproduce either sexually or asexually. The presence of the hard silica-containing cell wall complicates the process of asexual reproduction. The diatom cell divides and forms two cells within the old cell wall, and then two new cell walls form back to back between the two cells. Thus each daughter cell ends up with two cell walls, one inherited from the parent, and a new one that fits inside the old one. Each successive generation, of course, gets a little smaller, because each new cell wall fits inside the old one. Finally, a special cell is formed, which discards both of the old cell walls, enlarges and forms a pair of new, large cell walls.

The other two classes of Chrysophyta are the yellow-green algae Xanthophyceae and the golden-brown algae Chrysophyseae, both of which have silica-impregnated, two-shelled cell walls, and chloroplasts that are rich in carotenes and xanthophylls. These pigments give them their characteristic yellow or brown color. Some members of each class have, in the course of evolution, lost their chlorophylls and other pigments, and have become colorless, flagellate or amoeboid forms that are heterotrophic rather than autotrophic.

SECTION 12–11

PYRROPHYTA

The **dinoflagellates,** which comprise the phylum Pyrrophyta, are single-celled algae, most of which are surrounded by a shell made of thick, interlocking plates (Fig. 12–17). They are motile, with two flagella, one projecting from one end, and the other running in a transverse groove. Like diatoms, they have fucoxanthin in addition to chlorophyll and food reserves stored as oils and as polysaccharides. Most dinoflagellates are marine and are important photosynthesizers in the ocean. Occasionally they accumulate in large numbers in some part of the sea, coloring the water red or brown. Some species of dinoflagellates are poisonous to vertebrates, and when these accumulate, large numbers of fish in that region of the ocean may be killed. Other species of dinoflagellates are taken up as food by mussels. The mussels are apparently unharmed by the dinoflagellates, but if a man eats some of these infected mussels, he may become seriously ill.

SECTION 12–12

THE BROWN ALGAE, PHYLUM PHAEOPHYTA

The brown algae include about 1000 species of multicellular forms ranging in size up to giant kelps 100 meters long. They are the prominent brownish-green seaweeds that usually cover the rocks in the tidal zone and extend out into water 15 or so meters deep. These plants contain the golden-brown pigment **fucoxanthin,** which tends to mask the chlorophyll present; the color of the plants ranges from light golden to dark brown or

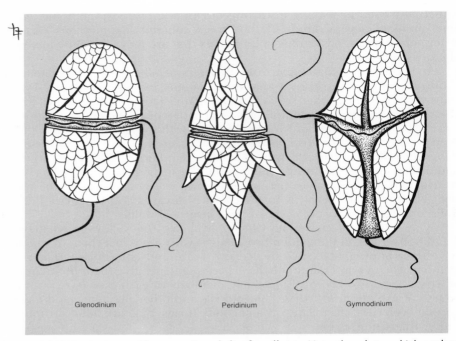

Glenodinium   Peridinium   Gymnodinium

FIGURE 12–17   *Three species of dinoflagellates. Note the plates which enclose and encase the single-celled body and the characteristic two flagella, one of which is located in the transverse groove.*

black. Some brown algae are large plants with complex body structures and parts that resemble the leaves, stems and roots of higher plants. In the algae, these parts are called **blade, stipe** and **holdfast,** respectively, to indicate that they are not homologous to the corresponding structures of higher plants. Brown algae are found in shallow waters along the coasts of all seas but are larger and more numerous in cool waters. The largest and most rugged of the algae, they are attached by their holdfasts to the rocks beneath the surface and usually have air bladders to buoy up the free ends.

The body, or thallus, of a brown alga may be a simple filament, such as the soft brown tufts of *Ectocarpus,* commonly found on pilings, or tough, ropelike, slimy strands, such as *Chorda,* the "Devil's shoelace," or thick, flattened, branching forms, such as *Fucus, Sargassum* or *Nereocystis* (Fig. 12–18). Phaeophytes have a well-defined alternation of generations. *Ectocarpus,* for example, consists of two kinds of plants that are similar in size and structure; however, one produces gametes and the other produces spores. The diploid form produces haploid spores (called zoospores) that divide and grow into mature haploid plants. These produce haploid gametes that fuse to produce a diploid zygote. This develops into the diploid plant, completing the life cycle. In certain other brown algae, the diploid sporophyte generation can be distinguished from the haploid gametophyte generation. In some, such as *Fucus,* the gametophyte generation is greatly reduced, as in the higher plants.

Brown algae furnish food and hiding places for many marine animals. Some kelps are used as food in oriental countries. Kelps such as *Laminaria* are processed commercially to yield a colloidal carbohydrate known as algin, a pectin-like component of the cell wall. This has the property of gelling and thickening mixtures and is widely used in making ice cream, for with it the ice cream manufacturer can use much less real cream and

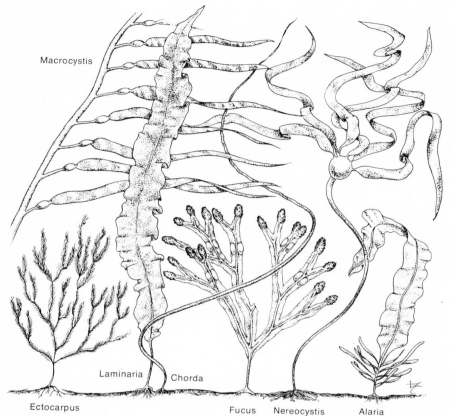

Macrocystis

Laminaria   Chorda

Ectocarpus   Fucus   Nereocystis   Alaria

FIGURE 12–18   *Some of the kinds of brown algae or kelps, all of which are multi-cellular marine plants. The sketches are not drawn to the same scale.*

still have a smooth creamy product. Algin is also used in making candy, toothpaste and cream cosmetics.

SECTION 12–13

THE RED ALGAE, PHYLUM RHODOPHYTA

The red algae, like the brown algae, are found almost entirely in the oceans. They are usually smaller and have more delicate bodies than the brown algae. They are unique in having the red pigment **phycoerythrin** in addition to chlorophyll, and are various shades of pink to purple. Red algae can grow at greater depths than other algae, and are found as deep as 100 meters. As sunlight penetrates water, first the red, then the orange, yellow and green rays are filtered out and only the blue and violet rays remain. Chlorophyll does not absorb light energy in the blue or violet region very efficiently, and plants that have only chlorophyll cannot carry on photosynthesis at these depths. Phycoerythrin can utilize these blue rays, and hence red algae can live at greater depths than other plants. Although red algae occur as far up as the low tide line, they reach their greatest development in the deeper tropical waters. Some 3000 species of Rhodophytes are known.

FIGURE 12–19   A photograph of a red alga. Note the lacy, delicately branched body. (Courtesy of the New York Botanical Gardens.)

Many rhodophytes have lacy, delicately branched bodies that are not as well adapted to survive in the intertidal zone as the tough, leathery brown algae, but they do well in the quieter deep waters (Fig. 12–19). The coralline red algae accumulate calcium from the sea water and deposit it in their bodies as calcium carbonate. Coralline algae are abundant in tropical waters and are even more important in the formation of coral atolls than are coral animals. Red algae have complex life cycles, with a marked alternation of sexual and asexual generations and specialized sex organs.

Several kinds of red algae are used as food. *Porphyra* is considered a great delicacy by the Japanese and is widely cultivated in submarine gardens. Dulse (*Rhodymenia*) is boiled in milk and eaten by the Scottish. **Agar,** used in making culture media for bacteria, is extracted from the red algae *Gelidium* and *Gracilaria.* Agar is extensively used in baking and canning. **Carrageenin,** extracted from Irish moss, is used in the preparation of chocolate milk to keep the chocolate from settling out.

SECTION 12–14

THE FUNGI: SLIME MOLDS

The three main groups of nongreen plants, the bacteria, slime molds and "true" fungi, have little in common except the heterotrophic nutrition necessitated by the absence of chlorophyll. Because of this, and because of the fact that each group has evolved independently, they are classified as separate phyla. The slime molds, peculiar organisms which in certain respects resemble true fungi, exist as slimy masses on decaying leaves or wood and move by extending pseudopodia, as amoebas do. At successive stages in their life cycle, they are single-celled flagellates, single-celled amoebas, multinucleate masses of cytoplasm (the colonial mass is called a *plasmodium*), and finally plantlike bodies with a stalk and fruiting body. The plasmodium has no definite shape, may be brightly colored and flows over the substratum, engulfing organic substances which are ingested in food vacuoles.

Ordinarily these cells reproduce asexually, by fission, but one of the more curious biological phenomena is the aggregation and fusing of individual, amoeba-like slime molds to form a fruiting (spore-producing)

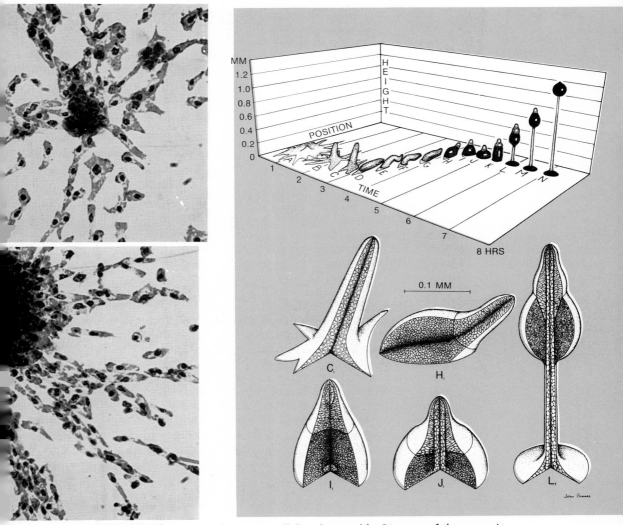

FIGURE 12–20   Reproduction in cellular slime molds. Diagram of the successive stages in the formation of a fruiting body. A—C, the fruiting body is formed by the aggregation of hundreds of individual amoeba-like slime molds. D—H, it crawls along the substrate surface for some time. I—N, it grows a stalk which lifts the spore-producing part off the surface. Finally the spores are released. (Courtesy of Dr. John Bonner.)

body (Fig. 12–20). The signal for the aggregation process appears to be cyclic 3′, 5′ adenosine monophosphate, which is involved as an intermediate in the mechanism of action of several hormones (p. 695). If a spore released from the fruiting body falls on some moist surface it will absorb water, split out of its wall and divide to form amoeboid cells. Eventually some of these amoeboid cells act as gametes and fuse to form a zygote, which then divides and grows to become the many-nucleate slimy mass, thereby completing the life cycle.

The spores of other species of slime molds germinate and produce one to four naked motile cells, each bearing two flagella. These **swarm cells** may divide mitotically or become amoeboid but eventually fuse in pairs to form a zygote. The spores are produced in the fruiting body by a meiotic division; thus the spores and swarm cells are haploid, but the plasmodium has diploid nuclei.

A few slime molds are parasites of plants; one causes a disease known as clubroot in cabbages. The slime molds have many characteristics in common with animals such as amoebas and flagellates, and whether they should be called animals, plants or protists is a matter of opinion. They may have evolved from flagellates; however, the formation of a fruiting body is characteristic of other fungi and for this reason they are usually classified with them.

SECTION 12–15

THE EUMYCOPHYTA, TRUE FUNGI

The 80,000 or more species of true fungi have a number of characteristics in common with the algae and are believed to have arisen from one or more of the algal phyla. Among the true fungi are the yeasts, molds, mildews, rusts, smuts and mushrooms. A few of the true fungi are unicellular, but most have many-celled bodies made of tubular branching filaments called **hyphae.** The outer wall of the body of the fungus may be composed of cellulose, chitin or a combination of the two. In some species the hyphae are subdivided by cross walls between successive nuclei and the fungus is multicellular; in other species there are no cross walls between adjacent nuclei and the fungus is multinucleate.

One of the distinguishing characters of the Eumycophyta is the presence of a **mycelium,** a mass of branching hyphae. In the common bread mold, this mycelium is visible as the cobwebby mass of fibers on the surface of the bread and penetrating into its interior. In a fungus such as the mushroom, much of the mycelium is below ground; the mushroom cap that we eat is a fruiting body, a specialized reproductive structure that grows out from the underground, nutritive mycelium.

Fungi grow best in dark, moist habitats; they are either saprophytic or parasitic and are found universally wherever organic material is available. Some fungi can grow under what are apparently very unfavorable conditions. They have a strong resistance to plasmolysis, for they can grow in salt or sugar solutions of high molarity (e.g., on jelly). As the mycelium branches and comes in contact with organic material, it secretes enzymes which hydrolyze proteins, carbohydrates and fats, and then absorbs the split products. Fungi, although important members of the carbon and nitrogen cycles, cause medically and economically important diseases of man, his domestic animals and his crop plants, and are responsible in large measure for the deterioration of wood, leather, cloth and similar materials. Fungal reproduction occurs in a variety of ways: asexually by fission, budding or by spores, and sexually by means that are characteristic for the subgroups. The spores of aquatic fungi typically have flagella, whereas the spores of terrestrial fungi are nonmotile cells dispersed by the wind or by animals.

Four classes of Eumycophyta are distinguished on the basis of the means of sexual reproduction: Phycomycetes, Ascomycetes, Basidiomycetes and Fungi Imperfecti. The latter is a heterogeneous group of fungi whose status is not completely understood, and in which sexual reproduction has not been detected. Some fungi may actually have no sexual phase but others probably have one that has not yet been discovered. When the means of sexual reproduction become known, the fungus can be removed from the Fungi Imperfecti and assigned to the appropriate class. This does not represent simply indecision on the part of the mycologists (specialists in the study of fungi) but commendable scientific caution in assigning a form to one of the clear-cut classes of fungi only when sufficient evidence is available to justify the classification.

SECTION 12–16

THE PHYCOMYCETES

The phycomycetes are the smallest class of fungi, consisting of about 500 species. Because of the similarity of their filamentous bodies to those of algae, and because they reproduce asexually by motile spores and sexually by gametes similar to those of certain algae, they are sometimes called the alga-like fungi. Their hyphae have few or no cross walls, and the mycelium consists of many nuclei in a common mass of cytoplasm. Some common phycomycetes are the bread mold, downy mildews and white rusts.

Bread becomes "moldy" when a spore of the black bread mold, *Rhizopus nigricans*, lands on it, germinates and grows to form a tangled mass of threads, the mycelium. Some of the hyphae, termed **rhizoids**, penetrate the bread and obtain nutrients; others, termed **stolons**, spread horizontally with amazing speed. Eventually certain hyphae grow upward and develop a **sporangium**, or spore sac, at the tip. Within this sac develop clusters of black, spherical spores which are released when the delicate spore sac ruptures.

Sexual reproduction occurs when the hyphae of two different plants come to lie side by side. Each hypha forms a swelling which grows toward the other; the tip of the swelling then enlarges and pinches off to form a gamete. Two adjacent gametes finally fuse to form a zygote (Fig. 12–21) from which develops the new hypha of the next generation.

There are "plus" and "minus" strains of the bread mold; sexual reproduction can occur only between a member of the plus strain and one of the minus strain. This is a sort of physiological sex differentiation, even though there is no morphologic sex differentiation. There is no justification for calling the two strains "male" and "female." Only the zygote of the bread mold is diploid. Meiosis occurs when the zygote germinates and all the hyphae are haploid.

SECTION 12–17

THE ASCOMYCETES

The largest class of fungi (about 35,000 species), the Ascomycetes or sac fungi, are so called because their spores are produced in sacs called

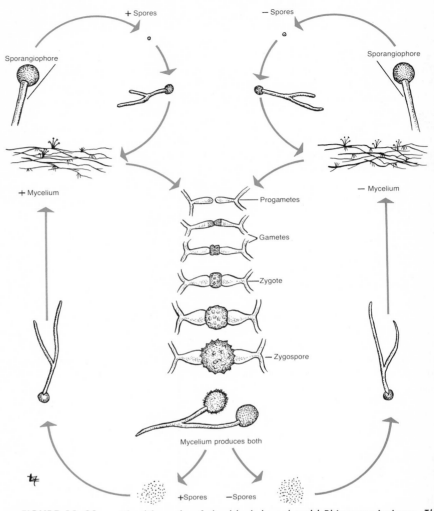

FIGURE 12–21   The life cycle of the black bread mold Rhizopus nigricans. The upper circles indicate the asexual production of mycelia from spores. In the center is a series of stages in sexual reproduction.

asci. Each ascus produces two to eight ascospores. Among the Ascomycetes are the yeasts, powdery mildews, the molds which appear on cheese, jelly and fruit, and the edible truffle. The ascomycete molds which appear on food may give it an unpleasant taste but they are not poisonous. The unique flavor of cheeses such as Roquefort and Camembert is produced by the action of ascomycetes. The ascomycete *Penicillium* produces the famous antibiotic **penicillin.**

The bodies of ascomycetes may be unicellular, as in yeasts; many-celled filamentous mycelia, as in powdery mildews; or thickened and fleshy, as in the truffle. Reproduction is accomplished asexually by budding (yeasts) or by spores called **conidia** that develop in sequence at the tips of certain hyphae, and sexually by gametes which unite to form a "fruit" containing the ascus. Although the structures in which asci are produced may be large and fleshy, superficially resembling true fruits, they have, of course, no relation to them.

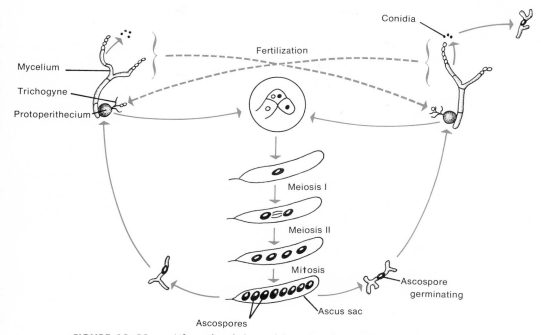

FIGURE 12–22   *Life cycle of the red bread mold* Neurospora crassa. *There are two mating types,* + *and* −, *indistinguishable in appearance. Sexual reproduction will occur only between two hyphae of opposite types. The resulting diploid cell divides immediately by meiosis and then by mitosis to produce eight haploid ascospores within the ascus sac. Under favorable conditions each ascospore germinates to form a minute mycelium. The haploid mycelium may also reproduce asexually by producing conidia, which will germinate to form additional haploid mycelia. In fertilization the nuclei of a conidium or mycelium passes into the protoperithecia of the opposite mating type through the trichogynes.*

The ascomycete *Neurospora crassa,* a saprophyte that appears on pies and cakes as a cottony white fluff at first, turning pink as it develops asexual pink spores, has been an important research tool in genetics and biochemistry. As in the black bread mold, *Rhizopus,* there are two mating types, indistinguishable in body form, and sexual reproduction will occur only between two hyphae of opposite types (Fig. 12–22). The diploid cell that results from sexual reproduction divides by meiosis and then by mitosis to produce eight haploid ascospores within the ascus. Under favorable conditions, each ascospore will germinate to produce a new mycelium. It is possible to dissect out the individual ascospores under a microscope and establish pure strains for use in genetic and biochemical research.

SECTION 12–18

THE BASIDIOMYCETES

Mushrooms, toadstools, puff balls, rusts, smuts and bracket fungi are among the 25,000 or more species of Basidiomycetes. They derive their name from the fact that they reproduce sexually by a **basidium,** a structure comparable in function to the ascus of ascomycetes. Each basidium is an

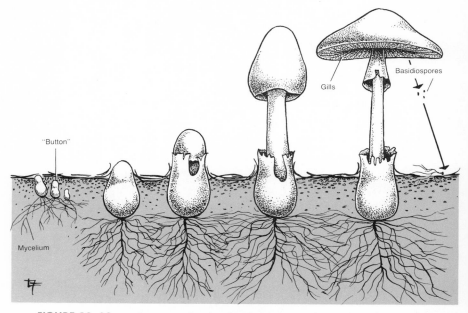

FIGURE 12–23   *Stages in the development of a mushroom from the mycelium, the mass of white, branching threads found underground. A compact mass, called a "button," appears and grows into the fruiting body or mushroom. On the under surface of the fruiting body are gills, thin perpendicular plates, extending radially from the stem. Basidia develop on the surface of these gills and produce basidiospores which are shed. If they reach a suitable environment, the basidiospores give rise to new mycelia.*

enlarged, clubshaped, hyphal cell, at the tip of which develop four **basidiospores.** Basidiospores develop on the *outside* of the basidium, whereas ascospores develop *within* the ascus. The basidiospores are released and will develop into new mycelia when they come in contact with the proper environment. The vegetative body of the plant consists of a mycelium made of many-celled hyphae. No motile cells are formed at any stage of the life cycle of basidiomycetes.

The cultivated mushroom, *Psalliota campestris,* has a vegetative body consisting of a mass of white, branching, threadlike hyphae that occur mostly below ground. Compact masses of hyphae, called "buttons," appear at intervals on the mycelium (Fig. 12–23). The button grows into the stalk and an umbrella we ordinarily call a mushroom. On the underside of the umbrella are many thin perpendicular plates called **gills,** which extend radially from the stalk to the edge of the cap. The basidia develop on the surface of these gills (Fig. 12–24). Each basidium contains two nuclei which fuse to form a diploid nucleus. This in turn divides by meiosis to form four haploid basidiospores. Each plant produces millions of basidiospores, each of which can, if it falls in the proper environment, give rise to a new mycelium.

The word "mushroom" does not refer to any particular species of basidiomycetes but simply to the fruiting body of a number of forms. There are some 200 edible kinds of mushrooms and about 25 poisonous ones (poisonous ones are sometimes called "toadstools"). There is no simple test that distinguishes edible and poisonous mushrooms; they must be identified by an expert.

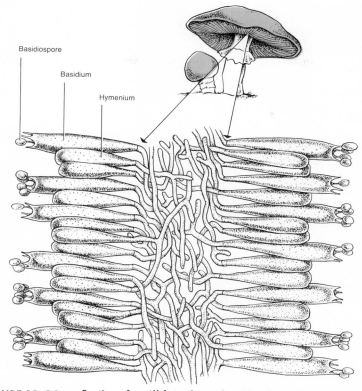

Basidiospore

Basidium

Hymenium

FIGURE 12–24   Section of a gill from the under side of a mushroom cap, magnified 500 × to show the basidia and their basidiospores.

SECTION 12–19

LICHENS

Although lichens look like individual plants, they are in fact an intimate combination of an alga and a fungus (Fig. 12–25), and provide a classic example of **mutualism.** The algal component is either a green or a blue-green alga, and the fungus is usually an ascomycete; in some lichens from tropical regions, the fungus partner is a basidiomycete. Lichens are resistant to extremes of temperature and moisture, and grow everywhere that life can be supported at all. Some kinds exist farther north than any other plants of the Arctic region, and others are found in steaming equatorial jungles. The alga, by photosynthesis, produces food for both, while the tough gelatinous mycelium of the fungus protects the alga and provides it with moisture and mineral salts. Lichens play an important role in the formation of soil, for they gradually dissolve and disintegrate the rocks to which they cling. The fungi of some lichens produce colored pigments. One of these, **orchil,** was used to dye woolens, and another, **litmus,** is widely used in chemistry laboratories as an acid-base indicator. The "reindeer moss" of the Arctic is a lichen.

There are some 10,000 species of lichens, and they present a nasty problem in classification—should they be classified according to their algal component, their fungal component or in some third way? Even though a lichen consists of an alga with one name and a fungus with another name,

FIGURE 12-25   Lichens. Above, leafy type growing on the bark of a tree. Lower left, the lichem known as "reindeer moss." Lower right, an encrusting type growing on the surface of a rock. Flat or cup-shaped fruiting bodies can be seen on some of the plants. (From Weatherwax, P.: Botany, 3rd ed. Philadelphia, W. B. Saunders Company, 1956.)

it is customarily given a third name and placed in a separate class of the phylum Eumycophyta.

SECTION 12-20

ECONOMIC IMPORTANCE OF THE FUNGI

Only a few fungi are used as food by man and only a few are human parasites. The only fungi poisonous to man are a few mushrooms and the ascomycete *Claviceps*, which causes a disease of rye plants known as **ergot.** If a man eats bread made with flour from diseased plants, he suffers ergot poisoning, characterized by hallucinations, insanity and death. A derivative of ergot, lysergic acid, commonly known as LSD, produces wild hallucinations like those of schizophrenia.

The plant diseases due to fungi cause tremendous economic losses. Phycomycetes cause "damping-off disease," which attacks young seedlings of corn, tobacco, peas, beans and even trees. The potato blight, caused by a phycomycete, destroyed almost the entire potato crop in Ireland in

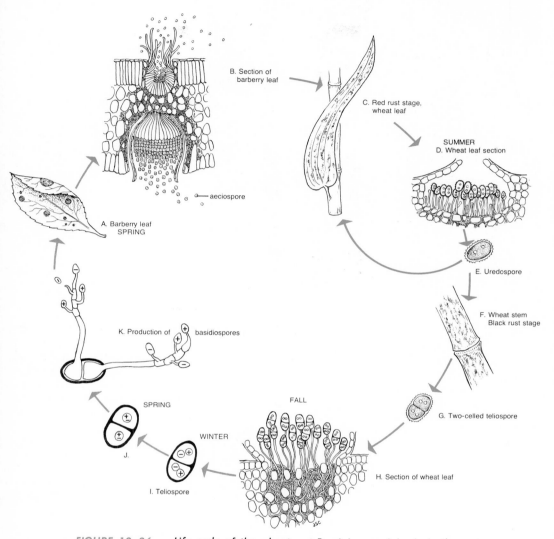

FIGURE 12–26   *Life cycle of the wheat rust Puccinia graminis. A, In the spring, basidiospores from wheat plants infected the previous year infect leaves of the barberry plant, forming pycnia-containing clusters of spermagonia on the upper surface and cluster cups of aecia on the lower surface of the leaf (B). Aeciospores are produced in the aecia; they are binucleate, containing N + N (not a single 2N nucleus) chromosomes. In early summer the aeciospores infect the leaves of young wheat plants (C). They develop into clusters of red single-celled uredospores (D) producing the "red rust" stage. Uredospores are released (E) and infect other wheat plants, producing more uredospores. In late summer, uredospores develop into dark-brown, two-celled teliospores (F) on the stems and leaf sheaths of wheat plants, forming the "black rust" stage (G). A section of the wheat stem (H) shows the N + N teliospores which are thick-walled and remain dormant over the winter (I). In the spring the N + N nuclei within each cell of the teliospore fuse to form a 2N nucleus (J). The teliospore, still attached to the wheat plant germinates, undergoes meiosis, and each teliospore cell produces four basidiospores (K). The haploid basidiospores then infect a barberry leaf to complete the cycle.*

FIGURE 12–27 *White spongy rot caused by the wood-decaying fungus Hydnum septentrionale, forming large clumps of soft, watery, cream-white brackets that rot the hardwood and sapwood of living maples. One tenth natural size. (Courtesy of the United States Department of Agriculture.)*

1845. The resulting famine led more than a million Irish to migrate to the United States.

A downy mildew which attacks grapes was introduced into France from the United States and almost destroyed the French vineyards before an effective fungicide, called Bordeaux mixture, was discovered. Some important plant diseases caused by ascomycetes are chestnut blight, Dutch elm disease, apple scab and brown rot, which attacks cherries, peaches, plums and apricots.

Basidiomycetes include smuts and rusts which attack corn, wheat, oats and other cereals. In general, each species of smut is restricted to a single host species. Some of these parasites, such as the stem rust of wheat and the white pine blister rust, have complicated life cycles passed in two or more different plants, and involving the production of several kinds of spores. The white pine blister rust must infect a gooseberry or a red currant plant before it can infect another pine.

The wheat rust must infect a barberry plant at one stage in its life cycle (Fig. 12–26). The eradication of barberry plants in wheat-growing regions has effectively reduced infection with wheat rust, but the eradication must be complete, for a single barberry bush can support enough wheat rust organisms to infect hundreds of acres of wheat. The basidiospores produced by the wheat rust in the fall have thick walls which enable them to survive a cold winter. These will grow only if they fall on a barberry leaf, whereas the usual thin-walled spores produced most of the summer can infect other wheat plants directly, and infection spreads from plant to plant in a wheat field in this way. If the winter is very mild, some of these thin-walled spores may survive and cause an infection the following year even in the absence of barberry plants. Thus, even the complete eradication of barberry plants does not provide a final solution to the wheat rust problem.

Bracket fungi (Fig. 12–27) cause enormous losses by bringing about the decay of wood, both in living trees and in stored lumber. The amount of timber destroyed each year by these basidiomycetes approaches in value that destroyed by forest fires.

Some of the Fungi Imperfecti cause important diseases of man. Monilia causes a disease of the throat and mouth called "thrush" and also infects the mucous membranes of the lungs and genital organs. The Tricho-phytoneae infect the skin of man and other animals, causing ringworm, athlete's foot and barber's itch. Other members of the Fungi Imperfecti are parasites of higher plants and cause important diseases of fruit trees and crop plants.

SECTION 12–21

UNICELLULAR ANIMALS: PROTOZOA

The final group of organisms included in the Protista are the single-celled animals, the Protozoa. These are functionally complex organisms, even though some may appear to be relatively simple morphologically. To carry out their life functions, many Protozoa have evolved specialized organelles, such as cilia or flagella, for movement, vacuoles, neurofibrils and eye spots. Although a very few live in damp soil, in the film of water surrounding each particle of soil, and others live parasitically in the blood and tissue fluids of animals or plants, most protozoa live in fresh or salt water. Almost all the 25,000 species of protozoa are microscopic, but a few are big enough to be seen with the naked eye. Some of the largest amoebas are 1 millimeter long. Some protozoa are shapeless blobs, others are elaborately and geometrically patterned. The cells may have internal skeletons, external skeletons or no skeleton at all. Some have protective coverings made of sand grains or other particles cemented together. The present-day flagellates constitute a bridge between plants and animals and are believed to be near the ancestral stem of both. The other classes of protozoa appear to have evolved from flagellate ancestors. Flagellates such as *Chlamydomonas* are clearly much more plantlike than animal-like, whereas the flagellate blood parasites, such as the trypanosomes causing African sleeping sickness, are clearly animals. The euglenoid flagellates have an exquisite blend of plant and animal characters and cannot reasonably be called one or the other. Most of the protozoa are typically animal-like in their motility, in their type of nutrition and in their behavior. Their intracellular organelles are more animal-like than plantlike, and they have either no cell wall or one composed of chitin rather than of cellulose.

The first class of Protozoa, the Flagellata, are named for the long whiplike cytoplasmic projections called **flagella,** which enable the organisms to move. Each flagellum is a long, supple fiber composed of 11 fibrils, a sheath of nine fibrils surrounding two central ones. The flagellate body is more or less oval with a definite front end from which the flagella project and a single, centrally located nucleus (Figs. 12–15 and 12–28). Some flagellates engulf food by forming pseudopods, others resemble paramecia and have a mouth and gullet. The flagellates with the largest number of flagella and the most specialized bodies are the ones living in the intestines of termites. Certain flagellates have some characteristics in common with the other classes of protozoa, and others in common with the sponges. These forms suggest some of the intermediate steps by which the other classes of protozoa and sponges may have evolved from flagellates.

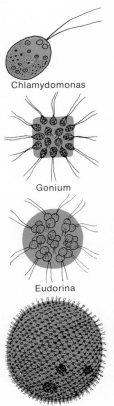

Chlamydomonas

Gonium

Eudorina

Volvox

FIGURE 12–28   Sketches of four representative flagellates.

Some flagellates are colonial; *Volvox,* for example, is a small hollow green sphere composed of thousands of individual cells arranged with their flagella outward. The colony of cells, moving by the beating of the flagella, rolls over and over but keeps one end directed forward. In such a colony there is incipient cellular specialization, for the cells at the anterior end have larger light sensitive spots than the others, and only those at the posterior end are capable of producing gametes.

Unlike other protozoa, members of the class Sarcodina (which includes the amoeba) have no definite body shape (Fig. 12–29). Some, such as the species causing amoebic dysentery in man, are parasitic. Other free-living species secrete hard shells around their body or cement sand grains together into a protective layer around their cells. The ocean contains untold trillions of amoeboid protozoa called **foraminifera,** which secrete chalky many-chambered shells with pores through which the animal extends its pseudopods. Dead foraminifera sink to the bottom of the ocean and are gradually transformed from a grey mud into chalk. Other sarcodinids, called **radiolaria,** secrete elaborate and beautiful glassy skeletons made of silica. The silica skeletons are durable and certain marine sediments are composed of fossil radiolarians.

The body of *Amoeba proteus* (Fig. 12–30) consists of a clear mass of shapeless, naked, gelatinous cytoplasm, containing a nucleus and cytoplasmic organelles. The nucleus occupies no fixed position in the cell, but is pushed around within the cytoplasm as the animal moves. An amoeba moves by pushing out temporary cytoplasmic projections, called pseudopods (false feet), from the surface of the cell. Additional cytoplasm flows into the pseudopods, enlarging them, until finally all the cytoplasm has entered and the animal as a whole has moved.

The pseudopods are used also to capture food (Fig. 12–31), two or more of them moving out to surround and engulf a bit of debris, another protozoan or a small, many-celled animal. The engulfed food is surrounded by a **food vacuole,** and the encircling cytoplasm secretes acids and enzymes to digest the food. The digested nutrients and water are absorbed from the food vacuole and the latter gradually shrinks. Any indigestible remnants are expelled from the body and left behind as the amoeba continues to move.

Gas exchange is an uncomplicated process in amoebas and other pro-

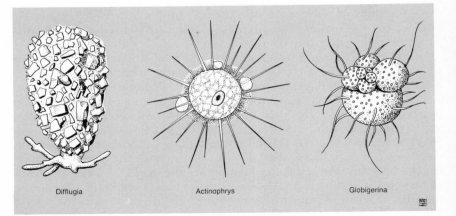

Difflugia   Actinophrys   Globigerina

FIGURE 12–29 *Some examples of the class Sarcodina. Difflugia is a free-living amoeba that builds a protective layer around itself by cementing together grains of sand. Actinophrys is a member of the order Heliozoa that lives in fresh water and Globigerina is a marine form, a member of the order Foraminifera.*

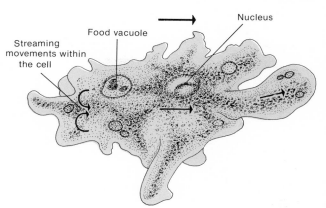

FIGURE 12–30   A diagram of the structure of Amoeba proteus.

tozoa, since they live in a liquid environment and can take in oxygen and give off carbon dioxide by diffusion. Excretion also occurs by this simple method. Many protozoa, amoebas as well as others, have a **contractile vacuole,** a cavity that regularly fills with water from the surrounding cytoplasm and empties into the environment. This structure is believed to be not an excretory organ but simply a pump to remove the excess water constantly entering by diffusion. The cytoplasm of a fresh water amoeba has a higher concentration of solutes than the surrounding water; hence, water tends to pass into it by osmosis. Without a pump to remove excess water, the amoeba would soon swell and burst, as our blood cells do when placed in fresh water. In contrast, most marine protozoa do not have or need a contractile vacuole, since the concentration of salts in the sea water is about the same as that in their cytoplasm.

A third class of protozoa, the Sporozoa (spore formers) have no special method of locomotion and are parasitic. Malaria, one of the great plagues of mankind, that formerly killed millions of people each year, is caused by a sporozoan, *Plasmodium*, with a complex life cycle involving man and the *Anopheles* mosquito.

A fourth group of protozoa, the Ciliata, is typified by *Paramecium* (Fig. 12–32). This animal, unlike the amoeba, has a definite, permanent shape, rounded in front and pointed in the rear. This shape is maintained by a sturdy though flexible outer covering (pellicle) of chitin secreted by the cell. The surface of the cell is covered by some 2500 **cilia,** fine, cytoplasmic hairs which extend through pores in the pellicle and move the animal along by their coordinated, rhythmic beating. The motion of the cilia is oblique, revolving the animal as it swims, and the coordination of ciliary beating is good enough to enable the ciliate to back up and turn around. A system of **neurofibrils** connects the rows of **basal bodies** at the

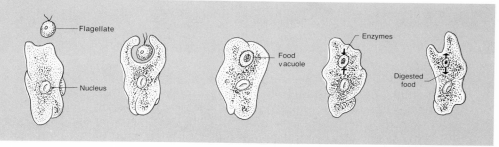

FIGURE 12–31   Diagram of an amoeba catching a small flagellate and digesting it in a food vacuole.

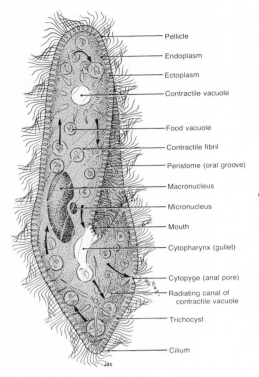

Pellicle

Endoplasm

Ectoplasm

Contractile vacuole

Food vacuole

Contractile fibril

Peristome (oral groove)

Macronucleus

Micronucleus

Mouth

Cytopharynx (gullet)

Cytopyge (anal pore)

Radiating canal of
   contractile vacuole

Trichocyst

Cilium

*FIGURE 12–32   Diagrammatic sketch of a typical
ciliate protozoan, the fresh-water Paramecium caudatum.*

bases of the cilia. If the neurofibrils are cut, the cilia can no longer beat in a coordinated fashion. Near the surface of the cell are many small bodies, called **trichocysts,** which can discharge filaments that presumably aid in trapping and holding the prey. Ciliates and suctorians are distinguished from other protozoa by the presence of two nuclei per cell: a **micronucleus,** which functions in sexual reproduction, and a **macronucleus,** which controls cellular metabolism and growth. A paramecium has a fixed gullet which ingests food and forms vacuoles, in which digestion occurs. Waste products leave the body at a fixed point, the anal pore.

A fifth class of protozoa, closely related to the ciliates, includes the **Suctorians.** Young individuals have cilia and swim, but adults lose their cilia and are sessile, attached to the substrate by a stalk (Fig. 12–33). They have delicate cytoplasmic **tentacles,** some of which are pointed to pierce the prey, while others are tipped with rounded adhesive knobs to catch and hold the prey.

SECTION 12–22

EVOLUTIONARY RELATIONSHIPS AMONG THE PROTISTS

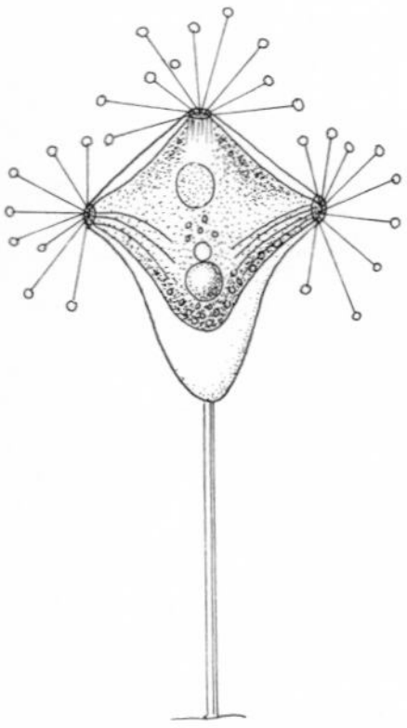

There is an incredible variety of relatively simple organisms, some autotrophs and some heterotrophs, and neither their evolutionary relationships nor the most appropriate way of classifying them has been resolved to everyone's satisfaction. The several basic similarities in the structure and chemical composition of bacteria and blue-green algae sug-

FIGURE 12–33   An example of a Suctorian, Acineta tuberosa.

gest that they lie somewhat more closely together in the evolutionary scheme than do other forms. Some taxonomists classify the blue-greens and the bacteria as members of the same phylum, the Schizophyta, or as a separate kingdom, the Monera. Others classify the bacteria in the phylum Schizomycophyta and the blue-green algae in a separate phylum, the Cyanophyta, in the subkingdom Thallophyta. It is widely held that bacteria evolved from the blue-green algae. They became adapted to a saprophytic or parasitic existence and subsequently lost chlorophyll. The fact that many bacteria have flagella leads other taxonomists to conclude that these organisms descended from simple flagellated forms, perhaps ones that also gave rise to the green algae. Alternatively the present-day heterotrophic bacteria may have evolved from early autotrophic bacteria similar to the present-day iron and sulfur bacteria. As a further possibility, the present-day bacteria could be direct descendents of the original bacteria-like heterotrophs that are generally believed to be the first types of living cells. It is even possible that different groups of bacteria have descended independently from different ancestors and that more than one of these explanations may be true. Perhaps it is more important at this time to realize the extent of this challenging problem in taxonomy rather than attempt to set up any rigid classification for these several disparate groups.

It also seems difficult at present to justify establishing a kingdom Protista, for although it may appear reasonable to put single-celled animal-like organisms and single-celled plantlike organisms together, the green, brown and red algae all include some forms that, though clearly related to the single-celled forms, are, in fact, relatively complex multicellular plants. The stoneworts, kelps and coralline algae are even more incongruous as members of the kingdom Protista than as members of the plant kingdom. The term protist is useful in referring in a general way to simple single-celled plants and animals without carrying the implication that it refers to a separate kingdom of organisms distinct from both plants and animals.

The evolutionary relationships of the slime molds and fungi to each other and to other living plants and animals are unclear, although ultimately they may have descended from some simple single-celled flagellated ancestor. The evolutionary relationships of the several classes of fungi have not been established. They may have evolved from one or another of the algae by taking up heterotrophic nutrition and losing chlorophyll. Alternatively they may have descended directly from primitive heterotrophs without ever having passed through an autotrophic stage. The phycomycetes are the most alga-like of the fungi, and may represent a more primitive descendent of some original alga, possibly the green algae. The characteristics of the ascomycetes are a curious mixture of those of the phycomycetes and the red algae. A number of mycologists believe that the ascomycetes evolved from red algae which became saprophytic and lost their photosynthetic pigments.

Finally, the evolutionary origin of the basidiomycetes is truly shrouded in mystery, for they show no relations with any of the algae. It is generally presumed that they are derived from some other fungi, perhaps from the ascomycetes.

SUGGESTIONS FOR FURTHER READING

Ahmadjian, V.: *The Lichen Symbiosis.* Waltham, Massachusetts, Blaisdell Publishing Company, 1967.
    This small but fascinating book presents some recent studies of the nature of the relationship between the fungal and algal components of a lichen.

Boney, A. D.: *A Biology of Marine Algae*. London, Hutchinson & Co., Ltd., 1966.
An excellent, ecologically and physiologically oriented treatise on marine algae which summarizes our current knowledge of the group.

Bonner, J. T.: *The Cellular Slime Molds*. 2nd ed. Princeton, New Jersey, Princeton University Press, 1968.
A record of experimental work with an unusually interesting group of organisms.

Bullock, William: *The History of Bacteriology*. New York, Oxford University Press, Inc., 1938.
A general history of the development of theories and knowledge about bacteria.

Burnett, J. H.: *Fundamentals of Mycology*. New York, St. Martin's Press, Inc., 1968.
A modern, thorough, physiologically and ecologically oriented account of the fungi.

Carson, Rachel: *The Sea Around Us*. New York, Oxford University Press, 1951.
This beautifully written book discusses the importance of algae as primary producers in the sea.

Curtis, H.: *The Viruses*. New York, The Natural History Press, 1966.
A popular account of the role of viruses as agents of disease and as tools for genetic research.

Davis, Kenneth S., and Day, John Arthur: *Water: The Mirror of Science*. Garden City, New York, Anchor Books, Doubleday & Company, Inc., 1961.
A discussion of the unusual and unique characteristics of water and its role in the origin and sustenance of life.

Dobell, Clifford: *Antonj von Leeuwenhoek and His Little Animals*. London, John Bale Medical Publishers, 1932.
This biography of Leeuwenhoek includes his description of the discovery of bacteria.

Fox, S. W. (ed.): *The Origins of Prebiological Systems and of Their Molecular Matrices*. New York, Academic Press, Inc., 1965.
The proceedings of a conference on problems of the origin of life held in October 1963. Included are 24 papers by leading authorities in the field and transcripts of the discussions that followed the presentation of each paper.

Gray, W. D. and Alexopoulos, C. J.: *Biology of the Myxomycetes*. New York, The Ronald Press Company, 1968.
An up-to-date account of the biology of the plasmodial slime molds, emphasizing ultrastructural, biochemical and physiological aspects.

Keosian, J.: *The Origin of Life*. New York, Reinhold Publishing Corporation, 1964.
A well rounded exposition of current theories and the evidence for each.

Luria, S. E., and Darnell, J. E.: *General Virology*. 2nd ed. New York, John Wiley & Sons, Inc., 1967.
A comprehensive account of the viruses, perhaps the best general account of the group.

Miller, S. L.: The Origin of Life. *In* Johnson, W. H., and Steere, W. C., (ed.): *This Is Life*. New York, Holt, Rinehart and Winston, Inc., 1962.
An interesting discussion of the author's theory of the origin of life and of his experiments bearing on this theory.

Oparin, A. I., (ed.): *Proceedings of the First International Symposium on the Origin of Life on the Earth* (International Union of Biochemistry, Symposium Series, Vol. 1). New York, Pergamon Press, Inc., 1959.
A collection of papers by authorities in the field.

Oparin, A. I.: *The Chemical Origin of Life*. Springfield, Illinois, Charles C Thomas, Publishers, 1964.
An up-to-date presentation of Oparin's classic theories of the origin of life.

Pollard, E. C.: *The Physics of Viruses*. Sci. Amer. *191*:62 December 1954.
The author's classic studies of viruses are summarized in this article.

Round, F. E.: *The Biology of the Algae*. London, Edward Arnold, Ltd., 1965.
A brief account of the algae, with special reference to their ecology and physiology; it introduces a number of interesting aspects of the biology of the group.

Stent, G. S.: *The Multiplication of Bacterial Viruses.* Sci. Amer. *188:*36, May 1953.
A summary of some of the early studies with bacteriophages.

Tiffany, L. H.: *Algae: The Grass of Many Waters.* Springfield, Illinois. Charles C Thomas, Publisher, 1939.
A nontechnical description of these important plants.

Vallery-Radot, R.: *Life of Pasteur.* New York, Doubleday, Doran & Company, 1928.
This book provides very interesting reading.

Wollman, E. L., and Jacob, F.: *Sexuality in Bacteria.* Sci. Amer. *195:*109, July 1956.
This text discusses the phenomenon of mating and the transfer of genetic material in bacterial organisms.

Zinsser, Hans: *Rats, Lice and History.* Boston, Little, Brown and Company, 1935.
A fascinating account of the great plagues and their influence on human history.

THE DIVERSITY OF LIFE: PLANTS

CHAPTER 13

Since it is generally believed that life began in the sea, it follows that land plants and animals evolved from aquatic ancestors. In tracing the evolutionary history of a given line of plants or animals we may find that, having once become adapted to terrestrial life, they may return to an aquatic habitat and perhaps reemerge later on and once again become terrestrial.

Aquatic plants can survive without many of the specialized structures found in terrestrial plants. The surrounding water keeps them supplied with nutrients, prevents the cells from drying out, buoys up and supports the plant body so that special supporting skeletal structures are unnecessary, and serves as a convenient medium for both the meeting of gametes in sexual reproduction and the dispersal of asexual spores. In migrating from water to the barren land, primitive plants had to become adapted by developing new structures to take over the many functions previously served by the water. It seems likely that the conquest of the land was a long and difficult process, perhaps fraught with many failures. The new forms of plants adapted to living in the soil had adequate salts, water and carbon dioxide but no light for photosynthesis. The ones that became adapted to living above ground had light and carbon dioxide but not enough water and no salts. The plants that became truly terrestrial were able to survive because they evolved specialized parts: **leaves** which extend into the air to absorb light and carry on photosynthesis, **roots** which extend into the soil to provide anchorage and absorb water and salts, and **stems** which support the leaves in the sunlight and connect them with the roots, providing a two-way connection for the transfer of water and nutrients in the xylem and phloem. Finally they evolved some appropriate means of reproduction—flowers, pollen and seeds—by which male and female gametes can unite in the absence of a watery medium and by which the zygote can begin development protected from desiccation.

THE BRYOPHYTES

Just as the present-day salamanders, newts, frogs and other members of the class Amphibia give us an idea of what the first land vertebrates may have looked like, the **mosses, liverworts** and **hornworts** — the Bryophytes — suggest the stages through which aquatic algae may have evolved before becoming fully terrestrial. Algae typically have body structures adapted to expose the maximum of surface for absorbing nutrients from the surrounding water. For survival on land, plants need a more compact body to decrease the water lost through the surface. Probably the first land plants laid flat and exposed only one surface to the air. Plants became successful on land only after they had developed a specialized epidermal tissue with thickened cell walls impregnated with a waxy, water-proof material. Most Bryophytes have an epidermis. It may be slightly thickened and waxy and provided with pores to permit the diffusion of gases.

The Bryophytes did not really solve the problem of reproduction in the absence of water — they simply got around the difficulty by evolving reproductive structures that would have a watery medium for the union of gametes. All land plants, including the Bryophyta, have evolved a life cycle in which the zygote is retained within the female sex organ. There it obtains nutrients and water from the surrounding parental tissues and is protected from drying while it develops into a multicellular **embryo.** The Bryophytes and Tracheophytes are classified together in a subkingdom, Embryophyta.

The name "moss" is erroneously applied to a number of plants that are not Bryophytes. The moss that grows on the bark of a tree may be an alga, **reindeer moss** is a lichen, and **Spanish moss** hanging from trees in the southern states is really a seed plant — a relative of the pineapple.

True mosses are all rather similar in structure, consisting of a filamentous green body or **protonema** on or in the soil from which grows an erect stem to which are attached a spiral whorl of leaves one cell thick. From the base of the stem extend many colorless, rootlike projections called **rhizoids.** Mosses are never more than 15 to 20 centimeters high, owing to the inefficiency of the rhizoids as water absorbers and the absence of true vascular and supporting tissues.

THE LIFE CYCLE OF A MOSS

The familiar, small, green, leafy plants called mosses are the haploid gametophyte generation of the plant. The gametophyte consists of a single, central stem, bearing "leaves" arranged spirally and held in place in the ground by a number of slender rootlets or **rhizoids,** which absorb water and salts from the soil. The leaf cells produce all the other compounds the plant needs for survival, so that each gametophyte is an independent organism. When the gametophyte has attained full growth, sex organs develop at the top of the stem in the middle of a circle of leaves (Fig. 13-1).

In some species the sexes are separate; in others, both male and female organs develop on the same plant. The male organs are sausage-shaped structures, **antheridia,** which produce a large number of slender, spirally coiled sperm or **antherozoids,** each of which has two flagella. After a rain or heavy dew, the sperm are released and swim through the film of water

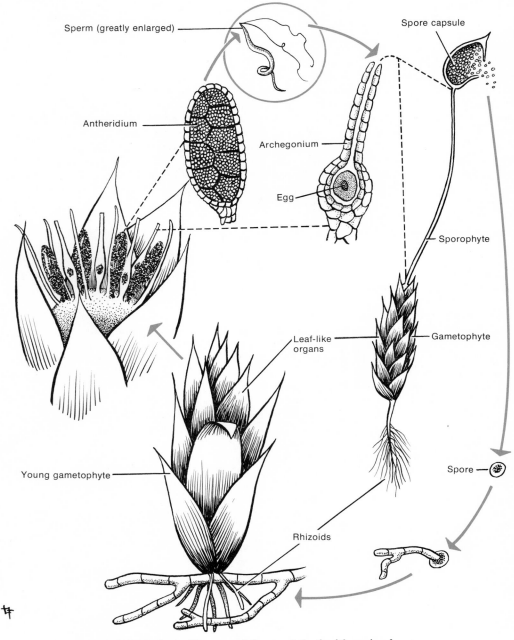

Sperm (greatly enlarged)

Spore capsule

Antheridium

Archegonium

Egg

Sporophyte

Leaf-like organs

Gametophyte

Young gametophyte

Spore

Rhizoids

FIGURE 13-1   *Diagram of the events in the life cycle of a moss.*

covering the plant to a neighboring female sex organ, either on the same plant or on another one. The female organ, the **archegonium,** is shaped like a flask and has one large egg in its broad base. This organ releases a chemical substance that attracts sperm and, guided by this, the sperm swim down the neck of the archegonium and into its base, where one sperm fertilizes the egg. The resulting zygote is the beginning of the diploid sporophyte generation.

In contrast to the independent green gametophyte, the sporophyte is a leafless, single stalk (**seta**) which lives as a parasite on the gametophyte, obtaining its nourishment by means of a **foot** which grows down into the gametophyte tissue. At the opposite, upper end of the sporophyte stalk, a **capsule** forms, the cells of which possess chloroplasts and produce some of their food photosynthetically. Within the cylindrical cavity of the capsule, each diploid **spore mother cell** undergoes meiotic divisions to form four haploid spores, which are the beginning of the gametophyte generation.

When the capsule matures, the upper end forms a **lid** which drops off. In some mosses the opening of the spore capsule is obstructed by one or two rings of wedge-shaped teeth. These bend inward in wet weather and prevent the escape of the spores, but bend outward in dry weather, permitting the liberation of spores when they are likely to be dispersed by the wind. When a spore drops in a suitable place it germinates and develops into a **protonema,** a green, branching filamentous structure which, by budding, produces several gametophytes, thereby completing the life cycle. The gametophyte may also undergo asexual reproduction by **gemmae** produced in gemma cups.

SECTION 13–3

LIVERWORTS AND HORNWORTS

The second class of bryophytes, the liverworts, are simpler and more primitive than mosses. Some of the 9000 species of liverworts are simply flat, sometimes branched, ribbon-like structures that lack a stem and lie on the ground attached to the soil by numerous rhizoids (Fig. 13–2). Other species tend to grow upright and have a leaflike gametophyte. A few others have gametophytes differentiated into stems, branches and leaves, but without vascular tissues. The upper surface of the gametophyte is an epidermis one cell thick, punctuated by many pores for the exchange of gases. The lower surface is an epidermis covered with many thin scales and from which grow many long, slender rhizoids. As in the mosses, the sporophyte of the liverwort grows as a parasite on the gametophyte plant. The upper surface of some liverwort gametophytes may bear gemma cups, within which are produced small, flattened, ovoid gemmae. These separate from the parent plant and grow into new gametophytes, a form of asexual reproduction.

Hornworts include some 300 species of plants—the small, leaflike, irregularly branched gametophytes. *Anthoceros*, a typical hornwort, has a sporophyte which grows upward as a slender, cylindrical capsule from the foot imbedded in the gametophyte.

Mosses and liverworts have many characteristics in common with the green algae and are generally believed to be derived from them. At one time it was thought that the higher vascular plants evolved from the bryophytes—from a liverwort or a hornwort. However, although there is fossil evidence of true vascular plants in the Silurian period, some 360,000,000 years ago, the first evidence of bryophytes dates from the Pennsylvanian, which began about 100,000,000 years later. For this, and other reasons, botanists are now inclined to believe that vascular plants evolved from green algae independently of the mosses. The mosses probably represent the end of a separate branch of the evolutionary tree.

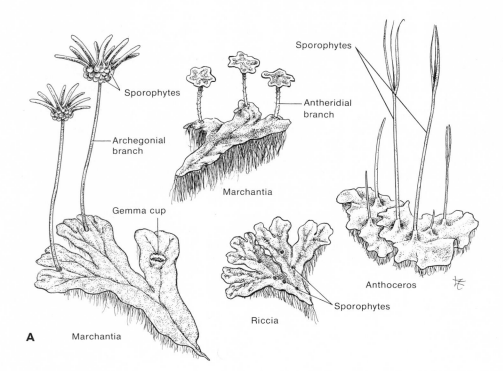

FIGURE 13–2   Sketches of some common liverworts, Marchantia and Riccia, and a hornwort, Anthoceros. The flat part of the body is the gametophyte upon which the sporophyte develops. Lower, photographs of Marchantia showing left an antheridial branch and right archegonial branches. (B and C, courtesy of the Carolina Biological Supply Company.)

SECTION 13–4

THE TRACHEOPHYTES

The phylum Tracheophyta is an ancient and diverse group which includes the dominant land plants of today, characterized by the presence of vascular tissues, **xylem** and **phloem,** in the sporophyte generation. The tracheophytes, like the mosses, have a life cycle with an alternation of gametophyte and sporophyte plants, but the sporophyte of these higher plants is a free-living, independent plant, whereas the gametophyte either is a small independent plant, or is contained within the sporophyte. It includes four subphyla: the **Psilopsida,** the most primitive vascular plants; the **Lycopsida,** the club mosses and quillworts; the **Sphenopsida** or horsetails; and the **Pteropsida,** an enormous group which includes all the ferns, conifers and flowering plants.

The most primitive vascular plants known are the Psilophytales which lived in the Devonian period and perhaps even earlier in the Silurian. These plants reached a height of 60 centimeters and had a creeping, horizontal stem or **rhizome,** from which grew branching erect green stems (Fig. 13–3). The smaller branches were coiled at the tip and probably unrolled as they grew, as present-day ferns do (fiddler heads). They lacked roots and either were leafless or had small scalelike leaves. Fossil remains of these plants found in Scotland were so well preserved that details of the internal structure were evident.

This subphylum is represented today by three species of living fossils. One of these, *Psilotum,* grows in southeastern United States. The sporophytes are rootless, but have an underground branching stem from which grow unicellular rhizoids and green photosynthetic upright branches bearing tiny, scalelike leaves. The gametophytes are small, underground,

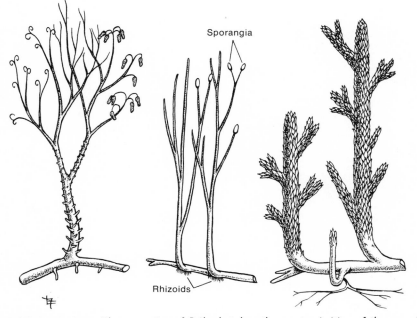

FIGURE 13–3   *Three species of Psilophytales, the most primitive of the vascular plants, which lived during the Devonian period.*

nongreen bodies bearing the sex organs. The Psilopsida are widely believed to be the ancestors of the other vascular plants. The several fossil species known each suggest the beginning of some sort of specialization found more fully developed in other subphyla.

The club mosses, quillworts and relatives which comprise the subphylum Lycopsida were widespread in the Devonian and Carboniferous periods and many of them were tall, treelike forms. Today only four genera remain, all of which are small, usually less than 30 centimeters high. These inconspicuous plants consist of a creeping stem which gives off true roots and upright stems with thin, flat, spirally arranged leaves (Fig. 13–4). The stem has a core of xylem surrounded by a cylinder of phloem. No cambium is present and therefore all the plant's growth is primary at the growing tip. The tip of the stem bears specialized leaves, **sporophylls,** arranged in a shape resembling a pine cone and bearing spore-producing structures, **sporangia.** In the genus *Lycopodium* the sporophyte produces spores, all of which are alike, and which germinate to form gametophytes. On the gametophytes develop sex organs in which are produced eggs and biflagellate sperm. After fertilization the developing embryo is, for a time, dependent upon the gametophyte for nutrition.

The **spike mosses** or little club mosses, members of the genus *Selaginella,* show an important evolutionary advance in having two types of spores, **megaspores** which germinate to produce female megagametophytes and **microspores** which germinate to produce male microgameto-

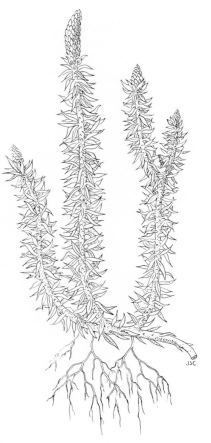

FIGURE 13–4   *Sketch of a club moss Lycopodium, about one half its natural size. This kind of club moss is sometimes used as a Christmas ornament. Note the spirally arranged leaves.*

phytes (Fig. 13–5). The gametophytes are greatly reduced in size and depend on the sporophyte for nourishment. When the haploid micro-spores are released from the sporophyte, they may drop near a megaspore. When moistened by dew or rain the microspore wall splits and the sperm are free to swim to the megaspore and fertilize the haploid egg. As in the higher seed plants, the embryo is produced in the female gametophyte while it is still within the sporophyte. Although the reproductive cycle of these plants foreshadows that of the seed plants, they are not their ancestors, but are a terminal group. The "resurrection plant" of the Southwest is a species of *Selaginella* that rolls up into a compact ball of apparently dead leaves during the dry season, then unrolls and carries on its normal activities when moisture is again present.

The **quillworts,** superficially quite different from the club mosses, are deciduous, perennial plants living in marshy places, with stems in the ground. Roots project down from the stem and slender, quill-like leaves resembling those of a bunch of garlic, project into the air, attached by

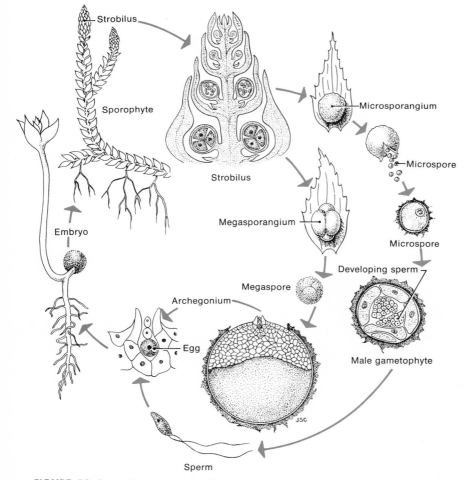

FIGURE 13–5   *Diagram of the life cycle of the "spike moss" Selaginella. Selag-inella produces megaspores in megasporangia which germinate to produce female gametophytes, and microspores produced in microsporangia which germinate to produce male gametophytes. Sperm develop within the male gametophyte and, when freed, swim to the archegonium on the female gametophyte and fertilize the egg within it. The fer-tilized egg then develops into the new sporophyte plant.*

their broad bases to the short stem (Fig. 13–6). The leaves have spore-bearing organs at the basal end.

Like the two previous subphyla, the Sphenopsida include many more fossil plants than living ones. Sphenopsida arose during the Devonian era and developed into a variety of species, some small and some that were gigantic treelike plants as much as 13 meters tall (Fig. 13–7). The latter flourished, especially during the Carboniferous period, and their dead bodies, together with those of certain other plants, are the source of our present coal deposits. The sphenopsids, or "horsetails," are widespread, from the tropics to the arctic on all continents except Australia. These plants, usually less than 40 centimeters tall, are found in both boggy and dry places. The name "horsetail" is appropriate because the multiple-branched, bushy structure of many species resembles a horse's tail. Some species are called scouring rushes, because they contain deposits of silica in the epidermis, which give the plants a harsh, abrasive quality. They were used to clean pots and pans before the invention of steel wool.

The sphenopsid sporophyte is made up of a horizontal, branching, underground rhizome from which grow slender branching roots and jointed aerial stems. The stem contains many vascular bundles arranged in a circle around a hollow center. The stems have conspicuous nodes that divide them into jointed sections; at each node is a whorl of smaller secondary branches and a whorl of small, scalelike leaves (Fig. 13–8).

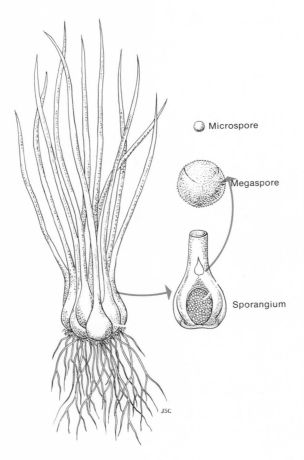

Microspore

Megaspore

Sporangium

FIGURE 13–6   *Left, sketch, natural size, of a quillwort, Isoetes. Right, sketch of the base of one of the leaves showing a sporangium.*

JSC

FIGURE 13-7   *Photograph of a calamite, a fossil sphenopsid. The whorls of long, linear leaves are clearly evident. (From Fuller, H. J., and Carothers, A. B.: The Plant World. 4th ed. New York, Holt, Rinehart and Winston, Inc., 1963.)*

Some branches develop a conelike structure at their tip, which contains numerous structures bearing spore sacs on their inner faces.

The spores released from these structures germinate into green gametophytes which have egg- and sperm-producing organs. The zygote formed after fertilization develops into the sporophyte plant; at first this is parasitic on the gametophyte, but it quickly develops its own stem and roots.

The subphylum Pteropsida is the largest group in the plant kingdom, composed of three classes: the **ferns,** class Filicinae; the **conifers,** class Gymnospermae; and the **flowering plants,** class Angiospermae. The sporophytes of these plants have well-differentiated roots, stems and leaves and are the important, complex generation. The gametophytes are small and are either independent, short-lived plants, as in the ferns, or minute structures, composed of only a few cells, found within the spores of the conifers and flowering plants. The leaves of the Pteropsida are typically large and have multiple, often branching, vascular bundles. It is believed that they evolved from stems by the flattening and transformation of the distal parts. Some of the primitive psilopsids have leaves formed in this way and may be the ancestors of the pteropsids. The specialized roots, stems and leaves of the pteropsids have enabled them to become truly adapted to terrestrial life, and the well-developed vascular and supporting tissues have enabled them to grow to a large size.

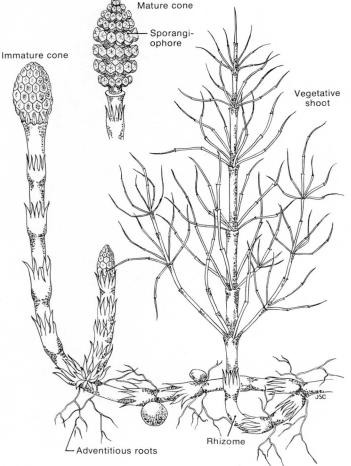

Mature cone

Sporangi-
ophore

Immature cone

Vegetative
shoot

FIGURE 13–8   Sketch, about
one half its natural size, of a horsetail, Equisetum.

Adventitious roots

Rhizome

SECTION 13–5

THE FERNS

Some 9000 species of ferns are widely distributed today in both temperate and tropical regions. Those in the temperate climate survive best
in cool, damp and shady places. The tropical rain forest has an abundance
of ferns, some of which are tall and superficially resemble palm trees,
having an erect, woody, unbranched stem, with a cluster of compound
leaves (fronds) at the top. Some tropical **tree ferns** reach a height of 16
meters and have leaves 4 meters long. Our common temperate zone ferns
have horizontal rhizomes growing at or just beneath the surface of the
soil from which grow hairlike roots. Each year from these perennial rhizomes grow new, erect fronds or compound leaves. These leaves characteristically are coiled in the bud and unroll and expand to form a mature leaf
(Fig. 13–9).

Ferns are remarkably like the seed plants, and are grouped with them

FIGURE 13–9   Photo-
graph of a young fern plant
showing the young leaves which
uncoil as they develop. (Weatherwax, P.: Botany. 3rd ed. Philadelphia, W. B. Saunders Company, 1956.)

in the subphylum Pteropsida. Their roots have a root cap, and mereste-
matic, elongation and mature zones, like the root of a seed plant. The fern
stem has a protective epidermis supporting the vascular tissues, and the
leaves have veins, chlorenchyma, a protective epidermis and stomata.
Ferns differ from the seed plants in that the xylem contains only tracheids —
no vessels — and the spores are all alike, and are produced in **sporangia** on
the under surfaces of certain leaves. The sporophyte plant may live for
several years and produce several yearly crops of haploid spores (Fig. 13–
10).

The spores are released at the proper time, fall to the ground, and
develop into flat, green, heart-shaped gametophytes 5 or 6 millimeters in
diameter. The gametophyte, termed a **prothallus,** grows in moist, shady
places such as decaying logs and on moist soil and rocks. The prothallus
develops rhizoids which grow into the soil and absorb water and salts. The
male and female sex organs, **antheridia** and **archegonia,** develop on the
under surface of the gametophyte. The sperm or antherozoid is in the
shape of a short spiral, and has many flagella on its slender anterior portion by means of which it swims. The archegonium, usually located near
the notch of the heart-shaped gametophyte, contains a single egg. The
antherozoids are released after a rain and, attracted by a chemical substance released by the archegonium, swim through the water to the under
surface of the gametophyte and reach the egg. The resulting zygote is the
beginning of a new sporophyte generation, the embryo of which is dependent upon the gametophyte for nourishment until it develops its own
roots and leaf, completing the life cycle. The conquest of land by the ferns
is incomplete, for the gametophyte generation can survive only where
there is plenty of moisture and shade, and the union of eggs and sperm
in fertilization requires a watery medium.

During the Carboniferous period there were great forests of tree ferns.
These had tall, slender trunks made of the stem plus an enveloping mass
of roots matted together by hairs. The bodies of these tree ferns also contributed to our present coal deposits. Another group of fossil plants, long
considered to be ferns, have been found to bear seeds, and these fossil
seed ferns are now classified with the gymnosperms.

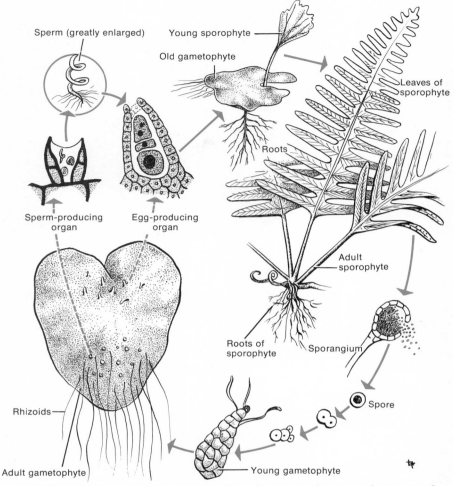

FIGURE 13–10   *The life cycle of a fern plant. As in the mosses there is an alter-*
nation of a sexual (gametophyte) and an asexual (sporophyte) generation; the large
familiar fern plant is the sporophyte.

SECTION 13–6

SEED PLANTS

The remaining two classes of the subphylum Pteropsida, the Gymnospermae and Angiospermae, differ from ferns in having no independent
gametophyte generation. Their two key characteristics are the formation
of **seeds,** which enclose the embryo during a resting stage, and the union
of male gametes with the eggs by **pollination,** by the growth of a pollen
tube. Seeds provide for the wide and rapid dissemination of a species and
are resistant to desiccation and high or low temperatures. The pollen tube
eliminates the need for an external supply of water through which sperm
can swim to the egg. These two traits have undoubtedly been responsible
in large measure for the success of seed plants as terrestrial organisms.

Seed plants characteristically produce two types of spores, **megaspores**
which develop into female gametophytes, and **microspores** which develop

into male gametophytes or **pollen.** The female gametophyte is retained within the megaspore and gives rise to an egg that is fertilized there. The resulting zygote develops into an embryo with the rudiments of leaves, stems and roots while still within the seed coat derived from the previous sporophyte. The **seed** thus consists of structures belonging to three distinct generations—the **embryo,** the new sporophyte; the **endosperm,** nutritive tissue derived from the female gametophyte; and the **seed coat,** derived from the old sporophyte.

There are over 250,000 species of seed plants, adapted to survive in a variety of terrestrial environments, and varying in size from the minute duckweed a few millimeters in diameter to giant redwood trees. They also are the plants of greatest value to man as sources of food, shelter, drugs and industrial products. The Gymnospermae ("naked seeds") and Angiospermae ("enclosed seeds") differ in the relationship of the seeds to the structures producing them. The seeds of angiosperms are formed inside a **fruit,** and the seed covering is developed from the wall of the ovule of the flower. The angiosperms which apparently violate this rule and have exposed seeds—wheat, corn, sunflowers and maples, for example—really do have enclosed seeds, for the structure commonly called the "seed" actually is a fruit enclosing the true seed. The seeds of gymnosperms are usually borne on **cones,** but they are never really enclosed as are angiosperm seeds. The seed habit has contributed immensely to the success of the seed plants; the stored food nourishes the embryo until it can lead an independent life; the tough outer coat protects the embryo from heat, cold, drying and parasites; and seeds provide a means for the dispersal of the species.

SECTION 13–7

THE GYMNOSPERMS

The gymnosperms are woody plants which do not have flowers, but bear their seeds on the inner sides of scalelike leaves that are usually arranged spirally to form a **cone.** There are only about 700 species living today, but there were many more in former geologic times. The four orders of gymnosperms with members living at present are the **Coniferales,** the **Cycadales,** the **Ginkgoales** and the **Gnetales.** Three other orders, the Cycadofilicales or seed ferns, the Bennettitales and Cordaitales, are known only from fossil remains. In addition to differences in methods of reproduction, the xylem of gymnosperms is composed only of tracheids and wood parenchyma; no wood fibers or vessels are present. Thus, the gymnosperms are softwoods rather than hardwoods.

Of the living gymnosperms the conifers—pine, cedar, spruce, fir and redwood trees—are the most successful biologically. The needle-like leaves of these evergreens are well adapted to withstand hot summers, cold winters and the mechanical abrasion of storms. Under the thick, heavily cutinized epidermis layer is a layer of thick-walled sclerenchyma. The stomata are set in deep pits that penetrate the sclerenchyma. The conifers bear two kinds of cones, one producing pollen and one producing eggs. The male cones release pollen which is carried by the wind to the female cones, where the eggs are fertilized. In pines, as much as a year may elapse between pollination and fertilization, and several more years may elapse between fertilization and the shedding of the seeds.

Conifers are important economically as the source of more than 75 per cent of the wood used in construction and in the manufacture of paper and plastics. Some of them produce resins used in the production of turpentine, tar and oils. A few conifer seeds—pine "nuts," for example—are used as food, and the berries of juniper produce aromatic oils used for flavoring alcoholic beverages such as gin.

The **cycads,** found mainly in tropical and semitropical regions, have either short, tuberous, underground stems or erect, cylindrical stems above the ground. With their large, compound, divided leaves (Fig. 13–11), they resemble ferns or miniature palm trees, for which they are frequently mistaken. The only cycad native to the United States is the sago palm, *Zamia,* found in Florida. In contrast to the pine, which produces two kinds of cones on the same tree, the cycad population consists of two kinds of trees, one of which produces only cones yielding pollen and the other produces only cones yielding eggs. These are not "male" and "female" trees, for they are sporophytes, asexually producing male and female gametophytes respectively.

The **ginkgo,** or maidenhair tree, is the only living representative of a once numerous and widespread order. It was cultivated in China and Japan as an ornamental tree, because of its distinctive, fan-shaped leaves, which it sheds in the fall (Fig. 13–12). Like the cycads, there are two kinds of ginkgo trees: one produces only staminate (pollen-producing) cones and the other only ovulate (egg-producing) cones. The ginkgo and cycads are the only seed plants that produce a swimming sperm rather than pollen tube nuclei.

The Gnetales is a small order of gymnosperms that includes some

FIGURE 13–11   Photograph of cycads. Above, Cyas; below left, a male plant of Zamia; and below right, a mature fruiting plant of Zamia. This plant found from Florida to Mexico is the only cycad native to the United States. (From Weatherwax, P.: Botany. 3rd ed. Philadelphia, W. B. Saunders Company, 1956.)

FIGURE 13–12   Ginkgo twigs. Left, a cluster of mature seeds; center, a twig with leaves and young ovules; right, a twig with leaves and pollen cones. (Copyright, General Biological Supply House, Chicago.)

very peculiar plants. The only one found in the United States is *Ephedra*, a low, much branched shrub that superficially looks like one of the horse-tails with naked twigs and rudimentary scale leaves. *Welwitschia*, a plant that grows in the desert in southwest Africa, consists of a short thick trunk, partly embedded in the soil, from which grow two long, flat, leath-ery leaves. Although the plant may live for centuries, it never produces more than these two leaves, which, subjected as they are to the sand and windstorms of the desert, eventually become very tattered.

In a number of respects the gymnosperms are more advanced, better adapted for land life, than the ferns. They have become independent of water for reproduction by evolving wind-borne pollen to transfer the male gametophyte to the female gametophyte, by developing a pollen tube to replace motile sperm as a means of effecting fertilization, and by evolv-ing the seed habit. The development of a cambium and, from it, secondary wood, has made possible the large size of many of the seed plants.

SECTION 13–8

LIFE CYCLE OF A GYMNOSPERM

The life cycle of a gymnosperm represents, in several respects, a transition between those of the ferns and the angiosperms. A pine tree, a typical gymnosperm, produces two kinds of cones, **staminate,** which are small, less than 3 centimeters long, and **ovulate,** which are the large easily visible cones, as much as 45 centimeters long in some species (Fig. 13–13). The ovulate cone is composed of many scales, and on the surface of each scale are two ovules. Within each **ovule** is a diploid **megaspore mother cell,** which divides by meiosis to form four haploid **megaspores.** Only one of these is functional and grows into a multicellular **megagametophyte.** On each megagametophyte are two or three female sex organs (archegonia), in each of which is a large egg.

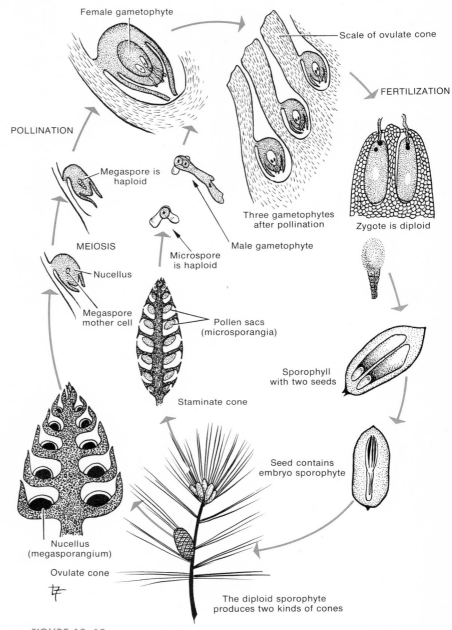

Female gametophyte

Scale of ovulate cone

FERTILIZATION

POLLINATION

Megaspore is haploid

Three gametophytes after pollination

Zygote is diploid

MEIOSIS

Nucellus

Microspore is haploid

Male gametophyte

Megaspore mother cell

Pollen sacs (microsporangia)

Sporophyll with two seeds

Staminate cone

Seed contains embryo sporophyte

Nucellus (megasporangium)

Ovulate cone

The diploid sporophyte produces two kinds of cones

FIGURE 13–13   *Diagram of the events in the life cycle of a pine tree (see text for discussion).*

On the underside of each scale of the staminate cone are two **microsporangia.** Within each microsporangium are many **microspore mother cells,** each of which divides by meiosis to form four **microspores.** While still within the microsporangium or pollen sac, the microspores divide mitotically to form a four-celled **microgametophyte** or **pollen grain.** These are released and carried by the wind. When a pollen grain reaches an ovulate cone it enters the ovule through an opening, the **micropyle,** and comes in contact with the megasporangium.

A year or more may elapse before one cell of the pollen grain elongates into a pollen tube which grows through the megasporangium to the megagametophyte. Another cell of the pollen grain divides to form two male **gamete nuclei,** not motile sperm as in lower plants. When the end of the pollen tube reaches the neck of the archegonium and bursts open, the two male nuclei are discharged near the egg. One fuses with the egg nucleus to form the diploid zygote, and the other disintegrates. After fertilization, the zygote divides and differentiates to produce a sporophyte embryo, surrounded by the tissues of the megagametophyte and those of the parent sporophyte. This entire structure is the **seed.**

The tissues of the megagametophyte provide nutrition for the developing embryo and are called **endosperm;** however, they are haploid cells and are quite different from the endosperm cells of angiosperms (p. 385) which are triploid ($3n$), though both serve the same nutritive function for their respective embryos. After a short period of growth during which several leaflike **cotyledons,** the **epicotyl** (which will become the stem) and the **hypocotyl** (from which the primary root grows) develop, the embryo remains quiescent until the seed is shed and drops to the ground. When conditions are favorable it germinates and develops into a mature sporophyte, the pine tree.

SECTION 13-9

THE ANGIOSPERMS

The **angiosperms,** or true flowering plants, are the largest class in the plant kingdom and include some 250,000 species of trees, shrubs, vines and herbs adapted to almost every kind of habitat. Some live completely under water, others in extremely arid regions. The vast majority are autotrophic, but some, such as the orchid, Indian pipe and mistletoe, have little or no chlorophyll and so are partly or wholly parasitic. A few angiosperms have evolved devices for catching insects and other small animals, and hence are holozoic and carnivorous. The angiosperms provide man with much of his food, clothing, shelter and drugs and liven the world with the beautiful colors and scents of their flowers.

Many angiosperms can complete an entire life cycle, from the germination of the seed to the production of new seeds, within a month; others require 20 or 30 years to reach sexual maturity. Some live for a single growing season; others live for centuries. The stems, leaves and roots present a bewildering variety of forms, but all angiosperms develop flowers that have a fundamentally similar pattern.

Angiosperms differ from gymnosperms in the abundance and prominence of the xylem vessels, in the formation of flowers and fruits, in the presence of sepals, petals or both in addition to the sporophylls, in the formation of a **pistil** through which the pollen tube grows to reach the ovule and egg (in gymnosperms the pollen lands on the surface of the ovule and the pollen tube grows in directly), and in the further reduction of the gametophyte generation to a few cells completely parasitic on the sporophyte.

Fossil remains of angiosperms have been found in rocks from the Cretaceous period, and, although they probably arose from some primitive gymnosperm, there are no intermediate plants in older rocks to indicate which group of gymnosperms might have been the ancestor of the angiosperms.

The angiosperms include some 175,000 species of Dicotyledones and 75,000 species of Monocotyledones. The two subclasses differ in the following ways:

1. The embryo in the monocot seed has only one seed-leaf or cotyledon; the dicot embryo has two. These are filled with starch and other foods which nourish the embryo and seedling until it is capable of making its own food by photosynthesis.

2. The leaves of monocots have parallel veins and smooth edges; those of dicots have veins which branch and rebranch and, usually, edges which are lobed or indented.

3. The flower parts of monocots — petals, sepals, stamens and pistils — exist in threes or multiples of three; dicot flower parts usually occur in fours or fives, or in a multiple of these.

4. Bundles of xylem and phloem are scattered throughout the stem of monocots; in dicots the xylem and phloem occur either as a single, solid mass, running up the center of the stem, or as a ring between the cortex and the pith.

There are a great many different families of both monocots and dicots, each of which usually takes its name from some conspicuous member. Some of the monocot families are the grasses, palms, lilies, orchids and irises. Some important dicot families are the buttercup, mustard, rose, maple, cactus, carnation, primrose, phlox, mint, pea, parsley and aster. The rose family, for example, includes, in addition to roses, the apple, pear, plum, cherry, apricot, peach, almond, strawberry, raspberry, hawthorn and other shrubs.

SECTION 13–10

THE LIFE CYCLE OF AN ANGIOSPERM

Angiosperms exhibit an alternation of gametophyte and sporophyte generations, but the gametophyte is reduced to a few cells lying within the tissues of the sporophyte flower. The sporophyte is the familiar tree, shrub or herb. Not all flowers are easily recognizable as such: Grasses and some trees have small green flowers which are quite different from the colorful blossoms we usually think of as flowers.

THE FLOWER. The flower of an angiosperm is a modified stem which bears, instead of ordinary foliage leaves, concentric circles of leaves modified for reproduction. A typical flower consists of four concentric rings of parts (Fig. 13–14) attached to the *receptacle,* the expanded end of the flower stem. The outermost parts, usually green and most like ordinary leaves, are called *sepals.* Within the circle of sepals are the *petals,* typically brilliantly colored to attract insects or birds and ensure pollination. Just inside the circle of petals are the *stamens,* the male parts of the flower. Each stamen consists of a slender filament with an *anther* at the tip. The anther is a group of *pollen sacs* (microsporangia), each of which contains a group of microspore mother cells — *pollen mother cells.* Each of these diploid cells undergoes meiotic division and gives rise to four haploid microspores which upon division of the nucleus become young microgametophytes or *pollen grains.*

In the very center of the flower is a ring of *pistils* (or a single fused one). Each pistil has a swollen, hollow, basal part, the *ovary,* a long slender portion above this, the *style,* and at the top a flattened part, the *stigma,*

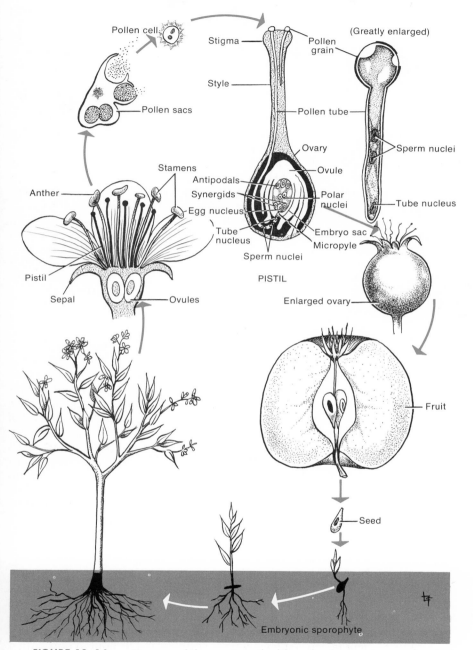

FIGURE 13–14   *Diagram of the events in the life cycle of an angiosperm (see text for discussion).*

which typically secretes a moist sticky substance to trap and hold the pollen grains that reach it. Different species show great variations in the number, position and shape of these various parts. A flower that has both stamens and pistils is called a **perfect flower;** one that lacks one or the other is called an imperfect flower. Flowers with stamens but no pistils are **staminate flowers;** those with pistils but no stamens are **pistillate flowers.** Willows, poplars and date palms are examples of plants which exist as two kinds of individuals, some bearing only staminate flowers, others bearing only pistillate flowers. The reproductive parts of the seed plants — stamen, pistil, stigma, style and so forth — were studied and named before the stages in the alternation of generations were understood and before the essential parallelism of the life cycles of mosses, ferns and seed plants was recognized.

Within the ovary, at the base of the pistils, are one or more **ovules.** An ovule is a megasporangium completely enclosed by one or two integuments. Each ovule typically contains one **megaspore mother cell** which divides by meiosis to form four haploid **megaspores.** One of the megaspores develops into the **megagametophyte;** the other three disintegrate. There is some variation in different species in the details of megagametophyte development, but typically the megaspore enlarges greatly and its nucleus divides. The two daughter nuclei migrate to opposite ends of the cell, then each divides and the daughter nuclei divide again. The resulting megagametophyte is an eight-nucleate cell called an embryo sac with four nuclei at each end. One nucleus from each end migrates toward the center; the two come to lie side by side in the center and are known as polar nuclei (Fig. 13–15). One of the three nuclei at one end of the megagametophyte becomes the **egg nucleus,** the other two and the three nuclei at the other end all disintegrate.

The haploid microspore develops into the pollen grain or young microgametophyte while still within the pollen sac. This occurs when the nucleus of the microspore divides into two, a larger **tube nucleus** and a smaller **generative nucleus.** Most pollen grains are released while in this state and are carried by the wind, insects or birds to the stigma of the same or a neighboring flower. The pollen grain then germinates; the pollen tube grows out of the pollen grain and down the style to the ovule. The tip of the pollen tube produces enzymes that dissolve the cells of the style, thus making room for the pollen tube to grow. The tube nucleus remains in the tip of the pollen tube as it grows. The generative nucleus migrates into the pollen tube and divides to form two nuclei, the **sperm nuclei.** The mature male gametophyte consists of the pollen grain and tube, the tube nucleus and two sperm nuclei, and some associated cytoplasm.

When the tip of the pollen tube penetrates the megagametophyte through the micropyle, it bursts and the two sperm nuclei are discharged into the megagametophyte. One of the sperm nuclei migrates to the egg nucleus and fuses with it. The resulting diploid cell is the **zygote,** the beginning of the new sporophyte generation. The other sperm nucleus migrates to the two polar nuclei and all three fuse to form an **endosperm nucleus,** made of three sets of chromosomes. The two polar nuclei may have fused to form a single one before the sperm nucleus arrives. This phenomenon of **double fertilization,** which results in a diploid zygote and a triploid endosperm, is peculiar to, and characteristic of, the flowering plants.

After fertilization the zygote undergoes a number of divisions and forms a multicellular embryo. The endosperm nucleus also undergoes a number of divisions and forms a mass of endosperm cells, gorged with

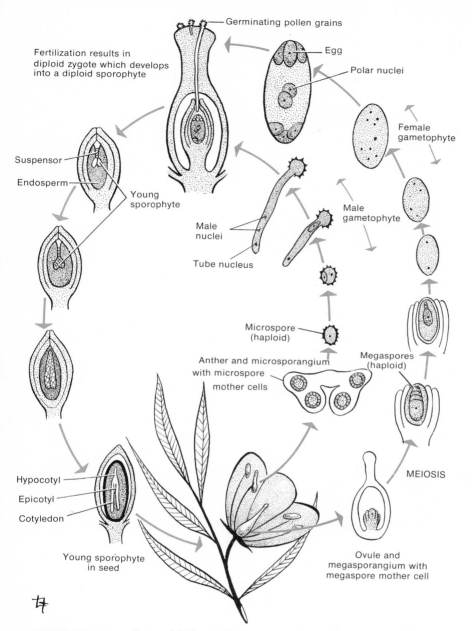

Germinating pollen grains

Egg

Polar nuclei

Fertilization results in
diploid zygote which develops
into a diploid sporophyte

Female
gametophyte

Suspensor

Endosperm

Young
sporophyte

Male
gametophyte

Male
nuclei

Tube nucleus

Microspore
(haploid)

Megaspores
(haploid)

Anther and microsporangium
with microspore
mother cells

Hypocotyl

Epicotyl

Cotyledon

MEIOSIS

Young sporophyte
in seed

Ovule and
megasporangium with
megaspore mother cell

*FIGURE 13-15   Some details of fertilization and the formation of seeds in a
dicotyledon (see text for discussion).*

nutrients, which fill the space around the embryo. The sepals, petals,
stamens, stigma and style usually wither and fall off after fertilization.
The ovule with its contained embryo becomes the seed; its walls become
thick and form the tough outer coverings of the seed. The seed consists
of the dormant sporophyte embryo plus stored food in the form of the
endosperm, all enclosed in a resistant covering derived from the wall of
the ovule. It serves in dispersing the species to new locations and in
enabling the species to survive periods of unfavorable environmental
conditions (such as winter!) which may kill the mature plants.

FRUITS.   The ovary, the basal part of the pistil containing the ovules, enlarges and forms the *fruit*. The fruit thus contains as many seeds as there were ovules in the ovary. In the strict botanical sense of the word, a fruit is a matured ovary containing seeds—the matured ovules. Although we usually think of only such sweet, pulpy things as grapes, berries, apples, peaches and cherries as fruits, bean and pea pods, corn kernels, tomatoes, cucumbers and watermelons are also fruits, as are nuts, burrs and the winged fruits of maple trees. A *true fruit* is one developed solely from the ovary. If the fruit develops from sepals, petals or receptacle as well as from the ovary it is known as an *accessory fruit.* The apple fruit consists mostly of an enlarged, fleshy receptacle; only the core is derived from the ovary.

True and accessory fruits are of three general types: simple fruits (e.g., cherries, dates, palms), which mature from a flower with a single pistil; aggregate fruits (raspberries and blackberries), which mature from a flower with several pistils; and multiple fruits (pineapples), derived from a cluster of flowers which unite to form a single fruit. Fruits are also classified as dry fruits if the mature fruit is composed of rather hard dry tissues and as fleshy fruits if the mature fruit is largely soft and pulpy (Fig. 13–16). Dry fruits are adapted primarily for being transported by the wind or by being attached to animal bodies by hooks. Fleshy fruits are adapted for dispersal by animals that eat them.

A *nut* is a dry fruit in which the ovary wall develops into a hard shell around the seed. The edible part of a chestnut is the seed within the fruit coat or shell. A Brazil nut is really a seed; there are about 20 such seeds borne within a single fruit. An almond is not a "nut" at all, but the seed or "stone" of a fleshy fruit related to the peach.

Grapes, tomatoes, bananas, oranges and watermelons are examples of fleshy fruits in which the entire wall of the ovary becomes pulpy; such

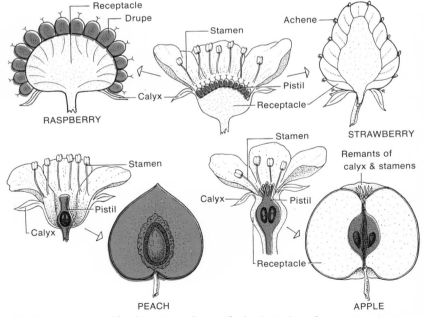

FIGURE 13–16   *The formation of some fleshy fruits from flowers. Raspberries and strawberries are derived from similar types of flowers, but the pistils of raspberries become fleshy drupes and the pistils of strawberries become dry achenes, the yellow seedlike spots scattered over the surface of the fruit.*

fruits are technically called **berries.** Peaches, plums, cherries and apricots are **drupes** or stone fruits in which the outer part of the ovary wall forms a skin, the middle part becomes fleshy and juicy, and the inner part forms a hard pit or stone around the seed. There are, then, many kinds of fruit, differing in the number of seeds present, in the part of the flower from which they are derived, as well as in color, shape, water and sugar content and consistency.

Fruits may form, or be induced to form, without the development of seeds. The banana, which has been cultivated for centuries, has only vestigial seeds (the black specks in the fruit) and must be propagated by vegetative means. Plant breeders have been able to develop seedless varieties of grapes, oranges and cucumbers. Many other plants have been induced to form seedless fruits by treatment with plant growth hormones.

SECTION 13–11

GERMINATION OF THE SEED AND EMBRYONIC DEVELOPMENT

A few seeds germinate shortly after being shed if conditions are suitable, but most seeds remain dormant during the cold or dry season and germinate only with the advent of the next favorable growing season. A prolonged period of dormancy usually occurs only in seeds with thick or waxy seed coats which render them impermeable to water and oxygen. The length of time that a seed will remain viable and capable of germination varies greatly. Willow and poplar seeds must germinate within a few days of being shed or they will not germinate at all. In contrast, seeds of the evening primrose and of yellow dock were able to germinate after 70 years. There are authentic records of lotus seeds germinating 200 years after being shed. The ability of a seed to retain its germinating power depends on the thickness of the seed coat, on a low water content, and on the presence of starch rather than fats as the stored food material. Dormant seeds are alive and do metabolize, though at a very low rate.

Germination is initiated by warmth and moisture and requires oxygen. The embryo and endosperm absorb water, swell and rupture the seed coats. This frees the embryo and enables it to resume development. Most seeds do not need soil nutrients to germinate; they will germinate equally well on moist paper.

The cell divisions that the zygote undergoes following fertilization first produce a filament of cells, called the **suspensor.** Most of the embryo forms from the end cell of this filament, which begins to divide in other planes to form a rounded mass of cells. From this grow (in dicotyledonous plants) two primary leaves or **cotyledons** and a central **axis.** The part of the axis below the point of attachment of the cotyledons is called the **hypocotyl** and the part above it, the **epicotyl** (Fig. 13–17). The embryo is in about this state of development when the seed becomes dormant.

After germination the hypocotyl elongates and emerges from the seed coat. The primitive root or **radicle** grows out of the hypocotyl and, since it is strongly and positively geotropic, it grows directly downward into the soil. The arching of the hypocotyl in a seed such as the bean pulls the cotyledons and epicotyl out of the seed coat, and the epicotyl, responding negatively to the pull of gravity, grows upward and forms the stem and leaves. The cotyledons digest, absorb and store food from the endosperm while within the seed. The cotyledons of some plants shrivel and drop off after germination; those of other plants become flat foliage leaves. The

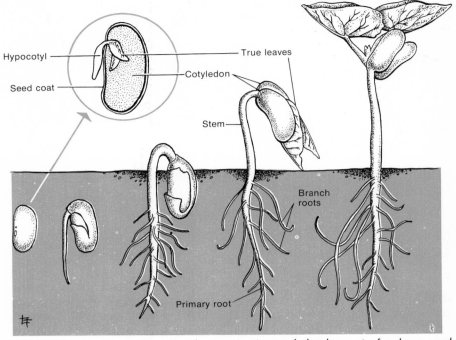

FIGURE 13–17   Stages in the germination and development of a bean seed. Inset: an enlarged view of an opened seed, showing the cotyledons, the hypocotyl, from which develops the root, and the epicotyl from which develop the stem and leaves.

cotyledons contain reserves of food that supply the growing seedling until it develops enough chlorophyll to become independent.

Seeds are used extensively by man as sources of food, beverages, textiles and oils. Almost all of man's carbohydrates are derived from seeds, the major exceptions being potato tubers, sugar cane and sugar beets. Wheat, rye, corn, rice, oats and barley are one-seeded fruits from members of the grass family; beans, peas and peanuts are the seeds of legumes and are rich in proteins as well as carbohydrates. Coffee and cocoa are beverages derived from seeds, and a variety of spices and seasonings are made from ground seeds. Cotton fibers are produced as epidermal hairs on the seed coats of the cotton plant. Oils derived from seeds may be important industrially, or as foods. Linseed and tung oils are used in the manufacture of paints and varnishes. Oils from peanuts, cottonseeds and soybeans are used to make salad oils and margarine. Coconut oil is used in making soaps and shampoos, as well as margarine.

SECTION 13–12

EVOLUTIONARY TRENDS IN THE PLANT KINGDOM

As we survey the many types of life cycles that are found from algae to angiosperms, a number of evolutionary trends are evident. One of these is a change from a population that is mostly haploid individuals to one that is almost entirely diploid (Fig. 13–18). In algae such as *Ulothrix* only one cell in each life cycle, the zygote, is diploid; all the rest are haploid.

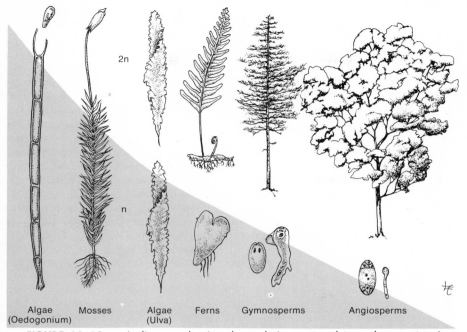

2n

n

Algae   Mosses   Algae   Ferns   Gymnosperms   Angiosperms
(Oedogonium)   (Ulva)

FIGURE 13–18   *A diagram showing the evolutionary trend toward a greater size and importance of the sporophyte (2n) and a reduction in the size of the gametophyte (n) generations of plants.*

The haploid phase of the moss is more conspicuous and longer-lived than the diploid phase, but the latter is a complex, multicellular plant. The relative importance of the two phases is reversed in the ferns: The diploid phase is the obvious, larger plant and the haploid gametophyte, though still an independent plant, is small and inconspicuous. The gymnosperms and angiosperms show progressive reductions of the haploid phase until, in the angiosperms, the male gametophyte consists of three cells and the female gametophyte of eight. The corollary of this trend toward diploidy is the trend toward the reduction of the gametophyte.

The trend toward immobile adult plants, anchored in place by roots, solved several physiological problems, such as the supply of water and salts, but raised the reproductive problem of the means of bringing the two kinds of gametes together. Aquatic plants, like aquatic sessile animals, may have motile sperm that swim to the egg. Such a system persisted in some of the early terrestrial plants, but the seed plants evolved pollen and the pollen tube. Pollen may be very light and carried by the wind to a height of several thousand meters and for distances of many kilometers. The pollen of other species may be carried by insects and other animals. Still other species are **monoecious,** having both sexes in the same plant, so that self-fertilization is possible; however, continued self-fertilization leads to the loss of the advantages of sexual reproduction, the opportunities for genetic recombination.

Immobile plants also have the problem of a method of dispersal by which the species can spread to occupy a larger territory. Algae have evolved motile aquatic spores, fungi and ferns have windborne spores, and seed plants have seeds that may be windborne or carried by "hooking" into the fur of an animal. Some seed plants have attractive fruits that are eaten by animals and seeds that resist digestion. The seeds eventually pass through the animal's digestive system, are excreted in the feces, and subsequently germinate.

There are a number of possible explanations for these evolutionary trends. As long as there was an independent gametophyte generation, the transfer of sperm to the egg required a film of water for the sperm to swim in. The evolution of a life cycle in which there is a reduction of the gametophyte to a small group of cells within the sporophyte, and sperm are transferred to the egg via a pollen tube, permitted reproduction to occur in the absence of moisture. The evolutionary advantages of this are obvious. There may be another, less obvious reason for this; a diploid individual can survive despite the presence of deleterious, recessive genes; a haploid individual would be much more susceptible to the effects of such genes. A third explanation is that, since terrestrial life required the development of conducting and supporting tissues, and since these have appeared only in sporophyte individuals, evolutionary processes on land favored those plants with longer sporophyte and shorter gametophyte generations.

SUGGESTIONS FOR FURTHER READING

Agrios, G. N.: *Plant Pathology.* New York, Academic Press, Inc., 1969.
     This discussion, based on a biochemical approach to the problems of parasitism, includes many detailed drawings of the disease cycles.

Bell, Peter R., and C. L. F. Woodcock: *The Diversity of Green Plants.* Reading, Massachusetts, Addison-Wesley Publishing Company Inc., 1968.
     A concise, up-to-date survey of the great variety of green plants.

Bold, H. C.: *Morphology of Plants.* 2nd ed. New York, Harper & Row, Publishers, Inc., 1967.
     A well illustrated treatment of the many variations on the theme of plant structure.

Corner, E. J. H.: *The Life of Plants.* New York, Mentor Brooks, New American Library, Inc., 1968.
     A renowned botanist presents his views of the evolution of plant life, indicating how plants modified their structures and functions to meet the challenge of new environments as they invaded the shore and spread across the land.

Delevoryas, T.: *Plant Diversification.* New York, Holt, Rinehart & Winston, Inc., 1966.
     A brief volume stressing evolutionary trends in the plants.

Doyle, W. T.: *Nonvascular Plants; Form and Function.* Belmont, California, Wadsworth Publishing Company Inc., 1965.
     A brief account of the lower plants which describes the important features of each type.

Esau, Katherine: *Plant Anatomy.* 2nd ed. New York, John Wiley & Sons, Inc., 1965.
     A beautifully illustrated book which provides an up-to-date treatment of all aspects of plant anatomy.

Watson, E. V.: *The Structure and Life of Bryophytes.* London, Hutchinson & Co., Ltd., 1964.
     A well written treatment of the entire group of mosses, hornworts and liverworts.

THE DIVERSITY OF LIFE: INVERTEBRATES

CHAPTER 14

There are well over 1,000,000 species of animals, catalogued in a hierarchy of species, genera, families, orders, classes and phyla. From this vast and diverse array we can sample only a small fraction, chosen because of their evolutionary or ecological significance, or their prevalence, or because they illustrate some general biological principle.

In order to survive, all animals have had to evolve solutions to the same basic biological problems. It follows that there is a basic unity of life and a discovery about one form may have broad, even universal, application. It seems obvious that rats, rabbits and guinea pigs are similar enough to man so that experiments on their digestive, circulatory and excretory systems will contribute to our understanding of the corresponding systems in man. It is not so obvious that much of our knowledge of nerve action has come from experiments on the nerves of earth worms and squid, that experiments on the heart of the horseshoe crab have given us information about the human heart, and that observations of the toad fish kidney have been useful in providing understanding of the human excretory system.

In determining relationships, biologists are careful to distinguish between homologous structures and analogous structures. **Homologous structures** are those which develop from similar embryonic rudiments, are similar in basic structural plan and development, and hence reflect a common genetic endowment and evolutionary relationship. In contrast, **analogous structures** are simply superficially similar and serve a similar function, but have quite different basic structures and developmental patterns. The presence of analogous structures does not imply an evolutionary relationship in the animals bearing them. For example, the arm of a man, the wing of a bird, and the pectoral or front fin of a whale are all homologous, with basically similar patterns of bones, muscles, nerves and blood vessels, and similar embryonic origins, though rather different functions (Fig. 14–1). The wing of a bird and the wing of a butterfly, in contrast, are simply analogous; both enable their possessors to fly, but they have no developmental processes in common. The wings of birds and the wings of bats have a similar structural plan and development and are anatomically homologous; however, they evolved independently as adaptations

391

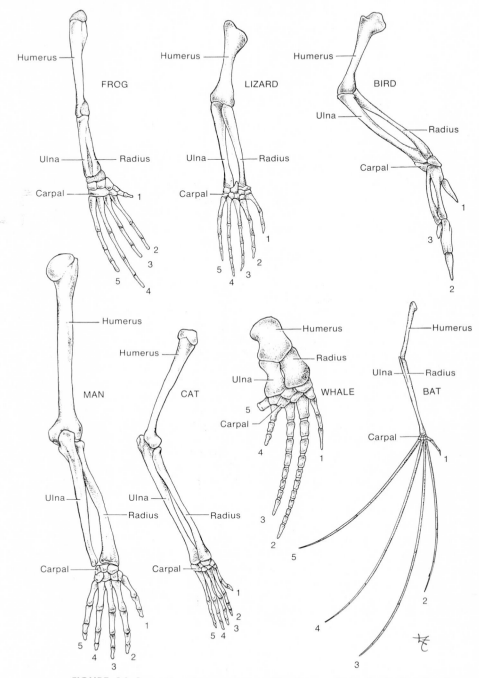

FIGURE 14-1   *Homologous organs. The bones of the forelimbs of a frog, lizard, bird, man, cat, whale and bat showing the arrangement of the homologous bones in these superficially different structures.*

for flying and are thus analogous in terms of their functions. As more has been learned about the molecular structure of cellular constituents, it has become clear that these terms can be applied at that level. The hemoglobins of different animals, the cytochrome c's present in different vertebrates, or the lactic dehydrogenases present in birds and mammals may be termed homologous proteins. The hemoglobins in different species, for example, have very similar sequences of amino acids, which again reflects a common genetic pattern and evolutionary relationship. In contrast, hemoglobin and hemocyanin may be termed analogous molecules, since they have similar functions (oxygen transport) but quite different molecular structures.

The differences that distinguish related species such as leopards and panthers are usually minor, superficial ones such as body color and proportion. In contrast, the major divisions of the animal kingdom, the phyla, are differentiated by basic body characteristics and patterns, which usually are not unique for a single phylum but occur in unique combinations. For example, the presence or absence of cellular differentiation distinguishes the single-celled animals, the protozoa, from the metazoa, composed of many kinds of cells specialized to perform some particular task in the body's economy.

Animals are distinguished by the type of symmetry present, whether their body plan shows spherical, radial or bilateral symmetry. Some of the metazoa have only two embryonic cell layers or germ layers—an outer **ectoderm** and an inner **endoderm** which lines the digestive tract. Others have these two plus a third, the **mesoderm,** which lies between the ectoderm and endoderm and makes up the rest of the body. In the simpler many-celled animals, the body is essentially a double-walled sac surrounding a single "gastrovascular" cavity with a single opening to the outside—the mouth. The more complex animals have two cavities and bodies constructed on a tube-within-a-tube plan. The inner tube, the digestive tract, is lined with endoderm and is open at both ends—the mouth and the anus. The outer tube or body wall is covered with ectoderm. Between the two tubes is a second cavity, termed a **coelom** if it lies within the mesoderm and is lined by it, or a **pseudocoelom** if it lies between the mesoderm and endoderm. The members of several phyla are characterized by bodies composed of a row of segments, each of which has the same fundamental plan, with or without some variation. In most of the vertebrates, the segmental character of the body is largely obscured. In man, the bones of the spinal column—the vertebrae—are among the few parts of the body that are still clearly segmented.

A few structures are found exclusively in members of a single phylum and help distinguish them from all other animals. For example, the coelenterates are unique in having stinging cells or **nematocysts;** echinoderms have a peculiar **water vascular system** found in no other phylum; and only the chordates have a dorsally located, hollow nerve cord. The appendix lists 21 animal phyla. Some systematists recognize more phyla, and some recognize fewer ones. Some 10 of these are major phyla that include most of the animals, both living and extinct.

SECTION 14–1

PROTOZOA

The single-celled animals, which make up the phylum **Protozoa,** are functionally complex, even though some may appear to be relatively simple structurally. Typically protozoa have a single nucleus and lead an independent existence, but some have multinucleate cells and in other species the cells are joined together to form a colony. A colonial protozoan can be distinguished from a multicellular animal, because its cells are quite similar and none is specialized for feeding. Most of the individuals in a population of protozoa are produced by simple cell division of the parent, although sexual reproduction by the mating of two individuals does occur. Protozoa are primarily aquatic and live in fresh or salt water, in small puddles or in the oceans. Some live in damp soils, crawling on the film of water that surrounds each dirt particle. Parasitic protozoa live in the body fluids of animals or in the saps of plants. Some species of protozoa can form inactive spores or cysts that can be dried and distributed with particles of dirt or dust from one habitat to another.

The 25,000 or so species of protozoa are divided into five classes, which differ in their means of locomotion as well as in other respects. The **Flagellata** have one or more long whiplike flagella; the **Sarcodina** move by forming pseudopods; the **Ciliata** are characterized by the presence of many short hairlike cilia which beat in a coordinated fashion and move the animal along; the young of the **Suctoria** have cilia but the mature animals have tentacles; and the **Sporozoa** are parasites lacking locomotor structures and reproducing by multiple fission.

Metazoan organisms are characterized by some division of labor between cells; certain cells are specialized to carry out nutritive, excretory, locomotor and other functions. In the protozoa these various activities are accomplished by specialized structures, termed **organelles.** Cilia and flagella are examples of locomotor organelles. The cilia beat with an oblique stroke so that the animal revolves as it swims. The coordination of the ciliary beating is good enough so that the animal not only can go forward but can back up and turn around. Coordination is achieved by a system of **neurofibrils** that connect the rows of basal bodies at the inner end of each cilium. If the neurofibrils are cut, the beating of the cilia is no longer coordinated. Near the surface of the cells of ciliates are many small **trichocysts,** organelles which can discharge filaments believed to aid the organism in trapping and holding its prey.

Flagellates move rapidly, pulling themselves forward by lashing one or more flagella located at the anterior end. Each **flagellum** is a long, supple filament containing an axial fiber, shown by electron microscopy to be composed of 11 filaments, a sheath of nine surrounding two in the center. The two central filaments give bilateral symmetry to the flagellum and influence its plane of motion. These 11 filaments have a chemical composition similar to that of actomyosin. Some protozoa can creep along a flat surface with wormlike movements that depend upon a layer of contractile fibers just beneath the surface of the cell which form an organelle comparable to the muscular body wall of a worm.

Although protozoa have no nervous system, they do have conductile organelles such as the basal body and system of neurofibrils just described. The **basal body** stimulates and controls the movement of the flagellum or cilium to which it is attached. The basal bodies may be joined to the centriole by filaments and appear to be produced by the division and differentiation of centrioles during development. Many flagellates have photosensitive organelles associated with the conductile and locomotor organelles. The photosensitive organelle or **eye spot** of *Euglena* consists of a patch of red pigment and a tiny light-sensitive photoreceptor beside the base of the flagellum. The shading of the photoreceptor by the pigment spot enables the animal to determine the direction of the source of light. In other species the photoreceptor may be set in a pigment cup with the opening of the cup directed anteriorly. In a few species the cuticle covering the animal is swollen over the cup to form an optic lens.

A prominent structure in many protozoans is the excretory organelle or **contractile vacuole.** This cavity regularly fills with water from the surrounding cytoplasm and then empties the water into the environment. It is not an excretory organ but a pump to remove the excess water that is constantly entering the cell by osmosis. The cytoplasm of a fresh-water protozoan has a higher concentration of dissolved materials — salts, sugars and organic acids — than the surrounding water. Water tends to pass into the cytoplasm by osmosis. Without a pump to remove the excess water, the amoeba or other protozoan would swell and burst, just as human blood cells do when they are placed in distilled water. In contrast, most marine protozoa do not have a contractile vacuole, since the concentration of salts in the sea water is about the same as that in their cytoplasm.

Flagellates

The largest class of protozoa, the flagellates, includes more than half of all living species of protozoa. Flagellates have spherical or elongate bodies, a single central nucleus, and one to many slender whiplike flagella at the anterior end which enable them to move. Some flagellates engulf food by forming pseudopods, others resemble paramecia and have a definite mouth and gullet. Although most flagellates are tiny and difficult to study, a few, such as members of the genus *Euglena*, are large and can be used as representatives of the class. Euglenas have chlorophyll contained in chloroplasts and synthesize much of their food, but no euglenoid is completely autotrophic. Healthy cultures of euglenas can be maintained only if amino acids are present in the medium. The flagellates with the largest number of flagella and the most specialized bodies are the ones living in the intestines of termites. Some species of flagellates have characteristics in common with other classes of protozoa or with the sponges and suggest some of the intermediate steps by which these forms might have evolved from them.

The body of *Euglena* is covered with a delicate pellicle with spiral thickenings. Beneath the pellicle is a layer of contractile fibrils which permit the organism to change its shape. Scattered inside the cytoplasm are green chloroplasts and transparent, colorless **paramylum bodies** containing stored polysaccharides. Protruding from the gullet is a single flagellum formed by the fusion of two (Fig. 14–2).

A *Euglena* swimming with a single flagellum may be likened to a one-armed man trying to swim. At each stroke the flagellum bends toward the side bearing the pigment spot, not in a simple backward lash but obliquely toward the long axis of the organism. The body not only turns toward one side but rotates a bit. Successive lashes of the flagellum thus move the organism forward in a spiral path with the pigment spot facing the outside of the spiral.

The more plantlike flagellates, dinoflagellates and phytomonads, such as the colonial flagellate *Volvox*, were described previously (p. 338). The cells of the *Volvox* colony are connected to each other by bridges of cytoplasm through which the activities of the different cells can be synchronized.

The more strictly animal-like flagellates are small and rather uncommon. Of special interest, because of their resemblance to the sponges, are the **choanoflagellates.** These sedentary flagellates are attached to the bottom by a stalk and their single flagellum is surrounded by a delicate collar

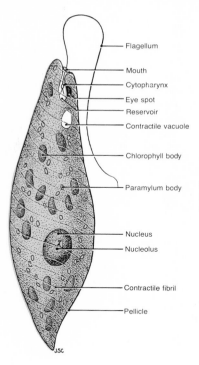

Flagellum

Mouth

Cytopharynx

Eye spot

Reservoir

Contractile vacuole

Chlorophyll body

Paramylum body

Nucleus

Nucleolus

Contractile fibril

Pellicle

FIGURE 14–2   *Diagram of the structure of Euglena. These single-celled organisms are classified by botanists as members of phylum Euglenophyta, whereas zoologists classify them as the order Euglenoidina within the class Flagellata.*

of cytoplasm. There are some parasitic flagellates of medical importance, such as the trypanosomes that produce sleeping sickness and are transmitted from man to man by the tsetse fly. The flagellates are generally considered to be the basic stock from which evolved not only other kinds of protozoa but also higher plants and animals.

Sarcodinids

The members of the class **Sarcodina,** unlike other protozoa, have no definite body shape. Their single cells are shapeless blobs that change form as they move. The nucleus, contractile vacuole and food vacuoles are shifted about within the cell as the animal moves.

An amoeba moves by pushing out temporary cytoplasmic projections called **pseudopods** (false feet) from the surface of the body. More cytoplasm flows into the pseudopods, enlarging them until all the cytoplasm has entered and the animal as a whole has moved. During amoeboid motion, a stable gel layer at the surface surrounds the central core of liquid, flowing cytoplasm. As a pseudopod forms, the outer layer of cytoplasm becomes a liquid sol momentarily and then returns to the semisolid gel state along the sides of the forming pseudopodial lobe. As the liquid cytoplasm reaches the tip of the pseudopod it is pushed to the side and converted into a gel to form that portion of the wall. At the posterior end of the amoeba, the gel walls are converted to a liquid to be pushed along. The molecular basis of amoeboid motion and the chemical reactions supplying energy for it appear to be fundamentally similar to those of muscular contraction. Human white blood cells also move by amoeboid motion.

The pseudopods are used to capture food, two or more of them moving out to surround and engulf a bit of debris—another protozoan, or even a small metazoan. The food that has been engulfed is surrounded by

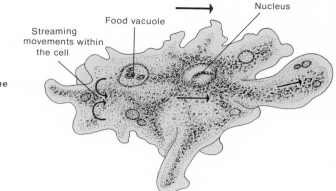

Streaming
movements within
the cell

Food vacuole

Nucleus

FIGURE 14–3   A diagram of the
structure of Amoeba proteus.

a **food vacuole** and acids and enzymes are secreted from the surrounding cytoplasm into the food vacuole to begin the process of digestion. The digested materials are absorbed from the food vacuole and the latter gradually shrinks as it becomes empty. Any indigestible remnants are expelled from the body and left behind as the amoeba moves along.

The parasitic members of the class Sarcodina include the species causing amoebic dysentery in man. Certain free-living species secrete shells around the body or cement sand grains together into a protective layer around their cells. The ocean contains untold trillions of amoeboid protozoa, the **foraminifera**, which secrete chalky, many-chambered shells with pores through which the animal extends its pseudopods. The dead foraminifera sink to the bottom of the ocean and form a grey mud which is gradually transformed into chalk. Other amoeboid protozoa, called **radiolaria**, secrete elaborate and beautiful skeletons made of silica. These skeletons become mud on the ocean floor and eventually are compressed and converted into siliceous rock (Fig. 14–3).

During the latter part of the Paleozoic era, a group of foraminiferans, the **Fusulinidae**, flourished. Over a relatively brief period (75,000,000 years) many species of fusulinids developed and then became extinct. Some of these were large protozoa, as much as 2 centimeters in diameter, that lay on the bottom of the shallow seas. Their fossils are now found in deposits that have accumulated oil. As an oil well is drilled through the sedimentary rock, the bit passes in rapid succession through these successive species of fusulinids. By analyzing the species present in a specific portion of the core, the driller can estimate how far into the paleozoic deposit he has drilled.

Ciliates

Members of the class Ciliata, typified by *Paramecium*, have a definite, permanent shape due to the presence of a sturdy flexible outer covering of chitin. The surface of the cell is covered by several thousand fine cytoplasmic hairs, cilia, which extend through pores in the covering and move the animal along. Ciliates and Suctorians differ from other protozoa in having two nuclei per cell, a **micronucleus** which functions in sexual reproduction and a **macronucleus** which controls cell metabolism and growth. Both nuclei divide at each mitosis, but at sexual reproduction the macronucleus disintegrates and the micronucleus gives rise to both nuclei of the offspring. The macronucleus appears to be a compound structure formed by the amalgamation of many sets of chromosomes. It is simply pulled in half during asexual reproduction, without any mitotic phenomena (Fig. 14–4).

Paramecia have two contractile vacuoles, which together can remove a volume of water equal to the total volume of the animal's body within half an hour. In contrast a man excretes an amount of water equal to his body volume in about three weeks. Well fed paramecia reproduce by division two or three times a day and are ideal subjects for studies of the laws governing population growth. Paramecia and other ciliates may have more than two, indeed as many as eight "sexes" or mating types. All the sexes look alike, but an individual of one sex will mate only with an individual of some other sex. The plantlike flagellates, the phytomonads, are haploid organisms, whereas ciliates, like animals, are diploid organisms. During sexual reproduction, two individuals of different sexes conjugate and press together by their oral surfaces. Within each individual the macronucleus disintegrates and the micronucleus undergoes meiosis to form four daughter nuclei. Three of these four degenerate, which is comparable to the degeneration of the polar bodies during oögenesis, leaving one viable haploid nucleus. This divides once mitotically, and one of the two identical haploid nuclei remains within the cell. The other crosses through the oral region into the other individual and fuses with its haploid nucleus. Thus each conjugation yields two fertilizations, and the two new diploid nuclei are identical.

One of the interesting features of certain ciliates is the presence of traits that are transmitted to offspring through the cytoplasm rather than through nuclear chromosomes as is usually the case. Some strains of paramecia produce and secrete into the medium **killer particles** which will

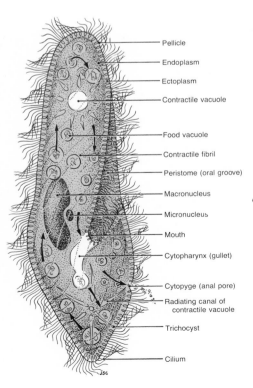

Pellicle

Endoplasm

Ectoplasm

Contractile vacuole

Food vacuole

Contractile fibril

Peristome (oral groove)

Macronucleus

Micronucleus

Mouth

Cytopharynx (gullet)

Cytopyge (anal pore)

Radiating canal of
contractile vacuole

Trichocyst

Cilium

FIGURE 14–4   *Diagrammatic sketch of a typical ciliate protozoan, the freshwater Paramecium caudatum.*

cause the death of a "sensitive" individual if they come in contact with it. All individuals that are unable to produce killer particles are sensitive, whereas all individuals that do produce these particles are resistant to their effect and do not get killed. These killer particles are manufactured in the cytoplasm from **kappa particles,** granules present in killer animals but not in sensitive ones. A killer paramecium may have in its cytoplasm some 800 kappa particles and will secrete a killer particle every five hours. The kappa particles are nucleoproteins which multiply in the cytoplasm independently of the division of the cell. At each division of the paramecium, the kappa particles are divided randomly between the daughter cells. As long as each daughter cell receives at least one kappa particle it will remain a killer. Under certain conditions the paramecium will divide more rapidly than the kappa particles do, and the number of particles per cell gradually decreases. Ultimately some cells will be formed that lack killer particles and they are then sensitive to them. Sensitive individuals, though unable to produce kappa particles or killer particles, occasionally mate with a killer before encountering a killer particle. During the mating process, kappa particles may be transferred into the sensitive cell. These will subsequently survive, divide and transform the sensitive cell into a killer. The trait will again be transmitted to the offspring as long as the reproduction of the particles keeps pace with the reproduction of the paramecium. Certain strains lack the ability to acquire the trait, because they lack a specific nuclear gene. Hence, even this kind of cytoplasmic inheritance is ultimately determined by nuclear genes.

Another ciliate of general biological interest is the genus *Tetrahymena*, which can be cultured in a chemically defined medium, i.e., one in which the exact amounts and kinds of all chemicals are known. By varying the chemicals present in the medium and by using appropriate radioactive or heavy atoms as labels, biologists have found what materials are needed for growth and maintenance and to what extent they may be converted into other materials, the nature of the enzymes that catalyze these transformations, and how all these processes are under genic control. Since the metabolic pathways in all organisms have many basic similarities, the study of *Tetrahymena* (in which information can be obtained rapidly and easily) is shedding light on similar problems in other organisms, even man.

Suctorians

Suctorians, a fourth class of protozoa, are closely related to ciliates and appear to have been derived from them in evolution. The suctorians, like the ciliates, have both a macronucleus and a micronucleus (Fig. 14–5). Young individuals have cilia and swim about, but the adults are sedentary and have stalks by which they are attached to the substrate. The body bears a group of delicate cytoplasmic tentacles, some of which are pointed to pierce their prey whereas others are tipped with rounded adhesive knobs to catch and hold the prey. The tentacles secrete a toxic material which may paralyze the prey. Although adult suctorians lack cilia they possess basal bodies. During asexual reproduction a bud forms on the suctorian, the basal bodies multiply, become arranged in rows, and develop cilia. After the nucleus has divided, the bud separates from the parent and swims away. When it becomes attached to the bottom, the cilia disappear and tentacles develop.

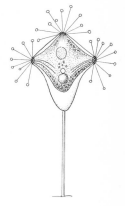

FIGURE 14–5   Diagram of a Suctorian, Acineta tuburosa.

Sporozoa

The sporozoans comprise a large group of parasitic protozoa, among which are the agents causing serious diseases such as **coccidiosis** in poultry and **malaria** in man. Sporozoa have neither locomotor organelles nor contractile vacuoles. Most sporozoans live as intracellular parasites in the host cells during the growth phase of their life cycle and absorb nutrients through their cell wall. The sporozoan causing malaria, *Plasmodium*, enters the human blood stream when an infected mosquito bites a man (Fig. 14–6). The plasmodia enter the red cells and each divides into 12 to 24 spores, which are released when the red cell bursts. The released spores infect new red cells and the process is repeated. The simultaneous burst-

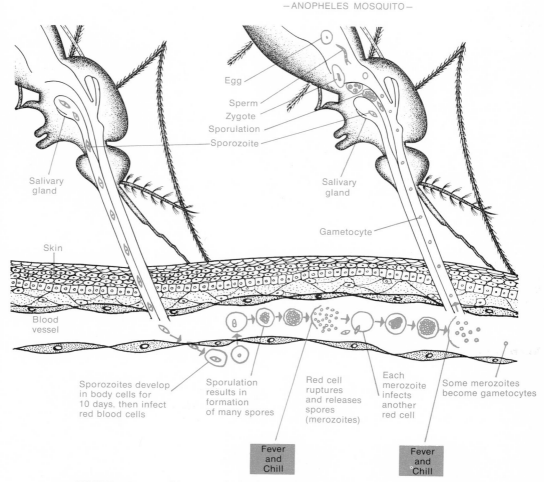

FIGURE 14–6   Diagram of the life cycle of the sporozoan Plasmodium which causes malaria in man and other mammals. An infected mosquito (left) bites a man and injects sporozoites of Plasmodium into the man's blood stream. These undergo sporulation and reproduce asexually within the red blood cells of the host. Periodically the infected red cells rupture and the new crop of merozoites released then infect other red cells. The bursting of the red cells releases toxic substances which cause the periodic fever and chill. Some merozoites develop into gametocytes which can infect another mosquito when one bites the man. The gametocytes develop into eggs and sperm (right) within the mosquito and undergo sexual reproduction. The resulting zygote by sporulation produces a host of sporozoites which migrate to the salivary glands and are ready to be injected when the mosquito bites the next man.

ing of millions of red cells causes the malarial chill followed by fever as toxic substances are released and penetrate other organs of the body. If a second, uninfected mosquito bites the infected man it will suck up some plasmodian spores along with its drink of blood. A complicated process of sexual reproduction occurs within the mosquito's stomach and new spores are formed, some of which migrate into the mosquito's salivary glands and are ready to infect the next man bitten. This process of sexual reproduction does not occur when the organisms are present in man.

SECTION 14–2

RELATIONSHIPS AMONG THE PROTOZOA

The flagellates are generally considered to be the basic stock of organisms from which other protozoa and indeed the higher animals and plants arose. Botanists usually claim all the flagellates in which a photosynthetic pigment occurs, including related forms such as some of the euglenas and dinoflagellates that have lost the pigment. Zoologists usually claim all the flagellates, even those that are completely autotrophic. This is perhaps not defensible, but there is a good argument for keeping all the flagellates together, since the transition from autotrophic to holozoic nutrition appears to have occurred independently in different groups.

The sarcodinids are related to the flagellates through several genera of amoeboid organisms that have flagella and through several forms that resemble typical flagellates when in open water but lose the flagella and creep like amoebas when next to a solid substrate. The existence of so many intergrades suggests that sarcodinids may have evolved several different times from the flagellates.

The ciliates are a distinct group and probably arose from flagellates only once. Suctorians are clearly related to the ciliates. The evolutionary origin of the macronucleus is unknown. During the conjugation of most ciliates, a bit of cytoplasm is transferred along with the migrating nucleus and its bit of cytoplasm separates in the mouth cavity as a *gamete* with a long tail. The two gametes move past each other to the opposite organism.

The sporozoa may well be a composite group. Some species show affinities with flagellates while others more nearly resemble sarcodinids. The phenomenon of multiple fission may be regarded as an adaptation to parasitism. The number of evolutionary changes needed to develop one class of protozoa from another is not very great, and the possibility that in the course of evolution, the change from one group to another occurred repeatedly cannot be ruled out.

SECTION 14–3

THE PHYLUM PORIFERA

The Porifera, the phylum of animals commonly called sponges, have porous body walls and internal cavities lined with **choanocytes** (Fig. 14–7). Much of the body is composed of a jelly-like matrix, containing a skeleton made of protein, calcium carbonate or silica. There seems to be no nervous system present. Sponges are organized on a cellular level. Instead of a single cell carrying on all the life activities as in protozoa, there is a division of labor, with certain cells specialized to perform particular functions such

FIGURE 14–7 Photograph of the skeleton of the glass sponge Euplectella. *The hexagons are fused to form intersecting girders. (Courtesy of the American Museum of Natural History.)*

as nutrition, support or reproduction. The sponge shows cellular differentiation but little or no coordination of cells to form tissues. The cells are very loosely organized and cell relations can be disrupted by passing the sponge through silk bolting cloth without damaging cellular integrity. The cells reaggregate and form a structure similar to the initial one. Because of their many distinctive morphologic features, the sponges are usually considered to be a side branch in the evolution of the metazoa. They evolved from flagellates independently of the other metazoa and have not given rise to any other phylum.

Living sponges resemble a piece of raw liver. They are drab colored, slimy to the touch, and usually have an unpleasant odor. Sponges are sedentary organisms ranging in size from 1 to 200 centimeters in height and varying in shape from flat, encrusting growths to balls, cups, fans and vases. Most sponges are marine; only one family occurs in fresh water.

Sponges make their living by filtering water, straining out microscopic organisms which they use for food. Characteristic of sponges is the presence of choanocytes or collar cells with flagella that beat, creating the currents of water necessary for bringing in food and oxygen and for carrying away carbon dioxide and wastes. The choanocytes of some complex sponges can pump a volume of water equal to the volume of the sponge each minute! Wandering through the gelatinous matrix of the sponge body are numerous amoebocytes, which collect food from the cells lining the pores, secrete the matrix and the protein, calcium carbonate or silica of the skeleton, collect wastes and become converted to epidermal cells as needed. Each cell of the body is irritable and can react to stimuli, but there are no sense cells or nerve cells which would enable the animal to react as a whole.

There are three classes of sponges, distinguished on the basis of the kind of skeleton present: **Calcarea** have a skeleton of calcium carbonate spicules, **Hexactinellida** have a skeleton of six-rayed siliceous spicules, and the **Demospongia** have a skeleton of spongin fibers or of siliceous spicules that do not have six rays (Fig. 14–8).

The bath sponges are found in warm shallow waters with a rocky bottom. Sponge fishermen hook them from the ocean bottom by poles with a pronged fork on the end. They are kept out of water until they die, then are left lying in shallow water until the flesh is decayed. After being beaten, washed and bleached in the sun they are ready for the market. All that remains is the spongin network whose many interstices enable it to soak up a large amount of water.

All sponges appear to be diploid and have the usual metazoan processes of oögenesis and spermatogenesis. The eggs are retained just be-

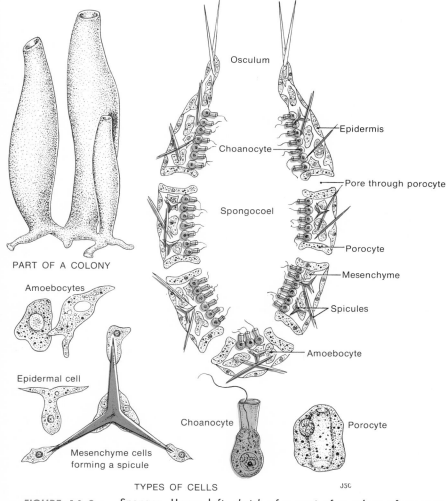

PART OF A COLONY

Osculum

Epidermis

Choanocyte

Pore through porocyte

Spongocoel

Porocyte

Mesenchyme

Spicules

Amoebocyte

Amoebocytes

Epidermal cell

Choanocyte

Porocyte

Mesenchyme cells forming a spicule

TYPES OF CELLS

JSC

FIGURE 14–8   Sponges. Upper left, sketch of a part of a colony of sponges. Upper right, diagram of a section through a simple sponge showing its cellular organization. Lower, sketches of the type of cells found in a sponge. The choanocytes are shown in light tint and the spicules in dark tint.

neath the choanocytes where they are fertilized by sperm from another sponge brought in with the current.

The pattern of development in the sponge is quite different from that in any other metazoan. The fertilized egg cleaves to form a blastula-like structure that is inside out compared to the blastulas of other animals. The nuclei lie toward the inner ends of the cells, and the flagella that appear on the cells project inward instead of outward. The embryo is also peculiar in having a mouth at the vegetal pole, through which food is taken from the parent. When fully developed the embryo turns inside out through its mouth, penetrates through the maternal choanocyte layer and escapes into the channels of the parent sponge and finally to the open ocean as a free swimming larva. The eversion process brings the flagella to the outside of the larva and by means of these the larva swims. It finally becomes attached to the bottom by its anterior end and the flagellated half invaginates into the posterior half to form a two-layered structure. The flagellated cells become the inner choanocytes, whereas the outer layer forms the remainder of the sponge.

The remarkable ability of the sponges to reorganize was demonstrated by H. V. Wilson in 1907. He squeezed sponges through fine silk cloth into a dish, thus disaggregating the sponge cells into minute clumps. The choanocytes liberated in this fashion swim about on the bottom and the amoebocytes crawl. When the cells come in contact they remain together. The bottom of the dish becomes covered with balls of cells, each of which develops into a tiny sponge if it includes both choanocytes and amoebocytes. The cells are species specific and if two kinds of sponges, one yellow and one orange, are disaggregated in the same dish, the clumps of cells that aggregate will be either all yellow or all orange but not a mixture of the two.

COELENTERATES AND CTENOPHORES

In addition to fishes and whales which swim actively, open water contains many organisms floating passively with the water currents. They may swim, but cannot swim strongly enough to travel in a horizontal direction or to stay in one place against the current. These organisms are the **plankton.** Radiolaria and foraminifera are planktonic protozoans, and a variety of algae are planktonic plants. The largest of the plankton are jellyfish, often seen from shipboard as vast swarms in the upper meter or so of water. The common name, **jellyfish,** is applied to a heterogeneous group of organisms with bodies having a jelly-like consistency which belong to the phylum Coelenterata or the phylum Ctenophora. Coelenterate jellyfish have numerous tentacles with stinging cells and swim weakly by muscular contractions of the umbrella-shaped body. Ctenophores have two tentacles and move by the beating of numerous combs, each of which is a row of fused cilia. In both phyla a simple epithelium, the epidermis, covers the body and another epithelium, the gastrodermis, lines the gut. Between the two epithelia is a jelly-like **mesoglea** which forms the bulk of the body and may contain a few cells. Both groups are primarily carnivorous and catch worms, shrimp and fish. A single pelagic coelenterate is called a **medusa** and a ctenophore is called a **comb jelly.** The coelenterate phylum also includes a number of forms living on the bottom—hydras, sea anemones and corals—and floating colonies such as the Portuguese man-of-war.

Coelenterates are believed to have evolved from the same stock as all the higher animals, because, like the latter, they have a central digestive cavity connected to the outside by a mouth (the sponges, in contrast, lack it). The tissues of coelenterates fall into roughly the same categories as do those of higher animals: epithelial, connective, muscular, nervous and reproductive.

The body plan of the coelenterates is typified by *Hydra*, a tiny animal, which lives in ponds and looks to the naked eye like a bit of frayed string (Fig. 14–9). It takes its name from the multiheaded monster of Greek mythology that had the remarkable ability to grow two new heads for each head cut off. The coelenterate hydra has a comparable ability to regenerate. When it is cut into several pieces, each one may grow all the missing parts and become a whole animal. The body, seldom longer than 1 centimeter, consists of two layers of cells enclosing a central gastrovascular cavity, which performs both digestive and circulatory functions. The outer epidermis serves as a protective layer; the inner gastrodermis is primarily

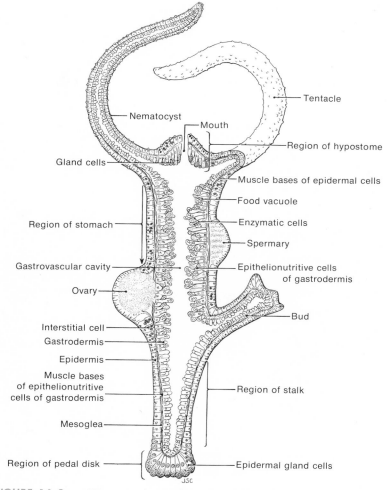

FIGURE 14–9   *Diagram of a Hydra cut longitudinally to show its internal structure. The sketch indicates both a spermary and an ovary but they occur on separate individuals; no single Hydra will have both an ovary and a spermary. Asexual reproduction by budding is represented on the right.*

a digestive epithelium. The bases of the cells of both layers are elongated into contractile muscle fibers; those of the epidermis run lengthwise, and those in the gastrodermis run circularly. By the contraction of one or the other, the hydra can shorten, lengthen or bend its body. Throughout its life the animal lives attached to a rock, twig or leaf by a disc of cells at its base. The mouth connects the gastrovascular cavity with the outside, and is surrounded by a circlet of tentacles, each of which may be as much as one and a half times as long as the animal's body. The tentacles, composed of an outer epidermis and an inner gastrodermis, may be hollow or solid.

Coelenterates are unique in producing ovoid "thread capsules" or **nematocysts** (Fig. 14–10) within stinging cells in the epidermis. When the nematocyst is stimulated it releases a coiled, hollow thread containing poison to paralyze the small animals which serve as prey. The nematocyst is shaped like a balloon, with a long tubular neck that develops tightly coiled and inside out within the cavity of the balloon. Each stinging cell has a small projecting trigger on its outer surface which responds to touch or to chemicals dissolved in the water ("taste") and causes the nematocyst to fire its thread. A nematocyst can be used only once; when it has been discharged, it is discarded and replaced by a new one, produced by the stinging cells.

The tentacles encircle the prey and stuff it through the mouth into the gastrovascular cavity, where digestion begins. The partially digested fragments are taken up by pseudopods of the gastrodermis cells, and digestion is completed within food vacuoles in those cells.

Respiration and excretion occur by diffusion, for the body of a hydra is small enough that no cell is far from the surface. The motion of the body, as it stretches and shortens, stirs the contents of the gastrovascular

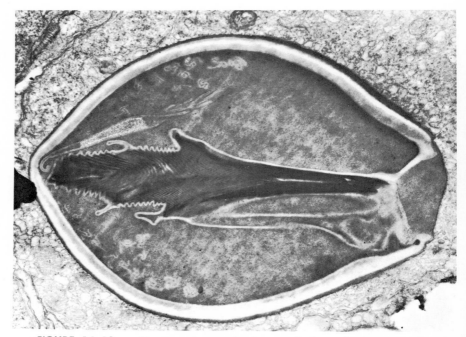

FIGURE 14–10   *Electron micrograph of an undischarged nematocyst of Hydra (sagittal section). (Courtesy of G. B. Chapman, Cornell University Medical College; from Lenhoff, H. M., and Loomis, M. F.: The Biology of Hydra. Coral Gables, Florida, University of Miami Press, 1961.)*

cavity. Some of the gastrodermis cells have flagella whose beating aids in circulation. The hydra has no other circulatory device.

The first true nerve cells in the animal kingdom are found in the coelenterates. These animals have many nerve cells, which form an irregular network and connect the sensory cells in the body wall with muscle and gland cells. The coordination achieved thereby is of the simplest sort; there is no aggregation of nerve cells to form a "brain" or spinal cord, and an impulse set up in one part of the body passes in all directions more or less equally.

Hydra reproduce asexually by budding when environmental conditions are optimal, but are stimulated to form sexual forms, males and females, when the pond water becomes stagnant. W. L. Loomis showed that the stimulus for the formation of sexual forms is an increased carbon dioxide tension in the water. Loomis showed further that males always bud to form other males, and females give rise by budding only to other females.

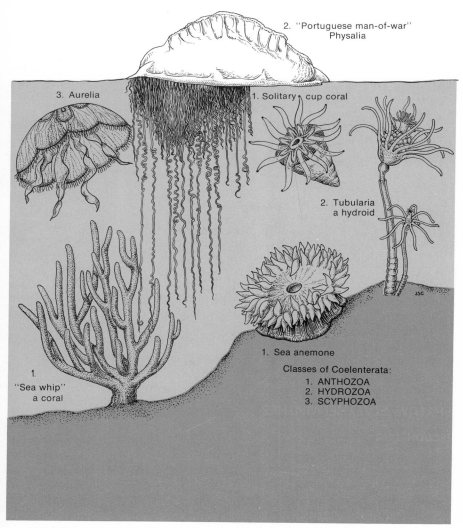

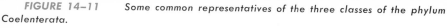

FIGURE 14–11   Some common representatives of the three classes of the phylum Coelenterata.

Besides the hydra and hydra-like organisms, the phylum Coelenterata includes such superficially different forms as jellyfish (Fig. 14–11), corals and sea anemones among its 10,000 different species. Both jellyfish and hydroids have bodies composed of an outer epidermis and inner gastrodermis, with a nonliving jelly (mesoglea) layer between them; in the hydra, the mesoglea layer is thin, whereas in the jellyfish it is thick and viscous, giving firmness to the body. The fundamental similarity between the two is illustrated in Figure 14–12. A jellyfish is like a hydra which has been turned upside down and whose mesoglea layer has been greatly increased. The hydra and jellyfish are, then, two ramifications of the same fundamental plan, one adapted for an attached life, the other for a free-swimming life.

Some of the marine coelenterates are remarkable for an alternation of sexual and asexual generations analogous to that in plants; however, both sexual and asexual generations of coelenterates are diploid organisms. Many species of jellyfish reproduce sexually to give rise to **planula** larvae which develop into sessile, sac-shaped, hydra-like animals (**polyps**). These, in turn, reproduce asexually to form new free-swimming jellyfishes, **medusae,** shaped like inverted bowls (Fig. 14–13). Many of the marine coelenterates form colonial organizations of hundreds or thousands of individuals. A colony begins with a single individual that reproduces by budding, but the buds, instead of separating from the parent, remain attached and continue to bud themselves. Several types of individuals may arise in the same colony, some specialized for feeding, others for reproduction.

The Portuguese man-of-war (*Physalia*, Fig. 14–11), which superficially looks like a jellyfish, is really a colony of hydroids and jellyfish. Its long tentacles are equipped with stinging capsules that can paralyze a large

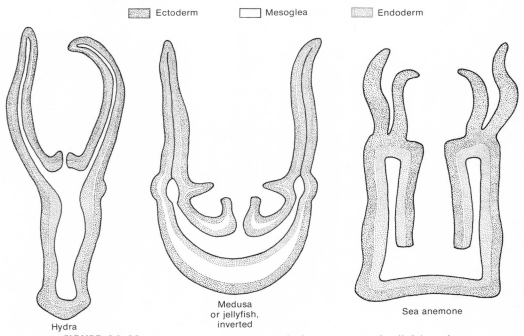

Ectoderm   Mesoglea   Endoderm

Hydra

Medusa
or jellyfish,
inverted

Sea anemone

FIGURE 14–12   *Diagrams comparing a hydra, an inverted jellyfish and a sea anemone to show the fundamental similarity of their structure.*

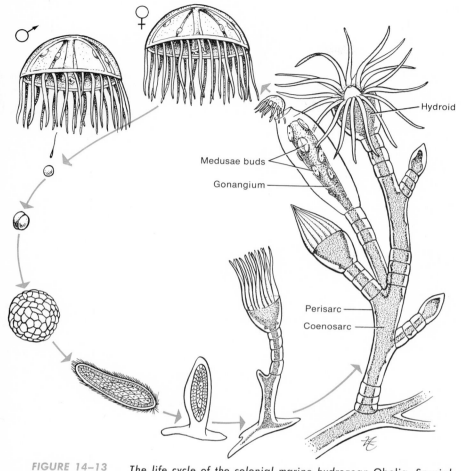

FIGURE 14–13 *The life cycle of the colonial marine hydrozoan Obelia. Special reproductive members of the colony produce medusae by asexual budding. Two types of medusae, one bearing testes and the other bearing ovaries are produced. Eggs and sperm are shed into the water where fertilization occurs. The zygote divides to produce a blastula, then a planula larva which swims about and finally becomes attached to some solid surface and develops into a new colony of polyps.*

fish and wound a man severely. The colony is supported by a gas-filled float of vivid, iridescent purplish-green.

The largest known jellyfish, *Cyanea,* may be 4 meters in diameter, and have tentacles 30 meters long. These orange and blue monsters, among the largest of the invertebrate animals, are a real danger to swimmers in the North Atlantic Ocean.

Sea anemones and corals have no jellyfish stage, and the polyps may be either individual or colonial forms. They differ from hydras in that the gastrovascular cavity is divided by a series of vertical partitions into a number of chambers, and the surface ectoderm is turned in at the mouth to line a gullet (Fig. 14–12). The partitions in the gastrovascular cavity increase the digestive surface, and an anemone can digest an animal as large as a crab or fish.

In warm shallow seas almost every square foot of the bottom is covered with coral or anemones, most of them brightly colored. The extravagant reefs and atolls of the South Seas are the remains of billions of microscopic, cup-shaped calcareous structures, secreted during past ages by

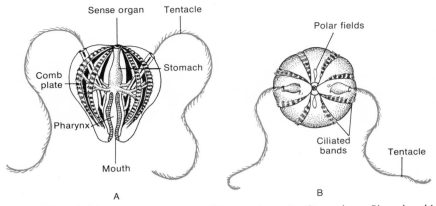

FIGURE 14–14   A, *Side view, and* B, *top view of a Ctenophore, Pleurobrachia.*

coral colonies and by coralline plants. Living colonies occur only in the uppermost part of such reefs, adding their own secretions to the mass.

The coelenterates are grouped in three classes: the **Hydrozoa,** which includes hydras, hydroids such as *Obelia,* and the Portuguese man-of-war; the **Scyphozoa,** which includes the jellyfish; and the **Anthozoa,** the sea anemones, true corals and alcyonarians, which include the precious corals.

CTENOPHORA.   The Ctenophores or comb jellies are similar in many ways to coelenterates, although the 100 or so species of the group are usually placed in a separate phylum. Their bodies, about the size and shape of an English walnut, consist of two layers of cells enclosing a mass of jelly. The outer surface is covered with eight rows of cilia, resembling combs, by which the animal moves through the water (Fig. 14–14). At the upper pole of the body is a sense organ, containing a mass of limestone particles balanced on four tufts of cilia connected to sense cells. When the body turns, these particles bear more heavily on the lower cilia, stimulating the sense cells, which cause the cilia to beat faster and bring the body back to its normal position. Nerve fibers running from the sense organ to the cilia control the beating. If they are cut, the beating of the cilia below the incision is disorganized. Ctenophores differ from coelenterates in lacking stinging capsules and in having only two branched tentacles instead of many.

Both coelenterates and ctenophores have remarkable powers of regeneration; a half, quarter or even smaller piece is able to grow into a whole animal. These animals also have a marked ability to return disarranged structures to their normal relationships. It is possible to turn a hydra inside out by pulling the base out through the mouth. The hydra, though unable to turn itself "inside in," does restore the normal relations of epidermis and gastrodermis by the migration of the individual cells to their proper position.

SECTION 14–5

PHYLUM PLATYHELMINTHES

Flatworms live in fresh and salt water, creeping over rocks, debris and leaves. Like the hydra, flatworms have a single gastrovascular cavity (Fig. 14–15)—sometimes extensively branched—connected to the outside by a single opening, the mouth, on the middle of the ventral surface. In addition to an outer ectoderm (epidermis) and an inner endoderm (gastrodermis), the flatworm has a third layer, the mesoderm, which comprises most of the body. The Platyhelminthes are the simplest animals that have well developed **organs,** functional units made of two or more kinds of tissue, such as the muscular pharynx for taking in food, the eyes and the complex reproductive organs. The worm is bilaterally symmetrical and has definite anterior and posterior ends. It keeps one surface (the back or dorsal one) always upward as it crawls along. Locomotion is achieved partly by means of cilia on its under or ventral surface, and partly by undulatory muscular contractions, similar to those of the earthworm. Gland cells in the ventral epidermis secrete a slime which facilitates movement.

The commonest free-living flatworms are the planarians, found in ponds and quiet streams all over the world. The common American planarian is *Dugesia*, about 15 millimeters long, with what appear to be crossed eyes and flapping ears. Planarians are carnivorous and feed on living and dead small animals (Fig. 14–16). Flatworms can survive without food for months, gradually digesting their own tissue and growing smaller as time passes. As in the coelenterates, respiration takes place by diffusion. To secrete waste products, the flatworm has a branching network of fine tubes opening to the surface by pores, and ending in branches known as **flame cells** or protonephridia (Fig. 19–1). Each of the latter is a single, hollow cell containing a tuft of cilia which, in beating, resembles a flame. The motion of the cilia drives the excreted fluid along the tubes and out the pores. Planarians living in fresh water have the same problem of get-

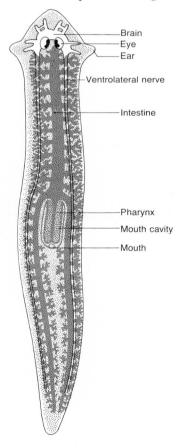

FIGURE 14–15   The common American planarian Dugesia. *The parts of the digestive system are shown in tint.*

Brain
Eye
Ear

Ventrolateral nerve

Intestine

Pharynx
Mouth cavity
Mouth

FIGURE 14–16   *Hunting and feeding in* Dugesia. *A small crustacean (*Daphnia*) is captured and eaten; its tough exoskeleton remains as an empty shell.*

ting rid of excess water faced by the fresh water protozoa, and the flame cells, like the contractile vacuole, solve it. Some flatworms confiscate intact nematocysts from the hydras they eat, incorporate them in their own epithelium and use them for defense.

Besides the free-living flatworms like *Dugesia*, which comprise the class **Turbellaria**, there are two classes of parasitic platyhelminthes, the **Trematoda**, or flukes, and the **Cestoda**, or tapeworms, both of which lack a cilated epidermis.

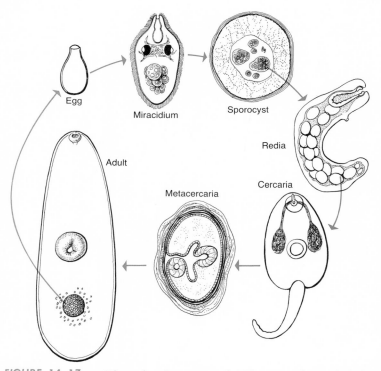

FIGURE 14–17   *Life cycle of a trematode. The miracidium which forms from a fertilized egg burrows into a snail, rounds up and becomes a sporocyst. Embryos within the sporocyst develop into redia which escape the sporocyst and feed on the tissue of the host snail. From reproductive tissue within the redia develop another set of embryos which grow into redia or into cercariae, which escape from the redia and are miniature flukes complete with a tail. The cercaria leaves the snail, swims to the next host, a crayfish, clam or fish, and bores into it, becoming surrounded by a cyst. There it develops into a meta-cercaria. When this is eaten by the final host it develops into an adult fluke and migrates to the liver or lungs of the final host. There it lays eggs which are fertilized and develop into the ciliated larval form, the miracidium, which completes the cycle.*

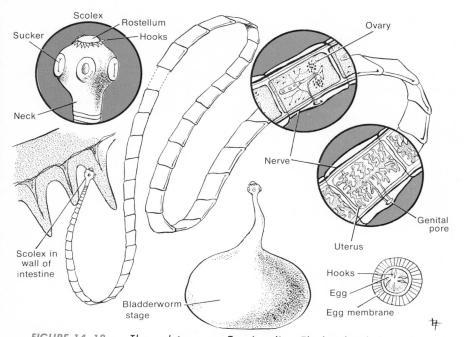

FIGURE 14–18   *The pork tapeworm* Taenia solium. *The head end of the tapeworm is equipped with suckers and hooks by means of which it is attached to the lining of the intestine (insert). Behind the head is a growing region which gives rise by budding to a succession of body sections or proglottids. The proglottids contain essentially nothing but a complete set of reproductive organs. Tapeworms have no mouth or digestive system but take up digestive materials from the intestinal cavity of their host.*

The flukes are structurally like the free-living flatworms, but differ in having one or more suckers with which to cling to the host and a thick outer layer, the cuticle, in place of the epidermis. The organs of digestion, excretion and coordination are like those of the other flatworms, but the reproductive organs are extremely complex. The flukes parasitic in human beings are the blood flukes, widespread in China, Japan and Egypt, and the liver flukes, common in China, Japan and Korea. Both of these parasites go through complicated life cycles, involving a number of different forms, alternation of sexual and asexual generations, and parasitism on one or more intermediate hosts such as snails and fishes (Fig. 14–17).

Tapeworms are long, flat, ribbon-like animals, some species of which live as adults in the intestines of probably every kind of vertebrate, including man. The head end of a tapeworm (Fig. 14–18) is equipped with suckers, and some species have a circle of hooks by means of which they may attach to the lining of the intestine. Behind the head a growing region constantly gives rise to new body sections, called **proglottids,** by budding. The rest of the body consists of a series of these sections, which contain little more than a complete set of reproductive organs. Tapeworms have no mouth and no trace of a digestive system; they live by soaking up the digested materials present in the intestine of their host.

Each proglottid mates with itself, or with an adjacent proglottid, becomes filled with fertilized eggs, each in its own capsule, breaks off and passes out of the host. The fertilized egg is eaten by another host, the larva hatches out of its capsule and continues development. Most cestodes must invade two or three species of hosts in succession to complete the life cycle.

SECTION 14–6

THE ORGAN SYSTEM LEVEL OF ORGANIZATION

THE NEMERTEA OR PROBOSCIS WORMS. This relatively small group of animals (550 species) is important as an evolutionary landmark, for the proboscis worms are the simplest living animals which illustrate the organ system level of organization (Fig. 14–19). None of them is parasitic and none is of economic importance. Almost all are marine, although a few species inhabit fresh water or damp soil. They have long narrow bodies, either cylindrical or flattened, varying in length from 20 centimeters to 20 meters. Some of them are a vivid orange, red or green with black or colored stripes. Their most remarkable organ—the **proboscis,** from which they take their name—is a long, hollow, muscular tube which is everted from the anterior end of the body and used for seizing food. It secretes mucus to catch and retain the prey. The proboscis of certain species is equipped with a hard point at the tip and poison-secreting glands at the base of this point. The pressure of the surrounding muscular walls on the contained fluid thrusts the proboscis outward; a separate muscle inside the proboscis retracts it.

Among the important evolutionary advances displayed by the nemertean is a complete digestive tract, with a mouth at one end for taking in food, an anus at the other for eliminating feces, and an esophagus and intestine in between. This is in contrast to the coelenterates and planarians, whose food enters and wastes leave by the same opening. In the proboscis worm, water and metabolic wastes are eliminated from the body by flame cells (protonephridia), as they are in the flatworms.

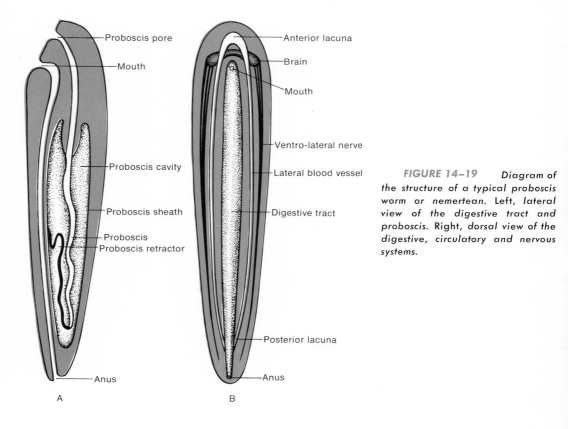

FIGURE 14–19   *Diagram of the structure of a typical proboscis worm or nemertean. Left, lateral view of the digestive tract and proboscis. Right, dorsal view of the digestive, circulatory and nervous systems.*

A second advance exhibited by the nemerteans is the separation of digestive and circulatory functions. These animals are the most primitive organisms to have a separate circulatory system. It is rudimentary, consisting of only three muscular tubes — the blood vessels — which run the length of the body and are connected by transverse vessels. Surprisingly, these primitive forms have red blood cells filled with **hemoglobin,** the same red pigment that transports oxygen in human blood. Nemerteans have no heart, and blood is circulated through the vessels by the movements of the body and the contractions of the muscular blood vessels. There are no capillaries. The nervous system is more highly developed than it is in the flatworm; there is a "brain" * at the anterior end of the body, consisting of two groups of nerve cells (ganglia) connected by a ring of nerves extending around the sheath of the proboscis; two nerve cords extend posteriorly from the brain.

THE NEMATODA.   The phylum **Nematoda,** the roundworms, has a great many members (about 8000 species), all of them remarkably similar in general body pattern. Some live in the sea, others in fresh water, in the soil, or in other plants or animals as parasites. About 50 species are human parasites, of which the most detrimental are the **hookworm, ascaris** (Fig. 14–20), **trichina** (Fig. 14–21), **filaria worm** and **guinea worm.** A microscopic examination of a shovelful of earth from almost anywhere in the world will reveal a number of tiny white worms which thrash around, coiling and uncoiling. Their long, cylindrical, threadlike bodies, pointed at both ends, are covered with a tough cuticle. A feature of nematode anatomy is the presence of a primitive body cavity, the **pseudocoelom,** between the body wall and gut wall. It is a derivative of the embryonic blastocoele and lies between the endoderm and mesoderm. A true coelom is a body cavity lying within tissues of mesodermal origin, and lined with a simple epithelium of mesodermal origin. In contrast to the nemerteans, which have cilia all over the epithelium and the lining of the digestive tract, none of the nematodes has any cilia at all. Nematodes have no circularly arranged muscle fibers, only longitudinal ones. They can only bend, and they swim poorly despite vigorous thrashing movements.

With the evolution of a complete digestive system, a separate circulatory system, and a nervous system composed of a "brain" and nerve cords, as illustrated in the proboscis worm, the essential structures of higher animals were established. The proboscis worm is not the actual ancestor of the higher forms, but is thought to be like the now extinct ancestor common to both the higher animals and itself. Evolution beyond this point branched out in a great many different directions, and the more advanced animals cannot be arranged in a single series of progressively higher and more complex forms. One main branch of evolution led to the vertebrates, another to the insects and other arthropods, and another to the clams, squids and other mollusks.

THE ROTIFERA.   Among the more obscure invertebrates are the "wheel animals" of the phylum Rotifera. These aquatic, microscopic worms, although no larger than many of the protozoa, have many-celled bodies with a complete digestive tract including a **mastax,** a muscular organ for grinding food; a pseudocoelom; an excretory system made up of flame cells

---

*The word "brain" is loosely applied to the aggregation of nerve cells at the anterior end of the nerve cord, which acts as a reflex center. It should not be inferred that anything like thought processes occurs in any of the lower animals.

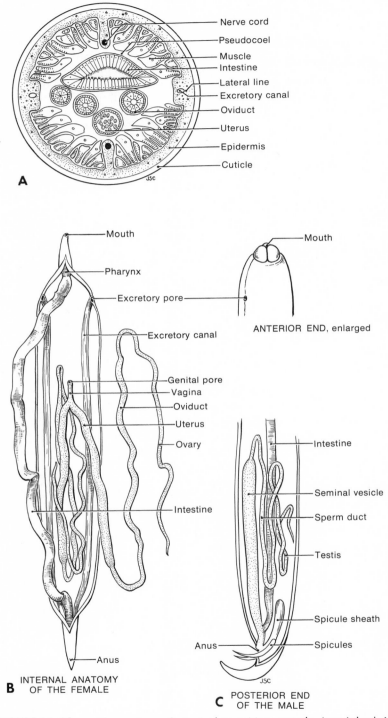

FIGURE 14–20   The anatomy of a typical parasitic nematode, Ascaris lumbricoides.
A, Transverse section showing arrangement of internal organs. B, Female worm dissected
open to show reproductive tract. C, Posterior end of male worm dissected open to show
reproductive organs.

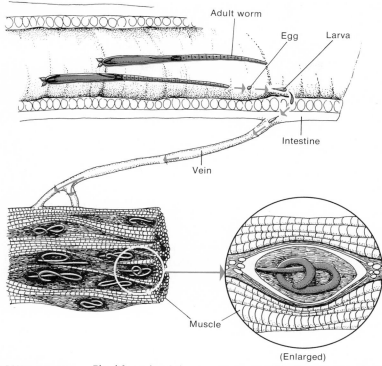

FIGURE 14-21   *The life cycle of the round worm Trichinella spiralis, which causes trichinosis. When a piece of pork infected with Trichinella is eaten and digested, the larvae are released and grow rapidly to maturity in the intestine. After fertilization, the females produce tiny larvae which burrow into the blood vessels and are carried to the muscles where they encyst (insert). When that muscle is eaten by the next host the cycle is repeated. Usually the pig becomes infected by eating a rat that had eaten scraps of infected pork.*

and a bladder; a nervous system with a "brain" and sense organs; and a characteristic crown of cilia on the head end which gives the appearance of a spinning wheel (Fig. 14–22).

Rotifers and gastrotrichs are "cell constant" animals: Each member of a given species is composed of exactly the same number of cells as every other one; indeed, each part of the body is made of a precisely fixed number of cells arranged in a characteristic pattern. Cell division ceases with embryonic development and mitosis cannot subsequently be induced; growth and repair are impossible. One of the challenging problems of biology is the

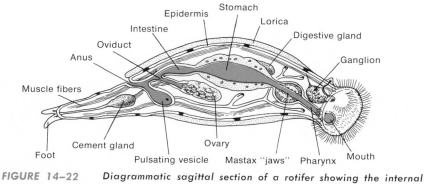

FIGURE 14-22   *Diagrammatic sagittal section of a rotifer showing the internal structures.*

nature of the difference between such nondividing cells and the dividing cells of other animals.

THE GASTROTRICHA.   The gastrotrichs are another phylum of microscopic, aquatic, cell-constant worms, like rotifers in many respects, but without a crown of cilia. Gastrotrichs move quickly, propelled by two longitudinal bands of cilia on their ventral surface. The fresh-water gastrotrichs are peculiar in that they consist entirely of females which reproduce by parthenogenesis; no males of this group have ever been found.

THE BRYOZOA.   The Bryozoa, or moss animals, live in colonies that superficially resemble those of the coelenterates (Fig. 14–23). The colonies of some species, delicately branched and beautiful, are sometimes mistaken for seaweed; other species form colonies which appear as thin, lacy incrustations on rocks. Each minute animal secretes about itself a protective case of calcium carbonate or of a horny, protein material, into which it can withdraw when danger threatens. This case may be shaped like a vase, a box or a tube. Around the animal's mouth is a circular or horseshoe-shaped ridge, called a **lophophore,** with a set of ciliated tentacles. An adaptation to living in a "vase" is the U-shaped digestive system. In one phylum, the **Entoprocta** (the anus lies within the ring of tentacles), the body cavity is a pseudocoelom and wastes are excreted by protonephridia. In the other phylum of bryozoa, the **Ectoprocta** (the anus lies outside of the ring of tentacles), the body cavity is a true coelom, and no excretory system is present. In both phyla, new members of the colony arise by bud-

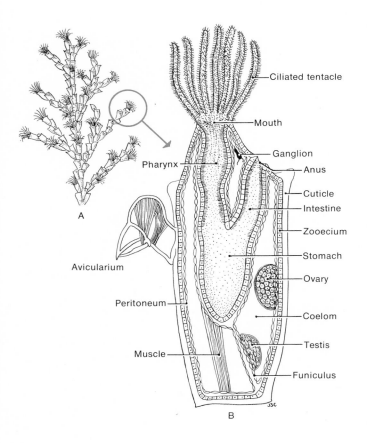

FIGURE 14–23   *A colonial ectoproct bryozoan, Bugula. The inset shows one animal cut open longitudinally to reveal the internal structures.*

ding from the older ones, and new colonies arise as the result of sexual reproduction during certain seasons.

The ectoproct bryozoan colony has specialized members, called **avicularia,** some of which resemble the head of a bird. These organisms are in constant motion from side to side and, as they move, a peculiar organ, shaped like a bird's lower beak and operated by muscles, frequently snaps open and shut. The purpose of the avicularia is not to catch food—other members do that—but to keep small animals from settling on the colony. The evolutionary relationships of both ectoproct and entoproct bryozoa are uncertain.

THE BRACHIOPODA.   Another phylum characterized by a lophophore includes the animals commonly known as lampshells, which superficially resemble clams. Like the latter, they have two shells, usually calcareous, that can be opened and closed by muscles. But unlike the clam, whose two shells are on the right and left sides of the body, the brachiopod shells lie above and below it, the larger ventral shell being attached to a rock or other object by a sturdy, muscular stalk. All brachiopods live in the sea. Although only about 200 species of this extremely ancient phylum exist today, there once were more than 3000 species. Because of their great age and well preserved hard shells, the fossil brachiopods are useful to geologists in determining the age of rocks. Fossils obtained from rocks more than 500,000,000 years old are almost exactly like the brachiopods living today. The genus *Lingula* is represented by both fossil and living forms; it is the oldest known genus that has living members.

SECTION 14–7

ADAPTING TO TERRESTRIAL LIFE

The "higher" invertebrates—the annelids, arthropods, mollusks and echinoderms—all have a separate mouth and anus, a muscular gut, a well developed circulatory system and a true **coelom,** a cavity within the mesoderm lined by peritoneum. The coelom is formed during development either by a splitting within originally solid masses of mesoderm (a **schizocoelom,** typically found in mollusks, annelids and arthropods) or by pouches that bud off from the original gut cavity (an **enterocoelom,** typically found in echinoderms and chordates), but there are notable exceptions to both of these generalizations.

Of these four phyla, only the arthropods are very successful terrestrial animals. The earthworm is terrestrial but most annelids are marine; there are a few land snails, but most mollusks live in the sea; all the echinoderms are marine. Of the five classes of arthropods, one, the Crustacea—crabs, lobsters and so on—includes largely marine forms, but the other four—insects, spiders, centipedes and millipedes—are mostly terrestrial animals. From the fossil record we know that the first air-breathing land animals were scorpion-like arachnids that came ashore in the Silurian, some 410,000,000 years ago. The first land vertebrates, the amphibians, did not appear until the latter part of the Devonian, some 60,000,000 years later.

In evolving to become adapted to terrestrial life, animals, like plants (Chap. 13), had certain problems to solve for survival in the absence of a surrounding watery medium. The chief problem for all land organisms is that of preventing desiccation. Reproduction provides a second problem: Aquatic forms can shed their gametes into the water and fertilization will occur there; the delicate embryos that result are protected by the surrounding water as they begin development. Land plants, by pollination, transfer sperm nuclei to the egg in the absence of a watery medium, and the developing embryo is protected by the tissues of the parent gametophyte or by seed coverings. Some land animals — most amphibians — return to the water for reproduction and the young forms — larvae or tadpoles — develop in the water. Earthworms, insects, snails, reptiles, birds and mammals transfer sperm from the body of the male directly to the body of the female by **copulation;** the sperm are surrounded by a watery medium or semen. The fertilized egg either is covered by some sort of tough, protective shell secreted around it by the female or it develops within the body of the mother. The problem of supporting a body against the pull of gravity in the absence of the buoyant effect of water is not too acute for small animals such as earthworms that burrow in the ground, but the larger ones and the above-ground ones need some sort of skeleton. Arthropods and mollusks evolved one on the outside of the body (an **exoskeleton**) and vertebrates have one within the body (an **endoskeleton**). Land forms are subject to much wider variations in temperature than marine organisms, for the ocean acts as a great, constant temperature bath. The deep water varies only a few degrees from summer to winter and even a small lake has less drastic temperature fluctuations than the air over it does. Land animals are exposed to both higher and lower temperatures than aquatic ones and had to evolve suitable adaptations to survive.

With all these disadvantages it might seem incredible that any land forms *did* evolve. However, one of the major tendencies in evolution is for organisms to become diversified and to spread into new types of environment. Wherever the environment can support life at all, some form of life, suitably adapted for survival there, will eventually evolve. A land environment is not without some advantages, however. Once land plants had evolved, the land offered the first animals an environment with a plentiful supply of food, no predators and few competitors.

The tough exoskeleton that evolved in arthropods and mollusks serves several purposes — it provides stiffening, enabling the body to stand against the pull of gravity; it serves as a point of attachment for muscles; it provides protection against desiccation; and it serves as a coat of armor to protect the animal against predators. Thus the evolution of an exoskeleton solved many of the problems of survival on land.

SECTION 14–8

THE ANNELIDA

One of the more familiar invertebrate animals is the earthworm, a member of the phylum Annelida (Fig. 14–24). This word, which means "ringed," refers to the fact that the body of the worm consists of a series of rings or **segments.** Both the internal organs and the body wall are segmented. The body is a bilaterally symmetrical tube composed of about 100 more or less similar units, each of which contains one or a pair of organs of each system. The segments are separated from each other by transverse, bulkhead-like partitions, the **septa.** The chief evolutionary advance shown by the earthworms over the lower forms is this development of segmentation, for each segment constitutes a subunit of the body that may be specialized to carry on a particular function. The dividing of the body into segments is thus similar, on a larger scale, to the original division of the

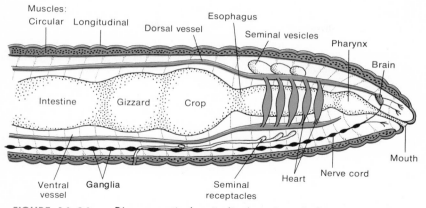

Muscles:
Circular   Longitudinal   Dorsal vessel   Esophagus   Seminal vesicles   Pharynx   Brain

Intestine   Gizzard   Crop

Ventral vessel   Ganglia   Seminal receptacles   Heart   Nerve cord   Mouth

FIGURE 14-24   *Diagrammatic longitudinal section of the anterior part of an earthworm showing internal structures. The circulatory system is indicated by tint.*

animal body into cells to provide for local specialization. In the earthworm the individual segments are almost all alike, but in many of the other segmented animals—the arthropods and chordates—the specialization of different segments reaches a point where the segmentation of the body plan is obscured.

The earthworm's body is protected from desiccation by a thin, transparent cuticle, secreted by the cells of the epidermis. The glandular cells of the epidermis secrete mucus which forms an additional protective layer over the skin. The body wall contains an outer layer of circular muscles and an inner layer of longitudinal muscles. Each segment, except the first, bears four pairs of bristles, **chaetae,** supplied with small muscles that can move the chaeta in and out and change its angle. The earthworm moves forward by contracting its circular muscles to elongate the body, grasping the ground or walls of the burrow with its chaetae, and then contracting its longitudinal muscles to draw the posterior end forward; locomotion occurs in waves.

The body cavity of the annelids is a large well developed true **coelom.** The body consists of two tubes; the outer tube is the body wall, and the inner tube, the wall of the digestive tract. The coelom is filled with a fluid which bathes the internal organs and is an intermediate, in transporting gases, nutrients and wastes, between the circulatory system and the individual cells of the body.

The digestive system of the earthworm shows several advances over that of the proboscis worm: there is a muscular pharynx for swallowing food, an esophagus and a stomach of two parts—a thin-walled **crop,** where food is stored, and behind it, a thick-walled muscular **gizzard,** where it is ground to bits. The rest of the digestive system is a long, straight intestine, where digestion and absorption take place, terminating in an anus which opens to the outside at the posterior end.

The circulatory system, more complex and efficient than that of the proboscis worm, consists of two main vessels. One, just dorsal to the digestive tract, collects blood from numerous segmental vessels. It is contractile and pumps the blood anteriorly. The other, in which blood flows posteriorly, lies just below the digestive tract and distributes blood to the various organs. In the region of the esophagus, the dorsal and ventral vessels are connected by five pairs of muscular tubes, called "hearts," which propel the blood to the ventral vessel. There are also smaller lateral and ventral distributing vessels, and tiny capillaries in all the organs as well as in the body wall.

The excretory system is composed of paired organs repeated in almost every segment of the body. Each individual organ, called a **metanephridium,** consists of a ciliated funnel opening into the next anterior coelomic cavity and connected by a tube to the outside of the body. Wastes are removed from the coelomic cavity partly by the beating of the cilia and partly by currents set up by the contraction of muscles in the body wall. The tube of the excretory organ is surrounded by a capillary network, so that wastes are removed from the bloodstream as well as from the coelomic cavity. The metanephridia, open at both ends, are quite different from the protonephridia of the lower invertebrates, which are blind tubules opening only to the exterior. Although the adults of the higher invertebrates usually have metanephridia, their larval forms usually have protonephridia as excretory organs, typically with a single long flagellum in place of the tuft of cilia. This is consistent with the theory that the higher invertebrates evolved from the lower ones.

Earthworms (and all oligochaetes) are **hermaphroditic;** each individual contains both male and female reproductive organs. Segments 10 and 11 each contain a pair of **testes** located in isolated coelomic cavities, sperm reservoirs. These have three pairs of prominent lateral pouches, the **seminal vesicles** (Fig. 14–24) that extend into segments 9, 10 and 11. Sperm produced in the testes are stored in the reservoirs and vesicles. Two pairs of **sperm funnels** collect sperm from the reservoirs and pass them through a pair of sperm ducts to the male pores on the ventral side of segment 15.

A single pair of tiny **ovaries,** in segment 13, shed their eggs into the coelomic cavity. They are collected by a pair of **egg funnels** into short **oviducts** that open via female pores on the ventral side of segment 14. Two pairs of **seminal receptacles,** in segments 9 and 10, store sperm received during copulation.

During copulation two worms, heading in opposite directions, press their ventral surfaces together (Fig. 14–25) and are held by thick mucous

FIGURE 14–25   Two earthworms copulating. (Photograph of living animals made at night, courtesy of General Biological Supply House, Chicago, Illinois.)

secretions of the **clitellum,** a thickened ring of epidermis in segments 32 to 37. Sperm from one worm pass posteriorly to its clitellum and are stored in the seminal receptacles of the second worm. The worms separate and the clitellum secretes a membranous cocoon containing an albuminous fluid. As the cocoon is slipped over the worm's head, eggs are laid into it from the female pores and sperm are added as the cocoon passes the seminal receptacles. When the cocoon is free, its openings constrict so that a spindle-shaped capsule is formed, and the eggs develop into tiny worms within the cocoon. This complex reproductive pattern is an adaptation to terrestrial life.

The nervous system, too, is more advanced than that of the proboscis worm, for it consists of a large, two-lobed aggregation of nerve cells located just above the pharynx in the third segment, and another ganglion just below the pharynx in the fourth segment. A ring of nerves around the pharynx connects the two ganglia. From the lower ganglion a nerve cord (actually, two closely united cords) extends the entire length of the body, beneath the digestive tract. In each segment there is a swelling of the nerve cord, a **segmental ganglion,** from which nerves extend laterally to the muscles and organs of that segment. The segmental ganglia coordinate the contraction of the muscles of the body wall, so that the worm can creep along. The nerve cord contains a few **giant axons** which transmit nerve impulses more rapidly than ordinary fibers. These, when danger threatens, stimulate the muscles to contract and draw the worm back into its burrow. Giant axons from annelids, squids and certain arthropods have been extensively used in studies of the mechanism of nerve conduction (p. 493). Living a subterranean life, the earthworm has no well developed sense organs, but some of its sea-dwelling relatives, such as the clamworm, *Nereis,* have two pairs of eyes and organs sensitive to touch and to chemicals in the water. The activities of the earthworm are governed by the two ganglia above and below the pharynx, the **brain** and **subpharyngeal ganglion.** Removal of the brain results in increased bodily activities and removal of the subpharyngeal ganglion eliminates all spontaneous movements. This is evidence of functional specialization of the nervous system; the brain is in part an **inhibitory center** and the subpharyngeal ganglion is a **stimulatory center.**

Besides serving as bait and as food for birds, earthworms turn over the soil, making it more porous and allowing air to penetrate freely and water to drain properly. They facilitate the growth of roots and hence increase the agricultural yield. In their burrowings through the earth, the worms swallow bits of soil, grind them to smaller bits, absorb some of the organic matter as food, and discharge the residue at the surface. The amount of earth moved in this way is remarkable; Charles Darwin calculated that every 10 years earthworms turn over enough soil to form a layer 5 centimeters thick over the entire land surface of the world.

The phylum Annelida contains some 10,000 species divided into four classes. One of these is the **Polychaeta** (many bristles), made up of marine worms which swim freely in the sea, burrow in the sand and mud near shore, or live in tubes formed by secretions from the body wall. Each segment of their bodies has a pair of thickly bristled paddles (called **parapodia**) extending laterally (Fig. 14–26). The anterior end of the body is a well developed "head" or **prostomium,** bearing eyes, antennae and a pair of lateral palps. Most species of polychaetes are predators, and all have separate sexes.

The eggs and sperm of polychaetes are released into the sea water where fertilization occurs. Many of these have evolved behavioral mechanisms that ensure fertilization. By responding to certain rhythmic variations in the environment, nearly all the males and females of a given spe-

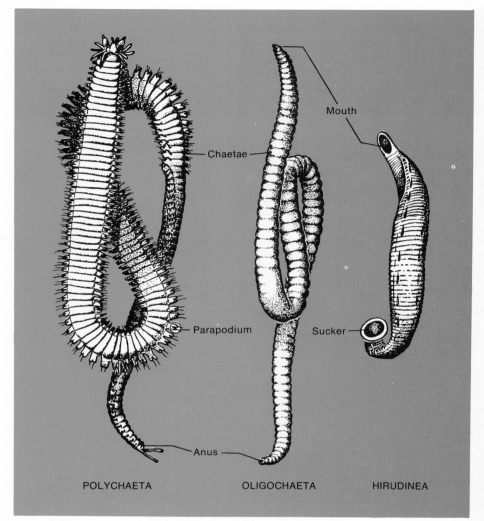

Chaetae

Mouth

Parapodium

Sucker

Anus

POLYCHAETA   OLIGOCHAETA   HIRUDINEA

FIGURE 14–26   *Representatives of three classes of the phylum Annelida. Left, a polychaete Nereis virens, the clamworm. Center, an oligochaete, Lumbricus terrestris, the earthworm. (From Villee, Walker and Smith: General Zoology. 3rd ed. Philadelphia, W. B. Saunders Company, 1968. After Lawson et al., 1955.) Right, Hirudo medicinalis, the medicinal leech, an example of the class Hirudinea. (From Villee, Walker and Smith: General Zoology. 3rd ed. Philadelphia, W. B. Saunders Company, 1968. After Hegner, 1935.)*

cies release their gametes at the same time. Seasonal cycles produce variations in temperature, length of day and amount of food; lunar cycles produce variations in the height of tides, strength of currents, the relation between time of tide and hour of day, and the amount of light at night; and diurnal cycles produce great variations in light from day to night. Over 99 per cent of the individuals of the population of **Palolo worms,** a species of polychaete living on coral reefs in the South Pacific, shed their eggs and sperm within a single two-hour period on one night of the year. The seasonal rhythm limits the reproductive period to November, the lunar rhythm to a day during the last quarter of the moon when the tide is unusually low, and the diurnal rhythm to a few hours just after complete darkness. The posterior half of the Palolo worm, loaded with gametes, actually breaks off from the rest, swims backward to the surface and eventually bursts, releasing the eggs or sperm so that fertilization may occur.

The 2000 species of the class **Oligochaeta** (to which the earthworm belongs) have few bristles per segment and are found almost exclusively in fresh water and in moist terrestrial habitats.

The **Archiannelida,** a small group of simple marine worms, are not segmented externally and do not have bristles.

Hirudinea, the leeches (Fig. 14–26), are provided with stout muscular suckers at the anterior and posterior ends for clinging to their prey. They differ from other annelids in having neither chaetae nor appendages. Most leeches feed by sucking the blood of vertebrates; they attach themselves by their suckers, bite through the skin of the host, and suck out a quantity of blood, which is stored in pouches in the digestive tract. An anticoagulant (**hirudin**), secreted by glands in its crop, ensures the leech a full meal of blood. Their meals may be infrequent, but they can store enough food from one meal to last a long time. The so-called "medicinal leech" is a fresh-water worm about 10 centimeters long which physicians used for bloodletting when the humoral theory of disease was in vogue. The archiannelids and polychaetes appear to represent one branch of annelid evolution and the oligochaetes and leeches another. The development of polychaetes and archiannelids is characterized by a larval form, called a **trochophore** (Fig. 14–27), which is very similar to the larva of mollusks.

It is generally believed that the annelids and arthropods developed from a common segmented ancestor, a theory substantiated by the existence of a curious animal called **peripatus** (Fig. 14–28), found in the moist, tropical forests of Africa, Australia, Asia and South America. This caterpillar-like creature, 5 to 8 centimeters in length, appears to be a connecting link between the two phyla. It is believed, however, not to be the ancestor of the present-day arthropods, but rather a relatively unchanged descendant of the original ancestor of both annelids and arthropods. Its anatomy is a mixture of the characteristics of both. It has many pairs of legs, each of which has a pair of claws at the tip. Its excretory, reproductive and nervous systems are similar to those of annelids, but its circulatory system and its respiratory system, which consists of air tubes (tracheal tubes), are arthropod-like. Most zoologists classify peripatus and related species as a separate phylum (Onychophora); others classify them with the annelids or with the arthropods.

Peripatus provides a possible link between annelids and the terrestrial

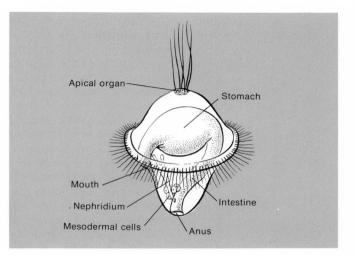

FIGURE 14–27   The trochophore larva of a polychaete.

FIGURE 14–28   *Photograph of Peripatus, a member of the phylum Onychophora, with structural features intermediate between those of the Annelida and the Arthropoda. (Courtesy of Ward's National Science Establishment.)*

arthropods but is quite unlike the trilobites. In 1930, 11 well preserved fossils from Cambrian deposits were found of a marine animal named *Aysheaia* similar in many respects to peripatus. Probably peripatus and *Aysheaia* are two representatives of a varied and widespread ancient group that had developed many arthropod-like characteristics before the arthropods arose.

SECTION 14–9

THE ARTHROPODA

The animals that make up this phylum are, without doubt, the most successful, biologically, of all animals. There are more of them, about 1,000,000 species, of which some 800,000 are insects. They live in a greater variety of habitats and eat a greater variety of food than the members of any other phylum. The word "arthropod" refers to the paired, jointed appendages characteristic of these animals. These are used as swimming paddles, walking legs, mouth parts or accessory reproductive organs for transferring sperm. All arthropods have segmented bodies covered by a hard chitinous exoskeleton secreted by the underlying epithelium. The cuticle has an outer, waterproof, waxy layer; a rigid, middle layer; and a flexible, inner layer of **chitin** (a polysaccharide composed of units of acetyl glucosamine). The rigid layer is thin in certain regions, such as the joints of the legs and between the body segments, allowing the cuticle to be bent. Such a layer provides protection against excessive loss of moisture and predators, and gives support to the underlying soft tissues. But it has disadvantages, too: body movement is somewhat restricted, and, in order to grow, the arthropod must shed the outer shell periodically and grow another larger one, a process which leaves him temporarily vulnerable.

GENERAL BODY PLAN OF ARTHROPODS

The bodies of most arthropods are divided into three regions: the **head,** always composed of exactly six segments, the **thorax** and the **abdomen,** both of which are composed of a variable number of segments. In contrast to most annelids, each arthropod has a fixed number of segments which remains constant throughout life. The variations in body plan and in the shape of the jointed appendages in the numerous species are almost indescribable! The nervous system of the more primitive arthropods, like that of the annelids, consists of a ventral nerve cord connecting segmental ganglia, but in the more complex arthropods the successive ganglia usually fuse together. Arthropods have a variety of well developed sense organs: complicated eyes, such as the compound eyes of insects; organs located in the antennas, which are sensitive to touch and to chemicals; organs of hearing; and touch cells on the surface of the body.

The true coelom is small and is made up chiefly of the cavities of the reproductive system. The large body cavity is not a coelom, but a **hemocoel,** a blood cavity—part of the circulatory system. The latter includes, besides the enclosed vessels, open spaces throughout the body by means of which the organs are bathed. There is a pumping organ, or "heart," in the dorsal part of the body which stirs the blood around in these spaces. Most of the aquatic arthropods have a system of gills for external respiration, whereas the land forms usually have a system of fine, branching air tubes or **tracheae,** which conduct air to the internal organs. The digestive system typically is a simple tube like that of the earthworm, lined in part with a cuticle similar to the outer covering of the body. In insects and some other forms, the excretory system consists of tubules which empty into the digestive tube. These metabolic wastes then pass out of the body with the feces through the anus.

CLASSES OF ARTHROPODS

The most primitive arthropods, the **Trilobita,** were marine arthropods, abundant in the Paleozoic era; they became extinct about 225,000,000 years ago. From fossil remains, some 3900 species of trilobites have been described; most of these lived on the sea bottom and walked on or dug into the sand and mud. They ranged in length from a millimeter to nearly a meter, but most were between three and 10 centimeters long. Their body was a flattened oval divided into three parts (Fig. 14–29)—an anterior **cephalon** of four fused segments bearing a pair of antennae and a pair of compound eyes; a **thorax,** consisting of a varying number of segments; and a posterior **pygidium** composed of several fused segments. The body was divided further into a median lobe and two lateral lobes by two furrows which extended from the anterior to the posterior end of the animal body. Each segment of the body had a pair of biramous (two-branched), segmented appendages. The inner branch of each appendage was used for walking and the outer branch served as a gill.

It is remarkable that fossil evidence has yielded information not only about the structure of the adult but also about the developmental stages of the trilobites. The trilobites passed through three larval periods, during each of which the larvae underwent several molts. As the successive molts

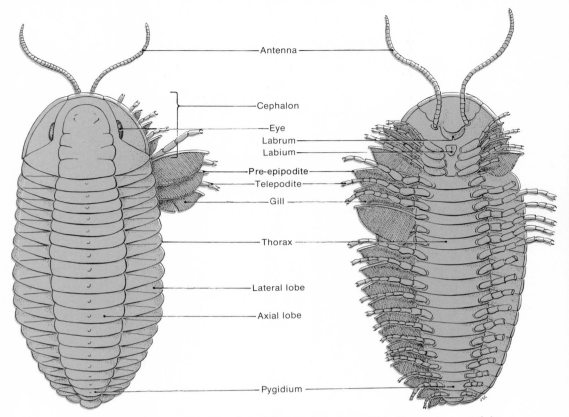

Antenna

Cephalon

Eye
Labrum
Labium

Pre-epipodite
Telepodite

Gill

Thorax

Lateral lobe

Axial lobe

Pygidium

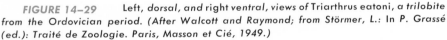

FIGURE 14–29   Left, *dorsal, and right ventral, views of Triarthrus eatoni, a trilobite from the Ordovician period. (After Walcott and Raymond; from Störmer, L.: In P. Grassé (ed.): Traité de Zoologie. Paris, Masson et Cié, 1949.)*

occurred, additional segments were added to the body, and the body structure became more complex. The trilobites have some characteristics in common with the crustacea and others in common with the arachnids and horseshoe crabs.

The centipedes, **Chilopoda,** and the millipedes, **Diplopoda,** are similar in having a head and an elongated trunk with many segments, each bearing legs (Fig. 14–30). All of them are terrestrial; they are typically found beneath stones or wood or in the soil in both temperate and tropical regions. The centipedes have one pair of legs on each segment behind the head. Most centipedes have many fewer than a hundred legs—the common numbers being in the thirties—although there are a few species with enough legs to merit the term "centipede." The legs of centipedes are rather long and enable them to run at a good pace. Centipedes are carnivorous and feed upon other animals, mostly insects; but the larger centipedes have been known to eat snakes, mice and frogs. The prey is captured and killed with poison claws located just behind the head on the first trunk segment. At the base of the claws is a pair of poison glands which empty into ducts that open at the tip of the pointed, fanglike claw. Centipedes breathe by means of a series of air tubes or **tracheae** that open to the exterior through openings called **spiracles.**

The millipedes or "thousand-leggers" are herbivorous scavengers that live beneath leaves, stones and logs. The distinguishing feature of the class is the presence of doubled-trunk segments, resulting from the fusion

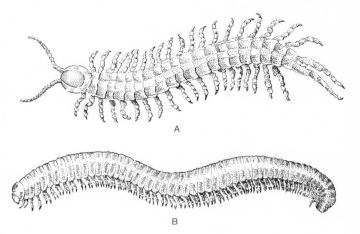

FIGURE 14–30   A, A centipede, a member of the class Chilopoda. Centipedes have one pair of
appendages per segment. B, A millipede, a member of the class Diplopoda. Millipedes have two pairs of
appendages per segment. (From
Hunter, G. W., and Hunter, F. R.:
College Zoology. Philadelphia, W. B.
Saunders Company, 1949. After
Koch.)

of two original somites. Each double segment has two pairs of legs and two
pairs of ganglia. The body of the millipede tends to be cylindrical, whereas
the body of the centipede tends to be flattened. Diplopods are not as agile
as chilopods, and most species can crawl only slowly over the ground. Millipedes also breathe by tracheae that open through spiracles. In both chilopods and diplopods, eyes may be completely lacking or the animals may
have simple eyes (ocelli). A few species of centipedes have eyes that are
similar to the compound eyes of insects, composed of a group of up to
100 optical units on each side of the head.

The **Crustacea** differ from other arthropods in having two pairs of
antennae or sensory feelers, a pair of **mandibles,** and two pairs of **maxillae**
on their heads. They usually have compound eyes. The 26,000 species of
crustacea include some familiar animals such as crabs, shrimp, lobsters
and crayfish (Fig. 14–31), plus thousands of species of less familiar ones.
The crustacea are the only class of arthropods that are primarily aquatic.
Most crustaceans are marine, but some live in fresh water. There are a few
species like the hermit crab of the Caribbean that survive for extended periods of time on land in a moist environment. Crustaceans are either

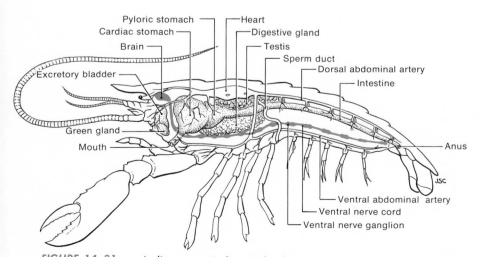

FIGURE 14–31   A diagrammatic longitudinal section of a male crayfish, a fresh-
water animal similar to a lobster. The brain and nervous system are shown in tint.

carnivores, scavengers or filter feeders. Certain appendages in the latter have fine hairs (setae) that function as a filter to collect small particles of food, which are then removed from the setae by other hairs and transferred to the mouth parts.

One of the more familiar crustaceans is the lobster, a decapod or "ten-legged" form. The six segments of the lobster's head and the eight segments of the thorax are fused into a **cephalothorax** which is covered on the top and sides by a shield, the **carapace,** composed of chitin impregnated with calcium salts. The two pairs of antennae are the sites of chemoreceptors and tactile sense organs; the second pair of antennae are especially long. The mandibles are short and heavy with opposing surfaces for grinding and biting. Behind the mandibles are two pairs of accessory feeding appendages, the first and second maxillae. The appendages of the first three segments of the thorax are **maxillipeds,** which aid in chopping up food and passing it to the mouth. The fourth segment of the thorax has a pair of large **chelipeds** or pinching claws, and segments 5 through 8 have pairs of **walking legs.** The appendages of the first abdominal segment are part of the reproductive system and function in the male to transfer sperm. On the following four segments of the abdomen are paired **swimmerets,** small paddle-like structures used for swimming. The last two segments bear the **uropods** and the **telson** which make up a fan-shaped tail used for swimming backwards.

Crustacean respiration is carried out generally by gills, usually attached to the proximal segment of most of the appendages. Crustaceans have an open circulatory system with a heart and arteries that end in the hemocoel, large blood-filled spaces that ramify through most parts of the body. The blood of the lobster contains a bluish pigment, **hemocyanin,** for the transport of oxygen.

Although lobsters, crabs and shrimps are the most familiar of the crustaceans, they are not the most important in the overall economy of nature. There are countless billions of microscopic crustaceans that swarm in the ocean, lakes and ponds, and form the food of many fish. The principal food of some of the largest whales is "krill," marine crustacea less than 25 millimeters long. The **Branchiopoda** include a variety of small shrimp-like forms—fairy shrimps, tadpole shrimps, water fleas—found mostly in fresh water. The **Ostracoda,** or mussel shrimps, are other minute crustaceans found in the sea and in fresh water. The ostracods are characterized by the presence of two round or elliptical protective shells, in addition to the usual cuticle, which look like miniature clam shells and are impregnated with calcium carbonate.

Another group of very small crustaceans, the **Copepoda,** are marine or fresh-water dwellers. In addition the group contains many species that are parasitic on other marine or fresh-water animals. Copepods are also important in the diet of whales and fishes. The free-living copepods have bodies that are typically short and cylindrical.

The **Cirripedia** or barnacles are the only sessile crustaceans. They differ quite markedly in their external anatomy from other crustacea, and it was only in 1830, when the larval stages were investigated, that the relationship between the barnacles and other crustaceans was recognized. The barnacles are exclusively marine and secrete complex calcareous cups within which the animal remains. The larvae of barnacles are free-swimming forms that molt several times and eventually become sessile and develop into the adult form. Barnacles were described many years ago by Louis Agassiz as "Nothing more than a little shrimplike animal standing on its head in a limestone house and kicking food into its mouth."

Two numerous groups of crustaceans are small animals that look rather like bugs and live either in the ocean, in fresh water or in damp places on land. The **Isopoda** and **Amphipoda** are commonly called "pill bugs," "wood lice," "beach flies," "sow bugs" and so on. Although they look superficially like insects, they are crustaceans. The isopods have bodies that are flattened from top to bottom and the amphipods have bodies that are flattened from side to side.

The order **Decapoda** is the largest order of crustaceans and contains some 8500 species of lobsters, crabs, crayfish and shrimp. Most decapods are marine but a few, such as the crayfish, certain shrimp and a few crabs, live in fresh water. The crustaceans in general and the decapods in particular show in a striking way the specialization and differentiation of parts in the various regions of the animal. The segments of the trilobites and perhaps of the earliest crustaceans bore appendages all of which were very similar. In the lobster, no two of the 19 pairs of appendages are identical, and the appendages in the different parts of the body are quite different in form and function.

The class **Insecta** is the largest, most successful and most diverse of all animals (Fig. 14–32). Insects are primarily terrestrial animals; some species live in fresh water and a few have become adapted to living along the shore between the tides. In contrast to the crustaceans, the insect head, composed of six completely fused segments, is clearly separated from the thorax, and the thorax is separated from the abdomen. The appendages of one of the head segments are the sensory antennae, and the other appendages are the complex mouthparts. In various species they are adapted for biting, for sucking or for piercing. The thorax consists of three segments fused together, each of which has a pair of legs. Insects typically have two pairs of wings on the last two thoracic segments. The abdomen has up to 11 segments, all of which usually lack appendages. Respiration is carried out by tracheae opening to the outside by spiracles; the circulatory system is open, without capillaries or veins. Insects have a variety of sense organs, including simple and compound eyes, chemoreceptors and receptors for sound waves. Insects are classified into 20 to 25 orders, each of which represents adaptations to a wide range of habitats and environments.

Another large and important group of arthropods is the chelicerates, members of the class **Arachnida** and the class **Merostomata.** The chelicerate body consists of a fused cephalothorax and an abdomen. The first pair of appendages on the head are the **chelicerae,** which are pinching claws in the Merostomata and are typically poison-injecting claws in spiders. The Arachnida have four pairs of walking legs and the Merostomata have five pairs of walking legs.

The class Arachnida includes the spiders, scorpions, mites and ticks. Except for a few groups that are secondarily aquatic, arachnids are terrestrial chelicerates. They are believed to have evolved from water scorpions, the second, now completely extinct, group of Merostomata. Arachnids have, in addition to the chelicerae, a pair of **pedipalps** and four pairs of legs, but no antennae. Most arachnids are carnivorous and prey upon insects and other small arthropods. Arachnids respire either by book lungs or by tracheae or both. **Book lungs** are internal and occur in pairs on the ventral side of the abdomen. An arachnid may have as many as four pairs. The eyes are simple rather than compound. In addition, arachnids have tactile hairs and slit sense organs that may serve as sensory organs of olfaction.

Horseshoe crabs or king crabs (Fig. 14–33), relics of a formerly numerous class, Merostomata, are living fossils clearly related to the arachnids, although superficially like crabs. The present species of horseshoe crabs

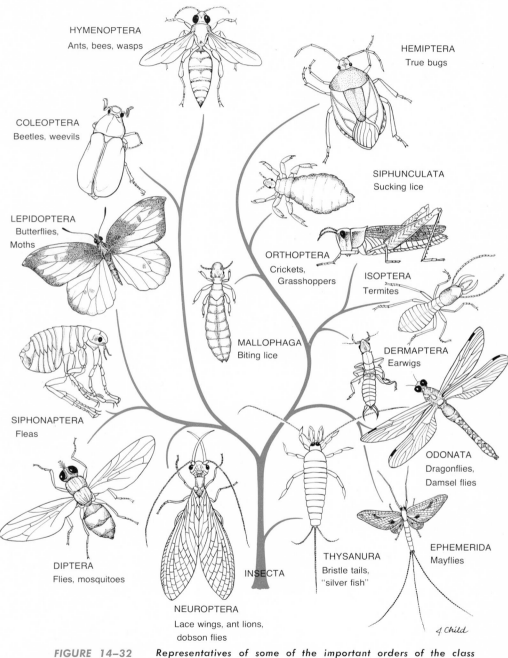

HYMENOPTERA
Ants, bees, wasps

HEMIPTERA
True bugs

COLEOPTERA
Beetles, weevils

SIPHUNCULATA
Sucking lice

LEPIDOPTERA
Butterflies,
Moths

ORTHOPTERA
Crickets,
Grasshoppers

ISOPTERA
Termites

MALLOPHAGA
Biting lice

DERMAPTERA
Earwigs

SIPHONAPTERA
Fleas

ODONATA
Dragonflies,
Damsel flies

DIPTERA
Flies, mosquitoes

INSECTA

THYSANURA
Bristle tails,
"silver fish"

EPHEMERIDA
Mayflies

NEUROPTERA
Lace wings, ant lions,
dobson flies

J Child

FIGURE 14–32   *Representatives of some of the important orders of the class*
Insecta.

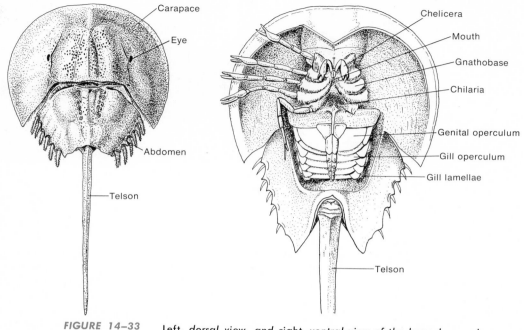

FIGURE 14–33   Left, dorsal view, and right, ventral view of the horseshoe crab
Xiphosura.

have survived essentially unchanged for 350,000,000 years or more. Other
fossil chelicerates, the **eurypterids,** were abundant in the Paleozoic era and
included some species as much as 3 meters long. These giant water scorpions had compound eyes, a pair of chelicerae, four pairs of walking legs
and a pair of large paddles for swimming. They were carnivorous, preying
on the earliest vertebrates, the ostracoderms (p. 452).

SECTION 14–12

ENDOCRINE CONTROL OF MOLTING IN CRUSTACEA

Crabs, lobsters, crayfish and the other crustacea molt many times during development from larval to adult forms. The lobster, for example,
molts seven times during the first summer; at each molt it gets larger and
resembles the adult more. After it reaches the stage of a small adult, additional molts provide for growth. Just before molting the glands in the epidermis secrete a **molting fluid** which contains enzymes to digest the chitin
and proteins of the inner layers of the cuticle. A soft, flexible, new cuticle
is formed under the old one, folded to allow for growth. The digested remains of the old cuticle are absorbed by the body; some substances, e.g.,
the calcium salts, are stored for reuse. The animal may swallow air or water
to aid in swelling up and bursting the old cuticle. It extricates itself from
the old cuticle, swells to stretch the folded new cuticle to its full size, and
then the epidermis secretes enzymes that harden the cuticle by oxidizing
certain compounds and by adding calcium salts to the chitin. Additional
layers of cuticle are secreted subsequently.

Molting is under the control of a hormone that is accumulated in the
sinus gland in the eyestalk. This normally prevents molting; molting occurs

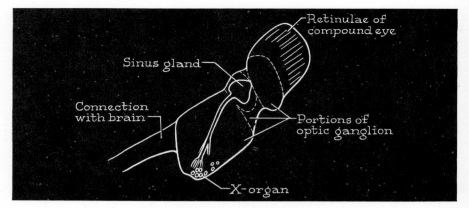

FIGURE 14–34   *Diagram of the eye stalk of the crab with the skeleton removed, showing the sinus gland and x organ.*

when the amount of this hormone falls below a certain threshold. Surgical removal of the sinus gland induces the animal to molt repeatedly, with no resting stage between molts. The sinus "gland" is not glandular tissue but the expanded tips of bundles of axons. The cell bodies of the axons are located in the **X organ,** farther up the eyestalk. The "antimolting" hormone is produced in the cell bodies of the X organ and travels along the axons to be stored and released at their tips in the sinus gland (Fig. 14–34). The hormone of the X organ inhibits the secretion of a molt-initiating hormone (apparently similar to ecdysone) secreted by the **Y organ,** a gland composed of strands of ectodermal cells located at the base of the mandibular muscles. The X organ secretes, in addition to the antimolting hormone, hormones that affect the distribution of pigment in the compound eyes and the control of pigmentation of the body, and several that influence metabolism and reproduction. The production of hormones by nerve cells is termed **neurosecretion.** A similar phenomenon occurs in vertebrates: The hormones vasopressin and oxytocin are produced in the hypothalamic region of the brain and accumulated in the posterior lobe of the pituitary.

SECTION 14–13

INSECT METAMORPHOSIS

Some insects develop, like the crustacea, by passing through a series of successive molts. With each molt they get to look more like the adult, but there is no striking change in appearance at any one molt. The grasshopper matures in this fashion. In contrast, moths, butterflies, flies and many other insects pass through successive stages that are quite unlike one another. From the egg hatches a wormlike **larva**—called a caterpillar (moths), maggot or grub (flies, bees)—which crawls about, eats voraciously, and molts several times, each time becoming a larger larva. The last larval molt forms a **pupa,** a form which neither moves nor feeds. Moth and butterfly larvae spin a cocoon and pupate (molt to form a pupa) within that. During pupation all the structures of the larva are broken down and used as raw materials in the development of the adult animal. Each part of the adult (legs, wings, eyes and so on) develops from a group of cells called a **disc,** which develop directly from the egg. They have never been a functional part of the larva but remain more or less quiescent during the larval period. During the pupal stage these discs grow and differentiate into the adult structures but remain collapsed and folded. When the adult hatches out of the pupa case, blood is pumped into these collapsed structures, they unfold and inflate, and chitin is deposited to make them hard. This striking change from larva to adult is called **metamorphosis** (Greek, change of form). The grasshopper is said to exhibit incomplete metamorphosis since it undergoes a gradual change from larva to adult (Fig. 14–35).

The larva and adult insect not only have a different appearance, they have quite different modes of life. The butterfly larva eats leaves; the adult drinks nectar from flowers. The mosquito larva lives in ponds and eats algae and protozoa; the adult sucks the blood of man and other mammals.

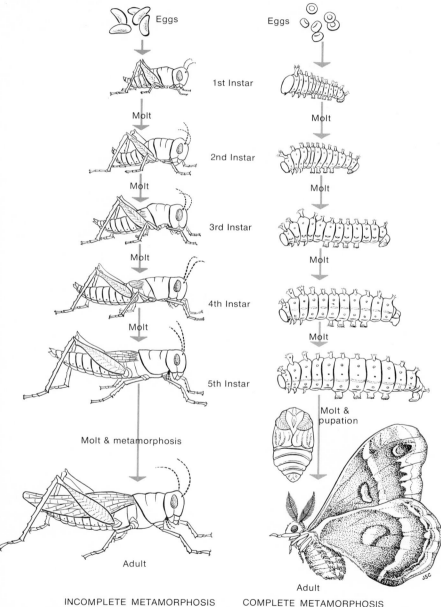

INCOMPLETE METAMORPHOSIS   COMPLETE METAMORPHOSIS

FIGURE 14–35   *Comparison of the life histories of an insect with incomplete metamorphosis, the grasshopper, left, and an insect with complete metamorphosis, the giant silkworm, Platysamia cecropia, right.*

In some species, such as the mayfly, the adult lives only a few hours, just long enough to mate and lay eggs.

The hormonal control of insect metamorphosis has been elucidated by the experiments of Wigglesworth in England and Carroll Williams at Harvard (Fig. 14–36). Williams has worked with the giant Cecropia silkworm, which undergoes a long period of dormancy (diapause) during the pupal stage. When the pupa has been chilled for about six weeks and returned to normal temperature it begins to develop again and completes its metamorphosis in about a month. Williams has found that the chilling stimulates neurosecretory cells in the **intercerebral gland,** a part of the brain, to secrete the hormone. Only the brain need be chilled to induce secretion of hormone. The hormone passes along the axons of the neurosecretory cells to their expanded tips, which comprise the **corpus cardiacum** (Fig. 14–37). This secretes the prothoracicotropic hormone (PTTH), which in turn stimulates the prothoracic glands to secrete **ecdysone,** a steroid with the formula $C_{27}H_{44}O_6$. Karlson and Butenandt in Germany extracted 500 kilograms of silkworms and obtained 25 milligrams of ecdysone. This was used to establish its chemical structure. Ecdysone stimu-

FIGURE 14–36   *Experiments illustrating the hormonal control of insect metamorphosis. A, A brainless, diapausing Telea polyphemus pupa (right) is grafted to a chilled Platysamina cecropia pupa (left). The chilling stimulates certain brain gland cells to secrete a hormone which in turn stimulates the prothoracic glands to secrete ecdysone. B, Ecdysone causes metamorphosis of both animals. C, The posterior portion of a diapausing Cecropia pupa is implanted with both chilled brain glands and prothoracic glands. D, Ecdysone produced by the implanted prothoracic glands, which were stimulated by the implanted brain glands, brings about the metamorphosis of the pupal abdomen. (After C. M. Williams; from Prosser, C. L., and Brown, F. A., Jr.: Comparative Animal Physiology. Philadelphia, W. B. Saunders Company, 1961.)*

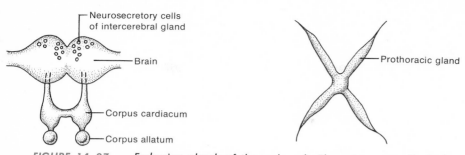

FIGURE 14–37   *Endocrine glands of the cockroach. The upper group lie in the head dorsal to the esophagus. The prothoracic gland is in the ventral part of the pro-thorax, among the muscle cells (After Bodenstein, D.: Recent Progress Hormone Research. Vol. 10. New York, Academic Press, Inc., 1954.)*

lates the epidermis to secrete molting fluid, thus leading to molt and meta-morphosis. The same system of brain hormone and ecdysone initiates the change from larva to pupal stage.

In 1960 Clever and Karlson found that injecting minute amounts of ecdysone into the larvae of the midge *Chironomus* caused, within 15 min-utes, the puffing or swelling of a specific region of a particular chromo-some. Certain tissues of dipteran insects have giant chromosomes composed of many chromatids. In these the individual bands of the chromosomes, corresponding to specific genetic loci, are visible under the microscope and can be identified. Puffing of specific chromosomal bands occurs at specific times in the course of normal development and is known to repre-sent the synthesis of RNA. Injecting ecdysone into larvae causes puffing at a specific band (I-18C) and the subsequent production in epidermal cells of the enzyme dopa decarboxylase, which catalyzes the conversion of DOPA to N-acetyl dihydroxy-phenylethylamine, an agent involved in the hardening of the cuticle. This suggestion that hormones may control the transcription of specific genes is being investigated by many biologists.

An additional hormone, **juvenile hormone,** is secreted by the **corpora allata** (Fig. 14–37), glands lying near the brain. This inhibits metamorpho-sis, but permits molts to occur, and ensures that the larva will molt several times and reach a large size before pupating. This hormone is not pro-duced during the last larval stage and so pupation can occur at the ensuing molt. If extra glands producing juvenile hormone are transplanted into mature larvae, they can be induced to undergo additional larval molts and grow to giant size before pupating and eventually forming giant adults. Removal of the glands early in larval life results in premature metamor-phosis in some insects.

SECTION 14–14

INSECT FLIGHT

Insects are the only invertebrates to have developed wings (though not all species have done so), but these structures are simply analogous, not homologous, to the wings of the vertebrates. Insects usually have two pairs of wings; flies and other Diptera have one pair of wings plus a pair of bal-ancers **(halteres)** which evolved from the second pair of wings. These beat up and down rapidly and apparently serve as gyroscopes during flight. In most insects, both sets of wings are functional as flying organs, but in grasshoppers and beetles the anterior pair are simply stiffened protective devices for the functional posterior pair.

Unlike birds, most insects do not have flight muscles attached to the wings. Instead the wings are attached to the body wall over a fulcrum in such a way that slight changes in the shape of the thorax cause the wings to beat up and down (Fig. 14–38). The contraction of the vertical muscles pulls down the **tergum,** a plate on the upper surface of the thorax, but raises the wings, on the opposite side of the fulcrum. The contraction of the longitudinal muscles causes the tergum to bulge upward and the wing is pulled downward. The movements of the body wall are barely perceptible, but the length of the lever on the two sides of the fulcrum is very different, and the distance moved by the tips of the wings is several hundred times as great.

In many insects—butterflies and moths, among them—the frequency of the wing beat is correlated with the frequency of the nerve impulses to the flight muscles. The impulses to the two sets of muscles are staggered so that rhythmic up and down movements of the wings occur. The rate of these movements ranges from about eight beats per second in large moths to about 75 beats per second in some smaller insects. In other insects—flies and bees, for instance—wing beats are not correlated with the number of nerve impulses. When nerve impulses reach the muscles with a frequency greater than a certain minimal threshold, the wings beat, but at a higher frequency, one set by the muscles themselves, of several hundred beats per second. The frequency is a function of the tension in the two sets of opposing muscles.

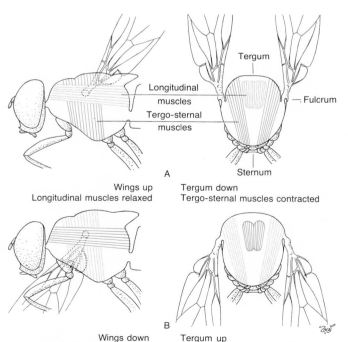

A
Wings up
Longitudinal muscles relaxed

Tergum down
Tergo-sternal muscles contracted

B
Wings down
Longitudinal muscles contracted

Tergum up
Tergo-sternal muscles relaxed

FIGURE 14–38   Diagram of the arrangement of the flight muscles of an insect. The contraction of the longitudinal muscles forces the tergum up and the wings down. Contraction of the tergosternal muscles forces the tergum down and the wings up.

SECTION 14–15

INSECT SOCIETIES

In a number of species of insects—bees, ants and termites—the population consists not of single individuals but of colonies or societies made of several different types of individuals, each adapted for some particular function. In this they resemble a coelenterate or bryozoan colony, but differ in that they are not joined together anatomically as are these lower forms; they constitute a **social colony.** A termite colony (Fig. 14–39), for example, contains "reproductives"—the king and queen—which give rise to all the other members of the colony; "soldiers," strong-jawed, heavily armored termites which protect the colony from enemies; and "workers," which gather food, build the nest and care for the young. Both soldiers and workers are sterile, and neither reproductives nor soldiers can feed themselves. Thus the members of the colony are completely dependent on each other. Each year new reproductives develop in a colony as winged forms that leave the group, mate and form a new colony. A queen termite may lay as many as 6000 eggs per day every day in the year for years. She is simply a specialized egg-laying machine and must be fed and cared for by the workers.

A honey bee colony consists of a single queen, a few hundred drones or males, and thousands of workers, sterile females. The queen bee mates just once, during her "nuptial flight," and stores the sperm in a sperm sac in her body. She returns to the hive and thereafter can lay either unfertilized eggs which develop parthenogenetically into haploid male drones, or fertilized eggs which develop into diploid females. If female larvae are fed "royal jelly" for about six days they develop into queens; if fed this for two or three days and then a mixture of nectar and pollen for three days they develop into workers. Young, adult workers serve as "nurse

FIGURE 14–39   *Model of a royal cell of the termite* Constrictotermes cavifrons *from British Guiana. The queen, with an enormously enlarged abdomen, occupies the center of the chamber with her head toward the right. The king, second in size, is at the lower left. Most of the individuals are workers. A few soldiers with "squirt gun" heads and reduced mandibles can be seen at the left. (Courtesy of the Buffalo Society of Natural Sciences.)*

bees" to feed the larvae and prepare brood cells. Older adults are "house bees" that stand guard at the entrance of the hive, receive and store nectar and pollen, secrete wax for new cells and keep the hive clean. The oldest adults are "field bees" that fly from the hive and forage for water, pollen and nectar.

SECTION 14–16

THE MOLLUSCA

This phylum, with its 80,000 living species and 35,000 fossil species, is the second largest of all the animal phyla. It includes the oysters, clams, octopuses, snails, slugs and the largest of all the invertebrates—the giant squid, which achieves a length of 16 meters, a circumference of 6 meters and a weight of several tons.

The adult body plan of these animals is quite different from that of any other group of invertebrates, but the more primitive mollusks have a characteristic *trochophore* larva, similar to the larval form of certain marine annelids. This suggests that mollusks and annelid worms arose from a common ancestral type; the worms, however, evolved a segmented body plan, while the mollusks evolved a unique body plan without segmentation. The sluggish marine animals, the **chitons** (Fig. 14–40), members of the class **Amphineura**, live by scraping algae off the rocks of the seashore. Their relatively simple structural characteristics provide clear illustrations of the basic molluscan traits: a broad, flat muscular **foot** for creeping across

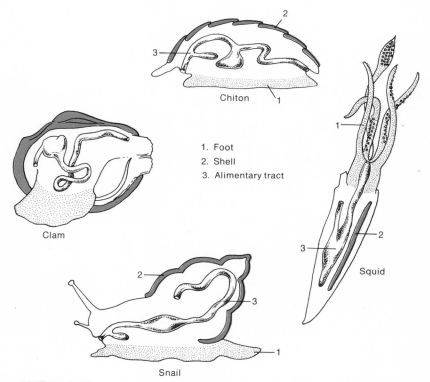

1. Foot
2. Shell
3. Alimentary tract

FIGURE 14–40   *Variations in the basic molluscan body plan in the chiton, snail, clam and squid. Note how the foot (1), shell (2) and alimentary tract (3) have changed their relative positions in the evolution of the several classes of mollusks.*

rocks; a **visceral mass** above the foot, containing most of the organs of the body; a **mantle,** or fold of tissue which covers the visceral mass and projects laterally over the edges of the foot; and a hard, calcareous **shell,** secreted by the upper surface of the mantle as eight separate plates. Like the outer covering of the arthropods, this shell gives protection but has the disadvantage of making locomotion difficult.

The molluscan digestive system is a single tube, sometimes coiled, consisting of a mouth, pharynx, esophagus, stomach, intestine and anus. The pharynx characteristically contains a rasplike structure, the **radula,** which, operated by a set of muscles, can drill a hole in another animal's shell or break off pieces of a plant. The bivalves are the only mollusks that lack a radula; they get their food by straining sea water. The circulatory system is well developed and consists of a pumping organ which sends blood through a system of branched vessels and open spaces containing the body organs. Two "kidneys," lying just below the heart, extract metabolic wastes from the blood, and discharge them through pores located near the anus. The nervous system consists of two pairs of nerve cords, one going to the foot, another to the mantle. The ganglia of these are connected around the esophagus, at the anterior end of the body, by a ring of nervous tissue, thus forming the "brain." Many mollusks have no well developed sense organs, but snails have a pair of simple eyes, usually located on stalks extending from the head, and squids and octopuses have well developed, image-forming eyes.

One usually thinks of snails as having a spirally coiled shell, and many of them do; yet many members of the same class **(Gastropoda),** such as the limpets and abalones, have shells like flattened dunce caps, and others, such as the garden slugs and some marine snails, the **nudibranchs,** have no shell at all. At a particular stage in the development of each gastropod there occurs a unique, sudden, permanent twisting of the body so that the anus is brought around and comes to lie above the head. Subsequent growth is dorsal and usually in a spiral coil. The twist limits space in the body and typically the gill, heart, kidney and gonad on one side are absent. The shell-less slugs and nudibranchs have viscera that undergo the same twisting during development.

Another class of mollusks, **Pelecypoda** (meaning hatchet-foot), commonly called **bivalves,** developed two shells, hinged on the dorsal side and opening ventrally. This arrangement allows the hatchet-shaped foot to protrude for locomotion, and the long, muscular siphon, containing two tubes for the intake and output of water, to be extended. Some bivalves, such as oysters, are permanently attached to the substrate; others, like clams and mussels, burrow rather slowly through sand or mud by means of the foot. A third type burrows through rock or wood, seeking protected dwellings (the shipworm, *Teredo,* which damages rock pilings and other marine installations, is just looking for a home). Finally, some bivalves, such as scallops, swim with amazing speed by contracting their large adductor muscle (the only part of the scallop that we eat) and thus clapping their shells together.

Clams (Fig. 14–41) and oysters obtain food by straining the sea water brought in over their gills by the siphon. The water is kept in motion by the beating of cilia on the surface of the gills, and food particles trapped in the mucus secreted by the gills are carried to the mouth. An average oyster filters about 3 liters of sea water per hour.

The innermost, pearly layer of the bivalve shell, made of calcium carbonate (mother-of-pearl), is secreted in thin sheets by the epithelial cells of the mantle. If a bit of foreign matter gets between the shell and the epithelium, the epithelial cells, in order to protect the animal, secrete concen-

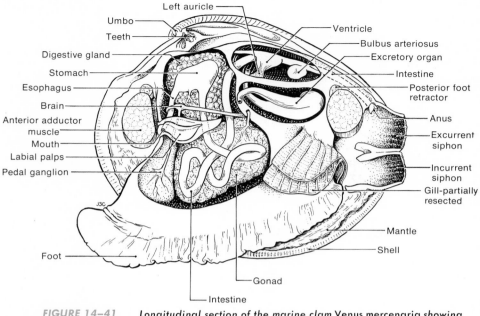

FIGURE 14–41   *Longitudinal section of the marine clam* Venus mercenaria *showing major organ systems.*

tric layers of this substance around the intruding particle; in this way, a pearl is formed.

In contrast to other mollusks, squids (Fig. 14–37), nautiluses and octopuses, comprising the class **Cephalopoda,** are active, predatory animals. They have evolved a specialized, complex head-foot with a large, well developed "brain" and two big eyes. These are strikingly like vertebrate eyes in structure, but develop quite differently as a folding of the skin, rather than as an outgrowth of the brain. This type of independent evolution of similar structures which carry on similar functions in two different, unrelated animals is known as **convergent evolution.** The foot of the squid and octopus is divided into 10 and eight long tentacles, respectively, covered with suckers for seizing and holding the prey. Besides having a radula, the animals have (in the mouth) two strong, horny beaks for killing the prey and tearing it to bits. The mantle is thick, muscular and fitted with a funnel. By filling the mantle cavity with water and ejecting it through this funnel, the animals attain rapid jet propulsion in a direction opposite to that in which the funnel is pointed.

Cephalopods are equipped with an **ink sac** which produces a thick, black liquid. This is released when the squid or octopus is alarmed. The ink distracts the pursuer; MacGinitie has shown that octopus ink paralyzes the chemoreceptors of the animals that pursue the octopus.

The shell of a nautilus is a flat, coiled structure, consisting of many chambers, built up year by year; each year the animal lives in the latest and largest chamber of the series. By secreting a gas resembling air into the other chambers, the nautilus is enabled to float. The shell of the squid is reduced to a small "pen" in the mantle, and the octopus has no shell at all.

Small octopuses survive well in aquaria and they are proving to have a surprising degree of intelligence. They can make associations among stimuli and generally show an adaptability of behavior that more closely resembles that of the vertebrates than the more stereotyped patterns seen in other invertebrates. Octopuses feed on crabs and other arthropods which they catch and kill by a poisonous secretion of their salivary glands. They live among rocks, taking shelter in small caves. Their motion is incredibly fluid and gives little hint of the considerable strength in their eight arms. They usually hide during the day and hunt for food in the evening.

The striking similarities in the development of mollusks and annelids — the process of spiral cleavage and the appearance of a trochophore larva — had suggested that these two phyla were related in evolutionary origin and had a common coelomate ancestor. This view was supported by the discovery in 1952 and subsequently of specimens of a primitive mollusk, *Neopilina*, in material dredged from a deep trench in the Pacific off Costa Rica. These animals are about 2.5 centimeters long and have some characteristics in common with gastropods and amphineurans. Their most remarkable feature is the segmental arrangement of certain internal organs — they have five pairs of retractor muscles, six pairs of nephridia and five pairs of gills. This has been interpreted by some zoologists as evidence of the segmental character of their ancestors, evidence that the mollusks, like the annelids, have a basically metameric body plan.

SECTION 14–17

THE ECHINODERMATA

The echinoderms (spiny-skinned) include the **sea stars** (Asteroidea), **sea urchins** (Echinoidea), **sea cucumbers** (Holothuroidea), **serpent stars** (Ophiuroidea) and **sea lilies** (Crinoidea) (Fig. 14–42) — a group of animals radically different from all other invertebrates. Curiously enough, they appear to be related to the chordates. The echinoderm larvae and those of the primitive hemichordates have many features in common (Fig. 14–46). All the 6000 or so species of this phylum are marine, and all have radial, rather than bilateral, symmetry.

A starfish or sea star (Fig. 14–43) consists of a central **disc** from which radiate five to 20 or more **arms.** On the under side of the disc, in the center, is the mouth. The skin of the entire animal is embedded with tiny, flat bits of calcium carbonate, some of which gives rise to spines. A number of these spines are movable. Around the base of some, especially near the delicate skin gills used in respiration, are still tinier, specialized spines, in the form of pincers; operated by muscles, these keep the surface of the animal free of debris. In order to move and attack their prey, echinoderms are equipped with a unique hydraulic arrangement called a **water vascular system.** The under surface of each arm is equipped with hundreds of pairs of tube feet — hollow, thin-walled, sucker-tipped muscular cylinders — at the base of which are round muscular sacs, called **ampullas.** To extend the feet, these sacs contract, forcing water into the feet. The feet are withdrawn by the contraction of muscles in their walls which forces the water back into the ampullas. The cavities of the tube feet are all connected by radial canals in the arms, and these, in turn, are connected by a circular canal in the central disc. The circular canal is connected by the stone canal to a button-shaped **madreporite** on the upper (aboral) surface of the central disc. The madreporite has many pores, as many as 250, by which the cavity of the stone canal opens to the exterior.

To attack a clam or oyster, the sea star mounts it, assuming a humped position as it straddles the edge opposite the hinge. Then, with its tube

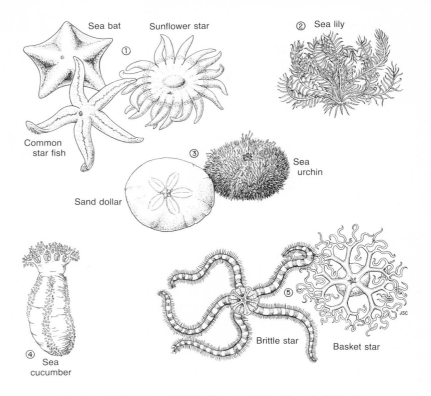

1. Asteroidea   2. Crinoidea   3. Echinoidea   4. Holothuroidea   5. Ophiuroidea

FIGURE 14–42   *Sketches of some representatives of the five classes of Echino-dermata.*

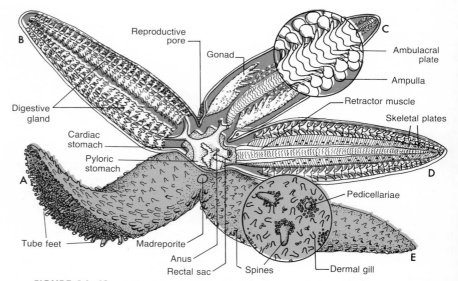

FIGURE 14–43   *The starfish Asterias viewed from above with the arms in various stages of dissection. A, Arm turned to show the lower side. B, The upper body wall removed. C, The upper body wall and the digestive glands removed, and a magnified detail of the ampullas and ambulacral plates. D, All the internal organs removed except the retractor muscles, showing the inner surface of the lower body wall. E, The upper surface, with a magnified detail showing the features of the surface.*

feet attached to the two shells, it begins to pull. The clam, of course, reacts by closing its shell tightly. But by using its tube feet in relays, the sea star can outlast the clam, which, becoming exhausted, relaxes, and opens its shell. The sea star then turns its own stomach inside out through the mouth and inserts it into the clam to digest it. The partly digested animal is later taken inside the sea star for further digestion in glands located in each arm. The water vascular system does not enable the starfish to move rapidly, but since it usually preys upon slow-moving or stationary clams and oysters, speed of attack is not necessary as it is for most predators. Starfish occasionally catch and eat small fish (Fig. 14–44).

There are no special respiratory or circulatory systems in these animals; both functions are accomplished by the fluid which fills the large coelomic cavity and bathes the internal organs. Nor is there any special excretory system—wastes pass to the outside by diffusion. The nervous system consists of a ring of nervous tissue encircling the mouth, and a nerve cord extending from this into each arm. There is no aggregation of nerve cells which could be called a brain.

A second class of echinoderms, the brittle stars or serpent stars, also have a central disc, but their arms are long and slender, enabling them to move rapidly. The arms are discarded and replaced when injured.

Sea urchins, a third class of echinoderms, look like animated pincushions, bearing on their spherical bodies long, movable spines between which the tube feet protrude. In these creatures, the calcareous plates have fused, forming a spherical shell, and in the center of the under surface of this sphere is the mouth. The tube feet, arranged in five bands on the surface of the shell, are longer and more slender than those of sea stars, but the water vascular system is otherwise similar.

Sea cucumbers, another class of the spiny-skinned phylum, are appropriately named, for many of them are green and about the size and shape

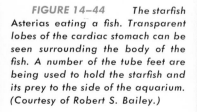

FIGURE 14–44   *The starfish* Asterias *eating a fish. Transparent lobes of the cardiac stomach can be seen surrounding the body of the fish. A number of the tube feet are being used to hold the starfish and its prey to the side of the aquarium. (Courtesy of Robert S. Bailey.)*

of a gherkin. Like the members of several other phyla, these animals have a circle of tentacles around the mouth, and, in common with the starfish, they have a water vascular system; some species have external tube feet. The bodies of sea cucumbers are flexible, hollow muscular sacs. Whenever environmental conditions are unfavorable, because of high temperatures, lack of oxygen or excessive irritation, the sea cucumber contracts violently and ejects its entire digestive tract. When conditions are again favorable, the cucumber grows a new digestive tract.

The crinoids, or sea lilies, are sessile creatures rather like a starfish turned mouth side up, with a number of arms extending upward and a stalk which attaches the animal to the sea bottom. There are many more fossil crinoids than living species.

SECTION 14-18

PHYLUM HEMICHORDATA: THE ACORN WORMS

The hemichordates are a small group of wormlike marine animals that burrow in sand or mud. The anterior section is a short, conical, muscular **proboscis** (which apparently resembled an acorn in the eyes of early biologists), connected by a narrow stalk to the **collar,** behind which extends a long, cylindrical, rather flaccid, wormlike body (Fig. 14–45). The mouth is on the lower side of the body at the base of the collar, and just behind the collar, perforated by many gill slits, is the **pharynx,** through which water passes. As it burrows along, the animal feeds on organic matter in the sand. The nervous system is a diffuse network over most of the body, but concentrated into dorsal and ventral nerve cords in the anterior region. Only the dorsal nerve cord extends into the collar, where it becomes thick and hollow. There is a short, rodlike outgrowth from the anterior

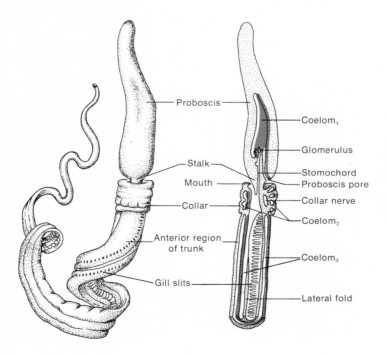

Proboscis
Coelom₁
Glomerulus
Stalk
Stomochord
Mouth
Proboscis pore
Collar
Collar nerve
Coelom₂
Anterior region of trunk
Coelom₃
Gill slits
Lateral fold

FIGURE 14–45   The acorn worm Saccoglossus, a member of the phylum Hemichordata. Left, external view showing external features. (After Bateson.) Right, a diagrammatic section through the anterior part of the body showing some of the internal organs. A lateral fold subdivides the pharynx into a ventral channel along which the sand passes and a dorsal channel containing the gill slits.

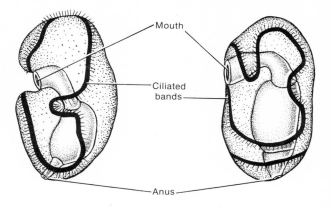

FIGURE 14–46   Left, the bipinnaria larva of a starfish, and right, the tornaria larva of an acorn worm. Note the striking similarities of the two. The digestive tracts are shown in color.

Mouth

Ciliated bands

Anus

end of the digestive tract called a "stomochord," which extends into the cavity of the proboscis. The larva of some acorn worms is much like that of some echinoderms (Fig. 14–46) and is often mistaken for it, but the later development of the two forms is quite different. This is taken as evidence of the evolutionary relationship of the two phyla. The hemichordates may also have some evolutionary affinity with the chordates (only these two phyla have pharyngeal clefts).

The foregoing description does not exhaust the great variety of animals. In addition to these phyla, there are other groups of less important invertebrates sometimes put in phyla of their own, sometimes classified under other phyla.

SUGGESTIONS FOR FURTHER READING

Barnes, R. D.: *Invertebrate Zoology.* 2nd ed. Philadelphia, W. B. Saunders Company, 1968.
   An excellent modern treatment that includes all the invertebrates.

Buchsbaum, R.: *Animals Without Backbones.* 2nd ed. Chicago, University of Chicago Press, 1948.
   A very popular, richly illustrated survey of the invertebrates.

Hyman, L.: *The Invertebrates.* New York, McGraw-Hill Book Co., Vols. 1 through 6, published 1940 through 1967.
   The definitive treatise of the invertebrates, covering all aspects of their structure, function, development and evolution.

Lee, D. L.: *The Physiology of Nematodes.* San Francisco, W. H. Freeman and Company, 1965.
   A readable synthesis of the biology of this important group of invertebrates.

Lenhoff, H., and Loomis, W. (eds.): *The Biology of Hydra.* Miami, University of Miami Press, 1961.
   A fascinating collection of reports of recent work on hydras and certain other coelenterates.

MacGinitie, G. E., and MacGinitie, N.: *Natural History of Marine Animals.* New York, McGraw-Hill Book Co., 1949.
   This book contains a wealth of general information on the habits, behavior and learning ability of a variety of marine animals.

Zim, H. S., and Ingle, L.: *Seashores.* New York, Simon and Schuster, Inc., 1955.
   A paperback field guide to the common invertebrates of the seashore.

THE PHYLUM CHORDATA

CHAPTER 15

HALLMARKS OF THE CHORDATES

The last great phylum of animals, that to which man belongs, is the phylum Chordata, whose members are distinctive in having a **notochord,** a dorsal, hollow **nerve cord** and paired, **pharyngeal pouches** and **gill slits.** The latter are present in all chordate embryos but are not evident in adult higher vertebrates. In addition to the fishes, amphibia, reptiles, birds and mammals, which make up the classes of the subphylum **Vertebrata** — characterized by a cartilaginous or bony vertebral column — the phylum Chordata includes two subphyla of curious soft-bodied marine animals which show the chordate characteristics to some extent, and are of interest as possible connecting links between vertebrates and invertebrates.

The notochord is a dorsal longitudinal rod composed of a fibrous sheath enclosing vacuolated cells. The turgidity of these cells makes the notochord firm yet flexible. The notochord, by preventing the body from shortening when the longitudinal muscles in the body wall contract, facilitates the lateral undulatory movements involved in the swimming motions of fish. The nerve cord dorsal to the notochord differs from that of invertebrates not only in its position but in its structure, for it is single rather than double and a hollow tube rather than a solid cord. Chordates also have pharyngeal pouches extending laterally from the anterior part of the digestive tract toward the body wall, perhaps breaking through as gill slits. The earliest chordates apparently were filter-feeders and this arrangement of pouches and slits permitted water to escape from the digestive system and concentrated the small food particles in the gut. Chordates share other characteristics with certain invertebrates. They are bilaterally symmetrical, they have three germ layers, and they have a tube-within-a-tube body plan, with a true coelom separating the gut from the body wall.

The vertebrates are less diverse and much less numerous and abundant than the insects but rival them in their adaptation to a variety of modes of existence and excel them in the ability to receive stimuli and react to them. Vertebrates are generally active animals and show a high degree of **cephalization,** accumulation of nerves and sense organs in the head.

SEA SQUIRTS OR TUNICATES: UROCHORDATA

The first chordate subphylum is composed of the sea squirts or Tunicates (Fig. 15–1), most of them barrel-shaped, sessile, marine animals unlike the other chordates; indeed, the primitive members are often mistaken for sponges or coelenterates. The larval form of the tunicates, however, is typically chordate, superficially like a tadpole. Its expanded body has a pharynx with gill slits, and its long muscular tail contains a notochord and dorsal nerve cord. The larva eventually becomes attached to the sea bottom, and loses its tail, notochord and most of its nervous system. In the adult, therefore, only the gill slits suggest that it is a chordate.

The adult develops a **tunic,** quite thick in most species and covering the entire animal, which is composed (curiously enough) principally of a kind of cellulose. The tunic has two openings, the **incurrent siphon,** through which water and food enter, and the **excurrent siphon,** through which water, waste products and gametes pass to the outside.

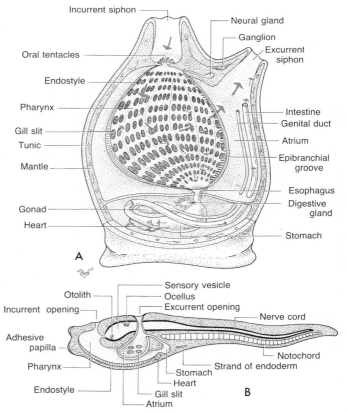

FIGURE 15–1   Diagrammatic lateral views of an adult (A) and a larval (B) tunicate showing major internal organs. The large arrows in A represent the course of the current of water and the small arrows represent the path of the food. The stomach, intestine and other visceral organs are embedded in the mantle.

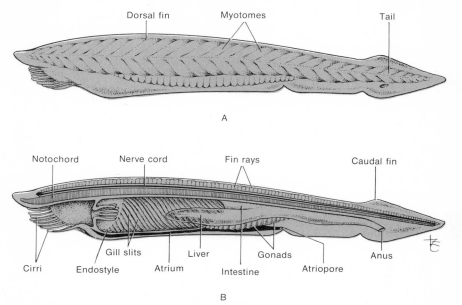

A

B

FIGURE 15–2   *External view and longitudinal section of* Amphioxus, *a member of the subphylum* Cephalochordata.

SECTION 15–3

CEPHALOCHORDATES

In the **Cephalochordata,** the second chordate subphylum, all three chordate characteristics are highly developed. The **notochord** extends from the tip of the head to the tip of the tail, a large pharyngeal region contains many pairs of **gill slits,** and the hollow, tubular, dorsal **nerve cord** stretches the entire length of the body (Fig. 15–2). The cephalochordates are small, translucent, fishlike segmented animals, 5 to 10 centimeters long and pointed at both ends. They are widely distributed in shallow seas, either swimming freely or burrowing in the sand near the low tide line. They feed by drawing a current of water into the mouth (by the beating of cilia) and straining out the microscopic plants and animals. The water passes through the gill slits into the **atrium,** a chamber lined with ecto-derm. This has a ventral opening, the **atriopore,** just anterior to the anus. Metabolic wastes are excreted by segmentally arranged, ciliated **protone-phridia** that open into the atrium. Although superficially similar to fishes, they are much more primitive, for they lack paired fins, jaws, sense organs and a brain. It is generally believed that the cephalochordate *Amphioxus* is rather similar to the primitive ancestor from which the vertebrates evolved. In contrast to the invertebrates, the blood of this animal flows anteriorly in the ventral vessel and posteriorly in the dorsal vessel.

SECTION 15–4

THE VERTEBRATES

The vertebrates are distinguished from these lower chordates by the presence of an internal skeleton of cartilage or bone that reinforces or replaces the notochord. The notochord is a flexible, unsegmented skeletal rod extending longitudinally in all chordates. In the vertebrates, segmental bony or cartilaginous **vertebrae** surround the notochord. In the higher vertebrates, the notochord is visible only early in development; later the vertebrae replace it completely. Vertebrates have a bony or cartilaginous brain case, the **cranium,** which encloses and protects the brain, the enlarged anterior end of the dorsal, hollow nerve cord.

Vertebrates have a pair of eyes that develop as lateral outgrowths of the brain. Invertebrate eyes, such as those of insects and cephalopods, may be highly developed and quite efficient, but they develop from a folding of the skin. Another vertebrate characteristic is a pair of ears, which in the lowest vertebrates are primarily organs of equilibrium. The **cochlea,** which contains the cells sensitive to sound vibrations, is a later evolutionary development.

The circulatory system of vertebrates is distinctive in that the blood is confined to blood vessels and is pumped by a ventral, muscular heart. The higher invertebrates such as arthropods and mollusks typically have

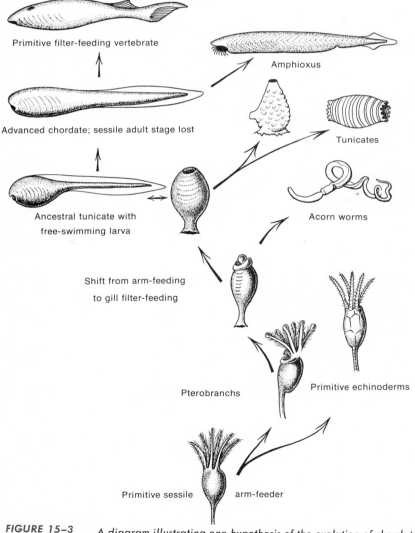

Primitive filter-feeding vertebrate

Amphioxus

Advanced chordate; sessile adult stage lost

Tunicates

Ancestral tunicate with
free-swimming larva

Acorn worms

Shift from arm-feeding
to gill filter-feeding

Pterobranchs

Primitive echinoderms

Primitive sessile   arm-feeder

FIGURE 15–3   *A diagram illustrating one hypothesis of the evolution of chordates.*
(After Romer, A. S.: The Vertebrate Body. 4th ed. Philadelphia, W. B. Saunders Company,
1970.)

hearts, but they are located on the dorsal side of the body and pump blood into open spaces in the body, called a **hemocoel.** Vertebrates are said to have a **closed circulatory system;** arthropods and mollusks have an open circulatory system, for the blood is not confined solely to tubular blood vessels.

The classes of the subphylum Vertebrata are as follows: the **Agnatha,** the jawless fishes such as the lamprey eels; the **Placodermi,** earliest of the jawed fishes, known only from fossils; the **Chondrichthyes,** the sharks and rays with cartilaginous skeletons; the **Osteichthyes,** the bony fishes; the **Amphibia,** frogs and salamanders; the **Reptilia,** lizards, snakes, turtles and alligators plus a host of fossil forms like the dinosaurs; the **Aves,** birds; and the **Mammalia,** the warm-blooded, fur-bearing animals that suckle their young. The Agnatha, Placodermi, Chondrichthyes and Osteichthyes comprise the superclass **Pisces,** and the Amphibia, Reptilia, Aves and Mammalia comprise the superclass **Tetrapoda.**

There is no clear fossil record of the ancestors of the chordates, for, whatever they were, they were undoubtedly small and soft bodied. An impression of an Amphioxus-like animal has been found in rocks of the Silurian period in Scotland. This animal, *Jamoytius*, has been interpreted by some paleontologists as a primitive vertebrate.

Theories of the origin of the chordates must depend on other types of evidence. The most widely held theory at present is that the echinoderms, hemichordates and chordates have a common evolutionary origin (Fig. 15–3). This is based on the striking similarity of the **tornaria** larva of the hemichordate and the **bipinnaria** larva of the starfish (Fig. 14–46), plus the generally similar modes of formation of the mesoderm and coelom in the three phyla.

SECTION 15–5

JAWLESS FISHES

The Agnatha, or jawless fishes, includes the **ostracoderms,** which are the earliest known fossil chordates (Fig. 15–4) and the living lamprey eels

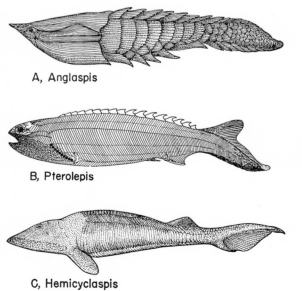

A, Anglaspis

B, Pterolepis

C, Hemicyclaspis

FIGURE 15–4   *Three examples of fossil ostracoderms — primitive, jawless, limbless fish. (From Romer, A. S.: The Vertebrate Body. 4th ed. Philadelphia, W. B. Saunders Company, 1970.)*

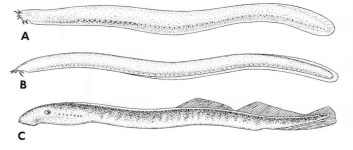

FIGURE 15–5   Three types of Cyclostomes. A, A slime hag; B, a hagfish; C, a lamprey. Note the absence of jaws and paired fins from all. (After Dean, B.: A Bibliography of Fishes. 3 vols. New York, American Museum of Natural History, 1916-1923.)

and hagfishes (Fig. 15–5). These have cylindrical bodies up to a meter long, with smooth scaleless skin and no jaws or paired fins. Lampreys and hagfishes have a circular sucking disc around the mouth, which is located on the ventral side of the anterior end. They attach themselves by this disc to other fish and, using the horny teeth on the disc and tongue, bore through the skin and get blood and soft tissues to eat. They are the only parasitic vertebrates; hagfishes may bore their way completely through the skin and come to lie within the body of the host. Both are of great economic importance because of their destruction of food fish such as cod, flounder, lake trout and whitefish. The trout of the Great Lakes have been killed off in great numbers by sea lampreys that apparently came up from the St. Lawrence via the Welland Canal. Lampreys leave the ocean or lakes and swim upstream to spawn. They build a nest, a shallow depression in the gravelly bed of the stream, into which eggs and sperm are shed. The fertilized eggs develop into **ammocoetes** larvae in about three weeks. These larvae, which probably resemble the ancestral primitive vertebrate more closely than any living adult vertebrate does, drift downstream to some pool and live as filter-feeders in burrows in the muddy bottom for several years. They then undergo a metamorphosis, become adult lampreys, and migrate back to the ocean or lake.

The earliest traces of the vertebrates are the fossil ostracoderms found in rocks of the Ordovician period. These were small, jawless, armored, bottom-dwelling, filter-feeding fishes that lacked fins. The head was covered with thick bony plates and the trunk and tail were covered with thick scales. They lived in fresh water and were probably preyed upon by eurypterids, giant water scorpions. They survived through the Silurian into the Devonian period and gave rise to the placoderms.

SECTION 15–6

THE EARLIEST JAWED FISHES: PLACODERMS

During the Silurian and Devonian periods, some descendants of the ostracoderms evolved jaws and paired appendages and changed from filter-feeding bottom dwellers to active predators. The earliest jawed fishes, the spiny-skinned sharks, are placed in the class Placodermi. These were generally small, armored, fresh water fish with a variable number (as many as seven) of paired fins (Fig. 15–6). One of the best known placoderms is *Dunkleosteus*, a monster that attained a length of 9 meters. The head and anterior part of the trunk had a bony armor, but the remainder of the body was naked. The evolution of jaws from a portion of the gill arch skeleton enabled the placoderms and their descendants to become adapted to new

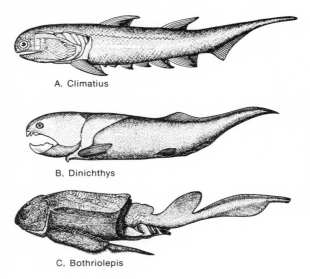

A, Climatius

B, Dinichthys

C, Bothriolepis

FIGURE 15–6   Three fossil placoderms from the Devonian Period. A, Climatius, a "spiny-skinned shark" with large fin spines and five pairs of accessory fins between the pectoral and pelvic pairs. B, Dinichthys, a giant arthrodire that reached a length of 10 meters. Its head and thorax were covered by bony armor but the rest of the body and tail were naked. C, Bothriolepis, a placoderm with a single pair of jointed flippers projecting from the body. (From Romer, A. S.: The Vertebrate Body. 4th ed. Philadelphia, W. B. Saunders Company, 1970.)

modes of life. The success of the jawed vertebrates undoubtedly contributed to the extinction of the ostracoderms.

SECTION 15–7

CARTILAGINOUS FISHES

The ostracoderms and placoderms were primarily fresh-water fish; only a few ventured into the oceans. The cartilaginous fishes evolved as successful marine forms in the Devonian and most have remained as ocean-dwellers; only a few have secondarily returned to a fresh-water habitat.

The Chondrichthyes—sharks, rays and skates—have a skeleton of cartilage which may or may not be calcified. The cartilaginous skeleton of these fishes represents the retention of an embryonic condition, not a primitive one, for the adult ancestors had bony skeletons. The dogfish is commonly used in biology classes because it demonstrates the basic vertebrate characteristics in a simple, uncomplicated form. All the Chondrichthyes have paired jaws and two pairs of fins. The skin contains scales composed of an outer enamel and an inner dentine layer. The lining of the mouth contains larger but essentially similar scales which serve as teeth. The teeth of the higher vertebrates are homologous with these shark scales.

All fish, from lampreys to the highest bony fish, have highly vascular **gills** which have a large surface for the transfer of oxygen and carbon dioxide. Cartilaginous fishes have five to seven pairs of gills. In some fish the gills also secrete salts to maintain osmotic equilibrium between the blood and surrounding water. A current of water enters the mouth, passes over the gills and out the gill slits, constantly providing the fish with a fresh supply of dissolved oxygen.

Cartilaginous fishes have no swim bladders and their bodies are denser than water, hence they tend to sink unless they swim actively. The large pectoral fins give a lift component, and the sculling action of the tail provides additional lift.

The whale shark, which reaches a length of 16 meters, is the largest fish known, but it feeds on microscopic crustacea and other plankton.

FIGURE 15–7   Above, the sawfish Pristis; below, the stingray Dasyatis. (The lower photograph courtesy of Marine Studios.)

Sharks are elongate, streamlined predators, that swim actively and catch other fish. Rays and skates are sluggish, flattened creatures, living partly buried in the sand and feeding on mussels and clams. They have enormous pectoral fins which undulate and propel the ray or skate along the bottom. The sting ray has a whiplike tail with a barbed spine at the tip, which can inflict a painful wound (Fig. 15–7). The electric ray has electric organs on either side of the head; these are modified muscles which can discharge enough electricity to stun fairly large fish. Shark skin is tanned and used in making shoes and handbags, and shark liver oil is an important source of vitamin A. Some sharks and rays are used for food.

SECTION 15–8

BONY FISHES

The Osteichthyes include some 20,000 species of fresh- and salt-water fishes, ranging in size from guppies to sturgeons, which may weigh over a ton. The fossil evidence now available indicates that bony fishes evolved from placoderms independently of, and at about the same time as, the cartilaginous fishes. They did not evolve from cartilaginous fishes.

The bony fish originally evolved in fresh water but subsequently entered the oceans and became dominant there, too. By the middle of the Devonian they had evolved into three major groups—lungfish, lobe-finned fish, and ray-finned fish—all of which had lungs and an armor of bony scales. A few lungfish have survived to the present, and the ray-finned fish,

FIGURE 15–8   *Photograph of the coelacanth Latimeria, 1.5 meters long, caught in the Indian Ocean off South Africa in 1952. Note the thick, lobe-shaped fins. (Life Magazine © Time, Inc.)*

after undergoing a slow evolution in the later Paleozoic and early Mesozoic eras, ramified greatly to give rise to the modern bony fish, the **teleosts.** The lobe-finned fishes, believed to be the ancestors of the land vertebrates, were almost extinct by the end of the Paleozoic. It was believed at one time that they had become extinct during the Mesozoic, but since 1939 several specimens of **coelacanths** nearly 2 meters long have been caught in the deep waters off the east coast of South Africa (Fig. 15–8).

Most of the bony fishes have beautifully streamlined bodies and swim by contracting the body and tail muscles which move the tail back and forth in a sculling motion. The fins are used chiefly for steering. Bony fish typically have a **swim bladder,** a gas-filled sac located in the dorsal part of the body cavity (Fig. 15–9). By secreting gases into the bladder or by absorbing them from it, the fish can change the density of its body and so hover at a given depth of water. The gills of bony fishes are covered by a hard, bony protective flap, the **operculum.** The skeleton is composed of bone rather than cartilage, and the head is encased in many bony plates which form a **skull.** Bony fish have protective, overlapping bony scales in the skin which differ from those found in sharks.

Many of the bony fishes, particularly those in tropical waters, are brightly and beautifully colored—red, orange, yellow, green, blue and

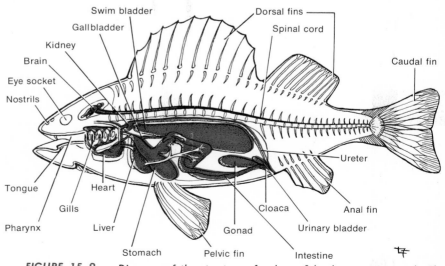

FIGURE 15–9   *Diagram of the structure of a bony fish, the common perch. The organs of the digestive tract are tinted.*

black. Some fish, such as flounders, can change color and pattern to conform to the color and pattern of the background, and thus render themselves inconspicuous to predators. Fish evolution has led to a tremendous variety of sizes, shapes and colors, and to a number of curious adaptations—the sea horse male that has a brood pouch in which eggs are carried until they hatch; the deep sea forms that have evolved luminescent structures as lures for their prey; the male stickleback which builds a nest of sticks held together by threads which he secretes, and then guards the eggs in the nest; the true eels which live as adults in streams in North America or Europe but which migrate to the Atlantic near Bermuda to spawn, and so on.

The lungfishes, **Dipnoi,** were once thought to be the ancestors of land vertebrates, but in the arrangement of the bones of the skull, the type of teeth present, the pattern of fin bones and type of vertebrae, the lobe-finned fishes resemble the primitive amphibians closely and the lungfishes do not.

THE AMPHIBIA

The four-legged land vertebrates, the amphibia, reptiles, birds and mammals, are sometimes placed together in the superclass Tetrapoda. Not all the tetrapods have four legs (e.g., the snakes), but they evolved from four-legged ancestors. Not all the tetrapods now live on land (e.g., whales, seals), but they evolved from terrestrial ancestors.

The first successful land vertebrates were ancient amphibians, the **labyrinthodonts** (Fig. 15–10), that closely resembled their ancestral lobe-finned fishes but had evolved a limb strong enough to support the weight of the body on land. These earliest arms and legs were five-fingered, a pattern that has generally been kept by the higher vertebrates. There were many different kinds of ancient amphibia, all of which became extinct in the first part of the Mesozoic. The labyrinthodonts, which ranged in size from fairly small animals up to ones as large as crocodiles, gave rise to other primitive amphibians, to the modern frogs and salamanders, and to the earliest reptiles, the **colylosaurs** or stem reptiles. The

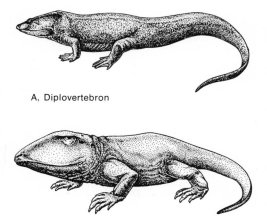

FIGURE 15–10   A, Diplovertebron, a primitive Paleozoic amphibian (labyrinthodont). B, Ophiacodon, an early Permian pelycosaur. Although the pelycosaurs were primitive reptiles, they had characteristics indicating that they represented a first stage in the evolution of the mammals. (From Romer, A. S.: The Vertebrate Body. 4th ed. Philadelphia, W. B. Saunders Company, 1970.)

A, Diplovertebron

B, Ophiacodon

modern amphibia, the frogs and salamanders, appeared in the latter part of the Mesozoic. The salamanders and water dogs more closely resemble the ancient amphibia; frogs and toads are highly specialized for hopping.

Although some adult amphibia are quite successful as land animals and can live in comparatively dry places, they must return to water to reproduce. Eggs and sperm are generally laid in water and the fertilized eggs, nourished at first by the yolk, develop into larvae or **tadpoles.** These breathe by means of gills and feed on aquatic plants. After a time, the larva undergoes *metamorphosis* and becomes a young adult frog or salamander, with lungs and legs. Metamorphosis is under the control of thyroxine, the hormone secreted by the thyroid gland, and can be prevented by removing the thyroid or the pituitary, which secretes a thyroid-stimulating hormone. The fore legs grow out of the fold of skin that had enveloped them, the gills and gill slits are lost, the tail is resorbed, the digestive tract shortens, the mouth widens, a tongue develops, the tympanic membranes and eyelids appear, and the shape of the lens changes. In addition, a host of biochemical changes occur to provide for the change from a completely aqueous form of life to one which is amphibious.

Adult amphibia do not depend solely on their primitive lungs for the exchange of respiratory gases; their moist skin, plentifully supplied with blood vessels, also serves as a respiratory surface. The skin of salamanders and frogs has no scales and may be brightly colored. Frogs especially have the ability to change color, from light to dark, by increasing or decreasing the size of the melanocytes, the pigment-containing cells of the skin (Fig. 15–11). The change in color is controlled by a hormone, called *melanocyte-stimulating hormone* (MSH), secreted by the intermediate lobe of the pituitary (p. 655). Salamanders, but not frogs, have a marked ability to regenerate lost legs and tails. In some species of frogs and toads, the fertilized eggs do not develop in the water but are kept on the back of the female, in the mouth of the male, or in a string wrapped around the male's hind legs.

A number of frogs, toads and salamanders have skin glands that secrete poisonous substances. These may serve as a means of protection for the species, by discouraging would-be predators.

A small group of tropical, legless, wormlike amphibia (the **Apoda**) burrow in moist earth.

FIGURE 15–11   A, A leopard frog adapted to a light background, and B, one adapted to a dark background. (From Turner, C. D.: General Endocrinology. 4th ed. Philadelphia, W. B. Saunders Company, 1970.)

SECTION 15–10

REPTILES

The class Reptilia has many more extinct than living species. They are true land forms and need not return to water to reproduce, as amphibians must. The embryo develops in a watery medium within the protective, leathery egg shell secreted by the female. Since a sperm could not penetrate this shell, fertilization must occur within the body of the female before the shell is added. This in turn necessitated the evolution of some means of transferring sperm from the body of the male to that of the female. Reptiles were the first to evolve a male copulatory organ, the **penis,** for this purpose.

The bodies of reptiles are covered with hard, dry, horny scales which protect the animal from desiccation and from predators. They breathe by means of lungs, for the dry scaly skin cannot serve as an organ of respiration. Like fish and amphibia, reptiles do not have a mechanism for regulating body temperature and therefore have the same temperature as their surroundings. In hot weather their body temperature is high, metabolism occurs rapidly, and they can be quite active. In cold weather their body temperature is low, their metabolic rates are low, and they are very sluggish. Because of this they are much more successful in warm than in cold climates. The reptiles living today are turtles (order Chelonia), alligators (order Crocodilia), snakes and lizards (order Squamata), and the tuatara of New Zealand (Fig. 15–12).

The first reptiles, the **cotylosaurs,** which appeared during the Pennsylvanian period, resembled their labyrinthodont ancestors. They were sluggish, lizard-like animals with short stubby legs extending laterally from the body. From these early stem reptiles evolved the **pelycosaurs** (Fig. 15–10), carnivorous, more slender and lizard-like than the cotylosaurs. These are believed to be in the direct line ancestral to the mammals. They gave rise, during the Permian period, to the **therapsids** (Fig. 15–13), another group of reptiles with a few more mammalian characteristics. *Cynognathus* was a slender, lightly built therapsid about 150 centimeters long, with teeth that were differentiated into incisors, canines and molars, rather than all alike, all conical-shaped, as reptilian teeth usually are.

During the Mesozoic era, many different kinds of reptiles arose, radiated into a great variety of descendants and finally became extinct. There were six major evolutionary lines of reptiles, the most primitive of which included the ancient stem reptiles and the turtles, originating in the Permian. The turtles have evolved the most complicated armor of any land animal, consisting of scales derived from the epidermis fused to the underlying ribs and breast bone. With this protection, both marine and land forms have survived with few structural changes since before the time of the dinosaurs. Their legs extend laterally, making locomotion difficult and slow, and their skulls are unpierced behind the eye sockets, a feature essentially unchanged from the old stem reptiles.

A second group of reptiles to survive with relatively few changes from the ancestral stem reptiles are the **lizards**—the most abundant of living reptiles—and the snakes. Most lizards have kept the primitive type of locomotion with the legs extended laterally, although many can run rapidly. Most of them are small, but the **monitor lizard** of the East Indies attains a length of 4 meters, and some fossil ones were 8 meters long. The **mosasaurs** (Fig. 15–14) of the Cretaceous were marine lizards which reached a length of 13 meters, and had a long tail useful in swimming.

F

FIGURE 15–12   Adaptive radiation among lizards. A, The Old World chameleon has grasping feet and a prehensile tail with which to climb about in the trees. B, The gecko climbs with the assistance of digital pads. C, The horned toad Phrynosoma is a ground-dwelling species that often burrows. D, The glass snake Ophisaurus also burrows. E, The Gila monster Heloderma and a related Mexican species are the only poisonous lizards in the world. F, The tuatara Sphenodon is one of the most primitive of living reptiles. (Courtesy of the New York Zoological Society.)

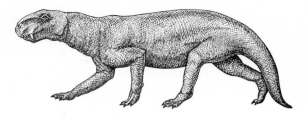

FIGURE 15–13   A mammal-like reptile (Lycaenops) from the late Permian Period in South Africa. (From Romer, A. S.: The Vertebrate Body. Philadelphia, W. B. Saunders Company, 1970; after Colbert.)

During the Cretaceous, snakes evolved from lizard-like ancestors. The important difference between snakes and lizards is not the loss of legs (some lizards are legless), but certain changes in the skull and jaws of the snake which enable it to open its mouth wide enough to swallow an animal larger than itself. A representative of an ancient line that has managed somehow to survive in New Zealand is the lizard-like **tuatara.** It shares several traits with the ancestral cotylosaurs, one of which is the presence of a third eye on the top of its head.

The main group of Mesozoic reptiles were the **archosaurs** or "ruling reptiles," of which the only living members are the alligators and crocodiles. At an early point in their evolution from stem reptiles, the ruling reptiles, which were then about 1 meter long, became adapted to two-legged locomotion—their front legs became short, while the hind legs became long, stout and considerably modified. These animals rested or walked on all fours, but in emergencies they reared up and ran on the two hind legs, assisted by their fairly long tail, which served as a balance. From the early archosaurs developed many different, specialized forms, some of which continued to use two-legged locomotion, others of which reverted to walking on all fours. These descendants include the **phytosaurs**—aquatic, alligator-like reptiles—common during the Triassic; the **crocodiles,** which evolved during the Jurassic and replaced the phytosaurs as aquatic forms; and the **pterosaurs,** or flying reptiles, which included animals the size of a robin, as well as the largest animal ever to fly—*Pteranodon*, with a wing-spread of 9 meters (Fig. 15–14). There were two types of flying reptiles, one with a long tail that had a steering rudder at the end, the other with a short tail. Both these types, apparently, were fish-eaters, and they probably flew long distances over the water in search of food. Their legs were not adapted for standing, and it is believed that, like bats, they rested by clinging to some support and hanging suspended.

Of all the reptilian branches, the most famous are the **dinosaurs** (meaning terrible reptiles). These were divided into two main types, one with a birdlike pelvis, the other with a reptilian pelvis.

The **Saurischia** (reptile pelvis) first evolved in the Triassic, and remained in existence until the Cretaceous. The early ones were fast, carnivorous, two-legged forms the size of a rooster, which probably preyed upon lizards and the archaic mammals then in existence. Throughout the Jurassic and Cretaceous there was a tendency in this group to grow larger, culminating in the gigantic carnivore of the Cretaceous, *Tyrannosaurus* (Fig. 15–15). Other Saurischia, beginning in the late Triassic, changed to a plant diet, reverted to a four-legged gait, and, during the Jurassic and Cretaceous, evolved into tremendous amphibious forms. Among these—the largest four-footed animals that ever lived—were *Brontosaurus*, with a length of 21 meters; *Diplodocus*, which attained a length of 29 meters, and *Brachiosaurus*, the biggest of them all, with an estimated weight of 50 tons.

The other group of dinosaurs, the **Ornithischia** (birdlike pelvis),

(Continued on Page 466)

FIGURE 15–14. A shallow inland sea covered the western half of the Mississippi Valley during much of the Cretaceous Period. Three reptiles characteristic of this time and place are shown. In the center is a mosasaur about 10 meters long; to the right is a giant marine turtle, 2.5 meters long; and flying in the left background are a number of reptiles of the genus Pteranodon, with short tails and a long crest extending back from the skull. (From the painting by Charles B. Knight, Copyright Chicago Natural History Museum.)

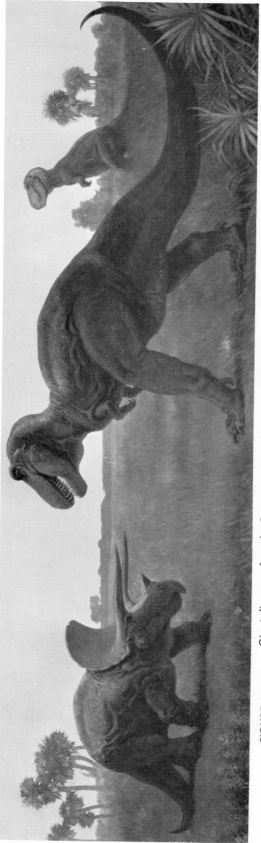

FIGURE 15-15   Giant dinosaurs from the Cretaceous Period of western North America. The largest flesh-eating dinosaur known was Tyrannosaurus, two of which are shown attacking the herbivorous, horned dinosaur Triceratops. Tyrannosaurus reached a length of 15 meters and a height of 6 meters. Its head, as much as 2 meters long, was equipped with many sharp teeth. The small front legs were completely useless. Tyrannosaurus walked on its powerful hind legs and balanced with its long tail. Triceratops was armed with a horn on its nose and a pair of horns over the eyes. A bony ruff covering the neck and shoulders protected that area. The rest of the body was covered with a leathery hide, thus the animal was vulnerable except when facing its enemies. (From the painting by Charles B. Knight, copyright Chicago Natural History Museum.)

FIGURE 15-16   Western Canada in the Cretaceous Period about 110,000,000 years ago. The land was low, well watered and covered with many swamps. Most of the dinosaurs were harmless, plant-eating forms of the Ornithischia group of reptiles, characterized by birdlike pelvic bones. Two types of duck-billed dinosaurs can be seen—three large, uncrested ones to the right and two types of crested ones in the left background. In the middle foreground is a heavily armored, four-footed dinosaur covered with bony plates and spines. In the center background are two "ostrich" dinosaurs—tall, slender animals with the general proportions of an ostrich, but with short forelegs and a long slender tail. (From the painting by Charles R. Knight, copyright Chicago Natural History Museum.)

FIGURE 15-17   Scene off the coast of North America in the Jurassic Period about 155,000,000 years ago. Two types of marine reptiles are shown: plesiosaurs with long necks, broad, flat bodies, and sturdy, paddle-shaped limbs; and ichthyosaurs with fishlike fins and tails. Both were fish-eaters. (From the painting by Charles R. Knight, copyright Chicago Natural History Museum.)

were vegetarians, probably from the beginning of their evolution. Although some of them walked upright, the majority had a four-legged gait. Having lost their front teeth, they developed a stout, horny, birdlike beak, which in some forms was broad and ducklike (hence the name "duck-billed" dinosaurs). Webbed feet were characteristic of this type; other species developed great armor plates as protection against the carnivorous saurischians. *Ankylosaurus* (Fig. 15–16) had a broad, flat body, covered with an armor plate and with large, laterally projecting spines. Still other ornithischians of the Cretaceous period developed bony plates around the head and neck. One of these, *Triceratops*, had two horns over its eyes and another over its nose—each 1 meter long.

Two other groups of Mesozoic reptiles, separate from each other and from dinosaurs, were the marine **plesiosaurs** and **ichthyosaurs** (Fig. 15–17). The former were characterized by an extremely long neck, which took up over half of their total length of 15 meters. The trunk was broad, flat and rather turtle-like, the tail was small, and the animal paddled along by means of finlike arms and legs. The ichthyosaurs (fish reptiles) had a body form superficially like that of a fish or a whale, with a short neck, a large dorsal fin and a sharklike tail. They swam by wiggling their tails, using their feet only for steering. The ichthyosaur young were apparently born alive, after having hatched from eggs within the mother, for the adults were too specialized to come out on land, and a reptilian egg will drown in water. The presence of skeletons of the young within the body cavity of adult fossils has strengthened this theory.

At the end of the Cretaceous, a great many reptiles became extinct; they were apparently unable to adapt to the marked changes brought about by the Rocky Mountain Revolution. As the climate became colder and drier, many of the plants which served as food for the herbivorous reptiles disappeared. Some of the herbivorous reptiles were too large to walk about on land when the swamps dried up. The smaller, warm-blooded mammals which had appeared were better able to compete for food, and many of them ate reptilian eggs. The demise of the many kinds of reptiles was probably the result of a combination of a whole host of factors, rather than any single one.

SECTION 15–11

BIRDS

The members of the class Aves are characterized by the presence of **feathers,** which are modified reptilian scales; these decrease the loss of water through the body surface, decrease the loss of body heat, and aid in flying by presenting a plane surface to the air. Birds and mammals are the only animals with a *constant* body temperature. They are sometimes called **warm-blooded**—other animals are called cold-blooded—but this is inaccurate, for a frog or snake may have a higher body temperature on a hot day than a bird or a mammal. Birds and mammals independently evolved mechanisms to keep body temperature constant despite wide fluctuations in the environmental temperature. The constant body temperature permits metabolic processes to proceed at constant rates and enables these animals to remain active in cold climates.

The earliest known species of bird, *Archaeopteryx*, was about the size of a crow, had rather feeble wings, jawbones armed with reptilian-like teeth, and a long reptilian tail covered with feathers (Fig. 15–18). Birds

FIGURE 15–18   A, A restoration of Archaeopteryx, the earliest known bird. B, A restoration of Hesperornis, a large diving bird of the Cretaceous Period. (Courtesy of the American Museum of Natural History.)

did not evolve from the flying reptiles, the pterosaurs, but from a group of primitive dinosaurs called *thecodonts*. Cretaceous rocks have yielded fossils of two other early birds—*Hesperornis*, an aquatic diving bird that had lost the ability to fly, and *Ichthyornis*, a powerful flying bird about the size of a tern. Like reptiles, birds lay eggs and have internal fertilization. Birds have reptilian scales on their legs, and the earliest birds, known only from fossils, had reptilian teeth. Adaptation to flight has involved the evolution of hollow bones and the presence of *air sacs*—extensions of the lungs that occupy the spaces between the internal organs. Not all birds fly; some, such as penguins, have small, flipper-like wings used in swimming (Fig. 15–19). Others, such as the ostrich and cassowary, have vestigial wings but well developed legs. Birds have become adapted to a variety of environments, and different species have very different types of beaks, feet, wings and tails.

Men have long been fascinated by the colors, songs and behavior of birds and these have been studied extensively. One of the most fascinating aspects of bird behavior is the annual migration that many birds make. Some birds such as the golden plover and arctic tern fly from Alaska to Patagonia and back each year, flying perhaps 25,000 miles en route. Others migrate only a few hundred miles south each winter and some, such as the bobwhite and great horned owl, do not migrate at all. The stimulus for the northward spring migration of certain birds that winter in California has been shown to be the increasing amount of daylight per day. This in some way stimulates the secretion of hypothalamic "releasing factors," which stimulate the pituitary gland to secrete gonadotropic hormones. These stimulate the growth of a testis or an ovary, and the increased amount of sex hormones circulating in the blood initiates the migration.

The services rendered to man by birds include the destruction of harmful rodents, insects and weed seeds and the dispersal of the pollen and seeds of many plants. The guano (excrement) deposited by sea birds in certain regions is a valuable fertilizer.

FIGURE 15–19   A group of neognathous birds. A, Penguins use their modified wings as flippers. B, Courtship of albatrosses. C, The young cormorant has to reach into the throat of its parent to get its food. D, An American egret or heron, a wading bird. E, a barn owl striking at its prey. F, two noddy terns on the nest. (A, Courtesy of Smithsonian Institute; B, Courtesy of Lieutenant Colonel N. Rankin; C, Courtesy of L. W. Walker from the National Audubon Society; D, and F, Courtesy of the American Museum of Natural History; E, Courtesy of David G. Allen.)

SECTION 15–12

MAMMALS

The distinguishing features of mammals are the presence of hair, **mammary glands** and sweat glands, and the differentiation of the teeth into **incisors, canines** and **molars.** Mammals have a constant body temperature and the covering of hair serves as insulation to aid in thermoregulation. Mammals evolved from a group of reptiles called **therapsids** (Fig. 15–13), probably during the Triassic period. At an early time in the evolution of mammals, one line of descent branched off from the major line and led to the egg-laying **monotremes.** Only two monotremes have survived to the present—the Australian duck-billed platypus and the spiny anteater (Fig. 15–20). The young, after hatching from the egg, are nourished by milk secreted by mammary glands.

The second subclass of mammals, the **marsupials** or pouched mammals, are also found largely in Australia—kangaroos, koalas and wombats. The opossum, which more closely resembles the primitive ancestral marsupials than the Australian marsupials do, is one of the few found outside of Australia. Marsupials do not lay eggs; the young are born alive in a very immature state and are transferred to a pouch on the mother's abdomen where they feed on milk secreted by the mammary glands and complete their development (Fig. 15–21).

The third subclass of mammals, the Eutheria or placental mammals, includes all the other mammals, all characterized by the formation of a **placenta** for the nourishment of the developing embryo while within the uterus (womb) of the mother. The placenta is a structure formed in part from tissues derived from the embryo and in part from maternal tissues by means of which the embryo receives nutrients and oxygen and eliminates wastes. The young are born alive in a more advanced state of development than the newborn marsupials.

FIGURE 15–20   *Two living examples of monotremes—mammals that lay eggs. Above, The duck-billed platypus with short fur, webbed feet, a horny, duck-shaped beak and an unusual tail. Below, The spiny anteater, about 45 centimeters long is covered with strong, pointed yellow spines with black tips. Its narrow black snout is cylindrical and it captures its food, mostly ants, with its long protrusible tongue, which is covered with sticky saliva. (Courtesy of the Australian News and Information Bureau.)*

FIGURE 15–21   A kangaroo mother with her young in her pouch. (Courtesy of Australian News and Information Bureau.)

The placental mammals evolved from a small, shrewlike, insect-eating, tree-dwelling ancestor during the Cretaceous. Some of these ancestral mammals remained in the trees and gave rise, through a series of intermediate forms, to the primates—monkeys, apes and men. Others lived on or under the ground, and during the Paleocene had conical, reptilian teeth, five digits on each foot, and a small brain; they walked on the soles of their feet instead of on their toes. During the Tertiary, the evolution of grasses which served as food, and forests which afforded protection, were important factors in leading to changes in the mammalian body pattern. Concomitant with a tendency toward increased size, the mammals all displayed tendencies toward an increase in the relative size of the brain, and toward changes in the teeth and feet. As modern forms better equipped for survival arose, the archaic mammals became extinct.

In the Paleocene and Eocene epochs the first carnivores, called **creodonts,** arose from the primitive insect-eating placental mammals (Fig. 15–22). They were replaced in the Eocene and Oligocene by more modern forms which eventually gave rise to the present-day carnivores, such as cats, dogs, bears and weasels, as well as to the web-footed, marine carnivores—the seals and walruses. One of the most famous fossil carnivores is the saber-toothed tiger (Fig. 15–23), which became extinct only recently in the Pleistocene. These animals had tremendously elongated, knifelike upper canine teeth, and a lower jaw that could be swung down and out of the way, allowing the teeth to be used as sabers for stabbing the prey.

The larger herbivorous mammals, most of which have hooves, are sometimes referred to as the **ungulates.** They do not form a single, natural group, but consist of several independent lines. Although both horses and cows have hooves, they are no more closely related than either one is to a tiger. The molar teeth of ungulates are flattened and enlarged to facilitate the chewing of leaves and grass. Their legs have become elongated and

FIGURE 15–22   Restoration of an archaic meat-eating mammal, a creodont from the Eocene Period, eating an Eohippus, a small ancestor of the horse. (Copyright by the American Museum of Natural History.)

adapted for the rapid movement necessary to escape predators. The earliest ungulates, called **condylarths,** appeared in the Paleocene; they had long bodies and tails, flat, grinding molars, and short legs ending in five toes, each of which bore a hoof. Corresponding to the archaic carnivores, or creodonts, were the archaic ungulates called **uintatheres.** During the Paleocene and Eocene, some of these were as large as elephants, and some had three large horns projecting from the top of the head.

The fossil records of several ungulate lines—the horse, the camel and the elephant—are complete, and it is possible to trace the evolution of these animals from small, primitive, five-toed creatures. The chief evolutionary tendencies in the ungulates have been toward an increase in the overall

FIGURE 15–23   Restoration of a scene at the Rancho La Brea tar pits (Los Angeles, California) during the Pleistocene epoch. Many well preserved specimens of animals now extinct have been found embedded in the asphalt. Left foreground, two saber-toothed tigers. Right foreground, three large ground sloths. The giant vultures, now extinct, had a wing spread of 3 meters. In the background are mastodons and dire wolves. (Copyright by the American Museum of Natural History; from a painting by Charles R. Knight.)

size of the body and a decrease in the number of toes. The ungulates were early divided into two groups—one characterized by an even number of toes, and including the cow, sheep, camel, deer, giraffe, pig and hippopotamus; the other characterized by an odd number of toes, and including the horse, zebra, tapir and rhinoceros. The elephants and their recently extinct relatives, the mammoths and mastodons, can be traced back to an Eocene ancestor the size of a hog which had no trunk. This primitive form, called *Moeritherium,* was close to the stem that also gave rise to such dissimilar creatures as the coney (a small, woodchuck-like animal found in Africa and Asia) and the sea cow.

The whales and porpoises descended from whalelike forms of the Eocene, called **zeuglodonts,** which in turn are believed to have evolved from the creodonts. The evolutionary history of the bats can be traced to ancestral, winged types, also of the Eocene, which descended from the primitive insectivores. The evolutionary history of some of the other mammals—the rodents, rabbits and edentates (anteaters, sloths and armadillos) —is less well known.

Some of the principal orders of placental mammals are the following:

(1) **Insectivora**—moles, hedgehogs and shrews. These are insect-eating animals, considered to be the most primitive placental mammals and the ones closest to the ancestors of all the placentals. The shrew is the smallest mammal alive; some weigh less than 5 grams.

FIGURE 15–24   *High speed stroboscopic photograph of a little brown bat, Myotis lucifugus, about to catch a falling meal worm. Note the large, extended ears used in detecting sounds reflected from the prey. (Courtesy of Mr. Frederic Webster.)*

(2) **Chiroptera**—bats. These mammals are adapted for flying; a fold of skin extends from the elongated fingers to the body and legs, forming a wing. They eat insects and fruit, or suck the blood of other mammals. Bats are guided in flight by a sort of biological sonar (Fig. 15–24): they emit high frequency squeaks and are guided by the echoes from obstructions. Blood-sucking bats may transmit diseases such as yellow fever and paralytic rabies.

(3) **Carnivora**—cats, dogs, wolves, foxes, bears, otters, mink, weasels, skunks, seals, walruses and sea lions. These are all flesh eaters, with sharp, pointed canine teeth and shearing molars.

(4) **Rodentia**—squirrels, beavers, rats, mice, porcupines, hamsters, chinchillas and guinea pigs. These numerous mammals have sharp, chisel-like incisor teeth.

(5) **Edentata**—sloths, anteaters and armadillos. Mammals with few or no teeth.

(6) **Primates**—lemurs, monkeys, apes and man. These mammals have highly developed brains and eyes, nails instead of claws, opposable great toes or thumbs and eyes directed forward.

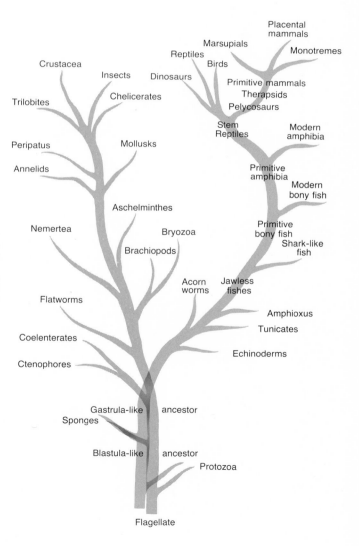

FIGURE 15–25   A diagram illustrating current theories of the evolutionary relationship in the animal kingdom. The branch including the Protostomia is indicated in black and the branch including the Deuterostomia is indicated in tint.

(7) **Artiodactyla**—cattle, sheep, pigs, giraffes and deer. Herbivorous hooved animals with an even number of digits per foot.

(8) **Perissodactyla**—horses, zebras, tapirs and rhinoceroses. Herbivorous hooved animals with an odd number of digits per foot.

(9) **Proboscidea**—elephants, mastodons and wooly mammoths. Animals with a long muscular proboscis or trunk, thick, loose skin, and incisors elongated as tusks. These are the largest land animals, weighing as much as 7 tons.

(10) **Sirenia**—sea cows, dugongs and manatees. These are herbivorous aquatic animals with finlike forelimbs and no hind limbs. They are probably the basis for most tales about mermaids.

(11) **Cetacea**—whales, dolphins and porpoises. These are marine mammals with fish-shaped bodies, finlike forelimbs, no hind limbs, and a thick layer of fat called blubber covering the body. Sulfur bottom whales are the largest animals ever known—up to 35 meters long and 150 tons.

The various members of the animal kingdom cannot be placed on a single scale ranging from lowest to highest, for evolution has occurred in the manner of a branching tree rather than in a single continuous series (Fig. 15–25). We cannot say, for example, that the starfish is "higher" or "lower" than the oyster; the two forms are simply representatives of the two main trunks of the evolutionary tree—one of which gave rise to echinoderms and chordates, the other to flatworms, nemerteans, annelids, arthropods and mollusks. Between the two groups are deep-lying differences of structure and development.

SUGGESTIONS FOR FURTHER READING

Allen, A. A.: *The Book of Bird Life.* 2nd ed. Princeton, New Jersey, D. Van Nostrand Company, Inc., 1961.
> A fine account of the ecologic aspects of birds, plus discussion of the methods used to observe and study birds in the field.

Barrington, E. J. W.: *The Biology of the Hemichordates and Protochordata.* Edinburgh, Oliver and Boyd, Ltd., 1965.
> A thorough, up-to-date account of the biological features of the lower chordates.

Berrell, N. J.: *The Origin of the Vertebrates.* New York, Oxford University Press, 1955.
> A review of the various theories regarding the evolutionary origin of the vertebrates.

Cochran, D. M.: *Living Amphibians of the World.* New York, Doubleday & Company, Inc., 1961.
> A fascinating and superbly illustrated description of the amphibians.

Florey, E.: *An Introduction to General and Comparative Animal Physiology.* Philadelphia, W. B. Saunders Company, 1966.
> An excellent text of the general physiology of animals.

Hartman, C. G.: *Possums.* Austin, University of Texas Press, 1952.
> A delightful account of the habits, natural history and reproduction of these unusual mammals.

Marshall, N. B.: *The Life of Fishes.* Cleveland, The World Publishing Company, 1966.
> A fine account of the biology of fishes written by a senior investigator at the British Museum of Natural History.

Orr, R. T.: *Vertebrate Biology.* 3rd ed. Philadelphia, W. B. Saunders Company, 1971.
> A valuable reference book covering many aspects of vertebrate life, such as territories, dormancy and population dynamics.

Romer, A. S.: *The Vertebrate Body.* 4th ed. Philadelphia, W. B. Saunders Company, 1970.
    A standard text of comparative anatomy of the vertebrates.

Schmidt, K. P., and Inger, R. F.: *Living Reptiles of the World.* New York, Doubleday & Company, Inc., 1957.
    A beautifully illustrated account of the reptiles.

Schmidt-Nielson, K.: *Desert Animals: Physiological Problems of Heat and Water.* New York, Oxford University Press, 1964.
    A careful analysis of the adaptations of camels, kangaroo rats, man and other animals to life in the desert.

Welty, J. C.: *The Life of Birds.* Philadelphia, W. B. Saunders Company, 1962.
    A remarkably complete discussion of birds and their ways.

Young, J. Z.: *The Life of Vertebrates.* 2nd ed. Clarendon, Texas, Clarendon Press, 1963.
    A fascinating account of the evolution and adaptation of vertebrates, interweaving discussions of structure and function.

PART
4

THE BIOLOGY
OF ORGANISMS

BIOLOGICAL MEMBRANES AND THE TRANSPORT OF MOLECULES

Living material, whether plant, animal or microbial, contains specific kinds of macromolecules arranged in a highly ordered fashion. In contrast, the nonliving environment of cells is characterized by the random distribution of nonspecific molecules. Between the two phases is an interface or boundary, the **plasma membrane.** The plasma membrane regulates, both actively and passively, the amounts and concentrations of water, inorganic ions and small organic molecules in the cell. In addition, by regulating the amounts of precursors and the ionic environment within the cell, the membrane may play a role in determining the macromolecular constituents of the cell.

A cell is not simply an unorganized sack of enzymes. Not only is the plasma membrane an active part of the cell that separates the "inside" of the cell from the "outside," but other membranes divide the space within the cell into specific compartments—nucleus, mitochondria, chloroplast, Golgi apparatus, lysosome, endoplasmic reticulum and so on—each of which contains specific enzymes and carries out specific functions. This intracellular compartmentalization permits the local accumulation of certain compounds and denies admittance to others, thus facilitating certain reactions and inhibiting others. It thus provides the morphological basis for the intracellular division of biochemical labor.

SECTION 16-1

THE PLASMA MEMBRANE: STRUCTURE AND FUNCTION

Early research on the properties of the plasma membrane showed that water can diffuse rapidly into most cells and that the cell surface acts as a semipermeable (differentially permeable) membrane. When cells are placed in solutions with a low concentration of salts (and a high concentration of water) water tends to diffuse into the cells and they swell and burst. When cells are placed in solutions with a high concentration of salts (and a lower concentration of water), water tends to pass out of the cell (water diffuses in each case from a region of high to one of lower concentration of water) and the cell shrinks. When cells are placed in an isosmotic solution, one with the same concentration of solutes as that inside the cells, there is no net transfer of water into or out of the cells, provided that the solute molecules on the outside cannot penetrate the plasma membrane.

Sucrose does not pass through the plasma membrane of the red blood cell but glycerol does. Red cells neither swell nor shrink when suspended in 0.3 M (isosmotic) sucrose and can be preserved for a long time in that solution. In contrast, when placed in 0.3 M glycerol (isosmotic at time zero), they gradually swell as glycerol diffuses into the cell, increasing the osmolarity of the cell contents and secondarily causing water to diffuse into the cell. Red cells placed in 0.15 M NaCl ("physiologic saline"), which is isosmotic at time zero, neither shrink nor swell, not because this is the exact concentration of Na^+ and Cl^- in the cell, but because they have a "sodium pump" which removes the Na^+ ions that diffuse in. If the red cell's sodium pump is blocked by an inhibitor, the cell will swell in isosmotic sodium chloride solution just as it does in isosmotic glycerol.

Research in the previous century established two generalizations regarding plasma membranes, sometimes called "Overton's rules." First, compounds soluble in lipids tend to diffuse through biological membranes faster than ones soluble in water. Stated another way, solutes penetrate membranes approximately in the order of their relative solubilities in olive oil and water. Second, small molecules tend to penetrate membranes more rapidly than larger ones. Water-soluble molecules of the size of sucrose (molecular weight 342) or larger generally do not penetrate plasma membranes at all. From these data it was inferred that the plasma membranes must contain a large amount of lipid and that it acts as a molecular sieve which regulates the size of molecules that may enter or leave the cell.

From mammalian red cell membranes (called red cell "ghosts") obtained by hemolyzing the cells and removing the contents, Gorter and Grendel extracted all the membrane lipid, spread it as a monolayer on water and compared its area with the calculated area of the red cell membranes from which it was derived. These calculations showed that each red cell membrane contained enough lipid to cover its surface with a layer two molecules thick. From these and other data, Gorter and Grendel concluded that the membrane is composed of two concentric layers of lipid molecules with their nonpolar ends between the layers and their polar, hydrophilic ends outermost.

Other analyses have shown that biological membranes of all kinds— Golgi, endoplasmic reticulum and mitochondrial, as well as plasma membranes—contain substantial amounts of lipids, especially phospholipids. The proportions of the several kinds of lipids present in different kinds of membranes may vary substantially, but they include **glycerophosphatides** such as phosphatidyl choline, **sphingolipids** such as sphingomyelin, and **sterols** such as cholesterol.

Biological membranes contain significant amounts of protein, indeed somewhat more protein than lipids by weight. When the values are corrected for their respective molecular weights, the data suggest that membranes have about 70 lipid molecules for each protein molecule. Most membranes have ATPase activity, which appears to play a role in the process of "active transport." In addition, red cell membranes have proteins resembling actin and myosin.

Electron micrographs agree with analytic data that biological membranes are a sandwich composed of a bimolecular layer of lipid with a layer of protein on both inner and outer surfaces (Fig. 16–1). Electron micrographs of the surface of a cell such as a red cell typically show two parallel, dark, osmiophilic lines separated by a light, osmiophobic space (Fig. 16–2). The lines vary somewhat in thickness in different cells, but the dark lines are usually 20 to 25 Å thick and the light space is 35 to 50 Å wide, with a total thickness of 75 to 100 Å. Robertson, in 1959, was the first to suggest that the three lines constitute one membrane, with the dark lines representing the polar ends of the phospholipid molecules and their associated proteins, whereas the light space between represents the hydrophobic core of the lipid bilayer.

To account for the passage of water and water-soluble molecules through the plasma membrane it was suggested initially that there must be "pores" in the lipid bilayer through which the aqueous phase on one side is continuous with the aqueous phase on the other (Fig. 16–3). Thompson has prepared synthetic lipid double layers made with purified lipids and has shown that they mimic many of the properties of natural membranes. These lipid membranes have a fairly large permeability coefficient for water, which suggests that it may be unnecessary to postulate pores for the passage of water molecules.

The results of many experiments led investigators in the past to suggest that biological membranes are mosaics of lipid and protein regions to account for the specific permeability effects observed. A variety of theories postulated the presence in the membrane of pores with permeability properties that differ from those of the rest of the membrane. From the rates at which various ions leak through biological membranes it has been possible to estimate the diameter that such pores should have. Several such studies suggest that the pores are about 8 Å in diameter (Fig. 16–3) and that there are about 10^{10} pores per cm². The total area of the pores is a very small fraction (0.06 per cent) of the total surface area of the cell.

The rate at which a substance passes by simple diffusion through a membrane is determined by its solubility in lipids and by the size of the diffusing particle. Water-soluble substances with a particle size greater

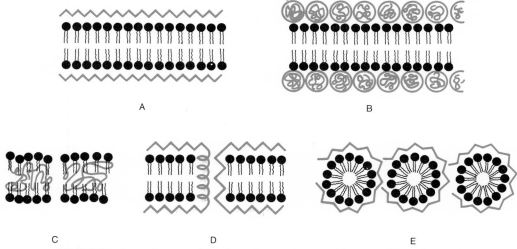

A   B

C   D   E

FIGURE 16–1   Some models of membrane structure. The several hypotheses about membrane structure differ in the way the polypeptide chains (tint) and the phospholipid molecules (black) are believed to be assembled. The diagrams demonstrate some of the more popular theories regarding membrane structure.

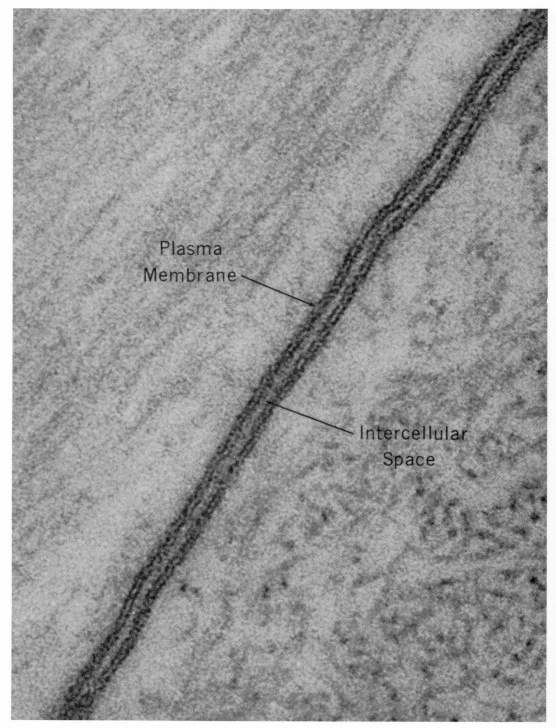

FIGURE 16–2   Electron micrograph of the boundary between two glial cells of the annelid Aphrodite illustrating the intercellular space between the two plasma membranes. Each plasma membrane is shown to consist of two dense lines and a less dense intermediate layer. The trilaminar structure is the unit membrane. × 260,000. (From Fawcett, D. W.: The Cell. Philadelphia, W. B. Saunders Company, 1966.)

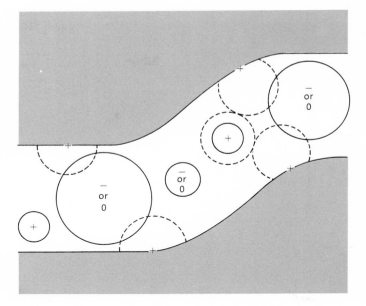

FIGURE 16–3   Diagram illustrating the postulated structure of the ultramicroscopic pores in cell membranes. The positive charges along the surface of the pore are indicated to show how they might influence the size and type of particle that could pass through the pore. (After Guyton, A. C.: Function of the Human Body. 3rd ed. Philadelphia, W. B. Saunders Company, 1969.)

than 8 Å, the diameter of the pores, are essentially unable to penetrate the membrane. Many charged particles are **hydrated,** surrounded by a shell of tightly bound water molecules, and the effective diameter of the diffusing particle is increased by this shell of water. The rate of diffusion of the particle is less than that predicted for the free, unhydrated ion. The pore behaves as though its surface were lined with positive charges, probably positively charged protein molecules. Each positive charge is surrounded by electrostatically charged space that protrudes into the pore. Any positively charged ion is also surrounded by a sphere of electrostatic charge, and the two positive charges tend to repel each other. This may explain the observation that positively charged particles pass through membranes only slowly and with difficulty, even when their hydrated diameter is less than 8 Å.

High-resolution electron micrographs indicate that the membrane has some sort of beaded substructure and that it may be made of hexagonal subunits with a central dense granule 25 Å in diameter. Similar subunits have been detected in the membranes of chloroplasts and mitochondria.

Other investigators have suggested that at least some biological membranes may be based not on lipid bilayers but on a skeleton of protein molecules assembled into a sheet with adherent lipids. It would appear that there is no single uniform structural plan that applies to all biological membranes. Rather, cells have a spectrum of membrane structures, ranging from pure protein in the head of the bacteriophage to the classic lipid bilayer with outer protein layers in the myelin sheath. Variations on the theme include membranes of protein polymers with interspersed lipids, and lipid bilayers with embedded globular enzymes (Fig. 16–1).

Cells rapidly frozen in liquid nitrogen, then fractured, stained with platinum-carbon and examined in the electron microscope showed that the fracture typically splits each unit membrane into two half-membranes. This confirms the reality of lipid bilayers in biological membranes and supports the idea that different kinds of membranes may have significant differences in their molecular architecture. There may be differences in the thickness and enzyme content (ATPase) of different regions of a given membrane within a single cell.

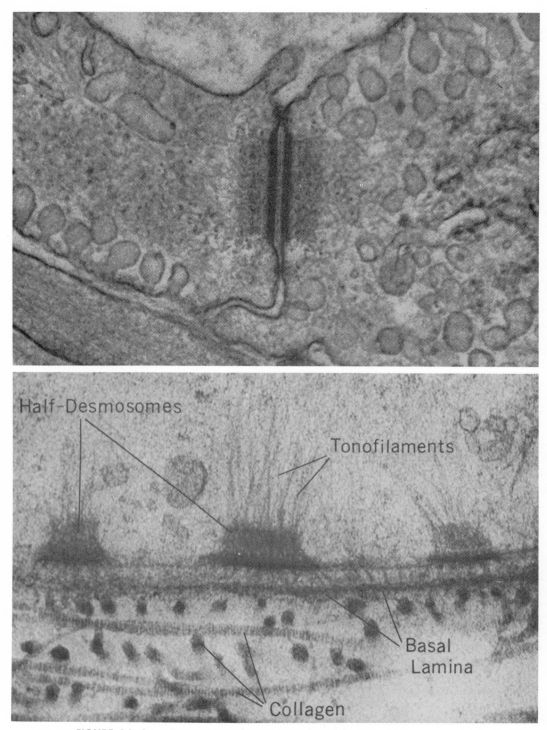

FIGURE 16–4   *Desmosomes, discontinuous button-like structures made up of two dense plaques which serve to hold opposing cell surfaces tightly together. The endothelial cell junction in the rete mirabile of the toadfish gas bladder. × 95,000. (From Fawcett, D. W.: The Cell. Philadelphia, W. B. Saunders and Company, 1966.)*

SECTION 16–2

DESMOSOMES

In certain areas where one cell is in close contact with an adjacent one, discontinuous, button-like plaques, termed **desmosomes,** are found on the two opposing cell surfaces, separated by an intercellular space about 250 Å wide (Fig. 16–4). Each plaque has a thin layer of dense material coating the inner leaf of the cell membrane to which are attached bundles of fine cytoplasmic filaments. The desmosomes of different tissues may differ slightly in their structure. Desmosomes are especially abundant in stratified squamous epithelium and apparently serve to hold the cells together.

Where membranes are synthesized and how this occurs are still challenging problems. In some cells, new plasma membrane seems to be formed by the coalescence of preexisting cytoplasmic vesicles. In other cells, the membrane appears to be formed spontaneously from precursor molecules by surface forces operating at the interface between the cell cytoplasm and the external medium.

SECTION 16–3

PHAGOCYTOSIS AND PINOCYTOSIS

The plasma membrane has the remarkable property of capturing materials in the external environment by forming small, membrane-lined indentations which fold to enclose droplets of external medium (Fig. 16–5). The vacuoles then sink into the deeper layers of the cytoplasm where they are processed appropriately. In this way, specific materials can be transferred from outside to inside the cell without any break in the continuity of the plasma membrane. This process of surface vacuolization, termed **phagocytosis** (cell-eating) if the vacuoles are large and contain particulate matter, or **pinocytosis** (cell-drinking) if the vacuoles are small and contain only dissolved substances, has been observed in many kinds of cells and studied especially in amoebae and in white blood cells (leukocytes). The reverse process, in which membrane-lined vesicles fuse with the plasma membrane and release their contents to the exterior of the cell, seen especially in secretory cells, has been termed **emeiocytosis** (cell-vomiting).

Pinocytosis can be induced in cells by the presence of specific solutes — proteins, acidic or basic amino acids — in the surrounding medium. In amoebae, short **pseudopodia** form, each with a narrow undulating channel extending from its tip to its base. Vacuoles are formed at the inner end of

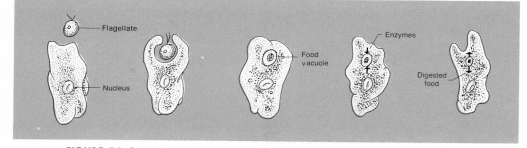

FIGURE 16–5   *Phagocytosis in an amoeba. The amoeba forms pseudopods which engulf a flagellate. The ingested flagellate is incorporated into a food vacuole within the amoeba.*

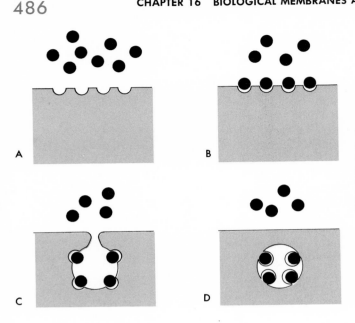

FIGURE 16–6   *Diagram illustrating the hypothetical mechanism of pinocytosis. Protein molecules in solution (dark circles) are bound to specific sites on the surface of the cell membrane which then folds in to bring the protein particles into a vesicle within the cytoplasmic substance of the cell.*

the channel, separate from it, and pass deep into the cytoplasm. The plasma membrane of cells of higher animals may exhibit undulations during pinocytosis, but no channels are evident. Pinocytosis is an active, energy-requiring process, and mitochondria are usually oriented near these active, undulating surfaces. Pinocytosis can be inhibited by appropriate metabolic inhibitors and is varied by changes in the temperature, pH and nutritive status of the cell. In a single cycle of pinocytosis an amoeba takes up a volume of fluid equal to from 1 to 10 per cent of its total volume. The process requires the synthesis of an area of new membrane equivalent to as much as 6 per cent of the initial surface area of the amoeba.

In amoebae, and possibly in other cells, protein molecules are specifically bound to the membrane of the vacuole and concentrated there, after which the membrane flows inward and forms a vesicle (Fig. 16–6). The amount of protein taken up in pinocytosis is much greater than could be accounted for if the proteins were simply taken in as part of the solution being "drunk" by the cell. For example, an amoeba can take up in five minutes an amount of protein equivalent to that present in 50 times its own volume of the protein-containing medium. Thus pinocytosis is a kind of membrane activity which brings about the selective uptake of materials from solutions. It can account for the rapid transport of large water-soluble molecules such as proteins and nucleic acids across the plasma membrane. Human cells in tissue culture have been shown to take up proteins from the medium; this is stimulated by the presence of polyamines such as polyornithine. Pinocytosis is an energy-requiring mechanism for the selective transport of substances across the plasma membrane and may account for some of the observed "active transport" of materials.

SECTION 16–4

ACTIVE TRANSPORT PHENOMENA

That living cells can regulate the concentration of ions in their interior has been known for more than 125 years, but it was only in the 1930's that it was suggested that cell membranes might use energy-requiring processes to transport specific molecules into or out of the cell. Since then abundant evidence has accrued to support the hypothesis that "active transport" is a universal property of the plasma membrane and the intracellular membranes of all cells. The plasma membrane is a **transducer** which uses chemical energy for the accumulation or transport of specific molecules. Associated with this and dependent on it are a variety of bioelectric phenomena, such as the propagation of the nerve impulse (p. 495).

Under "steady state" conditions, when there is no net change in the kinds of molecules or their concentrations in the cell, the interior of the cell contains many kinds of organic molecules, some small and others large, plus potassium, magnesium and phosphate ions and small amounts of sodium and chloride ions. Most of the larger organic molecules are present in ordered arrays; all of them are unable to leave the cell because they are too large to penetrate the plasma membrane. The fluid bathing the exterior of the cell, called the **extracellular fluid,** consists largely of water and inorganic ions, which in general are able to penetrate the plasma membrane. The macromolecules within the cell are bathed with an **intracellular fluid** containing water and inorganic ions. The composition of the intracellular fluid reflects, to a considerable extent, the composition of the extracellular fluid.

The distribution of uncharged molecules across the membrane is regulated by the relative concentrations of that molecule inside and outside the cell. The distribution of charged particles across the membrane reflects the electrical gradient in the system as well as the chemical concentration gradient of that species of ion. Thus, whether a given positively (or negatively) charged ion will enter or leave a cell is determined by a combination of the electrical and chemical gradients across the membrane. The electrical gradient itself may be partly or entirely the result of an asymmetric distribution of ions brought about by differences in the membrane's permeability to particles of different sizes or different charges. This is an important factor in the transport of oxygen and carbon dioxide by red blood cells and in their accomplishing this with a minimal change in the pH of the blood (p. 573).

It has been possible to introduce microelectrodes inside a number of cells — giant algae, the giant axons of the squid — and to record **potential differences** in the range of 50 to 200 mV, with the inside of the cell negative to the outside (Fig. 16–7). When the cell is made anoxic or is poisoned with metabolic inhibitors, the resting potential drops to 10 mV or less, because the intracellular potassium ions, K^+, leak out. It follows that the potential difference observed in normal cells is not the result of a simple physical equilibrium but rather is a state that requires the constant expenditure of energy by the cell to maintain. This potential difference results primarily from the difference in the concentration of potassium ions on the two sides of the membrane.*

Chemical analyses made more than a century ago by the German chemist Liebig showed that cells and tissues contain much less sodium than the extracellular fluid. Subsequent investigators had assumed that the cell

---

*The quantitative relationship of the membrane potential to the concentration of ions inside and outside the cell can be calculated from the Nernst equation, $E = \dfrac{RT}{z\,n\,F} \ln C_1/C_2$, where E is the electrical gradient (membrane potential) in volts, $\ln C_1/C_2$ is the natural logarithm of the concentrations of the ions on the two sides of the membrane, R is the universal gas constant, T is the absolute temperature, z is the valence of the ion, n is the number of moles and F is the Faraday constant.

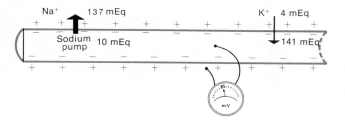

FIGURE 16–7   The concentration gradients of sodium and potassium at a nerve cell membrane. These differences in electrolyte concentrations account for the resting potential of the nerve membrane.

membrane is impermeable to sodium, but as soon as isotopic tracers were available, a constant flux of sodium ions in both directions across cell membranes was revealed. Thus, as Dean concluded in 1941, there must be some kind of **"sodium pump,"** some mechanism by which plasma membranes can selectively move sodium ions from inside the cell to outside the cell. Without this, the physical process of diffusion would cause sodium ions to enter, accompanied by water molecules that would ultimately lead to the swelling and bursting of the cell. There appear to be other selective ion pumps in some, if not in all, cells which are specific for potassium, hydrogen and chloride ions.

The distribution of ions across a cell membrane under steady state conditions is the resultant of two distinct processes: the physical forces causing ions to move down an electrochemical gradient, and energy-requiring processes that transport specific ions. Sodium ions are constantly being extruded from the cell by the energy-requiring sodium pump, and then leak back into the cell by diffusion. The differences in the concentrations of ions on the two sides of the membrane, and the **membrane potential** (the difference in voltage due to the differences in ionic composition) are determined by the rate of ion pumping and the rate of ionic leaking. A change in either process will change the membrane potential; both, as we shall see, play roles in the propagation of an impulse along a nerve.

Studies of active transport in membranes have revealed some striking examples of the 100-fold or more accumulation of specific ions by membranes, such as the accumulation of iodide by the vertebrate thyroid gland. All of these require the expenditure of metabolic energy by the cell and result in a decrease in the total entropy (p. 96) of the system. Isolated mitochondria, for example, can accumulate calcium and phosphate ions by metabolic processes so that the intramitochondrial concentration of calcium is 1000-fold greater than that outside, and calcium phosphate precipitates within the mitochondrion. The uptake of phosphate appears to compete with the process of oxidative phosphorylation within the mitochondria. This supports the suggestion that the active accumulation of ions by mitochondria is a modification of the process of oxidative phosphorylation, in which calcium ions instead of ADP act as phosphate acceptors.

The active transport of ions has been observed and analyzed in red cells, where transport occurs across a single plasma membrane, between the outside and inside of a cell. Ion transport has also been studied across a layer of cells such as the frog skin, the toad bladder or the kidney tubule, intestine or stomach of the rat. In these systems ions are taken up by cells on one side of the tissue and pumped out on the other. In a third type of system, ion transport has been studied in isolated subcellular organelles such as mitochondria, chloroplasts and nuclei. It seems likely that all biological membranes have selective mechanisms for the active transport of ions. The essence of active transport processes is that they require metabolic energy to drive them. This is usually demonstrated by inhibiting cell respiration by anoxia, by low temperature or by metabolic poisons, and observing a decreased rate of transport. Two drugs, **ouabain** and **strophanthidin,** appear to be specific inhibitors of active transport and are commonly used diagnostically, just as puromycin and actinomycin are inhibitors used diagnostically in studies of the synthesis of proteins and nucleic acids. The active transport of ions across a membrane can be counteracted by imposing an electric potential across the membrane.

Some of the earliest experiments to demonstrate the active transport of ions were carried out using pieces of frog skin stretched across a port between two compartments containing salt solutions. The transport of chloride ions and water from the outside of the skin to the inside, demonstrated by Huf in 1936, was later shown by Ussing to be a passive movement secondary to the active transport of sodium ions. When the two sides of the skin were "short circuited" so that the electrical charge as well as the concentrations of ions on the two sides were identical, only sodium ions were transferred from the outside to the inside of the skin. When the transfer of sodium was inhibited by adding carbon dioxide to the system, the current across the membrane also decreased. This indicated that the resting potential of the membrane is a result of the active transport of sodium.

A quantitative interdependence of the movements of sodium and potassium ions has been observed across the red cell membrane and across the membrane of the giant axon of the squid. This has led to the concept that a single pump, a single carrier mechanism, transports potassium ions in and sodium ions out. This theory appears to be true in red cells, neurons and yeast, but not in all cells; some appear to have separate mechanisms for the extrusion of sodium and the uptake of potassium.

It has been possible to calculate that each red cell has between 10^3 and 10^4 sites for the exchange of potassium and sodium. Since the exchange of sodium and potassium is inhibited by any factor which inhibits glycolysis or oxidative phosphorylation, it has been inferred that ATP is the direct source of energy for the sodium pump. Even red cell ghosts are able to accumulate potassium and extrude sodium if provided with ATP. An ATPase in red cell membranes requires both Na^+ and K^+ for activation and shares many of the properties of the sodium pump. It is inhibited by ouabain and its kinetic properties are quantitatively similar to those of the sodium transport system. The ATPase of intact red cell ghosts is stimulated by extracellular, but not by intracellular, potassium and by intracellular, but not by extracellular, sodium, as one might expect if the ATPase were an integral part of the sodium-potassium pump. Although the actual mechanism of the sodium pump is not yet clear, most investigators postulate that the membrane has special carrier molecules whose binding to, and separation from, the ion transported is coupled in some way to the hydrolysis of ATP and to metabolic processes (Fig. 16–8). Several models have been proposed for the mechanism of transport via carrier molecules, but none has been fully established.

This mechanism for the extrusion of sodium must have developed at a very early stage in the evolution of cells to control the amount of water within the cell and its total volume. Indeed, as soon as a cell membrane had evolved, a pump was necessary to keep the cell from swelling and bursting. The action of the sodium pump results in a voltage difference across the membrane and influences the distribution of other ions between cell and environment. In plant cells this mechanism is involved in turgor pressure and is supplemented by pumps for other kinds of ions. Beginning with primitive cells, changes in membrane voltage can be propagated along

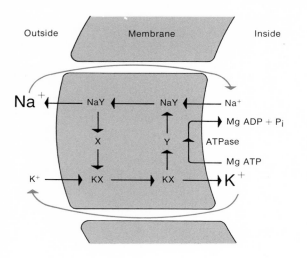

FIGURE 16–8   Suggested mechanisms for the active transport of sodium and potassium through the membrane. The two transport mechanisms are shown coupled to an energy yielding ATPase in the inner surface of the membrane. (After Guyton, A. C.: Function of the Human Body. 3rd ed. Philadelphia, W. B. Saunders Company, 1969.)

the surface of both plant and animal cells to transmit information about changes in the adjacent environment. This feature of ion transport is highly developed in neurons and is the basis of the propagated nerve impulse.

SECTION 16–5

ACTIVE TRANSPORT OF SUGARS

The energy-dependent uptake of uncharged molecules, such as glucose and galactose, by cells was shown in the twenties and early thirties, before the active transport of ions was recognized. The rate of transport of glucose into the red cell is decreased much more by lowering the temperature of the system than would be expected for simple physical diffusion, in which the rate would be proportional to the absolute temperature expressed in °K. The uptake of sugars by intestinal cells is inhibited when glycolysis is inhibited by iodoacetate.

Many of the more recent studies of the transport of substances across the intestinal wall have utilized a short length of intestine removed from a rat or other animal which is turned inside out and its ends tied shut. When placed in an appropriate solution, the intestinal cells lining the gut (now on the outside) take up materials from the solution and extrude them to the "outside" of the intestinal wall, now inside the little sausage-shaped bit of intestine. Such isolated loops of intestine can concentrate sugars as much as 100-fold into the central cavity. The process is inhibited by anoxia, by lowering the temperature or by any of a host of metabolic inhibitors such as cyanide or fluoroacetate. The cells lining the intestine are known to carry out pinocytosis very actively, and it is possible that some of the glucose is transported by this mechanism. Autoradiographs of intestinal cells taking up glucose labeled with tritium show a high concentration of the labeled molecules at the brush border of the intestinal cell, where the pinocytotic vesicles are concentrated.

Although it was originally believed that sugars are phosphorylated when taken up by the cell and dephosphorylated when extruded on the serosal side, this theory has now been abandoned. Other experiments have ruled out a variety of other possible chemical alterations of the glucose molecule during the transfer process, and it now appears that the transport of glucose across the cell membrane does not involve any change in the chemical structure of the molecule. Instead, glucose combines with a carrier molecule in the plasma membrane and is transported as a glucose-carrier complex.

SECTION 16–6

FACILITATED DIFFUSION

Substances such as glucose, which are very poorly soluble in lipids and have diameters greater than 8 Å, still manage to penetrate plasma membranes such as those of red blood cells at a respectable rate. This appears to occur by "facilitated diffusion," by the combination of the molecule with a specific carrier molecule, which is a peptide or protein (Fig. 16–9). The glucose molecule, G, combines with the carrier molecule, X, at the outer surface of the membrane, and the glucose-carrier complex, GX, which is lipid-soluble, can then diffuse across the membrane to the inner side, where the complex dissociates and glucose is released to the interior of the cell. The carrier molecule then diffuses back to the outer surface and is ready to take up another glucose molecule. The rate of glucose transport by such a system will be limited by the total number of carrier molecules per cell and by the rate at which the **glucose-carrier complex** can be formed and cleaved. It will show "saturation kinetics," i.e., when the extracellular concentration of glucose is low, the rate of penetration is proportional to that concentration. However, at higher extracellular concentrations of glucose, the rate is no longer proportional to that concentration; all the carrier molecules are saturated with glucose. The carrier molecules are specific and will accept glucose and certain closely similar sugars but not other sugars. One kind of sugar molecule will compete with others having a similar molecular structure for the binding sites on the carriers.

Facilitated diffusion does not require the addition of energy to the system for transport to occur when the extracellular concentration of glucose is greater than the intracellular concentration and glucose is moving down a chemical gradient. In other cells, such as those in the intestinal epithelium and the lining of the kidney tubule, where glucose is accumulated to a higher concentration and is moving up a chemical gradient, energy is required to drive the reactions. The hormone **insulin** greatly increases the rate of uptake of glucose by skeletal muscle and certain other cells in man. It is not clear whether insulin increases the number of ef-

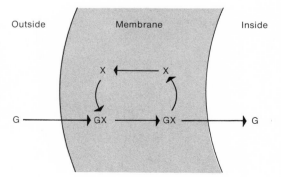

FIGURE 16–9   *The facilitated diffusion of glucose through a membrane. The theory suggests that the membrane contains specific carriers (X) to which the glucose is bound and transported through the membranes. A complex of carrier and glucose is formed in the outer membrane and cleaved at the inner membrane surface.*

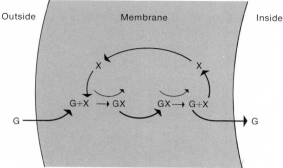

FIGURE 16–10   A mechanism which has been postulated to account for the active transport of glucose through membranes. The theory suggests that the formation or cleavage of the glucose-carrier complex may be coupled to an enzymatic system (small arrows) which supplies biologically useful energy and drives the transport system.

fective carrier molecules or whether it increases the rate of the reactions involved in loading or unloading the carriers. It is believed that the active transport of glucose also involves its combination with specific carrier molecules and that energy is used to drive the reactions involved in the formation or cleavage of the glucose-carrier molecule complex (Fig. 16–10).

A lipoprotein that has been isolated and partially purified by E. P. Kennedy from the cell membranes of *Escherichia coli* appears to be the transport factor, the ***"permease"*** for lactose. There are about 9000 molecules of permease per coli cell and it is believed that they operate by combining with lactose at the outer surface of the membrane. The protein-lactose complex diffuses across the membrane to the inner surface where lactose is released from the carrier. This hypothesis is formally similar to the mechanism proposed for the sodium pump.

SECTION 16–7

ACTIVE TRANSPORT OF AMINO ACIDS

In addition to the sodium pump, all cells studied so far have mechanisms for the active transport of amino acids. The transport of amino acids through the intestinal epithelium and kidney tubules is especially important in the overall economy of man and other higher animals. There is evidence that there are at least four different systems for the transport of different kinds of amino acids, all of which require as a cofactor a derivative of the vitamin ***pyridoxine*** (vitamin B_6). The hormones insulin, growth hormone, cortisol, testosterone and estradiol play a role in regulating the rate of transport of amino acids into one or another kind of cell.

SECTION 16–8

ACID SECRETION BY THE GASTRIC MUCOSA

One of the more remarkable cellular pumps is the one in the gastric mucosa, the cells lining the stomach, which secretes hydrochloric acid into the lumen of the stomach. ***Gastric juice*** is very acidic with a pH of nearly 1, corresponding to a hydrogen ion concentration of 0.1 M. The cells of the gastric mucosa produce gastric juice from extracellular fluid and plasma, which have a pH of about 7 (and a H^+ concentration of 10^{-7} M). The concentration of H^+ ions in gastric juice is 10^6 times greater than the concentration in plasma. A normal man secretes about 2.5 liters of gastric juice each day; the amount and its acidity vary with what he eats and how much he eats. The H$^+$ ions are actively transported from plasma to mucosal surface by a specific, energy-requiring process. Several models have been advanced to account for this, but our understanding of this process is even more obscure than of other active transport systems.

SECTION 16–9

ACTION POTENTIALS

To demonstrate an **action potential,** two electrodes are placed at different points on the surface of a neuron, a muscle fiber, a giant algal cell or some other large cell, and the two electrodes are connected through a voltmeter. When the cell is in the resting state, the potential on the outside of the cell is the same all over the surface; no voltage difference is evident between the two electrodes. When the cell is stimulated near one electrode, a voltage difference does appear, and the electrode near the stimulus becomes negative relative to the one farther away. After a brief flow of current, the voltage difference quickly returns to zero. The more distant electrode then becomes negative, relative to the one near the point of stimulation. Finally this second voltage difference disappears and the two electrodes once again have zero potential difference between them. If the electrodes are connected to a recording voltmeter or to an oscilloscope, the changes in voltage are seen as a single sharp spike followed by an inverted spike of the same amplitude (Fig. 16–11). The stimulus sets off a wave of **"negative potential"** which travels along the membrane, reaching first the near electrode (and it becomes negative) and then reaching the second electrode (and it becomes negative). The fact that the amplitude of the spike does not decrease with the distance traveled by the wave of negative potential is a clue that the wave is regenerated by energy-requiring processes as it passes each point on the membrane. Such propagated voltage changes in response to stimuli, called action potentials, were first recognized by Galvani in 1791.

Although action potentials have been studied most intensively in neurons and muscle fibers, they have been detected in plant cells, fertilized eggs and a variety of other cells. The height of the **spike potential,** its duration and its rate of propagation—all may vary quite a bit in different kinds of cells. In the giant axon of the squid, the action potential is about 90 mV, its duration is about 1 millisecond and it progresses at about 1 meter per second. The rate of propagation of the impulse is 1 centimeter or less per second in plant cells and as much as 120 meters per second in vertebrate neurons. Even the latter is slow, however, compared to the velocity of an electric current in a solution of electrolytes. The velocity of propagation of the action potential is proportional to the distance across which one local response, one change in membrane permeability, can trigger the next adjacent one.

The resting membrane potential of a nerve cell is about −50 mV, i.e., the inner side of the membrane is *negatively* charged relative to the outside of the cell. When the neuron is stimulated, the negative potential rapidly disappears, rising to zero, and then overshoots until the inside of the membrane is positively charged relative to the outside by some 40 mV. The action potential thus represents the momentary disappearance of the resting potential (the membrane is said to be "depolarized"), followed by a reversal of the usual voltage gradient. It has been shown experimentally that changing the concentration of sodium ions on the outside of the mem-

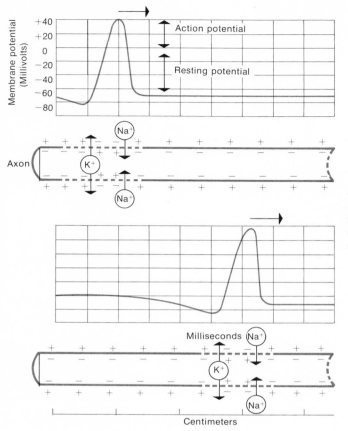

FIGURE 16–11   *The movement of an action potential along a nerve membrane. The wave of depolarization in the membrane is the initial event as the nerve impulse is transmitted along the neuronal membrane.*

brane leads to changes in the action potential. A decrease in the concentration of sodium ions on the outside of the membrane is accompanied by a decrease in the height of the spike potential. To explain these observations, Hodgkin suggested that the change in membrane potential is due to the sudden penetration of sodium ions into the interior of the cell. The action potential is related to some physical change in the plasma membrane that results in an increased permeability to sodium ions. Direct measurements of the flux of Na^+ and K^+, using isotopically labeled ions, showed that the **sodium flux** inward increases some 20-fold during stimulation and the **potassium flux** outward increases about ninefold. Thus the nerve impulse is accompanied by a sudden influx of sodium, down an electrochemical gradient, followed by an efflux of potassium, down an electrochemical gradient, which abolishes or reverses the resting potential. The actual number of ions that move through the membrane during a 1-millisecond pulse is very small and does not change significantly the distribution of Na^+ or K^+ ions on the two sides of the membrane. The change is detected primarily by its effect on the membrane potential.

By removing the entire cytoplasmic content of a squid axon and filling the axon with solutions of known ionic composition. Baker showed that these axons would respond to stimuli with action potentials of 90 to 130 mV for as long as five hours, during which they transmitted several hundred thousand impulses! The only requirement was for potassium salts in the internal medium and sodium salts in the external medium. Further studies showed that as a nerve impulse passed a given point there is first an inward flow of **current** due to the inward penetration of sodium ions. This is followed by a second, more gradual flow of current outward through the membrane due to the outward diffusion of potassium ions.

To summarize, the spike of an action potential results from the following sequence of events: (1) A sudden increase in the permeability of the membrane to sodium, (2) an inflow of sodium ions driven by the electrochemical gradient (the concentration of sodium outside a human neuron is about 137 mEq and the concentration of sodium inside it is 10 mEq); (3) this inflow causes the inside of the membrane to become positively charged relative to the outside and appears as the ascending portion of the spike, and (4) the membrane becomes less permeable to sodium and more permeable to potassium; an outflow of potassium ions along an electrochemical gradient (the concentration of potassium inside a human neuron is 141 mEq and the concentration of potassium outside is 4 mEq) follows, which appears as the descending curve of the spike potential.

SECTION 16–10

THE MEMBRANE THEORY OF NERVE CONDUCTION

The theory of nerve conduction presently accepted, the so-called **membrane theory,** states that the electrical events in the nerve fiber are governed by the **differential permeability** of the plasma membrane of the nerve cell to sodium and potassium and that these permeabilities are regulated by the **electric field** across the membrane. The interaction of these two factors, differential permeability and electric field, leads to the requirement for a critical threshold of change for excitation to occur. Excitation is a regenerative release of electrical energy from the nerve membrane, and the propagation of this change along the fiber is the brief, all-or-none electrical impulse, the **action potential.**

The resting nerve fiber is a long, cylindrical tube whose plasma membrane separates two solutions of different chemical composition. In the external medium, sodium and chloride ions predominate, whereas within the cell, potassium ions and organic anions predominate. The concentrations of sodium outside and inside the cell are in a ratio of about 10:1 and the concentrations of potassium inside and outside the cell are in a ratio of about 30:1. These concentrations of Na^+ and K^+ are kept relatively constant by the sodium pump, despite a steady, slow flux of ions in and out of the cell. As a result of this differential distribution of ions on the two sides of the membrane, there is a potential difference of 60 to 90 mV across the membrane, the **"resting" membrane potential.** The differential distribution of ions and the resulting membrane potential are maintained by the active extrusion of sodium ions from the interior of the cell by the sodium pump. The extrusion of sodium ions is accompanied by the entrance of potassium ions. Even when the sodium pump is turned off by a metabolic poison, many hours pass before the concentration gradients of sodium and potassium across the membrane disappear.

Electrical studies of the **cable properties** of the nerve fiber show that the axon could hardly serve as a passive transmission line (like a copper wire), because its cable losses are enormous. When a weak signal is applied to the fiber, one too small to excite its usual relay mechanism, the signal fades out within a few millimeters of its origin. The nerve impulse could not be propagated over the long distances in the nerves unless there were some process to boost the signal. The excitatory process regenerates and reamplifies the signal at each point along the nerve fiber. The cable properties of the nerve allow a change in electrical potential to spread along the nerve fiber for a short distance and stimulate the excitatory process in the adjacent portion of the nerve.

The generation of a nerve impulse involves a momentary change in the permeability of the nerve membrane which permits Na$^+$ ions to enter. This leads to a *depolarization* of the membrane in that region; it becomes positively charged on the inside. Although the permeability of the membrane to sodium is very low at the usual resting membrane potential, the permeability increases as the membrane potential decreases (Fig. 16–12). This permits the leakage of sodium ions down an electrochemical gradient into the interior of the nerve. This further decreases the membrane potential and further increases the permeability to sodium. The process is self-reinforcing and results in the upward deflection of the action potential. When the inside of the fiber becomes positively charged with respect to the outside (the membrane potential may reach +40 mV), the further net influx of sodium ions is prevented. The permeability of the membrane to potassium increases, and K$^+$ ions pass from inside the cell to outside the cell, down their electrochemical gradient.

The return of the membrane potential to its initial state, after the impulse has passed, with a negative charge inside and positive charge outside, is not brought about by a simple reversal of the movement of ions. There is no expulsion of the sodium ions that entered during the rising phase of

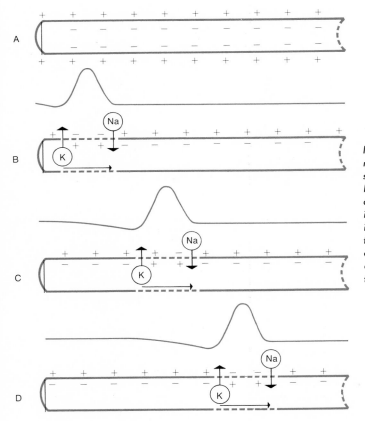

FIGURE 16–12   Diagram illustrating the membrane theory of nerve transmission. A, A resting nerve showing the polarization of the membrane with positive charges on the outside and negative charges on the inside. B, C, and D, Successive stages in the conduction of a nerve impulse from left to right showing the wave of depolarization of the membrane and the accompanying action potential propagated along the nerve.

the action potential. Instead, an equivalent quantity of potassium ions leaks out through the membrane. The actual quantities of Na^+ and K^+ that enter and leave a nerve during the passage of an impulse are so minute that they cannot be detected. The squid axon could conduct several hundred thousand impulses, even if its sodium pump were turned off, before its store of accumulated ions would be exhausted.

Because of the changes in permeability that accompany the depolarization of the nerve membrane, the fiber cannot immediately transmit another impulse. This period of inexcitability, called the **absolute refractory period,** is brief and lasts until normal permeability relations have been restored. The nerve impulse is, in essence, a wave of depolarization that passes along the nerve fiber. The change in membrane potential in one region renders the adjacent region more permeable, and the wave of depolarization is transmitted along the fiber. The entire cycle of depolarization and repolarization requires only a few thousandths of a second.

SECTION 16–11

NEURONAL FUNCTIONS

The idea that nerve fibers are simply some sort of cytoplasmic telephone wires has rapidly changed as we have come to appreciate that neurons are very dynamic and metabolically active cells. These cells, many of which are very long, carry out the intracellular transport of materials over long distances by active cytoplasmic streaming and undulations of the axons. Neurons may, in addition, be true secretory cells, secreting hormonal **neurohumors** such as **vasopressin, oxytocin** and the hypothalamic **releasing factors.** The regeneration of amputated limbs in salamanders requires specific neuronal influences and will not occur if the nerves are removed. In man and other mammals, when an axon is severed, a new cytoplasmic process grows out of its stump and advances by an amoeboid tip at the rate of 3 or 4 millimeters per day. This tip carries on pinocytosis and shows a remarkable ability to find and recognize its proper connection. After this tip has made contact with its proper distal portion, the mass of the axon is brought back to normal by a complex synthetic process which is accompanied by striking morphological changes in the cell body of the neuron. The **cell body** contains an abundant accumulation of endoplasmic reticulum and ribosomes that were detected by light microscopy in the nineteenth century and termed *"Nissl substance."* The Golgi apparatus of neurons is also abundant and well developed. The cytoplasm of the axon contains longitudinally oriented **neurofibrils** and **neurotubules,** which are associated with the intracellular transport of molecules.

SECTION 16–12

SYNAPTIC TRANSMISSION

The process by which a nerve impulse is transmitted across a **synapse** from one nerve to the next, or from a nerve to a muscle fiber, is not completely understood. In a few special situations where there is actual contact of the two plasma membranes, transmission proceeds by the direct electrical conduction of the action potential from one neuron to the next. At most synapses, however, a gap of some 200 Å separates the two plasma membranes and the impulse is transmitted across this gap by a special **chemical transmitter.** A specific chemical is synthesized by the neuron and released from the tip of its axon when a nerve impulse reaches it. This diffuses across the synaptic space and attaches to a special **chemoreceptor** on the surface of the dendrite of the adjacent cell (Fig. 16–13). The combination of this chemical with the chemoreceptor leads to a change in the membrane, depolarizing it and establishing a new action potential which passes along the length of that neuron to the next synapse, and so on.

Transmission at the neuromuscular junction is via **acetyl choline,** released by the tip of the motor nerve. It causes a depolarization of the membrane of the muscle cell which leads to the contraction of the muscle fiber. **Curare** prevents the transmission of impulses from nerve to muscle specifically at this type of synapse, apparently by combining with the receptors for acetylcholine and preventing their usual reaction with it. DeRobertis has shown that acetylcholine is released by motor nerves in discrete **packets,** each of which contains about 1000 molecules of acetylcholine. Calcium ions are essential for the operation of the mechanism which releases acetylcholine, whereas magnesium ions inhibit it. Electron micrographs of the tips of the neuron at the synapse have revealed masses of **synaptic vesicles** 200 to 650 Å in diameter, in which acetylcholine is stored (Fig. 16–14). How an action potential induces the release of a packet of acetylcholine molecules from its vesicle is still unknown, but their existence provides a satisfactory explanation for the polarity of the synapse,

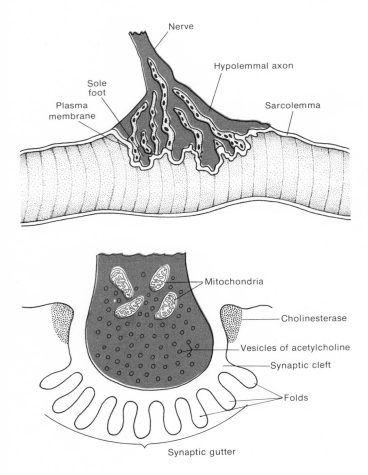

FIGURE 16–13   Diagram of a neuromuscular junction showing (below) the vesicles of acetylcholine in the sole foot of the nerve which penetrates into the membrane of the muscle fiber. This membrane becomes extensively folded to form the "synaptic gutter" in the plasma membrane of the muscle cell. The transmission of a nerve impulse from one neuron to a second neuron across a synapse is believed to involve a similar structural and functional junction.

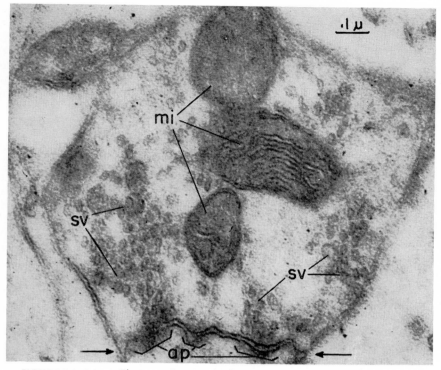

FIGURE 16-14   *Electron micrograph of a synaptic ending in the stimulated olfactory bulb of the rat. The synaptic ending contains three mitochondria (mi) and several synaptic vesicles (sv). The zone of contact between the two neurons is indicated by the two arrows. The nerve membranes appear to be thickened at active points (ap) in this zone of contact. (From DeRobertis, E., and Pellegrino de Iraldi, A.: Anat. Rec., 139:299, 1961).*

for the fact that a nerve impulse will travel in one direction but not in the reverse direction between axon and dendrite. Other neurons secrete **norepinephrine** at their tip instead of acetylcholine (and are termed **adrenergic** in contrast to the **cholinergic** fibers that secrete acetylcholine) Norepinephrine is also concentrated in synaptic vesicles and released by the arrival of an action potential. Other neurons may have still other neurochemical transmitters. Thus, information transmitted along the axon as a spike potential is coded and decoded at the synapse by processes involving graded thresholds, and the facilitation and summation of inhibitory and excitatory impulses. These are the substrates of animal behavior and the basis of the complex human processes of learning, memory and intelligence.

SECTION 16-13

TURGOR PRESSURE AND PLASMOLYSIS

It may seem a far cry from the transmission of nerve impulses, such as those involved as we listen to a great symphony, to the wilting of a stalk of celery, yet both processes have a common basis in biological membranes and their pumps. Some plants are supported by the thick cellulose walls secreted by each cell, which may be further fortified with a complex chemical called **lignin** present in woody plants. Other plants are supported by **turgor pressure.** The plant cell, inside its cellulose wall, has one or more large vacuoles filled with **cell sap.** This sap is an aqueous solution of salts, sugars, amino acids and other organic molecules. The plasma and vacuolar membranes, which separate the cell sap from the fluid outside the cell, are differentially permeable, and water molecules pass readily through their pores. Inorganic ions and organic molecules pass much less readily. When the concentration of solutes in the cell sap is greater than the concentration outside the cell, as it usually is, water tends to enter, moving by diffusion from a region of higher concentration (of water) to a region of lower concentration. This additional water distends the vacuole and presses the cytoplasm against the cellulose wall (Fig. 16–15). The cellulose wall is slightly elastic and is stretched by the internal pressure. After a certain amount of water has entered, an equilibrium is reached in which the pressure exerted by the stretched cell wall equals the pressure of the cell sap. After this, the number of water molecules entering the vacuole is equaled by the number leaving and the volume of the cell sap remains constant. **Turgor pressure,** the pressure exerted by the contents of the cell against the cell wall, should not be confused with the **osmotic pressure** of the cell sap, which is the pressure that could be developed if the cell sap were separated from pure water by a membrane completely impermeable to all the solutes in the cell sap. Turgor pressure is less than the osmotic pressure of the cell sap, because the fluid outside the cell is usually a dilute salt solution, not pure water, and because the cell membranes are permeable to the salts and other solutes in the cell sap. These in time would diffuse through

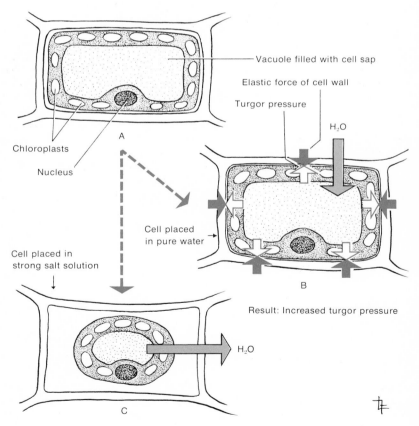

FIGURE 16–15   *Diagram illustrating the osmotic forces and the movement of water molecules which result in increased turgor pressure (B) or in plasmolysis (C).*

the membrane and reduce turgor pressure if it were not for the active processes of the living cell which can pump certain substances selectively into the cell and extrude others. In addition, the cell by photosynthesis produces new organic molecules, increasing the concentration of solutes in the cell sap and increasing turgor pressure. Turgor pressure provides the force to stretch the cell walls of young plant cells and provide for growth.

If the fluid outside the cell has a higher salt concentration than the cell sap, as when a stalk of celery is placed in a concentrated salt solution, water from the cell sap diffuses out of the cell, passing down a concentration gradient to the outside of the cell where the concentration of water is less. This decreases the turgor pressure within the cell and the stalk of celery wilts.

As water passes out of the cell, the volume of cell sap decreases and the cell contents are no longer pressed against the cellulose cell wall. Instead, it shrinks away from the cell wall, a process termed *plasmolysis* (Fig. 16–15 C). Plant cells exposed to hypertonic solutions eventually die, but if replaced in pure water after a brief exposure, they can regain their turgidity.

SECTION 16–14

CONTRACTILE VACUOLES AND MEMBRANE PUMPS

A plant or animal cell, placed in a solution that is not isotonic with it, may adjust to the changed environment by undergoing a change in its water content or by actively pumping certain solutes into or out of the cell. Amoebas, paramecia and other protozoa living in pond water, which is very hypotonic to the contents of the cell, have evolved *contractile vacuoles* which collect water from the interior of the cell and pump it to the outside (Fig. 12–30).

Some organisms living in the ocean have adapted to its high salt content by evolving an intracellular fluid with the same osmolarity as the sea water. Others, such as some of the bony fish, have evolved pumps in their gills by which they secrete excess salt. Fish such as salmon which migrate from fresh to salt water and back must adapt to each. The sea gull, which drinks salt water, gets rid of its excess salt by a special salt-excreting gland in its nose. Many seaweeds have developed membrane pumps by which they can accumulate *iodine* from the sea water so that the concentration within the cell may be 2×10^6 that in the sea water. The tunicates apparently can accumulate *vanadium* from sea water to a comparable extent.

SECTION 16–15

RESPIRATION: THE EXCHANGE OF GASES THROUGH MEMBRANES

The term *respiration* is generally used to refer to those processes by which animal and plant cells utilize oxygen, produce carbon dioxide and convert energy into biologically useful forms such as ATP (p. 100). Initially respiration was synonymous with breathing and meant inhaling and exhaling. Subsequently it was applied to the exchange of gases between the cell and its environment, and now it refers to the enzymatic processes

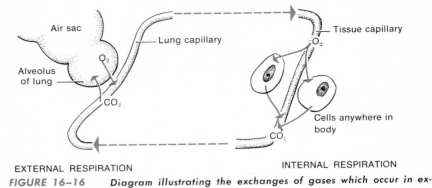

EXTERNAL RESPIRATION   INTERNAL RESPIRATION

FIGURE 16–16   *Diagram illustrating the exchanges of gases which occur in external and internal respiration.*

by which oxygen is utilized within the cell. Both oxygen and carbon dioxide
have a high solubility in lipids and pass readily by simple physical diffusion through the lipid matrix of plasma and intracellular membranes.
The exchange of gases is a simple process in small aquatic plants and animals such as paramecia or hydra. Dissolved oxygen from the surrounding
pond water diffuses into the cell, passing through the lipid matrix, and
carbon dioxide diffuses out. Each gas is moving down a chemical gradient
by simple physical diffusion. Such gas exchange may be called **direct respiration,** because the cells of the organism exchange oxygen and carbon
dioxide directly with the surrounding environment; no specialized respiratory system is required.

As animals evolved into larger, more complex forms, it was no longer
possible for each cell to exchange gases directly with the external environment, and some form of **indirect respiration,** by means of a structure
specialized for gas exchange, evolved (Fig. 16–16). This structure had to
be thin-walled to facilitate diffusion, it had to be kept moist so that oxygen
and carbon dioxide were dissolved in water, and it required a good blood
supply or connection with the extracellular fluids bathing the cells. These
structures for gas exchange include **lungs, gills, tracheal tubes** and even
the moist **skin.** The transfer of oxygen and carbon dioxide across membranes stands out as the prime example of the transfer of considerable
amounts of material through membranes by physical diffusion.

SECTION 16–16

CELL-CELL RECOGNITION

The development of systems for growing isolated cells in culture has
made possible experiments to analyze what makes cells move and how they
recognize other cells. The primordial germ cells in higher vertebrates first
appear in the **yolk sac** and then migrate a considerable distance into the
definitive **gonads.** Other kinds of cell migrations occur in the processes of
gastrulation and organogenesis. To a considerable extent, tissues are
formed by the selective adhesion of individual cells one to another. The
specific recognition of cells, and their adhesion, seems to involve yet another property of the plasma membrane.

An early demonstration of cell recognition was the experiment of
H. V. Wilson, who dissociated sponge cells by straining them through
bolting cloth and then observed the reaggregation of choanocytes and
amoebocytes to form complete sponges (p. 404). The specificity of this

reaction was shown by Spiegel in 1954. He dissociated two species of sponges, one cream-colored and the other orange, and mixed the cells together. The first cell aggregates formed were mixtures of cream- and orange-colored cells; however, the cells continued to move and reaggregate, and the aggregates ultimately formed contained either all cream-colored or all orange-colored cells. The reaggregation process requires Ca^{++} and Mg^{++} ions and a soluble material which is released when the sponges are dissociated chemically. It appears to be a protein, since its activity is destroyed by boiling it briefly; however, it is not inactivated by treatment with DNase, RNase or with collagenase. The cells can synthesize the protein material for reaggregation if they are incubated at room temperature (22°) but not if they are kept in the cold (5°). The aggregation proteins are species specific, causing aggregation and adhesion only of cells from the same species. Evidence on this point was supplied by experiments of Humphreys, who dissociated two kinds of sponges, *Haliclona* and *Microciona*, and kept the dissociated cells at 5° so they could not synthesize the reaggregation protein. Adding the aggregation protein from *Haliclona* led to the formation of blue-purple aggregates of *Haliclona* cells from the mixture, while the orange-red *Microciona* cells remained dissociated. The reciprocal experiment of adding *Microciona* aggregation factor to the mixture of cells led to the formation of orange-red *Microciona* aggregates, while the blue-purple *Haliclona* cells remained dissociated.

Comparable morphogenetic movements and specific cell recognitions have been demonstrated with the cells of dissociated amphibian embryos by Holtfreter. A mixture of presumptive epidermal cells from *Amblystoma* embryos and presumptive neural cells from *Triturus* embryos aggregated to form a blastula-like sphere; the epidermal cells subsequently migrated to the surface of the sphere and the neural cells migrated to the interior, forming what was identified as a **neural tube** surrounded by a double layer of epidermis (Fig. 16–17). The cells of the two species have different degrees of pigmentation which was helpful in distinguishing the cell types.

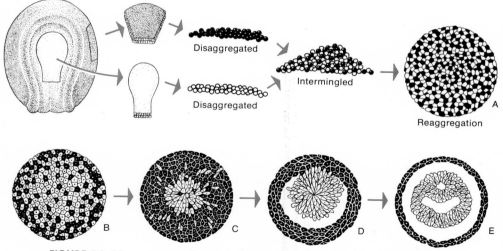

FIGURE 16–17   *A mass of neural plate cells and epidermal cells from the ventral surface were removed from an amphibian embryo and disaggregated. The two types of cells subsequently reaggregated in tissue culture to form a heterogeneous mass. Finally each type of cell migrated in an appropriate fashion so that the medullary plate cells rounded up and formed a neural tubelike structure in the center, whereas the epidermal cells migrated to the periphery of the mass and joined to form a new epidermis.*

Moscona has carried out comparable experiments with cells from chick and mouse embryos. The embryonic cells are dissociated by removing Ca^{++} and Mg^{++} ions and by treating the cells briefly with trypsin. Moscona's experiments showed that cells can recognize other cells of the same tissue type and aggregate with them. From a mixture of cartilage cells and liver cells, aggregates were formed which contained either liver cells or cartilage cells but not both. These aggregations are inhibited by lowering the temperature, which suggests that the process requires energy. In further experiments, Moscona incubated a mixture of chick kidney and cartilage cells and mouse kidney and cartilage cells. In marked contrast to the species-specific segregation of cells seen in the sponge experiments, these cells aggregated by *tissue* type, not by species. Chick and mouse kidney cells aggregated and formed kidney tubules (Fig. 16–18). Chick and mouse cartilage cells aggregated to form a bit of cartilage and secreted matrix.

A tissue formed of cells from two individuals or two species and, hence, of diverse genetic constitution, is said to be **chimeric.** The chimera was another monster of Greek mythology, an animal with a lion's head, a goat's body and a serpent's tail. The nature of the factors involved in cell recognition and cohesion is not at all clear and is another fascinating facet of the cell membrane.

SECTION 16–17

CELLULAR HOMEOSTASIS

Cells are remarkably conservative creatures with a variety of means of maintaining the *status quo ante.* Each cell can continue to function only as long as its internal constituents and the composition of the fluid bathing its plasma membrane are kept within certain limits. Typical values for the composition of the intracellular and extracellular fluids in man are given in Figure 16–19. Cells can continue to live indefinitely after being removed from the body if they are placed in a fluid that contains the appropriate constituents and has the same physical conditions as those of the body fluids. Organisms have evolved a remarkable variety of mechanisms to keep their internal environment constant. For man and the vertebrates in general, this internal environment is the **extracellular fluid** which bathes essentially all cells. Claude Bernard, more than a century ago, described the remarkable stability of the constituents of the extracellular fluid and noted several of the mechanisms by which mammals and birds maintain the constancy of their extracellular fluids. Lower animals have some mechanisms for regulating the extracellular fluid but are less capable than mammals in this respect. Walter Cannon, professor of physiology at Harvard early in the twentieth century, studied several other mechanisms that maintain constancy of the internal environment and coined the term **"homeostasis"** to describe this tendency of the extracellular fluid to remain constant, the tendency of animals to resist the effects of changes in external conditions and to keep the internal conditions constant. The mechanisms involved in homeostasis will be discussed in greater detail later, but the basic processes are in large part functions of biological membranes and involve physical diffusion, facilitated diffusion and active transport. In man, one of the major organs maintaining the constancy of the extracellular fluid is the kidney, but the lungs, gastrointestinal system and skin also play roles. Indeed, nearly every organ plays some role in controlling one or more of the constituents of the extracellular fluid.

The circulatory system transports blood and the substances dissolved

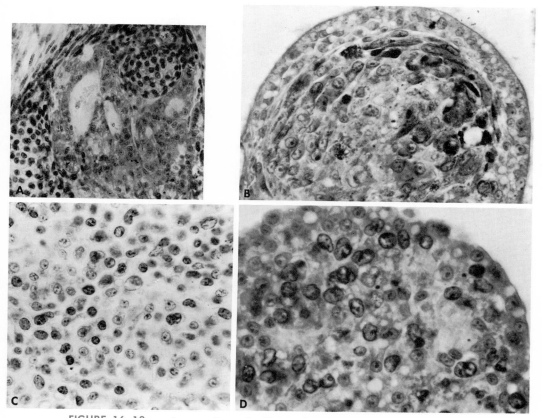

FIGURE 16–18   Aggregations of cells in culture by cell type. A, An aggregate of mouse cartilage cells (dark nuclei) and chick kidney cells. B, An aggregate of chick liver cells (the outside layer of lighter cells) and mouse melanoma cells. C, In contrast, mouse cartilage cells and chick cartilage cells become intermingled in a common matrix. The mouse cells can be distinguished by their darker nuclei. D, Rat and chick liver cells have aggregated and intermingled in a common tissue. (Courtesy of Dr. Aaron Moscona.)

in it throughout the body. Water and the dissolved materials pass from the blood to the fluids bathing each cell. This system provides for the constant exchange of materials between different parts of the body, and the constant mixing of the fluids aids in maintaining their composition constant. The respiratory system—lungs, gills or other organs—transfers oxygen from the external environment to the blood and the blood transports it to the fluids bathing each cell. The carbon dioxide produced during cellular metabolism passes via the extracellular fluid and blood to the respiratory surface. The gastrointestinal system takes in and processess nutrients, which are then absorbed into the blood or lymph and transported to the cells via the extracellular fluid. The liver, muscles and other tissues, by the enzymatic reactions termed intermediary metabolism, convert some of these nutrients into other compounds and release biologically useful energy in the form of ATP and other energy-rich compounds in the process. The kidneys regulate the concentrations of many substances in the blood and remove waste products from it. Each of these processes is in turn regulated by devices that tend to keep them constant. Many of these devices involve the principle of **feedback control,** in which the accumulation of the product of a reaction leads to a decrease in its rate of production, or a deficiency of the product leads to an increase in its rate of production. The functions

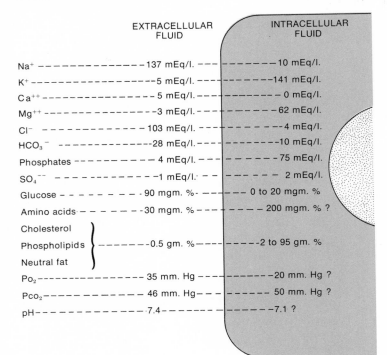

	EXTRACELLULAR FLUID	INTRACELLULAR FLUID
Na$^+$	137 mEq/l.	10 mEq/l.
K$^+$	5 mEq/l.	141 mEq/l.
Ca^{++}	5 mEq/l.	0 mEq/l.
Mg^{++}	3 mEq/l.	62 mEq/l.
Cl$^-$	103 mEq/l.	4 mEq/l.
HCO$_3^-$	28 mEq/l.	10 mEq/l.
Phosphates	4 mEq/l.	75 mEq/l.
SO$_4^{--}$	1 mEq/l.	2 mEq/l.
Glucose	90 mgm. %	0 to 20 mgm. %
Amino acids	30 mgm. %	200 mgm. % ?
Cholesterol Phospholipids Neutral fat	0.5 gm. %	2 to 95 gm. %
Po$_2$	35 mm. Hg	20 mm. Hg ?
Pco$_2$	46 mm. Hg	50 mm. Hg ?
pH	7.4	7.1 ?

FIGURE 16–19   Diagram illustrating the chemical composition of extracellular fluids and intracellular fluid. (After Guyton, A. C.: Function of the Human Body, 3rd ed. Philadelphia, W. B. Saunders Company, 1969.)

of the organs of the body are the integrated sums of the functions of its cells, and these in turn depend upon the properties of the biological membranes of those cells.

SUGGESTIONS FOR FURTHER READING

Eccles, J. C.: *The Physiology of Synapses.* New York, Academic Press, Inc., 1963.
   An advanced source book of the functional aspects of synaptic transmission.
Hokin, L. E., and Hokin, M. R.: *The Chemistry of Cell Membranes.* Sci. Amer., October (Offprint 1022) 1965.
   A presentation of the author's views of the chemical and physical nature of cellular membranes.

Katz, B.: *Nerve, Muscle and Synapse.* New York, McGraw-Hill Book Co., 1968.
   An up-to-date discussion of the role of membrane potentials in a variety of biological phenomena.

Locke, M. (ed): *Cellular Membranes in Development.* New York, Academic Press, Inc., 1964.
   A series of papers presented by experts at a symposium; an excellent source book for information on the development of membranes and the role of cell membranes in the process of development.

Solomon, A. K.: *Pores in the Cell Membrane.* Sci. Amer., December (Offprint 76) 1960.
   An entertaining account of the studies which provided evidence about the dimensions of these pores.

Stein, W. D.: *The Movement of Molecules Across Cell Membranes.* New York, Academic Press, Inc., 1967.
   A complete and up-to-date account at an advanced level of the mechanism by which various kinds of molecules pass across membranes.

Wilson, T. H.: *Intestinal Absorption.* Philadelphia, W. B. Saunders Company, 1962.
   A clear exposition of the role of the intestinal lining in the active uptake of nutrients.

PROCURING NUTRIENTS

CHAPTER 17

To carry out the biosynthesis of macromolecules, each cell in each kind of plant or animal needs a similar array of raw materials—sugars, amino acids, purines, pyrimidines, fatty acids and so on—which it must either make itself or obtain by some means from the environment. Cells by and large do not take in macromolecules in nutritional amounts; each one assembles its own out of raw materials—glycogen or starch out of simple sugars, proteins out of amino acids, nucleic acids out of purines and pyrimidines, and complex lipids out of fatty acids. The autotrophs of the world can make their own sugars and other raw materials from simple inorganic raw materials, carbon dioxide, water and minerals. The simplest autotrophs, certain algae, are single-celled, and everything needed for the synthesis of even the most complex macromolecules is made on the site. The large "higher" plants, the trees and shrubs, carry on photosynthesis in some parts, the leaves, but not in all cells. They thus must have some means of transporting the simple organic molecules made by photosynthesis in the leaves to the cells in the stem and root which generally lack chlorophyll and cannot carry out photosynthesis. The cells of the stem and root, in synthesizing their macromolecules, utilize small organic molecules made in the leaves and carried to them by a transport system, the phloem.

The heterotrophs of the world, the fungi and other nongreen plants and the animals, cannot use the energy of light to synthesize sugars and other small organic molecules. However, if supplied with these precursors, they have all the enzymes needed to make a variety of other organic compounds and to synthesize their own macromolecules. Man needs sugars as a source of energy, and certain organic compounds he cannot synthesize, "essential" amino acids, "essential" unsaturated fatty acids and vitamins. Given these raw materials, the cells of man and other animals can synthesize purines and pyrimidines and put them together as nucleic acids; they can synthesize glycogen and other polysaccharides from glucose, complex lipids and sterols from acetyl coenzyme A and proteins from the 20 different kinds of amino acids, some of which they can make out of other raw materials.

The food that we and other animals eat is not simply a mixture of the glucose, amino acids, fatty acids and other compounds that we need as raw materials, but is made up of proteins, fats, polysaccharides, polynucleotides and other complex molecules that must first be taken apart—digested, hydrolyzed to simpler subunits—before they can be transferred to the cells of the body and taken up by them. The molecules of proteins, nucleic acids, fats and carbohydrates are held together largely by **anhydro bonds,** formed by the removal of a molecule of water or by an equivalent reaction. These bonds are split by hydrolytic cleavage, by the addition of a molecule of water across the bond to split it. The H^+ from the water is incorporated into one of the split products and the OH^- into the other. The **peptide bonds** of proteins, the **glycosidic bonds** of polysaccharides and the **ester bonds** of nucleic acids and lipids are all anhydro bonds. Enzymes are highly specific for the substrate they will attack, the type of bond they will make or cleave, and thus different enzymes are required for the hydrolytic cleavage of each type of bond. The process of **digestion,** from amoeba to man, involves the same, or a very similar, array of enzymes, but differs in where the enzymes are produced, where they act and how the process is controlled. Digestion may be **intracellular**—food particles may be taken into the cell by phagocytosis and the digestive enzymes may act in food vacuoles within the cell—or **extracellular**—the enzymes may be secreted by cells into some cavity, typically the gut cavity, where they come in contact with the food and hydrolytic cleavage takes place.

THE DIGESTIVE SYSTEMS OF ANIMALS

Some animals (sponges, clams, tunicates and even some whales) are **filter feeders,** making a living by filtering out of sea water or pond water the microscopic plants and animals that are present. The choanocytes of sponges take up these small particles by phagocytosis and digest them intracellularly. Certain parasites, such as tapeworms, live in the gut cavity of man and other vertebrates and absorb "predigested" nutrients from the surroundings. Protozoa take food, either living microscopic plants or animals, or dead bits of organic matter into their cells by **phagocytosis.** An amoeba engulfs a ciliate or flagellate and forms a food vacuole around it. Hydrolytic enzymes synthesized in the cytoplasm are secreted through the vacuolar membrane into the cavity and digestion occurs within the vacuole as it circulates in the cell. The products of digestion are absorbed through the vacuolar membrane and utilized for the production of biologically useful energy or as substrates for the synthesis of macromolecules. Any undigested remnants are expelled and left behind as the animal moves on.

In coelenterates and platyhelminthes there has evolved a **gastrovascular cavity,** lined with a **gastrodermis.** Small animals are taken in through the mouth and digestion begins within the gastrovascular cavity. The gastrodermis secretes digestive enzymes into the cavity and the prey is partially digested there. The bits are absorbed into the gastrodermis cells where digestion is completed within food vacuoles. Digestion is partly extracellular, occurring in the gastrovascular cavity, and partly intracellular, within food vacuoles in the gastrodermis. The gastrovascular cavity of the flatworms may be greatly branched, ramifying through most of the body and facilitating the distribution of digested food. Neither coelenterates nor flatworms have an anal aperture; undigested wastes are excreted through the mouth.

In most of the rest of the invertebrates, and in all the vertebrates, the digestive tract is a tube with two apertures; food enters by the mouth and undigested residues leave by the **anus** (Fig. 17–1). The digestive tract may be short or long, straight or coiled, and may be subdivided into specialized organs. These organs, even though they may have similar names in different kinds of animals, may be quite different, and may have different functions. The digestive system of the earthworm, for example, includes a **mouth;** a muscular **pharynx** which secretes a mucous material to lubricate food particles; an **esophagus;** a thin-walled **crop** where food is stored; a thick, muscular **gizzard** where food is ground against small stones; and a long, straight **intestine** in which extracellular digestion occurs. The products of digestion are absorbed through the intestinal wall by diffusion, facilitated diffusion or by active transport, and the undigested residues pass out through the anus. Some invertebrates—worms, squid, crustacea and sea urchins—have hard, toothed mouthparts which can tear off and chew bits of food.

As the vertebrates have evolved, the digestive system was gradually elaborated and organs were added. The digestive systems of the vertebrates from fish to man are basically similar (Fig. 17–2). The first portion of the intestine, connected to the stomach, is the **small intestine,** in which most foods are digested and most absorption occurs. This is followed by a **large intestine,** in which digestion and absorption—especially the absorption of water—are completed. The **liver** and **pancreas** are large digestive glands, arising in development as outgrowths of the digestive tract, con-

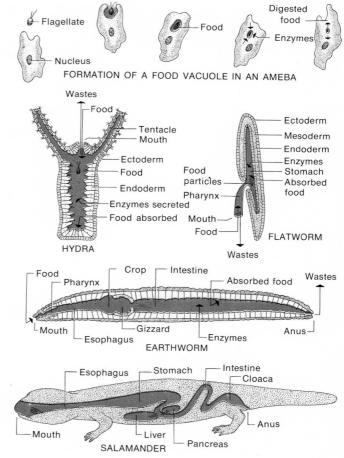

FIGURE 17–1   The structural basis of the process of digestion in amoeba, hydra, flatworm, earthworm and salamander, illustrating the similarities and differences in the digestive systems of these widely different animal forms.

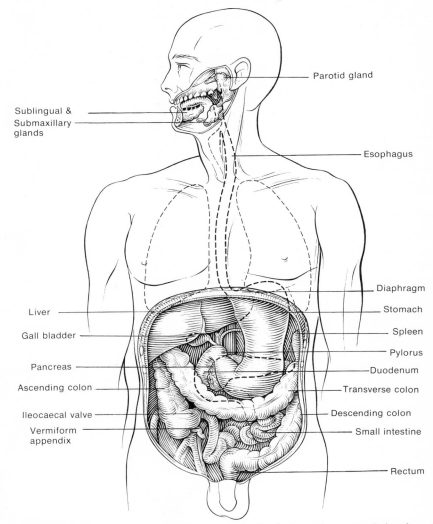

FIGURE 17-2   *Diagram of the human body showing the parts of the digestive system. The liver, which normally covers part of the stomach and duodenum, has been folded back to reveal these and the gallbladder on its undersurface.*

nected to the small intestine by ducts, and secreting **bile** and **pancreatic juice** respectively. These contain enzymes and other substances required for digestion.

SECTION 17-2

DIGESTIVE ENZYMES

Digestion, whether in amoeba or man, involves the cleaving of complex macromolecules into simpler subunits by hydrolysis, by the addition of water, catalyzed by enzymes called **hydrolases.** There are, of course, specific hydrolases for the splitting of proteins, fats and polysaccharides. Polysaccharides such as starch and glycogen form an important part of the food ingested by man and most animals. The glucose units of these large molecules are joined by **glycosidic bonds,** anhydro bonds linking carbon 4 or carbon 6 of one glucose molecule with carbon 1 of the adjacent one. These bonds are hydrolyzed by **amylases,** which will split polysaccharides to the disaccharide, maltose, but will not split the bond between the two glucose units of maltose. The amylases will split the α-glycosidic bonds present in starch and glycogen but not the β-glycosidic ones present in cellulose. The garden snail's digestive juice contains β-glycosidases that can hydrolyze cellulose. In most vertebrates, amylase is secreted only by the pancreas; in man and certain other mammals, amylase is secreted by the salivary glands as well. Each kind of disaccharide is cleaved by a specific enzyme: maltose by **maltase,** present in saliva and intestinal juice; sucrose by **sucrase** and lactose by **lactase** (β-galactosidase), both present in intestinal juice. The ultimate products of the digestion of carbohydrates are the hexoses, glucose, fructose and galactose, which are absorbed into the blood stream through the intestinal wall.

There are several kinds of hydrolases that attack the peptide bonds of proteins; each is specific for peptide bonds in a specific location in a polypeptide chain (Fig. 17–3). **Exopeptidases** cleave the peptide bond joining the terminal amino acids to the peptide chain. **Carboxypeptidase** splits the peptide bond joining the amino acid with the free terminal carboxyl group to the chain, and **aminopeptidase** removes the amino acid with a free terminal α-amino group. Other hydrolases, the **endopeptidases,** will cleave only peptide bonds within a peptide chain. **Pepsin,** secreted by certain cells in the gastric mucosa, and **trypsin** and **chymotrypsin,** secreted by the pancreas, are endopeptidases but differ in their requirements for specific amino acids adjacent to the peptide bond to be cleaved. Pepsin requires tyrosine or phenylalanine adjacent to the bond to be split; trypsin requires lysine or arginine; and chymotrypsin requires tyrosine, phenylalanine, tryptophan, methionine or leucine at the site of cleavage. These endopeptidases split peptide chains into smaller fragments which are then cleaved further by exopeptidases. The combined action of the endopeptidases and exopeptidases results in splitting the protein molecules completely to free amino acids, which are then absorbed through the intestinal wall into the blood stream by active transport.

These powerful proteolytic enzymes would constitute a serious threat to the tissues secreting them. However, pepsin, trypsin and chymotrypsin are not secreted as such, but are secreted in the form of inactive precursors — pepsinogen, trypsinogen and chymotrypsinogen. This prevents their digesting the proteins of the cells that produce them. In the gut each is activated by the removal of part of the precursor molecule to yield the active enzyme and an inactive fragment. **Pepsinogen,** with a molecular weight of 42,500, is converted to pepsin, with a molecular weight of 34,500, by the high concentration of H^+ in the gastric juice and by pepsin itself. **Trypsinogen** is converted to trypsin by **enterokinase,** an enzyme secreted by glands in the wall of the intestine, or by trypsin itself. The conversion of **chymotrypsinogen** to chymotrypsin is mediated by trypsin but not by chymotrypsin. As a further protection to the pancreas, which secretes trypsinogen and chymotrypsinogen, it also secretes a small protein called "trypsin inhibitor," which will combine with and inactivate any molecules of free trypsin that may be formed accidentally in the pancreas.

The amount of protein in the various digestive juices is not inconse-

FIGURE 17–3   Formula of a peptide indicating the points of attack of the enzymes pepsin (P), trypsin (T), chymotrypsin (C), aminopeptidase (AP) and carboxypeptidase (CP).

H_2N—gly—ala—leu—tyr—ala—asp—lys—val—glu—gly—COOH

AP   C   C or P   T   CP

quential, but comprises a significant fraction of the total protein hydrolyzed in the gut and is the source of a significant fraction of the amino acids absorbed through the intestine. The pancreas alone secretes about one quarter of its total protein content each day. It is estimated that about 100 grams of protein is secreted each day in the digestive juices of man. These proteins are subsequently hydrolyzed and their amino acids are reabsorbed. This amount is comparable to the total dietary intake of proteins by many people!

The digestion of fats is catalyzed by **esterases** that split the ester bond between glycerol and fatty acid. The principal mammalian esterase is **lipase,** secreted by the pancreas. Like other proteins, lipase is water-soluble, but its substrates are not. Thus the enzyme can attack only those molecules of fat at the *surface* of a fat droplet. The **bile salts** are surface-active agents that reduce the surface tension of fats, breaking the large droplets of fat into very fine ones. This greatly increases the surface area of fat exposed to the action of lipase and increases the rate of digestion of lipids. Conditions in the intestine are usually not optimal for the complete hydrolysis of lipids to glycerol and fatty acids. The products of digestion include glycerol and free fatty acids plus monoglycerides, diglycerides and some triglycerides, undigested fats.

The pancreas also secretes **ribonuclease,** an esterase which splits the phosphate ester bonds linking adjacent nucleotides in ribonucleic acids, and **deoxyribonuclease,** which splits the phosphate ester bonds linking adjacent nucleotides in deoxyribonucleic acids. Enzymes that complete the cleavage of nucleic acids are secreted by the intestinal mucosa. **Phosphodiesterase** removes nucleotides one at a time from the end of a polynucleotide chain. The nucleotides in turn are attacked by **phosphatases** which remove the phosphate group and leave the nucleosides, which are absorbed.

SECTION 17–3

REGULATION OF SECRETION OF ENZYMES BY THE DIGESTIVE GLANDS

Each of the glands secreting digestive enzymes must be stimulated to release its products at the proper time and in the proper amount. We salivate when we eat, gastric juice is produced when food reaches the stomach, and pancreatic juice and bile are secreted when food enters the duodenum, the first portion of the small intestine. It would be wasteful, indeed harmful, if the glands constantly secreted enzymes. The coordination of the flow of digestive juices with the presence of food is achieved in part by nervous coordination and in part by hormones, chemicals secreted in one part of the body and carried by the blood stream to another part where they produce a specific effect.

The smell of food or its presence in the mouth stimulates sensory nerves that carry impulses to a **salivation center** in the medulla of the brain. From there, impulses are relayed along motor nerves to the salivary glands, stimulating them to secrete. The mere presence of tasteless, odorless objects, such as pebbles, in the mouth stimulates certain cells lining the oral cavity to send impulses to the salivation center and initiates salivation. Impulses may also come to the salivation center from higher centers in the brain—just thinking about food can bring about salivation.

In 1822 a trapper, Alexis St. Martin, was shot in the stomach and treated by an Army surgeon, William Beaumont. St. Martin recovered nicely but the wound healed in such a way that there was an opening from the abdominal wall to the lumen of the stomach. Beaumont made a series of classic observations of the effects of dietary and emotional factors on the secretion of gastric juice and on other activities of the gastric mucosa.

Much of our knowledge of the control of gastric juice secretion was derived from experiments by the Russian physiologist Pavlov, who devised experimental techniques such as the Pavlov pouch and carried out many critical experiments. Although more famous in psychological circles for showing conditioned reflexes by getting dogs to salivate in response to the ringing of a bell, his contributions to gastric physiology are equally fundamental. Pavlov severed the esophagus of a dog, bringing the two cut ends to the surface of the neck. When the dog was fed, the food went out the hole in the neck instead of passing to the stomach. Although the food never reached the stomach, such a "sham-feeding" stimulated the flow of gastric juice, about one quarter of the normal amount, provided that the **vagus nerve,** which carries motor fibers to the stomach and other internal organs, was left intact. If the vagus nerve was cut, there was no flow of gastric juice in response to sham-feeding. Thus the flow of gastric juice is at least in part under nervous control. When Pavlov put food into the cut end of the esophagus leading to the stomach and prevented the dog from seeing, smelling or tasting it, gastric juice was produced—about half the normal amount. This flow occurred even when the vagus nerve was cut, showing that gastric secretion is also under some nonnervous control.

Other experiments showed that when partly digested food reaches the pyloric end of the stomach, certain mucosal cells produce the hormone **gastrin** (Fig. 17–4). This is absorbed into the blood through the wall of the stomach and finally reaches the gastric glands, stimulating them to secrete. Gastrin is a peptide hormone, containing about 30 amino acids. It was isolated in 1966 by Gregory at the University of Liverpool, who determined the sequence of its constituent amino acids. The final proof of the existence and action of this hormone was provided by experiments in which the circulatory system of one dog was connected by tubes to the circulatory system of another dog. When food was placed in the pyloric region of one dog, the gastric glands of the other began to secrete. Since there were no ner-

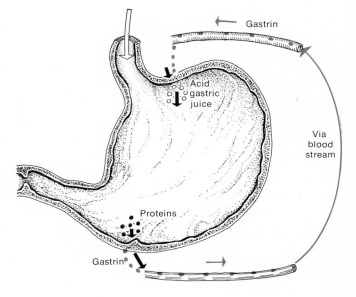

FIGURE 17–4   *The secretion of gastrin in response to the presence of acid food in the fundus of the stomach.*

vous connections between the two dogs, the secretion of gastric juice in the second dog must have been caused by a blood-borne substance, the hormone gastrin.

When fatty food reaches the duodenum it stimulates the duodenal mucosa to produce a different hormone, **enterogastrone,** which, on reaching the stomach, inhibits the secretion of the gastric glands and decreases the motility of the stomach. The presence of several different control mechanisms, some nervous and some hormonal, enables the stomach to provide the proper amount of gastric juice for the type and amount of food eaten: a meal rich in protein causes a copious flow and a meal rich in fats causes a small flow of gastric juice.

The hormone **secretin,** secreted by the intestinal mucosa, initiates the secretion of pancreatic juice (Fig. 17–5). The pancreas secretes its juice when the acidic contents of the stomach enter the small intestine even if the nerves to the pancreas are severed. The intestinal mucosa can be stimulated to secrete secretin by introducing dilute acid into the cavity of the small intestine. The hormone is normally secreted when the acid chyme enters the small intestine from the stomach. The hormone passes via the blood stream to all the cells of the body; the pancreas responds by secreting its enzymes and the liver responds with an increased production of

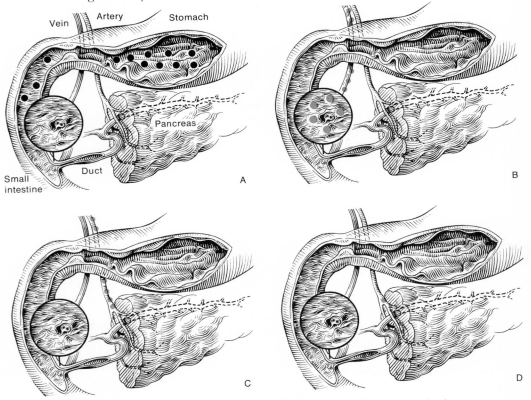

FIGURE 17–5   The control of the release of pancreatic juice by the hormone secretin. A, Hydrochloric acid is secreted by the glands in the wall of the stomach and passes through the pylorus to the duodenum. B, Some of the hydrochloric acid diffuses into the wall of the duodenum and causes the cells there to secrete the hormone secretin (tinted dots) which passes into the adjacent capillaries. C, Secretin is distributed by the blood vessels to all parts of the body, some of this passes via the pancreatic artery to the pancreas. D, The secretin stimulates the cells of the pancreas to secrete pancreatic juice which is visible in the pancreatic duct. The duct carries the pancreatic juice to the small intestine where its enzymes play an important role in the digestive process.

bile. When the nerves to the pancreas are stimulated, little secretion results; when they are cut there is little or no suppression of the flow. Its secretions are controlled almost entirely by secretin. Another hormone, **cholecystokinin,** also produced by the duodenal mucosa when acid food enters the duodenum, stimulates the gallbladder to contract and release bile.

SECTION 17-4

THE LIVER AND PANCREAS

The liver and pancreas are large glandular outgrowths from the small intestine. The liver, one of the largest organs of the human body, continually secretes **bile** which passes by a system of ducts to the **gallbladder.** Bile does not enter the intestine immediately, for a sphincter at the intestinal end of the bile duct remains closed until food enters the intestine. Stimulated by cholecystokinin, the sphincter relaxes, the wall of the gallbladder contracts and bile is forced out into the intestine. In the gallbladder, bile is concentrated by the removal of water and salts (Fig. 17-6).

Bile contains no digestive enzymes but is alkaline and aids in digestion by neutralizing the acid chyme from the stomach. The enzymes secreted by the pancreas and intestinal glands have pH optima in the neutral or slightly alkaline range. In addition, its **bile salts** emulsify the fats in the intestine and increase the surface area of the fat droplets, promoting the action of lipase. When the bile duct is obstructed and bile salts are absent from the intestine, both the digestion and absorption of fats are impaired and much of the fat eaten is excreted in the feces. The bile salts are carefully conserved by the body; they are reabsorbed in the lower part of the intestine and transported back to the liver in the blood stream to be secreted again. Another constituent of the bile, cholesterol, is very sparingly soluble in water. The removal of water from the bile in the gallbladder may concentrate the cholesterol to the point where it precipitates, producing hard little pellets called **gall stones.** These may obstruct the bile duct and stop the flow of bile.

The color of bile results from the presence of **bile pigments,** green, yellow, orange or red in different species of animals, which are excretory products derived from the degradation of hemoglobin in the liver. The bile pigments undergo further chemical reactions by the intestinal bacteria and are converted to the brown pigments responsible for the color of feces. If their excretion is prevented by a gall stone or some other obstruction of the bile duct, the bile pigments are reabsorbed by the liver and gallbladder and accumulate in the blood and tissues, giving a yellowish tinge to the skin, a condition called **jaundice.** The absence of the pigments from the intestinal contents gives the feces a clay-colored appearance.

The pancreas is an important digestive gland, producing quantities of enzymes that hydrolyze carbohydrates, proteins and fats. These enzymes pass from the pancreas to the intestine via a **pancreatic duct.** The pancreatic juice is a clear watery fluid with a pH of about 8.5. It is also an important factor in neutralizing the acid chyme. The average human being secretes from 1 to 1.5 liters of pancreatic juice per day.

The pancreas contains patches of endocrine tissue, the **islets of Langerhans,** which secrete the hormones **insulin** and **glucagon.** The two secretions, of enzymes into the pancreatic duct and of hormones into the blood stream, are entirely separate and unrelated. In man and most vertebrates

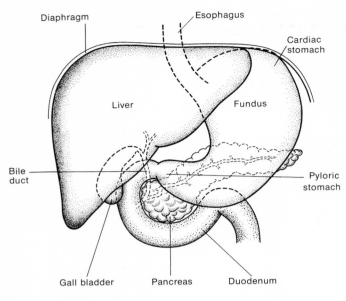

FIGURE 17–6   *Diagram of the anatomic relations of the stomach, liver, pancreas and duodenum.*

the two types of tissue occur together in the same gland. In certain species of fish, however, the two types are spatially separated into two different glands.

THE ABSORPTION OF NUTRIENTS

After the digestive enzymes have cleaved the large molecules of proteins, polysaccharides, nucleic acids and lipids into their constituent subunits, the products are absorbed through the wall of the intestine, especially the small intestine. The intestines of man and other vertebrates are greatly folded to increase the surface area through which absorption may occur. In addition, countless small, fingerlike projections, called **villi,** cover the entire surface of the intestinal mucosa (Fig. 17–7). Each villus contains a network of blood capillaries and a lymph capillary in its center, into which the nutrients are transferred. A third adaptation for increasing the surface area is the presence of countless, closely packed, cylindrical processes called **microvilli** on the surface of each epithelial cell of the intestine (Fig. 17–8). The folds, villi and microvilli together provide an enormous area through which absorption may occur. Some vertebrates have one or more blind pouches, **caeca,** which are attached to the intestine and increase the area available for absorption. The area of the mucosal surface of the intestine of sharks is increased by an epithelial fold (the spiral valve) attached to the side in a spiral fashion and shaped like a corkscrew. Food passing through the intestine must follow the spiral fold and thus is exposed to a greater area.

Absorption is a complex process, occurring in part by simple physical diffusion, in part by facilitated diffusion and in part by active transport. The various hexoses are absorbed by active transport, by a process which requires the expenditure of energy to move the molecules against a chemical gradient. The several hexoses, glucose, fructose and galactose, are absorbed at quite different rates. Galactose is absorbed more rapidly than

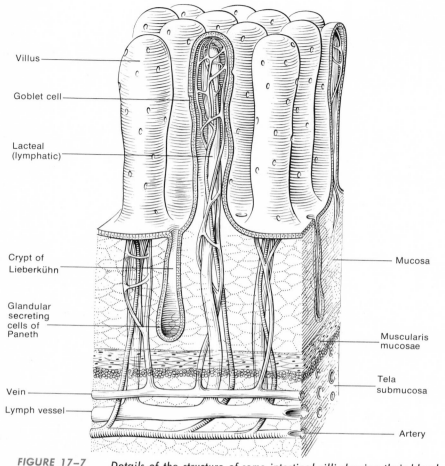

Villus

Goblet cell

Lacteal
(lymphatic)

Crypt of
Lieberkühn

Mucosa

Glandular
secreting
cells of
Paneth

Muscularis
mucosae

Tela
submucosa

Vein

Lymph vessel

Artery

FIGURE 17–7   *Details of the structure of some intestinal villi showing their blood
and lymph supply. Semi-schematic representation of jejunal intestinal villi, × 35 (approx.).*

glucose and glucose more rapidly than fructose. The hexoses pass through
the intestinal wall by an energy-requiring process which is essentially
unidirectional and which occurs without phosphorylation or any other
chemical change of the hexose molecule.

The amino acids are also absorbed by a process of active transport into
the blood capillaries and transported to the liver for short-term storage
and subsequent distribution to the rest of the body.

The products of the complete or partial hydrolysis of lipids enter the
body by a different process and a different route. Bile salts play an important role in enhancing the absorption of fatty acids, monoglycerides,
and diglycerides and of other lipid-soluble substances, such as the fat-
soluble vitamins. The lipids need not be completely hydrolyzed to glycerol
and fatty acids to be absorbed. The short-chain fatty acids are absorbed
into the blood capillaries, whereas the longer chain fatty acids are taken up
into the lymph capillaries. As the products of lipid hydrolysis pass through
the epithelial cells of the mucosa, they are resynthesized into molecules of
fat. These subsequently aggregate into **chylomicrons,** fine globules of fat
with a thin coat of protein, and enter the lymph. During the absorption
of a meal rich in fat, the lymphatic vessels of the intestine have a milky

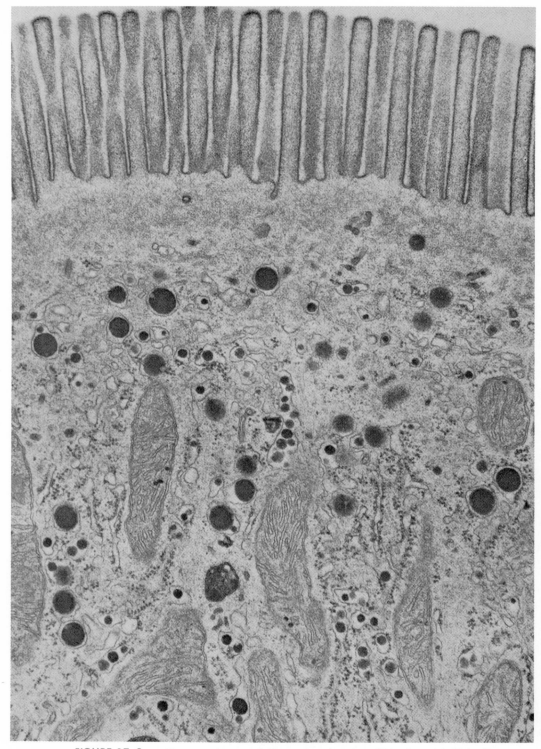

FIGURE 17–8   *Electron micrograph illustrating the microvilli in the epithelial cells of the intestine.*

color owing to this fat emulsion. The resynthesis of the fat occurs in the intestinal mucosal cells and requires the conversion of the free fatty acids to fatty acyl-coenzyme A compounds, which can then react with monoglycerides and diglycerides to form triglycerides. The intestinal lymph vessels empty into the great thoracic duct and then into the subclavian vein. The chylomicrons appear in the bloodstream and may give the blood itself a turbid, "milky" appearance after a meal rich in fat. Cholesterol is absorbed through the wall of the intestine in the free state, then is esterified in the mucosal cells and passes to the lymph capillaries.

A portion of the water in the intestinal contents is absorbed in the small intestine, but most of it is absorbed in the large intestine or **colon.** The colon absorbs water and converts the wastes to a semisolid state for defecation. The feces contain an enormous number of bacteria; half of the mass of feces may be bacteria. The intestinal bacteria synthesize a variety of vitamins and other nutrients that are absorbed and used by their vertebrate host. Bacteria have enzymes for digesting cellulose walls of plant cells and play an important role in the digestive processes of herbivores such as cattle and rabbits.

SECTION 17–6

METABOLISM: THE BASAL METABOLIC RATE

After entering the body, the molecules of nutrients participate in a tremendous variety of enzymatic reactions and these plus all the chemical activities of the organism are termed **metabolism.** The presence of metabolic processes is one of the fundamental characteristics of living things. The nutrients may be used as raw materials for the synthesis of new macromolecules or they may be oxidized to provide energy. Some of this energy is required for the continual synthesis of new tissue components, some for the functioning of the cells—the contraction of muscles, the transmission of nerve impulses, the secretion of cell products, the uptake of materials by active transport or pinocytosis—and the remainder is released as heat.

We may speak of "liver metabolism" if we are studying all the reactions being carried out within a single organ, or we may study all the chemical reactions undergone by a particular kind of molecule or ion (e.g., "carbohydrate metabolism" or "iron metabolism"). Carbohydrate metabolism would include all the reactions undergone by starches or sugars from the time they are eaten until, after digestion and absorption, they are stored, converted into other cellular constituents, or oxidized for energy and leave the body as carbon dioxide and water.

Human beings and many but not all animals can adapt to a variety of diets. Man can survive on a diet composed mainly of protein, with small amounts of fats and carbohydrates, on one composed mainly of carbohydrates, with small amounts of proteins and fats, or even on one consisting primarily of fats, with small amounts of the others. Cells can obtain biologically useful energy from any of these three types of fuels. The complete oxidation of a gram of carbohydrate or protein yields about 4 Calories and the oxidation of a gram of fat yields about 9 Calories. Whipped cream, butter and other foods with a high fat content have more calories per unit weight, and hence are more "fattening" than fruits, meat or bread.

The daily expenditure of energy varies widely from one person to another, depending on his activity, age, sex, weight, body proportions and hormonal state. Metabolic rates are measured under standardized conditions, with the subject reclining at complete rest at least 12 hours after his last meal. Under these conditions he uses energy to keep his heart beating, to breathe, to conduct a wealth of nerve impulses, to maintain the constancy of his body fluids and to maintain the body temperature. The amount of energy expended by the body just to "keep alive," when no food is being digested and no muscular work is being done, is termed the **basal metabolic rate.** For young adult males this is about 1600 Calories per day, for women it is about 5 per cent less. In other words, if a young adult remained in bed for 24 hours without eating or moving, he would expend about 1600 Calories in keeping alive. Chemical reactions occur more rapidly at higher temperatures and the basal metabolic rate increases about 5 per cent for each degree the body temperature rises. This is one reason why weight is lost during feverish illnesses (another is that we tend to eat less when we don't feel well).

A person remaining in bed for 24 hours but eating meals would expend about 1800 Calories. The additional 200 Calories are required for the movements of the muscles of the digestive tract, the synthesis and secretion of the digestive juices, and the active uptake of the products of digestion. A person leading a sedentary life uses some 2500 Calories per day and one doing heavy physical work may expend 6000 or more Calories per day. Most adults achieve a balance between the intake and utilization of calories and their weight remains remarkably constant for years.

Sugars and starches are the principal sources of energy in the diets of most people, but they are not essential to the body. We could obtain energy as well from a mixture of proteins and fats. Foods rich in carbohydrates are usually the cheapest ones available and this economic factor plays a role in determining the percentage of carbohydrates in the diet.

Fats and oils are the most concentrated foods, supplying not only more calories per gram than proteins and carbohydrates but also containing less water than other foods. Man and other vertebrates can synthesize most fatty acids but not the "polyunsaturated" ones containing two or more double bonds. These, termed "essential" fatty acids, must be present in the diet.

Foods rich in proteins are usually the most expensive and the amount of protein in the diet is determined in part by economic factors. Since all the protein constituents of the body are constantly undergoing degradation and replacement, there is a continued requirement for a certain minimum of protein in the diet, even for adults whose overall growth has ceased. Growing children, pregnant women and patients recovering from wasting diseases have increased requirements for protein in the diet. Proteins differ widely in the number and kinds of amino acids they contain. When a cell is synthesizing some particular protein it obviously must have available all the individual amino acids that comprise it. If even a single one is missing, no protein at all can be made. Animals can synthesize some, but not all, of the amino acids from other materials, and the ones that cannot be synthesized must be supplied in the diet. Such "essential" amino acids are no more essential for protein synthesis than other amino acids, but since they cannot be synthesized they are essential *in the diet.* Proteins that contain adequate amounts of all of these essential amino acids are termed "adequate proteins." Milk, meat and eggs contain nutritionally adequate proteins, but the major protein of corn kernels, **zein,** lacks two of the essential amino acids. An experimental animal raised on a diet of which corn is the sole protein source would lose weight and eventually die if the diet were not supplemented with tryptophan and lysine, which are missing in zein. Most people eat a mixture of different kinds of proteins and are in no danger of suffering from a deficiency of any of the essential amino acids. It is possible for an adult to maintain health with a dietary intake as little as 25 grams of protein per day, but larger amounts, 70 to 100 grams, are desirable.

Shortly after a meal rich in protein is eaten, the metabolic rate increases for a short time to as much as 30 per cent above the basal level. This appears to represent the energy necessary to convert some of the amino acids into carbohydrate or fat and is termed the **specific dynamic action** of proteins. It is not energy required for the digestion or absorption of proteins, for it also occurs if a solution of amino acids is injected intravenously into a patient.

METABOLISM OF CARBOHYDRATES, FATS AND PROTEINS

The amino acids and simple sugars absorbed from the intestine are carried to the liver by the **hepatic portal vein.** Perhaps, originally, the liver was important in digestion only, but during the evolutionary process it has assumed many other functions and is now a chemical jack-of-all-trades. It protects other cells of the body by detoxifying certain harmful substances; it is active in the storage and interconversion of carbohydrates, fats and proteins; it is important in the metabolism of hemoglobin; it stores certain vitamins; it manufactures substances necessary for the coagulation of the blood; and it converts some of the harmful waste products produced by the metabolism of other body cells to less harmful, more soluble ones which can be excreted by the kidneys.

Carbohydrate Metabolism

Three kinds of simple sugars—glucose, fructose and galactose—are derived from the hydrolytic cleavage of polysaccharides and double sugars and are absorbed from the digestive tract. They pass to the liver, which converts the other simple sugars to glucose and stores them all as **glycogen.** The metabolic pathways of these interconversions are outlined in Figure 17–9. Glycogen is a highly branched polysaccharide of high molecular weight, composed of glucose units linked by α-glycosidic bonds.

The role of the liver in storing carbohydrates was discovered over a century ago by the French physiologist Claude Bernard. He analyzed the glucose content of blood entering and leaving the liver just after a meal and found a much higher concentration of sugar in the blood entering the liver than in that leaving it. Analysis of the liver showed that new glycogen appeared simultaneously. Between meals, liver glycogen is reconverted to glucose and the concentration of glucose in the blood leaving the liver is greater than in that entering it. In this way Bernard discovered that the liver maintains the glucose concentration of the blood more or less constant throughout the day.

The liver can store enough glycogen to supply glucose from about 12 to 24 hours; after that the normal concentration of glucose in the blood is maintained by the conversion of other substances, principally amino acids, into glucose.

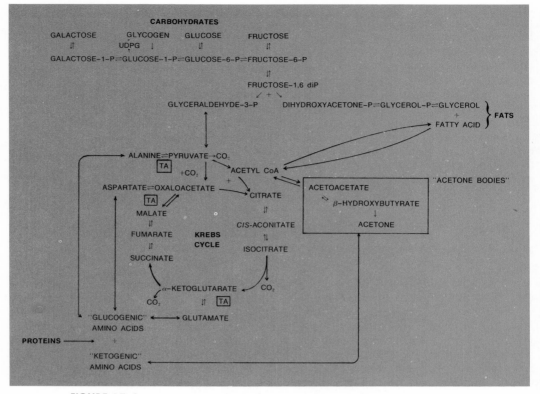

FIGURE 17–9   Diagram of the metabolic pathways by which the carbon chains of carbohydrates, fats and proteins can be interconverted. Transamination reactions, by which amino acids are converted to keto-acids, are indicated by TA.

Glucose is a primary source of energy for all cells, and its concentration in the blood must be maintained above a certain minimal level, about 60 milligrams per 100 milliliters of blood. The brain is the first organ to suffer when the concentration falls below this. In contrast to most other cells of the body, brain cells are unable to store any appreciable quantity of glucose as glycogen and they have only a limited ability to use fats or amino acids as sources of energy. When the glucose level is low, and the brain is not adequately supplied with fuel, symptoms appear which are similar to those accompanying a lack of oxygen—mental confusion, convulsions, unconsciousness and death. Whenever the brain cells are deprived of either glucose or oxygen, they cannot carry on the metabolic processes which yield energy for their normal functioning. Other tissues normally obtain glucose from the blood, but they can use other substances when necessary.

Muscle cells can change glucose into glycogen for storage, but muscle glycogen serves only as a local fuel deposit, available for muscular work, and is not available for regulating the concentration of glucose in the blood. Liver cells, but not muscle cells, contain the enzyme **glucose-6-phosphatase,** which converts glucose-6-phosphate to free glucose to be secreted into the bloodstream (Fig. 17–10).

In addition to being stored as glycogen or oxidized for energy, glucose may be transformed into fat for storage. Whenever the supply of glucose exceeds the immediate needs, the liver and other tissues convert it into fat to be used for energy at some later time. It has been known for many

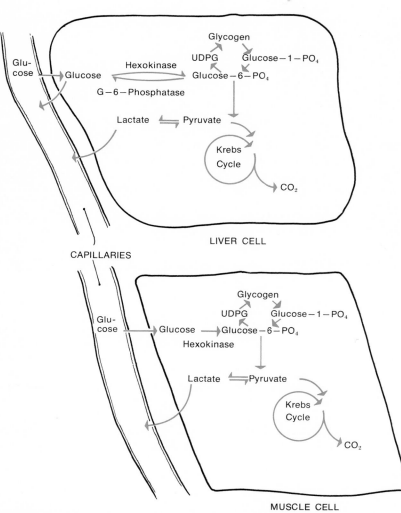

FIGURE 17–10   Diagram comparing the metabolism of glucose in liver cells and in skeletal muscle cells. Liver cells, but not muscle cells, have the enzyme glucose-6-phosphatase which enables them to secrete glucose into the bloodstream.

years that eating large amounts of starches or sugars is fattening; the starch of corn or wheat eaten by cattle and pigs is converted into the fat of butter and bacon. With the use of radioactive or stable isotopes, it is possible to demonstrate that a particular carbon or hydrogen atom that enters the body as carbohydrate can be recovered as fat in adipose tissue or liver. The metabolic pathways by which carbohydrates are converted to fats are outlined in Figure 17–9. Both the glycerol and the fatty acids of the lipid molecule can be synthesized from the carbon chain of glucose.

The functioning of the liver in carbohydrate metabolism is regulated by a complex interaction of four hormones—**insulin** from the pancreas, **epinephrine** from the adrenal medulla, **cortisol** from the adrenal cortex and **growth hormone** from the pituitary.

Lipid Metabolism

Each species of animal or plant deposits fat containing a certain proportion of the different kinds of fatty acids. When we eat beef fat or olive oil, it must be changed, largely by the liver, into the type of fat characteristic of human beings. The fat in adipose tissue, besides being available as a source of energy when needed, serves as a supporting cushion for certain internal organs and as an insulating layer under the skin, preventing too rapid heat loss. The role of adipose tissue in thermal insulation is especially evident in aquatic mammals such as whales, which have a thick layer of fat-laden cells, **blubber,** just under the skin.

The oxidation of fatty acids does not proceed properly unless oxaloacetic acid, derived primarily from carbohydrate metabolism, is available to condense with the acetyl coenzyme A formed from the fatty acids (Figs. 5–2 and 17–9). Diabetics, whose carbohydrate metabolism is interfered with, have abnormal lipid metabolism also, and certain injurious intermediate products (called ketone bodies) tend to accumulate in their blood and be excreted in the urine. In addition, large amounts of lipid collect in the liver, a symptom which occurs in certain other abnormalities of liver function.

Lipids, as well as proteins, are important structural components of the nuclear, mitochondrial and plasma membranes.

The metabolism of fats is controlled partly by hormones from the pituitary and adrenals, and partly by sex hormones, but the details of this regulation are not clear. Any severe disturbance of liver function results in the almost complete absence of fat from the usual adipose tissues, indicating that fat must be acted upon in some way by the liver before it can be stored or metabolized.

Protein Metabolism

The amino acids entering the liver from the hepatic portal vein may be removed from the blood and used or stored or carried to other cells to be incorporated into new proteins. Experiments using amino acids labeled with ^{15}N, "heavy" nitrogen, have shown that the body proteins are constantly and rapidly being torn down and rebuilt.

If there are more amino acids in the diet than are necessary for the synthesis of cell proteins, enzymes in the liver remove the amino group from amino acids, a process called **deamination.** Other enzymes combine the split-off amino groups with carbon dioxide to form a waste product,

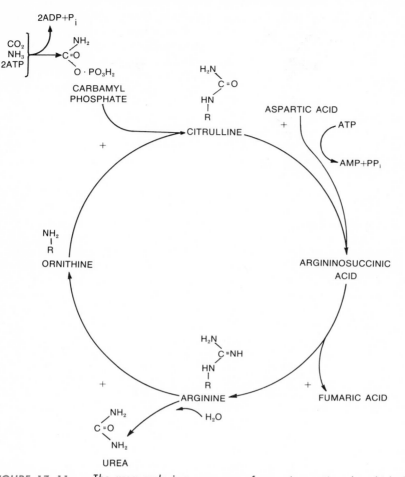

FIGURE 17-11   The urea cycle is a sequence of enzymic reactions by which the urea molecule is assembled from carbon dioxide and ammonia. It involves the cyclic utilization of the amino acids ornithine, citrulline and arginine; however, neither ammonia nor carbon dioxide reacts as such with these amino acids. First, carbamyl phosphate is synthesized from ammonia and carbon dioxide by a complex reaction which requires two ATPs. The carbamyl phosphate condenses with the terminal amino group of ornithine to form citrulline. This in turn reacts with aspartic acid, another amino acid, to form the intermediate argininosuccinic acid, a reaction which requires another ATP to drive it. The argininosuccinic acid is cleaved to yield arginine and fumaric acid; thus the amino group of aspartic acid is transferred to arginine. Arginine is hydrolyzed by the enzyme arginase to yield urea and ornithine, which can be used in the next cycle. The energy to drive the cycle and to synthesize urea is provided by the two $\sim$ P used in the synthesis of carbamyl phosphate and the ATP converted to AMP (adenosine monophosphate) and PP$_i$ (inorganic pyrophosphate) in the synthesis of argininosuccinic acid.

urea (Fig. 17–11), which is carried by the bloodstream to the kidneys and eliminated in the urine.

The parts of the amino acids remaining after deamination are simple organic acids. The carbon skeletons of certain amino acids, called "gluco-genic" amino acids, can be converted (Fig. 17–9) into glucose or glycogen. The carbon skeletons of the other amino acids, termed "ketogenic" amino acids, yield ketone bodies. There is little or no storage of proteins as such in the body; the proteins the body draws upon when carbohydrates and fats are exhausted are not stored proteins but are the actual enzymes and structural proteins of the cells.

The hormonal control of the metabolism of proteins and amino acids is even more obscure than that of lipids. Since growth is essentially the deposition of new protein, the **growth hormone** of the pituitary plays some part in it, but its nature is unknown. Insulin, the sex hormones and cortisol from the adrenal cortex are also involved in the control of protein metabolism.

SECTION 17-8

MINERALS AND VITAMINS

In addition to proteins, fats and carbohydrates, animals require water, minerals and vitamins to grow and maintain health. The constant loss of minerals from the body in urine, sweat and feces must be balanced by the intake of equivalent amounts in the food. Most foods contain adequate amounts of minerals and mineral deficiencies are comparatively rare. The deaths of entire herds of sheep in one region of Australia were finally traced to a deficiency of **cobalt** in the soil in that part of the world. The grass growing in that soil was, of course, also deficient in cobalt, and this mineral is required as a "trace element," one needed in very small amounts, for normal metabolism. Human diseases resulting from deficiencies of iron, copper, calcium, phosphorus and iodine are known.

The extracellular fluids of man and other vertebrates are about 0.9 per cent salt, mostly sodium chloride; the sodium and chloride ions play a major role in maintaining osmotic balance and acid-base balance in the body fluids. They are also present and play a functional role in the secretions of the digestive tract—the hydrochloric acid of the gastric juice and the slightly alkaline pancreatic and intestinal juices. The salts in these secretions are reabsorbed and used over again. Although large amounts of salts are secreted in the course of a day, they are reabsorbed almost completely and the loss via the digestive tract is negligible.

Calcium and phosphorus are the chief constituents of bones and teeth and a childhood deficiency of either one—or of **vitamin D**, needed for their absorption and metabolism—will result in **rickets**. Phosphorus has an important role in metabolism as a constituent of nucleotides and nucleic acids. In addition, sugars and fatty acids must be phosphorylated before they can be utilized as sources of biologically useful energy. **Iodine** is a component of the hormone thyroxine, synthesized in the thyroid gland. **Iron** is a constituent of the red blood pigment hemoglobin and of the respiratory enzymes, the **cytochromes,** present in the electron transmitter system. This iron is used and reused; as long as there is no loss of blood, the amount of iron required in the daily diet is minimal. Because women lose a considerable amount of blood each month by menstruation, their iron reserves are typically very small and they are more likely than men to become anemic owing to an iron deficiency. Traces of **copper** are needed as a component of certain enzyme systems and for the proper utilization of iron in hemoglobin. Traces of manganese, molybdenum, cobalt and zinc are required as components of certain enzyme systems; **zinc,** for example, is a component of carbonic anhydrase, alcohol dehydrogenase and a number of other enzymes. The role of **fluoride** ion in the prevention of dental caries is well known.

Water is required by every plant and animal and comprises some two thirds of the human body. Aquatic animals have no problem about obtaining water; indeed, their problem is to prevent the osmotic inflow of water and the consequent bursting of their cells. Most land animals drink water but others—certain desert animals—obtain all they require from the food they eat and from the water formed when the foods are oxidized. Water plays a major role in temperature regulation.

Vitamins

Vitamins are relatively simple organic compounds, required in small amounts in the diet. They differ widely in their chemical structure but have in common the fact that they cannot be synthesized in adequate amounts by the animal and hence must be present in the diet. It is clear that all plants and animals require these substances to carry out specific metabolic functions, but organisms differ in their ability to synthesize them. Thus, what is a "vitamin" for one animal or plant is not necessarily one for another. Only man, monkeys and guinea pigs require vitamin C, **ascorbic acid,** in their diets; other animals can synthesize it from glucose. The mold *Neurospora* requires the vitamin **biotin.** Insects cannot synthesize cholesterol and for insects it might be argued that cholesterol is a vitamin.

The vitamins whose role in metabolism is known—niacin, thiamine, riboflavin, pyridoxine, pantothenic acid, biotin, folic acid and cobalamin—have been found to be constituent parts of one or more **coenzymes.** Two major groups of vitamins can be distinguished: those that are soluble in lipid solvents, the **fat-soluble vitamins** A, D, E and K; and those that are readily soluble in water, the **water-soluble vitamins** C and the B complex. The lack of any one of these vitamins produces a particular deficiency disease with characteristic symptoms, e.g., **scurvy** (lack of ascorbic acid), **beriberi** (lack of thiamine), **pellagra** (lack of niacin) and **rickets** (lack of vitamin D) (Fig. 17–12).

Thiamine pyrophosphate is the coenzyme for the oxidative decarboxylation of α-keto acids such as pyruvate and α-ketoglutarate. It is also the coenzyme for transketolase. Riboflavin, as riboflavin monophosphate and **flavin adenine dinucleotide,** is required for the reactions of electron transport in the mitochondria and for certain oxidations in the endoplasmic reticulum. Riboflavin occurs in most foods and is synthesized by the intestinal bacteria, so that deficiencies of riboflavin are quite rare. Pyridoxine, as **pyridoxal phosphate,** is the coenzyme for many different reac-

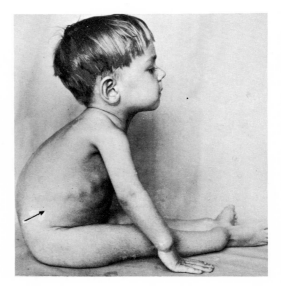

FIGURE 17–12   A child with rickets. A deficiency of vitamin D decreases the body's ability to absorb and use calcium and phosphorus. This results in the production of soft malformed bones. The malformation is most clearly evident in the ribs (arrow) and in the wrists and ankles. (Courtesy of Dr. Niilo Hallman.)

tions involving amino acids—transamination, decarboxylation to amines and so forth—and for glycogen synthetase. **Biotin** is a coenzyme for the reactions in which carbon dioxide is added to an organic molecule, such as the conversion of pyruvate to oxaloacetate, and the carboxylation of acetyl coenzyme A to form malonyl coenzyme A (the first step in the biosynthesis of fatty acids). It is widely distributed in foods and only individuals who eat raw eggs in large quantities are likely to become deficient in biotin. Egg white contains a protein, **avidin**, which forms a tight complex with biotin and prevents its functioning.

Niacin is part of the coenzymes **nicotinamide adenine dinucleotide** (NAD) and **nicotinamide adenine dinucleotide phosphate** (NADP) (also called DPN and TPN, respectively), which are coenzymes for many dehydrogenases and serve as hydrogen acceptors or donors. The vitamin folic acid appears in the coenzyme **tetrahydrofolic acid,** a coenzyme for many reactions involving one-carbon transfers, and in **biopterin,** the coenzyme for the conversion of phenylalanine to tyrosine. **Cobalamin** (vitamin B_{12}) serves as the coenzyme in certain reactions involving one-carbon transfers such as the transfer of methyl groups, CH_3. Cobalamin is synthesized by bacteria but not by higher plants or animals—thus this is a "vitamin" for the green plants as well as for animals. **Ascorbic acid** is known to play a role in the hydroxylation of proline to form hydroxyproline during collagen formation, but the overall function of this vitamin is still unknown.

Vitamin A, or **retinol,** is converted in the retina to retinal, a component of the light-sensitive pigment, **visual purple.** A deficiency of retinol may lead to "night-blindness" and a severe deficiency will lead to **xerophthalmia,** a blindness due to the abnormal deposition of keratin as a film over the cornea. Keratin is also deposited in other mucous membranes of individuals suffering from xerophthalmia. Vitamin A undoubtedly has metabolic functions other than its conversion to retinal in the visual process but these are unknown. Plants synthesize β-**carotene**, an orange-yellow pigment, which can be split to give two molecules of retinol. Large amounts of retinol are stored in the liver, enough to supply a person for several years.

Vitamin D, **cholecalciferol,** plays a role in the movement of calcium ions through membranes, perhaps by stimulating the synthesis of a specific protein required in the transport process. Cholecalciferol can be formed in the skin from a precursor, cholesta-5,7-dienol, by the action of ultraviolet light, which cleaves the B ring of the precursor molecule. Thus cholecalciferol is a "vitamin" only if the person isn't exposed to an adequate amount of sunlight. Some ten structurally related compounds have varying amounts of vitamin D activity; one of the more potent ones is **calciferol,** the usual commercial form of the vitamin which is manufactured from a plant sterol, ergosterol, by irradiation. Excessive doses of vitamin D are toxic, causing hypercalcemia and the deposition of calcium in soft tissues. An excess of vitamin A is also toxic. Acute poisoning with vitamin A can occur by eating the livers of polar bears, a pound of which contains a 20-year supply of retinol for a man. Chronic toxicity is more common and results from prolonged overdosage with vitamin preparations.

A number of similar substances, referred to as **vitamin K,** play a role in the normal coagulation of blood by promoting the synthesis of prothrombin in the liver. These compounds are found in many kinds of food and are synthesized by intestinal bacteria; thus a deficiency of vitamin K is more often associated with some abnormality of its absorption than with a lack in the diet. Vitamin E, or α-**tocopherol,** plays some unknown role in metabolism. Male animals made deficient in tocopherol undergo degenerative changes in the testes and become sterile, and eggs from vitamin E-deficient hens fail to hatch. The skeletal muscles of vitamin E-deficient animals eventually undergo degeneration. Tocopherol may serve as an antioxidant, protecting certain labile cellular components from being oxidized by removing intermediate free radicals. Some investigators believe that the effects of tocopherol deficiency can be traced to the accumulation of fatty acid peroxides, which react with and destroy other cellular components. There is some evidence that vitamin E plays a role in the mitochondrial electron transport system, but the nature of that role is unclear.

SECTION 17–9

NUTRIENT PROCUREMENT IN PLANTS

Autotrophic plants, whether algae or seed plants, can synthesize all the organic molecules they need from simple inorganic precursors and hence have no need for a digestive system to obtain nutrients. The simpler aquatic plants, such as algae, absorb all their nutrients from the surrounding water without the aid of any specialized structures. Land plants obtain carbon dioxide from the air, primarily through their leaves, but obtain water and minerals from the soil through their roots. There are a few insect-eating plants, all rather small, which, though lacking a digestive "system," do secrete digestive enzymes similar to those secreted by animals. When photosynthesis occurs, plants accumulate reserves of carbohydrates and fats that can be used subsequently when photosynthesis is impossible, at night or over the winter. An embryo plant cannot make its own food until the seed has sprouted and the embryo has developed a functional root, leaf and stem. Seeds must have a rich reserve of carbohydrates and fats to supply energy for the initial stages of growth. When these reserves are being made available, enzymatic processes akin to those in the intestine of man digest the starches, proteins and lipids of the seed and make their subunits available for cellular metabolism.

In addition to carbon dioxide and water as raw materials for the synthesis of their cell constituents, plants need nitrogen, sulfur and phosphorus from which to synthesize proteins and nucleic acids, and they need a number of minerals which are essential parts of certain enzyme systems in plant cells just as they are in animal cells. The nitrates, phosphates and sulfates and the potassium, magnesium, iron and other cations are obtained from the soil by the roots. A plant growing in a soil deficient in any of these nutrients will be stunted and may be unable to survive (Fig. 17–13). It is now known that in addition to the above, plants require trace amounts of boron, manganese, copper, zinc, molybdenum, cobalt, sodium and chloride. Perhaps research in the future will demonstrate the need for other trace elements.

One obvious function of the root system is to anchor the plant and hold the stem in an upright position. To do this it branches and rebranches extensively through the soil. Although the depth of the root seldom equals the height of the stem, the roots frequently extend laterally farther than the stem's branches do, and the total surface of the root usually exceeds that of the stem. In addition to anchoring the plant and absorbing minerals and water, the roots of some plants (e.g., carrots and beets) function for the storage of food reserves.

To understand how roots take up minerals and water, we must first learn a bit about the structure of the root system. The tip of each root is covered by a protective **root cap,** a thimble-shaped covering of cells which fits over the rapidly growing meristematic region (Fig. 17–14). The outer part of the root cap is constantly worn away as the root pushes through the

FIGURE 17–13   Tobacco plants showing mineral deficiency symptoms. The plant in
the center received all essential elements; the others were supplied with all essential
elements except the one indicated on the label. All plants are the same age and variety.
Note the marked growth inhibition resulting from lack of N and K, the green but distorted
leaves of the plant lacking Ca, the general chlorosis (except along the veins) of the
—Mg plant, and the chlorosis in older leaves only when P was lacking. The other leaves of
the —P plant are dark blue-green. The leaves of the —S plant show different degrees of
chlorosis, but more commonly the younger leaves are the most chlorotic. Note also the
necrosis in the —N, —P, and —K plants. (Courtesy of W. R. Robbins, Rutgers University.)

soil, thus its surface is rough and uneven. From this growing point of actively dividing meristematic cells are formed all the other tissues of the
root and the root cap. Just behind the growing point is the **zone of elongation,** where the cells remain undifferentiated but grow rapidly in length
by taking up large quantities of water. The growing point is about 1 millimeter in length and the zone of elongation is from 3 to 5 millimeters long;
these two regions alone account for the increase in length of the root.

Above the zone of elongation lies the **zone of maturation,** which can be
distinguished externally by a downy covering of whitish **root hairs.** In this
zone, the cells differentiate into the specific tissues of the root. Each root
hair is a slender, elongated lateral projection from a single epidermal cell,
through which much of the minerals and water are absorbed. Substances
may enter the root through any epidermal cell, but root hairs expose a
much greater surface and hence are more effective in absorbing nutrients
than ordinary epidermal cells. The delicate, short-lived root hairs are
formed just behind the zone of elongation and wither and die as the root
elongates. Only a short section of the root, perhaps from 1 to 6 centimeters
long, has root hairs.

Some measurements made by Dittmer will give an idea of the aston-

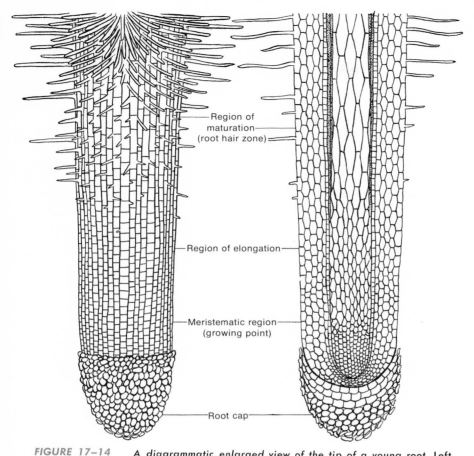

FIGURE 17–14   *A diagrammatic enlarged view of the tip of a young root. Left,
the surface of the root; right, a longitudinal section of the root showing the internal structure.*

ishing amount of root surface available for absorbing water and minerals.
A single rye plant, grown in a box 30 centimeters square and 55 centimeters deep, reached a height of 50 centimeters in four weeks. Dittmer
then carefully washed away the soil and measured the number of roots.
This single plant had 625 kilometers of roots with 285 square meters of
surface! Its root hairs, some 14 billion of them, were estimated to have a
total length of 10,500 kilometers and a surface area of 480 square meters.
The total area of roots plus root hairs was 765 square meters, about 6875
square feet (the building lots in some subdivisions are smaller than this).

The number and total area of the roots on any given plant depend on
many factors, one of which is how close each plant is to its neighbors.
Wheat plants grown 3 meters apart each had 70 kilometers of roots, but
the same variety grown 15 centimeters apart had less than 1 kilometer
of roots. Weed plants in your garden decrease the number and area of the
roots on your peas, beans and tomatoes and thus decrease their productivity.

The first root formed by a young seedling is called the **primary root;**
its branches are called **secondary roots.** Some species have a single large
tap root, usually growing straight down, from which branch many, much
smaller, secondary rootlets (Fig. 17–15). Other species have **diffuse roots,**
a system of many slender-branched roots, all of which are about the same
size.

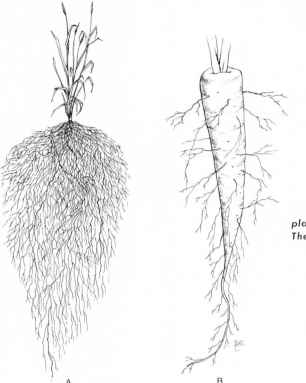

A   B

FIGURE 17–15   *Types of root systems in plants. A, The diffuse root system of a grass. B, The tap root system of a carrot.*

Additional roots that grow from the stem or leaf, or any structure other than the primary root or one of its branches are termed **adventitious roots.** The prop roots of corn plants and the aerial roots of ivy and other vines which attach the vine to a wall or tree are adventitious roots. Many kinds of flowering plants can be propagated by cutting off bits of their stems and producing adventitious roots at the base of these stem cuttings.

A cross section of a root in the zone of maturation reveals a complex, highly organized structure (Fig. 17–16). The outer surface, the **epidermis,** is a layer of rectangular, thin-walled cells one cell thick. The epidermis of the root, unlike the epidermis of the stem, usually has no waxy cuticle on its outer surface. A waxy cuticle would undoubtedly interfere with the

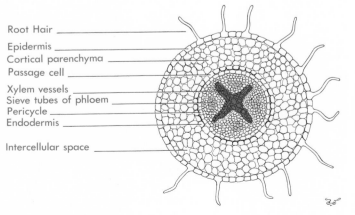

Root Hair

Epidermis
Cortical parenchyma
Passage cell

Xylem vessels
Sieve tubes of phloem
Pericycle
Endodermis

Intercellular space

FIGURE 17–16   *A cross section of a root near its tip where root hairs are present.*

absorption of water. Each root hair originates as a swelling on an epidermal cell which increases in size and becomes a hairlike projection as much as 8 millimeters long. There are hundreds of root hairs on each square millimeter of root surface. A branch of the vacuole of the epidermal cell fills most of the volume of the root hair, leaving just a thin film of cytoplasm between the vacuole and the cell wall.

Just inside the epidermis is the **cortex,** a wide area composed of many layers of large, thin-walled, nearly spherical parenchymal cells between which are many intercellular spaces. These cells serve as avenues for the conduction of water and minerals, and as storage places for starch and other foods. At the inner edge of the cortex, a single layer of cells, the **endodermis,** separates the loosely packed cortical parenchyma from the central core of vascular tissue, the **stele.** In the older parts of the root, the endodermis is thick, waxy and lignified, which decreases the diffusion of water out of the stele into the cortex. At various points in the endodermis are thin-walled **passage cells,** through which water and minerals pass to the stele. Just inside the endodermis is a single layer of parenchymal cells, the **pericycle,** which can be transformed into meristem to give rise to part of the root's cambium and to lateral roots. That secondary roots originate deep within the primary root ensures a good connection between the vascular tissues of branch and main roots.

The central portion of the stele, surrounded by endodermis and pericycle, is composed of the two vascular tissues, **xylem** and **phloem.** The cells of the xylem, **tracheids** and **xylem vessels,** are usually arranged in the form of a star, or like the spokes of a wheel. They are thick-walled, rounded and elongated—tubular—cells which conduct water. Between adjacent points of the xylem star are bundles of phloem cells, smaller and thinner-walled than the xylem. Some of the phloem cells are specialized **sieve tubes.** The roots of trees, shrubs and other perennial plants typically have a single-celled **cambium** layer between the xylem and phloem. This divides and gives rise to additional layers of xylem and phloem to provide for increased thickness of the root.

The transfer of water from the soil into the root hairs and across epidermis, cortex and endodermis into the xylem in the stele can be explained on purely physical principles (Fig. 17–17). The water available to plants is present as a thin film loosely held to the soil particles and is called **capillary water.** The roots, especially the root hairs, are in contact with the films of capillary water. The capillary water usually contains some dissolved inorganic salts and perhaps some organic compounds, but the concentration of solutes in capillary water is low and the solution is hypotonic to the fluid within the root hairs. The cell sap in the epidermal cells has a fairly high concentration of glucose and other organic compounds. Water molecules diffuse from a region of higher concentration (the capillary water) to a region of lower concentration inside the root hairs. Water is passing through a differentially permeable membrane, the plasma membrane of the root hair cell, by the process of osmosis. The plasma membrane is permeable to water but not to the glucose and other organic compounds in the cell sap.

As the root hair cells take in water, their contents become more dilute and their cell sap becomes hypotonic to that of the adjacent cell in the cortex. Water tends to pass from the epidermal cell to the cortical cell. The outermost cortical cells then have a lower osmotic pressure than the ones farther in and water tends to diffuse in toward the center. In this way water continues to diffuse inward, finally reaching the ducts of the xylem and then rises in the xylem to the stem and other parts of the plant body. The removal of water from the stele via the xylem maintains the concen-

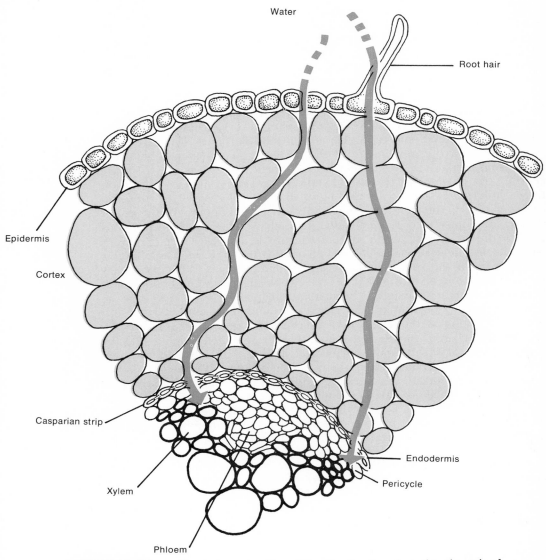

FIGURE 17–17   *The pathway for the transfer of water from the soil to the stele of the root.*

tration gradient of water from epidermis to xylem and permits the continued inflow of water. There is some evidence that the roots of some species are capable of an active energy-requiring secretion of water inwards toward the stele, which may come into play when the concentration gradient between soil and root cells is small or reversed.

Water may pass from epidermis to stele without passing through the intervening cells. The cell walls of all these cells are made of cellulose, which has a strong tendency to imbibe water (the "blotting paper" principle), and water may move inward along these cellulose cell walls to the endodermis. There it enters the stele through the thin-walled passage cells in the endodermis. The cell walls of many plant cells are interrupted by very small holes through which the cytoplasm of one cell may be connected to the cytoplasm of the adjacent one. These connections, called **plasmodesmata,** may be important as a further avenue for the transport of water, ions, sugars and amino acids from one cell to the next.

The water entering the xylem vessels increases the osmotic pressure within those ducts, just as water entering a semipermeable cellophane membrane containing a glucose solution does (p. 41). This **root pressure** is one of the factors that move sap upward through the root and stem to the leaves. Water vapor is constantly escaping from the leaves by **transpiration** (p. 550). New molecules of glucose and other organic substances are formed in the leaves by photosynthesis and sent via the phloem to the roots. These two processes maintain the hypertonicity of the sap in the roots and the absorption of water continues.

The entrance of mineral nutrients, inorganic salts, occurs in part by simple diffusion, for any ion present in greater concentration in the capillary water of the soil than in the cell sap of the root hair will tend to diffuse into the root hair. The absorption of inorganic ions is largely independent of the entrance of water, and the absorption of each kind of ion is independent of all others, for each diffuses down its own concentration gradient. The rate at which each type of ion enters is determined by the difference in the concentrations of that ion inside and outside the root and by factors such as the diameter of its hydrated form which determine its rate of penetration of cell membranes.

Some roots, and perhaps all, take in inorganic ions by processes of active transport, which require the expenditure of energy. An increase in the rate of cellular metabolism can be detected in such roots when they are absorbing inorganic ions against a concentration gradient. The concentration of inorganic ions in plant sap is rather low and large volumes of sap must flow to the stem and leaves to provide the required nutrients. The inorganic ions that enter the root are carried to other parts of the plant and utilized rapidly, thereby decreasing their concentration and maintaining the concentration gradient from soil to root to stem to leaf. The nitrates absorbed from the soil are reduced to amino groups and other nitrogenous compounds. Some of the amino groups are incorporated into amino acids which are used for protein synthesis and others become constituents of one of a host of nitrogenous organic molecules peculiar to one or another species of plant. These include flavorings, perfumes and drugs such as quinine and morphine.

SUGGESTIONS FOR FURTHER READING

Beaumont, W.: *Experiments and Observations on the Gastric Juice and the Physiology of Digestion.* Cambridge, Massachusetts, Harvard University Press, 1929.
> This text describes Beaumont's experiments on the trapper, Alexis St. Martin whose stomach was opened to the exterior by a gunshot wound; classic description of gastric function.

Carlson, A. J., Johnson, V., and Cavert, H. M.: *The Machinery of the Body.* 5th ed. Chicago, University of Chicago Press, 1961.
> An excellent elementary text of human physiology with good chapters on digestion and metabolism.

Guyton, A. C.: *Function of the Human Body.* 3rd ed. Philadelphia, W. B. Saunders Company, 1969.
> A clear and concise exposition of human physiology.

Marshall, P. T., and Hughes, G. M.: *The Physiology of Mammals and Other Vertebrates.* New York, Cambridge University Press, 1965.
> A useful textbook of mammalian physiology which gives the reader an appreciation of the structural and functional unity of organ systems.

Neurath, H.: *Protein-Digesting Enzymes.* Sci. Amer., December (Offprint 198) 1964.
> An interesting description of the structure of the enzymes and the basis of their specificity.

DISTRIBUTION SYSTEMS

CHAPTER 18

The energy required for all the myriad activities of plants and animals is derived from the reactions of biological oxidation, described in Chapter 5. The essential feature of these reactions is the transfer of hydrogen atoms from **hydrogen donors** to **hydrogen acceptors,** accompanied by the transfer of energy to phosphate ester bonds. Both animal and plant cells contain a series of compounds, each of which accepts a hydrogen atom (or its electrons) from the preceding one and donates it to the subsequent one. The ultimate hydrogen acceptor in the metabolism of most plants and animals is **oxygen,** which is converted to water: $2 H^+ + 2 e + 1/2 O_2 \rightarrow H_2O$. Only small amounts of oxygen can be stored as oxyhemoglobin in blood or oxymyoglobin in muscle; hence, the continuation of cellular metabolism requires an uninterrupted supply of oxygen.

Carbon dioxide is removed from substrate molecules by **decarboxylation** reactions, which can proceed independently of the reactions in which oxygen is utilized. Yeasts, for example, can metabolize glucose to ethanol and carbon dioxide without utilizing oxygen: $C_6H_{12}O_6 \rightarrow 2 C_2H_5OH + 2 CO_2$. In most animal and plant cells, the utilization of oxygen and the production of carbon dioxide proceed together. The carbon dioxide formed must be removed from the cells and body fluids, for it reacts with water to form carbonic acid, H_2CO_3. The sum of the processes by which animal and plant cells utilize oxygen, produce carbon dioxide and convert energy into biologically useful forms such as ATP is termed **respiration.** To carry on the process of respiration, cells need a continued supply of oxygen and a continued removal of carbon dioxide and water. For photosynthesis, green plant cells must have a continued supply of carbon dioxide and water and a continued removal of oxygen. Thus the survival of almost all cells requires the continuous exchange of gases with the environment.

SECTION 18–1

THE EXCHANGE OF GASES IN PROTISTS

The exchange of gases is a simple process in the protists and in small aquatic plants and animals. Cellular respiration in the mitochondria of a paramecium swimming in a pond uses oxygen and lowers the concentration of oxygen in the cytoplasm. Oxygen diffuses into the paramecium from the pond water, where the concentration of oxygen is greater, maintained by the diffusion of oxygen from the air into the water and by oxygen produced in photosynthesis by green plants in the pond. Carbon dioxide produced in metabolic processes within the paramecium diffuses down a concentration gradient out into the pond water. Some of the carbon dioxide diffuses out of the water into the air, some is used by plants in the pond in photosynthesis, and some is converted into carbonate ions, CO_3^{2-}, which react with calcium, magnesium and other cations to form insoluble compounds.

Carbon dioxide passes from the pond water into the cells of aquatic plants, and oxygen passes out of them by simple diffusion down a chemical gradient. Both simple animals and plants have a large enough ratio of surface to volume so that the rate of diffusion of gases through the surface of the body is not the factor that limits either the rate of respiration or the rate of photosynthesis. In larger animals, the ratio of surface to volume is smaller and cells located deep in the body cannot exchange oxygen and carbon dioxide with the environment rapidly enough by diffusion. Instead, cells exchange their gases with the **extracellular fluid** that bathes them, and this in turn exchanges gases with the environment. Most higher animals have evolved some specialized organ for gas exchange with the environment—**gills, lungs** or **tracheal tubes** (Fig. 18–1). The respiratory organ must be thin-walled to facilitate diffusion, it must be kept moist so that the gases are dissolved in water on both sides of the membrane, and some means of circulating the fluid bathing the cells must be provided. Thus in higher animals the systems for gas exchange and for internal transport are closely interrelated in function.

The higher animals living on land and breathing air have ready access to oxygen, for it comprises about 21 per cent of the air. They may have a problem in keeping their respiratory surface moist, for the air to which it is exposed is usually not saturated with water vapor and thus tends

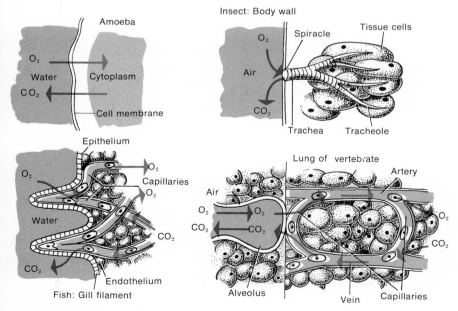

FIGURE 18–1   A diagram of some of the types of respiratory organs present in animals.

to dry out the respiratory surface. Animals living in water and exchanging gases with it have no problem of desiccation, but even well-aerated water has only about 1 ml of oxygen in every 100 ml of water, and the organism may require some mechanism to keep a continuous supply of fresh, oxygen-rich water flowing over its gills or other respiratory surface. The air we breathe contains about 210 ml of oxygen in each liter, whereas fresh pond water contains, dissolved in it, only about 7 ml of oxygen and sea water has about 5 ml of oxygen per liter. An earthworm can obtain from the air in its burrow enough oxygen by diffusion through its moist skin and does not have to stir up the air. In contrast, a marine worm living in a burrow or tube must undulate its body constantly and provide a current of water through the burrow if it is to obtain an adequate supply of oxygen from that dissolved in the sea water. The shore crab can live either in air or in water and obtains oxygen through a set of gills located in a gill chamber between the upper shell and the point of attachment of its legs. A paddle-shaped part, the **scaphognathite,** of the second maxilla moves back and forth in the gill chamber to keep a current of fresh water flowing over the gills. If the scaphognathite is paralyzed, the crab will soon die if placed in sea water; however, it will live indefinitely in air, for the rate of diffusion from air is rapid enough to supply all the oxygen the animal needs.

SECTION 18–2

GAS EXCHANGE IN TERRESTRIAL PLANTS

The higher plants have no special organ for gas exchange and there is little or no transport of gases from one part of a tree or shrub to another. Each cell in the root, stem and leaf exchanges oxygen and carbon dioxide with its environment by diffusion independently of the other cells.

The rates of gas exchange in roots and stems are not very large. The rate of cellular respiration in plant cells is usually considerably lower than in animal cells, and most roots and stems are not green and do not carry on photosynthesis. In fact, many of the cells in both root and stem are dead. Oxygen readily diffuses from the air in the spaces between the small particles of soil into the film of moisture surrounding the soil particles and into the root hairs (Fig. 18–2). The oxygen then diffuses into the cells deeper in the cortex and stele. The carbon dioxide produced in the cells passes by diffusion in the reverse direction and exits through the root hairs. In older, thicker roots that lack root hairs, there are innumerable small holes, called **lenticels,** in the protective layer of dead cells, the **cork,** through which gases diffuse in and out. When soil is thoroughly soaked with water, there are no air spaces between the soil particles and many plants are killed if the soil around their roots is completely water-logged.

The stems of trees and shrubs are also covered with a thick layer of dead cells, the cork, through which diffusion cannot readily occur. Stems, like roots, have an abundance of lenticels through which gases can pass. The green stems of annual plants have thin walls through which gas exchange occurs.

A leaf that is actively photosynthesizing has a high rate of gas exchange with its environment. The inflow of carbon dioxide and outflow of oxygen is accomplished by diffusion through pores, **stomata,** in the surface of the leaf (Fig. 18–3). Stomata are channels lying between two cells, called **guard cells,** which regulate the size of the pore. The interior

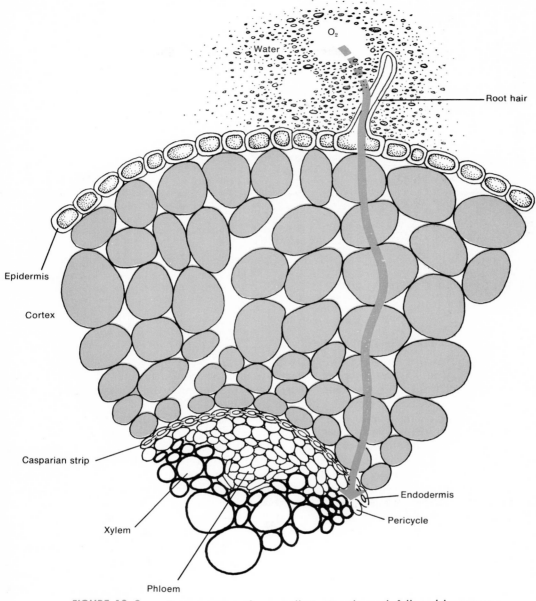

FIGURE 18–2   *A cross section of a root illustrating the path followed by oxygen as it diffuses into a root hair.*

of the leaf has many large intercellular spaces through which the gases may pass; these spaces are connected to the stomata. As carbon dioxide is utilized in photosynthesis in a cell within a leaf its concentration falls and carbon dioxide diffuses into the cell from the film of water surrounding it. Other molecules of carbon dioxide pass from the air space within the leaf and dissolve in the water film. In turn, carbon dioxide diffuses from the air outside the leaf (its concentration in the atmosphere is about 0.03 per cent), through the stomata and into the air spaces within the leaf. Throughout this course it is moving from a region of higher concentration to one of lower concentration. The molecules of carbon dioxide are kept moving

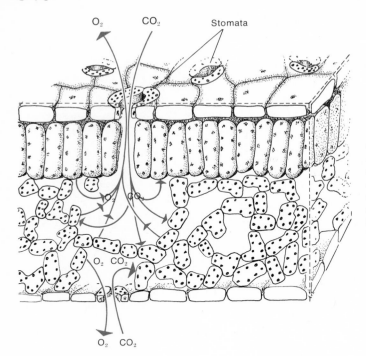

FIGURE 18–3   An enlarged cross section of a leaf showing its structure. The diffusion of carbon dioxide through the stomata to the interior of the leaf and the diffusion of oxygen from the photosynthetic cells through the stomata to the exterior are indicated by arrows.

because they are utilized within the cell and converted to something other than carbon dioxide. Oxygen produced within the cells in the leaf passes from cell to water film to air space and through the stomata to the exterior by diffusion down a chemical gradient.

The rate of entry of carbon dioxide and the rate of exit of oxygen is controlled by the size of opening in the stomata, which is regulated in turn by the **turgor pressure** in the guard cells. The stomata tend to open in the morning when exposed to light and to close at night or when exposed to hot dry conditions. The mechanism regulating the guard cells will be discussed in detail in connection with the movement of water into and out of the leaf (p. 550). In various species of plants, stomata may occur on upper or lower surfaces of the leaves, or on both, and may range in number from a few thousand to more than 100,000 per cm^2.

The leaves of plants have a problem comparable to that of the cells in the respiratory surfaces of terrestrial animals: how to ensure an adequate rate of gas exchange without losing too much water. Plants growing in a dry habitat usually have thick, fleshy leaves with a thick cuticle and stomata that are sunken below the plane of the leaf's surface.

SECTION 18–3

GAS EXCHANGE IN ANIMALS

Animals differ enormously in their rates of utilization of, and hence their requirements for, oxygen. A resting mouse uses some 2500 mm^3 of oxygen per hour for each gram of its body weight and as much as 20,000 mm^3 per hour when active. Each gram of earthworm uses perhaps 60 mm^3 and each gram of sea anemone uses only 13 mm^3 per hour. The earthworm and some amphibia use their moist skin as a surface across which gases are exchanged. The worm (and the frog out of water) secretes mucus and fluid to keep the skin moist and to permit gas exchange. Occasionally when a worm attempts to cross a rock or a cement sidewalk, the rate of evaporation from its skin caused by the sun and dry air exceeds the ability of its mucous glands and liquid secreting apparatus to replace the moisture. The respiratory surface, the skin, becomes dry and can no longer exchange gases, and the worm suffocates and dies. The frog must constantly seek a cool, moist environment to avoid a similar fate. Some amphibia use the moist membrane lining the mouth cavity as a respiratory organ.

Fishes, some amphibia, mollusks, crustacea and some worms have evolved **gills** as gas-exchange organs (Fig. 18–4). These fine filaments of tissue are covered with a thin epithelium and contain blood channels. Gases diffuse from the surrounding water through the thin moist epithelium into the blood vessels. The amount of oxygen dissolved in sea water is relatively constant, but the amount in fresh-water ponds and rivers may fluctuate widely. Each animal with gills has some arrangement to keep a current of water flowing over them. The fish opens its mouth, takes in a gulp of water, then closes the mouth and forces the water out past the gills by contracting its mouth cavity (Fig. 18–5).

The spiracles of insects and certain other arthropods represent a completely different solution to the problem of getting oxygen to the cells. Each segment of the body has a pair of openings, **spiracles**, through which air passes via a system of branched air ducts, **tracheal tubes**, to all the internal organs (Fig. 18–6). The ducts terminate in microscopic fluid-filled **tracheoles;** oxygen and carbon dioxide diffuse through these into the adjacent cells. The larger insects can pump air through the tracheal tubes by contracting the muscles in their abdominal walls. Grasshoppers draw air into the body through the first four pairs of spiracles when the abdomen expands and expel it through the last six pairs of spiracles when the abdomen contracts. This tracheal system is efficient; oxygen reaches the cells and carbon dioxide is removed by diffusion without the need of maintaining a rapid flow of blood as vertebrates must to supply the cells with oxygen.

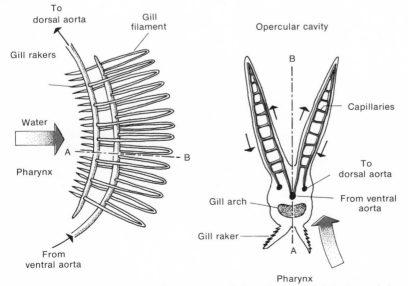

FIGURE 18–4   Diagram of the structure of the gill in a fish. Left, lateral view of a gill. Water enters from the mouth cavity (left), passes over the gill filament with its rich network of blood vessels and exits right. Right, horizontal section at the plane A-B. The direction of the water flow is indicated by the large arrow; the blood vessels are tinted.

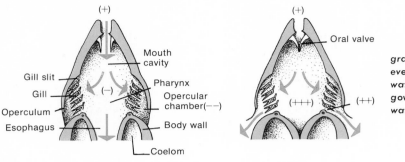

FIGURE 18–5   Diagram of the sequence of events when a fish gulps water. The relative pressures governing the movement of water are shown by + and −.

The respiratory system of man and other air breathing vertebrates includes the **lungs** for the exchange of gases with the environment, together with a system of tubes by which the air reaches the lungs (Fig. 18–7). Lungs are hollow spaces, usually greatly subdivided into thousands of small hollow pockets **(alveoli)** which are kept moist and are richly supplied with capillaries. The walls of the alveoli are very thin and gases readily diffuse across them from the cavity of the alveolus (which is continuous with the air outside the body) to the rich bed of capillaries on the other side (Fig. 18–8). Between the alveoli is a network of elastic fibers which support them and make the lung very pliable. The arrangement of the lung alveoli as pockets tends to minimize the loss of water and keeps the alveolar surface moist. The total alveolar surface in an adult man, across which gases may diffuse, has been estimated to be more than 100 m², more than 50 times the area of the skin.

There is no evidence to support the hypothesis that the cells of the lung actively secrete oxygen into the blood stream. Diffusion is more than rapid enough to supply the oxygen required. Oxygen moves down a steep diffusion gradient from a region of high concentration to a region of low concentration. The concentrations of oxygen, expressed as the partial pressure of oxygen (mm of mercury) in a gas phase or as the oxygen tension (mm of mercury) in a fluid, are 150 in the air we breathe, 105 in the air in the lungs, 100 in the blood leaving the lungs, and about 40 in the blood leaving the tissues (Fig. 18–8). The oxygen tension in tissues ranges from 40 mm Hg down to nearly zero.

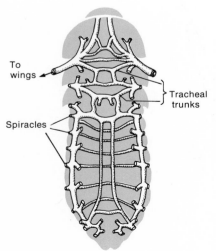

FIGURE 18–6   Diagram of the body of an insect showing the tracheal tubes by which oxygen is passed to the cells of the interior and carbon dioxide is removed. These are connected by spiracles, holes in the body wall, through which the gases are passed to the exterior.

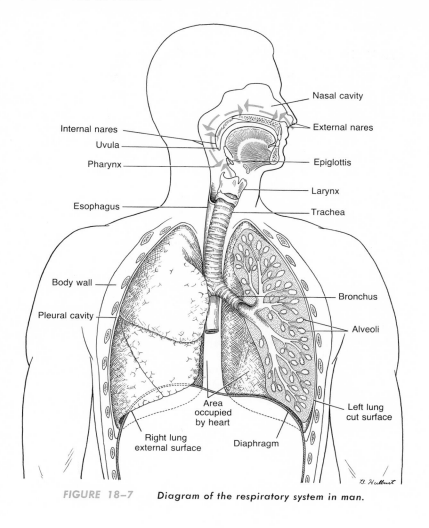

FIGURE 18–7   *Diagram of the respiratory system in man.*

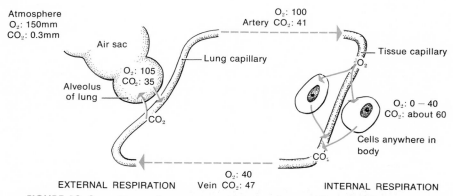

FIGURE 18–8   *Gas exchanges across the alveolus and tissue capillaries. Diagram illustrating the diffusion gradients responsible for the transfer of oxygen from lungs to tissues and for the transfer of carbon dioxide from tissues to lungs. The oxygen and carbon dioxide tensions are expressed in millimeters of mercury.*

SECTION 18-4

TRANSPORT SYSTEMS IN PLANTS

The simpler plants, consisting of single cells or small groups of cells, have no circulatory system. Simple physical diffusion, augmented in certain instances by facilitated diffusion or by active transport, suffices to supply each cell with the raw materials it requires. Although the large and more complex tracheophytes do have circulatory systems, they are simpler and constructed on an entirely different plan from those in animals. As we have seen, most cells in the higher plants exchange gases directly with their environment through intercellular spaces. However, tracheophytes have had to evolve a transport system to bring water from their roots in the soil to the leaves, where photosynthesis occurs, and a system to carry the organic materials synthesized in the leaves to the cells in the stems and roots to supply their requirements for metabolism and growth.

In contrast to the complex systems of heart and blood vessels which transport materials in higher animals, the higher plants have a simpler transport system made up of xylem and phloem. Some plants have a third system containing **latex,** a milky material, rich in carbohydrates, fats and proteins, from which are derived commercially valuable products such as rubber, chicle and opium. The system of **xylem** tubes transports water and minerals from the roots up the stem to the leaves and the **phloem** tubes transport food manufactured in the leaves down the stem for storage and use in the stems and roots. Phloem transports nutrients up the stem as well as down. In the spring, before leaves are available to synthesize food, the stored food in the root or stem is carried up to the buds to supply energy for their development and growth. The transport of water in the xylem and of nutrients in the phloem is termed **translocation.** The two processes have a somewhat different molecular basis: water rises in the vessels and tracheids of the xylem, which are not living cells, by the combined forces of **transpiration** and **root pressure.** Nutrients are transported in the living phloem cells with remarkable rapidity and by mechanisms not yet fully understood.

The arteries and veins of animals are large tubes which branch successively into smaller and smaller tubes. In contrast, all the xylem and phloem tubes are small and occur in bunches called **vascular bundles.** In the lower part of the stem, there are many vessels per bundle; in the upper part there are fewer per bundle as some enter each branch of the stem.

The material in the xylem and phloem, called **plant sap,** is a complex mixture of many substances, organic and inorganic, the composition of which varies greatly from one plant to another, from one part of the plant to another and from season to season. As much as 98 per cent is water. Other constituents include salts, sugars, amino acids, hormones such as indole acetic acid, enzymes and other proteins, and organic acids such as citric and malic acids. Plant saps, in contrast to the blood plasma in animals, are usually somewhat acid, with pHs ranging from 7 down to 4.6.

SECTION 18-5

THE STEM AND ITS FUNCTIONS

The stem, which in a tree includes the trunk, branches and twigs, is the connecting link between the roots, where water and minerals enter the plant, and the leaves, where foodstuffs are synthesized. The vascular tissues of the stem are continuous with those of root and leaf and provide a pathway for the exchange of materials. The cells of some stems contain chlorophyll and carry out photosynthesis; others are specialized for the storage of starches and other nutrients. The stem and its branches support the leaves so that each leaf is exposed to as much sunlight as possible. Stems also support flowers and fruits in the proper position for reproduction to occur. The growing points of the stem produce the primordia of the leaves and flowers.

The novice may have difficulty in distinguishing roots from stems, for many kinds of stems grow underground and some roots grow in the air. Ferns and grasses are examples of plants that have underground stems, called **rhizomes.** These grow just beneath the surface of the ground and give rise to leaves above ground. Thickened underground stems, adapted for food storage, called **tubers,** are found in plants such as the potato. An onion bulb is an underground stem surrounded by overlapping, tightly packed scale leaves. Roots and stems are structurally quite different: stems, but not roots, have nodes which give rise to leaves. The tip of a root is always covered by a root cap, whereas the tip of a stem is naked unless it terminates in a bud. The stem typically contains separate rings of phloem and xylem, with the xylem inside the phloem, whereas in the root the bundles of phloem tubes lie between the points of the star-shaped masses of xylem.

Plant stems are either herbaceous or woody. The soft, green, rather thin **herbaceous** stems are typical of plants called **annuals.** Such plants start from seed, develop, flower and produce seeds within a single growing season, dying before the following winter. Other herbaceous plants are **biennials,** having a two season growing cycle. During the first season, while the plant is growing, food is stored in the root. Then the top of the plant dies and is replaced in the second growing season by a second top which produces seeds. Carrots and beets are examples of biennials. Plants whose stems are soft and perishable and supported chiefly by turgor pressure are called **herbs.**

Quite different from the herbaceous annuals and biennials are the **woody perennials,** which live longer than two years and have a thick tough stem, or trunk, covered with a layer of cork. A **tree** is a woody-stemmed perennial that grows some distance above ground before branching and so has a main stem or trunk. **Shrubs** such as lilacs, oleanders and sagebrush are woody perennials with several stems of roughly equal size above the ground line.

It is generally believed that woody-stemmed plants are more primitive than the herbaceous ones, for the available evidence indicates that the first true seed plants were woody-stemmed perennials. In past geologic ages, such plants grew as far north as Greenland, but with the change in climate toward the end of the Mesozoic era, some of these plants were killed by the advancing cold and others were forced to retreat toward the equator. Still other woody plants adapted to the cold by evolving a life cycle in which growth and flowering were completed in the warm summer of a single year and the rigors of the winter were withstood by cold-resistant seeds; i.e., they became herbs.

The tissues of herbaceous stems are arranged around the bundles of xylem and phloem, but the details are different in the two main groups of Angiosperms or flowering plants, the Dicotyledoneae and the Monocotyledoneae. In the stem of dicots, such as the sunflower or clover, the circular arrangement of the xylem and phloem bundles subdivides the stem into three concentric regions: the outer **cortex,** the **vascular bundles** and the central core or **pith** composed of colorless parenchyma cells which serve as storage places (Fig. 18–9).

Each vascular bundle has an outer cluster of phloem cells and an inner cluster of xylem cells, separated by a layer of meristematic tissue, the **cambium** (Fig. 18–9). Continued mitotic activity in the cambium produces new phloem cells on its outer margin and new xylem on its inner margin. On the lateral border of the phloem is the **pericycle,** a layer of thick-walled supporting cells. Between the vascular bundles lie groups of cells known as **medullary rays,** which extend radially from the vascular region to both pith and cortex, and distribute materials from the xylem and phloem to these inner and outer parts.

The cortex consists of an inner layer, one cell thick, called the **endodermis,** which is immediately adjacent to the pericycle, then a layer of loosely packed, thin-walled parenchyma cells, and finally a layer of thick-walled collenchyma cells, which are supporting tissues. Immediately surrounding the collenchyma of the cortex is the outermost layer, the **epidermis.** The outer walls of the epidermal cells are thickened and contain cutin.

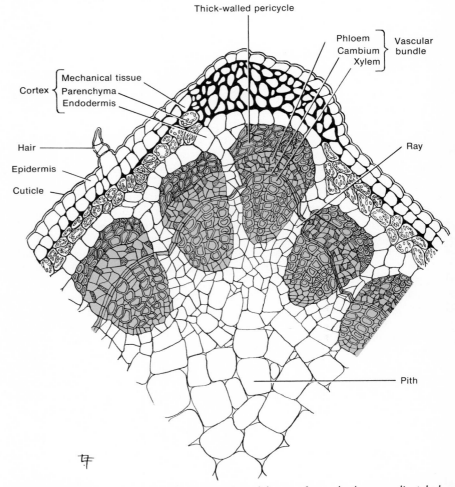

FIGURE 18–9   *A sector of a cross section of the stem from a herbaceous dicotyledon,* alfalfa.

The stems of woody plants resemble herbaceous ones during their first year of growth; but by the end of the first growing season, additional cambium has formed in the medullary rays so that a continuous circle of cambium extends between the vascular bundles as well as through them. In each successive year the cambium forms an additional layer of xylem and phloem. The phloem formed in this way eventually replaces the primary phloem and forms a continuous thin sheath of food-conducting tissue just outside the cambium. The yearly deposits of xylem form the **annual rings** (Fig. 18–10). These can be distinguished because the tracheids and vessels formed in the spring of the year are larger, and hence appear lighter than those elements of the xylem formed in the summer. Only the youngest, outermost layers of xylem, the **sapwood,** carry sap to the leaves; the inner layers of hard nonconducting xylem cells and fibers, known as the **heartwood,** increase the strength of the stem and accommodate the increasing load of foliage as the tree grows.

The width of the annual rings varies according to the climatic conditions prevailing when the ring was formed, so that it is possible to infer what the climate was at a particular time, several hundred or even thousands of years ago, by examining the rings of old trees.

The cambium plays an important role in the healing of wounds. When the stem's outer layer is removed through injury, the cambium grows over the exposed area and differentiates into new xylem, phloem and cambium, each of which is continuous with the same type of tissue in the uninjured part of the plant. Certain cells in the outer cortex of most woody plants become meristematic and form a second, or **cork, cam-**

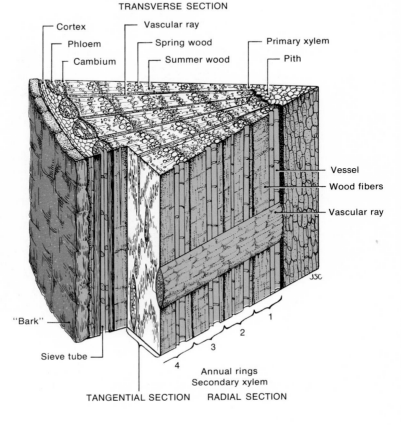

FIGURE 18–10   Diagram of a four-year-old woody stem showing transverse, radial and tangential sections and the annual rings of secondary xylem.

TRANSVERSE SECTION

Cortex — Vascular ray
Phloem — Spring wood — Primary xylem
Cambium — Summer wood — Pith

Vessel
Wood fibers
Vascular ray

"Bark"

Sieve tube

1
2
3
4

Annual rings
Secondary xylem

TANGENTIAL SECTION   RADIAL SECTION

bium. These outer cork cells become impregnated with a waterproof, waxy material and eventually die and fall off, partly under the stress of wind and rain, partly because of the outward pressure of the growing tissues within.

The stem of a monocot, such as corn, has an outer epidermis made of thick-walled cells and pierced by stomata similar to the ones in leaves. The epidermis and the cells of the cortex just beneath the epidermis become thick-walled and lignified and serve as supporting tissues. The vascular bundles are scattered throughout the stem (Fig. 18–11) instead of being arranged in a ring as in the dicots. The bundles are smaller and more numerous in the outer part of the stem. Each bundle contains xylem and phloem, but has no cambium; it is usually enclosed in a sheath of sclerenchyma cells which provide support. In some monocots, such as wheat and bamboo, the parenchyma cells of the center of the stem disintegrate, leaving a central pith cavity.

The epidermis of a young woody twig has stomata through which gases may enter and leave. Later, certain cells of the cork cambium divide repeatedly and form masses of thin-walled, loosely arranged cells that rupture the epidermis and form swellings, called **lenticels.** The intercellular spaces in the lenticels are continuous with those of the tissues within and permit the diffusion of gases in and out of the stem. Lenticels are visible as slightly elevated dots or streaks on the bark (Fig. 18–12).

The point on a stem where a leaf or bud develops is called a **node,** and the section of the stem between two nodes is called an **internode** (Fig. 18–12). In the upper angle of the point of junction of a leaf with the stem, a

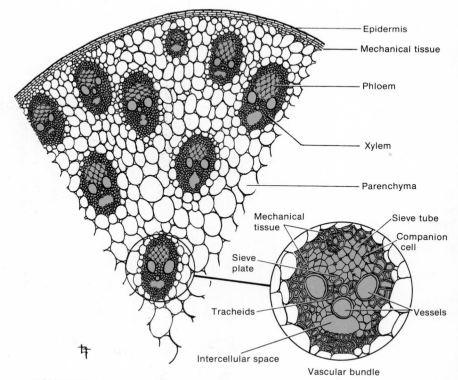

FIGURE 18–11   A sector of a cross section of a stem from a monocot, corn. Inset: an enlarged view of a vascular bundle containing phloem (sieve tubes and companion cells) and xylem (tracheids and vessels).

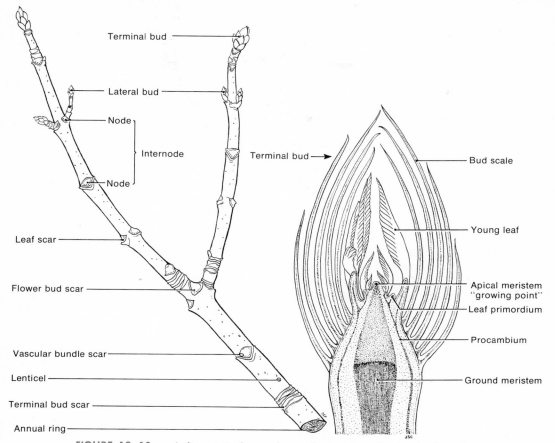

FIGURE 18–12   Left, a twig from a horse chestnut tree showing buds and scars.
Right, an enlarged longitudinal section through a terminal bud from a hickory tree.

bud usually appears, called a lateral or axillary bud to distinguish it from
the terminal bud at the tip of the stem. Terminal buds continue the growth
of the main stem; lateral buds give rise to branches. A bud consists of a
number of embryonic leaves, a growing point, and (in woody plants) a
ring of outer protective scales (Fig. 18–12). In some species these scales,
which are modified leaves, are coated with a waxy secretion or have a dense
covering of hairs to increase their protective value. The leaves within the
buds may be fairly well developed so that their ultimate shape can be distinguished or they may be shapeless rudiments.

When a terminal bud begins to grow in the spring, its covering scales
are forced apart and fall off, leaving a ring of scars (Fig. 18–12). These
scars, which mark the position of the end of the stem at the completion of
the season, may remain visible for several years, so that it is possible to
determine the age of a twig by counting the number of terminal bud scars.

The overall shape of a tree or shrub is determined by the position,
arrangement and relative activity of the terminal and axillary buds. In a
tree with a strong terminal bud, such as a cedar or poplar, the twig produced by the terminal bud is much more vigorous than those produced by
lateral buds, and a single, strong, straight main trunk results. Plants with
vigorous lateral buds have strong horizontal branches and a spreading
shape. Other factors influencing the shape of a tree are the direction and
strength of the prevailing wind and the presence of other trees nearby.

In addition to terminal and axillary buds, lenticels and bud scars, the surface of a twig may show **leaf scars** (Fig. 18–12) left when the stalk of a leaf breaks away from the twig, and **fruit scars,** produced by the breaking off of fruit.

The leaf is a specialized nutritive organ, adapted to carry on photosynthesis; the details of its structure were described earlier (p. 121). Bundles of xylem and phloem pass from the stem through the petiole to the blade where they branch and rebranch, forming a fine network of "veins." Each vein contains some xylem vessels and tracheids on its upper side and some phloem sieve tubes on its lower.

TRANSPIRATION

The leaves of a plant exposed to the air will lose moisture by evaporation unless the air is saturated with water vapor. The sun's heat vaporizes the water from the surfaces of the mesophyll cells, and the resulting water vapor passes through the stomata and escapes. This loss of water, called **transpiration,** may occur in all parts of the plant exposed to the air, but most of it occurs in the leaves. The rate of transpiration is very low during the night when the stomata are usually closed and the lower temperature decreases the rate of evaporation of water from the surface of the mesophyll cells. The stomata also tend to close in the latter part of the afternoon of a hot, sunny day. This greatly decreases the rate of transpiration and conserves the plant's supply of water. If the plant has an adequate supply of water, the stomata remain open and an amazing amount of water is transpired. Only a small fraction—1 or 2 per cent—of all the water absorbed by the roots is used in photosynthesis. All the rest passes through the stomata as water vapor in the process of transpiration. If the plant is not getting sufficient water from its roots, the guard cells around the stomata will become less turgid and the stomata will close, thereby conserving water.

The many small holes of the stomata provide a remarkably effective pathway for the diffusion of water vapor, oxygen and carbon dioxide. Although the total area of the pores is only 1 to 3 per cent of the total area of the leaf's surface, the rate of diffusion through the stomata is from 50 to 75 per cent of the rate through an open surface equal to the area of the leaf. In sunlight, an average plant will transpire about 50 ml of water per square meter of leaf surface per hour. An average corn plant uses more than 200 liters of water in the course of a growing season and a medium-sized tree will transpire that much in a single day. The amount transpired varies widely in different plants; for example, it is estimated that an acre of corn will transpire 1,400,000 liters of water in a growing season and an acre of mature maple trees may transpire twice that. In contrast, an acre of cactus in the Arizona desert will transpire no more than 1100 liters in a whole year. The amount of water vaporized from the leaves of trees in a forest is enough to influence significantly the rainfall, humidity and temperature of the region.

Transpiration contributes to the economy of the plant by assisting the upward movement of water through the stem, by concentrating in the leaves the dilute solutions of minerals absorbed by the roots and needed for synthesis of new cellular constituents, and by cooling the leaves in a manner analogous to the evaporation of sweat in animals. Although the leaf absorbs some 75 per cent of the sunlight reaching it, only about 3 per cent is utilized in photosynthesis. The rest is transformed into heat which must be removed or it would cause the death of the cells in the leaf. Some of its heat is removed by the vaporization of water, for 540 Calories are required to convert a liter of water to water vapor. The rest of the heat is lost by reradiation and by convection.

When water is lost by evaporation from the surface of a mesophyll cell, the concentration of solutes in the cell water increases and the cell becomes slightly hypertonic. Water thus tends to pass into it from neighboring cells that contain more water. These cells in turn receive water from the tracheids and vessels of the leaf veins. During transpiration then, water passes by the purely physical process of osmosis from the xylem vessels of the veins, through the intervening cells, to the mesophyll cells next to the air spaces of the leaf, where it is vaporized. In fact, a continuous stream of water passes from the soil into the vascular system of the roots, up through the stem and petiole to the veins of the leaf blade.

SECTION 18–7

THE MOVEMENT OF WATER

It had been shown experimentally many years ago that water and salts absorbed by the roots move upward in the stem primarily in the tracheids and vessels of the xylem and that sugars and other organic materials are carried primarily in the sieve tubes of the phloem. If a cut is made entirely around a stem, deep enough to penetrate the phloem and cambium but not the xylem, the leaves remain turgid and in good condition for a long time. They must be getting water via the xylem since the phloem has been completely cut. By special techniques it is possible to cut the inner xylem and leave the outer phloem relatively intact. When this is done the leaves wilt and die almost immediately, again showing that water reaches them primarily via the xylem. Although the route of water transport has been known for some time, the mechanism by which this occurs is still not completely clear. Any acceptable explanation must account for the high rates of water flow, as much as 75 to 100 ml per minute, observed in some plants, and for the fact that water can be moved to the tops of redwoods and Douglas firs that may be as much as 125 meters high. Although a pressure of about 12 atmospheres would support a column of water 125 meters high, additional pressure would be required to move the water upward against the frictional resistance of the small tubes. Estimates of the pressure required to move water to the top of a tall tree range up to 30 atmospheres. The pressure might be generated at the base of the plant and *push* water upwards, or it might be generated at the top of the plant and *pull* water up. Water might be moved by these two forces acting jointly.

Root Pressure

If the stem of a well-watered tomato plant is cut, sap will flow from the stump for some time. If a piece of glass tubing is attached to the stump with a watertight junction, water will rise to a height of 1 meter or more. This is evidence for a positive pressure, termed **root pressure,** at the junction of root and stem, generated by forces operating in the root. A plant growing in well-watered soil under conditions of high humidity so that little

FIGURE 18–13   Guttation. A photograph of a leaf of a strawberry plant showing the drops of water expressed from the veins at the tip of the leaf under pressure, a process termed *guttation.* (Courtesy of J. Arthur Herrick, Kent State University, Kent, Ohio.)

water is lost by evaporation from the leaves may force water under pressure out at the ends of the leaf veins, forming droplets along the edges of the leaves (Fig. 18–13). This phenomenon, termed **guttation,** is further evidence that under certain conditions the sap in the xylem may be under pressure generated by the roots.

The sap in the roots is hypertonic to the water in the surrounding soil; this may account at least in part for the generation of root pressure. The movement of water from the soil through the epidermis, cortex, endodermis and pericycle of the root to the xylem, and up the xylem to the stem and leaf, is down a concentration gradient and occurs at least in part by simple diffusion. If roots are killed or deprived of oxygen, root pressure falls to zero, indicating that some active, energy-requiring process is involved. Many plant physiologists have concluded that the cells of the endodermis of the root secrete water by an active process inward toward the stele and that this is responsible for much of the root pressure.

Although early measurements of root pressures indicated that they were small, newer methods of cutting off the stem and fitting a leakproof connection without injuring the stump have demonstrated root pressures of 6 to 10 or more atmospheres even in tomato plants that are raising water less than 1 meter.

In the spring, before leaves have been formed, root pressure is probably the sole force bringing about the rise of sap. Attempts to measure root pressures in conifers have been unsuccessful and perhaps these plants cannot generate this force. If root pressure were the major force causing the rise of sap, you would expect sap to flow out under pressure when a xylem vessel is punctured. Instead you may hear the hissing sound of air being taken in when the vessel is punctured. Thus root pressure may contribute to the rise of sap in some plants under certain conditions, but it probably is not the principal force causing water to rise in the xylem of most plants most of the time.

Transpiration and the Cohesion Theory

The alternative force that might raise water in a stem is a pull from above rather than a push from below. Water is lost by transpiration or used in photosynthesis in the leaves, and osmotically active substances are produced by photosynthesis; these processes keep the leaf cells hypertopic to the sap in the veins. They constantly draw water from the upper ends of the xylem vessels in the leaves and stem and this tends to lift the column of sap upward in each duct. This pull from above can be demonstrated by attaching a cut branch to a water-filled glass tube by a watertight connection and putting the other end of the tube in a beaker of water (Fig. 18–14). If a small air bubble is introduced into the tube, the rate of movement of the water can be measured by the rate at which the bubble moves.

The water columns in the xylem vessels, being under tension from above, are slightly stretched, but water molecules are joined by hydrogen bonds and have a very strong tendency to cling together. The slender column of water in the xylem vessel has a high tensile strength. Transpiration is the major process providing the pull at the top of the column. The tendency of water molecules to stick together transmits this force through the length of the stem and roots and results in the elevation of the whole column of sap. The idea that water rises in plants owing to a pull at the top resulting from transpiration was first stated by Stephen Hales early in the eighteenth century. The concept that the cohesive properties of water molecules play a role in the ascent of water in the xylem was formulated by Dixon and Joly in 1894, who predicted that a column of water would have great tensile strength. Some experimental measurements of it have given values considerably greater than the 30 atmospheres necessary to account for the rise of water to the top of a redwood. If water is evaporated from a porous clay tube at the top of a column of water immersed in a beaker of mercury (Fig. 18–15), the mercury can be drawn to a height of 100 centimeters or more, much higher than barometric pressure. For a successful

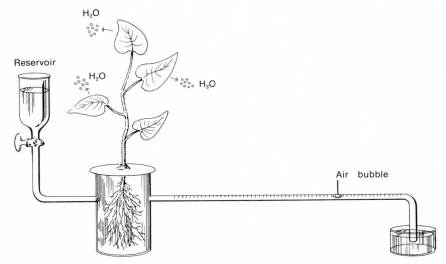

FIGURE 18–14   *A device for measuring the rate of transpiration in a stem. As water is transpired and evaporated from the leaves it is pulled from the reservoir on the right. The rate of transpiration is measured by the movement of the air bubble through the graduated tube. The reservoir on the right can then be refilled from the reservoir on the left and a new measurement begun.*

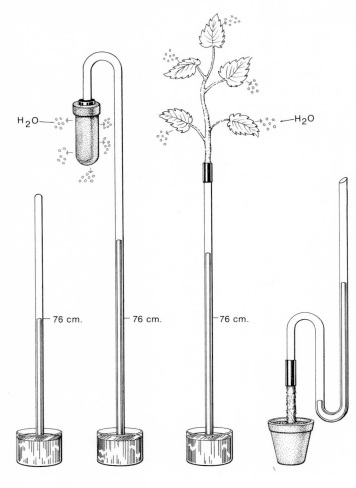

H_2O

H_2O

76 cm.   76 cm.   76 cm.

FIGURE 18–15   A model of the forces operating in transpiration. A porous clay cup (left) will draw water up in a glass tube as water is evaporated from its surface. In a similar fashion, water is moved upward in a stem as water molecules evaporate from the leaf (center). The measurement of root pressure is illustrated in the diagram on the right.

demonstration, the water must be carefully boiled first to remove dissolved gases that might form bubbles when the water column is subjected to tension.

This theory would predict that the diameter of a tree would decrease as transpiration occurs. If the water in the xylem ducts is under tension, there should be an inward pull on the wall of each duct. The inward pulls on the walls of all the ducts in a band of sapwood should be great enough to produce a detectable *decrease* in the diameter of a tree during active transpiration. This predicted result was found by D. T. MacDougal, who showed that the diameter of the Monterey pine fluctuates daily and is minimal shortly after noon, when transpiration is maximal.

Experiments analyzing the uptake of water by jungle vines also support the Dixon-Joly hypothesis. Some of these vines extend 50 meters or more up into the trees of the jungle. When the base of the vine is severed and placed in a pail of water, water is taken up by the plant at a high rate. If the stem is placed in a sealed container of water it again takes up water rapidly even though it produces a high vacuum in the container. This transpiration-cohesion theory is widely accepted at present and accounts for most of the rise of water by most plants under most conditions. Among the problems yet unsolved is how the column of water becomes established initially.

SECTION 18–8

THE TRANSLOCATION AND STORAGE OF NUTRIENTS

On a bright sunny day a green plant may produce more than 20 times as much food as it is using at that moment. During the night and over the winter season its consumption of nutrients continues but little or none is made. Each plant accumulates reserves of nutrients in leaves, stems or roots which tide it over periods when photosynthesis cannot occur. Leaves serve as temporary depots of food but are not suitable for long term storage for they are readily and rapidly lost. The stems of woody perennials serve as storage places for large amounts of food; potatoes and other plants use fleshy underground stems for this purpose. Perhaps the most common storage organs are roots; their underground location protects them from climatic changes and from animals looking for food.

The products of photosynthesis in the leaf must be transported to stem and root to be utilized or stored. Much of the glucose made in a day is converted into starch and stored in the leaves. The starch is subsequently reconverted to glucose and translocated in the phloem to the stem and root. In the root it may pass out of the phloem through the pericycle and endodermis to the cortical parenchyma where it is reconverted to starch and stored. The flow of liquid in the phloem, though not as great nor as rapid as the flow in xylem, is nonetheless sizeable and may attain 1000 cm per hour.

Translocation occurs in the sieve tubes of the phloem and is dependent on their metabolic activity, for the rate is greatly reduced when metabolism is decreased by low temperature, by lack of oxygen or by metabolic poisons. Dissolved nutrients, sugars and amino acids are carried both down and up in the phloem; indeed, two substances may be carried in opposite directions simultaneously. Very little, if any, translocation of solutes occurs in any system other than the phloem. Although the total amount of phloem sieve tubes that are functional in the trunk of a large tree is small, a variety of experimental evidence shows that the phloem is the only system involved in sugar transport and that it does this at a remarkable rate. Not all organic materials are synthesized in the leaves. For example, nitrates taken in through the roots are converted there to amino groups and incorporated into amino acids and other nitrogenous compounds.

Transport in the phloem is through living, active cells, and contrasts sharply with transport in the xylem which occurs through hollow tubes, the dead remains of cell walls without cytoplasm. The phloem sieve tubes contain cytoplasm and are connected end-to-end by cytoplasmic threads that penetrate the small pores *(sieve plates)* in the cell walls.

According to one hypothesis, materials pass into one end of a sieve tube through the sieve plate and are picked up by the cytoplasm which streams up one side of the tube and down the other (Fig. 18–16). At the other end of the sieve tube, the material passes across the sieve plate to the next adjacent tube by physical diffusion, by facilitated diffusion or by active transport. The *cyclosis* or cytoplasmic streaming within successive cells, and diffusion or active transport between cells, could move sugars and other materials over long distances. The system could account for the simultaneous transport of two substances in opposite directions.

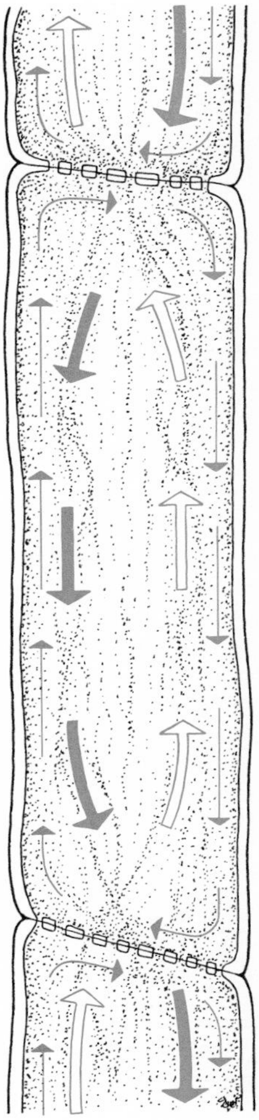

FIGURE 18–16   *Cytoplasmic streaming in a phloem tube. The light and dark arrows illustrate that cytoplasmic streaming could account for the simultaneous transport of two different substances in opposite directions.*

556

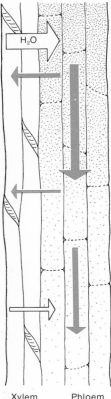

Sugar concentration
high

H₂O

Xylem   Phloem

Sugar concentration

Another widely held theory, termed the "pressure-flow" theory, suggests that water containing the solutes flows through the phloem under pressure along a gradient of turgor or osmotic pressure. Cells in the phloem in the leaf contain a high concentration of sugars and other products of photosynthesis and water thus tends to pass into them from the xylem ducts (Fig. 18-17). This increases the pressure within the phloem cells and tends to push fluid from one cell to the adjacent one down the phloem tube. As the fluid passes down the stem and root, sugars are removed and stored as starch, which is insoluble and exerts no osmotic effect. This withdrawal of solutes lowers the osmotic pressure in the phloem, water passes out of the phloem and the turgor pressure falls. The difference in turgor pressure is believed to bring about the mass flow of the contents of the sieve tubes from a region of high turgor pressure—such as the leaves when photosynthesis is occurring—to regions of lower turgor pressure—such as the stem or root where the materials are stored or used. This theory predicts that the contents of the sieve tubes should be under pressure, and this can be shown experimentally to be true. When a sap-sucking insect such as an aphid punctures a phloem tube with its stylet, the sap flows into the insect without assistance. If the body of the insect is removed but the stylet is left in place, the sugar-laden sap will continue to exude through the stylet for days. Although this theory is highly regarded by some botanists it does not readily account for the simultaneous transport of materials in both directions in the phloem. In addition, since it is based on physical phenomena it should be relatively unaffected by changes in the rate of metabolism in the phloem sieve tubes, but translocation is markedly affected by changes in metabolic rate.

The evolution of xylem and phloem systems has enabled tracheophytes to adapt to a land environment and to grow to great heights. A land plant exposing broad, chlorophyll-containing surfaces to the sun is subject to marked loss of water to the dry air. It must procure a supply of water not only as a raw material in photosynthesis but to replace the water lost by transpiration. The structure of the xylem adapts it to carry water absorbed from the soil to the leaf. The roots and stems in turn require a supply of nutrients to remain alive, to keep the plant anchored in the soil and to hold the leaves so they can be exposed to sunlight. The sieve tubes of the phloem transport these nutrients from the leaves to the stem and root and back again as required.

SECTION 18-9

TRANSPORT SYSTEMS IN ANIMALS

The metabolic processes of all cells require a constant supply of nutrients and oxygen and a constant removal of metabolic wastes, and all organisms have solved in one way or another the problem of transporting substances from one part of the body to another. The transport of hormones from the endocrine glands to their target organs and the equalizing of body temperature are functions added in some of the higher animals. In protozoans the transport of substances is effected by the diffusion of molecules, aided usually by the streaming movements of the cytoplasm. The flowing of cytoplasm from rear to front as the amoeba moves, and the circular streaming of the cytoplasm in paramecia and other protozoa with a fixed shape are effective transport systems in these tiny animals (Fig. 18–18). Transport from cell to cell in simple multicellular animals such as sponges, coelenterates and flatworms also occurs by diffusion, aided in some animals by the stirring of the fluids in the body cavity brought about by the contraction of muscles in the body wall or by the beating of cilia. The rate of diffusion is directly proportional to the difference in concentration in the two regions and inversely proportional to the square of the distance separating them. From this it follows that an adequate supply of oxygen and nutrients can be maintained by diffusion alone only if an animal is small and the distance over which diffusion occurs is short. In large animals the slower diffusion rate over the greater distance would not suffice. Such animals must develop some system of internal transport, a **circulatory system,** to supply their cells. The shape and activity of an animal, as well as its absolute size, may determine the need for a circulatory system.

The proboscis worms, nemerteans, are the simplest living animals to have a distinct and separate circulatory system; it consists of a dorsal and two lateral blood vessels which extend the entire length of the body and are connected by transverse vessels. The earthworm and other annelids have a more complex system: a dorsal vessel in which blood flows anteriorly, a ventral and a subneural vessel in which blood flows posteriorly, and five pairs of pulsating tubes ("hearts") at the anterior end which drive blood from the dorsal to the ventral vessel. In other segments of the body a network of fine, capillary vessels connects dorsal and ventral vessels and ramifies through the body wall and the wall of the intestine. The major advance of the annelid system over the nemertean one is the addition of capillary networks in the body wall, intestine and the glandular region of the metanephridia.

The circulatory systems of the larger and more complex animals typically include blood vessels, the blood in these vessels composed of a liquid **plasma** in which float one or more kinds of **blood cells,** and a **heart** to pump the blood. In most circulatory systems oxygen is not simply dissolved in the plasma but is combined with a heme protein. In man and the earthworm and many intermediate forms, the heme protein is **hemoglobin,** a red, iron porphyrin protein. The hemoglobin of vertebrate blood is located within the red blood cells, but the hemoglobin or other pigment in the bloods of many invertebrates is dissolved in the plasma, and any cells in the blood are colorless. The respiratory pigment of crab blood is a dif-

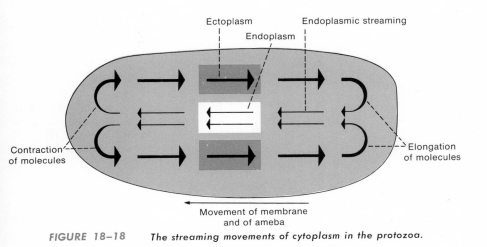

FIGURE 18–18   The streaming movements of cytoplasm in the protozoa.

ferent heme protein, blue-green **hemocyanin,** which contains copper rather than iron.

The circulatory systems of annelids and vertebrates are said to be "closed"; i.e., blood in the course of circulation remains within blood vessels (Fig. 18–19). In contrast, the circulatory systems of arthropods and mollusks are "open"; the blood vessels open to the body cavity, the **hemocoel,** and blood circulates partly within blood vessels and partly through the cavity of the hemocoel in making a complete circuit. In an arthropod such as a lobster, the heart and other organs lie free in the hemocoel and are bathed in blood. In the annelids and vertebrates the internal organs lie in the coelom and are supplied by blood which reaches them in closed vessels. The arthropod heart is typically a single, elongate, muscular tube lying in the dorsal midline. Blood enters the lobster heart from the **pericardial sinus,** a part of the hemocoel, through openings called **ostia,** and is pumped anteriorly, ventrally and posteriorly through arteries that branch to all parts of the body, emptying into the hemocoel. It drains ventrally into the **perineural sinus** from which veins carry it to the gills, where it is oxygenated, then other veins return it to the pericardial sinus. The details of the system may be quite different from one kind of animal to another but in all animals the circulatory system supplies oxygen and nutrients and removes metabolic wastes.

The hearts of most invertebrates are single muscular tubes which develop only very low pressures, a few millimeters of mercury, as they pump blood. In the closed circulatory systems of vertebrates, a higher pressure, as much as 100 to 200 mm Hg, is required to drive the blood against the resistance of the tremendous number of narrow capillaries. This has led to the evolution of powerful, thick-walled, muscular hearts. The chamber of the vertebrate heart called the **ventricle** has especially thick, muscular walls. A certain amount of pressure is required to distend this muscular wall and cause blood to flow in and fill the chamber during the relaxation phase **(diastole).** The low pressure in vertebrate veins is not sufficient to do this. The vertebrate heart has a second chamber, the **atrium,** with walls thin enough so that it can be filled by the low venous pressure, yet strong enough to pump blood into the ventricle and distend it. The octopus, whose heart is similarly arranged with two different chambers, has the highest blood pressure, 35 to 45 mm. Hg. of all the invertebrates.

The vertebrate heart is enclosed in a special cavity of the coelom, the

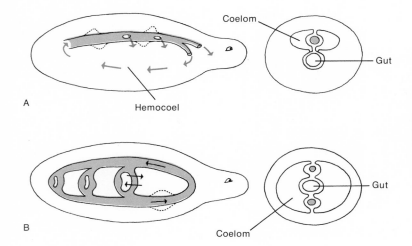

FIGURE 18–19 Diagrams illustrating an open (A) and a closed (B) circulatory system.

pericardial cavity, separated from the rest of the body by a thin, strong sheet of connective tissue, the **pericardium.** This cavity provides space for the heart to change in volume as it beats.

The circulatory systems of all vertebrates from fish and frogs through lizards to birds and man are essentially similar: a closed system composed of heart, aorta, arteries, capillaries and veins arranged in a basically similar plan (Fig. 18–20). Because of this similarity, students can learn much about the human circulatory system from the dissection of a dogfish or frog. **Arteries** carry blood away from the heart to the tissues, **veins** carry blood back to the heart from the tissues and **capillaries** are minute, thin-walled vessels connecting arteries and veins and completing the circuit from heart to heart. Only capillaries have walls thin enough to permit the exchange of nutrients, gases and wastes between blood and tissues.

The principal evolutionary changes in the vertebrate circulatory system have been associated with the change from gills to lungs as respiratory organs. The changes in the pattern of circulation permit the delivery of oxygen-rich blood to the brain and muscles. The fish heart consists of four chambers in a row: **sinus venosus, atrium, ventricle** and **conus** (Fig. 18–20A). Blood from the veins drains into the sinus venosus, and blood from the conus is pumped through the ventral aorta to the gills where it takes up oxygen. It then passes to the dorsal aorta and is distributed to all parts of the body. Blood passes through the fish heart only once each time it makes a circuit around the body.

In the group of fish from which the land vertebrates evolved, a number of changes occurred in the heart and blood vessels which are reflected in their amphibian descendents (Fig. 18–20B). A partition developed down the middle of the atrium, dividing it into right and left halves. The sinus venosus shifted its connection so that it emptied into the right atrium, and veins from the lungs emptied into the left atrium. Pulmonary arteries to the lungs grew out of the vessels which originally served the most posterior pair of gills.

The pattern of circulation in amphibia permits some mixing of oxygen-rich and oxygen-poor blood in the heart. Blood should pass from the veins to the sinus venosus, right atrium, ventricle, pulmonary arteries, lungs, pulmonary veins, left atrium, ventricle, aorta and then to the somatic circulation. There is some mixing of blood in the ventricle and blood from the sinus venosus may go to the aorta instead of to the pulmonary arteries. Blood from the right atrium enters the ventricle ahead of that from the left atrium and hence lies nearer the exit. As the ventricle contracts, non-aerated blood from the right atrium leaves first and enters the arteries branching off from the aorta, the pulmonary arteries to the lungs. The aerated blood from the left atrium leaves the ventricle toward the end of the contraction, is unable to enter the pulmonary arteries because they are already full of blood and so passes via the aorta to the body cells. A given blood cell might pass through the heart once, twice or even more times for each circuit around the body.

As reptiles evolved from their amphibian ancestors, one partition developed down the middle of the ventricle and one down the middle of the conus (Fig. 18–20C. The ventricular septum is incomplete in all the reptiles (except the alligators and crocodiles) and there is some mixing of aerated and nonaerated blood, though less than in the frog heart. The sinus venosus is small, foreshadowing its disappearance in the mammalian heart.

The hearts of birds and mammals (Fig. 18–20D) have completely separate right and left sides and aerated blood from the lungs is kept separate from the blood entering from the rest of the body. The sinus

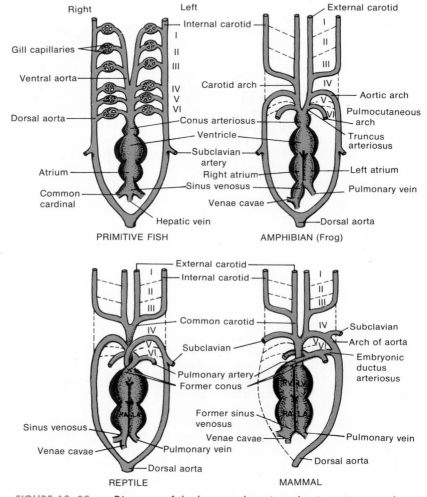

FIGURE 18–20   Diagrams of the heart and aortic arches in various vertebrates to
show the changes that have occurred during evolution from primitive fishes to mammals.
All of these are ventral views and the tube of the heart has been straightened so that the
atrium lies posterior to the ventricle. The roman numerals indicate the aortic arches of the
primitive fish and their remnants in higher vertebrates.

venosus has disappeared as a separate chamber but a vestige remains as the **sinoatrial node,** located at the junction of the vena cava and the right atrium. The sinoatrial node initiates the contraction of the heart muscle and regulates the rate of the heart beat. The separation of right and left hearts requires the blood to pass through the heart *twice* for each circuit of the body. As a result the blood in the aorta of birds and mammals contains more oxygen than that in the aorta of lower vertebrates. With this greater supply of oxygen, the body tissues of birds and mammals can maintain a higher metabolic rate and the "warm-blooded" condition is possible, i.e., birds and mammals can maintain a constant, high body temperature in cold surroundings.

SECTION 18–10

THE HEART AS A PUMP

For its function as a pump, the heart is equipped with valves that close automatically and prevent blood from flowing in the wrong direction. The valve between the right atrium and ventricle, with three flaps or cusps, is called the **tricuspid valve** and that between the left atrium and ventricle, with two cusps, is called the **bicuspid valve** (Fig. 18–21). These valves are held in place by stout cords or "heart-strings" attached to the valves and to the walls of the ventricles. These prevent the valves from being pushed back into the atria when the ventricles contract. At the bases of the pul-

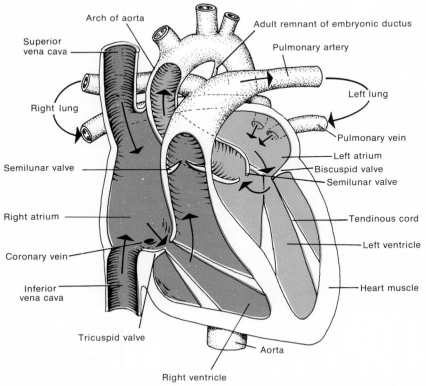

FIGURE 18–21   *A diagram of an adult mammalian heart showing its chambers, valves and connecting vessels. The course of the blood through the heart is indicated by the arrows.*

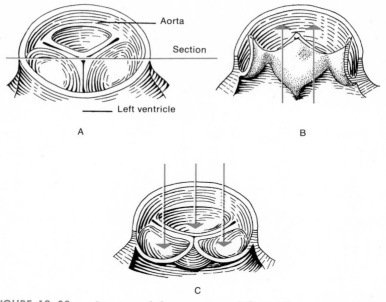

FIGURE 18–22   Diagram of the operation of the semilunar valves. A, Arrangement of the three pouches of the semilunar valves. The aorta has been cut across just above
its point of attachment to the ventricle to expose the valve. B, When the ventricle contracts,
the expelled blood (indicated by the arrows) pushes the pouches aside and passes into the
aorta. C, When the ventricle relaxes, blood from the aorta fills the pouches causing them
to extend across the cavity and prevent the leakage of blood back into the heart.

monary artery and aorta, which leave right and left ventricles respectively,
are the two half-moon shaped *semilunar valves,* pouches that open away
from the heart. When blood passes out of the ventricles the pouches are
pushed aside and offer no resistance. But when the ventricles are relaxing
and filling with blood from the atria, and the blood pressure in the arteries
is greater than that in the ventricles, blood fills the pouches, stretching
them across the lumen of the pulmonary artery and aorta and preventing
the flow of blood back into the ventricles (Fig. 18–22). There are no valves
at the opening of the veins into the atria and some blood is forced back
into the veins when the atria contract.

During a heart cycle, the atria and ventricles contract and relax in
succession. Contraction of the chamber is termed *systole* and relaxation is
termed *diastole.* The powerful ventricular systole drives blood past the
semilunar valves into the pulmonary artery and aorta under high pressure.
The muscle fibers in the ventricle are spirally arranged and their contraction "wrings" the blood out of the ventricular cavity. The elastic recoil
of ventricular relaxation reduces the pressure within the cavity and blood
enters from the atria.

Beating is an inherent property of the heart, beginning early in
embryonic development and continuing without pause throughout life.
A bit of cardiac muscle will continue to beat when excised and grown in
culture medium. The heart of a man at rest pumps about 75 ml per beat
and about 5 liters per minute. Thus the quantity of blood passing through
the heart each minute equals the total amount present in the body. Some
of the blood, on relatively short circuits through the body, will return to
the heart in less than a minute, whereas that on longer circuits will take
more than a minute to complete the trip. In a man's three score and ten
years, his heart beats some 2.6 billion times and pumps some 155,000,000

liters of blood. During exercise both the number of beats per minute and the amount of blood pumped per beat are increased.

The beat is initiated and regulated by **nodal tissue,** made of specialized cardiac muscle fibers called **Purkinje tissue** (Fig. 18–23). At appropriate intervals a heart impulse or action potential is initiated in the sinoatrial node and passes through the atrial muscles, causing them to contract. It reaches the **atrioventricular node** and is transmitted to the ventricles and throughout the ventricular tissue by bundles of Purkinje fibers. Impulses are conducted nearly 10 times faster in Purkinje tissue than in ordinary cardiac muscle fibers. The rapid conduction of the impulse through the nodal tissue ensures that all parts of the ventricle will contract nearly simultaneously.

The cyclic sequence of events that occurs during a single heart beat, termed the **heart cycle** (Fig. 18–24), begins with the initiation of an impulse, an action potential, in the sinoatrial node. This spreads over the atria, causing the atrial muscle to contract and force blood from the atria into the ventricles. The ventricles are partly filled with blood because the pressure is less there than in the atria and the tricuspid and bicuspid valves are open. The conduction of the impulse through the atrioventricular node is slower than in other parts of the nodal tissue; its fibers are very small and have a very slow rate of conduction. This accounts for the brief pause, about 0.1 second, after atrial systole before ventricular systole begins.

Stimulated by the action potentials streaming down the bundles of Purkinje fibers, the ventricular muscles begin to contract, causing a rapid increase in the intraventricular pressure. The bicuspid and tricuspid valves close quickly, producing the **first heart sound.** The pressure in the ventricle mounts rapidly, but until it equals the pressure within the arteries, the

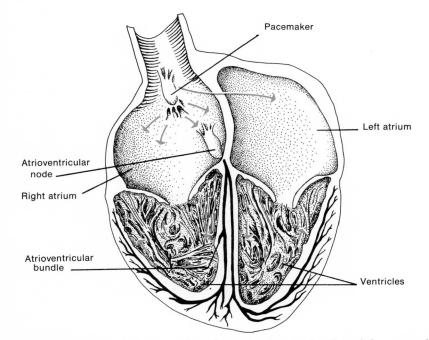

FIGURE 18–23   A diagram of the heart showing the location of the pacemaker (the sinoatrial node), the atrioventricular node and the atrioventricular bundle which regulates and coordinates the beating of the parts of the heart. The arrows indicate the direction in which action potentials are conducted.

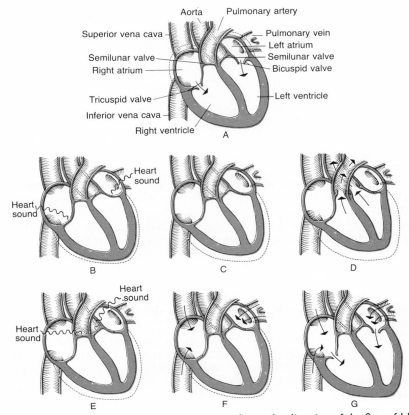

FIGURE 18-24   *The heart cycle. Arrows indicate the direction of the flow of blood; dotted lines indicate the change in size as contraction occurs. A, Atrial systole; atria contract, blood is pushed through the open tricuspid and bicuspid valves into the ventricles. The semilunar valves are closed. B, Beginning of ventricular systole; ventricles begin to contract, pressure within the ventricles increases and closes the tricuspid and bicuspid valves, causing the first heart sound. C, Period of rising pressure. Until the pressure within the ventricles equals that in the arteries, the semilunar valves remain closed and no blood flows in or out of the ventricles. D, The semilunar valve opens when the pressure within the ventricles exceeds that in the arteries and blood spurts into the aorta and pulmonary artery. E, The beginning of ventricular diastole. When the pressure in the relaxing ventricles drops below that in the arteries, the semilunar valve snaps shut, causing the second heart sound. F, Period of falling pressure; ventricles continue to relax and the pressure within them decreases. The tricuspid and bicuspid valves remain closed because the pressure within the ventricles is still higher than the pressure within the atria. Blood flows from the veins into the relaxed atria. G, The tricuspid and bicuspid valves open when the pressure in the ventricles falls below that in the atria, and blood flows into the ventricles.*

semilunar valves remain closed and no blood flows into or out of the ventricles. When the intraventricular pressure exceeds that in the arteries, the semilunar valves open and blood spurts into the pulmonary artery and aorta. As the ventricles complete their contraction, blood is ejected more slowly until it finally stops. As ventricular diastole begins, the pressure within the ventricles decreases until it is less than the pressure within the arteries and the semilunar valves snap shut, causing the **second heart sound.**

After the semilunar valves close, the muscles in the ventricular wall continue to relax and the pressure within the ventricle continues to fall.

When this becomes less than the pressure in the atria, the tricuspid and bicuspid valves open and blood flows from the atria into the ventricles. This is not caused by the contraction of any part of the heart but simply by the fact that the pressure within the relaxed ventricles is less than that in the atria and veins. The ventricles may be half filled with blood before the atria undergo systole, beginning a new cycle.

Each heart beat is accompanied normally by two **heart sounds,** described by the syllables "lubb dup." The first sound, **lubb,** low-pitched, not very loud and of long duration, accompanies the closing of the tri- and bicuspid valves. This is followed quickly by the second **(dup)** sound, which is higher pitched, louder, sharper and of shorter duration. The second sound results from the closure of the semilunar valves and marks the end of ventricular systole. Each heart beat is also accompanied by electrical phenomena, action currents, which can be detected even at the surface of the body by appropriate electrodes in the skin and recorded with an **electrocardiograph.** Since malfunctioning of the heart causes aberrant action currents which can be detected by an electrocardiogram, this is an important tool in assessing the kind and degree of abnormality that may be present.

SECTION 18–11

THE BLOOD VASCULAR SYSTEM

The blood vascular system of man and other vertebrates includes arteries, veins and capillaries. Arteries and veins are large vessels, distinguished by the direction of the flow of blood within them and by the structure of their walls (Fig. 18–25). **Arteries** carry blood from the heart

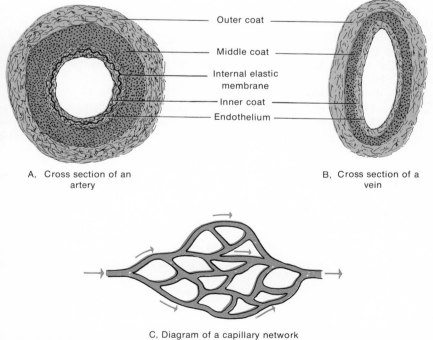

Outer coat

Middle coat

Internal elastic membrane

Inner coat

Endothelium

A, Cross section of an artery

B, Cross section of a vein

C, Diagram of a capillary network

FIGURE 18–25   · A comparison of a cross section of an artery (A) and a vein (B). The diagram of the capillary network (C) is drawn at a much greater magnification.

to the tissues and *veins* return blood from the tissues to the heart. *Capillaries* are microscopic thin-walled vessels located in the tissues and connecting arteries and veins. The walls of capillaries consist of a single layer of flattened cells, the *endothelium,* which is continuous with the endothelial lining of the artery and vein on either side. Only the walls of the capillaries are thin enough to permit the exchange of nutrients, wastes and gases between blood and the cells of the body.

Arterial walls have an outer coat of connective tissue, a middle coat of smooth muscle and an inner coat of endothelium and connective tissue (Fig. 18–25). The outer coat contains strong fibrous connective tissue which makes the artery resistant to internal pressure while permitting it to expand and contract with each heart beat. The smooth muscle in the middle layer, by contracting or relaxing and thus regulating the size of the lumen of the artery, controls the amount of blood going to any particular organ (p. 589). The inner coat of most arteries contains, in addition to the endothelial lining, a strong *internal elastic membrane* which strengthens the wall. The walls of veins have the same three coats present in arteries but each is thinner. The outer coat has fewer elastic fibers, the middle muscular coat is thinner than the corresponding layer in an artery of the same diameter and most veins have no internal elastic membrane. Veins, but not arteries are supplied with valves along their length to prevent the retrograde flow of blood.

Blood does not come in direct contact with the cells of the body; instead the cells are surrounded by and bathed with *interstitial fluid,* part of the extracellular fluid of the body. Substances must pass through the wall of a capillary from the blood and cross the space filled with interstitial fluid to get to the cells (Fig. 18–26). Professor A. Baird Hastings characterized the interstitial fluid, lymph and blood plasma as "the sea within us." An adult man has some 10^{15} cells, bathed by only 14 liters of extracellular fluid. An equivalent number of protozoan cells living in the sea would require 10,000,000 liters of sea water to provide them with the gases and nutrients they require. The remarkably efficient lungs, liver, intestines and kidneys, which continually replenish the oxygen and nutrients of the extracellular fluids and remove their wastes, enable the cells of the human body to survive even though they have relatively little extracellular fluid.

Each drop of blood passing through a capillary network is exposed to a large surface area through which substances may diffuse or be transported to the interstitial fluid bathing the cells. One estimate states that each milliliter of blood is exposed to 7000 cm² of capillary surface. The

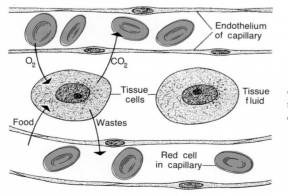

FIGURE 18–26   *A diagram of the diffusion of materials between capillaries and the cells of the body by way of the tissue fluid which bathes each of the cells.*

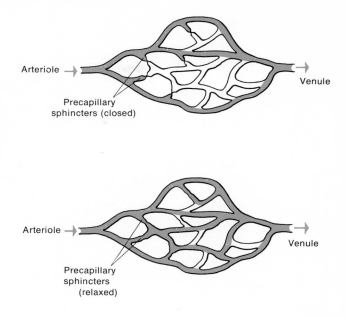

FIGURE 18–27 Changes in blood flow through a capillary bed as the tissue becomes active. A, The tissue at rest; only the thoroughfare channels are open in the capillary bed. B, In an active tissue the decreased oxygen tension in the tissue brings about a relaxation of the precapillary sphincters and more capillaries become open. This increases the blood supply and the delivery of oxygen to the active tissue.

number of capillaries in the human body is incredibly large. In tissues such as skeletal muscle, with a high metabolic rate, the capillaries are close together and the distance between adjacent capillaries is only twice the diameter of a capillary. Careful studies of serial sections of muscle suggested that there are some 240,000 capillaries per cm². Less active tissues have fewer capillaries and some, such as the lens of the eye, have none. Ordinarily only a fraction of the capillaries of any organ are filled with blood and functioning but during periods of intense activity all or nearly all of them are full (Fig. 18–27). Each capillary bed has certain "thoroughfare" channels through which some blood flows from **arterioles** (the smallest arteries) to the **venules** (smallest veins) all the time. Small muscular **precapillary sphincters** located at the arteriolar end of capillaries branching off the thoroughfare channels, open or close other parts of the capillary bed to meet the varying metabolic requirements of the tissue. These sphincters, plus the smooth muscle in the walls of the arteries and arterioles, regulate the supply of blood to each organ and its subdivisions.

The contraction and relaxation of the muscles in the wall of the arteriole are primarily under nervous control, responding to stimuli reaching them through the sympathetic nervous system. In contrast, the muscles of the precapillary sphincters are controlled by local conditions in the tissues. For example, decreased oxygen tension in the tissues causes these muscles to relax, which increases the flow of blood through the capillary bed and increases the oxygen supply to the tissue.

SECTION 18–12

THE EXCHANGE OF MATERIALS ACROSS THE CAPILLARY WALL

The capillary membrane behaves as though it had much larger pores than those in the plasma membranes of cells. The pores in the capillary wall are about 80 Å in diameter (in contrast to the 8 Å diameter of the pores in the plasma membrane) and glucose, amino acids and urea, and sodium, chloride and other ions readily pass through the capillary membrane by diffusion. Blood plasma contains about 7 gm of protein per 100 ml and these protein molecules are too large to pass through these pores. The interstitial fluid bathing the cells has a much smaller protein content, about 1.5 g per 100 ml. As a result, the blood plasma has a higher osmotic pressure (termed **colloid osmotic pressure**) than the fluid bathing the tissues, and water tends to pass from the interstitial fluid into the capillaries. If two solutions are separated by a semipermeable membrane that is permeable to water but not to the solute molecules, water will tend to pass by osmosis from the solution with the greater concentration of water (and lesser concentration of solute molecules) to the other. The osmotic pressure of the 7 gm per cent protein solution in the blood plasma is about 28 mm Hg and the osmotic pressure of the 1.5 gm per cent protein solution in the interstitial fluid is about 4 mm Hg. The difference, $28 - 4 = 24$ mm Hg, is the force tending to move water molecules into the plasma (Fig. 18–28).

This is countered by the **hydrostatic pressure** in the capillary that results from the beating of the heart. This is difficult to measure directly but is estimated to be 18 mm Hg on the average. It has been possible to obtain estimates of the pressure in the interstitial fluid, i.e., the pressure in the fluid just outside the capillaries. The values obtained are about −6 mm Hg, just slightly less than atmospheric pressure. The algebraic sum of these two, +18 on the inside of the capillary and −6 on the outside, gives 24 mm Hg as the total pressure difference between the two sides of the capillary membrane, which tends to push water out of the capillaries. This just equals the colloid osmotic pressure tending to move water into the capillaries. The balance between these two forces keeps the blood volume remarkably constant despite the higher hydrostatic pressure within the capillaries than in the interstitial fluid.

Although the colloid osmotic pressure is the same, or nearly the same in both the arterial and venous ends of the capillary, the hydrostatic pressure within the capillaries is higher (about 30 mm Hg) at the end next to the arteriole and lower (about 10 mm Hg) at the end adjacent to the venule. These differences in pressure bring about an actual *flow* of water molecules out of the capillary at the arterial end and back into the capillary at the venous end (Fig. 18–29). The sum of the forces acting at the arterial end gives a net **filtration pressure** of some 12 mm Hg moving water molecules out of the capillary. At the venous end, the sum of the forces gives a net **absorption pressure** of about 8 mm Hg moving water molecules back into the capillaries. The two forces are not equal, but part of the difference is counterbalanced by the fact that the venous ends of the capillaries are larger and there is half again as much surface area as at the arterial end. Some of the fluid that passes out of the arterial end of the

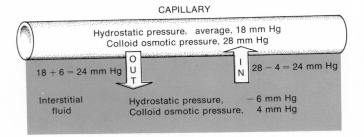

FIGURE 18–28   *Diagram of the hydrostatic and osmotic pressures which are responsible for the exchange of materials between capillaries and tissue fluid.*

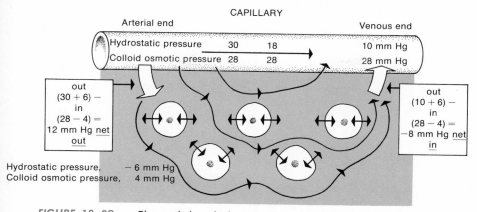

FIGURE 18-29   The path by which water flows in and out of capillaries to the interstitial fluid.

capillary, some 1 to 10 per cent of the total, is not reabsorbed at the venous end but passes into the lymphatic system and returns to the circulation by that route. Thus under normal conditions the blood volume remains constant as equal quantities of water pass in and out of the capillaries, yet the flow of water in and out is important in supplying nutrients to the tissues.

After the loss of blood in a hemorrhage, the decreased blood volume causes a decrease in blood pressure and hence in the filtration pressure. However, the amount of protein per ml of plasma remains the same and the colloid osmotic pressure is not reduced. This increased net pressure returning water to the capillaries rapidly restores blood volume at the expense of the interstitial fluid (and intracellular fluid). In "shock" which may occur after burns or accidents, the permeability of the capillary walls is increased, some of the plasma proteins escape into the interstitial fluid and the colloid osmotic pressure is decreased. Fluids tend to escape from the capillaries into the tissues, causing *edema* (accumulation of fluid in the tissue spaces).

SECTION 18-13

THE LYMPH VASCULAR SYSTEM

In addition to the blood vascular system vertebrates have a second, independent group of vessels, the **lymphatic system.** These small, thin-walled vessels originate as minute, dead-end **terminal lymphatic capillaries,** present in nearly all tissue spaces (Fig. 18-30). These join and form successively larger lymph vessels or veins, which, like the veins of the blood vascular system, have valves to prevent backflow. The final large lymphatic vessels drain into the blood vascular system at the junction of the internal jugular and subclavian (shoulder) veins.

The lymphatic vessels are an auxiliary system for the *return* of fluid from the tissue spaces to the circulation; there are no lymphatic arteries. The lymph capillaries, closed at one end, are very permeable and proteins and other large particles readily pass into them along with the tissue fluid. The fluid in the lymph capillaries, called **lymph,** has essentially the same composition as the interstitial fluid and, like it, has much less protein than

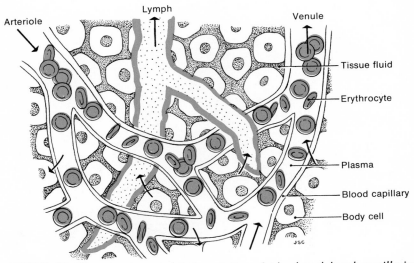

FIGURE 18–30   *A diagram of the relation of blood and lymph capillaries to tissue cells. Note that blood capillaries are connected at both ends, whereas lymph capillaries (shown in color) are "dead-end streets" and contain no erythrocytes. The arrows indicate the direction of flow.*

the blood. Thus, tissue fluid passes into the lymph capillaries, becomes lymph, and then is carried to the junction with the blood vascular system and is mixed with the blood.

At the junctions of lymph vessels are aggregations of cells, the **lymph nodes**, which produce one kind of white cell, the **lymphocytes,** and filter out bacteria and other particulate matter so that they do not enter the blood vascular system. Lymph flows very sluggishly through the minute, tortuous channels in the lymph nodes, and invading bacteria are trapped and phagocytized by the cells of the lymph node. Some bacteria may get past the first node and be caught in the second or third. In a massive infection the bacteria may penetrate all the lymph nodes and invade the blood stream. The presence of these phagocytized bacteria causes the lymph nodes to become swollen and tender—the lymph nodes of the neck may become noticeably swollen in individuals with sore throats. The lymph nodes in the lungs of heavy smokers are filled with particles of smoke and become a dark grey or black. These particles may eventually interfere with the functioning of the lymph nodes and reduce resistance to lung infections.

Although some of the lower animals, such as the frog, have four lymph "hearts" which pulsate and squeeze lymph along, the human lymph is moved by contractions of the adjacent skeletal muscles which compress the lymph vessels (the valves prevent backflow) and by the breathing movements of the chest. The lymphatics play an important role in the body's economy by returning fluid and especially proteins to the blood vascular system. There is a continual slow loss of protein from the circulation to the interstitial fluid; about 4 per cent of the total protein in the plasma passes out each hour. The lymphatics are the only means by which proteins can be returned to the circulation. Without lymphatics the concentration of protein in the interstitial fluid would soon equal the concentration in the blood capillaries with a consequent fall in colloidal osmotic pressure and decrease in blood volume. In addition, the lymph system produces lymphocytes, filters out bacteria and particulate matter, and plays a role in transferring fats absorbed from the intestinal villi into the blood vascular system (p. 517).

The rate at which lymph flows is quite slow and variable, but the total lymph flow is about 100 ml per hour (compared to the 5 liters per minute in the blood vascular system).

SECTION 18–14

THE BLOOD: PLASMA AND BLOOD CELLS

Contained within the arteries, veins and capillaries of the blood vascular system is a liquid tissue, the **blood,** composed of several kinds of cells suspended in an aqueous medium, the **plasma.** Plasma is a complex mixture that is in dynamic equilibrium with the interstitial fluid bathing the cells and the intracellular fluid within the cells. Plasma constantly takes up and loses substances as it flows through the capillaries, yet its composition is remarkably constant. It contains proteins, amino acids, carbohydrates, lipids, salts, hormones, enzymes, antibodies and dissolved gases. It is slightly alkaline, pH 7.4. The two chief constituents of plasma are water, 90 to 92 per cent, and proteins, 7 to 8 per cent. The concentrations of glucose, 0.1 per cent, and salts, 0.9 per cent, are small but kept nearly constant by a variety of regulatory devices.

Plasma contains several kinds of proteins, each with specific properties and functions—**fibrinogen,** involved in the clotting of blood, **albumin** and **globulins,** major contributors to the colloid osmotic pressure which regulates the water content of plasma; and **gamma globulins,** rich in antibodies which provide immunity to certain infectious diseases. The plasma proteins and hemoglobin in the red cells are important acid-base buffers that maintain the hydrogen ion concentration constant.

Floating in the plasma are red blood cells, **erythrocytes,** which are biconcave discs 7 to 8 microns in diameter and 1 to 2 microns thick. Mature mammalian erythrocytes have no nucleus; it is lost in the course of development from the precursor cell, the erythroblast. Each cubic millimeter of blood in an adult man contains about 5,400,000 red cells; the blood of a grown woman has about 5,000,000 red cells per cubic millimeter. Newborn infants have a greater number, 6,000,000 to 7,000,000 per cubic millimeter. Each red cell contains some 265,000,000 molecules of **hemoglobin,** the red pigment responsible for the transport of oxygen and carbon dioxide in the blood.

In the adult, red cells originate in the **red bone marrow,** which lies in the central hollow spaces of certain bones. The marrow contains unspecialized cells (hemocytoblasts) with a nucleus and no hemoglobin which divide to produce more precursor cells. After its last division, each precursor cell is gradually transformed into a mature red cell by a process which includes the loss of the nucleus, the manufacture of hemoglobin and the conversion from a spherical to the biconcave disc shape. The average life span of the human red cell is about 127 days, demonstrated by labeling the cells with radioactive iron or with some other isotope. The **spleen,** an oval organ some 12 centimeters long, lying to the left of the stomach, serves as a reservoir for red cells and, together with the liver, removes old red cells from the circulation. The hemoglobin molecules recovered from old red cells are dismantled in the spleen and liver; the iron atoms are recovered and reused and the heme portion is degraded and excreted by the liver as bilirubin and other bile pigments.

From the total number of red cells in the human body and their average life span, it is simple to calculate that about 2,500,000 red cells are made, and an equal number destroyed, each *second* throughout the day and night. And since there are some 265×10^6 molecules of hemoglobin per cell, some 650×10^{12} molecules of hemoglobin are synthesized each second. Experiments by Howard Dintzis indicated that about 90 seconds are required for the assembly on the ribosomes of the 574 amino acids that comprise each molecule of hemoglobin. Thus, at any moment, $90 \times 650 \times 10^{12}$, or 6×10^{16}, molecules of hemoglobin are in the process of being assembled!

The blood contains five kinds of white cells, **leukocytes,** all of which have a nucleus but no hemoglobin. All move actively by amoeboid movement. There are far fewer white than red cells, some 7000 per cubic millimeter on the average. The number fluctuates from 5000 to 9000 or 10,000 in different individuals and even in the same individual at different times of day, being lowest early in the morning and highest in the afternoon.

Two of the types of white cells, **lymphocytes** and **monocytes,** are produced in lymphoid tissue such as the spleen, thymus and lymph nodes. The other three, **neutrophils, eosinophils** and **basophils,** are produced in the bone marrow along with the red cells. All three contain cytoplasmic granules which differ in their size and staining properties. Neutrophils normally make up 60 to 70 per cent of the population of circulating white cells; they can move actively by amoeboid motion, slip through the wall of the capillary between adjacent cells and are important in taking up bacteria and other infective agents by phagocytosis. The neutrophils also phagocytize the remains of dead tissue cells. Neutrophils and other white cells are guided to points of infection by chemicals released by the inflamed and infected tissues.

Monocytes are the largest white cells, reaching 20 microns in diameter. Like the smaller lymphocytes, they have no cytoplasmic granules. Monocytes also move actively by amoeboid motion and phagocytize bacteria. After several hours in the tissue spaces, monocytes tend to enlarge and become **macrophages,** which can move quite rapidly and engulf 100 or more bacteria. Neutrophils are of prime importance in resisting acute bacterial infections; monocytes become of greater importance in countering long-term infections. Macrophages can ingest large particles of cellular debris and are important in cleaning up an infected area after the bacteria have been eliminated. Eosinophils, which have large granules that stain with eosin and other acidic dyes, are amoeboid and phagocytic. They increase greatly in number during allergic reactions and during infections with parasites such as trichina worms.

Lymphocytes are remarkably interesting cells which have the potential of being converted into many other types of cells in the body. The lymphocyte can swell and become a monocyte, then enter connective tissues and other spaces and become a macrophage. Lymphocytes can also enter the bone marrow and develop into either the precursors of red cells or the precursors of the granulocytic white cells such as neutrophils. Lymphocytes in tissues can develop into **fibroblasts** and secrete collagen fibers, elastic fibers and other elements of connective tissue. Lymphocytes may also become **plasma cells** which produce and secrete antibodies, of prime importance in the immune process (p. 579).

One final group of formed elements circulating in the blood are **thrombocytes,** important in initiating the clotting of the blood (p. 577). In most vertebrates other than mammals the thrombocytes are small, oval, pointed cells with a nucleus. The mammalian thrombocytes are tiny, spherical- or disc-shaped **blood platelets,** lacking a nucleus, which are formed by the fragmentation of giant cells, **megakaryocytes,** in the red bone marrow. Some platelets are formed from phagocytic cells in the lungs.

TRANSPORT IN THE BLOOD

Materials are transported in the blood in one of three states—in simple solution in the plasma, bound to one of the plasma proteins or carried in the red cell in combination with hemoglobin. A small amount of the oxygen and carbon dioxide is simply dissolved in the plasma water, but most of the oxygen and much of the carbon dioxide are bound to hemoglobin in the red cell. The inorganic ions, chloride, bicarbonate and phosphate anions and the cations sodium, potassium, magnesium and about half of the calcium are carried in plasma in solution. The remaining half of the calcium is bound to a plasma protein. Glucose, amino acids and organic acids such as lactic and citric are present in solution in the plasma, as are a variety of metabolic waste products such as urea, uric acid, ammonia and creatinine and very small amounts of certain vitamins.

Many other compounds in the plasma are bound to proteins. Neither free fatty acids nor triglycerides are soluble enough in water to dissolve in the plasma. Free fatty acids are carried in the blood largely in combination with serum albumin, and triglycerides are carried from the intestine to the liver and adipose tissue as **chylomicrons,** droplets of fat that are stabilized by a thin coat of proteins, phospholipids and cholesterol esters. The liver synthesizes and secretes several kinds of **lipoproteins,** which are about half peptide chain and half lipids—phospholipids, cholesterol and its esters and triglycerides. One type of lipoprotein is especially rich in triglycerides and is the form in which these fats are transported from the liver to skeletal muscle, adipose tissue and other tissues.

Some of the hormones, those of the anterior and posterior lobes of the pituitary, for example, are carried in the blood dissolved in the plasma. Others are bound to plasma proteins, either to albumin or to some unique protein that specifically binds that hormone. Thyroxine is bound either to a specific **thyroxine-binding globulin** or to albumin, from which it is slowly released over a period of days. This helps ensure a steady, slow supply of thyroxine to the tissues. Most of the insulin present in plasma is bound to globulins. Some of the steroid hormones, such as cortisol, are carried in plasma bound to specific proteins, e.g., **transcortin.**

The transport of oxygen and carbon dioxide in the blood depends largely on the hemoglobin present in the red cell. If blood were simply water, it could carry only about 0.2 ml of oxygen and 0.3 ml of carbon dioxide in each 100 ml. The properties of hemoglobin enable whole blood to carry some 20 ml of oxygen and 30 to 60 ml of carbon dioxide per 100 ml and serve simultaneously as an acid-base buffer to minimize changes in the pH of the blood. The protein portion of hemoglobin is composed of four peptide chains, typically two α chains and two β chains, to which are attached four heme (porphyrin) rings. An iron atom lies in the center of each heme ring (Fig. 18–31). Hemoglobin is found in all the major groups of animals above the flatworms; mollusks, crustacea and certain other animals have other heme pigments such as **hemocyanin.** Human blood contains about 15 gm of hemoglobin per 100 ml. Hemoglobin has the remarkable property of forming a loose chemical union with oxygen; the oxygen atoms are attached to the iron atoms in the hemoglobin molecule.

In the respiratory organ, the lung or gill, oxygen diffuses into the red cells from the plasma and combines with hemoglobin (Hb) to form oxyhemoglobin (HbO_2): $Hb + O_2 \rightleftarrows HbO_2$.

The reaction is reversible and hemoglobin releases the oxygen when it reaches a region where the oxygen tension is low, in the capillaries of

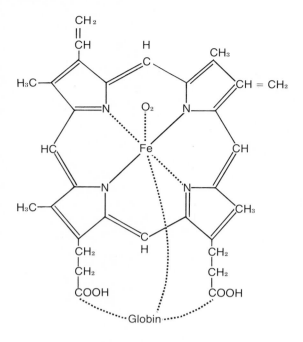

FIGURE 18–31 One of the subunits of the hemoglobin molecule, a combination of a porphyrin ring, ferrous iron, (Fe^{++}), and a peptide chain, globin. Four subunits, two α chains and two β chains, comprise the hemoglobin molecule.

the tissues. The combination of oxygen with hemoglobin and its release from oxyhemoglobin are controlled by the concentration of oxygen and, to a lesser extent, by the concentration of carbon dioxide. Carbon dioxide reacts with water to form carbonic acid, H_2CO_3; hence an increase in the concentration of carbon dioxide results in an increased acidity of the blood. The oxygen-carrying capacity of hemoglobin decreases as blood becomes more acid. Thus the combination of hemoglobin with oxygen is controlled indirectly by the amount of carbon dioxide present. This results in an extremely efficient system: In the capillaries of the tissues, the concentration of carbon dioxide is high and a large amount of oxygen is released from hemoglobin by the combined action of the low oxygen tension and the high carbon dioxide tension. In the capillaries of the lung or gill, carbon dioxide tension is lower and a large amount of oxygen is taken up by hemoglobin owing to the combined effect of the high oxygen tension and low carbon dioxide tension.

At the oxygen tension of arterial blood, 100 mm Hg, each 100 ml of blood contains about 19 ml of oxygen. At the oxygen tension of venous blood, 40 mm Hg, each 100 ml contains 12 ml of oxygen. The difference, 7 ml, represents the amount of oxygen delivered to the tissues by each 100 ml of blood. Since some 5 liters of blood are delivered to the tissues each minute, this means that 350 ml of oxygen can be supplied per minute. At rest, the cells of the body need about 250 ml of oxygen per minute and with exercise this requirement may increase 10- or 15-fold.

The cells of the body produce, at rest, about 200 ml of carbon dioxide per minute. If this were simply dissolved in plasma (which can transport in solution only 4.3 ml of carbon dioxide per liter), blood would have to circulate at a rate of 47 liters per minute instead of 4 or 5. This amount of carbon dioxide dissolved in water would yield a *p*H of 4.5. The unique properties of hemoglobin enable each liter of blood to transport some 50 ml of carbon dioxide from tissue to lung with only a few hundredths of a unit difference in the *p*H of arterial and venous blood. Some carbon dioxide is carried in a loose chemical union with hemoglobin as ***carbamino-***

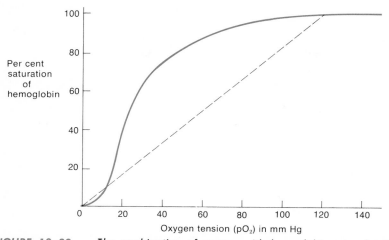

FIGURE 18–32   . The combination of oxygen with hemoglobin as a function of oxygen tension. The dotted line shows the relation that would exist if hemoglobin bound oxygen as a linear function of oxygen tension. Because of the interactions between the four heme groups in a single hemoglobin molecule, the binding of oxygen to hemoglobin follows the S-shaped curve shown in color.

hemoglobin, and a small amount is present as carbonic acid, but most of it is transported as bicarbonate ion, HCO_3^-. The CO_2 produced by cells dissolves in the tissue fluid to form H_2CO_3, a reaction catalyzed by **carbonic anhydrase,** and the carbonic acid is neutralized to bicarbonate by the sodium and potassium ions released when oxyhemoglobin is converted to hemoglobin. Oxyhemoglobin is a stronger acid than reduced hemoglobin, hence some cations are released when HbO_2 is converted to Hb. In the course of evolution this one molecule has been endowed with all the properties needed for the transport of large amounts of oxygen and carbon dioxide, with a minimal change in the pH of the blood while these gases are being transported.

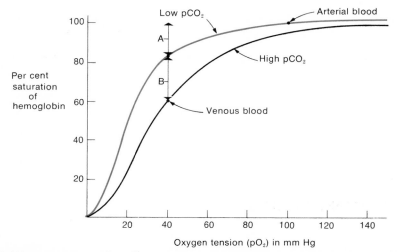

FIGURE 18–33   The effect of carbon dioxide tension ($_pCO_2$) on the delivery of oxygen to the tissues. The arrow A indicates the amount of oxygen released from hemoglobin as the $_pO_2$ falls from that of arterial blood (100 mm Hg) to that of venous blood (40 mm Hg). The arrow B indicates the additional amount of oxygen released because of the greater $_pCO_2$ in venous blood.

The properties of the heme pigments are such that the amount of oxygen taken up by the pigment is not directly proportional to the oxygen tension. A graph of the relationship gives an **S**-shaped curve (Fig. 18–32). The blood is a more effective transporter of oxygen than it would be if the oxygen content were a simple linear function of oxygen tension. The effect of carbon dioxide (really the change in *p*H brought about by the change in the carbon dioxide content) on the combination of oxygen with hemoglobin is shown in Figure 18–33. The oxygen dissociation curves for arterial blood, with low carbon dioxide tension, and for venous blood, with high carbon dioxide tension, illustrate how much more oxygen is delivered to the tissue by a given amount of blood as carbon dioxide is taken up in the tissue capillaries. The properties of the heme proteins in different species of animals are quite different and are generally adapted to the amount of carbon dioxide present. This is generally low in water-breathing animals and high in air-breathing animals. This points up the generalization that the evolution of air-breathing animals from water-breathing ones involved marked changes not only in the structure of the respiratory organs but also in the chemical properties of the heme proteins serving as blood pigments.

SECTION 18–16

THE PLASMA PROTEINS

There are about 7 gm of plasma proteins per 100 ml of plasma, of which 5 gm are **albumin** and 2 gm are **globulins.** The many other kinds of proteins are present in very small amounts. Albumin is primarily responsible for the colloid osmotic pressure which regulates the water content of cells and body fluids (p. 568). The globulins are a heterogeneous mixture of proteins, some of which, the γ **globulins,** are rich in antibodies. **Fibrinogen** is an essential component of the system involved in the clotting of blood. The **immunoglobulins** are synthesized by plasma cells and large lymphocytes, but the other plasma proteins are synthesized in the liver. When the concentration of proteins in the plasma decreases and the maintenance of colloid osmotic pressure is endangered, the liver responds by increasing the rate of production of plasma proteins. The control mechanism involved in this response has not yet been clarified. The role of specific plasma proteins in t ansport was discussed previously. In many acute infections, the synthesis of fibrinogen and of globulins is increased, and the increased amount of these two in the plasma increases the rate of sedimentation of red cells, a test commonly used to follow the course of a disease and the patient's recovery from it.

SECTION 18–17

THE CLOTTING OF BLOOD

Animals have evolved elaborate mechanisms to prevent the accidental loss of blood. Some animals, such as crabs, prevent blood loss by the powerful contraction of muscles in the wall of the blood vessel when the vessel is severed. In the vertebrates and in many invertebrates as well, blood loss is prevented by a series of chemical reactions in which a solid **clot** is formed to plug the broken vessel. Clotting is a function of the plasma, not of the blood cells, and involves the conversion of **fibrinogen,** a soluble plasma protein, to **fibrin,** an insoluble protein. This reaction involves the enzymatic removal of two peptides from the fibrinogen molecule to yield "activated" fibrin, which then polymerizes into long threads which form a network in which blood cells are trapped. The clotting mechanism is quite complex, involving many different substances in the plasma which interact in three sequential reactions. Each of the first two reactions produces an enzyme required to catalyze the succeeding reaction (Fig. 18–34).

The first step, the production of **thromboplastin,** is initiated when a blood vessel is cut. Traumatized tissues release a lipoprotein called thromboplastin which interacts with calcium ions and several protein factors in the blood plasma (proaccelerin, proconvertin) to produce **prothrombinase,** the enzyme that catalyzes the second step. Prothrombinase can also be synthesized by the interaction of factors released from platelets, calcium ions and other plasma globulins. One of these, termed **antihemophilic factor,** is present in normal plasma but is absent from the plasma of individuals with **hemophilia,** or "bleeder's disease." The prothrombinase, made either by the system involving thromboplastin released from traumatized tissue or by a comparable factor released by platelets, catalyzes a reaction in which **prothrombin,** a plasma globulin made in the liver, is split into several fragments, one of which is **thrombin.** This reaction also requires calcium ions. Finally, the thrombin acts as a proteolytic enzyme to cleave two peptides from fibrinogen and form an active fibrin monomer, which polymerizes to form long threads of insoluble fibrin. The network of fibrin threads traps red cells, white cells and platelets to form a clot.

This mechanism is admirably adapted to provide for rapid clotting when a blood vessel is injured and yet prevent clotting in the intact blood

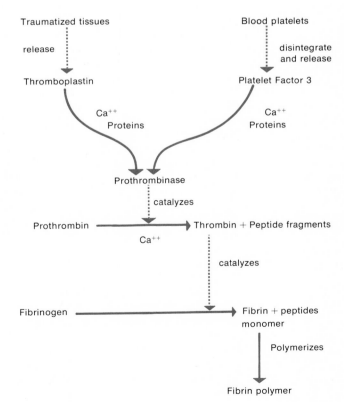

FIGURE 18–34   The sequence of reactions by which the clotting of blood occurs. Each reaction is catalyzed by an enzyme that is produced by the previous reaction. The process is initiated by the release of thromboplastin from damaged tissue or by the release of a specific factor when platelets disintegrate.

vessel. Although normal blood may contain a small amount of thrombo-plastin it also contains a strong anticoagulant, **heparin,** produced in mast cells in the lungs and liver. Heparin inhibits the conversion of prothrombin to thrombin. The synthesis of prothrombin in the liver requires an adequate supply of vitamin K, and anything that interferes with the dietary supply of vitamin K or its absorption from the intestine may lead to deficient clotting. Muscular contraction of the wall of the damaged artery or vein and the resulting closing of the hole may be important in stopping the loss of blood in man as in other animals.

SECTION 18–18

IMMUNOGLOBULINS AND IMMUNITY

The phenomenon of immunity was recognized centuries ago, when it was realized that a man who has had one of certain diseases was unlikely to have it again—he was "immune" to it. In the late eighteenth century Edward Jenner noticed that dairy workers handling cows with **cowpox** never had **smallpox.** When he scratched some serum from the pustules on a cow's udder into human skin, the individual had a mild disease with a single localized pox at the site of injection. Anyone so vaccinated never acquired smallpox. Cowpox and smallpox are caused by two distinct but closely related viruses; inoculation with cowpox virus **(vaccinia)** stimulates the production of antibodies that will react with the smallpox virus **(variola).** Suitable vaccination procedures for many other diseases have been worked out since then. These have been remarkably effective in improving the general level of health of the world's population and have protected untold millions of people against a variety of debilitating and lethal diseases, such as poliomyelitis, measles and whooping cough.

Actively acquired immunity, such as that produced in response to vaccination with cowpox, depends upon the production of specific proteins, **antibodies,** and their release into the blood and tissue fluids in response to some foreign protein, an **antigen.** To be an antigen and induce the formation of antibodies, the substance must somehow be recognized as "alien" by the antibody-producing cells. The basis for this recognition process is unknown.

The γ-globulin fraction of the plasma proteins contains three classes of **immunoglobulins,** antibodies, which differ in size and chemical composition. Some 85 per cent of the total is made of **γG immunoglobulin** (also called **IgG**), which has a molecular weight of 150,000. IgG is made of four subunits, two **"heavy" chains,** with molecular weights of 50,000, and two **"light" chains,** with molecular weights of 25,000 (Fig. 18–35). The next largest class of immunoglobulins includes the **γM immunoglobulins** or **IgM,** with a molecular weight of about 10^6. These larger molecules which sediment much faster (19S) than the IgG molecules (which are 7S) are not simply some polymer of the IgG molecules, although they are also composed of light and heavy chains. They have a different content of carbohydrate (12 per cent vs 3 per cent carbohydrate in IgG) and different heavy chains. The third, and rarest, class of immunoglobulins, the **γA immunoglobulins, IgA,** are also made of light and heavy chains and have molecular weights somewhat greater than the IgG molecules, with sedimentation constants of 7S, 9S and 11S. All three of these types of immunoglobulins have antigenically identical light chains but antigenically distinct heavy chains.

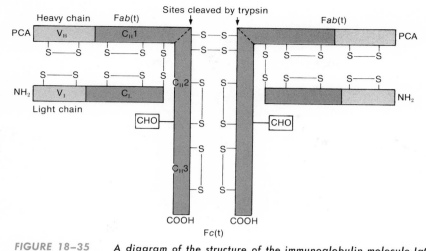

FIGURE 18–35   A diagram of the structure of the immunoglobulin molecule IgG, composed of two light chains and two heavy chains joined together by disulfide bonds. The constant (C) and variable (V) regions of the chains are indicated.

The IgM immunoglobulins are among the earliest antibodies to appear in response to an antigen; somewhat later in the immune response they are replaced by IgG antibodies. It is remarkable that the antibodies produced in response to a given antigen are not necessarily homogeneous but may differ in their specificity, their avidity for the antigen and their physiochemical properties—size, shape, net charge and amino acid sequence. This makes the problem of relating their structure to their specificity as antibodies even more difficult. The differences between different antibodies must be quite subtle, probably even more subtle than the differences between different enzymes.

Antibodies are produced in **plasma cells** and large lymphocytes located in lymph nodes and other lymphoid tissue. The plasma cells have an extensively developed endoplasmic reticulum (Fig. 18–36) for the synthesis of proteins.

There is evidence to suggest that only a small part of the large immunoglobulin molecule, some 15 to 30 amino acid residues, is involved in the immunologically active site. The differences between different antibodies appear to reside in small differences in the shape of the protein which are due to differences in the sequence of amino acids. The antigen and antibody are believed to have some sort of complementary geometrical configuration so that they fit together like a lock and key.

Following the injection of an antigen there is a latent period of nearly a week before any antibody appears in the blood. The titer of antibody rises slowly to a low peak, the **primary response,** then decreases (Fig. 18–37). A second injection of antigen several days, weeks or even months later induces a rapid production of antibodies after a shorter latent period, the **secondary response.** The titer of antibody reaches a higher level and decreases more slowly. Further injections induce additional secondary responses until a maximal titer is achieved. This usually declines with time, and periodic reimmunizations (booster shots) help maintain a satisfactory level of immunity. A secondary response may also be induced in a previously immunized individual by his exposure to the natural infectious agent, and antibody is usually produced rapidly enough to prevent the appearance of the disease symptoms.

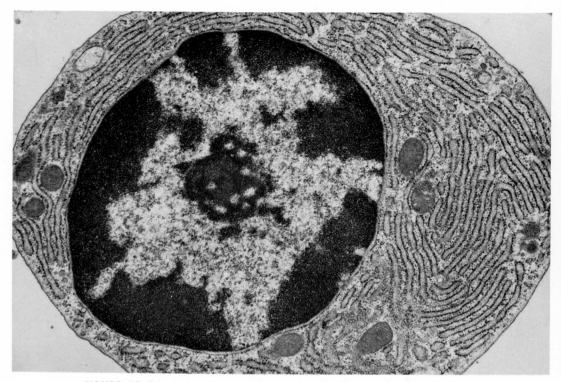

FIGURE 18–36   An electron micrograph of a human plasma cell showing the greatly hypertrophied endoplasmic reticulum. (From Bloom, W., and Fawcett, D.: Textbook of Histology. 9th ed. Philadelphia, W. B. Saunders Company, 1968.)

Although any foreign protein and many smaller molecules may act as an antigen and activate the plasma cells to produce a specific antibody, the more familiar antigens are the proteins of certain bacteria and viruses that produce infectious diseases. Antibodies can subsequently combat these bacteria or viruses by combining with them and neutralizing them and by causing the microorganisms to clump or agglutinate, thereby effectively preventing further penetration of the host; they may cause the microorganisms to undergo **lysis,** to break up and dissolve, or they may make invaders more susceptible to phagocytosis.

All of us have a **natural immunity** to certain infectious diseases that affect other organisms but not man. The virus for canine distemper, which kills about half of the dogs infected with it, does not infect human beings. The virus called **herpes simplex,** which is lethal to rabbits, usually causes only a "fever blister" in man. This natural immunity is due in large part to the presence of antibodies in the plasma that can combine with the infective agent. Most human beings, and most vertebrates in general, can form antibodies to a variety of infectious agents and develop **specific immunity** to that agent. A person who has contracted a disease and needs antibodies immediately to combat its antigens may be given a **serum,** a solution of antibodies produced by an animal that had been infected with that disease. This can tide the patient over until his own plasma cells can develop enough antibodies to the infective agent to protect him. This method of passively acquiring immunity is immediately effective but the immunity disappears in a few weeks.

To prepare a serum to a given disease, a pure strain of the infective agent is grown and injected in increasing doses into an animal such as a

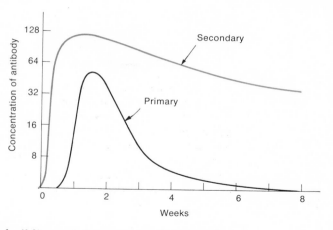

FIGURE 18–37   *Primary and secondary responses of antibody formation to successive doses of antigens.*

horse, which responds by gradually building up the amount of the specific antibody in its blood. Blood is drawn from the horse at intervals and processed to remove blood cells and the proteins other than the desired antibody.

The removal of the **thymus gland** of a newborn animal causes it to lose its ability to make antibodies. The thymus is a large organ in the fetus and newborn, full of **thymic cells** which resemble lymphocytes. These thymic cells are produced rapidly and pass out into the blood stream to all parts of the body, eventually becoming trapped in the lymph nodes and spleen. These lymphocytes can be converted into plasma cells at any time during life. If the thymus is removed at or before birth, the lymph nodes remain small, the level of lymphocytes circulating in the blood is very low and the animal cannot make antibodies to many antigens. Furthermore it is deficient in **cellular immunity** as well as in circulating antibodies or **humoral immunity** and will accept tissue grafts from other animals that differ from it genetically. These defects are alleviated by grafting a bit of thymus into the thymectomized animals. Apparently the thymus serves as a reservoir in which lymphocytes multiply and from which they migrate out and populate the lymph nodes. In addition, the thymus appears to elaborate a hormone that regulates the maturation of the lymphocytes.

Among the questions facing immunologists at present are two: (1) How can plasma cells make so many different kinds of antibodies—so many proteins with different amino acid sequences—and make a specific antibody on demand, when a specific antigen is present? (2) Why doesn't an animal's plasma cells make antibodies to that animal's own, native proteins?

A theory favored in the thirties and forties was that the antigen in some way supplies the information necessary for the production of its specific antibody. This **"instructive" theory** held that the antigen acts in some way as a template, and the globulin molecule as it is formed is molded around the antigen and thus develops a complementary conformation at its active sites. This theory, which assumed that the specificity of the antibody was due not to its specific amino acid sequence but to its conformation, was ruled out when it was found that different antibodies do have different amino acid sequences and that the specificity of the antibody is a function of its amino acid sequence.

An alternative theory, that of **"clonal selection,"** was advanced by Burnet in the late fifties. According to this theory the cells of the antibody-forming system of each individual have among them all the information they need to make any of a wide variety of antibodies even before they encounter the antigen. The antigen does not provide information to the antibody-forming cells; it simply selects cells that are making antibodies that fit it and stimulates those cells to proliferate and make large quantities of those antibodies. It is generally believed that there are many different kinds of plasma cells, each of which synthesizes a single or a few kinds of antibodies.

The cells synthesizing a given kind of antibody are believed to comprise a **clone,** which includes all the descendants of a single progenitor cell that had acquired by some random process the genetic capacity to synthesize an immunoglobulin with a specific amino acid sequence and the ability to react with a specific antigen. Until the specific antigen appears, the clone is believed to consist of relatively few cells. The presence of the antigen stimulates the proliferation of the clone that can make the appropriate antibody and in this way stimulates the production of the appropriate, specific immunoglobulins. This theory is satisfying in that it supposes that the synthesis of immunoglobulins involves the production on ribosomes of peptides with specific amino acid sequences by a mechanism identical to that for the synthesis of other proteins.

It is not yet clear how the specific antigen and the cells of the appropriate clone of plasma cells recognize each other and how this recognition leads to the proliferation of the cells of that clone.

Analyses of the amino acid sequences of immunoglobulins have shown that both light and heavy chains have "constant" and "variable" regions (Fig. 18–35). Certain regions are identical in all IgG immunoglobulins and other regions differ, are specific, in each kind of antibody. The clonal selection theory would require only a few structural genes for the immunoglobulin to be transmitted by the germ cells. Subsequent somatic mutations in different clones would provide the genetic diversity needed to code for all the different antibodies a given individual can make. This theory would also account for the presence of natural immunity, but it does not readily account for the fact that the ability to synthesize antibodies of different specificities appears during development in a fixed time sequence and not at random. The clonal selection hypothesis implies that any single plasma cell would make only one kind of antibody. Several experiments have shown that a single plasma cell may make more than one kind of antibody. Thus the idea that there are different clones of cells, each of which responds to only one or a few antigens, is not yet established. A related hypothesis suggests that there is one gene C that regulates the synthesis of the portion of the molecule that is common to all immunoglobulin chains, and one or a relatively few V genes that code for the part of the chain that varies in different antibodies. The recombination of a specific V gene with the C gene yields a VC gene that is read out and forms the specific immunoglobulin.

The antibodies described so far are all **humoral antibodies** that circulate in the plasma. In addition some antibodies are found within cells, apparently made there, and provide the basis for **tissue immunity,** which is involved in the phenomena of hypersensitivity and immunologic tolerance.

When cells from one adult animal, the donor, are injected into or grafted onto a second adult (the host), the transplanted cells may grow for a short time but will be rejected and slough off in a few weeks. If a second graft is transplanted from the same donor to the same host, it will be rejected in a much shorter time. This accelerated rejection, the **"second set" reaction,** is specific for the donor. If a skin graft is made from another genetically unrelated individual into this host it undergoes a first set reaction and is rejected only after some weeks. The immunologic basis of the second set reaction, the capacity to reject a transplant from another member of the same species, is acquired when the host tissue is exposed to the donor's tissue. It can be gained not only by a previous skin graft, but by the injection of a suspension of other cells, such as spleen cells, from the donor. Only if transplants are made between identical twins, or between members of a highly inbred, genetically homogeneous strain, will the grafts be successful and survive in the host's body.

If **embryonic** cells from a strain A mouse are infused into a newborn mouse of strain B, the mouse will develop normally and, when adult, will be able to accept transplants of skin and other tissues from strain A mice. It is said to have developed **immunologic tolerance.** If an animal is exposed to an antigen before it has developed the capacity to react to it, to produce antibodies to it, then the development of that capacity is delayed. In the continued presence of antigen, the development of that capacity can be postponed indefinitely. This same animal will respond quite normally to the presence of other antigens and will form antibodies to them. A new-born rabbit injected with bovine serum albumin does not develop anti-bodies to it. Several months later it can be injected with more bovine serum albumin and again will not form antibodies to it. Immunologic tolerance can be induced only in fetal or neonatal animals.

Transplantation immunity is determined by the genetic makeup of the host and donor; the genes that determine the antigens responsible for the response to transplanted tissue are termed **histocompatibility genes.** Genetic experiments with mice indicate that there are at least 14 inde-pendently segregating loci for these transplantation antigens, and the num-ber in man is undoubtedly at least as large.

Basically similar to the immune reaction is the phenomenon of **hyper-sensitivity** or **allergy,** evident in asthma, hay fever, skin rashes and so on. When an antigen-antibody reaction occurs in contact with cells, the lat-ter may be damaged and become swollen. The permeability of their plasma membranes becomes altered and they release histamine and other sub-stances. The histamine may secondarily affect other tissues and damage them. It causes contraction of the muscles in the walls of the bronchioles and relaxation of the muscles in the walls of the blood vessels. The latter results in local edema. In hay fever, the antigen-antibody reaction occurs in the nasal mucosa; the resulting local edema causes excessive secretion by mucosal glands, which results in the stuffiness and drippiness of hay fever.

Hypersensitivity originates when an antigen **(allergen)** enters the body and stimulates the production of antibodies specific for it. Exposure of the body cells to a second dose of allergen at any time after about two weeks after the first one results in the reaction of these antigens with the antibodies produced in response to the first dose of antigen. This may lead to secondary effects if the allergic reaction is generalized, such as the devastating symptoms of **anaphylaxis**—nausea, weakness, lowered tem-perature, convulsions and death. These symptoms are the outward signs of the body's reaction to the antigen. Allergies, like immunities, are highly specific for specific proteins.

SECTION 18-19

REGULATORY PROCESSES: RED CELL NUMBER

Each aspect of the circulatory system has one or more regulatory mechanisms that operate to keep it relatively constant under normal con-ditions and to bring it back to normal when it has been increased or decreased by some change in the environment. The constancy of the number of red cells circulating in the blood provides us with an excellent example of a **dynamic equilibrium.** Under normal circumstances the rate of formation of new cells just equals the rate of destruction of old ones and the total number of circulating red cells remains constant. The rate of production of red cells is increased by any factor that decreases the amount of oxygen delivered to the tissues. The loss of red cells by hemorrhage decreases the capacity of the blood to transport oxygen to the tissues and this leads to increased red cell production (Fig. 18–38). The stimulus is not the decreased number of red cells per se, for if a person with a normal number of red cells moves to a very high altitude for a few weeks, his red cell count will rise to 6,000,000 or 7,000,000 per cubic mm. At high altitudes there is less oxygen in the air and consequently less oxygen is delivered to the tissues. The same increase can be produced at sea level by keeping a man or experimental animal in a chamber the atmosphere of which has a low oxygen content but a total pressure equal to that of air at sea level.

Oxygen deficiency does not stimulate the red bone marrow directly. Instead, in response to lowered oxygen tension, the kidneys and perhaps the liver and other tissues secrete **erythropoetin,** which passes in the blood to the red bone marrow where it stimulates the initial stage in red cell production, the formation of hemocytoblasts from primordial stem cells in the marrow. When the number of red cells has returned to normal,

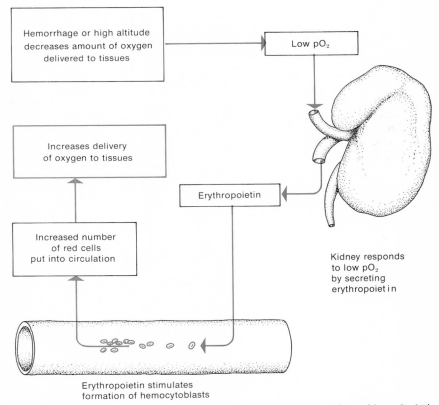

FIGURE 18–38   *The mechanism by which erythropoietin is released from the kidney in response to low $_pO_2$ and passes to the bone marrow where it stimulates the formation of hemocytoblasts, thereby increasing the number of red blood cells in the circulation and the delivery of oxygen to the tissues.*

the oxygenation of the tissues is also returned to normal and the stimulus for the production of erythropoetin is removed. The lower amount of erythropoetin results in a decreased rate of production of red cells and an excess of red cells is prevented. The number of circulating red cells is regulated automatically by the requirements of the tissues for oxygen via the production of erythropoetin.

SECTION 18–20

REGULATORY PROCESSES: THE HEART BEAT

An active tissue requires several times as much oxygen and nutrients as the same tissue at rest, and both the heart and blood vessels participate in the adjustments necessary to provide the required amounts. During periods of intense exercise the heart can pump seven or eight times as much blood as normal by increasing the number of beats per minute, by increasing the volume of blood pumped per beat, or a combination of the two. The heart of an adult normally pumps about 75 ml per beat, but it can pump as much as 200 ml per beat. The amount of blood pumped by the heart is normally regulated by the amount of blood flowing from the veins into the right atrium. This principle, first stated by the English physiologist E. H. Starling, is known as Starling's *Law of the Heart.* The regulation of the heart is achieved in part by factors operating within the heart and in part by extrinsic factors. When cardiac muscle is stretched before it contracts it will contract with greater force than when it has not been stretched. When a small quantity of blood enters the atrium, the muscle fibers are not stretched very much and they contract with a relatively weak force. When large quantities of blood enter the atrium the chamber is dilated, its muscle fibers are stretched and they contract with great force, emptying the chamber.

The heart is under nervous control and a *vasomotor center* in the medulla of the brain receives impulses from a variety of receptors and sends out impulses to the heart to regulate its rate and strength of contraction (Fig. 18–39). Impulses pass to the heart through two different sets of nerves, the sympathetic and parasympathetic systems (p. 646). Impulses passing through the parasympathetic system decrease the activities of the heart, whereas impulses through the sympathetic system increase its activities and cause increased heart rate, increased strength of contraction and increased blood supply through the *coronary vessels* that supply the muscles in the walls of the heart. Both sets of nerves end in the sinoatrial node and increase or decrease the frequency with which impulses are initiated there and pass down the Purkinje fibers. Sensory receptors in the wall of the vena cava and right atrium are stimulated when the vessels are distended with blood and the wall is stretched. They initiate impulses that pass to the vasomotor center and result in a faster heart beat. Sensory nerves in the walls of the aorta and carotid arteries, stimulated when these vessels are distended, conduct impulses to the vasomotor center that result in a slower heart rate.

The action of these feedback controls increases heart rate during exercise. The contractions of the skeletal muscles during exercise increase the return of venous blood and distend the vena cava and right atrium, stimulating the stretch receptors there. Impulses from these pass to the vasomotor center and lead to a stimulation of the heart rate. As the heart rate increases, the amount of blood in the aorta and carotid arteries increases, and the distension of their walls stimulates the stretch receptors

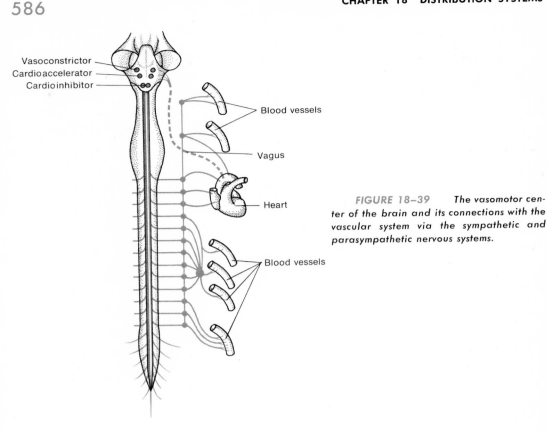

Vasoconstrictor
Cardioaccelerator
Cardioinhibitor

Blood vessels

Vagus

Heart

Blood vessels

FIGURE 18–39　　　*The vasomotor cen-ter of the brain and its connections with the vascular system via the sympathetic and parasympathetic nervous systems.*

there. Impulses from these receptors increase the flow of impulses to the sinoatrial node via the ***vagus nerves*** (parasympathetic) and result in a slowing of the heart rate. This control system nicely adjusts the rate of the heart to the metabolic demands of the body yet prevents any serious over-response, for an increased heart rate leads to the stimulation of the stretch receptors in the aortic arch and this in turn leads to a slowing of the heart rate.

Sensory receptors in the ***carotid body*** detect any reduction in the concentration of oxygen in the arteries and this leads to an increased heart rate. Other receptors detect an increased carbon dioxide content, or an increased temperature of the blood, and lead to an increased heart rate.

SECTION 18–21

REGULATORY PROCESSES: THE RATE OF BLOOD FLOW

The rate of blood flow to any given organ at any given moment is remarkably well adjusted to the requirements of that organ for oxygen and nutrients. A set of control mechanisms regulates the flow of blood to each tissue in proportion to its needs from moment to moment. The rate of blood flow depends primarily on the ***pressure,*** the force that drives the blood through the vessels, and the ***resistance*** offered by the smaller vessels to the flow of blood through them.

The amount of blood pumped by the heart, the ***cardiac output,*** is about 5 liters per minute at rest and as much as 30 liters per minute during violent exercise. The amounts of blood flowing to the various parts of the body are summarized in Table 18–1. Under basal conditions the liver, kidneys and brain receive 27, 22 and 14 per cent of the cardiac output, respectively, whereas the skeletal muscles get only 15 per cent. During exercise almost all the increased cardiac output goes to the skeletal muscles and they receive as much as 75 per cent of the total cardiac output.

In its course through the body, blood does not flow at a constant speed. The flow is rapid in the arteries (about 500 mm or 20 inches per second in the larger ones), a little less rapid in the veins (about 150 mm per second in the larger ones), and slow in the capillaries (less than 1 mm per second). The differences in the rate of flow depend upon the total cross-sectional area of the vessels. If a fluid passes from one tube into another of larger size, the rate of flow is less in the larger tube. When blood flows through a series of tubes of different sizes, connected together end-to-end, its velocity is always inversely proportional to the cross-sectional area of whatever tube it happens to be in.

The circulatory system is constructed in such a way that one large artery (the aorta) branches into many, intermediate-sized arteries. These in turn branch into thousands of small arteries, **arterioles,** each of which gives rise to many capillaries. Although the individual branches of the aorta are smaller than the vessel itself, there are so many of them that the *total* cross-sectional area is greater, and the rate of flow correspondingly less. It has been estimated that the total cross-sectional area of all the capillaries of the body is about 800 times that of the aorta. Therefore, the rate of flow in the capillaries is about 1/800 as great as in the aorta. At the other end of the capillary network, the capillaries join to form small veins, **venules,** which combine to form increasingly larger veins. As this occurs, the total cross-sectional area decreases and the rate of flow increases.

Since the heart pushes blood into the arteries only during ventricular systole, arterial blood moves spasmodically, rapidly when the ventricles contract, slowly at other times. When the semilunar valves are closed, the blood in that part of the aorta nearest the heart is stationary, but the blood in the arteries farther away from the heart does not stop completely

TABLE 18–1 The Rate of Blood Flow to Various Regions of the Body Under Basal Conditions with a Total Cardiac Output of 5 Liters Per Minute*

	ml per min	*per cent of total*
Brain	700	14
Heart	200	4
Bronchi	100	2
Kidneys	1100	22
Liver	1350	27
via the portal vein	(1050)	(21)
via the hepatic artery	(300)	(6)
Skeletal muscle	750	15
Bone	250	5
Skin, during cool weather	300	6
Thyroid gland	50	1
Adrenal glands	25	0.5
Other tissues	175	3.5
Sum	5000	100.0

*(From Guyton, A. C.: *Function of the Human Body.* Philadelphia: W. B. Saunders Company, 1969. Based on data compiled by Dr. L. A. Sapirstein.)

between systoles. In the arterioles the alternation in speeds is less marked; in the capillaries the flow of blood is almost constant, and the transfer of materials can occur continuously. This conversion of the intermittent flow in the arteries to the steady flow in the capillaries is made possible by the elasticity of the arterial wall. The force of the contracting ventricles pushes the blood forward, and distends and elongates the walls of the arteries (Fig. 18–40). During diastole the stretched walls contract and squeeze the blood along. Blood is prevented from flowing backward by the closure of the semilunar valves. The contraction of the arterial wall next to the heart distends the next section of the aorta or pulmonary artery, which, in turn, contracts and distends the next section, and so on. This alternate stretching and contracting passes along the arterial wall at the rate of 7 to 8 meters per second and is known as the **pulse.** The blood inside the artery flows at a much slower rate, about 50 cm per second.

The heart is assisted in moving blood through the veins by the movements of the skeletal muscles and the motion of the body in breathing. Most of the veins are surrounded by skeletal muscles, which, when they contract, cause the veins to collapse (Fig. 18–41). As the muscles relax, the collapsed section again fills with blood, which must come from the direction of the capillaries. This mechanism by which muscle contractions "milk" blood along the veins is especially important in returning blood to the heart from the legs against the pull of gravity. If one stands upright quietly for a time, tissue fluid tends to collect in the legs, and swelling **(edema)** results. In walking, the contractions of the leg muscles force the blood along the veins, and the feet and ankles are less likely to swell. In breathing, the chest muscles and diaphragm contract, increasing the space inside the chest and lowering the pressure within the chest cavity below that outside the body, causing air to flow into the lungs. The pressure within the veins in the chest also is lowered as one inhales. Blood

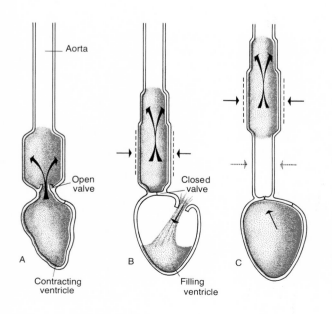

FIGURE 18–40   *Diagram of the movement of blood from the ventricle through the elastic arteries. For simplicity only one ventricle and artery are shown and the amount of stretching of the arterial wall is exaggerated. A, As the ventricle contracts, blood is forced through the semilunar valves and the adjacent wall of the aorta is stretched. B, As the ventricle relaxes and begins to fill for the next stroke, the semilunar valve closes and the expanded part of the aorta contracts, causing the next adjacent part of the aorta to expand as it is filled with blood. C, The pulse wave of expansion and contraction is transmitted to the next adjoining section of the aorta.*

Aorta

Open valve

Closed valve

A

B

C

Contracting ventricle

Filling ventricle

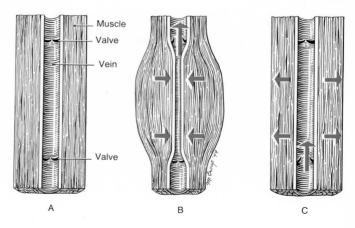

FIGURE 18–41   The action of skeletal muscles in moving blood through the veins. A, Resting condition; B, muscles contract and bulge, compressing veins and forcing blood toward the heart. The lower valve prevents backflow of blood. C, Muscles relax and the vein expands and fills with blood from below. The upper valve prevents backflow of blood.

A   B   C

moves into the chest veins and atria for the same reason that air moves into the lungs.

These two factors are important in enabling the circulatory system to respond to the increased need for blood during exercise. At that time both the "milking" action of the muscles on the veins and the breathing movements are greatly increased, sending more blood to the atria. The Law of the Heart ensures that the greater the volume of blood entering the heart, the more the heart muscle is stretched, the more forcibly the heart beats, and the greater will be the volume of blood ejected per beat. Because of this, the muscular contractions during exercise, which bring about increased requirements for food and oxygen, are in part instrumental in causing the circulatory system to satisfy the increased requirements.

The rate at which blood is delivered to each part of the body is regulated by smooth muscle fibers present in the walls of arteries and arterioles (Fig. 18–42). By contracting or relaxing they can change the diameter of the artery over a three- to fivefold range. Since the resistance to blood flow is inversely proportional to the fourth power of a vessel's diameter, this can be changed 100-fold to 1000-fold by the change in diameter.

These smooth muscles are innervated by two sets of nerves, one of which causes the muscles to contract, decreasing the size of the arterioles and lessening the supply of blood to that organ or to part of the body. An increase in the number of nerve impulses in the other set causes the muscles to relax, increasing both the size of the arterioles and the flow of blood to that organ. These muscles usually are partially contracted because of a balance between the two nerve impulses. The arterioles thus act as valves to regulate the amount of blood each organ in the body receives. The smooth muscles of the walls of the arterioles are regulated by the local concentration of carbon dioxide and by the hormone of the adrenal medulla, **epinephrine.** When the brain is metabolizing at a high rate, the increased amount of carbon dioxide acts directly on the smooth muscles of the arteries, causing them to relax, and thus increases the delivery of blood to the active tissue. In most other tissues it is the decreased amount of oxygen rather than the increased amount of carbon dioxide that brings about an increased blood flow. Epinephrine causes a relaxation of the walls of the arterioles serving the skeletal muscles, but causes a contraction of the walls of the arterioles serving the internal organs—the stomach, intestines and liver—which results in a greatly increased flow of blood to the skeletal muscles. These chemical effects are independent of the nerves, and act

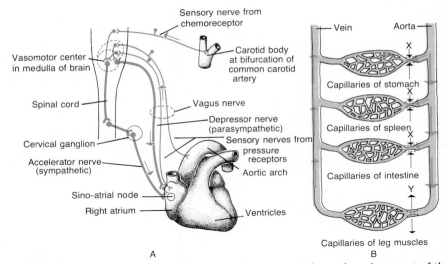

FIGURE 18–42   *Diagram of the reflexes systems which regulate the output of the heart and the circulation of blood to different parts of the body during exercise. A, The neuronal connections which enable the heart to increase cardiac output in response to exercise. B, Contraction of smooth muscles in the walls of the arterioles of the stomach, spleen and intestine (X) decreases the diameter of the arterioles and the flow of blood to those organs. Relaxation of the smooth muscles in the walls of the arterioles of the legs and other muscles (Y) increases the diameter of these vessels and the flow of blood to the skeletal muscles.*

equally well on normal arterioles and on those with severed nervous connections.

REGULATORY PROCESSES: BODY TEMPERATURE

The circulatory system has an additional function in birds and mammals which maintain a constant body temperature. Water has a high heat capacity which assists in equalizing the temperature of the whole body by transmitting heat between places of high metabolic rate where heat is released—the liver, kidneys, brain and muscles during exercise—and places where heat can be lost to the environment, the skin.

Just below the surface of the skin is an extensive **venous plexus** and the rate of flow through this can be varied as much as 100-fold by the constriction and dilation of the arteries supplying this plexus. When the flow of blood to the skin is great, a large amount of heat is transferred from the internal organs to the surface where it is lost by radiation and sweating. Certain areas of the skin have special **arteriovenous anastomoses,** short cuts, that connect the arteries and the venous plexuses. These, found especially in the hands, feet and ears, can open up and permit a large volume of blood to bypass the capillary network in the skin and go directly to the venous plexus. Such a vast increase in the blood flow to this region when it is exposed to cold brings in a large amount of heat which protects the tissues from freezing.

Nervous centers which regulate body temperature are found in the anterior part of the hypothalamus (p. 637). A group of neurons there can respond directly to temperature, giving off more impulses when the temperature of the blood rises and fewer impulses when the temperature of the blood falls (Fig. 18–43). From this temperature center impulses

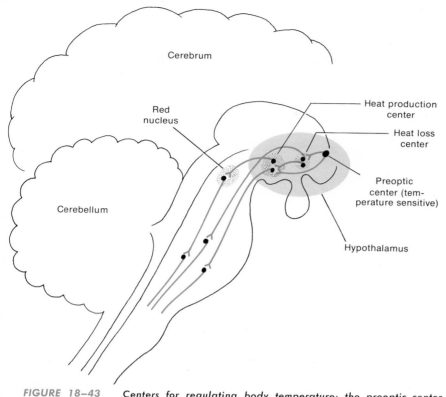

FIGURE 18–43   *Centers for regulating body temperature: the preoptic center contains neurons that respond to temperature changes; the heat loss center which causes vasodilation, sweating and lessened muscle tonus; and the heat production center which causes vasoconstriction, shivering, increased metabolism and decreased sweating. Impulses causing muscle tonus and shivering pass through the red nucleus in the pons.*

radiate to other parts of the hypothalamus and control the rates of heat production and heat loss. Two centers have been identified in the hypothalamus, an anterior one which regulates the rate of heat loss and a posterior one which regulates the rate of heat production.

When the temperature of the blood falls below normal the posterior heat-promoting center is stimulated and in turn activates several mechanisms which increase the body heat. The blood vessels of the skin are constricted, decreasing the transfer of heat to the skin to be lost by radiation. The rate of cellular metabolism is increased, thereby increasing the rate of heat production. Rapid cyclic contraction of certain muscles ("shivering") is stimulated by a complex mechanism. The small muscles located at the bases of the hairs are stimulated to contract, causing the hair to stand on end, increasing the effective insulation against the cold. When the temperature of blood reaching the hypothalamus rises above normal, the heat-losing center in the anterior part of the hypothalamus is stimulated and impulses radiate out to counteract the effects of the heat-production center. These include vasodilation, decreased metabolism and decreased muscular tone. In addition it promotes **sweating** and **panting,** two mechanisms effective in increasing heat loss.

The secretion of sweat by the sweat glands is under the control of the heat-losing center. By putting large amounts of water on the surface of the skin, the latter increases heat loss by the amount of heat required to convert liquid water to water vapor. Lower animals, but not man, respond to increased body temperature by breathing rapidly and shallowly (panting), which flushes a large volume of air into and out of the respiratory passages, increasing the evaporation of water and facilitating heat loss. Dogs and many other animals have very poorly developed sweat mechanisms and panting is their major means of keeping cool in a warm environment.

REGULATORY PROCESSES: BREATHING

During exercise oxygen consumption by muscles and other tissues may increase four or five times and the rate and depth of breathing must change automatically in response to these changing needs.

Breathing requires the coordinated contraction of a great many separate muscles, which is achieved by the **respiratory center,** a special group of cells in the medulla of the brain. From this center, volleys of nervous impulses pass out rhythmically to the diaphragm and rib muscles, resulting in their regular and coordinated contraction every four or five seconds. Under ordinary conditions the breathing movements are automatic and occur without our voluntary control. When the nerves to the diaphragm (the **phrenic** nerves) and the rib muscles are cut or destroyed, as in infantile paralysis, breathing movements stop at once. Of course, we can voluntarily change the rate and depth of breathing. We can even hold our breath for a while, but we cannot hold it very long – the automatic mechanism takes over and brings about an inspiration.

The question naturally occurs: Why does the respiratory center give off this volley of impulses periodically? Through a series of experiments it has been determined that if the connections of the respiratory center with all other parts of the brain are cut – i.e., if the sensory nerves and those from the higher brain centers are severed – the center sends out a constant stream of impulses, and the breathing muscles contract and remain contracted. The respiratory center, then, if left to its own devices, causes a complete contraction of the breathing muscles. If, however, either the sensory nerves or the nerves from the higher centers of the brain are left intact, the breathing movements continue in normal fashion. This means that for normal breathing to occur, the respiratory center must be *inhibited* periodically, so that it stops sending out impulses which cause contraction of the muscles. Further experiments revealed that the **pneumotaxic center** in the anterior part of the pons (Fig. 18–44) together with the medullary respiratory center form a "reverberating circuit" which provides for the basic control of the respiratory rate. The stretching of the walls of the alveoli during inspiration stimulates pressure-sensitive nerve cells in their walls which send impulses to the brain to inhibit the respiratory center and bring about the following expiration.

Many other nervous pathways connected with the respiratory center carry either stimulating or inhibiting impulses. Severe pain in any part of the body causes a reflex acceleration of breathing. Also, both the larynx and the pharynx have receptors in their linings which, when stimulated, send impulses to the respiratory center to inhibit breathing. These are valuable protective devices. When an irritating gas, such as ammonia or acid fumes, passes down the respiratory tract, it stimulates the receptors in the larynx, which send impulses to the respiratory center to inhibit breathing, and bring about an involuntary "catching of the breath." This prevents the harmful substance from entering the lungs. Similarly, when food acci-

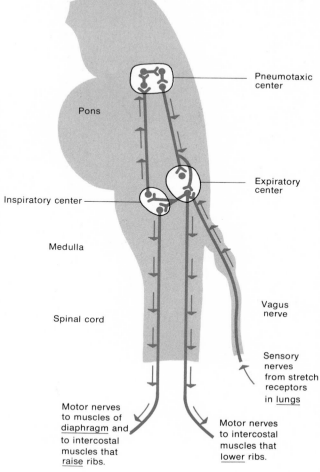

FIGURE 18–44   Diagram of the respiratory center of the brain. (1) Nerve impulses from the inspiratory center in the medulla stimulate the muscles of the diaphragm to contract and the intercostal muscles to raise the ribs. (2) Other impulses pass to the pneumotaxic center in the pons, go around its neuronal circuits and eventually (3) to the expiratory center in the medulla. The expiratory center is excited and sends impulses (4) to the intercostal muscles that lower the ribs. Other impulses (5) pass to the inspiratory center to inhibit it momentarily. When the impulses from the pneumotaxic center die out, inspiration begins again and the cycle is repeated. Sensory nerve endings in the lungs are stimulated by stretching during inspiration and send impulses via the vagus nerves (6) which stimulate the expiratory center and inhibit the inspiratory center. This reflex from the stretch receptors in the lungs provides for a second feedback mechanism to regulate the respiratory cycle.

dentally passes into the larynx, it stimulates receptors in the lining of that organ to send inhibitory impulses to the respiratory center. This momentarily stops breathing so that the food does not enter the lungs and injure the delicate lining epithelium.

During exercise both the rate and depth of breathing increase to meet the increased needs of the body for oxygen and to prevent the accumulation of carbon dioxide. The concentration of carbon dioxide in the blood is the prime factor controlling respiration. An increased concentration of carbon dioxide in the blood flowing to the brain increases the excitability of both the respiratory center and the pneumotaxic center. Increased activity of the first increases the strength of contraction of the muscles of respiration and increased activity of the second increases the rate of respiration. When the concentration of carbon dioxide returns to normal, these centers are no longer stimulated and the rate and depth of breathing return to normal.

This mechanism also works in reverse. If a person voluntarily takes a series of deep inhalations and exhalations, he reduces the carbon dioxide content of his alveolar air and blood to such a degree that when he stops breathing deeply, all breathing movements cease until the carbon dioxide in the blood again builds up to normal. The first breath of a newborn child is initiated largely by this mechanism. Immediately after a baby is born and separated from the placenta, the carbon dioxide content of its blood increases, stimulating the respiratory center to send nerve impulses to the diaphragm and rib muscles to contract in the first breath. When a newborn infant has difficulty in taking its first breath, air containing 10 per cent carbon dioxide may be blown into its lungs to set off this mechanism.

Experiments have shown that an increase in the carbon dioxide content, rather than a decrease in the oxygen content, of the blood is the factor primarily effective in stimulating the respiratory center. If a man is placed in a small, airtight chamber so that he breathes and rebreathes the same air, the oxygen in the air gradually decreases. If a chemical is placed in the chamber to absorb the carbon dioxide as fast as it is given off, so that its concentration in the lungs and blood does not increase, the man's breathing accelerates only slightly, even if the experiment is continued until the oxygen content is greatly reduced. If, however, the carbon dioxide is not absorbed, but is allowed to accumulate, breathing will be greatly accelerated, causing discomfort and a choking sensation in the subject. When he is supplied with air containing the normal amount of oxygen, but with an increased carbon dioxide content, there is again an acceleration of breathing. Obviously it is primarily the accumulation of carbon dioxide, not a deficiency of oxygen, that stimulates the respiratory center.

As additional protection against the failure of the body to respond properly to changes in the carbon dioxide and oxygen content of the blood, still another control has evolved. At the base of each internal carotid artery is a small swelling, called a **carotid sinus,** containing receptors sensitive to changes in the chemicals of the blood. If the carbon dioxide increases, or the oxygen decreases, these receptors are stimulated to send nerve impulses to the respiratory center in the medulla, increasing its activity.

The Effects of Training

The exercises and practice which an athlete performs in training increase his ability to do a certain task. First, the muscles increase in size with use and become stronger (owing to growth in the individual muscle fibers, not to an increase in the number of fibers). Second, as one performs a certain act repeatedly, he learns to coordinate muscles and to contract each one just enough and no more to produce the desired result, with consequent savings in energy expended. Third, changes occur in the circulatory and respiratory systems. The heart of a trained athlete is somewhat larger than normal and beats more slowly at rest. During exercise it pumps a greater volume of blood, more by increasing the strength of the beat than by increasing its rate. In addition, an athlete breathes more slowly and more deeply than the average person, and during exercise he increases the amount of air breathed by increasing the depth of his breathing, rather than by stepping up the rate. This is the more efficient way of doing it.

SUGGESTIONS FOR FURTHER READING

Brown, J. H. V.: *Physiology of Man in Space.* New York, Academic Press, Inc., 1963.
    A discussion of some of the physiological problems that had to be solved before man could accomplish space flight.

Comroe, J. H., Jr.: *The Lung.* Sci. Amer. *214*:57, February 1966.
    An entertaining and informative summary of pulmonary structure and function.

Fulton, John: *Selected Readings in the History of Physiology*. Springfield, Illinois, Charles C Thomas, Publisher, 1930.
A remarkable collection of 87 passages arranged to point up the landmarks in the history of physiological thought.

Harris, J. W.: *The Red Cell—Production, Metabolism, Destruction: Normal and Abnormal*. Cambridge, Massachusetts, Harvard University Press, 1963.
An excellent survey of the physiology of erythrocytes.

Perutz, M. F.: *The Hemoglobin Molecule*. Sci. Amer., November (Offprint 196) 1964.
The description of the hemoglobin molecule by the x-ray crystallographer who contributed a great deal to our understanding of its structure.

Richardson, Michael: *Translocation in Plants*. New York, St. Martin's Press, Inc., 1968.
A useful, brief review of experimental work on the movement of water in plants, with a judicious review of current theories.

Steward, F. C.: *Plants at Work*. Reading, Massachusetts, Addison-Wesley Publishing Company, Inc., 1964.
A well written exposition of many facets of plant function.

Sutcliffe, James: *Plants and Water*. New York, St. Martin's Press, Inc., 1968.
An excellent short account of the uptake and utilization of water by plants, including a review of experimental work in this area.

Waggoner, P. E., and Zelitch, I.: *Transpiration and the Stomata of Leaves*. Science, *150*:1413–1420, 1965.
An excellent discussion and summary of the mechanism of water movement in plants.

Zimmermann, M. H.: *How Sap Moves in Trees*. Sci. Amer., March (Offprint 154) 1963.
A readable, well illustrated, nontechnical article discussing the movement of water and solutes in trees.

HOMEOSTASIS AND THE REMOVAL OF WASTE PRODUCTS

CHAPTER 19

In the course of the metabolic processes by which nutrients are utilized for the production of energy and the growth and maintenance of cells, a variety of waste products are produced which must be removed. The constant synthesis and degradation of proteins, nucleic acids and other nitrogen-containing substances result in the production of ammonia, urea, uric acid and creatinine, nitrogenous wastes that are toxic and would interfere with metabolic processes unless removed. These wastes are removed from the blood and body fluids of vertebrates by **kidneys** and from the body fluids of other animals by other excretory devices. Although the removal of nitrogenous wastes is certainly an important function of the kidney, its major role is that of regulating the volume and composition of the blood and body fluids. The range of conditions which permit cells to survive and function is remarkably small. Evolution has been marked by the development of a host of mechanisms to maintain within each kind of organism a fluid environment with a composition appropriate for the special needs of its cells. This principle of the need for constancy was concisely stated in the much quoted aphorism of Claude Bernard, "La fixité du milieu intérieur est la condition de la vie libre." The capacity to regulate the internal fluid environment is essential for an organism to be able to live in a variety of environments. The term **homeostasis** was devised by Walter Cannon to refer to the tendency of organisms to maintain constant the conditions in their internal environment.

The vertebrate kidney and the excretory organs of most other animals not only eliminate nitrogenous wastes but regulate the volume of the body fluids (i.e., the water content of the body) and regulate the concentrations of salts, acids, bases and organic substances in the body fluids. By excreting certain substances and conserving others, the kidneys maintain the constant environment in the blood and body fluids required by cells for continued normal functioning.

The skin, lungs and digestive tract have excretory and regulatory functions, too. Carbon dioxide and water are eliminated by the lungs, bile pigments are excreted by the liver into the digestive tract, and certain cations such as iron and calcium are excreted through the digestive tract. The sweat glands of the skin are primarily concerned with the secretion of water for the regulation of body temperature but also excrete, more or less incidentally, some 5 to 10 per cent of the metabolic wastes such as urea. The composition of sweat and urine are qualitatively similar, but sweat is much more dilute and has only about one eighth as much solid matter as urine. The volume of perspiration in man ranges from some 500 ml. on a cool day to 2 or 3 liters on a hot one. In doing hard work in high temperature men have been found to secrete as much as 4 liters of sweat per hour!

SECTION 19–1

EXCRETION AND HOMEOSTASIS IN PLANTS

Plants, in contrast to animals, excrete only small amounts of nitrogenous wastes. Since plants neither ingest proteins (except for the few insectivorous ones) nor carry on muscular activity, which are the two largest sources of metabolic wastes in animals, the total amount of nitrogenous waste is small and can be eliminated by diffusion as ammonia through the stomata. Plants use many of the compounds that are animal wastes in the synthesis of new proteins and nucleic acids.

Plants do have some metabolic wastes and aquatic plants eliminate these by diffusion into the surrounding water. Land plants can store metabolic wastes such as salts and organic acids in leaves and eliminate them when the leaves are shed. Some herbaceous plants accumulate wastes in leaves and stems and these are lost when the top of the plant dies in the autumn. Spinach leaves, for example, accumulate crystals of oxalic acid which make up as much as 1 per cent of the mass of the leaf. Plant cells have a somewhat lesser ability than animal cells to regulate the composition of the fluids bathing them but a somewhat greater ability to withstand changes in those fluids. Plant cells can withstand, by increasing their **turgor pressure,** even marked shifts in the osmolar concentration of solutes in the surrounding fluid as long as it is less than the concentration within the cells; however, an increase in the concentration of solutes in the surrounding fluid will result in plasmolysis and death of the cell.

SECTION 19–2

EXCRETION AND HOMEOSTASIS IN PROTISTS

The removal of wastes is accomplished in most protozoa by simple diffusion through the cell membrane into the surrounding water where the concentration of the solute is lower. A major nitrogenous waste is **ammonia,** which is very toxic and inhibits many enzymatic processes but diffuses readily out of the cell before reaching a dangerous concentration. The protozoa living in fresh water have the additional problem of ridding the body of the water which constantly enters by osmosis, because the concentration of solutes in the cell is greater than in the surrounding environment. They have no firm cell wall like the algae and cannot use turgor pressure to counteract the osmotic pressure of the surrounding fluid. These forms have evolved **contractile vacuoles** which fill with fluid from the surrounding cytoplasm and then empty the fluid to the exterior of the cell. A protozoan in fresh water is like a leaky boat that must be bailed constantly to stay afloat. The contractile vacuole probably plays no significant role in ridding the cell of nitrogenous wastes, for these pass out readily by diffusion. Marine protozoa that live in an environment that is isotonic to their cell contents usually have no contractile vacuole.

EXCRETION AND HOMEOSTASIS IN THE INVERTEBRATES

Sponges and coelenterates have no specialized excretory organs and their wastes pass by diffusion from the intracellular fluid to the external environment. The vast majority of both sponges and coelenterates are marine organisms living in an isotonic environment and have no special problems of an excess of water intake. There are a very few fresh water sponges and a few fresh water coelenterates such as *Hydra* which live in a medium that is very hypotonic to their intracellular fluid. No contractile vacuoles have been observed in these animals and the means by which they prevent the inflow of water or pump it back out again remain a mystery.

The simplest animals with specialized excretory organs are the flatworms and nemerteans, which have **protonephridia** or **flame cells** (Fig. 19–1) equipped with flagella. A branching system of excretory ducts connects the protonephridia with the outside. The flame cells lie in the fluid which bathes the cells of the body, and wastes diffuse into the flame cells and from there into the excretory ducts. The beating of the flagella, which suggests a flickering flame when seen under the microscope, presumably moves fluid in the ducts out through the excretory pores. Like the con-

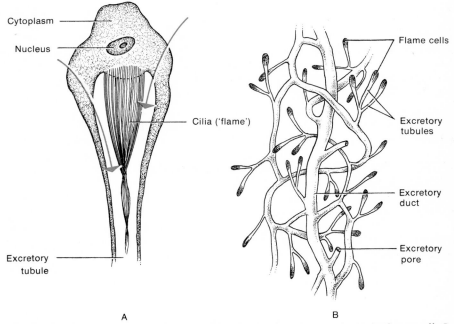

A   B

FIGURE 19–1   *The excretory organs of the flatworm. A, A single flame cell. B, flame cells are connected by excretory tubules and ducts to the excretory pore which opens to the exterior.*

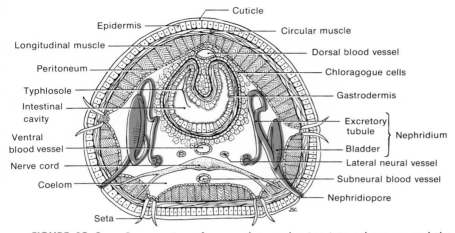

FIGURE 19–2   *Cross section of an earthworm showing internal organs and the excretory organs, the metanephridia.*

tractile vacuoles of the protozoa, the chief role of the flame cell is to regulate the water content of the animal. Metabolic wastes pass by diffusion through the skin or through the lining of the gastrovascular cavity. The number of protonephridia in a planarian is adjusted to the salinity of the environment. Planaria grown in slightly salty water develop few flame cells, but the number quickly increases if the concentration of salt in the environment is reduced.

Each segment of the body of an earthworm or marine worm contains a pair of specialized excretory organs, **metanephridia** (Fig. 19–2). The metanephridium is a tubule open at both ends, the inner end opening to the coelomic cavity of the next anterior segment by a ciliated funnel. The other end of the tubule opens to the outside of the body via an **excretory pore.** Around each tubule is a coil of capillaries which permit the removal of wastes from the blood stream. As fluid passes through the nephridium, moved by the beating of the cilia in the funnel and the contraction of the muscles in the body wall, water and substances such as glucose are reabsorbed by the capillaries and wastes are secreted from the capillaries into the fluid in the metanephridia. Thus wastes are concentrated in the urine passing out of the excretory pores. The earthworm excretes a very dilute, copius urine, at a rate of about 60 per cent of its total body weight each day. An accessory excretory method in annelids is provided by phagocytic **chloragen cells** that wander through the coelom, ingest solid particles of wastes by phagocytosis, and eventually deposit the wastes in the skin as pigment.

The excretory organs of crustacea are the **green glands,** a pair of large structures located at the base of the antennae and supplied with blood vessels. Each gland consists of a coelomic sac, a greenish glandular chamber with folded walls, and a canal which leads to a muscular bladder. Wastes from the blood pass to the coelomic sac and glandular chamber; the fluid in them is isotonic with the blood. Urine collects in the bladder and then is voided to the outside through a pore at the base of the second antenna.

The excretory organs of insects, the **malpighian tubules** (Fig. 19–3), are quite different from those of the crustaceans. They lie within the hemocoel and empty into the digestive tract. Each tubule has a muscular coat and its slow writhing assists the movement of the wastes down its lumen to the gut. The tubules are bathed in blood in the hemocoel and

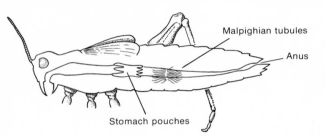

FIGURE 19-3   *Diagram of an insect, a grasshopper, showing its excretory organs, the Malpighian tubules.*

their cells transfer wastes by diffusion or active transport from the blood to the cavity of the tubule. Water is reabsorbed into the hemocoel both from the tubule and from the digestive tract. The major waste product, **uric acid,** is very sparingly soluble in water and as the water is reabsorbed the uric acid precipitates and is excreted as a dry paste. This is an adaptation which conserves body water.

SECTION 19-4

EXCRETIONS AND HOMEOSTASIS IN VERTEBRATES

The urinary systems of all the vertebrates are essentially similar. Each is composed of units called **kidney tubules** or **nephrons** which remove wastes from the blood, but the number and arrangement of the nephrons differ in different vertebrates. In the lower vertebrates the kidney tubules open into the body cavity instead of into the hollow ball of cells called **Bowman's capsule,** the arrangement found in the higher vertebrates. These **pronephric tubules** represent a type of excretory organ intermediate between the metanephridia found in annelids and the mesonephric and metanephric tubules of higher vertebrates.

Vertebrates living in or on the sea have evolved special means for coping with salt. The marine bony fishes have blood and body fluids that are hypotonic to the sea water. They tend to lose water osmotically and like Coleridge's "Ancient Mariner," are in danger of "drying up" even though surrounded by water. Some bony fish have, in the course of evolution, lost their kidney glomeruli and the resulting **aglomerular kidney** filters very little water from the blood. They compensate further by drinking sea water constantly, retaining the water and secreting the salts by specialized salt secreting glands in their gills (Fig. 19-4).

The elasmobranchs have evolved a different solution to this problem. These animals convert the ammonia from their nitrogenous wastes into **urea** and retain the urea in their blood and tissues in a concentration high enough to render these fluids slightly hypertonic to sea water. Their body fluids may take up some water osmotically and they are able to excrete a hypotonic urine. Vertebrates evolved originally in fresh water and the ancestral pronephric kidney was adapted for filtration so that it could eliminate the water that kept pouring in from the hypotonic environment. The freshwater fish seldom drink, they absorb salts by active transport across the gills and excrete a copious, dilute urine to eliminate the water taken in osmotically through the gills and lining of the mouth.

Marine turtles and sea gulls have specialized salt secreting glands in the head which can excrete the salts from the sea water they drink. The ducts from these salt glands empty either into the nasal cavity or onto the

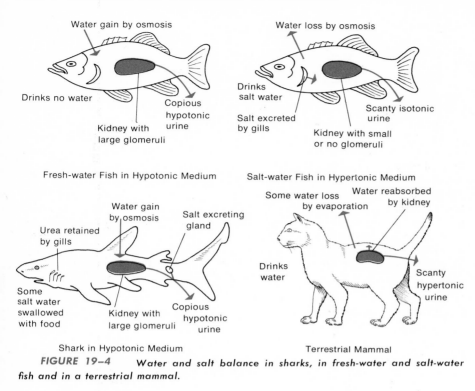

FIGURE 19–4 *Water and salt balance in sharks, in fresh-water and salt-water fish and in a terrestrial mammal.*

surface of the head. Marine mammals apparently eliminate their excess salt through their kidneys.

Amphibians have kept a primitive kind of kidney tubule with large renal corpuscles that produce a copious, dilute urine. A frog can lose through its skin and urine an amount of water equivalent to one third of its body weight in a day. Reptiles conserve water by having a dry horny skin and kidneys with small glomeruli. Less water is removed from the blood by this type of glomerulus than by the larger glomeruli of freshwater fish and amphibia. Birds and mammals have moderate-sized glomeruli and have evolved **loops of Henle** in which water is reabsorbed. This makes possible the excretion of a hypertonic urine. Mammals that live in the desert and must operate on very limited water supplies have evolved exceptionally long loops of Henle and can remove a greater fraction of water from the urine than other animals can. Toads and some reptiles can reabsorb water from the urinary bladder, but in most animals urine is not changed after it leaves the kidney.

The land vertebrates, reptiles, birds and mammals developed a third kind of kidney, a **metanephros,** with tubules that have two highly coiled regions and a long loop of Henle that extends deep into the medulla of the kidney. These long portions of the tubule function in the reabsorption of water and their ability to produce a concentrated hypertonic urine was a major factor in enabling their possessors to become efficient land animals.

The evolution of the urinary system has been complicated by the fact that in many animals the reproductive system has come to share some of its structures and several organs play a dual role. This relationship is so close that the two systems are frequently considered together as the **urogenital system.**

THE HUMAN KIDNEY AND ITS DUCTS

The human kidneys are a pair of bean-shaped structures about 10 centimeters long located on either side of the dorsal aorta and vena cava, just below the level of the stomach (Fig. 19–5). On the medial, concave side of each kidney is a funnel-shaped chamber, the **pelvis,** in which urine excreted by the kidney collects. From there it passes down the **ureter,** moved along by peristaltic waves of contraction in the ureteral walls, to the **urinary bladder.** This hollow muscular organ in the lower ventral part of the abdominal cavity distends as its muscular walls relax to make room for the accumulating urine. Valvular flaps of tissue at the openings of the ureters prevent the backflow of urine. The distension of the walls brought about by the increasing volume of urine stimulates nerve endings in the bladder walls to send impulses to the brain, producing the sensation of fullness. **Micturition,** the expulsion of urine from the bladder, requires nervous impulses to the bladder to cause its walls to contract, and impulses to the **sphincter** guarding the opening from the bladder to the urethra, causing it to relax.

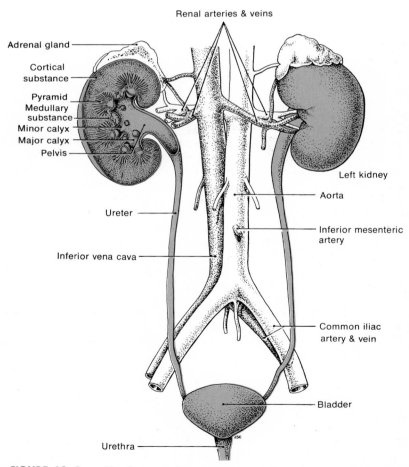

Renal arteries & veins

Adrenal gland

Cortical substance

Pyramid
Medullary substance
Minor calyx
Major calyx
Pelvis

Ureter

Inferior vena cava

Left kidney

Aorta

Inferior mesenteric artery

Common iliac artery & vein

Bladder

Urethra

FIGURE 19–5   *The human urinary system seen from the ventral side. The right kidney is shown cut open to reveal the internal structures.*

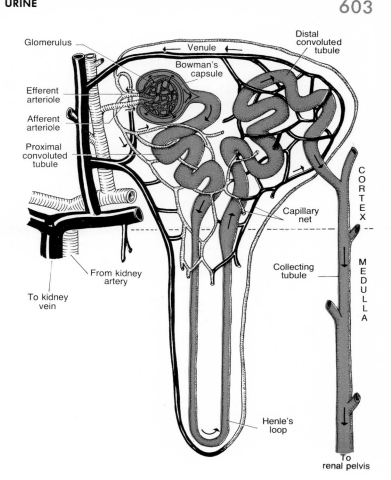

FIGURE 19–6   Diagram of a single kidney tubule and its blood vessels.

The functional unit of the mammalian kidney, the kidney tubule or **nephron** consists of a double-walled hollow sac of cells, **Bowman's capsule,** which surrounds a spherical tuft of capillaries, a **glomerulus** (Fig. 19–6), and the coiled, looped **tubules** which reabsorb into the blood some substances but not others. Branches of the renal artery ramify to all parts of the kidney; each ultimate arteriole passes to the end of one kidney tubule and supplies its glomerulus. The inner wall of Bowman's capsule consists of flat epithelial cells which adhere closely to the capillaries of the glomerulus, permitting ready diffusion of substances from the capillaries into the cavity of Bowman's capsule. Each kidney contains about 10^6 nephrons, each a separate, independent unit for excreting wastes and regulating the composition of the blood. Each filters the blood and then reabsorbs certain substances and not others as the filtrate passes through the tubule.

SECTION 19–6

THE FORMATION OF URINE

The three processes of **filtration, reabsorption** and **tubular secretion** enable the kidney to remove wastes but to conserve the useful components of the blood. Filtration occurs at the junction between the glomerular capillaries and the wall of Bowman's capsule. The blood is "filtered" as it passes through the capillaries so that water, salts, glucose, urea and all the substances in the blood, except the blood cells and the plasma proteins, pass into the cavity of Bowman's capsule to become the **glomerular filtrate** (Fig. 19–7). The total volume of blood passing through the kidneys is about 1200 ml per minute, or about one quarter of the entire cardiac output. The plasma passing through the glomerulus loses about 20 per cent of its volume as the glomerular filtrate; the rest leaves the glomerulus in the efferent arteriole. The mechanism underlying this process is the purely physical one of **pressure filtration**, and results from the fact that the arteriole entering the glomerulus, the **afferent arteriole**, is larger than the vessel leaving it, the **efferent arteriole**. The blood pressure in the glomerular capillaries is relatively high, about 70 mm Hg, and a fraction of the plasma fluid filters across into the capsule.

The pressure driving fluid out of the glomerulus and into Bowman's capsule is the pressure of the blood in the glomerular capillaries, 70 mm Hg. The pressure tending to move fluid in the reverse direction is the sum of the hydrostatic pressure in Bowman's capsule, about 14 mm Hg, and the colloid osmotic pressure of the plasma in the glomerular capillaries, about 32 mm Hg. Why is the colloid osmotic pressure greater here than in the capillaries elsewhere in the body? Thus the net force moving fluid out of the glomerulus, the **filtration pressure**, is $70 - (32 + 14)$ or 24 mm Hg. Most of the fluid filtered across the glomerular membrane is subsequently reabsorbed from the tubules into the capillaries surrounding them.

By introducing a fine glass syringe into the Bowman's capsule of a frog's kidney and collecting and analyzing some of the glomerular filtrate, A. N. Richards of the University of Pennsylvania showed that it has the same concentration of urea, salts, glucose and so forth as the plasma but lacks its proteins. The cells of Bowman's capsule are thin and unable to move materials from the capillaries; the work of pushing the filtrate from the plasma into the capsule is done by the heart. It can be shown experimentally that the rate at which fluid passes from the glomerulus into Bow-

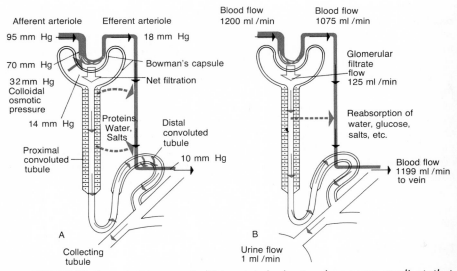

FIGURE 19–7   A, Diagram of a kidney tubule showing the pressure gradients that move fluids from blood to glomerular filtrate (filtration pressure). Substances are reabsorbed, largely in the proximal convoluted tubules, by processes of active transport. B, Diagram illustrating the total fluid movement in all the tubules of the kidneys.

man's capsule, the *glomerular filtration rate,* rises and falls with the blood pressure and consequently the filtration pressure. The normal glomerular filtration rate is about 125 ml per minute, which amounts to 180 liters per day. This is four and one half times the amount of fluid in the entire body!

The amount filtered is also regulated by the constriction or dilation of the arterioles leading to and from the glomerulus. The amount filtered is increased by the constriction of the efferent arterioles and dilation of the afferent arterioles. A rise in arterial blood pressure increases, in turn, glomerular pressure, glomerular filtration rate and the total amount of urine excreted. The increased loss of fluid from the blood decreases the blood volume and consequently the blood pressure. A fall in arterial blood pressure leads, by a comparable sequence of events, to a decrease in the amount of urine excreted. The decreased loss of fluid from the blood increases the blood volume and the blood pressure. The kidney in this way provides a mechanism by which the blood pressure is automatically regulated.

If the composition of the urine ultimately excreted were like that of the glomerular filtrate, excretion would be a wasteful process, and a great deal of water, glucose, amino acids and other useful substances would be lost; however, the concentrations of substances in the urine are quite different from those in the plasma and glomerular filtrate. From each Bowman's capsule, located in the cortex, the filtrate passes first through a *proximal convoluted tubule* (also in the cortex), then through a long *loop of Henle* passing deep into the medulla and back into the cortex, then through a *distal convoluted tubule* and empties at last into a *collecting tubule,* through which it passes again through the medulla into the pelvis (Fig. 19–6). There is no further change in the composition of the urine as it passes from the pelvis of the kidney through the ureters, bladder and urethra to be voided; the changes in composition occur when the filtrate passes from the Bowman's capsule through the long, coiled tubules and collecting tubule to the pelvis.

The walls of the kidney tubules are made of a single layer of cuboidal or flat epithelial cells. The cells making up the walls of the proximal convoluted tubules are richly endowed with mitochondria and their inner border is a *brush border* (Fig. 19–8) composed of many fine, hairlike processes which extend from the cells into the lumen of the tubule. As the filtrate passes through, these reabsorb much of the water and virtually all the glucose, amino acids and other substances needed by the body and secrete them back into the blood stream.

The efferent arteriole does not pass directly to a vein but connects with a second network of capillaries around the proximal and distal convoluted tubules (Fig. 19–6). Thus the route of blood in the kidney is unique —it passes through *two* sets of capillaries in sequence in passing from the renal artery to the renal vein. The ability of the kidney to regulate the composition of the blood depends upon this structural feature.

Substances are reabsorbed into the blood stream selectively and the rate is regulated in part by the momentary requirements of the body. The cells lining the tubules must expend energy, utilize ATP and do work to secrete these substances by the process of *active transport* back into the blood, usually against a diffusion gradient. A given amount of kidney tissue consumes more oxygen per hour than an equivalent weight of heart muscle, indicating that the kidneys work harder than the heart. The energy for this work is derived from biological oxidations within the mitochondria in their cells; when the kidney is deprived of oxygen, reabsorption, but not filtration, ceases. The substance reabsorbed in greatest amount is sodium chloride. Our kidney tubules reabsorb each day about 1200 grams of so-

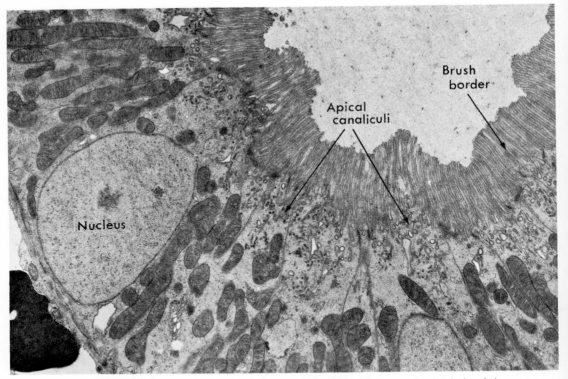

FIGURE 19-8   *An electron micrograph of the cells of the proximal tubule of the kidney showing the brush border at the lumen ×6000. (From Bloom, W., and Fawcett, D. W.: Histology. 9th ed. Philadelphia, W. B. Saunders Company, 1968.)*

dium chloride—a bit more than 2.5 pounds! Sodium ions are actively reabsorbed by a **sodium pump** (p. 488) and glucose and amino acids are reabsorbed by selective active transport mechanisms. These result in a decreased concentration of solutes in the tubular fluid and an increased concentration of solutes in the interstitial fluid surrounding the tubule. Water is reabsorbed osmotically, moved by the concentration gradient of water.

The human kidney produces about 125 liters of filtrate for every liter of urine formed; the other 124 liters of water are reabsorbed. In this way the waste products, urea, uric acid and creatinine, are greatly concentrated as the filtrate passes down the tubules. The concentration of urea in the urine is about 65 times that in the glomerular filtrate and would be even higher but for the fact that some urea is reabsorbed in the tubules. Urea, uric acid and creatinine are not actively reabsorbed by the tubules, but small amounts of them pass by diffusion from the lumen of the tubule back into the capillaries surrounding the tubules. The quantity of water reabsorbed depends on the body's current need for it and is regulated by the **antidiuretic hormone,** ADH, secreted by the posterior lobe of the pituitary (p. 655). If a large quantity of water or beer is drunk, less water is reabsorbed and a copious, dilute urine is excreted. If water intake is restricted, a maximum amount of water is reabsorbed by the cells of the tubules, conserving water, and a scanty, concentrated urine is excreted.

The cells of the kidney tubules not only remove substances from the filtrate and secrete them into the capillaries but also excrete additional wastes from the bloodstream into the filtrate by active transport mechanisms. This process of **tubular secretion** probably plays only a minor role in the function of human kidneys, but in animals like the toadfish, whose kidneys lack glomeruli and Bowman's capsules, secretion by the tubules is the only method available. When the blood pressure and consequently the filtration pressure drop below a certain level, filtration ceases, although urine is still formed by tubular secretion. Dyes injected into experimental animals can be seen to pass from the bloodstream into the urine through the cells of the tubules. Drugs such as penicillin and atabrine are removed from the blood and excreted by the process of tubular secretion. There is no doubt that tubular secretion can occur in the kidneys of man and other animals, but how large a role it normally plays in the process of excretion is unclear.

When the fluid reaches the end of the collecting tubule, and some substances have been reabsorbed and others added, the glomerular filtrate becomes *urine.*

THE REGULATION OF THE GLOMERULAR FILTRATION RATE

The amount of water, salts and sugars reabsorbed from the tubules depends to a large extent on the rate at which the glomerular filtrate passes through the tubules. If the rate is too great, important quantities of essential materials are lost in the urine because the fluid passes through the tubules before the reabsorption process is completed. If the rate is too slow, nearly everything, including urea and other waste products, is reabsorbed. There is then an optimal glomerular filtration rate which will ensure the reabsorption of water and salts but not of urea and other wastes.

The glomerular filtration rate in each nephron is regulated automatically by the concentration of certain substances in the distal convoluted tubule. The distal convoluted tubule comes to lie very close to the afferent

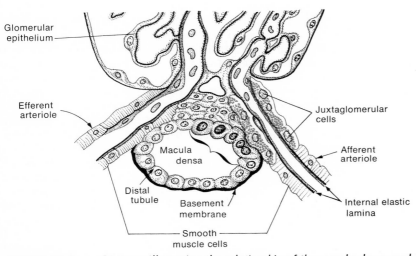

FIGURE 19–9   *Diagram illustrating the relationship of the macula densa and the juxtaglomerular cells, muscle cells in the wall of the afferent arterioles of the kidney. The cells of the macula densa monitor the concentration of substances in the glomerular filtrate and stimulate the juxtaglomerular cells to contract or relax, thereby controlling the glomerular filtration rate.*

arteriole serving the glomerulus of that tubule. At the point of contact the cells of the distal tubule become dense and increased in number, forming a structure called the **macula densa** (Fig. 19–9). The smooth muscle cells in the wall of the arteriole adjacent to the macula densa, called the **juxta-glomerular cells,** are swollen and filled with granules. By microinjection under the microscope it is possible to inject sodium chloride solutions directly into the distal convoluted tubule and show that the afferent arteriole immediately becomes constricted. Thus the composition of the fluid in the distal tubule is monitored, presumably by the cells in the macula densa, and this regulates the degree of constriction of the afferent arteriole, presumably via the juxtaglomerular cells. The glomerular filtration rate is of course controlled by the filtration pressure which depends on the degree of constriction of the walls of the afferent arteriole. In this way, each nephron, by continuously assaying the concentration of salts in the tubular fluid entering the distal tubule from the loop of Henle, regulates its glomerular filtration rate to the optimal value.

SECTION 19–8

RENAL THRESHOLDS AND RENAL CLEARANCES

Although glucose is present in the glomerular filtrate, there normally is little or none in the urine because it has been reabsorbed by the cells of the tubules. But if the concentration of glucose in the blood, and consequently in the glomerular filtrate, is very high, not all of it can be reabsorbed as the filtrate passes through the tubules and some glucose will appear in the urine. The concentration *in the blood* of a substance such as glucose at the point where it just begins to appear in the urine is termed its **"renal threshold."** The threshold for glucose is about 150 mg glucose per 100 ml blood. When this value is exceeded, glucose begins to "spill" into the urine. There are comparable "renal thresholds" for many other substances; the concentration at which the substance begins to appear in the urine is different for each.

Since a major function of the kidney is to "clear" the extracellular fluids of the body, renal physiologists have adopted the concept of **renal clearance** to express quantitatively the kidney's ability to eliminate any given substance from the blood. As a given volume of plasma passes through the glomerulus and glomerular filtrate is formed and passes through the tubules, a certain amount of the substance appears in the filtrate and some of this may be reabsorbed into the cells of the tubules. The relationship between the amount of fluid reabsorbed and the amount of the substance reabsorbed is expressed as the number of milliliters of plasma that is "cleared," completely freed of that substance, per minute. By taking simultaneous samples of blood and urine and measuring the concentration in each of the substance in question, you can calculate the quantity of that substance in each milliliter of blood and the quantity of that substance appearing in the urine each minute. Plasma clearance is defined as

$$\frac{\text{amount secreted in urine per minute}}{\text{amount in each milliliter of plasma}}$$

If the concentration of urea proved to be 0.2 mg per ml and the amount appearing in the urine is 12 mg per minute, the "clearance" is 12 mg per minute/0.2 mg per ml = 60 ml per minute. Thus, 60 ml of plasma are said to be "cleared" of urea per minute. Measures of renal clearance are commonly made to assess general kidney function.

HOW THE KIDNEY EXCRETES A CONCENTRATED URINE: THE COUNTER-CURRENT MODEL

To become successful on land, man and the other higher vertebrates had to evolve a mechanism to excrete a concentrated urine and thus conserve body water. This appears to depend on certain properties of the **loop of Henle.** The glomeruli and proximal and distal tubules are located in the outer part of the kidney, the **cortex,** whereas the loop of Henle extends deeply into the central **medulla** (Fig. 19–6). The peritubular capillaries also form long loops, **vasa recta,** that extend down into the medulla. Blood passes down into the medulla and then back up to the cortex in these vasa recta before emptying into the renal veins. The **collecting tubules,** into which the distal tubules empty, pass through the medulla and empty into the pelvis of the kidney.

This anatomic arrangement permits the kidney to excrete a urine that is *hypertonic* to the blood. As tubular fluid passes through the ascending loop of Henle, sodium is actively pumped into the interstitial fluid, chloride ions go along passively, and the concentration in the interstitial fluid increases. Some of the sodium and chloride ions diffuse passively back into the descending loop, and the cycling of sodium from ascending limb to interstitial fluid to descending limb results in the establishing of a concentration gradient of sodium and chloride in the tissue fluid surrounding the loop, with the lowest concentration near the cortex and the highest concentration deep in the medulla (Fig. 19–10).

As blood flows into the medulla in the vasa recta, sodium and chloride diffuse into it, but as the blood flows back up and out of the medulla, sodium and chloride diffuse out of the blood into the interstitial fluid. This "counter current" flow of blood prevents the loss of sodium and chloride from the medulla and permits the concentration gradient in the intersti-

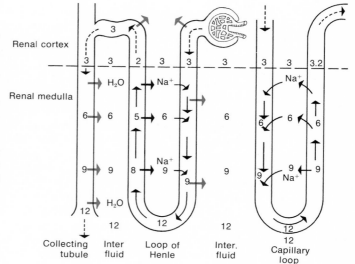

FIGURE 19–10   *Diagram of countercurrent flow mechanism of the mammalian kidney. The general direction of fluid movement is shown by dotted arrows; active sodium transport by heavy arrows; passive sodium transport by small arrows and the movement of water by tinted arrows. The numerals refer to the relative concentrations of osmotically active solutes. When two zeros are added they refer to the concentration of solutes in milliosmoles per liter.*

tial fluid to be maintained. The active transport of sodium out of the tubular fluid in the ascending loop of Henle is so powerful that the fluid reaching the distal convoluted tubule actually has a lower concentration of sodium than the fluid in the glomerular filtrate. The walls of the ascending limb of the loop of Henle appear to be impermeable to water, for water does not diffuse out as the sodium is pumped out. As the filtrate passes through the loop of Henle it loses a lot of sodium but very little water. Then the urine enters the collecting tubules and flows down through the medulla, through an ever-increasing concentration of sodium and chloride in the interstitial fluid. The walls of the collecting tubule are permeable to water and water moves by osmosis from the dilute urine in the collecting tubule to the interstitial fluid with a high concentration of solutes. The urine finally passing into the pelvis is nearly as concentrated as the interstitial fluid deep in the medulla and is quite a bit more concentrated than the initial glomerular filtrate.

The steroid hormone **aldosterone,** secreted by the adrenal cortex, acts on the cells in the ascending loop of Henle and increases the active reabsorption of sodium ions. As you might guess, the rate at which the adrenal secretes aldosterone is in turn regulated by the concentration of sodium ions in the blood.

The permeability of the walls of the collecting tubules to water can be varied and is controlled by a hormone from the posterior lobe of the pituitary, **antidiuretic hormone.** This, by a mechanism mediated by **cyclic AMP** (p. 156), greatly increases the permeability of the collecting tubules to water so that water is removed from the urine and a concentrated urine is excreted. In the absence of antidiuretic hormone, the permeability of the collecting tubules to water is very low, and the dilute fluid entering the collecting tubule from the loop of Henle passes through the collecting tubule nearly unchanged and is excreted as a very dilute urine. In **diabetes insipidus,** in which there is a deficiency of ADH, the output of urine may reach 30 to 40 liters per day instead of the normal 1.2 to 1.5 liters.

SECTION 19-10

OSMORECEPTORS

The complex counter current flow mechanism in the loop of Henle, by which the kidney can vary the amount of water reabsorbed from the tubular filtrate, is controlled by the amount of antidiuretic hormone secreted by the posterior lobe of the pituitary. This in turn is under the control of **osmoreceptors** (Fig. 19–11) in the supraoptic nuclei of the hypothalamus (p. 637). These monitor the concentration of solutes in the blood and increase or decrease the secretion of ADH to correct any change in the **osmolarity,** the total concentration of solutes, in the blood.

Osmoreceptors are specialized neurons that are believed to have small fluid chambers which swell when the concentration of solutes in the blood falls and decrease in size when the concentration of solutes in the blood rises. When contracted by the increased osmolarity of the blood they initiate impulses which pass through nerve fibers to the posterior lobe of the pituitary and cause the release of ADH (Fig. 19–11). This passes in the blood to the kidneys where it increases the permeability of the collecting tubules to water, and a greater fraction of the water passing through the collecting tubules is reabsorbed. This leads to the retention of water, while solutes continue to pass into the urine and the osmolarity of the body fluids

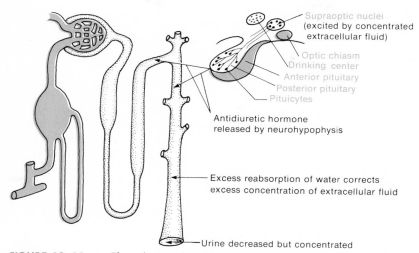

FIGURE 19-11   *The release of antidiuretic hormone in response to stimulation of osmoreceptors in the supraoptic nuclei by increased concentration of solutes in the extra-cellular fluid. The antidiuretic hormone passes to the collecting tubules of the kidney and causes increased reabsorption of water, which reduces the concentration of solutes in extracellular fluid.*

decreases. If the osmolarity of the body fluids decreases, the osmoreceptors emit fewer or no impulses, the secretion of ADH is decreased or stopped, and the kidneys excrete a dilute urine because less water is reabsorbed in the collecting tubules.

When you drink several glasses of water or beer it is absorbed from the gut in 15 or 20 minutes and dilutes the blood. The osmoreceptors detect the decreased osmolarity of the blood, decrease the secretion of ADH and the collecting tubules of the kidney become less permeable to water, less water is reabsorbed from the urine and a large quantity of dilute urine is passed from the kidneys. The rate of urine flow will rise from a normal value of about 2 ml per minute to 7 or 8 ml per minute until the osmolarity of the extracellular fluids is returned to normal.

SECTION 19-11

KEEPING THE *p*H CONSTANT

The cells of man and most animals have become adapted to surviving only within a relatively small range of hydrogen ion concentrations and several mechanisms have evolved to maintain the *p*H within these limits. The role of hemoglobin in transporting carbon dioxide and its role with the plasma proteins in serving as acid-base buffers was described in the previous chapter (p. 574). The lungs, by removing carbon dioxide from the blood as fast as it is formed, play a role in maintaining the *p*H constant, for CO_2 combines with water to form carbonic acid.

The kidneys have a primary role in regulating the hydrogen ion concentration of the intra- and extracellular fluids by excreting acidic or basic constituents when these deviate from normal and restoring the normal balance of the two. It does this by exchanging hydrogen ions derived from carbonic acid within the tubular cells for sodium ions in the tubular fluid.

The hydrogen ions pass into the urine and combine with the bicarbonate there to form carbonic acid; this dissociates to form CO_2 and water. The CO_2 is reabsorbed into the blood from the tubules and is eliminated through the lungs. The decrease in the concentration of bicarbonate ions in the tubular fluid represents a net excretion of hydrogen ions. Ordinarily the amount of hydrogen ions exchanged for sodium in the distal tubules is just equivalent to the amount of bicarbonate ions in the tubular fluid. However, if the extracellular fluids become very acidic, the quantity of bicarbonate ions in the filtrate decreases and hydrogen ions combine with other buffers in the tubular fluid such as phosphate and are excreted.

If the extracellular fluids become too alkaline, the amount of bicarbonate in the glomerular filtrate exceeds the amount of hydrogen ions secreted by the tubules. The bicarbonate that has not reacted with hydrogen ions to form carbonic acid is simply excreted as bicarbonate along with the excess sodium ions, and the loss of the sodium bicarbonate makes the body fluids more acidic, returning the acid-base balance to normal.

The kidney has an additional mechanism for coping with an excess of acids: it can secrete ammonium ions, NH_4^+, derived from the hydrolysis of glutamine in the kidney by the enzyme **glutaminase.** The NH_4^+ is exchanged for Na^+ ions in the tubular filtrate, conserving sodium ions and excreting an acid anion, such as Cl^- in combination with the NH_4^+ ions.

An individual who is fasting, or eating a high fat diet, or is suffering from diabetes mellitus has an increased content of acid **ketone bodies** in his blood and extracellular fluids. The kidney compensates for this by increasing the exchange of hydrogen ions for sodium ions in the distal convoluted tubules, conserving sodium, and by increasing the output of ammonium ions to be exchanged for additional sodium ions.

SECTION 19–12

REGULATING THE VOLUMES OF BODY FLUIDS

The volumes of the blood, extracellular fluid and intracellular fluid are all regulated by automatic feedback controls. The volume of blood in an adult man is kept within a few hundred milliliters of the normal total of 5 liters by two mechanisms, both of which depend on the fact that an increase in blood volume increases the blood pressure. One operates in the capillaries of all tissues and the other in the glomeruli of the kidneys.

An increase in blood volume raises the blood pressure and will increase the pressure in the capillaries above the normal value of about 18 mm Hg. This, the pressure that tends to move fluid out of the capillaries into the interstitial fluid, becomes greater than the colloid osmotic pressure, the force that tends to move fluid into the capillaries from the interstitial fluid. The result is a net movement of fluid from the capillaries into tissue spaces until the blood pressure and blood volume are returned to normal. A decrease in blood volume and hence of blood pressure has the opposite effect, and fluid tends to move from the tissue spaces into the capillaries to restore normal pressures and volumes.

An increased blood volume increases the pressure within the glomeruli of the kidney, filtration pressure is increased and the volume of urine formed increases. This is reinforced by another mechanism that is based on the presence of **baroreceptors** in the walls of the arteries in the chest and neck (Fig. 19–12). These are sense organs that detect the degree of stretching of the walls of the arteries. An increase in the blood pressure stretches the walls of the arteries, and stimulates the baroreceptors to send impulses

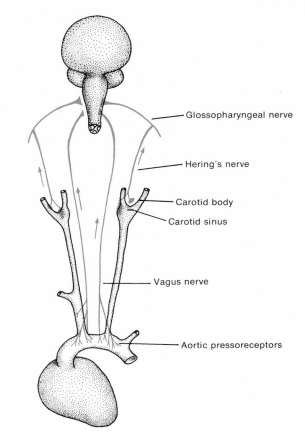

Glossopharyngeal nerve

Hering's nerve

Carotid body

Carotid sinus

Vagus nerve

Aortic pressoreceptors

FIGURE 19–12   Diagram of the barore-ceptors involved in regulating pressures within the circulatory system.

to the vasomotor center in the medulla of the brain. The **vasomotor center** initiates impulses to the smooth muscles in the walls of the renal afferent arterioles, causing them to relax, allowing the arterioles to dilate and increase the flow of blood into the glomeruli.

An increase or decrease in the volume of interstitial fluid will lead to a corresponding change in its pressure and will alter the pressure relations between capillaries and interstitial fluid. The interstitial fluid normally has a slight negative pressure (p. 568) which is one of the forces moving fluid out of the capillaries. An increase in volume will change this pressure toward zero, or make it positive, moving fluid into the blood. The increased blood volume in turn would increase the output of urine and the excess interstitial fluid would be removed from the body via the blood into the urine.

Each cell maintains its own fluid volume, largely by transporting in certain electrolytes and excluding others; it maintains an osmotic equilibrium with the extracellular fluid. The kidneys do not directly regulate intracellular fluid volume but do this indirectly, by maintaining constant the volume and constituents of the extracellular fluids.

One further center that plays a role in regulating fluid volumes is the **drinking center,** located in the hypothalamus, immediately posterior to the osmoreceptors in the supraoptic nuclei (Fig. 19–11) Stimulation of this center, when the body fluids are too concentrated, leads to the sensation of thirst and the individual is stimulated to find and drink water, which will dilute the body fluids and turn off the stimulation of the drinking center.

SUGGESTIONS FOR FURTHER READING

Christensen, H. N.: *Body Fluids and Acid-Base Balance*. Philadelphia, W. B. Saunders Company, 1964.
> A helpful discussion of the biochemical mechanisms involved in maintaining the constancy of the compositions of body fluids.

Pitts, R. F.: *Physiology of the Kidney and Body Fluids*. Chicago, Year Book Medical Publishers, 1968.
> An up-to-date reference book for students interested in kidney function.

Smith, H. W.: *The Kidney*. Sci. Amer. *188*:40, January 1953.
> An excellent discussion of the role of the kidney in maintaining a constant salt and water content of the body.

Smith, H. W.: *Principles of Renal Physiology*. New York, Oxford University Press, 1956.
> Smith's definitive account of his lifelong study of kidney function.

Wolf, A. V., and Crowder, N. A.: *Introduction to Body Fluid Metabolism*. Baltimore, The Williams and Wilkins Company, 1964.
> An advanced account of the function of the kidney and other organs in regulating body fluids.

BEHAVIOR AND ITS
BIOLOGICAL BASIS

CONTROL SYSTEMS: NEURAL INTEGRATION

CHAPTER 20

Orderly and efficient functioning of a complex multicellular organism is dependent upon the different parts operating in concert. To this end there must be means for monitoring the activities of the different parts, whether they be cells, tissues or organs, and for providing a flow of information among parts or between parts and centers that collate, integrate, store and retrieve information and issue appropriate commands. Two systems work in intimate association in providing these functions. The **endocrine system** controls relatively slow and long-lasting events. The **nervous system** is concerned with rapid events of the order of millisecond duration.

On gross inspection the nervous system appears as a network of semi-transparent to whitish or grayish cords of various diameters emanating from a central whitish-gray mass. As recently as the beginning of this century, most biologists believed that the system consisted of a continuous network of cells in which there was uninterrupted continuity of material. This **reticular theory** was finally disproved by the great Spanish histologist Ramón y Cajal who demonstrated beyond doubt that the nervous system was constructed of discrete cells, the **neurons**. It is interesting that he proved this point by employing a silver staining technique that had been perfected by a staunch supporter of the reticular theory, the Italian, Golgi. Both men received the Nobel Prize in 1906 for their studies of the nervous system.

THE NEURON

The structural and functional unit of the nervous system of all multicellular animals is the **neuron** (Fig. 20–1). The average neuron is slightly less than 0.1 millimeter in diameter. Traditionally the neuron is pictured as having three parts: a long **axon** emerging from one end of the cell body

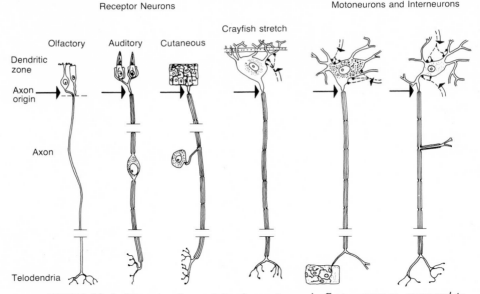

FIGURE 20–1 *Diagram of a variety of receptor and effector neurons, arranged to illustrate the idea that impulse origin, rather than cell-body position, is the most reasonable focal point for the analysis of neuron structure in functional terms. Thus, the impulse conductor or axon may arise from any response generator structure, whether transducing receptor terminals or synapse-bearing surfaces (dendrites, cell body surface, or axon hillock). The interior of the cell body (chromidial neuroplasm, or perikaryon) is conceived of as related primarily to the outgrowth of axon and dendrites and to metabolic functions other than membrane activity. Thus, the position of the perikaryon in the neuron is not critical with respect to the "neural aspects" of neuron function—namely, response generation, conduction and synpatic transmission. Except for the stretch neuron of the crayfish, the neurons shown are those of vertebrates. (After Bodian, D.: The Generalized Vertebrate Neuron. Science, 137:323–326, 3 August 1962. Copyright 1962 by the American Association for the Advancement of Science.)*

and bushy **dendrites** emerging from the other; however, there are so many exceptions that the usefulness of this anatomical subdivision is limited. Functional distinctions are more accurate; thus there are three functional parts. The dendrites constitute that part of the neuron specialized for receiving excitation, whether it be from environmental stimuli or from another cell. The axon is that part specialized to distribute or conduct excitation away from the dendritic zone. It is generally long and smooth but may give off an occasional collateral. Inside the central nervous system it is surrounded by nonnervous cells called **neuroglia**. Outside the central nervous system it is wrapped in **Schwann cells**. It ends in a distribution or emissive apparatus, the **telodendria**. The cell body is concerned with metabolic maintenance and growth and may be situated anywhere with respect to the other parts. In the vertebrate sensory nerves, from the skin, for example, it is situated on an offshoot of the axon. The variety in structure of neurons is enormous (Figs. 20–2 and 20–3). Different functional and anatomical types of neurons characterize different parts of the nervous system.

It is the axon that is responsible for the tremendous length of some neurons. The axon of a **sensory** cell barely 0.1 millimeter in diameter located in the toe of a giraffe traverses a distance of many feet before ending in the spinal cord. It is the bundling together of many axons that makes the nerves and nerve trunks observed in gross anatomical dissection. A common connective tissue sheath surrounds the nerves.

Neurons are classified as **sensory, motor** or **interneurons** (connector

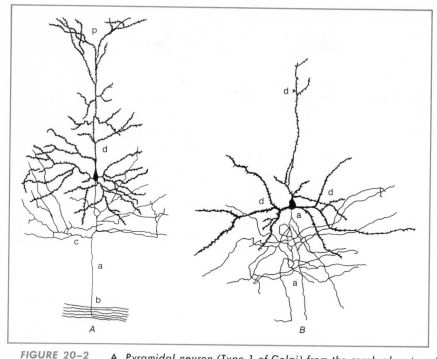

FIGURE 20–2   A, Pyramidal neuron (Type 1 of Golgi) from the cerebral cortex of a rabbit. The axon gives off numerous collateral branches close to the cell body and then enters the white substance within which it extends for a long distance. Only a small part of the axon is included in the drawing: a, Axon; b, white substance; c, collateral branches of axon; d, ascending or apical dendrite; p, its terminal branches at the outer surface of the brain. (After Ramón y Cajal.) B, Neuron of Type II from the cerebral cortex of a cat. The entire neuron is included in the drawing: a, Axon whose branches terminate close to the cell body; d, dendrites. (After Kolliker; from Maximow, A. A., and Bloom, W.: A Textbook of Histology. Philadelphia, W. B. Saunders Company, 1941.)

neurons) on the basis of their functions. Sensory neurons or **afferent neurons** are either receptors (olfactory receptors) or connect to receptors (taste receptors) and conduct information to the central nervous system. Motor neurons **(efferent neurons)** conduct information away from the central nervous system to the **effectors** (muscles, glands, electric organs, light organs). **Interneurons** are those that connect two or more neurons. They usually lie wholly within the central nervous system, whereas the sensory and motor neurons have one of their endings in the central nervous system and the other close to the external environment or to the animal's internal environment.

In contrast to axons and dendrites that are scattered and extended throughout the body, the cell bodies of neurons are usually grouped together in masses called **ganglia**. In the simplest sense a ganglion is any aggregation of neural cell bodies. Examples of ganglia are the **dorsal root ganglia** of vertebrates (Fig. 20–4) which are merely collections of cell bodies of sensory neurons, and the **autonomic ganglia** of vertebrates which are groups of cell bodies of motor neurons. More commonly a ganglion is an aggregation of neural cell bodies plus interneurons. It is a place where different neurons connect with one another and where much integration may occur. The brains of animals are fusions of many ganglia.

In the central nervous systems of all animals the cellular part of the neurons and the fibrous part are separated into two zones. In vertebrates the **gray matter** (usually inside, but also outside in higher brain centers)

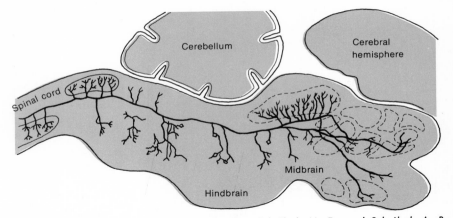

FIGURE 20–3   *A central neuron. (After Scheibel, M. E., and Scheibel, A. B.: Structural Substrates for Integrative Patterns in the Brain Stem Reticular Core. In Jasper, M. M., et al. (eds.): Reticular Formation of the Brain. Boston, Little, Brown and Company, 1958; by permission of the publisher.)*

contains cell bodies plus axons and dendrites. The **white matter** consists exclusively of axons. In invertebrates the outside or rind consists solely of cell bodies while the inside (core) consists of fibers.

A neuron consists of the usual cellular components: a nucleus, cytoplasm that extends to the outermost branches, and a cell membrane enclosing all. Enveloping the axon that is outside the central nervous system is a cellular sheath, the **neurilemma,** composed of the Schwann cells. These cells, migrating from **mesenchyme,** line up along axons and wrap around

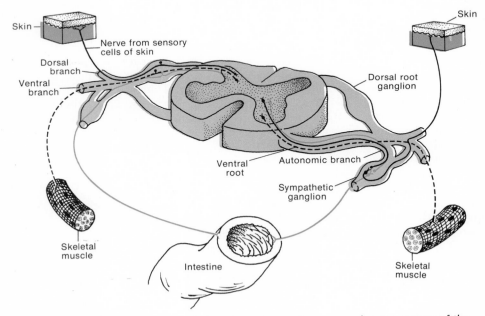

FIGURE 20–4   *Diagram of the primary types of sensory and motor neurons of the spinal nerves, and their connections with the spinal cord. For convenience, the sensory neurons are shown on the left and the motor neurons on the right, though both kinds are found on each side of the body.*

them. On some axons the Schwann cell lays down within its folds a spiral wrapping of insulating fatty material called **myelin** (Fig. 20–5). Between adjacent cells there are gaps. At these gaps, or **nodes**, the axon is free of myelin. Axons lying within the brain and spinal cord have myelin provided by satellite cells **(oligodendrocytes)** rather than by Schwann cells. Nerves consisting of heavily myelinated fibers (e.g., those in the brain, spinal cord; those to skin and skeletal muscle) are white in appearance; those with little or no myelin are gray.

The roles of the neurilemma and myelin are not completely understood; however, there is clearly an interdependence between the neuron and its sheath cells. When an axon is separated from its cell body by a cut it soon degenerates. A hollow tube of Schwann cells remains, but myelin eventually disappears. As long as the cell body of the neuron has not been injured it is capable of regenerating a new axon. Sprouting begins within a few days following cutting (Fig. 20–6). The growing axon enters the old sheath tube and proceeds along it to its final destination in the central nervous system or periphery. Axons can grow in the absence of sheaths if some conduit is provided for them. They can, for example, be made to grow within sections of blood vessels or extremely fine plastic tubes. The length of time required for regeneration depends on how far the nerve has to grow and may require as long as two years. When cuts occur within the spinal cord or brain, regeneration is very feeble or usually totally absent.

It is a remarkable fact that each regenerating axon of a cut nerve finds its way back to its former point of termination, whether this be a specific connection in the central nervous system or a specific muscle or sense organ in the periphery. If, during the early stages of development of an amphibian, one transplants an extra limb bud next to the normally developing limb, both will grow to maturity. The extra limb then moves synchronously with the normal one. Anatomical examination reveals that

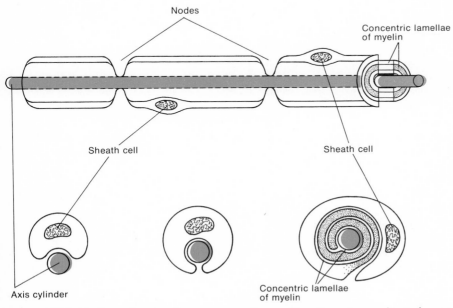

FIGURE 20–5   Sheath cells on neuron. Upper: Dissection of a myelinated nerve fiber. Lower: Envelopment of axis cylinder by a sheath cell. (After B. B. Geren, 1954; William W. Ballard: Comparative Anatomy and Embryology. Copyright © 1964. The Ronald Press Company, New York.)

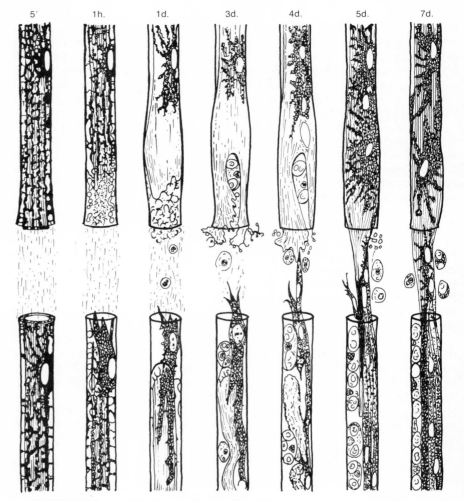

FIGURE 20–6   Distal part of nerve fiber five minutes after cutting shown at the left. Retraction of the central end of the cut axon one hour later. Budding at third and fourth day. Slender axon grows into the distal degenerated stumps at days 5 and 7. (Courtesy of J. Z. Young; from Young, J. F.: Factors Influencing the Regeneration of Nerves. Advances Surg. 1:165, 1949.)

the nerve that innervates the normal limb sent out branches to the extra one. Clearly the extra limb exerted some stimulating influence on the growing nerves to produce more branches and some directive influence as well.

Similarly, in the optic system after the optic nerve is cut, the axons of retinal cells regenerate and make proper connections within the central nervous system. Behavioral observations confirm this. When a frog with a regenerated optic nerve sees a fly, it strikes at it accurately (Fig. 20–7, *Top A*). This visual-motor performance indicates that there is topological correspondence between points of the retinal area and points in the visual area of the CNS. When an excised eye is rotated 180 degrees before re-implantation, the frog strikes 180 degrees off target. This aberration is explained by the fact that the retinal cells still regenerated to their proper destinations in the CNS but the retina itself is upside down with respect to the CNS (Fig. 20–7, *Botton A*).

The principal known function of the myelin sheath is to provide for

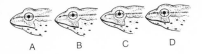

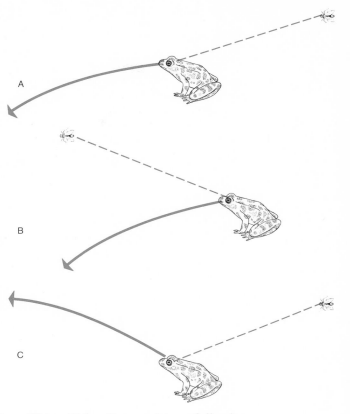

FIGURE 20–7 Top: Optic nerve cut and eye rotated. A, Normal; B, 180° rotation, C, dorsoventral inversion; D, nasotemporal inversion. Bottom: Eye rotated 180° and frog strikes at a point diametrically opposite to the position of the fly used as a lure (A). Dorsoventral inversion of the eye and frog strikes inversely in the dorsoventral direction (A). Nasotemporal inversion of the eye and frog strikes inversely with respect to the nasotemporal direction (C). (From Sperry, R. W.: Mechanisms of Neural Maturation. In Stevens, S. S. (ed.). Handbook of Experimental Psychology. New York, John Wiley & Sons, Inc., 1951.)

a special kind of nerve conduction. This will be discussed in a following section. It may have other functions as well. The neuron clearly has an effect on it. When the distant nerve cell is cut off from its axon, the myelin in the axon begins degenerating within a few minutes. The neuron obviously has some trophic function that is necessary for the well-being of the Schwann cell. As a matter of fact it is now realized that the nervous system has, in addition to its role of transmitting impulses, an important trophic relation with all the organs it innervates. The presence of nerves is essential to the regeneration of amputated amphibian limbs, the normal maintenance of taste buds and the continued functional integrity of muscles. Something other than impulses is obviously transported along the long cellular extensions of the neurons, and it has been demonstrated that there is active flow of cytoplasm away from the cell body (Fig. 20–8).

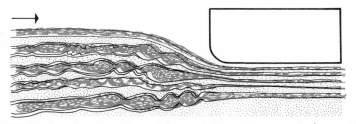

FIGURE 20–8 A region of partial constriction of the nerve is shown at the right by the white block, with the fibers proximal to the constricted region at the left. After several weeks of damming, the fibers proximal to the dam are swollen, tortuous, convoluted, and beaded. The fibers in the constricted region are thinner than normal. (Courtesy of P. Weiss; from Weiss, P., and Hiscoe, M. B.: Experiments on the Mechanism of Nerve Growth. J. Exp. Zool. 107:315, 1948.)

SECTION 20–2

SYNAPSES

Since the nervous system is composed of discontinuous units, the neurons, but behaves like a continuous transmission system, there are obviously functional connections between neurons. These functional junctions were termed **synapses** by Sherrington. A **synapse** is a region where one cell (the **presynaptic**) comes into contact or near contact with, and influences, another cell (the **postsynaptic**) (Fig. 20–9). When there is a space it seldom exceeds 500 Å in width (Fig. 20–10). The synaptic connection involves only limited areas of the neurons involved. In the vertebrate nervous system many synapses are between the telodendria of axons and the cell body of the postsynaptic neuron. In invertebrates the majority of synapses are between axon telodendria and dendrite arborizations. Most synapses transmit excitation in one direction only, i.e., from the pre- to the postsynaptic cell. Characteristically the synaptic region of the presynaptic axon is packed with small rounded bodies **(synaptic vesicles)** 100 to 200 Å in diameter in vertebrates (Fig. 20–11). The vesicles are believed to contain a specific chemical which is released when the axon is excited and which transmits the excitation to the postsynaptic cell (see p. 632).

The variety and complexity of synapses is almost beyond description (Fig. 20–12). The different kinds and their geometric arrangement obviously have functional meaning, although our knowledge of this is still rudimentary.

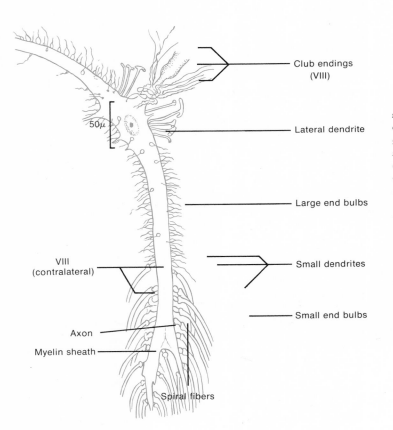

FIGURE 20–9 *Variety of synapses on one neuron. Part of the Mauthner's neuron of the goldfish, drawn from protargol stained sections. Club end. (VIII), club endings of ipsilateral vestibular nerve fibers; lat. dend., main lateral dendrite; myelin sh., the first internode of the myelin sheath on the axon; spiral f., spiral fibers in the region of the axon cap; VIII (contralat.), contralateral vestibular nerve fibers giving rise to collaterals that terminate as small club endings. (From Bodian, D.: Introductory Survey of Neurons. Sympos. Quant. Biol. 17:1–13, 1952.)*

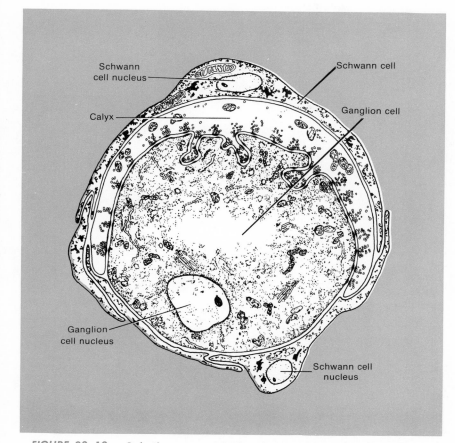

FIGURE 20–10   Calyciform synapse in the ciliary ganglion of the chick, as revealed by the electron microscope. Note the locally dense regions of the opposed synaptic membranes, the clusters of synaptic vesicles at these sites on the presynaptic side, the uniform cleft width (300 to 400 Å). (From DeLorenzo, A. J.: The Fine Structure of Synapses in the Ciliary Ganglion of the Chick. J. Biophys. Biochem. Cyt. 7:31–36, 1960.)

FIGURE 20–11   A common form of synapse in the mammalian brain, according to de Robertis (1962). The axonal (presynaptic) side above; the dendritic (postsynaptic) side below. (From Ultrastructure and chemical organization of synapses in the central nervous system. In Brazier, M. A. B. (ed.): Brain and Behavior. Amer. Ass. Adv. Sci., 1962.)

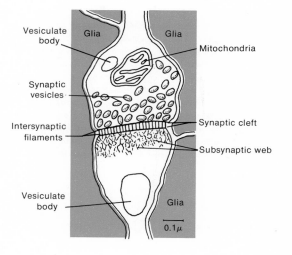

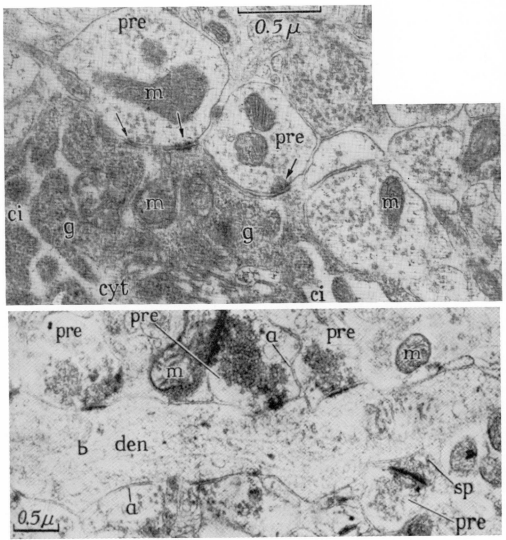

FIGURE 20–12   *Two types of synapse in the cerebral cortex, as seen by electron microscopy. Visual cortex of the rat. A, A group of synaptic contacts of Type 1 (arrows) on a small dendrite (den.). This type exhibits increased thickness and density of the apposed membranes, especially of the postsynaptic side, over a large percentage of the length of close contact; the thickened regions lie farther apart than unthickened apposed surfaces; an intermediate band of material can be seen at high magnification in the cleft. B, Type 2 synapses (arrow) on a nerve cell body. The cytoplasm of the postsynaptic cell body (post) contains characteristic granules (g) and cisternae (ci) of the endoplasmic reticulum. This type of synapse has a small percentage of the length of apposed membranes thickened, both surfaces are equally thickened, the spacing is little different from adjacent regions, and an intermediate band is not clearly visible. m, mitochondria; pre, presynaptic ending with cluster of vesicles; sp. spine of the small dendrite with a spine synapse. (From Gray, E. G.: Axosomatic and axo-dendritic synapses of the cerebral cortex: An electron microscope study. J. Anat.*

SECTION 20–3

THE NERVE IMPULSE

In the last half of the eighteenth century, the Italian Galvani discovered that a frog's leg would contract if the exposed sciatic nerve touched the muscle. Although he did not realize it at the time, Galvani had demonstrated that nerves are a source of electricity and that this electricity causes muscles to contract. In the years that followed, the idea grew that the nerve impulse was an electric current, that nerves transmitted like wires. Helmholz showed, however, that conduction was too slow to be electric. He measured the speed of impulse conduction in a nerve muscle preparation by stimulating the nerve at different distances from the muscle and recording the elapsed time between stimulation and contraction. In frog nerves the impulse traveled about 30 meters per second. Furthermore, dead or crushed nerves still conduct electricity but not nerve impulses. We now know that the nerve impulse is not an electric current but a propagated **electrochemical reaction.**

Let us consider the electrical aspects first. In all cells there is an electrical potential difference between the inside and the outside. This can be measured by placing an electrode, insulated except at the tip, *inside* the cell, another on the *outside* surface, and a suitable recording instrument **(galvanometer** or **oscilloscope)** connected between the two. The potential difference in a neuron is about 60 millivolts (the inside is negative with respect to the outside). This potential difference is called the **resting potential.** If the two electrodes are placed on the *outside* surface of the neuron, they do not register any potential difference, because all points on the outside are at equal potential.

If a neuron is stimulated by any means whatsoever (electrically, by touch, by injury and so on), the resting potential changes. It may decrease, increase or drop to zero. When the neuron is unstimulated it is said to be **polarized;** at stimulation it becomes **depolarized.** When the neuron is fully depolarized, a recording instrument would detect no difference between the *inside* and the *outside.* If, however, one electrode was placed on the outside of the neuron at the point of stimulation and a second electrode placed outside at a point that was not stimulated, a recording instrument would show that the first electrode was negative with respect to the second (Fig. 20–13). This is so because the second electrode is on a portion of the neuron that is still polarized (the outside is positive with respect to the inside), while the first electrode is on a spot where the inside and the outside of the neuron are more nearly electrically equal, thus negative *with respect* to the *unstimulated* area. This **local state of depolarization** is the starting point of the **nerve impulse.** It is a local potential and a slow one.

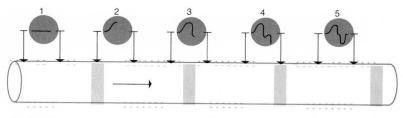

| Resting state | Impulse reaches first electrode | Impulse passes first electrode | Impulse reaches second electrode | Impulse passes second electrode |

FIGURE 20–13   *Diagram to illustrate the electrical changes in a nerve fiber as an action potential passes along, in this case toward the right. For actual recording only one oscilloscope is employed. Five oscilloscope faces are drawn here to illustrate the successive changes that occur when the impulse travels past the two externally placed electrodes. 1, Resting state, both electrodes recording same potential, no deflection of line. 2, Impulse reaches first electrode, depolarization at this point, this electrode negative with respect to other. 3, Impulse passes, area under first electrode repolarized, second electrode not yet depolarized, line on oscilloscope returns to zero since both electrodes are at equal potential. 4, Impulse reaches second electrode, this electrode now negative with respect to first since current is flowing in opposite direction. 5, Areas under both electrodes now repolarized so line returns to zero. This is a double or diphasic action potential.*

If it is not of critical magnitude, restorative processes in the neuron, working toward repolarization, cause the potential differences to disappear slowly, and nothing is transmitted along the nerve.

If, on the other hand, the depolarization attains a critical magnitude within a short enough time, it stimulates the axon on either side of itself. As a consequence of this local stimulation current flows from the two inactive loci to the point of depolarization. The two newly depolarized areas now stimulate inactive areas adjacent to them and so on in a chain reaction. Thus, a self-generating condition of depolarization passes along the neuron.

This propagated depolarization is the **action potential.** Unlike the **generator potential,** which is slow and graded, the action potential is fast (about 2 milliseconds in duration) and **all-or-none.** It "fires" at maximum voltage or not at all. After it has occurred, a finite recovery time must elapse before a second one can be generated. This interval, the **absolute refractory period,** lasts from one half to two milliseconds. Under most circumstances the action potential is repetitive, i.e., whole trains of impulses are generated and transmitted along a nerve. Within a neuron the action potential can travel in either direction from point of origin, but the normal direction of travel is from the dendrite toward the cell body and away from the cell body in the axon.

The electrical and chemical processes involved in the transmission of a nerve impulse are similar in many ways to those involved in muscle contraction. Compared to a contracting muscle, however, a transmitting nerve expends little energy; the heat produced by 1 gram of nerve, stimulated for one minute is equivalent to the energy liberated by the oxidation of 0.000001 gm of glycogen. This means that if a nerve contained only 1 per cent of glycogen to serve as fuel, it could be stimulated continuously for a week without exhausting the supply. Nerve fibers are practically incapable of being fatigued as long as an adequate supply of oxygen is available. Whatever "mental fatigue" may be, it is not a real fatigue of the nerve fibers.

SECTION 20–4

THE MEMBRANE THEORY OF NERVE CONDUCTION

The **membrane theory of nerve conduction** states that the electrical events in the nerve fiber are governed by the differential permeability of the neuron membrane to **sodium** and **potassium ions** and that these permeabilities in turn are regulated by the electrical field across the surface.

The neuron is essentially a long cylindrical tube whose surface membrane separates two solutions of different chemical composition though they have the same total number of ions. In the external medium sodium and chloride ions predominate, whereas within the cell potassium and various organic ions predominate. In the giant axon of the squid, on which most experiments have been done and which is the prototype for all axons, sodium is nearly 10 times more concentrated outside the membrane than inside, chloride is five to 10 times more concentrated outside than in, and potassium is about 40 times more concentrated inside than out. The membrane, which is about 50 A thick, has high electrical resistance, low selective ionic permeability, and high electrical capacity. Potassium and chloride diffuse relatively freely across this membrane, but permeability to sodium is low. Potassium tends to leak out of the neuron and sodium tends to leak in. Because of the selective permeability of the membrane, potassium tends to leak out faster than sodium leaks in and this, plus the fact that the negative organic ions inside cannot get out, causes increasing negativity on the inside. As the inside becomes more negative it impedes the exit of potassium. Ionic conditions would eventually change and come to a new equilibrium if something were not done to counteract leakage. A steady state is maintained by what has been called the **sodium pump.** Sodium ions are actively transported from the inside to the outside against a concentration and an electrochemical gradient at the expense of energy derived from the usual metabolic processes involving **ATP** within the nerve cell. The sodium pump can be turned off experimentally by poisoning with such metabolic inhibitors as cyanide.

The extrusion of sodium ions is accompanied by the entrance of potassium ions and, although the details of the process are unknown, there is probably an exchange of cations at the cell surface, with a potassium ion entering for each sodium ion extruded. The relatively low ionic permeability of the membrane is such that even when the pump is poisoned many hours elapse before the concentration gradients of sodium and potassium ions across the membrane disappear. The resting potential of the unstimulated nerve arises as a consequence of this ionic difference and is maintained metabolically.

The electrochemical potential energy of the resting potential forms the basis for the generation and transmission of the action potential (Fig. 20–14). The permeability of the membrane depends upon the magnitude of the transmembrane electrical potential. Excitation of a nerve involves

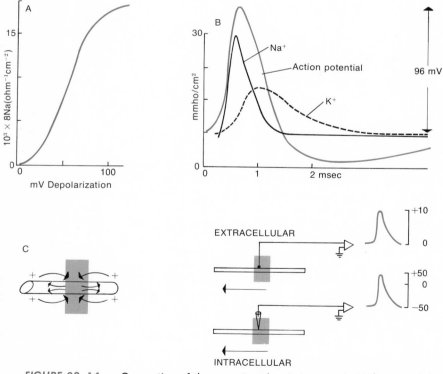

FIGURE 20–14   Generation of the nerve impulse. A, Increase in Na$^+$ permeability as membrane depolarizes. B, Na$^+$ and K$^+$ movement across membrane during action potential. C, Diagram illustrating flow of electrotonic current in vicinity of action potential (colored), and examples of extracellular and intracellular recording of action potentials. With a penetrating microelectrode the entire action potential is recorded; whereas with an extracellular electrode only the positive overshoot is defected (A and B after Katz, B.: The Croonian Lecture: The Transmission of Impulses from Nerve to Muscle, and the Subcellular Unit of Synaptic Action. Proc. Roy. Soc. Biol. 155:455, 1962.)

a momentary change in the permeability of the membrane to sodium. As sodium enters, the membrane depolarizes. As the membrane depolarizes, more sodium enters. This process becomes progressive; it is self-reinforcing. The entering sodium drops the potential to zero and beyond by about + 50 millivolts (Fig. 20–14). After 1 to 2 milliseconds, permeability to sodium decreases and potassium begins to move out. This movement leads to the restoration of the resting potential, i.e., to repolarization. When repolarization is complete, the permeability to potassium has become normal. Excess sodium that entered during the process is now slowly removed by the sodium pump. All the other events described require only a few milliseconds. During the process, both sodium and potassium ions are moving down their respective concentration gradients. The actual quantities of ions involved are so small that there is no detectable change in the concentration of either ion in the fiber during one impulse. The giant axon of the squid could conduct several hundred thousand impulses even if its sodium pump was turned off, before its store of accumulated ions would be exhausted.

Propagation of the action potential depends upon **electrotonic currents** flowing ahead of the nerve impulse. The process may be likened to the charging and discharging of a condenser. At any point where an action potential has been generated, current flows in the directions indicated in Figure 20–15. Notice that the circuit is completed by current flowing in the solution that bathes the nerve. If the nerve is immersed in mineral oil, only a thin film of conductive fluid bathes the nerve, and the external resistance (Re) is greatly increased. When this is done, the rate of conduction of impulses is decreased. The effectiveness of the flowing current in propagating the impulse depends on the magnitude of the current and the resistances of the neuron membrane, the cytoplasm and the surrounding medium. These factors determine how far away from the active site the membrane permeability will be sufficiently increased to start regenerative sodium entry. The concept that the action potential is propagated by electrotonic currents moving ahead of it is termed the **local circuit theory of propagation.**

The **rate of conduction** of impulses increases as the diameter of the axon increases because the internal resistance (R) decreases. Thus, large nerve fibers conduct faster than small ones. Sporadically throughout the animal kingdom (higher vertebrates excepted), **giant nerve fibers** have evolved. The giant fibers of the squid have already been mentioned. These fibers conduct very rapidly. They generally serve the purpose of conducting danger signals. In these cases, speed rather than detailed information is critical.

High rates of conduction are achieved in vertebrates by a different evolutionary development (Fig. 20–16). Here the myelin sheath plays an important role. It is highly insulating; consequently current flow between the fluid external to the sheath and the fluid in the axon is not possible.

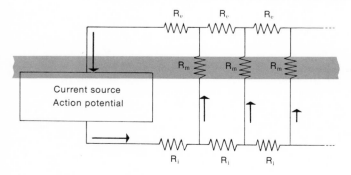

FIGURE 20–15   *Cellular electrical properties affecting electrotonic current flow induced by action potential. Shaded area is axonal membrane; Re, extracellular resistance; Rm, membrane resistance; Ri, axoplasmic resistance. Lengths of arrows indicate relative magnitudes of current.*

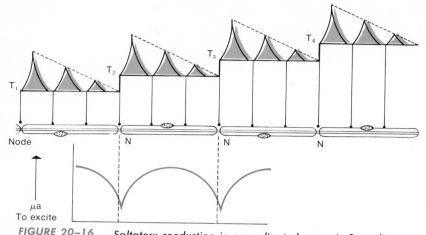

FIGURE 20–16   *Saltatory conduction in a myelinated axon. A, Recording at and between successive nodes of Ranvier demonstrates electrotonic transmission between nodes, as impulse appears instantaneously at all points within a node while diminishing in magnitude. At the node a time delay occurs and the impulse regains initial magnitude, showing the node to be the site of the active, impulse-generating process. B, Support to the idea that electrotonic current flowing from a previous active site exits and excites at the next node comes from a demonstration that current (μA = microamperes) required to excite the axon is least at the nodes. (B after Tasaki, I.: Nervous Transmission. Springfield, Illinois, Charles C Thomas, Publisher, 1953, p. 6; from Case, J.: Sensory Mechanisms. New York, The MacMillan Company, 1966.)*

At the **nodes of Ranvier,** however, where there are gaps in the sheath, there is no insulation. Only here is free ionic communication possible. Impulses are generated only at the nodes. As a result, nerve impulses "leap" from one node to the next. This leaping mode is known as **saltatory conduction**. A myelinated nerve that is only microns in diameter can conduct at velocities up to 100 meters per second as compared with velocities of 20 to 50 meters per second in the largest (1 mm in diameter) unmyelinated axons.

SECTION 20–5

SYNAPTIC TRANSMISSION

Since the nervous system is made up of individual neurons that are discontinuous, there must be some mechanism for the neural message to jump the gap from the axon of one neuron to the **dendrites** or **soma** of the other neuron or, in the case of **neuromuscular** junctions, to the muscle. At some specialized **synapses,** transmission is accomplished electrically. The arrival of the action potential at the end of the axon of the **presynaptic cell** sets up electrical currents in the external fluid of the synaptic gap, and these currents stimulate the **postsynaptic cell** to generate an action potential. Transmission and the generation of the action potential are basically the same as in the nerve fiber itself. **Electrical synapses** of this sort are found in parts of the nervous system of crayfish and of fish. Experiments of Furshpan and Potter in 1957 showed that transmission across the giant synapse in an abdominal ganglion of the cord of the crayfish is by electrical means. The membrane contact of this special synapse is able to act as a **rectifier** and allows current to pass easily in one direction, from the axon of a **connector** neuron to the dendrite of a motor neuron. Thus impulses are transmitted one way, in the normal direction.

The most common means of synaptic transmission, however, is

through the agency of a *chemical transmitter substance.* The classical paired-heart experiment of Loewi provided a striking demonstration of the fact that stimulated nerves can produce a diffusible substance that excites muscle. When two isolated hearts were arranged in such a manner that the blood leaving one heart entered the other (Fig. 20–17), stimulation of the *vagus nerve* of one heart arrested the beat of both hearts. The diffusible substance was subsequently shown to be *acetylcholine.* Acetylcholine is a potent stimulant and causes a local depolarization of the membrane of the muscle cell which sets up propagated impulses in the membrane and causes the contraction of the muscle fiber. *Curare* prevents the transmission of impulses from nerve to muscle specifically at this type of synapse, apparently by combining with the receptors for acetylcholine and preventing their normal reaction with it.

By similar experiments it has been shown that the *sympathetic post-ganglionic fibers* accelerating the heart rate release a chemical called *sympathin,* which is similar to, or identical with, the hormone *epinephrine.* You may wonder why, if the tip of the axon secretes a substance such as acetylcholine, that substance does not stimulate the next dendrite or the muscle continuously. Histochemical studies have shown that there is a high local concentration at the synaptic area of a potent enzyme, *acetylcholinesterase,* which specifically hydrolyzes acetylcholine and renders it inactive. A different enzyme oxidizes sympathin.

Chemical transmission at the synapse involves two processes: (1) the release, by the arrival of a nerve impulse, of the specific chemical from its storage place in the tip of the axon into the narrow space between the adjacent neurons, and (2) the process by which the specific transmitter substance is attached to specific molecular sites in the dendrite and produces a change in the properties of its cell membrane so that a new nerve impulse is set up. The first is a special example of *neurosecretion* and the second is a special example of *chemoreception,* resembling the process which occurs in the chemical sense organs such as those of taste and smell. The passage of the chemical transmitter from axon to dendrite can be accounted for by simple diffusion. Over the short distance involved, diffusion would be rapid enough to account for the speed of transmission observed at synapses. It has been shown that acetylcholine is released by motor nerves in discrete packets which contain a large number of molecules. *Calcium ions* are essential for the operation of the mechanism which releases acetylcholine, and *magnesium ions* inhibit it. This suggests that the transmitter substance is stored within the nerve endings in tiny intracellular structures and that these discharge their entire contents to the

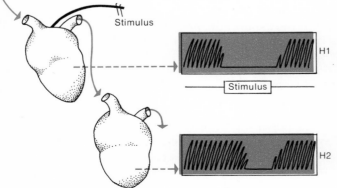

Stimulus

Stimulus

H1

H2

FIGURE 20–17   A demonstration of junctional chemical transmission. Paired heart experiment showing stimulation of vagus nerve of one heart arrests its beat, as shown in the kymograph recording. Action of the vagus nerve must have involved a diffusible substance (subsequently determined to be acetylcholine), because the beat of a second heart is arrested even though its only connection is by way of the perfusion fluid flowing.

surface. Electron micrographs (Fig. 20–18) of the tips of the neuron at the synapse have revealed masses of **synaptic vesicles** which might be sites of storage of acetylcholine. It is conceivable that the arrival of a nerve impulse would lead to the liberation of the contents of one of these vesicles into the synaptic space.

It is not yet definitely known whether transmission across the synapses in the brain and spinal cord is by means of acetylcholine, sympathin or some other chemical, or by an electrical mechanism.

The synaptic junction is a point of resistance to the flow of impulses in the nervous system, and not every impulse reaching the synapse is transmitted to the next neuron. The resistance varies in different synapses, so that they are important in determining the route of impulses through the nervous system and the response of the organism to a specific stimulus.

The entire nervous system is a functional unit, and an impulse arising in any receptor can be transmitted to every effector in the body. Consider, for example, the effects of burning a finger: The muscles of the arm contract to pull the finger away from the heat, a sensation of pain is produced in the brain, a cry may be emitted, the heart beat, digestion and breathing may be altered—in fact, it is conceivable that every muscle and gland in the body may be affected temporarily. Our sense organs receive a constant stream of stimuli, but selective resistance at the synaptic junction prevents the uncontrolled, continuous contraction of muscles and secretion by glands. The drug **strychnine** decreases synaptic resistance; the slightest

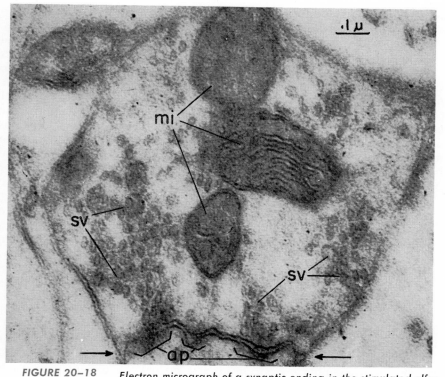

FIGURE 20–18   *Electron micrograph of a synaptic ending in the stimulated olfactory bulb of the rat. The synaptic ending contains three mitochondria (mi) and several synaptic vesicles (sv). The zone of contact between the two neurons is indicated by the two of contact. (From De Robertis, E., and Pellegrino De Iraldi, A.: A plurivesicular component in adrenergic nerve endings. Anat. Rec. 139:299, 1961.)*

stimulus to a person suffering from strychnine poisoning sets off the secretion of all glands and the convulsive contraction of all the muscles of the body.

The amount of synaptic resistance can be modified by nerve impulses. One impulse might cancel out the effect of another, a process known as **inhibition.** The opposite condition, whereby one impulse strengthens another, is called **facilitation.** These two processes are of prime importance in effecting integration of body activities. We have seen that all the muscles of the body are in a state of constant, slight contraction, known as tonus, owing to a constant volley of nerve impulses reaching them. But when one muscle, such as the triceps, is to contract, its antagonist, the biceps, must relax. This is achieved by impulses that inhibit the volley of impulses to the biceps and impulses that reinforce the volley to the triceps. Inhibition and facilitation can occur only at the synapse, since once an impulse starts along a neuron, it can be neither stopped nor accelerated. Whether or not a given impulse crosses a synapse depends on whether it is inhibited or facilitated by other impulses.

SECTION 20–6

THE CENTRAL NERVOUS SYSTEM: SPINAL CORD

The 10 billion or so neurons that make up the nervous system of man are divided into two main parts: those belonging to the **central nervous system,** which make up the brain and spinal cord, and those belonging to the **peripheral nervous system,** which make up the cranial and spinal nerves.

The tubular **spinal cord,** surrounded and protected by the neural arches of the vertebrae, has two important functions: to transmit impulses to and from the brain and to act as a reflex center. In cross section two regions are evident, an inner, butterfly-shaped mass of **gray matter,** made up of nerve cell bodies, and an outer mass of **white matter,** made up of bundles of axons and dendrites (Fig. 20–19). The whiteness of these bundles is due to the myelin sheaths of the axons and dendrites; the ends of the axons and dendrites, present in the central gray matter, have no myelin sheath. The "wings" of the gray matter are divided into two dorsal horns and two ventral horns. The latter contain the cell bodies of motor neurons whose axons pass out through the spinal nerves to the muscles; all the other neurons in the spinal cord are connector neurons.

The axons and dendrites of the white matter are segregated into bundles with similar functions: the **ascending tracts,** which carry impulses to the brain, and the **descending tracts,** which carry impulses from the brain to the effectors. Neurologists have carefully noted the symptoms of persons with injured spinal cords and then have correlated these with the particular tracts found to be destroyed when the patient's nervous system was examined after death. From these observations they have been able to map out the location and functions of the various tracts (Fig. 20–19). For example, the dorsal columns of the white matter transmit impulses originating in the sense organs of muscles, tendons and joints, by means of which we are aware of the position of the parts of the body. In advanced syphilis these columns may be destroyed so that the patient cannot tell where his arms and legs are unless he looks at them, and he must watch his feet in order to walk.

One curious fact, still not satisfactorily explained, emerged from these studies of the location and function of the fiber tracts. All the fibers in the

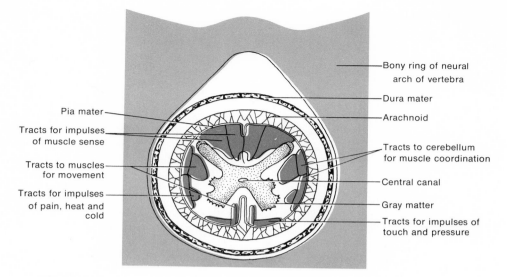

Pia mater

Tracts for impulses of muscle sense

Tracts to muscles for movement

Tracts for impulses of pain, heat and cold

Bony ring of neural arch of vertebra

Dura mater

Arachnoid

Tracts to cerebellum for muscle coordination

Central canal

Gray matter

Tracts for impulses of touch and pressure

FIGURE 20–19   *Cross section of the spinal cord surrounded by the bony vertebra, showing the meninges (dura mater, pia mater and arachnoid), the gray matter and some of the important nerve tracts of the white matter.*

spinal cord cross over from one side of the body to the other somewhere along their path from sense organ to brain, or from brain to muscle. Thus the right side of the brain controls the left side of the body and receives impressions from the sense organs of the left side. Some fibers cross in the spinal cord itself; others cross in the brain.

In the center of the gray matter is a small canal, running the entire length of the neural tube, filled with *cerebrospinal fluid,* which is similar to plasma but contains much less protein. The spinal cord and brain are wrapped in three sheets of connective tissue known as *meninges.* Meningitis is a disease in which these wrappings become infected and inflamed. One of these sheets *(dura mater)* is fastened against the bony neural arches of the vertebrae; another *(pia mater)* is located on the surface of the spinal cord, and the third *(arachnoid)* lies between. The spaces between the meninges are filled with more cerebrospinal fluid so that the spinal cord (and the brain) floats in this liquid and is protected from bouncing against the bone of the vertebrae (or skull) with every movement.

SECTION 20–7

THE CENTRAL NERVOUS SYSTEM: THE BRAIN

The *brain* is the enlarged, anterior end of the spinal cord. In man the enlargement is so great that much of the resemblance to the spinal cord is obscured, but in the lower animals, the relationship of brain to spinal cord is clear. The detailed anatomy of the brain is exceedingly complex, and we shall consider only six main regions: *medulla, pons, cerebellum, midbrain, thalamus* and *cerebrum* (Fig. 20–20).

The most posterior part of the brain, lying next to the spinal cord, is the *medulla.* Here the central canal of the spinal cord enlarges to form a cavity called the *fourth ventricle* (three other ventricles lie farther up in the brain). The roof of the fourth ventricle is thin and contains a cluster of blood vessels which secrete part of the cerebrospinal fluid; the rest of

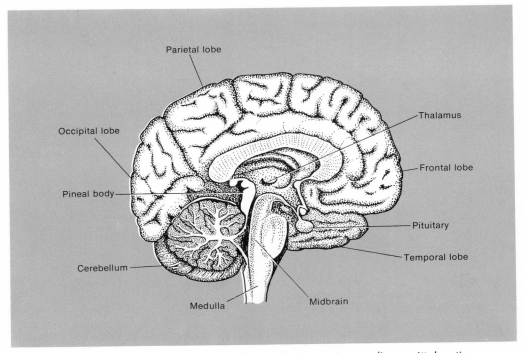

FIGURE 20-20   *The parts of the human brain seen in a median sagittal section.*

this fluid is secreted by similar clusters of blood vessels in the other ventricles. In the roof of the fourth ventricle are three tiny pores through which cerebrospinal fluid escapes into the meningeal spaces. The walls of the medulla are thick and made up largely of nerve tracts connecting with the higher parts of the brain. The medulla also contains a number of clusters of nerve cell bodies, the nerve centers, which are reflex centers controlling respiration, heart rate, the dilatation and constriction of blood vessels, swallowing and vomiting.

Above the medulla is the **cerebellum,** consisting of a central part and two hemispheres extending laterally which resemble pine cones in shape. Its gray surface is made up of the cell bodies of neurons, and beneath is a mass of white tissue composed of fiber tracts connecting with the medulla and with the higher parts of the brain. The size of the cerebellum in different animals is roughly correlated with the amount of their muscular activity. It regulates and coordinates muscle contraction and is proportionately large in extremely active animals, such as birds. Removal or injury of the cerebellum is not accompanied by paralysis but by impairment of muscle coordination. When the cerebellum of a bird is removed it is unable to fly and the wings thrash about jerkily. When the human cerebellum is injured by a blow or disease, all muscular movements are uncoordinated, and any activity requiring delicate coordination, such as threading a needle, is impossible.

Running crosswise on the ventral side of the brain just below the cerebellum is a thick bundle of fibers known as the **pons** or bridge, which carries impulses from one hemisphere of the cerebellum to the other, thus coordinating muscle movements on the two sides of the body.

In front of the cerebellum and pons lies the **midbrain,** which has thick walls and a small central canal connecting the fourth ventricle of the medulla with the third ventricle in the thalamus. The thick walls of the midbrain contain certain reflex centers and the main fiber tracts leading to the thalamus and cerebrum. On the upper side of the midbrain are four low, rounded protuberances *(corpora quadrigemina)* in which are centers for certain visual and auditory reflexes. The reflex constriction of the pupil when light shines on the eye and the pricking up of a dog's ears in response to a sound are controlled by reflex centers there. The midbrain also contains a cluster of nerve cells regulating muscle tonus and posture.

In front of the **midbrain** the central canal again widens and becomes the third ventricle, the roof of which contains another cluster of blood vessels secreting cerebrospinal fluid. The thick walls of the third ventricle are called the **thalamus.** This is a relay center for sensory impulses; fibers from the spinal cord and lower parts of the brain synapse here with other neurons going to the various sensory areas of the cerebrum. The thalamus appears to regulate and coordinate the external manifestations of emotions; thus, by stimulating the thalamus, a sham rage can be produced in a cat—the hair stands on end, the claws protrude, the back becomes humped and other signs of anger are evinced. But as soon as the stimulation stops, the appearance of rage ceases.

In the floor of the **third ventricle** (the **hypothalamus**) are centers regulating body temperature, appetite, water balance, carbohydrate and fat metabolism, blood pressure and sleep. Curiously the front part of the hypothalamus prevents a rise in body temperature and the rear part prevents a fall. The hypothalamus controls certain functions of the anterior lobe of the pituitary, such as the secretion of gonadotropins, and produces certain hormones, those released in the posterior lobe of the pituitary.

The parts of the brain considered so far have to do with unlearned, automatic behavior which is determined by the fundamental structure of these parts, a structure which is essentially the same from fish to man. The **cerebral hemispheres,** the most anterior and the largest part of the human brain, have a basically different function, that of controlling learned behavior. The complex psychological phenomena of consciousness, intelligence, memory, insight and the interpretation of sensations have their physiological basis in the activities of the neurons of the cerebral hemispheres. The importance of the cerebrum to different animals can be investigated by removing it surgically. A cerebrumless frog behaves almost exactly like a normal one, and a pigeon whose cerebral cortex has been removed can fly and balance on a perch, but tends to remain quiet for hours. When stimulated it moves about, though in a random, purposeless way, and it fails to eat when given food. A dog whose cerebral cortex has been removed can walk and will swallow food if it is placed in the mouth, but shows no signs of fear or excitement. Human infants occasionally are born whose cerebral cortex fails to develop, and although they can carry out the vegetative functions of breathing and swallowing, they are incapable of learning and make no voluntary movements. They usually die soon after birth.

The **cerebrum** contains slightly more than half of 10 billion neurons of the human nervous system. The **cerebral hemispheres** develop as outgrowths of the anterior end of the brain and in man and other mammals they grow back over the rest of the brain and hide it from view. Each hemisphere contains a cavity, the first and second ventricles, each of which is connected to the third ventricle of the thalamus by a canal. These ventricles, like the others, contain clusters of blood vessels which secrete cerebrospinal fluid. The cerebrum is made of both gray and white matter; the latter, composed of tracts of nerve fibers, is on the inside of the cerebrum, while the gray matter, made of nerve cell bodies, lies on the surface or **cortex** of the cerebrum. Deep in the substance of the cerebral hemispheres lie other masses of gray matter, nerve centers which act as relay stations to and from the cortex. The lower vertebrates, with little gray matter, have smooth cerebral cortices, but in man and other mammals, the surface of the cerebral hemispheres is convoluted. Ridges separated by furrows increase the amount of space available for the cortical gray matter. The pattern of these convolutions is quite constant even in men of widely different degrees of intelligence, and the geography of the cerebral cortex has been carefully studied (Fig. 20–21). The idea that certain parts of the brain have special functions is an old one, the "science" of phrenology having been based on the premise that functions were localized in the brain. A specially gifted person would have an enlargement in a particular area of the brain and a corresponding bump on the skull. It was believed that an analysis of such bumps would indicate what a man was best fitted for.

Experimental evidence has established that there is a considerable amount of localization of function in the cortex. By surgical removal of particular regions of the cortex from experimental animals it has been possible to localize many functions exactly; and by observing the paralysis or loss of sensation in a man with a brain injury or tumor, and then examining the brain after death to see where the injury was located, it has been possible to map the human brain. During operations on the brain, surgeons have electrically stimulated small regions and observed which muscles contracted, and, since brain surgery can be carried on under local anesthesia, the patient could be asked what sensations he felt when a particular region was stimulated. Curiously the brain itself has no nerve endings for pain so that stimulation of the cortex is not painful. Brain activity may be studied by measuring and recording the electrical potentials or "brain waves" given off by various parts of the brain when active.

By combining the data obtained in several ways, investigators have been able to locate many functions of the brain (Fig. 20–21). The posterior

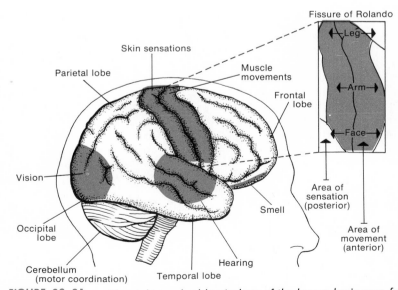

FIGURE 20–21   *The right cerebral hemisphere of the human brain seen from the side. The shaded areas are regions of special function; the light areas are "association areas." Inset: Enlarged view of the sensory and motor areas adjacent to the fissure of Rolando, showing the location of the nerve cells supplying the various parts of the body.*

part contains the visual center; its removal causes blindness, and stimulation of it, even by a blow on the back of the head, causes the sensation of light. Removal of the region from one side of the brain causes blindness in half of each eye, for the nerves from each eye split, and half go to each side of the brain. The center for hearing is located on the side of the brain above the ear. Its stimulation by a blow causes a sensation of noise. Although removal of both auditory areas causes deafness, removal of one does not cause deafness in one ear, but a decrease in the auditory acuity of both ears.

Running down the side of the cerebral cortex is an easily recognizable, deep furrow called the **fissure of Rolando,** which separates the motor area, controlling the skeletal muscles, from the area just behind the furrow which is responsible for the sensations of heat, cold, touch and pressure from stimulation of sense organs in the skin. In both there is a further specialization along the furrow from the top of the brain to the side — neurons at the top of the cortex control the muscles of the feet; the neurons next in line control those of the shank, thigh, abdomen and so on; and the neurons farthest around the side control the muscles of the face. The size of the motor area in the brain for any given part of the body is proportional not to the amount of muscle but to the elaborateness and intricacies of movement; thus there are large areas for the control of the hand and face. There is a similar relationship between the parts of the sensory area and the region of the skin from which it receives impulses. Thus, in the connections between the body and the brain, there is not only a twisting of the fibers so that one side of the brain controls the opposite side of the body, but a further "reversal" which makes the uppermost part of the cortex control the lower extremities of the body.

When all the areas of known function are plotted, they cover almost all of the rat's cortex, a large part of the dog's, a moderate amount of the monkey's, but only a small part of the total surface of man's cortex. The rest, known as **association areas,** is made up of neurons that are not directly connected to sense organs or muscles, but which supply interconnections between the other areas. These regions are responsible for the higher intellectual faculties of memory, reasoning, learning, imagination, and for personality. In some way, the association regions integrate all the diverse impulses constantly reaching the brain into a meaningful unit, so that the proper response is made. They interpret and manipulate the symbols and words by means of which our thought processes are carried on. When disease or accident destroys the functioning of one or more association areas, the condition known as **aphasia** may result. In this the ability to recognize certain kinds of symbols is lost. For example, the names of objects may be forgotten, although their functions are remembered and understood.

SECTION 20–8

BRAIN WAVES

Metabolism is invariably accompanied by electrical changes, and the electrical activity of the brain can be recorded by a device known as the **electroencephalograph**. To obtain recordings, electrodes are taped to different parts of the scalp and the activity of the underlying parts of the cortex is measured. The electroencephalograph has shown that the brain is continuously active. The most regular manifestations of activity, called **alpha**

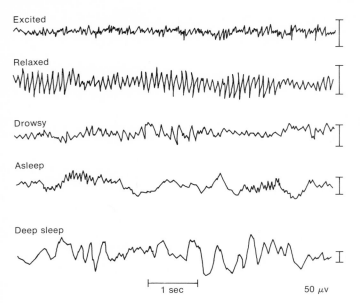

Excited

Relaxed

Drowsy

Asleep

Deep sleep

1 sec   50 μv

FIGURE 20–22 Electroencephalograms made while the subject was excited, relaxed and in various stages of sleep. Recordings made during excitement show brain waves which are rapid and of small amplitude, whereas in sleep the waves are much slower and of greater amplitude. The regular waves characteristic of the relaxed state are called alpha waves. (From Jasper; Epilepsy and Cerebral Localization, by Penfield and Erickson.)

waves, come from the visual areas at the back of the brain when the subject is resting quietly with his eyes shut. These waves occur rhythmically at the rate of nine or 10 per second and have a potential of about 45 microvolts (Fig. 20–22). When the eyes are opened, the waves disappear and are replaced by more rapid, irregular waves. That the latter are produced by objects seen can be demonstrated by presenting the eyes with some regular stimulus, such as a light blinking at regular intervals, and observing that brain waves of a similar rhythm appear. Sleep is the only normal condition in which the brain waves are drastically altered. During sleep, the waves become slower and larger (have a greater potential) as the subject falls into deeper and deeper unconsciousness. The dreams of a sleeping subject are mirrored in flurries of irregular waves.

Certain brain diseases alter the character of the waves; epileptics, for example, exhibit a distinctive, readily recognizable wave pattern, and even people who have never had an epileptic attack, but might under certain conditions, show similar abnormalities. The location of brain tumors can be detected by noting the part of the brain showing abnormal waves.

SECTION 20–9

SLEEP

The neural mechanisms involved in sleep are unknown, and investigators are still trying to discover why sleep is necessary. Sleep is characterized by decreased electrical activity of the cerebral cortex (Fig. 20–22), and this might be correlated with its recuperative effect on the nervous system. Only the higher vertebrates with fairly well developed cerebral cortices sleep, and those with larger hemispheres seem to require more sleep than others. Fatigue is popularly considered the cause of sleep, but there is no experimental evidence to verify this belief. An important sleep-inducing factor is the absence of stimuli; it is easy to go to sleep, even when one is not particularly tired, if there is nothing interesting to occupy the mind. But although we tend to be wakeful in the presence of attention-holding stimuli, there is a limit beyond which sleep is inevitable. For all the higher animals, life is characterized by a basic rhythm of sleep alternating with wakefulness, a pattern regulated by the hypothalamus. There is a sleep center in the anterior part, and a wakefulness center in the posterior part of the hypothalamus. It is believed that the change from wakefulness to sleep and back is controlled by "feedback" circuits involving these two centers.

INSANITY AND NEUROSES

Certain types of brain derangements have an easily understood basis in damage to the brain tissue produced by disease or a wound. If the pores in the roof of the fourth ventricle become clogged, trapping the cerebrospinal fluid within the ventricles, the pressure of this fluid inside the brain will gradually destroy the tissue there. Or a blood vessel in the meninges covering the brain may rupture and the pressure of the accumulated blood destroys parts of the brain. Tumors and infectious diseases such as syphilis can damage tissue; the actual symptoms—paralysis, loss of sensation or other functions—depend on which part of the brain is affected.

The causes of other types of disorder, the so-called functional disorders—psychoneuroses and psychoses—are more baffling, for they occur without any structural or chemical change in the brain which pathologists have so far been able to detect. Typically these involve emotional disturbances rather than changes in intelligence.

Psychoneuroses are comparatively mild and common disorders with a great variety of symptoms: anxiety, fear, shyness and oversensitiveness. The emotional upsets may actually produce organic disorders such as irregular heart beat or digestive disturbances. The cause of this type of mental disorder is not positively known, and there is evidence for believing that it differs from person to person, and is complex in every instance. The most widely accepted theory at present states that psychoneuroses are due to deep-seated emotional conflicts. Heredity, present environment, past experience and general health may play a part in causing the condition. The patient is usually completely unaware of the cause or causes of his unhappiness. There is no single cure for the various pyschoneuroses; many of them respond to psychiatric treatment, by means of which the reason for the sense of anxiety, guilt, conflict or fear is brought to the patient's attention. Other psychoneuroses disappear gradually for no apparent reason, others become increasingly worse, and a few develop into the more serious psychoses.

Psychoses are the severe mental diseases which usually require hospitalization because the patient may be a danger to himself or to other members of the community. There are three main types of psychoses, each of which represents an exaggeration of normal tendencies. **Manic-depressive** psychoses are characterized by an alternation of excessive elation and depression, sometimes accompanied by delusions and hallucinations. Most manic-depressives are normal for most of their lives, but suffer recurrent episodes of insanity. **Paranoia** is a psychosis characterized by delusions typically of grandeur or persecution. **Dementia praecox** or **schizophrenia** is marked by a withdrawal from the everyday world into a world of daydreams which becomes the world of reality. Most psychotic patients have symptoms which represent combinations of these types.

Psychoses are most difficult to cure than psychoneuroses, but encouraging results have been obtained in recent years with psychotherapy. One of the most drastic therapeutic methods is shock treatment, based on the theory that psychotics actually can be jolted back into sanity. Violent fits or comas are produced by the injection of insulin or by the application of electric currents. The drawbacks to such treatments are many, and the neural mechanisms underlying the results are not clearly understood, but a number of cases have been cured by some variation of shock treatment. Treatment with tranquilizing drugs such as chlorpromazine has been successful in many cases and is gradually supplanting shock treatment.

SECTION 20–11

THE VERTEBRATE PERIPHERAL NERVOUS SYSTEM

Emerging from the brain and spinal cord and connecting them with every receptor and effector in the body are the paired *cranial* and *spinal nerves;* these make up the *peripheral nervous system.* Cranial and spinal nerves are made of bundles of nerve fibers. The only nerve cell bodies present in the peripheral nervous system are those of the sensory neurons, aggregated into clusters (ganglia) near the brain or spinal cord, and of certain motor neurons of the *autonomic system* which will be discussed later.

TABLE 20–1   The Cranial Nerves of Man*

Number	Name	Origin of Sensory Fibers	Effector Innervated by Motor Fibers
I	Olfactory	Olfactory mucosa of nose (smell)	None
II	Optic	Retina of eye (vision)	None
III	Oculomotor	Proprioceptors of eyeball muscles (muscle sense)	Muscles which move eyeball (with IV and VI); muscles which change shape of lens; muscles which constrict pupil
IV	Trochlear	Proprioceptors of eyeball muscles (muscle sense)	Other muscles which move eyeball
V	Trigeminal	Teeth and skin of face	Some of muscles used in chewing
VI	Abducens	Proprioceptors of eyeball muscles (muscle sense)	Other muscles which move eyeball
VII	Facial	Taste buds of anterior part of tongue	Muscles of the face; submaxillary and sublingual glands
VIII	Auditory	Cochlea (hearing) and semicircular canals (senses of movement, balance and rotation)	None
IX	Glossopharyngeal	Taste buds of posterior third of tongue, lining of pharynx	Parotid gland; muscles of pharynx used in swallowing
X	Vagus	Nerve endings in many of the internal organs—lungs, stomach, aorta, larynx	Parasympathetic fibers to heart, stomach, small intestine, larynx, esophagus
XI	Spinal accessory	Muscles of shoulder (muscle sense)	Muscles of shoulder
XII	Hypoglossal	Muscles of tongue (muscle sense)	Muscles of tongue

*(From Villee, C. A.: *Biology.* 5th ed. Philadelphia, W. B. Saunders Company, 1967.)

CRANIAL NERVES.   Twelve pairs of nerves originate in different parts of the brain and innervate primarily the sense organs, muscles and glands of the head. The same twelve pairs, innervating similar structures, are found in all the higher vertebrates—reptiles, birds and mammals; fish and amphibia have only the first 10 pairs. Like all nerves, these are composed of neurons; some have only sensory neurons (nerves I, II and VIII), some are composed almost completely of motor neurons (III, IV, VI, XI and XII), and the others are made up of both sensory and motor neurons. The names and structures innervated by the cranial nerves are given in Table 20–1. One of the most important cranial nerves is the vagus, which forms part of the autonomic system and innervates the internal organs of the chest and the upper abdomen.

SPINAL NERVES.   All the spinal nerves are mixed nerves, having motor and sensory components in roughly equal amounts. In man they originate from the spinal cord in 31 symmetrical pairs, each of which innervates the receptors and effectors of one region of the body. Each nerve emerges from the spinal cord as two strands or roots which unite shortly to form the spinal nerve. All the sensory neurons enter the cord through the **dorsal root** and all motor fibers leave the cord through the **ventral root** (Fig. 20–23). If the dorsal root is severed, the part of the body innervated by that nerve suffers complete loss of sensation without any paralysis of the muscles. If the ventral root is cut, there is complete paralysis of the muscles innervated by that nerve, but the senses of touch, pressure, temperature, kinesthesis and pain are unimpaired. The size of each spinal nerve is related to the size of the body area it innervates; the largest in man is one of the pairs supplying the legs. Each spinal nerve, shortly beyond the junction of the dorsal and ventral root, divides into three branches: the **dorsal branch,** serving the skin and muscles of the back; the **ventral branch,** serving the skin and muscles of the sides and belly; and the **autonomic branch,** serving the viscera (Fig. 20–4).

SECTION 20–12

REFLEXES AND REFLEX ARCS

One of the simplest actions of the nervous system involving stimulation and response is the **reflex.** A reflex is a relatively stereotyped, automatic and innate response that typically involves part of the body rather than the whole. Flexion of the leg in response to painful stimulus and constriction of the pupil in bright light are examples. When we step on something sharp or come in contact with something hot, we do not wait until pain is experienced by the brain and then, after deliberation, decide what to do. Our responses are immediate and automatic. The foot or hand is withdrawn reflexly *before* pain is experienced. Many of the more complicated activities of our daily lives, such as walking, are regulated to a large extent by reflexes. We have already discussed how important reflexes control heart rate, blood pressure, breathing, salivation, movements of the digestive tract and so forth. The pioneering studies on reflexes were conducted by Sir Charles Sheerington around the turn of the century and were presented in 1906 in a book that is now one of the classics of biology, *The Integrative Action of the Nervous System.*

The minimum anatomical requirements for reflex behavior are a sensory neuron with a receptor to detect the stimulus, connected synap-

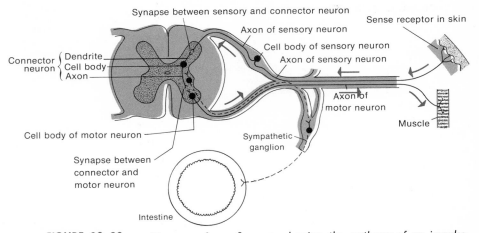

FIGURE 20–23   *Diagram of a reflex arc, showing the pathway of an impulse,*
indicated by arrows.

tically with a motor neuron attached to an effector (e.g., a muscle). This
circuit is a **reflex arc.** It is the simplest type of arc and is described as **monosynaptic** because there is only one synapse between the sensory and the
motor neuron. Most reflex arcs consist, in addition, of one or more interneurons between the sensory and motor neuron (Fig. 20–23).

As a rule reflexes involve the connections of many interneurons in the
spinal cord. The functioning of the spinal cord is thus not merely the conduction of messages to and from the brain; its other important function
is the integration of reflex behavior. Its importance in this respect is easily
demonstrated in spinal animals, i.e., animals from which the brain has been
removed. In these animals, all spinal reflexes remain intact. An experiment
demonstrating this consists in removing the brain of a frog while leaving
the spinal cord intact and then applying a piece of acid-soaked paper to the
animal's back. No matter how many times the piece of paper is placed on
the skin, one leg will invariably come up and flick it away. This response,
involving many muscles working in a coordinated fashion, is purely reflex,
and clearly demonstrates one of the chief characteristics of a reflex: fidelity
of repetition. A frog with a brain might make the response two or three
times, but eventually it would do something else—perhaps hop away.

The degree of complexity of the reflexes depends upon how many of
the spinal segments are involved. In some of the simple reflexes already
mentioned, the entire reflex arc is contained in a single spinal segment.
Many reflexes, however, require interneuronal connections with one or
more spinal segments. For example the scratch reflex, righting reflex and
walking reflex require the organization of a great deal of sensory information over many segments of the spinal cord. Thus reflexes may involve
various levels of the nervous system and be of different degrees of complexity. Some reflexes, as, for example, walking reflexes, even require some influence from the midbrain. Some of the details about specific reflexes,
including conditioned reflexes, are discussed in the chapter on behavior.

SECTION 20–13

THOUGHTS, MEMORY AND THE LEARNING PROCESS

It is not yet possible to give a complete physiological explanation for
the psychological phenomena of thoughts, memories and learning.
Thoughts are believed to have a physiological basis in specific patterns of

nerve impulses traveling in certain neuronal pathways in the brain. The nerve impulses related to one particular thought may travel over a particular group of neurons arranged in a closed circuit. Such a reverberating circuit might include from half a dozen to several hundred or more neurons, and it might be initiated either by incoming sensory impulses or by spontaneous impulses originating in the brain. On the basis of this theory, thoughts would change as the nerve impulses pass over a different reverberating circuit, one involving a different group of neurons.

Memories are thoughts that occur some time—minutes, months or years—after the original thought and are presumably caused by the reestablishing of the original reverberating circuit of impulses. It has been suggested that the continued passage of impulses across the synapses of a given circuit leads to a decrease in the resistance of those synapses and an increase in the ease with which succeeding nerve impulses can pass over that pathway. The learning process may involve the repeated use of a particular neuronal pathway until its synaptic resistance is markedly decreased.

It has also been postulated in recent years that the memory involves the synthesis of RNA and that specific memories are coded in the central nervous system by specific sequences of nucleotides in RNA. There are some interesting correlations of RNA metabolism and neural function, but the clarification of the physiological and biochemical basis of memory remains for the future.

Certain experiments, such as those in which animals learn to run a maze or labyrinth to get food or to avoid electric shocks, have emphasized the role of "trial and error" in learning. The higher mammals, apes and men show, in addition, the phenomenon of "insight" or getting the idea. After a few random trials, the subject of the experiment sees the point of the problem ("puts two and two together") and thereafter performs with a high frequency of success.

MOODS AND EMOTIONS.   Such phenomena as moods, emotions and personality as a whole also depend on the activities of the cerebral cortex, the thalamus and the hypothalamus, but the neural mechanisms underlying them are not understood. These, as well as other activities of the higher centers of the brain, are influenced greatly by the conditions of the body; the state of mind can be markedly affected by the state of the stomach. The secretions of several endocrine glands also affect the functioning of the brain; e.g., many women have periods of mental depression just before and during menstruation, and it is quite common for the menopause (the period from 40 to 50 years of age when the recurring menstrual cycle ceases) to be accompanied by profound emotional and mental disturbances.

SECTION 20–14

THE AUTONOMIC NERVOUS SYSTEM

The heart, lungs, digestive tract and other internal organs are innervated by a special set of peripheral nerves, collectively called the **autonomic nervous system.** This system in turn is composed of two parts: the **sympathetic** and the **parasympathetic** nerves.

The autonomic system contains only motor nerves, and is distinguished from the rest of the nervous system by several features. There is no willful control by the cerebrum over these nerves; we cannot voluntarily speed up or slow down the heart beat or the action of the muscles of the

TABLE 20-2   Actions of the Autonomic System*

Organ Innervated	Action of Sympathetic System	Action of Parasympathetic System
Heart	Strengthens and accelerates heart beat	Weakens and slows heart beat
Arteries	Constricts arteries and raises blood pressure	Dilates arteries and lowers blood pressure
Digestive tract	Slows peristalsis, decreases activity	Speeds peristalsis, increases activity
Urinary bladder	Relaxes bladder	Constricts bladder
Muscles in bronchi	Dilates passages, makes for easier breathing	Constricts passages
Muscles of iris	Dilates pupil	Constricts pupil
Muscles attached to hair	Causes erection of hair	Causes hair to lie flat
Sweat glands	Increases secretion	Decreases secretion

*(From Villee, C. A.: *Biology.* 5th ed. Philadelphia, W. B. Saunders Company, 1967.)

stomach or intestines. Another important characteristic of the autonomic system is that each internal organ receives a *double* set of fibers, one set coming via the sympathetic nerves and one set via the parasympathetic nerves. Impulses from the sympathetic and parasympathetic nerves always have antagonistic effects on the organ innervated. Thus, if one speeds up an activity, the other decreases it. These effects are summarized in Table 20-2.

Still another peculiarity of the autonomic system is that the motor impulses reach the effector organ from the brain or spinal cord, not by a single neuron, as do those to all other parts of the body, but by a relay of two or more neurons. The cell body of the first neuron in the chain, termed the **preganglionic** neuron, is located in the brain or spinal cord; that of the second neuron (the **postganglionic** neuron) is located in a ganglion somewhere outside the central nervous system (Fig. 20-23). The cell bodies of the postganglionic neurons of sympathetic nerves are close to the spinal cord; those of the parasympathetic nerves are close to, or actually within, the walls of the organs they innervate. Afferent fibers from the internal organs enter the central nervous system along with the somatic nerve fibers.

THE SYMPATHETIC SYSTEM.   The **sympathetic system** consists of nerve fibers whose preganglionic cell bodies are located in the lateral portions of the gray matter of the spinal cord. Their axons pass out through the ventral roots of the spinal nerves in company with the motor neurons to the skeletal muscles and then separate from these and become the autonomic branch of the spinal nerve going to the **sympathetic ganglion.** These ganglia are paired, and there is a chain of 18 of them on each side of the spinal cord from the neck to the abdomen (Fig. 20-24). In each ganglion, the axon of the preganglionic neuron synapses with the dendrite of the postganglionic neuron. The cell body of this neuron is located within the ganglion, and its axon passes to the organ innervated.

In addition to the fibers going from each spinal nerve to each ganglion, there are fibers passing from one ganglion to the next. The axons of some of the postganglionic neurons pass from the sympathetic ganglion back to the spinal nerve and through it to innervate sweat glands, the muscles that make the hair stand erect, and the muscles in the walls of blood vessels. The axons of other postganglionic neurons pass from the sympathetic ganglia of the neck up to the salivary glands and the iris of the eye. The sensory neurons (fibers) of the sympathetic system are located within the

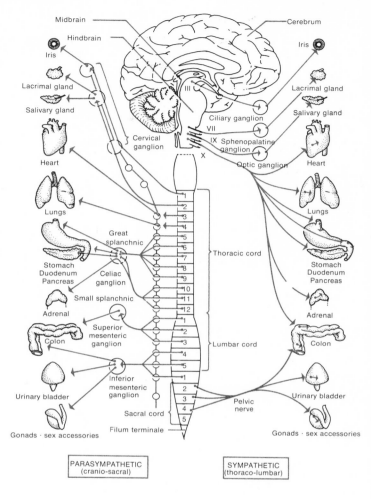

FIGURE 20-24 *Diagram of the autonomic nervous system. The parasympathetic system is shown on the left, the sympathetic system on the right. Roman numerals refer to the numbers of the cranial nerves.*

same nerve trunks as the motor neurons, but enter the spinal cord by way of the dorsal root, together with other sensory nerves of the nonautonomic system.

THE PARASYMPATHETIC SYSTEM. The **parasympathetic system** consists of fibers originating in the brain and emerging via the third, seventh, ninth and especially the tenth or vagus nerves, and of fibers originating in the pelvic region of the spinal cord and emerging by way of the spinal nerves in that region (Fig. 20-24). The **vagus nerve** arises from the medulla and passes down the neck to the chest and abdomen, innervating the heart, respiratory system and digestive tract as far as the small intestine. The large intestine and the urinary and reproductive systems are innervated by parasympathetic nerves from the pelvic spinal nerves. The iris of the eye, the sublingual and submaxillary glands and the parotid gland are innervated by the third, seventh and ninth cranial nerves, respectively. These nerves all contain the axons of preganglionic neurons in the chain; the ganglia of the parasympathetic system are located in or near the organs innervated, so that the axons of the postganglionic neurons are all relatively short.

SECTION 20-15

THE NERVOUS SYSTEMS OF LOWER ANIMALS

The unicellular animals such as the amoeba and paramecium have no neurons, since their entire body consists of but one cell. The amoeba manages to exhibit some simple responses, such as movement toward food or away from a needle point, without any obvious specialized structure for integration. The paramecium, whose body is covered with thousands of tiny cilia, coordinates the beating of these cilia so that it can move by a system of tiny neuromotor fibers which stretch from the anterior end of the animal to all the cilia. These fibers can be severed by a surgical operation performed under the microscope; thereafter the cilia no longer beat in a coordinated fashion, but at random.

Insofar as multicellular organisms are concerned, nervous tissue evolved early in the history of animals. Neurons, synapses and neuromuscular junctions are found in coelenterates and have changed comparatively little in the course of evolution. The neurons and synapses found in the various invertebrates are essentially identical with those in man. Evolutionary development of the nervous system has taken the form of enormous **multiplication of units** (neurons), **diversification of form** and specialized function of neurons, increasing **complexity** of interneural connections, a gathering together into groups of neurons with common or associated functions, and a **centralization** of neural tissue.

In the coelenterates, for example, there is no central nervous system. Neurons tend to be more evenly scattered throughout the body, usually in the form of a net; however, there is some tendency for neurons to be gathered together in cords in the jellyfish and sea anemones. Synapses are the rule although in hydras it appears that the nerve net is a syncitium—the neurons are fused. Generally speaking, all the neurons are similar; however, in the more complicated coelenterates there is some differentiation into sensory, ganglion and motor neurons.

Among the invertebrates, the nervous system attains its highest development in the arthropods (insects, spiders, crabs, lobsters) and ceph-

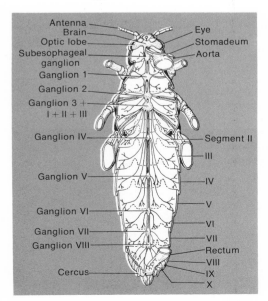

FIGURE 20-25   Ventral nervous system and brain of a grasshopper, Dissosteira carolina.

alopods (octopuses and squids). In these animals there is a brain and a longitudinal cord analogous to the spinal cord of vertebrates. In the invertebrates, the brain lies in a dorsal position and the solid cord in a ventral position. The two are connected by a pair of cords that pass around the esophagus. In the annelids and the more primitive arthropods there is a pair of large ganglia in every segment of the body. In some of the more highly evolved species, for example, spiders and flies, the ganglia become fused. A comparison of the cockroach or locust with a fly reveals that the first two have paired ganglia in every segment whereas the last has, in addition to the brain and subesophageal ganglion, only one large fused, ganglionic mass in the thorax (Fig. 20–25).

Despite its small size, the invertebrate nervous system is enormously complex. The crayfish system, for example, contains 97,722 neurons. Sense organs such as the compound eye, the taste organs, and all the mechanical sense organs concerned with postural relations, tactile discrimination and tactile manipulation of the environment are highly developed. Synaptic connections are so complex as to defy analysis. The brain and ganglia are literally jammed with integrating interneurons. There is also an analogue of the vertebrate autonomic nervous system that innervates the heart, parts of the alimentary canal and the principal endocrine organs. The relation between the nervous system and the endocrine organs is similar to the situation in the vertebrates where there is a close relation between the hypothalamus and the pituitary gland.

SUGGESTIONS FOR FURTHER READING

Cannon, W. B.: *The Wisdom of the Body*. New York, W. W. Norton & Company, Inc., 1932.
Sherrington, C. S.: *The Intergrative Action of the Nervous System*. New Haven, Yale University Press, 1947.
> These two books discuss the role of the nervous system in maintaining constant body conditions.

Pavlov, I. P.: *Conditioned Reflexes*. New York, International Publishers, 1941.
> A comprehensive account of Pavlov's experiments is given in his book.

Zilboorg, G.: *A History of Medical Psychology*. New York, W. W. Norton & Company, Inc., 1941.
> This book describes the development of psychology and psychiatry.

Garrett, H. E.: *Great Experiments in Psychology*. 3rd ed. New York, Appleton-Century-Crofts, 1951.
> Some of the classic experiments in the field are described in this text.

Ranson, S. W., and Clark, S. L.: *The Anatomy of the Nervous System*. 10th ed. Philadelphia, W. B. Saunders Company, 1959.
> A presentation of the details of the structure of the brain and nerves.

Cobb, Stanley: *Foundations of Neuropsychiatry*. 5th ed. Baltimore, The Williams & Wilkins Company, 1952.
> An interesting discussion, at not too technical a level, of the anatomical and physiological bases of normal and abnormal mental activity.

Wiener, Norbert: *Cybernetics*. New York, John Wiley & Sons, Inc., 1948.
> A development of the analogy between the regulation of body activities by the nervous system and the regulation of industrial processes by feedback controls.

Young, J. Z.: *Doubt and Certainty in Science*. New York, Oxford Press University, 1951.
> A discussion of the biological basis of memory.

Fulton, John: *Selected Readings in the History of Physiology*. Springfield, Illinois, Charles C Thomas, Publisher, 1930.
> In Chapters 6 and 7 there are some classic experiments in neurophysiology.

Katz, B.: *The Nerve Impulse.* Sci. Amer. *187*:55, November 1952.

Katz, B.: *How Cells Communicate.* Sci. Amer. *205*:78, September 1961.

Keynes, R. D.: *The Nerve Impulse and the Squid.* Sci. Amer. *199*:83, December 1958.
    Discussions of the nature of the nerve impulse and some of the experiments that have
    led to our present theories regarding the nerve impulse are presented.

Gray, G. W.: *The Great Ravelled Knot.* Sci. Amer. *179*:28, October 1948.

Olds, J.: *Pleasure Centers in the Brain.* Sci. Amer. *195*:105, October 1956.

Snider, R. S.: *The Cerebellum.* Sci. Amer. *199*:84, August 1958.
    Popular discussions of the functions of the brain and its several parts.

Walter, W. G.: *The Electrical Activity of the Brain.* Sci. Amer. *190*:54, June 1954.
    A discussion of the technique of electroencephalography.

Gerard, R. W.: *What Is Memory?* Sci. Amer. *189*:118, September 1953.
    Some experimental attempts to reach an understanding of the memory process are
    described by this author.

Bullock, T. H., and Horridge, G. A.: *Structure and Function in the Nervous Systems of Invertebrates.* 2 Vols. San Francisco, W. H. Freeman & Company, 1969.
    A comprehensive account of invertebrate nervous systems is given in this two-volume
    work.

CONTROL SYSTEMS: HORMONAL INTEGRATION

CHAPTER 21

The activities of the various parts of the bodies of animals and plants are integrated, at least in part, by the hormones of the endocrine system. The rapid responses of muscles and glands, measured in milliseconds, are typically mediated by impulses transmitted by the neurons of the nervous system. The hormones secreted by the endocrine glands diffuse or are transported by the bloodstream to cells in other parts of the body and regulate their activities. The responses controlled by hormones are generally somewhat slower and measured in minutes, hours or even weeks but tend to persist for a longer time than those under nervous control. The long-term adjustments of metabolism, growth and reproduction are typically under endocrine control. The concentrations of glucose, sodium, potassium, calcium and many other substances in the blood and extracellular fluids are regulated by hormones.

Endocrine glands by definition secrete substances into the blood stream rather than into a duct leading to the outside of the body or to a cavity in one of the internal organs. Because of this they are sometimes called ductless glands or glands of internal secretion. The thyroid, parathyroid, pituitary and adrenal glands function only in the secretion of hormones and are strictly ductless glands. The pancreas, ovaries and testes are glands with dual functions; they have external secretions via ducts and internal secretions carried by the bloodstream.

METHODS OF INVESTIGATING ENDOCRINE SYSTEMS

To determine whether a gland suspected of producing a hormone is in fact an endocrine gland, an investigator usually begins by removing the gland surgically and observing the effect on the animal. Next he replaces the gland with one transplanted from another animal and determines whether the changes induced by removing the gland can be reversed by replacing it. In replacing the gland he is careful to ensure that the new gland becomes connected only with the vascular system of the recipient and not with a duct of some sort. He might next try feeding an animal from which the gland has been removed with dried glands to determine whether the active substance, the hormone, can be replaced in this way. Finally he extracts the gland with a variety of solvents to determine the solubility characteristics of the active material and gain a clue as to its chemical nature. By making an extract of the gland, or perhaps by making an extract of blood or urine, and purifying the extract by a sequence of appropriate chemical and physical methods, he finally obtains the pure compound and can determine its exact chemical structure. At each step of his extraction and purification procedure he tests his material by injecting it into a test animal from which the gland has been removed. In this way he finally determines what specific chemical in the original extract will reverse the changes caused by removing the gland.

Hormones are remarkably effective substances and a very small quantity produces a marked effect in the structure and function of one or another part of the body. Only small amounts of a hormone are secreted at any one time by the endocrine gland and the amount circulating in the blood is very small. Even the amount excreted in the urine is not very large. Because of this the isolation of a pure hormone can be a difficult job indeed. To obtain a few milligrams of pure estradiol, one of the female sex hormones, more than two tons of pig ovaries were extracted! When the hormone secreted by a given gland has been extracted and identified and its chemical structure determined it can then be prepared synthetically to be used in treating diseases which result from a deficiency of that hormone.

SECTION 21–2

WHAT IS A HORMONE?

The term "hormone" is not a chemical term—it does not describe some particular class of chemical compounds. It is an operational term and in E. H. Starling's classic definition "a hormone is a substance secreted by cells in one part of the body that passes to another part where it is effective in a very small concentration in regulating the growth or activity of the cells." Hormones are usually carried in the blood from their site of production to their site of action but neurohumors may pass down an axon and prostaglandins are transferred in the seminal fluid. Hormones play an integrative role in vertebrates, insects, crustacea and plants. Other animals may have endocrine systems, too; sex hormones have been isolated from many of the lower animals.

The range of chemical structures that have hormonal activity is remarkable. Some hormones are **amino acids** or derivatives of amino acids (thyroxine, epinephrine, melatonin, indole acetic acid); others are **purines** (cytokinin), **fatty acid** derivatives (prostaglandins, juvenile hormone), **short peptides** (oxytocin, antidiuretic hormone), **long peptides** (ACTH, glucagon, insulin), **proteins** (gonadotropins, growth hormone), **steroids** (sex hormones, adrenal cortical hormones, ecdysone) or **gibberellins,** which are diterpene compounds with a complex ring structure. It does not seem likely that all of these diverse compounds would affect cell function by the same or similar mechanisms. Indeed it is becoming apparent that some and perhaps many hormones have several independent mechanisms of action by which they regulate cellular activities.

AMINO ACID DERIVATIVES.   *Epinephrine* and *norepinephrine* (Fig. 21–1) are amines derived by decarboxylation of the amino acid **tyrosine.** They are synthesized and stored in the **chromaffin cells** of the adrenal medulla and released when the cells are stimulated by nerve impulses transmitted by the sympathetic nervous system. Norepinephrine is also a **neurotransmitter** produced by the tips of the axons of certain nerves and transmitting the impulse to the adjoining dendrites.

Thyroxine is another derivative of tyrosine that is synthesized in the thyroid gland. The thyroid has a remarkably effective iodide pump that can accumulate iodide from the blood stream and concentrate it many fold. Tyrosine residues are iodinated to form diiodotyrosine and two of these are coupled to form thyroxine (Fig. 21–2). The iodination of the tyrosine and the joining of the iodinated tyrosines occur while the amino acids are part of a large protein molecule, **thyroglobulin.** Thyroxine contains four atoms of iodine and the related compound, **triiodothyronine,** which is also hormonally active, has three atoms of iodine. A hormone secreted by the anterior lobe of the pituitary, **thyrotropin** or thyroid stimulating hormone **(TSH),** enhances all the reactions involved in the production of thyroxine—the uptake of iodide, the addition of iodine to the tyrosine and the release of thyroxine from thyroglobulin.

Indole acetic acid, the primary growth hormone or **auxin** of plants, is synthesized from the amino acid **tryptophan** by transamination and decarboxylation (Fig. 21–3). Only very small amounts of auxin are present in plants at any moment for it is converted enzymatically to the inactive compound indole formaldehyde.

PURINE DERIVATIVES.   A second plant growth hormone, originally termed "coconut milk factor," has been identified as a derivative of adenine. It has a five-carbon isoprenoid chain attached to the amino group at carbon 6 in the purine ring (Fig. 21–4). **Zeatin,** found in young corn seeds, is 6-(4-hydroxy-3-methyl-trans-2-butenyl amino) purine. A purine very similar to this has been found to be a constituent of both serine transfer RNA and tyrosine transfer RNA (p. 191). This is a 6-(γ,γ-dimethyl allyl amino) purine, which differs from zeatin only in lacking a hydroxyl group on one of the methyl groups in the side chain.

FATTY ACID DERIVATIVES.   The **prostaglandins,** derivatives of 20-carbon unsaturated fatty acids, have a five-carbon ring in the middle of the chain and hydroxyl or ketone groups on certain carbons in the chain (Fig. 21–5). They are secreted by the seminal vesicle and transferred in the semen to the female reproductive tract where they regulate the activity of uterine muscles. They are synthesized from polyunsaturated fatty acids such as **arachidonic,** a 20-carbon fatty acid with four double bonds.

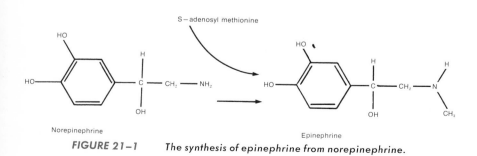

FIGURE 21–1   *The synthesis of epinephrine from norepinephrine.*

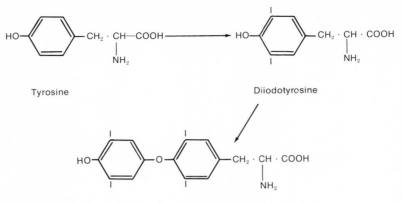

Tyrosine   Diiodotyrosine

Thyroxine

FIGURE 21-2   *Diagram of the reactions by which thyroxine is synthesized from tyrosine.*

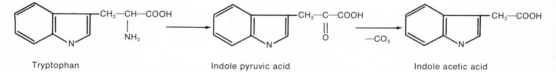

Tryptophan   Indole pyruvic acid   Indole acetic acid

FIGURE 21-3   *Diagram of the reactions by which indole acetic acid (auxin) is synthesized from tryptophan.*

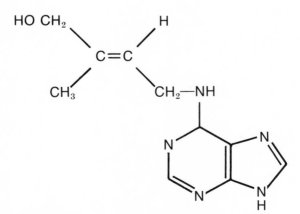

FIGURE 21-4   *The structure of zeatin, a cytokinin.*

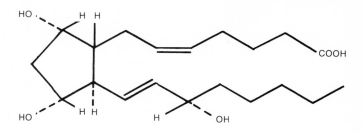

FIGURE 21–5   One of the prostaglandins, prostaglandin F₂α.

Prostaglandins affect tissues other than the uterus and perhaps are se-creted by tissues other than the seminal vesicle.

SHORT PEPTIDES.   **Oxytocin** and vasopressin or **antidiuretic hormone** (Fig. 21–6) are both short peptides, composed of nine amino acids, seven of which are identical in the two hormones. The two substances have quite different physiological properties even though they differ in only two amino acids. Both are synthesized in neurosecretory cells in the supraoptic and paraventricular nuclei of the hypothalamus and then pass along the axons of these cells to the posterior lobe of the pituitary where they are stored and subsequently released.

Melanocyte stimulating hormone (MSH), a peptide containing 13 amino acids, is secreted by the anterior lobe of the pituitary. In frogs and other amphibians, MSH seems to have a physiological role in bringing about the darkening of the skin by causing the melanocytes in the skin to expand. The role of MSH in man and other mammals is not clear, but the human pituitary produces MSH and skin does respond somewhat to MSH.

LONG PEPTIDES.   Insulin, glucagon, ACTH and calcitonin are some-what longer peptides, with some 30 amino acids in the chain. **Insulin,** secreted by the β cells of the pancreas, consists of two peptide chains joined by disulfide bonds. One chain contains 21 amino acids and the other contains 30 amino acids. It was recently found that insulin is produced in the β cells of the pancreas as a single peptide chain composed of 84 amino acids. This, termed **proinsulin,** undergoes folding, three disulfide bonds are formed and a peptide chain containing 33 amino acids is removed from the center of the original chain by hydrolysis, leaving two peptide chains joined by the disulfide bridges (Fig. 21–7).

Glucagon, secreted by the α cells of the pancreatic islets, is a short peptide containing 29 amino acids. **Secretin,** a hormone secreted by the duodenal mucosa, is a peptide containing 27 amino acids, 16 of which are identical in sequence with those in glucagon.

Adrenocorticotropic hormone (ACTH), secreted by the anterior lobe of the pituitary, has 39 amino acids in a single peptide chain, 10 of which

FIGURE 21–6   The structure of oxyto-cin and vasopressin (antidiuretic hormone). Note that the two hormones have amino acid sequences which differ only in two amino acids.

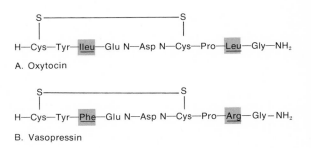

A. Oxytocin

B. Vasopressin

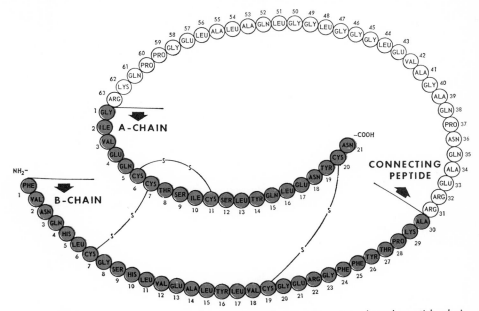

FIGURE 21-7   *The molecular structure of proinsulin, a single polypeptide chain, and its conversion to insulin, composed of two peptide chains. This is achieved by the formation of three disulfide bonds and the removal of a section of the peptide chain between peptide A and peptide B. (From McGilvery, R. W.: Biochemistry. Philadelphia, W. B. Saunders Company, 1970.)*

form a sequence similar to a portion of the peptide chain in MSH. ACTH is synthesized rapidly, little is stored in the pituitary, and it is rapidly removed from the plasma. Its *biological half-life,* the time required for half of a given amount produced to be removed from the plasma, is 20 minutes or less. It stimulates the growth of the adrenal cortex and its production of adrenal cortical steroids.

An even longer peptide, *parathormone,* is secreted by the parathyroid glands and involved in the regulation of calcium metabolism. Parathormone is a single peptide chain containing 84 amino acids. *Calcitonin,* secreted by the *ultimobranchial bodies* embedded in the thyroid, has an action antagonistic to that of parathormone. It is a peptide chain containing 32 amino acids and a 23-membered ring formed by an intrachain disulfide bridge between two cysteines.

PROTEINS.   The gonadotropins, thyrotropin and growth hormone, secreted by the anterior pituitary, are all true proteins with molecular weights of 25,000 or more. The *growth hormones* of different mammals differ markedly in molecular weight. Man and other primates have growth hormones with molecular weights of about 25,000, whereas those of sheep and ox have molecular weights of about 45,000 or 48,000. Growth hormone is a single peptide chain containing 84 amino acids. *Calcitonin,* sehuman and beef growth hormones will stimulate growth in the rat. Human growth hormone has been found to consist of 188 amino acids in a single chain. As much as 10 per cent of the dry weight of the human pituitary may be growth hormone. It is curious but a man in his eighties has about as much growth hormone in his pituitary as a rapidly growing child does. The growth hormone is rapidly degraded after it has been secreted into the plasma; its biological half-life is about 25 minutes.

Thyrotropin or TSH is a basic glycoprotein with a molecular weight of

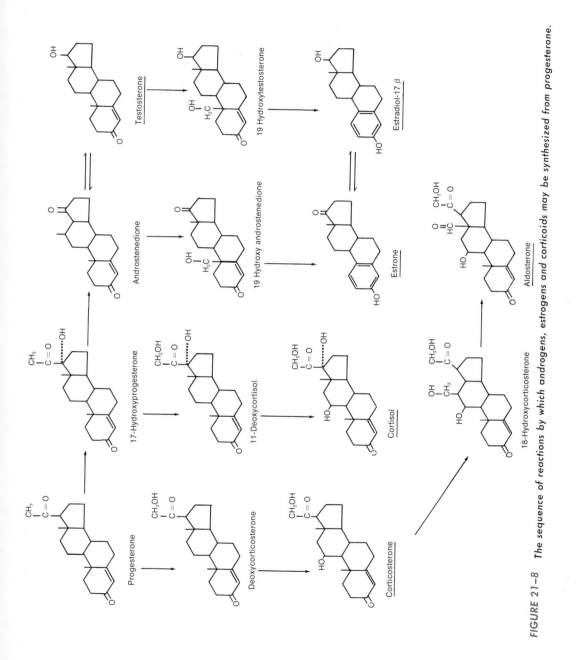

FIGURE 21-8   The sequence of reactions by which androgens, estrogens and corticoids may be synthesized from progesterone.

about 25,000. It contains several kinds of carbohydrates — N-acetyl galactosamine, mannose and fucose.

The gonadotropins, **follicle-stimulating hormone,** FSH, and **luteinizing hormone,** LH, secreted by the anterior lobe of the pituitary, are both glycoproteins. FSH, with a molecular weight of about 31,000, contains about 8 per cent carbohydrate, including some **sialic acid.** FSH loses its biological activity when the sialic acid is split by the enzyme **neuraminidase** or when the other carbohydrate components are removed by treating with amylase. LH is a slightly smaller protein, with a molecular weight of about 26,000. LH's from different species have carbohydrate contents ranging from 4.5 per cent in human LH to 11 per cent in sheep LH. A third gonadotropin, **prolactin** or **luteotropic hormone,** is a protein composed of a single peptide chain of 205 amino acids, and has a molecular weight of about 25,000.

STEROIDS.   The adrenal cortex, testis and ovary secrete steroids, synthesized from cholesterol and composed of 18, 19 and 21 carbons arranged in four connected rings. Three of the rings have six carbons and the fourth has five. It is a curious fact that the primary female sex hormone **estradiol,** with 18 carbons, is synthesized from the male sex hormone **testosterone** which has 19 carbons (Fig. 21–8). This in turn is synthesized from the second type of female sex hormone, **progesterone,** which has 21 carbons. Progesterone is also a precursor in the adrenal cortex of both glucocorticoids and mineralocorticoids. **Glucocorticoids** such as cortisol stimulate the conversion of proteins to carbohydrates, and **mineralocorticoids** such as aldosterone regulate sodium and potassium metabolism. The adrenal cortex of both men and women produces dehydroepiandrosterone and adrenosterone, 19-carbon steroids with slight male sex hormone activity, and a small amount of the potent male sex hormone testosterone.

The hormone stimulating molting in insects, **ecdysone,** has also been found to be a steroid with 27 carbons arranged in four rings and a tail like the cholesterol molecule (Fig. 21–9). The cholesterol molecule is assembled enzymatically by putting together five-carbon units termed **isoprenoid units.** Six of these are put together to make **lanosterol** with 30 carbons, then three methyl groups are removed enzymatically to yield

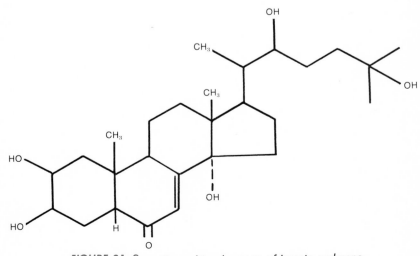

FIGURE 21–9   The molting hormone of insects, ecdysone.

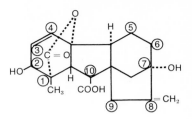

FIGURE 21-10   The structure of one of the gibberellins, a plant hormone with strong growth-promoting properties. The system of numbering the carbon atoms in the molecule is indicated.

cholesterol with 27 carbons. Insects synthesize ecdysone from cholesterol but cannot synthesize cholesterol; it must be a constitutent of their diet.

GIBBERELLINS.   The final group of compounds with hormonal activity includes substances with specific growth promoting effects in plants. In investigating a disease of rice plants that caused excessive elongation of the shoot, Kurosawa found that the active substance was secreted by an infecting fungus named *Gibberella fujikuroi* and named the substance **gibberellin**. Some nine or more kinds of gibberellins have been identified and found to be normal constituents of green plants. Gibberellins are composed of two six-membered carbon rings joined in an intricate fashion to two five-membered carbon rings (Fig. 21-10). They are, like the steroids, derived from isoprenoid units via 20 carbon intermediates, **diterpenes**. The growth-promoting effect of these compounds was strikingly shown in 1956 by Radley, who found he could make dwarf pea plants grow tall by providing them with a gibberellin extracted from normal tall pea plants.

SECTION 21-3

THE ENDOCRINE GLANDS OF MAN

The location of the glands of the human body known to secrete hormones is shown in Figure 21-11. The source and major physiological effects of the principal hormones of man are summarized in Table 21-1.

The **thyroids** are a pair of glands located in the neck, joined by a narrow isthmus of tissue which passes in front of the trachea just below the larynx. The thyroids consist of groups of cuboidal epithelial cells arranged in follicles, hollow spheres one cell thick. The cavity of the sphere contains a gelatinous colloid. secreted by the epithelial cells (Fig. 21-12).

The **parathyroids** are four masses of tissue about the size of a small pea, which are attached to or embedded in the substance of the thyroid gland. The cells of the parathyroids are arranged in a compact mass and not in follicles like those of the thyroid (Fig. 21-12).

Scattered among the acinar cells of the pancreas that secrete the digestive enzymes are a million or more islands of endocrine tissue called the **islets of Langerhans**. These contain two types of cells which can be readily distinguished in histological sections. The β cells secrete insulin and the α cells secrete glucagon (Fig. 21-12).

The paired **adrenal glands** lie at the upper end of each kidney and are combinations of two entirely independent glands, the adrenal medulla which secretes epinephrine and norepinephrine and the adrenal cortex which secretes the adrenal cortical steroids. The two parts have different embryonic origins and different cellular structures. In some of the lower

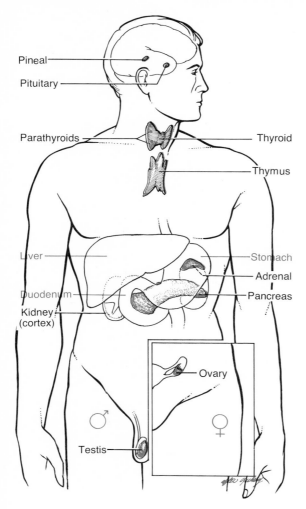

Pineal

Pituitary

Parathyroids

Thyroid

Thymus

Liver

Stomach

Adrenal

Duodenum

Pancreas

Kidney (cortex)

Ovary

Testis

FIGURE 21–11   *The approximate locations of the major endocrine glands in man.*

vertebrates the cells corresponding to the medulla and cortex are spatially separated. The adrenal medulla is derived embryologically from the same source as the nervous system and its cells look like modified nerve cells. The medulla secretes large quantities of epinephrine when the man or animal is frightened or angered. The epinephrine promotes several responses, all of which are helpful in coping with emergencies—the blood pressure rises, the heart rate increases, the glucose content of the blood rises, the spleen contracts and squeezes out a reserve store of blood, the clotting time of blood is decreased, the pupils dilate and the muscles which erect the hairs contract, providing a thicker protective mat in those mammals with fur and gooseflesh in man.

The **pituitary** (also called the hypophysis), which lies in a small depression on the floor of the skull just below the hypothalamus, is a double gland about the size of a pea. Its anterior lobe forms in the embryo as an outgrowth of the roof of the mouth. The posterior lobe grows down from the floor of the brain. The two parts meet and the anterior lobe grows partly around the posterior one. The anterior lobe loses its connection with the mouth but the posterior lobe keep its connection with the hypothalamus. Certain cells in the hypothalamus secrete **releasing factors** which bring about the secretion of the hormone of the anterior pituitary. The

TABLE 21–1. Vertebrate Hormones and Their Physiologic Effects

Hormone	Source	Physiologic Effect
Thyroxine	Thyroid gland	Increases basal metabolic rate
Parathormone	Parathyroid glands	Regulate calcium and phosphorus metabolism
Calcitonin	Ultimobranchial bodies	
Insulin	Beta cells of islets in pancreas	Increases glucose utilization by muscle and other cells, decreases blood sugar concentration, increases glycogen storage and metabolism of glucose
Glucagon	Alpha cells of islets in pancreas	Stimulates conversion of liver glycogen to blood glucose
Secretin	Duodenal mucosa	Stimulates secretion of pancreatic juice
Cholecystokinin	Duodenal mucosa	Stimulates release of bile by gallbladder
Epinephrine	Adrenal medulla	Reinforces action of sympathetic nerves; stimulates breakdown of liver and muscle glycogen
Norepinephrine	Adrenal medulla	Constricts blood vessels
Cortisol	Adrenal cortex	Stimulates conversion of proteins to carbohydrates
Aldosterone	Adrenal cortex	Regulates metabolism of sodium and potassium
Dehydroepiandrosterone	Adrenal cortex	Androgen, stimulates development of male sex characters
Growth hormone	Anterior pituitary	Controls bone growth and general body growth; affects protein, fat and carbohydrate metabolism
Thyrotropin	Anterior pituitary	Stimulates growth of thyroid and production of thyroxine
Adrenocorticotropin (ACTH)	Anterior pituitary	Stimulates adrenal cortex to grow and produce cortical hormones
Follicle-stimulating hormone (FSH)	Anterior pituitary	Stimulates growth of graafian follicles in ovary and of seminiferous tubules in testis
Luteinizing hormone (LH)	Anterior pituitary	Controls production and release of estrogens and progesterone by ovary and of testosterone by testis
Prolactin (LTH)	Anterior pituitary	Maintains secretion of estrogens and progesterone by ovary; stimulates milk production by breast; controls "maternal instinct"
Oxytocin	Hypothalamus, via posterior pituitary	Stimulates contraction of uterine muscles and secretion of milk
Vasopressin	Hypothalamus, via posterior pituitary	Stimulates contraction of smooth muscles; antidiuretic action on kidney tubules
Melanocyte stimulating hormone	Anterior lobe of pituitary	Stimulates dispersal of pigment in chromatophores
Testosterone	Interstitial cells of testis	Androgen; stimulates development and maintenance of male sex characters
Estradiol	Cells lining follicle of ovary	Estrogen; stimulates development and maintenance of female sex characters
Progesterone	Corpus luteum of ovary	Acts with estradiol to regulate estrous and menstrual cycles
Chorionic gonadotropin	Placenta	Acts together with other hormones to maintain pregnancy
Placental lactogen	Placenta	Has effects like prolactin and growth hormone
Relaxin	Ovary and placenta	Relaxes pelvic ligaments

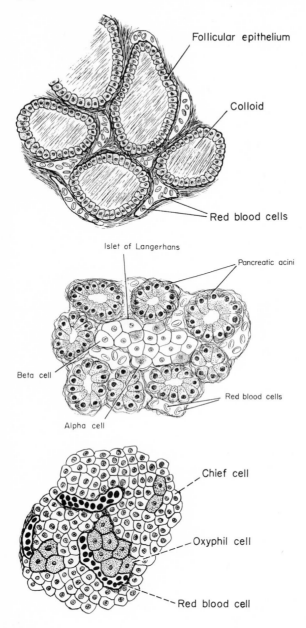

Follicular epithelium

Colloid

Red blood cells

Islet of Langerhans

Pancreatic acini

Beta cell

Red blood cells

Alpha cell

Chief cell

Oxyphil cell

Red blood cell

FIGURE 21–12   *Diagrams of the microscopic structure of the thyroid (above), the islets of Langerhans in the pancreas (center) and the parathyroid (below). (From Guyton, A. C.: Function of the Human Body. Philadelphia, W. B. Saunders Company, 1969.)*

first of these to be chemically characterized was **thyrotropin-releasing factor** (TRF) shown to be a tripeptide, pyroglutamyl-histidyl-proline amide.

The anterior lobe contains at least five different types of cells that differ in their shape, size, staining properties and the kind of granules present in the cytoplasm. It seems likely that each type produces and secretes a different kind of hormone. The cell secreting growth hormone has been identified as a rounded cell whose cytoplasm is packed with dense, round, acidophilic granules. These have been isolated and shown to have a high content of growth hormone. The cells secreting prolactin stain deeply with carmine stain and contain granules that are larger and more ovoid than the granules in the growth hormone-secreting cells.

In between the seminiferous tubules that produce the sperm are *interstitial cells* (Fig. 21–13) which produce and secrete the male sex hormones (androgens) such as *testosterone.* If the testes remain in the abdominal cavity instead of descending into the scrotal sac the seminiferous tubules degenerate and the man is sterile, but his interstitial cells are normal and secrete a normal amount of testosterone. The sperm-forming cells are particularly susceptible to heat and the scrotal sac, which is about three degrees cooler than the abdominal cavity, provides an environment in which the cells can develop.

The cells of the ovary that produce and secrete the steroid sex hormones are the cells lining the cavity of the follicle and the cells of the corpus luteum, formed from these cells after ovulation has occurred. The ovaries secrete two types of female sex hormones, estrogens such as *estradiol,* and *progesterone.*

The *pineal gland*, a small round structure on the upper surface of the thalamus, lying between the cerebral hemispheres, is derived embryologically as an outgrowth of the brain. It secretes *melatonin*, a methoxy indole synthesized from tryptophan, which is converted to 5-hydroxy tryptamine (serotonin) (Fig. 21–14). Serotonin is acetylated at the amine nitrogen and a methyl group is added to the 5-hydroxy group by *hydroxy indole-O-methyl transferase,* an enzyme found uniquely in the pineal. The enzyme is stimulated by norepinephrine, released by the tips of the sympathetic nerves that extend to the pineal from a cervical sympathetic ganglion. Light falling on the retina of the eye increases the synthesis of melatonin by the pineal. A small nerve, the *inferior accessory optic tract,* passes from the optic nerve through the medial forebrain and connects with the sympathetic nervous system. The pineal secretion inhibits ovarian functions either directly or via an effect on the pituitary. Girls blind from birth undergo puberty earlier than normal, apparently because they lack the inhibitory effect of melatonin on ovarian function.

The *placenta* is primarily an organ for the support and nourishment of the developing fetus but also is an endocrine organ, secreting the steroids estradiol and progesterone and two protein hormones. *Chorionic gonadotropin* is similar in many respects to the luteinizing hormone secreted by the pituitary, and *placental lactogen* has some properties similar to growth hormone and some similar to prolactin. During pregnancy large amounts of chorionic gonadotropin are produced and pass into the urine of the mother. This is the basis for one of the classic pregnancy tests: an extract of the urine of a pregnant woman will have characteristic effects

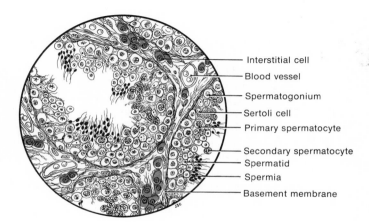

FIGURE 21–13   Section of a human testis showing parts of several seminiferous tubules containing cells in various stages of spermatogenesis. Between the seminiferous tubules are the interstitial cells that synthesize and secrete testosterone.

— Interstitial cell
— Blood vessel
— Spermatogonium
— Sertoli cell
— Primary spermatocyte
— Secondary spermatocyte
— Spermatid
— Spermia
— Basement membrane

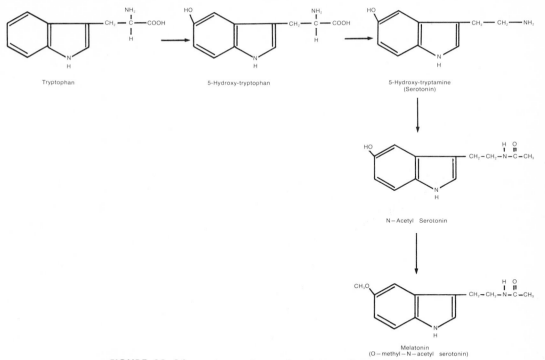

Tryptophan

5-Hydroxy-tryptophan

5-Hydroxy-tryptamine
(Serotonin)

N—Acetyl Serotonin

Melatonin
(O—methyl—N—acetyl serotonin)

FIGURE 21-14   *The synthesis of melatonin from tryptophan.*

on the ovaries of a test animal, rabbit or rat, or will cause sperm production in a frog or toad. This test makes possible the diagnosis of pregnancy a few weeks after conception. Much more sensitive immunoassays for HCG, which can diagnose pregnancy just a few days after implantation, have now largely replaced the bioassays. The placenta produces large amounts of the steroid hormones, especially in the last month or two of gestation—some 250 mg of progesterone and 30 mg of estrogens each day. The placentas of some animals, such as the rabbit, produce another protein hormone, **relaxin**, which is also produced by the ovary. This relaxes the ligaments of the pelvis just before birth to facilitate the passage of the young through the birth canal. Relaxin is effective only after the connective tissue of the pubic symphysis has been sensitized by prior treatment with estradiol.

The lining of the digestive tract produces hormones that stimulate or inhibit the secretion of digestive juices—**gastrin**, secreted by the mucosal cells in the pyloric region of the stomach, and **secretin, cholecystokinin** and **enterogastrone**, secreted by the mucosal cells of the duodenum.

TARGET ORGANS.   All the hormones secreted by the endocrine glands in man and other vertebrates (with the possible exception of the prostaglandins) are secreted into the blood stream and carried by the blood to all parts of the body. A few hormones, such as thyroxine or growth hormone, affect the metabolic conditions of every cell in the body; every cell responds to the presence of the hormone and every cell shows an altered metabolic state when deprived of the hormone. Most hormones, however, affect only certain cells in the body, despite the fact that the blood stream carries them to all parts of the organism. Only the pancreas responds to the secretin circulating in the blood. The cells that respond to a given hormone are called the "target organ" of that hormone. The thyroid gland is the target organ of thyrotropin or TSH secreted by the pituitary and the ovary or testis is the target organ of the gonadotropins FSH and LH from the pituitary. Some hormones, such as estradiol, have marked effects on their primary target organs, the uterus, vagina and mammary glands, lesser effects on other characters, the voice, distribution of hair on the body, bone growth and so forth and still smaller effects on other tissues.

It is now clear that the ability of a tissue to respond to estradiol is correlated with the presence in the tissue of a protein that specifically takes up and binds estradiol. The uterus, vagina, pituitary and hypothalamus contain this protein and can accumulate estradiol multifold from the blood. It is assumed that this protein plays a role in transporting the estradiol from outside the cell into the nucleus where it produces its effect. There is evidence for comparable protein receptors in the oviduct which are specific for progesterone and receptors in the prostate specific for testosterone. The specificity of the effects of the peptide and protein hormones is also a property of the cells responding to them, although the mechanism involved may be something other than a "receptor" molecule.

As a result of the process of cellular differentiation (about which we know very little, p. 877), certain cells become competent to respond to a certain hormone whereas others do not. Other cells develop the ability to secrete the hormone, others do not. Thus both the *secreting* cells and the *responding* cells must develop their special potentialities to have an effective chemical coordination system. A remarkably striking example of this is seen in an inherited human condition called **"testicular feminization."** Individuals with this condition are genetic males, XY, and secrete normal male levels of testosterone, yet they are phenotypically female, with typical female breast development and female external genitalia. The vagina is a short blind pouch and the rest of the internal genitalia are rudimentary or completely absent. The gonads are typical undescended testes. After puberty these individuals develop typical feminine proportions, voice and habitus, and are usually quite attractive. Some could readily qualify for a *Playboy* centerfold. For some reason not yet clear the cells of these individuals are unable to respond to androgens (perhaps they lack the specific protein androgen receptors), and they develop in part as females under the stimulus of the estrogens secreted by the testis and adrenal.

SECTION 21–4

ENDOCRINES AS SIMPLE TRANSMITTERS OF INFORMATION

Perhaps the simplest model of hormone action is the kind of "information transmitter' exemplified by secretin (Fig. 21–15). The presence of acidic food in the duodenum stimulates the release of secretin from the cells of the duodenal mucosa which passes via the blood to all cells in the body. The specific arrangement of amino acids in the 27-amino acid peptide means little or nothing to any cell except the acinar cells of the pancreas, and to them it means "secrete your enzymes." The duodenal mucosal cells continue to secrete secretin as long as acid food is present. They also secrete cholecystokinin, another transmitter of information. This hormone stimulates the gallbladder to contract and release bile, which passes down the bile duct to the duodenum and plays a role in neutralizing the contents of the duodenum. It thus turns off the stimulus for the secretion of both hormones.

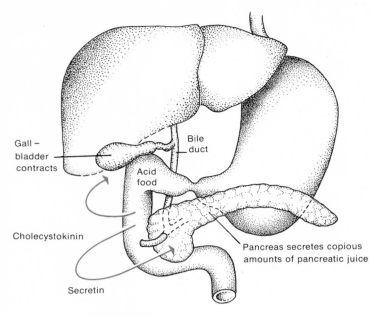

Gall-
bladder
contracts

Bile
duct

Acid
food

Cholecystokinin

Pancreas secretes copious
amounts of pancreatic juice

Secretin

FIGURE 21–15 *The con-*
trol of the secretion of pancreatic
juice and bile by secretin and
cholecystokinin. The hormones
are released from the lining of
the upper duodenum when the
cells are stimulated by acidic
food present in the lumen of the
duodenum.

SECTION 21–5

ENDOCRINES AS LIMIT CONTROLS

A second, somewhat more complex model of endocrine function is one in which two hormones act together to set upper and lower limits for some physiological function. As an example, let us consider the regulation of the amount of glucose in the blood by the hormones insulin and glucagon (Fig. 21–16). Following a meal rich in carbohydrates or following an injection of glucose the concentration of glucose in the blood rises (1). Some glucose is taken up by the liver and stored as glycogen but the rise in the concentration of glucose is the signal for the secretion of insulin (2). A rise in the concentration of insulin in the blood can be detected within two minutes after an experimentally induced rise in blood glucose. A major effect of insulin is its dramatic increase in the rate of transport of glucose into skeletal muscle and adipose tissue. This leads to a decrease in the concentration of glucose in the blood (3). The secretion of glucagon by the alpha cells of the pancreas is also controlled by the concentration of glucose in the blood but is inhibited by high concentrations of glucose and stimulated by low concentrations of glucose. Thus as glucose concentrations fall below the optimal level (4) the release of glucagon is stimulated (5). This, by activating the **glycogen phosphorylase** system in the liver, stimulates the conversion of glycogen to glucose-1-phosphate, which is converted in turn to free glucose via glucose-6-phosphate and is secreted. The glucose secreted by the liver returns the blood concentration of glucose to normal and serves to define the "lower limit."

The pancreas continually monitors the concentration of glucose in the blood passing through it and secretes either insulin or glucagon as required. An increased concentration of glucose in the blood serves as the stimulus to a system—the secretion of insulin by the β cells of the islets in the pancreas—which will return it to normal. A decreased concentration of glucose in the blood serves as a stimulus to the opposing system—the secretion of glucagon by the α cells of the pancreatic islets—which will return it to normal.

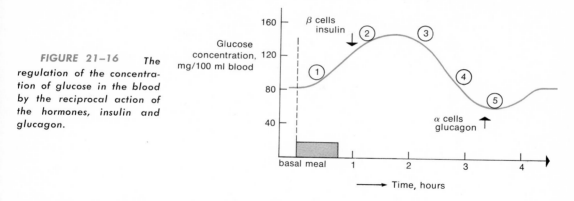

FIGURE 21–16   *The regulation of the concentration of glucose in the blood by the reciprocal action of the hormones, insulin and glucagon.*

Another example of a system of limit controls is the regulation of the concentration of calcium in the blood by the hormones parathormone and calcitonin, which tend to increase and decrease respectively the concentration of calcium in the blood. **Parathormone**, secreted by the parathyroid glands, causes the release of calcium from the bones and teeth, and an increased concentration of calcium in the blood. Parathormone is secreted in response to a decrease in the concentration of calcium in the blood and, by causing the dissolution of bone mineral, brings about a release of calcium (and phosphate) and an increased concentration of calcium in the blood. Calcitonin is secreted by the **ultimobranchial bodies** embedded in the thyroid in response to an increase in calcium concentration in the blood above normal and stimulates the deposition of calcium phosphate in the bones. Thus calcitonin regulates the upper limit of calcium concentration and parathormone regulates its lower limit.

SECTION 21–6

ENDOCRINES AS RECIPROCAL OR NEGATIVE FEEDBACK CONTROLS

The major glucocorticoid produced by the human adrenal cortex is **cortisol**. This hormone promotes the mobilization of amino acids from skeletal muscle and other peripheral tissues and increases the conversion, in the liver, of the carbon chains of these amino acids into glucose and glycogen, a process termed **gluconeogenesis** (the new formation of glucose). It decreases the utilization of glucose in peripheral tissues and favors the mobilization of lipids and the formation of ketone bodies. Its effects in general are opposite to those of insulin.

The synthesis of cortisol in the adrenal cortex is stimulated by ACTH. This pituitary hormone, which increases the size of the adrenal cortex by stimulating the synthesis of RNA and protein, specifically stimulates the production of cortisol from cholesterol by increasing the activity of one or more of the enzymes concerned in this conversion. Thus stimulation of the adrenal cortex by ACTH leads to an increased production of cortisol and an increased concentration of cortisol in the blood (Fig. 21–17 A). This in turn controls, by a negative feedback relationship, the secretion of ACTH by the pituitary. Cortisol may do this by directly inhibiting the synthesis of ACTH in the pituitary or indirectly by decreasing the production of corticotropin-releasing factor, CRF, by the hypothalamus.

The primary stimulus that elicits the secretion of cortisol is some

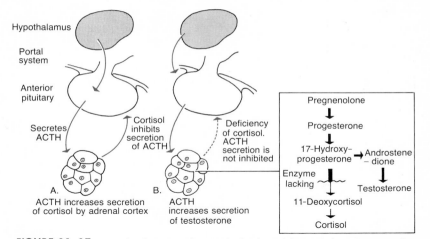

FIGURE 21–17   *Left, the normal control of adrenal function by ACTH. A releasing factor from the hypothalamus stimulates the pituitary to secrete ACTH, which increases the synthesis of cortisol in the adrenal cortex. The cortisol then inhibits the secretion of ACTH by the pituitary. Right, in one type of adrenal cortical virilism, the adrenals lack an enzyme and the conversion of progesterone to cortisol is greatly reduced. The production of cortisol cannot be increased by ACTH and ACTH secretion is not inhibited. The secretion of ACTH remains high, stimulating an increased size of the adrenal and the secretion of androgens such as testosterone instead of cortisol.*

sort of physical stress—an injury, a burn, a painful disease, exposure to heat or cold—that sends impulses to the brain that are forwarded to the hypothalamus. The hypothalamus secretes the corticotropin-releasing factor which passes along a special portal system of blood vessels directly to the anterior lobe of the pituitary. This stimulates the appropriate cells in the pituitary to secrete ACTH, which is carried in the blood to the adrenal cortex, and stimulates the production and release of cortisol. The mobilization of amino acids and lipids from peripheral tissues and the gluconeogenesis in the liver provide substrates for the repair of the damage and decrease the stimulus which led to the production of the releasing factor. The increased concentration of cortisol in the blood acts in the hypothalamus or pituitary or both to decrease the production and release of ACTH.

An inherited defect of any one of the several enzymes involved in the synthesis of cortisol from pregnenolone and cholesterol may lead to enlargement of the adrenal cortex. The commonest defect is a deficiency of the enzyme adding a hydroxyl group at carbon 21 of the steroid: such a defect leads to an accumulation of intermediates in the biosynthetic pathway; some of which, since they cannot be converted to cortisol, are converted to androgens such as **androstenedione** (Fig. 21–17 B). This may be converted either in the adrenal or elsewhere in the body into testosterone, the most potent androgen. The failure of the adrenal cortex to secrete cortisol results in the oversecretion of ACTH, since there is nothing to shut it off. The adrenal grows larger and secretes even more androgens and the individual becomes virilized. A female fetus lacking the enzyme has external genitalia that are masculinized to varying degrees but a male fetus with the enzymatic defect may show no abnormality at birth. After birth virilization progresses in both males and females with enlargement of the phallus, early development of pubic and axillary hair, lowering of the voice and other effects of androgens. Patients with this condition, **adrenal cortical hyperplasia**, can be treated by injecting cortisol to turn off the production of ACTH by the pituitary.

A similar feedback system regulates the synthesis of TSH by the pituitary and of thyroxine by the thyroid gland. The anterior pituitary secretes TSH which stimulates several processes in the thyroid gland involved in the production and secretion of thyroxine. An increased concentration of thyroxine in the blood leads to a decreased secretion of TSH, whereas a decreased concentration of thyroxine in the blood leads to an increased secretion of TSH. This reciprocal feedback relationship maintains a fairly constant concentration of both in the blood.

The release of TSH from the pituitary can be increased by a TSH-releasing factor secreted by the hypothalamus in response to neural stimulation triggered by lowered body temperature. The increased releasing factor raises the output of TSH which in turn increases the secretion of thyroxine, which increases the metabolic rate in all body cells and raises body temperature.

The existence of these feedback controls raises certain problems in the treatment of a patient with a hormone. Giving thyroxine depresses the output of TSH just as endogenous thyroxine does and this will decrease the output of thyroxine by the patient's own thyroid gland. Similarly giving cortisol to a patient decreases the secretion of ACTH by his pituitary which decreases the size of his adrenal cortex and its secretion of cortisol. When a patient has been treated for a long time with thyroxine and then the treatment is stopped abruptly, the sudden decrease in the concentration of the thyroxine in the blood may cause a greater than normal secretion of TSH by the pituitary and result in undue stimulation of the thyroid.

SECTION 21–7

COMPLEX CONTROL SYSTEMS: REGULATION OF THE ESTROUS AND MENSTRUAL CYCLES

To ensure that fertilization and conception occur, an egg must be released from the ovary at a time when sperm are likely to be present in the oviduct to fertilize it and when the lining of the uterus, the **endometrium**, is in the proper condition to permit the implantation and development of the fertilized egg. The timing and coordinating of these events is controlled by a battery of eight hormones and a series of events in the central nervous system. Three of the hormones are specific releasing factors secreted by the hypothalamus which pass down the **hypophyseal portal system** of blood vessels to the anterior pituitary (Fig. 21–18). The portal system begins with a plexus of capillaries in the median eminence at the base of the hypothalamus and forms vessels which pass down the pituitary stalk to the anterior lobe of the pituitary where they connect with a second network of sinusoids and capillaries. Follicle-stimulating hormone-releasing factor, FRF, luteinizing hormone-releasing factor, LRF, and luteotropic hormone-releasing factor, LTRF, are each small peptides that specifically cause the release of the corresponding protein hormone from the anterior pituitary.

Follicle-stimulating hormone acts on the ovary and causes a few of its primary follicles to begin development (Fig. 21–19). The cells surrounding the ovum proliferate rapidly and begin to secrete estradiol. The second pituitary hormone, **luteinizing hormone,** causes further growth of the follicle and a surge of LH, perhaps coupled with a surge of FSH, causes the follicle to rupture, releasing the egg, the process termed **ovulation.** The remaining follicular cells, under the stimulation of LH and luteotropic

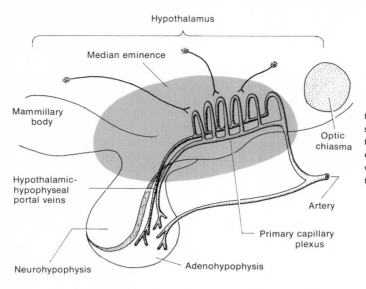

FIGURE 21–18 *Diagram of the relations of the blood vessels supplying the pituitary and hypothalamus, indicating the hypothalamic-hypophyseal portal veins by which hormones are transferred from the hypothalamus to the pituitary.*

hormone, LTH, increase in size, develop a yellow fatty appearance, and form a **corpus luteum** in the space previously occupied by the follicle. The cells of the corpus luteum secrete progesterone and estrogens. The human corpus luteum persists for about two weeks if fertilization does not occur, then degenerates.

The primary estrogenic hormone secreted by the ovary, estradiol, stimulates the development of the characteristic features of the female and plays a role in regulating the cyclic changes of the estrous or men-

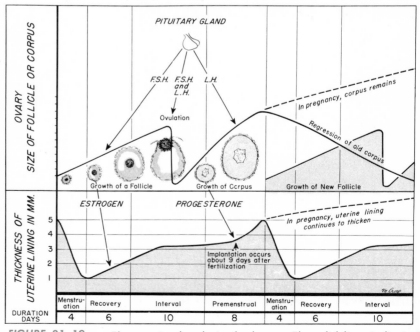

FIGURE 21–19 *The menstrual cycle in the human. The solid lines indicate the course of events if the egg is not fertilized; the dotted lines indicate the course of events when pregnancy occurs. The actions of the hormones of the pituitary and ovary in regulating the cycle are indicated by arrows.*

strual cycle. Estradiol stimulates the growth of the uterus at puberty by causing the muscle cells to increase in number and size. The growth of the vagina, the development of the labia, clitoris and other external genitalia, the growth of pubic hair, the broadening of the hips and changes in the structure of the pelvic bones to the female type, the growth of the breasts, the proliferation of the glandular cells in the breasts, and the deposition of fat in hips and thighs characteristic of adult females are all brought about by estradiol. The growth of the lining of the uterus, the endometrium, during the first part of the proliferative phase of each menstrual cycle is controlled by estradiol (Fig. 21–19).

Progesterone, the second type of female sex hormone, has little effect on the development of the female sexual characteristics but plays a major role in stimulating the development of the endometrium during the second or secretory phase of the menstrual cycle to a condition suitable for the implantation of a fertilized egg. The cells of the endometrium accumulate glycogen and fats, and glands develop which secrete a nutrient fluid. The blood vessels in the endometrium grow, becoming long and coiled. Progesterone also prepares the breasts for the secretion of milk by increasing the secretory activity of the cells in the glands.

The females of most species of mammals show cyclic periods of the sex urge and will permit copulation only at certain times, known as periods of **estrus** or "heat," when conditions are optimal for the fertilization of the egg. Most wild animals have one estrous period per year, the dog and cat have two or three and rats and mice have periods of estrus every five days. Estrus is characterized by heightened sex urge, ovulation and changes in the lining of the uterus and vagina. Following estrus the endometrium thickens and its glands and blood vessels develop to provide an optimal environment for the implanting embryo.

The menstrual cycle of the primates is characterized by periods of vaginal bleeding called **menstruation** (Latin, *menstrualis*, monthly) resulting from the degeneration and sloughing of the endometrial lining of the uterus. Ovulation occurs about midway between two successive menstruations. Primates, unlike other mammals, show little or no cyclic change in the sex urge and permit copulation at any time in the menstrual cycle.

A key point in both estrous and menstrual cycles is **ovulation,** the release of an ovum from the ovary. One group of mammals—the rabbit, cat, ferret, mink—are "reflex ovulators" and the nervous stimulation of mating acts reflexly to bring about ovulation. This direct coupling of copulation and conception may account for the well known rapid procreation of rabbits. Direct electrical stimulation of the appropriate region of the hypothalamus, the **tuber cinereum** and **preoptic regions,** can cause ovulation in the rabbit, cat or monkey. In the rabbit, the neural stimulation of coitus causes the production of LRF, luteinizing hormone-releasing factor, from the hypothalamus, which passes via the portal system to the anterior pituitary and causes a marked rise, a **surge**, in the release of LH from the pituitary. This passes to the ovary and causes the follicle to rupture and release the ovum (Fig. 21–20).

The second and larger group of mammals are "spontaneous ovulators" in which ovulation is not stimulated by coitus but the timing and frequency of ovulation may be influenced by environmental factors. The laboratory rat, kept under normal conditions of day and night lighting, ovulates very early in the morning, between 1:00 and 2:30 A.M. If rats are kept for two weeks or more under artificial conditions in which the periods of light and dark are reversed, the time of ovulation is shifted 12 hours. Rats exposed to continuous light for 24 hours a day eventually stop ovulating and develop persistent vaginal cornification. The primates with men-

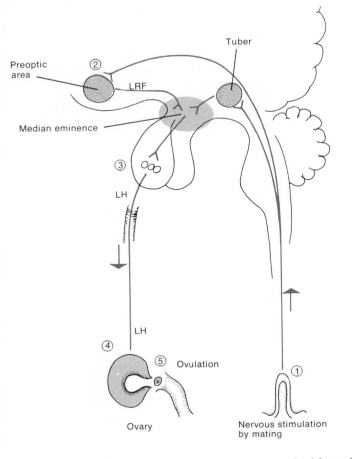

Tuber

Preoptic
area

LRF

Median eminence

LH

LH

Ovulation

Ovary

Nervous stimulation
by mating

FIGURE 21–20   *Diagram of
the nervous pathways involved in the
reflex ovulation in an animal such as
the rabbit. The stimulation of receptors in the vagina by the mating act
generates impulses which pass to the
hypothalamus and bring about the
secretion of releasing factors. These
pass to the pituitary and cause the
release of luteinizing hormone which
goes via the blood stream to the
ovary and initiates ovulation, the
release of the egg.*

strual cycles — women, apes and old-world monkeys — undergo spontaneous ovulation. There is ample evidence that the rhythm of the human
menstrual cycle, and probably of ovulation as well, is influenced by environmental factors. Nurses on night duty and airline hostesses who travel long
distances east and west to different time zones frequently report changes
in their menstrual cycles. The spontaneous ovulators appear to have some
sort of light-dependent hypothalamic clock which provides the neural
stimulation for the release of the hypothalamic-releasing factors. The
releasing factors are produced by neurons that end in the median eminence where they are released and pass into the hypophyseal portal vessels
through which they reach the pituitary. This stimulates the release of a
surge of LH which initiates ovulation in the ovary (Fig. 21–21).

The pituitaries of males as well as of females produce and secrete
FSH and LH, which control the development and function of the testis.
The pituitary of a male transplanted into a hypophysectomized female
will support a normal estrous or menstrual cycle. An ovary transplanted
into a castrate male will develop ripe follicles but will not undergo ovulation — there is no cyclic release of an LH surge from the male pituitary.
The difference between the two sexes — the development of the hypothalamic clock — appears at a critical stage early in the development of the
central nervous system. In the developing male testosterone inhibits the
development of the cyclic center. Young females lacking testosterone do
develop a hypothalamic clock which regulates the rhythmic release of
gonadotropins controlling the sexual cycles. If testosterone is injected

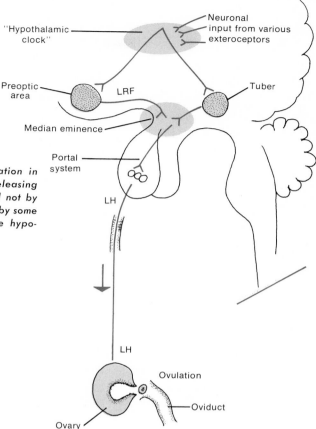

FIGURE 21-21 *Spontaneous ovulation in man and other primates. The secretion of releasing factors from the hypothalamus is stimulated not by nerve impulses coming from the vagina but by some sort of endogenous biological clock in the hypothalamus.*

into a female rat anytime between the second and fifth day after she is born, the activity of the cyclic center in the hypothalamus is permanently abolished. She never ovulates but remains in a state of constant estrus, similar to a rat kept under constant illumination.

Two types of feedback mechanisms are involved in regulating the estrous and menstrual cycles. Just as thyroxine or cortisol decreases the secretion of TSH and ACTH respectively by the pituitary, the concentrations of estradiol and progesterone in the blood affect the output of gonadotropins by the pituitary either directly or by some effect in the hypothalamus. In addition, there is some evidence that the pituitary gonadotropins have a negative feedback effect on the hypothalamus and decrease the output of releasing factors.

The development of sensitive, precise methods for measuring FSH and LH by radioimmunoassays and for measuring estradiol and progesterone by specific protein-binding methods has made possible the measurement of each of these in a small blood sample taken each day in a menstrual cycle (Fig. 21-22). Such assays show that the concentrations of LH and FSH rise abruptly and then fall over a period of two to three days in the middle of the menstrual cycle. The concentration of FSH in the blood is also elevated during the first week or so of the proliferative phase, beginning during the previous menstruation and continuing after the menstrual flow stops. The concentration of estradiol in the blood is low

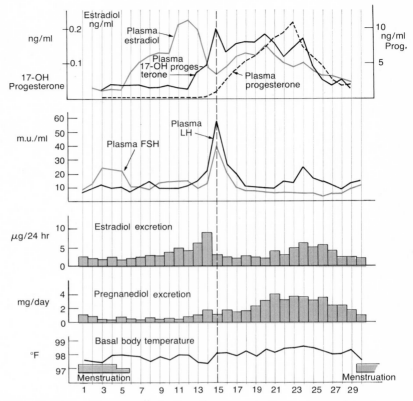

FIGURE 21–22   *Diagram illustrating the concentrations of gonadotropins, estrogens and progestins in the plasma during a single human menstrual cycle. The urinary excretion of estradiol and pregnanediol and the changes in basal body temperature are also shown.*

during the first ten days or so of the proliferative phase but rises sharply and reaches a peak about the time when the concentration of LH begins to rise. The concentration of estradiol then falls and has nearly reached the low basal level at the time the LH concentration reaches its peak. A second, broader and lower peak of estradiol concentration is typically seen during the secretory phase of the cycle. The concentration of progesterone in the blood begins to rise at about the time of the LH peak and reaches a peak about 6 days later. It remains high until nearly the end of the cycle, then falls and the withdrawal of the progesterone precipitates a decreased blood flow to the endometrium, the death and sloughing of the cells in the endometrial lining and the menstrual flow. The menstrual flow, made up of necrotic endometrial tissue, blood that oozes from the ruptured ends of the endometrial blood vessels and tissue fluid from the uterine surface, is gradually expelled from the uterus over a three- to five-day period. The basal portion of the endometrium remains intact during menstruation and is the source of the new epithelium and glands that develop under the influence of estradiol following menstruation.

The regulation of the menstrual cycle appears to involve the following sequence of events (Fig. 21–23).

1. Following the previous menstruation, the withdrawal of progesterone removes its inhibitory influence on the hypothalamus and FRF is released, stimulating the release of FSH from the pituitary. This accounts for the increased level of FSH circulating in the blood early in the cycle.

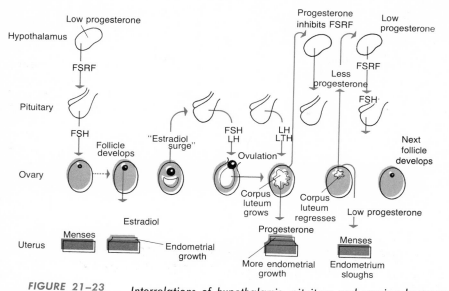

FIGURE 21-23   *Interrelations of hypothalamic, pituitary and ovarian hormones in regulating the events of the menstrual cycle in women.*

2. The FSH causes one or more of the follicles in the ovary to enlarge rapidly and begin to secrete estradiol. The estradiol secreted by the follicular cells causes the proliferation of the endometrial cells. When the follicle reaches a certain size it secretes a surge of estradiol which rises and falls and triggers in turn a surge of FSH and LH which causes ovulation.

3. The peak of LH together with luteotropic hormone secreted by the pituitary causes the follicular cells to undergo luteinization, forming a corpus luteum and increasing the secretion of progesterone.

4. The concentration of progesterone in the blood rises and remains high during most of the secretory phase of the cycle. Progesterone causes the continued growth of the endometrial lining and stimulates the endometrial glands to secrete a nutrient fluid.

5. Progesterone also has an important function in inhibiting FSH release and preventing the development of any additional follicles and eggs. Eventually the corpus luteum begins to regress, the concentration of progesterone in the blood falls and the hypothalamus, freed of its inhibitory effects, releases FRF and a new cycle begins.

If the egg has been fertilized and implants in the endometrium, the cells of the trophoblast in the developing placenta secrete **chorionic gonadotropin**. This has strong luteinizing and luteotropic activities which maintain the corpus luteum and stimulate the continued secretion of progesterone. By the sixteenth week or so of pregnancy the placenta itself produces enough progesterone so that the corpus luteum is no longer needed and undergoes involution.

Progesterone blocks ovulation not by a direct effect on the ovary but by preventing the secretion of the releasing factors FRF and LRF by the hypothalamus. The **oral contraceptives** contain synthetic estrogens and progestins which block ovulation in a similar manner. The natural hormones, estradiol and progesterone, are rapidly metabolized in the body but the synthetic hormones have slight changes in the molecular structure that markedly decrease the rate at which they are destroyed. The oral contraceptives, like the natural hormones, inhibit the release of LH by the pituitary and thus prevent ovulation. A woman taking "the pill" has no midcycle surge of LH and FSH and does not ovulate. Some of the synthetic progestins alter the character of the mucus secreted by the cervix of the uterus and make it less readily penetrated by sperm, thus decreasing the probability of fertilization.

After an egg has been released from the ovary and is passing down the oviduct it retains the ability to be fertilized for a relatively short time, about 24 hours. The sperm deposited in the female reproductive tract during intercourse retain their ability to fertilize an egg for 48 hours or less. The period of maximum fertility in human beings thus narrows down to a few days midway between two successive menstrual periods, at the time of ovulation. It is difficult to determine precisely when this period of maximum fertility occurs because of the variability in the time of ovulation in the menstrual cycle.

COMPLEX CONTROL SYSTEMS: REGULATING THE DEVELOPMENT AND FUNCTION OF THE MAMMARY GLANDS

The growth and development of the mammary glands after puberty and the production and secretion of milk after parturition are controlled by a complex sequence of hormonal events. After puberty the breasts develop and their glandular elements proliferate under the control of estradiol. Progesterone causes further development of glands that have been stimulated previously by estradiol. During pregnancy the breasts undergo further development, stimulated by the large quantities of estradiol and progesterone secreted by the placenta. The gland cells develop and become secreting cells and the ducts and alveoli of the mammary glands develop. Despite their marked effects on the development of the breasts, estradiol and progesterone together *inhibit* the actual production of milk. After parturition there is a marked decrease in the concentration of progesterone circulating in maternal blood and the inhibition of milk production is released. Throughout pregnancy the production of pituitary gonadotropins has been inhibited by the placental gonadotropins. With the disappearance of the placenta and its hormones the pituitary begins to secrete large quantities of luteotropic hormone or prolactin, which stimulates the breast to produce milk.

If the breasts are emptied of milk regularly the production of prolactin by the pituitary continues. If milk is not withdrawn from the breast the production of prolactin in the pituitary ceases and milk production in the breasts halts. The mechanism for this (Fig. 21–24) appears to involve nerve impulses from the nipples which pass via the spinal cord to the hypothalamus, which produces LTRF, *luteotropic hormone-releasing factor*. This passes via the hypophyseal portal system to the pituitary and stimulates the secretion of prolactin. If the nipples are not stimulated by suckling, neither LTRF nor prolactin is secreted and milk production stops.

The release of milk from the mammary gland is under yet another hormonal control system. When a baby sucks on a nipple it usually gets no milk for the first minute or so. Then milk appears in the ducts of both breasts, though suckling has occurred on only one of them. The fact that milk appears in both breasts indicates that some general stimulatory process, rather than some local process confined to the suckled breast, is concerned. Nerve impulses initiated by the suckling of the breast pass via the

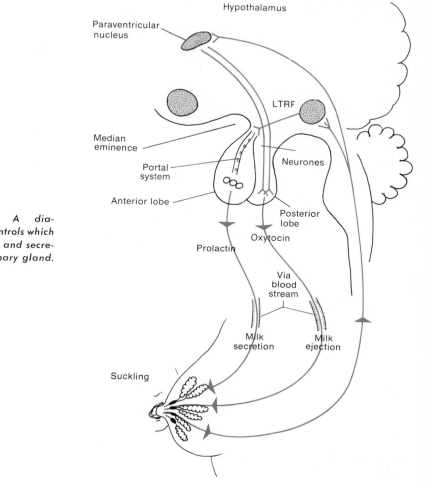

FIGURE 21–24 A diagram of the hormonal controls which stimulate the production and secretion of milk in the mammary gland.

spinal cord to the hypothalamus and stimulate the release of both **vasopressin** and **oxytocin** (Fig. 21–24). Oxytocin is much more effective than vasopressin in producing milk ejection. The oxytocin circulates in the blood to the breasts and stimulates the contraction of myoepithelial cells surrounding the alveoli, raising the pressure within the breast and expressing the milk in the alveoli into the ducts leading to the nipples. Insulin and cortisol play minor roles, not yet clearly defined, in regulating the development of the breasts and the secretion of milk.

SECTION 21–9

THE REGULATION OF GROWTH IN PLANTS

Plant hormones, like animal hormones, are organic compounds which can produce striking effects on cell metabolism and growth even though present in extremely small amounts. The plant hormones are produced primarily in actively growing tissue, especially by meristem tissue in the growing points at the tip of stems and roots. Like animal hormones, the plant hormones usually exert their effects on parts somewhat removed from the site of production. The plant hormones have many different types of effects on metabolism and cell division: (1) they stimulate the longitudinal growth of individual cells in the growing part of the plant, (2) they initiate the formation of new roots, especially adventitious roots, (3) they initiate the development of flowers and the development of fruit from the flower parts, (4) they stimulate cell division in the cambium, (5) they inhibit the development of lateral buds, and (6) they inhibit the formation of abscission regions and hence prevent the fall of leaves and fruit. Three groups of chemically defined substances that occur naturally and control growth and development in flowering plants are the **indole auxins**, the **cytokinins** and **gibberellins.**

Auxins

Some of the first experiments with growth-promoting substances were made by Charles Darwin and his son Francis. It had been known for many years that plants would grow toward the light (were positively phototropic). To see what part of the plant received the light stimulus, Darwin grew a number of canary grass seedlings, covered the tips of some with black paper caps and covered everything but the tips of others with black paper cylinders (Fig. 21–25). He then put all the seedlings near a window so that they received light from one direction only. By the next day the seedlings with no cover at all and the seedlings with everything but the tip covered were both bent very strongly toward the light, but the seedlings

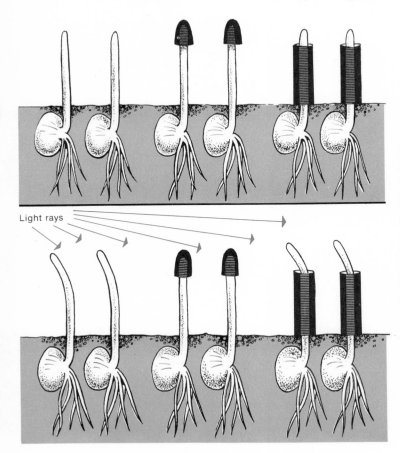

Light rays

FIGURE 21–25 *Darwin's experiment with canary grass seedlings. Upper row: some plants were uncovered, some were covered only at the tip and others were covered everywhere but at the tip. After exposure to light coming from one direction (lower row), the uncovered plants and the plants with uncovered tips bent toward the light; the plants whose tips were covered (center) grew straight up. Darwin's conclusion: the tip of the seedling is sensitive to light and gives off some "influence" which passes down the stem and causes it to bend.*

which had been capped had grown straight up. From these experiments Darwin concluded that the light was received by the tip of the plant and that some "influence" moved down the stem from the tip to cause the plant to bend.

The investigations of Boysen-Jensen of Denmark, Frits Went of Holland and others from 1910 until about 1930 clarified the mechanism underlying tropistic responses and showed that this "influence" was a plant growth hormone. These classic experiments were performed using the **coleoptile** of the oat seedling. The coleoptile is a hollow, practically cylindrical organ that envelops the unexpanded leaves like a sheath. After a certain stage of growth, its further increase in length is due almost entirely to cell elongation. In 1910 Boysen-Jensen discovered that if the coleoptile tip is cut off at this stage, the decapitated coleoptile immediately stops elongating. If the tip is replaced, the coleoptile again begins to grow (Fig. 21–26).

Placing a thin sheet of mica between the tip and the rest of the coleoptile also prevented cell elongation; however, a thin sheet of agar gel could be placed between the tip and the rest of the coleoptile without interfering with the growth of the coleoptile. Indeed, the decapitated tip could be placed for a time on a block of agar, then removed, and the bit of agar, when placed on the coleoptile, would stimulate it to grow (Fig. 21–26 D). These experiments showed that the growth of the coleoptile is controlled by some substance produced in the tip that normally passes downward and stimulates the coleoptile cells to elongate. By leaving a tip on an agar block for varying lengths of time it was shown that the amount of growth

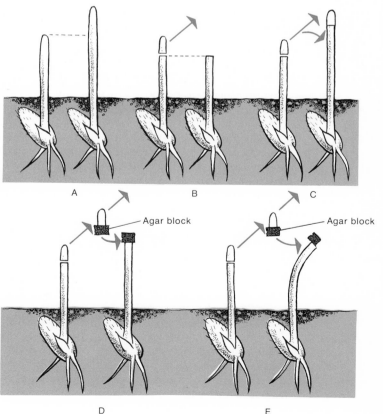

FIGURE 21–26   *A series of experiments which demonstrate the existence and mode of action of plant growth hormones in oat coleoptiles. In each pair of drawings, the figure on the left indicates the experiment performed and the figure on the right the growth of the coleoptile after a period of time. A, Control, no operation performed and normal growth results. B, If the tip of the coleoptile is removed, no growth occurs. C, If the tip is cut off and then replaced, normal growth ensues. D, If the tip is cut off, placed on a block of agar for a time and the agar block but not the coleoptile tip is then placed on the seedlings, growth occurs. E, If the tip is placed on an agar block for a time and the agar block is then placed asymmetrically on the seedling, curved growth results.*

A   B   C

Agar block   Agar block

D   E

substance that diffused into the agar was proportional to the length of time they were in contact. The amount of growth substance was measured by its effect on the elongation of the coleoptile.

Further experiments by Frits Went showed that if the agar block is placed on one side of the decapitated coleoptile, growth is asymmetrical; the coleoptile bends *away* from the side on which the block was placed, i.e., growth is more rapid in the cells directly under the agar block (Fig. 21–26 E). This test is extremely sensitive and Went could measure the growth-promoting substance, called **auxin,** in terms of "curvature units," the number of degrees of bending produced in the coleoptile. This is an example of a **bioassay,** a test in which some chemical substance is measured or assayed in terms of its effect on some biological system.

Auxins are present only in minute quantities even in actively growing tissues. The growing shoot of the pineapple plant contains about six micrograms of indole acetic acid per kilogram of plant material. As J. P. Nitsch concluded, this is comparable to the weight of one needle in a 22-ton haystack!

Indole acetic acid, the primary natural auxin, is rapidly metabolized by indole acetic acid oxidase. This enzyme is inhibited by certain orthodiphenols. The growth-promoting effects of orthodiphenols had been recognized for some time and they were initially believed to be auxins themselves; however, they promote growth by inhibiting indole acetic acid oxidase and thus raising the effective concentration of endogenous indole acetic acid.

Indole acetic acid is synthesized from tryptophan (Fig. 21–3), but the mechanism by which light falling on the growing tip of a plant stimulates the conversion of tryptophan to indole acetic acid remains unclear. Auxin is transported down the stem at speeds ranging from 0.5 to 1.5 cm per hour. The kinetics of auxin transport suggest that this is an active process, driven by metabolic processes, which requires that the substance transported have a certain molecular configuration. The synthetic auxin **2, 4–D** (2,4-dichlorophenoxy acetic acid) is a potent stimulator of plant metabolism but is only poorly transported down the stem and thus does not produce correlated growth of the several parts of the plant as indole acetic acid does.

The growth response of the coleoptile sheath is a metabolic process, for it will occur only in the presence of oxygen and growth is improved if glucose is supplied as a source of energy. The roots, buds and stems each have growth responses that occur at different concentrations of auxin. Each shows an increasing growth response at low concentrations of auxin and then an inhibition of growth at higher concentrations (Fig. 21–27). The auxin concentration which produces optimal growth of stems is much higher than the concentration for optimal growth of roots or buds.

The bending of a plant shoot in response to light, or the bending of the root in response to gravity, is due to differential growth resulting from the differential distribution of auxin. Such tropistic responses can be separated into (1) the perception of the stimulus, light or gravity; (2) the induction of a lateral physiological difference and (3) a lateral response expressed as differential growth. The perception of light involves the absorption of light energy by pigments such as carotenoids or flavins. Unilateral illumination sets up a transverse gradient of absorbed energy which induces differential physiological activity, the production of auxin. If the tip of a coleoptile is illuminated unilaterally, then cut off and placed on the stump of another coleoptile, the latter will respond with differential growth and will bend toward the direction from which light originally

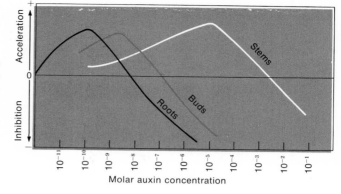

FIGURE 21-27   *The variations in the concentration of indole acetic acid required to induce a growth response in the root, bud and stem of a plant.*

reached the coleoptile tip. Since the side toward the light should absorb more energy, and since the production of auxin is an energy-requiring process, this, at first glance, would seem to account for the bending toward the light. However, a bending toward the light means that the *shaded* side of the plant must grow faster and have more auxin. To account for this some have suggested that light causes greater destruction of auxin on the lighted side of the stem or that light produces oxidation products of auxin which inhibit auxin transport down the lighted side. Others postulate that the plant redistributes the auxin that is synthesized and transports it laterally to the shaded side. Experiments with ^{14}C-labeled auxin did show lateral transport of the auxin following unilateral stimulation by light or gravity.

Auxins may bring about the differentiation of tissues in the parts of the plant through which it is transported or to which it is taken. The differentiation of xylem in a polar fashion is a process which results in the production of an integrated system of pipes that pass vertically through the plant from the apical meristem and expanding leaves down through the stem to the root tips. This differentiation is directed in the terminal regions of the stem by the apical meristem. In an interesting series of experiments, Wetmore and Sorokin showed that a bit of lilac callus grown in tissue culture would not form xylem cells unless a piece of meristem was grafted onto it. However, if auxin was applied in a localized position in place of a meristem graft, some xylem cells were formed in the callus. The apical meristem appears to be a source of auxin which passes down into the tissues below and participates in directing them toward differentiation as xylem.

The differentiation of roots is also controlled by auxins. It has long been known that the lower tip of a cut stem placed in water may form roots. Cuttings from lower in the stem or from older wood show a lesser ability to form roots, which agrees with the decreasing amount of auxin toward the base of the stem. By placing a cutting in a dilute solution of either natural or synthetic auxin, roots can be readily produced.

Auxins and other plant hormones determine the growth correlations of the several parts of a plant. The terminal bud of the stem normally inhibits the development of lateral buds. If the terminal bud is cut off, the lateral buds, freed of the inhibition induced by the auxin produced in the apical meristem of the terminal bud, may begin to develop. That this inhibition is caused by auxin can be demonstrated by removing the terminal bud and replacing it with an appropriate amount of indole acetic acid held in place by a suitable material such as fat. The lateral buds of a stem treated in this fashion remain inhibited.

The abscission or shedding of leaves, flowers, fruits and stems from the parent plant is another process controlled by auxin. Auxin produced in leaves passes down the petiole and inhibits the development of the abscission zone. As long as the leaf continues to produce auxin its abscission or shedding is inhibited. Abscission is a natural indicator of the decreased auxin formation that normally accompanies aging. Low concentrations of auxin promote abscission and high concentrations inhibit abscission.

Auxins have a variety of practical uses and are of tremendous economic importance in stimulating the growth of roots from cuttings, in producing **parthenocarpic fruits** (ones formed without pollination and hence without seeds), in hastening the ripening of fruit and in preventing their dropping from the tree before harvest. The synthetic auxin 2,4-D is widely used as a weed killer. Most of the common weeds are dicotyledonous plants which are much more sensitive to stimulation by auxins than are the monocots such as grasses. A lawn sprayed with the proper concentration of 2,4-D, enough to stimulate the weeds but not the grass, will be freed of weeds. They take up the 2,4-D and are stimulated to metabolize at a high rate, consuming their cellular constituents and finally dying. Many of the important crop plants—corn, oats, rye, barley and wheat—are monocotyledons and millions of acres of these crop plants are treated each year with 2,4-D to eliminate weeds and to increase the yield of these plants.

The molecular mechanism by which auxins produce their remarkably diverse effects is not yet clear. The elongation of a plant cell in response to auxin requires the uptake of water by the cell and a softening of the cell wall to permit the cell to swell as it takes up water osmotically. Auxin does indeed soften the cell wall and increase its plastic bending in response to a standard force. It might do this by altering the chemical structure of the pectins in the cell wall, and some experimental evidence supports this hypothesis. The finding that actinomycin D, which inhibits the synthesis of RNA, also inhibits the elongation of plant cells induced by auxin indicates that the synthesis of RNA is required for at least some aspects of the cell elongation induced by auxin. An enlarging cell would have more plasma membrane and the RNA synthesis might be involved in producing the structural components of the plasma membrane.

Auxin may be regarded as the most important of the plant hormones, since it has the most marked effects in correlating growth and differentiation so that the normal pattern of development results. Auxin, together with certain other chemical messengers, brings about the differentiation of the dividing plant cells into a true organism instead of into a simple multicellular colony.

Gibberellins

Gibberellins increase the length of the stem of some species of plants and increase the size of fruits in others. A unique effect of gibberellins is the stimulation of the germination of seeds. The seeds of wheat, oats, barley and other cereals have two major parts (p. 387), the **embryo** and the reserve food supply, the **endosperm** (Fig. 21–28). The storage cells of the endosperm appear to be dead but are surrounded by the **aleurone,** a three-layered coat of living cells. During germination the starch in the storage cells is hydrolyzed by α-**amylase,** secreted by the aleurone layer; however, the embryo must be present before the aleurone layer will secrete α-amylase. If a grain of wheat is cut in half and the half with the embryo is removed, the starch in the remaining half will not undergo hydrolysis. Gibberellin, a chemical messenger secreted by the embryo, activates the

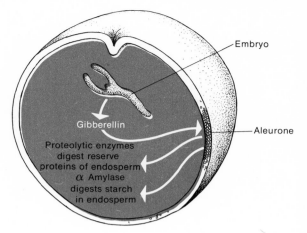

FIGURE 21–28   A diagram of the production of amylase and other enzymes by the aleurone layer of the seed in response to gibberellin secreted by the embryo.

cells in the aleurone layer to produce and secrete α-amylase. Gibberellins activate other enzymes that break down the material in the cells of the seed coats, weakening them so that the growing embryo can burst through. The formation of proteolytic enzymes in the aleurone layer of the endosperm is also stimulated by gibberellin. The resulting hydrolysis of proteins liberates tryptophan which serves as the precursor of indole acetic acid in the tip of the coleoptile of the young embryo.

The synthesis of α-amylase by the aleurone layer in response to gibberellin is completely inhibited by actinomycin D. **Actinomycin D** is a peptide antibiotic which complexes with DNA through the amino group of guanine and displaces DNA-dependent RNA polymerase. Thus the DNA-dependent synthesis of RNA is inhibited. This evidence suggests that gibberellins regulate in some fashion the expression of the genetic information contained in the DNA, perhaps by uncovering a portion of the DNA so that it can be transcribed to produce a specific RNA which in turn results in the formation of the specific enzymes.

More than a dozen different gibberellins have been isolated from plants, each with some slight difference in structure and some difference in biological activity. Some will induce flower formation, others will not. One will produce male sex organs, **antheridia,** on fern gametophytes, but the others will not.

Cytokinins

A third type of plant hormone, the **cytokinins,** stimulate growth of cells in tissue culture or organ culture and have a marked effect in increasing the rate of cell division. **Zeatin**, isolated from young corn seeds, was shown to be a derivative of adenine, 6-(4-hydroxy-3-methyl-*trans*-2-butenyl amino) purine (Fig. 21–4). Cytokinins occur in plants in such small amounts that they were identified only by the technique of mass spectrometry. The side chain attached to the amino group at position 6 of the purine is an isoprenoid and, like the gibberellins, is derived from mevalonic acid. A substance very closely related to zeatin, 6-(γ,γ-dimethylallyl amino) purine was isolated as the ribonucleoside from both serine and tyrosine **transfer RNAs** from yeast, calf liver, peas and spinach. In both kinds of tRNA this odd base is located just adjacent to the anticodon. Five other transfer RNAs have been tested and shown to lack this curious base.

Cytokinins not only promote cell division but can change the structure of plant cells growing in culture. When the concentration of cytokinin in the culture medium is very low (10^{-9} M or less) only loose, friable tissues appear. At somewhat higher concentrations, 10^{-8} to 10^{-7} M, roots develop on the mass of cells in culture and finally at somewhat higher concentrations, 3×10^{-6}M, shoots are induced. This experimental demonstration that cytokinins may control the relative production of shoot and root in tissue culture preparations suggests that they may perform a similar function in the intact plant.

It is not yet possible to link cytokinins with any specific biochemical reaction and the mechanism by which they stimulate cell division is unknown. DNA synthesis in dividing tobacco cells is stimulated by the addition of cytokinin.

In the plant the three major types of plant hormones interact in pairs or all three may interact to regulate specific biological phenomena. The optimal growth of tobacco callus tissue in culture requires specific concentrations of all three factors. Cytokinins and gibberellins have dominant roles in controlling the early phases of growth and development and auxins become dominant later in controlling cell elongation.

Photoperiodism in Plants

It had, of course, been known for a long time that different kinds of plants flower at different seasons of the year and that the time of flowering can be related to the number of hours of daylight per day, the **photoperiod.**

FIGURE 21–29 *Photograph of an experiment suggesting the existence of a flowering hormone in plants. All the petunias received eight hours of daylight each day. The plant on the left with small buds received in addition eight hours of fluorescent light which contains flower-suppressing red light but no infrared. The flowering plant in the center was given an additional eight hours of incandescent light which contains both red light and flower-stimulating infrared light. (Courtesy of the United States Department of Agriculture.)*

Some species of plants (e.g., asters, cosmos, chrysanthemums, dahlias, poinsettias and potatoes) will produce flowers only when the photoperiod is less than about fourteen hours per day. Such plants were termed **"short-day" plants** and normally flower in the early spring or late summer or fall. "Short-day" plants are actually "long-night" ones for the controlling factor is the length of the period of uninterrupted darkness. "Long-night" plants will flower only when exposed to darkness for nine or more hours. They can be made to flower earlier than usual by decreasing their daily exposure to light—by covering them—or they can be kept from flowering by giving them artificial illumination (Fig. 21–29).

Other species of plants (e.g., beets, clover, coreopsis, corn, delphinium and gladiolus) will produce flowers only when the photoperiod is more than 14 hours per day. These **"long-day"** (actually "short-night") **plants** normally flower in the later spring and early summer. If they are covered part of each day, so that their daily exposure to light is less than 13 to 14 hours, flowering is greatly delayed or prevented entirely.

Carnations, cotton, dandelions, sunflowers and tomatoes are examples of indeterminate plants which flower at a given time and are relatively unaffected by the amount of daylight per day. The time of flowering is not controlled solely by the photoperiod, for temperature, moisture, soil nutrients and the amount of crowding may also play a role.

Florigens and Phytochromes

The mechanism by which the length of darkness affects time of flowering is not known in detail, but the results of some experiments suggest that a flower-producing hormone which has been named **florigen** is involved. A typical experiment, using cocklebur, a "long-night" plant, is as follows (Fig. 21–30): One plant is exposed to 12 hours of light per day

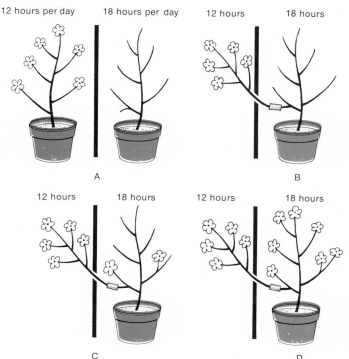

FIGURE   21–30   A diagram of an experiment to demonstrate the existence of a flower-inducing hormone. A, Two cocklebur plants were grown in pots separated by a light-tight partition and exposed to 12 and 18 hours of light respectively per day. The plant exposed to a 12-hour daily photoperiod flowered; the plant exposed to an 18 hour daily photoperiod did not. B, The "12-hour" plant was cut off, inserted through a light, tight hole in the partition and grafted to the "18-hour" plant. The two plants continued to receive 12 and 18 hours of light respectively. The 18-hour plant gradually developed flowers, first on the twigs nearest to the graft (C) and eventually on all the twigs (D). If no graft had been made, the 18-hour plant would not have developed flowers.

12 hours per day   18 hours per day   12 hours   18 hours

A   B

12 hours   18 hours   12 hours   18 hours

C   D

until it is producing flowers. It is then grafted to another plant that had been exposed to 18 hours of light per day (and thus had been inhibited from producing flowers). The two parts, though grafted, are separated by a light-tight partition and the first part continued to receive 12 hours and the second 18 hours of daylight. The "long-night" part of the plant continues to produce flowers and, in time, the "short-night" part of the plant also produces flowers, usually beginning at the point nearest the graft. This is taken as evidence for a diffusible, flower-inducing hormone produced in the leaves and transported in the phloem to the buds. Nothing is known of the chemical composition of this hormone nor of how it might act to induce flowering.

Red light inhibits flowering in "short-day" plants but induces flowering in "long-day" plants. Infrared light induces flowering in "short-day" plants and inhibits flowering in "long-day" plants. A light-sensitive protein pigment, **phytochrome**, appears to have a basic role in photoperiodism. This exists in two forms, phytochrome$_{660}$, sensitive to red light (660 mμ), and phytochrome$_{735}$, sensitive in infrared light (735 mμ). P$_{660}$ appears to be the quiescent form in which the plant stores the potentially active compound and P$_{735}$ is the active material. Light converts inactive P$_{660}$ to active P$_{735}$ and infrared light converts P$_{735}$ to P$_{660}$; these are photochemical reactions that occur as rapidly at 0° C as at 35° C. P$_{735}$ is also converted to P$_{660}$ in the dark by a reaction mediated by enzymes and requiring oxygen. During the day phytochrome exists predominantly in the P$_{735}$ form and during the night it gradually is converted enzymatically into the P$_{660}$ form. This could provide the plant with a means of detecting whether it is light or dark. The rate at which P$_{735}$ is converted to P$_{660}$ might provide the plant with a "clock" for measuring the duration of darkness.

A number of plant functions other than flowering may also be affected by the daily photoperiod: the formation of tubers by Irish potatoes is accelerated when the daily exposure to light is shortened. Since the growth of the tuber (the part we know as the potato) involves the deposition of starch, the photoperiod must in some way stimulate the transfer of carbohydrates from the leaves to the tuber.

SECTION 21–10

THE HORMONAL CONTROL OF MOLTING IN INSECTS

Some insects, such as the grasshopper, pass through a series of successive molts during development. With each molt they look more like the adult but there is no striking change in appearance with any one molt (Fig. 21–31). In contrast, moths, butterflies, flies and many other insects pass through successive stages that are quite unlike one another. From the egg hatches a wormlike *larva*—called a caterpillar (moths) or maggot (flies)—which crawls about, eats voraciously and molts several times, each time becoming a larger larva. The last larval molt forms a *pupa*, which neither moves nor feeds. Moth and butterfly larvae spin a cocoon and pupate within that. During pupation the structures of the larva are broken down and used as raw materials for the formation of parts of the adult animal. Each part of the adult—legs, wings, eyes, antennae—develops from a group of cells called a *disc*, which develops directly from the egg. They have never been a functional part of the larva but remain quiescent during the larval period. During the pupal stage these discs grow and differentiate into the adult structures, but remain collapsed and folded.

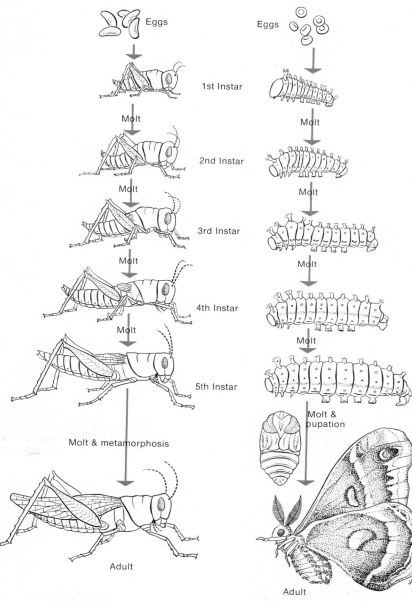

FIGURE 21–31   *Comparison of the life histories of an insect with incomplete meta-morphosis, the grasshopper (left) and an insect with complete metamorphosis, the giant silkworm, Platysamia cecropia (right).*

When the adult hatches out of the pupal case, blood is pumped into these collapsed structures, they unfold and inflate, and chitin is deposited to make them hard. This striking change in appearance from larva to adult is termed **metamorphosis**. The grasshopper, which undergoes a gradual change in form from larva to adult, is said to exhibit incomplete metamorphosis.

The larva and adult insect not only have a different appearance, they have quite different modes of life. The butterfly larva eats leaves; the adult drinks nectar from flowers. The mosquito larva lives in ponds and eats algae and protozoa; the adult sucks the blood of man and other mammals. The adults of some species, such as the mayfly, live only a few hours, just long enough to mate and lay eggs.

Because its exoskeleton is firm and rigid, an insect can grow and change its appearance only by **molting**, by periodically shedding its chitinous exoskeleton and growing a new one. Before shedding the old exoskeleton the insect develops a new one underneath the old (Fig. 21–32); some of the materials of the old exoskeleton are salvaged and used in the synthesis of the new one. A line appears along which the old skeleton will break and the insect swells up by swallowing water or air, bursting the old exoskeleton and stretching the new one to its full size. There is a period of time when the insect is likely to dry out or to be eaten by a predator while it lacks its hard exoskeleton.

The molting of insects and other arthropods is under hormonal control. The major organ initiating molting is the **intercerebral gland** on the surface of the brain (Fig. 21–33). Axons from these neurosecretory cells pass posteriorly to the **corpora cardiaca**, composed of the expanded tips

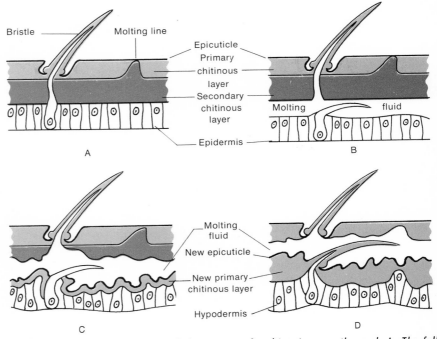

FIGURE 21–32   *Diagram of the process of molting in an arthropod. A, The fully formed exoskeleton and underlying epidermis between molts. B, Separation of the epidermis and the secretion of molting fluid and a new epicuticle. C, Digestion of the old secondary chitinous layer and secretion of a new primary chitinous layer. D, The situation just before the old exoskeleton is shed.*

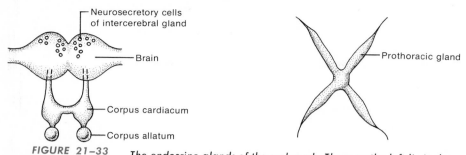

FIGURE 21–33   *The endocrine glands of the cockroach. Those on the left lie in the head dorsal to the esophagus; the prothroracic gland, right, is in the ventral part of the prothorax, among the muscle cells. (After Bodenstein, D.: Recent Progress Hormone Research. Vol. 10. New York, Academic Press, Inc., 1954.)*

of these axons. The intercerebral gland secretes several hormones which regulate various aspects of body activity. One of these, **prothoracicotropic hormone**, is released from the corpora cardiaca and initiates the molting process by stimulating a second endocrine gland, the **prothoracic gland**, to secrete **ecdysone**. The conditions that stimulate the activity of the intercerebral glands have been studied in the blood-sucking bug, *Rhodnius*. Stretch receptors in its abdomen are stimulated when the bug has a large meal of blood and these stimulate the intercerebral glands to activity. *Rhodnius* feeds only infrequently, but one large meal is enough to support the metabolic activity associated with a molt.

The prothoracic gland is a diffuse set of strands of large ectodermal cells in the ventral part of the prothorax. It secretes **ecdysone**, a steroid synthesized from cholesterol. Ecdysone, $C_{27}H_{44}O_6$, differs from the steroids secreted by vertebrates in retaining the side chain of cholesterol (Fig. 21–9); it has five hydroxyl groups, two of which, at positions 14 and 25, are quite unusual. It has a ketone group adjacent to a double bond but in ring B instead of in ring A as in the vertebrate steroids. More than 500 kilograms of silk worms were extracted to obtain 25 mg of ecdysone. There is a striking parallelism between the stimulation of production of ecdysone by prothoracicotropic hormone and the stimulation of production of adrenal steroids by ACTH. Ecdysone stimulates the epidermis to secrete molting fluid, thus leading to molt and metamorphosis.

In 1960 Clever and Karlson found that injecting minute amounts of ecdysone into the larva of the midge, *Chironomus*, caused within 15 minutes the puffing or swelling of a specific region of a particular chromosome. Dipteran insects may have in certain tissues **giant chromosomes** composed of many chromatids. In these giant chromosomes individual bands, corresponding to specific genetic loci, are visible under the microscope and can be identified. The puffing of certain bands in the chromosomes occurs at specific times in the course of normal development and is known to represent the synthesis of RNA at those sites (Fig. 21–34). Injecting ecdysone causes puffing at band I-18 C and the subsequent production of the enzyme **dopa decarboxylase** in epidermal cells. Dopa decarboxylase catalyzes the conversion of DOPA to N-acetyl dihydroxyphenylethyl amine, a substance involved in the hardening of the cuticle. This suggestion that hormones may control the transcription of specific genes is being investigated by many biologists.

The control of metamorphosis involves a third set of endocrine organs, the **corpora allata**, which are small glands in the head, just behind the corpora cardiaca. If these glands are removed from a young larva, it will undergo metamorphosis at the next molt, even though this may be one or more molts too soon. Conversely, if corpora allata from young larvae are transplanted into older larvae due to undergo metamorphosis at the next

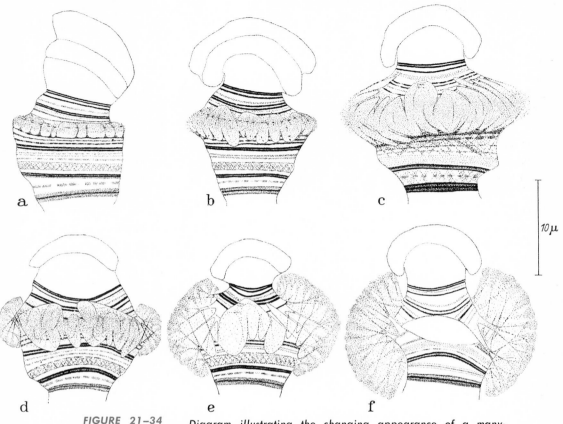

10 μ

FIGURE 21-34   *Diagram illustrating the changing appearance of a many-stranded (polytene) chromosome in a salivary gland of Chironomus tentans as a chromosome puff gradually appears. The material that makes up the puff has been shown by histochemical tests and by autoradiography with tritium labeled uridine to consist largely of RNA. (From Beermann: Chromosoma 5:139–198, 1952.)*

molt, the juvenile form is retained at the molt. The corpora allata secrete **juvenile hormone**, $C_{18}H_{30}O_3$, which has been characterized as a derivative of a fatty acid (Fig. 21–35). It is probably an isoprenoid derivative of mevalonic acid and is unusual in having ethyl side chains. A substance extracted from certain American paper products (e.g., *The New York Times*) which are prepared from the woods of certain trees has marked juvenile hormone activity. It is conjectured that the trees protect themselves from certain insects by secreting a juvenile hormone-like material that prevents the larvae from becoming adults and reproducing. It has also been suggested that we use juvenile hormone as an insecticide to prevent the multiplication of insects. Under test conditions, spraying caterpillars or the

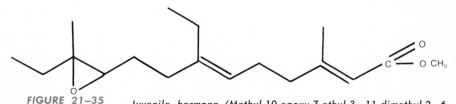

FIGURE 21-35   *Juvenile hormone (Methyl-10-epoxy-7-ethyl-3, 11-dimethyl-2, 6-tridecadienoate).*

foliage on which they are feeding with solutions containing juvenile hormone has prevented their pupating normally. They do not molt into giant larvae but die.

Juvenile hormone inhibits metamorphosis but permits molting to occur, thus ensuring that the larva will molt several times and reach a large size before pupating. Juvenile hormone is not produced during the last larval stage, hence pupation can occur at the ensuing molt.

The corpora allata secrete hormones that affect other phenomena in certain groups of insects. The deposition of yolk in eggs and the secretions of the accessory sex glands are under the control of hormones from the corpora allata. In some cockroaches the male is attracted to the female by a pheromone (p. 697) that she produces and secretes under the control of a hormone from the corpora allata.

Some pupae, such as those of the giant silkworm *Platysamia cecropia*, remain in a state of arrested development or dormancy (called **diapause**) over the winter. If a newly formed pupa is kept at 24° C it will remain inactive for five or six months but eventually will begin to develop. If a newly formed pupa is chilled to 4° C for six weeks and then kept at 24° it begins at once to develop. Thus chilling initiates further development. It does this by stimulating the release of prothoracicotropic hormone from the intercerebral gland. If intercerebral glands are removed and chilled and then transplanted into a diapausing pupa that has not been chilled, the pupa begins development.

The molting process in crustaceans is under the control of a hormone that is accumulated in the **sinus glands** in the eyestalk (Fig. 21–36). These glands contain several hormones, one of which initiates molting. The sinus glands are composed of the expanded tips of axons surrounding a blood sinus. The cell bodies of these axons are located some distance away in the eyestalk in the **X organ**. The hormone is produced in the cell bodies of the X organ and passes along the axons to the sinus gland where it is stored and released. When the eyestalk of a crab or other crustacean is removed a molt usually occurs, indicating that the hormone made in the X organ and secreted by the sinus gland tends to *inhibit* molting. The **Y organ**, composed of diffuse strands of ectodermal cells at the base of the large muscles of the mandibles, is an endocrine gland that produces the hormone that induces molting. Ecdysone prepared from insects will induce molting when injected into crustaceans and it is likely that the molting hormone produced by the Y organ is ecdysone or a very similar steroid. The hormone of the X organ inhibits molting by preventing the secretion of molting hormone (ecdysone) by the Y organ.

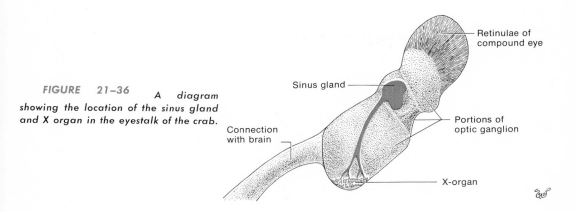

FIGURE   21–36   A   diagram showing the location of the sinus gland and X organ in the eyestalk of the crab.

Retinulae of compound eye

Sinus gland

Portions of optic ganglion

Connection with brain

X-organ

SECTION 21–11

HORMONAL REGULATION OF METABOLIC RATES

The survival of a complex organism such as man requires, among other things, that order be maintained among the multitude of possible metabolic reactions. Each of these reactions must increase or decrease in an appropriate fashion in response to specific changes in the environment. In part these changes result from the kinetic properties of the enzymes; the rates increase and decrease as the concentrations of substrates and cofactors rise and fall. For example, glucose-6-phosphate can be converted in the liver to glucose-1-phosphate, to fructose-6-phosphate, to 6-phosphogluconate or to free glucose (Fig. 21–37). Each reaction is catalyzed by a specific enzyme and the rate of each is controlled by the **affinity** of the enzyme for glucose-6-phosphate and by the **maximum rate** of the enzyme when it is fully saturated with substrate. Merely changing the concentration of glucose-6-phosphate present will change the direction in which metabolism flows and not simply the rate at which metabolism occurs. At low concentrations of glucose-6-phosphate, **glucose-6-phosphate dehydrogenase**, which has the highest affinity for the substrate, is the only enzyme saturated with substrate and most of the small amount of glucose-6-phosphate present will be converted to 6-phosphogluconic acid. At this low concentration of substrate, **glucose-6-phosphatase**, which has a very low affinity for substrate, will hardly operate at all and little or none of the glucose-6-phosphate will be converted to free glucose. At high concentrations of glucose-6-phosphate, when all the enzymes are saturated with substrate, the factors determining the direction of metabolic flow are the relative maximum rates of the four enzymes. Glucose-6-phosphatase has a high maximum rate and at high concentrations of glucose-6-phosphate most of the substrate is converted to glucose.

In addition to the effects of the concentrations of substrate and cofactors, the activity of many enzymes is **modulated** by the presence of other

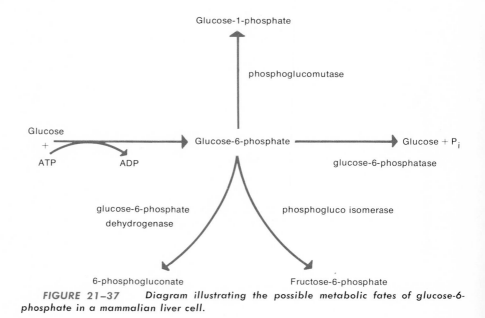

FIGURE 21–37   *Diagram illustrating the possible metabolic fates of glucose-6-phosphate in a mammalian liver cell.*

molecules. The activity of **phosphofructokinase**, for example, which converts fructose-6-phosphate to fructose-1.6-diphosphate, is increased by adenylic acid, AMP, but is inhibited by ATP (Fig. 21–38). The enzyme which catalyzes the conversion of fructose-1,6-diphosphate back to fructose-6-phosphate, **fructose diphosphatase**, is inhibited by AMP. The opposite effects of AMP on the two enzymes prevent both of them from being active at the same time which would result in a wasteful cyclic hydrolysis of ATP. The subject of enzyme modulation and the details of these changes in enzyme activity are quite complex and beyond the scope of this book, but have been introduced to help the student appreciate the many kinds of metabolic controls that may operate within cells.

Superimposed on the controls that depend on the innate properties of the enzymes and their responses to metabolites that modulate their activity are other changes induced by specific hormones. Several hormones—thyroxine, insulin, growth hormone and cortisol—have broad effects on the metabolism of a wide variety of tissues. These controls also help maintain order among the many possible metabolic reactions in those tissues.

Most vertebrates, for example, will not attain their normal adult form and dimensions in the absence of thyroxine. Thyroxine and growth hormone appear to act synergistically in promoting the normal growth of the skeleton. The effect of thyroxine in controlling amphibian metamorphosis is well known and thyroxine affects the growth and differentiation of other vertebrates though not in such a marked fashion. Thyroxine and triiodothyronine accelerate the general metabolic rate and oxygen consumption of nearly every organ and tissue in the body. The activities of more than 100 enzymes have been reported to be increased following the administration of thyroxine. Since thyroxine increases oxygen consumption and oxygen is utilized primarily in the mitochondria it was natural to look for a primary effect of thyroxine on mitochondrial structure and function. Isolated mitochondria do indeed undergo swelling when exposed to thyroxine. Although very high concentrations of thyroxine can "uncouple" oxidative phosphorylation so that energy is released as heat rather than being retained in energy-rich phosphate bonds ($\sim$P), this

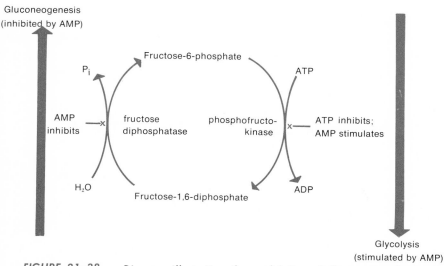

Gluconeogenesis
(inhibited by AMP)

Fructose-6-phosphate

P_i

ATP

AMP inhibits —x— fructose diphosphatase

phosphofructo-kinase —x— ATP inhibits; AMP stimulates

H_2O

Fructose-1,6-diphosphate

ADP

Glycolysis
(stimulated by AMP)

FIGURE 21–38   Diagram illustrating the modulation of the enzymes fructose-1, 6-diphosphatase and phosphofructokinase by adenylic acid (AMP) and adenosine triphosphate (ATP).

phenomenon probably bears no relation to the physiologic effects of thyroxine. The increased basal metabolic rate of a hyperthyroid individual cannot be explained in terms of an effect on any single metabolic process.

Insulin regulates the metabolism of proteins and lipids as well as of carbohydrates. Insulin facilitates the entry of glucose into skeletal muscle and adipose tissue; it also increases the activity of a number of enzymes such as glucokinase and glycogen synthetase. The uptake of amino acids into cells and the incorporation of amino acids into proteins are both increased by insulin. Insulin increases the conversion of glucose to fatty acids and inhibits the hydrolysis of triglycerides in adipose tissue. In insulin deficiency, as in **diabetes**, the peripheral tissues are unable to utilize glucose and mobilize fatty acids for metabolism. These are incompletely metabolized and lead to an accumulation of metabolic acids, **ketone bodies**, which cause acidosis. In acute diabetes the production of acidic ketone bodies may exceed the capacity of the body to handle them and the acidosis may result in coma and death.

Cortisol has metabolic effects that oppose those of insulin. It brings about a mobilization of amino acids from peripheral tissues and accelerates **gluconeogenesis**, the conversion of the carbon chains of the amino acids to glucose and glycogen. Cortisol inhibits the utilization of glucose by skeletal muscle and other peripheral tissues and accelerates the mobilization of fatty acids, increasing the rate of production of ketone bodies. If the adrenals (or the pituitary) are removed from a diabetic animal it is much less likely to develop a lethal acidosis. Cortisol has a remarkable effect in suppressing inflammation but how this is related to its metabolic effects is not clear.

Growth hormone has a host of effects on the tissues of the body, promoting the transfer of amino acids into cells and increasing protein synthesis. It increases the synthesis of DNA and RNA in liver and muscle and the synthesis of collagen and mucopolysaccharides in connective tissue and skin. A deficiency of growth hormone increases the sensitivity of the individual to insulin and a given dose of insulin produces a greater than normal decrease in the concentration of glucose in the blood. The concentrations of both urea and amino acids in the blood are decreased by growth hormone, reflecting the greater uptake of amino acids and their utilization in protein synthesis. Growth hormone decreases the rate of conversion of amino acid nitrogen to urea. It stimulates the mobilization of fat from adipose tissue and increases the concentration of fatty acids in the plasma. Many of the effects of growth hormone counter those of insulin.

SECTION 21–12

MECHANISMS OF HORMONE ACTION

Any theory of the molecular mechanism by which a given hormone produces its specific effects in specific tissues must account for the high degree of **specificity** of many hormones and for the remarkable biological **amplification** inherent in hormonal processes. Hormones circulate in the blood in very low concentrations—steroid hormones at concentrations of 10^{-9} M or less and peptide and protein hormones at concentrations of 10^{-12} M. The several current theories regarding the mechanism of hormone action are alike in suggesting that the hormone goes to or into a cell and combines with some specific **receptor**. Many cells do not recognize certain hormones, presumably because they have no receptor for that hormone and because the receptors are highly specific. The hypotheses differ as to what the receptor is, where it is located and what happens after the hormone is bound to its receptor. Appropriate experiments with highly labeled steroid hormones have shown the presence of specific protein receptors for estradiol in the uterus, for progesterone in the oviduct, for aldosterone in the kidney and for testosterone in the prostate gland. They indicate that these receptors are in the nuclei of the cells.

The effects of hormones in facilitating the entrance of certain substances in the cell, such as the uptake of glucose by muscle cells stimulated by insulin, has suggested that the hormone combines with a protein or some other substance in the cell membrane. This leads to a change in the molecular architecture of the membrane and hence in its permeability to specific substrates.

Another hypothesis states that the receptor is a protein whose enzymatic activity is altered by the combination. One variation of this hypothesis currently popular is that the hormone combines with a specific **adenyl cyclase** and stimulates the production of **cyclic 3′,5′-adenylic acid** from ATP (p. 156). The cyclic AMP is regarded as a "second messenger" that mediates the effect of the hormone. Epinephrine, for example, has been shown to stimulate the adenyl cyclase of liver cells (Fig. 21–39). The resulting cyclic AMP is an activator of another enzyme, a protein kinase, that transfers a phosphate group from ATP to a third enzyme, phosphorylase kinase, and activates it so that it can convert an inactive fourth enzyme, phosphorylase b, to active phosphorylase a. The latter then catalyzes the production of glucose-1-phosphate from glycogen. At each of these successive steps there is an amplification of 10- to 100-fold, so that a very small amount of epinephrine will lead to the production of a very large amount of glucose-1-phosphate.

A third general hypothesis suggests that the hormone enters the nucleus and activates specific genes that were previously repressed so that

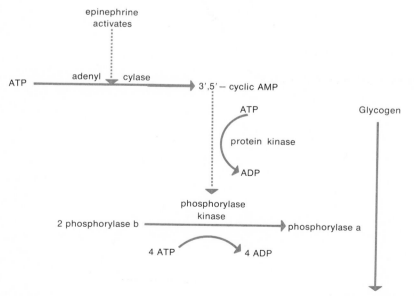

FIGURE 21–39   *The sequence of enzymatic events by which epinephrine or glucagon stimulates adenyl cyclase and brings about the synthesis of 3′,5′-adenosine monophosphate (cyclic AMP). This in turn activates a protein kinase that phosphorylates phosphorylase kinase which in turn phosphorylates and activates phosphorylase. This finally brings about the cleavage of glycogen and the secretion of glucose.*

they are transcribed (Fig. 21–40). This leads to the production of new kinds of messenger RNA which code for the synthesis of new specific proteins. This theory also accounts for the marked amplification of hormonal effects, for a very small amount of hormone, by turning on the transcription of a specific gene, could result in many molecules of messenger RNA and many more molecules of protein. Evidence supporting this hypothesis comes from studies of the effect of estradiol on the uterus, of estradiol and progesterone on the oviduct and of testosterone on the seminal vesicle. The stimulation of molting by ecdysone also appears to involve this type of mechanism.

Whether hormones are typically used up as they regulate metabolism in their target tissues is not clear. Estradiol is not used up or changed chemically as it stimulates the growth of the uterus. The hormones bound to their receptors appear to be relatively stable, but hormones circulating in the blood have relatively short biological half-lives. They are inactivated and eliminated from the body and must be replaced by new hormone molecules synthesized in the appropriate endocrine gland.

It seems unlikely that all hormones have a common molecular mechanism by which their effects are produced. Indeed there is evidence that certain hormones produce their effects not by a single mechanism but by several different mechanisms acting in parallel. The theories current at any given moment tend to reflect our general knowledge of cellular and molecular biology. Theories implicating effects of hormones on cell membranes have given way to theories implicating an effect on altering the activity of an enzyme and these in turn have been replaced by theories

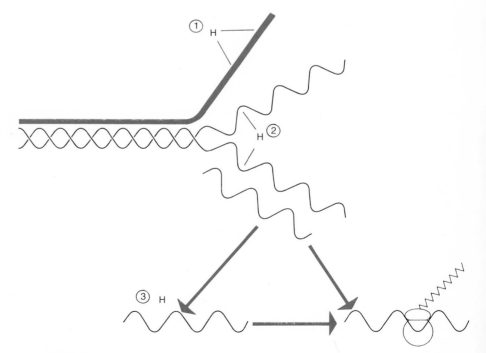

FIGURE 21–40   *A diagram illustrating some of the possible sites at which a hormone may regulate genetic transcription, either by combining with a histone (1) or other protein which is bound to the DNA or by interacting with the DNA itself (2). The hormone might act by combining with a specific messenger RNA (3) and protecting it from being inactivated.*

involving effects of the hormone on the genetic mechanism which result in the synthesis of specific kinds of RNA and proteins.

PHEROMONES

In recent years it has been appreciated that the behavior of animals may be influenced not only by hormones—chemicals released into the internal environment by endocrine glands and which regulate and co-ordinate the activities of other tissues—but also by **pheromones**—substances secreted by *exocrine* glands, released into the *external* environment and which influence the behavior of other members of the same species. We are used to thinking that information can be transferred from one animal to another by sight or sound; pheromones represent a means of communication, a means of transferring information, by smell or taste.

Some pheromones act in some way on the recipient's central nervous system and produce an immediate effect on its behavior. Among these are the **sex attractants** of moths and the **trail pheromones** and **alarm substances** secreted by ants. Other pheromones act more slowly and trigger a chain of physiological events in the recipient which affect its growth and differentiation. These include the regulation of the growth of locusts and control of the numbers of reproductives and soldiers in termite colonies.

The sex attractants of moths provide some of the more spectacular examples of pheromones. Among the ones that have been isolated and identified are **bombykol**, a 16-carbon alcohol with two double bonds, secreted by female silkworms, and **gyplure**, 10-acetoxy-$\triangle^7$-hexadecenol, secreted by female gypsy moths. The male silk moth has an extremely sensitive device in his antennae for sensing the attractant. It is possible to record the nerve impulses coming from the antennae. These electro-antennagrams show specific responses to bombykol and not to other substances. He cannot determine the direction of the source by flying up a concentration gradient because the molecules are nearly uniformly dispersed except within a few meters of the source. Instead he responds by flying *upwind* to the source. With a gentle wind the bombykol given off by a single female moth covers an area several thousand meters long and as much as 200 meters wide. An average silkworm contains some 0.01 mg of bombykol. It can be shown experimentally that when as little as 10,000 molecules of attractant are allowed to diffuse from a source 1 centimeter from a male he responds appropriately. He can have received only a few hundred of these molecules, perhaps less. Thus the amount of attractant in one female could stimulate more than one billion males! The attractants, generally hydrocarbons, contain 10 to 17 carbons in the chain, which provides for the specificity of the several kinds of attractants.

The sex attractant of the American cockroach is not a long chain alcohol like bombykol and gyplure but has a central three-carbon ring to which are attached methyl groups and a propanoxy group. Sex attractants have been tested as possible specific insecticides. By putting sex attractant on stakes placed every 10 meters in a large field, investigators could blanket the air with sex attractant. This confused the males and greatly decreased the probability of their finding females and mating with them.

The fire ants, when returning to the nest after finding food, secrete a "trail pheromone" which marks the trail so that the other ants can find their way to the food. The trail pheromone is volatile and evaporates within two minutes, so that there is little danger of ants being misled by old trails. Ants also release alarm substances when disturbed and this (rather like ringing the bell in a firehouse) in turn transmits the alarm to ants in the vicinity. These alarm substances have a lower molecular weight than the sex attractants and are less specific, so that members of several different species respond to the same alarm substance.

Worker bees, on finding food, secrete **geraniol**, a 10-carbon, branched chain alcohol, to attract other worker bees to the food. This supplements the information conveyed by their waggle dance (p. 811). Queen bees secrete 9-ketodecenoic acid which, when ingested by worker bees, inhibits the development of their ovaries and their ability to make royal cells in which new queens might be reared. This substance also serves as a sex attractant to male bees during the queen's nuptial flight.

In colonial insects, such as ants, bees and termites, pheromones play an important role in regulating and coordinating the composition and activities of the population. A termite colony includes morphologically distinct queen, king, soldiers and nymphs or workers. All develop from fertilized eggs; however, queens, kings and soldiers each secrete inhibitory substances, pheromones, that act on the corpus allatum of the nymphs and prevent their developing into the more specialized types. If the queen dies there is no longer any "antiqueen" pheromone released and one or more of the nymphs develop into queens. The members of each colony will permit only one queen to survive and will eat up any excess ones. Similarly the loss of the king termite or a reduction in the number of soldiers permits other nymphs to develop into the specialized castes to replace them. Males of migratory locusts secrete a substance from the surfaces of their skin which accelerates the growth of young locusts.

There are examples of pheromones in mammals as well as in insects. When female mice are placed four or more per cage there is a greatly increased frequency of **pseudopregnancy**. If their olfactory bulbs are removed this effect disappears. When more females are placed together in a cage their estrous cycles become very erratic; however, if one male mouse is placed in the cage his odor can initiate and synchronize the estrous cycles of all the females (the "Whitten effect") and reduce the frequency of reproductive abnormalities. Even more curious is the finding (the "Bruce effect") that the odor of a strange male will block pregnancy in a newly impregnated female mouse. The nerve impulses from the nose pass to the hypothalamus and block the output of prolactin-releasing factor. The subsequent lack of prolactin leads to regression of the corpora lutea and the failure of the fertilized ova to implant.

The question of whether there are human pheromones remains unanswered, but of interest in this respect is the observation of the French biologist J. LeMagnen that the odor of 14-hydroxytetradecenoic acid is perceived clearly only by sexually mature females and that it is perceived most sharply at about the time of ovulation! Males and young girls are relatively insensitive to this substance, but male subjects became more sensitive to it after an injection of estrogen.

SUGGESTIONS FOR FURTHER READING

Barrington, E. J. W.: *Textbook of Comparative Endocrinology.* London, Oxford University Press, 1963.
    A clear statement of endocrine principles from the comparative viewpoint.

Frieden, E.: *The Chemistry of Amphibian Metamorphosis.* Sci. Amer., November (Offprint 170) 1963.
> An interesting article about the many changes accompanying the change from tadpole to adult.

Gorbman, A., and Bern, H. A.: *A Textbook of Comparative Endocrinology.* New York, John Wiley & Sons, Inc., 1962.
> An excellent treatment of the evolutionary aspects of endocrine systems in both vertebrate and invertebrate animals.

McKerns, K. W. (ed.): *The Gonads.* New York, Appleton-Century-Crofts, 1969.
> A fine source book, with chapters by 46 authorities. The papers were presented at a symposium and the transcript of the discussion that follows provides further insight into the hormonal aspects of reproduction.

Pincus, G. (ed.): *Recent Progress in Hormone Research.* New York, Academic Press, Inc., published annually.
> Each volume of this series of books contains the papers given at an annual symposium and a transcript of the discussion following each presentation. A mine of information on the latest discoveries in endocrinology.

Scharrer, E., and Scharrer, B.: *Neuroendocrinology.* New York, Columbia University Press, 1963.
> An excellent account of the process of neurosecretion, with examples from both insects and vertebrates.

Turner, C. D.: *General Endocrinology.* 4th ed. Philadelphia, W. B. Saunders Company, 1966.
> An excellent introductory text covering the basic biological aspects of endocrinology.

Williams, R. H. (ed.): *Textbook of Endocrinology.* 4th ed. Philadelphia, W. B. Saunders Company, 1967.
> A standard reference text of endocrinology, which deals primarily with the human and medical aspects of endocrine function.

Wilson, E. O.: *Pheromones.* Sci. Amer., *208*:95, May 1963.
> A fascinating description of the many facets of animal behavior controlled by the sense of smell.

Wurtman, R. J., and Axelrod, J.: *The Pineal Gland.* Sci. Amer., *213*:60, 1965.
> A fascinating brief discussion of the endocrine functions of this hitherto obscure gland.

Young, W. C. (ed.): *Sex and Internal Secretions.* (2 Vols.) 3rd ed. Baltimore, The Williams & Wilkins Company, 1961.
> A series of papers by experts in the many fields relating to the hormonal control of sex cycles and behavior.

SENSE ORGANS AND RECEPTORS

CHAPTER 22

SECTION 22-1

THE FUNCTION OF SENSE ORGANS

In order to regulate in response to changes in its internal and external environments, an organism must have means of detecting such changes. The sense organs have the dual function of detecting changes and transmitting information concerning the nature of the change to the central nervous system.

A sense organ would not make much information available to an organism if it responded indiscriminately to all environmental change. The nature of the event must be stated; whether, for example, it is a change in blood sugar level, an increase in temperature or the sudden appearance of a predator. No single sense organ could possibly assess all of these diverse events, consequently there has evolved a specificity whereby one kind of organ detects light, another chemicals and so on.

No sense organ would be very useful if it responded to gross changes only. On the other hand, if it were so sensitive as to respond to every moving molecule or electron it would transmit only noise. Imagine, for example, the chaos if the ear were so sensitive as to respond to the Brownian movement of molecules hitting the ear drum. Any sense organ must maintain an optimum signal to noise ratio.

In addition to specificity and optimal (not maximal) sensitivity, a sense organ must have a capacity for discrimination and for recording not merely "on" and "off" but also rate, magnitude and direction of change.

Every **sense organ** is a specialized structure consisting of one or more **receptor cells** and **accessory tissues.** For example, the receptors of the vertebrate eye are the **rods** and **cones;** the accessory structures are the **cornea, lens, iris** and **ciliary muscles** (Fig. 22–23). While some of the capacities enumerated above are resident in the receptor itself, others are conferred upon the sense organ by the accessory structures. Without a lens, an eye would be incapable of detecting many events. In this case the lens enhances the versatility of the organ. Accessory structures may also act as filters that limit the performance of receptors. For example, ultraviolet is not seen by man because it is filtered out before reaching the retina which itself is not insensitive to light of this wavelength. In contrast the eyes of insects have no UV filters, and the detection of UV is important in the lives of these organisms.

The importance of accessory structures is especially evident in the mechanical senses where they modify sensitivity and determine the parameter of the stimulus. Among invertebrates, for example, one mechanoreceptor may be sensitive only to gross touch because it is associated with a stout rigid spine, while another may actually be an auditory receptor because it is connected to a long filamentous hair that is moved by sound waves. The hairs on the cerci (paired posterior appendages) of crickets are examples of this type of auditory organ.

Receptors are generally nerve cells, the axons of which extend directly into the central nervous system or connect synaptically with one or more interneurons which in turn connect with the central nervous system. Most receptors are of this type. Some receptors (e.g., the receptors in the mammalian taste bud) are modified epithelial cells connected to one or more nerve cells (Figs. 22–17 and 22–18).

SECTION 22–2

RECEPTOR PROCESSES

The sense organ performs two functions: it **detects** and it **transmits** information to the central nervous system. In those sense organs where the receptor is a **primary neuron** this neuron detects and transmits. Where the receptor is an **epithelial cell,** it detects, but the transmission of information is the function of the associated neuron.

In its capacity as detector or sensor a receptor receives a small amount of energy from the environment. Each kind of receptor is specialized to receive one particular form of energy more efficiently than another. Rods and cones absorb the energy of photons of certain specific energies. Temperature receptors respond to energy of radiation transferred by radiation, conduction or convection. Electricity is detected by the energy of electrons. Tastes and smells are detected by the potential energy in the mutual attraction and repulsion of atomic particles. Each receptor clearly possesses some degree of intrinsic specificity.

The various kinds of environmental energy act as triggers which cause the receptors to perform biological work. This work transforms metabolic energy into electrical energy. These relationships are best explained by referring to a very simple sense organ, the tactile hair of an insect. This hair plus its associated cells is a complete sense organ (Fig. 22–1). The **bipolar neuron** at its base is the receptor. Its dendrite is attached to the base of the hair near the socket; its axon passes directly to the central nervous system without synapsing. In its unstimulated state this neuron maintains a steady resting potential; i.e., there is a potential difference between the inside and the outside. This potential difference exists because the ionic composition of the fluids on each side of the semipermeable cell membrane are different. The difference is maintained by metabolic work performed by the cell. When the hair is touched, the shaft moves in the socket and mechanically deforms the dendrite. Deformation (the stimulus) renders the membrane permeable to ions with the result that the potential difference between the two sides of the membrane decreases, disappears or increases. If it decreases or disappears, the cell is said to have **depolarized;** if it increases, the cell is said to have **hyperpolarized.** The state of depolarization caused by the stimulus is called the **receptor potential.** It spreads relatively slowly down the dendrite, decaying exponentially as

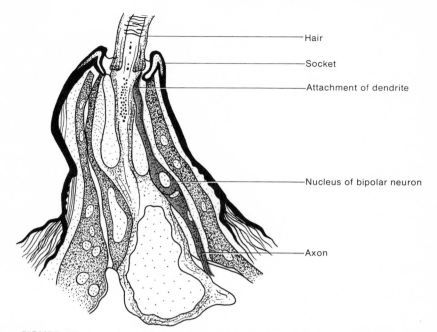

Hair
Socket
Attachment of dendrite
Nucleus of bipolar neuron
Axon

FIGURE 22–1   A tactile hair from a caterpillar showing the attachment of the dendrite of the bipolar neuron (the mechanoreceptor) at the point where the shaft of the hair enters the socket. (From Hsu, F.: Étude cytologique et comparée sur les sensilla des insectes. La Cellule 47:1–60, 1938.)

it goes. When a special area of the cell near the axon (the **axon hillock**) becomes depolarized, **action potentials** are generated. The action potentials then travel along the axon to the central nervous system. The primary receptor thus performs all the essential functions of a sense organ: it detects an event in the environment (a force acting on the hair); it generates electrical energy at the expense of its metabolic energy (the receptor potential); it transmits information (action potentials) to the central nervous system. With minor variations this is the mode of operation of all receptors.

The relations between the stimulus, the receptor potential and the action potentials are summarized in Figure 22–2. The amplitude and

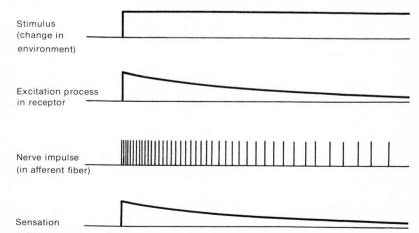

Stimulus
(change in
environment)

Excitation process
in receptor

Nerve impulse
(in afferent fiber)

Sensation

FIGURE 22–2   A diagram showing the relations among the stimulus, the receptor potential, the action potential and sensation. (From Adrian, E. D.: The Basis of Sensation. London, Chatto & Windus Ltd., 1949.)

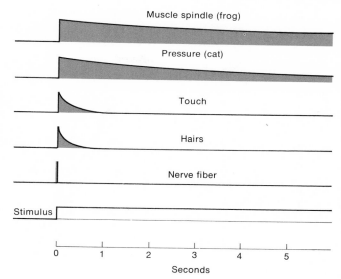

FIGURE 22–3   A diagram showing the relation between the stimulus and the different rates of adaptation for different receptors and a nerve fiber. The heights of the curves indicate the rates of discharge of action potentials. (From Adrian, E. D.: The Basis of Sensation. London, Chatto & Windus Ltd., 1949.)

duration of the receptor potential are related to the **strength** and **duration** of the stimulus. Thus a strong stimulus causes a greater depolarization of the receptor membrane than a weak one. The action potentials are **repetitive,** and the frequency at which they are generated is related to the magnitude of the receptor potential. The frequency is higher with a large potential than a small one. Accordingly the strength of a stimulus is reflected in the *frequency* of the action potentials. The amplitude of each action potential bears no relation to the stimulus; it is characteristic of the particular neuron under the usual recording conditions.

It must be remembered that the action potential is an all-or-none phenomenon. The receptor potential by contrast is a **graded response.** Once a stimulus has triggered a receptor to generate action potentials, the stimulus has no further control over the action potentials. The situation is analogous to lighting a fuse. The heat of the match is the stimulus. When it raises the end of the fuse to the combustion point, the fuse begins to burn, and utilizing its own energy it ignites adjacent parts of itself. In this way the "message" travels the length of the fuse and bears no direct relation to the match.

Even though the stimulus may continue unabated, neither the receptor potential nor the action potentials continue unchanged. The receptor potential gradually falls, and the frequency of action potentials decreases (Figs. 22–2 and 22–3). This is the phenomenon of **adaptation.** Some receptors adapt very rapidly and completely, but others, more slowly (Fig. 22–3). It is reasonable that receptors like those recording a change in posture should adapt slowly and incompletely, because a continued flow of information is required.

SECTION 22–3

SENSORY CODING AND SENSATION

The action potentials are the code by which sense organs "inform" the central nervous system of some event occurring in the internal or external environment. All events are described in the same code. Thus light of the wavelength 400 mμ (blue), molecules of sugar (sweet) and sound waves of 440 vibrations per second (note A above middle C) all cause action potentials to be sent to the brain via the appropriate nerves, and these action potentials are identical. Compare, for example, the patterns of action potentials from the pressure receptor of the carotid sinus and the optic receptor of the horseshoe crab (Fig. 22–4). How then can the organism accurately assess its environment?

As already indicated the frequency of the repetitive action potential codes the intensity of the stimulus. Since each receptor *normally* responds to but one category of stimuli (i.e., light, sound, taste and so forth), a message arriving in the central nervous system along this nerve is interpreted as meaning that a particular stimulus occurred. That is to say, one method of coding makes use of "name lines" that enter the brain at specific points. Interpretation of the message and, in the case of human beings, the quality of sensation depend upon which central interneurons receive the message. Sensation, when it occurs, occurs in the brain. Rods and cones do not see; only the combination of rods, cones and centers in the brain see. Furthermore, many sensory messages never give rise to sensations. For example, chemoreceptors in the carotid sinus and the hypothalamus sense internal changes in the body but never stir our consciousness. The situation with respect to name lines is analogous to a person having two telephones in his house, a red one that was a direct line to the fire station and a blue one that was a direct line to the police station. Whenever the red telephone rang in the house the message would be interpreted as originating in the fire station regardless of who did the talking there.

Spatial localization of stimuli impinging on the body, especially mechanical and pain stimuli, also depends upon the destination of specific nerves in the brain. The importance of the brain in localization and in making sensations possible is emphasized by the phenomenon of "misreference," which occurs occasionally in connection with pain. A well

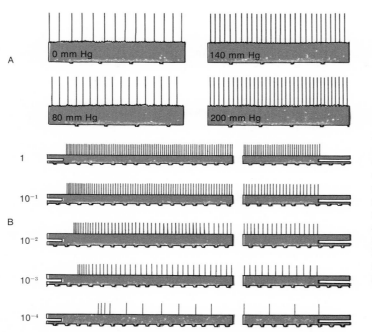

FIGURE 22–4   The rate of discharge of a nerve fiber from a carotid sinus pressure receptor (A) is shown at different levels of intercarotid pressure. (After Bronk, D. W., and Stella, G.: Am. J. Physiol., 110:708, 1935.) The discharge from a single optic receptor nerve fiber of the Limulus eye (B) is shown at different intensities of illumination. The filled bar underneath signifies the duration of stimulation. (After MacNichol, E. F. In Grenell, R. G., and Mullins, L. J. (eds.): Molecular Structure and Functional Activity of Nerve Cells, American Institute of Biological Sciences, 1956.)

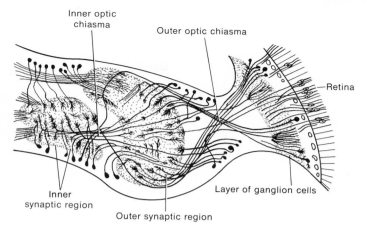

FIGURE 22–5   An example of neural circuitry in the optic lobes of insects. This is a diagrammatic representation of the optic lobe of the blowfly Calliphora vomitoria. (From Cajal, S. R., and Sanchez, D.: Contribucion al conocimiento de los centros nerviosos de los insectos. Trabajas del Lat. de Investig. Biologicas (University of Madrid) 13:1–167, 1915.)

Inner optic chiasma

Outer optic chiasma

Retina

Layer of ganglion cells

Inner synaptic region

Outer synaptic region

known example of this is the experience of people suffering from heart pains who complain of pain in the shoulder, upper chest or medial side of the left arm. Actually the stimuli originate in the heart, but the nerve impulses terminate in the same part of the brain as impulses genuinely originating in the shoulder, chest or arm.

Cross-fiber patterning is another method of coding information (see Fig. 22–19). This is probably the method used in olfactory organs. It is unlikely that the olfactory organ contains a specific receptor for each of the thousands of individual odors that can be recognized. There is, in fact, evidence that there are a limited number of categories of receptors, each of which responds to a spectrum of odors. There is not rigid specificity because the spectra overlap. Perception of different characteristic odors probably depends, therefore, on the pattern of response of all fibers responding together.

There is also some evidence that the **temporal pattern** of action potentials generated in a single neuron may serve as a code for different stimuli. Single taste receptors of flies, for example, generate action potentials at an even regular frequency when the stimulus is salt, but generate irregular frequencies when the stimulus is acid.

In invertebrates it is usual for the axon of a sensory neuron to travel all the way to the central nervous system without synapsing. In these circumstances the message generated at the periphery arrives unaltered. The compound eye is an exception. In it, as in most vertebrate sense organs, many interneurons are interposed between the receptor and the central nervous system (Fig. 22–5). A glance at the neuroanatomy of the vertebrate retina or olfactory bulb reveals how complicated the neural circuitry is. As a consequence of all these synaptic connections, the original message is altered and may lose or gain some of its information. The message that finally arrives at the brain has been well censored by interneurons and bears even less resemblance to the original stimulus than did the action potentials from the receptor.

SECTION 22–4

MECHANORECEPTORS

It is customary to classify receptors according to the nature of their effective stimuli. Most animals are equipped with **mechanoreceptors, chemo-**

receptors, photoreceptors and *temperature receptors.* Some fish are equipped with *electroceptors.*

Mechanoreceptors respond in the final analysis to energy derived from the surface gravitation of the earth. They are sensitive to stretch, compression or torque imparted to tissues by the weight of the body, the relative movement of parts, the gyroscopic effects of moving parts and impact of the substrate or the surrounding medium (air or water). Mechanoreceptors are concerned with enabling an organism to maintain its primary body attitude with respect to gravity (for us, anterior end up and posterior end down; for a dog, dorsal side up and ventral side down; for a tree sloth, ventral side up and dorsal side down). They are also concerned with maintaining postural relations, the position of one part of the body with respect to another, information that is essential for all forms of locomotion and for all coordinated and skilled movements from spinning a cocoon to writing a book. Mechanoreceptors provide, in addition, information about the shape, texture, weight and topographical relations of objects in the external environment. Finally mechanoreception is necessary for operation of some of the internal organs. They supply, for example, information about the presence of food in the stomach, feces in the rectum, urine in the bladder, a fetus in the uterus. For purpose of further discussion it is convenient to divide the mechanoreceptors into the following categories: tactile, proprioceptive and auditory.

The Tactile Sense

Among the simplest tactile receptors are the **tactile hairs** of invertebrates. The tactile hair of an insect has already been described (Fig. 22–1). It is a **phasic** receptor; i.e., it responds only when the hair is moving—it responds to velocity (compare Fig. 22–3). Figure 22–6 shows a record of electrical events in such a hair. When the hair is displaced, a receptor potential develops and a few action potentials are generated, but all activity ceases when motion ceases, even though the hair is maintained in the displaced position.

The remarkable tactile sensitivity of man, especially in the finger tips and lips, is due to a large and diverse number of sense organs in the skin (Fig. 22–7). By making a careful survey of a small area of skin, point by point, using a stiff bristle to test for touch, a hot or cold metal stylus to test for temperature and a needle to test for pain, it has been found that receptors for each of these sensations are located at different spots. By compar-

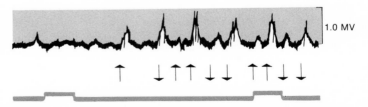

FIGURE 22–6   *Electrical responses to repeated mechanical stimulation of a hair on the wing of the flesh fly Sarcophaga. The arrow indicates the onset of a rapid one-directional displacement, and the direction of the displacement with reference to the preceding stimulus. Negative potential at the recording electrode is up. Time marks recur at 0.2-second intervals. Notice that with each displacement the baseline goes up indicating brief depolarization (the receptor potential). Action potentials (thin line superimposed on the baseline) are seen to occur at this time. Activity ceases even though the hair remains at the displaced position until the next stimulus. This is a phasic receptor. (Courtesy of M. L. Wolbarsht.)*

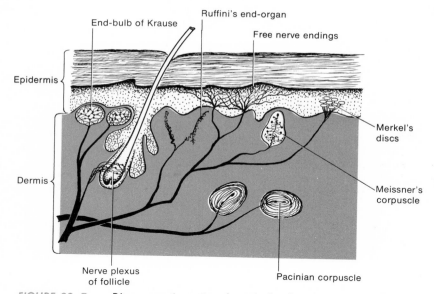

FIGURE 22-7   *Diagrammatic section through the skin showing the types of sense organs present. The sense organs respond to the following stimuli: cold—end-bulb of Krause; warmth—Ruffini's end-organs; touch—Meissner's corpuscles and Merkel's discs; deep pressure—pacinian corpuscles; and pain—free nerve endings.*

ing the distribution of the different types of sense organs and the types of sensations produced, it has been found that the free nerve endings are responsible for pain perception, that basket nerve endings around hair bulbs, **Meissner's corpuscles** and **Merkel's discs** are responsible for touch, that the **end-bulbs of Krause** and **Ruffini's endings** are responsible for cold and warmth and that **Pacinian corpuscles** are responsible for deep pressure.

The Pacinian corpuscle has been particularly well studied. The bare axon is surrounded by lamellae interspersed with fluid. Compression causes displacement of the lamellae which provides the deformation stimulating the axon. Even though the displacement is maintained under steady compression, the receptor potential rapidly falls to zero and action potentials cease. This is a phasic receptor responding to velocity.

Proprioception (Kinesthesis)

Among invertebrates the sense organs most commonly concerned with relaying postural information are hairs, plates (campaniform organs) and other modified cuticular structures. These are **tonic** (static) sense organs. Unlike phasic receptors (compare Figs. 22–6 and 22–8), the receptor potential is maintained (though not at constant magnitude) as long as the stimulus is present, and action potentials continue to be generated. Thus there is continued information about the position of the organ concerned.

In man each muscle, tendon and joint is equipped with **proprioceptors** sensitive to muscle tension and stretch. By reason of these sense organs we can, even with our eyes closed, perform manual acts such as dressing or tying knots. Impulses from the proprioceptors are also extremely important in ensuring the harmonious contraction of different muscles involved in a single movement; without them, complicated skillful acts would

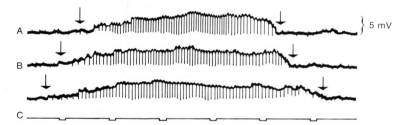

FIGURE 22–8   *Electrical responses to mechanical stimulation of a tactile hair on the outer clasper (external genitalia) of a male blowfly, Phormia, to mechanical stimulation. Conditions are the same as in Figure 22–6. The amplitude of the response in each record is approximately proportional to the degree of displacement of the hair. Notice that the receptor potential and action potentials continue as long as the hair remains displaced. This is a tonic receptor. (Courtesy of M. L. Wolbarsht.)*

be impossible. Impulses from these organs are also important in the maintenance of balance. Proprioceptors are probably more numerous and more continuously active than any of the other senses, although we are less aware of them than of any of the others. The existence of this sense was discovered only a little more than one hundred years ago. One obtains some idea of what life without proprioceptors would be like when a leg or arm "goes to sleep"—a feeling of numbness results from the lack of proprioceptors.

The Mammalian Muscle Spindle

Of the various stretch receptors concerned with postural behavior, the mammalian muscle spindle is certainly one of the more versatile. It illustrates very beautifully how sensory performance may be modified by the consequences of its own action. It is an example of a **feedback** mechanism. In the muscles of higher vertebrates there are in addition to the regular striated muscle fibers (**extrafusal** fibers) special fibers (**intrafusal** fibers) associated with sensory nerve endings. A bundle of intrafusal fibers together with their sensory endings is called a **muscle spindle** (Fig. 22–9). The intrafusal fiber is striated except in the region of the nucleus. Here there are two kinds of nerve endings: flower-spray endings belonging to a thin sensory (afferent) nerve and annulospiral endings belonging to a thick sensory (afferent) nerve. In the region of attachment of the extrafusal muscle fibers to the tendon there is another sense organ, the **Golgi** or **tendon organ**. There are two sets of motor neurons to the muscle: the **alpha efferents** innervate the ordinary (extrafusal) muscle fibers; the **gamma efferents** innervate the intrafusal fibers. The neural circuitry is shown in Figure 22–10.

For violent muscular contraction, commands from the central nervous system come mostly via the alpha efferents (the "emergency" pathway). If, as a consequence, a muscle is stretched excessively in a damaging way, the Golgi organ and the muscle spindle are stimulated. Messages from the Golgi organ pass up the sensory nerve to a point where they synapse with the alpha efferent. They inhibit the alpha efferent and the muscle stops contracting. Thus tension is kept within bearable limits and the muscle is kept at constant length under a specific load.

Under "ordinary" circumstances, i.e., in the production of slow voluntary movements, commands from the central nervous system descend the gamma efferents to the intrafusal fibers. These begin a slow graded contraction. As a consequence the muscle spindle is stimulated, sends

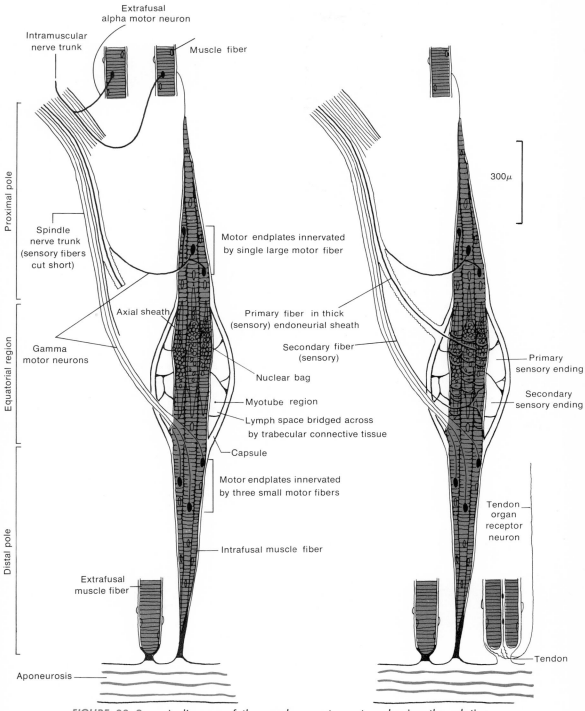

FIGURE 22-9   A diagram of the muscle receptor system showing the relations between the intrafusal and extrafusal fibers, the tendon organ, the spindle receptors and the motor innervation.

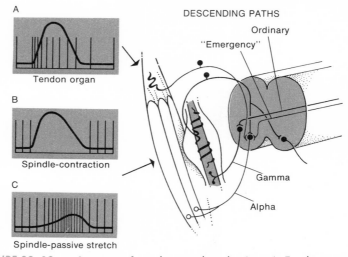

FIGURE 22–10   *Summary of muscle control mechanisms. A, Tendon organ responds as tension increases during contraction (indicated by heavy line). Tendon organ shown terminating in inhibitory junction on alpha motorneuron. B, Sustained activity of spindle receptors ceases as contraction of extrafusal muscle elements shortens spindle. Spindle receptor neuron terminates in excitatory synapse with alpha motorneuron of muscle, in which receptor lies and in inhibitory synapse on alpha motorneuron of antagonist muscle. Normal excitatory path from brain terminates on gamma motorneuron and "emergency" pathway terminates on alpha motorneuron. C, Spindle receptor becomes active upon stretching, signaling for increased activity in surrounding muscle fibers. (From Case, J.: Sensory Mechanisms. The Macmillan Company, 1966.)*

impulses to the synapse with the alpha efferent which is excited to cause the extrafusal fibers to contract. Since the intrafusal fibers are hitched in parallel with the other fibers rather than in series as are the Golgi organs, they become slack (Fig. 22–11). The spindle then stops exciting the alpha efferent, and the muscle ceases contracting. The net result is to cause the muscle to come to a new state of tension. Thus is muscle tone maintained and precise voluntary movements made.

While the muscle has been contracting it has been stretching its an-

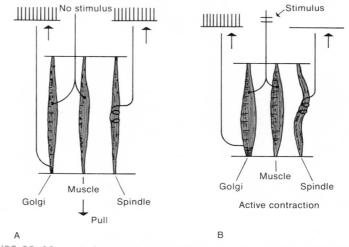

FIGURE 22–11   *Golgi spindle receptors have a series relationship to the muscle fibers. Spindles are in parallel. Pull on the muscle increases the rate of firing of both receptors (A). Active contraction of the muscles (B) will cause an increase in discharge of the Golgi tendon organ and a decrease in rate of discharge from the spindle. (From Ochs, S.: Elements of Neurophysiology. John Wiley & Sons, Inc., 1965.)*

tagonist. Naturally the spindle of the antagonist excites it to contract. If it continues to do so in the face of the pull being exerted upon it, the excessive strain stimulates its Golgi organ. This inhibits its contraction. Thus are antagonistic muscles prevented from "fighting" each other.

SECTION 22–5

THE EAR OF MAN

Two different sense organs equipped with mechanoreceptors are located in the ear. They are concerned with hearing and with equilibrium. These organs are buried deep in the bone of the skull, and a number of accessory structures are needed to transmit sound waves from the outside to the deep-lying sensory cells. The ear may be divided into outer, middle and inner parts; by reference to the diagram (Fig. 22–12) the path of the sound waves may be followed.

The outer ear consists of two parts, the skin-covered cartilaginous flap or *pinna* and the *auditory canal* leading from it to the middle ear.

The pinnas, or visible ears, are of some slight use in man for directing sound waves into the canal, but in other animals such as the cat, the larger, movable pinnas are very important. At the junction of the auditory canal and the middle ear is stretched a thin, connective tissue membrane, the *ear drum*, which the sound waves set vibrating.

The middle ear is a small chamber containing three tiny bones connected in a series, the *hammer, anvil* and *stirrup* (so called because of their shapes), which transmit the sound waves across the middle ear cavity. The hammer is in contact with the ear drum, and the stirrup is in contact with the membrane of the opening into the inner ear called the *oval window.* The middle ear is connected to the pharynx by the narrow *eustachian tube,* which serves to equalize the pressure on the two sides of the ear drum. If the middle ear were completely closed, any variation in atmospheric pressure would cause a pronounced and painful bulging or caving in of the ear drum. At the pharyngeal end of the eustachian tube is a valve, normally closed, which prevents one from becoming unpleasantly aware of his own voice. This valve is opened during yawning or swallowing, and during an abrupt ascent or descent in an elevator or airplane such acts help prevent the cracking sensation of the ear drums produced by the changes in atmospheric pressure accompanying changes in altitude. Unfortunately, the

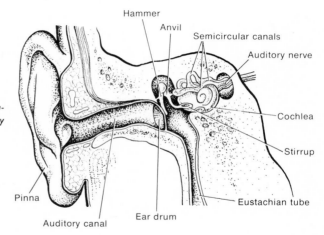

FIGURE 22–12   *Structure of the human right ear, cut open to show schematically the outer, middle and inner ear.*

eustachian tube also provides a path for organisms which sometimes cause infections resulting in the fusing of the middle ear bones and loss of hearing.

The inner ear consists of a complicated group of interconnected canals and sacs, often referred to, most appropriately, as the *labyrinth*. The part of the labyrinth concerned with hearing is a spirally coiled tube of two and a half turns, resembling a snail's shell, called the *cochlea*. If the cochlea were uncoiled, as in Figure 22–13, it could be seen to consist of three canals separated from each other by thin membranes and coming almost to a point at the apex. The oval window is connected to the base of one of these tubes, the *vestibular canal*. At the base of the *tympanic canal* is another opening covered by a membrane, the *round window*, which also leads to the middle ear. These two canals are connected with each other at the apex of the cochlea and are filled with a fluid known as the *perilymph*. Between the two lies a third, the *cochlear canal*, filled with a fluid called *endolymph* and containing the actual organ of hearing, the *organ of Corti*. This structure consists of five rows of cells with projecting hairs which extend the entire length of the coiled cochlea. Each organ of Corti contains about 24,000 of them. These cells rest upon the *basilar membrane*, which separates the cochlea from the tympanic canal. Overhanging the hair cells is another membrane, the *roof* (or tectorial) *membrane*, attached

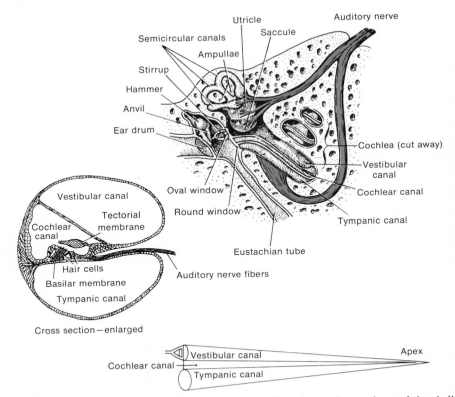

FIGURE 22–13   Upper right, *The coiled cochlea shown dissected out of the skull and cut open to reveal the vestibular and tympanic canals. Lower right, A diagram of the cochlea as though it were uncoiled and drawn out in a straight line. Lower left, A cross section through the cochlea to show the organ of Corti resting on the basilar membrane and covered by the tectorial membrane. Vibrations transmitted by the hammer, anvil and stirrup set the fluid in the vestibular canal in motion; these vibrations are transmitted to the basilar membrane and the organ of Corti. The hair cells of the organ of Corti are the receptor cells for hearing and are innervated by branches of the auditory nerve.*

along one edge to the membrane on which the hair cells rest, and with the other edge free. The hair cells initiate impulses in the fibers of the auditory nerve.

For a sound to be heard, sound waves must first pass down the auditory canal and set the ear drum vibrating. These vibrations are transmitted across the middle ear by the hammer, anvil and stirrup, which are so arranged that they decrease the amplitude, but increase the force, of the vibrations. The stirrup transmits the vibrations via the oval window to the fluid in the vestibular canal. Since fluids are incompressible, the oval window could not cause a movement of the fluid in the vestibular canal unless there were an escape valve for the pressure. This is provided by the round window at the end of the tympanic canal. The pressure wave presses upon the membranes separating the three canals, is transmitted to the tympanic canal, and causes a bulging of the round window. The movements of the basilar membrane produced by these pulsations are believed to rub the hair cells of the organ of Corti against the overlying roof membrane, thus stimulating them and initiating nerve impulses in the dendrites of the auditory nerve, lying at the base of each hair cell.

Since sounds differ in pitch, intensity and quality, any theory of hearing must account for the ability to discriminate such differences. Microscopic examination of the organ of Corti reveals that the fibers of the basilar membrane are of different lengths along the coiled cochlea, being longer at the apex and shorter at the base of the coil, thus resembling the strings of a harp or piano. Sounds of a given frequency (and pitch) set up resonance waves in the fluid in the cochlea that cause a particular section of the basilar membrane to vibrate. The vibration stimulates the particular group of hair cells in that section. Thus the pitch of a sound is sensed by the particular hair cells stimulated. Loud sounds cause resonance waves of greater amplitude and lead to a more intense stimulation of the hair cells and to the initiation of a greater number of impulses per second which pass over the auditory nerve to the brain.

When the ear is subjected to intense, continuous sound, the organ of Corti is injured. This was demonstrated by an experiment in which guinea pigs were exposed to continuous pure tones for a period of several weeks. When their cochleas were examined microscopically after death, it was found that the guinea pigs subjected to high-pitched tones suffered injury only in the upper part of the cochlea. Workers such as boilermakers, subjected to loud, high-pitched noises over a period of years, frequently become deaf to high tones because of injury of the cells toward the base of the organ of Corti. Recent research indicates that the nerve impulses produced by particular sounds have the same frequency as those sounds, so that the brain may recognize particular pitches by the frequency of the nerve impulses reaching it, as well as by the identity of the nerve fibers conducting the impulses.

The auditory nerves transmit two kinds of nerve impulses: ordinary nerve impulses like those of any other nerve, and a different type called **microphonic**. The energy for the latter is not derived from the metabolism of the nerve fiber, as is the energy for the former; instead, the cochlea acts as a microphone to convert the mechanical energy of the sound vibrations into electrial energy. For this reason the wave form of the electrical potential from the cochlea closely resembles that of the stimulating sound wave. In fact, Wever and Bray placed electrodes on the auditory nerve of a decerebrated cat, and then, listening with a telephone receiver to the amplified signals of the nerve, were able to hear not only musical tones, but actual words spoken to the cat. The hair cells of the organ of Corti are believed to be responsible for this conversion of mechanical to electrical energy, the upper and lower ends of the cochlea responding to low and high tones, respectively. It is still a disputed question, however, whether these microphonics have anything to do with the actual sensation of hearing in the normal animal.

Variations in the quality of sound, such as are produced when an oboe, a cornet and a violin play the same note, depend upon the number and kinds of **overtones** or **harmonics** present, which stimulate different hair cells in addition to the main stimulation common to all three; thus, differences in quality are recognized by the pattern of the hair cells stimulated. Careful histological work has shown that the nerve fibers from each particular part of the cochlea are connected to particular parts of the auditory area of the brain, so that certain brain cells are responsible for the perception of sensations of high tones, others for low tones.

The human ear is equipped to register sounds of frequencies between about 20 and 20,000 cycles per seond, although there are great individual differences. Some animals — dogs, for example — can hear sounds of much higher frequencies. The human ear is more sensitive to sounds between 1000 and 2000 cycles per second than to higher or lower ones. Within this range the ear is extremely sensitive; in fact, when compared with the energy of light waves necessary to produce a sensation, the ear is 10 times more sensitive than the eye.

The normal human ear is just about as efficient a hearing device as anything could possibly be, for, like the eye, it has evolved to the point where any further increase in sensitivity would be useless. If it were more sensitive, it would pick up the random movement of the air molecules, which would result in a constant hiss or buzzing. If the eye were more sensitive, a steady light would appear to flicker because the eye would be sensitive to the individual photons (light particles) impinging on it.

There is little fatigue connected with hearing. Even though it is constantly assailed by noises, the ear retains its acuity and fatigue disappears after a few minutes. When one ear is stimulated for some time by a loud noise, the other ear also shows fatigue — loses acuity — indicating, not unexpectedly, that some of the fatigue is in the brain rather than in the ear itself.

Deafness may be caused by injuries or malformations of either the sound-transmitting mechanisms of the outer, middle or inner ears, or of the sound-perceiving mechanism of the latter. The external ear may become obstructed by wax secreted by the glands in its wall; the middle ear bones may become fused after an infection; or, more rarely, the inner ear or auditory nerve may be injured by a local inflammation or the fever accompanying some disease.

Relatively few animals have a sense of hearing. The vertebrate ear began as an organ of equilibrium, the cochlea being a later evolutionary outgrowth of the saccule which reaches full development only in mammals. The human ear is indeed a curious evolutionary hodge-podge: the cells sensitive to sound are apparently adaptations of cells sensitive to the motion of liquids; the middle ear and eustachian tube were originally part of the respiratory apparatus of fish; the stirrup was originally a structure which attached the jaws of primitive fishes to the cranium, and the hammer and anvil are the remnants of the lower and upper jaws, respectively, of our ancestral fish. In the jawless fish ancestral to these the structures were part of the support for the gills. Thus respiratory organs became, first, eating organs, and then organs for hearing. This is an example of one of the fundamental patterns of evolution — the reshaping of old organs to perform new functions, rather than the setting up of completely new structures.

SECTION 22-6

EQUILIBRIUM

Besides the cochlea, the labyrinth of the inner ear consists of two small sacs—the saccule and the utricle—and three semicircular canals (Fig. 22–14). These structures are filled with endolymph and float in a pool of perilymph. Destruction of them causes a considerable loss of the sense of equilibrium, and a pigeon whose organs have been destroyed is unable to fly. In time it can relearn to maintain equilibrium using visual stimuli.

Equilibrium in man depends upon the sense of vision, stimuli from the proprioceptors, and stimuli from cells sensitive to pressure in the soles of the feet, as well as upon stimuli from these organs in the inner ear. In certain types of deafness the equilibrium organs of the inner ear as well as the cochlea are inoperative, yet the sense of equilibrium remains unimpaired.

The utricle and the saccule are small, hollow sacs lined with sensitive hair cells and containing small ear stones or **otoliths** made of calcium carbonate. Normally the pull of gravity causes the otoliths to press against particular hair cells, stimulating them to initiate impulses to the brain via sensory nerve fibers at their bases. When the head is tipped, the otoliths press upon the hairs of other cells and stimulate them.

Many invertebrates, such as the crayfish and lobster, have similar organs. An ingenious experiment was performed to demonstrate the action of these organs in the crayfish; it depended on the fact that as the crayfish molts—sheds its skin and grows another, larger one—it also develops new organs of equilibrium and supplies them with grains of sand picked up from the environment. By supplying the molting crayfish with particles of iron, the experimenters could subsequently cause the animals to respond to a magnet. When the magnet was placed directly over the animal, pulling the iron filling against the hair cells on the top of its equilibrium organ, the crayfish thought that "up" was "down" and responded by turning over and swimming on its back.

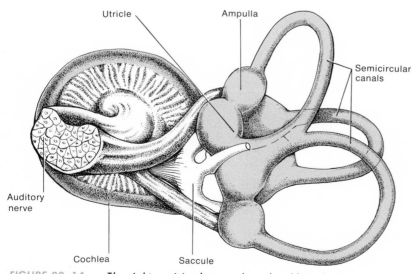

FIGURE 22-14   *The right semicircular canals and cochlea of an adult man, shown dissected free of surrounding bone and enlarged about five times, seen from the inner and posterior side. Note that the plane of each semicircular canal is perpendicular to those of the other two.*

The labyrinth of each ear has three semicircular canals, each of which consists of a semicircular tube connected at both ends to the utricle. The canals are so arranged that each is at right angles to the other two. At one of the openings of each canal into the utricle is a small, bulblike enlargement (the *ampulla*) containing a clump of hair cells similar to those in the utricle and the saccule, but lacking otoliths. These cells are stimulated by movements of the fluid (endolymph) which fills the canals. When the head is turned, there is a lag in the movement of the fluid within the canals, so that the hair cells in effect move in relation to the fluid, and are stimulated by its flow. This stimulation produces not only the consciousness of rotation, but certain reflex movements in response to it, movements of the eyes and head in a direction opposite to the original rotation. Since the three canals are located in three different planes, a movement of the head in any direction will stimulate the movement of the fluid in at least one of the canals. By irrigating the canal of the outer ear with warm or cold water, convection currents can be set up causing movements in the fluid of the canals without movement of the head. Sensations of rotation and dizziness result. Man has become used to movement in the horizontal plane, which stimulates certain semicircular canals, but he is unused to vertical movements parallel to the long axis of the body. Such movements—the motion of an elevator or of a ship pitching in a rough sea—stimulate the semicircular canals in an unusual way and may produce the sensation of nausea and the vomiting of sea or motion sickness. When one lies down, the movement stimulates the semicircular canals in a different way, and nausea is less likely to occur.

SECTION 22-7

THE HALTERES OF FLIES

Millions of years before man invented the gyroscope, flies had evolved a gyroscope for use as a balancing organ to stabilize flight. A flying machine must maintain stability if it is to be controllable in the air, and flies are no exception. They must be able to control lift and to stabilize in all three planes of rotation; i.e., they must correct for pitch, roll and yaw. They accomplish this with information derived from the *halteres*, a pair of marvelously modified hind wings. Each looks lika a dumbbell. It is a heavy mass of tissue on a thin stalk (Fig. 22–15). The folding and articulation of the base is very complicated and is equipped with about 418 mechanoreceptors that respond to strains produced in the cuticle by *gyroscopic torque*. This comes about because the halteres beat actively like wings. They are oscillating masses that generate forces at the base of the stalk as the whole fly rotates. They are clearly gyroscopes; however, they probably do not act as stabilizing gyroscopes of the sort placed in ships to offset their movement. Their action is indirect in that their mechanoreceptors signal the central nervous system to make the necessary corrections in flight.

SECTION 22-8

CHEMORECEPTION: TASTE AND SMELL

Throughout the animal kingdom many sexual, reproductive, social and feeding activities are initiated, regulated or in some way influenced

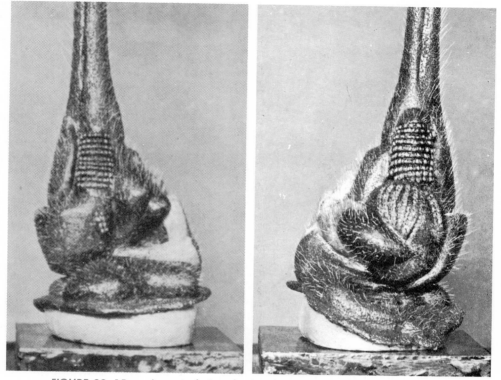

FIGURE 22-15   A, ventral view; B, dorsal view, of a model of the left haltere of the blowfly Lucilla sericata. Rows of sense organs are visible at the base. Magnified 130×. (Courtesy of J. W. S. Pringle, 1948.)

by specific chemical aspects of the environment. Insects, for example, rely upon a great number of chemicals for communication, for defense from predators, for the recognition of specific foodstuffs. Many vertebrates employ chemical secretions to mark territory, to attract their sexual partners, to defend themselves. Anyone who has ever encountered a skunk appreciates the last mentioned role. Chemoreception is also involved in the tracking and location of prey in carnivores and in the detection of carnivores by intended prey.

Sensitivity to chemicals may be very specific in that only certain compounds act as stimuli and then in low concentrations. On the other hand the sensitivity of some receptors may be gross and nonspecific. In this category fall the receptors in the skin of frogs. As most beginning physiology students know, a frog will scratch its back when dilute acid or high concentrations of inorganic salts are applied to the skin. Free nerve endings are the receptors in this case. This **common chemical sense** is widely distributed among aquatic animals. Among vertebrates it is restricted to moist areas of the body. Recall how the eye smarts and waters in the presence of ammonia fumes or how a broken blister smarts if a nonphysiologic solution touches it.

The specific and highly sensitive chemoreceptive systems comprise the senses of **taste** and **smell** (olfaction). These are easily distinguishable in ourselves and other terrestrial organisms. As one examines aquatic organisms, and especially those lower on the phylogenetic scale, it becomes increasingly difficult to decide what is taste and what is olfaction.

Because man is primarily a visual animal and employs his chemical senses to a far lesser degree than other animals he tends to deemphasize the importance of the chemical senses. Probably for this reason we still know comparatively little about their mode of action.

THE SENSE OF TASTE IN MAMMALS

The organs of taste are budlike structures located predominantly on the tongue and soft palate. They are situated in papillae of which there are four kinds: **circumvallate, foliate, fungiform** and **filiform** (Fig. 22–16). The cells of the **taste buds** were originally classified on the basis of size and histology as gustatory cells and supporting cells (Fig. 22–17). Recent studies with the electron microscope reveal that this concept is too simple, that there are all gradations between the two. There is also a very rapid substitution of cells. Every 10 to 30 hours the cells are completely replaced. Each cell has at its free surface a border of microvilli, many of which project into a tiny pore connecting with the fluids bathing the surface of the tongue. There are no taste hairs as once believed.

Each taste cell is an epithelial cell and is the receptor. The connections with the nerve cells are complicated. Each taste cell is innervated by more than one neuron. Furthermore, some neurons may connect with one taste cell and others with many. This complexity of connections renders interpretation of the physiology of taste difficult.

Traditionally there are four basic tastes: sweet, salt, sour and bitter. To this must now be added water. While it is true that the greatest sensitivity to each of these tastes is restricted to a given area of the tongue (in man especially) (Fig. 22–18), not all papillae are restricted in their sensitivity to a single taste modality. Some indeed are specific to salt, acid or sugar, but the majority respond to two or more categories of taste solutions. Nor is a single taste bud restricted in its sensitivity to different chemicals. Furthermore, a single receptor (epithelial taste cell) may respond to more than one category of taste. Thus the detection and processing of information in the taste organs of the tongue are very complex indeed. Taste discrimination probably depends on a code that consists of cross-fiber patterning; i.e., each receptor responds to more than one

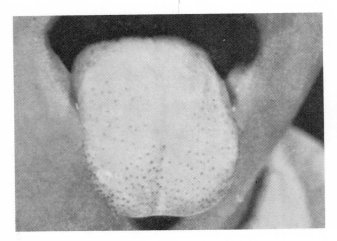

FIGURE 22–16   Tongue of a four-year-old boy showing numerous fungiform papillae. Milk applied to tongue for contrast. (From Beidler, L.: Taste Receptor Stimulation. In Progress in Biophysics and Biophysical Chemistry. New York, Pergamon Press, Inc., 1961.)

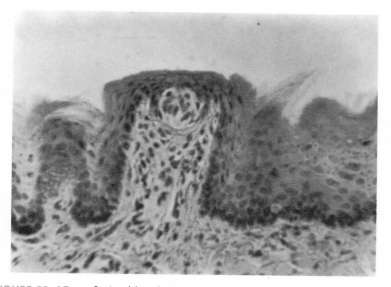

FIGURE 22–17   *Stained histological section of the tongue of a rat showing fungiform papilla with a taste bud, ×286. (From Beidler, L.: Taste Receptor Stimulation. In Progress in Biophysics and Biophysical Chemistry. New York, Pergamon Press, Inc., 1961.)*

kind of chemical, but no two respond exactly alike, so that the total pattern of messages going to the brain is different for different solutions (Fig. 22–19).

Flavor does not depend on the perception of taste alone. It is compounded of taste, smell, texture and temperature. Smell affects flavor because odors pass from the mouth to the nasal chamber via the internal nares.

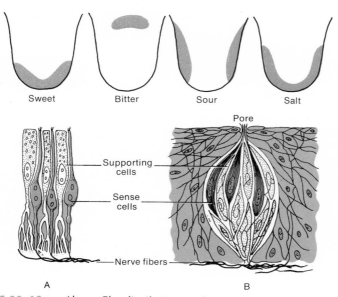

FIGURE 22–18   *Above: The distribution on the surface of the tongue of taste buds sensitive to sweet, bitter, sour and salt. Below: A, Cells of the olfactory epithelium of the human nose. B, Cells of a taste bud in the epithelium of the tongue.*

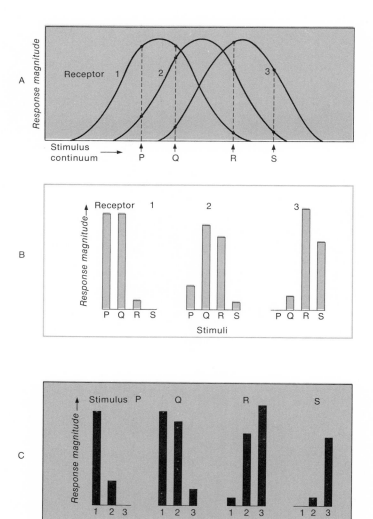

FIGURE 22-19   Afferent fiber types and patterns of neural activity. A, Afferent
fiber types (or receptor types). Curves 1, 2 and 3 represent the responsiveness of three
hypothetical afferent fiber types (or receptor types) along a hypothetical stimulus continuum. P, Q, R and S represent four stimuli along this stimulus continuum. The responsiveness of a fiber type to one of these stimuli is indicated by the intersection of the response curve and the ordinate erected at the stimulus. B, Responsiveness of the three fiber
types to the four stimuli in A. In each of the bar graphs is shown the responsiveness of one
of the fiber types to each of the stimuli in A. If recordings were obtained from one of the
fiber types shown in A using these stimuli, one of these three "response profiles" would
be obtained, depending upon which fiber type was being sampled. There would be as
many "response profiles" as fiber types. C, Across-fiber patterns. In these bar graphs are
shown the patterns of activity across the three fiber types produced by the four stimuli in
A. Each stimulus produces a characteristic pattern across the three fiber types. There would
be as many across-fiber patterns as stimuli. (After Erickson, R. P.: Sensory Neural Patterns
and Gustation. In Zotterman, Y., (ed.): Olfaction and Taste. New York, Pergamon Press, Inc.,
1963.)

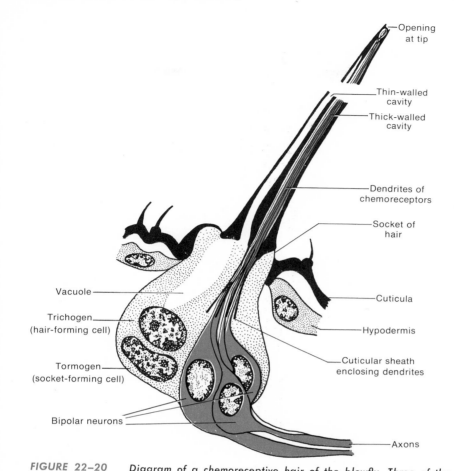

Opening
at tip

Thin-walled
cavity

Thick-walled
cavity

Dendrites of
chemoreceptors

Socket of
hair

Vacuole

Trichogen
(hair-forming cell)

Tormogen
(socket-forming cell)

Bipolar neurons

Cuticula

Hypodermis

Cuticular sheath
enclosing dendrites

Axons

FIGURE 22–20   Diagram of a chemoreceptive hair of the blowfly. Three of the
five neurons are shown. (From Dethier, V. G.: Insects and the Concept of Motivation. In
Levine, D., (ed.): Nebraska Symposium on Motivation. Lincoln, Nebraska, University of
Nebraska Press, 1966.)

SECTION 22–10

THE SENSE OF TASTE OF INSECTS

One of the most thoroughly studied organs of taste is the taste hair
of the fly (Fig. 22–20). The terminal segments of the legs and the mouthparts of flies, moths, butterflies and a number of other insects are equipped
with very sensitive hairs and pegs. In the fly each one of these contains
four taste receptors and a tactile receptor. All are primary neurons. One
taste receptor is more or less specific to sugars, one to water and two to
salts. If water is placed on one hair of a thirsty fly, the water cell generates
action potentials which pass directly to the central nervous system and
cause the fly to respond by extending its retractible proboscis and drinking. Similarly, sugar on one hair stimulates the sugar receptor and causes
feeding. Salt causes the fly to reject the solution.

SECTION 22–11

THE SENSE OF SMELL (OLFACTION)

The sense of smell of terrestrial vertebrates is served by primary neurons located in the **nasal epithelium** (Fig. 22–21). Each of these neurons has

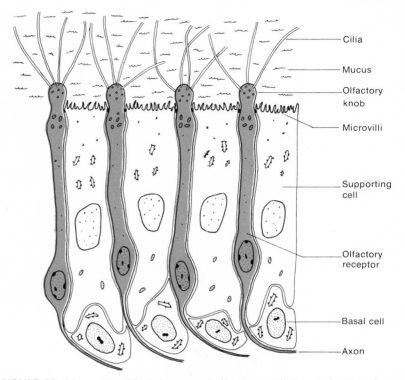

Cilia

Mucus

Olfactory
knob

Microvilli

Supporting
cell

Olfactory
receptor

Basal cell

Axon

FIGURE 22–21   *Simplified diagram of olfactory epithelium indicating the various
cellular components. (From Moulton, D. G., and Beidler, L.: Structure and Function in the
Peripheral Olfactory System. Physiol. Rev. 47:4, 1967.)*

a short axon that passes through the **cribriform plate** and immediately
synapses with other neurons. The complexity of this neural circuitry is
illustrated in Figure 22–22. In the rabbit, for example, there are 10^8
receptors. Twenty-six thousand of these synapse with each **glomerulus.**
Each glomerulus connects with 24 **mitral cells** and 68 **tufted cells.** The
possibilities for processing olfactory data generated by the receptors before
it even reaches the brain are enormous.

Little is known at present about the way in which odors stimulate the
primary receptor or how the many different odor qualities are distinguished. There are numerous hypotheses but none have yet stood the test
of time.

SECTION 22–12

PHOTORECEPTION — VISION

Light-sensitive cells exist in almost all living matter. Even protozoa
respond to changes in light intensity, usually moving away from the source
of light. Most plants orient their leaves and flowers toward the sun, although they have no special light-sensitive structures. In most of the higher
animals this light-sensitivity is localized in certain cells and is highly
developed. The human eye is an excellent example of an extremely sensitive, specialized organ for perceiving light. The late **Selig Hecht** and his
collaborators at Columbia University showed that an eye, when fully dark-

adapted, can detect as little as 6 to 10 quanta of light. Just as matter consists of tiny particles known as atoms, light consists of units called **photons,** and, by definition, the energy of 1 photon is 1 **quantum.** The light reaching the eye from a candle 14 miles away is just at the limit of visibility of a normal dark-adapted eye and is about 6 or 7 quanta of light.

Some protozoa have "eye spots" which are more sensitive to light than the rest of the cell, but the most primitive light-sensitive organs in the evolutionary scale are those of flatworms. They are **photoreceptors** rather than eyes because they are incapable of forming images. They are bowl-shaped structures containing black pigment, at the bottom of which are clusters of light-sensitive cells. These are shaded by the pigment from light coming from all directions except above and slightly to the front. This arrangement enables the planarian to detect the direction of the source of light.

The acquisition of a **lens** which could concentrate light on a group of photoreceptors was the first step in the evolution from photoreceptors to true eyes. As better lens systems evolved, the formation of images became possible. This marked the advent of eyes in the truest sense. The most highly developed eyes are found in arthropods (insects, crabs, lob-

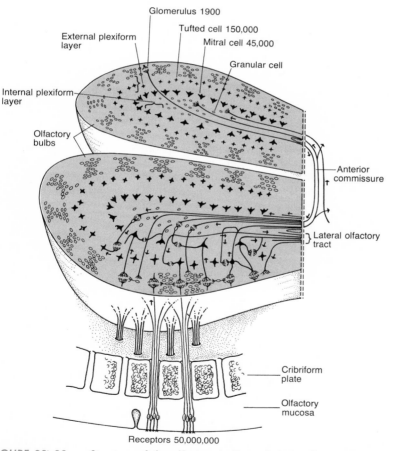

FIGURE 22–22   *Structure of the olfactory bulbs and their relations to the nerves and mucosa. (Modified after Gastaut & Lammers, 1961). A.C., anterior commissure. The figures are estimates of the numbers of each type of cell in an olfactory bulb and in the olfactory mucosa lining one nasal cavity of the rabbit. (After Allison, A. C., and Warwick, R. T. T.: Quantitative Observations on the Olfactory System of the Rabbit. Brain 72:186–197, 1949.) (From Moulton, D. G., and Tucker, D.: Electrophysiology of the Olfactory System. Ann. N.Y. Acad. Sci. 116:380–428, 1964.)*

sters and so on), in cephalopods (octopuses, squids and so forth) and in vertebrates. There are two basic types: the **camera-type eye** of the cephalopods and vertebrates and the **compound eye** of the arthropods.

SECTION 22–13

THE HUMAN EYE

The squid or octopus eye is rather like a simple Brownie camera, equipped with slow, black and white film, whereas the human eye is like a de luxe Leica loaded with extremely sensitive color film.

The analogy between the human eye and a camera is complete: the eye (Fig. 22–23) has a **lens** which can be focused for different distances, a diaphragm (the **iris**) which regulates the size of the light opening (the **pupil**) and a light-sensitive **retina** located at the rear of the eye, corresponding to the film of the camera. Next to the retina is a sheet of cells filled with black pigment which absorbs extra light and prevents internally reflected light from blurring the image (cameras are also painted black on the inside). This sheet, called the **choroid coat,** also contains the blood vessels which nourish the retina.

The outer coat of the eyeball, called the **sclera,** is a tough, opaque, curved sheet of connective tissue which protects the inner structures and helps maintain the rigidity of the eyeball. On the front surface of the eye this sheet becomes the thinner, transparent **cornea,** through which light enters.

Immediately behind the iris is a transparent, elastic ball, the **lens,** which bends the light rays coming in, bringing them to a focus on the retina. It is aided by the curved surface of the cornea and by the refractive properties of the liquids inside the eyeball. The cavity between the cornea and the lens is filled with a watery substance, the **aqueous humor;** the larger chamber between the lens and the retina is filled with a more viscous fluid, the **vitreous humor;** both fluids are important in maintaining the shape of the eyeball. They are secreted by the **ciliary body,** a doughnut-shaped structure which attaches the ligament holding the lens to the eyeball.

The eye accommodates, or changes focus for near or far vision, by

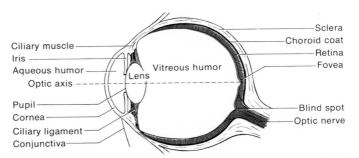

FIGURE 22–23   *Diagrammatic section of the human eye illustrating the location of the retina, which contains the light-sensitive rods and cones; the lens and cornea, which focus light rays on the retina; and the iris, which regulates the amount of light entering the eye by changing the diameter of the pupil. The human eye changes focus for near and far vision by changing the shape of the lens; certain birds change the curvature of the cornea; fish change the position of the lens in the eye (cameras are focused by changing the position of the lens); and mollusks shorten the eye, bringing the retina nearer the lens for distant vision.*

changing the curvature of the lens. This is made possible by the stretching
and relaxing of the lens by the **ciliary ligament,** which attaches the lens to
the ciliary body. Because of the pressure of the fluids within, the eyeball
is under tension, which is transmitted by the ciliary ligament to the lens.
Tension on the ligament flattens the lens and focuses the eye for far vision,
the condition of the eye at rest. Just in front of the ciliary body, and attached to the ciliary ligament, are ciliary muscles which, when contracted,
take up the strain on the ligament and lens, leaving the latter free to assume the more spherical shape for near vision.

   As people grow older, the lens becomes less elastic and thereby less
able to accommodate for near vision. When this occurs, spectacles with
one part ground for far vision and one for near vision are worn to accomplish what the eye can no longer do.

   The amount of light entering the eye is regulated by the **iris,** a ring
of muscle which appears as blue, green or brown, depending on the
amount and nature of pigment present. The structure is composed of two
sets of muscle fibers, one arranged circularly, which contracts to decrease
the size of the pupil, and one arranged radially, which contracts to increase

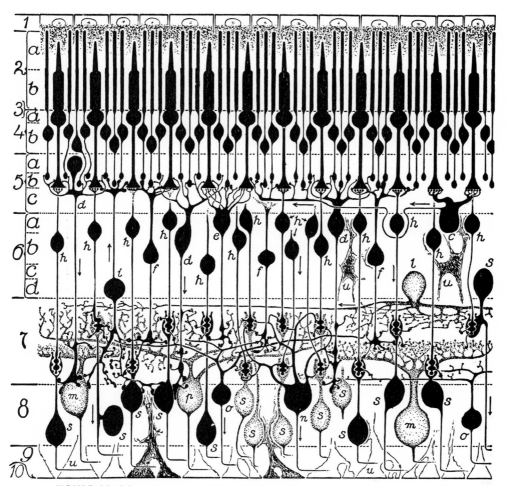

FIGURE 22–24   *Diagram of the primate retina composed from numerous Golgi-
stained preparations of man, chimpanzee and macaque. (From Maximow, A. A., and Bloom,
W.: Textbook of Histology. Philadelphia, W. B. Saunders Company, 1941.)*

the size of the pupil. The response of these muscles to changes in light intensity is not instantaneous, but requires 10 to 30 seconds; thus when one steps from a light to a dark area, some time is needed for the eyes to adapt to the dark, and when one steps from a dark room to a brightly lighted street, the eyes are dazzled until the size of the pupil is decreased.

Each eye has six muscles stretching from the surface of the eyeball to various points in the bony socket which enable the eye as a whole to move and be oriented in a given direction. These muscles are innervated in such a way that the eyes normally move together and focus on the same area.

The only part of the human eye which is light-sensitive is the **retina,** a hemisphere made up of an abundance of receptor cells, called, according to their shape, **rods** and **cones.** There are about 125,000,000 rods and 6,500,000 cones. In addition, the retina contains many sensory and connector neurons and their axons. Curiously enough, the sensitive cells are at the back of the retina; to reach them, light must pass through several layers of neurons (Fig. 22–24). The eye develops as an outgrowth of the brain, and folds in such a way that the sensitive cells eventually lie on the farthermost side of the retina (Fig. 22–25). At a point in the back of the eye, the individual axons of the sensory neurons unite to form the optic nerve and pass out of the eyeball. Here there are no rods and cones. This area is called the "blind spot," since images falling on it cannot be per-

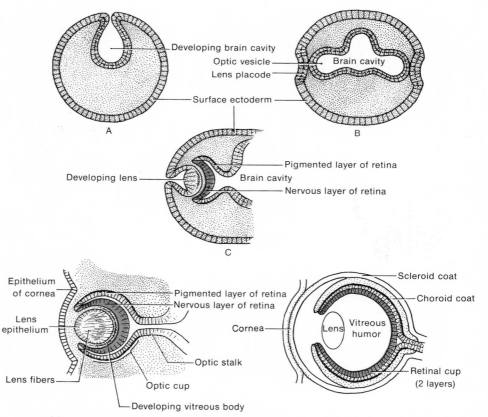

FIGURE 22–25   *Diagrams illustrating some of the steps in the development of the eye in man. The optic vesicles grow laterally from the brain (B) to meet the lens placode. This folds in (C), pushing part of the optic vesicle in with it to form the optic cup (lower left).*

FIGURE 22–26 Demonstration of the blind spot on the retina. See text for details.

ceived. Its existence can be demonstrated by closing the left eye and focusing the right one on the + in Figure 22–26. Starting with the page about 13 centimeters from the eye, move it away until the circle disappears. At that position the image of the circle is falling on the blind spot and is not perceived.

In the center of the retina, directly in line with the center of the cornea and lens, is the region of keenest vision, a small depressed area called the **fovea.** Here the light-sensitive cones, responsible for bright light vision, for the perception of detail and for color vision, are concentrated.

In normal vision the eyes are constantly in motion; there are small involuntary movements even when the eye is fixed on a stationary object. Thus the image on the retina is in constant motion, drifting away from the center of the fovea and flicking back to it. Superimposed on these movements is a high speed tremor of the eye. When, by some suitable device, the image is fixed immovably on the retina it fades and disappears, later reappearing completely or in part. Such experiments by Donald Hebb and his colleagues have shown that what a subject "sees" is determined at least in part by whether or not it makes sense to him.

The other light-sensitive cells, the rods, are more numerous in the periphery of the retina away from the fovea. These function in twilight or dim light and are insensitive to colors. One is not ordinarily aware that color can be perceived only in those objects which are more or less directly in front of the eyes, but the fact can be demonstrated by a simple experiment. Close one eye and focus the other on some point straight ahead. As a colored object is gradually brought into view from the side, you will be aware of its presence and of its size and shape before you are aware of its color. Only when the object is brought closer to the direct line of vision, so that its image falls on a part of the retina containing cones, can its color be determined. The rods are actually more sensitive in dim light than are the cones. Since the rods are located not in the center but in the periphery of the retina, it is a curious fact that you can see an object better in dim light if you do not look at it directly (for then its image will fall on the cones in the center of the retina), but slightly to one side of it, so that its image falls on the rods in the periphery of the retina.

SECTION 22–14

THE CHEMISTRY OF VISION

All **visual pigments** have a common plan — a **chromophore (retinal$_1$)** bound to a **protein (opsin).** The combination is known as **rhodopsin** or **visual purple.** It is the visual pigment in the rods of most vertebrates and in the retinal cells of insects. The cones contain **iodopsin,** a visual pigment consisting of the same chromophore (retinal$_1$) but a different protein. It is found in man, chickens, cats, snakes, frogs and crayfish. Fresh-water fish have different visual pigments. Their rod pigment **(porphyropsin)** consists of **retinal$_2$** plus opsin. The cone pigment consists of retinal$_2$ plus a different protein. It is **cyanopsin.** Cyanopsin is found in the eyes of fresh-water fish, tortoises and tadpoles. It is interesting that **euryhaline** eels, salmon and trout have both rhodopsin and porphyropsin, but the one (porphyropsin) commonly associated with spawning (in fresh water) predominates. The sea lamprey has mostly rhodopsin on its downstream migration to the sea and mostly porphyropsin when going upstream as a sexually mature adult. Amphibians change from porphyropsin to rhodopsin when they change from tadpole to adult.

Retinal is the aldehyde of Vitamin A **(retinol).** It is formed from retinol by an oxidation catalyzed by **alcohol dehydrogenase.** Retinal combines with the different opsins, as mentioned above, to form the various kinds of visual pigments.

The eyes of mollusks, arthropods and vertebrates, which arose independently in the course of evolution, all utilize the same basic chemical reaction, the conversion by light of the *cis* form of retinal to the *trans* form. This is the only action of the light. All else consists of dark reactions. The light reaction involves a simple rearrangement of the molecular structure which can occur very rapidly. Light energy converts rhodopsin (with the *cis* form of retinal) into **lumirhodopsin,** an unstable compound containing the *trans* form of retinal which decays first into **metarhodopsin** and then into free retinal and opsin (Fig. 22–27). The initiation of nerve impulses by the rods occurs when light strikes rhodopsin and triggers the **isomerization** of *cis*-retinal into the all-*trans* form. The all-*trans* form of retinal does not fit onto the opsin molecule as neatly as does the bent, *cis* form and hence the lumi- and metarhodopsin are more easily hydrolyzed. The cleavage of retinal from opsin is a much slower reaction than the isomerization, and is too slow to account for the high speed of the visual process.

The resynthesis of rhodopsin from retinal and opsin requires that the retinal first be **reisomerized** to the *cis* form. The retinal can then be recombined with opsin to form rhodopsin which will provide for further excitation of the rods. Rhodopsin is involved in a cyclic process: it is con-

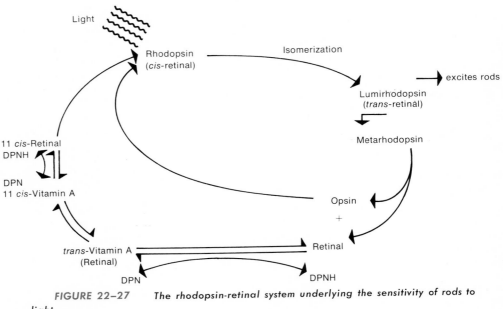

FIGURE 22–27   *The rhodopsin-retinal system underlying the sensitivity of rods to light.*

tinually synthesized and, after the light-induced isomerization of *cis*-retinal to the *trans* form of retinal in lumi- and metarhodopsin, it is hydrolyzed to yield free retinal and opsin. The free all-*trans*-retinal is then converted to vitamin A, reisomerized to the *cis* form, and reoxidized to *cis*-retinal before it can be combined with opsin to yield new rhodopsin.

It has been shown that a single quantum of light can be absorbed by a single molecule of rhodopsin and lead to the excitation of a single rod. When the eye is exposed to a flash of light lasting only one millionth of a second, the eye sees an image of light that persists for nearly one tenth of a second. This is the length of time that the retina remains stimulated following a flash and this presumably reflects the length of time that lumirhodopsin persists in the rods. This persistence of images in the retina enables the eye to fuse the successive flickering images on a moving picture or television screen, and one has the impression of seeing a continuous picture.

The ability to see an exceedingly faint light depends on the amount of rhodopsin present in the retinal rods and this in turn depends on the relative rates of synthesis and breakdown of rhodopsin. In bright light, much of the rhodopsin is broken down to free retinal and opsin. The synthesis of rhodopsin is a relatively slow process and the concentration of rhodopsin in the retina is never very great as long as the eye is exposed to bright light. When the eye is suitably shielded from light, the breakdown of rhodopsin is prevented and its concentration gradually builds up until essentially all the opsin has been converted to rhodopsin. The sensitivity of the eye to light, which is a function of the amount of rhodopsin present, can increase 1000-fold if the eye is dark-adapted for a few minutes and can increase about 100,000-fold if the eye is dark-adapted for as much as an hour.

SECTION 22–15

COLOR VISION

The chemistry of the cones and of color vision is less well understood, but the cones contain an analogous light-sensitive pigment, **iodopsin,** composed of retinal and a different opsin. The cones are considerably less sensitive to light than are rods and cannot provide vision in dim light. The prime function of the cones is to perceive colors. The evidence from certain psychologial tests is consistent with the hypothesis that there are three different types of cones which respond respectively to blue, green and red light. This has been substantiated recently by the demonstration that from human and monkey retinas can be extracted three kinds of color receptors—red, green and blue. Each type can respond to light with a considerable range of wavelengths; the green cones, for example, can respond to light of any wavelength from 450 to 675 millimicrons (i.e., blue, green, yellow, orange and red light), but they respond to green light more strongly than to any of the others. Intermediate colors, other than blue, green and red, are perceived by the simultaneous stimulation of two or more types of cones. Yellow light (i.e., light with a wavelength of 550 millimicrons) stimulates, according to that theory, green and red cones to an approximately equal extent and this is interpreted by the brain as "yellow color." Colorblindness results when one or more of the three types of cones is absent because of the absence of the gene which is necessary for the formation of that type of cone.

SECTION 22–16

DEFECTS IN VISION

The defects of the eye most common in man are nearsightedness **(myopia),** farsightedness **(hypermetropia)** and **astigmatism.** In the normal eye (Fig. 22–28 *A*) the shape of the eyeball is such that the retina is the proper distance behind the lens for the light rays to converge on the fovea. In a near-sighted eye (Fig. 22–28 *B*), the eyeball is too long and the retina is too far from the lens, so that the light rays converge at a point in front of the retina and are again diverging when they reach it, resulting in a blurred image. In a far-sighted eye (Fig. 22–28 *C*), the eyeball is too short and the retina too close to the lens, causing the light rays to strike the retina before they have converged again, resulting in a blurred image. Concave lenses correct for the near-sighted condition by bringing the light rays to a focus at a point farther back, and convex lenses correct for the far-sighted condition by causing the light rays to converge farther forward.

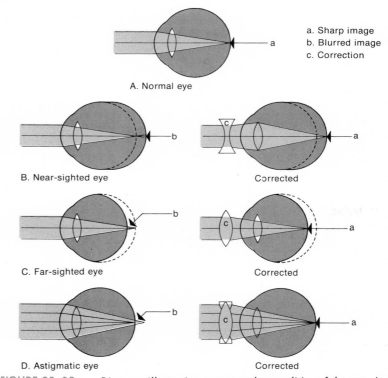

a. Sharp image
b. Blurred image
c. Correction

A. Normal eye

B. Near-sighted eye   Corrected

C. Far-sighted eye   Corrected

D. Astigmatic eye   Corrected

FIGURE 22–28   *Diagram illustrating common abnormalities of the eye. A, Normal eye, in which parallel light rays coming from a point in space are focused as a point on the retina. B, Near-sighted eye, in which the eyeball is elongated so that parallel light rays are brought to a focus in front of the retina (on dotted line, which represents the position of the retina in the normal eye) and so form a blurred image on the retina. This situation is corrected by placing a concave lens in front of the eye. This diverges the light rays, making it possible for the eye to focus these rays on the retina. C, Far-sighted eye, in which the eyeball if shortened and light rays are focused behind the retina. A convex lens coverges the light rays so that the eye focuses them on the retina. D, Astigmatic eye, in which light rays passing through one part of the eye are focused on the retina, while light rays passing through another area of the lens are not focused on the retina, owing to unequal curvature of the lens or cornea. A cylindrical lens will correct this by bending light rays going through only certain parts of the eye.*

Astigmatism is a condition in which the cornea is curved unequally in different planes, so that the light rays in one plane are focused at a different point from those in another plane (Fig. 22–28 D). To correct for astigmatism, lenses must be ground unequally to compensate for the unequal curvature of the cornea.

In old age the lens may lose its transparency, become opaque and interfere with the transmission of light to the retina, causing blindness. The only cure for this is surgical removal of the lens. This restores sight, but removes the ability to focus, so that a lensless person must wear special spectacles as a substitute for the lens.

SECTION 22–17

BINOCULAR VISION AND DEPTH PERCEPTION

The position of the eyes in the head of man and certain other higher vertebrates permits both of them to be focused on the same object. This **binocular vision** is an important factor in judging distance and depth. To focus on a near object, the eyes must converge (become slightly cross-eyed). In the eye muscles causing this convergence are proprioceptors, stimulated by this contraction to send impulses to the brain; hence, part of our judgment of distance and depth depends upon impulses which result when the sensory fibers in those muscles are stimulated. In addition, the eyes, being a little over 5 centimeters apart, see things from slightly different angles and thus get slightly different views of a close object. Depth perception is also made possible by the differential size of near and far objects on the retina, by perspective, by overlap and shadow, by distance over the horizon and by aerial perspective (i.e., dimness increasing with distance).

SECTION 22–18

THE COMPOUND EYE

The compound eye of arthropods is made up of hundreds to thousands of **monopolar neurons** that are gathered together into groups of seven or eight, each group being provided with its own cornea and lens (Fig. 22–29). The **cornea** is a transparent thickened area of the body cuticle, and the **lens** is formed by special epithelial cells. Each unit consisting of cornea, lens and neurons is called a **facet** or **ommatidium.** It has a sheath of pigmented, enveloping cells. In nocturnal and crepuscular insects and many crustacea, this pigment is capable of migrating proximally and distally (Fig. 22–29). When the pigment is in the proximal position, each ommatidium is shielded from its neighbor and only light entering directly along its axis can stimulate the receptors. When the pigment is in the distal position, light striking at an angle may pass through several ommatidia and stimulate many retinal units. Thus in dim light sensitivity of the eye is increased, and in bright light the eye is protected from excessive stimulation. Pigment migration is under neural control in insects and under hormonal control in crustacea. In some species it follows a circadian rhythm.

Form perception by compound eyes is poor. Even though the lens system of each ommatidium is adequate enough to focus a small inverted image on the retinal cells, these images are physiologically unimportant.

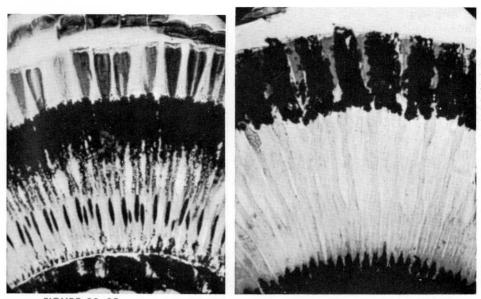

FIGURE 22-29   *Histological sections of two compound eyes of the codling moth showing the individual ommatidia, each with a cornea and a lens. In picture A the eye is light-adapted and the shielding pigment surrounds the retinal cells. In picture B the eye is dark-adapted and the pigment has moved up around the lenses leaving the retinal cells free to be stimulated by light entering at an angle. (From Collins, D. L.: Iris—Pigment Migration and Its Relation to Behavior in the Codling Moth. J. Exp. Zool. 69:164–198, 1934; in Roeder, K. D. (ed.) Insect Physiology. New York, John Wiley & Sons, Inc., 1953.)*

Each ommatidium in gathering light from a narrow sector of the visual field is in fact sampling a mean intensity from that sector and projecting it as a point of light on the retinal field. All of these points of light taken together form a **mosaic** picture. To appreciate the nature of this mosaic picture one need only look at a newspaper photograph through a magnifying glass. It is a mosaic of many dots of different intensities. The clearness and definition of the picture will depend upon how many spots there are per unit area—the more spots the better the picture. So it is with the compound eye.

Although the compound eye forms only coarse images, it compensates for this by being able to follow **flicker** to high frequencies. Flies are able to detect flickers up to about 265 per second as compared to man's value of 45 to 53 per second. Because flickering lights fuse above these values, man sees motion pictures as smooth movement and the ordinary 60-cycle light in the room as steady. To an insect, both motion picture and light flicker horribly. Because the insect has such a high critical flicker fusion rate it detects motion very well, and motion is of more significance to animals with compound eyes than is form.

Compound eyes are superior to man's eyes in two other respects. They are sensitive to different wavelengths of light from the red into the ultraviolet and they are able to analyze the **plane of polarization** of light. Accordingly an insect can see well in the ultraviolet, and its world of color is much different than ours. Since different flowers reflect ultraviolet to different degrees, two flowers that appear identically white to us may appear strikingly different to insects. How the world actually appears to an insect insofar as ultraviolet light is concerned can be appreciated by viewing the landscape through a television camera with an ultraviolet-transmitting lens (Fig. 22–30). Similarly a sky that appears equally blue to us

FIGURE 22–30   A, Ultraviolet video-viewing of marsh marigolds, Caltha palustris, in the field. Video-camera and video-tape recorder (Sony VCK 2400) are battery-operated and portable. B—E, Marsh marigolds, as seen by man, are uniformly yellow (B, D); in the ultraviolet (C, E) they are shown to have darkly absorbent centers. F-G, Sympatric, yellow-petaled Compositae from Florida (clockwise from upper left: Helenium tenuifolium, Rudbeckia sp., Heterotheca subaxillaris, Coreopsis leavenworthii, Bidens mitis) as seen in visible light (F) and in the ultraviolet (G). H—I, Yellow crab spider Misumenoides formo-sipes on yellow flower Viguiera dentata in visible (H) and ultraviolet light (I). Panels D to I are actual video-images as they appear on a monitor screen. Direct ultraviolet photography yields a sharper image (C), but the technique is much more laborious. (From Eisner, T. et al.: Science 166:1172, November 1969. Copyright 1969 by the American Association for the Advancement of Science.)

in all quadrants reveals quite different patterns to an insect, because the plane of polarization of the light is not the same in all parts of the sky, and the insect's eye can detect the difference. Honeybees and some other arthropods employ this ability as a navigational aid.

SUGGESTIONS FOR FURTHER READING

Adrian, E. D.: *The Physical Background of Perception.* New York, Oxford University Press, 1947.
   A classic account of the relation between stimulus, electrical events and sensation.

Dennis, W.: *Readings in the History of Psychology.* New York, Appleton-Century-Crofts, 1948.
   A number of original papers describing the perception of sensory stimuli.

Bartley, S. H.: *Vision.* New York, D. Van Nostrand Company Inc., 1941.
Wald, George: *Eye and Camera.* Sci. Amer. *183*:32, August 1950.
Rushton, W. A. H.: *Visual Pigments in Man.* Sci. Amer. *207*:102, November 1962.
   Presentations of the structure and functioning of the eye.

Stevens, S. S., and Davis, H.: *Hearing: Its Psychology and Physiology.* New York, John Wiley & Sons, Inc., 1938.
Bekesy, G. von: *The Ear.* Sci. Amer. *197*:66, August 1957.
   Discussions in detail about the sense of hearing.

Lowenstein, W. R.: *Biological Transducers.* Sci. Amer. *203*:98, August 1960.
Miller, W. H., Ratliff, F., and Hartline, H. K.: *How Cells Receive Stimuli.* Sci. Amer. *205*:222, September 1961.
   The question of how stimuli such as light and pressure can be converted into nerve impulses is discussed in these two texts.

Dethier, V. G.: *The Physiology of Insect Senses.* New York, Barnes and Noble, 1963.
   A comprehensive account of the senses of insects.

EFFECTORS

CHAPTER 23

To survive, each organism has evolved systems whereby it makes appropriate, meaningful, adaptive responses to specific changes in the environment. This requires systems of receptors (sense organs) to detect the changes, integrative systems of nerves and endocrines to coordinate the information received and trigger the response, and a system of *effectors*, organs which carry out the responses. The effectors of man and other mammals are primarily muscles and glands, but other animals may change color, produce light or electricity or release stinging cells in response to certain stimuli.

Each organism receives information about its external and internal environment via its receptors—eyes, ears, nose, nerve endings in the skin, muscles and joints and so forth—and coordinates and integrates these sensory inputs by its nervous and endocrine systems. One or another of its effectors is stimulated to respond in some way appropriate to the kind of information received. The presence of food in the stomach and duodenum leads to the secretion of gastric and pancreatic enzymes. A mammal sees an enemy or predator and the central nervous system brings about an outpouring of epinephrine by the adrenal medulla. An infant suckles at the breast and the hypothalamus is stimulated to release oxytocin; this stimulates the myoepithelial cells around the alveoli of the mammary glands to contract, and milk is forced out of the alveoli into the ducts leading to the nipples. A cat falls from a height and certain muscles contract so that it turns and lands on its feet. Each effector provides an adaptive response to the stimulus.

The major effector systems in most animals are the skeletal, smooth and cardiac muscles. Other effectors include the glands—digestive glands, sweat glands, accessory sex glands, mammary glands and so on—pigment cells, electric organs and luminescent structures.

SECTION 23–1

EFFECTORS IN PLANTS

The tropistic responses of plants, the differential growth of cells in response to light or gravity (p. 678), occur very slowly and do not involve special effector cells. Some plants exhibit much more rapid movements which involve specialized effector cells. Perhaps the most striking of these is the response to the "sensitive plant," *Mimosa pudica*, to touch (Fig. 23–1). The leaves of *Mimosa* are normally held in a horizontal plane, but if one is lightly touched all the leaflets fold within two or three seconds. Touching one leaf causes not only the stimulated leaf but also the neighboring leaves to fold and droop. After a few minutes the leaves return to their original position. The folding of the leaves results from a sudden decrease in the turgor pressure of specialized cells in the **pulvinus** at the base of the petiole (Fig. 23–2). The excitation is transmitted along the sieve tubes of the phloem in the leaves and stems at a rate of about 5 cm per second. The excitation is accompanied by electrical phenomena and increased permeability of the excited cells. The molecular mechanism involved is unclear but the excitation may turn off the sodium pump, decreasing the influx of sodium and secondarily decreasing the influx of water.

The **Venus flytrap** is another plant with a rapid response; its leaves close quickly and trap any insects that have landed on them. The closing of the leaves is accomplished by changes in the turgor pressure of special effector cells located along the hinge of the leaf.

The flowers of many species of plants open or close regularly at certain times of day or in response to environmental stimuli, again owing to changes in the turgor pressure of special cells in the petals.

The streaming of the cytoplasm in certain plant cells, in protozoa and in slime molds, and the movements of cilia and flagella represent types of movements that may occur as responses to specific stimuli and which do not involve the contraction of muscles. As we learn more about the chemical basis of these phenomena it is becoming clear that they have many features in common with muscular contraction.

FIGURE 23–1   Photograph of the sensitive plant Mimosa pudica. Left, the plant before being disturbed. Right, the plant five seconds after being touched. Note how the leaves have folded and drooped. (Courtesy of the General Biological Supply House, Chicago.)

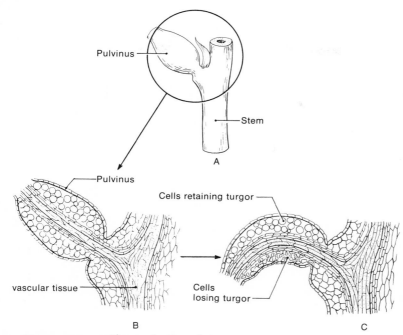

FIGURE 23–2   *The mechanism of the response of Mimosa to touch. A, The base of the petiole showing the pulvinus. B, Section through the pulvinus showing the conditions of the cells when the leaf is extended horizontally. C, Section through the pulvinus showing the cells losing turgor to produce folding of leaves.*

SECTION 23–2

GLANDULAR SECRETION

An effector system of importance in many animals is the secretion of specific substances by their glands. Some glandular cells secrete their products continuously, others secrete intermittently in response to specific stimuli. The secretion of epinephrine by the adrenal medulla and the secretion of enzymes by the pancreas have been studied in some detail.

In the cells of the adrenal medulla, small vesicles that are part of the Golgi complex near the nuclear membrane become filled with dense material which has been shown to be epinephrine. The vesicles, surrounded by membranes, migrate to the surface while increasing in size and density (Fig. 23–3). A large number of such vesicles filled with epinephrine may accumulate in the cytoplasm of the cell. The cells are innervated by splanchnic nerves whose tips are filled with synaptic vesicles which contain **acetylcholine.** Stimulation of the nerve releases acetylcholine from the synaptic vesicles and this triggers the expulsion of epinephrine from the vesicles in the cells of the adrenal medulla. The droplets become attached to the plasma membrane of the cell, increase in size and evacuate their contents by a process resembling pinocytosis in reverse.

The protein enzymes secreted by the pancreatic acinar cells are produced on the ribosomes of the cell's endoplasmic reticulum as inactive precursors, e.g., **trypsinogen**. The endoplasmic reticulum is concentrated in the base of the cell. The apex of the cell, bordering on the lumen of the gland, is full of **zymogen granules** (Fig. 23–4). These have been isolated and found to contain the enzymes secreted by the pancreas; they are the secretory products of the cell. The proteins synthesized on the ribosomes

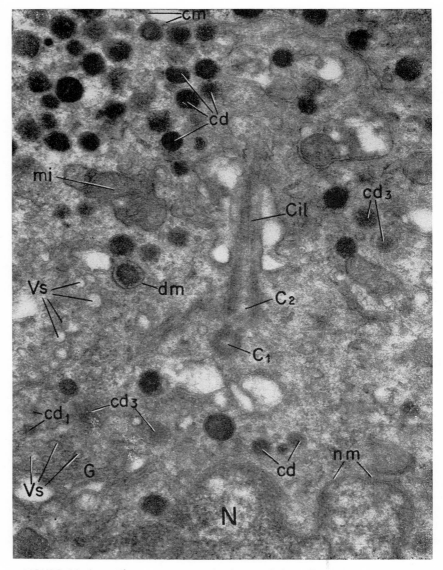

FIGURE 23–3   Electron micrograph of one of the cells in the adrenal medulla in which epinephrine is produced and secreted. The nucleus (N) and the folded nuclear membrane (nm) can be seen at the bottom. The cytoplasm, above, contains a Golgi complex (G), two centrioles (C_1 and C_2), a cilium (cil) arising from one of the centrioles, and mitochondria (mi). The epinephrine collects in droplets (cd) which enlarge (cd_1, cd_2, cd_3) and are accumulated in the peripheral part of the cytoplasm near the cell membrane (cm), top. ×51,500. (From DeRobertis, E., and Sabatini, D. D.: Submicroscopic analysis of the secretory process in the adrenal medulla. Fed. Proc. 19:71, 1960.)

pass into the cavity of the endoplasmic reticulum, forming intracisternal granules. These pass to the Golgi apparatus and are packaged as zymogen granules which move to the apex of the cell. These granules discharge their contents into the lumen of the gland by a fusion of the membrane surrounding the granule with the plasma membrane. How this discharge may be triggered by secretin remains unknown.

SECTION 23–3

SKELETAL MUSCLES

The principal effectors of all the multicellular animals, which provide for movements in response to stimuli, are muscles, composed of specialized contractile cells. A typical skeletal muscle of a vertebrate is an elongated mass of tissue composed of millions of individual **muscle fibers** bound together by connective tissue fibers. The entire structure is surrounded by a tough, smooth sheet of connective tissue so that it can move freely over adjacent muscles and other structures with a minimum of friction. The two ends of the muscle in vertebrates are typically attached to two different bones and the contraction of the muscle draws one bone toward the other. The joint between the two acts as the fulcrum of the lever system. A few muscles pass from a bone to the skin or, as in the muscles of facial expression, from one part of the skin to another. The end of the muscle that remains relatively fixed when the muscle contracts is called the **origin**; the end that moves is called the **insertion**; and the thick part between the two is called the **belly** (Fig. 23–5). The origin of the **biceps** is on the shoulder and its insertion is on the bone in the forearm called the radius; when the biceps contracts the shoulder remains fixed and the elbow is bent.

Muscles typically contract in groups rather than singly. You cannot, for example, contract the biceps alone; you can only bend the elbow, which involves the contraction of a number of other muscles in addition to the biceps. Muscles can exert a pull but not a push, hence muscles are typically arranged in **antagonistic pairs**, one of which pulls a bone in one direction and the other pulls it in the reverse. The biceps, for example, which bends or flexes the arm is termed a **flexor** and its antagonist, the **triceps**, which straightens or extends the arm is termed an **extensor.** Similar pairs of opposing flexors and extensors are found at the wrist, knee, ankle and other joints. When a flexor contracts, the opposing extensor must relax to permit the bone to move; this requires proper coordination of the nerve impulses going to the two sets of muscles. Other antagonistic pairs of muscles are **adductors** and **abductors** that move parts of the body toward or away from the central axis of the body; **levators** and **depressors** that raise and lower parts of the body; **pronators** that rotate body parts downward and backward and **supinators** that rotate them upward and forward; and **sphincters** and **dilators** that decrease and enlarge the size of an opening.

The vertebrate skeleton is an **endoskeleton**, a bony or cartilaginous framework lying within the body and surrounded by muscles. The arthropod skeleton is an **exoskeleton,** a chitinous framework on the outside of the body surrounding the muscles. The protective advantages of an exoskeleton are counterbalanced by the problems it raises in the growth of the organism. The basic differences in the mechanical arrangement of vertebrate and arthropod joints are illustrated in Figure 23–6. The muscles of the vertebrate surround the bones; one end of each is attached to one

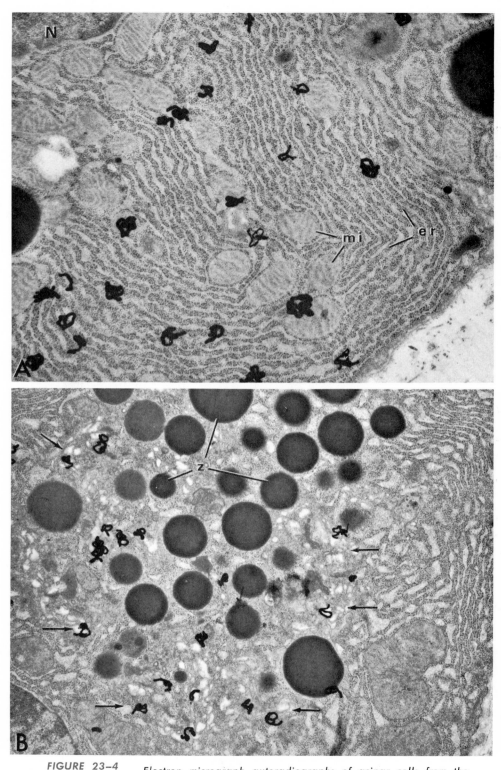

FIGURE 23–4   Electron micrograph autoradiographs of acinar cells from the pancreas of a guinea pig illustrating the sequence of events by which enzymes are secreted. A, Three minutes after labeling with ³H-leucine. The autoradiographic grains are located almost exclusively on the granular endoplasmic reticulum (er). B, Seven minutes after labeling with ³H-leucine, the autoradiographic grains are in the region of the Golgi complex (arrows); Z, zymogen granules.

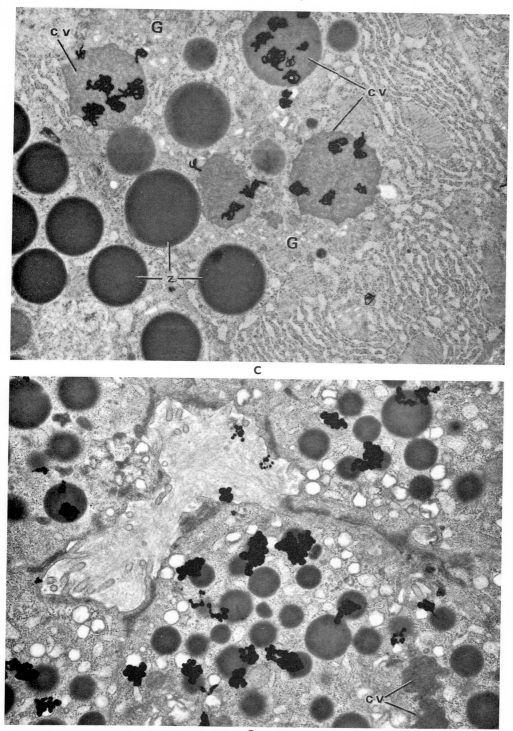

FIGURE 23–4 (Continued)   C, Thirty-seven minutes after labeling with ³H-leucine, the autoradiographic grains are concentrated in the condensing vacuoles (cv) of the Golgi complex (G). The zymogen granules are unlabeled. D, One hundred seventeen minutes after labeling with ³H-leucine, the autoradiographic grains are concentrated primarily over the zymogen granules, while the condensing vacuoles are unlabeled. ×17,000. (Courtesy of J. D. Jamieson and G. E. Palade; From DeRobertis, E. D., Nowinski, N. W., and Saez, F. A.: Cell Biology. 5th ed. Philadelphia, W. B. Saunders Company, 1970.)

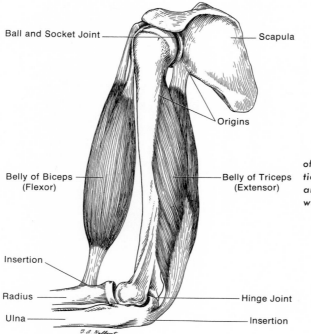

Ball and Socket Joint

Scapula

Origins

Belly of Biceps
(Flexor)

Belly of Triceps
(Extensor)

Insertion

Radius

Ulna

Hinge Joint

Insertion

FIGURE 23–5   The muscles and bones
of the upper arm illustrating the origin, insertion and belly of a muscle and the antagonistic
arrangement of the biceps and triceps by
which the forearm may be bent or extended.

bone and the other end to the second bone. The contraction of the muscle
moves one bone with respect to the other. The muscles of the arthropod
lie within the skeleton and are attached to its inner surface. The arthropod
exoskeleton has certain regions, *joints*, in which the exoskeleton is thin and
flexible so that it can bend. A muscle may stretch across the joint so that
its contraction will move one part on the next. Or, a muscle may be located
entirely within one section of the body or appendage and be attached at
one end to a tough *apodeme*—a long, thin, firm part of the exoskeleton
extending into that section from the adjoining one.

The hydra and other coelenterates, the planarian and other flatworms,
and the earthworm and other annelids move by the same basic principle
of antagonistic muscles even though they have no hard exo- or endoskeleton to anchor the ends of the muscles. Instead, the fluid contents of
the body cavity, which are noncompressible, serve as a *hydrostatic skeleton*.
Such animals typically have a set of circular muscles which decrease the
diameter and increase the length of the animal, and a set of antagonistic
longitudinal muscles which decrease the length and increase the diameter
of the animal (Fig. 23–7). Some of the marine worms have additional

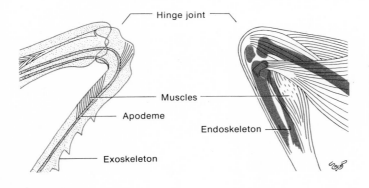

Hinge joint

Muscles

Apodeme

Endoskeleton

Exoskeleton

FIGURE 23–6   A comparison
of the vertebrate endoskeleton and
joint with the arthropod exoskeleton and joint. Note the difference in
the arrangement of the muscles and
the skeletal elements at the two types
of joints.

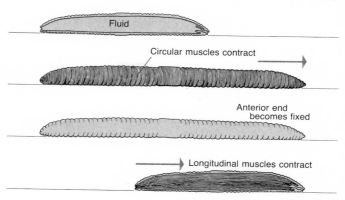

FIGURE 23–7 A diagram illustrating the hydrostatic skeleton of an earthworm. The fluid within the animal's body enables the alternate contraction of the circular and longitudinal muscles to move the animal along.

diagonally arranged muscles that permit more complex movements of the body and of the paddle-like **parapodia** that extend laterally from the body wall. Many marine worms live in tubes and the movements of the parapodia are important not only in locomotion but in moving currents of water laden with oxygen and nutrients through these tubes.

When a muscle is not contracting to effect movement it is not completely relaxed. As long as an individual is conscious, all his muscles are contracted slightly, a phenomenon termed **tonus**. Posture is mainta.ned by the partial contraction of the muscles of the back and neck and of the flexors and extensors of the legs. When a person stands, both the flexors and extensors of the thigh must contract simultaneously so that the body neither sways forward nor backward on the legs. The simultaneous contraction of the flexors and extensors of the shank locks the knee in place and holds the leg rigid to support the body. When movement is added to posture, as in walking, a complex coordination of the contraction and relaxation of the leg muscles is required. It is not surprising that learning to walk is such a long and tedious process.

Some of the larger muscles of the human body are remarkably strong. The **gastrocnemius** muscle in the calf of the leg has its origin at the knee and its insertion, by the tendon of Achilles, is on the heel bone. Because the distance from the toes to the ankle joint is at least six times that from the ankle joint to the heel, the gastrocnemius is working against an adverse lever ratio of 6:1. Thus when a man weighing 70 kg stands on one leg and rises on his toes, the one gastrocnemius muscle must exert a force of 420 kg. When one ballet dancer holds another in his arms and rises on the toes of one leg, his gastrocnemius is exerting a force of nearly 1000 kg.

Normal nervous coordination prevents muscles from contracting maximally, but in certain diseases in which nervous control is impaired, a muscle may contract forcefully enough to rip tendons and break bones.

SECTION 23–4

THE PHYSIOLOGY OF MUSCLE ACTIVITY

The functional unit of vertebrate muscles, called the **motor unit**, consists of a single motor neuron and the group of muscle cells innervated by its axon, all of which will contract when an impulse travels down the motor neuron to the **motor end plate**. In man it is estimated that there are some 250,000,000 muscle cells, but only some 420,000 motor neurons in spinal nerves. Obviously, some motor neurons must innervate more than one muscle fiber. The degree of fine control of a muscle is inversely proportional to the number of muscle fibers in the motor unit. The muscles of the eyeball, for example, have as few as three to six fibers per motor unit, whereas the leg muscles have as many as 650 fibers per unit.

If a single motor unit is isolated and stimulated with brief electric shocks of increasing intensity beginning with stimuli too weak to cause contraction, there will be no response until a certain intensity is reached, then the response is maximal. This phenomenon is termed the "all-or-none" effect. An entire muscle, in contrast, composed of many individual motor units can respond in a graded fashion depending on the number of motor units contracting at any given time. Although an entire muscle cannot contract maximally, a single motor unit can contract *only* maximally or not at all. The strength of contraction of an entire muscle composed of thousands of motor units depends upon the number of its constituent motor units that are contracting and upon whether the motor units are contracting simultaneously or alternately.

Muscles retain their ability to contract after they have been removed from the body. The muscle usually used for experimental purposes is the **gastrocnemius** muscle of the frog, and if care is taken to keep it moist it will contract for hours. To make a record of these contractions, the muscle is mounted with its origin attached to a fixed hook and its insertion connected by means of another hook to a lever with a pointed stylus at its tip (Fig. 23–8). This stylus is in contact with a cylinder, covered with recording paper and revolved by a motor. Each contraction of the muscle raises the stylus and its vigor and duration are recorded. Additional styluses can be used to record an appropriate time scale and to mark when the muscle is stimulated.

THE SINGLE TWITCH.   When a muscle is given a single stimulus, a single electric shock, it responds with a single, quick twitch which lasts about 0.1 second in a frog's muscle and about 0.05 second in a human muscle. A record of a single twitch (Fig. 23–9) reveals that it consists of three separate phases: (1) the **latent period**, lasting less than 0.005 second, an interval between the application of the stimulus and the beginning of the visible shortening of the muscle; (2) the **contraction period**, about 0.04 second in duration, during which the muscle shortens and does work; and (3) the **relaxation period**, the longest of the three, lasting 0.05 second, during which the muscle returns to its original length.

Skeletal muscle fibers, like nerve fibers, have a **refractory period**, a very short period of time immediately after one stimulus during which they will not respond to a second stimulus. The refractory period in skeletal muscle is so short (about 0.002 second) that muscle can respond to a second stimulus while still contracting in response to the first. The superposition of the second contraction on the first results in a greater than normal shortening of the muscle fiber, an effect known as **summation**.

The first event after the stimulation of a muscle is the initiation and propagation of an **action potential** (Fig. 23–9), followed by the changes in the structure of the contractile proteins **actin** and **myosin** which are evident as a change in the birefringence of the muscle. After a twitch the muscle consumes oxygen and gives off carbon dioxide and heat at a rate greater than during rest, marking a **recovery period** in which the muscle is restored to its original state. This recovery period lasts for several seconds, and if a muscle is stimulated repeatedly so that successive contractions occur before the muscle has recovered from the previous ones, the muscle becomes fatigued and the twitches grow feebler and finally stop. If the fatigued muscle is allowed to rest for a time, it regains its ability to contract.

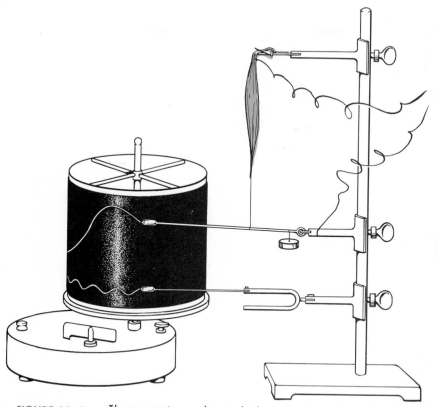

FIGURE 23–8   *The apparatus used to study the contraction of an isolated muscle. The stylus attached to the insertion of the muscle writes a record of the contraction on the rotating cylinder, the kymograph. The contraction is timed by the vibrating tuning fork whose stylus draws a wavy line on the kymograph record.*

TETANUS.   Muscles do not normally contract in single twitches, but in sustained contractions evoked by volleys of nerve impulses reaching them in rapid succession. Such a sustained contraction is called **tetanus**,* and during a tetanic contraction the stimuli occur so rapidly (several hundred per second) that relaxation cannot occur between the contractions of successive twitches. In most tetanic contractions the individual muscle fibers are stimulated in rotation rather than simultaneously, so that although individual muscle fibers contract and relax, the muscle as a whole remains partly contracted. From personal experience you know that any muscle of your body can contract to different degrees. This gradation of contraction is controlled through the nervous system: in a weak contraction only a small percentage of the muscle fibers is stimulated at one time; for a stronger contraction a larger percentage of muscle fibers contracts simultaneously.

TONUS.   The term tonus or "tone" refers to the state of sustained partial contraction present in all normal skeletal muscles as long as the nerves to the muscle are intact. Cardiac and smooth muscles exhibit tonus

*This term should not be confused with tetany, the muscular spasms occurring in deficiencies of the parathyroid hormone, or with the disease tetanus ("lockjaw"), characterized by abnormal muscular contractions and caused by the tetanus bacillus.

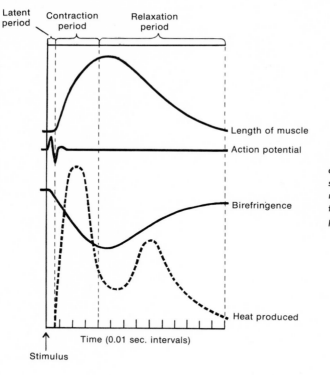

FIGURE 23–9   *A diagram of the changes that occur in a muscle during a single muscle twitch. Note the temporal relationship between the action potential, the shortening of the muscle and the production of heat.*

even after their nerves are cut. Each muscle is normally stimulated by a continuous series of nerve impulses which cause a constant, slight contraction or tonus. Severing the nerve to a skeletal muscle eliminates tonus immediately. Tonus is a mild state of tetanus, present at all times and involving only a small fraction of the fibers of a muscle at any moment. It is believed that the individual fibers contract in turn, working in relays, so that each fiber has a chance to recover completely, while other fibers are contracting, before it is called upon to contract again. A muscle under slight tension can react more rapidly and contract more strongly than one that is completely relaxed.

SECTION 23–5

THE BIOCHEMISTRY OF MUSCULAR CONTRACTION

A steam engine can convert only about 10 per cent of the heat energy of its fuel into useful work; the rest is wasted as heat. But muscles are able to use between 20 and 40 per cent of the chemical energy of glucose in the mechanical work of contraction. The remainder is converted into heat, but is not wholly wasted, since it is used to maintain the body temperature. If one refrains from contracting the muscles, the heat produced elsewhere in the body is insufficient to keep it warm in a cold place. In these circumstances the muscles contract involuntarily (one "shivers"), and heat is thereby produced to restore and maintain normal body temperature.

Physiologists and biochemists have been attempting for many years to solve the problem of how a muscle can exert a pull, but the actual chemical and physical events that occur in muscle contraction are still a matter of conjecture rather than established fact. Muscle is about 80 per cent water, the rest being mostly protein, with small amounts of fat and glycogen, and two phosphorus-containing substances, **phosphocreatine** and **adenosine triphosphate**. The actual contractile part of a muscle fiber is a protein chain which apparently shortens by the sliding together of its parts. Two proteins, **myosin** and **actin**, have been extracted from muscle, neither of which is capable of contracting alone. When they are combined to form a thread of actomyosin and calcium and potassium and adenosine triphosphate are added, the thread undergoes contraction.

As a first step toward analyzing the biochemical mechanism and the sequence of events involved in muscle contraction we can measure what substances are used up in the process. Glycogen, oxygen, phosphocreatine and adenosine triphosphate decrease in amount during contraction, and carbon dioxide, lactic acid, creatine, adenosine diphosphate and inorganic phosphate increase. The fact that oxygen is used up and carbon dioxide is formed suggests that muscular contraction is an oxidative process. But this oxidation is not essential, for a muscle can twitch a good many times even when completely deprived of oxygen—e.g., when it is removed from the body and placed in an atmosphere of nitrogen. Such a muscle becomes fatigued, however, more rapidly than one contracting in an atmosphere of oxygen. Furthermore, although we breathe faster during exertion, the accelerated breathing continues for some time *after* the physical work has ceased. This suggests that oxidative processes are involved, not in muscular contraction, but in the *recovery* from contraction.

The disappearance of glycogen and the formation of lactic acid are related, for in the absence of oxygen, the amount of lactic acid formed is just equivalent to the glycogen that disappears. Since the breakdown of glycogen to lactic acid requires no oxygen, and since it liberates energy rapidly, it was once thought that this reaction was directly responsible for muscle contraction. When oxygen is present, the muscle oxidizes about one fifth of the lactic acid to carbon dioxide and water, and the energy released by this oxidation is used to reconvert the other four fifths of the lactic acid to glycogen. This explains why lactic acid does not accumulate as long as muscle has sufficient oxygen, and why a muscle becomes fatigued more rapidly (uses up its glycogen and accumulates lactic acid) when it contracts in the absence of oxygen.

About 1930 it was found that a muscle poisoned with iodoacetate, which inhibits the chemical reactions by which glycogen breaks down to lactic acid, can still contract, although it is capable of twitching only 60 to 70 times instead of the 200 or more times achieved by a muscle deprived of oxygen. But the fact that it can twitch at all when the breakdown of glycogen is prevented shows that this is not the primary source of energy for contraction.

The other chemical change that can be detected during contraction is a splitting off of inorganic phosphate from phosphocreatine and adenosine triphosphate, accompanied by the release of energy. It is now believed that this is the immediate source of energy for contraction. The reactions whereby glucose and other substances are metabolized to yield energy-rich phosphate compounds such as adenosine triphosphate were described in Chapter 5. Phosphocreatine serves as a reservoir of ~P in the muscle; its energy-rich phosphate must be transferred to ADP to form ATP before it can be utilized in contraction. After a muscle has contracted, the breakdown of glycogen to lactic acid and the oxidation of lactic acid in the Krebs citric acid cycle provide energy for the resynthesis of adenosine triphosphate and phosphocreatine.

In summary, muscle contraction involves the following chemical reactions:

(1) Adenosine triphosphate $\rightarrow$ Inorganic phosphate + Adenosine diphosphate + Energy (used in actual contraction)

(2) Phosphocreatine + ADP $\rightleftarrows$ Creatine + ATP

(3) Glycogen $\rightleftarrows$ Intermediates $\rightleftarrows$ Lactic Acid + Energy (used in resynthesis of organic phosphates)

(4) Part of lactic acid + $O_2 \rightarrow CO_2 + H_2O$ + Energy (used in resynthesis of rest of lactic acid to glycogen and in resynthesis of ATP and phosphocreatine)

Myosin is an enzyme as well as a contractile protein and can catalyze the cleavage of ATP to ADP and inorganic phosphate. The enzyme **creatine kinase** catalyzes the transfer of ~P from ATP to creatine to form ADP and phosphocreatine (reaction 2).

It is estimated that the energy from organic phosphates alone could sustain maximal muscular contraction for only a few seconds. A man might run a 50-meter dash with the ATP and phosphocreatine in his muscles. By calling upon all the sources of energy available in the absence of oxygen, a man might be able to continue maximal contractions for 30 to 60 seconds.

THE OXYGEN DEBT.   The fact that the contraction and part of the recovery from contraction occur without oxygen is extremely important. Our muscles are often called upon to do great spurts of work, and although both the rate of breathing and the heart rate increase during exertion, oxygen cannot be supplied in sufficient quantities to permit these exertions. During violent exercise, such as running the 100-meter dash, glycogen breaks down to lactic acid faster than the lactic acid can be oxidized so that the latter accumulates. In such circumstances the muscle is said to have incurred an **oxygen debt**, which is afterward repaid by rapidly breathing enough extra oxygen to oxidize part of the lactic acid, which furnishes energy for resynthesizing the rest to glycogen. In other words, during short spurts of extreme muscular activity, muscles use energy from sources that do not require the utilization of oxygen. After the activity has ceased, the muscles and other tissues pay off the "oxygen debt" by utilizing an extra amount of oxygen to restore the energy-rich phosphate compounds and glycogen to their original condition. During a long race a runner may reach an equilibrium in which he gets a "second wind," and, because of the increase in his breathing and heart rate, he takes in enough oxygen to oxidize the lactic acid formed at that moment so that his oxygen debt is not increased.

FATIGUE.   A muscle that has contracted many times, exhausted its stores of organic phosphates and glycogen and accumulated lactic acid, is unable to contract any more and is said to be **fatigued**. Fatigue is primarily induced by this accumulation of lactic acid, although animals feel fatigue before the muscle reaches the exhausted condition.

The exact spot most susceptible to fatigue can be demonstrated ex-

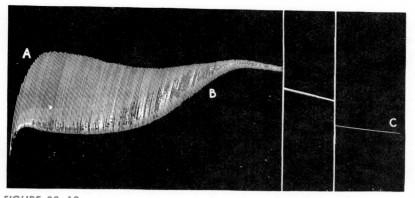

FIGURE 23-10   *Kymograph records of a fatigue curve showing (A) the staircase effect, (B) contracture and (C) complete fatigue in a muscle. (From Tuttle, W. W., and Schottelius, B. A.: Textbook of Physiology. 15th ed. St. Louis, C. V. Mosby Company, 1965.)*

perimentally if a muscle and its attached nerve are dissected out and the nerve stimulated repeatedly by electric shocks until the muscle no longer contracts (Fig. 23–10). If the muscle is then stimulated directly by placing the electrodes on the muscle tissue it will respond vigorously. With the proper apparatus for detecting the passage of nerve impulses it can be shown that the nerve leading to the muscle is not fatigued; it is still capable of conduction. The point of fatigue, then, is the *junction* between the nerve and the muscle, where nerve impulses instigate muscle contraction.

SECTION 23–6

THE BIOPHYSICS OF MUSCLE CONTRACTION

Electron micrographs show that muscle myofibrils are made of longitudinal filaments, the **myofilaments**. These are of two kinds: thick **primary filaments** (100 Å thick and 1.5 μ long) and thin **secondary filaments** (50 Å thick and 2 μ long) (Fig. 23–11). By selectively extracting the proteins and by histochemical and immunochemical staining it has been possible to show that the primary filaments consist of **myosin** and the secondary filaments of **actin**. The primary and secondary filaments are arranged in such a fashion that, seen in cross section, each primary filament is surrounded by six secondary filaments, which in turn are shared with six surrounding primary filaments.

The alternating light and dark bands seen in the light microscope (Fig. 3–28) consist of dense A bands and light I bands (Fig. 23–12). Each unit consists of an A band bounded on each side by an I band and separated from the adjacent unit by a Z line, a thin, dense, rigid band of protein extending through the center of the I band. The central portion of the A band is somewhat less dense and is called the H zone. Electron micrographs reveal that the thick primary filaments are found only in the A band and that the I band contains only thin secondary filaments. These are rigid rods projecting from both sides of the Z band. The secondary filaments, however, are not limited to the I band but extend for some distance into the A band, interdigitating with the primary filaments. Thus at either end of the A band are both primary and secondary filaments interdigitating, but the central part of the A band (the H zone) contains only primary filaments. The secondary filaments appear to be smooth but the primary filaments appear to have minute spines every 60 to 70 Å along their length which project toward the adjacent secondary filament. These

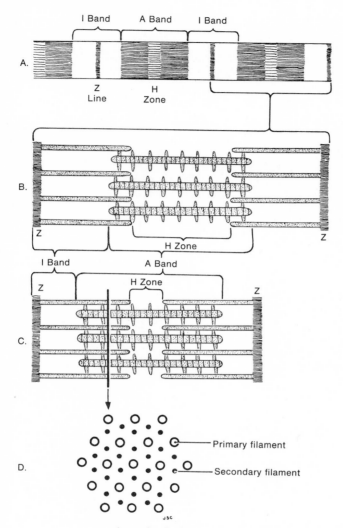

FIGURE 23–11   Diagrams illustrating the sliding filament hypothesis of the mechanism of muscle contraction. A, Diagram of part of a single myofibril showing the pattern of light (I) and dark (A) bands. B, Longitudinal view of the arrangement of thick and thin filaments within a myofibril in the relaxed state. C, Longitudinal view of the arrangement of thick and thin filaments in a contracted myofibril showing that the I band decreases in thickness. Note that the two types of filaments appear to slide past one another during contraction. D, Transverse view through C at the arrow, showing each thick primary filament surrounded by six thinner secondary filaments. (After Huxley, H. E.: The mechanism of muscular control. Science 164: 1357, 1969.)

spines look like bridges connecting the two sets of filaments. Chemical analyses show that myosin is concentrated in the A bands and actin is concentrated in the I bands. Thus the primary filaments are myosin and the secondary filaments are actin.

During contraction the length of the A band remains constant but the I band shortens and the length of the H zone within the A band decreases. Huxley and others have proposed that during contraction the filaments maintain their initial length, but the primary and secondary filaments *slide* past each other. During contraction the thin secondary filaments of actin are believed to extend farther into the A band, decreasing its central H zone and narrowing the I band as the ends of the primary myosin filaments approach the Z band.

The physicochemical mechanism by which this sliding may occur is not yet clear; perhaps the bridges are broken and then reformed further along. The energy of the ~P may be utilized in forming new bridges, cross-linkages, between the primary and secondary filaments. If we assume, as Huxley did, that the formation of one new crossbridge requires the energy of one ~P bond, we can calculate from the rate of contraction and the total number of crossbridges per fiber how much ATP would be utilized per second by the fiber. The resulting figure is of the same order of magnitude as the rate of ATP utilization determined experimentally.

FIGURE 23–12   *Electron micrograph of one muscle unit (sarcomere) extending from one dense Z line above to the next Z line below. The central A band of medium density is composed of myosin filaments 100 Å thick. The less dense I bands above and below the A band are made of thinner (60 Å) actin filaments extending from the Z band. The actin filaments extend into the A band and occupy the spaces between the myosin filaments. Actin and myosin filaments interdigitate in the region indicated by X. The region marked Y, the central portion of the A band, contains only myosin filaments. Papillary muscle of the cat heart ×78,000. (From Fawcett, D. W.: The Cell. Philadelphia, W. B. Saunders Company, 1966.)*

The contraction of a muscle is initiated by nervous stimuli that arrive at the specialized ends of motor nerves, the **motor end plates,** and cause the release of acetylcholine. The acetylcholine presumably depolarizes the membrane of the muscle fiber, triggering an action potential which changes ion concentrations (increases calcium ion concentration) and causes myosin to hydrolyze ATP. The energy released must in some way cause the thin filaments to move with respect to the thick filaments, perhaps by repeatedly forming new crossbridges one "bridge" further along. Clearly there is much yet to be learned about the details of this sequence of events.

When a muscle contracts it becomes shorter and fatter but there is no change in its total volume. This has been shown experimentally by dissecting out a muscle, placing it in a glass vessel with a narrow neck, and filling the vessel with water. The muscle is then stimulated electrically, and as it contracts and relaxes there is no change in the water level in the neck of the vessel.

SECTION 23–7

CARDIAC AND SMOOTH MUSCLE

The muscles of the heart and internal organs, though resembling skeletal muscle in a general way, have certain distinctive characteristics. They both contract much less rapidly than skeletal muscle: skeletal muscle fibers contract and relax in 0.1 second, but cardiac muscle requires from one to five seconds, and smooth muscle needs from three to 180 seconds. All the phases of contraction are prolonged.

Smooth muscle exhibits wide variations in tonus; it may remain almost relaxed or tightly contracted. It apparently can maintain the shortened condition of tonus without the expenditure of energy, perhaps owing to a reorganization of the protein chains making up the fibers.

Each beat of the heart represents a single twitch. Cardiac muscle has a long **refractory period,** the period following one stimulus when it is unable to respond to any other. Consequently it is unable to contract tetanically, since one twitch cannot follow another quickly enough to maintain a contracted state.

The basic mechanism of the contraction of cardiac and smooth muscles is probably very similar to the sliding filament mechanism that operates in skeletal muscle. Both cardiac and smooth muscles contain actin and myosin and the contraction process involves the hydrolysis of ATP and an interaction between actin and myosin initiated by calcium ions.

A unique feature of cardiac muscle is its inherent rhythmicity; it contracts at a rate of about 72 times per minute even when denervated and removed from the body. Cardiac muscle discharges its membrane potential each time after it has built up to a certain level. After each impulse has passed, the membrane becomes repolarized but then suddenly becomes permeable again, initiating the transmission of the next action potential.

SECTION 23–8

THE MUSCLES OF LOWER ANIMALS

From the flatworm to man all animals have muscles that are similar in their elongate, cylindrical or spindle shape and in their content of contractile protein filaments. Even the coelenterates, which lack true muscle fibers, have cells which can contract. Most of the invertebrates have only smooth muscle whereas arthropods have only striated muscle.

In contrast to the pattern in the vertebrates of a single neuron innervating relatively few muscle fibers to form a motor unit, a single axon in an arthropod may not only innervate *all* the fibers of one muscle but may innervate those of another muscle as well. Furthermore, most muscles receive just a few axons, perhaps only two, but each axon has a different effect upon muscular contraction. In a three-axon system, one axon produces a strong brief contraction, another a weak sustained contraction and the third inhibits the action of the other two (Fig. 23–13). By varying the frequency of stimulation among the axons, the strength and duration of muscular contraction can be varied considerably. This relatively simple pattern of connections between nerves and muscles permits remarkably fine control of activity.

The flight muscles of insects are not attached to the wings; instead, the wings are articulated with the wall of the thorax in such a way that slight changes in the shape of the thorax cause the wings to beat up and down. The flight muscles are located entirely within the thorax, which is like a box with an undersized cover (Fig. 23–14). The inner end of each wing is attached by a movable joint to the upper edge of the sides of the box. When the vertical muscles of the thorax contract, the upper part of the thorax (the lid of the box), called the **notum**, is pulled down and the wings flip upward. When muscles extending longitudinally in the thorax contract, the notum is arched upwards and the wings flip down. The stout flight muscles undergo little change in length during contraction. The two sets of opposing muscles contract alternately against each other. There are other muscles that govern the angle of the wing beat and enable the insect to turn, hover or even back up in flight.

In butterflies and moths the frequency of the wing beat is correlated with the frequency of the nerve impulses to the muscles. The impulses are rhythmic and result in a similarly rhythmic up and down movement of the wing. The rate of the wing beat ranges from 8 beats per second in large moths to 75 or more beats per second in some smaller insects.

In flies, bees and certain other insects, the wing beats are not correlated with the frequency of the nerve impulses. Impulses of less than about 100 per second cause no wing beats at all, but more frequent im-

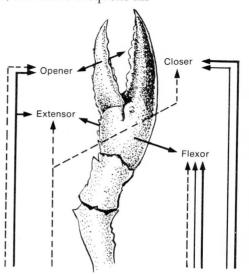

FIGURE 23–13   *The nerve supply to the last four muscles of the crayfish pincers. Each vertical line represents a single axon which supplies all the fibers of the indicated muscles. Dotted lines indicate inhibitory axons; heavy lines indicate rapid strong excitors; and thin lines indicate slow sustained excitors. The arrows on the pincers indicate the directions of movement.*

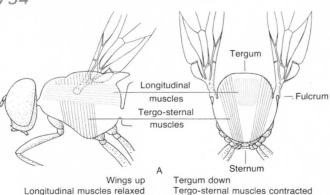

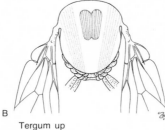

A
Wings up   Tergum down
Longitudinal muscles relaxed   Tergo-sternal muscles contracted

B
Wings down   Tergum up
Longitudinal muscles contracted   Tergo-sternal muscles relaxed

FIGURE 23–14   The flight muscles of the insect. A, The longitudinal muscles relax and the tergosternal muscles contract pulling the tergum down and the wings up. B, The longitudinal muscles contract and the tergosternal muscles relax, bringing the tergum up and the wings down.

pulses lead to rapid contraction of the flight muscles. Once the frequency of the wing beats exceeds a certain threshold, the rhythm of the muscular contraction originates within the muscle and is termed a **myogenic rhythm** in contrast to the **neurogenic rhythm** in other insects. Insects with myogenic rhythms may achieve wing beat frequencies of 300 or 400 wing beats per second. A critical feature of myogenic rhythms is the tension in the system. The two muscles must act together, each alternately stretching the other. The frequency of contraction depends on the tension in the flight muscles and the tension can be increased by the contraction of other muscles in the thorax. Thus in the evolution of the small fast-flying insects, the coordination of the flight muscles has passed from the nervous system and has been built into the muscles themselves. The ability to contract repeatedly under tension results from some adjustment in the physiology of the muscle cells, perhaps at the level of the interaction of actin and myosin molecules.

The tip of the wing of many insects moves evenly from the up or down position toward the horizontal and then "clicks" suddenly the rest of the way up or down. Certain small muscles are attached to the inner upper edge of the elastic thoracic box, just below the point where the wings articulate. The contraction of these muscles tends to pull the sides of the elastic thoracic box together. The distance between the upper edge of the box is least when the wings are up or down and greatest when the wings are horizontal. As the wings begin their beat they move *against* the force of these small muscles until they reach the halfway point, when they move *with* their force and are suddenly accelerated. This ensures full amplitude to the wing beat and a sudden stretching of the relaxing set of muscles just before they are to contract. Stretching improves the strength of contraction of both arthropod and vertebrate muscles.

Electrical phenomena, action potentials, are associated with all types of muscle contraction (Fig. 23–9). In general, muscles are arranged with their fibers in parallel so that the voltage difference in a large muscle is no greater than that of a single fiber. In the electric organ of the electric eel, however, the electric plates are modified muscle cells, motor end plates, arranged in series (Fig. 23–15). Although each plate has a potential difference of about 0.1 volt, the discharge of the entire organ, made of several thousand plates, may amount to several hundred volts. Records of the electric eel show that it can produce a potential of 400 volts or more, enough to stun or kill the fish on which it preys and to give quite a jolt to a man.

NEMATOCYSTS, CHROMATOPHORES AND LUMINESCENT ORGANS AS EFFECTORS

The members of the phylum Coelenterata are unique in having special stinging cells termed **cnidoblasts** which produce thread capsules or

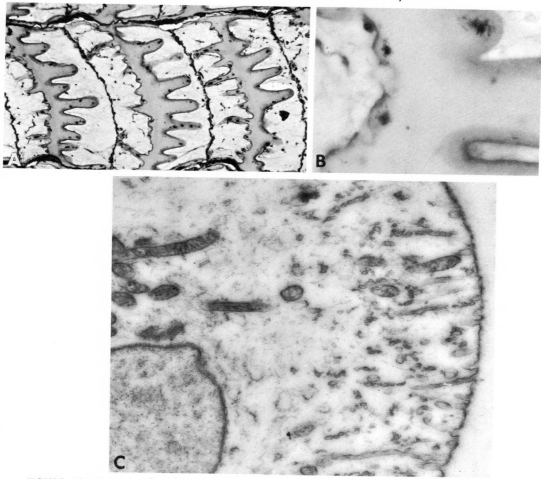

FIGURE 23–15   A, The electric organ of the eel. Photomicrograph of a small portion of the electric organ showing six to eight electroplaxes (cells specialized for the generation of electricity) within their connective tissue compartments. Magnified 250×. B, A photomicrograph of a single electroplax magnified 1500×. C, Electron micrograph of the noninnervated surface of an electroplax magnified 25,000×. A nucleus (the round structure in the lower left corner) and mitochondria (the round or elongate dark structures near the nucleus) are shown. The plasma membrane, right, is invaginated by numerous tubules which dip into the interior of the cell. (Courtesy of Dr. John Luft.)

nematocysts (Fig. 14–10). The nematocyst contains a coiled, poison-filled, barbed thread which it can release on stimulation. This traps and paralyzes the small animals which serve as prey for the coelenterates. The release of the thread occurs when the trigger of the nematocyst is touched. A nematocyst can be used only once; after it has been discharged it is discarded and replaced by a new one produced by the cnidoblast. At the time of discharge the coiled thread rapidly turns inside out and uncoils, extending its full length out from the nematocyst, much as the proboscis of the nemertean is extended (p. 414). The nemertean, however, has a muscle attached to its proboscis and can pull it back and use it over again.

Squid, crabs, fish, frogs and many other animals can change the color of their skin by expanding and contracting cells in the skin, **chromatophores**, which contain particles of pigment such as melanin. By changing color to blend with the background, the animal increases its chances of escaping detection by possible predators. In most animals the contraction or expansion of the chromatophores is under hormonal control. Information received by the eyes and other light-sensitive organs is integrated by the nervous system and the release of the hormone is stimulated.

The luminescent organs of the firefly, glowworm, squid, fish and other animals are a final group of effectors by which certain animals make adaptive responses to certain inputs. The chemical reactions involved in luminescence were described in Chapter 6. Luminescent organs, such as those of the firefly and deep-sea fish, serve to attract mates and assure the continuation of the species. Others use light as a lure to attract prey.

The combination of receptors and effectors, linked and coordinated by the integrative activities of the nervous and endocrine systems, enables the animal to detect changes in the external or internal environment and respond in some appropriate fashion. The responses are the basis for the myriad types of behavior exhibited by animals which are considered in the next chapter.

SUGGESTIONS FOR FURTHER READING

Bourne, C. (ed.): *Structure and Function of Muscles.* New York, Academic Press, Inc., 1960.
    An extensive collection of papers relating to the anatomy and physiology of muscles.

De Robertis, E., Nowinski, W. W., and Saez, T. A.: *Cell Biology.* 5th ed. Philadelphia, W. B. Saunders Company, 1970.
    This text has excellent chapters on both muscle contraction and cellular secretion.

Gergely, J.: *Biochemistry of Muscle Contraction.* Boston, Little, Brown and Company, 1964.
    An excellent discussion of the biochemical basis of muscle function.

Gray, James: *How Animals Move.* Baltimore, Penguin Books, 1964.
    A simplified account of the various kinds of locomotor devices found in animals.

Huxley, H. E.: *The Mechanism of Muscular Contraction.* Science *164*:1356, 1969.
    An interesting summary of the sliding filament theory by its chief proponent.

Porter, K. R., and Fanzani-Armstrong, C.: *The Sarcoplasmic Reticulum.* Sci. Amer. March (Offprint 1007) 1965.
    A brief, readable description of the subcellular organelles of muscles.

Rivera, J. A.: *Cilia, Ciliated Epithelium, and Ciliary Activity.* New York, Pergamon Press, Inc., 1962.
    A readable account of the structure and function of cilia.

Smith, D. S.: *The Flight Muscles of Insects.* Sci. Amer., June (Offprint 1014) 1965.
    An interesting brief account of the operation of those remarkable muscles.

BEHAVIOR

CHAPTER 24

The metabolic processes of living organisms operate most efficiently under uniform conditions. The environments of the world are not uniform either in space or in time. Since all organisms constantly exchange material and energy with the external environment, it stands to reason that fluctuations in this environment will cause fluctuations in internal states of organisms unless some sort of regulation occurs. The general nature of the regulation is set in part by the period, i.e., the frequency of environmental fluctuations relative to the life span of the animal. If the period of change is much greater than a generation time, the regulation is genetic. Climatic oscillations that span millions of years, as, for example, glacial and interglacial periods, are meaningless insofar as the regulation of individuals is concerned. In these cases, the animals that cannot tolerate the changes die; those that can, survive and reproduce. Eventually new species evolve.

Environmental variations that oscillate within the life span of an individual are countered by changes in the individual itself. Slow, relatively long-lasting variations are met by acclimatization, i.e., a gradual physiological change. Everyone is familiar with the fact that mountain climbers usually acclimatize at a high base camp for days before attempting the ascent to high altitudes. The acquisition of a tan in the summertime is another example of a physiological change brought about slowly, if not painfully. A tanned person obviously can tolerate exposure to sunlight better than an untanned one. There are comparable gradual adjustments to hot and cold climates.

When environmental changes are more abrupt, animals regulate by rapid changes which may be physiological or behavioral. When the external temperature rises rapidly, a man may sweat; when it drops, he may shiver. When a situation angers a man, the adrenalin level in his blood rises, his heart pumps faster, his face may get red. Alternatively when a man is hot, he may remove some clothing, fan himself or depart for a cooler place. When cold, he may beat his arms, blow on his hands, put on more clothing or seek a warmer place. There are behavioral responses to environmental variation. Behavior is one means of coping with different environments and with changes in environment.

WHAT BEHAVIOR IS

It is difficult to draw a hard and fast line between physiological responses and behavioral responses, since the latter inevitably involve physiological change. We usually think of behavior as that which an animal does. Of course, growing is also something that an animal does, so we refine our description of behavior by saying that it is most often a coordinated series of muscular activities. These may result in movement of only part of the body. A dog may wag its tail, a bird may sing a song, a butterfly may release a volatile sex attractant. They may also result in more complicated postural or locomotory movement whereby the whole animal is involved. On the other hand, they can result in an animal not moving, as when a pointer "freezes" upon detecting a pheasant, a possum plays dead or a gull chick hugs the ground motionless and disappears into the background.

Although behavior is most commonly a muscular response (or ciliary, flagellar or pseudopodial response) it can also involve other effectors. Fireflies, marine plankton and some abyssal fish can flash light organs; electric eels can deliver electrical shocks.

Behavior is thus most often a response to environmental variation. The variation may be that occurring in the external physical environment (light, temperature, humidity, oxygen, carbon dioxide, pH, texture and so on). It may be variation that the organisms themselves brought about in the physical environment. Man is the horrible example of an animal that brings about changes in the physical environment. Any animal that depletes the environment of oxygen or pollutes it with waste products is an example of organismic-induced variation. Beetles living in stored grain alter the environment by raising the temperature and by secreting quinones. In plant succession, as we have seen, organisms alter the physical environment in a number of ways.

Animals also respond to changes in the living environment, i.e., to other organisms, as we shall see. The mouse flees the cat, the mosquito seeks the man, the bee visits the flower, the mallard drake courts the duck.

Changes in an animal's internal environment also initiate behavior. The most familiar example involves hunger. When an animal has been deprived of food for some time, a number of changes occur in its internal environment. The blood sugar level drops and the stomach increases its motility. In response to these and other changes an animal becomes restless, moves about and "seeks" food. When the animal has eaten, internal conditions change, restlessness ceases, and grooming or sleep may ensue.

Behavioral responses are adaptive either from the point of view of survival of the individual or of the species. The result of any behavior is to remove the animal from a source of danger or in one way or another to alleviate an actual or potential threat to its well-being by initiating some form of regulation. Many times, however, a behavioral response may be one that is maladaptive to the individual but maximizes the likelihood of survival of the offspring, hence of the species. In those species that die after producing eggs, the reproductive behavior leading to egg production is obviously maladaptive insofar as the female was concerned but is essential to the survival of the species.

The problems arising from a variable environment are basically the same for all animals and throughout geologic time. Each animal must live long enough to reproduce. This means that it must avoid inhospitable environments, predators, parasites and competition from its own species. It must ensure itself of an adequate supply of raw materials and energy for its metabolic machinery. It must find another of its kind (unless it reproduces asexually or is a self-sufficient hermaphrodite), mate with it successfully and, in some cases, guard or educate its young. No two species accomplish these ends the same way. Behavior serving these ends is just as diverse as is structure; it is just as characteristic of a species as size, form, color, smell or song. Indeed some species of animals resemble each other so closely that they can be told apart only on the basis of some distinctive behavior. There are, for example, two species (or subspecies) of ground crickets (*Nemobius*) that are indistinguishable except by their songs which are absolutely distinctive.

Similar situations call forth different behavioral responses in different species. Also, different situations confronting any one animal call forth different and appropriate behavior. Both categories of difference, interspecific ones and different behavioral responses in an individual, reflect different patterning of muscular contraction. A blood-sucking mosquito is displaying a different pattern of muscular action from a leaf-chewing grasshopper; a predatory bat catching mosquitoes is displaying a pattern different from that exhibited by a mantis stalking a fly. Similarly an oriole weaving its nest employs different motions than a swallow constructing a mud nest. And a winter wren singing its incredibly rapid song has a different sequence of muscular action than the hermit thrush singing its bell-like notes.

Correspondingly the different behaviors of an individual from one minute to the next reflect different patterns of muscular activity. Fighting involves one pattern, fleeing another, courtship another and nest building still another. These differences may be topographic, i.e., a matter of which groups of muscles contract, or temporal i.e., a matter of frequency and sequence of contraction. More often both types of change are involved.

Since behavior is usually some movement, lack of movement or secretion in response to an external or internal environmental change, it depends not only upon the effector system but also upon the sensory capacities and the capabilities of the nervous system. An animal cannot respond to temperature change if it lacks temperature receptors. It cannot fly if it lacks wings nor vocalize if it lacks vocal cords nor walk upright if its structure prohibits it. It cannot write words even if it has a hand and opposable thumb if it lacks the neural equipment to direct the hand in the formation of letters. Accordingly behavior is expressed within the limitations imposed by the receptor, effector and nervous systems. Each one of these contributes to and shapes behavior. The role of the environmental variable, the stimulus, depends upon how these various systems are interconnected and integrated. Their organization may be such that the stimulus triggers a simple set of muscle actions that proceed automatically once started. On the other hand, as the muscular action proceeds it may be constantly monitored by stimuli which it itself has generated in the body and therefore steer and guide it. Or the stimulus may trigger a complex response that is genetically completely programed in the central nervous system. Examples of these alternatives follow.

SECTION 24–2

REFLEXES

In the very simplest cases, the stimulus triggers a muscle or set of muscles to respond in a predictable unvarying fashion, and behavior is determined only by the presence of an appropriate sense organ connected with a particular muscle or set of muscles. The knee jerk, or **stretch reflex,** is a classic example. A sharp tap on the tendon below the knee cap stretches the attached muscle. This stretching stimulates the spindle organ (a stretch receptor) (Fig. 24–1). Messages from this receptor neuron are conducted along its axon to the spinal cord where they pass by way of a synapse to a motor neuron that stimulates the muscle to contract, hence the knee jerk. This behavior is clearly dependent upon a stretch receptor and upon its being connected with a specific muscle. This is an example of a **mono-synaptic reflex arc.**

In complex animals there are many simple bits of behavior that arise from the use of simple **neural circuits.** The reflex arcs upon which these bits of behavior depend are not, however, simple on/off switches, because the synapses by which the incoming and outgoing segments of the circuit are connected possess a number of characteristics that alter the incoming message. There is delay at the synapse. The stronger the stimulus, the shorter the delay. Thus the onset of behavior can be directly related to the strength of the stimulus. There is **afterdischarge;** i.e., the synapse may continue to be active after the stimulus has ceased. A cockroach that is stimulated to run in response to a touch on its anal cercus does not cease running the second the stimulus ceases. The continuance is due in part to afterdischarge at a synapse in the ventral nerve cord. Synapses exhibit **temporal summation,** i.e., an incoming message may be too weak to activate the motor fiber, but subsequent weak messages, provided they arrive soon enough, add to produce a response. This phenomenon can be observed in the proboscis response of flies. One can apply to the mouthparts of a fly a sugar solution that is concentrated enough to stimulate the taste hairs but too weak to evoke a behavioral response, the characteristic extension of the proboscis. If this stimulation is followed quickly enough by a second one of equal strength, a proboscis extension does occur.

When more than two neurons meet at a synapse there may be **spatial summation.** In other words, an incoming message from one fiber that is too weak to cause a response may, when added to a message coming in on another channel, be of sufficient strength to be effective.

Both temporal and spatial summation are found in the **scratch reflex** of the dog. When any part of a saddle-shaped area of skin on a dog's body is shocked, rubbed or otherwise mechanically stimulated, the dog raises the hind leg on the appropriate side and scratches. The leg moves

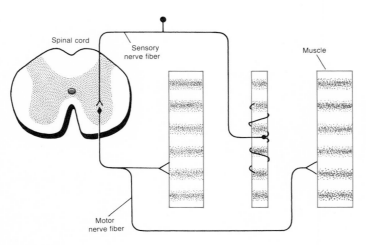

FIGURE 24–1   A diagram of the anatomy of the stretch reflex, a monosynaptic reflex arc. (After Van der Kloot, W. G.: Behavior. New York, Holt, Rinehart and Winston, Inc., 1968.)

Spinal cord

Sensory nerve fiber

Muscle

Motor nerve fiber

back and forth rhythmically four times per second, regardless of the frequency of stimulation. If a weak electrical shock is repeated 18 times per second, there is no response. After 44 stimuli have been received, there is a response. This is a result of temporal summation. If a shock too weak **(subthreshold)** to elicit a response is given simultaneously with several other similarly weak shocks applied at different places on the skin, a response occurs. This is an example of spatial summation.

The knee jerk is a monosynaptic reflex (see Fig. 24–1). Other reflexes in the vertebrate involve more than one synapse. The **flexion reflex** involves two synapses, because an internuncial neuron is interposed between the sensory and the motor fiber (Fig. 24–2). Here the interneuron is **inhibitory.** When it is stimulated by sensory input from one muscle, it inhibits activity in the **antagonistic** muscle.

By increasing the number of synapses in reflex arcs, considerable complexity of behavior can be developed. For example, in the flexion reflex, different behavioral responses can be elicited as the strength of stimulus is increased. In experiments with spinal dogs, dogs in which the connection between the spinal cord and brain is cut, a weak noxious stimulus applied to the foot pad results in withdrawal of the foot. An increase in stimulus strength causes strong flexion of the lower leg. A still stronger stimulus elicits flexion of the upper leg as well. This spreading of response with increase of stimulus strength is termed **irradiation.** Figure 24–3 illustrates the complexities that can be introduced into reflex behavior by adding chains of reflex arcs and interneurons.

Conceivably one could construct a behaving animal completely out of reflexes. In so-called higher animals reflexes account for only a small part of the total behavior. In some of the more simple animals, however, re-

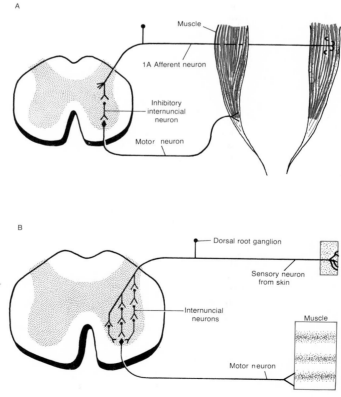

FIGURE 24–2   A, The most direct nerve pathway between a stretch receptor and a motor neuron running to the antagonist muscle. B, A polysynaptic reflex arc between sensory neurons in the skin and motor neurons. (After Van der Kloot, W. G.: Behavior. New York, Holt, Rinehart and Winston, Inc., 1968.)

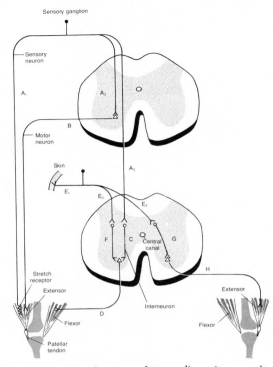

FIGURE 24-3   *Schematic diagram of two adjacent segments of the spinal cord, showing examples of connections of sensory neurons, interneurons, and motor neurons, making up various reflex arcs.*

The monosynaptic arc involved in the knee jerk is activated when the extensor muscle is stretched by the patellar tap. This involves sensory neuron A_1 leading from the stretch receptor through its central branch A_2 to the motor neuron B which activates contractions in the extensor, lifting the leg. At the same time, there is reciprocal inhibition of the antagonistic flexor muscles, probably through the polysynaptic arc, involving another branch of the sensory neuron A, and the interneuron C, resulting in inhibition of the motor neuron D and relaxation of the flexors.

The polysynaptic arc in the withdrawal reflex to painful stimulation of the skin on the distal part of a leg involves sensory neuron E, and its central branch E_2, interneuron F, and the excitation of motor neuron D, leading to a contraction of the flexors.

At the same time, a crossed-extension reflex occurs over another branch of the sensory neuron, E_3, through the interneuron G and the motor neuron H, resulting in excitation of the extensor muscles on the opposite side. Not shown in these last two cases are the re-ciprocal inhibitory influences on the muscles antagonistic to those contracting and the feedback from the stretch receptors that are present in all the muscles. Also, it should be pointed out that while only one neuron is portrayed in each component of the reflex arc, actually many, many neurons typically work in concert. (After Lloyd, D. P. C.: Synaptic mechanisms. In Fulton, J. F., (ed.): A Textbook of Physiology. Philadelphia: W. B. Saunders Company, 1955.)

flexes play a much more important role and as a consequence the behavior of these animals tends to be more rigid, stereotyped and simple. Many of the activities of starfish can be understood in terms of reflex arcs. A great deal of their behavior is based on the movements of the tube feet. They can extend, retract and make postural movements associated with stepping. Extension and retraction are unoriented reflex responses. Apparent co-ordination of all feet, which retract in unison when a wave washes over the animal, are in fact individual responses to a common stimulus. In crustacea there are numerous aspects of behavior that are reflex—e.g., withdrawal of the eyestalks, closing and opening of claws and movements associated with escape, defense, feeding and copulation.

By hooking reflex arcs together a capacity for constant monitoring of the behavioral response is introduced. More finely graded complex behavior now becomes possible. A description of prey catching by the praying mantis illustrates the remarkable degree of coordination that can be achieved by providing appropriate feedback mechanisms. A **feedback mechanism,** of which the ordinary thermostat controlling a home furnace is an example, is an arrangement whereby the effect produced by a stimulus is "fed back" into the circuit to stop the effect or to augment it. The behavior of the mantis also illustrates how component units of a series of reflex actions may depend on a succession of different stimuli.

The mantis lies in ambush waiting for its victim to appear. When the prey, a fly, for example, appears in the mantis' visual field, the mantis, whose eyes are immovable, turns its head to face the fly. The rest of the body is then usually brought into line with the head, although this alignment is not necessary for a successful strike. When the fly is brought appropriately close, either through its own action or by the careful stalking of it by the mantis, the mantis makes a lightning strike with its front legs. The strike is completed in 10 to 30 milliseconds; therefore, all the information required to direct the strike must be acquired before the action. The mantis requires information about the position of the prey relative to the head and the head relative to the prothorax. The first is provided by the eyes and the second by **proprioceptors** in the neck (Fig. 24–4). Normal mantids hit about 85 per cent of the flies at which they strike. If information from the proprioceptors of the neck is eliminated by cutting the sensory nerve, hitting accuracy drops to between 20 and 30 percent. If the head is immovably cemented to the thorax in the median position, performance remains normal. If the head is turned to the right and fixed, the prey is missed to the left, and vice versa. If the proprioceptors are eliminated on one side and the head fixed in a turned position, the hitting errors are compounded.

Thus the direction of the stroke is determined by feedback processes which control the position of the head (Fig. 24–5). The alignment of the head preceding the strike is steered by the difference between the **optic-center** message (a function of the angle between prey and fixation line) and the **proprioceptive-center** message (a function of the angle between the head and the body axis). When fixation movements cease, the direction of the strike is set principally by the optic-center messages, and to some

FIGURE 24–4   Proprioceptors of the neck region of the mantis (left side). K = sternocervical hair plate, situated on the anterior end of the laterocervical sclerite (L). N = tergocervical plate; O = ventral border of the laterocervical sclerite. (After Mittelstaedt, H.: Recent Advances in Invertebrate Physiology, Scheer, B. T. (ed.). Eugene, University of Oregon Books, 1957.)

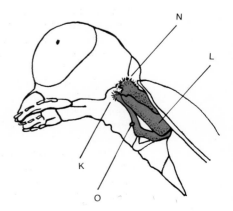

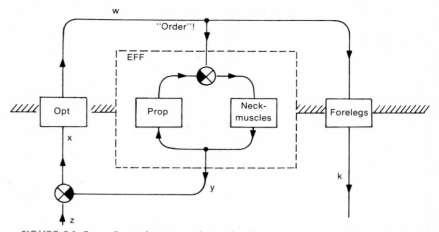

FIGURE 24–5   *Control pattern of prey localization in mantids. The control pattern of subsystem EFF is shown to be a loop, which turns the head into a position determined by the order (w). (From Mittelstaedt, H.: Ann. Rev. Entom. 7:183, 1962.)*

extent by the proprioceptive-center messages. Mantids that have been hand-fed from hatching and have never had to catch a fly are able to perform accurately at the first opportunity.

SPONTANEOUS ACTIVITY

The direct stimulus-response type of activity represented by reflexes is limited in its performance. It requires rather elaborate neural circuitry and it depends upon the stepwise intervention of a series of different stimuli. Behavior is also served by **spontaneous activity** of the nervous system and by a system of programing activity centrally. In the first instance, patterns of behavior occur independently of stimuli from the external environment. In the second instance, external stimuli serve simply to release a neural program. Examples of both follow.

Spontaneous activity constitutes a large part of behavior in many simple organisms in which there are recognizable ganglia. In jellyfish there are concentrations of grouped neurons and a **statocyst** (a gravity-detecting sense organ) at various locations around the margin of the bell. Rhythmic waves of excitation originate in these bodies. They are the genesis of the smooth swimming movements which consist of rhythmic contractions of the bell forcing out water. Coordination is lost if the circumferential nerve of the bell is cut; however, an isolated small section of bell with a marginal body will continue to contract rhythmically.

The behavior of the lugworm (*Arenicola marina*) is another example of spontaneously generated behavior. This marine worm lives in a **U**-shaped burrow in the mud. It keeps the burrow open by rhythmic respiratory, locomotory and feeding movements. These movements, which do not arise in direct response to environmental change, are rhythmic muscular contractions driven by a series of spontaneously active "clocks" in the various ganglia. The brain is not necessary for these activities. Isolated fragments of the worm perform as well as the whole.

SECTION 24–4

CENTRALLY DETERMINED SEQUENCES

Whereas prey-capture by the mantis requires continual monitoring of all preparatory movements preceding the actual strike, and the various movements of the lugworm are performed independently of information from the outside, the behavior of most invertebrate animals is a consequence of centrally determined sequences of motor impulses that need not depend on any feedback from the periphery. In other words, there are behavior programs built into the central nervous system and these need only be triggered or triggered and guided by specific stimuli. Flight by locusts is a well analyzed case. Wind blowing on special receptors on the head causes impulses to travel from the brain to the thoracic ganglion, which in turn sends impulses to the motor nerves driving the wing muscles. A complicated patterning of muscular activity is necessary to produce flight. The pattern of impulses from the thoracic ganglion determines the muscular sequence. Even when all the muscles have been removed, and when there is no stimulation fed back from the wings (no reflex feedback from proprioceptors), normal neural patterns are delivered by the ganglion. Under normal circumstances proprioceptive feedback alters some of the characteristics of flight but is not necessary for basic performance. Reflexes from eyes and **halteres** (balancing organs) control fine wing movements to prevent random roll, yaw and pitch.

The singing of crickets is a fine example of patterning being determined by the central nervous system. The complicated series of muscular contractions required to produce the different kinds of song, courtship, aggression and so forth are partly determined in the brain. Electrical stim-

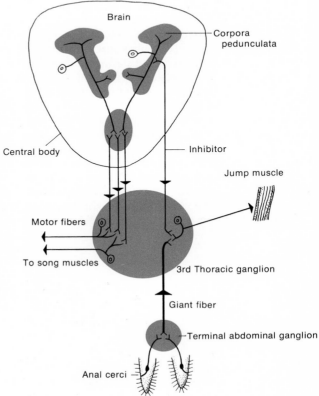

FIGURE 24–6   *The main interactions between ganglia that control singing outbursts in grasshoppers and crickets. Electrical stimulation of a nucleus called the corpora pedunculata in the brain causes a burst of normal song and simultaneously inhibits the motoneurons involved in jumping. The central body in the brain organizes the patterns of the song, and stimulation here produces abnormal noises. Stimulation in the 3rd thoracic ganglion excites motoneurons in a meaningless pattern. The alerting mechanism of the anal cerci, which are sensitive to sounds and air movements, is inhibited during song, thus preventing the animal's alarming itself by its own noise. (After Huber, F.: Invertebrate Nervous Systems, Wiersma, C. A. G. (ed.). Chicago, University of Chicago Press, 1967.)*

FIGURE 24–7   *Male cricket (Gryllus campestris) with two chronically implanted
electrodes in the brain (1) and two grounded ones inserted into the prothoracic segment
(2). 1 = steel wires insulated to the tip, diameter 20μ. 2 = steel wires having a loop
within the prothoracic segment, diameter 100 μ. (Courtesy of D. Otto, unpublished).*

ulation of one area of the brain triggers normal song rhythms by sending
a command to another part of the brain which organizes the patterns to be
produced by the motor fibers in the thoracic ganglion. Stimulation in the
area of the brain that organizes pattern triggers atypical songs never heard
in the males's normal vocal repertoire (Figs. 24–6 and 24–7). The singing
of crickets is an example of a ***fixed-action pattern.*** It is characteristic of these
patterns that they require environmental stimuli only to start them going.
Frequently, however, there are fixed-action patterns that are oriented by
environmental cues. The egg-rolling response of the greylag goose is one
of these. This goose and many other birds that nest on the ground retrieve
eggs that happen to roll out of the nest by placing the bill on the far side
of the egg and rolling it back with side-to-side movements to prevent the
egg from slipping away (Fig. 24–8). It was shown by replacing the egg with
a cylinder which did not wobble that the side-to-side compensatory movements ceased. On the other hand, if egg or cylinder was removed once the
retrieving movement had commenced, the bill was still drawn to the breast.
In other words, this part of the behavior is a fixed-action pattern that,
once elicited by the stimulus (sight of the egg outside of nest), continues
independently. The side-to-side movements are constantly steered by
stimuli coming from the wobbly egg. These steering stimuli are modifying
the orientation of the basic pattern. From these selected examples it can
be seen that the mechanisms underlying the behavior of animals are of

FIGURE 24–8   *A, Greylag goose retrieving an egg which is outside the nest. This
movement is very stereotyped in form and used by many ground-nesting birds. B, The goose
attempts to retrieve a giant egg in precisely the same fashion. (From Lorenz, K. Z., and
Tinbergen, N.: Z. Tierpsychol. 2, 1, 1938.)*

many kinds. *Reflex activity, feedback mechanisms, endogenous activity* and *centrally patterned activity* triggered by specific external stimuli all work in concert to provide smoothly coordinated muscular activities appropriate to the environmental situation at hand.

SECTION 24–5

STIMULI

The role of the stimulus in these processes is as varied as the processes themselves. It is important to remember that the sensed world is not the same for all animals. The honeybee sees a far different world than we do. Her eye is sensitive to ultraviolet (which we cannot see). For her the primary colors are: ultraviolet, blue-violet and yellow. She can distinguish white light with ultraviolet from that without ultraviolet; therefore, "bee white" is different from "human white." For us a mixture of the terminal ends of the spectrum, i.e., red and violet, gives purple. For a bee, a mixture of the terminal end, i.e., yellow and ultraviolet, gives a new fundamental color, "bee purple." These and other relations are summarized in the color circles for man and bee (Fig. 24–9). Additionally bees can detect differences in the plane of polarization of light in the blue sky. A sky that appears uniform to us is patterned to the bee.

Other differing sensory capacities are found throughout the animal kingdom. Bats and moths are able to hear their own sounds and the sounds of each other. The bat emits high frequency sounds far above the human

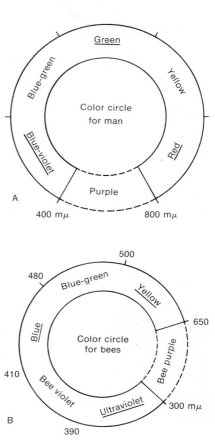

FIGURE 24–9 *Color circle (A) for man; (B) for the bee. The three primary colors are underlined; by mixing them the intermediate colors can be produced. Complementary colors are opposite each other in the circles. (After Daumer, K.: Reizmetrische Untersuchungen des Farbensehens der Biene. Z. Vergl. Physiol. 38:413–478, 1956.)*

audible range. These sounds bounce off any insect flying in the vicinity, and the bat is able by echolocation to home-in on the insect. A moth can hear the ultrafrequency cries of the bat and take evasive action. Sometimes the bat wins, sometimes the moth wins.

In the field of chemical senses the olfactory superiority of many animals is well known. Dogs are famous for their tracking abilities, male moths for their ability to detect minute quantities of sex attractants elaborated by females.

Still other sensory capacities are developed in other animals: rattlesnakes can sense heat well enough in the dark to strike unerringly, electric fish can detect obstacles by sensing disturbances produced in the electric field with which they surround themselves (Fig. 24–10), porpoises utilize sonar for underwater navigation, spiders "read" the vibrations imparted to their webs by victims blundering in. Even more subtle sensory filtering is represented by the "bug detectors" of frogs. Frogs, whose principal item of diet is flies, respond more readily to small, dark, circular moving objects near them than to larger more diffuse or distant movements. This filtering of the visual field is accomplished by segments of the retina that are specialized to react to small, dark, convex objects.

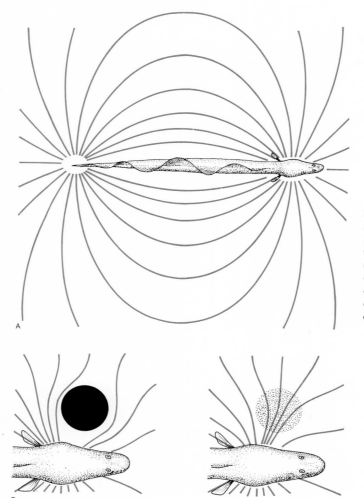

FIGURE 24–10   Electrolocation in the fish Gymnarchus. The fish generates an electric field, its tail negative to the head (A). Objects that conduct less (B) or more (C) electricity than the surrounding water distort the electric field and this is detected by special sense organs along the sides of the fish. (Courtesy of H. W. Lissmann.)

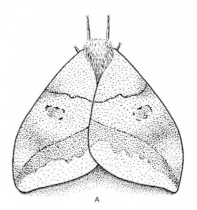

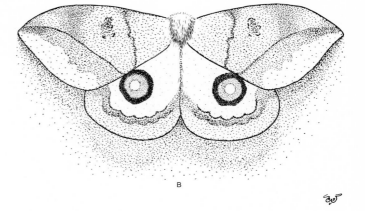

FIGURE 24–11 The moth Automeris coresus (A) at rest, and (B) displaying the vivid eye-spots on its hind wings in response to a light touch. (After Blest: Behaviour 11:257, 1957.)

It is clear that different parts of the world have different meaning for different animals. No animal detects all changes in the environment nor are all the changes that are detected effective. Furthermore, a stimulus that might be effective at one time or under one set of conditions may not be effective at other times under other conditions. Some of the filtering of environmental information is effected by the sense organs, other by the central nervous system. As already implied, the capacities of the sense organ are the first determinants of what parts of the environment will be significant behaviorally.

If an animal were to process all the sensory information that it was capable of receiving, the central nervous system would undoubtedly be swamped. The animal would not be able to "see the forest for the trees." It is not surprising, therefore, to discover a selective responsiveness, responsiveness to a particular element or elements of the environment that are a "sign" of an appropriate situation. *Sign stimuli* are regular features of fixed-action patterns of behavior. Where quick action is essential, as in fleeing a potential predator, a sign of danger is more important than a

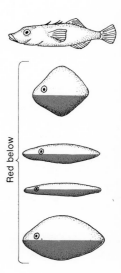

FIGURE 24–12 A series of models used in tests of aggression with male sticklebacks. The four crude imitations colored red below released more attacks than the accurate model which lacked red. (After Tinbergen, N.: The Study of Instinct. London, Oxford University Press, 1951.)

FIGURE 24–13   *Two models used for identifying the stimulus characters important in eliciting the zig-zag courtship dance of the male three-spined stickleback. The lower and cruder model has a swollen abdomen which is absent in the upper one. (After Tinbergen, N.: The Study of Instinct. London, Oxford University Press, 1951).*

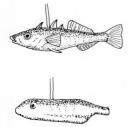

detailed description of the danger. Alarm signs are usually simple and contrasting, whether they be sounds or sights. Small birds show a violent flight reaction to animals with large eyes. This is understandable considering the ever-present danger from owls and hawks. It is not surprising, therefore, that small birds flee any staring eye. Many moths, the prey of birds, have evolved "eyes" on their hind wings. When the moth is at rest, the "eyes" are concealed by the forewings. When the moth is disturbed, it "flashes" the "eyes" by suddenly extending its forewings (Fig. 24–11). Many caterpillars also have false eyes.

By employing models to vary systematically the characters of a stimulus it is possible to discover the "sign." This approach was employed in discovering how male sticklebacks are able to distinguish sexes. In the springtime the males establish territories from which they drive other males. They ignore or court females. Presentation of various models (Fig. 24–12) revealed that the most important single feature eliciting attack is a red belly (characteristic of males this time of year). Other morphological details are unimportant. Courtship, expressed as a zigzag dance, is elicited by a swollen belly (Fig. 24–13).

Sometimes more than one characteristic is important, as in the case of young gulls begging food from the parent. The chick pecks at the tip of the parent's bill where there is a red spot. At this moment, the parent regurgitates food on the ground, picks some up and presents it to the chick. Presentation of cardboard models from a series with spots of different shades revealed that the important sign stimuli are a red patch and a long, thin, low, moving form (Fig. 24–14). The relation of parts to one another

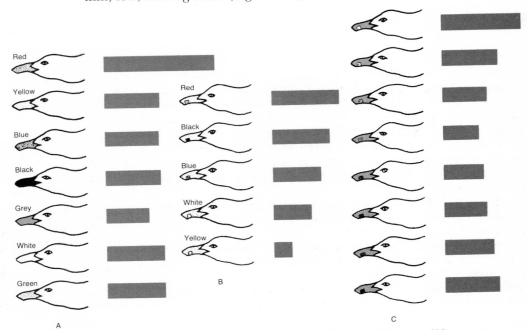

FIGURE 24–14   *Three series of model heads used in the pecking tests: (A) measures the effect of bill color, (B) that of patch color (all the bills were yellow), and (C) the effect of varying the contrast between patch and bill color (all the bills were grey). The length of the bar beside each model is proportional to the number of pecks it received. (From Tinbergen, N., and Perdeck, A. C.: Behaviour 3:1, 1950.)*

is also frequently important, as illustrated by experiments with nestling thrushes. At age eight days, these small birds respond to the sight of the parent's head over the edge of the nest by gaping in that direction. When the models shown in Figure 24–15 were presented, the young gaped to the smaller head in the combination on the left and the larger head in the combination on the right.

Curiously enough nature does not always provide the best stimulus. An experimenter, by exaggerating the best features of a stimulus, can construct **"superoptimal" stimuli.** Two examples are shown in Figures 24–16 and 24–17.

The fact that animals can be conditioned to stimuli to which they do not ordinarily respond indicates very clearly that the selective action of many stimuli is due not to sensory filtering but to filtering elsewhere in the nervous system. How this is brought about is not known.

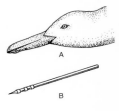

SECTION 24–6

BEHAVIORAL CHANGES IN TIME

So far we have considered behavior as though it were constant and unvarying in time, as if once the neuromuscular system was built it performed routinely according to specifications. We know, however, that this is not the case, that the animal is not a rigid input/output, stimulus/response, machine. Although simple reflex systems seem to operate with an unvarying monotony, few other systems do. A stimulus that evokes one response at one time may be totally ineffective another or may even evoke an opposite response. Clearly an organism is not the same from one moment to the next. Somewhere between the stimulus and the response there are variables. They may cause rapid, slow, short-term, long-term, reversible or irreversible changes in behavior. Among those commonly rec-

FIGURE 24–15   Cardboard models used to release gaping responses from nestling thrushes. (After Tinbergen, N.: The Study of Instinct. London, Oxford University Press, 1951.)

FIGURE 24–16   (A) Accurate, three-dimensional model of a herring-gull's head, and (B) "supernormal" bill. The latter received 26 per cent more pecks from young chicks. (After Tinbergen, N., and Perdeck, A. C.: Behaviour 3:1, 1950.)

FIGURE 24–17   An oyster-catcher attempting to brood a giant egg in preference to its own egg (foreground) or a herring-gull's egg (left foreground). The bird's original nest site was equidistant between the three "test" eggs. (From Tinbergen, N.: The Study of Instinct. London, Oxford University Press, 1951.)

ognized are: **sensory adaptation, fatigue, central excitatory and inhibitory states, mood, motivation, endogenous rhythms, biological clocks, maturation, experience** and **learning.** Some of these terms describe processes, some relate to the kind of effect produced on behavior, some refer to hypothetical states presumed to exist in the animal, some are concepts. As a step toward understanding these variables, a case history will be presented followed by a more detailed discussion of each variable.

The case history concerns the feeding behavior of the black blowfly. The proboscis of the fly is equipped with a moustache of sensory hairs. Associated with each hair are five bipolar neurons. The dendrite of the neuron is the end affected by the stimulus. The axon transmits sensory messages directly into the central nervous system. One neuron is a mechanoreceptor responsive to touch; one neuron is sensitive to water, one to sugars and two to salts. If one hair of a hungry, thirsty fly is bent, the fly responds by extending its proboscis; this is the initial act in the feeding sequence. After repeated stimulation of the hair (no feeding is permitted), the fly ceases responding. Clearly the behavior has changed even though the stimulus has not. If a neighboring, previously unstimulated hair is now touched, the proboscis is extended; i.e., responsiveness resumes. A return of responsiveness can also be brought about by changing the stimulus. If the first hair is stimulated with water instead of by bending, responsiveness returns. After awhile water too fails as a stimulus. Now the application of sugar causes a renewal of proboscis extension. It is clear in these instances that the waning and return of responsiveness of the fly was a consequence of successive sensory adaptation and restimulation via other sensory neurons.

A reduction in responsiveness of the fly can also be brought about by repeated massive stimulation of all the sensory hairs simultaneously (again feeding is prevented in order to keep the nutritional state constant). In this instance, the waning of sensitivity is not due to sensory adaptation, because stimulation of unadapted taste hairs on other parts of the body (legs) fails to reawaken the response. Refractoriness has occurred somewhere else in the central nervous system.

There are many patterns of behavior that fit the case just described, but in most cases the mechanism is not known. The effect of **sensory adaptation** is a matter of common human experience. A person who has just entered a darkened theatre from a sunny exterior does not at first see or respond to objects in the theatre. Conversely, upon returning to the sunlight he is not able to respond to visual stimuli until he has become light-adapted. In some instances of response waning, interneurons have been incriminated as the sites of change. This is true of the leg-wiping reflex of the frog. If a spot on the back of a frog is lightly touched, one hind leg sweeps up and forward over the spot. The response wanes after repeated touches. If a new spot is touched, wiping resumes.

Most often we have only a description of the behavioral change and no analysis of the mechanism. When a male parasitic wasp (*Mormoniella*) courts a nonreceptive female, his response wanes rapidly. Chaffinches mob an owl at first sight, but with repeated appearances of the owl they react less vigorously (Fig. 24–18). Female mice pay less attention to their litter the more they are exposed to them. Virgin females presented with young once a day may actually give more maternal responses than mothers in continuous company of their own young.

As with the fly, changing a stimulus frequently reawakens a response. Nestling passerine birds will respond to the sight of a dark object appearing over the edge of the nest or to vibration of the nest by extending their

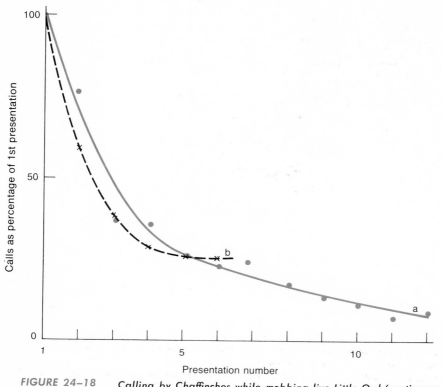

FIGURE 24–18   *Calling by Chaffinches while mobbing live Little Owl (continuous line) presented for 20 min a day, or stuffed Tawny Owl (broken line) presented for 3 min a day. The number of calls given by each bird were expressed as a percentage of the number given on the first day: the ordinate represents the mean. (After Hinde, R. A.: Proc. Roy. Soc. (Biol.) 142:331, 1954; by permission of the Royal Society.)*

necks and gaping. If the visual response is given repeatedly, the nestlings eventually stop responding. They resume if the nest is jarred.

In some instances, repeated application of a stimulus may not only cause waning of response but may actually elicit a new one in its place. A finger rubbed gently near the mouth of a newborn baby at first causes the head to turn toward the finger. With repeated stimulation, the head no longer turns. With still more persistent stimulation, the head now turns away from the finger.

Repeated stimulation may enhance rather than decrease responsiveness of an animal. Returning to the fly as a case history, it can be shown that some change in the central nervous system is responsible. If one taste hair of a nonthirsty fly is stimulated with water, there is no proboscis extension, even though the water receptor in the hair is sending action potentials into the central nervous system. If a different hair is stimulated briefly with sugar, there is a proboscis extension. Now, after the sugar is removed, the first hair is once again stimulated with water. This time the proboscis extends. Since there has been no change in the sensitivity of the water receptor following stimulation of the sugar receptor in the other hair, the change in sensitivity must have occurred somewhere in the central nervous system.

There is considerable indirect evidence that the receptivity of the central nervous system to incoming information changes as a consequence of previous sensory excitation. This state of change has been referred to

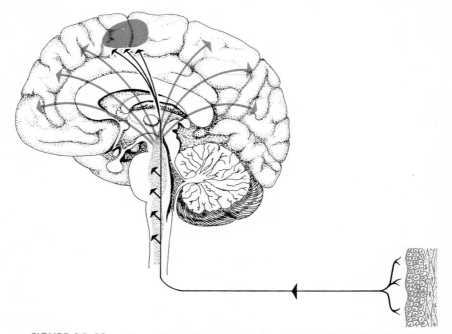

FIGURE 24–19   The reticular formation *is the area stippled with red in this cross section of the brain. A sense organ (lower right) is connected to a sensory area in the brain (upper left) by a pathway extending up the spinal cord. This pathway branches into the reticular formation. When a stimulus travels along the pathway, the reticular formation may "awaken" the entire brain (black arrows). (From The Reticular Formation by J. D. French. Copyright © May 1957 by Scientific American, Inc. All rights reserved.)*

as **central excitatory state, arousal, alertness.** It varies with the general level of sensory input to the central nervous system. Sense organs have a dual function: they transmit specific information and they help maintain a general level of excitability. The total amount of stimulation to which an animal is subject at any time influences its sensitivity to specific stimulation. Animals deprived of sensory input (if blinded, for example) are more lethargic than normal. Electrophysiological findings are in agreement. In cats, the electrophysical activity evoked in the brain-stem by holding a ticking watch near the ear are larger when light is shone in the eyes than when the cat is in darkness.

In vertebrates, general excitatory input and arousal have been associated with the **ascending reticular activating system,** columns of nerve cells extending through the lower brain to the thalamus and having diffuse connections with the cortex (Fig. 24–19). Specific sensory pathways give off collaterals to these columns as they pass through the brain-stem. Activity in any sensory system thus has a diffuse effect on the cortex in addition to its specific cortical effect. Sensory stimulation or electrical stimulation of the reticular formation of a sleeping animal can produce behavioral arousal (Fig. 24–20) and a change in the **electroencephalogram** (EEG). Although the system serves the purpose of general arousal, it is also selective. The urbanite in the country is aroused by all the night songs of frogs, crickets and so forth through which the countryman sleeps soundly unaware. In short, general sensory stimulation has an activating effect on behavior and alters the state of the brain.

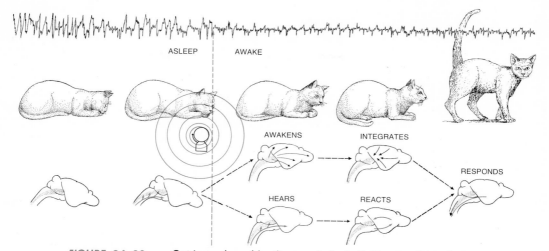

FIGURE 24–20 *Cat is awakened by the sound of a bell. The sound stimuli (incoming red arrows) reach the reticular activating system, or RAS, and the auditory area of the brain. The RAS acts (black arrows) to awaken the cortex so that it can "hear" signals arriving in the auditory area. The brain waves at the top change from a pattern of sleep to one of wakefulness. The RAS then integrates the brain's activity so that the brain can react as a whole. The cat finally responds with a motor impulse (outgoing red arrow) that is regulated by RAS. The cat then jumps to its feet and runs away. The entire process takes place in a matter of a few seconds. (From The Reticular Formation by J. D. French. Copyright © May 1957 by Scientific American, Inc. All rights reserved.)*

SECTION 24-7

MOTIVATED BEHAVIOR

Changes in behavior that can be ascribed neither to the factors just discussed nor to growth and development, hormonal levels or learning, and that are associated with some deprivation, are termed **motivational.** Some examples of motivated behavior are feeding, drinking, courting, copulating, nest building, defense of territory, care of young. The most striking feature of motivational behavior is that it is highly specific to a particular state and particular stimuli. Normally a hungry animal will not court nor will a highly sexually receptive animal feed. Characteristically a motivated animal begins **appetitive behavior,** i.e., searching for a goal. This does not mean that the animal is consciously "searching" and that it necessarily "knows" what the goal is. It merely means that there is a phase of undirected behavior that in the normal course of events leads to a particular source of further stimulation. A hungry fox commences hunting. When the sight or odor of food is encountered, the fox now switches to an oriented type of behavior. When the food is seized, eating, the **consummatory act,** ensues. After the meal the fox no longer exhibits appetitive, goal-oriented, or consummatory behavior, even though the appropriate stimuli are present.

The level of motivation can be gauged by the threshold of response to relevant stimuli, the level of general activity, the amount of work an animal will perform to achieve the goal, the degree of noxious stimulation it will tolerate to achieve the goal.

The most thoroughly studied form of motivated behavior is feeding. A discussion of feeding will illustrate the characteristics of motivated behavior and the concepts built around it. We return first to our case history, the fly.

A fly that has been deprived of food is more active than it is when satiated. Its behavioral threshold to sugar is more than a thousand times lower; i.e., it will extend its proboscis to 10^{-6} M sucrose when hungry as compared with a satiety threshold of 1 M. Additionally the longer the fly is deprived, the more noxious stimulation (salt in the sugar) it will tolerate. It flies about more or less randomly until it encounters the odor of food. This causes its flight to be oriented. Upon landing it steps in the food, the taste hairs on its feet are stimulated, the proboscis is extended, the taste hairs on the proboscis are stimulated and feeding commences. After a minute or so sensory adaptation occurs, no more sensory information reaches the central nervous system and feeding is terminated. It does not resume again, even after the sense organs of taste have unadapted, and sugar is still present. Now the presence of food in the gut stimulates stretch receptors in the wall of the esophagus and the abdominal body wall. Messages from these receptors inhibit excitatory messages from the organs of taste, and the stimulus of food is ignored. The **behavioral threshold** has risen. This rise and fall of responsiveness associated with food deprivation and satiation fits our idea of motivated behavior. It can be explained satisfactorily in terms of competing excitatory and inhibitory systems (Fig. 24–21).

The feeding behavior of rats is similar in many respects to that of the fly. The longer a rat has been deprived of food, the more restless it becomes, the lower its threshold to food stimuli, the more food it eats and the more aversive stimulation (quinine) it will tolerate in the food. After

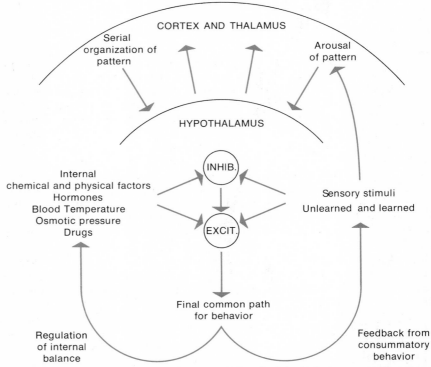

FIGURE 24–21   *Schematic diagram of the physiologic factors controlling the excitatory and inhibitory hypothalamic mechanisms that govern motivated behavior. (From Dethier, V. G., and Stellar, E.: Animal Behavior. Englewood Cliffs, N.J., Prentice-Hall, Inc., 1961.)*

satiation, the stimuli that were previously effective are ignored. Appetitive behavior ceases.

The rat's ingestion is under the control of **oral** and **pharyngeal** stimuli and postingestive stimuli that act as a feedback to shut off eating. The taste of food is instrumental in determining what food will be eaten. It also assists in stopping feeding. In rats with an esophageal fistula (the esophagus cut and both ends sewn to the outside skin so that ingested food drops out again), the passage of saccharin (sweet but nonnutritious) through the mouth brings eating to an end. **Gastric** (stomach) factors also play a role. Rats fitted with stomach tubes that bypass all the oral and pharyngeal receptors (Fig. 24–22) are able to feed themselves by pressing a bar that delivers food directly into the stomach. These animals can regulate their food intake and body weight for weeks. They can make necessary adjustments when food is diluted or concentrated. The relation between oral and gastric effects is complex because, as was shown in rats with both esophageal fistulae and gastric cannulae (allowing one substance to pass through the mouth and a different one to be pumped into the stomach), the influence of any given oral substance varies with the nature of the substance actually pumped into the stomach.

In rats, and other mammals, there is an additional neural structure that exerts great influence on feeding behavior. This is the **hypothalamus** (Fig. 24–23). It is also concerned with thirst, sexual behavior, emotional behavior, sleep, maternal behavior and behavioral responses to temperature change. It is considered to be the controlling center for much motivated behavior.

Experiments with feeding behavior illustrate its role. Destruction of the ventromedial regions on both sides result in a prodigious increase in eating. A rat might triple its weight. It is inferred that these regions constitute an **inhibitory center.** Bilateral destruction of the lateral hypothalamus causes an animal to starve to death even in the presence of abundant food. The lateral area is an **excitatory center.** If electrodes are chronically implanted into these areas (Fig. 24–24), stimulation of the lateral area induces eating even when the animal is normally satiated and stimulation of the ventromedial area inhibits eating.

The relation between the hypothalamus and motivated behavior is also beautifully demonstrated by experiments on thirst in goats. When a goat is in need of water, a deficit of water in the blood stimulates special cells in the lateral area of the hypothalamus. Messages from these cells act through the pituitary to initiate secretion of an **antidiuretic hormone.** This hormone causes the kidneys to resorb more water. At the same time

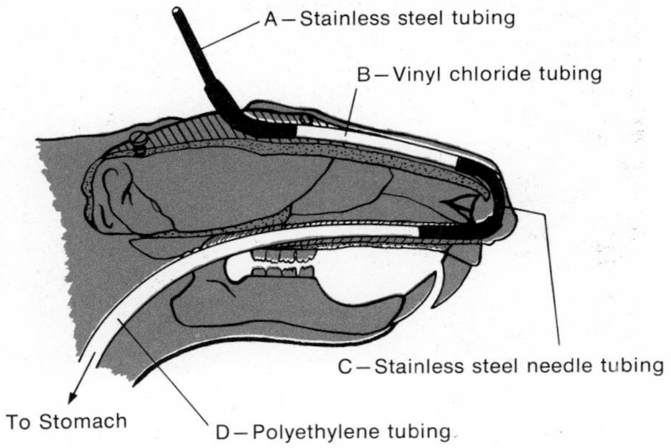

FIGURE 24–22   *Nasophyaryngeal gastric tube used for delivering liquids to rat's stomach without stimulation of the oropharyngeal receptors. (After Epstein, A. N., and Teitelbaum, P.: J. Comp. Physiol. Psychol. 55:753, 1962.)*

A—Stainless steel tubing

B—Vinyl chloride tubing

C—Stainless steel needle tubing

To Stomach

D—Polyethylene tubing

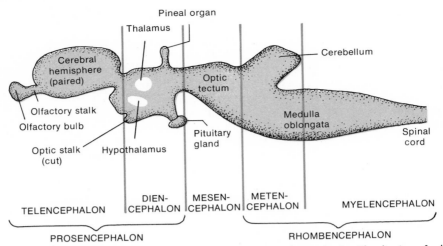

FIGURE 24–23   *The basic divisions of the vertebrate brain. The brains of all vertebrates pass through a stage rather like this during development, but in mammals and birds in particular, the adult brain is dominated by the enormous growth of the cerebral hemispheres and cerebellum. These come to overlie all the rest and obscure the original layout. (After Romer, A. S.: The Vertebrate Body, 3rd ed. Philadelphia, W. B. Saunders Company, 1962.)*

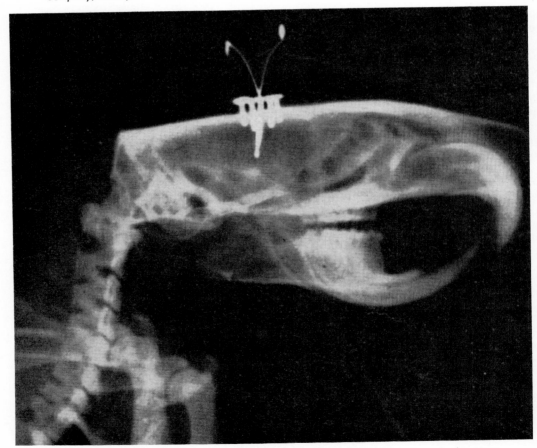

FIGURE 24–24   Implanted electrodes in the brain of a rat are shown in this x-ray photograph. The electrodes are held in a plastic carrier screwed to the skull. They can be used to give an electrical stimulus to the brain or to record electrical impulses generated by the brain. (From Pleasure Centers in the Brain by J. Olds. Copyright © October 1956 by Scientific American, Inc. All rights reserved.)

the animal looks for water and drinks it. If minute amounts of hypertonic salt solution are injected through fine needles directly into the specific areas of the hypothalamus, the goats consume enormous quantities of water even though they had previously satisfied their thirst and are in no need of water. Even if the water is adulterated with unpalatable concentrations of salt or quinine, the goats drink furiously. They exhibit all the normal aspects of appetitive and consummatory behavior.

Additional evidence that the hypothalamus is important insofar as motivational behavior is concerned derives from the observation that rats will work hard to stimulate certain areas. If an electrode is implanted into certain areas of the **hypothalamus, septum** or **tegmentum** and connected to a circuit that enables the animal to stimulate itself by pressing a bar, the animal will stimulate incessantly. It will even cross an electrified grid to reach the bar. This is a dramatic example of motivated behavior.

The hypothalamus is influenced by sensory input from the periphery and by changes in the animal's internal environment. It has been demonstrated, for example, that genital stimulation causes enhanced electrical activity in only that area of the hypothalamus whose destruction causes sexual arousal to be abolished. With respect to the internal environment, the injection of minute amounts of hypertonic saline solutions has just been mentioned. Injection of sex hormones into other areas incites mating behavior. There are also areas where heating by implanted wires causes the animal to sweat and pant and cooling causes the animal to shiver.

The hypothalamus is not the only area of the brain that influences motivated behavior. The septum and tegmentum have already been mentioned in connection with self-stimulation. The **cerebral cortex** also plays a role, especially with respect to emotional behavior. The **neocortex** is probably important in arousing aggressive behavior because its destruction causes cats to be extremely placid. Destruction of the old cortex produces fantastically ferocious cats.

Numerous measures of the level of motivation have been employed. In the case of feeding we have already mentioned general activity, threshold, amount of food eaten and amount of adulterants tolerated. For vertebrates we can add the strength of electric shock that an animal will tolerate in crossing a grid to get food and the amount of work, such as bar-pressing, that an animal will perform. Not all of these have the same

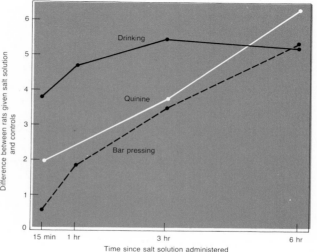

FIGURE 24–25   How three differ-ent measures of thirst change in the period following placing 5 cc of strong salt solu-tion directly into the stomach of a pre-viously water-satiated rat. The units on the vertical axis are arbitrary; they are simply the difference between control and experi-mental rats on the various measures of thirst. (From Annals of the New York Acad-emy of Sciences, Volume 65, Article 4, Figure 2, page 321, N. E. Miller.)

FIGURE 24–26   Schematic representation of the sexual behavior of the three-spined stickleback.

The female, with a swollen belly (top right) enters the territory of the male, swimming in a special posture. The male courts her, and when she responds he leads her to the nest. If she follows, he adopts a special posture by the nest entrance. When she enters (inset), a trembling movement by the male elicits spawning. Subsequently the male enters the nest and fertilizes the eggs (not shown). (After Tinbergen, N.: The Study of Instinct. London, Oxford University Press, 1951.)

time course and there is a question as to whether they are indeed measuring the same thing (Fig. 24–25). It is essential, therefore, to employ more than one measure. Furthermore, different kinds of motivated behavior require different yardsticks. This is exemplified by studies of the mating behavior of the stickleback.

The stickleback has a complex pattern of reproductive behavior (Fig. 24–26). After the male has migrated and staked out his territory, he builds a nest and courts any egg-laden female who enters his territory by performing a "zigzag dance." She then follows him, and he leads her toward the nest at which point he directs his head to the entrance. The pointing stimulates the female to enter the nest. She in turn stimulates him to "tremble." His trembling stimulates her to spawn. He then fertilizes the eggs, and stimuli from the eggs terminate his sexual behavior. His aggressiveness now returns. The strength or tendency toward the different components of this behavior are susceptible to measure. Sex and aggression can be measured by presenting to a male on his home territory another fish confined in a glass tube. If it is a rival male in nuptial colors, the resident male tries to bite him. If it is a receptive female, he performs the zigzag dance. Counts of the bites or zigzags reveal the respective strengths of the male's tendencies (Fig. 24–27).

In short, levels of motivation are related to changes in states in the central nervous system which are brought about by such nonneural factors as hormones, blood composition and blood temperature. Specific environmental stimuli influence or trigger motivational behavior. Motivational behavior is a response to a need to restore **homeostasis**. When balance has been restored, there is generally sensory feedback signaling that the need has been satisfied.

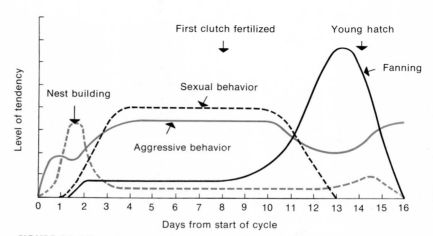

FIGURE 24–27   Long-term fluctuations in the tendencies to perform nest building, sexual behavior, aggressive behavior and nest ventilation ("fanning") during the reproductive cycle of the male stickleback. (From Sevenster, P.: Behavior (Suppl.) 9:1, 1961.)

SECTION 24-8

CYCLIC FLUCTUATIONS

There are other powerful fluctuations in behavior, most often cyclic, that are still not clearly understood. Many animals exhibit **circadial** (approximately daily), **monthly** or **seasonal** changes that are associated with specific environmental events but, once set, can persist for long periods of

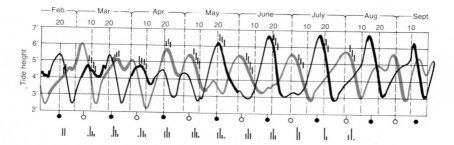

FIGURE 24–28   *Tidal and lunar periodicity. Bottom, A photograph (by flash) of a remarkable phenonenon exhibited by the grunion (Leuresthes tenuis) on the southern coasts of California. The grunion wriggle out of the waves at the highest points during the highest (night) high tides of the lunar cycle. Momentarily stranded on the wet beach between waves, the females wriggle down into the sand tail-first. (Note the females with only their heads protruding from the sand.) In this position the female lays her eggs while the male, curled around her body, ejaculates sperm which travel over the female's wet flank to reach the eggs. Both male and female return to sea with the next wave that covers them. The entire grunion population's egg laying is thus synchronized, being restricted to the two periods of highest tides in the lunar cycle. Furthermore, although there are two high tides each 24 hours, the grunion limit their activity to the night high tide. The eggs buried in the sand, hatch as young fish in time to go to sea with the next set of very high tides about two weeks later. Top, A chart showing the occurrence of grunion egg laying in relation to tidal and lunar cycles at La Jolla, California. The heights of high tides (in feet) about 24 hours apart have been connected by smooth lines. The two tides each day yield the two series of curves; the night tide is indicated by the heavier line. Grunion activity is represented by short vertical bars above the curves. Moon phases are given below (the black circles indicate new moon, and the white circles full moon). The vertical bars under each set of high tides show the relative intensities of individual runs on successive nights. (From Simpson, G. G., and Beck, W. S.: Life. New York, Harcourt, Brace & World, Inc., 1957.)*

time in the absence of environmental cues. Circadial rhythms are common-place. Some animals are nocturnal, some *diurnal* and some *crepuscular* (twilight). Even parasites within the bodies of their hosts have persistant rhythms. The nematode worm that causes elephantiasis in man in Africa resides in the circulation of the deep tissues during the day and migrates to the peripheral circulation at night. This behavior is synchronized well with that of the nocturnal mosquito that is an intermediate host and trans-mits the parasite from one man to another. A similar nematode parasite, *Loa loa*, infecting natives on Pacific islands, lives in the deep circulation at night and in the peripheral circulation in the daytime. It is transmitted by a day-flying fly.

A number of animals have bimonthly or monthly *lunar cycles.* The pre-cision of these cycles can be astounding. The Atlantic fireworm swarms in the surface waters around Bermuda for three or four months each summer at the full moon and for 55 minutes after sunset. The Palolo worm of the Pacific swarms and breeds at the break of dawn at the third quarter of the October and November moons. The grunion, a small pelagic fish of the Pacific coast of the United States, swarms from April through June on those three or four nights when a spring tide occurs. At precisely the high point of the tide the fish squirm onto the beach, deposit eggs and sperm in the sand, then rush back to the sea (Fig. 24–28). By the time the next tide reaches the spot 15 days later, the young fish have hatched and are ready to go to sea.

The *menstrual cycle* of man and apes is another example of a lunar cycle. Other animals have different breeding cycles. Dogs typically come into estrus twice a year. Domesticated cattle, chickens, rabbits and laboratory rodents are polyestrous; they produce many successive batches of eggs throughout the year. Among these various animals some breed in the spring (most fishes, birds and mammals), while others breed in the fall (deer, sheep and so forth).

While some external stimulus may set the biological clock for many behavioral rhythms, it does not control the tempo. The rhythm is endo-genous. When animals are placed in a constant environment, their rhyth-mic behavior continues. Crabs that normally become active at low tide continue to do so in time with the tide at their home territory even though they are no longer there. They act as though they possessed within their bodies tide charts giving local latitude, longitude and time of year. Animals that are normally active during daylight continue to mark off night and day appropriately even though they are being kept in constant light or darkness.

The keying of these endogenous rhythms to the local environment is obviously of supreme adaptive importance. Although the cycles are set and persistent under constant conditions, they can be reset if necessary. If spring breeding animals are transported from one hemisphere to the other, their cycle eventually shifts over to coincide with the occurrence of spring in their new home. In short, all cycles are set by local time.

SECTION 24-9

MATURATION OF BEHAVIOR

The variations and fluctuations in behavior that have been considered thus far are of short duration. Other changes reflecting the maturational process develop more slowly and tend to be irreversible. The behavior of the baby is different from that of the adolescent, which in turn differs from that of the adult. The behavior of a skilled person is different from that of an unskilled person. Although it is difficult to make clear dichotomies, there is no doubt that some changes in behavior are due to **maturational processes** and some to **experience**. Some maturational processes clearly involve developments in the nervous system. Consider the **Babinski reflex** in humans and chimpanzees. When the sole of the foot of a normal adult human is stroked, *flexion* of the toes results. In babies less than eight or nine months old, the same kind of stimulation causes *extension* of the great toe. The difference in response is correlated with the level of maturation of the nervous system, especially with development of the myelin sheath in the **pyramidal tract.** The sheath has not yet developed in young babies. The **moro reflex** is another example, If a young baby is placed flat on its back on a pillow and the pillow suddenly hit, the baby responds by throwing out both arms and spreading its fingers in a gesture resembling an embrace. The reflex occurs in response to a **vestibular reaction** caused by the head being shaken. After the first three months of life it disappears. The disappearance is correlated with maturation of the nervous system.

These reflexes have even been observed in fetuses. They persist until maturation, and practice enables the baby to control movement and position. They may appear in adults when there is a pathological loss of function of higher inhibitory centers.

Even more striking changes associated with the development of the nervous system are seen in insects. A caterpillar obviously behaves as a caterpillar, but when the nervous system and muscle system are reorganized at metamorphosis, the behavior becomes that of a moth or butterfly. The butterfly is able to fly as soon as it has emerged from the chrysalis and its wings have hardened. Performance of the new pattern of behavior, flying, is strictly a matter of development.

SECTION 24-10

HORMONAL REGULATION OF BEHAVIOR

Another category of maturational changes involves the growth and development of endocrine glands whose products, the **hormones,** alter the internal milieu in which the nervous system works. Hormones may alter behavior by enhancing the development of organs employed in behavior, by affecting the early development of the nervous system, by producing changes in peripheral organs concerned with the generation of sensory input to the central nervous system, by influencing special centers in the central nervous system and by exerting rather general nonspecific effects on the animal as a whole. Exemplifying the first category is the development at sexual maturity of the comb employed by cockerels in display.

Hormonally induced neural growth occurs preeminently in antenatal or prehatching life and is correlated with subsequent behavior. Hormones often cause rapidly accelerated growth in "target" organs, the organs upon which their potency is specifically registered. **Estrogen** causes growth of the uterus and mammary glands and vaginal epithelial changes in female mammals. In birds **estrogen,** often plus **progesterone,** causes the oviduct to enlarge, and, in some species, ventral feathers to be shed while the underlying skin vascularizes to form a brood pouch. Notice of some of these changes is undoubtedly fed back to the central nervous system from sensory nerves in the respective organs.

One well studied example of hormonally induced changes in peripheral organs affecting behavior is the effect of **androgens** on the sensitivity of the rat's penis. Ejaculation by male rats depends upon stimuli received through the penis. One way in which androgens increase the sensitivity is by increasing the density of cornified papillae between the folds of the glans penis. Another example is the effect of **prolactin** on parental feeding by doves. "Pigeon milk," on which doves feed their young, is actually sloughed epithelium of the crop. Prolactin causes the crop to enlarge just before the young hatch. Sensory stimulation from the crop at this time is an important factor in initiating parental feeding. Support for this interpretation has been derived from experiments in which the application of local anesthesia to the crop has reduced parental feeding.

Effects on behavior resulting from the influence of hormones on general body conditions are exemplified by experiments in which **thyroidectomy, gonadectomy, adrenalectomy** and **hypophysectomy** each enhance the nest-building behavior of rats. Each of the operations mentioned reduces body temperature, and nest-building is known to be enhanced by low temperature.

Generally hormones influence behavior by acting directly on some specific area of the central nervous system. In mammals the effect is primarily on the hypothalamus. Normal sexually mature female cats have three estrous cycles a year. When in estrus, a female cat in the presence of a male will elevate the rump, deflect the tail to one side and make treading movements with the hind legs. A cat in **anestrus** will fight if a male approaches too close. Mating behavior of the cat in heat is accompanied by morphological changes in the reproductive system. Small doses of estrogen injected into the blood stream of a spayed cat over a period of time will induce the expected morphological changes but will not cause the cat to behave as if in heat. These results suggest that under normal circumstances the hormonal effect on behavior is not mediated via **peripheral feedback.** If, on the other hand, solid **stilbestrol dibutyrate**, a chemical that mimics the action of the normal hormone, is implanted into their posterior hypothalamus, full development of sexual behavior follows even though the genital tract retains its anestrous state. Similarly sexual behavior is possible in spayed rats when estradiol is implanted into the **medial-basal preoptic** and anterior **hypothalamus.** Figure 24–29 diagrams some of the relationships between hormones and behavior.

Thus the role of hormones in the development and expression of behavior has been demonstrated in many vertebrates by castrating young animals and observing the absence of certain adult or sexual behavior later on, by injecting hormones into sexually immature animals and inducing sexual behavior prematurely and by restoring sexual behavior in spayed animals by injecting or implanting the appropriate hormones.

The emphasis thus far has been on the hormonal effect on the development of behavior from the young to the adult. Everything that has been said applies as well to recurrent patterns of behavior that occur in animals having cyclic reproductive periods. This refers to the **seasonal reproductive cycles** of fishes and birds and the estrous cycle of mammals. The periodic changes in behavior are related to these reproductive cycles in which hormones play an important role.

The interaction between hormonal states and environmental stimuli is very complex. Hormones influence animals' responses to environmental stimuli, and environmental stimuli influence the production of hormones. For example, the **migration** up rivers in the spring to breeding grounds by male three-spined sticklebacks is due to changes in pituitary and thyroid secretion. The influence of environment on hormones is seen also in the reproductive behavior of ring doves. In these birds egg laying can be stimulated by the injection of hormones. If a female is isolated from all

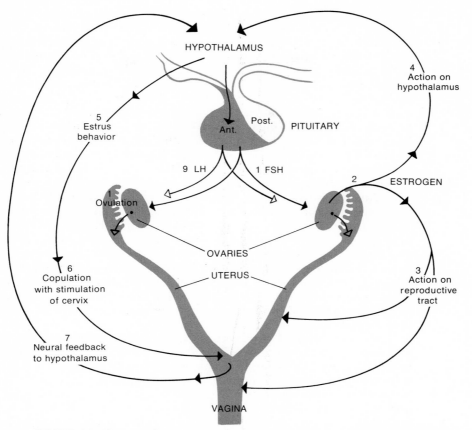

FIGURE 24–29   A diagrammatic representation of some of the hormonal and behavioral events during the reproductive cycle of an "induced ovulator," such as a cat or rabbit. The numbering of arrows indicates the approximate sequence of events—there is probably considerable overlapping in some cases. (From Manning, A.: An Introduction to Animal Behavior. Reading, Massachusetts, Addison-Wesley Publishing Company, Inc., 1967.)

other birds, her hormonal level remains low and she will not lay. If, however, she is permitted only to see a male, the sight activates centers in the hypothalamus which control pituitary secretion. This effects the release of **FSH** and **LH** which stimulate growth of the ovary which then secretes estrogen. Egg laying follows. Incubation behavior follows a similar pattern. A female alone in her cage with a nest full of eggs will not incubate unless she is injected with progesterone. If, instead of receiving an injection she is given the companionship of a male, the two begin courting. Courting apparently stimulates the production of progesterone. Incubation, by both birds, soon follows.

SECTION 24-11

ROLE OF EXPERIENCE

All of the foregoing examples describe changes in behavior related to structural and physiological changes that proceed in an orderly and predictable manner as the organism grows or cycles. At the same time it must be remembered that the organism is practicing with its behavioral machinery and acquiring experience as it ages. It is extraordinarily difficult, therefore, to separate the contributions of strictly maturational processes and of experience to the development of behavior. Indeed the two are so intimately interwoven that there may be little profit in attempting a separation.

Many **species-specific patterns** of muscle activity appear fully perfected at birth, and it is tempting to consider that their appearance represents only the coming of age of neuromuscular mechanisms and neural integrative circuits. Newly hatched nidicolous birds beg before they have ever experienced parental feeding and when they hear their parents' alarm call for the first time, "freeze" in the posture characteristic of their species. The dromedary performs characteristic movements without practice at specified times (in minutes) after birth: chewing, 10, rolling on the ground, 74, sucking, 100, jerking the head, 120, kicking, 156, yawning, 160 and tail beating, 198. The Babinski and Moro reflexes of babies have been mentioned. On the perceptual side babies show preferences for facelike shapes (Fig. 24–30). Baby chicks hatched in darkness and tested immediately upon exposure to light on objects graded in angularity from spherical to pyramidal pecked 10 times more often at the spheres. Apparently the ability to perceive shape, size and three dimensions is already present.

There have been numerous experimental attempts to assess the importance of maturation, as opposed to practice, in the development of behavior. One classical experiment consisted of rearing larvae of salamanders under continuous anesthesia from before the time the peripheral nervous system had developed until the time that a control group had hatched and was swimming well. As soon as the effects of the anesthetic had disappeared, the experimentals swam as well as the controls even though they had not had the benefit of all the prehatching practice in the egg capsule that the controls had.

A similar experiment concerned young swallows that had been reared in cages so narrow that they could not practice their wing move-

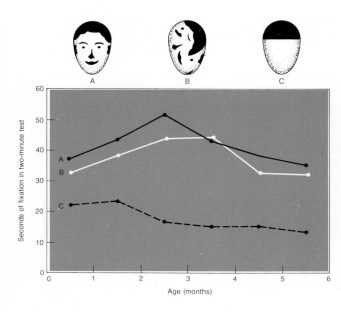

FIGURE 24–30 Adaptive significance of form perception was indicated by the preference that infants showed for a "real" face (A) over a scrambled face (B), and for both over a control (C). The results charted here show the average time scores for infants at various ages when presented with the three face-shaped objects paired in all the possible combinations. (From The Origin of Form Perception by R. L. Fantz. Copyright © May 1961 by Scientific American, Inc. All rights reserved.)

ments. When released they flew just as expertly as normally reared young birds of the same age.

Two other experimental approaches also reveal the importance of the developmental process. The first involves surgical manipulation of neuromuscular circuits. When the right and left forelimb buds of embryo salamanders were exchanged, the resultant adult limbs faced backwards. Since the correct nervous connections developed between the grafted limbs and the spinal cord, these limbs moved backwards while the normal hind limbs moved forward. The animal was never able to compensate for this artificial developmental defect.

The other experimental approach bearing on this question involves the raising of hybrids. Some species of lovebirds collect nesting material by cutting long thin strips from leaves, bark or paper. One species carries the strips to its nesting hole in its beak; the other species tucks the strips among the feathers of its rump. Hybrids attempt to tuck but invariably fail because they tuck in the wrong place or forget to let go of the strip after tucking it. It takes the hybrids nearly two years to learn to cope with this inherited pattern (Fig. 24–31).

As convincing as these examples seem it is still difficult to rule out practice completely, except in the case of the insects, because the existence

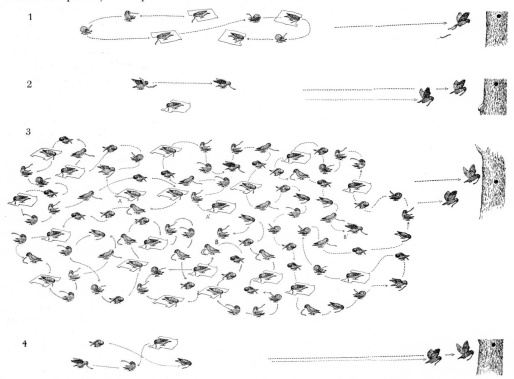

FIGURE 24–31   Hybrid lovebird inherits patterns for two different ways of carrying nest-building materials. From the peach-faced lovebird (1) it inherits patterns for carrying strips several at a time, in feathers. From Fischer's lovebird (2) it inherits patterns for carrying strips one at a time, in the bill. When the hybrid first begins to build a nest (3), it acts completely confused. Colored lines from A to B and black lines from A' to B' indicate the number of activities necessary for it to get two strips to the nest site, a feat achieved only when the strips are carried singly, in the bill. It takes three years before the bird perfects its bill-carrying behavior (4), and even then it makes efforts to tuck its nest materials in its feathers. As the bird gains experience it becomes more and more proficient in this activity, which, however, never results in successful carrying. (From Behavior of Lovebirds by W. C. Dilger. Copyright © Jan. 1962 by Scientific American, Inc. All rights reserved.)

of intraegg and intrauterine practice cannot be excluded. Behavior does not begin at hatching or at birth. Developing embryos behave. Chicks in the egg respond to warning cries of the cock by stopping movement. Human babies undergo experiences in utero that influence behavior after birth. There is, for example, a correlation between the position of the fetus during the last week preceding delivery and its postnatal behavior and leg reflex patterns. A baby that developed in a normal position flexes its legs if it receives a pin prick on the sole of its foot, but extends its legs if the sole is touched gently. Breech babies whose legs were extended, extend in response to a pin prick; breech babies whose legs were flexed, flex in response to a pin prick. If a breech baby is turned during the last few weeks of uterine life and born in the normal position, its reflex patterns are normal.

The importance of practice and experience in the development of behavior is strikingly illustrated by studies of *sensory-motor coordination.* When an animal is subjected to prolonged sensory or motor deprivation, performance of motor and perceptual skills declines. Some of the decline may be traced to pathology, but a large measure results from lack of practice. The same mechanisms are involved in the development of normal performance in young, the maintenance of normal performance once it has developed and adaptation to marked changes (e.g., learning motor skills while wearing prisms that turn the world upside down). For normal development there must be visual stimulation correlated with movement. This dependence was strikingly demonstrated by arranging two kittens that had been raised from birth to 8 weeks in total darkness as shown in Figure 24–32. One was carried about in a little cart pulled by his litter mate who could move more or less freely. Both kittens received the same visual stimulation and the riding kitten received some bodily stimulation from the movements of his partner; however, he could not walk. After a total of 30 nonconsecutive hours of training both kittens were tested. The walking one fended off objects with his paw, avoided little cliffs, and in general performed normally. The riding one was totally inept; however, when finally allowed to walk for himself he eventually performed normally.

The necessity for practice varies from one species of animal to the next and from one type of task to the next. Whereas the development of distance perception depends on visual experience in kittens and primates, it does not in chicks. Pattern discrimination is not markedly impaired by visual deprivation in birds and rodents but is in primates.

Quite apart from practice or self-learning is the business of *imitation,* learning from others. In nature, growing animals are constantly exposed to the behavior of their parents and siblings. One way to investigate the part played by imitation in the development of behavior is to raise animals in isolation. Difficulties in interpreting results arise, however, because isolation frequently involves debilitating sensory deprivation. Nevertheless, the results do demonstrate the great importance of learning.

The development of singing in birds is another fine example. Each species of bird has a repertoire of songs (often 15 to 18), so characteristic of the species that an accomplished bird watcher need never see a bird to identify it. The different sounds that a bird produces are employed in distinct functional contexts, i.e., in territorial defense, as alarm calls and so on. In most cases they develop without being learned from other individuals. Birds raised in isolation sing the songs characteristic of their species. Male whitethroats (*Sylvia communis*) in isolation develop their songs in the usual way by adding more and more notes as time goes on. Even though each male is auditorially isolated from all others, the songs of all individuals

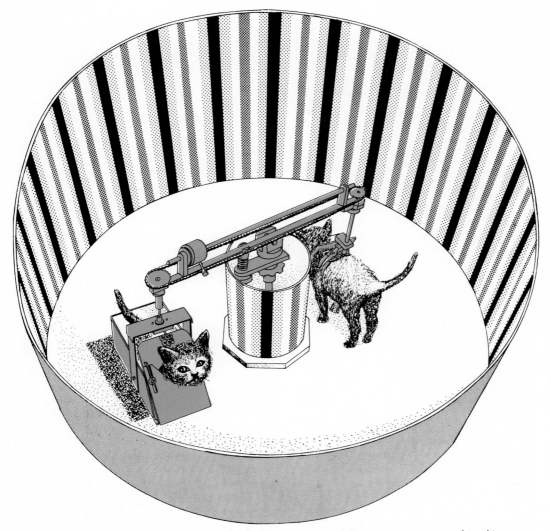

FIGURE 24–32   Active and passive movements of kittens were compared in this apparatus. The active kitten walked about more or less freely, its gross movements were transmitted to the passive kitten by the chain and bar. The passive kitten, carried in a gondola, received essentially the same visual stimulation as the active kitten because of the unvarying pattern on the wall and on the center post. Active kittens developed normal sensory-motor coordination; passive kittens failed to do so until after being freed for several days. (From Plasticity in Sensory-Motor Systems by R. Held. Copyright © Nov. 1965 by Scientific American, Inc. All rights reserved.)

develop in synchrony. Some species of birds depend upon imitation for the normal development of their song. Chaffinches reared in isolation sing only a simple song. If a number of young are reared together but isolated from adults, their songs are more complex probably because they counter-sing with one another. If they can hear an adult, they learn a complete repertoire. They also profit by being able to hear themselves sing. Both self-learning and imitation are important.

Rearing monkeys in social isolation has also revealed the importance of *environment* in shaping behavior. Rhesus monkeys deprived of rela-tionship with mothers or siblings never develop normal sexual and

FIGURE 24–33   Cloth and wire mother-surrogates were used to test the preferences of infant monkeys. The infants spend most of their time clinging to the soft cloth "mother" (foreground) even when nursing bottles were attached to the wire mother (background). (From Love in Infant Monkeys by H. F. Harlow. Copyright © June 1959 by Scientific American, Inc. All rights reserved.)

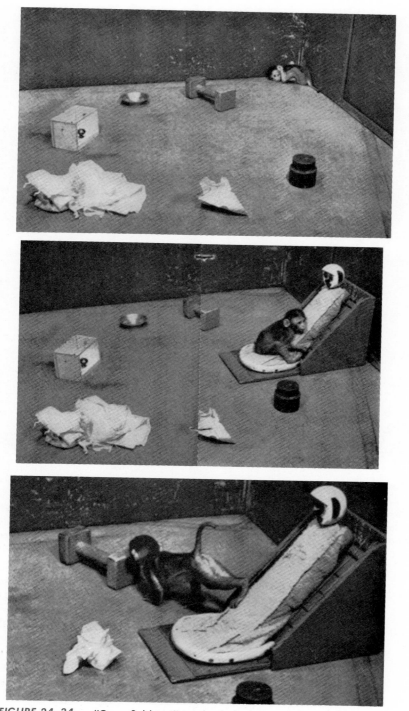

FIGURE 24–34   "Open field test" involved placing a monkey in a room far larger than its accustomed cage; unfamiliar objects added an additional disturbing element. If no mother was present, the infant would typically huddle in a corner (top). The wire mother did not alter this pattern of fearful behavior, but the cloth mother provided quick reassurance. The infant would first cling to her (center) and then set out to explore the room and play with the objects (bottom), returning from time to time for more reassurance. (From Love in Infant Monkeys by H. F. Harlow. Copyright © June 1959 by Scientific American, Inc. All rights reserved.)

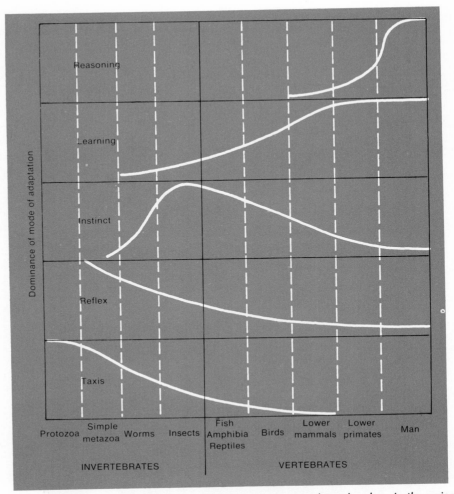

FIGURE 24-35   *Schematic portrayal of the changes that take place in the major modes of adaptive behavior in phylogeny. Reading from left to right, it shows the relative development of different modes of adaptation; reading up and down, it shows the relative pattern of modes of adaptation at different levels of the phylogenetic scale. (From Dethier, V. G., and Stellar, E.: Animal Behavior. Englewood Cliffs, N.J., Prentice-Hall, Inc., 1961.)*

maternal relationships. They do not enter into normal social relationships with other monkeys, they tend toward homosexuality, and, if they eventually have families themselves, they neglect their own young. Similar abnormalities of behavior have been observed in other animals raised in isolation.

Studies of the relationship between baby monkeys and cloth and wire surrogate mothers (Fig. 24-33) revealed that bodily contact and visual stimulation are as important psychologically as milk is physiologically. Young monkeys formed strong emotional attachments to cloth mothers, ran to them when frightened by a mechanical teddy bear beating a drum, and readily explored strange objects in their room as long as the "mother" was there as a psychological refuge. Monkeys that had no "mother" fled to a corner of the room, clasped their heads and bodies and rocked convulsively back and forth (Fig. 24-34). Their actions resembled the autistic behavior of neglected or institutionalized children.

From all the considerations so far it is clear that the ability to modify behavior as a result of experience has great adaptive value. This capacity assumed an ever greater role as animals evolved. The behavior of most invertebrates is built around reflexes, fixed-action patterns (instincts) and stereotyped orientation (phototaxis, geotaxis and so on). These modes are seen to be replaced to an ever greater extent by learning as one ascends the vertebrate scale (Fig. 24–35).

SECTION 24–12

LEARNING

Learning generally refers to the process of instituting relatively permanent changes in behavior and involves the modification of behavior by experience and its retention and its recall. It is usually defined by exclusion, i.e., it is modification of behavior not accountable by sensory adaptation, central excitatory states, endogenous rhythms, motivational states and maturation. It should not be thought of, however, as a unitary phenomenon. It exhibits many forms. It undoubtedly has different characteristics and involves different mechanisms in different species of animals. It is convenient to divide learning into several categories: habituation, classical conditioning, instrumental (operant) conditioning, trial and error.

Habituation is the loss of old responses rather than the acquisition of new ones. An animal gradually stops responding to stimuli that have no relevance to life, that are neither rewarding nor punishing. The taming of animals represents a most commonly encountered form of habituation. Animals that were once wild may become "accustomed" to the presence of human beings who do not harm them. Crows in a field will eventually ignore a scarecrow. Habituation is distinguished from fatigue and sensory adaptation by its relatively long persistence. The waning may never reverse itself. On the other hand, it may be rapid. The alarm calls given by chaffinches when mobbing an owl cease almost completely after 30 minutes of stimulation, i.e., the birds are completely habituated; however, if the owl is removed, recovery is only about 55 per cent. This observation suggests that recovery may have a rapid and a slow phase and that habituation is not really a "simple" type of learning. The nature of the processes underlying it is obscure.

All other types of learning consist of strengthening responses that are "significant" to the animal. The simplest kind is *Pavlovian* or *classical reflex conditioning.* It is *associative learning*—the animal learns to associate a reward or a punishment with a heretofore irrelevant stimulus. Pavlov trained dogs to associate the sound of a bell with meat powder placed in the mouth. When meat powder was placed in the dog's mouth, the dog salivated. When this was done a bell was rung. Eventually the dog salivated at the sound of the bell even when no meat powder was presented. Pavlov termed the salivation to meat (the *unconditioned stimulus,* UCS) the *unconditioned reflex* (UCR), and the salivation to the bell (the *conditioned stimulus,* CS) the *conditioned reflex* (CR).

Another kind of associative learning is *instrumental learning* (also called *trial and error, conditioned reflex type II* and *operant conditioning.* In this kind of learning animals have some control over the stimuli they receive, and sometimes the responses they employ. Their behavior controls the situation. In the simplest case a rat presses a bar to obtain food. Food is the reward for pressing the bar. The rat determines how often he is to be rewarded. He may determine what response he will employ; i.e., he can press the bar with his nose or his foot. In the usual experiment the animal begins by responding in a variety of ways that are natural to him. The experimenter then decides which of these various responses will be rewarded. Thus the rat may be rewarded only if he presses the bar with his leg. As this response is rewarded, the others eventually cease. The **reinforcement** or strengthening of the response can be in the form of a **reward** (reward training), escape from an uncomfortable stimulus **(escape training)** or the possibility of avoiding a noxious stimulus entirely **(avoidance training).**

Maze running is a more complicated form of instrumental conditioning, because the animal is required to make successive discriminations at the numerous choice points. Here the animal has the possibility of making errors. Correct responses are rewarded or incorrect responses are punished. Other types of trial and error experiments consist of placing animals in boxes where they have to manipulate a variety of bars or locks in order to escape (Fig. 24–36). Trial and error is probably one of the most common forms of learning experienced by animals in nature.

In the many conditioning experiments that have been conducted on many different species of animals from worms to man, certain general relationships have been observed that enable us to characterize more accurately the nature of this behavior. First, time relations are critical. If the UCS precedes the CS, there is little or no conditioning. If the CS precedes the UCS by more than a second, conditioning is established with difficulty (Fig. 24–37). Thus **contiguity** is one requirement for associative learning.

Repetition is another relevant feature. The more often the CS and UCS are paired, the stronger is the CR acquired. The amount of saliva produced by Pavlov's dogs increased with each successive trial. A rat learning to run a maze makes fewer and fewer errors with each successive trial. The rate at which an animal learns is described by a learning curve generated by plotting the time to completion of task at each trial or by the number of errors committed at each trial (Fig. 24–38). This is one way of measuring learning capacity under different sets of conditions or of comparing the performance of different animals.

Repetition finally produces a degree of learning beyond which there is no improvement. While training beyond this point (overtraining)

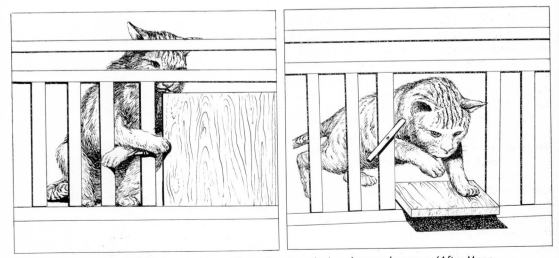

FIGURE 24–36   A cat operating a bar to unlock a door and escape. (After Munn, N. L.: Evolution and Growth of Human Behavior. 2nd ed. Boston, Houghton Mifflin Company, 1965.)

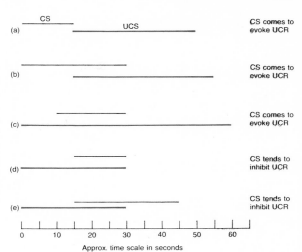

FIGURE 24–37   *The effect of the sequence of stimuli upon the formation of a conditioned reflex. In each case the upper, thin line denotes the duration of the conditioned stimulus (CS), the lower, thick line that of the unconditioned, reinforcing stimulus (UCS). The results are given on the right; note that the CS must not end with or persist beyond the UCS if a positive conditioned reflex is to be established. If it does so, then the CS will tend to inhibit the response to the UCS. (From Konorski, J.: Conditioned Reflexes and Neuron Organisation. New York, Cambridge, University Press, 1948.)*

causes no improvement, it does make the response more resistant to extinction. **Extinction** is the decay of learning in the absence of reinforcement. Without reinforcement the conditioned response may disappear entirely; however, if the animal is left alone and later presented with the CS, the conditioned response returns. This spontaneous recovery is only partial and is more easily extinguished than before. Pavlov believed that extinction is new learning that interferes with CR. It is certainly true that a distracting stimulus presented during early training causes a reduction in CR. It inhibits it. If the distracting stimulus is given during extinction, the CR is enhanced. Presumably it removes the inhibiting effects of the new learning by inhibiting that new learning. Pavlov termed this disinhibition.

Two other concepts associated with conditioning are generalization and discrimination. **Generalization** refers to the fact that an animal conditioned to one stimulus will also be conditioned to closely related stimuli. The closer the two resemble each other, the better the response to the new one. If, for example, a dog is conditioned to a 1000-cycle-per-second tone, it will respond somewhat to a 500- or 1500-cycle-per-second tone. Response to a 100- or 2000-cycle-per-second tone would be poorer. **Discrimination** refers to the fact that the dog can be made to respond to one particular tone (e.g., a 100-cycle-per-second tone) by rewarding it for that and punishing it for other tones. This technique has been employed in investigations of the fineness of sensory discrimination. This was the technique employed by von Frisch to measure honeybees' ability to discriminate different colors.

A type of learning whereby an animal learns without reinforcement and later puts the information to good use has been called **latent learning.** Exploration falls into this category. Most animals spend a lot of time exploring new situations and environments. In this way they "familiarize" themselves with their surroundings and use the information later. Even hymenopterous insects (ants, bees, wasps) make orientation flights around a nest they have just constructed and get a fix on its position. Some wasps can learn essential landmarks around the nest in an orientation flight lasting only nine seconds.

Insight learning, or **reasoning,** is considered to be the highest form of learning. It has been defined as the ability to combine two or more isolated experiences to form a new experience tailored to a desired end. Detour

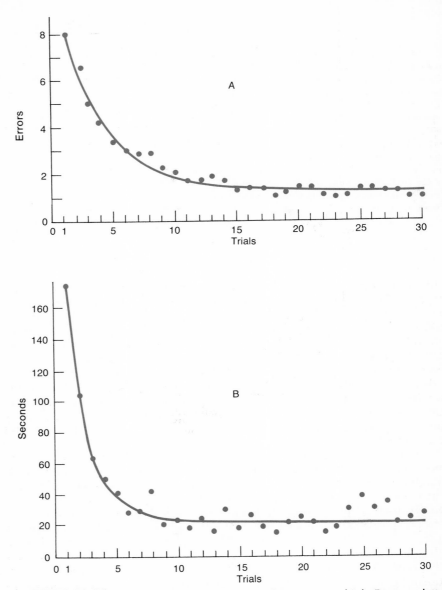

FIGURE 24-38   *The learning curves of rats learning a multiple T maze plotted (A) by errors, and (B) by time to reach the goal box. Each circle represents the mean value for 47 animals. (From Woodrow, H.: Psychol. Bull. 39:1, 1942.)*

problems have been the device most frequently employed to test an animal's ability to utilize insight. If an animal blocked from food that he can see selects a detour correctly on the first try, he probably has "sized up the situation." Of all animals only monkeys and chimpanzees succeed on the first exposure; all others, even smart dogs and raccoons, fail at the beginning. A classic example of insight learning is that of the chimpanzee that would pile up boxes or fit two bamboo sticks together to obtain bananas hung out of reach. The chimpanzee "figured out" the solution without being taught (Fig. 24-39).

Other tests of problem-solving include *"oddity-principle learning,"* *"delayed reaction"* and *"learning set."* In the first, the animal must pick out

FIGURE 24–39   *Chimpanzee solves the problem of getting a banana that is out of reach by stacking boxes on top of one another. (From Maier, N. R. F., and Schneirla, T. C.: Principles of Animal Psychology. New York, Dover Publications, Inc., 1964.)*

the odd member in a set of three regardless of what the objects are (Fig. 24–40). In the second, the animal is allowed to see which of two cups has food hidden under it. Later, after a delay, he has to lift the correct cup. In learning set, an animal is presented with two dissimilar objects, one of which always has food under it. Repeated trials are necessary before the animal goes immediately on first try to the object with the reward. Now the experiment is repeated with two different objects. Again a number of trials are required before the animal goes directly to the reward. This identical discrimination problem is repeated with many different pairs of objects. Each problem is just as difficult as the first, but the animal performs better and better. Eventually presented with two objects the animal lifts one. If food is there, he chooses that one on every presentation. If food is not there, he lifts the other object and then always chooses that one.

FIGURE 24-40   *A rhesus monkey performing an oddity-principle test. (From Stone,*
K. P. (ed.): Comparative Psychology. Englewood Cliffs, New Jersey, Prentice-Hall, Inc.,
1951.)

In other words he finally "gets the idea" (formed a learning set) that whatever object the food is under it will always be so. If he chooses correctly
the first time, there is no need to look elsewhere. If he misses the first time,
there is only one other alternative and that is the one he chooses from
then on. The rate at which various mammals can form discrimination
learning sets are shown in Figure 24–41.

    Imprinting is a form of learning that is considered by some to belong
to a distinct category. It was first described in birds, but is now known to
occur in sheep, goats, deer, buffalo and other animals whose young are
mobile at birth. It is a phenomenon whereby young become "attached" to

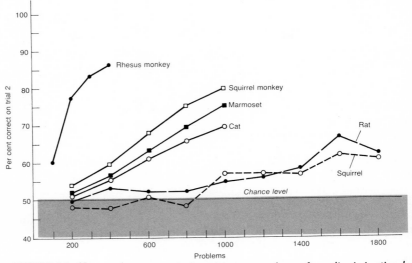

FIGURE 24-41   *The rate at which various mammals can form discrimination learn-*
ing sets. With each new problem the animal's choice on the first trial has to be random,
but if it has learned the principle behind the problems, trial 2 should be correct. Note how
long it is before the scores of rats or squirrels on trial 2 become better than chance or 50
per cent. Many monkeys reach almost 100 per cent within 400 problems. (From Warren:
The Behavior of Non-human Primates. Vol. 1. New York, Academic Press, 1965.)

the first moving object they see and react to it as they would toward their actual mother. Their reaction usually takes the form of following, although following is not an essential part of imprinting (Fig. 24–42). There is almost no limit to the kinds of visual stimuli that will induce imprinting. Sound alone suffices in some species; however, sound and movement together usually constitute the most effective stimulus.

It is characteristic of imprinting that stimuli are effective only if presented within a limited sensitive or **critical period.** The duration of the critical period depends upon which of several behavioral criteria are measured. If the criterion is simply following, the critical period lasts about 70 hours; if the criterion is the formation of a lasting attachment, how close a duckling will discriminate and stay close to a decoy, the period lasts only about 30 hours. Mallard ducklings begin to imprint immediately upon hatching, but the strength of the response declines with age (Fig. 24–43). By varying the rate at which a decoy travels and by placing hurdles in the path of the young, it was discerned that the strength of imprinting increases with the effort expended.

As young birds become older the decoy ceases to be a "mother"; however, at sexual maturity the decoy becomes an object of courting and copulation. Thus Lorenz became a sex symbol to the geese imprinted upon him. Young birds reared with foster parents always tried to mate at maturity with the foster species.

Despite the great research efforts that have been expended in the search for the **"engram"** or **"memory trace,"** the problem remains unsolved. Learning in some way involves the storage of information in the nervous system and its retrieval on demand. Different processes are undoubtedly involved. As every student knows, it is possible to have learned something very well indeed but be unable to recall it at the time of an examination, although later it may be recalled without difficulty. We frequently have things "at the tip of our tongue." There is also evidence that memory, at least for mammals, may be of two varieties, **short-term** and **long-term.** This is strikingly demonstrated in various types of **amnesia.** A person may be

FIGURE 24–42   Remote-controlled decoy was used by Hess and his colleagues in other imprinting experiments. Here both decoy and duckling move about freely rather than on a runway. (From "Imprinting" in Animals by E. H. Hess. Copyright © March 1958 by Scientific American, Inc. All rights reserved.)

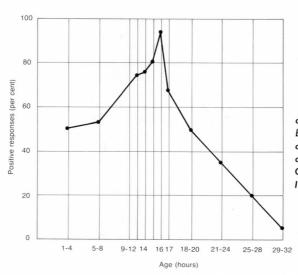

able to remember accurately something that happened to him 50 years earlier but not remember what occurred within the previous five minutes. Conversely he can remember recent events but not past events. Somewhere in the nervous system changes are occurring which are concerned with learning and memory, storage and recall. Whether they directly involve DNA, RNA and proteins as some theorists believe remains to be seen.

SECTION 24–13

SOCIAL BEHAVIOR

The environment to which animals must adjust includes not only the inanimate but also other individuals, both of the same species and of other species. In the broadest sense, any interaction between two or more individuals constitutes **social behavior.** Usually, however, social relationships imply interactions among members of the same species.

Mere presence of more than one individual does not mean that the behavior is social. Many factors of the physical environment bring animals together in **aggregations,** but interactions, if they occur here, are apt to be circumstantial. A light at night is a stimulus that causes large numbers of moths and other insects to collect around it, humidity beneath a log is a common stimulus causing aggregations of wood lice, and a water hole in the dry savannah is a common focus for large numbers of birds and mammals. None of these groups are social.

The basis of social life is fundamentally **mating behavior.** It may also be family life, group life and, strangely enough, fighting. Except in the case of parthenogenetic species, male and female must come together in a cooperative fashion. Although this accomplishment might appear at first glance a simple business, it does in fact face a number of obstacles. Elaborate patterns of behavior have evolved to overcome these.

First of all, eggs and sperm must be produced at the same time. This means that male and female must find each other even if fertilization is external, as in fish. It also means that both individuals must be in a reproductive state. The synchrony of time and place is accomplished by common stimuli from the external environment and by the stimulatory effect of one sex upon the other. Many fish, birds and mammals are brought into synchronous reproductive readiness through the agency of increasing day length, which acts on the pituitary to release in turn gonadal hormones affecting behavior. In the stickleback, the hormonal change predisposes the fish to migrate into shallow water where the rise in temperature and visual stimulation provided by vegetation releases reproductive behavior.

Orientation of the sexes in response to common environmental stimuli results in some remarkable feats of *migration* and *navigation.* Some of the most extensive studies have been of the salmon. These fish live in the sea until the time for spawning approaches. The five species of Pacific salmon migrate from the sea to the headwaters of fresh-water streams once in their life. They spawn and die. The Atlantic salmon returns year after year. At spawning time each salmon swims to the mouth of the same stream in which it was born and swims up, always making the correct turns at the forks of tributaries, until it arrives at the spawning ground. Elaborate experiments have shown that different streams have different odors derived from the country through which they flow. Salmon respond to these odors. The young salmon "remember" the characteristic odor of the stream in which they were hatched, and respond to this stimulus at spawning time. Similar but less well understood stimuli presumably guide eels across the Atlantic Ocean from the streams of Europe and America to their breeding grounds in the Sargasso Sea, whence the young return via the Gulf Stream to fresh water.

For many species of animals, however, the meeting of the sexes comes about as a result of responding to signals sent out by one of the partners. Chemical, auditory and visual stimuli are widely employed. The chemical attractants of female moths are legendary. Special abdominal glands secrete highly volatile compounds (*pheromones,* secretions employed in intraspecific communication) that are more or less species specific. The pheromone of the commercial silk worm moth (*Bombyx mori*) is an unsaturated, straight chain alcohol with 16 carbon atoms. Bombykol, as it is called, attracts males from downwind. Only one molecule hitting each olfactory cell every two seconds is needed to attract.

Vertebrates also deploy pheromones, either in the excrement or from glands on the feet, near the eyes or horns, near the tail or around the anus (Fig. 24–44). In addition to being *sex attractants* these may serve as *territorial markers* and *orientation signals.*

FIGURE 24–44   Male *Antelope cervicapra* marking a tree with the secretion of the scent-gland located in front of the eye. (After Hediger, H.: Bÿdr. tot de Dierk 28:172, 1949.)

Auditory signals are of prime importance among insects, frogs and birds. Crickets and long-horned grasshoppers have highly specific court-ship songs. Males produce the sounds by rubbing a scraper on one wing across a file on the other or similarly by rubbing the legs against the closed forewings (Fig. 24–45). Females respond only to songs of their own species. Similarly among frogs and toads there are highly distinctive songs that attract females to the breeding place in springtime. The songs of all these animals not only bring the sexes together but ensure that individuals meet others of their own species, thus preventing hybridization.

Visual signals are particularly characteristic of fish and birds. The male stickleback does a zigzag dance to cause the female to approach and

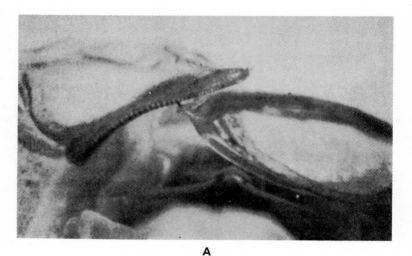

A

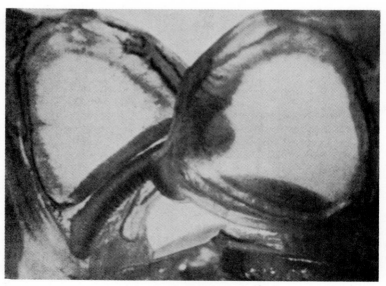

B

FIGURE 24–45   A, Pterophylla camellifolia *(Katydid). Enlarged rear view of* file in contact with scraper. B, Pterophylla camellifolia. *Enlarged view of file and scraper showing wing civer discs. (From Pierce, G. W.: The Songs of Insects. Cambridge, Mass-achusetts, Harvard University Press, 1948.)*

then leads her down to the waiting nest. Birds of the open spaces, tundra and prairies, have conspicuous specialized movements, such as lifting their wings and revealing bright under surfaces.

Once the sexes are together, persuasion and receptivity are necessary, especially when copulation is required. Copulation involves bodily contact and contact in all other contexts means capture by predators. During copulation animals are extremely vulnerable to attack and of the two sexes the female is usually the more vulnerable and the male the more aggressive. It is not surprising, therefore, that males usually do most of the courting, i.e., most of the persuading. Nor is it surprising that the females do most of the appeasing to assuage the pugnacity of the male. How behavior operates to achieve these ends is described in the following examples.

Mallard ducks, in common with certain other surface-feeding ducks, have 10 **courtship** poses (Fig. 24–46). These are performed by the male in his brilliant nuptial colors. Other ducks employ the same poses but perform them in different sequences characteristic of the species. Herring gulls signal readiness to copulate by tossing their heads up and down (Fig. 24–47). Black-headed gulls "appease" by head-flagging (Fig. 24–48), a movement by which the threatening brown face mask is turned away. **Appeasement** by the female is especially important in species in which the plumage of both sexes is nearly identical. The male tends to be very **aggressive** during the mating season and in some way the female must identify herself in order not to provoke attack. Females of some species accomplish this by acting infantile. The response of the male is frequently to feed the female. Feeding is a common feature of courtship in these species.

Most display features of fish and birds (threat and courtship movements) are presumed to have evolved from intention movements, displacement activities and autonomic responses. **Intention movements** are incomplete and slightly modified segments of patterns that occur complete in other contexts. For example, the **display movements** shown in Figure 24–49 are derived from flight take-off leaps. A **displacement activity** is a normal activity performed out of context. In conflict situations many animals do something that has no relevance to the situation. A person may bite his nails, a gull may preen when it is not necessary and a bird may drink when engaged in an unrelated activity. Preening has evolved into complex displays in some birds. Displays have also evolved from autonomic responses such as defecation and urination. In short, animals have evolved rituals that function as social signals.

In many species of animals, family life, a relationship between male, female and young, is a natural consequence of sexual relations. The basis of family relationships is the provision of shelter, food and defense for the young. Within this relationship there must be synchrony. Behavior must change as family relationships do. A bird must be broody when there are eggs in the nest, and must feed the young when they hatch. Broodiness in the gull depends not only on the presence of prolactin but also on the visual stimulus of eggs. During the early part of this period, a gull will refuse to incubate a pipped egg. Later, however, about the time when the young are ready to hatch, she accepts it. The change in behavior is related to a hormonal change plus the auditory stimulus of the young in the egg. Still later, the broken egg shells stimulate the gull to remove them.

If parents and young are to get along, certain tendencies must be suppressed. The young equal food from an adult point of view. To the young the parents equal predators. Some of the most interesting examples of suppression of inappropriate responses are observed among fish. Some species of cichlids carry the young about in the mouth. While the

FIGURE 24–46   Ten courtship poses *which belong to the common genetic heritage of surface-feeding ducks are here shown as exemplified in the mallard: (1) initial bill-shake, (2) head-flick, (3) tail-shake, (4) grunt-whistle, (5) head-up — tail-up, (6) turn toward the female. (7) nod-swimming, (8) turning the back of the head, (9) bridling, (10) down-up. (From The Evolution of Behavior by K. Lorenz. Copyright © Dec. 1958 by Scientific American, Inc. All rights reserved.)*

young are in the mouth, the adult does not feed. A most amusing incident involved a cichlid male of the species that carries the young each evening to a "bedroom," a pit dug in the sand. This male was collecting young, snapping up stragglers, when it spied a worm. After several moments of indecision, it spat out the young, snapped up and swallowed the worm, and then retrieved its young for transport to the pit.

Parents and young must react to one another in other ways, too. Many birds will not feed the young unless they gape or beg. Young are stimulated by the sight of the parent's head over the nest, vibration of the nest by the

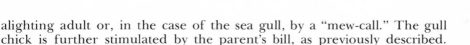

alighting adult or, in the case of the sea gull, by a "mew-call." The gull
chick is further stimulated by the parent's bill, as previously described.

Groups larger than families may consist of families or may be made up
of individuals no longer tied to the family. Among the common kinds of
groups are flocks of birds, schools of fish and herds of animals. The
adaptive value of grouping are several: A group is more alert to danger
than a single individual. What one member may miss, another sees and
the whole group takes flight. Often several unrelated species group together. In mixed groups of baboons and impala, the keen-sighted baboons
spot visible danger and the keen-scented impala detect hidden danger. A
group offers protection against attack. A classic example is that of the ring
which male musk oxen make around the young and female musk oxen.
A group is capable of communal attack. The most frequently observed
form of communal attack is the mobbing of crows, owls or hawks by
groups of smaller birds. Communal hunting has been developed to a fine
art by wolves and lions. The origin of a group and its continued cohesive-

FIGURE 24-49   *Some examples of bird displays derived from the take-off leap.*
(From Hinde, R. A.: Animal Behavior. New York, McGraw-Hill, Inc., 1966.)

ness depends upon a constant flow of information within the group. Mere
aggregation is effected by visual, auditory or chemical signals.

Reproductive fighting and **dominance fighting** are also aspects of group
behavior that are of adaptive value. The weapons employed range all the
way from biting, butting and kicking to mere bluffing (Fig. 24-50). Seldom
are the combats mortal. Threat and threat display, therefore, serve the
same purpose. Threat takes the form of visual bluffing in fish and birds
especially and in olfactory warning in mammals. In visual threat animals
assume aggressive postures and display brightly colored or contrasting
parts of their bodies. Robins display their red breast; some cichlids display
black gill covers ringed with gold (Fig. 24-51). Dogs urinate, brown bears
rub their backs on trees, some antelopes mark trees with scent from a
gland near the eye (Fig. 24-44).

The result of this fighting, which is directed toward reproductive
rivals, is to space out the breeding pairs. The end result is a more even
distribution of food, more nesting and breeding space, and the prevention
of multiple copulations with their attendant waste.

FIGURE 24-50   *Fighting red deer. (From Tinbergen, N.: Social Behaviour in Ani-*
mals. London, Methuen & Co., Ltd., 1953.)

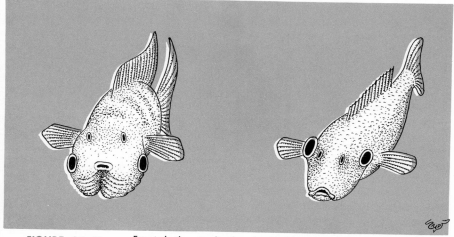

FIGURE 24–51   Frontal threat display in Cychlasoma meeki (left) and Hemichromis bimaculatus (right). (After Tinbergen, N.: The Study of Instinct. London, Oxford University Press, 1951.)

This kind of spacing provides each pair with its own **territory** from which it drives trespassers. The size of a territory may vary from the very small domain of a nesting gull to the extensive hunting preserve of a large carnivore. With gulls, as with many colonially nesting birds, the size of the territory is the diameter of a circle from within which one bird can reach out to peck another without having to stir from its nest. This type of territory is nest-centered.

In most cases the territory is more than the breeding area proper; it is the area which supplies food. Size is then partly a function of the nature of the food. Large carnivores may have to range far and wide to get enough food to sustain them in contrast to a herbivore that may require only a limited grazing area. Size may also be determined by topography, i.e., the actual amount of space available. It does not appear to change with changes in the abundance of individuals. There are "psychological" factors that determine the absolute minimum to which territory may be reduced.

Animals with highly organized societies have a **colony territory,** or range, rather than a pair territory. Troops of baboons may claim a territory of three to six square miles. The location and extent of ranges is set by group tradition. Territorial boundaries are learned by each generation. As long as food and water are available, troops may never meet. In arboreal species in which foliage prevents visual observation, troops are kept apart by vocal signals. Howler monkeys, for example, begin each day by howling for about 30 minutes and so advertise their position to other troops. When there are shortages, as, for example, when only one water hole is available for baboons, the troops may come together, but they do not mix.

Other social animals, as exemplified by prairie dogs, inhabit territories that are subdivided into fixed and stable family units. The resemblance to human social life in this respect is striking. Whereas the baboons resemble human nomadic tribal life, the prairie dogs resemble urban apartment living.

Although the size of territories for each species is a fixed characteristic, the location and boundaries of a territory are learned. Once staked out, the territory is marked, as already mentioned, and may be vigorously defended.

Another way of sharing the available environment may be called **temporal spacing.** It is achieved by the establishment of a dominance

TABLE 24–1 Behavioral and Ecological Comparisons of Apes and Men

	Ecology				Economic System
	Group Size, Density and Range	Home Base	Population Structure	Food Habits	Economic Dependence
(Men)	Groups of 50–60 common but vary widely. One individual per 5–10 square miles. Range 200–600 square miles. Territorial rights; defend boundaries against strangers.	Occupy improved sites for variable times where sick are cared for and stores kept.	Tribal organization of local, exogamous groups.	Omnivorous. Food sharing. Men specialize in hunting, women and children in gathering.	Infants are dependent on adults for many years. Maturity of male delayed biologically and culturally. Hunting, storage and sharing of food.
(Apes)	10–200 in group. 10 individuals per square mile. Range 3–6 square miles; no territorial defense.	None: Sick and injured must keep up with troop.	Small, inbreeding groups.	Almost entirely vegetarian. No food sharing, no division of labor.	Infant economically independent after weaning. Full maturity biologically delayed. No hunting, storage or sharing of food.

	Social System					Communication
	Organization	Social Control	Sexual Behavior	Mother-Child Relationship	Play	Communication
(Men)	Bands are dependent on and affiliated with one another in a semiopen system. Subgroups based on kinship.	Based on custom.	Female continuously receptive. Family based on prolonged male-female relationship and incest taboos.	Prolonged; infant helpless and entirely dependent on adults.	Interpersonal but also considerable use of inanimate objects.	Linguistic community. Language crucial in the evolution of religion, art, technology and the co-operation of many individuals.
(Apes)	Troop self-sufficient, closed to outsiders. Temporary subgroups are formed based on age and individual preferences.	Based on physical dominance.	Female estrus. Multiple mates. No prolonged male-female relationship.	Intense but brief; infant well developed and in partial control.	Mainly interpersonal and exploratory.	Species-specific, largely gestural and concerned with immediate situations.

(From The Social Life of Baboons by S. L. Washburn, and I. DeVore. Copyright © June 1961 by Scientific American, Inc. All rights reserved.)

APES AND MEN are contrasted in this chart, which indicates that although apes often seem remarkably "human," there are fundamental differences in behavior. Baboon characteristics, which may be taken as representative of ape and monkey behavior in general, are based on laboratory and field studies; human characteristics are what is known of preagricultural Homo sapiens. The chart suggests that there was a considerable gap between primate behavior and the behavior of the most primitive men, a

hierarchy in a group that occupies common territory. No individual calls an area his own or occupies it more than briefly; however, not all spots are equally available to all individuals. The best of everything comes first to dominant members of a group, while members lower on the social ladder must wait their turn.

Among the earliest studies of social order were those of chickens. In any flock of hens one individual usually dominated; she could peck at any member of the flock without being pecked in return. Next in order came a hen that could peck all but number one; below her was number three, and so on—hence the term "peck-order." **Peck-order** is established gradually in a flock of growing chicks by a series of individual **dominance fights** that establish who can beat whom. Once the order has been established, little fighting occurs and individuals learn their place. Any strange bird entering the flock must fight each member and so establish her social place.

Social order is generally the rule in animals that live in groups. Among vertebrates it attains its highest development in troops of primates. A male achieves a dominant position in the group by physical condition, aggressiveness and fighting ability. He then commands the immediate space around him. Once achieved, his position is held by symbolic threats rather than by actual fighting. These threats are ritualized behavior patterns developed from preparatory movements for fighting, just as when a man shakes a clenched fist at his enemy.

The **hierarchy** is further complicated in baboons by the fact that a number of males may join together in a governing clique. There may be an outside male who can outfight any individual member, but the group combines to stand against him.

Each animal comes to know his rank in the social order and if challenged by his superior makes **ritualized submission** by raising his rump, and the dominant asserts his dominance by mounting. **Grooming,** which is basically a hygienic measure, has acquired social significance in the more aggressive species of primates. The dominant member is more often groomed than grooming.

The cohesiveness of a primate group and the intensity of the dominance hierarchy are directly related to the level of potential danger to which the species is subjected. Gorillas, for example, do not face the same dangers as baboons, are much less aggressive animals, legend and appearance to the contrary notwithstanding, and lead a considerably more relaxed life. There are dominant males who lead, but there is little violence. Baboons and macaques, on the other hand, are subjected to many dangers in their environment; they are, accordingly, very aggressive. Aggressiveness would tend to disrupt troops if some order and discipline were not maintained. This is the end served by dominance hierarchies. The basis of the society is not, therefore, sex, as once believed, but the need for protection. Dominant males clearly play an important role in defense of the troop. As a troop moves these males lead, less dominant males bring up the rear and females and juveniles occupy the protected center. When danger threatens, the males present a united front.

There are many similiarities between the social arrangements of non-human primates and humans, but as an examination of Table 24–1 will reveal there is actually a wide gap between the two.

SECTION 24–14

SOCIAL INSECTS

Among the oldest and most highly developed societies are those of insects. Three hundred million years ago these societies were already in existence. In the course of evolution social habits arose 24 different times. The establishment of social life in these usually short-lived animals was made possible by the prolongation of parental life. First the progeny depended on the parents, then the parents on the progeny.

Various gradations between solitary and social life are reflected in existing species. Three levels of parent-progeny relations are seen among solitary insects. Some such as mosquitoes and dragonflies drop their eggs anywhere and go about their business. Others such as butterflies and flesh-flies lay their eggs on food suitable for the young. Still others like the solitary digger wasps dig a hole, provision it with food (paralyzed spiders, caterpillars and so on), lay an egg, seal the entrance and depart. Subsocial insects also **mass provision** but remain to guard the nest or young. Examples of some of these relationships are to be found among the dung beetles including the sacred scarab of Egypt (*Scarabaeus sacer*). In some species the female collects and rolls a ball of dung, excavates a burrow, drops the ball in, lays eggs and departs. In other species the male assists to the point of guarding the dung ball while the female excavates. In still another species both sexes dig chambers, stock them with dung on which the female lays eggs, then guard them until the young hatch. At this time all disperse.

True social insects **provision progressively** and then the young cooperate in caring for the next generation. The most highly developed and complex of these are the social termites, ants and bees. They share in common the fact of being matriarchies, all of whose members are the offspring of a single female, hence identical genotypes. Most diversity is found among ants. There are the army ants that have no permanent nest but travel in long marauding columns and rest in temporary bivouacs. There are ants that kill off other species, carry off their young and are thenceforth fully dependent upon the "slaves" that develop from these larvae. Other ants collect nectar and feed it to special members of the colony that become living honey casks hanging from the roofs of underground chambers. There are ants that feed on the honeydew produced by aphids, carry these "ant cows" into overwintering quarters and protect them from predators. There are the leaf-cutter or fungus-growing ants that cut leaves and then

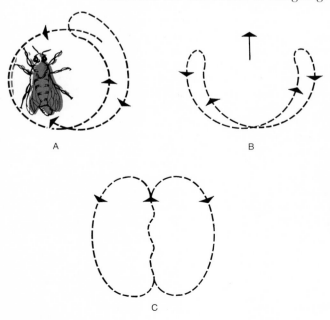

A   B

C

FIGURE 24–52   *Three kinds of communication dances performed by honey bees. A, The round dance, B, the sickle dance, C, waggle dance. (From von Frisch, K., and Lindauer, M.: In McGill, T. (ed.): Readings on Animal Behavior. New York, Holt, Rinehart and Winston, Inc., 1965.)*

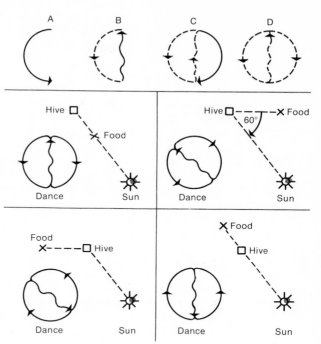

FIGURE 24–53   *Indication of direction by the waggle dance. (From Villee, C. A., Walker, W. F., Jr., and Smith, F. E.: General Zoology. 3rd ed. Philadelphia, W. B. Saunders Company, 1968.)*

carry them underground to serve as a substrate for growing pure strains of fungus which serves as food.

The honeybee colony surpasses all, and is unique among invertebrate societies in that its members can communicate navigational information to one another. In the normally established bee colony, the queen is usually the only reproductive individual. As long as she is healthy and functioning, the colony functions. Members of the hive are kept informed of "queen rightness" by a chemical signal in the form of a secretion from the queen that is passed from those individuals attending her bee by bee to all in the hive. **Chemical signals** in general are instrumental in maintaining cohesiveness of the hive. Scent-fanning bees at the entrance of the hive blow hive odor outward to guide foragers home; bees signal alarm by extending the sting and fanning to disperse a specific odor; queens, during the nuptial flight, attract drones by odors from their mandibular glands.

One of the most astounding features of the hive is the system of **communication** that is employed to direct foragers to sources of food or sites for a new hive. When a forager locates a source of food less than 50 to 100 meters distant from the hive, she returns to the hive and performs a round dance (Fig. 24–52). Receptive hive mates follow the dancing forager as she dances and pick up from her the scent of the flower she had visited. They are thus informed of the presence of food nearby and its nature.

When the source of food lies at a distance greater than 50 to 100 meters, the returning forager performs a waggle dance that indicates the direction in which the food lies and its distance from the hive. The waggle dance is in the form of an eight (Fig. 24–53). As the bee traverses the waist of the eight, the straight run, her abdomen vibrates from side to side. Distance is indicated by the number of straight runs per unit time and the number of waggles. As distance increases, the number of runs decreases while the number of waggles increases. Distances up to 11 kilometers can be recorded.

Direction is indicated by the orientation of the dance. In normal circumstances the dance is performed on a vertical comb in a dark hive. If the source of food lies on a direct line from the hive to the sun so that the bees have to fly toward the sun, the dance is oriented so that the bee is traveling *up* the straight run; if the food lies in a direction directly opposite to the sun, the dance is oriented so that the bee is traveling *down* the straight run. When the course of flight to the food lies at some angle to the line between the hive and the sun, the dance is oriented so that the straight run is at that angle with respect to gravity (Fig. 24–53). In other words, the visual angle that guides flight is transformed to a gravitational angle in the dance.

The cohesiveness and organization of insect societies often prompts comparison with human societies, and indeed there are similarities. It is revealing to make the following six-point comparison: (1) Insect societies arise from discrete families, i.e., one female and her offspring, and frequently send off new colonies (the swarming of ants and bees). Human society originally began as discrete families, but this is not repeated phylogenetically. (2) Man has tradition and social heredity; insects, in this case ants, bequeath fungi and real estate (hunting grounds). (3) Man uses tools; insects, with the rare exception of the ant that uses its larvae as spinning shuttles to fasten leaves together with larval silk, do not. (4) Man controls the environment of his dwelling; bees and termites also control the temperature of their nests. (5) Man domesticates animals; ants "domesticate" aphids. (6) Man has a language; bees and ants communicate by dances and pheromones.

A basic difference between human and insect societies is in the manner in which the two adapt to different requirements. Man has evolved learning and abstract intelligence whereby he meets different situations. This adaptation provides him with a flexible society. Insects evolve different genotypes and different castes (soldiers, workers and so on), each of which meets a specific need. Their society is, therefore, a rigid one.

SUGGESTIONS FOR FURTHER READING

Manning, A.: *An Introduction to Animal Behavior.* Reading, Massachusetts, Addison-Wesley Publishing Company, Inc., 1968.
Dethier, V. G., and Stellar, E.: *Animal Behavior.* 3rd ed. Englewood Cliffs, New Jersey, Prentice-Hall, Inc., 1970.
  Two concise introductions to animal behavior, written from different points of view.

Hinde, R.: *Animal Behavior:* 2nd ed. A Synthesis of Ethology and Comparative Psychology. New York, McGraw-Hill, Inc., 1970.
Marler, P., and Hamilton, W. J.: *Mechanisms of Animal Behavior.* New York, John Wiley & Sons, Inc., 1966.
  More detailed accounts of animal behavior.

McGill, T. E. (ed.): *Readings in Animal Behavior.* Holt, Rinehart & Winston, 1965.
McGaugh, J. L., Weinberger, N. M., and Whalen, R. E. (eds.): *Psychobiology.* W. M. Freeman, 1967.
  Samples of exciting areas of research and contemporary topics of discussion.

Tinbergen, N.: *Social Behavior of Animals.* New York, John Wiley & Sons, Inc., 1953.
  A readable introduction to social behavior.

REPRODUCTIVE BIOLOGY

THE REPRODUCTIVE PROCESS

CHAPTER 25

If there is any one feature of living systems that qualifies as the "essence of life" it is the ability to reproduce one's kind and perpetuate the species. Even the smallest viruses, which may have none of the other characteristics of living things, are able to reproduce or to be reproduced by their host cells. At the molecular level, reproduction involves the unique capacity of the nucleic acids for self-replication which depends on the specificity of the relatively weak hydrogen bonds between pairs of nucleotides. Reproduction at the level of the whole organism ranges from the simple fission of bacterial and other unicellular organisms (a process which does not involve sex at all) to the incredibly complex structural, functional and behavioral processes of reproduction in higher plants and animals. Reproduction in higher animals involves not only genetic phenomena, the transfer of biological information from one generation to the next, but the endocrine regulation of oögenesis, ovulation and spermatogenesis, intricate behavior patterns that ensure that male and female gametes are released at the same time and meet to form a fertilized egg or zygote, and the complex sequence of processes in development and differentiation by which a zygote becomes an adult organism.

The survival of each species of plant or animal requires that its individual members multiply, that they produce new individuals to replace the ones killed by predators, parasites or old age. The details of the process vary tremendously from one kind of organism to another but we can distinguish sexual and asexual reproduction. **Asexual reproduction** involves only a *single* parent, which splits, buds or fragments to give rise to two or more offspring, all of which have hereditary traits identical to those of the parent. Even the higher animals may reproduce asexually; the production of identical twins in man by the splitting of a single fertilized egg is a kind of asexual reproduction. **Sexual reproduction** involves *two* parents, each of which contributes a specialized cell, a gamete (eggs and sperm) which fuse to form the **zygote**, or fertilized egg. Eggs are typically large, nonmotile cells containing a store of nutrients which will support the development of the zygote until the resulting embryo can obtain food in some other way. The typical sperm is small, motile and adapted to swim actively to the egg by the beating of its long whiplike tail. Sexual reproduction has the biological advantage of making possible the recombination, of the best inherited traits of the two parents; thus the offspring may be better able to survive than either parent. Evolution can proceed much more rapidly and effectively with sexual reproduction then with asexual reproduction.

SECTION 25–1

ASEXUAL REPRODUCTION

Asexual reproduction occurs commonly in all kinds of plants—bacteria, algae, fungi, mosses and tracheophytes—and in protozoa, coelenterates, bryozoa and tunicates, but may occur even in mammals. Perhaps the simplest form of asexual reproduction is the splitting of the body of the parent into two more or less equal daughter parts, each of which becomes a new, independent, whole organism (Fig. 25–1). This form of reproduction, termed **fission**, occurs chiefly among the protists, the single-celled plants and animals. The cell division involved is mitotic.

Hydras and yeasts reproduce by **budding**, in which a small part of the parent's body separates from the rest and develops into a new individual. It may split away from the parent and take up an independent existence or it may remain attached and form a more or less independent member of the colony.

Salamanders, lizards, starfish and crabs can grow a new tail, leg, arm or certain other organs if the original one is lost. When this ability to regenerate a part is carried to an extreme, it becomes a method of reproduction. The body of the parent may break into several pieces and each piece then regenerates the missing parts and develops into a whole animal. This reproduction by **fragmentation** is common among the flatworms; starfish have the ability to regenerate an entire new starfish from a single arm. You cannot kill starfish by chopping them in half and throwing the pieces back in the sea, you simply double the number of starfish preying on your oysterbeds.

Almost all plants reproduce asexually at one stage in their life cycle by **spores**, special cells that usually have some sort of capsule or coat to protect the cell from excessive heat, cold, desiccation or other unfavorable environmental conditions. The class of parasitic protozoa, the Sporozoa, typically live as intracellular parasites within the cells of the host during the growth phase of their life cycle and reproduce by spores. Sporozoa cause some serious diseases such as **malaria** in man and other animals and **coccidiosis** in poultry. The infective spore matures as a feeding form or trophozoite and divides by multiple fission into a number of young. Each

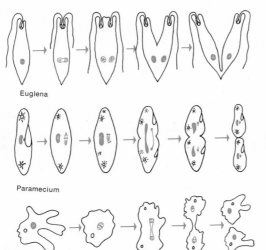

Euglena

Paramecium

Amoeba

FIGURE 25–1   *Asexual reproduction in Euglena, Paramecium and Amoeba. In each, one cell divides mitotically to give rise to two cells.*

of these infects a new cell of the same host and matures as another trophozoite. Eventually some trophozoites undergo metamorphosis into sexual forms; the females become eggs and the males divide repeatedly to form sperm. After fertilization the new individual grows and divides by multiple fission to form infective spores able to infect new hosts. The spores of most sporozoa are encapsulated to withstand desiccation, but the spores of the blood parasites (such as plasmodia which cause malaria) are naked and must be transferred directly into the blood stream of the new host. The formation of eggs and sperm, fertilization and the formation of infective spores usually occur in a different kind of host, a mosquito, from that in which the trophozoite stages are found, in man or monkey.

SEXUAL REPRODUCTION IN PLANTS

The many variations on the theme of sexual reproduction in plants were discussed in Chapters 12 and 13. From the bacteria and blue-green algae, in which sexual reproduction is rare or nonexistent, through the green algae, mosses, and seed plants, sexual reproduction has played an increasingly important role in the life cycle of these forms, and the diploid phase occupies an increasingly large portion of the life cycle. Even the seed plants reproduce asexually by **stolons** (horizontal stems growing along the surface of the ground), by **rhizomes** (stems growing laterally underground) or by **tubers**. The florist and farmer use **cuttings** of stems to propagate many varieties of cultivated trees, shrubs and plants. Among the evolutionary trends discernible in plants are the shrinking of the gametophyte generation and the expansion of the sporophyte. These permit the land plants to reproduce without the necessity of a film of water for swimming sperm. The sperm are transferred to the egg via a pollen tube. The fertilized egg can develop into an embryo nourished by the supply of food in the endosperm until the embryo develops its own leaves and becomes an independent plant.

SEXUAL REPRODUCTION IN ANIMALS

Certain animals such as the coelenterates have alternate sexual and asexual generations superficially reminiscent of the life cycles of plants, but both generations of coelenterates are diploid. Most animals typically reproduce sexually and have permanent sex organs. Paramecia and certain other protozoa have a complicated process of sexual reproduction in which two diploid individuals come together, fuse on their oral surfaces and exchange nuclear material (Fig. 25–2). The original micronucleus of each divides meiotically and three of the four nuclei degenerate. The remaining haploid nucleus divides mitotically and one of the two migrates across the cytoplasmic bridge to the other animal and fuses with one of its haploid nuclei. Thus, two fertilizations result and the two new diploid nuclei are identical. The animals subsequently separate and further nuclear and cytoplasmic divisions result in four individuals, each with a micronucleus and a macronucleus.

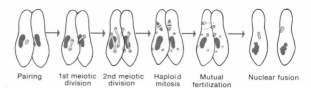

Pairing   1st meiotic 2nd meiotic Haploid Mutual   Nuclear fusion
        division   division   mitosis fertilization

FIGURE 25–2   *Sexual reproduction in paramecium. Two individuals with diploid micronuclei undergo conjugation (left). After meiosis (second and third figures) three of the nuclei degenerate and the fourth nucleus divides by mitosis (fourth figure). The organisms undergo mutual fertilization; one nucleus passes from each organism to the other. This is followed by fusion of the haploid nuclei to form a new diploid nucleus in each of the two organisms (last figure). The original macronuclei disappear. The two new diploid nuclei divide several times by mitosis and eventually establish both the new micronuclei and the new macronuclei.*

All sponges appear to be diploid and undergo the usual processes of oögenesis and spermatogenesis. The eggs are retained just beneath the choanocytes and are fertilized by sperm brought in with the current. Sponges have no special gonads and the sex cells may arise from choanocytes, from amoebocytes or from certain embryonic cells in the matrix.

HERMAPHRODITISM. Many of the lower animals are hermaphroditic (from Hermes, the Greek god of male beauty, and Aphrodite, the goddess of female beauty); both ovaries and testes are present in the same individual and it produces both eggs and sperm. Some hermaphroditic animals, the parasitic tapeworms, for example, are capable of self-fertilization. (Does this violate the generalization that sexual reproduction involves two different individuals?) Since a host may be infected with but one parasite this ability is an important adaptation for the survival of the species. Most hermaphroditic animals, however, do not reproduce by self-fertilization; rather, two individuals copulate and each inseminates the other, as in the earthworm (p. 422). In other species self-fertilization is prevented because the ovaries and testes develop at different times. Oysters, though hermaphroditic, cannot undergo self-fertilization, for the ovaries and testes produce gametes at different times.

PARTHENOGENESIS. A rather rare modification of sexual reproduction found among honeybees, wasps and certain other arthropods is the development of unfertilized eggs into adult animals, a process called **parthenogenesis** (virgin birth). Some species of arthropods, representing the ultimate in women's liberation, perhaps, consist entirely of females that reproduce by parthenogenesis. In other species parthenogenesis occurs for several generations; then some males develop, produce sperm and mate with the females to fertilize the eggs. The queen honeybee is fertilized by a male just once during her lifetime, during her "nuptial flight." The sperm are stored in a pouch connected with the genital tract and closed by a muscular valve. As the queen subsequently lays eggs she can either open the valve, permitting some sperm to escape and fertilize the eggs or keep the valve closed so that the eggs develop parthenogenetically. Fertilization usually occurs in the fall and the fertilized eggs are quiescent during the winter. The fertilized eggs become females, queens and worker bees; unfertilized eggs become males (drones). Some species of wasps give birth alternately to a parthenogenetic generation and one which develops from fertilized eggs.

The eggs of species which do not normally undergo parthenogenesis may be stimulated artificially to develop without fertilization by changing the temperature, pH or salinity of the surrounding water, or by chemical or mechanical stimulation of the egg. Frog eggs can be stimulated to undergo parthenogenetic development by pricking them with a fine needle. A variety of species of marine invertebrates as well as frogs and salamanders have been produced parthenogenetically. It has been possible to stimulate rabbit eggs to develop parthenogenetically; the resulting cleaving egg must be placed in the uterus of a hormonally primed female to complete development.

FERTILIZATION

Most aquatic animals simply liberate their eggs and sperm into the water and their union occurs by chance. This primitive and rather uncertain method of gamete union is called **external fertilization**. These animals usually have no accessory sex structures except the ducts that transport the gametes to the exterior.

Other animals, especially those living on land, have accessory sex organs for transferring the sperm from the body of the male to that of the female and fertilization occurs within her body. This method of **internal fertilization** requires the cooperation of the two sexes and many species have evolved elaborate behavior patterns to ensure that the two sexes are brought together, mate at the appropriate time and care for the resulting offspring. A variety of secondary sex characteristics have evolved and serve as stimuli to attract the opposite sex and elicit a mating response.

The male salamander mounts and clasps the female, stroking her chin with his nose. He then dismounts in front of her and deposits a **spermatophore**, a packet of sperm. She picks up the spermatophore and stuffs it into her cloaca, where the packet breaks and releases the sperm so that fertilization may ensue.

The breeding habits of some animals are dangerous or even fatal to the individuals performing them. Salmon swim hundreds of miles upstream to spawn and die. The male spider is frequently eaten by the female after he has inseminated her. The survival of the species, however, requires that the individual sacrifice his own interests to perform these acts.

Fertilization involves not only the penetration of the egg by the sperm but the union of egg and sperm nuclei and the activation of the egg to undergo cleavage and development (Fig. 25-3). The egg may be in any stage from primary oöcyte to mature ovum when the sperm enters it, but the fusion of egg and sperm nuclei can occur only after the egg nucleus has completed the two meiotic divisions and has become a mature **ovum**. The eggs of some species secrete **fertilizin,** an important constituent of the jelly coat surrounding the egg. Fertilizin causes sperm to clump together and stick to the surface of the egg. Other substances extracted from the jelly coat (which may be identical with fertilizin) stimulate sperm motility and respiration and prolong sperm viability.

After one sperm enters the egg a **fertilization membrane** forms around the eggs of some species which prevents the entrance of other sperm. This prevents polyspermy and the possibility that more than one sperm nucleus will fuse with the egg nucleus. The presence of two sperm nuclei may lead to the formation of tripolar spindles and abnormal development.

The females of all birds, most insects and many aquatic invertebrates

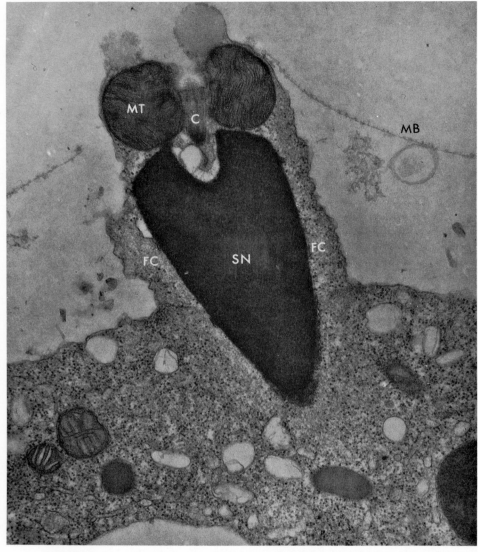

FIGURE 25-3   *Spermatozoon of the sea urchin Arbacia punctulata being drawn onto the egg cytoplasm by the fertilization cone. FC, fertilization cone, C, centriole of spermatozoon, MB, fertilization membrane, MT, mitochondrial body, SN, spermatozoon nucleus. (Courtesy of Professor E. Anderson.)*

lay eggs from which the young eventually hatch; such animals are said to by **oviparous** (egg-bearing). The small eggs of mammals are kept in the uterus and provided with nutrients from the mother's blood until development has proceeded to the stage where they can live independently. Such animals are termed **viviparous** (live-bearing). The females of sharks, lizards and some insects and snakes are **ovoviviparous**; they produce large, yolk-filled eggs which remain in the female reproductive tract for considerable periods after fertilization. The developing embryo usually forms no close connection with the wall of the oviduct or uterus and receives no nourishment from maternal blood.

MATING BEHAVIOR AND SYNCHRONIZATION OF SEXUAL ACTIVITY

Since most animals breed only during relatively brief seasons of the year, the production and release of eggs and sperm must be synchronized if fertilization is to occur. Typically the males and females are triggered by some environmental cue such as a change in the photoperiod, a change in the ambient temperature, seasonal rainfall or by specific relations of tidal and lunar cycles (cf. the Palolo worm, p. 424). In some species the males and females must not only be brought to full sexual activity at the same time, but must be induced to move or migrate to specific mating and breeding grounds. The migration of salmon upstream to breed, the migration of eels to a specific breeding ground in the Central Atlantic, the migration of the gray whale to Baja, California, and the migration of turtles and birds are examples of such movements. Animals that live a solitary life during most of the year — seals, penguins, certain sea birds — come together for brief periods of mating at specific times and places.

The specific reproductive synchronization of a particular male for a particular female frequently involves some sort of **courtship behavior** (Fig. 24–26). The courtship, usually initiated by the male, may be a very brief ceremony, or in certain species of birds may last for many days. The courtship behavior serves two additional roles: it tends to decrease aggressive tendencies and it establishes species and sexual identification, i.e., it identifies a member of the same species but of the opposite sex. The members of many species normally fight whenever they meet and some special cue is needed if two animals are to avoid fighting long enough to mate. Special structural and functional adaptations have evolved in some species in which aggression seems to be especially difficult to control. The male praying mantis is smaller than the female and is usually attacked by her. Mating in this species can continue even after the male's head has been bitten off by the female, since the nervous activities controlling copulation are centered in an abdominal ganglion. The males of certain species of flies present their predatory females with little packages of food wrapped in silk thread. The male cockroach secretes a special substance on his back which is especially attractive to the female and attracts her attention during copulation. The males of other insects present the female with an empty silk balloon which diverts her from attacking him. The nature of the balloon aids the female to identify the species and sex of the male presenting it.

The song patterns of birds and the mating calls of certain fishes, frogs and insects provide effective cues for the discrimination of the species. Female frogs are attracted to calling males of their own species but not to calling males of related species.

CARE OF THE YOUNG

The evolution of more efficient methods for bringing about fertilization has been accompanied by the evolution of other behavior patterns for the care of the young. In general, the number of eggs produced by a female of a given species and the chance that any particular egg will survive to maturity are inversely related. In the evolution of the vertebrates from fish to mammal, the trend has been toward the production of fewer eggs and the development of better parental care of the young. Fish such as the cod or salmon produce millions of eggs each year but only a small number of these ever become adult fish. At the other end of the spectrum mammals produce relatively few eggs but take good care of their offspring so that a large fraction of the total attain maturity. The eggs of reptiles are usually laid in the sand or mud where they develop without parental care, warmed by the sun. Birds lay relatively few eggs in nests and incubate them by sitting on them. The parents take care of the newly hatched, helpless chicks for several weeks until they are able to survive independently. The mammalian egg develops within the mother's uterus, where it is safe from predators and from most harmful environmental factors. Most mammalian females have a strong "maternal instinct" to take care of the newborn until they can shift for themselves. This pattern of behavior is regulated largely by hormones secreted by the pituitary and ovary.

SECTION 25–7

ANIMAL LIFE CYCLES

The typical life cycle of an animal is a simple one of growth of a fertilized egg into an adult which produces more eggs for fertilization. The life cycles of other animals may include two or more forms that intervene between one fertilized egg and the fertilized egg of the next generation.

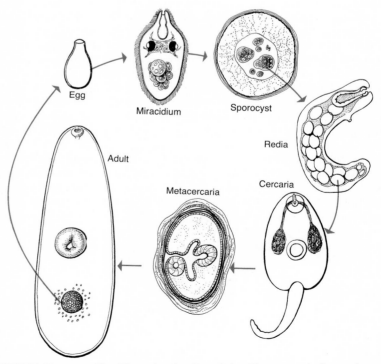

FIGURE 25–4   *The life cycle of a liver fluke. The arrows indicate that one stage either becomes the next or produces the next by a reproductive process.*

Amphibians and cyclostomes have young which do not resemble the parents but develop as *larvae*. At the proper time the frog larva, or *tadpole*, undergoes a relatively rapid change, a *metamorphosis*, to become an adult. Insect larvae hatch out of the egg and molt several times to become successively larger larvae. The final larval stage undergoes metamorphosis to a pupa and the pupa undergoes a second metamorphosis to become an adult.

The life cycles of certain parasitic worms involve two or three different larval forms which infect different hosts (Fig. 25–4). The larvae of both amphibians and insects may be adapted to quite different environments from the adults of the same species and may occupy quite different ecologic niches and places in the food chains. The frog tadpole is completely aquatic and feeds on plant material, whereas the adult frog is amphibious and carnivorous.

SECTION 25–8

DEVELOPMENT OF THE GONADS AND REPRODUCTIVE PASSAGES

The reproductive systems of different kinds of animals have a fundamentally similar plan with many variations on the theme. The gonads and their ducts may be single, paired or multiple. Sperm, produced in the testes, are transported in ducts to the exterior, frequently suspended in seminal fluid secreted by glands associated with the reproductive tract.

In many vertebrates a number of accessory structures have developed which facilitate the transfer of sperm from the male to the female reproductive tract and provide a place for the fertilized egg to develop. These structures have evolved either from, or in close association with, the urinary system and the two are frequently referred to as the *urogenital system*. The vasa deferentia of the male mammal empty into the *urethra*, which also carries urine from the bladder to the exterior of the body.

Eggs produced in the ovaries of the female are released at ovulation and pass from the ovary into the abdominal cavity, then into the funnel-shaped end of the oviduct, the *ostium*. Eggs are moved along the oviduct by peristaltic contractions of its muscular wall and by the beating of the cilia lining the lumen of the duct. The yolk of the bird's egg is formed while the egg is still in the ovary but the egg white and shell are added by glands in the wall of the oviduct. The oviducts may open to the exterior directly or they may expand to form a terminal duct, the *uterus*, a thickened muscular pouch in which the embryo develops. In mammals the uterus is connected to the exterior by the *vagina*, adapted to receive the penis of the male in copulation.

The sex of an embryo is determined at the time of fertilization by the XX-XY mechanism, but early in development the embryo has the developmental potential of differentiating into either a male or female, for the primordia of both male and female duct systems are present (Fig. 25–5). The paired *pronephric tubules* and ducts appear briefly at the anterior end of the coelom. The tubules degenerate and the ducts grow posteriorly as the paired *wolffian ducts*. A second group of tubules, the *mesonephros*, appears and connects with the wolffian duct. The definitive kidney of the higher vertebrates, the *metanephros*, develops from the posterior part of the nephrogenic cord. A bud develops at the posterior end of the wolffian

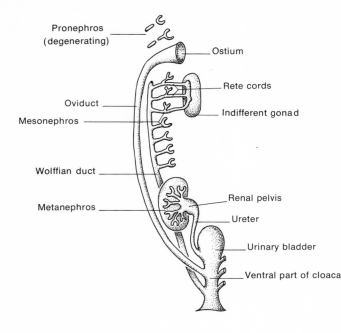

Pronephros
(degenerating)

Ostium

Rete cords

Oviduct

Indifferent gonad

Mesonephros

Wolffian duct

Metanephros

Renal pelvis

Ureter

Urinary bladder

Ventral part of cloaca

FIGURE 25–5   A ventral view of the urogenital organs of the sexually indifferent stage of an embryo.

duct, near the junction of the duct with the cloaca, and grows toward the nephrogenic cord forming a duct, the **ureter**. The end of the ureteric bud branches, giving rise to the collecting tubules of the kidney. Nephrogenic mesenchyme accumulates around the tips of the collecting tubules and differentiates into renal tubules and glomeruli.

The paired paramesonephric or **mullerian ducts** develop just lateral to the wolffian ducts (Fig. 25–6) and are the precursors of the fallopian tubes, uterus and part of the vagina. The gonads develop as thickenings, **germinal ridges**, on the mesodermal epithelium lining the coelom. The ridge becomes composed of several layers of cells and protrudes into the coelomic cavity, suspended from the peritoneal wall by a double layer of peritoneum, the **mesorchium** or **mesovarium**. The cells in the germinal ridge are of two types—one is very similar to other cells of the peritoneal epithelium and the second is a very different type, the **primordial germ cells** (Fig. 25–7). These are larger cells, with large, vesicular nuclei and clear cytoplasm that stains deeply with an alkaline phosphatase reaction. These primordial germ cells appear in the endodermal epithelium of the yolk sac in the vicinity of the allantoic stalk (p. 836). They migrate by amoeboid motion into the mesenchyme and eventually reach the germinal ridge. There are less than 100 primordial germ cells when they are first evident in the yolk sac but they multiply during their migration and some 5000 or more reach the germinal ridge. The mesenchyme of the germinal ridge is gradually replaced by finger-like accumulations of cells, the **sex cords**, which migrate in from the mesonephric cord or the epithelium of the germinal ridge. The epithelium of the germinal ridge forms the cortex of the gonad and the sex cords comprise its medulla. Development to this stage, the **indifferent gonad**, is similar in ovary and testis.

In male embryos the primordial germ cells migrate from the cortex into the primitive sex cords of the medulla which become organized into the **seminiferous tubules**. The primordial germ cells become the spermatogonia and the **Sertoli cells** develop from the sex cords. The medulla of the testis becomes its functional part while the cortex is reduced and con-

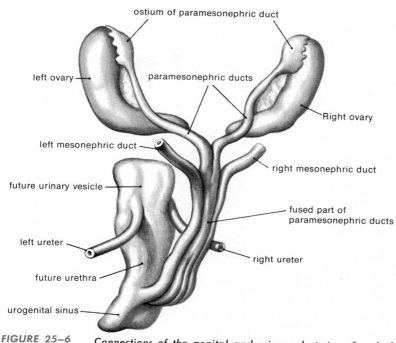

ostium of paramesonephric duct

left ovary

paramesonephric ducts

Right ovary

left mesonephric duct

right mesonephric duct

future urinary vesicle

fused part of paramesonephric ducts

left ureter

right ureter

future urethra

urogenital sinus

FIGURE 25–6   *Connections of the genital and urinary ducts in a female human embryo of 29 mm (beginning of third month) in which the genital ducts are still in the indifferent stage. (After Balinsky, B. I.: An Introduction to Embryology. 3d ed. Philadelphia, W. B. Saunders Company, 1970.)*

verted into a thin epithelial layer covering the coelomic surface of the testis.

In female embryos the medulla is reduced, the primitive sex cords are resorbed, and the central part of the gonad becomes filled with loose mesenchyme permeated by blood vessels. The primordial germ cells remain in the cortex, which increases in thickness. Masses of cortical cells form primary follicles surrounding one or several primordial germ cells, which swell and become oögonia. These undergo rapid mitotic divisions and after 20 weeks of development in the human fetus, some 7,000,000 oögonia are present. Thereafter mitosis ceases and no further ova are produced. The number is reduced by **atresia** (cell degeneration) to about 1,000,000 at the time of birth and to about 400,000 at the time of puberty. Only about 450 of these are destined to undergo ovulation during a woman's reproductive period, all the rest undergo atresia.

In male embryos the seminiferous tubules become connected to the **rete testis**, a system of thin tubules which develop from the dorsal part of the gonad and form connections with the adjoining mesonephric tubules (which become the **epididymis**) and through them to the wolffian duct, which becomes the **vas deferens**. This development proceeds under the stimulation of testosterone secreted by the interstitial cells of the developing testis. The testis appears to secrete a second hormone, possibly a peptide, which inhibits the development of the mullerian ducts so that no oviduct or uterus appears.

In female embryos the wolffian ducts degenerate when the mesonephros stops functioning as an excretory organ. The mullerian ducts open at the anterior end by a funnel-shaped **ostium** into the coelomic cavity and grow posteriorly and connect with the cloaca. The cloaca is subsequently divided by a **cloacal septum** into a ventral **urogenital sinus** and a

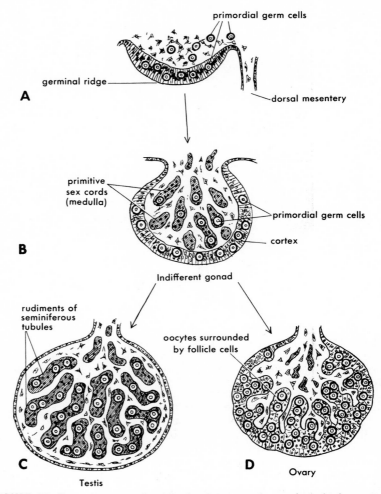

FIGURE 25–7   *Diagram showing development of gonads in higher vertebrates. A, Germinal ridge stage; primordial germ cells partly embedded in epithelium of the ridge and located partly in the adjacent mesenchyme. B, Indifferent gonad, germ cells in the cortex and in primary sex cords. C, Gonad differentiating as testis; cortex reduced; germ cells in sex cords (future seminiferous tubules). D, Gonad differentiating as ovary; primary sex cords reduced; proliferating cortex contains the germ cells.*

dorsal **rectum**. The latter opens to the exterior by the anus. The mullerian ducts are lateral to the wolffian ducts initially but the posterior ends of the mullerian ducts pass over the wolffian ducts. They fuse to form the uterus and cervix and combine with part of the urogenital sinus to form the vagina. The ventral part of the urogenital sinus into which the ureters open, together with the adjoining part of the allantoic stalk, expand to form the **urinary bladder**. The caudal part of the ventral urogenital sinus becomes a narrow tube, the **urethra**.

The external opening of the urogenital sinus is flanked on both sides by elongated thickenings, the **genital folds**, which meet anteriorly in front of the sinus and form a median outgrowth, the **genital tubercle,** the rudiment of the **phallus** (Fig. 25–8). From this sexually indifferent condition, under the stimulation of the sex hormones, development occurs toward the male or female condition. In the male the rudiment of the phallus grows and becomes the penis. The genital folds fuse on the posterior surface of the penis forming a tube, the penile urethra, continuous with the

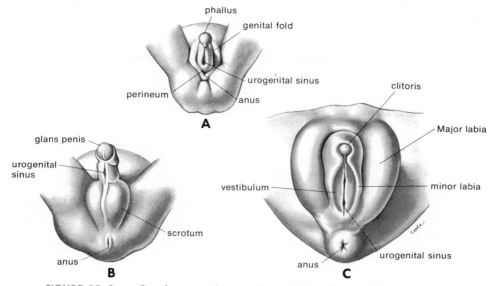

FIGURE 25–8   *Development of external genitalia in human embryos. A, Indifferent stage in an embryo of 29 mm (beginning third month). B, Male embryo of 145 mm (16
weeks). C, Female embryo of 150 mm (16 weeks). (After Balinsky, B. I.: An Introduction to
Embryology. 3rd ed. Philadelphia, W. B. Saunders Company, 1970.)*

urethra leading from the urinary bladder. In the female the phallus grows
slightly and becomes the **clitoris** and the genital folds do not fuse but remain as the **labia minora**. The outer folds become the **labia majora**, comparable to the skin of the scrotum.

SECTION 25–9

THE HUMAN REPRODUCTIVE SYSTEM: MALE

The pair of testes develop within the abdominal cavity but in man and
some other mammals they descend shortly before or after birth into the
scrotal sac, an outpocketing of the body wall covered by a loose pouch
of skin. The cavity of the scrotal sac is part of the abdominal cavity and
is connected with it by the **inguinal canal**. After the testes have descended
this canal is usually closed by the growth of connective tissue. The normal
descent of the testes into the scrotal sac is necessary for the production
of sperm. If the testes remain in the abdominal cavity the slightly higher
temperature there prevents the formation of sperm.

Each testis consists of about 1000 highly coiled **seminiferous tubules**
in which sperm are produced, and the **interstitial cells**, lying between the
tubules, which produce male sex hormones (Fig. 21–13). The lining of
the seminiferous tubules consists of **spermatogonia**, derived from the primordial sex cells, and **nurse cells** (Sertoli cells) which nourish the sperm
as they develop from rounded cells into mature, tailed forms. The formation of sperm proceeds in waves along the tubules. The seminiferous
tubules are connected via fine tubes, the **vasa efferentia**, derived from the
rete testis, to the **epididymis**, a single, complexly coiled tube as much as
6 meters long in man in which sperm are stored. From each epididymis
a duct, the **vas deferens**, passes from the scrotum through the inguinal
canal into the abdominal cavity and over the urinary bladder to the lower
part of the abdominal cavity where it joins the urethra.

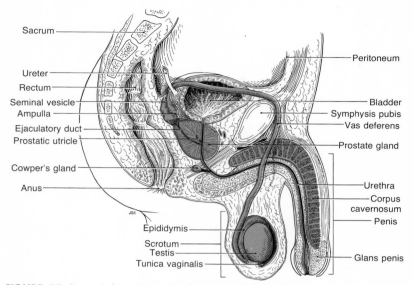

FIGURE 25–9   *Schematic sagittal section of the pelvic region of a male human being showing the organs of the reproductive tract in color.*

The **urethra** is a tube connecting the urinary bladder to the exterior. In the male it passes through the **penis,** flanked by three columns of **erectile tissue** (Fig. 25–9). These spongy masses can be engorged with blood during sexual excitement by the dilation of the arteries supplying them. The spaces, filled with blood under pressure, cause the penis to change from its usual flaccid condition and become enlarged, firm, erect and able to serve as a copulatory organ.

The sperm, suspended in seminal fluid, are transferred to the vagina of the female during copulation. The seminal fluid, amounting from 2 to 5 ml per ejaculation, is produced by three different glands. The paired **seminal vesicles** empty into the vasa deferentia just before they join the urethra. Around the urethra, near its source in the urinary bladder, are the paired **prostate** glands (which in man are fused to form a single prostate), which secrete their contribution to the seminal fluid into the urethra via two sets of short, thin ducts. Farther along the urethra, at the base of the erectile tissue of the penis, lies a third pair of glands, **Cowper's glands**, which contribute the final component of the seminal fluid. Mucous alkaline secretions are provided by the seminal vesicles and Cowper's glands and a thin milky fluid with a characteristic odor is added by the prostate. Seminal fluid may contain glucose and fructose which are metabolized by the sperm, acid-base buffers, and mucous materials which lubricate the passages through which sperm travel.

SECTION 25–10

THE HUMAN REPRODUCTIVE SYSTEM: FEMALE

The **ovaries**, each about 3 centimeters long and the shape of a shelled almond, are held in place by ligaments within the lower part of the abdominal cavity. The egg is released by ovulation from the ovary into the abdominal cavity whence it passes into one of the two *fallopian tubes* through the funnel-shaped ostium at its end (Fig. 25–10). The egg is directed into the ostium by the beating of cilia in the epithelial lining of the fallopian tubes. Very rarely an egg may be fertilized within the abdominal cavity and begin to develop attached to an organ such as the liver. Usually when this occurs development cannot be completed and the embryo must be removed; there have been instances reported of such embryos completing development and being "born" by surgical removal.

The two fallopian tubes empty into the upper corners of the pear-shaped *uterus*, in which the embryo develops until the time of birth. The uterus is in the central part of the lower abdominal cavity, just behind the urinary bladder. About the size of a clenched fist, it has thick walls of smooth muscles and a mucous lining richly supplied with blood vessels. The uterus terminates in a muscular ring, the *cervix*, which projects a short distance into the vagina. The *vagina*, a single muscular tube, extends from the uterus to the exterior and serves both as a receptacle for sperm during coitus and as the birth canal when the fetus completes development.

The external female sex organs, collectively known as the *vulva*, include the *labia majora*, two folds of fatty tissue covered by skin richly endowed with hair and sebaceous glands, which extend back and down enclosing the opening of the urethra and vagina and merging behind it. The *labia minora*, thin, pink folds of tissue devoid of hair, lie within the folds of the labia majora and are usually concealed by them. At the junction of these two in front is the *clitoris*, a sensitive erectile organ about the size of a pea. In most women the clitoris, which is homologous to the male penis, is completely covered by the prepuce. Like the penis, the clitoris contains spongy tissue which becomes engorged with blood during sexual excitement; nerve endings in the clitoris and the labia minora respond to erotic stimulation. Behind the clitoris is the opening of the urethra, which in the human female has only a urinary function, and behind the urethra lies the opening of the vagina. The vaginal opening is partly occluded by the *hymen*, a thin membrane composed of elastic and collagenous connective tissue which is ruptured by the first sexual intercourse. At the junction of the thighs and torso, just above the clitoris, is a small mound of fatty tissue, the *mons veneris*, covered in the adult with pubic hair.

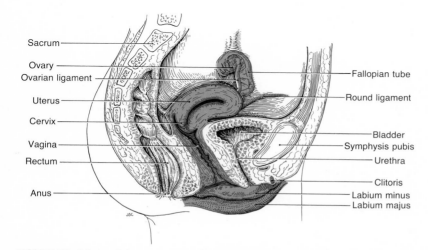

FIGURE 25–10   *Schematic sagittal section of the pelvic region of the female human being showing the organs of the reproductive tract in color.*

HUMAN REPRODUCTION: THE SEX ACT

Human reproduction is accomplished sexually by the union of ova and sperm, by the introduction of the erect penis of the male into the vagina of the female. At the height of the male's sexual excitement seminal fluid is ejected into the upper end of the vagina, around the cervix of the uterus, from which sperm are carried through the cervical canal and uterus to the upper part of the fallopian tubes, where fertilization occurs. The complex structures of the reproductive systems in male and female and the complex physiological, endocrine and psychological phenomena associated with sex have just one purpose: to ensure the successful union of the egg and sperm and the subsequent development of the fertilized egg into a new individual.

The first phase of sexual intercourse, termed **excitement**, may last from a few minutes to a few hours. The essential feature of this phase in the male is erection of the penis and psychic tension. Erection of the penis results when the corpora cavernosa become engorged with blood. The excitement phase in the female is characterized by erection of the clitoris and labia minora. The several phases of intercourse merge into each other and the erection of the clitoris and labia may not occur until the second or **plateau** phase. During this second phase the breasts may enlarge as much as 20 per cent in volume and the nipples become erect. During this time there is typically a flushing of the skin from the breast and neck up into the face. During the excitement phase, fluid passes through the walls of the vagina and moistens the vaginal canal and entrance. The tissues surrounding the entrance to the vaginal canal undergo tumescence to form what has been termed the "orgasmic platform." (Fig. 25–11).

The movement of the penis in and out of the vagina, lubricated by mucoid secretions from the male urethra and from the paired Bartholin's glands just inside the vagina, massages the clitoris and labia causing psychic stimulation of both female and male and leading to the climax or **orgasm**. The male orgasm is marked by ejaculation of seminal fluid, brought about by the contraction of the muscles in the walls of the epididymis, vasa deferentia and seminal vesicles and by the contraction of the bulbocavernosus

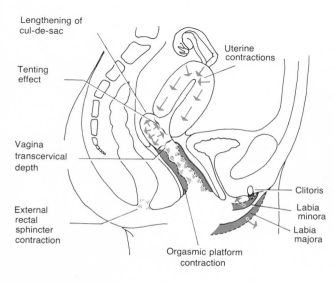

Lengthening of
cul-de-sac

Tenting
effect

Vagina
transcervical
depth

External
rectal
sphincter
contraction

Orgasmic platform
contraction

Uterine
contractions

Clitoris

Labia
minora

Labia
majora

FIGURE 25–11   *Sagittal section of the pelvic region of the human female showing the structural and functional changes that occur during the orgasmic phase. (After Masters, W. H., and Johnson, V. E.: Human Sexual Response. Boston, Little, Brown and Company, 1966.)*

muscle at the proximal end of the urethra. During the female climax the orgasmic platform contracts rhythmically, accompanied by contractions of the uterus and fallopian tubes.

Following orgasm there is a detumescence of penis, labia and clitoris, a decrease in the size of the breasts and a general decrease in muscle tone in both male and female. Not every woman achieves orgasm at every intercourse but orgasm is not a prerequisite for fertilization. There is intense physical and psychological tension during intercourse which is relieved by the orgasm. During the excitement and orgasmic phases the respiratory rate may reach 40 per minute, the pulse rate may reach 170 per minute and the blood pressure may increase 30 to 80 mm Hg during systole and 20 to 40 mm Hg during diastole. Orgasm is followed by a *resolution* phase of muscular relaxation and a feeling of physical and mental lassitude. The male cannot immediately return to another orgasm but requires a latent period of some minutes before beginning again. The female, in contrast, requires no such latent period between climaxes.

Sperm become motile only after they have been in contact with the secretions of the epididymis; sperm removed directly from the testis are nonmotile. Sperm can survive only briefly in the acidic environment of the vagina (*p*H about 4.5), but the alkaline seminal fluid aids in overcoming this acidity. Sperm appear in the fallopian tubes within a very brief period, 20 minutes or less, after copulation and apparently are transported through the cervix and uterus and up into the fallopian tube by muscular contractions of the uterus and tubes. Since female orgasm is not at all essential for fertilization either the uterine contractions occur independently of orgasm, or else sperm can swim through the uterus without the aid of myometrial contractions.

Prostaglandins, secreted by the seminal vesicles into the seminal fluid, increase the activity of uterine muscle and may play a role in furthering the union of egg and sperm.

The inability of a man to have an erection of the penis is termed *impotence*. This is to be contrasted to male *sterility*, which is present in 3 to 4 per cent of men and results from the deficiency or complete absence of normal sperm from the seminal fluid. Some sperm have abnormalities readily seen under the microscope, such as two tails or two heads. Male sterility usually occurs when the total number of sperm in a single ejaculation falls below 150,000,000. Some 14 per cent of women are sterile either because their fallopian tubes have been blocked by infection, the ovary is covered with a thick capsule, or because their pituitary secretes inadequate amounts of gonadotropins. The cervical mucus secreted by some women resists the passage of sperm (some times just the sperm of some men but not others) apparently by some sort of antigen-antibody reaction.

SECTION 25 12

HUMAN   EPRODUCTION: FERTILIZATION

Sperm deposited in the vagina during intercourse travel up the vagina and into the uterus partly under their own power but mostly by the force of the muscular contractions of the walls of these organs. Most of the sperm become lost on the journey, but a few find their way to the openings of the fallopian tubes and swim up them. Sperm swim against a current and the same current which draws the egg into the fallopian tube probably assists the sperm in finding their way. If the egg is fertilized this

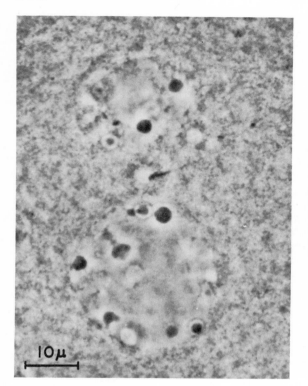

usually occurs in the upper third of the fallopian tube. Only one of the hundreds of millions of sperm deposited at each ejaculation fertilizes a single egg (Fig. 25–12).

Each human egg is surrounded by a layer of cells derived from the follicle and termed the **corona radiata**. By the time the ovum has reached the fallopian tube it has completed the first meiotic division and extruded the first polar body. It is probable, though not proven, that hyaluronidase from the seminal plasma and the hydrolytic enzymes of the lysosomes in the sperm head play a role in the penetration of the sperm through the corona radiata and the vitelline membrane. Once one sperm has entered the egg there is some sort of change in the surface layer of the ovum which prevents the entrance of other sperm. The sperm either leaves its tail outside the egg or sheds it shortly after entering the egg's cytoplasm; only the nuclear material of the head and the centriole remain (Fig. 25–13).

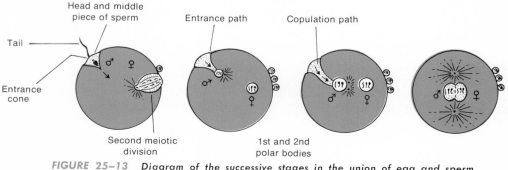

FIGURE 25–13   *Diagram of the successive stages in the union of egg and sperm during fertilization.*

The ovum completes the second meiotic division and extrudes the second polar body. The head of the sperm swells to form the **male pronucleus** and the nucleus of the ovum becomes the **female pronucleus.** The fusion of the haploid male pronucleus with the haploid female pronucleus forms the nucleus of the fertilized egg or zygote and restores the diploid number of chromosomes.

In view of the many factors working against fertilization it may seem remarkable that it ever does occur, and indeed, man is a relatively infertile animal (yet fertile enough so that overpopulation is the major problem facing mankind today). Sperm remain alive and retain their ability to fertilize an egg from 24 to 48 hours at most after having been deposited in the female tract and the egg loses its ability to be fertilized about 24 hours after ovulation. Sperm cells are delicate, their cytoplasm contains meager resources of food, and they are sensitive to heat, slight changes of pH and so on. The leukocytes of the vaginal epithelium which engulf millions of sperm are another hazard. However, the frequency of copulation and the large number of sperm deposited at each ejaculation enable the human race not only to maintain itself but to increase at an alarming rate.

SECTION 25–13

HUMAN REPRODUCTION: DEVELOPMENT AND IMPLANTATION OF THE BLASTOCYST

The first cleavage of the fertilized egg occurs about 30 hours after insemination and the succeeding mitoses occur every 10 hours or so. By the time the developing ovum reaches the uterus, perhaps 3 to 7 days after fertilization, it is a tight ball of some 32 cells, called a **morula**. If the fertilized egg passes through the fallopian tube too rapidly and reaches the uterus prematurely it cannot be embedded in the wall. One type of contraceptive device, the **intrauterine coil**, may stimulate muscular contractions of the fallopian tube and uterus so that the fertilized ovum reaches the uterine cavity prematurely and dies before it can undergo implantation.

When the developing ovum reaches the uterine cavity it begins to differentiate into a **blastocyst** (Fig. 25–14) composed of an outer envelope

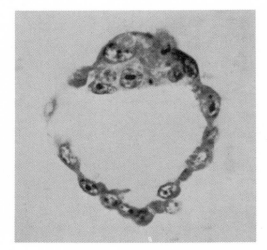

FIGURE 25–14   Photomicrograph of a human blastocyst. (From Reid, D. E.: Principles of Obstetrics. Philadelphia, W. B. Saunders Company, 1962.)

of cells, the **trophoblast,** and an **inner cell mass,** a ball of cells at one pole within the trophoblast which is the precursor of the embryo. This stage implants in the endometrial lining of the uterus by secreting enzymes which erode the cells of the endometrium, permitting the embedded blastocyst to establish close contact with the maternal blood stream (Fig. 25–15). The cells of the trophoblast grow and divide rapidly; they and the adjacent cells of the uterine lining, the **decidua,** form the placenta and fetal membranes. The cells of the endometrium heal over the site of entry of the blastocyst so that it lies wholly within the endometrium and out of the uterine lumen.

The embedding reaction of the endometrium can be elicited by pricking with a glass needle the uterine lining of a female suitably primed with estradiol and progesterone. This stimulus leads to **"pseudopregnancy"** and the uterus develops for a short period just as though an embryo were present.

The trophoblast initially consists of two layers of cells, an inner **cytotrophoblast** composed of individual cells and an outer **syncytiotrophoblast** composed of a multinucleate syncytium. The trophoblastic cells digest and phagocytize materials in the endometrium that were stored prior to implantation. The trophoblast soon is bathed in and nourished by the maternal blood. Normally menstruation occurs about 14 days after ovulation in the nonpregnant woman. To prevent this the embedded embryo must signal the maternal organism in some way. Because of the time required for traversing the oviduct, and because the fertilized egg remains in the lumen of the uterus for some days, about 11 days elapse between ovulation and implantation. Thus the embryo has only a very few days to provide the signal which will prevent menstruation. Fairly frequently the signal does not arrive in time and menstruation sweeps out the fertilized egg. The woman has been pregnant in the sense that she had a fertilized egg in her reproductive tract, but was never aware of it and

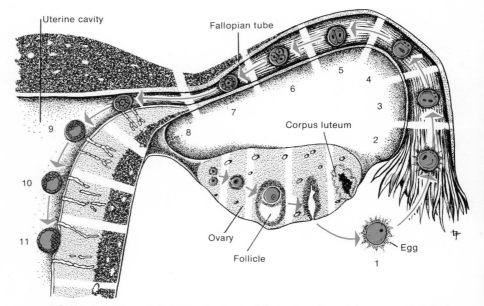

FIGURE 25–15   *Schematic diagram of the maturation of the egg in a follicle in the ovary, its release (ovulation) (1), fertilization in the upper part of the oviduct (2), cleavage of the egg as it descends the oviduct or fallopian tube (3–7), stages in the development of the embryo in the uterus before implantation (8–10) and implantation of the embryo in the wall of the uterus (11).*

menstruated at her usual time. One of the major contributions of the trophoblast is its secretion of **chorionic gonadotropin**, probably by the cells of the cytotrophoblast. Chorionic gonadotropin has properties similar to those of luteinizing hormone and luteotropic hormone of the pituitary; it prevents the corpus luteum from involuting. The secretion of chorionic gonadotropin begins the day the trophoblast is embedded in the endometrial lining.

The process of implantation has implicit in it two questions of general biological interest. Why does the trophoblast generally cease invading the endometrial lining once it has formed a connection with the maternal blood? Why doesn't it continue to invade as a group of cancer cells would? And why, since the cells of the trophoblast have the genotype of the developing fetus, a genotype different from that of the mother, do the maternal cells not react as though the trophoblast were a transplant and reject it, as an animal rejects a skin graft from another, genetically different member of the same species (p. 581)?

SECTION 25–14

HUMAN REPRODUCTION: NUTRITION OF THE EMBRYO

After implantation in the uterine lining, the embryo continues to develop, at first obtaining its nourishment by enzymatically breaking down the cells of the uterine lining immediately around it, and later by extracting the nutritional essentials from the blood stream of the mother via the blood vessels of the placenta.

The new human being develops only from the inner cell mass, the cells which lie along one side of the hollow ball originally implanted in the uterus; the other cells form membranes which nourish and protect the developing child. The problem of supplying the embryo with food during development has been solved in different ways by the several groups of vertebrates.

Fish and amphibia produce relatively large eggs, containing yolk to supply the necessary proteins, fats and carbohydrates. These eggs are laid and develop in water, whence they obtain oxygen, salts and water itself. The embryos of these animals have a pouchlike outgrowth of the digestive tract, the **yolk sac**, which grows around the yolk, digests it and makes it available to the rest of the organism.

The eggs of reptiles and birds are usually laid on land and are enclosed within a shell which protects them from excessive drying. They have additional membranes for the protection and nutrition of the embryo. The familiar "white" of the hen's egg is an extra store of protein and water to support the embryo until the time of hatching. Both the shell and the white of eggs of reptiles and birds are secreted by glands located in the wall of the oviducts while the egg is passing down these tubes.

SECTION 25–15

THE EMBRYONIC MEMBRANES

Several embryonic membranes have evolved to enfold, protect, support and nourish the embryos of reptiles, birds and mammals. These membranes, the **amnion**, the **chorion** and the **allantois**, are sheets of living tissue growing out of the embryo itself—the amnion and chorion out of the body wall to enfold the embryo (Fig. 25–16) and the allantois out of the digestive tract to function in the absorption of food.

The formation of the amnion is a complex process, differing in detail in different species, but it is essentially an outfolding of the body wall of the embryo which grows around the embryo to meet and fuse above it (Fig. 25–16). The space between the embryo and the amnion, known as the **amniotic cavity**, becomes filled with a clear, watery fluid secreted by both the embryo and the membrane. Embryos of the higher vertebrates reach the birth stage enclosed in a small pool within the shell or uterus. The amniotic fluid prevents desiccation of the embryo and acts as a protective cushion which absorbs shocks and prevents the amniotic membrane from sticking to the developing embryo, while permitting the organism a certain freedom of motion. During the birth process of man and other mammals, pressure of the contracting uterus is transmitted via the amniotic fluid and helps to dilate the neck of the uterus; later, the amnion normally ruptures, releasing shortly before the fetus is born about a liter

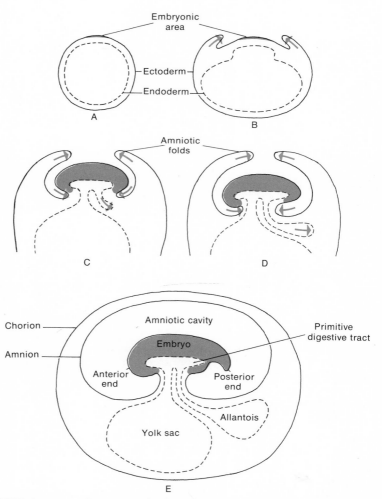

FIGURE 25–16   Diagram indicating the steps in the formation of the embryonic membranes—the amnion, chorion, yolk sac and allantois—in a typical mammal such as a pig. The arrows indicate the direction of growth and folding.

of amniotic fluid, the so-called "waters." Sometimes the amnion fails to burst and the child is born with it still enveloping its head. The amnion is then popularly known as a "caul," and is the source of many odd superstitions.

The amnion develops from the inner part of the original fold from the body wall; the outer part forms a second membrane, the **chorion.** This membrane, in the eggs of reptiles and birds, rests in contact with the inner surface of the shell, and in mammals it is established next to the cells of the uterine wall.

The **allantois**, like the yolk sac, is an outgrowth of the digestive tract. It grows between the amnion and chorion and, in animals like the chick, where it is a large and functional membrane, fills most of the space between these two. The allantois of the reptilian or avian egg serves as a depot for nitrogenous wastes. The products of nitrogen metabolism are excreted as uric acid by the kidney of the developing embryo. The relatively insoluble uric acid is deposited as crystals in the cavity of the allantois and discarded along with the allantois when the young hatch out of the egg shell.

The allantois fuses with the chorion to form a compound chorioallantoic membrane, full of blood vessels, by means of which the embryo takes in oxygen, gives off carbon dioxide and excretes waste products. Since the embryo "breathes" through the shell, it will suffocate if the shell is coated with wax. In the human, the allantois is small and nonfunctional, except for furnishing blood vessels to the placenta, and the yolk sac is completely nonfunctional. When the chick hatches from the shell, or when the child is born, most of the allantois and all the other membranes are discarded. But the base of the allantois, the part connected to the digestive tract originally, remains within the body and is converted into part of the urinary bladder.

As the embryo grows, the region on the ventral side from which the folds of the amnion, the yolk sac and the allantois grew becomes relatively smaller, and the edges of the amniotic folds come together to form a tube which encloses the other membranes. This tube, the **umbilical cord**, contains in addition to the yolk sac and allantois the large blood vessels through which the embryo obtains nourishment from the wall of the uterus. The human umbilical cord, about 1 centimeter in diameter and about 70 centimeters long at birth, is composed chiefly of a peculiar jelly-like material found nowhere else. It is usually twisted spirally, and in its contortions before birth, the fetus* sometimes passes through a loop of the cord and actually ties a knot in it.

SECTION 25-16

THE PLACENTA

The outer surface of the chorion in man and the higher mammals is thin over most of its surface, but at the outer extremity of the umbilical cord it develops a number of finger-like projections, known as **chorionic villi**, which grow into the tissue of the uterus. These villi, plus the tissues of the uterine wall in which they are embedded, make up the organ known

---

*As soon as the zygote or fertilized egg begins to divide it is called an embryo. After the embryo has begun to resemble a human being (some two months after fertilization), it is referred to as a fetus until the time of birth.

as the *placenta*, by means of which the developing embryo obtains nutrients and oxygen and gets rid of carbon dioxide and metabolic wastes
(Fig. 25–17). There are many capillaries in the villi which receive blood
from the embryo by a branch of one of the two **umbilical arteries**, and
return it to the embryo by way of the **umbilical vein**. The uterine lining
becomes thickened and highly vascularized, forming a mass of spongy
tissue filled with blood from the mother's bloodstream. The blood of the
mother and fetus do not mix at all in the placenta or any other place. The
blood of the fetus in the capillaries of the chorionic villi comes in close contact with the mother's blood in the tissues between the villi, but they are
always separated by a membrane through which substances must diffuse
or be transported by some active, energy-requiring process. As the embryo
develops, the placenta necessarily grows too. At the time of birth, it is a
thick, circular disc 15 to 20 centimeters in diameter and 2 to 3 centimeters
thick, and weighing about 500 grams. The placenta is a very active tissue
with a high rate of blood flow and a high rate of oxygen consumption. At
term about 600 milliliters of maternal blood passes each minute through
placental spaces that total about 140 milliliters and have about 11 square
meters of surface area. The fetal blood flow, entering by the two umbilical
arteries, is about 300 ml per minute. The oxygen consumption of the
placenta, about 10 μl per g tissue and minute, is twice that of the fetus.
Besides serving as the nutritive, respiratory and excretory organ of the
fetus, the placenta is an important endocrine gland (see p. 663).

The uterus increases in size as the fetus grows, and by the end of nine
months its mass is 24 times as great as at the beginning of pregnancy.
After six months of fetal development the upper end of the uterus is on a
level with the navel; by eight months it is as high as the lower edge of the
breastbone. Within the uterus the fetus assumes a characteristic "fetal"
position with elbows, hips and knees bent, arms and legs crossed, back
curved, and the head bowed and turned to one side. At birth the fetus
usually is turned head downward so that its head emerges first, but occasionally the buttocks or feet are presented first, making delivery more
difficult.

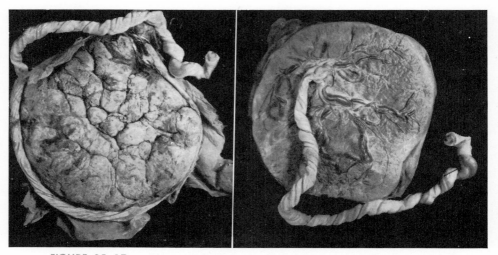

FIGURE 25–17   *Photograph of the maternal (left) and fetal (right) surfaces of
a human placenta at term, the end of a normal pregnancy. (From Greenhill, J. P.: Obstetrics. 13th ed. Philadelphia, W. B. Saunders Company, 1965.)*

SECTION 25-17

THE BIRTH PROCESS

The human **gestation period**, the duration of pregnancy, is normally 280 days from the time of last menstrual period to the birth of the baby. Babies born as early as 28 weeks or as late as 45 weeks after the last menstrual period may survive. The factors which actually initiate the process of birth or **parturition**, after the period of gestation is complete, are unknown. Childbirth begins with a long series of strong contractions of the uterus, experienced as "labor pains." Labor may be divided into three periods. During the first period, which lasts about 12 hours, the contractions of the uterus move the fetus down toward the cervix and cause the latter to dilate, enabling the fetus to pass through. At the end of this period the amnion usually ruptures, releasing the amniotic fluid, which flows out through the vagina. In the second period, which normally lasts between 20 minutes and an hour, the fetus passes through the cervix and vagina and is born or "delivered" (Fig. 25-18). The fetus is expelled from the uterus by the combined forces of uterine contraction plus the contractions of the muscles of the abdominal walls. With each uterine contraction the woman holds her breath and bears down.

After the baby is born and before the umbilical cord is cut the contractions of the uterus may squeeze much of the fetal blood from the placenta back into the infant. After the pulsations in the umbilical cord cease the cord is tied and cut, severing the child from the mother. The stump of the cord gradually shrivels until nothing remains but the depressed scar, the **navel**. During the third stage of labor, which lasts perhaps 10 or 15 minutes after the birth of the child, the placenta and fetal membranes are loosened from the lining of the uterus by another series of contractions and expelled. In man and certain other mammals in which the placenta forms a very tight connection with the lining of the uterus, the expulsion of the placenta is accompanied by some loss of blood. In other mammals, where the connection between the fetal membranes and uterine lining is not so close, the placenta can pull away from the uterine wall without causing

FIGURE 25-18   Models showing the process of birth. (From the Dickinson-Belski Series, The Maternity Center Association.) Left, Head passing through the dilated cervix of the uterus into the upper part of the vagina. Right, Head passing through the external opening of the vagina. (From Dickinson-Belskie Series in the Birth Atlas, published by Maternity Center Association, New York.)

bleeding. After birth the size of the uterus decreases and its lining is rapidly restored.

The women of civilized countries appear to have more difficulty during childbirth than the women of primitive tribes do, and they frequently require help in being delivered. The obstetrician may administer drugs such as oxytocin to increase the contractions of the uterus or he may have to assist the expulsion of the child by using a pair of forceps to draw it out. In some women the aperture between the pelvic bones, through which the vagina passes, is too small to permit the passage of the baby. In such cases the child must be delivered by an operation in which the abdominal wall and uterus are cut open from the front. This operation is now called a "cesarean delivery." It seems unlikely that Julius Caesar was born this way (his mother was still alive when he was a grown man). The term is probably derived from the latin *caedere,* to cut, or from the Roman law, *lex caesarea,* that governed the cutting open of a dying woman in an attempt to save her unborn child.

In a little more than 20 per cent of all known pregnancies the infant is born before it is prepared to carry on an independent existence. When this occurs, the birth is called an **abortion** or miscarriage. Such premature births may be caused by improper implantation of the embryo, by faulty functioning of the placenta or by injury to the mother.

Among mammalian species there are great differences as to the condition of the young at birth. The newborn of some species, such as the rat, are blind, hairless and helpless, while others, such as the guinea pig, are well developed and able to walk and eat solid food. There is also great variation in the weight of the newborn in comparison to the mother's weight: a newborn polar bear weighs 0.1 per cent of its mother's weight; a newborn human weighs about 5 per cent as much as its mother; and a newborn bat may weigh as much as 33 per cent of its mother's weight.

SECTION 25–18

NUTRITION OF THE INFANT

During pregnancy both the glands and ducts of the mammary glands grow, stimulated by estradiol and progesterone. The volume of the breast also increases because of the deposition of fat and the engorgement with blood. Each breast increases in volume by about 200 milliliters on the average, but the increased volume may be as much as 800 milliliters. The onset of milk secretion occurs under the stimulus of hormones from the ovary and the anterior and posterior pituitary, but the continued secretion of milk depends upon the presence of a suckling child. If the newborn infant is not breast fed, the breasts stop secreting milk after a few days and decrease in size. Otherwise, they continue to yield milk from six to nine months or even longer. Formerly it was not unusual for women to nurse their children for as long as a year and a half.

As everyone knows, milk is an excellent food, containing proteins, fats and carbohydrates. But it is deficient in some things, especially iron and vitamins C and D, so that at the present time an infant's diet is usually supplemented after the first month or so with eggs to supply iron, and orange juice and cod liver oil to supply the vitamins. The milks of various species differ in their nutrient content, and to raise a human infant on cow's milk, the latter must be diluted and sugar must be added to approximate the content of human milk.

The widely prevalent assumption that women are infertile after the delivery of a baby is only relatively true. A woman who does not breast feed her child may ovulate within six weeks after delivery. Lactation does not necessarily inhibit either ovulation or menstruation, but the return of ovulation tends to occur later in women who are breast feeding. Even in the latter, ovulation may occur by the twelfth week postpartum.

SECTION 25-19

CONTROLLING THE SIZE OF THE HUMAN POPULATION

The rapidly rising world population has made clear the danger that the human population will outrun the resources of the world needed to maintain a reasonable standard of life. Many of the other problems facing mankind at present, such as the pollution of the environment, stem directly from the vastly increased number of people on this planet. Even if ways are found to increase the food supply to feed the seven billion people expected on the earth in the year 2000 compared to the 3.5 billion living now, man may well be limited by the supply of air pure enough to breath, the supply of water pure enough to drink, or the availability of places to put his accumulated trash and garbage.

The methods now available for restricting the number of infants born include: (1) suppressing the formation and release of gametes, (2) preventing fertilization, (3) preventing the implantation of a fertilized egg and (4) abortion, premature delivery of the implanted embryo. There are several types of contraceptive pills now in use but each typically consists of a small amount of estrogen combined with a larger amount of a progesterone-like compound. The pills are taken daily from the fifth to the twenty-fifth day of the cycle and act on the pituitary to suppress the formation and secretion of FSH and LH (p. 672). The pills may decrease somewhat the amount of glandular tissue in the endometrium and change the nature of the cervical mucus so that it is more difficult for sperm to enter the uterus. No comparably safe and effective method for suppressing spermatogenesis in the male has been found.

Another method of contraception is to prevent fertilization by interposing a barrier between the gametes. A permanent barrier can be raised by tying or removing a section of the fallopian tubes or of the vasa deferentia. Temporary barriers are provided by a condom on the penis or a diaphragm which stretches across the vault of the vagina and occludes the opening of the cervix. Spermicidal jellies and foams, used alone or in combination with a diaphragm, also reduce the probability of fertilization.

The intrauterine coil, of which there are several designs, is placed in the lumen of the uterus and kept there for months. These probably increase the motility of the fallopian tubes so that the fertilized egg reaches the endometrium before it can undergo implantation.

The fourth method of birth control, abortion, has been widely practiced in all parts of the world. Emptying the contents of the uterus may be brought about in any one of several ways and is a safe procedure when carried out by qualified physicians under sterile conditions in a hospital. It can be a very dangerous procedure under any other condition. The cervix can be dilated and the lining of the uterus scraped with a special curette or the implanted embryo can be removed with a special suction device. A needle can be inserted through the abdominal wall and the wall of the uterus to the amniotic cavity and some of the amniotic fluid is withdrawn and replaced by hypertonic saline. This kills the fetus and the uterus gradually goes into labor and empties its contents during the next two or three days.

The demographic unit for measuring the efficacy of any method of birth control is the number of pregnancies that occur per "100 woman years," i.e., in a group of 100 women using that method of contraception for one year, or for 1200 to 1300 ovulations, assuming that all the women ovulate about once every 28 days. With no attempt to restrict conception, between 50 and 80 pregnancies per 100 woman years will result. Restricting intercourse to the "safe period" results in about 24 pregnancies per 100 woman years. Using the condom or diaphragm results in 14 and 12, respectively, pregnancies per 100 woman years. The intrauterine coil reduces the number of pregnancies to about 2 per 100 woman years and women using the contraceptive pills have between 0 and 1 pregnancies per 100 woman years. The failures of the last two methods probably result not from any defect inherent in the method but in the way the method is used.

SUGGESTIONS FOR FURTHER READING

Asdell, S. A.: *Patterns of Mammalian Reproduction.* 2nd ed. Ithaca, New York, Comstock Publishing Associates, 1964.
> An excellent source book on differences in reproductive cycles and processes in mammals ranging from the aardvark to the zebu.

Austin, C. R.: *Fertilization.* Englewood Cliffs, New Jersey, Prentice-Hall, Inc., 1965.
> A brief text summarizing our knowledge of this field.

Diamond, M.: *Perspectives in Reproduction and Sexual Behavior.* Bloomington, Indiana, Indiana University Press, 1968.
> A collection of papers analyzing a wide variety of problems dealing with the endocrine control of reproduction and sexual behavior.

Masters, W. H., and Johnson, V. F.: Human Sexual Response, Boston, Little, Brown and Company, 1966.
> An excellent summary of the authors' investigations of the physiological aspects of human mating.

Metz, C. B., and Monroy, A. (eds.): *Fertilization.* New York, Academic Press, Inc., 1967.
> A series of up-to-date papers covering a variety of aspects of the fertilization process in invertebrate and vertebrate forms.

Milne, L. J., and Milne, M. J.: *The Mating Instinct.* Boston, Little, Brown and Company, 1954.
> Interesting accounts of mating behavior in a variety of animals.

Villee, C. A. (ed.): *The Control of Ovulation.* New York, Pergamon Press, Inc. 1961.
> A series of papers dealing with the process of ovulation and its control presented at a symposium, plus a transcript of the discussion following each.

Young, W. C. (ed.): *Sex and Internal Secretions.* 3rd ed. Baltimore, The Williams & Wilkins Company, 1961.
> This two-volume work by many experts in the field covers a wide variety of the endocrine control of reproduction and behavior.

DEVELOPMENT

CHAPTER 26

The division, growth and differentiation of a fertilized egg into the remarkably complex and interdependent system of organs which is the adult animal or plant is certainly one of the most fascinating of all biological phenomena. Not only are the adult organs that result complicated and reproduced in each new individual with extreme fidelity of pattern, but many of the organs begin to function while they are still developing. The human heart begins to beat, for example, during the fourth week of gestation, long before its development is complete.

The pattern of cleavage, blastula formation and gastrulation appears, with various modifications, in the development of nearly all multicellular animals. The details of later development may be quite different in animals of different phyla but are similar in more closely related forms.

The development of a new individual begins with the formation of eggs and sperm in members of the parental generation. During gametogenesis (p. 227), meiosis reduces the number of chromosomes from the diploid to the haploid condition and there is a random selection of the specific genes which will be united in the new individual and guide its development. The egg cell undergoes growth and accumulates in its cytoplasm a variety of substances which will be used in the early stages of development of the fertilized egg. The sperm matures and changes from a spherical cell to one adapted to move actively to the egg and fertilize it.

The fertilization of the egg, the second stage of development, begins with the various processes which bring the parents together and stimulate them to release their gametes simultaneously, and includes the penetration of the egg by the sperm followed by the activation of the fertilized egg so that it begins to develop. During the activation process the several kinds of organ-forming material within the egg's cytoplasm may undergo marked rearrangements of their positions.

During the third stage, the fertilized egg undergoes a rapid succession of mitotic divisions termed *cleavage;* these result in a compact mass of cells or in a hollow ball of cells, a *blastula,* with a single layer of cells, the *blastoderm,* surrounding a cavity, the *blastocoele.* During the cleavage divisions the size of the embryo does not increase, and the cleavage cells or *blastomeres* become successively smaller with each division. Since glycogen and other nutrients stored in the cytoplasm of the egg are metabolized to provide energy for the cleavage divisions, the mass of the blastula (dry weight) may actually be less than the mass of the unfertilized egg.

The fourth phase of development, **gastrulation,** results in the appearance of two or more layers of cells, the **germinal layers,** from the single-layered blastoderm. The bodies of higher animals consist of several layers of tissues and organs all of which can be traced back to the three germinal layers, **ectoderm, mesoderm** and **endoderm.** The outer ectoderm forms the epidermis of the skin and the nervous system; the innermost layer, the endoderm, forms the lining of the gut and the digestive glands; and the middle layer, the mesoderm, is the source of the muscles, the vascular system, the lining of the coelomic cavity, the reproductive organs, the excretory organs and the bones and cartilages of the internal skeleton.

Part of the blastoderm disappears from the surface and is enclosed by the remainder of the blastoderm. The portion remaining on the exterior becomes the ectoderm and the part in the interior of the embryo becomes endoderm and mesoderm. The endoderm and mesoderm cells may reach the interior by the folding-in of one part of the blastoderm. The simple spherical ball of cells becomes a double-walled cup by a process termed **invagination.** The double-walled cup of cells is called a **gastrula,** hence the term **gastrulation** for the process by which it is formed. The mesoderm separates from the endoderm in the interior of the gastrula by quite different processes in different kinds of animals. The cavity of the double-walled gastrula formed by invagination is called the **archenteron** or primitive gut, and the opening from the archenteron to the exterior is the **blastopore.**

The archenteron, or a portion of it, eventually gives rise to the cavity of the digestive tract. The blastopore develops into the oral opening in coelenterates and becomes the anal opening in echinoderms and chordates. In the mollusks, annelids and arthropods the blastopore becomes divided into two openings, one of which becomes the mouth and the other the anus. The central portion of the digestive tract is formed from endoderm but typically the ectoderm undergoes invaginations at both oral and anal ends which fuse with the endoderm. Thus the most anterior portion of the digestive tract, the **stomodeum** (mouth cavity), and the most posterior portion of the digestive tract, the **proctodeum,** adjoining the anus, are lined with ectoderm rather than endoderm.

The next stage of development is **organogenesis,** during which the three germ layers split into smaller masses of cells, each destined to form a specific organ or portion of the adult. Each organ is first discernible as a group of cells, a **rudiment,** which becomes segregated from the other cells of the embryo. Organogenesis is followed by growth and differentiation of the cells of the organ rudiments. Eventually the cells acquire the specific structure and the chemical and physiological properties which enable them to perform their functions so that the developing animal can take up an independent existence.

The young emerging from the egg may be a miniature copy of the adult, or it may lack certain organs and differ to some extent from the adult in form as well as in size. In other animals the form emerging from the egg has special organs, not present in the adult, which are needed for the survival of the young in its particular mode of existence. Such a young animal, termed a **larva,** may lead a mode of life quite different from the adult. Eventually the larva undergoes a relatively rapid and marked change of form, **metamorphosis,** and achieves the adult form. During metamorphosis the special larval characteristics are lost and the adult characteristics appear. The young and adult animals continue to undergo changes with time, which we call **aging,** and which must be considered part of the development process.

TYPES OF EGGS

The amount of yolk, or stored nutrients, present in an egg plays an important role in determining the pattern of development that ensues following fertilization. Yolk granules vary in composition from one species to another but usually contain varying amounts of protein, phospholipids and neutral fats. Most invertebrates and the lower chordates have small eggs with relatively small amounts of yolk. The yolk granules in the egg of the sea urchin, *Arbacia*, for example, make up about 27 per cent of the total volume of the egg. Eggs such as these, with small amounts of yolk present in fine granules distributed fairly uniformly throughout the cytoplasm, are termed **isolecithal** (equal yolk) (Fig. 26–1).

The eggs produced by fish, amphibia, reptiles and birds have a large amount of yolk, usually present in large granules called **yolk platelets.** The yolk is massed toward the lower or **vegetal pole** of the egg while the cytoplasm is concentrated toward the upper or **animal pole** of the egg. Such eggs are termed **telolecithal** (end yolk). Two proteins found in the yolks of both amphibians and avian eggs are phosvitin and lipovitellin. **Phosvitin** is a curious protein; about half of its amino acid residues are phosphorylated serine. Lipovitellin is a much larger protein molecule which contains about 17 per cent lipid. In amphibians and birds phosvitin is synthesized in the maternal liver and transported in the blood to the ovary. Once taken up by the egg it may be phosphorylated further, which makes it less soluble and causes it to precipitate in the oöcyte. A protein kinase has been found in the frog ovary which can catalyze the incorporation of phosphate from ATP into partially phosphorylated phosvitin. Yolk makes up about 45 per cent of the weight of an amphibian egg and as much as 90 per cent or more of the eggs of birds, reptiles and bony fish. In the latter three the cytoplasm is restricted to a thin layer on the surface of the egg; a thicker cap of cytoplasm at the animal pole contains the nucleus. The entire center of the egg is filled with yolk platelets.

The eggs of arthropods, especially insects, have a different pattern of distribution of yolk, termed **centrolecithal.** The yolk is concentrated in the center of the egg and the cytoplasm is present as a thin layer on the surface.

FIGURE 26–1   Photographs of eggs of the sea urchin Arbacia punctulata subjected to centrifugation to separate the lighter nucleate half from the heavier yolk-laden anucleate half. Upper left, mild gravitational force stratifies constituents of the egg, in order form top to botton: oil cap (dark) circular nucleus, clear area, mitochrondria, dark yolk granules. Further centrifugation first elongates egg (lower left), then separates it into nucleate half (upper right) and anucleate half (lower right). (Courtesy of Grant Patton and Lawrence Mets.)

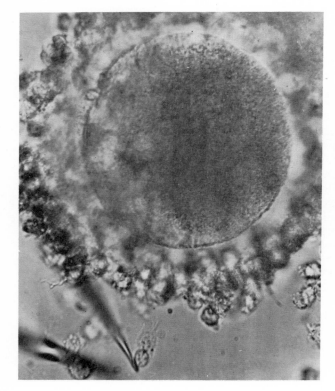

FIGURE 26–2   A human egg recovered from an ovarian follicle. The large egg cell is surrounded by the many small cells which make up the corona radiata, cells which were originally part of the follicle. The photograph was made in the course of an experiment in which microneedles were used to dissect away the corona radiata cells and to dissect the egg itself. The two needles can be seen in the lower part of the picture dissecting the nucleus out of one of the corona radiata cells. ×600. (Courtesy of Dr. William R. Duryee.)

In addition there is an island of cytoplasm in the center of the egg which contains the nucleus.

The duck-billed platypus and the echidna have small, yolk-filled eggs comparable in size to those of a small lizard, which are laid and develop outside the mother's body. The eggs of other mammals are small and relatively free of yolk (Fig. 26–2). They superficially resemble isolecithal eggs but their pattern of development is more nearly like that of telolecithal eggs, probably because mammals evolved from reptiles which laid the latter kind of egg.

SECTION 26–2

CLEAVAGE AND GASTRULATION

The entrance of the sperm into the egg initiates a rapid series of changes—the completion of the meiotic divisions, the fusion of male and female pronuclei, complex movements of the egg's cytoplasmic constituents and the beginning of the cleavage divisions. There is a sharp rise after fertilization in the rates of oxygen consumption and protein synthesis in the eggs of certain species. There is some evidence to support the hypothesis that the RNA present in the fertilized egg, which directs protein synthesis during the early stages of cleavage, was synthesized in the oöcyte before fertilization and hence is a product of the maternal genotype rather than of the genotype of the embryo. The RNA is "masked" in the unfertilized egg, perhaps by combination with a protein, and is unavailable for translation into a peptide chain. The ribosomes of the unfertilized egg are generally single and polyribosomes appear only after fertilization has occurred. One of the effects of fertilization thus appears to be an "unmasking" of the maternal messenger RNA so that it can be translated in protein synthesis. It is only later, at about the time of gastrulation, that the genome of the embryo is transcribed to form "embryonic" messenger RNA which codes for proteins that play a role in the gastrulation process and in later development. The problem of whether fertilization does lead to the "unmasking" of previously masked, inactive, "maternal" messenger RNA, and how this is achieved, remains unsolved, but the hypothesis has attracted widespread interest and has stimulated many types of experimental approaches.

CLEAVAGE IN ISOLECITHAL EGGS.　　The first cleavage division of an isolecithal egg passes through both animal and vegetal poles and splits the egg into two equal cells (Fig. 26–3). The second cleavage division also passes through both poles of the egg but at right angles to the first, and separates the two cells into four equal cells. The third division is horizontal, at right angles to the other two, and separates the four cells into eight—four above and four below the line of cleavage. Further divisions result in embryos containing 16, 32, 64, 128 cells, and so on until a hollow ball of cells, the **blastula,** is formed. The wall of the blastula is a single layer of cells, the **blastoderm,** surrounding the **blastocoele,** the cavity in the center.

The single-layered blastula is converted into a double-layered sphere, a **gastrula,** by the invagination of a section of one wall of the blastula. This eventually meets the opposite wall and obliterates the original blastocoele. The new cavity of the gastrula is the **archenteron,** and the opening of the archenteron to the exterior is the **blastopore,** which marks the site of the

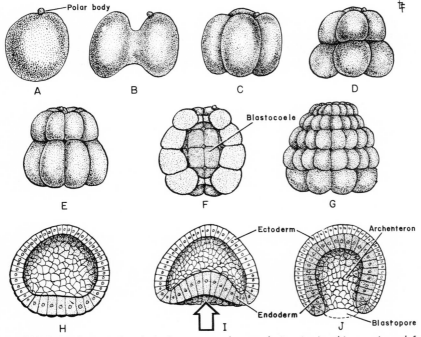

FIGURE 26–3　Isolecithal cleavage and gastrulation in Amphioxus viewed from the side. A, Mature egg with polar body. B—E, two-, four-, eight- and 16-cell stages. F, Thirty-two-cell stage cut open to show the blastocoele. G, Blastula and H, blastula cut open. I, Early gastrula showing the beginning of invagination at the vegetal pole. J, Late gastrula, invagination completed and blastopore formed.

invagination which produced the gastrula. The outer of the two walls of the gastrula is the **ectoderm,** which eventually forms the skin and nervous system. The inner layer of cells, lining the archenteron, is mainly the presumptive endoderm, which will form the digestive tract and its outgrowths such as liver, pancreas and lungs, plus the presumptive notochord and the presumptive mesoderm, which will form the remaining organs of the body. As the gastrula elongates in the anteroposterior axis the presumptive notochord is stretched into a longitudinal band of cells occupying the mid-dorsal part of the inner layer. The presumptive mesoderm forms two longitudinal bands of cells, one on each side of the presumptive notochord. The remainder of the lateral, ventral and anterior parts of the inner layer are presumptive endoderm cells.

CLEAVAGE IN TELOLECITHAL EGGS.   The processes of cleavage and gastrulation are markedly modified in telolecithal eggs by the large amount of yolk present. The cleavage divisions of the cells originating from the lower part of the frog's egg are slowed by the inert yolk so that the blastula consists of many small cells at the animal pole and a few large cells at the vegetal pole (Fig. 26–4). The lower wall is much thicker than the upper one and the blastocoele is displaced upward.

In eggs with a larger amount of yolk, such as a hen's egg, cleavage occurs only in the small disc of cytoplasm at the animal pole (Fig. 26–5). At first, all the cleavage planes are vertical and all the blastomeres lie in a single plane. The cleavage furrows separate the blastomeres from each other but not from the yolk; the central blastomeres are continuous with the yolk at their lower ends and the blastomeres at the circumference of the disc are continuous both with the yolk beneath them and with the uncleaved cytoplasm at their outer edge. As cleavage continues more cells become cut off to join the ones in the center but the new blastomeres are also continuous with the uncleaved underlying yolk. The central blastomeres eventually become separated from the underlying yolk either by cell divisions with horizontal or tangential cleavage planes or by the appearance of slits beneath the nucleated portions of the cells. Horizontal cleavages separate an upper blastomere which is a cell with a complete plasma membrane, separated from its neighbors and from the yolk, and a lower blastomere which remains connected with the yolk. The blastomeres at the margin of the disc and the lower cells in contact with the yolk eventually

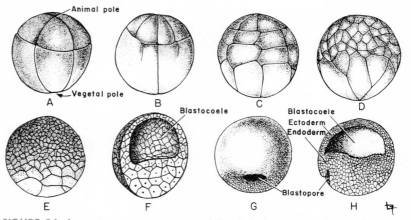

FIGURE 26–4   Successive stages in telolecithal cleavage and gastrulation in the frog, viewed from the side. A–D, Cleavage. E, Blastula. F, Blastula cut open. G, Early gastrula. H, Early gastrula cut open.

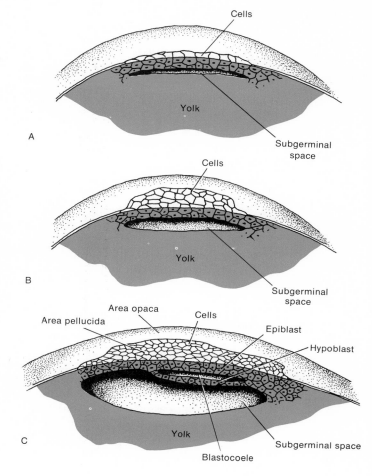

FIGURE 26–5 *Successive stages in the cleavage of a hen's egg. Cleavage is restricted to a small disc of cytoplasm on the upper surface of the egg yolk called the blastodermic disc. A subgerminal space appears beneath the blastodermic disc separating it from the unsegmented yolk. The blastodermic disc cleaves into an upper epiblast and a lower hypoblast separated by the blastocoele.*

lose the furrows that partially separated them and fuse into a continuous syncytium with many nuclei, termed the **periblast,** which does not participate in the formation of the embryo but is believed to break down the yolk and make its nutrients available for the growing embryo.

In birds and some reptiles the free blastomeres with complete plasma membranes become incorporated into two layers, an upper **epiblast,** and below that a thin layer of flat epithelial cells, the **hypoblast.** The hypoblast is separated from the epiblast by a cavity, the blastocoele, and is separated from the underlying yolk by another cavity, the **subgerminal space.** The subgerminal space appears only under the central portion of the blastoderm. The area of the blastoderm over the subgerminal space is more transparent and called the **area pellucida,** whereas the more opaque part of the blastoderm which rests directly on the yolk is called the **area opaca.**

CLEAVAGE IN CENTROLECITHAL EGGS.   Cleavage in centrolecithal eggs such as those of insects begins with the division of the nucleus in the central island of cytoplasm (Fig. 26–6). After several nuclear divisions without cytoplasmic division the nuclei migrate out from the center of the egg. Each nucleus is surrounded by a bit of the original central cytoplasm. When the nuclei reach the surface of the egg the cytoplasm surrounding them fuses with the superficial layer of cytoplasm, resulting in a syncytium

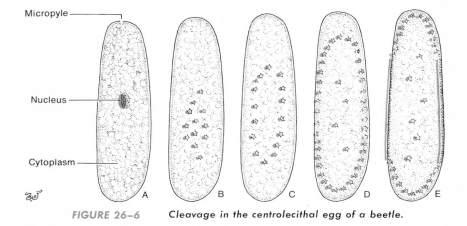

Micropyle

Nucleus

Cytoplasm

A   B   C   D   E

FIGURE 26–6   *Cleavage in the centrolecithal egg of a beetle.*

covering the surface of the egg. The cytoplasm subsequently becomes sub-divided by furrows that extend in from the surface. These blastomeres are connected to the yolk mass for a time but eventually become separated from it; the constituents of the yolk are gradually used up as nutrients for the developing embryo. This stage can be compared to the formation of the blastula even though there is no cavity comparable to the blastocoele. The blastoderm surrounds a mass of uncleaved yolk rather than a cavity.

SYNTHESIS OF NUCLEIC ACIDS IN CLEAVAGE.   There are no qualita-tive changes in the chemical composition of the embryo during cleavage — no new substances appear. The amount of DNA increases, since the number of nuclei is increasing rapidly and the amount of DNA per nucleus re-mains constant (Fig. 26–7). Limited amounts of messenger RNA and trans-fer RNA are synthesized during cleavage but little or no ribosomal RNA is produced until the onset of gastrulation. The synthesis of RNA does not appear to be necessary for cleavage, since eggs treated with large doses of **actinomycin D,** which inhibits DNA-dependent RNA synthesis, continue to cleave normally. However, if fertilized eggs are treated with **puromycin,** which inhibits RNA-dependent protein synthesis, cleavage is halted or proceeds in a very abnormal fashion, indicating that protein synthesis is necessary for normal cleavage.

SEGREGATION OF GENETIC POTENTIALITIES IN CLEAVAGE.   An im-portant question is whether the daughter nuclei formed during cleavage di-visions are exactly equivalent or whether some sort of parceling out of potentialities occurs. It is conceivable that the different fates of the dif-ferent blastomeres may be due to differences in the genetic information present in the different cleavage cells. There is now a wealth of experi-mental evidence indicating that there is no such segregation of genetic potentialities during cleavage. Isolated blastomeres, one of the first four or eight, can develop into an entire embryo with all of its parts but simply reduced in size. The limiting factor in such experiments is not the genetic potential of the nucleus but the quantity of cytoplasm needed for development.

Another type of experiment avoids this by transplanting nuclei from blastomeres or from cells of later stages of development into an enucle-ated uncleaved egg. A ripe frog's egg is prepared by removing its nucleus. A cell from an advanced stage of embryonic development is then separated from its neighbors and sucked up into a micropipette (Fig. 26–8). The

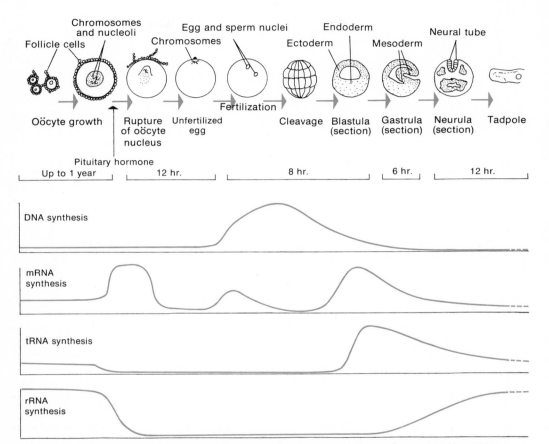

FIGURE 26–7   Diagram illustrating the changes in the synthesis of DNA, mRNA, tRNA and rRNA during oögenesis, at fertilization and during early development in the frog embryo. (After Gurdon, J. B.: Nucleic acid synthesis in embryos and its bearing on cell differentiation. Essays in Biochemistry, 4:25–68, 1968.)

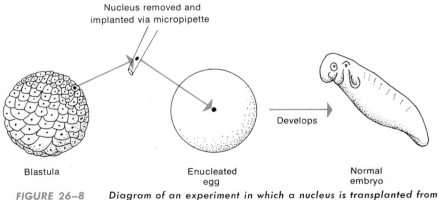

FIGURE 26–8   Diagram of an experiment in which a nucleus is transplanted from a cell of a gastrula to an enucleated egg. (See text for discussion.)

plasma membrane of the cell is ruptured by the process and the nucleus plus some cytoplasmic debris is injected deep into the enucleated egg, after which the pipette is carefully withdrawn. The operated eggs begin cleaving and some develop normally and undergo metamorphosis. Nuclei obtained from late blastula or early gastrula stages, when there are as many as 16,000 cells, can, when transplanted in this fashion into an uncleaved egg, lead to the development of a normal embryo. Even nuclei from later stages of development, from the neural plate or from ciliated cells in the digestive tract of a swimming tadpole, may lead to the development of normal embryos when transplanted into enucleated eggs. The nuclei of cells from malignant tissue (cancer cells) but not from normal adult cells supported development when transplanted into enucleated eggs. It appears that the chromosomes of nuclei from cells in advanced stages of embryonic development or from adult cells are unable to divide rapidly enough to keep up with the rate of cytoplasmic division early in embryonic development. Chromosomal duplication occurs too slowly and as a result the daughter cells receive incomplete sets of chromosomes and develop defectively.

   AMPHIBIAN GASTRULATION.   Gastrulation in amphibians is a complex process which differs from that in *Amphioxus* because of the large, yolk-laden cells in the vegetal half of the blastula. A groove appears on one side of the blastula and cells at the bottom of the groove stream into the interior of the embryo. The groove spreads transversely and its lateral ends extend along the margin between the vegetal region and the upper part of the blastula until they meet on the ventral side of the embryo (Fig. 26–9). This groove, the **blastopore**, is produced by the invagination of presumptive endoderm and mesoderm into the interior of the embryo. The process involves the **invagination** of cells at the slit-shaped blastopore, the growth of cells of the roof of the blastocoele down over the lower, yolk-filled cells **(epiboly)**, and a rolling in of these cells when they reach the blastopore **(involution)**. By the time the blastopore has become ring-shaped, the yolk-filled cells of the vegetal pole remain as a **yolk plug** filling the space enclosed by the lips of the blastopore. The rim of the blastopore continues to contract and eventually completely covers the yolk plug. A cavity leading from the groove on the surface of the embryo into the interior appears on the dorsal side, lined on all sides by cells that have invaginated from the surface. This cavity, the **archenteron,** is a narrow slit at first but gradually expands at the anterior end, encroaching on the blastocoele, which is eventually obliterated.

   The cells of the animal region of the blastula increase their surface greatly during gastrulation. By the end of gastrulation, after the presumptive mesoderm and endoderm have gone through the blastopore into the interior of the embryo, they cover the entire embryo. The outer layer of cells includes the presumptive epidermis and the presumptive nervous system. The latter expands in a longitudinal direction but contracts in the transverse direction and the presumptive nervous system becomes oval in shape, elongated in the anteroposterior axis. The presumptive notochord rolls over the dorsal lip of the blastopore into the interior of the embryo and becomes stretched longitudinally along the dorsal midline of the roof of the archenteron. The presumptive mesoderm rolls over the lateral and ventral lips of the blastopore, then moves anteriorly as a sheet of cells between the outer ectoderm and the inner endoderm, concentrated toward the dorsal side of the embryo. The mesodermal layer is thickest in the roof of the archenteron where it adjoins the notochord. More presumptive mesoderm rolls in over the lip of the blastopore, even after the

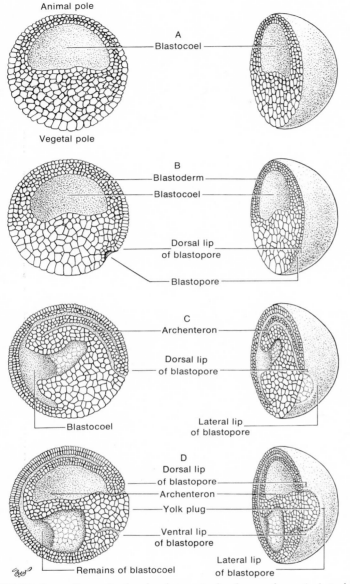

FIGURE 26-9   *Stages in the development of a frog embryo. A, Late blastula. B, Early gastrula. C, Middle gastrula. D, Late gastrula.*

yolk plug has disappeared from the surface. A portion of the presumptive endoderm lies at the equator of the blastula and is taken into the interior in the original invagination of the blastopore. The remainder of the presumptive endoderm, the cells of the vegetal region, are taken into the interior more or less passively and form the floor of the archenteron.

AVIAN GASTRULATION.   Gastrulation in the bird is accomplished by a form of cell migration that differs from that in the frog; the cells from the epiblast appear to move downward singly. Initially a strip of epiblast extending forward in the midline of the embryo from the posterior edge of the area pellucida becomes thickened as the **primitive streak,** with a narrow furrow, the **primitive groove,** in its center. At the anterior end of the primitive streak is a thickened knot of cells, **Hensen's node.** The thickening of the epiblast in the primitive streak is brought about by the migration of cells in from the lateral portion of the epiblast. The cells of the epiblast invaginate at the primitive streak, move into the space between the epiblast and hypoblast, and reach the latter, forming a mass of moving cells (Fig. 26–10). The cells continue to migrate, moving laterally and anteriorly from the primitive streak area. The primitive streak is a dynamic structure and persists even though the cells comprising it are constantly changing as they migrate in from the epiblast, sink down at the primitive streak, then move out laterally and anteriorly in the interior. The primitive streak, where invagination occurs, is considered to be homologous to the blastopore of the amphibian or Amphioxus egg, but there is no cavity in the embryo of the chick that is homologous to the archenteron.

The presumptive notochord cells become concentrated in Hensen's node and then grow anteriorly as a narrow process from Hensen's node just beneath the epiblast. Presumptive mesoderm grows laterally and anteriorly from the primitive streak between the epiblast, which becomes presumptive ectoderm, and the hypoblast. The original hypoblast cells plus other cells which migrate into the lower layer from the primitive streak form the presumptive endoderm, which forms the digestive tract and yolk sac.

DEVELOPMENT OF MAMMALIAN EGGS.   Cleavage of the mammalian egg typically results in a solid ball of cells, the **morula,** which then becomes subdivided into a mass of cells in the interior, the **inner cell mass** from which the embryo develops, and an enveloping hollow sphere of cells, the **trophoblast.** The cells of the inner cell mass have a more basophilic cytoplasm, which increases with development, which reflects the rapid synthesis of RNA occurring within them. The cavity of the blastocyst may be compared to the blastocoele but the embryo is not a blastula, for its cells are differentiated into two types. The cells of the inner cell mass differentiate further into a thin layer of flat cells, the **hypoblast,** on the interior surface of the mass, adjacent to the blastocoele, which represents presumptive endoderm (Fig. 26–11). The remaining cells of the inner cell mass become the epiblast. The cells of the hypoblast spread along the inner surface of the trophoblast and eventually surround the cavity of the blastocyst, forming a **yolk sac,** although the cavity is filled with fluid, not yolk.

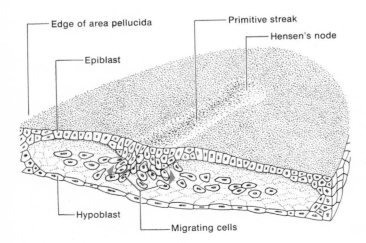

Edge of area pellucida

Primitive streak

Hensen's node

Epiblast

Hypoblast

Migrating cells

FIGURE 26–10   Gastrulation in the bird. The anterior half of the area pellucida of a chick embryo is cut transversely to show the migration of mesodermal and endodermal cells from the primitive streak.

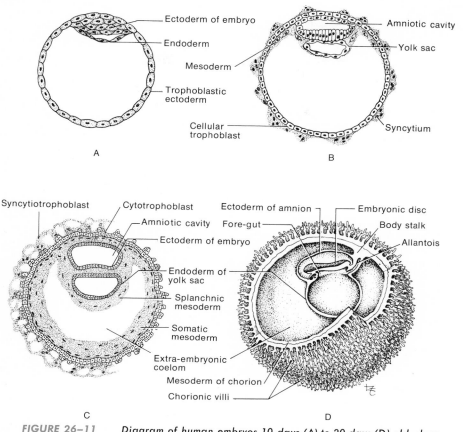

FIGURE 26–11   *Diagram of human embryos 10 days (A) to 20 days (D) old, showing the formation of the amniotic and yolk sac cavities and the origin of the embryonic disc.*

As the hypoblast spreads out, the inner cell mass also spreads and becomes a disc-shaped plate of cells similar to the blastodisc of avian and reptilian eggs. The **blastodisc,** composed of an epiblast of thick columnar cells and a hypoblast of more irregular flat cells, becomes delimited from the rest of the embryo. Gastrulation begins with the formation of a primitive streak and Hensen's node in which cells migrate downward, laterally and anteriorly between the epiblast and hypoblast. Some of these migrating cells join the hypoblast and become presumptive endoderm; others migrate laterally between the epiblast and hypoblast to become presumptive mesoderm. The amniotic cavity appears as a crevice between the cells of the inner cell mass which then enlarges. The cavity of the blastocyst becomes filled with mesenchymal cells, presumptive extraembryonic mesoderm, and a secondary small, yolk sac cavity appears just beneath the hypoblast. The bilaminar embryonic disc comes to lie as a plate between the two cavities, connected to the trophoblast at the posterior end by a group of extra-embryonic mesoderm cells, the **body stalk** or **allantoic stalk.** The nonfunctional endodermal part of the allantois, which develops as a tube from the yolk sac, is rudimentary and never reaches the trophoblast. Thus after two weeks of development the human embryo is a flat, two-layered disc of cells about 250 microns across, connected by a stalk to the trophoblast.

MORPHOGENETIC MOVEMENTS

Gastrulation in each of these types of embryo involves the movement and migration of cells which occur in specific ways and lead to specific arrangements of cells. These **morphogenetic movements** involve considerable parts of the embryo which stretch, fold, contract or expand. The movements are not, apparently, analogous to the contraction of muscles or to some sort of amoeboid movement of the whole embryo. The forces driving these movements and the factors controlling their direction are unknown. A section of presumptive ectoderm from the early gastrula of the frog will expand actively, just as it does *in situ* in contributing to the movements of gastrulation, if removed from the embryo and grown in organ culture. A bit of the blastopore lip transplanted from one embryo into another embryo will invaginate and form an archenteric cavity independent of the archenteron of the host embryo. Invagination involves not only the movement of the cells, but changes in their shape, contractions at certain ends of the cells and expansions at others.

The movements of the cells, and the positions they take up, are guided, at least in part, by **selective affinities** of certain cells which can be demonstrated if the cells of an early embryo are disaggregated and then incubated in various combinations (p. 503). Epidermis cells become concentrated on the exterior of the cell mass and mesodermal cells take up a position between the epidermis and the endoderm. Neural plate cells form hollow vesicles resembling a neural tube or brain vesicle and mesodermal cells tend to arrange themselves around coelomic cavities. This sorting out of mixtures of cells is believed to result from their selective affinities. When cells touch as a result of random movement, they may remain in contact if held together or they may move apart if not bound strongly. You could postulate that different kinds of cells have qualitatively different specific affinities, or you could account for many of the experimental results by postulating quantitative differences in the degree of adhesiveness of different kinds of cells.

The continued mitotic divisions during gastrulation lead to more cells and more nuclear material but little or no change in the total volume or mass of the embryo. The rate of metabolism, as measured by the rate of oxygen consumption, increases two- or threefold over the rate during cleavage, presumably to supply biologically useful energy for the morphogenetic movements, and for the sharply increased rate of synthesis of messenger RNA and of proteins. A considerable body of indirect evidence indicates that the genes undergo little or no transcription during cleavage but become active in this respect during gastrulation.

DIFFERENTIATION AND ORGANOGENESIS

In all multicellular animals, except sponges and coelenterates, a third layer of cells, the mesoderm, develops between ectoderm and endoderm. In annelids, mollusks and certain other invertebrates the mesoderm develops from special cells which are differentiated early in cleavage (Fig. 26–12). These migrate to the interior and come to lie between the ectoderm and endoderm. They multiply to form two longitudinal cords of cells

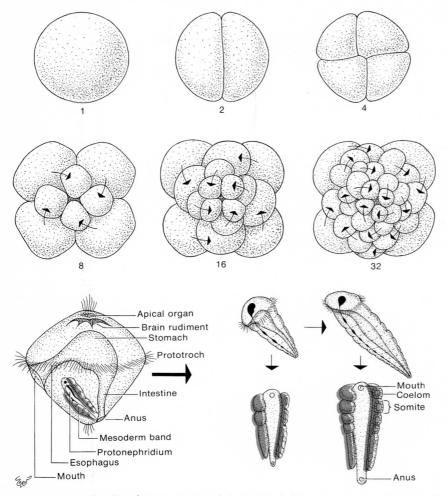

FIGURE 26-12 *Development in annelids. Upper half, The successive cleavage divisions occur in a spiral pattern as indicated. Lower left, A typical trochophore. The upper half of the trochophore develops into the extreme anterior end of the adult worm and all the rest of the adult body develops from the lower half. A series of cavities appears within each mesodermal somite (tinted, lower right), which coalesce to form the coelom.*

which develop into sheets of mesoderm. The coelomic cavity originates by the splitting of the sheets to form pockets and hence is called a **schizocoele.**

In primitive chordates (*Amphioxus*) the mesoderm arises as a series of bilateral pouches from the archenteron. (Fig. 26–13). These lose their connection with the gut and fuse one with another to form a connected layer. The cavity of the original pouches remains as the coelom, termed an **enterocoele** because it is derived indirectly from the archenteron. The mesoderm is formed in amphibia by the invagination of cells over the lateral and ventral lips of the blastopore; the cells then move forward as a sheet between ectoderm and endoderm. The presumptive mesoderm of the chick migrates down through the primitive streak area and then laterally and anteriorly between the presumptive ectoderm and endoderm.

However the mesoderm may originate in the various kinds of chordates, it splits into two sheets of cells which grow laterally and anteriorly between the ectoderm and endoderm. The cavity between the two sheets is the coelom and when the digestive tract becomes separated as a tube

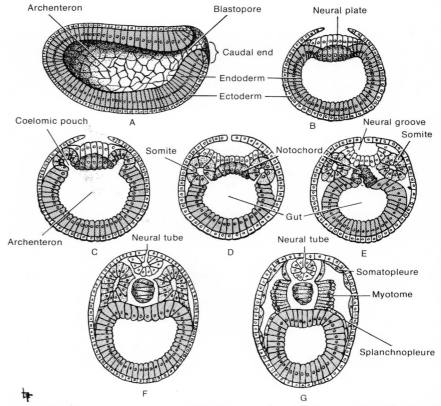

FIGURE 26–13   *Stages in the development of an Amphioxus embryo showing the formation of mesoderm by the budding of pouches from the archenteron and the formation of the neural tube. A is a sagittal section and B to G are cross sections.*

from the yolk sac cavity, the inner mesoderm grows around it forming the muscles of the digestive tract. The endoderm forms only the inner lining of the digestive system.

The **notochord,** a flexible, unsegmented, skeletal rod which extends longitudinally along the dorsal midline of all chordate embryos, is formed at the same time as the mesoderm. Its exact mode of formation, like that of the mesoderm, differs from species to species. In all vertebrates the notochord is a short-lived structure, eventually replaced by the vertebral column, but in some lower vertebrates remnants of it can be found between the vertebrae, even in adults.

Although the two-week-old human embryo is a simple flat disc, the two-month-old embryo has nearly all its structures present at least in rudimentary form. The brain and spinal cord are among the earliest organs to appear. At about the third week the ectoderm overlying the notochord in front of the primitive streak develops as a thickened plate of cells, the **neural plate** (Fig. 26–14). The center of this becomes depressed as a neural groove while the outer edges of the plate rise as two longitudinal neural folds, which meet at the anterior end and appear, when viewed from above, like a horseshoe. These folds gradually come together at the top, forming a hollow neural tube (Fig. 26–15). The cavity of the anterior part of the neural tube becomes the ventricles of the brain while the cavity at the posterior part becomes the neural canal, extending the length of the spinal cord.

The anterior part of the neural tube, which will become the brain, is

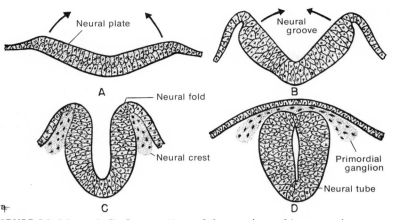

FIGURE 26–14   A–D, Cross sections of the ectoderm of human embryos at successive stages to illustrate the origin of the neural tube and the neural crest. The neural crest forms the dorsal root ganglia and the sympathetic nerve ganglia.

much larger than the posterior part and continues to grow so rapidly that the head region comes to bend down at the anterior end of the embryonic disc. All the regions of the brain—forebrain, midbrain and hindbrain—are established by the fifth week of development and a week or two later the outgrowths which will form the large cerebral hemispheres begin to grow.

The motor nerves grow out of the brain or spinal cord, but the sensory nerves have a separate origin. When the neural folds fuse to form the neural tube, bits of ectodermal tissue are pinched off near the apex of each neural fold and form a ridge, the neural crest, on each side of the neural tube (Fig. 26–14). These cells migrate down from their initial position and form the dorsal root ganglia of the spinal nerves and the postganglionic sympathetic fibers. Other neural crest cells migrate and form the medullary cells of the adrenal gland, the neurilemma sheath cells of the peripheral neurons and certain other structures.

A pair of saclike protrusions, the **optic vesicles,** appear on the lateral walls of the forebrain, grow laterally, and the base of the vesicles becomes constricted as the optic nerve. The optic vesicle comes in contact with the inner surface of the overlying epidermis, then flattens out and invaginates to form a double-walled optic cup. The inner, much thicker layer of the

FIGURE 26–15   A dorsal view of a human embryo showing a stage in the closure of the neural tube. Closure is completed between somites 4 and 7, but the neural folds anterior and posterior to this are still open.

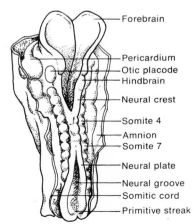

cup becomes the sensory *retina* of the eye and the outer, thin layer of the cup becomes the pigment layer of the retina. When the optic cup touches the overlying epidermis it stimulates the latter to develop into a *lens* rudiment. In birds and mammals the epidermis thickens and folds in to produce a pocket. This pinches off and forms a *lens vesicle* lying in the opening of the optic cup and surrounded by the *iris,* formed from the rim of the optic cup. The cells on the inner side of the lens vesicle become columnar and then are transformed into long fibers. Their nuclei degenerate and the cytoplasm becomes hard, transparent and able to refract light.

SECTION 26–5

DEVELOPMENT OF BODY FORM

The conversion of the two-week-old flat disc into a roughly cylindrical embryo is accomplished by three processes: (1) the growth of the embryonic disc, which is more rapid than the growth of the surrounding tissue; (2) the underfolding of the embryonic disc, especially at the front and rear ends; and (3) the constriction of the ventral body wall to form the future umbilical cord and to separate the embryo proper from the extraembryonic parts. In addition, the body begins to separate into head and trunk and the appendages begin to appear.

Growth is rapid at the anterior end of the embryonic disc, and soon the head region bulges forward from the original embryonic area. The tail, which even human embryos have at this stage, bulges to a lesser extent over the posterior end. The sides of the disc grow downward, eventually to form the sides of the body. The embryo becomes elongated, because growth is more rapid at the head and tail ends than laterally. The enlarging of the embryo has been compared to the increase in size of a soap bubble blown from a pipe, which, as it grows, swells out in all directions above the mouth of the pipe (the yolk sac). What is to become the mouth and heart originally lies in front of the embryonic disc, and as the disc grows and bulges over the tissues in front, the mouth and heart swing underneath to the ventral side. A similar underfolding occurs at the posterior end. By such growth and underfolding the lateral and eventually the ventral walls of the body are formed, and the embryo becomes more or less cylindrical in shape.

While the embryo is still a simple disc, its entire under surface is open to the yolk cavity. As the body wall folds, a *foregut* and a *hindgut* (which form, respectively, the anterior and posterior parts of the digestive tract) are cut off from the yolk sac but remain connected to it by the yolk stalk. As the embryo grows and folds, the amnion also grows to enclose it, finally constricting the yolk stalk and the body stalk (with its allantois and blood vessels) together into a single cylindrical tube, the *umbilical cord.* This takes place about four weeks after development has commenced, and allows the embryo to float free in the liquid-filled *amniotic cavity,* connected to the trophoblastic chorion and placenta only by the umbilical cord (Fig. 26–16).

The month-old embryo, about 5 millimeters long, is now recognizable as a vertebrate of some kind. It has become cylindrical, with a relatively large head region, and has prominent gills and a tail. Meanwhile, blocks of muscle, known as *somites,* are forming rapidly in the mesoderm on either side of the notochord and the beating heart is present as a large

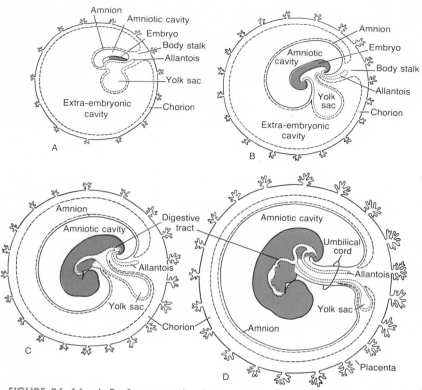

FIGURE 26–16   A–D, *Stages in the development of the umbilical cord and body form in the human embryo.*

bulge on the ventral surface behind the gills. The arms and legs are still mere buds on the sides of the body.

By the end of six weeks the embryo is about 12 millimeters long; the head begins to be differentiated; the arms and legs have grown out, but the tail and gills are still present.

At the end of two months of growth, when the embryo is 25 millimeters long, it begins to look definitely human. The face has begun to develop, showing the rudiments of eyes, ears and nose. The arms and legs have developed, at first resembling tiny paddles, but by this stage the beginnings of fingers and toes are evident (Fig. 26–17). The tail, which was prominent during the fifth week of development, has begun to shorten and to be concealed by the growing buttocks. As the heart moves posterior-

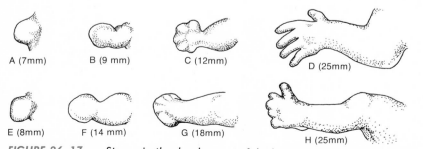

FIGURE 26–17   *Stages in the development of the human arm (upper row) and leg (lower row) between the fifth and eighth weeks.*

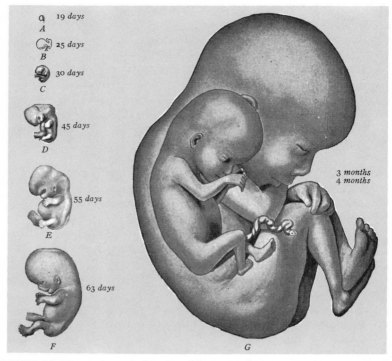

FIGURE 26-18   *A graded series of human embryos. Note the characteristic position of the arms and legs in the four-month fetus. (From Arey, L. B.: Developmental Anatomy. 7th ed. Philadelphia, W. B. Saunders Company, 1965.)*

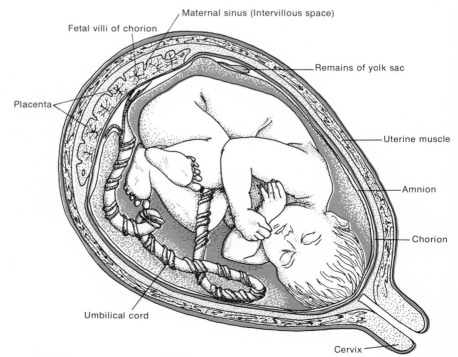

FIGURE 26-19   *Diagrammatic section through the uterus showing the placenta and fetus shortly before birth.*

ly on the ventral side, and the gill pouches become less conspicuous, a neck region appears. Now most of the internal organs are well laid out so that development in the remaining seven months consists mostly of an increase in size and the completion of some of the minor details of organ formation (Fig. 26–18).

The embryo is about 75 millimeters long after three months of development, 250 millimeters long after five months and 50 centimeters long after nine months. During the third month the nails begin forming and the sex of the fetus can be distinguished. By four months the face looks quite human; by five months, hair appears on the body and head. During the sixth month, eyebrows and eyelashes appear. After seven months the fetus resembles an old person with red wrinkled skin. During the eighth and ninth months, fat is deposited under the skin causing the wrinkles partially to smooth out; the limbs become rounded, the nails project at the finger tips, the original coat of hair is shed, and the fetus is "at full term," ready to be born (Fig. 26–19). The total **gestation period,** or time of development, for human beings is about 280 days from the beginning of the last menstrual period before conception until the time of birth.

SECTION 26–6

FORMATION OF THE HEART

In contrast to many organs which develop in the embryo without having to function at the same time, the heart and circulatory system must function while undergoing development. The heart forms first as a simple tube from the fusion of two thin-walled tubes beneath the developing head (Fig. 26–20). In this early condition it is essentially like a fish heart, consisting of four chambers arranged in a series: the **sinus venosus,** which re-

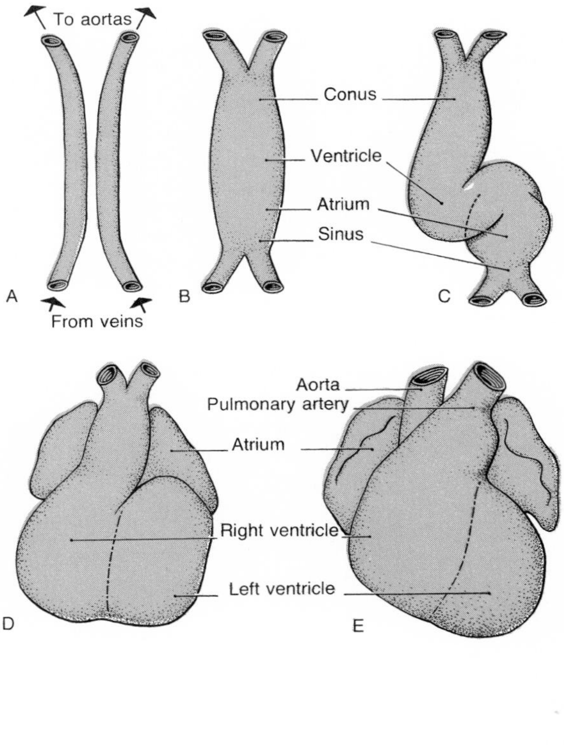

FIGURE 26–20   A–E, Ventral view of successive stages in the development of the mammalian heart. (See text for discussion.)

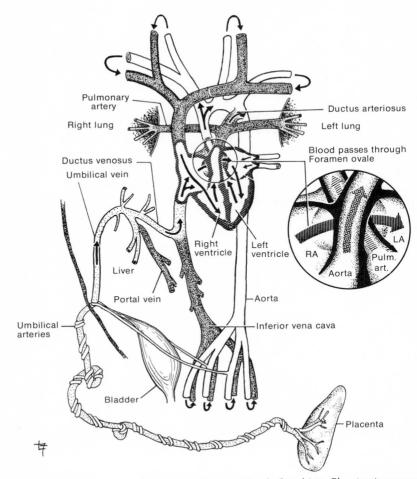

FIGURE 26-21   *The human circulatory system before birth. The structures peculiar to the fetal circulatory system are the umbilical arteries and veins to the placenta, the arterial duct (ductus arteriosus) which connects the pulmonary artery and aorta, and the venous duct (ductus venosus) which connects the umbilical vein and the inferior vena cava. The oval window (foramen ovale) connects the right and left atria.*

ceives blood from the veins; the single *atrium;* the single *ventricle;* and the *arterial cone,* which leads to the aortic arches.

Initially the heart is a fairly straight tube, with the atrium lying posterior to the ventricle; but since the tube grows faster than the points to which its front and rear ends are attached, it bulges out to one side (Fig. 26-20). The ventricle then twists in an **S**-shaped curve down and in front of the atrium, coming to lie posterior and ventral to it as it does in the adult. The sinus venosus gradually becomes incorporated into the atrium as the latter grows around it, and most of the arterial cone is merged with the wall of the ventricle.

The embryonic heart, when it first appears, is a single structure with only one of each chamber, whereas the adult heart is a double pump, with separate right and left atria and ventricles. This separation prevents the mixing of aerated blood from the lungs with nonaerated blood from the rest of the body. The lungs are nonfunctional in the embryo, and not much blood passes through them. The heart begins separating into four chambers at an early stage. The two ventricles are completely separated by the end of the second month. The atria are partly separated, but complete separation does not occur until after birth, when the **oval window** between them finally closes (Fig. 26–21). Before birth, this must be kept open to permit blood to get into the left side of the heart, for in the fetus only a small amount of blood passes through the lungs to the left atrium. Without this "window" the left side of the heart would be nearly empty, and most of the blood would pass through the right side only.

SECTION 26–7

DEVELOPMENT OF THE DIGESTIVE TRACT

The digestive tract is first formed as a separate foregut and hindgut by the growth and folding of the body wall, which cuts them off as two simple tubes from the original yolk sac (Fig. 26–22). These tubes grow as the rest of the embryo grows, becoming greatly elongated. The lungs, liver and pancreas originate as hollow, tubular outgrowths from the original

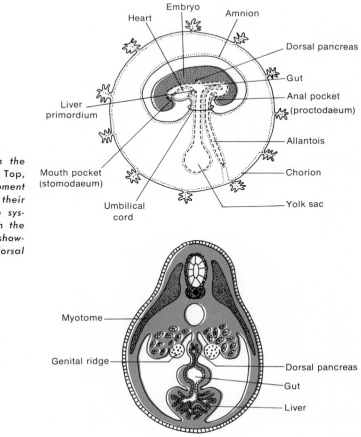

FIGURE 26–22   Stages in the development of the digestive tract. Top, A sagittal section showing development of the yolk sac and allantois and their relation to the primitive digestive system. Botton, Cross section through the anterior part of the digestive tract showing the formation of the liver and dorsal pancreas as buds from the foregut.

foregut, and hence are composed of endoderm. But these outgrowths always are associated with some mesodermal tissue, which forms the blood and lymph vessels, connective tissue and muscles of these organs. The endoderm forms only the internal epithelium of the digestive tract and lungs, and the actual secretory cells of the pancreas and liver. Both the enzyme-secreting (acinar) cells and those of the islets of Langerhans in the pancreas are derived from tubular outgrowths from the foregut. The lung first develops as a single median outgrowth from the ventral side of the foregut. This single tube, which is the forerunner of the trachea, soon branches into two, the rudiments of the bronchi, which in turn divide repeatedly and eventually grow into the complex structure which is the adult lung.

The most anterior part of the foregut flattens out to become, in cross section, a flattened oval rather than a circle, and develops into the **pharynx.** In the pharyngeal region a series of five paired pouches, the **gill pouches,** bud out laterally from the endoderm and meet a corresponding set of in-pocketings from the overlying ectoderm. In lower vertebrates such as fish, the two sets of pockets fuse to make a continuous passage from the pharynx to the outside—the **gill slits**—which function as respiratory organs.

In the higher vertebrates this normally does not occur; the pouches exist, but are nonfunctional vestiges which give rise to other structures or disappear. For example, the first pair of pouches become the cavities of the middle ear and their connection with the pharynx, the **eustachian tubes.** The second pair of pouches become a pair of tonsils, while parts of the third and fourth pouches become the thymus gland, and other parts of them become the parathyroids. The fifth pouches become the ultimobranchial bodies which secrete calcitonin. The thyroid develops from a separate outgrowth on the floor of the pharynx.

The mouth cavity arises as a shallow pocket of ectoderm which grows in to meet the anterior end of the foregut; the membrane between the two ruptures and disappears during the fifth week of development. Similarly the anus is formed from an ectodermal pocket which grows in to meet the hindgut; the membrane separating these two disappears early in the third month of development.

SECTION 26–8

THE DEVELOPMENT OF THE KIDNEY

The development of the kidney is one of the finest and most clear-cut examples of the principle of recapitulation. Within the subphylum of vertebrates are three different types of kidney. The earliest, or **pronephros,** is the adult kidney of certain primitive fishes. The second, or **mesonephros,** is the adult kidney of amphibians and the higher fishes. The third, the **metanephros,** is the adult kidney of reptiles, birds and mammals. But in development, each of the higher animals repeats the evolutionary sequence of this organ. Thus frog embryos first develop a pronephros, which functions during early embryonic life, before the permanent mesonephros develops. And man develops first a nonfunctional pronephros, then a mesonephros, which may be functional during fetal life, and finally the permanent metanephros (Fig. 26–23). The three kidneys develop one after another in both time and space, each new kidney lying posterior to the previous one.

The pronephros, which in the human embryo consists of about seven pairs of rudimentary kidney tubules, develops in the mesoderm and degenerates during the fourth week of embryonic life. From the tubules a

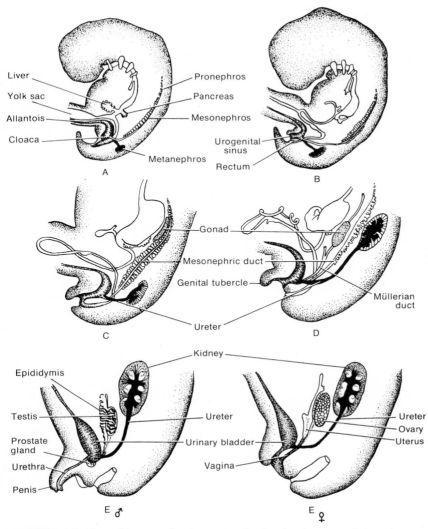

FIGURE 26-23 *Diagram showing stages in the development of the urinary and reproductive systems in man. A, Early in the fifth week of development; B, early in the sixth week; C, seventh week; D, eighth week; E♂, three-month-old male fetus; E♀, three-month-old female fetus.*

pair of wolffian ducts grows back to the hindgut and connects with it.

The tubules of the mesonephros originate during the fourth week, reach their height at the end of the seventh week, and degenerate by the sixteenth week. These tubules connect with the ducts left by the degenerated pronephros and empty into them. In the female the mesonephros and its ducts degenerate completely except for a few nonfunctional remnants, but in the male some of the tubules remain and are converted into the epididymides, while the wolffian ducts become the vas deferens.

The metanephros of reptiles, birds and mammals develops as a pair of buds from the ducts of the mesonephros. The ureter and collecting tubules of the kidney develop from these buds, while the Bowman's capsules and convoluted tubules develop from the same sort of mesoderm that formed the tubules of the pronephros and mesonephros at a more anterior point. Later, these two portions unite to form the kidney tubules of the adult. The metanephros begins forming during the fifth week and is practically complete by 16 weeks.

SECTION 26–9

MORPHOGENETIC MOVEMENTS OF CELLS IN ORGANOGENESIS

The various rudiments of the organs arise by morphogenetic movements of sheets of epithelial cells or loose mesenchymal cells which are similar to those involved in gastrulation. Epithelial cells may undergo thickening in localized areas to form elongated cells; this occurs in the thickening of the ectoderm to form the neural plate. Epithelial layers may undergo folding to form grooves or pockets which, by continued folding, yield tubes or vesicles. Crevices may appear between epithelial layers or between masses of cells, splitting a single layer apart into two separate layers. Vesicles and tubes may be formed by the thickening of an epithelial layer followed by the formation of a cavity in the middle of the solid thickening, rather than by the folding of an epithelium. Masses of cells that were previously separated may fuse to form a new organ rudiment; for example, the edges of the neural plate fuse after the neural plate has been folded into a tube. Epithelial layers may break up to form mesenchyme and mesenchyme cells may secondarily be rearranged into an epithelium. Mesenchyme cells may aggregate in a mass without forming an epithelial sheet, and later differentiate as cartilage, bone or muscle tissue. Mesenchyme cells are often accumulated near the surface of an epithelium or around a structure such as a vesicle or tube. They subsequently differentiate into cartilage or bone and produce a skeletal capsule around the organ developing from the vesicle or tube.

A sort of programed cell death may be a morphogenetic force. The death of localized cells in part of a rudiment while the adjacent cells remain healthy and continue to proliferate may lead to a change in the shape of an organ. For example, the death of cells in a certain pattern during the development of the hand and foot separates the digits from one another.

The cellular migrations which are the basis of the morphogenetic movements of organogenesis can be readily observed when isolated embryonic cells are kept in culture medium. The factors that direct cell migration along certain channels and to certain destinations in the embryo are not at all clear but probably involve specific properties of the cell surface plus, in certain instances, the properties of the substrate along which the cells are migrating. The intercellular spaces of the mesenchymal areas of the embryo are filled with a colloidal, partially gelated solution, which results in the stretching of molecular fibers in various directions through the space. These serve as a substratum for migrating mesenchyme cells and determine in part their direction of migration.

SECTION 26–10

MALFORMATIONS

In view of the extreme complexity of the developmental process it is indeed remarkable that it occurs so regularly and that so few malformations occur. The development of a human arm involves the formation of 29 bones, each of a specific size and shape and each forming a joint with the next in a very specific way. It involves the formation of some 40 or more muscles, each with the proper origin, insertion and size. It involves the development of a large number of motor and sensory nerves, each with the proper synaptic connections on the motor end plates of muscles or in the sensory receptors in the skin, tendons and joints. And it involves the development of a large number of arteries and veins arranged in a fairly specific pattern to supply each of the parts of the arm.

About one child in 100 is born with some major defect such as a cleft palate, club foot or spina bifida. Some of these are inherited; others result from environmental factors. Experiments with fruit flies, frogs and mice have shown that x-rays, ultraviolet rays, temperature changes and a variety of chemical substances will induce alterations in development. The kind of defect produced depends on the time in development at which the environmental agent is applied, and does not depend to any great extent on the kind of agent used. For example, x-rays, the administration of cortisone and the lack of oxygen will all produce similar defects in mice—harelip and cleft palate—if applied at comparable times in development. There are certain critical periods in development, during which particular organs are differentiating and growing most rapidly and are most susceptible to interference.

According to old superstitions, such abnormalities are caused by fright or injury to the mother, and it was even supposed that injury to a particular part of the mother's body resulted in the malformation of that part of the fetus. However, the injuries commonly believed to cause malformations usually occur later in pregnancy, long after the organs have completed differentiation.

SECTION 26–11

TWINNING

In man, apes and monkeys, as well as in many other species of mammals, offspring are usually produced singly, although in other animals, more than one (up to 25 in the pig) are produced in a single litter. About once in every 88 human births, two individuals are delivered at the same time. More rarely, three, four, five and even six children are born simultaneously. About three fourths of the twins (and triplets, quadruplets, quintuplets and so on) are the result of the simultaneous release of two eggs, one from each ovary, both of which are fertilized and develop. Such *fraternal twins* may be of the same or different sexes, and have only the same degree of family resemblance that brothers and sisters born at different times have. They are entirely independent individuals with different hereditary characteristics and result from the fertilization of two or more eggs ovulated simultaneously.

Recently, purified human FSH (follicle-stimulating hormone) has been used to induce ovulation in women whose infertility appears to be due to deficient production or release of gonadotropin. In the initial attempts many women apparently received too much FSH, had multiple ovulations and released from 2 to 10 or more ova, all of which were fertilized and began to develop. Most of the women with multiple fetuses have been unable to carry them to term but others with one or two fetuses have had normal births.

In contrast, *identical twins* (or triplets, and so on) are formed from a single fertilized egg which at some early stage of development divides into 2 (or more) independent parts, each of which develops into a separate fetus. Such twins, of course, are the same sex, have identical hereditary traits and are so similar that it is difficult to tell them apart.

Occasionally identical twins develop without separating completely and are born joined together *(Siamese twins)*. All grades of union have been known to occur, from almost complete separation to fusion throughout most of the body so that only the head or the legs are double. Sometimes the two twins are of different sizes and degrees of development, one being quite normal, while the second is only a partially formed parasite on the first. Such errors of development usually die during or shortly after birth.

SECTION 26–12

CHANGES AT BIRTH

Great changes take place within a short time after a child is born. Hitherto it has received both food and oxygen from the placenta; now its own digestive and respiratory systems must function. Correlated with these changes are major changes in the circulatory system: the umbilical arteries and veins are cut off, and the blood flow through the heart and lungs is altered.

It is believed that the first breath of the newborn infant is initiated by the accumulation of carbon dioxide in the blood after the umbilical cord

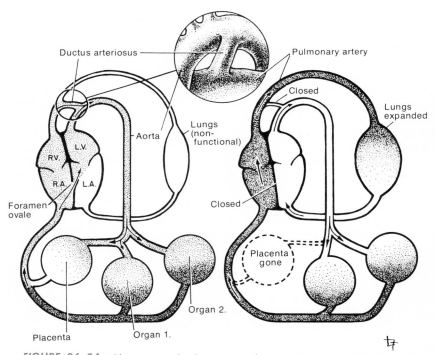

FIGURE 26–24 *Changes in the human circulatory system at birth. Left, the circulatory system of the fetus (compare with Figure 26–21). Right, the circulatory system of a newborn child. Aerated blood is shown in white, nonaerated blood in dark stippling, and a mixture of the two in light stippling. In the embryo, little of the blood goes through the lungs. Most of the blood entering the right atrium gets into the aorta either by way of the oval window (foramen ovale) between the pulmonary artery and aorta (inset). At birth (1) the placenta is removed, (2) the lungs expand, (3) the foramen ovale closes and (4) the arterial duct closes and gradually degenerates.*

is cut; this stimulates the respiratory center in the medulla. The resulting expansion of the lungs enlarges its blood vessels, which previously were partially collapsed, and blood from the right ventricle flows in ever-increasing amounts through the vessels of the lung, instead of through the arterial duct which connected the pulmonary artery and aorta during fetal life. The resulting increase in blood returning from the lungs to the left atrium results in the closing of the valvelike opening of the oval window (Fig. 26–24). These changes take place within a short period after birth, and eventually the flap of the oval window grows into place and the arterial duct degenerates, so that the adult pattern of circulation is established. Occasionally the oval window fails to close or the arterial duct fails to degenerate and there is a mixing of oxygenated and nonoxygenated blood. This results in a "blue baby," a child whose skin has a purplish hue because of the inadequate oxygenation of its blood. Delicate surgical procedures, developed after years of practice operations on dogs, now make it possible to operate on the heart itself and cure this condition.

SECTION 26–13

POSTNATAL DEVELOPMENT

Development does not, of course, cease at birth. At birth the teeth and genital organs of the human infant are only partly formed and the body proportions are quite different from those of the adult. The head is proportionately much larger early in development than in the adult (Fig. 26–25). The head comprises about half of the length of the two-month-old fetus but its growth terminates early in childhood so that the head of

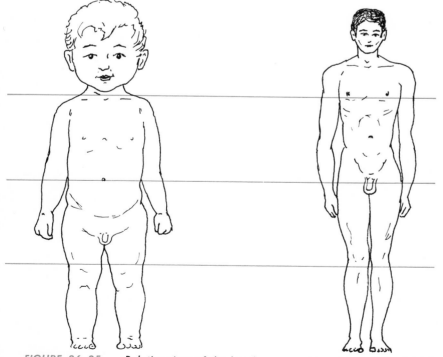

FIGURE 26–25   Relative sizes of the head, arms and legs in a young child (left and in an adult (right).

the adult is proportionately smaller than that of the newborn. The arms attain their proportionate size shortly after birth but the legs attain theirs only after some 10 years of growth. The last organs to mature in man are the genitals, which do not begin to grow rapidly until 12 to 14 years after the infant is born.

The degree of maturity and self-sufficiency of the newly hatched bird or newly born mammal varies widely from one species to another. Baby chicks and ducks can run around and eat solid food just after hatching, but baby robins are blind, have very few feathers and cannot stand. The newborn guinea pig has fur and teeth and can eat solid food. Newborn rats, mice and humans are quite helpless and require a lot of parental care to survive. The developmental processes that occur after birth involve some multiplication and differentiation of cells but in large part they involve the growth of cells formed earlier. The weight or size of an animal or plant as a function of time usually yields an **S**-shaped growth curve, one remarkably like the growth curve of a population of individuals (p. 964). Although biologists can identify some of the factors ("environmental resistance") that stabilize the size of a population at a certain level (p. 966), very little is known of the factors that lead to the cessation of cell multiplication and growth. Some plants and animals do not, in fact, stop growing but continue to grow, though perhaps at a slower pace, all through their life.

The human egg is about 100 microns in diameter, just barely visible to the naked eye. A baby at birth is about 50 centimeters long, roughly 5000 times as long as the egg. In developing from an infant to an adult the height increases only an additional three and one half times, to about 175 centimeters. The maximal rate of linear growth occurs before birth, in the fourth month of fetal life. There is a final growth spurt at the time of adolescence which reaches its peak at about age 12 years in girls and at about age 14 in boys.

Each structure and organ has its characteristic rate of growth. The growth rates of the various organs can be assigned to one of four types (Fig. 26–26). The growth curve of the skeleton follows that of the body as

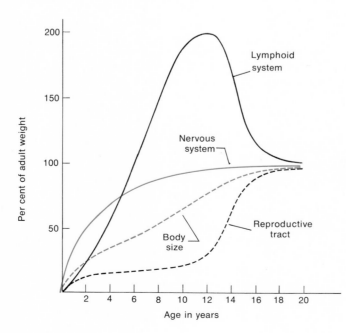

Figure 26–26   *Diagram showing the relative rates of growth of the several different organ systems of the human fetus and newborn.*

a whole. The brain and spinal cord grow relatively rapidly early in childhood and nearly reach their adult size by the age of nine. Lymphoid tissue, including the thymus, has a third type of growth curve, reaching a maximum at age 12 that exceeds the adult value. It then involutes until about age 20, when it attains adult values. The fourth pattern of growth is shown by the reproductive system which grows very slowly until age 12 or thereabouts and then undergoes rapid growth at puberty.

SECTION 26–14

THE AGING PROCESS

Since development in its broadest sense includes any biological change with time, it also includes those changes that result in the decreased functional capacities of the mature organism, the changes commonly called *aging.* The declining capacities of the various systems in the human body, though most apparent in the elderly, may begin much earlier in life, during childhood or even during prenatal life. The various systems of the body may begin their decline at quite different times; the aging process is far from uniform in the various parts of one's body. A 75-year-old man, for example, has lost 64 per cent of the taste buds, 44 per cent of the renal glomeruli and 37 per cent of the axons in his spinal nerves that he had at age 30. His nerve impulses are propagated at a rate 10 per cent slower, the blood supply to his brain is 20 per cent less, his glomerular filtration rate has decreased 31 per cent and the vital capacity of his lungs has declined 44 per cent. Relatively little is known about the aging process itself but this is now an active field of scientific investigation. The marked improvements in medicine and public health have led to a larger fraction of the total human population surviving to an advanced age, but have not increased the maximum life expectancy for man.

A remarkable model of the aging process in man is provided by a rare, inherited type of abnormal development called *progeria.* Individuals with this condition develop more or less normally until about one year old, then undergo changes which are considered typical of aging—they lose their hair, stop growing and develop the appearance of wizened old men. The collagen in the connective tissues of their skin becomes highly cross-linked as is seen in old age. They usually die at age 10 to 15 of coronary artery disease secondary to extensive atherosclerosis.

Cells that differentiate and stop dividing appear to be more subject to the changes of aging than are those that continue to divide throughout life. Nerve and muscle cells, which lose the capacity for cell division at an early age, show a decline in their respective functional capacities at an earlier age than tissues such as liver and spleen, which retain the capacity to undergo cell division.

Several theories have been advanced regarding the nature of the aging process—that it involves hormonal changes; that it involves the development of *autoimmune reactions* (allergies against certain components of the organism's own body that result in destruction of those components by antibodies); that it involves the accumulation of specific waste products within the cell (the "clinker" theory); that it involves changes in the molecular structure of macromolecules such as collagen (an increased cross-linkage between the helical chains); that there is a decrease in the elastic properties of connective tissues owing to an accumulation of calcium which results in stiffening of the joints and hardening of the arteries; that it involves the peroxidation of certain lipids by free radicals; or that cells are destroyed by hydrolases released by the breaking of lysosomes. There are several current theories regarding the cause of the aging process. Some believe that it involves the accumulation of somatic mutations caused by the continued exposure to cosmic radiation and x-radiation, mutations which decrease the ability of the cell to carry out its normal functions at the normal rate. In all likelihood aging is a part of and due to the same kinds of developmental processes that bring about the increasing functional capacities of the various systems of the body during earlier development. They may be part of the program of timed development built into the genome. Like other developmental processes, aging may be accelerated by certain environmental influences and may occur at different rates in different individuals because of inherited differences. The best guarantee of a long life is to have long-lived parents and grandparents. There is some experimental evidence that aging, at least in rats, can be delayed by dietary means, by caloric restriction. The thin rats, by and large, live longer than the fat rats!

SECTION 26–15

DEVELOPMENT IN PLANTS

The development of a seed plant is basically different from the development of an animal zygote. The egg cell is retained within the ovary and is fertilized by a sperm nucleus that reaches it via the pollen tube (Fig. 26–27). Seed plants have a unique double fertilization; another sperm nucleus fuses with two polar nuclei within the egg to give rise to the triploid endosperm (p. 384). The fertilized egg, or zygote, undergoes several divisions and forms a row of cells extending into the embryo sac. One of these, the last in the row, divides in other planes to form a cluster of eight cells which is the precursor of the embryo. The rest of the row of cells becomes the **suspensor** which grows and pushes the developing embryo into the embryo sac. The initial divisions of the endosperm nucleus yield a large number of endosperm nuclei which cluster around the developing embryo and the

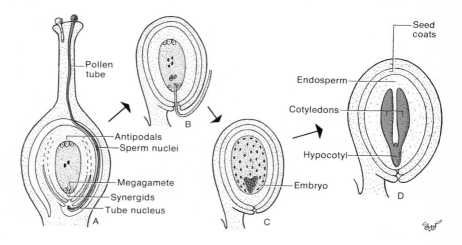

FIGURE 26–27   *The process of fertilization in an angiosperm which leads to the formation of a 2n (diploid) zygote and a 3n (triploid) endosperm. (See text for discussion.)*

walls of the embryo sac. Cell membranes subsequently appear between these nuclei so that separate endosperm cells are formed. These continue to divide and grow to form a large mass of endosperm. The embryo and endosperm are surrounded by one or two integuments or **seed coats,** derived from maternal tissue.

The embryo develops one or two **cotyledons** or seed leaves, which may become thick and fleshy, full of stored food, or remain thin. As the bean seed germinates, its cotyledons enlarge, develop chlorophyll and are active in photosynthesis. Above the point where the cotyledons are attached to the embryo develops the **epicotyl,** which consists of a growing point (the **plumule)** enclosed by a pair of folded, miniature leaves. Below the cotyledons extends the **hypocotyl.** The tip of the hypocotyl, the **radicle,** or embryonic root, forms the primary root of the young plant (Fig. 26–28).

The cotyledons of peas and beans absorb all the nutrients from the endosperm before the seed is released from the parent plant, whereas corn embryos absorb endosperm only after the seed begins to germinate. The mitotic divisions of the embryo cease and it becomes dormant within the seed. The ripe seed is eventually released from the parent plant and remains dry and dormant for months or even years. Eventually it germinates by taking up water which softens the seed coats and increases the activity of the enzymes in the embryonic cells. During germination the metabolic rate increases markedly, cell division resumes, and the increased size of the embryo due to the uptake of water and cell growth splits the seed coats. The radicle elongates, grows down into the soil and begins to produce root hairs and branch roots. In the bean, the upper portion of the hypocotyl elongates and arches, raising the cotyledons and epicotyl above the ground (Fig. 26–29). When exposed to the light the plumule expands and develops into the stem and the first true leaves of the plant. The nutrients stored in the cotyledons are rapidly used as the young plant grows and the cotyledons shrivel and are shed as the seedling derives its food from photosynthesis occurring in the stem and young leaves. In other plants, such as the pea, the hypocotyl does not elongate and the cotyledons are not raised above the ground. The epicotyl of the pea elongates, grows upward, emerges from the soil and the young leaves begin to unfold.

The embryo within the seed is really a miniature plant, with its root, shoot and leaves already differentiated. The germination of a seed is a resumption of growth that follows the greatly increased rate of metabolism which occurs when the seed is placed in a warm, moist, aerobic environment.

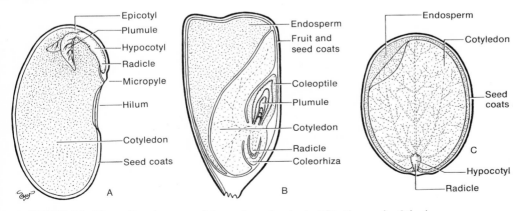

FIGURE 26–28   Development of the embryonic plant within the seeds of the bean (left) the corn (center) and the castor bean (right).

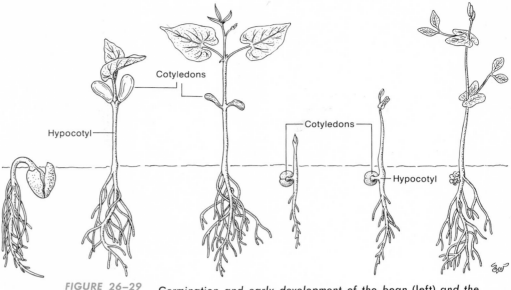

FIGURE 26-29   *Germination and early development of the bean (left) and the pea (right).*

The seedling plant grows slowly at first, then more rapidly, and finally more slowly, describing an **S**-shaped growth curve; it forms more branches and leaves, and eventually forms flower buds and flowers. The flowering process is initiated in some plants by changes in temperature or in the length of daylight, and perhaps by hormonal factors (p. 685). Shrubs and trees usually do not form flowers and seeds until after several years of growth. Trees and shrubs have a mass of cells, the **cambium,** that continues to divide and differentiate. The cells on the outer side of this cambium layer differentiate into phloem, whereas those on the inner side differentiate into xylem; the factors controlling this differentiation are unknown. Experimental analyses of the factors initiating and controlling morphogenesis in plants are underway in many laboratories. Methods for growing isolated cells, tissues or organs from plants in sterile cultures (Fig. 26-30) have provided clues about the processes involved and promise to provide more in the future. These somewhat simpler systems can be analyzed and controlled more readily than intact plants and have been used in studies of the polarity and symmetry of embryonic growth, the correlated growth of the several parts of the plant, such as the root and shoot, and the nature of differentiation and regeneration.

Most biologists believe that an important factor in differentiation is the production of new kinds of messenger RNA which code for new, specific proteins characteristic of the new kind of tissue. The new proteins, with specific enzymatic properties, catalyze specific biochemical processes which are reflected in the differentiation of the tissue. Some classic experiments by James Bonner and his colleagues have shown the appearance of a specific protein, a globulin, in pea seeds that is absent from all other tissues of the pea plant. Although a conceptual framework is available to explain the appearance of a new protein, we are not yet in a position to explain the remarkable process of morphogenesis, the development of a specific shape. The leaves that appear first on a plant such as ivy may have quite a different size and shape from those that appear later. This is controlled at least in part by hormones such as gibberellin, but other morphogenetic factors as yet unknown also are effective.

FIGURE 26–30   *A carrot plant which was grown in culture from a single cell isolated from a mature carrot. Thus the single adult cell contained all the genetic information to produce an entire plant.*

SECTION 26–16

DIFFERENTIATION

One of the important, unsolved problems of modern biology is the nature of the mechanisms that regulate developmental processes so that each organ appears at the proper time and in the proper spatial relations to the other organs. How can a single cell, a single fertilized egg, give rise to the many different types of cells that make up the adult organism, cells which differ so widely in their structure, functions and chemical properties? The advances in biochemical genetics in the past two decades have permitted new intellectual and experimental approaches to this problem.

We now have a detailed working hypothesis as to how biological information is transferred from one generation of cells to the next and how this information may be transcribed and translated in each cell so that specific enzymes and other proteins are synthesized. The operation of this system would produce a multicellular organism in which each cell would have the same assortment of enzymes as every other cell. Additional hypotheses are needed to account for: (1) the means by which the amount of any given enzyme produced in a cell is regulated, (2) the control of the time in the course of development when each kind of enzyme appears, and (3) the mechanism by which unique patterns of proteins are established in each of the several kinds of cells in a multicellular organism despite the fact that they all contain identical quotas of genetic information.

Any satisfactory model of the developmental process must be able to explain how genetic and nongenetic factors can interact to control the differentiation of cells and tissues. Why does the epicotyl of the developing pea seed become negatively geotropic and grow upwards whereas the closely adjacent hypocotyl, composed of cells with the same genetic information as those of the epicotyl, becomes positively geotropic and grows downwards? Why do the cells on the outer margin of the cambium differentiate into phloem while genetically identical cells on the inner margin of the cambium differentiate into xylem?

Biologists have speculated that differentiation might occur: (1) by some sort of segregation of genetic material during mitosis, (2) by the establishment of chemical gradients within the developing embryo, (3) by somatic mutations, (4) by the action of chemical organizers, or (5) by the induction of specific enzymes.

A theory widely held by early embryologists was that the egg or sperm contained a completely formed but minute germ which simply grew and expanded to form the adult. This **preformation theory** was gradually displaced by the contrasting theory of **epigenesis**, which stated that the unfertilized egg is structureless, not organized, and that development proceeds by the progressive differentiation of parts. However, development is not *simply* epigenetic. Certain potentialities, though not structures, may be localized in certain regions of the egg and early embryo; this restricts the developmental possibilities of that part. When the embryos of echinoderms or chordates are separated experimentally at the two- or four-cell stage, each of the separated cells will form an embryo complete in all details although smaller than normal. However, when embryos of annelids or mollusks are separated at the two-cell stage, neither cell can develop into a whole embryo. Each cell develops only into those structures it would have formed normally—half an embryo, perhaps, or some part of one. This localization of potentialities eventually occurs in the development of all eggs; it simply occurs earlier in some than others.

Some sort of chemical or physiological differentiation must be present before any structural differentiation is visible, but the basic problem of how the chemical differentiation arises remains unsolved. By appropriate experiments it has been possible to map out the location of these potentialities in the frog, chick and other embryos (Fig. 26–31).

Cellular differentiation might be explained if genetic material were parceled out differentially at cell division and the daughter cells received different kinds of genetic information. Although there are a very few clear

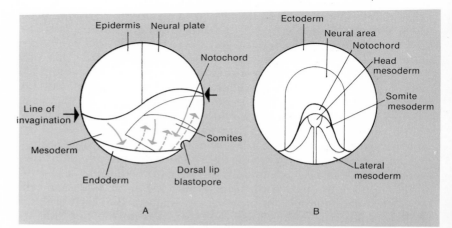

FIGURE 26–31   *Embryo maps. A, Lateral view of a frog gastrula showing the presumptive fates of its several regions. The solid arrows indicate cell movements on the surface and the dotted arrows indicate cell movements in the interior. B, Top view of a chick embryo showing the location in the primitive streak stage of the cells which will form particular structures in the adult.*

instances of differential nuclear division in animals such as *Ascaris* and *Sciara*, this does not appear to be a general mechanism of differentiation. By and large, genes are neither lost nor gained during developmental processes. The generalization that the mitotic process ensures the exact distribution of genes to each cell of the organism is a valid one. Thus the differences in the kinds of enzymes and other proteins found in different cells of the same organism must arise by differences in the *activity* of the same set of genes in different cells. The experiments of Briggs and King (p. 850) showed that even the nucleus from a differentiated cell taken from an advanced stage of embryonic development can, when placed in an enucleated egg, lead to the development of a normal embryo. Thus it clearly had retained a full set of genetic information.

Some striking evidence regarding the differential activity of genes comes from cytologic studies of insect tissues. In certain insect tissues the chromosomes undergo repeated duplication, and the daughter strands line up exactly in register, locus by locus, so that characteristic bands appear along the length of the giant chromosome. When these bands are examined carefully, either in the same tissue at different times or in different tissues at the same time, certain differences in appearance become evident. A particular section of a chromosome may have the appearance of a diffuse puff (Fig. 26–32). Histochemical tests and autoradiographic evidence from experiments using tritium-labeled precursors have shown that the puff consists of RNA. It has been inferred that genes show this puffing phenomenon when they become active and that the puff represents the messenger RNA produced by the active gene in that band. It has been possible to correlate the appearance of puffs at certain regions in the chromosome with specific cellular events such as the initiation of molting and pupation (p. 686).

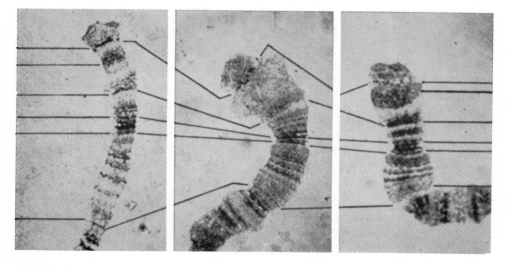

PRE-PUFF   **PUFF**   **POST-PUFF**

50 µ

FIGURE 26–32 *Photomicrographs of the changing appearance of a polytene chromosome in a salivary gland of Chironomus tentans as a puff gradually appears on the chromosome. The corresponding bands in the three photomicrographs are indicated by connecting lines. The material which makes up the chromosomal puff has been shown histochemically and autoradiographically to be largely ribonucleic acid. (Courtesy of G. Rudkin.)*

The turning on and off of the synthesis of specific proteins—differentiation at the molecular level—could occur by some process involving the genic DNA, the transcription of the DNA to form messenger RNA, the combination of the messenger RNA with the ribosome during protein synthesis or even some transformation of the ultimate protein product. In view of the tremendous number of kinds of DNA that are represented by the genic complement of a cell in a multicellular plant or animal, we might ask what prevents that cell from producing continuously all the tremendous variety of messenger RNAs and their corresponding protein products that are possible. What determines which molecules of DNA are to be transcribed at any given moment in a given cell?

A mechanism which controls the transcription of DNA to regulate the production of messenger RNA would probably be the most economical one biologically, for it would clearly be to the cell's advantage not to have its ribosomes encumbered with nonfunctional molecules of messenger RNA. A cell optimally should produce only those kinds of messenger RNA that will code for the specific proteins required at that moment.

The continued synthesis of any protein requires the continued synthesis of its corresponding messenger RNA. Each kind of messenger RNA has a half-life ranging from a few minutes in certain microorganisms to 12 or 24 hours or longer in man and other mammals. Although each molecule of RNA template can serve to direct the synthesis of many molecules of its protein, the RNA is eventually degraded and must be replaced. This provides a mechanism by which a cell can alter the kind of protein being synthesized as new types of messenger RNA replace the previous ones. Thus the cell can respond to exogenous stimuli with the production of new types of enzymes.

The induction of enzymes by environmental stimuli has been cited as a model for embryonic differentiation. Bacteria and, to some extent, animal cells, respond to the presence of certain substrate molecules by forming enzymes to metabolize them. Jacques Monod has suggested that extracellular or intracellular influences may initiate or suppress the synthesis of specific enzymes, thus affecting the chemical constitution of the cell and leading to differentiation. As the embryo develops, the gradients established as a result of growth and cell multiplication could result in quantitative and even qualitative differences in enzymes. The induction or inhibition of one enzyme could lead to the accumulation of another chemical product which would induce the synthesis of a new enzyme and confer a new functional activity on these cells.

Morphogenesis appears to be too complex a phenomenon to be explained in terms of any single process such as enzyme induction. Enzymes can be induced in an embryo by the injection of an appropriate substance. Adenosine deaminase, for example, has been induced in the chick embryo by the injection of adenosine; however, no enzyme has been induced which is not normally present to some extent in the embryo. Adult tissues such as the liver may show marked differences in their enzymatic activities, differences which might be the result of adaptations comparable to those seen in bacteria. Adaptive changes in enzymes, however, are temporary and reversible, whereas differentiation is a permanent, essentially irreversible process.

The DNA of the genes not being transcribed at any given moment may be bound to a histone or to some other kind of protein which makes the DNA unavailable for the transcription system. The hypothesis that in the nucleus some genes are free and can be transcribed whereas others are bound and not transcribable is supported by a variety of experimental evidence. The protein, pea seed globulin, is synthesized in the seed of the pea plant but not in any other part of the plant. James Bonner has provided evidence that the DNA which codes for the synthesis of the globulin is bound to histone in the cells elsewhere in the plant. In the seed the histone is removed; that particular segment of the DNA becomes free and can be transcribed, forming messenger RNA which codes for the synthesis of the globulin. The question of what controls the binding and release of a specific segment of DNA remains to be answered.

Other biologists, such as Albert Tyler, have postulated that DNA may be transcribed to form messenger RNA but the mRNA is "masked" and inactive as a template for protein synthesis until it is subsequently unmasked by a separate process. This form of control may operate especially during early embryonic development. The mRNA synthesized in the oöcyte from the maternal genome may remain masked and inactive until activated by the fertilization process and freed to undergo translation.

Studies of the synthesis of RNA in the nucleus have shown that a large fraction of it, perhaps 80 per cent of the total RNA synthesized, is destroyed without leaving the nucleus; only about 20 per cent of the nuclear RNA is identical with RNA present in the cytoplasm. Some have suggested that this "heterogeneous" RNA serves in some way to regulate genic action. There appears to be far more DNA in the cells of multicellular organisms than is necessary to serve as template for the mRNA used in directing the synthesis of protein. Is the rest of the DNA part of some enormous, complex regulatory system which controls the activity of the DNA that does produce mRNA? Others interpret the rapid turnover of nuclear RNA as indicating that all genic DNA is transcribed into RNA all the time, but much of the RNA produced is rapidly destroyed before leaving the nucleus. Only the RNA that is stabilized in some way survives and passes to the cytoplasm. This second interpretation implies that gene action is regulated not during transcription but subsequently, between transcription and translation, by some process which selectively stabilizes certain kinds of RNA.

The kinds of enzymes present in different cells and tissues of the same organism may show marked qualitative and quantitative differences. Mammalian liver cells, for example, have a glucose-6-phosphatase and can convert glycogen and other precursors to free glucose, whereas skeletal muscle cells lack this enzyme. Enzymes that catalyze the same reaction in different tissues may differ in molecular size, in their amino acid sequences, in their immunologic properties and in their responses to hormones and other control mechanisms.

Even within a single tissue or a single cell, multiple molecular forms of an enzyme, termed **isozymes,** may be found. All these proteins catalyze the same general reaction but have distinct chemical and physical properties. The different molecular forms may bear a different net charge and thus can be separated by electrophoresis.

The lactic dehydrogenase isozymes have been extensively studied. They appear to be formed of subunits, four of which combine to form the active enzyme. There are two major kinds of subunits, A and B, each of which is a polypeptide chain with a specific, gene-determined sequence of amino acids. The entire molecule is analogous to a hemoglobin molecule which is composed of two α and two β polypeptides; however, in the lactic dehydrogenase molecule, any combination of the two types of subunits is permissible. The combinations of two kinds of subunits taken four at a time (A_4, A_3B, A_2B_2, AB_3, B_4) add up to the five kinds of lactic dehydrogenases that are typically observed when the enzyme is extracted from a tissue and subjected to electrophoresis.

It appears that there are two genes, one for each of the subunits. Different types of tissues have characteristic ratios of the different chains in their tetramers, presumably reflecting differences in the relative activities of the two genes. What controls the relative activities of the two genes remains unknown. A very curious observation, one that requires explanation, is that the lactic dehydrogenase in the breast muscle of the chick changes during embryonic development from a pure B_4 isozyme through a series of intermediates to a pure A_4 type in the adult.

When a piece of the dorsal lip of the blastopore of a frog gastrula is excised and implanted beneath the ectoderm of a second gastrula, the tissue heals in place and causes the development of a second brain, spinal cord and other parts at that site, so that a double embryo, or closely joined Siamese twins, result (Fig. 26–33). Many tissues show similar abilities to organize the development of an adjoining structure. The eye cup will initiate the formation of a lens from overlying ectoderm even if it is transplanted to the belly region where the cells would normally form belly epidermis. Such experiments indicate that development involves a coordinated series of stimuli and responses, each step determining the succeeding one. The term "organizer" is applied to the region of the embryo with this property and also to the chemical substance released by that region which passes to the adjoining tissue and directs its development.

It had been widely accepted that organizers can transmit their inductive stimuli only when in direct contact with the reactive cells, but Niu and Twitty showed that induction may be mediated by diffusible substances which can operate without direct physical contact of the two tissues. Niu and Twitty grew small clusters of frog ectoderm, mesoderm and endoderm cells in tissue culture and found that ectoderm alone would never differentiate into nerve tissue. Ectoderm cells placed in a medium in which mesoderm cells had been grown for the previous week did differentiate into chromatophores and nerve fibers. No comparable differentiation occurred when ectoderm cells were placed in cultures that had contained endoderm. It appears that inductive tissues such as notochord and mesoderm cells contain and release diffusible substances, nucleoproteins, which can operate at a distance and induce the differentiation of ectoderm.

Further evidence that morphogenetic substances are diffusible is provided by Grobstein's experimental analysis of the differentiation of kidney tubules. Nephrogenic mesenchyme is normally induced to form renal tubules by the tip of the ureter (p. 867). If the two rudiments are separated by treatment with trypsin and the dissociated cells are grown in tissue cul-

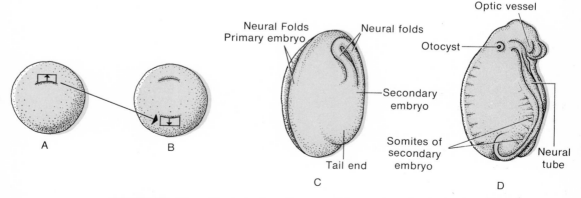

FIGURE 26–33   *The induction of a second frog embryo by the implantation of a dorsal lip of a blastopore from embryo A onto the belly region of embryo B. Embryo B then develops through stage C to a double embryo D.*

ture, they will reaggregate and form tubules. Certain other tissues, the ventral portion of the spinal cord, for example, also proved to be efficient inducers of renal tubules. Grobstein then separated the inducing and reacting tissues by cellophane membranes of varying thicknesses and porosities. The inducing substance can pass through membranes up to 60 microns thick and with pores 0.4 microns in diameter (Fig. 26–34). Electron micrographs revealed that induction occurred even when there was no cellular contact through the fine pores; the inducing principle is diffusible. It appears to be a large molecule and is at least in part protein for it is inactivated by trypsin.

Other experiments of Grobstein have reemphasized that extrinsic factors as well as nuclear factors may play a role in the process of differentiation. For example, embryonic pancreatic epithelium will continue to differentiate in organ culture only in the presence of mesenchyme cells.

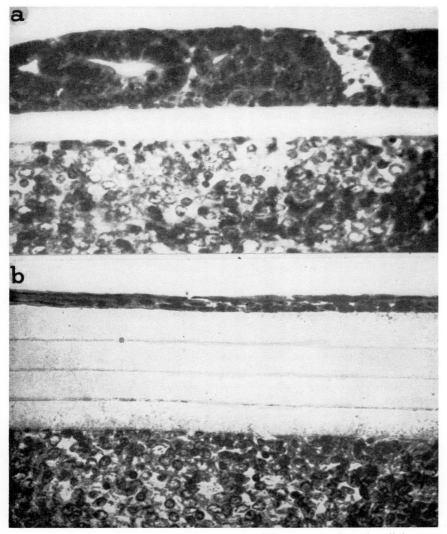

FIGURE 26–34   *Induction of kidney tubule differentiation through cellulose ester membrane filters. Below the filters is the inducer, spinal cord tissue; above the filters is the reacting nephrogenic mesenchyme. a, Successful induction through one layer of filter (20 μ thick). b, Four layers of filter preclude induction. (From Grobstein, C.: Some transmission characteristics of the tubule-inducing influence on mouse metanephrogenic mesenchyme. Experimental Cell Research 13:575-587, 1957.)*

This requirement can be met not only by pancreatic mesenchyme from the mouse but by mesenchyme from a variety of other sources, even embryonic chick mesenchyme. The mesenchyme can be replaced by a chick embryo juice and the active principle of the embryo juice appears to be a protein, for it is inactivated by trypsin but not by ribonuclease or deoxyribonuclease. It can be sedimented by high speed centrifugation and appears to be a large protein. This factor is only weakly effective in causing the differentiation of salivary gland epithelium in culture and is ineffective in inducing the formation of kidney tubules or of cartilage from their respective mesenchymes. Grobstein's experiments indicate that there is a spectrum of protein factors, each of which is more or less specific for the differentiation of one kind of cell.

The possible morphogenetic role of steroids has been explored by Dorothy Price of the University of Chicago. When the entire reproductive tract of a fetal male rat is dissected out and explanted it grows normally in culture if the testes are present and in their normal spatial relationship to the tract. If both testes are removed from the explanted reproductive tract, its development is inhibited and the accessory organs, the vas deferens, seminal vesicles and prostate do not differentiate. If the removed testes are replaced by pellets of testosterone implanted in their place, development of the explanted tract proceeds normally (Fig. 26–35). Thus testosterone can diffuse, at least over short distances, and induce the development of the male reproductive tract.

Suspensions of dissociated, individual healthy cells can be prepared by treating a tissue briefly with a dilute solution of trypsin. When the suspension of cells is placed in tissue culture medium the cells may re-

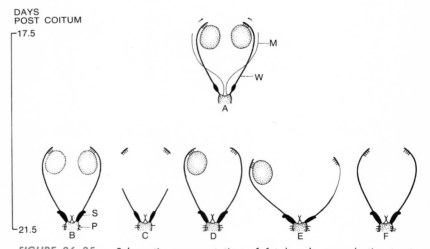

FIGURE 26–35   *Schematic representation of fetal male reproductive tracts explanted after about 17 days of development. A, The tract at 17 days, at the time of explantation. B--F, Tracts cultured for four days and hence at 21 days of development. B, Tract cultured with both testes present: the seminal vesicles (S) and prostate glands (P) have developed. C, Tract cultured without testes showed regression of the wolffian ducts, no seminal vesicles and only a few prostatic buds appeared. D, Tract cultured with one testis in place developed normally. E, Tract cultured with one testis in place, but spread apart showed some regression of the wolffian duct of the side without the testis, and the seminal vesicle on that side was small or absent. F, Tracts cultured with no testes but with testosterone micropellets developed with normal wolffian ducts, seminal vesicles and prostatic buds. M, Mullerian duct; W, wolffian duct. (From Price, D., and Pannabecker, R.: Ciba Foundation Colloquia on Aging in Transient Tissues, Wolstenholme, G. E. W. and Millar, S. C. P. (eds.) London, J. and A. Churchill, 1956)*

aggregate and continue to differentiate in conformity with their previous pattern. The cells in tissue culture reaggregate not in a chaotic mass but in an ordered fashion, forming recognizable morphogenetic units. The cells appear to have specific affinities, for epidermal cells join with each other to form a sheet, disaggregated kidney cells join to form kidney tubules and so on (p. 504). The question of how one cell recognizes another cell and joins with it to form a tubule, sheet or other structural unit is indeed a fascinating one.

Disaggregated cells growing in culture can be subjected to a variety of experimental procedures to analyze the factors controlling their differentiation. Dissociated epidermal cells grown in the usual culture medium will reaggregate and differentiate as a stratified squamous epithelium. But if vitamin A is added to the culture medium or if the dissociated cells are exposed to vitamin A for as short a time as 15 minutes, the cells reaggregate and differentiate into a columnar epithelium with mucus-secreting goblet cells (Fig. 26–36). Vitamin A under these circumstances has a strong morphogenetic effect.

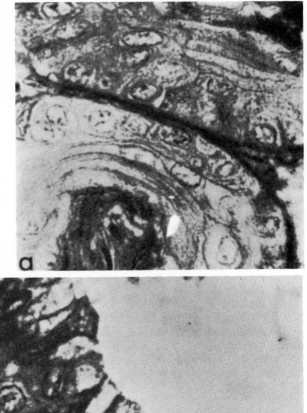

FIGURE 26–36   Influence of a short treatment with vitamin A on the differentiation of dissociated and reaggregated epidermal cells. a, Control culture, differentiating as typical stratified squamous epithelium. b, Treated culture, developing into columnar epithelium with goblet cells. (From Weiss, P., and James, R.: Experimental Cell Research, Supplement 3, 1955).

Embryonic tissues growing *in vivo* have differential sensitivities to changes in nutrients, to the presence of inhibitors and antimetabolites and to various environmental agents. Any of these factors, applied during the appropriate critical period in development, may change the course of development and differentiation and mimic the phenotype of a mutant gene, producing what is termed a **phenocopy.**

A striking demonstration that the same set of genes operating in dissimilar environments may have different morphological effects was provided by experiments with three races of frogs found in Florida, Pennsylvania and Vermont. Each of these races normally develops at a speed which is adapted to the length of the usual spring and summer season in its locale. Southern frogs develop slowly and northern frogs develop more rapidly. When northern frogs are raised under southern environmental conditions their development is overaccelerated, whereas when southern frogs are raised under northern conditions their development is over-retarded.

When an egg is fertilized with a sperm from a different race and the original egg nucleus is removed before the sperm nucleus can unite with it, it is possible to establish a cell with "northern" genes operating in "southern" cytoplasm or the reverse. Northern genes in southern cytoplasm resulted in poorly regulated development; the animal's head grew more rapidly than the posterior region and became disproportionately large. Southern genes introduced into northern cytoplasm led again to poorly regulated development but the head rather than the posterior region was retarded and disproportionately small. Genes from the Pennsylvania race of frog acted as "northern" with Florida cytoplasm but as "southern" with Vermont cytoplasm. The same set of genes had diverse morphological effects when they operated in different cytoplasmic environments.

Cellular differentiation may involve the differential activation of specific genes in different tissues; it may involve mechanisms affecting protein synthesis but operating at the ribosomal level, or operating at the cell surface where the transport of substances into or out of the cells is regulated.

Differentiation may be controlled at least in part by influences originating outside the cell, by "organizers" from neighboring cells in early differentiation, by the mesenchymal proteins studied by Grobstein, or by hormones from distant cells. Such systemic influences participate in the integration of the differentiation of individual cells into the larger pattern of differentiation of the tissues of the whole organism. Eventually it should be possible for developmental biologists to bridge the gap between studies of development at the level of the whole organism and studies at the molecular level and trace in detail the sequence of events from the initial action of the gene to the final expression of the phenotype.

Our descriptive knowledge of the phenomena of fertilization, cleavage, gastrulation, morphogenesis and organogenesis is quite extensive, but our understanding of the fundamental molecular mechanisms involved in each of these processes is indeed rudimentary. This is a fertile field for future investigation.

SECTION 26–17

REGENERATION

Animals have to varying degrees the ability to repair the damage caused by a wound, by the loss of an organ or by the severing of an appendage. The repair process involves a reinitiation of morphogenetic processes halted when development was completed. Hydras, planarians and earthworms have a high degree of ability to regenerate. If a planarian is cut in half, for example, each half will regenerate the missing part and two planaria are formed. Earthworms cut in half have a similar ability to regenerate the missing end. Echinoderms such as the starfish and crustaceans such as the crab or lobster can readily regenerate a lost arm or leg. Among vertebrates the salamanders and newts are outstanding in their regenerative capacity and can regenerate a severed arm or leg. Certain lizards can regenerate a lost tail; indeed some lizards, when threatened, can sever their tail near the base, leaving the tail to distract their predator while they escape and regenerate a new tail. The regenerated lizard tail differs from

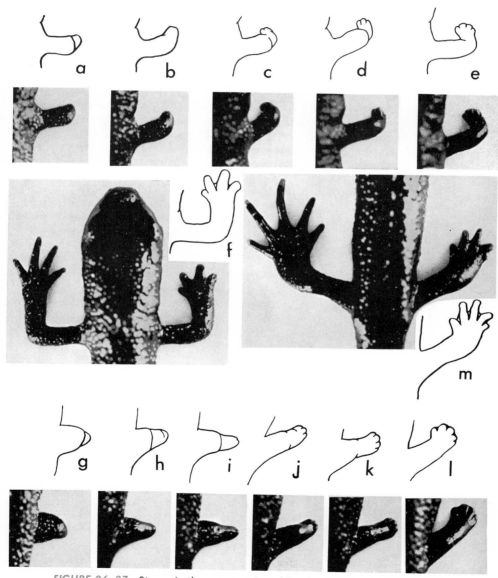

FIGURE 26–37   *Stages in the regeneration of fore limbs (a-f) and hind limbs (g-m) in salamanders after they have been removed. (From Balinsky, B. I.: An Introduction to Embryology. 3rd ed. Philadelphia, W. B. Saunders Company, 1970.)*

the original in the shape of the vertebrae and in the kind of scales cover-ing it. These regenerative phenomena, especially the regeneration of limbs in salamanders and the regeneration of parts in planaria, have been stud-ied extensively to learn about the regenerative process and about wound healing, and as a possible source of clues about normal differentiation.

Both adult and larval salamanders can regenerate limbs to a remark-able degree. After the limb has been amputated, the surrounding tissues close the wound and a mass of cells, a **blastema,** accumulates under the skin. The blastema grows rapidly by active proliferation of cells. At first it has a conical shape and then flattens dorsoventrally at the tip (Fig. 26–37). Rudiments of the digits appear and cell masses segregate in the proper pattern in the interior to become the rudiments of the bones and muscles. Organogenesis is followed by histological differentiation and the entire regenerating limb continues to grow until it attains the size of a normal limb. The regenerated limb can move and eventually becomes indistin-guishable from a normal limb. The cessation of growth when the normal size is reached is just as remarkable and unexplained as the initiation of growth when the limb is lost.

The presence of a nerve supply is required for regeneration to pro-ceed normally. If the nerves supplying the arm or leg of the salamander are destroyed when the limb is removed the blastema stops growing and the development of the regenerating limb is halted (Fig. 26–38). However, if the regenerating limb has begun to differentiate, the nerve can be cut with-out interfering with the regenerative process. The nerve supply is required for the initiation of the regenerative process but not for its completion. If a nerve to a limb is cut across and the proximal end of the nerve is placed in a cut in the skin a short distance from the limb, a blastema will form as the wound heals. The nerve will initiate the development of a limb rudi-ment or even of a complete new limb in an area that does not normally form a limb. It appears that the regenerative stimulus supplied by the nerve is not the ordinary nerve impulse but perhaps some sort of growth-promot-ing neurosecretory product.

Regeneration can be inhibited by irradiating the blastema with x-rays in an appropriate dose, about 7000 r. The irradiated blastema regresses, becomes filled with connective tissue cells and may be completely resorbed. If a normal leg of an adult newt is irradiated with 7000 r no effects are visi-ble; neither the appearance nor the function of the leg is affected. How-

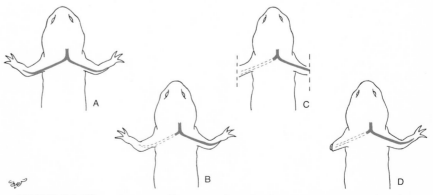

FIGURE 26–38   *Diagram illustrating the role of an intact nerve supply in bringing about regeneration of amputated limbs in salamanders. A, Salamander. B, The nerve supply to one forelimb is removed. C, Both forelimbs are amputated just above the elbow. D, The limb with a nerve supply regenerates; the other limb does not.*

ever, at any subsequent time, even months later, if the irradiated leg is amputated, no regeneration will occur. X-rays have a similar inhibitory effect on all regenerative systems tested and thus cause some marked change in a fundamental property of cells.

SUGGESTIONS FOR FURTHER READING

Arey, L. B.: *Developmental Anatomy.* 7th ed. Philadelphia, W. B. Saunders Company, 1965.
   A fine description of the development of man and other mammals. One of the classic texts which has been kept up to date in successive editions.

Balinsky, B. I.: *An Introduction to Embryology.* 3rd ed. Philadelphia, W. B. Saunders Company, 1970.
   An excellent general text which includes the details of organogenesis in each of the vertebrate types, plus a synthesis of the experimental analysis of cellular differentiation.

Bell, Eugene: *Molecular and Cellular Aspects of Development.* New York, Harper & Row, Publishers, 1965.
   Reprints of many original papers dealing with the various facets of experimental embryology.

Comfort, A.: *Aging, The Biology of Senescence.* New York, Holt, Rinehart and Winston, Inc., 1964.
   A general survey of the biological aspects of the phenomenon of aging.

DeHaan, R. L., and Ursprung, H. (eds.): *Organogenesis.* New York, Holt, Rinehart and Winston, Inc., 1965.
   A collection of papers by investigators in the field; an excellent source book.

Falkner, F. (ed.): *Human Development.* Philadelphia, W. B. Saunders Company, 1966.
   Chapters by 29 authorities will serve as a useful reference work for students interested in all aspects of human development.

Galston, A. W.: *The Green Plant.* Englewood Cliffs, New Jersey, Prentice-Hall, Inc., 1968.
   A convenient and concise summary of plant growth and development.

Hayflick, L.: *Human Cells and Aging.* Sci. Amer. *218*:32, March, 1968.
   A brief presentation of current theories of aging based on results of experiments with human cells kept in culture.

Konigsberg, I. R.: *The Embryological Origin of Muscle.* Sci. Amer., August (Offprint 191) 1964.
   A brief summary of the development of skeletal muscle in vertebrates.

Leopold, A. C.: *Plant Growth and Development.* New York, McGraw-Hill Book Company, 1964.
   An important research work that provides material of interest in connection with this chapter.

Monroy, A., and Moscona, A. A. (eds.): *Current Topics in Developmental Biology.* New York, Academic Press, Inc., Published annually.
   Each volume contains up to date summaries by experts in the field of their particular areas of research in developmental biology.

Steward, F. C.: *Growth and Organization in Plants.* Reading, Massachusetts, Addison-Wesley Publishing Company, Inc., 1968.
   Treatment of the problems of growth and development which synthesizes all aspects of knowledge of plants—their physiology, biochemistry and morphology.

Strehler, B.: *Time, Cells, and Aging.* New York, Academic Press, Inc., 1962.
   An interesting account of our knowledge of the process of aging and of several theories to account for it.

Talland, G. A. (ed.): *Human Aging and Behavior,* New York, Academic Press, 1968.
   An interesting series of papers relating to the effects of aging on behavior.

Torrey, J. G.: *Development in Flowering Plants.* New York, The Macmillan Company, 1967.
   The morphogenetic approach to plant growth and development is here presented in a
   brief and readable book.

Torrey, T. W.: *Morphogenesis of the Vertebrates.* 2nd ed. New York, John Wiley & Sons, Inc.,
1967.
   A good account of vertebrate embryology.

Willier, B. H., Weiss, P. A., and Hamburger, V.: *Analysis of Development.* Philadelphia, W. B.
Saunders Company, 1955.
   A classic book by many authors summarizing the methods and findings of experimental
   embryology.

Willier, B. H., and Oppenheimer, J. M. (eds.): *Foundations of Experimental Embryology.* Englewood Cliffs, New Jersey, Prentice-Hall, Inc., 1964.
   A collection of 11 important papers.

POPULATION BIOLOGY: ECOLOGY

ECOLOGY

If a biologist from another planet arrived on earth, the first thing that he would notice is the great diversity of species here—well over a million. Next he would observe that different species live in different places but also that more than one species lives in the same place. He might also notice that few places on the planet are uninhabited. No life would be found in the craters of active volcanoes, on recent lava flows, in the Dead Sea nor at the bottom of the Black Sea. Life would be very rare at the poles. There would be an increase in kinds and absolute numbers as one moved closer to the equator. Then he would notice that no species lives alone; the populations are mixed. Finally he might notice that in any given space different species are present at different times (winter versus summer, night versus day, present versus past). He might also observe that the abundance of any given species is never constant. Ecology is concerned with studying these phenomena, understanding them and making predictions about future changes. The application of ecology is the management of species and their environments.

Since all the factors bearing on these phenomena are intertwined, there is no correct place to begin an ecological investigation. *Climatic* and *edaphic* (soil) factors influence the numbers and distribution of plants; plants and climatic and edaphic factors influence the number and distribution of animals, which in turn may influence plants. It is convenient to begin our discussion with an examination of the kind of environment that the earth provides for living organisms.

SECTION 27–1

DISTRIBUTION OF SOLAR ENERGY

The earth is a remarkably nonuniform, patchy planet where conditions vary in space and time. It derives nearly all its energy from the sun. This energy is unevenly distributed over the face of the globe. The incoming solar radiation *(insolation)* varies with the length of the path of the sun's rays through the atmosphere, the area of horizontal surface over which a "bundle" of rays of a given cross section spreads the distance of the sun from the earth, which changes seasonally owing to the elliptical orbit of the earth around the sun, the amount of water vapor and dust in the atmosphere and the length of day. Because the angle of incidence is lower at high latitudes the energy is spread thinner and passes through a thicker layer of atmosphere there (Fig. 27–1); consequently, the polar regions receive less radiant energy than equatorial regions even though the length of day decreases with distance toward the equator.

Major variations in the incoming solar energy are caused by the various movements of the earth with respect to the sun (Fig. 27–2). The earth completes an orbit around the sun every 365¼ days. Because the axis of the earth is tilted 23½ degrees relative to its plane of orbit, the distribution of energy varies throughout the year. Twice a year, at the **vernal equinox** (March 21) and the **autumnal equinox** (September 23), the sun is at zenith at the equator at noon and its circle of illumination reaches the North and South Poles simultaneously. After March 21 the north pole moves nearer to the sun until June 21 **(summer solstice)**. Then the movement of the pole is reversed; it moves away from the sun until December 21 **(winter solstice)**. When the North Pole of the axis is inclined away from the sun, winter comes to the Northern Hemisphere. Days get shorter and temperatures drop. The exact reverse occurs in the Southern Hemisphere. These seasonal differences become more pronounced with distance from the equator.

The rotation of the earth around its own axis every 24 hours gives night and day and all the energy changes associated with it. Temperature changes, however, lag behind changes in insolation. The daily maximum is usually in midafternoon and the minimum just before sunrise. Soil temperatures lag even more.

The presence of an atmosphere imposes further variations on the pattern of energy. Short wavelengths of radiant energy from the sun are not absorbed by water or water vapor; hence, they pass through the atmosphere with little absorption. Some energy is absorbed by the earth, the rest is reflected back as long wavelengths (heat). Snow, water and white-colored soils reflect much heat, while bare ground and black soils absorb. Man's

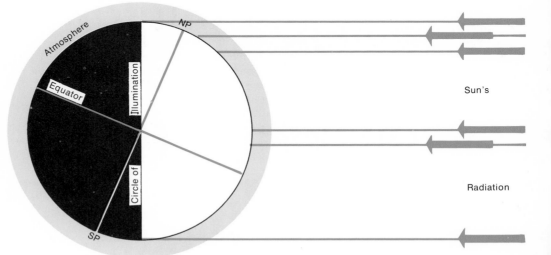

FIGURE 27–1   Circle of illumination, areas of daylight and darkness, angles of sun's rays at different latitudes and differences in areas affected and thickness of atmosphere penetrated at time of summer solstice. (After Ward, H. B., and Powers, W. E.: Introduction to Weather and Climate. Evanston, Illinois, Northwestern University Press, 1942.)

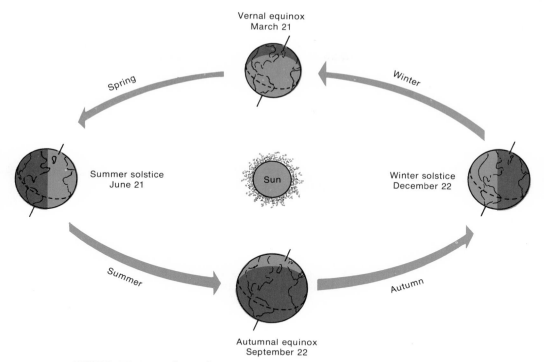

Vernal equinox
March 21

Spring

Winter

Summer solstice
June 21

Sun

Winter solstice
December 22

Summer

Autumn

Autumnal equinox
September 22

FIGURE 27–2   The sunlit portions of the Northern Hemisphere are seen to vary from greater than one half in summer to less than one half in winter. The proportion of any latitude that is sunlit is also the proportion of the 24-hour day between sunrise and sunset. (From MacArthur, R. H., and Connell, J. H.: The Biology of Populations. New York, John Wiley & Sons, Inc., 1966.)

alteration of the earth's surface also contributes to temperature differences. Paving, building, plowing, timber removal and air pollution all affect the radiation of insolation. Oceans receive or lose energy by radiation, heat exchange with the atmosphere, evaporation and condensation. Probably as much as 40 per cent of the heat of the atmosphere is derived from condensation of water vapor derived principally from evaporation from the surface of the ocean. The moisture-laden air rises, moves to higher latitudes where it is cooled and gives up its moisture as clouds or rain.

The heat is then absorbed by the wet atmosphere. Some earth radiation also ascends by convection. Thus the atmosphere is heated from below and radiates back to earth. The atmosphere is a heat trap working on the same principle as a greenhouse where glass substitutes for clouds and water vapor.

SECTION 27–2

CLIMATE

This nonuniform heating of the atmosphere from below affects climate. Heat and pressure differences around the world cause the circulation of air. These wind systems are primarily responsible for the climatic differences in the world. The primary circulation involves an exchange of air between high and low latitudes. At the equator air is hotter, hence, lighter. As it rises it is replaced by heavier, colder air flowing in from the north and the south. The equatorial air cools at the top so that it falls to the surface near the poles (Fig. 27–3). Since the earth is not stationary, but rotates, the Coriolis force skews the flow clockwise north of the equator and counterclockwise in the south. Furthermore, since the force is 0 at the equator and maximum at the poles, the prevailing currents vary with the latitude. This accounts for the equatorial doldrums, the east trade winds and the westerlies (Fig. 27–3). The doldrums near the equator is a calm hot area with frequent showers. The horse latitudes (30 degrees north and south of the equator), the region between the westerlies and the trade winds are regions of calms and little rain. In the belt of the trades there are steady winds but little rain. In the belt of the westerlies lie regions of changeable and stormy weather. Around the poles are circulating winds that are extremely cold in winter.

This **primary atmospheric circulation** is complicated by a number of factors. For example, temperature differences between oceans and land masses cause circulation to and from summer-heated and winter-cooled continents. Perturbations in primary circulation are also associated with meanders in the high altitude jet stream. Further perturbations in atmospheric circulation are caused by barriers to flow (e.g., mountain ranges) and by instabilities of fronts. For example, alternating cyclones (ascending air, low pressure) and anticyclones (descending air, high pressure) give the middle latitudes the maximum weather change.

In addition to causing patterns of temperature, circulation of the atmosphere produces a heterogeneous distribution of moisture. As warm air rises it is cooled. If it contains moisture, this condenses as clouds or rain because cold air holds less moisture than warm. Falling air is warmed, and any liquid water that it contains is vaporized. Accordingly, trade winds blowing toward the equator get warmer and gather moisture from the oceans. At the equator they rise, cool and cause heavy rainfall in the tropics. Now dry, the air returns to higher latitudes. En route to higher latitudes winds begin to fall and pick up moisture in the region of latitude 30 de-

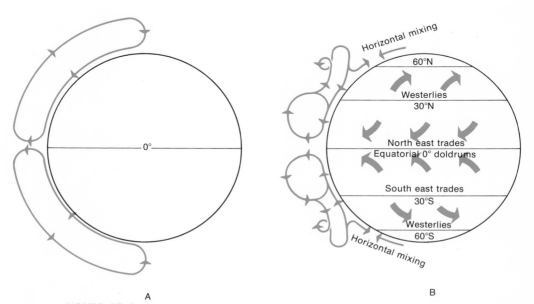

A   B

FIGURE 27–3   *The circulation of the earth's atmosphere: (A) on a hypothetical nonrotating earth, (B) incorporating the effects of the earth's rotation. (After Byers, M. G.: The Atmosphere Up to Thirty Kilometers. In G. P. Kuiper (ed.): The Earth As A Planet. Chicago, University of Chicago Press, 1954.)*

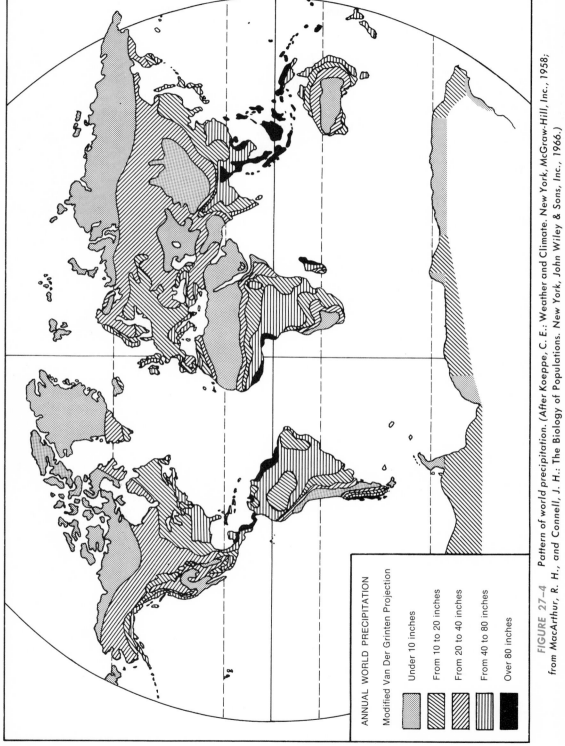

FIGURE 27–4   *Pattern of world precipitation. (After Koeppe, C. E.: Weather and Climate. New York, McGraw-Hill, Inc., 1958; from MacArthur, R. H., and Connell, J. H.: The Biology of Populations. New York, John Wiley & Sons, Inc., 1966.)*

ANNUAL WORLD PRECIPITATION

Modified Van Der Grinten Projection

Under 10 inches

From 10 to 20 inches

From 20 to 40 inches

From 40 to 80 inches

Over 80 inches

grees. Most of the world's deserts occur near these latitudes. The temperature of various ocean currents also affects the moisture pattern. For example, where there are cold currents off western coasts the air cannot pick up moisture, so western coasts tend to be dry. Here the air at the surface of the oceans is cold and the falling warmer air cannot pick up moisture; however, fogs occur because of the meeting of the cold and warm air. At a higher latitude (from 40 to 60 degrees), the westerlies are blowing in off a warm ocean and another zone of heavy rainfall occurs (Fig. 27–4).

Temperature and moisture patterns are also affected by the relation of land masses to seas and the altitudinal configuration of land masses. Land tends to heat up more rapidly than water and also to cool more rapidly. Thus continents have great seasonal temperature extremes and diurnal extremes. These conditions, in conjunction with temperature conditions in oceans, cause important seasonal winds (monsoons). In Asia, for example, the continent heats up in the summer, thus creating a huge continental low due to hot air rising. It is replaced by cool winds blowing in from cool oceans and the summer monsoon rains result. Conversely, in the winter the coldness of the continent creates a huge high pressure area from which cold winds blow out over the warm ocean.

By contrast, islands and coastal regions have a more even climate, because in the winter time the ocean slowly releases the heat accumulated during the summer. In the summer the ocean has a cooling effect due to its lag in picking up summer heat.

Finally, high mountain chains act as barriers to air flow. Air is forced to rise as it encounters a mountain chain. As it rises it cools, its moisture condenses, and rain falls on the side of the mountain. On the lee side as the air falls it warms and picks up moisture so that this area is dry.

Altitude causes mountain and plateau climates to differ greatly from those of the lowlands. With increasing altitude, temperature and absolute humidity drop, though not to the same degree at all latitudes. Insolation, relative humidity and precipitation increase with altitude. Most mountains have a rainy side and a dry side and usually more precipitation than plateaus (Fig. 27–5). Because high mountains act as barriers to winds, they cause different climates in adjacent areas on each side. The Himalayas protect India from the cold of Tibet, keep the north side dry and the south side wet. The Rocky Mountains stop the cold waves from Canada. The Andes prevent the southeastern trade winds from blowing over; the east sides are wet and the coast dry.

Thus we have **mountain climates** (see above), **marine climates** (moderate temperatures with small diurnal and annual ranges, high humidity, much cloudiness and rainfall), **continental climates** (wide temperature variation, low humidity except in winter), **littoral climates** (coast climates with features of both marine and continental), **desert climates** (great temperature extremes and little rainfall) and **monsoon climates** (much cloudiness and intense rainfall in summer, no clouds and little precipitation in winter) (Fig. 27–6).

All over the world there are local winds that cause regular predictable changes in temperature and humidity. Among the more famous warm winds are: the foehn of the Alps, the chinook of the Great Plains of North America, the simoom of the Sahara, the harmattan of West Africa and the Santa Ana of California. Cold winds include: the mistral of the Rhone Valley, the pampero of Argentina and the norther of Texas.

In summary, the factors that affect climate are: **latitude, altitude, land and water masses, mountains as barriers, ocean currents, general atmospheric circulation, local winds** and **cyclonic storms**. Additional factors are: local topography, land control, surface covering (presence and type

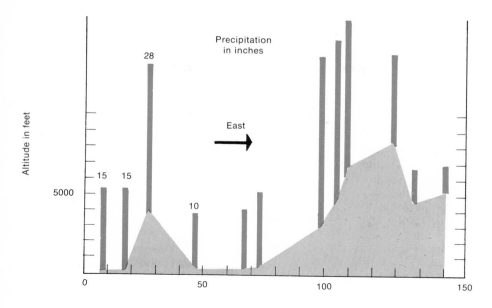

FIGURE 27–5   *Mean annual rainfall (vertical columns) in relation to altitude (ornamented line) at a series of stations extending from Palo Alto in the Pacific coast eastward across the Coast Range and the Sierra Nevada to Oasis Ranch in the Nevada desert. The diagram shows (1) the approach effect on the west edge of the Sierra Nevada, (2) the zone of maximum rainfall on the middle western slope of the Sierra Nevada, and (3) rain shadows to the landward of the two mountain ranges. (After Daubenmire, R. F.: Plants and Environment. New York, John Wiley & Sons, Inc., 1947.)*

of vegetation), color, wetness, dryness and snow cover. Thus, as we shall see later, climate is primarily responsible for vegetational types, but vegetation in turn affects local climate.

SECTION 27–3

PATTERNS IN THE EARTH'S CRUST

The patchiness of the earth also extends to its crust. The greatest amount of the crust is igneous rock and relatively uniform; however, **weathering, erosion** and **sedimentation** lead to great geochemical differentiation. There tends to be a segregation of chemicals. Quartz sand and sandstone represent the segregation of silicon. Clay muds are predominantly aluminum and other bases. Calcium is separated as limestone. Iron is separated as ferric hydroxide. Sedimentation involves not only the mechanical and physical action of the atmosphere and hydrosphere but also the incorporation of H_2O and CO_2 in sedimentary minerals and the addition of soluble ions, especially sodium, to the hydrosphere. The biosphere is also involved in geochemical differentiation. Photosynthesis has, until the industrial revolution, probably been responsible for the composition of the atmosphere. Organisms are greatly responsible for the balance of dissolved material in the oceans. Organisms also make a major contribution to soil. They assist in the deposit of calcium and sulfur. They are the source of the great deposits of fossil fuels—peat, coal and oil.

All the processes of weathering of the earth's crust result in the deposition on the surface of an unconsolidated coat whose characteristics

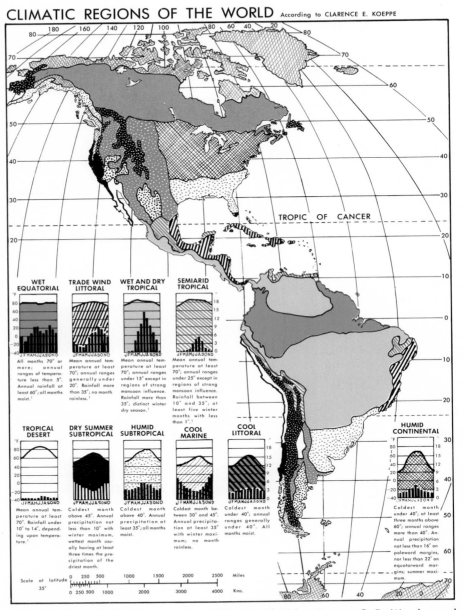

FIGURE 27-6   Climatic regions of the world. (After Koeppe, C. E.: Weather and Climate. New York, McGraw-Hill, Inc., 1958.)

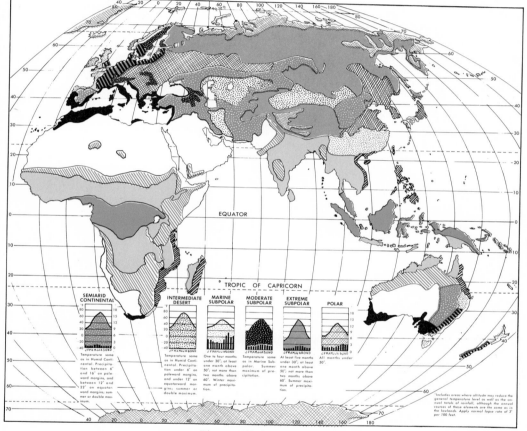

Figure 27–6 *Continued.*

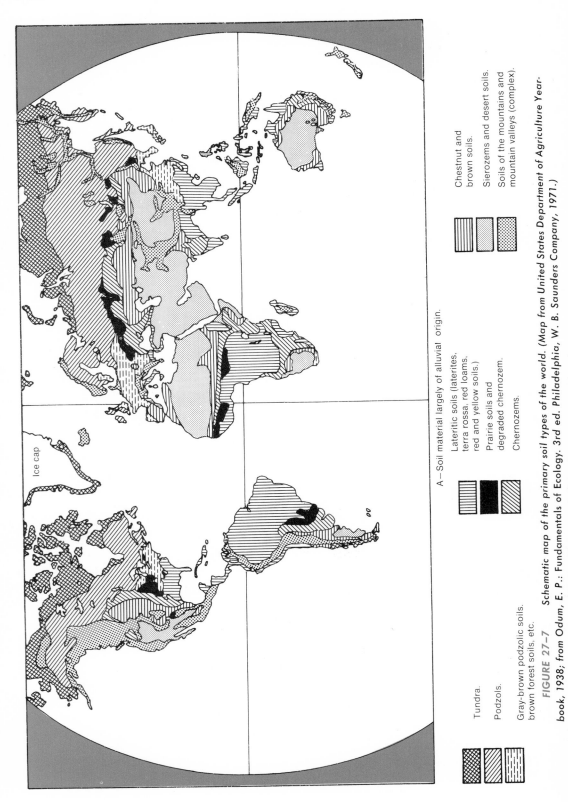

A – Soil material largely of alluvial origin.

Lateritic soils (laterites, terra rossa, red loams, red and yellow soils.)

Prairie soils and degraded chernozem.

Chernozems.

Chestnut and brown soils.

Sierozems and desert soils.

Soils of the mountains and mountain valleys (complex).

Tundra.

Podzols.

Gray-brown podzolic soils. brown forest soils, etc.

Ice cap

FIGURE 27-7   Schematic map of the primary soil types of the world. (Map from United States Department of Agriculture Yearbook, 1938; from Odum, E. P.: Fundamentals of Ecology. 3rd ed. Philadelphia, W. B. Saunders Company, 1971.)

depend on the nature of the parent rock, the kind of weathering to which it has been subjected and its age. This is soil, although strictly speaking the term is reserved for this material after organic debris has been added to it. Most soils remain where they are formed from the parent rock or from deposition of organic material (peat bogs). Other soils may have been transported from their place of origin to another location by wind (sand dunes, loess [Iowa, Nebraska and Kansas farm soils] [dust from flood plains of glacial rivers of long ago]), by water (alluvial deposits, deltas and so forth and by glaciers (most soils of Canada and Northeastern United States are glacial).

From the time of its origin, soil undergoes progressive development. **Soil development** is controlled directly and indirectly by climate—directly in that climate, especially temperature and rainfall, determines the rate at which materials in solution and suspension are transported by percolating water to levels of accumulation; indirectly in that climate determines the kind of vegetation, which in turn determines what organic materials are available for incorporation into the soil. Three processes are involved in soil development: **podzolization**, in humid cold regions where rainfall exceeds evaporation and vegetation produces acid humus giving ash gray soils; **laterization,** in tropical regions where temperatures are high, rainfall heavy and acidity from the decay of tropical vegetation low, all giving red soils high in iron; and **calcification,** where there is low rainfall unevenly over the year and high rates of evaporation. As soils develop, a **horizontal stratification** appears which may be seen in a soil profile. **Profiles** show three general layers or horizons: (1) the upper layer from which substances have been removed by draining water, (2) a layer of accumulated materials below this and (3) a bottom layer of unweathered parent material.

As an example of the different types of soils developed in different climates, Figure 27–8 illustrates conditions in the United States. **Pedalfers** are those soils without a carbonate layer; **pedocals** are those with. The types of pedalfers are: **podsol soils,** with a thick white or gray-leached layer over a brown layer with accumulated aluminum and iron; **gray-brown podsol soils,** with less leaching, prairie soils, black or dark brown with brown subsoils; **red and yellow soils** in which both podsolization and laterization occurs. Pedocals include: **chernozem soils,** black and rich in organic matter with brown or reddish calcareous subsoils; **chestnut soils,** with less organic matter, lighter brown; **gray desert** and **semidesert soils,** with little organic content and largely physical weathering.

All of these soils have different characteristics of importance to plants, because soils supply anchorage, water, mineral nutrients and aeration of roots. Among the important characteristics are: texture (whether it is gravel, sand, silt or clay), organic content, soil water, soil atmosphere, acidity, alkalinity and salinity.

SECTION 27–4

ENVIRONMENTAL VARIATION AND THE REQUIREMENTS OF LIFE

This **patchiness** of the earth, in terms of climate (temperature, moisture, clouds, fogs, snow, rain and winds), edaphic factors (soils) and physiography (continents, islands, oceans, rivers, lakes and so forth) extends from large areas to minute areas. That is, the difference may be macro, as, for example, the difference between the tropics and the poles, or micro, the difference between the surface of the soil and 4 centimeters below.

Between the extremes, the coarse-grained differences and the fine-grained differences, are all intermediates.

It is against the world's background of patchiness and fluctuating conditions that life developed and continues apparently in defiance of the Second Law of Thermodynamics, which states essentially that changes in the material of the universe are unidirectional and irreversible and result in increased entropy (disorganization). Living materials do not escape this law, so there is an inexorable cycle of death and decay. Life itself continues, however, because it functions as an open system constantly supplied by an external source of energy. In short, organisms maintain a steady state in a varied and fluctuating environment. To this end physiological processes must be maintained above certain minima. Water, oxygen, carbon, nitrogen, trace elements and a workable temperature are basic requirements that must be met irrespective of the environment. Organisms satisfy these requirements by adapting to the environment, moving out or becoming extinct. **Adaptation** may be behavioral or physiological, or both. The environment determines the mode and degree of regulation required. An organism is no stronger than the weakest link in the ecological chain of its requirements. This **Law of the Minimum** was originally stated by Liebig in terms of the growth requirements of plants. The growth of a plant is dependent upon the amount of that foodstuff which is presented in minimum quantity.

The most vulnerable stage of the species represents its limits. For each environmental factor affecting a species there is obviously a minimum and a maximum value. These values mark the limits of tolerance **(Shelford's Law of Tolerance)**. A species may have a wide tolerance for one factor, as, for example, temperature, but a narrow tolerance for another (e.g., salinity or pH). Thus an organism may not be living under optimum conditions insofar as one factor is concerned simply because it cannot tolerate another accompanying factor. Over long periods regulation assumes the form of a change in gene frequency, i.e., natural selection weeds out the unfit, and eventually new species come to occupy a particular area.

Conceivably life could be adapted to the total melange of worldly conditions but in fact this is much too inefficient. One cannot have an organism that can maintain its steady state on land, in the desert, at the poles, in the tropics. The best general strategy is to specialize. This is the basic reason why there are so many different forms of life and why they are not evenly distributed over the face of the earth.

SECTION 27–5

THE DISTRIBUTION OF PLANTS

The most basic relation between the patchiness of the environment and the forms and distribution of organisms is that of plants. Plants are the primary procedures. The environmental factors that control the distribution of different plant species are climatic and edaphic. In any given climate and soil condition, certain plant species can survive. The world distribution of various species of plants correlates remarkably well with world climate and soil conditions as a comparison of Figure 27–6, 27–7 and 27–9 will show. In addition, all the plants that can survive particular climatic and edaphic conditions, themselves contribute to local climatic and microclimatic conditions. All interact and form communities. Climate

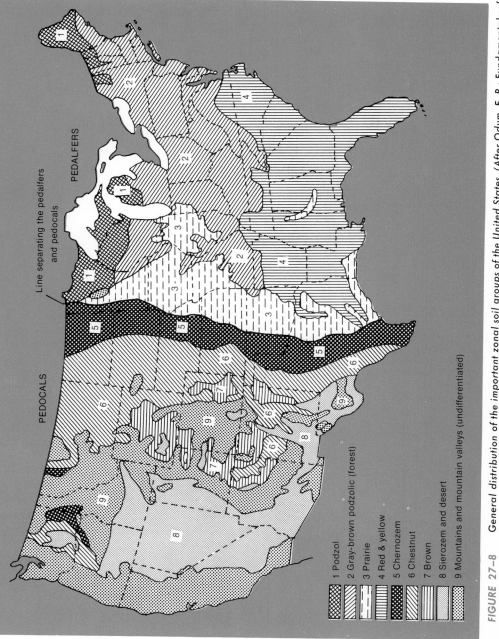

PEDALFERS

Line separating the pedalfers
and pedocals

PEDOCALS

1 Podzol
2 Gray-brown podzolic (forest)
3 Prairie
4 Red & yellow
5 Chernozem
6 Chestnut
7 Brown
8 Sierozem and desert
9 Mountains and mountain valleys (undifferentiated)

FIGURE 27-8   General distribution of the important zonal soil groups of the United States. (After Odum, E. P.: Fundamentals of Ecology. 3rd ed. Philadelphia, W. B. Saunders Company, 1971.)

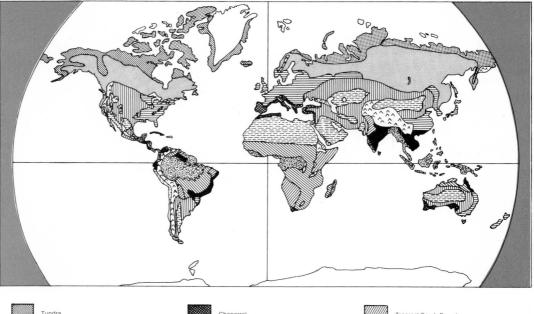

Tundra	Chaparral
Northern Conifer Forest (Taiga)	Desert
Temperate Deciduous & Rain Forest	Tropical Rain Forest
Temperate Grassland	Tropical Deciduous Forest

Tropical Scrub Forest
Tropical Grassland & Savanna
Mountains (complex zonation)

FIGURE 27–9   Schematic map of the major biomes of the world. Note that only the tundra and the northern conifer forest have some continuity throughout the world. Other biomes of the same type (temperate grassland or tropical rain forest, for example) are isolated in different biogeographical regions and, therefore, may be expected to have ecologically equivalent but often taxonomically unrelated species. The pattern of the major biomes is similar to but not identical with that of the primary soil groups as mapped in Figure 27–7. (Map prepared using Finch and Trewartha's (1949) map of original vegetation as a basis.) (From Odum, E. P.: Fundamentals of Ecology. 3rd ed. Philadelphia, W. B. Saunders Company, 1971.)

and vegetation types together regulate the distribution of animals. Thus species cannot live alone. Accordingly there are easily recognizable **communities** (biomes) of plants and animals.

The complex and intimate interactions between climates, organisms and substrates result in characteristic recognizable regional communities. The world-wide distribution of these extensive communities, called **biomes,** corresponds closely with the distribution of climates and of soils (Figs. 27–6 and 27–7). They are recognized, and frequently named for, the major fully developed (climax) plant community. They are best appreciated by flying by jet from pole to pole.

SECTION 27–6

THE MAJOR BIOMES

Tundra

Circling the north pole, on Greenland and the continents of North America and Eurasia, is a vast frozen treeless plain of which Marco Polo

FIGURE 27–10   The tundra biome. Above, view of the low tundra near Churchill, Manitoba, in July. Note the numerous ponds. Below, view of tundra vegetation showing "lumpy" nature of low tundra and a characteristic tundra bird, the willow ptarmigan. (Below, courtesy of C. Lynn Haywood.)

said, "A land inaccessible because of its quagmires and ice" (Fig. 27–10). This land, the tundra, has no counterpart in the Southern Hemisphere, because that hemisphere is largely ocean. The limiting environmental factors are low temperatures, low precipitation and a short growing season (± 60 days). From a few inches to a few feet below the surface, the ground is alternately freezing and thawing, and beneath this layer it is permanently frozen (permafrost). Consequently drainage is poor in low places and there is much wet ground and many bogs and ponds. The low areas are matted with thick spongy low and slowly decaying vegetation, slowly because low temperatures slow bacterial action. Higher places are bare and rocky. The extensive frost action keeps the soil in constant movement so that hummocks are formed, topsoil is sorted into polygonal shapes and thawed soil flows over the permafrost leaving ridges and terraces. Under these conditions plants have a poor foothold. They are constantly buffeted by winds, and small differences in topography (e.g., leeward versus windward side of a hummock) and snow cover become important environmental variables. The number of plant species is low, and the growth form is low. Characteristic plants are cotton grass, dryland sedges and rushes, mosses, lichens, dwarf heaths, dwarf birches and dwarf willows. The animals, which must be adapted to low temperatures, long winters, short summers and limited vegetation are usually circumpolar in distribution because of connecting land masses, although some Eurasian species are replaced by other but similar species in North America. Thus in the Old World are found the reindeer and glutton, replaced in the New World by the caribou and wolverine. Other animals are the musk ox, arctic hare, arctic fox, arctic ground squirrel, lemming, ptarmigan, pipit and many sandpipers and plovers in wet areas. Invertebrates are few; however, the **tundra** is plagued by immense swarms of mosquitoes and black flies.

The life cycle of invertebrates may be prolonged so that the short summer is compensated for by several summers, punctuated by hibernation. Several seasons may be required to complete development. Vertebrates tend to speed up their cycles. The northern robin feeds its young 21 hours a day so that the growing period is shortened from the 13 days of the southern forms to the eight days of the northern forms.

Coniferous Forests

Immediately south of the tundra lies an enormous belt of conifer forest **(taiga)** circling the Northern Hemisphere (Fig. 27–11). In Siberia alone it is 3600 miles long and 800 miles wide, the largest forest in the world. It occupies glaciated land with thin mineral soils, and is dotted with lakes, streams and bogs. It is a region of short-growing season and bitterly cold winters. The dominant plant forms are the needle-leafed evergreens, mostly spruces, firs and pines which do not grow to great stature but stand straight and dense in an immense monotonous sea. They provide

FIGURE 27–11 *The coniferous forest biome covers parts of Canada, northern Europe and Siberia and extends southward at higher altitudes on the larger mountain ranges (From Orr, R. T.: Vertebrate Biology. 3rd ed. Philadelphia, W. B. Saunders Company, 1971.)*

dense shade with resultant poor shrub and herb layers. Needle decay is slow; there is a podsol soil. Along the waterways are thickets of alders, willows and birches. The bogs (muskegs of Canada) are acid, moss and sphagnum-filled, and have blueberries, wild cranberries, bayberry and so forth. Forests of the taiga are dark, windless below the surface, cold, with no underbrush and apparently devoid of animals. This region is, however, the home of the moose, porcupine, beaver, snowshoe hare, lynx, sable, otter, bear, wolf, ermine, grouse, squirrel, loon, siskin, crossbill, titmouse, woodpecker, warbler in the summer and billions of mosquitoes.

The coniferous forest extends southward along the mountains—in North America, along the Appalachians into North Carolina and the Rockies and Sierras into Mexico. The character of the forest changes. The northern high Appalachians are essentially extensions of the taiga, but in the southern Appalachians the dominant conifers are red spruce and Frasier fir. In the west, south of Alaska, the situation is different because of the enormous rainfall (see Section 27–2) along the Coast Range. In the winter as rain or snow it amounts to as much as 635 centimeters (250 inches). In the summer, fogs collecting on the trees drip an additional 127 centimeters (50 inches). The forest here is called the "temperate rain forest" (Fig. 27–12). Mosses and lichens festoon the trees and cover the ground. The trees are western hemlock, western red cedar, Sitka spruce

FIGURE 27–12   *Douglas fir (Pseudotsuga) and western arborvitae (Thuja plicata) in the coastal montane forest—Snoqualmie National Forest. Washington. (Courtesy of the United States Forest Service.)*

and Douglas fir. The southern portion is famous for its greedily exploited Redwood Forest. Comparable, although less moist, conditions occur in the Rockies. At high elevations the dominant forms are Englemann spruce, alpine fir and bristle-cone pines. Douglas fir and ponderosa pine characterize lower elevations. In the Sierra Nevada and Cascade Range grow the mountain hemlock, red fir and lodgepole pine at high elevations and the sugar pine, incense cedar and giant sequoia at lower elevations. The trees of this forest tend to be huge and grow in thick stands. One of the big sequoias is 267 feet high, 40 feet in diameter and 3500 years old. Western cedars attain heights of 200 feet. Sitka spruce attain such girth that it requires 10 men with hands joined to encircle the base. Coast redwoods may reach 359 feet in height and be approximately 2000 years old (Fig. 27–13).

FIGURE 27–13   *Pacific coastal forest in California showing redwood (Sequoia sempervirens) predominating and Douglas fir in association. Conspicuous subordinate species are Lithocarpus densiflorus, Rhododendron californicum, Gaultheria shallon, Vaccinium spp., Polystichum munitum. (Courtesy of the United States Forest Service.)*

Deciduous Forests

South of the coniferous forests are, or were, the great **temperate deciduous forests** which at one time covered eastern North America, all of Europe, the tip of South America and parts of China, Japan and Australia (Fig. 27–14). In Europe, North America and the Far East civilization attained its greatest development in these areas. As a consequence little of the vast forests remain today. When English settlers reached America, the forest was so great and unbroken that it was said that a squirrel could travel from the Atlantic Ocean to the Mississippi River without ever coming to ground. And in the denuded Europe of today we simply have no concept of the extent and density of the forests of the Middle Ages and before.

This forest is rich and complex. The number of species of trees is very great, and the regional subdivisions are numerous. In North America, for example, where the taiga grades into the **eastern deciduous forests,** the dominant trees are white pine, eastern hemlock, jack pine, red pine, white cedar, beech, sugar maple, basswood, yellow birch, black cherry and red oak. The stately white pine was a dominant species especially around the Great Lakes where it was destroyed by indiscriminate logging in the 1880's and 1890's. Various important subdivisions of the deciduous forest are: the beech-maple forest (North Central), maple-basswood (Wisconsin, Minnesota), oak-hickory (the West and the South), oak-chestnut (before the chestnut blight) (Appalachian Mountains), mixed mesophytic (a huge number of species dominated by yellow poplar) (Appalachian plateau) and southern pine forests. The last is not a final stage; it is kept by cutting and fire. If left alone it would go to oak, hickory and magnolia. Live oak and magnolia dominate in Florida and the southern Gulf States.

FIGURE 27–14   An example of a temperate deciduous forest is in Noble County, Ohio. The dominant trees are white and red oaks with an understory of hickory. (Courtesy of the United States Forest Service.)

912

The eastern deciduous forest is a region of deep rich soil, evenly distributed rainfall and no temperature extremes. Seasons are marked. Herb and shrub layers are luxuriant. There are very large bird and mammal populations so that it is difficult to single any particular ones out for attention.

Rain Forests

The most elaborate of all forests are the great **tropical rain forests,** the home of teak, ebony and mahogany, that stretch in a belt around the equator (Fig. 27–15). These are regions of high even temperatures and heavy rainfall (250 to 400 centimeters). In parts of the Cameroons the rainfall may total 1017 centimeters and in Assam it reaches the prodigous total of 1161 centimeters. The most extensive continuous expanse of rain forest is in the Amazon Basin, second only to the forest in the Congo Basin, West Africa and the Indo-Malayan region. From the air the rain forest is seen as a vast monotonous undulating carpet of green. The traveler on foot does not encounter the proverbial inpenetrable steaming jungle. Instead he moves in a gloomy, moderately warm world, free of brush and herbs underfoot. The closed canopy overhead lets little light through to the impoverished soil below. There are no seasons; it is forever midsummer. Where rainfall is seasonal there are changes, but they are inconspicuous. The trees themselves may be seasonal, but their cycles overlap so no overall change is seen. In New Hebrides natives plant their various crops at regular intervals signaled by the fruiting or flowering of certain trees. The overwhelming majority of plants are woody and of the dimensions of trees. Plant families that in temperate regions are low herbs are trees in the rain forest. Vervains become spreading trees, grasses are represented by 60-foot-high bamboos and violets are represented by a

FIGURE 27–15   *The rain forest biome—border of a clearing in the Ituri Forest of Nala, the Congo. (Courtesy of Herbert Lang and The American Museum of Natural History.)*

species the size of apple trees. Trees, woody lianas and epiphytes (orchids, mistletoes, mosses, lichens and stranglers) fill the upper storys of the forest. The trees, while not so high as redwoods nor so big in girth as trees of the temperate zone, give the impression of stately height because their trunks of smooth thin bark rise straight to great height before branching. The crowns are relatively small. The average height is from 150 to 180 feet (from 46 to 55 meters), although some grow to a height of 200 feet (60 meters). Many have huge buttressed roots. Others have stilt roots and others, enormous surface roots snaking over the jungle floor. Commonly rain forest trees bear flowers directly on their trunks. Hanging and looping among the trunks and tangled among the crowns are the lianas, woody cables sometimes as long as 240 meters, thick as a man's thigh and strong enough and dense enough to hold up a tree. Ten elevenths of all climbing species of plants occur in the tropics. Common, too, are the stranglers, trees that begin their lives as epiphytes then send shoots to the ground after which they wrap around the host tree literally strangling it till it dies and rots away leaving only a hollow core. Among the epiphytes are some that have large honeycombed tubers in which ants may live, and the bromeliads which have water tanks in which mosquitoes and frogs breed. Characteristic of a tropical rain forest is a stratification that consists of five **storys** or **canopies** (Fig. 27–16). Each stratum is composed of the crowns of different species; however, the young individuals of a high canopy temporarily compose part of a lower canopy. The general appearance is one of the entire space being filled with greenery. The topmost canopy

FIGURE 27–16 *Profile diagram of primary mixed forest, Shasha Forest Reserve, Nigeria. The diagram represents a strip of forest 200 ft (61 m) long and 25 ft (7.6 m) wide. All trees, 15 ft (4.6 m) and over, are shown. Ab, Pausinystalia or Corynanthe sp.; Ak, Ako ombe (unidentified); Eb, Casearia bridelioides; Ek, Lophira procera; Ep, Rinorea sp. (cf. dentata); Er, Picralima umbellata; Es, Diospyros 'confertiflora'; Ip, Strombosia sp.; It, Strombosia pustulata; Od, Scottellia kamerunensis; Om, Rinorea sp. (cf. oblongifolia); Op, Xylopia quintasii; Os, Diospyros insculpta; Te, Casearia sp.; Y, Parinari sp. (cf. excelsa). (From Richards, P. W.: The Tropical Rain Forest. New York, Cambridge University Press, 1952.)*

may be almost continuous. It is here that most of the animals that live in the rain forest are found. Many of them never reach the ground. It is nearly impossible to appreciate the **stratification** of the forest either from above or below except where man makes a clearing. Where rivers cut through they let in so much light that a great profusion of secondary growth obscures the true structure of the forest.

The number of plant species is enormous. The Indo-Malayan forest is the richest; the African, the poorest. A given area in Malaya may have from 2500 to 10,000 species. In contrast to other forests the population is one of individuals. Individuals of a given species may be a mile apart.

There are modified rain forests in other parts of the world. These include the monsoon forest which is leafless in the dry season and the montane rain forests at higher altitudes.

By day the rain forest is a quiet place and few animals are to be seen. Monkeys and a few birds can be heard from time to time, but most birds are silent. Insects and frogs are noisy at night. Mammals are few, the exceptions being arboreal primates in the canopy and bush pigs and a few small antelopes on the ground.

Savannah

Where the forest grades into grasslands one encounters the **savannah** (Fig. 27–17). The largest true tropical savannah is that of Africa. This is the grassland dotted with scattered trees and clumps of trees — the parkland big game country. This warm region with its prolonged dry season supports the largest, most varied ungulate population in the world. Here roam the great herds of impala, gnus (wildebeests), buffalo, zebra, all the fleet of foot, and their predators the lions, jackals and so on. Where there are rivers the forest extends into the savannah as the gallery forests. Elsewhere the trees are few in species and number. They are adapted to sea-

FIGURE 27–17   The savanna biome; characteristic animals of the African Grasslands, Zebra and wildebeest. Kruger National Park, Transvaal. (Courtesy of Herbert Lang.)

sons regulated by rainfall. They include the baobab, candlearbre, euphorbias and thorny flat-topped acacias.

Other savannahs are in Australia and South America. In North America they once extended from Kentucky and Texas to the Gulf of Mexico
and the Eastern seaboard. Elsewhere savannah has been produced by burning and grazing.

At one time 42 per cent of the land mass of the world was grassland
which has now been reduced by overgrazing and poor farming practices.
These areas, the steppes of the Old World, the prairies of North America,
llanos and pampas of South America and veld of Africa, are rolling plains
watered intermittently with a small amount of rain (from 10 to 20 inches)
(Fig. 27–18). Through some a few great rivers roll. The largest continuous grassland is the *steppe* of Russia and Siberia. Like the others it is a
land of grasses and, in the spring, a profusion of wild flowers. There are
fierce extremes of climate, a monotony of landscape and frequent drought
and famine. Animal life is represented by herbivorous rodents (mostly
burrowing) and a predominance of ungulates. Here live gerbilles, moles,
pikas, marmots, hamsters, quail, cranes, larks, swallows, buzzards and
kestrels. Perching birds are few. Amphibians are rare, but reptiles are well
represented. They include lizards, burrowing snakes, vipers and cobras.
The Eurasian steppe favors nomadic human cultures, and this area is the
cradle of the great human migrations that periodically swept into Europe.

The North American *temperate grasslands* are the New World
counterpart. Today they are dominated by introduced grasses from
Europe (timothy, bluegrass and so forth). Originally they were covered
with native bunch grasses, sod-formers, and forbs (composites and legumes, as, for example, aster, goldenrod, snakeroot). There is a marked
stratification in the vegetation. The root layer comprises more than half
of the grasses and may extend 5 feet into the soil. Above this is an accumulated mulch. There may be 5000 pounds of fresh mulch and 9000 pounds
of humus per acre. Next is a layer of ground-hugging herbs (dandelions,
wild strawberries, cinquefoil, violets). A middle layer consists of short
grass, wild mustard and cone flowers. The top layer is made up of tall

FIGURE 27–18   A region of shortgrass grassland with a herd of bison, originally
one of the major grazing animals in the grassland biome of the western United States and
Canada. The bison in the center is wallowing. (From Odum, E. P.: Fundamentals of Ecology.
3rd ed. Philadelphia, W. B. Saunders Company, 1971.)

grasses and forbs. Animals include bison, prairie chickens, antelopes, and an incredible number of insects, especially grasshoppers and leaf hoppers.

Well marked categories of grasslands are recognizable. Present-day eastern grasslands are largely tame; nonetheless, they support sparrows, bobolinks, meadowlarks, deer mice and cottontail rabbits. **Tall-grass prairies** once extended north-south adjacent to the deciduous forest. As long as the Indians kept them burned, they remained stable. When the settlers moved in and stopped burning, the forest began to encroach. In general this prairie consists of grasses and forbs with continually changing series of species as local conditions change. Originally there were vast expanses of 8-foot-tall grasses—big bluestem, switch grass, Indian grass and others. The **mixed prairie**, consisting of grasses from 2 to 4 feet tall, was made up of little bluestem, needle grass, western wheat grass and Indian rice grass. The **short-grass prairie** consisted of buffalo grass and species of grama. It was this kind of prairie that became the Dust Bowl through poor land management. To the west is an area of desert grassland that continues only as long as there are fires. Without fire, mesquite, cacti and low trees invade from the west.

Chaparral

In parts of California and Mexico, around the shores of the Mediterranean, on the southern coast of Australia and in the middle of Chile is a type of vegetation called **chaparral** in America, **maqui** in Europe and **mellee scrub** in Australia (Fig. 27–19). In areas where it flourishes, winters are mild and rainy and summers are long, hot and dry. The plants have hard, thick, evergreen leaves, are shrublike and grow as low dense thickets. Familiar species in California are: chamiso and manzanita; in Australia,

FIGURE 27–19   An example of the chaparral biome in California. The shrubs in the picture are Eriodictyon tomentosum (Courtesy of the United States Forest Service.)

eucalyptus. This type of plant community is perpetuated by fire. Seeds of some of the species actually germinate better after exposure to fire. In particularly dry seasons the shrubs actually explode when ignited. Too frequent firing causes the death of chaparral. It is then replaced by grass. When fire is too rigorously controlled, a sclerophyll forest of live oak invades. In the coastal ranges of California where this flora attains its best development, the sclerophyll forest characteristically grows on northern slopes and chaparral on southern ones.

Deserts

One seventh of the land surface of the world consists of areas where evaporation exceeds rainfall. These areas, the **deserts**, may differ in other respects but have in common a lack of available water. Only in northern Chile and in the central Sahara are there deserts with absolutely no rain. Most other deserts have some rainfall, usually less than 10 inches a year. The large deserts of the world form two discontinuous belts around the globe, roughly on the Tropic of Cancer and the Tropic of Capricorn. The Sahara is the largest, being 3,500,000 square miles; however, it is a varied area. Only 10 per cent has sand dunes. In the western part of it there are mountains with snow. Continuous with it is the Arabian Desert, one third of which is sand-covered. It is distinguished by having no rivers originating in it or flowing through it. Continuing east one encounters the Turkestan, Iranian, Indian, Takla Makan and Gobi deserts. In North America deserts occur in the western United States and extend down into Mexico. They are: The Great Basin, Mohave, Sonoran and Chihauhuan. South of the equator lies the Australian desert which occupies 44 per cent of the continent, the Patagonian desert in Argentina, the Atacama Desert in Chile and the Kalahari Desert in Africa.

Most deserts are hot. Summer air temperatures reach 120° F. Ground temperatures may be from 30 to 50 degrees hotter. Altitude makes some deserts cold (e.g., the Gobi Desert). In general there is great contrast between day and night temperature and summer and winter. Deserts can be very cold as well as very hot. Organisms that live in deserts must adapt to these extremes, to paucity of water, to violent winds, to occasional torrential downpours and in some areas to salinity of soil.

Desert plants are either **drought-evaders** or **drought-resisters** (Fig. 27–20). Evaders are annuals that persist as seeds, sprout after sufficient rain has fallen, bloom quickly, produce seeds and then die. Resisters are perennials that send roots to deep sources of underground water, or shed leaves in times of drought, or store enormous quantities of water. Many of these plants grow in wadis or washes where underground sources of water exist. A mesquite may send its roots down 100 feet. The paloverde of American deserts avoids excessive evaporation by having minute leaves. Succulents are those resisters that store water in thick, fleshy leaves or in barrel-like stems. One cactus, the mature saguaro, may weigh 10 tons, of which four fifths is water. Many desert plants are also excessively spiny, thus being protected from some grazing animals. The limiting factor in the desert is water, not, as in the case of the forest, light. Some characteristic plants are: acacias, cassias, Joshua trees, yucca, agave, cacti, sagebrushes, creosote bushes, fleshy milkweeds and incense bushes.

Animals face problems of water shortage and extreme heat. Many are nocturnal or burrowing. Some utilize metabolic water. Actually the desert supports many species of animals. The kangaroo rat, gerbil, jerboa,

FIGURE 27–20   Two types of desert in western North America, a "cool" desert in Idaho dominated by sagebrush (above) and (below) a rather luxuriant "hot" desert in Arizona, with giant cactus (Saguaro) and palo verde trees, in addition to creosote bushes and other desert shrubs. In extensive areas of desert country the desert shrubs alone dot the landscape. (Upper, courtesy of the United States Forest Service; lower, courtesy of the United States Soil Conservation Service.)

skinks and other lizards are typical. Predators include the dingo (wild dog) of Australia, the wild dog of Africa, coyotes and desert foxes. There is also a fantastic number of insects.

The communities or biomes just described obviously grade into one another, either gradually or abruptly. In either case there is a zone of transition *(ecotone)*. In addition to an overlap of species, there are sometimes species that are restricted to these transition zones. They are often called "edge" species.

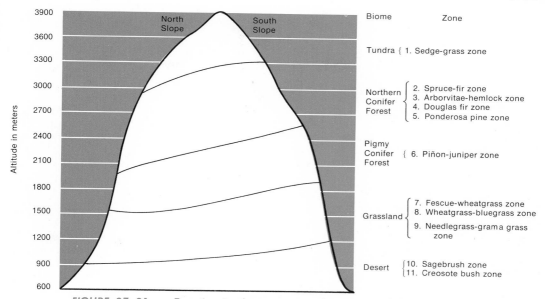

FIGURE 27-21   Zonation in the mountains of western North America. The dia-gram does not refer to a specific mountain but shows general conditions which might be expected in the central Rockies (so-called "intermountain" region) of Utah. North, south, east or west of this region, zonation may be expected to vary somewhat. Thus, the creosote bush and the pinon-juniper zone are absent northward and eastward. Eastward an oak-mountain mahogany zone or westward various zones in the chaparral biome occur between the grassland and the northern conifer forest. (After Woodbury, 1945; and Daubenmire, R. F.: The Life Zone Problem in the Northern Intermountain Region. Northwest Science, 20:28–38, 1946.)

SECTION 27–7

ALTITUDINAL DISTRIBUTION

As pointed out earlier, climate varies not only latitudinally and locally but also with altitude. It is to be expected then that the distribution of organisms will do the same. Consequently in any given mountain there may be from four to five narrow vegetational zones (Figs. 27–21 and 27–22), with a correspondence at successively higher altitudes at the same

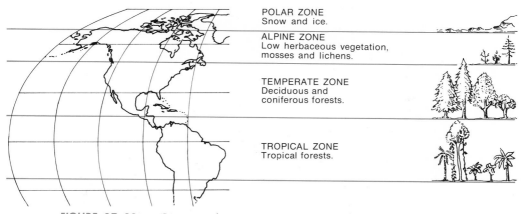

FIGURE 27-22   Diagram showing correspondence of life zones at successively higher altitudes at the same latitude (right), and at successively higher latitudes at the same altitude (left). (From Villee, C. A.: Biology. 5th ed. Philadelphia, W. B. Saunders Company, 1967.)

FIGURE 27–23   Giant lobelias and senecios at an altitude of 5000 meters in the Ruwenzori (Mountains of the Moon). (Ern. Thill, Bruxelles.)

latitude and successively higher latitudes at the same altitude. A striking example is to be found on the Mountains of the Moon in Africa. These are high mountains on the equator which bear perpetual ice and snow on their summits. They are islands of elevation rising out of a sea of tropical warmth. As one ascends these mountains he passes successively through tropical rain forest, a zone with species characteristic of the East African Steppe (the *Hagenia* zone), bamboo forests, giant treelike heaths draped in mosses and lichens, various temperate-zone plants which include lobelias and senecios that grow to tree size (normally these are small herbs) and live for as long as 100 years (Fig. 27–23), a tundra zone and finally snow and ice.

SECTION 27–8

DISTRIBUTION IN TIME

We have seen that the distribution of land plants in space is determined principally by climate and secondarily by soils. The distribution of animals is determined by climate and by vegetation. In both cases the organisms themselves impose further influences upon their distribution because they modify the environment simply by living in it. Plants affect the amount of light that will reach the ground, they alter the composition and structure of soil and they affect the climate near the surface of the ground, specifically the temperature and humidity. Animals by dying and by depositing their waste products on the ground affect the soil. Many animals, as, for example, earthworms and burrowing rodents, affect the texture of soil. Herbivorous animals affect the vegetation.

Because of these complex interrelations the distribution of vegetation and animals just described represents only an instant in time. The distribution of organisms is not contant. It has changed in geologic time and it changes before our eyes. Fossil evidence shows that at various times in the past different plants and animals existed and that survivors from past eras as well as their descendents occupy different geographical ranges. Many of these changes were responses to changes in climate. **Climatic changes** of sufficient magnitude and extent to affect the wide distribution and composition of organisms are usually slow and relatively long-lasting. They were associated with the emergence of mountains, the sinking and rising seas and glaciation. In the Tertiary, for example, the world's climate was generally tropical. It was constant and widespread. The climate of the Quaternary from the Pliocene to the present has been one of glaciation, resulting in steep climatic gradients between the poles and the equator. The flowering plants evolved during a time of genial, moist, more or less constant climate. The surface of the world was flatter and conditions more uniform. With the change in world climate conditions of stress developed between plant and environment, some species became extinct, new ones evolved and geographic distribution began to reflect nonuniform world climate.

The present distribution of organisms is also partly a reflection of **history**. This is especially true of the distribution of animals. Climate and vegetation alone do not explain why some animals are present in one area and absent from another that appears to be ecologically just as suitable. Nor does ecology explain which similar but separated habitats are occupied by similar but not identical animals. One can recognize different geographic faunal regions for which only an historical explanation suffices.

The great faunal regions and some of the animals that characterize them (Fig. 27–24 *A* and *B*) are: The **Nearctic** (wolf, moose, caribou, mountain goat, musk ox, bison, porcupine), the **Palearctic** (reindeer, bison,

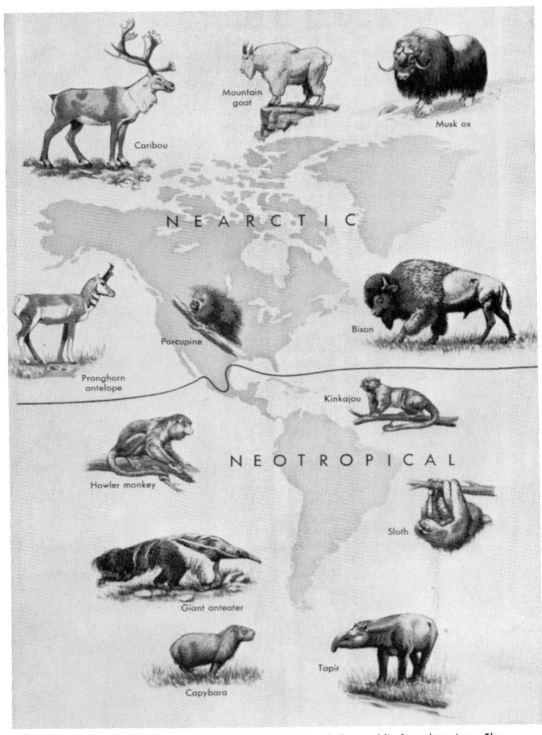

FIGURE 27–24   A, Mammals characteristic of the world's faunal regions. The New World. (From Simpson, G. G., and Beck, W. S.: Life. New York, Harcourt, Brace & World, Inc., 1957.)

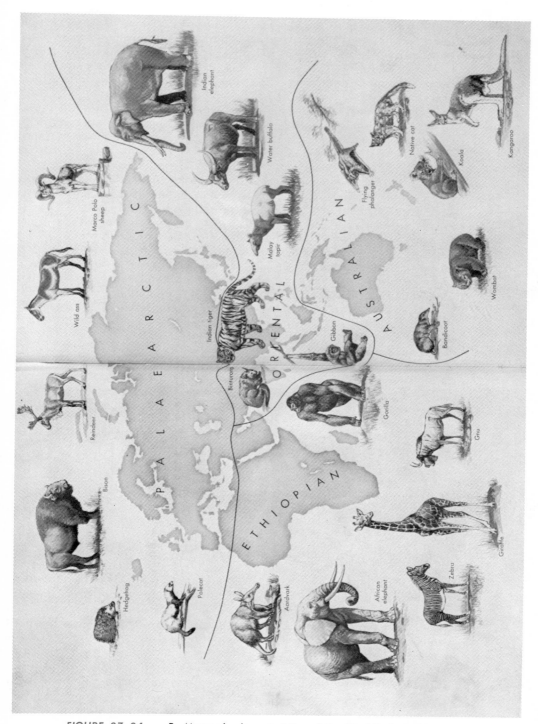

FIGURE 27–24   B, Mammals characteristics of the World's faunal regions. The
Old World. (From Simpson, G. G., and Beck, W. S.: Life. New York, Harcourt, Brace &
World, Inc., 1957.)

polecat, hedgehog), **Neotropical** (howler monkey, capybara, tapir, sloth, guinea pig, armadillo), the **Ethiopian** (African elephant, gorilla, giraffe, zebra, gnu, impala), the **Oriental** (Indian elephant, tiger, gibbon, water buffalo), the **Australian** (mostly marsupials—kangaroo, wombat, koala, platypus). Some of these distributions are the result of land bridges between the various continents changing through the ages so that the dispersal of animals was prevented during some periods but permitted during others.

The responses to these changes, which exceeded in period the life span of an organism, were genetic. Organisms that could not tolerate the changes became extinct. It was then the population that regulated. Natural selection altered the genetic composition of the population. When the climatic change was of sufficient duration, new species arose.

Local changes in habitat are occurring all the time, less as a direct consequence of climate and more as a consequence of the organisms themselves and such local and relatively rapid physiographic changes as the silting up of a river, the encroachment of a sand dune and the eruption of a volcano, or such drastic alteration of the environment by man as burning, overgrazing, poor farm management, lumbering, covering of large areas with buildings, concrete and asphalt to make cities, highways, parking lots and air strips, draining of swamps, landfill in estuaries and harbors, building of dams, irrigation of deserts and pollution of all sorts.

Modification of habitats by organisms themselves is a natural phenomenon. For example, under vegetation, light, air movement, humidity and temperature are altered. There is an accumulation of litter which affects the run-off of water, the temperature of the soil itself and the formation of humus. As a consequence of these changes soil development is affected; water relations, available nutrients, pH and aeration change. These changes affect organisms that depend directly on soil conditions, and they, in turn, affect other organisms. Thus, directly or indirectly, every organism affects the environment. Habitats generally become less favorable to organisms that produced the change and they are replaced by other organisms that can compete more successfully.

SECTION 27–9

SUCCESSION

Thus none of the communities previously described are completely stable. There is, for example, generation overlap so that individuals die and are constantly being replaced by new ones. When death of an individual makes space, light and soil available, different species may be able to invade the area. The changes that occur are not haphazard; within any particular climatic area they are predictable. In similar habitats there are similar communities and when change occurs, the various species usually succeed each other in the same order.

Thus there are two causes of **succession** operating simultaneously, that caused by the organisms themselves (biotic) and that caused by physiography.

Two kinds of succession are recognized. **Primary succession** occurs whenever some event causes barren ground without soil—bare rocks, cliffs, volcanic lava flows and sand dunes. Conditions are extreme in these habitats and unsuited to the growth of most plants and hence animals. Pioneer plants must be adapted to this dry soilless condition. Mosses and lichens are the first inhabitants. They are succeeded by hardy annual herbs, followed by biennials and perennials, shrubs and eventually trees.

The progress is dictated by the rate at which soil accumulates and moisture increases. One spectacular example began with the volcanic explosion of the island of Krakatau, between Java and Sumatra, on August 7, 1883. The greater part of the island disappeared. What little was left was covered with a 200-foot mantle of hot volcanic debris. All life there was obliterated. One year later some grass and one spider was found. In 1908, 202 species of animals were living there. In 1919, 621 species were found; in 1934, 880 species. There was now a young forest. Another classic example is the sand dune area of Lake Michigan. Sand is derived from pulverized rock deposited by wind or water. Lake Michigan originally exceeded its present boundaries. As it receded it exposed sand. The sequence of colonization of this virgin sand is summarized in Table 27–1 which shows the floral and faunal succession.

Table 27–1 *Ecological succession of plants and invertebrates on the Lake Michigan dunes**

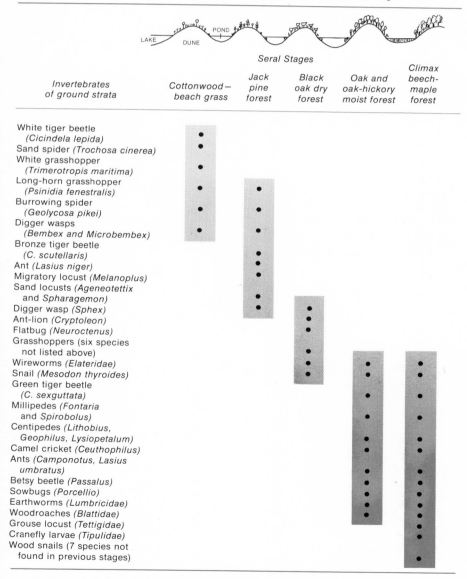

*From Shelford, V. E.: *Animal Communities in Temperate America.* Chicago, University of Chicago Press, 1913. A few species of invertebrates are listed to illustrate the general pattern of change: for more complete listing see his tables L to LV.

FIGURE 27–25   A diagram of the probable evolution of bogs on the Laurentian Shield, reading from top to bottom and showing the processes that operate in the case of closed drainage (A, B) and in situations where there is more seepage of water movement and wind action (C, D, E). Not shown is the accumulation of washed-in material in the main depression (E) and elsewhere, which would be filling the lake in its deeper parts before and after the mat advanced from the periphery and would begin as early as Stage II. (From Dansereau, P., and Segadas-Vianna, F.: Ecological Study of the Peat Bogs of Eastern North America. I. Structure and Evolution of Vegetation. Canad. J. Bot. 30:490–520, 1952.)

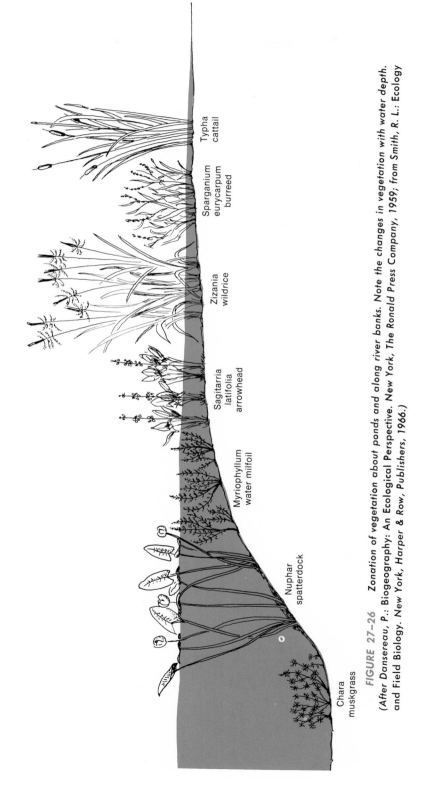

FIGURE 27–26   *Zonation of vegetation about ponds and along river banks. Note the changes in vegetation with water depth.*
(After Dansereau, P.: Biogeography: An Ecological Perspective. New York, The Ronald Press Company, 1959; from Smith, R. L.: Ecology
and Field Biology. New York, Harper & Row, Publishers, 1966.)

TABLE 27-2 Distribution of breeding passerine birds in a secondary upland sere, Piedmont region, Georgia*

Plant Dominants / Age in years of study area / Bird species (having a density of 5 or more in some stage)†	Forbs 1–2	Grass 2–3	Grass-shrub 15	Grass-shrub 20	25	Pine forest 35	Pine forest 60	100	Oak-hickory climax 150–200
Grasshopper sparrow	10	30	25						
Meadowlark	5	10	15	2					
Field sparrow			35	48	25	8	3		
Yellowthroat			15	18					
Yellow-breasted chat			5	16					
Cardinal			5	4	9	10	14	20	23
Towhee			5	8	13	10	15	15	
Bachman's sparrow				8	6	4			
Prairie warbler				6	6				
White-eyed vireo				8		4	5		
Pine warbler					16	34	43	55	
Summer tanager					6	13	13	15	10
Carolina wren						4	5	20	10
Carolina chickadee						2	5	5	5
Blue-gray gnatcatcher						2	13		13
Brown-headed nuthatch						2	5		
Wood pewee							10	1	3
Hummingbird							9	10	10
Tufted titmouse							6	10	15
Yellow-throated vireo							3	5	7
Hooded warbler							3	30	11
Red-eyed vireo							3	10	43
Hairy woodpecker							1	3	5
Downy woodpecker							1	2	5
Crested flycatcher							1	10	6
Wood thrush							1	5	23
Yellow-billed cuckoo								1	9
Black and white warbler									8
Kentucky warbler									5
Acadian flycatcher									5
Totals: (including rare species not listed above)	15	40	110	136	87	93	158	239	228

*After Johnston, D. W., and Odum, E. P.: Breeding Bird Populations in Relation to Plant Succession on the Piedmont of Georgia. Ecology 37:50, 1956.) Figures are occupied territories or estimated pairs per 100 acres.

†By density, the "dominant" species for each stage are as follows:

1. Forb and grass stage: Grasshopper sparrow and meadowlark.
2. Grass-shrub stage: Field sparrow, yellow throat, and meadowlark.
3. Young pine forest (25–60 years): Pine warbler, towhee, and summer tanager.
4. Old pine forest (with well developed deciduous understory): Pine warbler, Carolina wren, hooded warbler, and cardinal.
5. Oak-hickory climax: Red-eyed vireo, wood thrush, and cardinal.

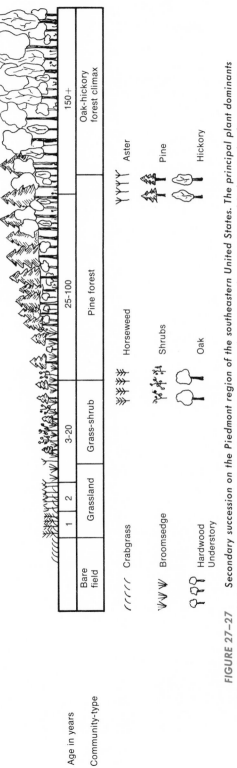

FIGURE 27-27 Secondary succession on the Piedmont region of the southeastern United States. The principal plant dominants of the upland sere which follows abandonment of crop land (cotton, corn and so on) are shown in pictorial fashion in this diagram. (After Odum, E. P.: Fundamentals of Ecology. 3rd ed. Philadelphia, W. B. Saunders Company, 1971.)

Primary succession also occurs in open water. Whereas the change on bare ground tends toward increased wetness, that in open water tends toward greater dryness. It progresses as depth is decreased. A typical succession is shown in Figures 27–25 and 27–26. Pioneer plants are submerged species. Nearer the shore where the water is more shallow, there are floating species. Where the water is still shallower, there are rushes, reeds and cattails. Beyond these shoreward are shrubs. Thus there are around a lake concentric rings of vegetation representing different successional stages, because each type tends to build up the bottom with accumulated debris, making the water shallower, and permitting the next stage in the succession to invade. The succession therefore moves toward the center of the pond changing it eventually from a pond to a peat-filled bog to eventual dryland forest.

Succession also occurs in previously inhabited areas that have been artificially disturbed and where soil and some organisms remain. Typical

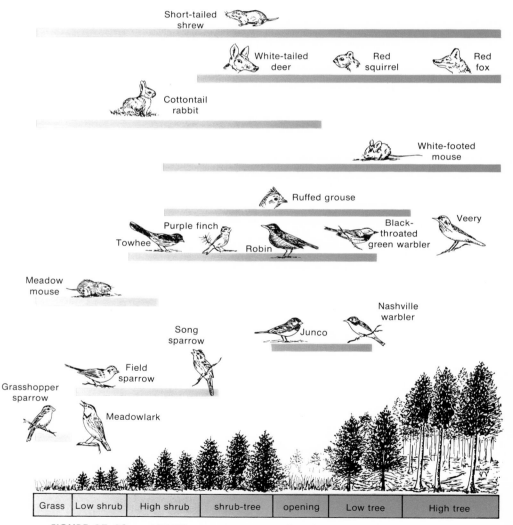

FIGURE 27–28   *Wildlife succession in conifer plantations in central New York. Note how some species appear and others disappear as vegetation density and height change. Other species are common to all stages. (From Smith, R. L.: Ecology and Field Biology. New York, Harper & Row, Publishers, 1966.)*

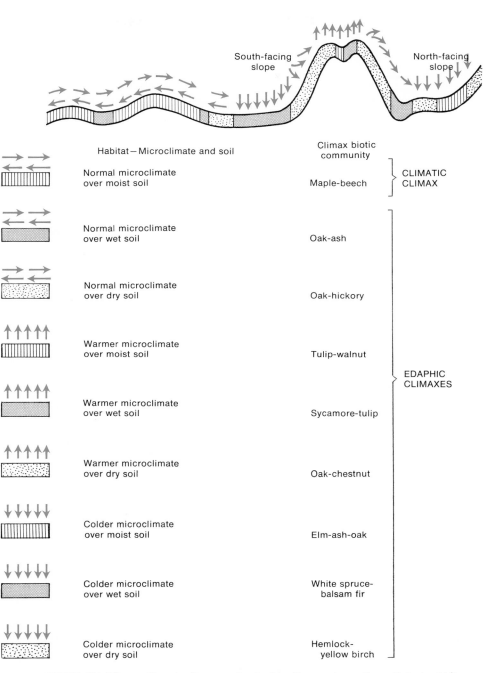

FIGURE 27–29   *Climatic climax and edaphic climaxes in southern Ontario. (After Hills, G. A.: The Classification and Evaluation of Site for Forestry. Res. Rpt. 24. Ontario, Department of Lands and Forests, 1952.)*

is that which occurs on abandoned farms. An example of this type of succession is given in Table 27–2 and Figure 27–27. Another example is depicted in Figure 27–28.

In actuality succession may be very complicated because stages may be skipped, telescoped or extended. And man is constantly setting succession back by fire, lumbering, grazing practice, farming and urbanization. Eventually, however, if the habitat is left undisturbed a more or less stable **climax** is reached.

When an area is not disturbed, changes in the community no longer occur. Succession comes to a halt and a steady state is achieved between the environment and the fauna and the flora. This **climax** is reached probably because there are no more successful species to replace those presently occupying the site. The community in residence is able to tolerate the conditions it has brought upon itself and continues permanently. Potentially the climax is determined by the regional climate. Unless other limiting factors intervene all successions will reach an end determined by the climate. Thus the tundra, taiga, savannah and so on are climatic climaxes. However, as pointed out earlier, soil conditions determine what plants can actually grow in a given climate. Consequently, although all successions in a given climatic region are tending toward one climax soil, topography and water may cause successions to end short of their goal. This kind of climax is called **edaphic**. Some people believe that all successions in a given region will eventually reach the same climax given enough time (the **monoclimax theory**), while others believe that this is theoretical and that in actuality successions stop at a number of edaphic climaxes (the **polyclimax theory**). An example is given in Figure 27–29.

SECTION 27–10

THE MARINE HABITAT

The atmosphere is not truly a habitat for any organism. Birds and insects fly in it; seeds, spores and small organisms are borne aloft for a time, but all of these are actually terrestrial organisms. The terrestrial habitat is a thin stratum covering the planet only to a thickness of about 60 meters at the most. The ocean, on the other hand, constitutes about 78 per cent of the earth's surface, is a layer of 4000 meters average thickness and offers for living nearly 300 times the space that land does.

This vast environment has profound effects on atmospheric conditions as already pointed out. Conversely, the atmospheric conditions that regulate climate on land greatly influence conditions in the ocean. The interactions between atmosphere and ocean are very complex. Here we are concerned primarily with patterns of circulation. The force of easterly trade winds on the ocean surface causes westerly equatorial currents north and south of the equator. When these encounter the east coasts of Asia and the New World, they are deflected north and south. The currents then turn eastward at high latitudes, strike the west coasts of the Americas, Europe and Africa, thence returning to the equator. Coriolis forces also act upon the moving currents. The effects of these currents are such that there are vast clockwise circulations in the north Pacific and Atlantic Oceans and counterclockwise circulations in the oceans of the Southern Hemisphere (Fig. 27–30).

Although surface currents in the oceans are maintained by wind acting on the surface and by the barriers imposed by continents, there are also deeper circulations, which though maintained by action of the wind at the surface are moved largely by differences in density caused by

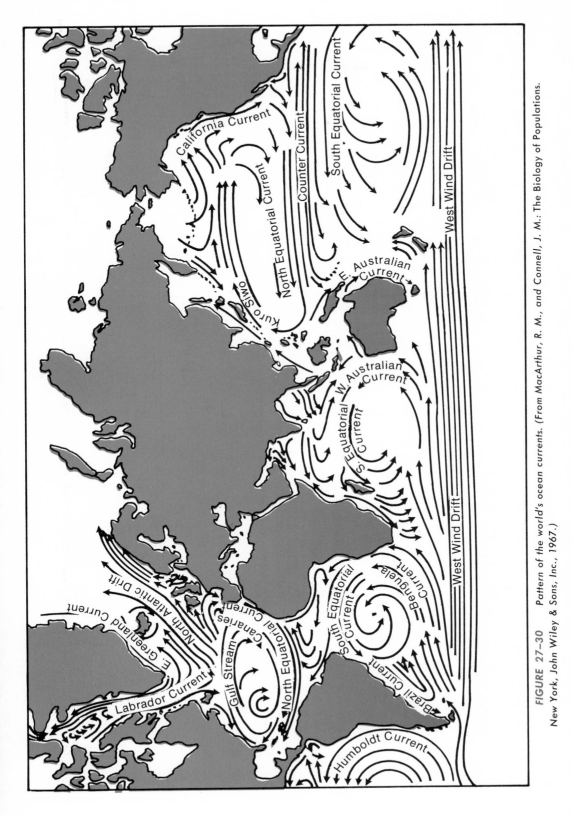

FIGURE 27–30   Pattern of the world's ocean currents. (From MacArthur, R. M., and Connell, J. M.: The Biology of Populations. New York, John Wiley & Sons, Inc., 1967.)

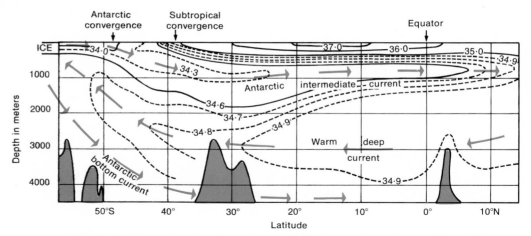

FIGURE 27–31   A section through the Atlantic Ocean, from latitude 55°S to 15°N
along the meridian 30° W, showing the water of varying saltness (34-00/00 to 37-00/00)
and the directions of the main water movements at different depths. It shows how the great
Antarctic icecap extends its influence into the Northern Hemisphere. (After Deacon, G. E. R.:
A General Account of the Hydrology of the Southern Ocean. Discovery Rpt. 7:171–238,
1933; from Hardy, A. C.: The Open Sea. Vol. 1. London, William Collins Sons & Co., Ltd.,
1956.)

salinity and temperature gradients. These currents account for temperature and salinity differences in various parts of the ocean. The ocean is
stratified with respect to density, the most dense water occurring at the
greatest depths. Water of greatest density originates at the surface, where
changes are brought about by heating, cooling, evaporation, precipitation,
ice formation and the addition of river water, and moves to the deeps. For
example, in the North Atlantic the Gulf Stream carries water of high
salinity to the Labrador Sea where it is cooled, becomes even more dense,
sinks and slowly flows south. It is also joined by the deep Mediterranean
Current flowing out of the Mediterranean Sea. There is an easterly surface
current through the Strait of Gibralter. In the Mediterranean, evaporation
increases salinity. When cooled in winter, the surface water sinks and flows
out along the bottom. This current crosses the equator and rises near
Antarctica. There is also a current of deep water flowing north of Antarctica. There is no comparable circulation in the Indian and Pacific
Oceans. The great ice cap on the continent of Antarctica is constantly
moving toward the coast where it juts out as a floating ice shelf and breaks
off to form icebergs. These, upon melting, form a cold but light (because
of low salinity) surface layer. This flows away to the north. Below it is
another layer, cool but not fresh. It sinks and flows north along the surface.
Between these two is a layer of warm water traveling south (Fig. 27–31).
It takes at least seven years for this antarctic water to reach the Northern
Hemisphere.

A glance at Figure 27–30 will show that currents moving from the
equator (e.g., the Gulf Stream) will carry heat to northern latitudes, while
currents moving toward the equator (e.g., the Labrador, California and
Humbolt Currents) will move colder water toward the equator.

SECTION 27–11

PATCHINESS OF OCEAN

Although the marine habitat is a vast continuous domain, it is by no
means uniform. It is a patchy environment. Just as the patchiness of land

prevents an even distribution of all terrestrial species so also does the patchiness of the sea prevent an even distribution of all marine species. Temperature is the major barrier to the uniform distribution of organisms. The ocean consists essentially of two parts, the Atlantic and the Arctic and the Pacific and the Indian. In the Southern Hemisphere these two parts are connected by wide stretches of water; however, in the North, the only connection is the narrow Bering Strait. The warm areas of the oceans are separated from each other by the cold waters at the poles. Thus the warm Indian and Pacific oceans are continuous but separated from the warm South Atlantic by cold areas.

   Temperature is of paramount importance in determining the distribution of organisms in the ocean. It varies with locations, seasons and depth. In polar and tropical seas it is remarkably constant. In temperate seas there is marked seasonal variation which is, however, restricted mostly to surface layers and to shallow coastal waters. Temperature drops with depth, and the deeps are universally cold. Organisms that can tolerate only warm water are restricted to the two isolated warm areas, the Atlantic-Arctic and the Indian-Pacific. They are different in the two areas. Animals that can tolerate cold and depth, on the other hand, have a continuous habitat, because below 2000 meters the ocean is rarely above 3° C anywhere. **Depth** is a barrier to the dispersion of animals that can tolerate only the pressures of the continental shelf. They can move around and inhabit all coasts unless they encounter a thermal barrier. Animals of the open sea that have a wide temperature tolerance may occur throughout the world. Among mammals, a world-wide distribution is found only among the sperm, killer and other whales. Land masses are **barriers** to all marine animals that are not strong enough swimmers to circumnavigate continents, and irregularities in the topography of the ocean floor may prevent the even distribution of bottom forms.

   Differences in **salinity** also contribute to the patchiness of the ocean. A striking illustration of the effect on salinity is seen in the salt pits at Bourg de Batz (Loire Inférieur), where seawater is channeled into a series of evaporation basins for the separation of salt. The channels themselves contain two nemertines, four lamellibranchs, nine snails, six annelids, one crab and nine amphipods and isopods; the first evaporation basin contains three tubellarians, three lamellibranchs, seven snails, six annelids, one crab, one shrimp and seven amphipods and isopods; the second series of tanks contain one turbellarian, two annelids, one crab, one isopod and one amphipod; the next series contain one turbellarian and one annelid; the salt beds themselves contain only the brine shrimp, *Artemia salina*.

   Although the marine habitat is quantitatively the largest on earth, it, together with freshwater, supports only one fifth of the animal species of the world; however, the diversity of form is exceedingly great. The major communities of the sea are distributed according to temperature, depth and salinity. The plainest distinction is between tropical marine communities and polar marine communities. The boundaries between the two are not sharply divided and they shift with seasons. The greatest numbers of organisms occur in tropical waters. Numbers decrease progressively as one proceeds north.

   While tropical marine communities are distinguished by a large number of species, they have a smaller number of individuals per species. Also, the **vertical stratification** of communities is different than elsewhere; surface species penetrate deeper, consequently the density of organisms at the surface is reduced. For example, there are fewer sea birds obtaining their food from the surface. There are no seals. Foraminifera and jellyfish are extremely abundant. Other groups of animals that are represented in cold water are numerous in warm. Two dominant communities of the tropical oceans are coral reefs and mangrove swamps. The distribution of coral reefs is rigorously controlled by temperature. They flourish where the mean annual temperature rises to at least 23.5° C; below 20° C no appreciable reefs are formed. Where upwelling brings cold water to the coast, as in the western North and South Americas and Africa, there are no reefs (Fig. 27–32). Of course, in the absence of suitable bedrock there are no reefs despite favorable temperature conditions. Clear water, high salinity and depths not in excess of 50 meters are also prerequisites for reef formation.

Coral reefs (which may be formed principally by coralline algae or Foraminifera as well as corals) are a habitat for a gigantic and truly amazing collection of plants and animals. The Great Barrier Reef of Australia is one of the richest marine areas in the world.

Wherever in the tropics there are swampy seacoasts with a rich, loose, muddy sediment there are mangrove swamps. The dominant plants are species of mangroves characterized by branching prop roots (Fig. 27–33). As the plant grows these roots sprout from the limbs and grow down into the water where they become anchored in the mud. By blocking tidal currents the roots cause debris to accumulate so that gradually soil is built up to high-tide level.

Polar communities are characterized by a small number of species. Where there is a freezing winter, algae and sessile animals are absent along shorelines. In the Arctic Sea drifting ice scrapes all life from the rocky shores. Polar plankton has a great scarcity of free-swimming larvae. There is a great abundance of diatoms. The copepods that do occur, occur in enormous numbers. Life in the surface layer of polar seas is very abundant. As a direct consequence there are great numbers of sea birds and mammal predators.

Where the ocean is partially shut off by land configuration to form

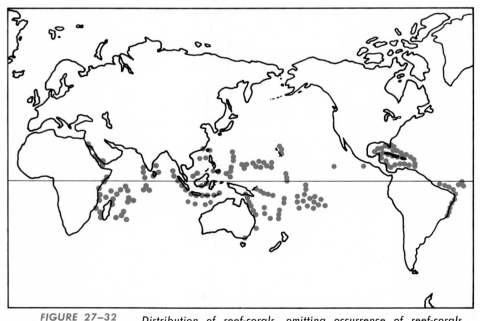

FIGURE 27–32   *Distribution of reef-corals, omitting occurrence of reef-corals without reef formation. (After Schott, G.: Geographie des Atlantischen Ozeans. Hamburg, Germany, Boysen & Maasch, 1926.)*

FIGURE 27–33   A, Red mangrove (Rhizophora mangle) at low tide on the west coast of Florida. Note prop roots which promote deposition of silt and provide a solid substrate for attachment of marine organisms. B, Zonation in a tropical mangrove swamp. The five zones between low and storm tide are: Rhizophora, Avicennia, Laguncularia, Hibiscus, and Acrostichum. (After Dansereau, P.: Biogeography: An Ecological Perspective. New York, The Ronald Press Company, 1957.)

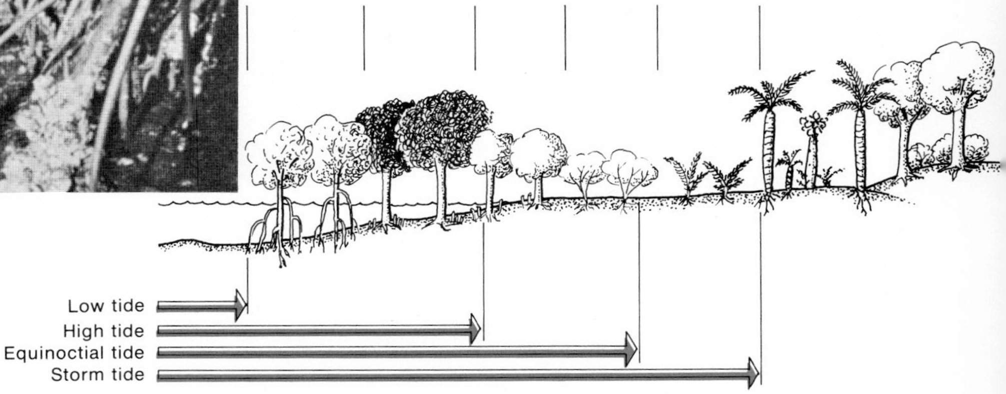

Low tide
High tide
Equinoctial tide
Storm tide

seas, bays and gulfs, there are physical conditions associated with restricted interchange with the main body of water. Salinity is the large variable. In tropical areas increased evaporation causes separated areas to be saltier. In polar areas the reverse is true because of the low evaporation and the addition of freshwater from melting ice. In the open sea at a 300-meter depth the salt content is uniformly 35 parts per thousand. In the western Baltic it is 12 parts per thousand, and in the Gulf of Finland, 0.6 because of dilution. The Mediterranean, with a low influx of freshwater, is 37 at its western end, 39.5 at Crete, and 40 on the Syrian coast. The Red Sea, with no source of freshwater, reaches 46.5 parts per thousand.

The major limiting factors for organisms in the ocean are **temperature, salinity, mineral salts, pressure,** and, for plants, **light**. With respect to each, different organisms possess varying degrees of tolerance. In general, marine organisms have a narrower temperature tolerance than terrestrial species. The restriction of corals to a narrow range of warm temperature is an example. Salinity poses a formidable problem for marine animals. Since many marine organisms have blood of the same osmotic concentration as that of seawater, they have no water problems as long as they remain in water of the correct concentration. If they move into areas of high salinity, they tend to lose water to the sea; if they move into areas of low salinity, they tend to be flooded by water entering from the sea. If they have physiological regulating mechanisms in the form of kidneys to expel excess water or gills or similar structures that can secrete excess salt, or if they just are able to tolerate lack of constancy in their blood, they can then survive in waters of different salinity. Such organisms are said to be **euryhaline**. Those living in brackish water and those that migrate periodically from salt- to freshwater (eels, salmon and so forth) have the widest tolerance. If animals lack osmoregulatory mechanisms or cannot tolerate inconstancy in their blood, they have a very narrow tolerance to changes in salinity. They are **stenohaline**. Most marine species, especially those of the open ocean, fall into this category.

Nutrient salts, especially nitrates and phosphates, are especially important. The concentration of these salts varies greatly in different parts of the sea and at different times of the year. They are constantly being washed

937

into the sea from water running off from the land where they occur either in native rock or are the products of terrestrial life. Ordinarily much of this material would sink to the bottom. In regions where there is upwelling (i.e., cold water from the deep being returned to the surface), as off the west coasts of California, Peru and Portugal, the essential nutrients are brought to the surface. Off the coast of Peru much of the nitrate and phosphate comes from the tons of guano deposited annually on rocky islands and headlands occupied by thousands of fish-eating birds. As a consequence of upwelling off these western coasts, this material is kept from sinking. It is, therefore, readily available to phytoplankton. These areas are regions of great marine productivity. The rich tuna and sardine fisheries are evidence of this.

Finally, **light** is a limiting factor of enormous importance for photosynthetic organisms. Their vertical distribution is sharply delimited by the depth to which effective intensities and wavelengths of light can penetrate. Ultraviolet and red are quickly absorbed at the surface; green and blue penetrate more deeply. The human eye can detect light at depths as great as 500 meters. Sensitive photographic plates record it at 1000 meters. Below this all is pitch dark.

The ocean consists essentially of three regions: the interface separating water and air; the interface separating water and land (shore and bottom); the vast area of water between the two interfaces. Because the ocean is a thick three-dimensional environment it has a greater potential for vertical variation than is offered on land. This potential is fully realized in a marked **zonation** of fauna and flora related to depth. Where the sea washes the edges of continents there is a zone of shallow water (to 200 meters). This is the region of the continental shelf. Between the high- and low-tide marks is the intertidal zone **(littoral),** periodically exposed to air. Beyond this is the **subtidal** zone. From here on out is the **open ocean**. That part of the open ocean through which enough light penetrates to permit photosynthesis is the **euphotic** zone. The deep dark regions below constitute the **bathyl** and **abyssal** zones. Communities within all of these zones may live on the bottom (benthic) or free of it (pelagic) (Fig. 27–34).

The seashore is one of the most prolific and fascinating habitats in the world. It is a world of great temperature extremes, of alternate ex-

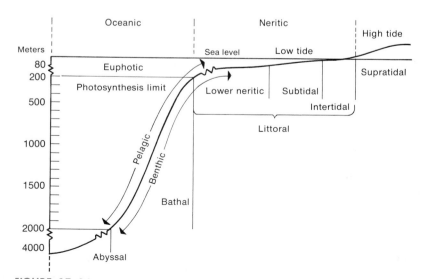

FIGURE 27–34   Zonation in the sea. (After Hedgpeth, J. W.: The Classification of Estuarine and Brackish Waters and the Hydrographic Climate. Rpt. 11. National Research Council Committee on a Treatise on Marine Ecology and Paleoecology, 1951.)

posure (in the intertidal zone) to air and water and of tidal action that may involve surf, swells and strong currents to a considerable depth. The gravitational pull of the sun and moon each cause two bulges on opposite sides of the earth. The land bulges are inconsequential, but the water bulges are profound. The bulge on the moon side results from gravitational attraction; that on the opposite side from the fact that gravitational force at that surface is less than at the center of the earth. The tides daily advance westward because the earth is rotating eastward on its axis. Since the earth rotates once daily there will be two high tides and two low tides within this period. Twice a month at the new and full moon the earth, sun and moon are in line; consequently the forces of the sun and moon are additive and the difference between high and low tides is greater than usual. These are the fortnightly spring (referring to brimming fullness rather than to season) tides. When the sun and moon pull at right angles (at the quarter moon), the difference between high and low tide is less than usual. These are the neap tides.

Tides are not uniform throughout the world because they are shaped in part by the configuration of the land, latitude, barometric pressure, winds, natural periods of oscillation and other factors. In the Gulf of Mexico, for example, there is only one daily tide because ebb and flood partially cancel each other. The difference in tidal range goes from less than 1 foot in the open sea to approximately 50 feet in places like the Bay of Fundy and near Mont St. Michel in France.

Rocky shores set off zonation to best advantage (Figs. 27–35, 27–36 and 27–37). The limits common to all rocky shores are shown in Figure 27–35. Details differ from one part of the world to another. One type may be described as follows. On the landward side there is usually a zone of bare rock marking the transition from land to ocean. Next is a spray zone marked by black patchlike lichens and algae. These are grazed on by periwinkles. Seaward of this is the littoral or intertidal zone characterized by barnacles, whelks, periwinkles and mussels. Next comes a zone of rock kelp and Irish moss and rock crabs, then kelp. Tidal pools are also characteristic

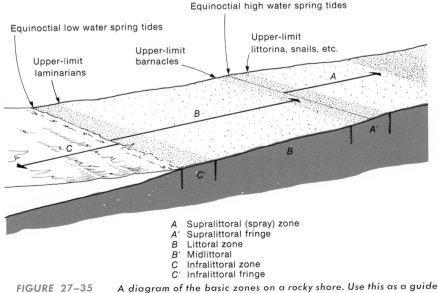

A Supralittoral (spray) zone
A' Supralittoral fringe
B Littoral zone
B' Midlittoral
C Infralittoral zone
C' Infralittoral fringe

FIGURE 27–35   A diagram of the basic zones on a rocky shore. Use this as a guide when studying the subsequent drawings of zonation on rocky shores. (After Stephenson, T. A., and Stephenson, A.: The Universal Features of Zonation Between Tide-Marks in Rocky Coasts. J. Ecol. 37:289–305; from Smith, R. L.: Ecology and Field Biology. New York, Harper & Row, Publishers, 1966.)

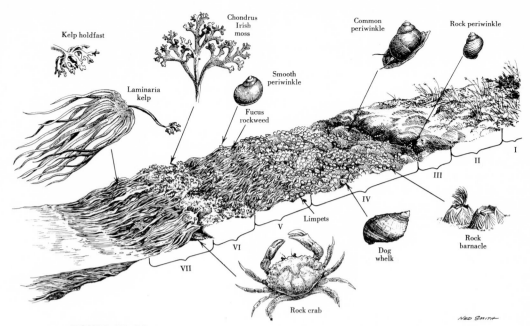

FIGURE 27–36   *Zonation on a rocky shore along the North Atlantic. (I) Land: lichens, herbs, grasses, and so forth; (II) bare rock; (III) zone of black algae and rock periwinkles, (IV) barnacle zone: barnacles, dog whelks, common periwinkles, mussels; (V) the fucoid zone: rockweed and smooth periwinkles; (VI) Chondrus zone: Irish moss; (VII) laminarian zone: kelp. (Zonation drawing based on the data from Stephenson, 1954, and author's photographs and observations. From Smith, R. L.: Ecology and Field Biology. New York, Harper & Row, Publishers, 1966.)*

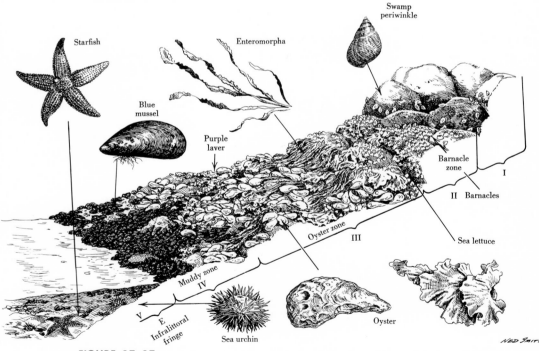

FIGURE 27–37   *Zonation along a rocky shore—mid-Atlantic coast line. (I) Bare rock with some black algae and swamp periwinkle; (II) barnacle zone; (III) oyster zone: oysters, Entromorpha, sea lettuce, and purple laver; (IV) muddy zone: mussel beds; (V) infralittoral fringe: starfish and so on. Note absence of kelps. (Zonation drawing based on Stephenson, 1952; sketches done from life or specimens. From Smith, R. L.: Ecology and Field Biology. New York, Harper & Row, Publishers, 1966.)*

of rocky coasts and are teeming microcosms with their own characteristic life.

The sandy shore may be even more harsh than the rocky shore. It is subject to all the extremes of the latter plus the inconvenience of a constantly shifting substrate. The last makes life on the surface almost impossible; life has retreated below the surface. Zonation on a sandy beach is illustrated in Figure 27–38. It does not, however, conform to a universal pattern as does that of rocky shores. In the example depicted, the supralittoral zone is inhabited by ghost crabs and beach hoppers. These animals spend most of the daytime hidden in damp burrows. They forage at night. Ghost crabs nightly go to the water to dampen their gill chambers. The intertidal zone is not as rich here as on the rocky shore, but it is the home of ghost shrimps, clams and bristle worms. Lower down the beach are lugworms, trumpet worms and other species of clams. Two interesting inhabitants of this zone are the mole crab and the coquina clam. As waves roll up the beach these two small creatures emerge from the sand, ride the waves up the beach and, as the velocity of the water decreases, burrow quickly into the sand as the waves retreat. Once settled the crab extends its antennae and the clam its siphon to extract particulate food from the receding waves. The sublittoral zone is the home of cockles, razor clams, moon shells and others. Still farther out are starfish, sand dollars, sea cucumbers, killifish, silversides and flounders.

In this area of the marine environment, as in all others, the **primary producer organisms**, equivalent to the flowering plants on land, are the **phytoplankton** (Fig. 27–39), consisting principally of diatoms and dinoflagellates. It is difficult to appreciate their importance because they are so small. An absolutely minimal estimate would place their density at

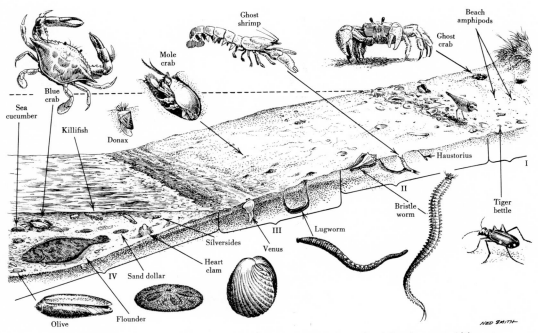

FIGURE 27–38   *Life on a sandy ocean beach along the Atlantic coast. Although strong zonation is absent, organisms still change on a gradient from land to sea. (I) Supratidial zone: ghost crabs and sand fleas; (II) flat beach zone: ghost shrimp, bristle worms, clams; (III) intratidal zone: clams, lugworms, mole crabs; (IV) subtidal zone: the dotted line indicates high tide. (From Smith, R. L.: Ecology and Field Biology. New York, Harper & Row, Publishers, 1966.)*

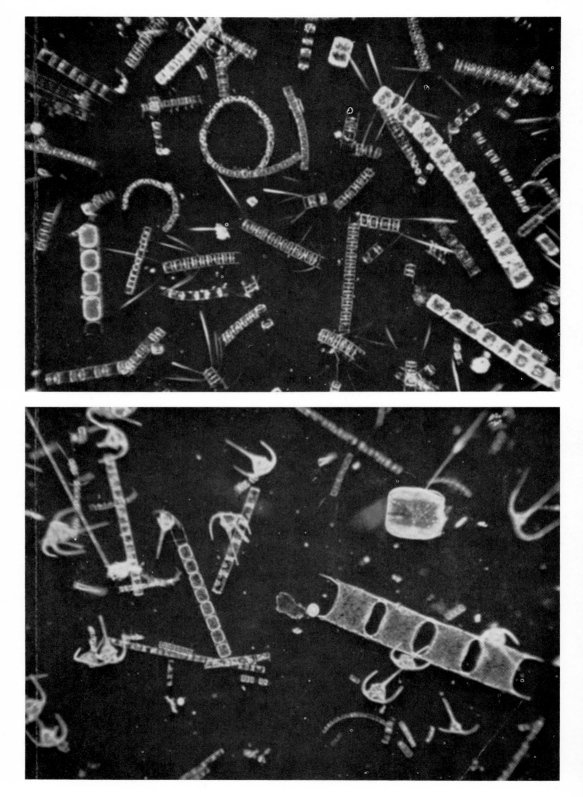

12,500,000 individuals per cubic foot. In temperate regions the phytoplankton undergoes two seasonal population explosions or "blooms," one in the spring, the other in late summer or fall. The mechanism is similar to that responsible for "blooms" in lakes. In the wintertime low temperatures and reduced light restrict photosynthesis to a low level; however, these factors do not prevent bacteria and other microorganisms from generating high concentrations of nitrogen, phosphorus and other nutrient elements. When spring brings higher temperatures and more light, photosynthesis accelerates, and there is an ample supply of nutrients. The nutrient supply is ample because the winter mixing of surface and deep water brings up nutrients that have fallen to and accumulated at the bottom. Within a fortnight, the diatoms multiply 10,000-fold. This prodigious growth accounts for the spring bloom. Soon, however, the nutrients are exhausted. Replacement from lower layers no longer occurs because warming of the surface water keeps it on top and prevents mixing. Nutrients are now locked in the bodies of animals that have eaten the phytoplankton or are slowly falling to the bottom in dead bodies. Whereas temperature and light were the limiting factors during the winter, nutrient level is the limiting factor during the summer, especially since existing phytoplankton is now being consumed by animals. Now nutrients begin to accumulate again in lower layers. As fall approaches the upper layers of water begin to cool again. The accompanying density change, together with the autumn equinoctial gales, begin mixing the water again. Water rich in phosphates and nitrates is brought up from below. Other forms of phytoplankton, especially nitrogen-fixing blue-green algae, now bloom until reduced nutrients or temperature again intercedes (Fig. 27–40). In general, where there is upwelling, as already mentioned in connection with the west coasts of continents, there is a rich phytoplankton. In the open tropical sea, plankton is sparse because the surface layers are continually warmed and mixing is minimal.

The other important population throughout the ocean is the **zooplankton** (Fig. 27–41). By definition zooplankton is those animals that are carried passively by moving water. Every major phylum of the animal kingdom is represented, if not as adults, at least as eggs or larval stages. All the organisms are small, but not so small as not to be visible through a 6X hand lens. The beauty and almost limitless variety of forms of these animals beggar description. A sample haul near the Isle of Man gave an average of 4500 animals per cubic meter. The same hauls gave about 727,000 planktonic plants per cubic meter.

The active surface dwellers of the **neritic** zone are many bony fish, large crustacea, turtles, seals, whales and a host of sea birds. All these **consumers** are restricted in their distribution by temperature, salinity and nutrients. There is horizontal zonation and vertical stratification. The greatest numbers of fish (individuals—not species) are found in northern

FIGURE 27–39   Top, *Living plants of the plankton (phytoplankton)*, ×110. *Chains of cells of several species of* Chaetoceros *(those with spines), a chain of* Thalassiosira condensata *(at and pointing to bottom right corner), and a chain of* Lauderia borealis *(above the last named). Bottom, More living phytoplankton from the English Channel*, ×60. *Diatoms:* Biddulphia sinensis *(the four large cells linked together),* Coscinodiscus conicinus *(the single large cell),* Melosira borreri *(the chain of eight small cells, pointing "N.N.W.") and* Rhizosolemia faeroense *(the chain of six small cells, pointing "N.N.E."), also the dinoflagellate* Ceratium tripos *and closely related species (the anchor-shaped cells). Both by electronic flash. (From Hardy, A.: The Open Sea. Vol. I. London, William Collins Sons & Co., Ltd., 1966.)*

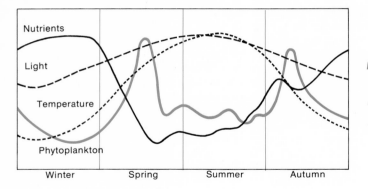

FIGURE 27–40   *The probable mechanisms for phytoplankton "blooms." See text for explanation. (From Odum, E. P.: Fundamentals of Ecology. 3rd ed. Philadelphia, W. B. Saunders Company, 1971.)*

waters and where there are cold upwellings. Only a few species make up the bulk of commercial fisheries. Three fourths of the world's catch belong to the following groups in decreasing order of abundance:

herring family
cod, haddock, pollock
salmon
flounder, sole, plaice, halibut
mackerel, tuna, bonito

The cod and flounder groups are bottom fish. Upon these and other marine organisms feed a great number of birds. Sandpipers, plovers, herons, curlews and others search the supra- and intertidal zones; cormorants, sea ducks, and pelicans, the subtidal zones; petrels and shearwaters, the lower neritic zone farther out to sea.

The **oceanic region** is less rich in species and numbers than the coastal areas, but it has its characteristic species. Many of these are transparent or bluish and since the sediment-free water of the open sea is marvelously transparent, these animals are nearly invisible. Animals that are too thick to be transparent frequently have smooth shiny and silvery bodies which make them invisible by mirroring the water in which they swim. Among the most characteristic animals of the open ocean are the baleen and toothed whales, the only mammals that are truly and absolutely marine. The baleen or whalebone whales live on phytoplankton which they strain from the water. The toothed whales live on nekton. Giant squids comprise part of their diet. The distribution of these whales is correlated directly or indirectly with the distribution of phytoplankton.

Above the surface of the open sea are the oceanic birds. They are not truly marine because they must come to land to breed, but apart from that they may spend all their lives at sea out of sight of land. They include petrels, albatrosses and frigate birds. Like the whales their distribution may be world-wide but not uniform. They are restricted by the distribution of fish which, in turn, are dependent on the phytoplankton (Fig. 27–42).

Eighty-eight per cent of the ocean is more than 1 mile deep. It is continuous throughout the world except for the deep water of the Arctic Ocean, which is cut off from the rest by a narrow submerged mountain range connecting Greenland, Iceland and Europe. This is the area of great pressure and of perpetual night. Since no photosynthesis is possible, the only source of energy is the constant rain of organic debris, the bodies and waste products of organisms in the surface layers, that falls toward the bottom. The other prerequisite for life, oxygen, gets to the bottom by means of the oceanic circulation discussed earlier.

Life in the deep is poorly known because of the enormous difficulty of observation. The **pelagic** animals are strong swimmers, not easily caught in nets. The scant knowledge available has been gleaned from studies of net hauls and by observation from special undersea craft or via underwater television. In general, the animals of this world are either filter-feeders which sieve out particles before they reach the bottom, grubbers which ingest sediment, or predators on the aforementioned. The sediment is generally an ooze consisting principally of the siliceous shells of diatoms in some regions and calcareous shells of protozoa in others. Animals living on these oozes characteristically possess long thin appendages, spines or stalks.

The deep pelagic animals are strikingly different from all others. The largest of these is the giant squid that attains a length of 55 feet (the dreaded Kraken of Norse legend). Many of the others are very small. Scarlet and black are the usual colors. Large eyes and luminous organs are common features. Shapes of many of the denizens of the deep are bizarre, to say the least.

So far, we have discussed only terrestrial and marine communities. Equally prolific and interesting communities occupy the waters of the land masses. They cover about 2 per cent of the land. Just as the land masses are discontinuous in the ocean, so the freshwaters of the world are discontinuous on land. They are separated by land or, as in the case of rivers that flow into the ocean, isolated from each other by the oceans's salinity.

On land the distribution of organisms is controlled by climatic, edaphic, vegetational and historical factors. In freshwater the limiting factors are temperature, turbidity, current, oxygen, carbon dioxide, salts and osmotic conditions. **Temperature variations** are smaller in water than in air, but aquatic organisms often have very narrow tolerances. Variations of temperature are also affected by circulation and stratification. **Turbidity** is important because it affects the penetration of light into water. **Oxygen** and **carbon dioxide** concentrations are frequently limiting.

It is convenient to recognize standing waters (lakes, ponds, bogs, swamps and marshes) and flowing waters (springs, streams [brooks, creeks] and rivers). Natural lakes tend to be of recent origin, geologically speaking. They are most numerous in the glaciated parts of northern America and Eurasia where they were formed about 10,000 to 12,000 years ago as the ice retreated. Lakes occur also in regions of recent volcanism either in the craters of volcanoes or where water was dammed up by lava flows. Cut off loops of meandering streams frequently become oxbow lakes. Ponds are formed by all of the aforementioned events. In addition, they are formed by silting of rivers, by beaver dams and by man. Actually sharp boundaries do not exist between these habitats because erosion and sedimentation are constantly changing them. Within freshwaters organisms are not evenly distributed. Different zones support different communities. In standing water the commonly recognized zones are: **littoral** (shallow regions, usually near the shore, where light penetrates to the bottom), **limnetic** (open water to the depth where there is still sufficient light for the rate of photosynthesis to equal respiration) and **profundal** (bottom and deep water below the level of effective light penetration—in shallow ponds this zone may be absent). In moving water the principal zones are: rapids (current keeps bottom clear of loose material) and pools (reduced velocity of current permits the accumulation on the bottom of loose material).

Aquatic life is probably most prolific in the littoral zone. Within this zone the plant communities form concentric rings around the pond or lake as the depth increases (Fig. 27–26). At the shore proper are the cattails, bulrushes, arrowheads and pickerelweeds—the emergent, firmly

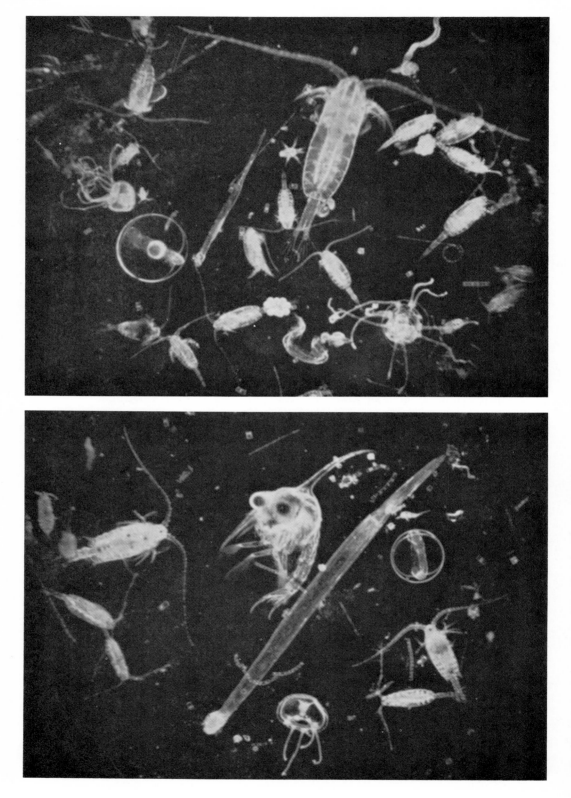

rooted vegetation linking water and land environments. Out slightly deeper are the rooted plants with floating leaves such as the water lilies. Still deeper are the fragile thin-stemmed water weeds, rooted but totally submerged. Here also are found diatoms, blue-green algae and green algae. Common green pond scum is one of the latter.

The littoral zone is also the scene of the greatest concentration of animals (Fig. 27–43). They too are distributed in recognizable communities. In or on the bottom are various dragonfly nymphs, crayfish, isopods, worms, snails and clams. Other animals live in or on plants and other objects projecting up from the bottom. These include the climbing dragonfly and damsal fly nymphs, rotifers, flatworms, bryozoa, hydra, snails and others. The zooplankton consists of water fleas such as Daphnia, rotifers and ostracods. The larger freely swimming fauna **(nekton)** includes diving beetles and bugs, dipterous larvae (e.g., mosquitoes) and large numbers of many other insects. Among the vertebrates are: frogs, salamanders, snakes and turtles. Floating **(neuston)** members of the community include whirligig beetles, water striders and numerous protozoa (Figs. 27–44 and 27–45). Many pond fish (sunfish, top minnows, bass, pike and gar) spend much of their time in the littoral zone.

The limnetic or open-water zone is occupied by many microscopic plants (dinoflagellates, *Euglena, Volvox*), many small crustaceans (copepods, cladocera and so on) and many fish.

Deep (profundal) life consists of bacteria, fungi, clams, blood worms (larvae of midges), annelids and other small animals capable of surviving in a region of little light and low oxygen.

As compared to ponds where the littoral zone is large, the water usually shallow and temperature stratification usually absent, lakes have large limnetic and profundal zones, a marked **thermal stratification** and a seasonal cycle of heat and oxygen distribution. In the summertime, the surface water **(epilimnion)** of lakes becomes heated while that below **(hypolimnion)** remains cold. There is no circulatory exchange between upper and lower layers with the result that the lower layers frequently become oxygen deprived. Between the two is a region of steep temperature decline **(thermocline)**. As the cooler weather of fall approaches, the surface water cools, the temperature is equal at all levels, the water of the whole lake begins to circulate and the deep is again oxygenated. This is the **"fall overturn."** In winter, as the surface temperature drops to 4°C the water becomes less dense. Remaining at the surface it impedes circulation. The bottom is now warmer than the top. Because bacterial decomposition and respiration are less at low temperatures and cold water holds more oxygen, there is usually no great winter stagnation. The formation of ice may, however, cause oxygen depletion and a heavy winter kill of fish. The

Text continued on page 954.

FIGURE 27–41   A, *Living animals of the plankton (zooplankton)*, ×16. *The copepods* Calanus finmarchicus *(the largest animal) and* Pseudocalanus elongatus *(similar in shape, but much smaller than Calanus, and the one with the cluster of eggs); two small anthomedusae with long tentacles; a fish egg (the circular object); a young arrow-worm* Sagitta *(to right of the fish egg); small nauplius (larval stage) of copepod (close to left side of Calanus) and the planktonic tunicate* Oikopleura *(curly objects top right and middle bottom).* B, *More living zooplankton*, ×16. *The transparent arrow-worm* Sagitta setosa *(diagonal); zoea, young stage of crab (above arrow-worm); fish egg, with developing fish, pilchard (spherical); anthomedusan (bottom center); and copepods; two specimens of* Centropages *(left of zoea and below fish egg) and the slightly smaller ones are* Acartia. *Both by electronic flash, but partially narcotized. (From Hardy, A.: The Open Sea. Vol. I. London, William Collins Sons & Co., Ltd., 1966.)*

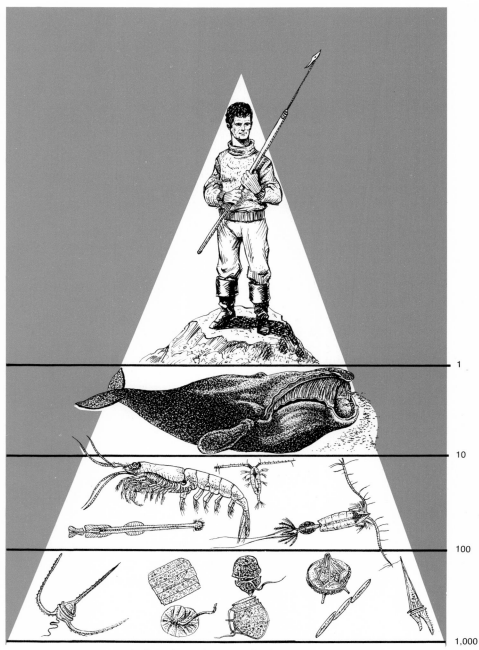

FIGURE 27–42   A hypothetical pyramid showing energy relations between whales and principal organisms in their food chain. The ratio of phytoplankton net production to production of zooplankton and whales is indicated by numbers to the right of the diagram. (From Whales, Plankton and Man by W. E. Pequegnat. Copyright © Jan. 1958 by Scientific American, Inc. All rights reserved.)

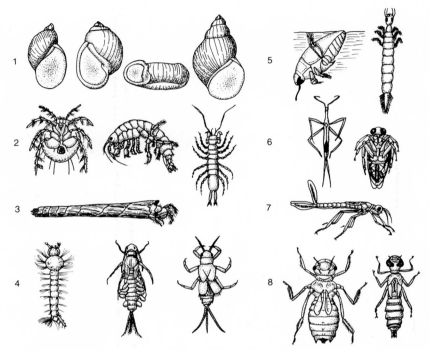

FIGURE 27-43   Some representative animals of the littoral zone of ponds and lakes. Series 1 to 4 are primarily herbivorous forms (primary consumers); series 5 to 8 are predators (secondary consumers). 1, Pond snails (left to right): Lymnaea (pseudosuccinea) columella; Physa gyrina; Helisoma trivolvis; Campeloma decisum. 2, Small arthropods living on or near the bottom or associated with plants or detritus (left to right): a water mite, or Hydracarina (Mideopsis); an amphipod (Gammarus); an isopod (Asellus). 3, A pond caddis fly larva (Triaenodes), with its thin, light portable case. 4, (Left to right) A mosquito larva (Culex pipiens); a clinging or periphytic mayfly nymph (Cloeon); a benthic mayfly nymph (Caenis)—note gill covers which protect gills from silt. 5, A predatory diving beetle, Dytiscus, adult and (right) larva. 6, Two predaceous Hemipterans, a water scorpion, Ranatra (Nepidae), and (right) a backswimmer, Notonecta. 7, A damsel fly nymph, Lestes (Odonata-Zygoptera); note three caudal gills. 8, Two dragonfly nymphs (Odonata-Anisoptera), Helocordulia, a long-legged sprawling type (benthos), and (right) Aeschna, a slender climbing type (periphyton). (After Pennak, Robert W.: Fresh-water Invertebrates of the United States. New York, The Ronald Press Company, 1953.)

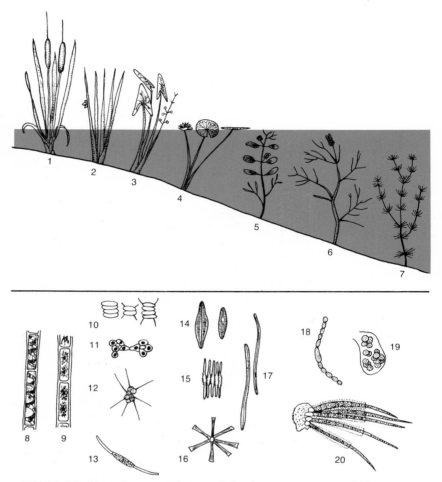

FIGURE 27–44   *Some producers of lentic communities, including emergent, floating and submergent rooted littoral plants (1–7), filamentous algae (8–9) and phytoplankton (10–20). The phytoplankton include representative green algae (10–13), diatoms (14–17) and blue-green algae (18–20). Note that phytoplankton exhibit "flotation" adaptations which enable them to remain suspended or at least decrease markedly the rate of sinking (these organisms, of course, have no power of movement of their own)— for example, reduction in integumentary material, flotation process and colonial life-habit, which increases surface area, and gas vacuoles indicated by black spots in the blue-green algae cells (18–20). Organisms diagrammed are: 1, cattail (Typha); 2, bulrush (Scirpus); 3, arrowhead (Sagittaria); 4, water lily (Nymphaea); 5 and 6, two species of pond weeds (Potamogeton diversifolia, P. pectinatus); 7, muskgrass (Chara); 8, Zygnema; 9, Spiroggra; 10, Scenedesumus; 11, Coelastrum; 12, Richteriella; 13, Closterium (a desmid); 14, Navicula; 15, Fragilaria; 16, Asterionella (which floats in the water like a parachute); 17, Nitzschia; 18, Anabaena; 19, Microcystis; 20, Gloeotrichia (19 and 20 represent parts of colonies enclosed in a gelatinous matrix). (8 to 17 after Needham, J. G., and Needham, P. R., A Guide to the Study of Fresh-Water Biology. Ithaca, New York, Comstock Publishing Associates, 1941; 18 to 20 after Ruttner, F.: Fundamentals of Limnology. Toronto, University of Toronto Press, 1953.)*

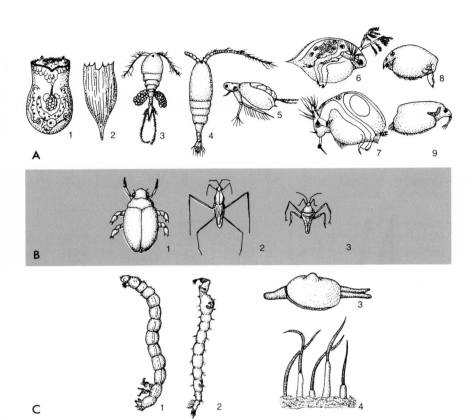

FIGURE 27–45 A, *Representative zooplankton. Rotifers: 1, Asplanchna; 2, Notholca (lorica only). Copepods: 3, a cyclopoid copepod, Macrocyclops; 4, a calanoid copepod, Senecella. Cladocera (representative of each of five families): 5, Diaphanosoma (Sididae); 6, Daphnia (Daphnidae); 7, Bosmina (Bosminidae); 8, Pleuroxus (Chydoridae); 9, Acantholeberis (Macrothricidae). B, Zooneuston. 1, a whirligig beetle, Dineutes (Gyrinidae); 2, a water strider, Gerris (Gerridae); 3, a broad-shouldered water strider, Rhagovelia (Veliidae). C, Some characteristic profundal zone types. 1, a chironomid larva, or blood worm, Tendipes (note prolegs and abdominal gills); 2, a "phantom larva: Chaoborus (note air sacs which apparently aid the animal in performing vertical migrations); 3, a "pea-shell" clam, Musculium (Sphaeriidae), with "foot" and two branchial siphons extended; 4, Tubifex, a red annelid which builds tubes on the botton and vigorously waves its posterior end around in the water. (After Pennak, Robert W.: Fresh-water Invertebrates of the United States. New York, The Ronald Press Company, 1953.)*

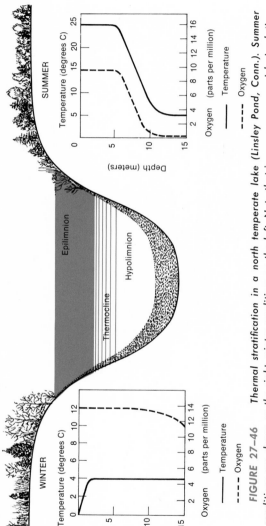

FIGURE 27-46 Thermal stratification in a north temperate lake (Linsley Pond, Conn.). Summer conditions are shown on the right, winter conditions on the left. Note that in summer oxygen-rich circulating layer of water, the epilimnion, is separated from the cold oxygen-poor hypolimnion waters by a broad zone, called the thermocline, which is characterized by a rapid change in temperature and oxygen with increasing depth. Two typical hypolimnion organisms are shown. (From Life in the Depths of a Pond by E. S. Deevey, Jr. Copyright © Oct. 1951 by Scientific American, Inc. All rights reserved.)

TABLE 27-3   A Classification of Wetlands

Fresh Areas

Inland Fresh Areas	Description
1. Seasonally flooded basins of flats	Soil covered with water or waterlogged during variable periods but well drained during much of the growing season. In upland depressions and bottomlands. Bottomland hardwoods to herbaceous growth.
2. Fresh meadows	Without standing water during growing season; waterlogged to within a few inches of surface. Grasses, sedges, rushes, broadleaf plants.
3. Shallow fresh marshes	Soil waterlogged during growing season; often covered with six or more inches of water. Grasses, bulrushes, spike rushes, cattails, arrowhead, smartweed, pickerelweed. A major waterfowl-production area.
4. Deep fresh marshes	Soil covered with six inches to three feet of water. Cattails, reeds, bulrushes, spike rushes, wild rice. Principal duck-breeding area.
5. Open fresh water	Water less than 10 feet deep. Bordered by emergent vegetation: pondweed, naiads, wild celery, water lily. Brooding, feeding, nesting area for ducks.
6. Shrub swamps	Soil waterlogged; often covered with 1 foot of water. Alder, willow, buttonbush, dogwoods. Ducks nesting and feeding to limited extent.
7. Wooded swamps	Soil waterlogged; often covered with 1 foot of water. Along sluggish streams, flat uplands, shallow lake basins. North: tamarack, arborvitae, spruce, red maple, silver maple; South: water oak, overcup oak, tupelo, swamp black gum, cypress.
8. Bogs	Soil waterlogged; spongy covering of mosses. Heath shrubs, sphagnum, sedges.

Coastal Fresh Areas

9. Shallow fresh marsh	Soil waterlogged during growing season; at high tides as much as six inches of water. On landward side, deep marshes along tidal rivers, sounds, deltas. Grasses and sedges. Important waterfowl areas.
10. Deep fresh marshes	At high tide covered with 6 inches to 3 feet of water. Along tidal rivers and bays. Cattails, wild rice, giant cutgrass.
11. Open fresh water	Shallow portions of open water along fresh tidal rivers and sounds. Vegetation scarce or absent. Important waterfowl areas.

Saline Areas

Inland Saline Areas	Description
12. Saline flats	Flooded after periods of heavy precipitation; waterlogged within few inches of surface during the growing season. Vegetation: seablite, salt grass, saltbush. Fall waterfowl-feeding areas.
13. Saline marshes	Soil waterlogged during growing season; often covered with 2 to 3 feet of water; shallow lake basins. Vegetation: alkali hard-stemmed bulrush, wigeon grass, sago pondweed. Valuable waterfowl areas.
14. Open saline water	Permanent areas of shallow saline water. Depth variable. Sago pondweed, muskgrasses. Important waterfowl-feeding areas.

Coastal Saline Areas

15. Salt flats	Soil waterlogged during growing season. Sites occasionally to fairly regularly covered by high tide. Landward sides or islands within salt meadows and marshes. Salt grass, seablite, saltwort.
16. Salt meadows	Soil waterlogged during growing season. Rarely covered with tide water; landward side of salt marshes. Cord grass, salt grass, black rush. Waterfowl-feeding areas.
17. Irregularly flooded salt marshes	Covered by wind tides at irregular intervals during the growing season. Along shores of nearly enclosed bays, sounds, etc. Needlerush. Waterfowl-cover area.
18. Regularly flooded salt marshes	Covered at average high tide with 6 or more inches of water; along open ocean and along sounds. Salt-marsh cord grass along Atlantic. Pacific: Alkali bulrush, glassworts. Feeding area for ducks and geese.
19. Sounds and bays	Portions of salt-water sounds and bays shallow enough to be diked and filled. All water landward from average low tide line. Wintering areas for waterfowl.
19. Mangrove swamps	Soil covered at average high tide with 6 inches to 3 feet of water. Along coast of southern Florida. Red and black mangroves.

(After S. P. Shaw and C. G. Fredine, 1956, *Wetlands of the United States,* U.S. Fish and Wildlife Serv. Circ. 39.)

"spring overturn" occurs when the ice melts and the heavier surface water sinks to the bottom (Fig. 27–46).

Moving waters differ in three major aspects from lakes and ponds: current is a controlling and limiting factor; land-water interchange is great because of the small size and depth of moving water systems as compared with lakes; oxygen is almost always in abundant supply except when there is pollution. Temperature extremes tend to be greater than in standing water. Plants and animals living in streams are usually attached to surfaces or, in the case of animals, are exceptionally strong swimmers. Characteristic stream organisms are: caddis fly larvae, blackfly larvae, attached green algae, encrusting diatoms and aquatic mosses.

There are numerous other fascinating kinds of fresh-water habitats. Table 27–3 gives a summary of these.

SECTION 27–12

COMMUNITIES

The distributions of organisms that we have just described in the three *habitats*, terrestrial, marine and freshwater, represent large, easily recognized assemblages *(biomes),* in each of which there is a characteristic life form of *climax* vegetation (tree, shrub, grass, forb and so on) with which certain animal life is correlated. Within a biome there may occur one or more naturally occurring assemblages of organisms *(communities).* Thus the deciduous forest of North America is beech-maple in some areas and oak-chestnut in others. Communities are most often named for their most conspicuous organism, usually a plant in the terrestrial habitat and an animal in the marine habitat (e.g., the barnacle community).

The composition of communities is determined by whatever organisms are available in the area and can tolerate the prevailing conditions, i.e., the habitat. The result is a mutually sustaining and interdependent population, fluctuating by reason of death, replacement, immigration, emigration and seasonal change. The community is, however, by no means haphazard. Each has a *taxonomic unity*; i.e., certain species occur together in an orderly manner. One species dominates; others are less common. Species are, nevertheless, replaceable in space and time so that functionally similar communities may have quite different species. Every community has a definite feeding (trophic) organization and metabolic pattern. Similar communities in different geographic areas have ecological equivalents. Thus, the cacti of the New World deserts are the equivalents of the similar-looking but unrelated euphorbias of the Old World. Kangaroos are the Australian equivalent of the North American bison and antelope. In each example, the equivalent organisms occupy similar niches in their respective communities; i.e., they do similar jobs. As has frequently been remarked, an organism's *niche* is its profession; its *habitat* is its address. And as in any community there can be only so many doctors, lawyers and chiefs. Each organism uses the community in such a way that it is unavailable to others; no two may occupy the same niche (*competitive exclusion principle*).

The community cannot be understood, however, solely in terms of the number of species and of individuals that comprise it (composition and size) or its *standing crop* (its living weight or biomass). To appreciate fully its functional aspects one must inquire also into its energy relationships.

All organisms and the nonliving world are inexorably locked in an inseparable relationship. Together they constitute the *ecosystem*. No organism can live without an environment. Together the living and the nonliving world carry on the circulation, transformation and accumulation of matter and the transformation of energy.

ENERGY AND THE ECOSYSTEM

Of all the sun's energy impinging on the earth most is eventually lost as heat. A small portion of the energy of light is absorbed by plants and of this only a small portion again is transformed into potential or food energy. The rest leaves the plant and becomes part of the earth's general heat loss. All life, except green plants, obtains its energy by taking in the products of photosynthesis or microorganism chemosynthesis. In this view all organisms are durable forms through which flow energy and matter. When the materials of the earth enter organisms they temporarily assume the form of the particular organisms and then are returned to the environment. Since the earth has a finite amount of material it must all be constantly recycled or life would cease. If all organisms eventually used up all the available nitrogen in the world and bound it in forms that rendered it no longer usable, there could be no further life. In the laboratory one could grow a pure culture of some organism by supplying the necessary molecules for its metabolism. When, however, one terminated the supply, the culture, after using up the last available amount, would then die. Man unwittingly brings about this state of affairs in nature in an alarming number of instances. Today, for example, there is very little virgin tropical rain forest left because of man's activities. In Africa the natives clear portions of the rain forest to grow crops of manioc. Without fertilization the cleared area may produce a crop for two or three years. The area is then abandoned and more forest is felled for new cropland. The abandoned area will probably never again produce a mature rain forest. The thin tropical soils have only a small supply of mineral nutrients because these are leached out by rain. The rain forest, in precarious balance with this soil, can maintain itself only so long as the balance is not disturbed. Once man upsets this balance the forest is irretrievably lost. Civilized man in the technologically advanced countries does the same thing in a more sophisticated manner much more efficiently but with equally devastating results. Many other tragic examples can be given which illustrate the point that no organism can live alone, that all must work together to recycle the limited supply of essential materials, and that man is the great destroyer upsetting critical ecological balances. The necessity of maintaining these balances, of recycling material helps explain why there is no one single uniform species of life on earth and why organisms living in complex interacting communities of many species are fitter than if they lived in isolation. Clearly then organisms evolve not solely in terms of the limitations and peculiarities of the abiotic environment (temperature, soil, pH, salinity), but also in relation to the other organisms with which they are associated. Organisms make the community, but at the same time the community makes the organism.

CYCLES

In surveying the energetics and other complex relationships of the ecosystem, it is convenient to think in terms of raw materials (water, carbon dioxide, oxygen, nitrogen, phosphorus and so on), **producers** (primarily photosynthesizing organisms) and **consumers** (primarily animals, but also fungi and many bacteria). Some ecologists consider fungi and bacteria in a class by themselves and designate them as decomposers. On land the **primary producers** are the flowering plants. The **primary consumers** are herbivorous vertebrates and insects. All the organisms that prey upon them as predators or parasites are **secondary consumers**. In the sea the primary producers are the phytoplankton, especially diatoms and dino-flagellates. The consumers are the zooplankton, marine vertebrates such as the baleen whales, the jellyfish, crustacea, mollusks, marine worms, bony fish, marine birds and all their predators and parasites. In freshwater the producer organisms are phytoplankton and rooted and floating plants in shallow water. Consumers are zooplankton, insect larvae, crustacea, fish and aquatic fungi and bacteria, usually decomposing bottom material. All these organisms play vital roles in recirculating essential elements of protoplasm.

Some cycles are predominantly **geochemical**, although organisms do play some role. The **sedimentary** cycle has already been referred to on page 899. The **water cycle,** illustrated by Figure 27–47, is a closed cycle that can operate in the absence of life but is influenced by it. The **carbon cycle** (Fig. 27–48) involves the use of CO_2 for photosynthesis. When the plant respires it returns CO_2 to the atmosphere. Animals eat plants, return more CO_2 to the air and by dying, together with plants, provide a substrate for bacteria and fungi to metabolize and eventually turn into CO_2. Erosion and dissolution of limestone liberates carbonates and eventually CO_2. Some dead organisms, by becoming fossilized, tie up large quantities of carbon in the ground in the form of oil, gas, coal and peat. When these are burned, CO_2 is released. Organisms that build calcareous shells also tem-porarily tie up carbon when they die by contributing to the formation of limestone or coral reefs.

Nitrogen is removed from the air by lightning and by nitrogen-fixing

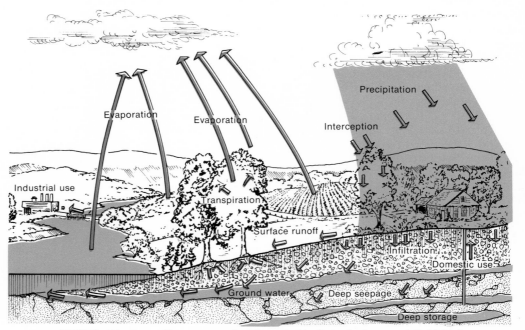

FIGURE 27–47   The water cycle. (From Smith, R. L.: Ecology and Field Biology. New York, Harper & Row, Publishers, 1966.)

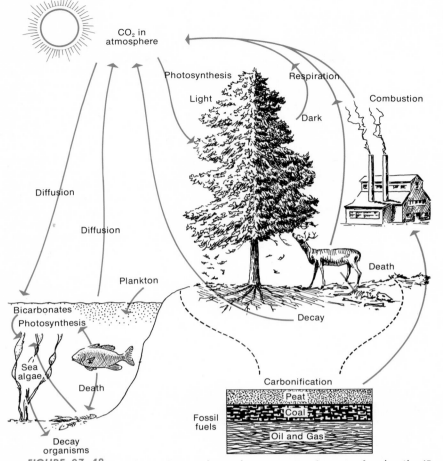

FIGURE 27–48   *The carbon cycle in the ecosystem. See text for details. (From Smith, R. L.: Ecology and Field Biology. New York, Harper & Row, Publishers, 1966.)*

bacteria and algae which turn it into soluble nitrates. Nitrates find their way into the soil or water where they become available for plants (Fig. 27–49). Some is secreted into the soil by plants and some by animals; the rest comes from the eventual degradation of plant and animal material by decomposing bacteria that convert it to ammonia. Ammonia is also produced by volcanic action. Nitrite bacteria convert ammonia into nitrites from which other bacteria produce nitrates. Denitrifying bacteria return nitrogen to the air. The cycle also operates in the marine habitat.

The **phosphorus cycle** is probably simpler (Fig. 27–50). Whereas the reservoir of nitrogen is the air, the reservoir of phosphorus is the rocks from which it is released by erosion. Most of this is lost by washing into the sea and becoming locked in marine sediments which will only become available when there is a major geologic uplift. Phosphorus in shallow marine sediments is available to fish from which it is acquired by marine birds. They in turn return it to use in their excreta (guano), which again washes into the sea and is utilized by plankton and by fish. There is some reason to believe that the return of phosphorus to the cycle is inadequate and that in the long run no more will be available. The loss of phosphorus

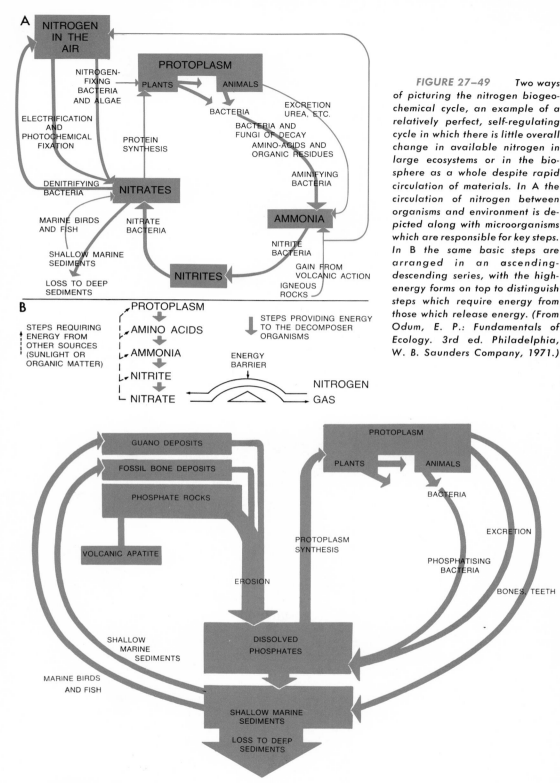

A

NITROGEN IN THE AIR

PROTOPLASM

PLANTS   ANIMALS

NITROGEN-FIXING BACTERIA AND ALGAE

BACTERIA

EXCRETION UREA, ETC.

ELECTRIFICATION AND PHOTOCHEMICAL FIXATION

PROTEIN SYNTHESIS

BACTERIA AND FUNGI OF DECAY AMINO-ACIDS AND ORGANIC RESIDUES

DENITRIFYING BACTERIA

NITRATES

AMINIFYING BACTERIA

MARINE BIRDS AND FISH

NITRATE BACTERIA

AMMONIA

SHALLOW MARINE SEDIMENTS

NITRITE BACTERIA

LOSS TO DEEP SEDIMENTS

NITRITES

GAIN FROM VOLCANIC ACTION IGNEOUS ROCKS

B

PROTOPLASM

AMINO ACIDS

AMMONIA

NITRITE

NITRATE

STEPS REQUIRING ENERGY FROM OTHER SOURCES (SUNLIGHT OR ORGANIC MATTER)

STEPS PROVIDING ENERGY TO THE DECOMPOSER ORGANISMS

ENERGY BARRIER

NITROGEN GAS

FIGURE 27–49   Two ways of picturing the nitrogen biogeochemical cycle, an example of a relatively perfect, self-regulating cycle in which there is little overall change in available nitrogen in large ecosystems or in the biosphere as a whole despite rapid circulation of materials. In A the circulation of nitrogen between organisms and environment is depicted along with microorganisms which are responsible for key steps. In B the same basic steps are arranged in an ascending-descending series, with the high-energy forms on top to distinguish steps which require energy from those which release energy. (From Odum, E. P.: Fundamentals of Ecology. 3rd ed. Philadelphia, W. B. Saunders Company, 1971.)

GUANO DEPOSITS

PROTOPLASM

PLANTS   ANIMALS

FOSSIL BONE DEPOSITS

BACTERIA

PHOSPHATE ROCKS

EXCRETION

PROTOPLASM SYNTHESIS

VOLCANIC APATITE

PHOSPHATISING BACTERIA

BONES, TEETH

EROSION

SHALLOW MARINE SEDIMENTS

DISSOLVED PHOSPHATES

MARINE BIRDS AND FISH

SHALLOW MARINE SEDIMENTS

LOSS TO DEEP SEDIMENTS

FIGURE 27–50   The phosphorus cycle. Phosphorus is a rare element compared with nitrogen. Its ratio to nitrogen in natural waters is about 1 to 23 (Hutchinson, G. E.: Nitrogen in the biogeochemistry of the atmosphere. Am. Sci. 32:178–195, 1944.) Chemical erosion in the United States has been estimated at 34 metric tons per square kilometer per year. Fifty-year cultivation of virgin soils of the Middle West reduced the P_2O_5 content by 36 per cent (Clarke, F. W.: The Data of Geochemistry. U.S. Geol. Surv. Bull. No. 228, 1924). As shown in the diagram, the evidence indicates that return of phosphorus to the land has not been keeping up with the loss to the ocean. (After Odum, E. P.: Fundamentals of Ecology. 3rd ed. Philadelphia, W. B. Saunders Company, 1971.)

is accelerated by man who mines and eventually loses irretrievably more than he returns to the cycle.

Since 1944 man has been adding radioactive materials to the bio-geochemical cycle. The fate of strontium 90 illustrates the importance of some of these. Strontium moves with calcium in the sedimentary cycle. Seven per cent of sedimentary material flowing down rivers is calcium. Strontium follows calcium into the biological system. In the Far North, for example, where there has been heavy radioactive fallout, lichens absorb nearly 100 per cent of the radioactive particles falling to earth. Caribou and reindeer grazing on the lichens concentrate the strontium. Men eating these animals further accumulate the element. Some now have one third to one half the permissible amounts. The problem is not confined to this region. In Europe and North America there has been a steady accumulation of strontium in the bones of children and adults who have acquired it from the milk of cows, which in turn have acquired it from the vegetation.

SECTION 27–15

THE FLOW OF ENERGY

Thus we see in the ecosystem that there is a conservation of usable matter by elaborate recycling processes. The amount of matter is limited, but the amount of energy available is unlimited. In contrast to the cyclic flow of matter, therefore, the flow of energy is one-way (Fig. 27–51). The amount of energy at any stage in the process can be estimated by burning the material in question in a calorimeter and measuring the amount of heat evolved. A study of energy flow sheds further light on the functional structure of communities. It also can tell us how much life a given area can support. That information in turn enables one to calculate how many individuals of each species the area can support. As this potential energy is transferred from the primary producers, plants, through the organisms that eat plants, and successively through their predators and parasites, and finally, at their death, through the microorganisms that decompose, a large proportion of it is lost at each step as heat. Because of this progressive loss as heat, the energy flow must be less and less at each succeeding level. The loss at each level is largely through respiration, but there are also losses due to poor utilization of food. The loss as heat is not immediate, because the undigested food passed as feces is eventually acted upon by decomposers, with a loss of heat then. It is amusing that among farm animals the pig is the most efficient converter of energy. Under the best management, about 20 per cent of the gross energy ingested by a pig is converted into pig biomass. The fact that the pig at 20 per cent is the best farm converter illustrates strikingly the great inefficiency of the energy transfer system in general. Since the successive transfer occurs through eating, the chain of energy transfer is referred to as the **food chain**. Descriptively three types of food chains exist: the predator chain, the parasite chain and the saprophytic chain. Many food chains operate simultaneously and interlock. The total pattern is referred to as the **food web**. One is illustrated in Figure 27–52.

The transfer of energy through a community begins with the fixation of sunlight by photosynthesis. The process is not very efficient. Only 8 per cent of the sun's energy reaching the planet strikes plants and of this only 2 per cent is utilized in photosynthesis. Part of this accumulation of energy (the primary production) is used by the plant itself; the amount

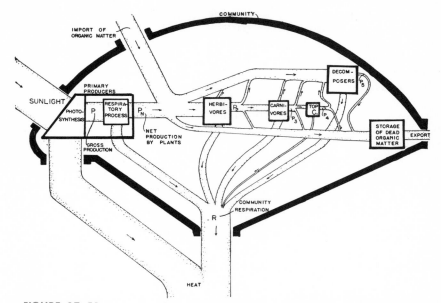

FIGURE 27–51   *Energy flow diagram of a community showing successive fixation and transfer by components, and the large respiratory losses at each transfer. Compare the one-way flow of energy through the community as shown in this diagram with the circulation of materials as shown in Figures 27–48 and 27–49. P = gross primary production, PN = net primary production and P_2, P_3, P_4 and P_5 = secondary production at the indicated levels. (After Odum, H. T.: Primary Production in Flowing Waters. Limnol. and Oceanogr. 1:102–117, 1956.)*

left over that is stored and expressed as growth represents the **net primary production.** The net primary production of a 55-year-old Scots pine plantation, for example, was about 442×10^{10} cal per hectare. This equaled a mean annual transformation of solar energy of about 8×10^{10} cal. The average annual insolation was about 770×10^{10} cal per hectare. Therefore, the efficiency of the Scots pine is about 1.3 per cent. Tropical forests have an efficiency of about 2 per cent; sugar cane fields, about 1.9 per cent.

The stored energy accumulates as growth or **biomass.** Part is cycled seasonally by death and decomposition; the rest, remaining alive, is the **standing crop biomass.** As is to be expected, this can vary with season. In grasslands there is an annual turnover, but in forests much energy is tied up in wood. On an energy basis the most productive ecosystems are coral reefs and estuaries; the least productive, deserts and the open ocean (Fig. 27–53). Available evidence suggests that the production of plant material in each given area has reached an optimum level limited only by climate and soil.

Consumers depend upon the net production of the green plants. Of the net production that is available to herbivores not all is assimilated. A grasshopper assimilates only 30 per cent of its food; some mice, nearly 90 per cent. In all cases most goes into maintenance and is eventually lost as heat and by respiration. It is lost to the ecosystem. The residue is stored in the form of new tissue and new individuals. It is this stored energy in the herbivore that is of interest to the carnivore.

Various kinds of graphic representations have been employed by ecologists to illustrate the trophic functional structure of communities. There are three general types of these ecological pyramids: the **pyramid of numbers,** the **pyramid of biomass** and the **pyramid of energy flow.** These are compared in Figure 27–54. The **pyramid of numbers** owes its geometry

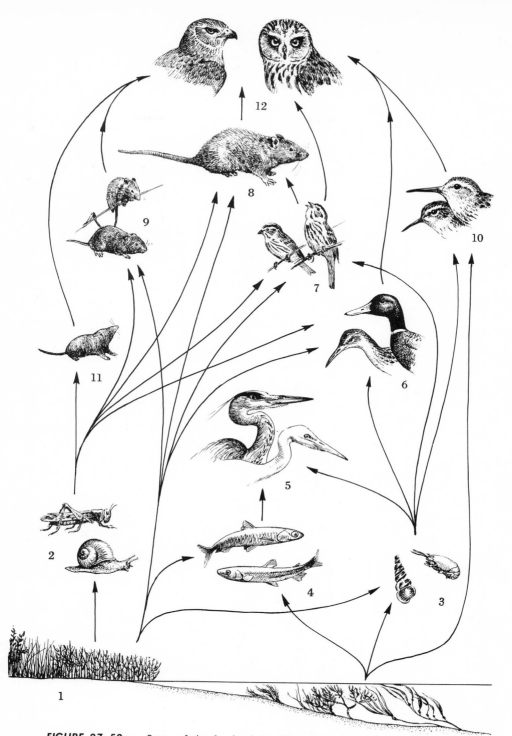

FIGURE 27–52   Some of the food relationships in a *Salicornia* salt marsh *(San Francisco Bay area). Producer organisms (1), terrestrial and salt marsh plants, are consumed by herbaceous terrestrial invertebrates, represented by the grasshopper and snail (2). The marine plants are consumed by herbivorous marine and intertidal invertebrates (3). Fish, represented by smelt and anchovy (4), feed on vegetative matter from both ecosystems. The fish in turn are eaten by first-level carnivores, represented by the great blue heron and common egret (5). Continuing through the web, we have the following omnivores: clapper rail and mallard duck (6); savanna and song sparrows (7); Norway rat (8); California vole and salt marsh harvest mouse (9); the least and western sandpipers (10). The vagrant shrew (11) is a first-level carnivore, while the top carnivores (second level) are the marsh hawk and shorteared owl (12). (Relationships after Johnston, R. F.: Predation by Shorteared Owls in a Salicornia Salt Marsh. Wilson Bull. 68:91–102, 1956; Ecology and Field Biology. New York, Harper & Row, Publishers, 1966.)*

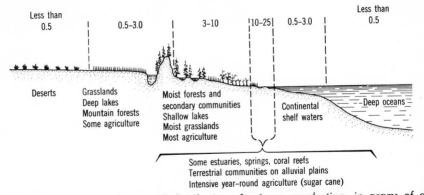

0.5–3.0   3–10 |10–25| 0.5–3.0   Less than 0.5

Deserts

Grasslands
Deep lakes
Mountain forests
Some agriculture

Moist forests and
secondary communities
Shallow lakes
Moist grasslands
Most agriculture

Continental
shelf waters

Deep oceans

Some estuaries, springs, coral reefs
Terrestrial communities on alluvial plains
Intensive year-round agriculture (sugar cane)

FIGURE 27–53   *The world distribution of primary production, in grams of dry matter per square meter day, as indicated by average daily rates of gross production in major ecosystems. (From Odum, E. P.: Fundamentals of Ecology. 3rd ed. Philadelphia, W. B. Saunders Company, 1971.)*

to the following facts: a great many small units are required to equal the mass of one big unit; energy is lost progressively at each link in the food chain so that much less is available to higher levels; small organisms metabolize much faster than do large ones. The **pyramid of biomass** illustrates better the quantitative relations of the standing crop. When the total weight of individuals at any one moment at successive trophic levels is plotted, a gradually sloping pyramid is found as long as there is no very great difference in size in the various organisms. If, however, organisms at the lower levels are very much smaller than those at the higher levels, the

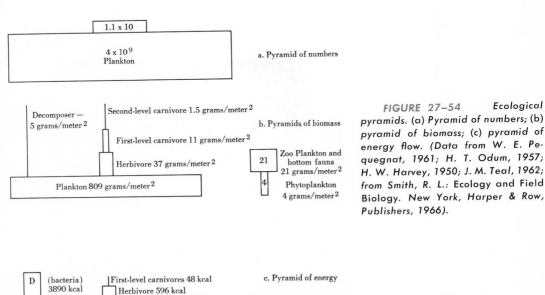

1.1 x 10

4 x 10^9
Plankton

a. Pyramid of numbers

Decomposer —
5 grams/meter^2

Second-level carnivore 1.5 grams/meter^2

First-level carnivore 11 grams/meter^2

Herbivore 37 grams/meter^2

Plankton 809 grams/meter^2

b. Pyramids of biomass

21   Zoo Plankton and bottom fauna 21 grams/meter^2

4   Phytoplankton 4 grams/meter^2

FIGURE 27–54   *Ecological pyramids. (a) Pyramid of numbers; (b) pyramid of biomass; (c) pyramid of energy flow. (Data from W. E. Pequegnat, 1961; H. T. Odum, 1957; H. W. Harvey, 1950; J. M. Teal, 1962; from Smith, R. L.: Ecology and Field Biology. New York, Harper & Row, Publishers, 1966).*

D   (bacteria) 3890 kcal

First-level carnivores 48 kcal

Herbivore 596 kcal

c. Pyramid of energy

Net production: 8763 kcal/meter^2/year
Gross production: 36,380 kcal/meter^2/year

pyramid may be inverted. The rapid metabolism and turnover of the smaller organisms accomplish a larger output of energy with a smaller standing crop biomass. In short, the number and weight of organisms supported at any level depends on the *rate* at which food is being produced. The **pyramid of energy** gives the best picture of the functional nature of communities. Whereas the other pyramids picture the situation at a given moment, the energy pyramid depicts rates of food mass through the food chain. From knowledge of the energy pattern it is possible to calculate the standing crop and the number of individuals the community will contain if one knows in addition the mean weight of individuals and their mean life span.

SECTION 27–16

POPULATIONS

Considerations of energy flow do not, however, give information about the sizes of individual populations comprising the community. Moreover, the sizes of populations normally are not constant. Some fluctuations take the form of orderly growth and decline; others are violent and erratic; still others are cyclic. These changes are attributable to a number of causes, among which are: birth, death, immigration and emigration. In certain instances a population change is due to birth rate alone. The spring and fall "blooms" of diatoms and dinoflagellates are population explosions due to birth rate. Each cell divides, then each daughter cell divides and so on in the succession 1, 2, 4, 8, 16. A plot of the numerical increase with each generation gives a logarithmic or exponential curve (Fig. 27–55). Another example of nearly **exponential growth** is provided by the results of planting 435 Atlantic striped bass in San Francisco Bay in 1879 and 1881. In 1899, 1,234,000 pounds of bass were fished out of the bay. Obviously there are limits to the duration of exponential growth in nature. Sooner or later some **limiting factor** intercedes rather suddenly. In the blooms of phytoplankton, the limiting factor is nutrient material. Birth rate declines, mortality takes effect and the size of the population is drastically reduced.

Another form of population growth is that described by a **sigmoid curve.** In this case the rate of increase gradually tapers off, and an **equilibrium** is reached. This upper leveling off or **asymptote** has sometimes been called the **carrying capacity**. For example, sheep were introduced into Tasmania about 1800 and careful records of their numbers kept over the ensuing years. By about 1850, the population stood at 1,700,000 and has remained more or less at that level ever since (until 1934), with mild irregular fluctuations due probably to climatic factors which limit the number that can be maintained with profit.

The course of mortality may be different for different species. There may be **exponential mortality**, i.e., increasing mortality independent of age. This straight line curve is characteristic of bacteria and some birds. Some organisms have higher mortalities at advanced ages; the curve is convex. Man is an example. Many small organisms experience the greatest mortality during youth; the curve is concave (Fig. 27–56).

Insects that have a single brood per year and that do not live more than one year have no generation overlap. Most other species with longer life spans or repeated reproductive cycles do have generation overlap; therefore, the population has an **age structure.** Each age group has its own characteristic birth and mortality rates which remain rather constant as

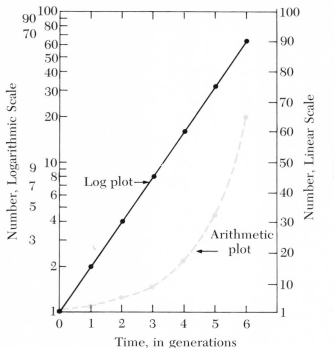

FIGURE 27–55   *Exponential population growth of an organism reproducing by binary fission, plotted in two ways.*

long as environmental conditions do not change. The proportion of different groups in the population influences birth and death rates and determines the reproductive status, or replacement rate, of the whole. This is because mortality varies with age, and because reproduction is usually associated with specific ages. Under any schedule of constant birth and death rates, a particular age structure will develop. Barring environmental interference, there will be a characteristic stable age distribution, and the population will grow smoothly. The stable age structure may temporarily be altered by unusual increases in births or deaths but eventually will return to the stable or normal distribution. Under ideal optimum

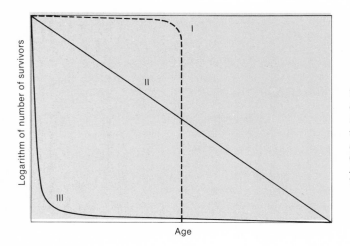

FIGURE 27–56   *Schematic representation of theoretical types of survivorship curve (after Deevey, 1947). The survival axis can be graduated either arithmetically or logarithmically, but the logarithmic scale is more instructive in that a straight line implies equal rates of mortality with respect to age. (From MacArthur, R. M., and Connell, J. M.: The Biology of Populations. New York, John Wiley & Sons, Inc., 1966.)*

conditions, when the environment is unlimited, the per cent growth rate, i.e.,) the population growth rate per individual, is constant and maximum. This rate is often called the **biotic potential.** It represents an intrinsic rate of natural increase, i.e., the inherent power of a population to grow. Some laboratory populations of insects and rodents approximate these optimum conditions. Some comparative values expressed as the rate of increase per female per year are: man (white, American, 1920), 0.0055; brown rat in the laboratory, 5.4; flour beetle, 36.8. If the exponential rate of growth continued for a year, each population would increase the following number of times: man, 1.0055; rat, 221; flour beetle, 1.06×10^{15}. In fact, the environment intercedes. The difference between the theoretical rate and the actual rate is a measure of environmental resistance.

So far we have been discussing populations as though they were isolated entities. In fact, birth and death rates are supplemented by emigration, immigration and migration. In most cases **emigration** and **immigration** balance each other; however, a mass dispersal in or out greatly affects the structure of a population. Conversely the status of the population determines the impact of **dispersal.** If a population is on the verge of extinction, a mass immigration may prevent the demise of the population. If a population is overcrowded, mass emigration can bring about a crash. This situation is seen, for example, when snowy owls or lemmings depart from an overcrowded area.

Migration is a special type of dispersal in which nearly total populations abandon an area temporarily. By migrating seasonally or daily, organisms can take advantage of optimum conditions in areas that otherwise would be uninhabitable on a permanent basis. Thus, by always moving with optimum conditions, these species can maintain high activity and density, while species that stay behind must remain inactive (by going into diapause or hibernating) until better times return. In all cases, dispersal is greatly influenced by barriers—water for nonflying terrestrial species, land for aquatic species, salinity for aquatic species, extremes of temperature, high mountain ranges and so on. It is probable, however, that the internal factors of a population are more significant in influencing the size of a population than is dispersal.

What does limit the size of populations? Some species are abundant; some are rare. The physical environment is a crucial factor for some species. This appears to be especially true of small organisms which have short life cycles and high rates of metabolism per gram. In general, numbers may be limited by a shortage of natural resources (e.g., food, places in which to breed), by inaccessibility of these resources relative to the animals' capacities for dispersal and searching (e.g., plants too widely dispersed), and by too little time for reproduction (e.g., limited wet season, limited day length as in the Arctic). Herbivorous insects in a desert are limited by the amount of food available during a short growing season. When drought sets in, the plant crop disappears. The size of the particular insect population depends on their intrinsic rate of increase and the duration of the growing season.

Populations of larger organisms have longer life cycles and their numbers and biomass more clearly reflect energy flow. These are not limited by the physical environment as much as they are regulated by interactions among their own members or by interactions with competitors, predators and parasites. This is not to say that these factors do not affect small organisms; they tend to be less critical there, except under laboratory conditions.

Intrinsic factors tending to regulate population size may be physiological or behavioral, or both. **Stress** is one factor that operates physiologically.

For example, when some rodent populations become too dense, the animals encounter one another more often. Fights ensue, conditions in general are more stressful and there is enlargement of the adrenals; the resulting hormonal imbalance adversely affects mating and reproduction. Mortality also increases.

Territoriality and *peck orders* are behavioral methods of regulation. Both of these patterns of behavior tend to prevent available food from being over exploited. They cause the food to be distributed among stronger members of the population and limit the number of breeding members of the population.

One of the more obvious factors regulating populations is predation. The negative effects of predation (also of parasitism) tend to be small when the two species involved have had a long evolutionary or historical association. It is when prey and predator or host and parasite first encounter one another that the effect is spectacular. Similarly when the ecosystem of which they are a part becomes unbalanced (e.g., by interference by man), predators and parasites are apt to have a profound effect on populations. A striking example is afforded by the sequence of events occurring on the Kaibab plateau in Arizona between 1907 and 1925. In 1907 the area had about 4000 deer and a large number of pumas and wolves. Between 1907 and 1923 man killed most of these predators. By 1925 the deer population increased to 1,000,000. The area could not support this great population. It was almost completely overgrazed. Forty per cent of the herd of deer died in two winters. The forest is still depleted. Clearly the predators had kept the deer in balance with the vegetation and themselves.

Many striking examples involving parasitism are known. The chestnut blight, an introduced fungus, has virtually wiped out the American chestnut; the Dutch elm disease is making short work of the American elm. In old associations, parasites have established a fine working balance with their hosts, because they obviously cannot afford to kill them. For examples, natives of Africa have built up a relative immunity to malaria so that both host and parasite can survive. The parasite of sleeping sickness, a trypanosome, lives in its reservoir hosts, the large grazing animals, without harming them. Tapeworms live in people generally without killing them. In each case, however, this fine balance can be upset by changes in the ecosystem or by changes which otherwise affect the health of the host.

Earlier we referred to cyclic oscillations in populations. Some of these are controlled by regular seasonal changes in the environment. Others are the result of intrinsic factors and of interactions among populations, as the following example will illustrate. Data are drawn from the records of the Hudson Bay Company of Canada which has kept a tally of the pelts of animals trapped since 1800. A plot of the abundance of lynx and snowshoe hare (Fig. 27–57) reveals that the lynx population peaked about every 9.6 years. In the intervening periods, the lynx became very rare. The snowshoe hare showed a similar cycle, but its peaks preceded those of the lynx by approximately one year. The correlation between the two cycles strongly suggests that the cycles represent interactions between prey and predator.

Another classic example of cyclic abundance involves lemmings, mice and voles, and the snow owl and foxes that prey upon them. Every three to four years the lemmings of the world become fantastically abundant. At some of these times they emigrate from crowded areas and march toward the sea (in Norway) and drown. As the lemmings increase, the foxes and owls do also. Suddenly, often within a single season, the lemming popu-

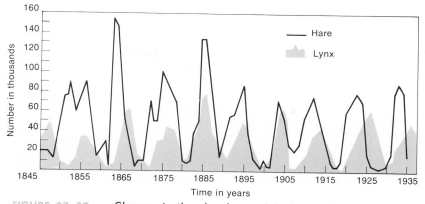

FIGURE 27–57   *Changes in the abundance of the lynx and the snowshoe hare,*
as indicated by the number of pelts received by the Hudson Bay Company. This is a classic
case of cyclic oscillation in population density. (After MacLulich, D. A.: Fluctuations in the
numbers of the varying hare (Lepus Americanus). Univ. Toronto Studies, Biol. Ser., No.
43., 1937.)

lation "crashes," the foxes starve and the owls migrate south. Like the
lemmings, the owl movement is one-way, an emigration.

It will be noted that these examples involve organisms living in the
tundra. This is one of the simpler ecosystems. Marked oscillations of the
sort described do not occur in more complex ecosystems, or, if they do,
are less apparent. It appears that complexity makes for stable prey-pred-
ator relations. It should be emphasized that fluctuations in natural populations depend upon complex interactions of all the phenomena so far
discussed: intrinsic rate of natural increase, environmental resistance,
behavioral and physiological effects of crowding, predation and parasitism.

Energy relationships determine how much life an area can support,
and the previously discussed factors regulating population density determine how many individuals of a given species will actually inhabit the
area. None of these, however, tell how many different species share the
area. It is generally true that a single species cannot exploit a wide range
of resources as well as a combination of species. There is no single herbivore that is equal to the cow at eating grass, the hippopotamus at eating water plants, the sloth at eating the leaves of the cecropia tree and
the camel at munching thorny desert bushes. On the other hand, there
cannot be a different species of herbivore for every species of plant. There

FIGURE 27–58   Notonecta (left) and Corixa (right), *two aquatic bugs (Hemiptera)*
which occupy the same general habitat but very different ecological niches because of
differences in food habits. (After Mellanby, H.: Animal Life in Fresh Water. London,
Methuen & Co., Ltd., 1938.)

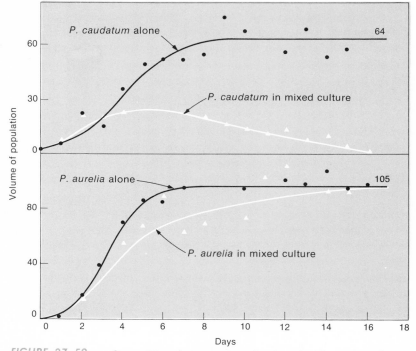

FIGURE 27–59   Competition between two closely related species of protozoa which have identical niches. When separate, Paramecium caudatum and Paramecium aurelia exhibit normal sigmoid growth in controlled cultures with constant food supply; when together, P. caudatum is eliminated. (From Alee, W. C., et al.: Principles of Animal Ecology. Philadelphia, W. B. Saunders Company, 1949; after Gause, G. F.: The Struggle for Existence. Baltimore, Maryland, The William & Wilkins Company, 1934.)

is some happy medium between extreme specialization and infinite generalization. It is clear that different parts of the world support different numbers of species. In general, the tropics supports more species of plants and animals than any other part of the world. The number thins out rapidly as one proceeds to the poles. Furthermore, islands have fewer species than continents, and the smaller and more remote the islands, the fewer the species. Any small uniform habitat has a limited capacity for species. The first species to arrive in a habitat will find a niche and establish itself successfully until the habitat is saturated. From then on there is no room for new species unless, of course, a new one can succeed in replacing an old one. In short, species compete for a niche in the environment.

Niche has already been described as the "profession" of an organism. The two aquatic bugs, the backswimmer (*Notonecta*) and the water boatman (*Corixa*), offer a striking example. The two look much alike (Fig. 27–58), but the backswimmer swims upside down. Both occupy the same habitat— the shallow, vegetation-choked area of ponds and lakes. Their niches, however, are different. *Notonecta* is a predator. *Corixa* feeds on decaying vegetation.

Interspecific competition is any interaction between two or more species populations which adversely affect their growth and survival. It may involve common space, food, light, mutual predation and so forth. A population will stop increasing when the amount of essential resources results in mortality-balancing reproduction. If another population were to use the same resources, both species would suffer a decrease in numbers. They are competing. In general, no two species can have exactly the same niche and remain two species. Sooner or later they will completely interbreed, if that is possible, or one species will be eliminated. A classic laboratory experiment that demonstrated competition involved two closely related species of *Paramecium* that have the same niche, *P. caudatum* and *P. aurelia*. When they are grown in separate cultures, each shows normal sigmoid growth; when grown together, *P. caudatum* is eventually eliminated (Fig. 27–59).

SUGGESTIONS FOR FURTHER READING

Odum, E. P.: *Fundamentals of Ecology.* 3rd ed. Philadelphia, W. B. Saunders Company, 1971.
A useful introduction to the subject matter and approach to the study of ecology.

MacArthur, R. H., and Connell, J. H.: *The Biology of Populations.* New York, John Wiley & Sons, Inc. 1969.
A concise treatment that emphasizes the population aspect.

Elton, C.: *The Ecology of Animals.* 4th ed. New York, Barnes & Noble, Inc., 1969.
Every student interested in the subject should read this small classic.

Oosting, H. J.: *The Study of Plant Communities.* 2nd ed. San Francisco, W. H. Freeman & Company, 1969.
A general treatment of plant ecology.

Good, R.: *The Geography of the Flowering Plants.* 3rd ed. New York, John Wiley & Sons, Inc, 1969.

Richard, P. W.: *The Tropical Rain Forest.* New York, Cambridge University Press, 1952.

Haviland, M. D.: *Forest, Steppe and Tundra.* New York, Cambridge University Press, 1926.

Bourliere, F.: *The Tropics.* New York, Alfred A. Knopf, Inc., 1957.
These books are readable accounts of specific biomes.

Hardy, A.: *The Open Sea.* Boston, Houghton Mifflin Company, 1969.
This fascinating and comprehensive account of the oceans is a two-volume work.

Carson, Rachel L.: *The Sea Around Us.* New York, Oxford University Press, 1951.
Carson, Rachel L.: *The Edge of the Sea.* Boston, Houghton Mifflin Company, 1955.
For sheer beauty these two are superlative.

Coker R. E.: *Lakes, Streams and Ponds.* Chapel Hill, University of North Carolina Press, 1954.

Reid, G.: *Ecology of Inland Waters and Estuaries.* New York, Reinhold Publishing Corporation, 1961.
Fine books for the ecology of freshwater.

THE CLASSIFICATION
OF
PLANTS AND ANIMALS

The system of classifying animals and plants—cataloguing by phylum, class, order, family, genus and species—was discussed in Chapter 11. In the synoptic survey given here the phyla within the kingdoms and the classes within the phyla are arranged in the order of increasing complexity as far as possible, but since evolution has proceeded in a branching fashion, it is impossible to arrange the animals and plants rigidly in order, from simple to complex. The numbers given are estimates of known species in the phylum; for many of the groups there are many additional species as yet undescribed.

I. THE PLANT KINGDOM

Organisms classified as plants usually have stiff cell walls and chlorophyll.

SUBKINGDOM THALLOPHYTA:

Plants not forming embryos. The simplest plants, without true roots, stems or leaves; the body is either a single cell or an aggregation of cells with little differentiation into tissues. (107,000)

PHYLUM CYANOPHYTA. The blue-green algae, with no distinct nuclei or chloroplasts; probably the most primitive of existing plants. (2500)
PHYLUM EUGLENOPHYTA. Euglenoids, with definite nuclei and chloroplasts; lack an outer cellulose wall; have a pigmented eye spot. (350)
PHYLUM CHLOROPHYTA. The green algae, with definite nuclei and chloroplasts. (6000)
PHYLUM CHRYSOPHYTA. The yellow-green algae, the golden-brown algae and the diatoms. (10,000)
PHYLUM PYRROPHYTA. The cryptomonads and dinoflagellates. (1100)

PHYLUM PHAEOPHYTA. The brown algae, with multicellular, often large bodies—the large seaweeds. (1000)

PHYLUM RHODOPHYTA. The red algae. Multicellular, usually marine plants, sometimes impregnated with calcium carbonate. (3000)

PHYLUM SCHIZOMYCOPHYTA. The bacteria. (3000)

PHYLUM MYXOMYCOPHYTA. The slime molds. The body consists of a mass of cytoplasm containing many nuclei, but not sharply divided into cells. Movement by amoeboid motion. (450)

PHYLUM EUMYCOPHYTA. The true fungi. (80,000)

Class Phycomycetes. The algal fungi—bread molds and leaf molds. (500)

Class Ascomycetes. The sac fungi—yeasts, mildews and cheese molds. (35,000)

Class Basidiomycetes. Mushrooms, toadstools, rusts and smuts. (25,000)

Class Fungi Imperfecti. A heterogeneous collection of fungi in which sexual reproduction is unknown, and which are not easily assigned to one of the other classes.

SUBKINGDOM EMBRYOPHYTA:

Plants forming embryos.

PHYLUM BRYOPHYTA. Embryophyte plants without conducting tissues. Multicellular plants, usually terrestrial, with a marked alternation of sexual and asexual generations. The prominent plant is the gametophyte (sexual generation), on which the sporophyte is dependent. (25,000)

Class Musci. Mosses. The gametophyte plant has an erect stem, and leaves arranged in a spiral. (15,000)

Class Hepaticae. The liverworts. Usually simple, flat plants, living in moist, shady places. (9000)

Class Anthocerotae. Hornworts. (300)

PHYLUM TRACHEOPHYTA. Vascular plants.

SUBPHYLUM PSILOPSIDA. Leafless and rootless vascular plants.

Class Psilophytinea.

    *Order Psilophytales.**

    Order Psilotales.

SUBPHYLUM LYCOPSIDA. The clubmosses with simple conducting systems and small green leaves. (900)

Class Lycopodineae. The clubmosses and quillworts.

    Order Lycopodiales. Clubmosses.

    Order Selaginellales. Small clubmosses.

    *Order Lepidodendrales.** Giant clubmosses.

    *Order Pleuromeiales.**

    Order Isoetales. Quillworts.

SUBPHYLUM SPHENOPSIDA. Horsetails with simple conducting systems, jointed stems and reduced, scalelike leaves.

Class Equisetineae.

    *Order Hyeniales.**

    Order Sphenophyllales.

    Order Equisetales. Horsetails.

SUBPHYLUM PTEROPSIDA. Complex conducting systems, large conspicuous leaves.

---

*Known only as fossils; no living representatives.

Class Filicineae.   Ferns. (9000)

*Order Coenopteridales.**

Order Ophioglossales.

Order Marattiales.

Order Filicales.

Class Gymnospermae.   Conifers, cycads and most other evergreen trees and shrubs. No true flowers or ovules are present; the seeds are borne naked on the surface of the cone scales. (640)

SUBCLASS CYCADOPHYTAE.

*Order Cycadofilicales.**   The seed ferns—the most primitive of the seed plants—known only from fossils (late Paleozoic era).

*Order Bennettitales.**

Order Cycadales.   The cycads, the most primitive living seed plants, found in tropical and subtropical regions.

SUBCLASS CONIFEROPHYTAE.

*Order Cordaitales.**   Extinct, large-leaved evergreen trees. Fossil remains of these have been found in deposits from the Devonian to the Permian.

Order Ginkgoales.   The ginkgo, or maidenhair tree, is the only living member of this group.

Order Coniferales.   The conifers, the common evergreen trees and shrubs, with needle-shaped leaves.

Order Gnetales.   Climbing shrubs or small trees found in tropical and semitropical regions, with many characteristics in common with the angiosperms.

Class Angiospermae.   Flowering plants, with seeds enclosed in an ovary. (250,000)

SUBCLASS DICOTYLEDONEAE.   Most flowering plants. Embryos with two cotyledons or seed-leaves; vascular bundles in a ring in the stem; leaves with netlike venation; flower parts (sepals, petals, stamens and carpels) in fives, fours or twos. (175,000)

SUBCLASS MONOCOTYLEDONEAE.   The grasses, lilies and orchids. Leaves with parallel veins, stems in which the vascular bundles are scattered, and flower parts in threes or sixes. The embryo has only one seed-leaf. (75,000)

II.   THE ANIMAL KINGDOM

Organisms classified as animals usually lack stiff cell walls and have no chlorophyll; nutrition is heterotrophic, either holozoic or parasitic.

PHYLUM PROTOZOA.   Microscopic, unicellular animals, which sometimes aggregate in colonies. Some are free-living; others are parasitic. (25,000)

Class Flagellata.   Protozoa which move by whiplike cytoplasmic protrusions called flagella. Primitive animals, probably the group most closely related to the one-celled plants.

Class Sarcodina.   Protozoa which move by pseudopodia.

---

*Known only as fossils; no living representatives.

Class Sporozoa. Parasitic protozoa which reproduce by spores and have no method of locomotion.

Class Ciliata. Protozoa which move by means of cilia.

Class Suctoria. Protozoa with cilia only in young stages, adults attached by stalk to substrate. Some parasitic.

PHYLUM PORIFERA. The sponges (sessile, aquatic animals), both fresh-water and marine. The simplest of the many-celled animals, resembling in many respects a protozoan colony. The body is perforated with many pores to admit water, from which food is strained. Three classes: Calcarea (calcareous spicules), Hexactinellida (siliceous spicules) and Demospongiae (protein spicules—bath sponges). (3000)

PHYLUM COELENTERATA. Radially symmetrical, aquatic animals with a central gastrovascular cavity. The body wall consists of two layers of cells, in the outer of which are stinging cells, nematocysts. (10,000)

Class Hydrozoa. Hydra-like animals, either single or colonial. There is usually an alteration of a hydra-like (asexual) generation with a jellyfish (sexual) generation.

Class Scyphozoa. True jellyfishes.

Class Anthozoa. The corals and sea anemones, which have no alteration of generations. The digestive cavities of these animals are divided by mesenteries to increase the effective surface.

PHYLUM CTENOPHORA. The comb jellies or sea walnuts. These animals lack the stinging capsules of coelenterates and move by means of eight comblike bands of cilia. (100)

PHYLUM PLATYHELMINTHES. The flat-worms, with flat, and either oval or elongated, bilaterally symmetrical bodies, and three cell layers. The excretory organs are flame cells, protonephridia. There is a true central nervous system. (6000)

Class Turbellaria. Nonparasitic flat-worms with a ciliated epidermis.

Class Trematoda. The flukes, parasitic nonciliated flatworms with one or more suckers. Many are internal parasites with complicated life cycles.

Class Cestoda. The tapeworms, parasitic flatworms with no digestive tract; the body consists of a head and a chain of "segments" or individuals which bud from the head.

PHYLUM NEMERTEA. The proboscis worms. Nonparasitic, usually marine animals with a complete digestive tract and a protrusible proboscis armed with a hook for capturing prey. The simplest animals with a blood vascular system. (550)

PHYLUM NEMATODA. The roundworms. An extremely numerous phylum. Characterized by elongated, cylindrical, bilaterally symmetrical bodies; they live as parasites in plants and animals, or are free-living in the soil or water. (8,000)

PHYLUM ACANTHOCEPHALA. The hook-headed worms. Parasitic worms with no digestive tract and a head armed with many recurved hooks. (100)

PHYLUM CHAETOGNATHA. Arrow worms, free-swimming marine worms, with a body cavity (coelom) which develops from pouches of the digestive tract (as in the echinoderms and lower chordates). (30)

PHYLUM NEMATOMORPHA. The horsehair worms. Extremely thin, brown or black worms about 15 centimeters long, resembling a horsehair. The adults are free-living, but the larvae are parasitic in insects. (200)

PHYLUM ROTIFERA. Small, wormlike animals, commonly called "wheel animalcules," with a complete digestive tract, flame cells and a circle of cilia on the head, the beating of which suggests a wheel. (1200)

PHYLUM GASTROTRICHA. Microscopic, wormlike animals resembling the rotifers, but lacking the crownlike circle of cilia. (100)

PHYLUM ENTOPROCTA. Bryozoa or "moss" animals with mouth and anus within the U-shaped circlet of ciliated tentacles. Stalked, sessile, microscopic animals, mostly marine forms. (60)

PHYLUM ECTOPROCTA. Sessile, colonial bryozoa with a lophophore of ciliated tentacles and an anus which opens outside of the lophophore. Have a true coelom. (5000)

PHYLUM BRACHIOPODA. The lamp shells. Marine animals with two hard shells (one dorsal and one ventral), superficially like a clam. They obtain food by means of a lophophore. (200 at present; 3000 extinct)

PHYLUM PHORONIDA. Wormlike marine forms which secrete and live in a leathery tube; they have a U-shaped digestive tract and a lophophore. (10)

PHYLUM ANNELIDA. The segmented worms. There is a distinct head, digestive tract, coelom, closed circulatory system and—in some—nonjointed appendages. The digestive system is divided into specialized regions. (10,000)

Class Polychaeta. Mostly marine worms. Each segment of their bodies has a pair of paddle-like structures (parapodia) for swimming and many bristles (chaetae). Some burrow in sand and mudflats; some live in calcareous tubes which they secrete; others swim freely in the ocean.

Class Oligochaeta. Fresh-water or terrestrial worms, with no parapodia and few bristles per segment.

Class Archiannelida. Primitive annelids without bristles or external segmentation.

Class Hirudinea. The leeches—flattened annelids lacking bristles and parapodia, but with suckers at anterior and posterior ends.

PHYLUM ONYCHOPHORA. Rare, tropical animals, structurally intermediate between annelids and arthropods, with an annelid-like excretory system and an insect-like respiratory system. Body segmented, with a hemocoel and one pair of unjointed appendages per segment. Only a few species known.

PHYLUM ARTHROPODA. Segmented animals with jointed appendages and a hard, chitinous skin, with a body divided into head, thorax and abdomen. Hemocoel present. (800,000)

Subphylum Trilobita. Trilobites, primitive marine arthropods that originated in the Cambrian and became extinct in the Permian. Segmented body divided by two longitudinal furrows into three lobes. All segments except last had a pair of biramous appendages.

Subphylum Chelicerata. Chelicerae on third segment, no antennae.

Class Xiphosura. Horseshoe crabs. Book gills on opisthosoma.

Class Eurypterida. Opisthosoma divided into mesosoma and metasoma. Mesosomal appendages gill-like.

Class Pycnogonida. Sea spiders. Body greatly reduced in size.

Class Arachnida. Spiders, scorpions, ticks and mites. Adults have no antennae; the first pair of appendages ends in pincers, the second pair is used as jaws, and the last four pairs are used for walking.

Subphylum Crustacea. Lobsters, crabs, barnacles, water fleas and sowbugs. Animals that are usually aquatic, have two pairs of antennae and respire by means of gills.

Subphylum Labiata. Antennae on second segment, nothing on third, mandibles on fourth. Second maxillae form lower lip.

Class Chilopoda. The centipedes. Each body segment, except the head and tail, has a pair of legs.

Class Diplopoda. The millipedes. Each external segment (really two segments fused) bears two pairs of legs.

Class Insecta. The largest group of animals, mostly terrestrial. The body is divided into a distinct head, with four pairs of appendages; the thorax has three pairs of legs and usually two pairs of wings; the abdomen has no appendages. Respiration by means of tracheae. There are about 24 different orders of insects, of which the following are common:

Order Orthoptera. Grasshoppers, crickets and praying mantids.

Order Isoptera. Termites.

Order Odonata. Dragonflies and damsel flies.

Order Anoplura. Lice.

Order Hemiptera. Water boatmen, bedbugs and backswimmers.

Order Homoptera. Cicadas, aphids and scale insects.

Order Coleoptera. Beetles, weevils and fireflies.

Order Lepidoptera. Butterflies and moths.

Order Diptera. Flies, mosquitoes and gnats.

Order Hymenoptera. Ants, wasps, bees and gallflies.

PHYLUM MOLLUSCA. Unsegmented, soft-bodied animals, usually covered by a shell, and with a ventral, muscular foot. Respiration is by means of gills, protected by a fold of the body wall—the mantle. (80,000)

Class Amphineura. Chitons, marine forms with a shell composed of eight plates.

Class Scaphopoda. Tooth shells, marine forms living in sand or mud; tubular shells open at both ends.

Class Gastropoda. Snails, slugs, whelks, abalones; asymmetrical animals with a single spiral shell, or no shell.

Class Pelecypoda. Clams, mussels, oysters, scallops. These lack a head and have a hatchet-shaped foot for burrowing. The shell consists of two plates or valves (the animals are called "bivalves"), one on each side of the body.

Class Cephalopoda. Squids, cuttlefish, octopuses. Marine animals having a well developed "head-foot," with eight or 10 tentacles, and well developed eyes and nervous system.

PHYLUM ECHINODERMATA. Marine animals which are radially symmetrical as adults, bilaterally symmetrical as larvae. The skin contains calcareous, spine-bearing plates. The animals have a unique water vascular system of canals, and tube feet for locomotion. Respiration is by skin gills or by out-pocketings of the digestive tract. (6000)

Class Asteroidea. The starfishes. The body is a central disc with broad arms (usually five) not sharply marked off from the disc.

Class Ophiuroidea. The brittle stars and serpent stars. The animals have long, narrow arms sharply differentiated from the central, disc-shaped body.

Class Echinoidea. The sea urchins and sand dollars. Spherical or flattened oval animals with many long spines.

Class Holothuroidea. Sea cucumbers. Long, ovoid, soft-bodied armless echinoderms, usually with a ring of tentacles around the mouth.

Class Crinoidea. Sea lilies and feather stars. The body is cup-shaped and attached by a stalk to the substrate. Most of these are known only as fossils; only a few species survive.

PHYLUM HEMICHORDATA. The acorn worms. Marine animals with an anterior muscular proboscis, connected by a collar region to a long worm-like body. The larval form resembles an echinoderm larva. (100)

PHYLUM CHORDATA. Bilaterally symmetrical animals with a noto-chord, gill clefts in pharynx and a dorsal, hollow neural tube. (70,000)

SUBPHYLUM UROCHORDATA. The tunicates or sea squirts. The adults are saclike, attached, filter-feeding animals with a tunic of cellulose; the adults often form colonies, but the larval forms are free-swimming and have a notochord in the tail region.

SUBPHYLUM CEPHALOCHORDATA. Amphioxus. Marine animals with a segmented, elongated fishlike body. They burrow in the sand and take in food by the beating of cilia on the anterior end. They have a noto-chord extending from the tip of the head to the tip of the tail.

SUBPHYLUM VERTEBRATA. Animals having a definite head, a backbone of vertebrae, a well developed brain and, usually, two pairs of limbs. They have a ventrally located heart, and a pair of well developed eyes and other sense organs.

Superclass Pisces. Aquatic vertebrates, fishes.

Class Agnatha. Lampreys, hagfishes and fossil ostracoderms. Vertebrates without jaws or paired fins.

Class Placodermi. The spiny-skinned sharks. The earliest fishes with jaws and paired appendages, known only from fossils.

Class Chondrichthyes. Sharks, rays, skates and chimaeras. Fishes with a cartilaginous skeleton and scales of dentin and enamel im-bedded in the skin.

Class Osteichthyes. The bony fishes. Lungs or swim bladder usually present and body usually covered with bony scales. The sturgeon, bowfin, salmon and lungfishes.

Class Amphibia. Frogs, toads, salamanders and the extinct forms, labyrinthodonts. As larvae these forms breathe by gills; as adults they breathe by lungs. There are two pairs of five-toed limbs; the skin is usually scaleless.

Class Reptilia. Lizards, snakes, turtles, crocodiles, the ex-tinct dinosaurs and other forms. The body is covered with horny scales derived from the epidermis of the skin. The animals breathe by means of lungs and have a three-chambered heart.

Class Aves. The birds. Warm-blooded, typically winged animals whose skin is covered with feathers. Present-day birds are tooth-less, but the primitive ones had reptilian teeth. The forelimbs are mod-ified as wings.

Class Mammalia. Warm-blooded animals whose skin is covered with hair. The females have mammary glands which secrete milk for the nourishment of the young.

SUBCLASS PROTOTHERIA. The monotremes, primitive forms that lay eggs. Most of them are extinct; only two species survive, the duckbilled platypus and the spiny anteater.

SUBCLASS METATHERIA. The pouched mammals or marsupials. The young are born alive, but in a very undeveloped state. They complete development at-tached to teats located in a pouch on the mother's abdomen. The subclass includes opossums, and a variety of forms found only in Australia: kangaroos, wallabies, koala bears, wombats and so on.

SUBCLASS EUTHERIA. The placental mammals. The young develop within the uterus of the mother, ob-taining nourishment via the placenta.

Order Insectivora. Primitive, insect-eating mammals; moles and shrews.

Order Chiroptera. Bats.

Order Carnivora. Dogs, cats, bears, sea lions and seals.

Order Rodentia. Rats, squirrels, beavers and porcupines.

Order Lagomorpha. Rabbits and hares.

Order Primates. Monkeys, apes and man.

Order Artiodactyla. Even-toed ungulates—cattle, deer, camels and hippopotamuses.

Order Perissodactyla. Odd-toed ungulates—horses, zebras and rhinoceroses.

Order Edentata. Armadillos, sloths and anteaters.

Order Proboscidea. Elephants.

Order Cetacea. Whales, dolphins and porpoises.

Order Sirenia. Sea cows—large, plant-eating aquatic mammals.